OCCUPATIONAL SAFETY AND HYGIENE II

SELECTED EXTENDED AND REVISED CONTRIBUTIONS FROM THE INTERNATIONAL SYMPOSIUM OCCUPATIONAL SAFETY AND HYGIENE, GUIMARÃES, PORTUGAL, 13–14 FEBRUARY 2014

Occupational Safety and Hygiene II

Editors

Pedro M. Arezes
University of Minho, Guimarães, Portugal

João Santos Baptista
University of Porto, Porto, Portugal

Mónica P. Barroso, Paula Carneiro, Patrício Cordeiro & Nélson Costa
University of Minho, Guimarães, Portugal

Rui B. Melo
University of Lisbon, Lisbon, Portugal

A. Sérgio Miguel
University of Minho, Guimarães, Portugal

Gonçalo Perestrelo
SPOSHO, Porto, Portugal

CRC Press
Taylor & Francis Group
Boca Raton London New York Leiden

CRC Press is an imprint of the
Taylor & Francis Group, an **informa** business
A BALKEMA BOOK

CRC Press/Balkema is an imprint of the Taylor & Francis Group, an informa business

Typeset by V Publishing Solutions Pvt Ltd., Chennai, India

Published by: CRC Press/Balkema
P.O. Box 11320, 2301 EH Leiden, The Netherlands
e-mail: Pub.NL@taylorandfrancis.com
www.crcpress.com – www.taylorandfrancis.com

ISBN: 978-1-138-00144-2 (Hbk)
ISBN: 978-1-315-77352-0 (eBook PDF)

Occupational Safety and Hygiene II – Arezes et al. (eds)
© 2014 Taylor & Francis Group, London, ISBN 978-1-138-00144-2

Table of contents

Occupational Safety and Hygiene II – Arezes et al. (eds)
© 2014 Taylor & Francis Group, London, ISBN 978-1-138-00144-2

Foreword

This book contains selected full papers presented at the annual congress of the Portuguese Society of Occupational Safety and Hygiene (SPOSHO), which was celebrating its 10th edition in 2014. The SHO2014-International Symposium on Occupational Safety and Hygiene, similarly to the past seven years, was held in the School of Engineering at University of Minho in Guimarães, Portugal.

The 140 papers included in this book were individually reviewed by the international Scientific Committee (SC) of the Symposium, which has involved more than 120 international specialists in the various scientific fields covered by the event.

We take this opportunity to thank our academic partners of the organisation of SHO2014, namely, the School of Engineering of the University of Minho, the Faculty of Engineering of the University of Porto, the Faculty of Human Kinetics of the University of Lisbon, the Polytechnic University of Catalonia and the Technical University of Delft. We also thank the scientific sponsorship of more than 20 academic and professional institutions and the official support of the Portuguese Authority for Working Conditions (ACT), as well as the valuable support of several Companies and Institutions, including the several media partners, which have contributed to the broad dissemination of the event.

Finally, we also hope that the papers included in this book will be a valuable contribution to improve the dissemination of the research results in the mentioned domains, by presenting the state-of-the-art of the considered topics, as well as to show the practical application of new research methodologies and the relevance of emergent topic within the domain of occupational health and safety.

The Editors

Pedro M. Arezes
J. Santos Baptista
Mónica P. Barroso
Paula Carneiro
Patrício Cordeiro
Nélson Costa
Rui B. Melo
A. Sérgio Miguel
Gonçalo Perestrelo

The editors wish also to thank all the reviewers, listed below, which were involved in the process of reviewing and editing the papers included in this book.

A. Sergio Miguel
Ana Colim
Anabela Simões
Andrew Hale
Angela Macedo Malcata
Anil R Kumar
Anna Sophia Moraes
Beata Mrugalska
Béda Barkókebas Júnior
Bianca Vasconcelos
Camilo Valverde
Carla Barros-Duarte
Catarina Silva
Celeste Jacinto
Celina P. Leão
Cristina Reis
Delfina Ramos
Denis A. Coelho
Divo Quintela
Eliane Lago
Ema Sacadura Leite
Enda Fallon
Enrico Cagno
Evaldo Valladao Pereira
Fernanda Rodrigues
Fernando Amaral
Fernando Moreira Da Silva
Filomena Carnide
Florentino Serranheira
Francisco Masculo
Francisco Rebelo
Francisco Silva
Gabriel Burdalo Salcedo
Guilherme Teodoro Buest
Hamilton Costa Junior
Hernani Veloso Neto
Ignacio Castellucci
Ignacio Pavon Garcia
Isabel Loureiro

Isabel L. Nunes
Isabel S. Silva
João Santos Baptista
Jacinta Renner
Javier Llaneza
Joana Guedes
Joana Madureira
João Areosa
João Quaresma Dias
João Paulo Rodrigues
João Porto
João Ventura
Joaquim Gois
Jorge A. Santos
Jorge Patricio
Jose Cardoso Teixeira
Jose Carvalhais
Jose Torres Da Costa
Jose Keating
Jose L. Meliá
Jose Miquel Cabeças
José Pedro Domingues
Juan Carlos Rubio Romero
Karen Jacobs
Ken Parsons
Laura Martins
Luis Antonio Franz
Luis Silva
Luisa Matos
Luiz Bueno Da Silva
Ma. Carmen Rubio
Mahmut Eksioglu
Marcelo M. Soares
Marcelo Pereira Da Silva
Maria Antónia Goncalves
Maria D. Martínez Aires
Maria Eugénia Pinho
Maria José Abreu
Marianne Lacomblez
Marino Menozzi
Mário A.P. Vaz
Marta Santos
Matilde Rodrigues
Miguel Tato Diogo
Mohammad Shahriari
Mónica Barroso
Monica Dias Teixeira
Nélson Costa
Olga Mayan
Paul Swuste
Paula Carneiro
Paulo Flores
Paulo Noriega
Paulo Sampaio
Paulo Vila Real
Pedro Arezes
Pedro Ferreira
Pedro Mondelo
Pere Sanz-Gallen
Raquel Santos
Ricardo Vasconcelos
Rui Azevedo
Rui Garganta
Rui Melo
Samir N.Y. Gerges
Santiago López
Sara Bragança
Sergio Sousa
Silvia A. Silva
Susana Costa
Susana Sousa
Susana Viegas
Tarcisio Saurin
Teresa Cotrim

Occupational Safety and Hygiene II – Arezes et al. (eds)
© 2014 Taylor & Francis Group, London, ISBN 978-1-138-00144-2

Safety of peak less tower cranes, an accident analysis conducted by the Dutch Safety Board

Paul Swuste
Safety Science group, Delft University of Technology, The Netherlands

ABSTRACT: Since cranes and tower cranes are complex installations they constitute critical aspects of safety at construction sites. The risks posed by cranes are specific and should be treated as such. Prior to assessing the impact of management and organizational factors, accident analysis should first start with an analysis of the actual accident process. The Dutch Safety Board conducted such an accident analysis involving a non-mobile, peak less, trolley tower crane. This tower crane collapsed at a Rotterdam building site on July 10th 2008. The results show that the flexibility of the configuration of the mast and the horizontal arm of the crane or the jib was greater than that calculated by the design engineer. While hoisting a heavy load, the crane collapsed. The defects in the design of the crane were not identified, so the accident was classified as a 'normal accident', one that was essentially integral to the design and could also thus occur in other tower cranes of the same make. Such tower crane design shortcomings emerge as process disturbances once the crane is operational. Despite its shortcomings, the collapsed crane did have a CE mark. Other officially required safety audits and crane inspections did not address possible defects in the design, production, or operation of the crane. Once on the market there appears to be no further effective safety net for the detection of structural weaknesses. Also the role of parties involved in construction and inspection of tower cranes will be discussed.

1 INTRODUCTION

Tower cranes are complex and impressive installations. The increased technical quality of the cranes is the main reason why scenarios like 'crane instability', 'jib instability' and 'hoisting equipment instability' contribute little to accidents. Nowadays 'load instability' is still the most dominant accident scenario in crane accidents (Beaver et al., 2006; Paas and Swuste, 2006; Tam and Fung, 2011). Tower cranes are therefore a critical component in a series of elements that make construction sites inherently dangerous (Parfitt, 2009; Sertyesilik et al., 2010). Oddly, scientific literature devotes relatively little attention to analysing the causes of crane accidents and considering how they could be prevented (Swuste, 2005; Shapira and Lyachin, 2009).

Crane accident causes are often seen as operator errors on the part of crane operators. Cranes are, however, robust installations and crane drivers are professionals. Lifting activities are risky and management time and finance is always limited. The risks attached to cranes are specific and planning must be linked to specific hazards and to crane-centred risks, like its load, its location and its environmental conditions (Schexnayder, 2003; Aneziris et al., 2008; McDonald et al., 2011). On July 10, 2008, a tower crane collapsed in Rotterdam, seemingly without cause. The Dutch Safety Board (Onderzoeksraad voor Veiligheid – OVV, 2009) extensively investigated this accident;

- To determine the cause of the collapse of the tower crane at Rotterdam;
- To bring light any possible shortcomings in the design of the tower crane;
- To identify a safeguard (or lack thereof) that could identify and help prevent accidents resulting from shortcomings in design or production;
- In addition the question was asked if the crane accident was a so-called 'normal accident'.

2 MATERIALS AND METHODS

2.1 *Data*

Immediately after the accident, an OVV research team conducted an on-site survey. In the technical study, three different scenarios were extensively studied:

1. Exceeding operational parameters.
   The electronic components and the data carrier were checked, as well as what is known as the crane's electronic compatibility. That kind of compatibility involves the possible influence of electromagnetic fields, such as power supply,

Figure 1. Peak less tower crane.

on the electronic data flow required to control crane movements.

2. Failing tower construction, due to failures or weaknesses in the construction or other equipment failure.
   For this scenario, destructive materials research was conducted on the crane construction.
3. Jib overload
   The accident might have been caused by jib overload, because of the trolley—the runner on the jib—(unintentionally) moving too far out. To investigate this scenario a similar, but stiffer crane was mounted on the same site to measure the results produced by computer models and to test the simulations made. In connection with the third objective, the European product and social guidelines for tower cranes were inventoried, including the relationship with Dutch national legislation and specific crane inspection regimes.

The last bullet point under the objectives of the investigation raises the question if the accident of the crane tower was a 'normal accident'. This term comes from Charles Perrow (1984), when he provided a sociological explanation for accidents in complex technological systems, such as air and marine traffic, chemical plants, dams, and nuclear power plants according to their riskiness. According to Perrow, multiple and unexpected failures were built into society's complex systems. These accidents were called normal accidents or system accidents, because such accidents are inevitable in complex systems. Given the characteristic of the system involved, multiple failures which interact with each other will occur, despite efforts to avoid them. Such events appear trivial to begin with before cascading through the system in unpredictable ways to cause a large event with severe consequences. Complex systems, like production processes, with system determinants as interactive complexity and tight coupling, fulfill system conditions where these normal accidents will occur.

3 RESULTS

3.1 *The circumstances of the accident*

The crane in question is a peak less trolley tower crane. This type of crane is easier to transport than other tower cranes and it is often used in construction projects with height restrictions and on sites where several cranes operate together in close proximity to each other (Figure 1). The crane was involved in the construction of a 24-storey, high rise flat in Rotterdam. It had a hook height of 59 meters and had been approved by an accredited inspection body. At the time of the accident the mast was extended to a height of 96 meters.

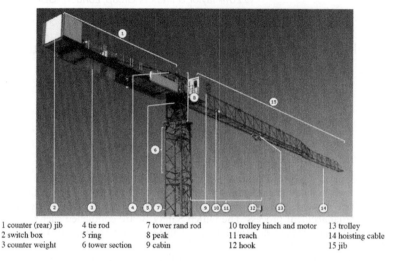

On the day of the accident, a 7-ton balcony slab had been lifted and, including the balancing device, the total burden had been 12.8 tons. The weather conditions were good. Because of the high weight of the load, it was initially decided to conduct the hoisting around the building instead of over it. However, the trolley could not reach far enough around the corner of the building, because the load moment protection stopped the trolley. The load was therefore hoisted over the building. When it was almost in place, the staff on the balcony reported to the crane driver not to move the load outwards but inwards. In his reply the driver denied he had given an outward moving control command. At that stage, the outreach was 27 meters. The men on the balcony saw the load sway away from them before the crane collapsed. The crane driver fell forty meters, landed on the top floor of the building but did not survive the fall. The devastation was great.

3.2 *Technical research, Scenario 1: Exceeding operational parameters*

The tower crane had a maximum load weight of 16 tons at an outreach 3 to 23 meters, and 8,7 tons at 40 meters outreach. If the load protecting device is not switched on, or if it fails, the crane would collapse when exceeding the load moment. The analysis of the crane wreckage showed the load moment protection device time was switched on at the time of hoisting and this device had prevented the load being hoisted around the building. For a 13 ton load the maximum outreach was 28 meters. With an actual load of 12.8 tons and a flight of 27 meters the crane had in fact reached the limit of its maximum load torque.

3.3 *Technical research, Scenario 2: Failing tower construction*

The mast was bent, broken and wrenched apart. The chemical analysis and tensile tests showed that the steel structure had been built according to design specifications. The fracture surfaces in the structure are described as 'tough breaks'. These breaks therefore indicate overload and not material defects.

3.4 *Technical research, Scenario 3: Jib overload*

Several factors can lead to jib overload, a load can catch, weather conditions, control malfunctioning, or a bending jib. Weather conditions were calm on the day of the accident. The starting position of the load, just before the accident, was furthermore virtually static.

Crane control malfunctioning can be the result of incorrect commands from the operator. The communication between crane driver and site personnel made it plausible to believe that just before the accident the crane driver did not commission an outward command of the trolley. Failure could also have been caused by the crane control cables located in the direct proximity of strong electromagnetic fields, like the antenna for the 'global system for mobile communications' (gsm); and the power cables from the motors. The gsm antenna was installed in the crane switch box. Researchers did observe that power and data cables were bundled together and could therefore have had an impact on data transport. In one test, such electromagnetic disturbances were detected. OVV did notice that the programmable operating system was insufficiently responding to adverse events. These events might have led to error messages, but not to actions.

The accident could also have been caused by the bending of the jib. The manufacturer had accounted for a maximum sag of the jib of 1° below horizontal. In a simulation model an unloaded crane with an outreach of 27 meters had a sag of 0.5° above horizontally level. In a stationary hanging load of 13 tons, the sag was 2.1° degrees below the level point. That means the flexibility of the crane exceeded the values calculated by the manufacturer, all of which had a significant impact on the functionality of the trolley. The trolley motor controls the trolley. If the motor is accessed, the settings will determine when maximum power can be delivered, as well as the moment the brakes should be released. Examination of the collapsed crane showed a 0 seconds the setting of the 'brake time', meaning the brake must have been released when the motor was started up. That in itself should not lead to problems. However, the setting of the 'magnetising time' was clearly insufficient to allow the motor to operate at maximum capacity. As a consequence the motor brake was released when the motor was not at full power thus causing the load to run down the sagged jib. The extensive analysis of this scenario is mentioned in the OVV report (OVV, 2009).

3.5 *Normal accident*

Following Perrow (1984), the crane accident is not as complex and tightly coupled as for instance nuclear power systems, because a crane, and its activities, is not a production process, basically not a transformation system. But there is an exception, the mass of the loads, the height of operation, and the quality of the controls. This has transformed the safety envelope of the crane. And combined with an over-bending of the jib the crane system has characteristics of transformation systems: poorly understood dynamics, unobservable processes and intrinsically poorly monitored and instrumented processes, and critically narrow limits of safety. Also similar to tight coupling, there was very limited time for recovery from failures, no time-buffers or redundancies.

4 DISCUSSION AND CONCLUSION

The reconstruction of the accident with the tower crane shows the specific crane risks. The proximal factors of the accident process are clear. The excessive inclination of the jib tested the other components of the crane to the limit. The trolley and its load could not be controlled when the jib bent and the trolley motor provided insufficient torque. The load itself was racked beyond the maximum load moment. Ultimately, it was the dynamic forces that led to the collapse of the crane. The flexibility of the jib made the load unstable. This, then, answers the first two objectives. The extensive bending had not been predicted by the manufacturer and the accident was classified as a 'normal accident' (Perrow, 1984). It was a fault in the design of the crane.

Once the accident process had become clear, its distal management and organizational factors could then be identified. The primary responsibility for good design lies with the manufacturer. It was hard to establish whether the manufacturer had carried out risk analysis during the crane's construction as all documentation on that topic was missing. A design review of the criteria, such as the bending of the jib during maximum load, was not executed. Such a review was not required for CE qualification. Error messages on tower cranes were resolved by the manufacturer on an ad-hoc basis and these reports have not so far led to adjustments to the design of the tower crane. Once on the market there appears to be no further effective safety net for the detection of structural weaknesses. Apart from the manufacturer no one tested or checked defects in the design or production of the crane (Swuste et al., 2012).

The conclusion therefore is that risks of tower cranes are documented and analysed insufficiently. The Safety Board recommends a hotline for accidents or near-accidents involving cranes and tower cranes. The contact point should provide information to all interested parties and warn crane manufacturers, owners, users and clients. The Safety Board has two recommendations when it comes to detecting defective design deficiencies. Firstly, tower cranes should be added to the list of dangerous machinery in the Machinery Directive, making independent inspections mandatory. The second recommendation involves extending existing certification scheme by having functional assessment criteria in addition to mechanical failure which also will relate to the electronics and the control crane movements.

A final question that remained after the Safety Board analysis had been completed concerned the matter of tower crane selection. Why was a peak less crane chosen? A crane with a peak has a more rigid jib and can make other bending moments. The arguments in favour of a peak-free crane, height restriction and reduced risk of accidents with other cranes, was not applicable to the Rotterdam construction site case because no other cranes were active in its vicinity and there was no evidence of a height restriction. This question is relevant for the contractor, who hires the crane. Unfortunately the accident analysis could not provide an answer to this question.

REFERENCES

Aneziris O, Papazouglou I, Mud M, Damen M, Kuiper J, Baksteen H, Ale B, Bellamy L, Hale A, Bloemhoff A, Post J, Oh J (2008). Towards risk assessment for crane activities. *Safety Science* 46:872–884.

Beaver J, Moore J, Rinehart R Schriver W (2006). Crane related fatalities in the construction industry. *Journal of Construction Engineering and Management* 132(9):901–910.

McDonald B, Ross B Carnahan R (2011). The Bellevue crane disaster. *Engineering Failure Analysis* 18:1621–1636.

OVV, De Onderzoeksraad voor Veiligheid - Dutch Safety Board (2009). *Collapse of a tower crane Rotterdam July 10th 2008*.

Paas C, Swuste P (2006). Mobile cranes, what goes wrong? A survey on dominant accident scenarios. *Journal of Applied Sciences* 19(3): 47–55 (in Dutch).

Parfitt M (2009). Cranes, structures under Construction, and temporary facilities: are we doing enough to ensure they are safe? *Journal of Architectural Engineering*, March 1–2.

Perrow C (1984). *Normal accidents*. Basic Books, New York. A reprint has been published in 1999: Perrow C (1999). Normal Accidents: living with high-risk technologies. Princeton University Press, Princeton, New Jersey.

Schexnayder C (2003). Crane accidents construction sessions roebling award books conferences. *Practical periodical on structural design in construction*. May p. 67–73.

Sertyesilisik B, Tunstall A, McLouglin J (2010). An investigation of lifting operations on UK construction sites. *Safety Science* 48:72–79.

Shapira A, Lyachin B (2009). Identification and analysis of factors affecting safety on construction sites with tower cranes. *Journal of Construction Engineering and Management* 135(1):24–33.

Swuste P (2005). *Safety Analysis vertical transport mobile cranes*. Safety Science Group, Delft University of Technology (in Dutch).

Swuste P Frijters A Guldenmund F (2012). Is it possible to influence safety in the building sector? A literature review extending from 1980 until the present. *Safety Science* 50:1333–1343.

Tam V, Fung I (2011). Tower crane safety in the construction industry: a Hong Kong study. *Safety Science* 49:208–215.

Occupational Safety and Hygiene II – Arezes et al. (eds)
© 2014 Taylor & Francis Group, London, ISBN 978-1-138-00144-2

A methodology to identify negative occupational well-being hazards

José Miquel Cabeças
Faculdade de Ciências e Tecnologia da Universidade Nova de Lisboa, Lisboa, Portugal

ABSTRACT: This article presents hazard taxonomy to negative occupational well-being, organized as a matrix or checklist (Negative Well-Being Matrix), to be applied during the systematic hazards identification at a specific workplace. The matrix serves as a quasi-exhaustive list of negative occupational well-being hazards to work related diseases (psychosomatic and physical as well as mental, social and behavioral diseases) and to negative mental or physical well-being (not directly related to injuries or diseases symptomatology neither to the exposition to accidents and occupational diseases hazards). The matrix allows safety and health technicians to analyze workplaces, questioning the presence of the hazards mentioned in the matrix. Individual risk factors, which may drive tendencies to negative emotional styles and work related diseases, are included in the matrix.

1 INTRODUCTION

The identification of hazards or risk factors (Eurostat 2010, BS 8800:2004) at workplace level is a crucial procedure to the risk identification, risk analysis and risk evaluation.

The relationship between work and disease may be identified in the following categories (Indulski 1993): (1) occupational diseases, having a specific or a strong relation to occupation; (2) work-related diseases, with multiple causal agents, where factors in the work environment may play a role, together with other risk factors, in the development of such diseases; (3) diseases without causal relationship with work but which may be aggravated by occupational hazards to health, eg. communicable diseases, parasitic diseases and malnutrition in developing countries (WHO 1985). Work-related diseases are also named as "multifactorial diseases" (el Batawi 1984, WHO 1985).

1.1 *Well-being at work*

The definition of well-being may be written in many ways. According to Anttonen & Räsänen (2008) and Räsänen (2011), well-being at work (or occupational well-being) refers to the experience of the worker that is influenced by how safe, healthy, well-led, and well-organized work is, how effectively the changes in work are managed, the level of community support to the individual, and how meaningful and rewarding a person finds work, accounting for the factors of competence and productivity. Employee well-being limits the term to those aspects of work which are perceived to influence well-being and are modifiable by the employer.

Employee well-being has a wide number of dimensions, varying according to sector and type of work (Briner 2000, Cooper et al. 2008, Warr 1990, Warr 2002, Danna & Griffin 1999): (1) the facilities, including the environmental and physical work context; (2) the work content; (3) the work context or the psychosocial work environment (Schnall et al. 2000, Benavides et al. 2002); (4) the work-life balance or the home-work interface; (5) the physical health and the psychological health (Danna & Griffin 1999).

To Danna & Griffin (1999) the term health generally appears to encompass both physiological and psychological symptomatology within a more medical context (e.g., reported symptomatology or diagnosis of illness or disease). This definition allowed these authors to distinguish between well-being and intrinsic health (Celia et al. 2009). Danna & Griffin (1999) proposed a definition of well-being, including work-related experiences such as job satisfaction. In this sense, occupational well being is a broader concept than occupational health and may be negatively affected by existing individual physical or psychological pathologies. Well-being is viewed as comprising the various life/non-work satisfactions enjoyed by individuals. Health, in turn, is seen as being a sub-component of well-being and comprises the combination of such mental/psychological indicators as affect, frustration, and anxiety and such physical/physiological indicators as blood pressure, heart condition, and general physical health.

1.2 *Causes of work-related diseases*

A 1985 report of a World Health Organization Expert Committee (WHO 1985) concluded that adverse occupational psychosocial factors have become increasingly important in responses and psychosomatic disease causation. Ample evidence shows that a relation exists between certain types of working conditions and behavioral and psychosomatic disorders among workers.

There is a reasonable consensus in the literature of the nature of psychosocial hazards (Leka & Jain 2010): job content, workload & work pace, work schedule, control, environment & equipment, organizational culture & function, interpersonal relationships at work, role in organization, career development and home-work interface. Psychosocial risks, work-related stress, violence, harassment, bullying (or mobbing) are recognized major challenges to occupational health and safety (EASHW 2007). A WHO review (Leka & Jain 2010) collates evidence across all WHO regions on exposures to selected psychosocial risk factors (where available), including job control and demand, work organization, working hours, and relative risks for major health outcomes, including coronary heart disease, depression, and back pain.

1.3 *Psychological well-being and work related diseases*

As stated by Huppert (2009), it has long been known that negative emotions are related to a higher prevalence of disease. Evidence supports the view that positive mental states can have direct effects on physiological, hormonal, and immune function which, in turn, influences health outcomes. The functioning of the immune system likes to be an important physiological factor in the relationship between positive emotions and health (see Danner et al. 2001, Huppert & Whittington 2003, Ostir et al. 2001, Cohen et al. 2003a, b, Marsland et al. 2006, Davidson et al. 2003, Fredrickson et al. 2000, Lai et al. 2005, Pressman & Cohen 2006). Positive mood has also been shown to influence the cardiovascular response to stress; prolonged reactivity to stress is harmful to immune function and to other physiological processes.

Several studies provide evidence linking mental health domains to physical conditions and the complex nature of their relationships are being increasingly explored (WHO 2005).

Robust evidence (ILO 2010) indicates that many of the most commonly experienced physical effects due to work related stress and psychosocial risks relate to hypertension, heart disease, wound-healing, musculoskeletal disorders, gastro-intestinal disorders, and impaired immuno-competence. Disorders usually cited as being stress-related include bronchitis, coronary heart disease (Eller et al. 2009, Peter & Siegrist 2000, Allesøe et al. 2010), mental illness, thyroid disorders, skin diseases, certain types of rheumatoid arthritis, obesity, tuberculosis, headaches and migraine, peptic ulcers and ulcerative colitis, and diabetes.

Psychosocial working conditions may have a detrimental impact on both affective and cognitive outcomes, namely anxiety, depression, distress and burnout (Cox et al. 2000, Stansfeld & Candy 2006). Burnout is the result of chronic stress (at the workplace) which has not been successfully dealt with, characterized by exhaustion and depersonalization (negativism/cynicism) (Maslach & Jackson 1981). Depression is one of the leading causes of disability and is projected by the WHO to become the second leading cause of the global burden of disease by 2020. Exposure to psychosocial risks has been linked to a wide array of unhealthy behaviors such as physical inactivity, excessive drinking and smoking, poor diet and sleep (Cox et al. 2000).

Increasing attention is being placed on the interactive effects of physical and psychosocial hazards in the etiology of work-related MSDs, particularly in the low back, neck and upper limbs (ILO 2010). The interactive role of psychosocial factors, like low social support, low job satisfaction, poor work organization and low job content and factors related to the physical aspects of work was highlighted in lower back pain (De Beeck & Hermans 2000).

The etiology of coronary heart disease (CHD) may include type A behavior, stressful life events, lack of social support, shift work and a sedentary lifestyle (Knutsson 1989). The evidence on the relationship between work-related psychosocial factors and the development of ischemic heart disease (IHD) has been increasing (ILO 2010).

The metabolic syndrome is a cluster of risk factors that increases the risk of heart disease and type 2 diabetes (Chandola et al. 2006). Depressive symptoms and severity of stressful life events were associated with the cumulative prevalence of the metabolic syndrome (Räikkönen et al. 2007).

Evidence from the Whitehall II study ($n = 5{,}895$) has indicated that psychosocial work stress was an independent predictor of type 2 diabetes among women after a 15-year follow-up (period 1991–2004) (Heraclides et al. 2009).

1.4 *Individual risk factors to occupational well-being*

The personality dimensions of extraversion and neuroticism are strong predictors of the usual emotional style (Huppert 2009). Extraversion (sociability) is strongly associated with a positive emotional

Table 1. Relevant negative well-being hazards (Negative Well-Being Matrix), adapted and developed from the Risk Factors-Disorders Matrix (Cabeças 2013, Cabeças & Paiva 2010).

Group of hazard	Sub-group of hazard	Code	Type of hazard	WD[1]	NW[1]
4. Radiations	4.5 Radiant heat	4.5.1	Radiant heat		
5. Noise	5.1 Noise	5.1.3	Disturbing noise		
6. Vibrations	6.1 Vibration	6.1.3	Whole body and hand-arm disturbing vibration		
7. Chemical	7.2 Solid moistures	7.2.1	Dust		
		7.2.3	Smoke		
9. In the physical work environment	9.1 Indoors climate	9.1.1	Indoors thermal discomfort (cold, hot)		
		9.1.2	Indoors humidity discomfort (wet or dry humidity)		
	9.2 Indoor ventilation (natural or mechanical)	9.2.1	Insufficient air renewal, insufficient air flow		
		9.2.2	Inadequate location/direction of air flow		
		9.2.3	Circumstances that cause odor nuisance, strong odor		
	9.3 Work area lighting	9.3.1	Illuminance or light level (light intensity in a work area)		
		9.3.2	Luminance or bright		
	9.4 Physical layout	9.4.1	Open workplaces (offices)		
	9.5 Outdoor work environment (climate, weather)	9.5.1	Extreme temperature (too hot, too cold)		
		9.5.2	Rain, wind, extreme humidity		
		9.5.3	Ice, snow, hailstorm		
	9.8 Confined space	9.8.1	Works in confined spaces		
10. Musculoskeletal (ergonomics)	10.1 Force	10.1.1	Lifting loads or moving people activities; carrying or moving		
	10.2 Repetition	10.1.2	Activities with repetitive hand or arm movements		
	10.3 Postures	10.1.3	Tiring or painful working positions; inadequate postures		
	10.4 PC work	10.1.4	Working regularly with computers		
	10.5 Standing, sitting	10.1.5	Standing jobs; long term sitting jobs		
11. Psychosocial	11.1 Job content	11.1.1	Lack of variety (monotonous) or short work cycles		
		11.1.2	Fragmented or meaningless work		
		11.1.3	Under use of skills		
		11.1.4	High uncertainty		
		11.1.5	Continuous exposure to people through work		
	11.2 Workload and work pace	11.2.1	Continually subjected to deadlines		
		11.2.2	Work overload or under load		
		11.2.3	Machine pacing		
		11.2.4	Insufficient working breaks		
		11.2.5	High levels of time pressure		
	11.3 Work schedule	11.3.1	Shift working		
		11.3.2	Night shifts		
		11.3.3	Unpredictable hours		
		11.3.4	Long or unsociable hours		

(*Continued*)

Table 1. (Continued).

Group of hazard	Sub-group of hazard	Code	Type of hazard	WD[1]	NW[1]
		11.3.5	Inflexible work schedules	■	■
	11.4 Control	11.4.1	Low participation in decision making (job influence)	■	■
		11.4.2	Lack of control over workload, pacing	■	■
	11.5 Organisational culture & function	11.5.1	Poor communication	■	■
		11.5.2	Low levels of support	■	■
		11.5.3	Lack of definition on organizational objectives	■	■
	11.6 Interpersonal relationships at work	11.6.1	Social or physical isolation	■	■
		11.6.2	Poor relationships with superiors	■	■
		11.6.3	Interpersonal conflict	■	■
		11.6.4	Lack of social support	■	■
		11.6.5	Bullying	■	■
		11.6.6	Harassment (moral, sexual)	■	■
		11.6.7	Discrimination/intolerance	■	■
		11.6.8	Disrespected to private property	■	■
		11.6.9	Physical and verbal violence	■	■
		11.6.10	Emotional labour	■	■
	11.7 Role in organisation Career development	11.7.1	Role ambiguity, role conflict	■	■
		11.7.2	Poor pay (precarious work)	■	■
		11.7.3	Career stagnation and uncertainty	■	■
		11.7.4	Under promotion or over promotion	■	■
		11.7.5	Job insecurity (fixed term contract)	■	■
		11.7.6	Low social value to work	■	■
		11.7.7	Responsibility for people	■	■
	11.8 Home-work interface	11.8.1	Conflicting demands of work and home	■	■
		11.8.2	Low support at home	■	■
		11.8.3	Dual career problems	■	■
12. Individual	12.1 Individual susceptibility	12.1.1	Obesity/overweight	■	
		12.1.2	Poor nutrition	■	■
		12.1.3	Physical inactivity/lack of exercise	■	
		12.1.4	Sleep disorders, smoking habits, consumption of alcohol	■	
		12.1.5	Type A behavior patterns	■	
		12.1.6	Locus of control	■	■
		12.1.7	Neuroticism/negative emotional style	■	■
	12.2 Vulnerable workers	12.2.4	Migrant worker	■	■
		12.2.6	Temporary worker	■	■
		12.2.7	Subcontracted worker	■	■
		12.2.8	Precarious worker	■	■
14. Others	14.2 Personal Protective Equip. (PPE)	14.2.1	Uncomfortable, temperature discomfort, hinder work		■

(1) WD: Work related diseases with multiple causal agents and not included in Annexes I and II of 2003/670/EC; NW—negative mental or physical well-being, like discomfort complaints and insatisfaction not directly related to injuries or diseases symptomatology, neither to the exposition to accidents or diseases hazards.

style, while neuroticism is associated with a negative emotional style. Neuroticism appears to drive negative mood and common mental disorders, whereas extraversion drives positive emotional characteristics. Cross-sectional studies have shown strong associations between psychological well-being and both extraversion and neuroticism (Huppert 2009, Cotton & Hart 2003, Danna & Griffin 1999).

The linkage between type A individuals to cardiovascular disease (coronary heart disease and coronary artery disease) is the elevated blood pressure. Substantial research has found that certain types of illnesses appear more frequently in workers with Type A tendencies (Danna & Griffin 1999).

Two main objectives may be identified in this paper: (1) to define a methodology to identify occupational hazards associated to negative well-being, based on the Risk Factors-Disorders Matrix (Cabeças 2013, Cabeças & Paiva 2010) and (2) to propose an organization of risk factors in terms of negative mental or physical well-being.

2 MATERIALS AND METHODS

The relevant negative well-being hazards defined in this article was adapted and developed from the general hazards or risk factors taxonomy presented in the Risk Factors-Disorders Matrix (Cabeças 2013, Cabeças & Paiva 2010), allowing the safety and health technician to observe the work environment with an extensive checklist of risk factors, checking the identified hazards, and questioning systematically all the remaining hazards existing in the matrix.

The individual consequences of exposure to negative well-being hazards may be classified in the following three levels: (1) Psychosomatic and physical work related diseases (WD in Table 1), as for example, heart disease, stroke, hypertension, ischemic heart disease, metabolic disease and diabetes, asthma, cancer, wound-healing, gastro-intestinal disorders, peptic ulcers and ulcerative colitis, impaired immunocompetence, thyroid disorders, skin diseases, certain types of rheumatoid arthritis, obesity, tuberculosis, headaches and migraine, musculoskeletal disorders, low back pain, shoulder and neck pain; (2) Mental, social and behavioral work related diseases (WD in Table 1), as for example, major depression, burnout, neurotic reactions, personality disturbances, psychotic illness, mental disorders and F00-F99 disorders (Karjalainen 1999); and (3) Negative mental or physical well-being (NW in Table 1) not directly related to injuries or diseases symptomatology and not related to the exposition to accidents and occupational diseases hazards, including, for example, discomfort complaints and insatisfaction, reactions to psychosocial stress, changes in the individual's mental state, anxiety or depressive reactions, diminution of interest in social activities and problems in interpersonal relations, psychogenic disorders (for example, symptoms of headache, persistent general fatigue and dizziness), lumbar fatigue with feeling of a "tight band" across the back, transitory musculoskeletal pain and mild hypertension or elevation of the cholesterol concentration.

Based on evidences from the literature, a correspondence was established in the Negative Well-Being Matrix (Table 1), between each type of hazard and work related diseases. To some types of hazard, two associations were established; some hazards may cause a work related disease (WD in Table 1) or a discomfort complaints and insatisfaction (NW in Table 1), depending on the exposition, intensity and multiple individual and environmental factors.

3 DISCUSSION AND CONCLUSIONS

This article presents hazard taxonomy to negative occupational well-being, organized as a matrix or checklist (Negative Well-Being Matrix), to be applied during the systematic hazards identification at a specific workplace. The matrix is an open matrix (different items may be added to the matrix) and serves as a quasi-exhaustive list of negative occupational well-being hazards to work related diseases (psychosomatic and physical as well as mental, social and behavioral diseases) and to negative mental or physical well-being (not directly related to injuries or diseases symptomatology neither to the exposition to accidents and occupational diseases hazards); according to the Risk Factors-Disorders Matrix methodology (Cabeças 2013, Cabeças & Paiva 2010), accidents and occupational diseases hazards must be previously identified. The matrix allows safety and health technicians to analyze the workplaces, questioning the presence of the hazards mentioned in the matrix. Individual risk factors, which may drive tendencies to negative emotional styles and work related diseases, are included in the matrix.

REFERENCES

Allesøe, K., Hundrup, Y.A., Thomsen, J.F. & Osler, M. 2010. Psychosocial work environment and risk of ischaemic heart disease in women: the Danish Nurse Cohort Study. *Occupational and Environmental Medicine* 67:318–322.

Anttonen, H. & Räsänen, T. (Eds.) 2008. *Well-being at Work—New Innovations and Good Practices (progress/application for programme VP/2007/005/371)*. Helsinki: Finnish Institute of Occupational Health.
Benavides, F., Benach, J. & Muntaner, C. 2002. Psychosocial risk factors at the workplace: is there enough evidence to establish reference values? *Journal of Epidemiology & Community Health* 56(4):244–245.
Briner, R.B. 2000. Relationships between work environments, psychological environments and psychological well-being. *Occupational Medicine* 50(5):299–303.
BS 8800:2004 Occupational health and safety management systems-Guide.
Cabeças, J.M. 2013. Taxonomy and procedures to characterize occupational hazards (risk factors) at workplace level: incorporating new knowledge in hazards identification. In *Proceedings from SHO2013* (60–62). Guimarães: Portuguese Society of Occupational Safety and Hygiene.
Cabeças, J.M. & Paiva, A. 2010. Taxonomia e estrutura dos procedimentos de análise de riscos ocupacionais. In *Proceedings from SHO2010* (133–137). Guimarães: SPOSHO—Portuguese Society of Occupational Safety and Hygiene.
Celia, C., Disley, E., Hunt, P., Marjanovic, S., Shehabi, A., Villalba-Van-Dijk, L. & Van Stolk, C. 2009. *Health and wellbeing at work in the United Kingdom*. London: The Work Foundation.
Chandola, T., Brunner, E. & Marmot, M.G. 2006. Chronic stress at work and the metabolic syndrome: prospective study. *BMJ (Clinical Research Ed.)* 332(7540):521–5.
Cohen, S., Doyle, W.J., Turner, R.B., Alper, C.M. & Skoner, D.P. 2003a. Emotional style and susceptibility to the common cold. *Psychosomatic Medicine 65*(4):652–657.
Cohen, S., Doyle, W.J., Turner, R., Alper, C.M. & Skoner, D.P. 2003b. Sociability and susceptibility to the common cold. *Psychological Science* 14(5):389–95.
Cooper, R., Boyko, C. & Codinhoto, R. 2008. The Effect of the Physical Environment on Mental and Capital Wellbeing, in: *Mental Capital and Wellbeing: Making the most of ourselves in the 21st century*. London: Government Office for Science.
Cotton, P. & Hart, P.M. 2003. Occupational wellbeing and performance: a review of organisational health research. *Australian Psychologist*, 38(2):118–127.
Cox, T., Griffiths, A. & Rial-Gonzalez, E. 2000. *Research on work related stress*. Luxembourg: Office for Official Publications of the European Communities.
Danner, D., Snowdon, D. & Friesen, W. 2001. Positive emotions in early life and longevity: Findings from the Nun Study. *Journal of Personality and Social Psychology* 80(5):804–13.
Danna, K., & Griffin, R.W. 1999. Health and Wellbeing in the Workplace: A Review and Synthesis of the Literature. *Journal of Management* 25(3):357–384.
Davidson, R.J., Kabat-Zinn, J., Schumacher, J., Rosenkrantz, M., Muller, D., Santorelli, S.F., Urbanowski, F., Harrington, A., Bonus, K. & Sheridan, J.F. 2003. Alterations in brain and immune function produced by mindfulness meditation. *Psychosomatic Medicine* 65(4):564–70.
De Beeck, R.O. & Hermans, V. 2000. *Research on work-related low back disorders (No. 204)*. Luxembourg: Office for Official Publications of the European Communities.
EASHW—European Agency for Safety and Health at Work 2007. *Expert Forecast on Emerging Psychosocial Risks Related to Occupational Safety And Health. Luxembourg*: Office for Official Publications of the European Communities.
el Batawi, M.A. 1984. Work-related diseases. A new program of the World Health Organization. *Scandinavian Journal of Work, Environment & Health* 10(6 Spec No):341–6.
Eller, N.H., Netterstrøm, B., Gyntelberg, F., Kristensen, T.S., Nielsen, F., Steptoe, A. & Theorell, T. 2009. Work-related psychosocial factors and the development of ischemic heart disease: a systematic review. *Cardiology in Review* 17(2):83–97.
Eurostat 2010. *Health and safety at work in Europe. (1999–2007): a statistical portrait*. Eurostat Statistical books. Publications Office of the European Commission, Directorate-General for Employment, Social Affairs and Inclusion.
Fredrickson, B.L., Mancuso, R.A., Branigan, C. & Tugade, M. 2000. The undoing effect of positive emotions. *Motivation and Emotion* 24(4): 237–258.
Heraclides, A., Chandola, T., Witte, D.R. & Brunner, E.J. 2009. Psychosocial Stress at Work Doubles the Risk of Type 2 Diabetes in Middle-Aged Women—Evidence from the Whitehall II Study. *Diabetes Care* 32(12): 2230–5.
Huppert, F.A. & Whittington, J.E. 2003. Evidence for the independence of positive and negative well-being: Implications for quality of life assessment. *British Journal of Health Psychology* 8(1):107–122.
ILO—International Labour Organisation 2010. List of occupational diseases (revised 2010). *Identification and recognition of occupational diseases: Criteria for incorporating diseases in the ILO list of occupational diseases (OSH 74)*.
Indulski J.A. 1993. International Symposium on Work-related Diseases—Prevention and Health Promotion (27–30 October, 1992, Linz, Austria). *Polish journal of occupational medicine and environmental health* 6(2):195–201.
Karjalainen, A. 1999. *International Statistical Classification of Diseases and Related Health Problems (ICD-10) in Occupational Health*. Geneva: World Health Organization.
Knutsson, A. 1989. Shiftwork and coronary heart disease. *Scandinavian Journal of Social Medicine*, Supp. 44:1–36.
Lai, J.C, Evans, P.D., Ng, S.H., Chong, A.M., Siu, O.T., Chan, C.L., Ho, S.M., Ho, R.T., Chan, P. & Chan, C.C. 2005. Optimism, positive affectivity, and salivary cortisol. *British Journal of Health Psychology* 10:467–484.
Leka, S. & Jain, A. 2010. *Health impact of psychosocial hazards at work: an overview*. Geneva: World Health Organization.
Marsland, A.L., Cohen, S., Rabin, B.S. & Manuck, S.B. 2006. Trait positive affect and antibody response to hepatitis B vaccination. *Brain, Behavior, and Immunity* 20(3):261–269.

Maslach, C. & Jackson, S.E. 1981. The measurement of experienced burnout. *Journal of Occupational Behaviour* 2(2):99–113.

Ostir, G.V., Markides, K.S., Peek, M.K. & Goodwin, J.S. 2001. The association between emotional well-being and the incidence of stroke in older adults. *Psychosomatic Medicine* 63(2):210–215.

Peter, R. & Siegrist, J. 2000. Psychosocial work environment and the risk of coronary heart disease. *International Archives of Occupational and Environmental Health* 73 Suppl:S41-S45.

Räsänen, T. 2011. *Well-being at work: innovation and good practice*. Finnish Institute of Occupational Health. Eurofound, European Working Conditions Observatory (EWCO).

Räikkönen, K., Matthews, K.A. & Kuller, L.H. 2007. Depressive symptoms and stressful life events predict metabolic syndrome among middle-aged women: A comparison of World Health Organization, Adult Treatment Panel III, and International Diabetes Foundation definitions. *Diabetes Care* 30(4):872–877.

Schnall P., Belkic K., Landsbergis P. & Baker, D. (Eds.) 2000. *The workplace and cardiovascular disease*. Philadelphia: Hanley and Belfus.

Stansfeld, S. & Candy, B. 2006. Psychosocial work environment and mental health—a metaanalytic review. *Scandinavian Journal of Work Environment & Health* 32(6):443–462.

Warr, P. 1990.The measurement of well-being and other aspects of mental health. *Journal of Occupational Psychology* 63(3):193–210

Warr, P. 2002. The study of well-being, behaviour and attitudes. In P. Warr (Ed.). *Psychology at work*. London: Penguin Books, pp. 1–25.

WHO Expert Committee on Identification and Control of Work-Related Diseases World Health Organization 1985. *Identification and control of work-related diseases*. Technical report Series, 714.

WHO World Health Organization 2005. *Promoting mental health: concepts, emerging evidence, practice*. Report of the World Health Organization, Department of Mental Health and Substance Abuse in collaboration with the Victorian Health Promotion Foundation and the University of Melbourne.

Occupational Safety and Hygiene II – Arezes et al. (eds)
 ISBN 978-1-138-00144-2

Causes of musculoskeletal accidents identified in official investigations

J.A. Carrillo-Castrillo & L. Onieva
University of Seville, Seville, Spain

J.C. Rubio-Romero & A. López-Arquillos
University of Malaga, Malaga, Spain

ABSTRACT: Accident causation involves multiple factors. Although musculoskeletal disorders have been studied deeply in the literature, very few studies have been published regarding their causes when they are considered accidents. Accident investigation can be used to research the mechanisms that explain associations between accident causation and risk factors. This study focuses on the association between causes of musculoskeletal disorders analyzing official investigation reports of 153 accidents in Andalusia from 2004 to 2012. Besides safety improvements to equipment, the most crucial areas to target are work organization, safety management and personal factors.

1 INTRODUCTION

1.1 *Musculo-skeletal accidents*

In Spain, musculoskeletal disorders (hereinafter MSD) can be reported both as accidents and as occupational diseases. If the injury is related to movements whereby the injured person's physical exertion exceeded what is normal, it is classified as a musculoskeletal accident. If the damage resulted from a long-term influence of working conditions it is reported as a musculoskeletal disease and it is not reported as accident but as occupational disease.

Most of the musculoskeletal accidents reported are strain and sprain injuries. It is important to note that around of 30% of total occupational accidents reported are MSD, although most of them are usually slight ones.

1.2 *The analysis of MSD accident causation*

Although the risk factors for MSD have been researched, most of studies concentrate in the exposure and not in the mechanism of the injuries, with a illness approach. In spite of the importance of those exposure factors, many of the MSD are related to a sudden event and can be modeled as an accident.

Epidemiologic analysis has identified collectives of workers at risk and risk factors of MSD (Malchaire *et al.*, 2001). In an extensive review, there is evidence of the causal relationship between physical work factors and MSD (National Institute for Occupational Safety and Health, 1997). In the case of Spain, physical demands at work are the leading cause of work injuries (Caicoya & Delclos 2010).

Regarding that period in Andalusia, in the manufacturing sector, previous studies have identified a number of individual worker characteristics can affect likelihood of occupational safety injuries (Carrillo *et al.*, 2012) and in case of injury they also affect the severity of injuries (Carrillo & Onieva 2012).

In addition, the relative risk of musculoskeletal injury has been estimated based in the Continuous Sample of Working Lives obtained from Social Security Office (Carrillo *et al.*, 2013). Worker collectives at risk in the manufacturing sector are male workers, young workers and low qualified manual workers.

However, from the preventive point of view, it is also important to identify the underlying mechanisms in order to design specific actions to fight the accident causes.

Accident causes should be classified (International Labour Organization 1998) as immediate causes (unsafe acts, unsafe conditions) and contributing causes (safety management performance, mental and physical conditions of workers). Contributing causes are part of the latent causes (Reason, 2000).

The analysis of accident investigations is one of the most used tools to identify those causes, and the barriers to preventing them. There is evidence of the association between the type of causes identified and the accident scenarios (Carrillo-Castrillo *et al.*, 2013).

1.3 *Scope of this research*

This research is based on a cross-sectional analysis of the causes identified in the musculoskeletal accident investigations performed by public officers in Andalusia.

It is important to consider that Andalusia is one of the biggest regions of Europe, and represents approximately 12% of the Spanish manufacturing sector and employs on average more than two hundred thousand workers (European Commission, 2012).

The main purpose of this study is to analyze the causes identified in the musculoskeletal accidents officially investigated in the Andalusia manufacturing sector.

2 MATERIALS AND METHOD

2.1 *Accident investigation reports*

In this study, all of the accidents are musculoskeletal and have been investigated by safety officers. The accidents included occurred from 2004 to 2012.

Data is also available for accident circumstances following the European Statistics on Accidents at Work Methodology, hereinafter ESAW (European Commission 2002). Musculoskeletal accidents are identified with the *Contact* (mode of injury) that is the ESAW variable that describes how the worker was hurt.

In 2004, the Labour Authority in Andalusia adopted a common codification system for the causes identified in official accident investigation.

In the coding system, approved by the National Institute for Safety and Health at Work in Spain, there are 255 possible causes (Fraile 2011).

Codes are defined with four digits, the first digit identifies the group of causes (which is a type of cause), the first two digits identify the subgroup of causes and the other last two digits are the different causes in each subgroup.

For example, cause 7206 is used for "inadequate training/information about risks of preventive measures", classified as group 7 (safety management) and the subgroup 72 (preventive activities).

Group of causes identify a type of causes. Active causes are usually classified in the groups with first digit 1 to 5 and most of the causes classified in the groups 6 to 8 are contributing (Reason 2000).

The number of causes identified per accident investigated is variable, reaching a maximum of nine, but in most cases the number of causes is between three and five. Note that all causes are considered equally except ones marked as the main cause, but in some cases no main cause is considered.

2.2 *Accident scenarios*

Accident scenario is defined by the main characteristics of the job ongoing and the event preceding the contact.

According to ESAW, the accident scenario can be defined using two variables of the sequence of events: *Specific Physical Activity* (activity being performed by the victim just before the accident) and *Deviation* (last event differing from the norm and leading to the accident).

2.3 *Causation patterns*

When there is a main cause, analysis of association is straight forward. For the rest of causes, which may have more than one cause in the same group, the concept of causation pattern is defined.

Causation pattern is the defined by the group of causes with at least one cause identified in the accident investigation. Thus, if at least one of the causes is included within a certain subgroup, then that group of causes is part of the causation pattern.

Code is formed as $G_1G_2G_3G_4G_5G_6G_7G_8G_9$ where each digit is 1 if at least one cause of a group is included and 0 if not. For example, code 001001000 is for an accident with causes identified in Group 3 (Machinery) and Group 6 (Work organization).

2.4 *Analysis at cell level of contingency tables: Phi coefficient*

The association between two categorical variables can be tested with Chi-square.

However, in terms of prevention, what really matters is the relationship between categories, thus the analysis should be performed at cell level.

Cell level relationships can be tested with Phi coefficient. Phi coefficients are calculated between each combination of categories of two variables (Chi *et al.*, 2009).

For dichotomous variables, Phi coefficient at cell level is equivalent to the Phi coefficient for the variable.

Only positive and significant ($p<0.05$) values of Phi coefficients are useful for identifying association between categories. Negative values would indicate that there is little contribution of the category to the accident causation.

3 RESULTS

3.1 *Cross-sectional analysis*

The most prevalent specific causes identified are presented in Table 1, with indication of the times they were identified as the main cause. Only causes identified in at least ten accidents are included.

The number of specific causes identified in each of the group of causes is presented in Table 2, with indication of the times they were identified as the main cause.

The most frequent circumstances in the accident investigated and those in all musculoskeletal accidents in the period are presented in Table 3.

3.2 *Causation patterns*

Most prevalent causation patterns are presented in Table 5.

Table 1. Most prevalent causes identified.

Cause (code)	No. cases	No. cases as main cause
Inadequate work methods (6102)	55	12
Inadequate training or information (7206)	53	4
Causes related to personal factors (8999)	40	17
Inadequate preventive measures (7202)	17	–
Worker overload (6108)	16	11
Overload regarding specifications (6107)	12	–
Hazards not identified (7201)	12	–

Table 2. Distribution of the identified causes in the groups of causes.

Group of Causes (first digit of the code of the cause)	N° of cases*	N° of causes	N° of cases as main cause
Workplace condition (1)	27	37	21
Installations (2)	0	0	0
Machinery (3)	13	17	3
Other equipment (4)	16	19	2
Materials/substances (5)	9	9	4
Work organization (6)	84	123	27
Safety management (7)	79	135	7
Personal factors (8)	58	75	26
Others (9)	18	21	0
Total	153	436	90

*Note that each case can have causes classified in different groups. Cases are all accidents with at least one cause classified in each type of cause.

3.3 *Associations*

All possible associations between categories of the variables have been tested. Only significant and positive Phi coefficients are presented (see Table 6, Table 7 and Table 8).

Most prevalent scenarios in the accident investigated are presented in Table 4. Scenarios are identified with the first digits of the ESAW codes for *Specific Physical Activity* and *Deviation.*

Table 3. Most frequent circumstances in the investigated accidents and in all musculoskeletal accidents in the period.

		No. Accidents	
Variable	Category	Investigated	All
Hours since the beginning of the shift	1–3	80	30,398
	4–5	45	14,837
	6–8	22	12,492
	>8	6	956
Day of the week	Monday	33	15,358
	Weekend	17	2,838
	Rest	103	40,487
Specific Physical Activity	Operating machine	5	3,153
	Working with tools	22	5,914
	Handling objects	50	24,316
	Carrying with hands	26	12,955
	Movement	42	10,725
Deviation	Loss of control	8	2,321
	Slipping, falling	17	1,489
	Movement without physical stress	28	10,068
	Movement with stress	83	41,163

Table 4. Most prevalent scenarios.

Code*	Description	No. of cases	Scenario
47	Body movement with stress when handling objects	32	S1
67	Body movement with stress when moving	12	S2
57	Body movement with stress when carrying by hand	10	S3

*Code is formed with the first digit of ESAW variables of *Specific Physical Activity* and *Deviation*.

Table 5. Most prevalent causation patterns.

Code*	No. of cases	Pattern
000001100	32	P1
000001000	12	P2
000000010	10	P3
000001110	10	P4

*Code is formed with $G_1G_2G_3G_4G_5G_6G_7G_8G_9$ where each digit is 1 if at least one cause of the group of causes is included and 0 if not.

Table 6. Significant associations: Circumstances of the accident and Type of Causes.

Variable	Category	Type of causes	Phi
Specific Physical Activity	Movement	Workplace	0.24**
	Operating machine	Machinery	0.21*
	Operating machine	Other equipment	0.30***

$* p < 0.05$ $** p < 0.01$ $*** p < 0.001$.

Table 7. Significant associations: Circumstances of the accident and Causation Patterns.

Variable	Category	Pattern #	Phi
Deviation	Movement with stress	P1	0.18**
Specific Physical Activity	Handling objects	P2	0.16**
	Carrying by hand	P3	0.16**
Day	Monday	P3	0.31***

$* p < 0.05$ $** p < 0.01$ $*** p < 0.001$.
See Table 5.

Table 8. Significant associations: Circumstances of the accident and Type of the Main Cause.

Variable	Category	Type of the main causes	Phi
Specific Physical Activity	Movement	Workplace	0.24**
	Operating machine	Machinery	0.24**
	Operating machine	Other equipment	0.30***
	Handling objects	Safety management	0.18*
Deviation	Slipping, Falling	Workplace	0.22**

$* p < 0.05$ $** p < 0.01$ $*** p < 0.001$.

4 DISCUSSION

Analysis of accident causes is a useful tool in occupational safety (Khanzode *et al.*, 2012). Learning from cases where prevention has already failed can show potential areas of improvement. Moreover, analysis of differential causation can give an insight in the risk factors and the potential preventive measures to reduce them.

The focus of this research is on the mechanism of injuries classified as musculoskeletal. Although there are risk factors affecting the progressive harm produced in the musculoskeletal system and eventually lead to an injury classified as an accident, there are also specific causes that intervene in sudden damages that can be prevented.

The prevalence of musculoskeletal accidents is so high that any measure to prevent these accidents is a priority for any safety program.

4.1 *Causes of Musculoskeletal Accidents*

The proportion of contributing causes (76%), even among those identified as the main cause (66%), is very high. This result has been found previously and reveals that root causes in manufacturing have a very important role (Jacinto *et al.*, 2009; Katsakiori *et al.*, 2012; Carrillo-Castrillo *et al.*, 2013).

Moreover, the most prevalent patterns only include contributing causes. The two first patterns, including organizational and safety management causes, explain 28% of the accidents investigated. The next two patterns, which include personal factors, explain another 10% of the accidents investigated.

Inadequate work methods and inadequate training are the two more prevalent causes, present in 52% of the accidents investigated showing the link between work methods and training (Katsakiori *et al.*, 2010). The first finding is that the most likely effective preventive measures should be those aimed to improve work methods and training of workers.

Musculoskeletal accidents investigated have certain prevalent scenarios that need to be investigated. Half of the accidents involved movement with or under stress in physical activities such as handling, carrying and moving.

4.2 *Preventive action identification*

One of the first applications of the results presented is the identification of preventive actions.

Association of physical activities and active causes has been found. Accidents when worker is moving or related to slips or falls have a significant association with workplace causes. Measures should focus in the housekeeping activities to prevent them. In the same way, accidents operating machinery are related to the condition of machines and equipment.

When the activity is handling, an association with safety management have been found. Risk assessment and training of workers are the most likely effective measures according to the accidents investigated.

Finally, human factors and worker causes are associated with certain activities carrying by hand. Also, a significant association was found with Mondays.

Monday is the day with higher prevalence of accidents in general and of musculoskeletal kind too. The analysis show that most of accidents on Mondays have only causes related to human factors. Although there are other explanations related to cultural issues that should be deeply research (Camino *et al.*, 2011), it seems that safety measures should focus on that human factors to reduce injury rates on Mondays.

5 CONCLUSIONS

Contributing causes are a serious concern for safety practitioners. According to the accident investigated, the real challenge for musculoskeletal accidents prevention is to fight organizational and personal risk factors. Although the safety improvements to equipment are still needed, the most crucial areas to target are work organization, safety management and personal factors.

The findings of differential causation in terms of the individual differences indicate that preventive actions should be oriented to the most likely causes.

The data suggests that there are areas of improvement such as workplaces and housekeeping, training and work methods.

Another important missing issue is the risk factors of MSD in the jobs of the injured workers and a clear differentiation of the injuries as a sudden event without previous processes and those accidents with a previous deterioration in the musculoskeletal system caused by the already known epidemiologic risk factors.

These results apply to Andalusian manufacturing sector. In other regions or sectors other areas of improvement based on the accident causation can be identified with the methodology used.

REFERENCES

Camino López, M.A., Fontaneda, I. González Alcántara, O.J. & Ritzel, D.O. 2011. The special severity of occupational accidents in the afternoon: "The lunch effect". *Accident Analysis and Prevention* 43(3): 1104–1116.

Carrillo, J.A., Gómez, M.A. & Onieva, L. 2012. Safety at work and worker profile: analysis of the manufacturing sector in Andalusia in 2008. *Occupational Safety and Hygiene – SHO*. Guimaraes: Portuguese Society of Occupational Safety and Hygiene.

Carrillo, J.A. & Onieva, L. 2012. Severity Factors of Accidents: Analysis of the Manufacturing Sector in Andalusia. *Occupational Safety and Hygiene – SHO*. Guimaraes: Portuguese Society of Occupational Safety and Hygiene.

Carrillo, J.A., Rubio-Romero, J.C., López-Arquillos, A. & Onieva, L. 2013. The role of worker individual characteristics in the severe injuries causation in the Andalusian manufacturing sector. *Occupational Safety and Hygiene – SHO*, Guimaraes: Portuguese Society of Occupational Safety and Hygiene.

Carrillo-Castrillo, J.A., Rubio-Romero, J.C. & Onieva, L. 2013. Causation of Severe and Fatal Accidents in the Manufacturing Sector. *International Journal of Occupational Safety and Ergonomics* 19(3): 423–434.

Caicoya, M. & Delclos, G.L. 2010. Work demands and musculoskeletal disorders from the Spanish National Survey. *Occupational Medicine* 60: 447–450.

Chi, C.-F., Yang, C.-C. & Chen, Z.-L. 2009. In-depth accident analysis of electrical fatalities in the construction industry. *International Journal of Industrial Ergonomics* 39(4): 635–644.

European Commission. 2002. *European Statistics on Accidents at Work (ESAW) – Methodology (ed. 2001)*. Luxembourg: Office for Official Publications of the European Communities.

European Commission. 2012. Eurostat: Your Key to European Statistics. http://epp.eurostat.ec.europa.eu/.

Fraile, A. 2011. NTP924: Causas de accidentes: clasificación y codificación. Madrid: Instituto Nacional de Seguridad e Higiene en el Trabajo.

International Labour Organization. *Encyclopaedia of Occupational Health and Safety: Part VIII, Chapter 56*. SW: International Labour Organization.

Jacinto, C., Canoa, M. & Guedes, C. 2009. Workplace and organizational factors in accident analysis within the Food Industry. *Safety Science* 47(5): 626–635.

Katsakiori, P., Kavvathas, A., Athanassiou, G., Goutsos, S. & Manatakis, E. 2010. Workplace and Organizational Accident Causation Factors in the Manufacturing Industry. *Human Factors and Ergonomics in Manufacturing and Service Industries* 20(1): 2–9.

Khanzode, V.V., Maiti, J. & Ray, P. 2012. Occupational injury and accident research: A comprehensive review. *Safety Science* 50(5): 1355–1367.

Malchaire, J., Cock, N. & Vergracht, S. (2001), "Review of the factors associated with musculoskeletal problems in epidemiological studies", International Archives of Occupational and Environmental Health, 74(2): 79–90.

National Institute of Occupational Safety and Health (1997). Musculoskeletal Disorders and Workplace Factors. A Critical Review of Epidemiologic Evidence for Work-Related Musculoskeletal Disorders of the Neck, Upper Extremity, and Low Back. Cincinnati: DHHS (NIOSH) Publication No. 97-141.

Reason, J. 2000. Human errors: models and management. *British Medicine Journal* 320: 768–770.

Occupational Safety and Hygiene II – Arezes et al. (eds)
© 2014 Taylor & Francis Group, London, ISBN 978-1-138-00144-2

Assessing the inter-analyst reliability of the RIAAT process

P.F. Cordeiro
Systems Department—Quality, Safety and Hygiene at work and Environment, Telcabo—Telecomunicações e Electricidade Lda, Lisboa, Portugal

C. Jacinto
UNIDEMI—R&D Unit in Mechanical and Industrial Engineering, Faculty of Sciences and Technology, Universidade Nova de Lisboa, Portugal

F.P. Santos
Centre for Marine Technology and Engineering (CENTEC), Instituto Superior Técnico, University of Lisbon, Portugal

ABSTRACT: This paper presents a preliminary assessment of the inter-analyst reliability of the RIAAT procedure, which is a recent tool developed for the Recording, Investigation and Analysis of Accidents at Work. The study involved 5 analysts who applied the RIAAT procedure to a set of 11 accidents at work occurred in a Portuguese company. The study focused on 8 nominal variables considered as key-variables within RIAAT. Reliability was measured with three coefficients for calculating the level of agreement between analysts. Overall, the results showed a low inter-analysts reliability for all variables tested; the first 4 (Eurostat variables included in RIAAT) held a reliability level almost acceptable. The remaining 4 variables (RIAAT-specific) led to lower agreements; this may be explained by their use in the analysis and coding of more "distant" causal factors. The authors argue for the benefit of having a well-trained team of investigators rather than one person, as it enhances reliability of information.

1 INTRODUCTION AND BACKGROUND

Valid information is essential for the prevention of accidents. In content analysis, the concepts of reliability and validity are interconnected; it is argued that reliability is always a precursor of validity (Potter & Levine-Donnerstein 1999). Coded data obtained from records, either textual or audible, are typically generated by analysts. Interpretation and conclusions from such data can be trusted only after establishing their reliability (Lombard *et al* 2002, Hayes & Krippendorff 2007).

Krippendorff (2004a) distinguishes three types of reliability: *stability*, *reproducibility* and *accuracy*; of these, the most frequently used is reproducibility, which is also called intercoder or inter-analyst reliability. The concept is defined by Krippendorff (2004a, p. 215) as "*the degree to which a process can be replicated by different analysts working under varying conditions, at different locations, or using different but functionally equivalent measuring instruments*"; in short, the inter-analysts reliability measures the *level of agreement* between different people performing the same analysis.

There is a wide variety of reliability coefficients and the specialty literature also provides discussions on the most appropriate to use (Craggs & Wood 2005, Hayes & Krippendorff 2007). Potter & Levine-Donnerstein (1999) discuss the differences between four of the most popular coefficients: percent agreement (%-agreement), Scott's π, Cohen's κ and Krippendorff's α. They encourage the use of coefficients that take into account the agreement that would be obtained by chance (i.e. chance-corrected).

This study used three reliability coefficients, namely, the %-agreement, Scott's π and Krippendorff's α, following the suggestion of using more than one coefficient (Lombard *et al* 2002, Taylor & Watkinson 2007). Unlike π and α, the %-agreement is not chance-corrected but it was included because it is easy to calculate.

The aim of this work was to make a preliminary inter-analysts reliability assessment of an investigation tool; the study object was the RIAAT process (Recording, Investigation and Analysis of Accidents at Work), in terms of its analytical procedure and embedded coding system (Jacinto *et al* 2010a, 2011). This process covers the complete cycle of accident information. The motivation lays on the fact that RIAAT is a recent approach, which still needs validation in terms of usability and reliability.

The study covered 11 accidents at work (coding units) that occurred in 2010–2011 in a Portuguese company, *Telcabo*, dealing with the construction, installation and operation/maintenance of telecommunications and energy networks.

2 RIAAT PROCESS

The RIAAT tool allows processing accident or incident information (input) into an organisational learning practice (output) as a result of the investigation and analysis of an event, in order to promote constant improvement of Occupational Safety and Health (OSH). This process is divided into four sequential parts covering the whole cycle of accident information in the following order: I) Recording; II) Investigation and Analysis; III) Plan of Action; IV) Organisational Learning (Jacinto *et al* 2010a, 2011).

Part I consists of recording the accident information and the active failures leading to it. This part is aligned with the European Statistics on Accidents at Work (ESAW) methodology (Eurostat, 2001) implemented in all European Union (EU) countries. The second part comprises a multi-layer investigation and analysis of the accident, to identify its causes and the relevant factors that have contributed to the occurrence. It is explicitly based on Reason's (1997) model of organisational accidents, which establishes three main levels of concern: the *person*, the *local workplace* and the *management*. Part III is designed to check previous risk assessment and to establish a plan of action that allows making corrections and improvements. Finally, Part IV is designed to create and disseminate organisational learning, ensuring that the relevant lessons and knowledge are used by the organisation (Jacinto *et al* 2010a, 2011).

3 METHODOLOGY

This section explains the study design. Fieldwork was carried out at *Telcabo* and it involved 5 participants (analysts) from its OSH staff.

3.1 *Data and sampling process*

The study base consisted of 11 accidents at work occurred in 2010–2011. They were selected by the first author upon taking into account their severity and learning potential. Thus, sample size was N = 11. The participants analysed these 11 accidents independently, and then coded the key information selected for this study; these eleven cases and their corresponding RIAAT records became the *coding units* (CU) of this inter-analyst reliability study. Each analysis produced coded information, and, depending on the factor analysed, three *coding instruments* (or protocols) were used for the coding task, namely:

- Classification schemes of the ESAW methodology (Eurostat, 2001), used to produce harmonised accident statistics within EU countries;
- Classification schemes from Reason (1997), who divides human failures into errors and violations;
- Classification schemes specific to RIAAT, for identifying underlying causal factors (Jacinto *et al* 2010a, 2010b).

3.2 *Study design*

A set of 8 nominal variables were chosen, as *key-variables*, to assess the level of reproducibility (inter-analysts reliability) of the RIAAT procedure.

- Four harmonised ESAW variables: Deviation, Material Agent of Deviation (MAD), Contact, and Material Agent of Contact (MAC).

The accident information pertaining to these variables was recorded in part I of each RIAAT form. They were chosen because they are fundamental for the characterisation of the accident itself, i.e., the last deviant event from normality and leading to the accident (Deviation); the mode of injury (Contact) which describes how the victim was injured; and the material agents (MAD, MAC) associated with both variables (Eurostat 2001).

The information related to the next four variables was recorded in part II of the RIAAT form.

- One variable for Human Failures (HF) adopted from Reason (1997). It divides human failure (unsafe acts) into two main categories: *Errors* and *Violations*. Additionally, errors are subdivided into slips and lapses, or mistakes. The former are non-intentional, whereas mistakes are intentional.
- Three variables specific to the RIAAT procedure: Individual Contributing Factors (ICF), Local Workplace Factors (LWF) and Organisational and Management Factors (OMF).

The ICF are *individual factors* that can influence errors and behaviours; they are divided into temporary (e.g. memory, fear and threats, distraction, fatigue, pain, discomfort, physical/mental stress) and permanent factors (permanent physical or psychological condition, personality). The LWF are *workplace characteristics* which can influence people's behaviour (e.g. physical environment, equipment, job and task related factors, competence, communication practices, and weather/external conditions). Finally, OMF are *organisational conditions*, i.e., weaknesses that might have facilitated the occurrence.

As mentioned, the study involved 5 analysts, four of which received a short training session (4hr) given by the most experienced member of the group. The study objective and the RIAAT procedure were explained to the participants. They received the RIAAT user's manual and form (Jacinto *et al* 2010b), plus a copy of the codes and definitions of all variables. One demonstration case, other than the 11 selected, was discussed within the group. The analysts were asked to read/study the support materials within a week, and were informed that the next step would be to interview each one of the 11 workers who suffered the accidents. The interviews took place in three non-consecutive mornings, and were conducted by the leading member. In methodological terms the main purpose was to guarantee that each analyst was "exposed" to the same "stimulus", i.e., exposed to the same information and means of communication. Afterwards, the analysts were asked to fill-in the forms (11 cases) and to code the 8 variables without sharing opinions with one another.

The variables were coded at their category level. In the case of the variables ICF, LWF and OMF, for which several categories (factors) may apply simultaneously, the participants were asked to select and code only the "two most important" factors in their own opinion (no order included).

3.3 *Reliability coefficients*

Three of the most often used reliability coefficients were selected to estimate the inter-analysts reliability (agreement level). These were the %-agreement, Scott's π and Krippendorff's α, which were calculated with the following software applications:

- PRAM (freeware application; no longer available on the internet) - Scott's π;
- ReCal3 (Freelon 2010) - %-agreement and Krippendorff's α.

3.4 *Demographics (analysts)*

The five participants from *Telcabo* included three Health & Safety Senior Technicians (two have engineering background and similar OSH experience, and the third has a different background and no experience on the field), one Medical doctor and one specialised Nurse. None of them had previous experience in accident investigation. They were 3 women and 2 men, aged from 34 to 46 years old and have been in the company from 1 to10 years. All participants analysed the 11 cases independently and coded the information (8 variables) into the RIAAT forms.

4 INTER-ANALYST RELIABILITY RESULTS

4.1 *Inter-analyst reliability of the ESAW variables*

The inter-analysts reliability of the four ESAW variables was estimated at the category level. Jacinto *et al* (2010c) refer that it is more important to code the ESAW variables at the category level—rather than subcategory—because only the category is used for the production of official statistics.

The coding of each variable resulted in 55 judgements (11 CU × 5 coders) and in 110 paired combinations of values (C_2^m × 11 CU; m = number of coders). In total, 220 judgments were made giving 440 pairs of values.

Table 1 shows three reliability values for the variable Material Agent of Deviation (MAD), which were estimated using the %-agreement, Scott's π and Krippendorff's α respectively. This procedure was performed for the eight variables.

Both %-agreement and Scott's π can only be calculated based on pairs of judgments. Thus, their values are the average of all pairwise results for each nominal variable (Lombard *et al* 2002, Krippendorff 2004a), as can be seen in Table 1. The %-agreement consists of the number of observed pairwise agreements divided by the total number of pairwise decisions. Unlike Scott' π, the %-agreement is not chance-corrected i.e., it does not take into account the agreement that would be expected by chance. Consequently, %-agreement overestimates reliability and can be misleading. This is the reason why its value is way higher than π. Although easy to compute, %-agreement is not considered suitable for assessing coding reliability in content analysis (Krippendorff 2004a, Craggs & Wood 2005).

Looking at the pairwise values of Scott's π in Table 1 there are differences between them. These are due to the number of categories selected by each pair of analysts and to the collective distribution of their judgments over such categories. Moreover, there are six pairs (in grey) which contributed to lower the average result of the coefficient; all pairs in which analyst 2 is present contributed to this situation. It must also be pointed out that analysts 4 and 5 reached an agreement of 100% i.e., perfect agreement. According to Krippendorff (2004b), this is as statistically unexpected as full disagreement is, hence, it must not occur. It may be possible that these two analysts did not comply with the criterion for not sharing information during the process. In contrast, high disagreement situations may unveil insufficient training.

Krippendorff's α is also chance-corrected, but unlike the two previous coefficients, it can be estimated for any number of coders simultaneously, any

Table 1. Intercoder reliability (%) of the MAD variable estimated using the %-agreement, Scott's π and Krippendorff's α.

		Pairs of analysts										
Software	Coefficients	1,2	1,3	1,4	1,5	2,3	2,4	2,5	3,4	3,5	4,5	Value
ReCal3	%-agreement	45.5	54.5	36.4	36.4	45.5	36.4	36.4	63.6	63.6	100	51.8
PRAM	Scott's π	29.4	39.2	11.5	11.5	30.9	18.1	18.1	51.1	51.1	100	36.1
ReCal3	Krippendorff's α					N.A.						37.7

Table 2. Acceptability criteria for Reliability (Krippendorff 2004a, Krippendorff & Bock 2009).

Krippendorff's α	Criteria
≥ 0.8	Reliable data. Reproducibility is guaranteed.
[0.667, 0.8]	Reliability non-acceptable to make conclusions. It does not guarantee reproducibility.
< 0.667	Unreliable data.

sample size and can deal with other levels of measurement than nominal, such as, ordinal, interval and ratio (Lombard *et al* 2002, Krippendorff 2004a).

One can see both α and π indices as being complementary; α holds the above mentioned unique merits (and a few others), making it eligible for "the most adequate" measure to estimate reliability (Craggs & Wood 2005, Hayes & Krippendorff 2007, Taylor & Watkinson 2007). In turn, π enables detecting coders with very different performances and, consequently, to perceive training needs.

The acceptable level of agreement depends on the phenomena being studied and purpose of the assessment; most importantly, it should minimise the risk of drawing conclusions out of unreliable data.

A proposal for acceptance criteria is given in Table 2 (Krippendorff 2004a, Krippendorff & Bock 2009).

Figure 1 depicts the inter-analysts reliability levels estimated for the four ESAW variables.

As expected, the %-agreement was always higher than the other two (because is not chance-corrected). With regard to the other coefficients (Krippendorff's α and Scott's π), their values were of similar magnitude and showed a low reliability level for all four variables under scrutiny (or unreliable if criterion ≤0.667 is applied); agreement was somewhat better for Deviation and Contact, as compared with their respective Material Agents. Despite the small sample size, these levels are corroborated by Jacinto *et al* (2010c) in which several coders were also OSH staff. The same variables held significantly higher values when the coding was performed by a "reference group" (government workers who do this job regularly). The 2010 study clearly demonstrated that training and experience was paramount.

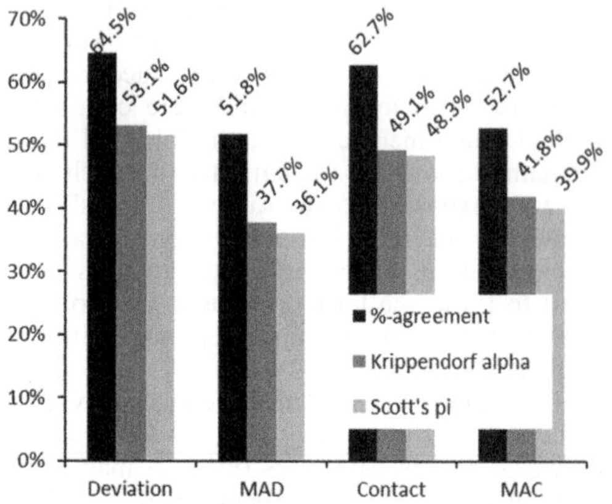

Figure 1. %-agreement, Krippendorff's α, and Scott's π for the 4 ESAW variables.

4.2 *Inter-analysts reliability of the variables HF, ICF, LWF and OMF*

The %-agreement, Krippendorff's α and Scott's π results of the variables, Human Failure (HF), Individual Contributing Factors (ICF), Local Workplace Factors (LWF) and Organisational and Management Factors (OMF) are shown in Figure 2.

The inter-analyst reliability of this second set of variables was much lower than those of the ESAW; this means higher levels of disagreement between analysts, which was someway expected.

This may have been influenced by the "latent" content of the data. As opposed to "manifest" content, in which data are on the "surface" (Potter & Levine-Donnerstein 1999) and are objective and easily identifiable (Taylor & Watkinson 2007), latent content demands focus on the meaning underlying the elements on the surface (Potter & Levine-Donnerstein 1999). Thus, latent data requires more interpretation from the analysts, which is indeed the case with Part II—RIAAT, which includes the variables of Figure 2.

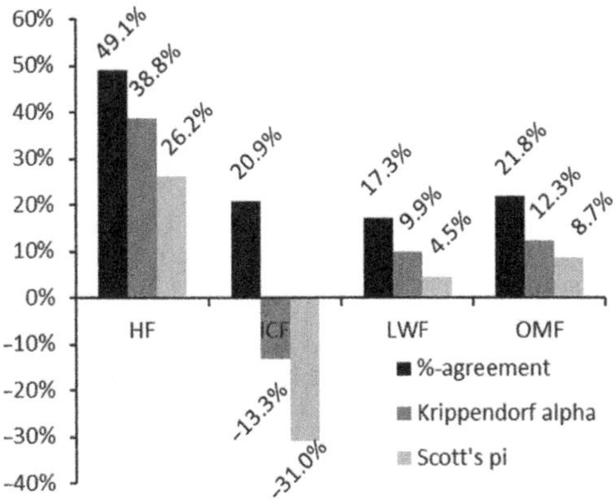

Figure 2. %-agreement, Krippendorff's α, and Scott's π for the variables HF, ICF, LWF and OMF.

Causal factors regarding these four variables were inferred from the interviews, and they incorporate perceptions and judgements, thus the subjectivity is always much higher at this stage of the analysis. Within these variables, the further "away" the analyst gets from the accident event, the higher the subjectivity. There are no reference values for comparison, but it is expected that a larger sample size and more training (and experience) in accident investigation would improve agreement.

However, the results for Individual Contributing Factors (ICF) are a paradox and there is no explanation for it, since the observed agreement is lower than the expected agreement due to chance. This is why both α and π returned negative values.

Both Scott's π and Krippendorff's α are chance-corrected. They express reliability values above what might be expected due to chance. However, they differ in the way they calculate the agreement, thus the difference between their values. In what concerns α, the expected agreement for a category *c* is estimated multiplying (n_c/n) *by* $(n_c - 1)/(n - 1)$, where n_c and *n* are the overall frequency of *c* judgments and the total number of judgments made by coders, respectively. Instead, Scott's π uses $(n_c/n)^2$. Moreover, Krippendorff's α is corrected for small sample sizes (Krippendorff 2004a), while π is not. Consequently, and in the case of nominal data and two coders, the π value tends asymptotically to the α value as the sample size increases (Krippendorff 2004a).

Overall, these results mean that coded data of the variables in Figure 2 are unreliable. Indirectly, it might also suggest that the "in-depth" analysis described in RIAAT (Part II) may lead to different conclusions, depending on the individual analyst performing the analysis. This is likely to happen with most methods for accident investigation, thus the recommendations in the literature for appropriate training and team investigation.

5 CONCLUSION

This paper has presented a study on the inter-analyst reliability of eight variables of the RIAAT process and discussed the results. It was not possible to derive significant conclusions about the reliability of this process, mainly due to some methodological limitations, such as:

- Small sample size (N = 11 accidents analysed);
- Insufficient training of four of the five participants/analysts;
- Analysts working individually, with different technical backgrounds (engineers, administrative personnel, medical doctor and specialised nurse);
- Analysts were asked to choose only two categories in some variables, which may have reduced the level of observed agreement.

Despite these limitations, the four ESAW variables reached reliability levels almost acceptable; moreover, when the analysts have low experience, as was the case here, these results are corroborated by a previous study in which the sample size was significantly larger (N = 100), therefore more robust.

In what concerns the other four variables of the RIAAT process, which are used in the analysis of the influencing factors and distant "latent" causes, the results show levels of inter-analyst reliability much lower than those of the ESAW variables. Considering the overall results, this kind of evolution is not surprising, since the analysis (inference process) "moves" away from the immediate (observable) causes of the accident; thus, the likelihood of different interpretations is greater as the subjectivity gets higher. The worse results (notorious disagreement) were found for the Individual Contributing Factors (ICF) and this may be due to the methodological limitations. This first attempt suggests that the study should continue, providing that the authors are able to solve the previously mentioned limitations, namely by comparing several well-trained teams (made of different backgrounds and experience) and by analyzing the effect of the number of categories in the level of agreement. Future assessments will determine to what extent, and under what conditions, RIAAT can promote reliable accident data. This would impact positively on the decision-making of safety professionals, managers and researchers.

ACKNOWLEDGMENTS

The authors are grateful to *Telcabo* and its personnel for participating in this study, especially to Systems Director Luís F. Coelho, for his helpful contribution.

REFERENCES

Craggs, R. & Wood, M.M. 2005. Squibs and Discussions—Evaluating Discourse and Dialogue Coding Schemes. *Computational Linguistics* 31(3):289–296.

Eurostat, 2001. European Statistics on Accidents at Work (ESAW)—Methodology. European Commi., Luxembourg.

Freelon, D.G. 2010. ReCal: Intercoder Reliability Calculation as a Web Service. *International Journal of Internet Science* 5(1):20–33.

Hayes, A.F. & Krippendorff, K. 2007. Answering the Call for a Standard Reliability Measure for Coding Data. *Communication Methods and Measures* 1(1):77–89.

Jacinto, C., Guedes Soares, C., Fialho, T. & Silva, S.A. 2010a. A new process for managing accident information and improving safety. In: Arezes *et al* (eds), *International Symposium on Occupational Safety and Hygiene—SHO*: 285–289. Guimarães, Portugal, 11–12 Feb. 2010.

Jacinto, C., Guedes Soares C., Fialho, T. & Silva, S.A. 2010b. RIAAT—Users Manual, Rev.1.1. Unpublished (http://www.mar.ist.utl.pt/captar/riaat. aspx).

Jacinto, C., Santos, F.P., Fialho, T., Guedes Soares, C. & Silva, S.A. 2010c. Reliability of coding of a set of ESAW variables within the Portuguese official system—Research Report, sub-task 2.3 of Project CAPTAR, ref: PTDC/SDE/71193/2006 (unpublished research report).

Jacinto, C., Guedes Soares, C., Fialho, T. & Silva, A.S. 2011. The Recording, Investigation and Analysis of Accidents at Work (RIAAT) process. Policy and Practice in Health and Safety 9(1):57–77.

Krippendorff, K. 2004a. *Content Analysis: An Introduction to Its Methodology*. 2nd Edition. Thousand Oaks, CA: Sage.

Krippendorff, K. 2004b. Reliability in Content Analysis: Some Common Misconceptions and Recommendations. *Human Communication Research* 30(3):411–433.

Krippendorff, K. & Bock, M.A. 2009. *The Content Analysis Reader*. Thousand Oaks, CA: Sage Publications.

Lombard, M., Snyder-Duch, J. & Bracken, C.C. 2002. Content Analysis in Mass Communication—Assessment and Reporting of Intercoder Reliability. *Human Communication Research* 28(4):587–604.

Potter, J.W. & Levine-Donnerstein, D. 1999. Rethinking validity and reliability in content analysis. *Journal of Applied Communication Research* 27(3):258–284.

Reason, J. 1997. *Managing the risks of organisational accidents*. Aldershot: Ashgate Publishing Ltd.

Taylor, J. & Watkinson, D. 2007. Indexing Reliability for Condition Survey Data. *The Conservator volume* 30: 49–62.

Occupational Safety and Hygiene II – Arezes et al. (eds)
 ISBN 978-1-138-00144-2

Polymeric nanocomposites production risk assessment using different qualitative analyses

S.P.B. Sousa & M.C.S. Ribeiro
Institute of Mechanical Engineering and Industrial Management (INEGI), Oporto, Portugal

J. Santos Baptista
Research Laboratory on Prevention of Occupational and Environmental Risks (PROA/LABIOMEP), Faculty of Engineering, University of Oporto (FEUP), Oporto, Portugal

ABSTRACT: Polymer nanocomposite global market is expected to reach $3000 billion, by 2015. The rising production and consumption inspire some concerns as the polymer nanocomposite production can expose workers to new risks associated to the nanomaterials. The present investigation is aimed at studying the risk assessment in the polymer nanocomposites production by using several qualitative risk methods. For this purpose, qualitative analyses based on the following methodologies ANSES, CB Nanotool, EPFL, GWSNN, ISPESL and PMSN were applied and compared. Divergent risk levels were obtained in the qualitative methods; however, all methods consider that the pre-production stage is the one with the higher risk. These risks assessment criteria must be reviewed and complemented with quantitative methods in order to fill the existing gaps.

1 INTRODUCTION

Polymer nanocomposite (NC) is a compound in which one or more constituent materials at nanoscale size are completely dispersed in the polymer (Aitken *et al.* 2006). Almost all NC materials have new and improved properties when compared with their equivalents, macro and micro composites (Šupová *et al.* 2011, Zou *et al.* 2008). It is expected, by 2015, that the NCs global market reaches $3000 billion, since nanomaterials (NMs) are increasingly playing a decisive role in diverse market sectors (Kiliaris and Papaspyrides 2010). However, the rising production and consumption of NCs based materials also lead to new concerns as the polymer NC production can expose workers to new risks associated to the NMs. Exposure can occur by inhalation, accidental ingestion and absorption through the skin during the work (Crosera *et al.* 2009, Gupta 2011). Thus, with the nanotechnology based products in rapid growth, researchers, manufacturers, regulators and consumers are increasingly concerned about the potential safety impacts that these products may possibly have (ISO 2011).

Until now there is no specific legislation for NMs, but recently, it started to appear some ISO, ASTM and BSI related standards. Most of these regulations can be interpreted as the transposition of the chemical procedures to the NMs (Amoabediny *et al.* 2009, INRS 2011, Kaluza *et al.* 2009, Lövestam *et al.* 2010, MESD 2006).

The present study aims to analyse the application of different qualitative risk assessment methods to the polymeric NCs production at laboratory scale. This study has been developed under a national Portuguese project (PTDC/ECM/110162/2009) in the Institute of Mechanical Engineering and Industrial Management (INEGI).

2 MATERIALS AND METHOD

The NCs were produced by mixing an unsaturated polyester polymer matrix with 5 wt% of Al_2O_3 (45 ηm) and $Mg(OH)_2$ (15 ηm) nanoparticles. Both manual stirring and ultrasound sonication techniques were used in the mixing process.

The production only involved a single researcher who used personal protective equipment (a pair of nitrile gloves, three latex gloves, a mask with A_2P_3 filters, protection glasses and tyvek protective cloth) and collective protection (fume hood and ventilation/general exhaust).

The risk assessment involved in the manufacturing process was made by using the qualitative analyses (Table 1) based on:

1. French Agency for Food, Environmental and Occupational Health & Safety method (ANSES);

Table 1. Risk assessment methods parameters.

Parameters/ method	Physical	Health	Chemical	Exposure
1) ANSES	X	X	X	X
2) CB Nanotool	X	X		X
3) EPFL	X			X
4) GWSNN	X	X		X
5) ISPESL	X	X	X	X
6) PMSN	X		X	X

2. Control Banding Nanotool (CB Nanotool),
3. Ecole Polytechnique Fédérale de Lausanne method (EPFL);
4. Guidance working safely with nanomaterials and nanoproducts (GWSNN);
5. Istituto Superiore per la Prevenzione e la Sicurezza del Lavoro, Italy method (ISPESL); and
6. Precautionary Matrix for Synthetic Nanomaterials (PMSN).

2.1 *Environmental and occupational health & safety method (ANSES)*

ANSES is a risk assessment method by control bands. The risk values are obtained by overlapping the hazard bands and the exposure bands (Riediker *et al.* 2012).

The hazard bands are defined according to the severity level of the hazard resulting from the analysis of the available information of similar chemicals. The hazard levels may assume five classifications from HB_1—very low (no significant risk to health) to HB_5—very high (severe hazard requiring a full hazard assessment by an expert). These classifications can be aggravated with increment factors which aim to mitigate uncertainties regarding the assumed toxicity of the NMs (Riediker *et al.* 2012).

The exposure bands are defined according to the NM emission potential, whether raw or included in a matrix. The physical form is a key parameter to be considered in order to assess the NMs emissivity from the product, and hence, the potential exposure level of the operator when handling the product. The exposure bands can assume four levels: EP_1—solid; EP_2—liquid; EP_3—powder; and EP_4—aerosol (Riediker *et al.* 2012).

From the resultant matrix, control level can be defined that corresponds to technical solutions for collective prevention to be implemented at the workplace. The suggested control levels correspond to: CL_1—natural or mechanical general ventilation; CL_2—local ventilation; CL_3—enclosed ventilation; CL_4—Full containment; and CL_5—full containment and review by a specialist required (Riediker *et al.* 2012).

2.2 *CB Nanotool*

CB Nanotool is a four by four factors matrix that relates severity parameters on one axis and probability parameters on the other (Zalk 2009).

The severity parameters consider that the physicochemical and general properties of NMs are often unknown. The problem is solved by adding information about the parent material (PM). This information is generally much more available and can include input factors in a more appropriate way for each weighing factor. The considered severity factors for NMs are: surface chemistry, particle shape, particle diameter, solubility, carcinogenicity, reproductive toxicity, mutagenicity, dermal toxicity and asthmagen; for PM are: toxicity, carcinogenicity, reproductive toxicity, mutagenicity, dermal hazard potential and asthmagen. The overall severity score is determined based on the sum of all the points from the severity factors (Zalk 2009).

The probability axis fits with traditional information. The probability scores are based on factors determining the extent to which employees may be potentially exposed to NMs, analysing: estimated amount of NMs used during operation, dustiness/mistiness, number of employees with similar exposure, frequency of operation and duration of operation (Zalk 2009).

The obtained control bands by risk level can be classified in: RL_1—general ventilation, RL_2—fume hoods or local exhaust ventilation, RL_3—containment, and RL_4—seek specialist advice (Zalk 2009).

2.3 *EPFL risk assessment method*

EPFL method consists in a decision tree for “nano-laboratories” with three risk classes, which correspond to similar approaches applied to other hazards types (biological, chemical or radiation). This decision tree analyses the established collective protection measures (*closed milieu*), NMs form/state (fibre, powder, suspension, and matrix), handling typology (production or use), NMs quantity use, possibility to release dust or aerosol and NMs agglomeration ability. The risk classification can be $Nano_1$ (low), $Nano_2$ (medium) and $Nano_3$ (high). With the risk classification it can be defined several safety measures: technical, organizational, personal and cleaning management (Groso *et al.* 2010).

2.4 *Guidance working safely with nanomaterials and nanoproducts*

GWSNN risk assessment method analyses different scenarios through a three by three decision matrix, informing the policy options and procedures to guarantee safe working conditions with NMs. The hazard category can be classified as:

1—soluble nanoparticles (solubility > 100 mg/l), 2—synthetic, persistent NMs (non-fibrous) and 3—fibrous, nonsoluble NMs for which asbestos-like effects cannot be ruled out. The exposure classification is made based on the NMs potential exposure in the different activities related with the polymeric NCs production: I—no emission of free nanoparticles due to working in full containment, II—emission of nanoparticles embedded in a matrix is possible, and III—emission of free nanoparticles is possible. The recommended control measures are: A—applying sufficient (room) ventilation, and when required, local exhaust ventilation and/or containment of the emission source and use appropriate personal protective equipment; B—according to the hierarchic Occupational Hygienic Strategy, the technical and organizational feasible protective measures are evaluated on their economic feasibility (control measures will be based on this evaluation); and C—the hierarchic Occupational Hygienic Strategy will be strictly applied and all protective measures that are both technically and organizationally feasible will be implemented (Cornelissen *et al.* 2011).

2.5 *ISPESL risk assessment method*

The ISPESL risk assessment method is based on 10 factors: A) numerousness of the exposed workers, B) frequency of exposure, C) frequency of direct manipulation, D) dimensions of the nanoparticles, E) nanoparticles behaviour (e.g. dispersion or agglomeration), F) effectiveness of Personal Protection Devices (PPD) used, G) work organization/procedures, H) toxicological characteristics of the substances, I) risk of fire and explosion, and J) suitability of workspaces and installations. The aforesaid factors are denominated "factors level risk" and each one of them may assume three increasing values: 1 (low), 2 (medium) and 3 (high), referred to as "risk levels" (Giacobbe *et al.* 2009).

Since the use of NMs presents uncertainness about danger level, the risk assessment takes into consideration these aspects through the index denominated "corrective factor". This index assumes a value within the range 0.5 and 2.0 in accordance to the established level of scientific knowledge. This assumes the following values: 0.5—good scientific knowledge; 1.0—enough scientific knowledge; 2.0—insufficient scientific knowledge (Giacobbe *et al.* 2009).

The evaluation risk (Eq. 1) is calculated through the factor level risk (flr) sum (from A to J) and multiplied by the corrective factor (cf). The evaluation result consists in several risk levels subdivided in an increasing way (risk level: "low" 5–15, "medium" 16–35, "high" 36 to 60) (Giacobbe *et al.* 2009).

$$Evaluation\ Risk = \sum_{i=A}^{J} (flr)_i \times cf \qquad (1)$$

2.6 *Precautionary matrix for synthetic nanomaterials*

The PMSN estimates the precautionary need that represents the relation between the parameters: "Nano-relevance according to the precautionary matrix" (N), potential effect (W), potential exposure of humans or inputs into the environment (E) and "Specific framework conditions" (S) (Höck *et al.* 2011).

The precautionary matrix should be completed according to the following procedures: 1) making an inventory of materials/products/applications; 2) checking the nano-relevance; 3) finding and dividing up (process) steps; 4) positioning each (process) step found in the value chain; 5) inserting general information in the relevant matrix; 6) completing the technical part of the precautionary matrix; 7) specifying information sources; 8) obtaining information; 9) finishing the matrix; and 10) clarifying any need for action (Höck *et al.* 2011).

The precautionary matrices are logically completed and evaluated in two iterative steps:

- 1st—a rapid evaluation to demonstrate knowledge gaps and uncertainties and lead to a preliminary precautionary matrix;
- 2nd—exact clarifications on the fundamentals of the results from 1st step and from the specific answers to knowledge gaps that afford a finished and definitive evaluation of precautionary matrix (Höck *et al.* 2011).

The potential risk can be classified into class A (the nanospecific need for action can be rated as low, even without further clarification) or class B (nanospecific action is needed; existing measures should be reviewed, further clarification undertaken and, if necessary, measures to reduce the risk associated with development, manufacturing, use and disposal implemented in the interest of precaution) (Höck *et al.* 2011).

3 RESULTS AND DISCUSSION

The NMs go through different "states" during the NCs production (Fig. 1): in the pre-production (stage 1), they are at powder state; in the production (stage 2), they are dispersed in a resin/solution; and, at last, in post-production (stage 3) they are inserted into the solidified resin matrix. The NMs (powder) can create potential exposure in air, in both the pre and the production stages, and with lower probability in the post-production phase. The final NC is unlikely to present a direct risk because the NMs are trapped in the solid resin. Though, it is still common practice incinerating the end-of-life composite materials, including NCs; thus, at the end of the life cycle, the NMs

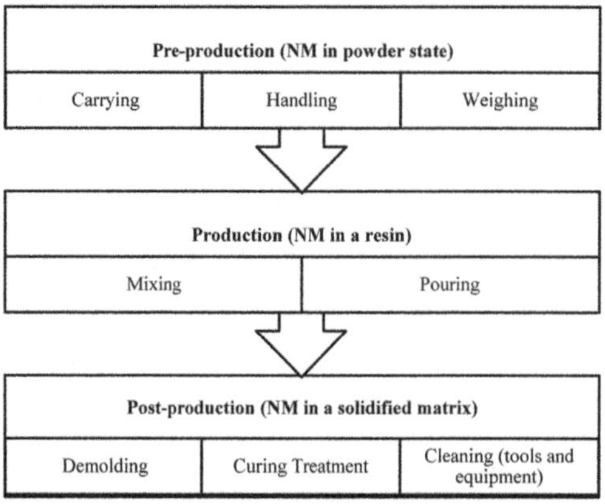

Figure 1. Polymeric NC production.

might still be released to the environment, and this risk could not be disregarded (Roes *et al.* 2012). However, the present analysis is only addressed and restricted to the risk assessment during the manufacturing process of NCs, namely during the processing stages described in Figure 1.

Analysing the results, it was found that the most critical operation in the polymeric NC processing is the pre-production phase, because it deals with NMs in powder state and in which a lack of care in the procedures may lead to typical scenarios of exposure.

According to the analysed risk assessment methods it was also verified that the preventive measures (collective protection measures and personal protective equipment) used during the polymer NCs production are quite suitable.

It was verified that the applied methods did not have fully convergent results (Table 2). However, they give the same useful indications relative to the control measures that must be taken to reduce the risk levels.

Despite the same useful indications gave by the applied methods (relative to the control measures that must be taken to reduce the risk levels), they did not result, however, in convergent outputs. This feature is likely due to the great variety of parameters and factors used in the different risk assessment methods (Table 1 and Fig. 2).

The chemical related parameters are less used. The ANSES and CB Nanotool methods focus in the health-related properties based on control of substances hazardous to health (COSHH), whereas the EPFL, GWSNN and PMSN analyse rather the physical properties. Under this point of view, the ISEPSL method seems to be the most complete risk assessment method as it makes use of all these parameters for the analysis.

The use of various parameters and/or differences in their interpretation can lead to differences

Table 2. Obtained risk levels for the production of both Al_2O_3 and $Mg(OH)_2$ based NCs.

Method	Pre-production	Production	Post-production
1) ANSES	Medium (CL_2)	Low (CL_1)	Low (CL_1)
2) CB Nanotool	Medium (RL_2)	Low (RL_1)	Low (RL_1)
3) EPFL	Medium ($Nano_2$)	Medium ($Nano_2$)	Medium ($Nano_2$)
4) GWSNN	High (C)	High (C)	Medium (B)
5) ISPESL	Medium (16–35)	Medium (16–35)	Medium (16–35)
6) PMSN	High (B)	High (B)	Low (A)

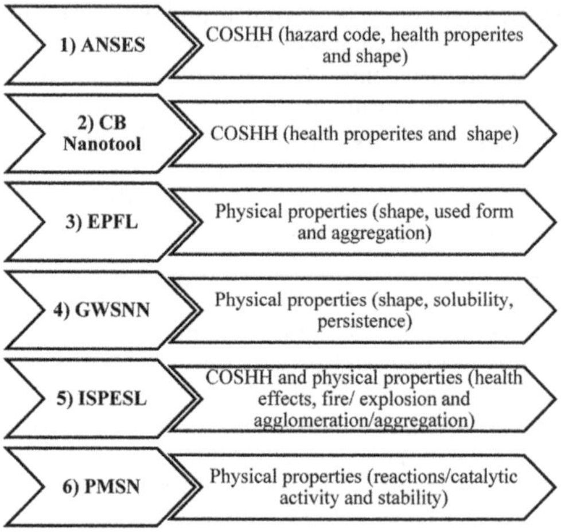

Figure 2. Main factors for hazard assessment.

in risk level results. It was verified that the methods derived from "professional" sources, like GWSNN and PMSN tend to give a higher risk degree in the first two phases of the product life cycle. This may be due the origin policy of these methods, which have the tendency to be more cautious.

These methods should be reviewed in order to give more convergent results avoiding unequal final outcomes.

4 CONCLUSIONS

The risks are a key issue to be considered especially in the early stages of any new material production. By studying proactively emerging risks, one can prevent future problems. The risks are inherent to any technology and nanotechnology is no exception.

Within this scope, in this study the inherent risks associated to the main production stages of two different polymer NCs were evaluated. The experimental study was conducted in a laboratory environment but provided an overview of the measures that could be applied to improve the safety levels during polymer NCs production.

The results obtained with the qualitative methods were divergent; however, all methods consider that the pre-production stage is the one with the higher risk. This highlights the need to create new risk assessment guidelines and standardized methodologies that allow obtaining more consistent and credible data about the risk levels and adequate measures that should be taken in the workplace. The qualitative risk assessment should also be complemented with appropriate quantitative methods.

The typical risk assessment process includes hazard identification, dose-response assessment, exposure assessment and risk characterization, but in NMs case, there is still a lack of information that hinders somehow the fully completion of these steps. Until a new evolution of risk assessment methods, relative to NMs, the qualitative methods should not be used in extreme conditions of exposure (as, for instance, large amounts of NMs). In these situations it will be necessary to perform an analysis with special expertise. However the use of qualitative methods can be a first step to reveal risk situations in the polymeric NCs production, reinforcing workplace safety and prevention.

With the current standards it is already possible to implement prevention and protection generic measures. While implementing these measures, it should be taken into account that:

- NMs must be regarded as new products and not as mere "minor" materials whose risks and hazards are already known.
- NMs should be systematically assessed by standard toxicity tests, according to the tests recommended on the regulations for the chemical agents.
- The regulations for the chemical agents should be applied at least in the pre-production stage (when NMs are in the powder state), because in this phase it is easier to contaminate the work environment and consequently the other production phases.
- All development programs should be linked to NMs research to assess the safety health and environment.
- It is important to ensure the traceability of NC residues with dispersal potential, using specific labelling.

These measures should be adopted as soon as possible to ensure a responsible development of NCs production. The current knowledge about the NM toxicity is insufficient and the scientific preliminary assessments show that there are enough suspicions that these materials may have harmful effects on human health. Thus more research work is needed to fill existing gaps. Until further information, NM should be considered as hazardous materials.

ACKNOWLEDGMENTS

The authors wish to acknowledge FCT, COMPETE and FEDER (under PTDC/ECM/110162/2009), for funding the research. The authors also wish to thank J. A. Rodrigues from INEGI, for his valuable help and assistance.

REFERENCES

Amoabediny, G.H., Naderi, A., Malakootikhah, J., Koohi, M.K., Mortazavi, S.A., Naderi, M. & Rashedi, H. 2009. Guidelines for safe handling, use and disposal of nanoparticles. *Journal of Physics: Conference Series*, 170, 012–037.

Cornelissen, R., Jongeneelen, F., Broekhuizen, P. & Broekhuizen, F. 2011. Guidance working safely with nanomaterials and—products, the guide for employers and employees. FNV, VNO-NCV and CNV.

Crosera, M., Bovenzi, M., Maina, G., Adami, G., Zanette, C., Florio, C. & Filon Larese, F. 2009. Nanoparticle dermal absorption and toxicity: a review of the literature. *International Archives of Occupational and Environmental Health*, 82, 1043–1055.

Giacobbe, F., Monica, L. & Geraci, D. 2009. Nanotechnologies: Risk assessment model. *Journal of Physics: Conference Series*, 170, 012–035.

Groso, A., Petri-Fink, A., Magrez, A., Riediker, M. & Meyer, T. 2010. Management of nanomaterials safety in research environment. *Particle and Fibre Toxicology*, 7, 40.

Gupta, N.S., A. 2011. Issues Associated with Safe Packaging and Transport of Nanoparticles. *ASME Pressure Vessels and Piping Conference.*

Höck, J., Epprecht, T., Furrer, E., Hofmann, H., Höhner, K., Krug, H., Lorenz, C., Limbach, L., Gehr, P., Nowack, B., Riediker, M., Schirmer, K., Schmid, B., Som, C., Stark, W., Studer, C., Ulrich, A., Von Götz, N., Weber, A., Wengert, S. & Wick, P. 2011. Guidelines on the Precautionary Matrix for Synthetic Nanomaterials. Berne: Federal Office of Public Health and Federal Office for the Environment.

INRS. Risks associated with nanoparticles and nanomaterials. INRS Nano 2011 Conference, 5–6–7 April 2011 Nancy—France. 1–131.

ISO. 2011. *How toxic are nanoparticles? New ISO standard helps find out* [Online]. Available: http://www.iso.org/iso/pressrelease.htm?refid = Ref1394 [Accessed 10/15 2011/].

Kaluza, S., Balderhaar, J., Orthen, B., Honnert, B., Jankowska, E., Pietrowski, P., Rosell, M., Tanarro, C., Tejedor, J. & Zugasti, A. 2009. Workplace exposure to nanoparticles. *Brussels, Belgium: European Agency for Safety and Health at Work (EU-OSHA)*, no. 2, 89.

Kiliaris, P. & Papaspyrides, C.D. 2010. Polymer/layered silicate (clay) nanocomposites: An overview of flame retardancy. *Progress in Polymer Science,* 35, 902–958.
Lövestam, G., Rauscher, H., Roebben, G., Klüttgen, B.S., Gibson, N., Putaud, J.-P. & Stamm, H. 2010. Considerations on a definition of nanomaterial for regulatory purposes. Luxembourg: European Union.
MESD 2006. Nanotechnologies, Nanoparticles: What Hazards—What Risks? *In:* DEVELOPMENT, M.O.E.A.S. (ed.) *Committee for prevention and precaution.* Paris.
Riediker, M., Ostiguy, C., Triolet, J., Troisfontaine, P., Vernez, D., Bourdel, G., Thieriet, N., Cad, A., #232 & NE 2012. Development of a control banding tool for nanomaterials. *Journal of Nanomaterials,* 2012, 8–8.
Roes, L., Patel, M.K., Worrell, E. & Ludwig, C. 2012. Preliminary evaluation of risks related to waste incineration of polymer nanocomposites. *Science of The Total Environment,* 417–418, 76–86.
Šupová, M., Martynková, G.S. & Barabaszová, K. 2011. Effect of Nanofillers Dispersion in Polymer Matrices: A Review. *Science of Advanced Materials,* 3, 1–25.
Zalk, D.P., S 2009. Control Banding and Nanotechnology Synergist *The Synergist,* 21, 26–29.
Zou, H., Wu, S. & Shen, J. 2008. Polymer/Silica Nanocomposites: Preparation, Characterization, Properties, and Applications. *Chemical Reviews,* 108, 3893–3957.

Occupational Safety and Hygiene II – Arezes et al. (eds)
© 2014 Taylor & Francis Group, London, ISBN 978-1-138-00144-2

Comparative study between productivity predictive models for work in hot environments

A. Sousa
Research Laboratory on Prevention of Occupational and Environmental Risks (PROA/LABIOMEP), Higher Engineering Institute, University of Algarve, Faro, Portugal

J. Santos Baptista
Research Laboratory on Prevention of Occupational and Environmental Risks (PROA/LABIOMEP), Faculty of Engineering, University of Porto, Portugal

ABSTRACT: Productivity is a key factor for the maintenance and success of the companies. Given the importance of the topic, several studies have focused on the different issues that influence work performance (e.g. technological, behavioural, training, skills or environmental). This article focuses on the question of the relationship between labour productivity and thermal environmental conditions (hot) in existing workspaces. The proposed approach is to identify, test and compare the main quantitative models available on the subject. As a result of this analysis, it was possible to identify advantages and limitations in the practical application of each method, and the need to develop further studies on the subject.

1 INTRODUCTION

The effect of the thermal environment in labour productivity, in qualitative terms, is recognized by the majority of studies on/about the subject. However, some authors report that their quantification is a topic still open (Parsons 2009), considering the need to develop further studies on this issue.

The generality of the available studies only uses air temperature to predict productivity. Some others use this parameter together with humidity, or else more complex indices as PMV or WBGT, as key variables to characterize the conditions of the existing thermal environment. In a qualitative perspective, several authors consider that the productivity decreases with the appearance of heat stress conditions (Lamberts & Xavier, 2002), (Kjellstrom, Holmer, & Lemke, 2009; Miller, Bates, Schneider, & Thomsen, 2011; Guedes, Baptista, & Diogo, 2011), usually measured by WBGT index, which includes, among others, the effect of the two variables in question.

The present study focuses on the analysis of quantitative models for predicting productivity using these variables. It aims at analyzing and comparing the answers of proposed quantitative models that relate the loss of labour productivity with hot thermal environmental conditions, as well as the possibility of a wider application of these models.

2 MATERIALS AND METHOD

A literature search was performed to identify the models that associate the variables: productivity and thermal environment. Then, the models were tested with values within the proposed domain for the respective input variables.

3 MODELS

In this section are described the quantitative studies that focus on the relationship "productivity—thermal environment", identified in the literature review. Among them, two distinct groups (1 and 2) can be highlighted:

- Group 1—Experimental studies with a focus on real work context that relate discrete values for the productivity loss as a function of the air temperature. For a given temperature value, it was indicated a productivity percentage reduction which was then compared with a standard value. Belong to this group the studies based on survey data in the following activities: office work (Witterseh, et al. 2004), heavy drilling mining (Eston 2005), call centers and textile industry (Seppänen, Fisk & Faulkner 2005) and school learning (Wyon 2010).
- Group 2—Studies that propose mathematical models to predict the productivity, based on the knowledge of environmental variables

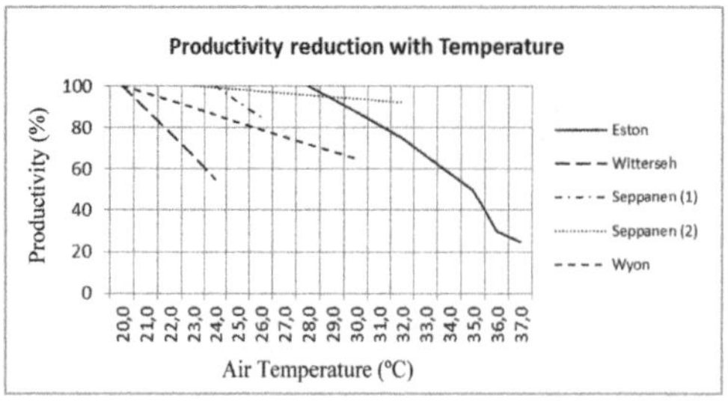

Figure 1. Relationship between temperature and productivity.
1)—*Call centers*; (2)—Textile industry.

and sometimes on effort level. In this group the works presented by Koehn & Brown (1985), Thomas & Yiakoumis (1987), Mohamed & Srinavin (2002, 2005) should be mentioned, as well as the one by Zhao, Zhu & Lu (2009) published more recently.

3.1 *Data collected in a real context*

Starting by the works of group 1, it is it was found that in these studies the productivity loss values are obtained by collecting data in different and specific activities (e.g. learning, industry), considering only 'air temperature' as a parameter. The graphical representation of their results can be seen in Figure 1, where can be easily verified that the slope of the functions is different for each one of the studies. This fact can be explained by the absence of 'metabolism' (induced effort inherent to the activity) and 'humidity' parameters, as independent variables, complementary to air temperature. Indeed, for several decades it has been recognized the key role of humidity and effort level in the metabolic response of human beings, as it is evident in the reference model for thermal comfort, developed by Fanger (1972).

In addition to the above mentioned factors there are many others (eg, acclimation, age, gender, ethnicity, medication, alcohol, coffee, etc.) which may also contribute to obtain different results for each situation.

Anyway, not considering the humidity factor, precludes by itself the comparison and generalization of the results. This is evident by analyzing the graph of Figure 1. From this observation it can be concluded that this group of proposals (group 1) only responds to situations with similar environmental conditions to those under which the data were collected.

3.2 *Predictive models*

Focusing attention on the four 'productivity models' that constitute group 2, it is worth noticing that they consist of a set of parameterized functions based on the collection of real data on activities carried out in hot and humid environments (Koehn & Brown 1985), (Thomas & Yiakoumis 1987) (Mohamed & Srinavin 2005) or on data collected in controlled laboratory trials for the same type of thermal environment (Zhao, Zhu & Lu 2009).

Regardless of the data source, all studies culminate in the proposal of a function set (model) that intends to achieve a analytical relationship between productivity and thermal environment. Each model was analyzed using the variables within the ranges of allowable values.

3.3 *Koehn & Brown model*

Koehn & Brown (1985) developed two functions: one for cold environments (–29°C to 10°C) and another for hot environments (21°C to 49°C). These functions relate productivity with air temperature and relative humidity. In the present study, the function for hot environments will be analyzed. The expression proposed by the authors is the following:

$$P = 5{,}17.10^{-2}.T_a + 1{,}73.10^{-2}.H_r - 3{,}20.10^{-4}.(T_a)^2 - 9{,}85.10^{-5}.(H_r)^2 - 9{,}11.10^{-5}.(T_a.H_r) - 1{,}459 \quad (1)$$

where: T_a—Air temperature (°F); H_r—Relative humidity (%); P—Productivity (%).

The results obtained by using this model can be seen in figure 2.

3.4 *Thomas & Yiakoumis model*

A similar model was proposed by Thomas and Yiakoumis (1987), based on the same set of independent variables, but using a fit with a different function:

$$P_r = 9{,}448 + 0{,}0518.Ta - 2{,}89.\ln(Ta) + 3{,}89.10^{-37}.e^{(Hr)} \quad (2)$$

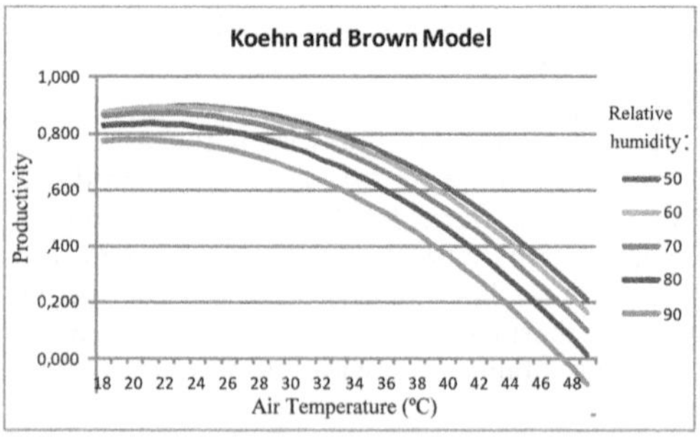

Figure 2. Koehn & Brown results.

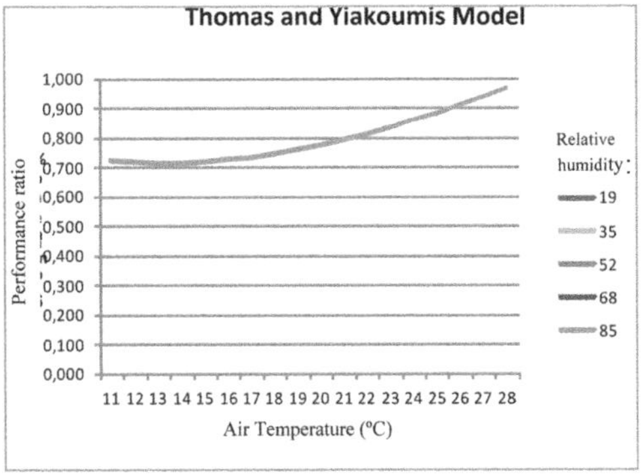

Figure 3. Thomas & Yiakoumis results.

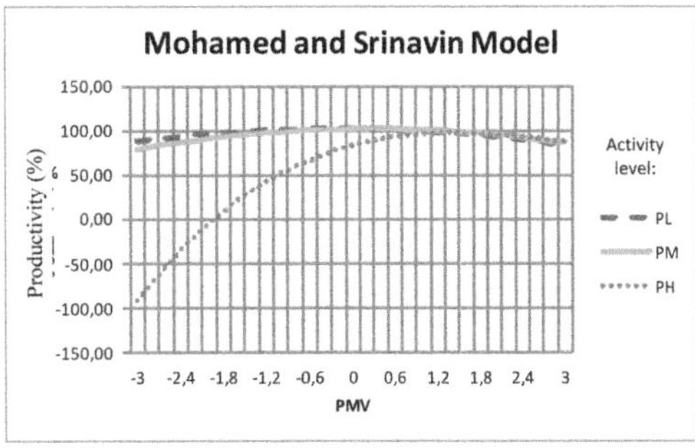

Figure 4. Mohamed & Srinavin results.

where: P_r—Performance ratio (observed/predicted) (%), T_a—Air temperature at 1 p.m. (°F); H_r-Relative humidity at 1 p.m. (%).

The authors indicate as validity range for the results an air temperature between 11°C and 28°C, and a relative humidity between 19% and 85%.

As can be seen in figure 3, the model shows the same responses for each temperature, regardless of the variation on humidity values.

The low sensitivity to humidity values of the model advises us not to generalize the application of the proposed function, given the importance of this parameter.

3.5 *Mohamed & Srnavin model*

Mohamed and Srinavin (2005) developed three functions that relate productivity with the PMV index (Predicted Mean Vote), in situations of light, moderate and heavy tasks, which have been designated, respectively, by P_L, P_M and P_H.

These three functions are applicable for a temperature range between 5°C and 45°C, being defined, respectively, by the following analytical expressions:

$$P_L = 102 - 0{,}80\ PMV - 1{,}84\ (PMV)^2 \quad (3)$$

$$P_M = 102 + 1{,}19\ PMV - 2{,}17\ (PMV)^2 \quad (4)$$

$$P_H = 83 + 21{,}64\ PMV - 9{,}53\ (PMV)^2 + 0{,}91\ (PMV)^3 \quad (5)$$

Knowing that the PMV values may vary in the range [−3, 3] (ISO 7730:2005), the results of applying this model can be represented as shown in the diagram of Figure 4.

For the situations of light and moderate tasks, the results show that productivity value is maximum for PMV values around zero (thermal neutral zone), which reflects an expected outcome (maximum performance under conditions of thermal comfort).

For heavy tasks, the same situation occurs for PMV = 1.4 (corresponding to moderately hot environments), an conclusion for which a reasonable explanation has not yet been found: in activities that require more effort, performed in warmer environments, it is generally agreed that productivity decreases, which contradicts the just mentioned results.

Still concerning the mentioned function for heavy tasks, it is noteworthy that for the PMV values between −2 and −3, productivity (%) is negative. Needless to say that there is no practical validity in this case.

This problem seems to be caused by an error in the abovementioned function. In fact, if the sign of the second term of the equation 5 changed from (+) to (−) keeping the same parameters values, the equation would be (6):

$$P_H = 83 - 21{,}64\ PMV - 9{,}53\ (PMV)^2 + 0{,}91\ (PMV)^3 \quad (6)$$

This change has the graphical illustration shown in Figure 5, which highlights an 'answer' more in line with known patterns.

It should also be noted that the authors did not validate the function for heavy tasks. They did it just for the expressions related to light and moderate tasks.

3.6 *Zhao, Zhu & Lu model*

This last model considers the individuals '*heat tolerance time*' (H_{tt}) when they are exposed to certain thermal environment (WBGT = x) in three physical work conditions: light (L), medium (M) and heavy (H). In this case, the proposed functions are respectively:

$$H_{ttL} = 0{,}0869\ x^3 - 9{,}3769\ x^2 + 336{,}24\ x - 4004{,}5 \quad (7)$$

$$H_{ttM} = 0{,}1508\ x^3 - 16{,}601\ x^2 + 608{,}11\ x - 7411{,}8 \quad (8)$$

$$H_{ttH} = 0{,}0519\ x^3 - 5{,}6694\ x^2 + 206{,}04\ x - 2490{,}3 \quad (9)$$

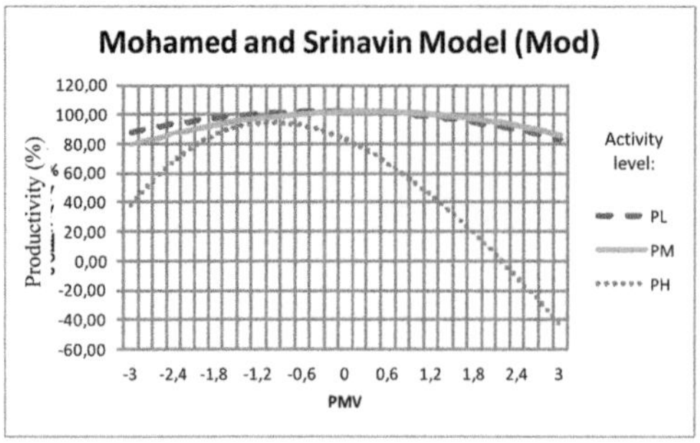

Figure 5. Mohamed & Srinavin modified.

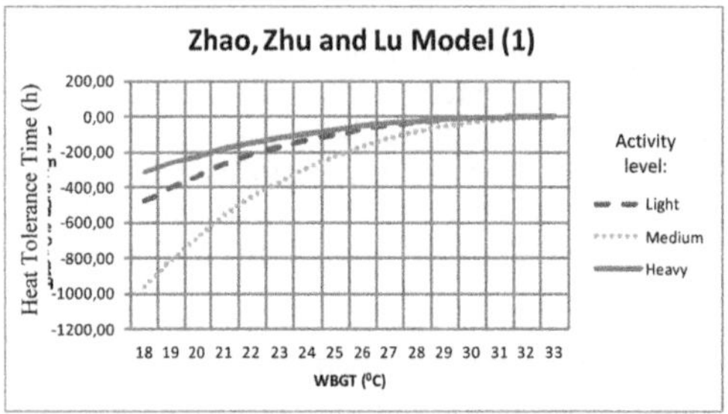

Figure 6. Zhao, Zhu & Lu results.

The WBGT index range used to estimate the equations was [34°C, 38.5°C]. Note that the application of the model to WBGT values below 34°C, as can be seen in Figure 6, leads to '*heat tolerance time*' negative, which has no physical counterpart.

It should be noticed that in figure 6 is shown the reference interval for the WBGT index ([18°C, 33°C] according to ISO 7243:1989). In that interval, the functions proposed by Zhao, Zhu & Lu (2009) are not valid.

Bypassing the component of inadequacy of the model for practical applications in the context of the abovementioned standard presented below, the results obtained for the three physical work conditions (L, M and H), within the domain of the model variables, namely, '*heat tolerance time*' (eq. 7, 8 and 9) (figure 7) and the corresponding '*productivity level*' (figure 8), obtained by using the following equations, proposed by the same authors:

$$P_L = -0,286\,t^2 + 0,6256\,t - 0,07\,x + 2,94\ (t \le H_{tt}) \quad (10)$$

$$P_M = -0,364\,t^2 + 0,7476\,t - 0,05301\,x + 2,09\ (t \le H_{tt}) \quad (11)$$

$$P_H = -0,5963\,t^2 + 0,9115\,t - 0,0676\,x + 2,44\ (t \le H_{tt}) \quad (12)$$

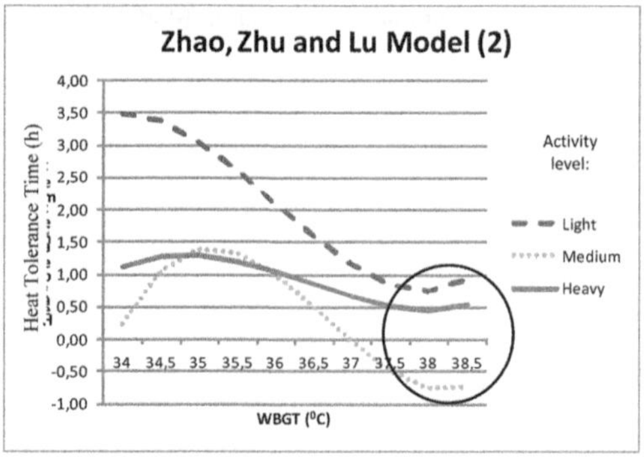

Figure 7. Heat tolerance time.

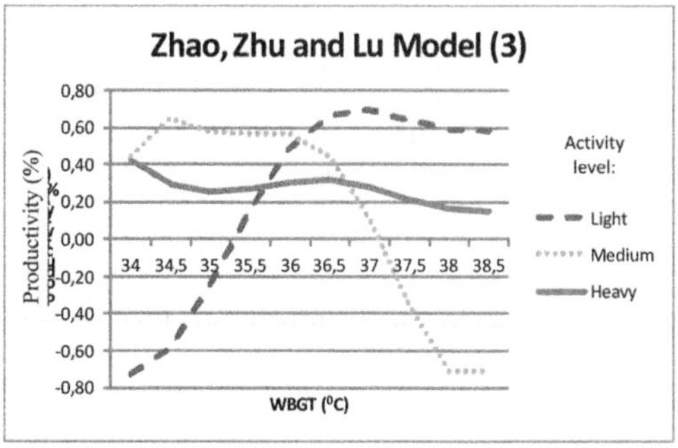

Figure 8. Productivity.

where: t is work time (h), H_{tt} is the *heat tolerance time* (h), x is the WBGT index (°C) and P_L, P_M and P_H are the productivity (%), respectively, to light, medium and heavy tasks.

Note that the *heat tolerance time* variation for medium work conditions should be in the intermediate region between those obtained for the light and heavy working conditions (figure 7). However, this does not occur.

It is also evident that for WBGT = 38°C, there is an inflection point (minimum) from which the values of heat tolerance time increase again (marked with a circle in figure 7). This means a longer or higher *heat tolerance time* from that point, an outcome that does not make sense. In fact, observing figure 8, the results indicate a decrease in productivity for the same WBGT values, which is, apparently, a contradictory result.

Regarding productivity levels, the simulations carried out for the three working conditions show different tendencies, opposed in some cases (Figure 8). Once again, it appears that the results obtained for the *medium* activity level differ in the corresponding interval from *light* and *heavy* activities, which is inconsistent in practice.

In the same way it can be observed as an example, that for light tasks productivity increase continuously between 34°C and 37°C, which is not corroborated by the consensually accepted information.

According to what has been said so far the ISO 7243:1989 standard defines increasing rest periods related with the increase of the WBGT index, whatever the metabolic activity level, and it is also not allowed the work for WBGT ≥ 33°C. Hence, productivity growth/increase between 34°C and 37°C indicated by the model does not seem realistic.

4 CONCLUSION

In summary from this research work can be drawn the following conclusions:

- The outcomes from the studies that focus on 'productivity—air temperature' relationship (group 1), wherein the humidity effect is not considered, show very significant differences (Figure 1), so, its generalization to other similar contexts is not desirable;
- The productivity forecasting models (group 2), within the respective application domains, can also lead to outcomes that cannot be generalized, as evidenced the conclusions drawn from the graphs shown in Figures 3, 4, 5 and 8;
- The Thomas & Yiakoumis model considers the humidity parameter as an independent input variable, but the results presented does not clearly show the effect of this variable,. This fact arouses the need for further quantitative studies on this subject;
- Finally, the Koehn & Brown model uses the same input variables as the Thomas & Yiakoumis proposal, but it provides results in line with known and accepted trends. From this point of view, the model is appropriate.

However, in the conclusion of their study, the authors became aware that the results vary with several factors, such as, work load, task complexity and duration or age, physical condition, motivation and acclimatization of the worker, among others, but they do not indicate how to adapt their model function to this kind of new situations.

Therefore, the establishment of a quantitative relationship between thermal environment conditions and productivity requires new approaches and further studies in order to define a relational model that can be used, in a generalized way, to other applications in the same thematic scope.

REFERENCES

Costa, E., J. Baptista & M. Diogo. 2011. "Adaptação climática, metabolismo e produtividade." CLME. Maputo, 3710A.

Eston, S. 2005. "Problemas de conforto termo-corporal em minas subterrâneas." Revista de Higiene Ocupacional 4, n.° 13: 15–17.

Fanger, P.O. 1972. Thermal comfort: analysis and applications in environmental engineering. New York: McGraw-Hill.

Guedes, J.C., J. Santos Baptista & M. Tato Diogo. 2011. "Factores condicionantes da tolerância ao calor." Ed. A.S. Miguel, G. Perestrelo, N. Costa, M. Barroso, P. Arezes, P. Carneiro, P. Cordeiro, Rui Melo J.S. Baptista. International Symposium on Occupational Safety and Hygiene—SHO2011. Guimarães: SPOSHO, 300–304.

ISO 7243:1989. "Hot environments—Estimation of the heat stress on working man, based on the WBGT-index (wet bulb globe temperature)", International Organization for Standardization, Genève, Switzerland.

ISO 7730:2005. "Ergonomics of the thermal environment—Analytical determination and interpretation of thermal comfort using calculation of the PMV and PPD indices and local thermal comfort criteria", International Organization for Standardization, Genève, Switzerland.

Kjellstrom, T., I. Holmer & B. Lemke. 2009. "Workplace heat stress, health and productivity—an increasing challenge for low and middle income countries during climate change." Global Health Action.

Koehn, Enno & Gerald Brown. 1985. "Climatic effects on construction." Journal of Construction Engineering and Management 111, n.° 2: 129–137.

Lamberts, R. & A. Xavier. 2002. "Conforto térmico e stress térmico." http://www.dec.ufms.br/lade/docs/cft/ap-labeee.pdf 19/11 (acedido em 20 de novembro de 2010).

Miller, Veronica, Graham Bates, John Schneider & Jens Thomsen. 2011. "Self-pacing as a protective mechanism against the effects of heat stress." Annals of Occupational Hygiene 55, n.° 5: 548–555.

Mohamed, Sherif & Korb Srinavin. 2002. "Thermal environment effects on construction workers' productivity." Work Study 51, n.° 6: 297–302.

Mohamed, Sherif & Korb Srinavin. 2005. "Forecasting labor productivity changes in construction using the PMV index" International Journal of Industrial Ergonomics 35: 345–351.

Parsons, K. 2009. "Maintaining health, comfort and productivity in heat waves." Department of Human Sciences, Loughborough University, Loughborough, UK.

Seppänen, O., W. Fisk & D. Faulkner. 2005. "Control of temperature for health and productivity in offices." ASHRAE. Vol. III. n.° Part 2. 680–686.

Thomas, Randolph & Iacovos Yiakoumis. 1987. "Factor model of construction productivity." Journal of Construction Engineering and Management 113, n.° 4: 632–639.

Witterseh, T., D. Clausen, P.P. Wyon & Geo. 2004. "The effects of moderate heat stress and open-plan office noise distraction on SBS symptoms and on the perfomance of office work." Indoor Air 14, n.° 8: 30–40.
Wyon, D. 2010. "Thermal and air quality effects on the performance of schoolwork by children." http://web1.swegon.com/upload/AirAcademy/Seminars/Documentation_ 2010/ Rotterdam/David%20 Wyon.PDF (acedido em 21 de Dezembro de 2011).
Zhao, J., N. Zhu & S. Lu. 2009. "Productivity model in hot and humid environment based on heat tolerance time analysis." Building and Environment 44, n.° 11: 2202–2207.

Occupational Safety and Hygiene II – Arezes et al. (eds)
© 2014 Taylor & Francis Group, London, ISBN 978-1-138-00144-2

Determinants of psychosocial risks

L. Duarte & L. Freitas
CIEG-Universidade Lusófona, Lisboa, Portugal

ABSTRACT: The concern with organizational and psychosocial factors and their relation to occupational health is not new. The importance and wider recognition, acquired in recent years as a result of major changes in organizations and processes—globalization—has point out consequences of psychosocial risks in the health of workers. This article aims to present the design of a study that is part of an assessment project on the determinants of psychosocial risks in health and education organizations. It is expected from its application to collect data that allow the organizations' heads of human resources to design and implement preventive measures regarding the workplaces. Also contributing to fulfill legal requirements about consulting and obtaining the involvement of workers in what concerns the discussion of issues related to Safety and Security at Work, and allows, through the use of relevant and comparable data, the design of new work policies.

1 INTRODUCTION

1.1 *General situation*

The significant changes that took place in the past decades in work environment resulted in emerging risks in the field of occupational health and safety and led to the identification of a new typology of risks in addition to the physical, chemical and biological hazards, which were named psychosocial risks.

The theoretical concept behind them was defined by the Joint Committee ILO/WHO in 1984 as the existing working conditions, related to the organization, content and amount of work, which may affect both the welfare and health (physical, psychological or social) of workers and work performance. Later, the literature reveals that this definition has been subject to several revisions, although preserving, in essence, the dynamic perspective of the concept, embedded in the interaction between the work and the worker and its risks to their health, derived from the psychosocial environment around them.

It is noted that the impact due to inadequate psychosocial situations, can affect the health of the person at various levels, triggering: physiological changes (e.g., cardiac disease, gastric disorders, skin diseases), mental changes (e.g., behavioral and cognitive disorders) or emotional. These changes can occur in a direct manner or be mediated by a stressful situation, or by interaction with other factors (e.g., physical background) (Bilbao & Cuixart, 2012) .

On the other hand, it is noted that the negative outcomes does not occur only at the individual level, but also on organizations as a whole, resulting in increased absenteeism or workplace conflict, voluntary abandonment of employment or low productivity, etc. However, it is noted that the interaction between the work and the person may also translate into positive results when people has the opportunity to develop their skills. Therefore, it appears that what distinguishes psychosocial factors of other types of work risks is its preventive purpose, consisting primarily on its disposal, although in reality it requires frequent resort to mitigation of adverse effects (Bilbao & Cuixart 2012).

In fact, psychosocial risks and work-related stress usually go together. Being sure that this is a pure answer to the misfit resulting from requests and pressures to which the person is subject while performing his tasks, bearing his skills and abilities. High demands and low control of work provision conditions are rightly considered risk factors for both physical and mental health (Melchior, 2007; Vahtera & Kivimäki, 2008). There is a consistent relation between health complaints made by workers and exposure to psychosocial risks, which translates into a set of consequences for the individual (physical, mental and social) and organizational (absenteeism, illness, satisfaction, productivity, etc.) (Cox *et al.*, 2000).

In general, the psychosocial risks involve work related factors, with causes focused on processes or methods of work as well as the psychosocial context in which they occur. That is the case of organizational labour conditions, working alone, supervision issues, quality requirements and violence associated with work. Other determinants to include in an assessment are the outcome measurement in terms of stress experienced by workers, resulting in physiological or psychological reactions such as depression, anxiety or hypertension. One must also consider the relationship between management/supervisors and workers or between peers, as well as businessexternal constraints, issues that do not derive neither directly nor indirectly from the work situation, including wages, pensions, promotions, influence on labour management decisions and job uncertainty as a result of market conditions (Rasmussen *et. al.*, 2010). We thus realize that psychosocial risks do not establish a new problem. Developing tools to identify and evaluate them in order to find effective solution proposals will be a priority to the elimination of problems in the workplace, such as stress, harassment, bullying andother types of violence.

In conclusion, psychosocial risks originate in poor work planning of and hence in the resulting negative interactions. In this context, line and top managers have an important role while holding the power to redesign initiatives at work. However, the psychosocial reality refers in fact not only to the conditions that objectively exist, but also as to the way they are received and experienced by the individual. Therefore, in analyzing a situation, rather than knowing the characteristics of work, it is necessary to identify the perceptions of workers in this field (Bilbao & Cuixart, 2012). Additionally, in legal terms, we observe that through the Directive—89/391/EEC and its special directives, the community legislation provides a framework that allows EU workers benefit from high levels of OSH in the workplace.

In Portugal, the assessment of psychosocial risks derives from imperatives of community and national rules, based on a set of solid and consistent commands: the requirement of technical and scientific research on new risk factors (subparagraph e) of n.º 2, Art. 5, the employer's commitment in assuring psychosocial risk factors do not affect OSH (subparagraph d), of n.º 2, Art. 15), the adaptation to new forms of organization work (ibid., subparagraph f), the obligation to consult workers or their elected representatives, paragraph 1, article 18, and also the responsibility of prevention of occupational hazards services to monitor the health of workers and develop activities to promote health (subparagraphs g) and h) paragraph 1, Art. 98). There are several legislative references in the Framework Law OSH to the significance of the employers intervention to anticipate the risks derived from the organization and content of work.

In 2007, the Danish Authority for the Working Environment launched a new strategy to strengthen and upgrade the primary prevention of work-related stress, with positive results. This has resulted in increasing the number of inspections of the psychosocial work environment (Rasmussen *et al.*, 2010), establishing a greater role of supervisory authorities in the assessment of exposure to very specific risks, not sensitive to classical models of risk assessment.

The Authority for Working Conditions (AWC) has developed, in consequence of their functional and also as of the campaigns undertaken since 2012 by the Committee of Senior Labour Inspectors (SLIC), with representatives of all EU Member States, a specific activity centered in the diagnosis of psychosocial issues, aiming at improving the quality improvement of existing risk assessments in this field, through the inspection action. In essence were addressed sources of risk, consequences and preventive and corrective measures.

1.2 *Purpose of the study*

The EU has consistently validated several models on the theme: HSE Indicator Tool (2004), FPSICO-Cuestionario de evaluación de riesgos psicosociales, the INSHT, the method Working Conditions and Control Questionnaire (WOCCQ) (Keyser and Hansez, 2001), COPSOC, Copenhagen Psychosocial Questionnaire (Kristensen *et al.*, 2005), General Nordic Questionnaire (QPS NORDIC) (Lindstrom, 2002), Occupational Stress Index (OSI) (Belkic, 2000).

In Portugal there are few studies on this issue and, in special, on organizations linked to the areas of health and education. This study aims to establish a major contribution to the analysis of psychosocial risk factors in these organizational contexts.

2 METHODOLOGY

2.1 *Study design*

It is a study in cross-type organizations, therefore, held in a single moment. Such studies are appropriate to identify characteristics in populations, in this case the perceptions of workers regarding certain variables associated with psychosocial risks.

The study is ongoing and targets workers from welfare institutions in the areas of early childhood education, elderly support and health. There are different aimed jobs (eg, assistants, technicians and superior technicians). Since the target is high-numbered, about 600 respondents, a questionnaire was chosen as the measuring instrument, in an electronic data collection model. The questionnaire aims to identify problems, measure the opinions/perceptions concerning psychosocial risks at study, and get the corresponding indicators.

2.2 *Measure instrument*

The use of questionnaires as an instrument for measuring psychosocial risks has been commonly referred to in the literature (e.g., Bilbao & Cuixart, 2012; Rasmussen *et al.,* 2010; Barón, 2008). The measure instrument chosen in this study is a self-report questionnaire, designed to identify problems, measure the opinions/perceptions of individuals concerning psychosocial risks and colect their indicators.

The scales used for measurement of the concepts addressed in this study were adapted from a questionnaire used to evaluate psychosocial risks at the Instituto Nacional de Seguridad e Higiene en el Trabajo—Spain (INSHT, 2013), and the contents of the method CoPsoQ-ISTAS 21, v1.5, of Instituto Sindical de Ambiente, Trabajo y Salud (ISTAS, 2010), designed following the approach of multiple indicators so that each variable was measured by multiple items. Based on the literature review related to the variables involved in our study, we tried to adapt these items to the reality of our field and target. The survey has a total of 113 items, four pages.

2.3 *Variables operationalization*

All variables are measured on a five levels response interval scale, *likert* type (agreement and frequency scales), where 1 corresponds to the most unfavorable opinion, 3 reflects a neutral opinion, and 5 corresponds to most favorable opinion. The variables operationalized from INSHT (2013) were:

Working time: This factor refers to the aspects associated with the management and work time structure throughout the week and during a working day. It evaluates the impact of working time from the resting periods the activity allows, its quantity and quality, and the effect of working time in social life. This variable was operationalized through four items.

Autonomy: Under this factor there are aspects related to working conditions, namely, the capacity and capability of the worker to manage and make decisions, both in the aspects of time structuring and on procedures and work organization. This factor is designed in two blocks: TimeAutonomy and Time Decision. This variable was operationalized using eleven items.

Workload: This is the level of the job requirement that the person is subject to, namely the degree of mobilization needed to solve the demands of the working activity, regardless the nature of the load (cognitive or emotional). It is understood that the workload is high when the quantitative component is associated with a qualitative component, namely the quantity and difficulty appear together. This factor assesses the workload from the following factors: Pressure of Time, Attention Effort and Quantity and Task Difficulty. This variable was operationalized using fourteen items.

Psychological requirements: Psychological job demands relate to the nature of the different requirements that the subject is asked to apply. Such requirements are usually of cognitive and emotional nature. Thus, the cognitive system is more or less committed, depending on the job requirements, while requiring the use of information and knowledge, demand planning, decision making, etc. Emotional demands occur in situations where performance demands withhold and manage feelings or emotion to correspond to the nature of work or its requirements, for example, in the case of patient treatment, customer service etc. The effort associated with emotional restraint and its management may also occur within the context/work environment with superiors, subordinates, peers, etc.This variable was operationalized through seventeen items.

Work variety/Content: This factor includes the recognition of work as meaningful and useful in itself, both for the worker and for the whole company and society at large, thus making it a target of recognition and appreciation for them, offering the employee a sense of purpose beyond economic considerations. This factor is evaluated by a series of items in order to study the extent to which the work is designed to integrate varied and meaningful tasks. If it is an important job and is subject to recognition by the worker's environment. This variable was operationalized using nine items.

Participation/Supervision: This factor contains two dimensions of control over work: what the worker exercises through participation in the various aspects of the job and what the organization has over the worker, through the supervision of tasks. This variable was operationalized using twelve items.

Concern with employee/Compensation: Concern for worker refers to the degree of individualized concern that the company shows the worker in the long term. These issues manifest themselves in the organization's approaches to the promotions, training, career development, the information path and content conveyed on these matters, as well as the perception of job security and the balance between the contributions of the employee to the company and remuneration. This variable was operationalized through nine items.

Role performance: This factor considers the problems that can arise from the definition of each job. Includes three aspects: Clarity of Role,Role Conflict and Role Overload. This variable was operationalized using eleven items.

Relationships and social support: This factor refers to the conditions of employment originated from the relationships established between people in the workplace. This factor resorts to the concept of *"social support"*, understood as a stress moderating factor, which the method materializesby studying the possibility of instrumental support or help obtain from others in the workplace (bosses, colleagues, ...) and, the quality of such relationships. This variable was operationalized considering 19 items.

The variables operationalized from ISTAS 21 (2010) were:

Insecurity about the future: This factor depicts the concerns about the future in terms of job loss and unwanted changes of basic working conditions (working hours, tasks, contract, salary, etc.). This is related on one hand with the stability of employment and employability of the labor market, and on the other, with the threat of worse working conditions, or because the externalcontext shows working conditions of greater insecurity and arbitrariness, or even because they identify possible restructuring, relocation of the company etc. This variable was operationalized using three items.

Double presence: This factor is associated with synchronic demands of the workplace and the domestic-family environment. They are high when job demands interfere with family. In the workplace they refer to quantitative requirements, organization, duration, extension, or modification of the working day and also with the level of autonomy over this, for example, working hours or days incompatible with caregiving or with social life. This variable was operationalized by three items.

2.4 *Psychometric tests*

To evaluate the psychometric properties of the scales it is planned to study their construct validity using factor analysis, an assessment of the dimensionality of the usedset of items and determine the values of test Kaiser-Meyer-Olkin (KMO), which is a measure of suitability factor analysis. Their reliability is evaluated by the level of internal consistency, i.e., the degree of uniformity and consistency between subjects' responses to each item by calculating Cronbach's alpha coefficient.

3 FINAL CONSIDERATIONS

It was found in the context of organizations in general and in the areas of education and health the existence of a gap in the evaluation of psychosocial risks and their prevalence, due to the scarcity of instruments with good psychometric qualities.

Hence, in theoretical terms, this study intends to contribute to this specific universe, using an adaptation of model INSHT (2013) to the Portuguese reality, adding knowledge to the subject in discussion.

In terms of practical implications, this study provides relevant data for human resource management in social organizations, guiding an intervention that is more focused on the most relevant psychosocial factors. On the other hand, the presentation of a valid instrument for assessing psychosocial risks in these institutions will allow to assess the effectiveness of measures taken in the workplace. Highlighting the following factors:

a. At the organizational level:
   - Identifying the prevalence of psychosocial risk factors in the organization;
   - Definition of the risk profile of the organization:
   1. Contributions aiming those responsible for human resources management:
   2. The design and implementation of prevention measures at the workplace and new policies for Occupational Safety and Health with favorable prognosis in terms of their effectiveness;
   3. The lower costs for the institution by reducing the number of sick leaves as a result of measures taken in terms of process control and risk reduction;
   4. The increase in labor productivity by increasing the working days.

b. At the individual level:
   - Increased personal welfare, in professional terms, with favorable implications in the family life by reducing identified psychosocial risks;
   - Increased productivity, with consequent absence of reduction in wage due illness related leaves and/or work related accident;
   - Auscultation of workers on the issues involved in this matter, promoting and reinforcing the mechanisms of organization identification and loyalty towards its principles.

REFERENCES

Barón, Francisco López; García, María Àngeles Carrión; Moreno, Manuel Pando; Salazar, Erika Mayté del Ángel (2008). DIAGNÓSTICO PSICOSOCIAL EN UNA EMPRESA SIDEROMETALÚRGICA ESPAÑOLA. RESULTADOS DEL FPSICO DEL INSHT Y DEL INVENTARIO DE VIOLENCIA Y ACOSO PSICOLÓGICO EN EL TRABAJO IVAPT-E-R. *Revista Cubana de Salud y Trabajo,* 9 (1), 54–61.

Bilbao, Jesús Pérez & Cuixart, Clotilde Nogareda (2012). Factores psicosociales: metodología de evaluación. Notas Técnicas de Prevención, Serie 27, NTP 926. Instituto Nacional de Seguridad e Higiene en el Trabajo. Madrid.

Cox, T.; Griffiths, A.; Rial-Gonzalez, E. (2000). Research on work related stress. Luxembourg: Office for Official Publications of the European Communities.
INSHT (Instituto Nacional de Seguridad e Higiene en el Trabajo), (2013). Factores Psicosociales: Método de evaluación. Versión 3.0. http://www.insht.es/portal/site/Insht/menuitem.1f1a3bc79ab34c578c2e8884060961ca/?vgnextoid = 886e58055a35f210VgnVCM1000008130110aRCRD&vgnextchannel = 25d44a7f8a651110VgnVCM100000dc0ca8c0RCRD.
ISTAS (Instituto Sindical de Ambiente, Trabajo y Salud), (2010). *Manual del método CoPsoQ-istas 21 (versión 1.5) para la evaluación y prevención de los riesgos psicosociales para empresas com 25 o más trabajadores y trabajadoras.* Centro de Referencia de Organización del Trabajo y Salud. Barcelona.
Keyser, V. De & Hansez, Isabelle (2001). Working Conditions and Control Questionnaire (WOCCQ). http://www.woccq.be/
Melchior, M.; Caspi, A.; Milne, B.J.; Danese, A.; Poulton, R.; Moffitt, T. E.. (2007). Work stress precipitates depression and anxiety in young, working women and men. *Psychological Medicine*, 37 (8), 1119–1129.
Rasmussen, Mette Bøgehus; Hansen, Tom; Nielsen, Klaus T. (2010). New tools and strategies for the inspection of the psychosocial working environment: The experience of the Danish Working Environment Authority. *Safety Science* 49 565–574.
Vahtera, Jussi & Kivimäki, Mika (2008). Reducing sickness absence in occupational settings. *Occupational & Environmental Medicine*, 65, 219–220.

Occupational Safety and Hygiene II – Arezes et al. (eds)
© 2014 Taylor & Francis Group, London, ISBN 978-1-138-00144-2

Occupational Health and Safety (OHS) at construction projects. A perspective from formworks & falseworks companies

J.C. Rubio-Romero & A. López-Arquillos
Universidad de Málaga, Málaga, Spain

J.A. Carrillo-Castrillo
Universidad de Sevilla, Sevilla, Spain

ABSTRACT: Construction sector has poor indicators about their Occupational safety levels. In civil construction projects, Occupational Health and Safety in the design and use of formwork/falsework is not properly addressed many times. Aim of current research is to analyze the opinion of the professionals from formwork/falsework manufacturing companies about the design and use of formwork/falsework. A Likert-scale survey was distributed between main international manufacturing formwork companies. Questions were divided in three different categories: a) Design phase b) Construction phase and c) Legal issues. Results showed that although safety and final cost are considered during the design of the formwork, customers prefer a cheap product before a safe product. In opinion of the manufacturers, users do not follow instructions about the product frequently. New standards about formwork/falseworks could improve the occupational safety of the civil construction projects.

1 INTRODUCTION

Studies about Occupational Health and Safety at civil construction are frequently focused on topics as the impact of the different variables on the severity of the accidents (Sawacha, Naoum and Fong 1999), (Salminen 2004), (Lopez-Arquillos et al., 2012) contributing factors in construction accidents (Haslam et al., 2005) or occupational safety risk assessment at construction activities (Hallowell and Gambatese 2009).

Researches focused on the concept of design for construction safety also known as Prevention through Design concept (PtD) are also important in the framework of occupational health and safety at construction. Authors like Behm (2005), and Gambatese et al., (2008) studied the influence of the project design in the construction worker's safety. Influence of contractors has been studied by authors like Wang et al., (2006), Saurin et al., (2008) or Fadier & De la Garza (2006). In others similar researches, influence of the designers or structural engineers has been analyzed by Gambatese and Hinze (1999), Behm (2005), or Gambatese et al., (2008).

In contrast of cited influence groups, although they are present in majority of civil construction works, literature about influence of the formwork manufacturing companies on the occupational health and safety of construction site has not been found. Among construction tasks, formwork activities are frequently associated with high rates of accidents and injuries. According to Huang and Hinze (2003), 5.83% of falls were attributed to the construction of formwork or to the construction of temporary structures, and around 21% of all accidents involved wood framing or formwork construction.

This study aims to analyze the opinion about occupational safety in the design and use of formwork/falsework from formwork/falsework manufacturing companies' professionals.

2 METHODOLOGY

A Likert-scale questionnaire (Likert 1932), was designed in order to collect the opinion from the expert selected. Likert scale questionnaire has been demonstrated as a very useful tool in previous papers about occupational health and safety in construction (Ismail et al., 2011, Gittleman et al., 2011, Melia et al., 2008) this is the main reason of the application of this methodology in the present work.

A total number of 70 questionnaires were delivered between the selected formwork/falsework manufacturing companies.

Average of experience in the group of respondents was 11.81 years. Only 13% of respondents from companies selected were female.

Table 1. Distribution of respondents in selected companies.

Company	Number of respondent	Percentage
ULMA	19	27%
PERI	19	27%
ALSINA	16	26%
DOKA	14	20%
TOTAL	70	100%

Due to the sensitivity of the data and in order to ensure the understanding of the instructions, in addition of the traditional surveys method as mailing or virtual surveys, possibility of interview was provided to the respondents. Face to face interaction also provided the interviewer with the opportunity to clarify questions about the content of the items.

Possible answers ranged from value 1 (strongly disagree) to value 5 (strongly agree). The questionnaires contained 17 items, grouped in three different categories: a) Design phase, b) Construction phase, and c) Legal issues. The questionnaires were designed to be simple and brief. They were checked previously by five different experts on Occupational Health and Safety in Construction for suitability and quality of the questions. Suggestions of the experts about the language level, comprehensiveness or item content were included in the final version of the questionnaire.

3 RESULTS

Results were divided according to the three different categories of the questionnaire.

3.1 *Design phase*

In the design phase were included items about the design of the project, and the design and manufacturing process of the formwork/falsework as product.

Questions included in the current category are the following:

Q1—The project designer would ask the formwork manufacturing company for advice while they are designing the structure.

Q2—The project designer designs the structure without consulting the formwork/falsework manufacturing company. When the design is finished the construction company asks the formwork/falsework manufacturer for constructive solutions to suit the structure as designed.

Q3—The majority of projects do not specify type of the formwork/falsework in the project's documentation. Formwork/falsework selection is up to the construction company.

Q4—When formwork/falsework is being designed safety is considered as a very important design factor.

Q5—When formwork/falsework is being designed productivity is considered as a very important design factor.

Q6—When a formwork/falsework is being designed final cost is considered as a very important design factor.

Figure 1 showed the percentage of respondents based on their level of agreement [From 1 (strongly disagree) to 5 (strongly agree)]. Results showed in figure 1 pointed that the project designer do not use to ask manufacturer about the best formwork solution when he is designing the construction project. It is remarkable that although safety is considered during the design phase of the product, final cost is the most important factor, between both.

Figure 2 showed the mean and mode values obtained for items related with design phase of the project and falseworks.

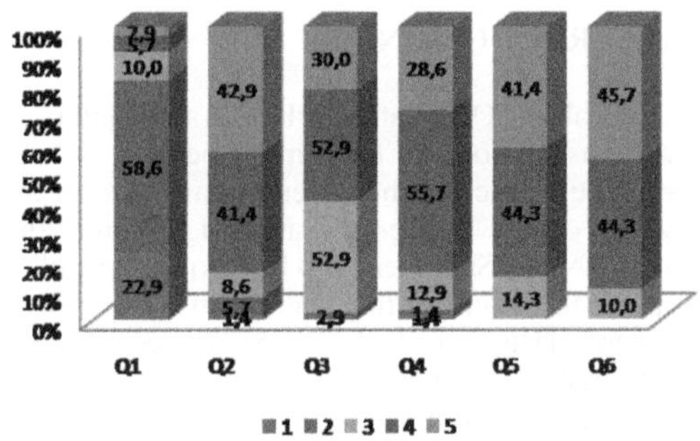

Figure 1. Level of agreement. Items Q1–Q6.

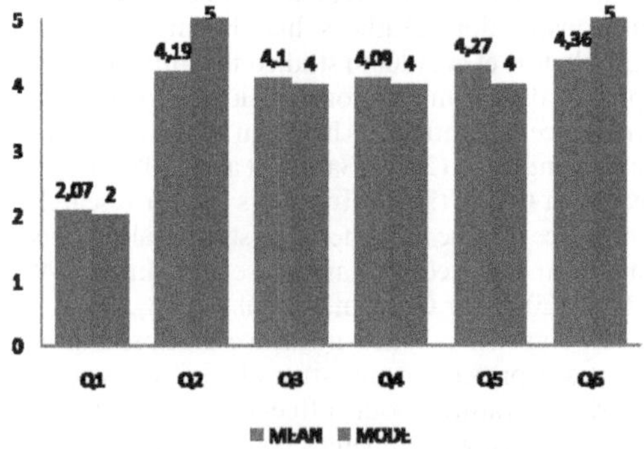

Figure 2. Mean and mode values. Items Q1–Q6.

3.2 *Construction phase*

In this subsection questions about the development of civil construction project on site are included. Items included in this category are:

Q7—The formwork/falsework customer chooses always the safest one.

Q8—The formwork/falsework customer chooses always the cheapest one.

Q9—The user always follows the manufacturer's instructions about the product.

Q10—Technical advice from formwork/falsework companies to users includes advice about safety issues related with use of the formwork/falsework.

Q11—Formwork/falsework suppliers are the same suppliers for the rest of temporary equipment (e.g. scaffolds or hoists).

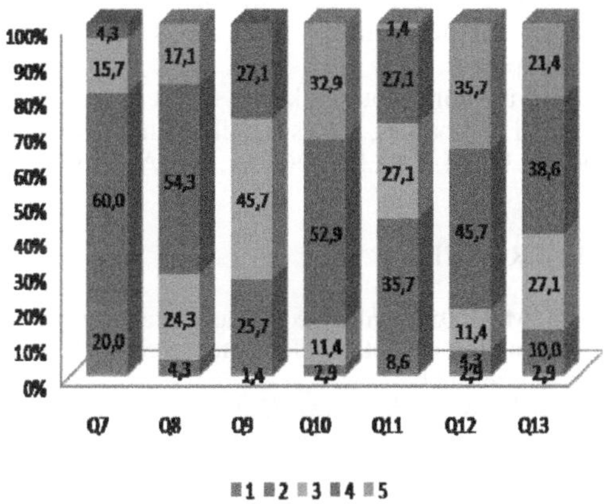

Figure 3. Level of agreement. Items Q7–Q13.

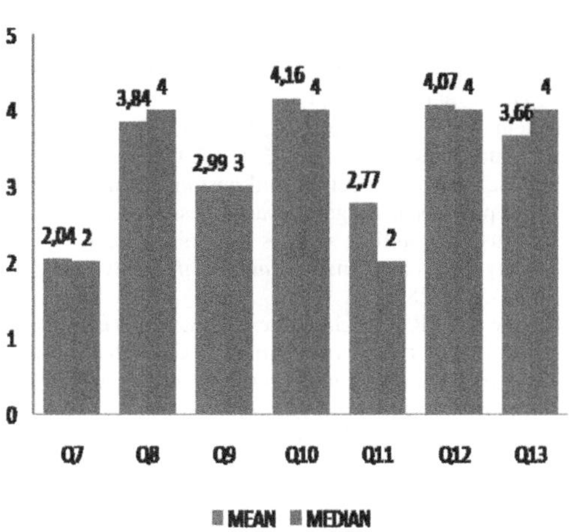

Figure 4. Mean and mode values. Items Q7–Q13.

Q12—Formwork/falsework manufacturers provide training in health and safety to their customers in the use of their products.

Q13—Formwork/falsework manufacturers provide the customer with qualified technicians to erect, use and dismantle the formwork/falsework and their auxiliary equipment.

Figure 3 showed the percentage of respondents based on their level of agreement, according to the Likert scale proposed. It is especially remarkable that in opinion of the experts, customers do not use to choose the safest product, because they choose the cheapest one. Item Q10 about the safety advices included in technical advices of the product obtained the higher level of agreement from respondents. In contrast item Q7 about the importance of the safety as influence factor when a formwork/falsework is selected obtained the lowest level of agreement.

Figure 4 include the mean and median values obtained by items related with construction phase. Similar values between means and medians were observed.

3.3 *Legal issues*

In the last category of the questionnaire, items about the influence of hypothetical new standards about formwork/falsework and Occupational Health and Safety issues were included:

Q14–A compulsory standard about formwork/falsework design and manufacture would improve health and safety in the final formwork/falsework as a product.

Q15–A non-compulsory specific standard (ISO, BS or similar) about formwork/falsework design and manufacture would improve health and safety in the final formwork/falsework as a product.

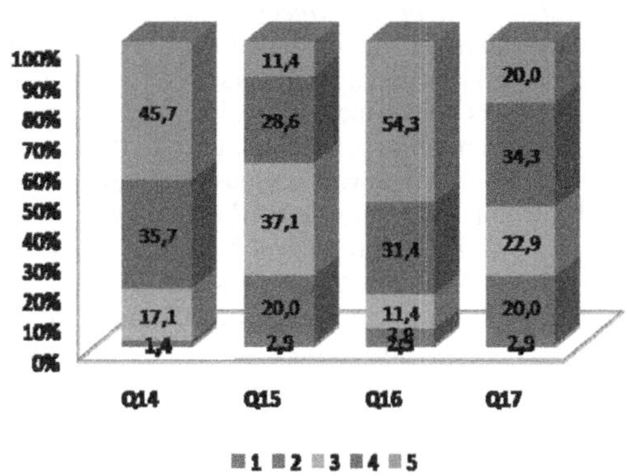

Figure 5. Level of agreement. Items Q14–Q17.

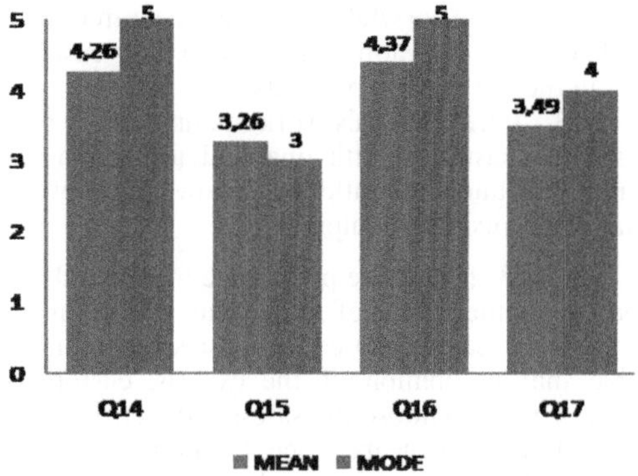

Figure 6. Mean and mode values. Items Q14–Q17.

Q16—A compulsory standard about formwork/falsework use would improve health and safety for the formwork/falsework workers.

Q17—A non-compulsory specific standard (ISO, BS or similar) about formwork/falsework use would improve health and safety for the formwork/falsework workers.

The percentage of respondents based on their level agreement, for each legal issue item, can be observed in figure 5. It must be highlighted that a big consensus were obtained about the positive influence of new standards about manufacturing and use of formworks. Compulsory standards obtained higher values of agreement than non-compulsory standards.

Mean and mode values from legal issues items are showed in figure 6.

4 CONCLUSIONS

Conclusions can be summarized according to the results as follows:

a. *Design phase of the project and product*: In the design phase of the construction project, lack of interaction between designers and formworks/falseworks companies was pointed by respondents as a common practice. Regard to the design phase of the formwork/falsework, final cost of the product was considered the most important factor while the product is being designed. In addition, productivity factor was considered an important factor too. In contrast, safety factor obtained the lowest score.
b. *Construction phase*: In opinion of the manufacturers, users do not follow instructions about the product frequently. Big consensus was obtained in the statement about training in health and safety from manufacturers to the customers in the use of their products.

   Another remarkable finding was that formwork/falsework suppliers are not usually the same suppliers for the rest of temporary equipment.
c. *Legal Issues*: New standards about design and use of formwork, would improve health and safety levels.

In future researches will be interesting to include in the sample studied other stakeholders as construction companies and project designers in order to contrast opinion from different groups of interest implied in the civil construction process.

ACKNOWLEDGMENTS

This study was financed by Spanish Government [Ministry of Science and Technology] through the project named Safety from Design in performance of formwork activities at civil works, referenced like BIA2011-27338 into the list of approved project in the National Plan of Research and Development.

The authors would like to thank the experts from all companies, but especially from companies such as ULMA, PERI, DOKA, and ALSINA.

REFERENCES

Behm, M. (2005). Linking construction fatalities to the design for construction safety concept. *Safety Science*, 43(8), 589–611.

Fadier, E., & De la Garza, C. (2006). Safety design: Towards a new philosophy. *Safety Science*, 44(1), 55–73.

Gambatese, J., & Hinze, J. (1999). Addressing construction worker safety in the design phase: Designing for construction worker safety. *Automation in Construction*, 8(6), 643–649.

Gambatese, J.A., Behm, M., & Rajendran, S. (2008). Design's role in construction accident causality and prevention: Perspectives from an expert panel. *Safety Science*, 46(4), 675–691.

Gittleman, J.L., Gardner, P.C., Haile, E., Sampson, J.M., Cigularov, K.P., Ermann, E.D., & Chen, P.Y. (2010). [Case study] CityCenter and cosmopolitan construction projects, las vegas, nevada: Lessons learned from the use of multiple sources and mixed methods in a safety needs assessment. *Journal of Safety Research*, 41(3), 263–281.

Hallowell, M.R., & Gambatese, J.A. (2009). Activity-based safety risk quantification for concrete formwork construction. *Journal of Construction Engineering and Management*, 135(10), 990–998.

Haslam, R., Hide, S., Gibb, A.G.F., Gyi, D.E., Pavitt, T., Atkinson, S., & Duff, A. (2005). Contributing factors in construction accidents. *Applied Ergonomics*, 36(4), 401–415.

Huang, X., & Hinze, J. (2003). Analysis of construction worker fall accidents. *Journal of Construction Engineering and Management*, 129(3), 262–271.

Ismail, Z., Doostdar, S., & Harun, Z. (2011). Factors influencing the implementation of a safety management system for construction sites. *Safety Science*.

Likert, R. (1932). A technique for the measurement of attitudes. *Archives of Psychology*.

López Arquillos, A., Rubio Romero, J.C., & Gibb, A. (2012). Analysis of construction accidents in Spain, 2003–2008. *Journal of Safety Research*.

Meliá, J.L., Mearns, K., Silva, S.A., & Lima, M.L. (2008). Safety climate responses and the perceived risk of accidents in the construction industry. *Safety Science*, 46(6), 949–958.

Salminen, S. (2004). Have young workers more injuries than older ones? an international literature review. *Journal of Safety Research*, 35(5), 513–521.

Saurin, T.A., Formoso, C.T., & Cambraia, F.B. (2008). An analysis of construction safety best practices from a cognitive systems engineering perspective. *Safety Science*, 46(8), 1169–1183.

Sawacha, E., Naoum, S., & Fong, D. (1999). Factors affecting safety performance on construction sites. *International Journal of Project Management*, 17(5), 309–315.

Wang, W., Liu, J., & Chou, S. (2006). Simulation-based safety evaluation model integrated with network schedule. *Automation in Construction*, 15(3), 341–354.

Occupational Safety and Hygiene II – Arezes et al. (eds)
© 2014 Taylor & Francis Group, London, ISBN 978-1-138-00144-2

Information and Communication Technologies (ICT) in Occupational Health and Safety (OHS) courses at engineering under/degree studies

J.C. Rubio-Romero, A. López-Arquillos & M. Suárez-Cebador
Universidad de Málaga, Málaga, Spain

J.A. Carrillo-Castrillo
Universidad de Sevilla, Sevilla, Spain

ABSTRACT: Effectiveness of information and communication technologies (ICT) in OHS courses at university engineering students was measured through the students' perception. A specific online questionnaire was designed for current research. A total of 308 students from engineering courses completed the questionnaire successfully. Results showed that ICT significantly contributed to the understanding of the topics. Combination of use of ICT and face-to face teaching sessions is the best way to learn OHS concepts.

1 INTRODUCTION

Nowadays, developments in Information and Communication Technologies (ICT) have provided new tools which have been progressively implemented in a wide range of educational contexts from primary schools (Hadjithoma & Karagiorgi, 2009) to universities (Brower, 2003) and masters programs (Arbaugh, 2002, 2004).

At university under/degrees, courses focused on occupational safety have been included in the academic curriculum for many years, as reflected in the extensive literature on the subject. For example, Heinrich (1956), Grossel (1992), and Phoon (1997) authored pioneering studies on the integration of safety and health training in graduate and undergraduate engineering programs. More recently, Arezes and Swuste (2012) give an overview of post-graduate courses on occupational health and safety in Europe, the majority of which are in engineering programs.

Previous researches (Chow et al, 1999; Fender, 2002) were focused on OHS in distance learning courses. For instance, Fender (2002) concluded that industries would be positively impacted by distance-education programs as new educational opportunities arose for working professionals. Within this context, and taking into account the ever expanding use of new technologies in education, this study collected data from engineering students with a view to discovering how they perceive the use of ICT in their OHS courses. It is hoped that the results obtained will contribute to a more effective use of information and communication technology in academic contexts, especially in studies related with OHS issues integrated in engineering courses.

2 METHODOLOGY

A total population of 461 students was surveyed, but only 308 of them completed the questionnaire. Questionnaires were answered during academic year 2012/2013. Students are member of three different Spanish Universities. The OHS courses in which the students were enrolled were subjects in the academic programs of engineering under/degrees. Questionnaires were uploaded to the servers and made available to the students who accessed them through the virtual campus.

A Likert-scale questionnaire was designed to collect data concerning the students' opinion and expectations of the ICT tools used in their OHS courses. The items were grouped in three categories:

a. Communication and interaction
Items in this category pertained to the impact of ICT on teacher-student and student-student communication in health and safety subjects and training at the university engineering courses studied.
b. Content and quality
Items in this category pertained to the instrumental application of ICT to academic contents and sought to discover whether ICT enhanced knowledge acquisition and facilitated the learning of course contents.

Table 1. Universities, technical schools, and subjects in the degree programs.

University	School	Subjects
Malaga	Industrial Engineering	Industrial Safety
		Quality Management
		Business Administration Economy
Granada	Civil Engineering	Prevention at Work
		Safety at Work
Seville	Industrial Engineering	Occupational Safety

c. Risk prevention

Items in this category pertained to the students' current habits and expectations regarding their present and future training in risk prevention in the context of their professional trajectories as future engineers.

Possible answers ranged from value 1 (strongly disagree) to value 5 (strongly agree). Some of the statements in this questionnaire were based on a previous study by Castro and Chirino (2010), a study focused on teachers' opinions and perceptions of ICT. The data thus collected were statistically processed in order to compare responses. The mean value was regarded as a measurement reflecting the general tendency of the answers. Data dispersion was evaluated by means of standard deviation values.

3 RESULTS AND DISCUSSION

In this section, results obtained from students perception are showed. Results were divided in different sections attending to their item categories.

3.1 *Communication and interaction*

Questions contained in this subsection are the following:

Q1—The use of ICT tools decreases face-to-face interaction between students.

Q2—The use of ICT tools decreases face-to-face interaction between students and teachers.

Q3—The use of ICT improves cooperation between students.

Q4—In all courses, ICT tools are more social than academic.

In current subsection the highest agreement mean value was obtained for the item stating that ICT improves cooperation between students.

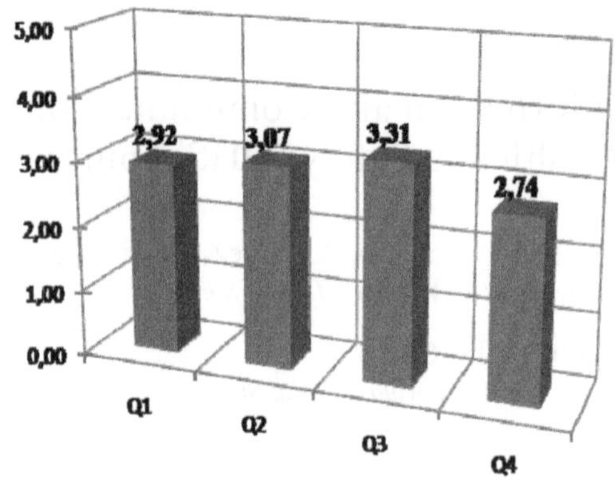

Figure 1. Results of level of agreement by communication and interaction items. Mean values.

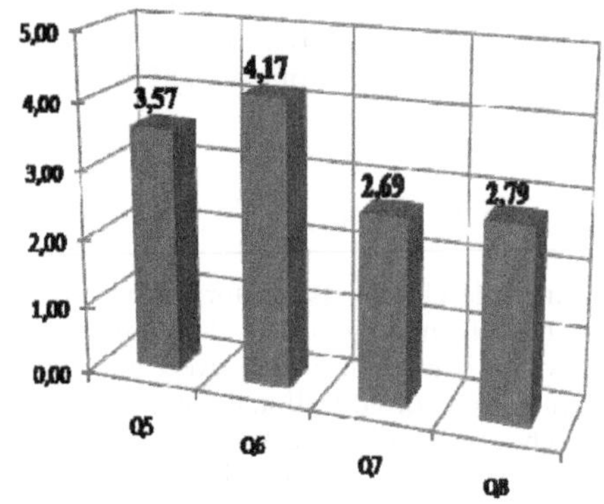

Figure 2. Results of level of agreement by content and quality items. Mean values.

In contrast the lowest level of agreement corresponded to the use of ICT more as a social than an academic tool. It is significant that according to the opinion of the students, the decrease of interactions face-to face between students and teachers is bigger than decrease of interactions between the students.

3.2 *Content and quality*

Item included in the questionnaire in this subsection are:

Q5—The use of ICT tools facilitates the understanding of course contents.

Q6—The use of ICT facilitates and improves access to course contents.

Q7—The use of ICT is confusing because it provides too much information.

Q8—The use of ICT makes face-to-face tutorial sessions less necessary.

The results obtained in current subsection showed the positive impact of ICT tools on teach-

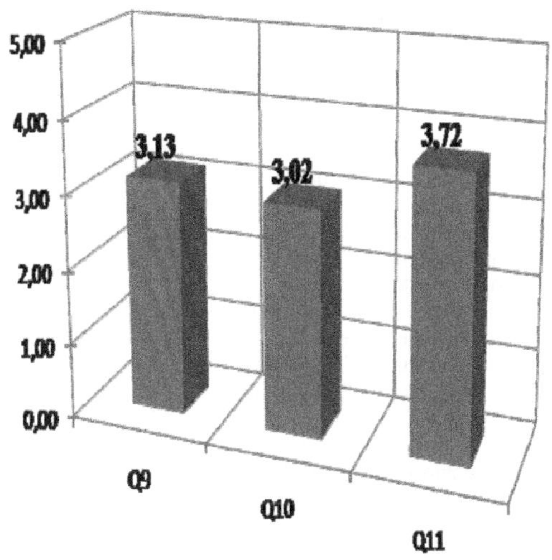

Figure 3. Level of agreement by risk prevention items. Mean values.

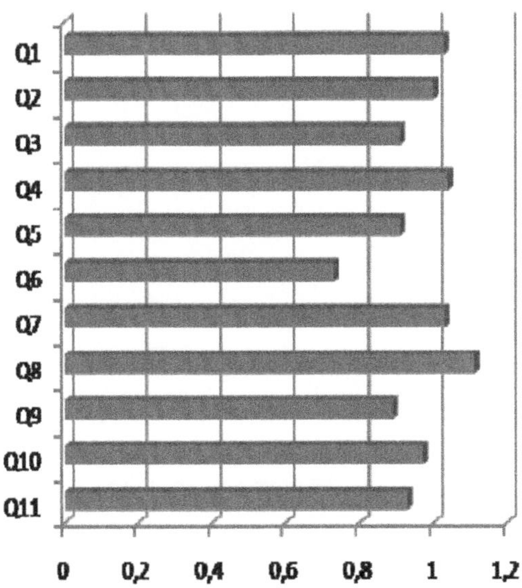

Figure 4. Standard deviation values of the questions.

ing course contents and the quality of learning this achieved. On the one hand, the highest agreement mean value were obtained for the item stating that ICT facilitated and improved the access to course contents.

On the other hand, the lowest level of agreement was obtained for the item stating that the use of ICT is confusing because it provides too much information. So according students opinion information provided by new technologies is a good quality information and not confusing.

3.3 *Risk prevention*

In current subsection the following questions were added:

Q9—The use of ICT has improved my knowledge of occupational risk prevention.
Q10—I often use ICT for self-learning in occupational risk prevention.
Q11—The best way to continue learning about occupational risk prevention is through the combined use of ICT and face-to-face support.

Results of the last three items included in the questionnaires obtained a level of agreement higher than three in their mean values. Item with the highest mean value affirmed that the combined use of ICT and face to face support was the best way to continue learning OHS issues.

In order to test the statistical quality of the results, standard deviation was calculated for all items studied. Majority of the items obtained a value of standard deviation lower than the unit. Results are showed in figure 4.

4 CONCLUSIONS

Results obtained in current research showed that students from OHS courses have a positive perception about the effectiveness of the use of ICT tools. Respondents indicated that the integration of ICT in OHS university courses could benefit the learning process in many ways.

Students can access to course content more easily and at the same time ICT tools could facilitate understanding of the contents. In addition cooperation between students would increase.

According to the results obtained, effectiveness of ICT tools would be even greater if they were used in addition to other learning tools.

ICT tools for OHS courses could be improved if they are adapted to new generation of application for mobile devices.

ACKNOWLEDGEMENTS

Authors would like to thank their financial support to Consejería de Economía Innovación Ciencia y Empleo de la Junta de Andalucía, and LIMASA III. Without their help, this study would not have been possible.

REFERENCES

Arbaugh, J.B. (2004). Learning to learn online: A study of perceptual changes between multiple online course experiences. The Internet and Higher Education, 7, pp. 169–182.

Arbaugh, J.B. (2002). Managing the on-line classroom: A study of technological and behavioral characteristics of web-based MBA courses. The Journal of High Technology Management Research, Volume 13, Issue 2, pp. 203–223.

Arezes, P.M. Swuste, P. (2012). Occupational health and safety post-graduation courses in Europe: A general overview. Safety Science, Volume 50, Issue 3, pp. 433–442.

Brower, H.H. (2003). On emulating classroom discussion in a distance-delivered OBHR course: Creating an online community. Academy of Management Learning and Education, 2(1), pp. 22–36.

Castro, J.J., Chirino, E. (2010). Teachers' opinion survey on the use of ICT tools to support attendance-based teaching. Computers & Education, Volume 56, Issue 3, pp. 911–915.

Chow, W., Wong, L., Chan, K., Fong, N., Ho, P. (1999). Fire safety engineering: comparison of a new degree programme with model curriculum. Fire Safety Journal, 32, pp. 1–15.

Fender, D.L. (2002). Student and faculty issues in distance education occupational safety and health graduate programs. Journal of safety research 33 (2): pp. 175–193.

Grossel, S., (1992). Current status of process safety/prevention education in the US. Journal of Loss Prevention in the Process Industry 5, 2.

Hadjithoma,C., Karagiorgi, Y. (2009). The use of ICT in primary schools within emerging communities of implementation. Computers & Education, Volume 52, Issue 1, pp. 83–91.

Heinrich, H., (1956). Recognition of safety as a profession, a challenge to colleges and universities. National Safety Council Transactions. In: Proceedings of the 44th National Safety Congress, October 22–26, Chicago, Ill, pp. 37–40.

Phoon, W.(1997) Education and training in occupational and environmental health. Environmental Management and Health 8(5), pp. 158–161.

Occupational Safety and Hygiene II – Arezes et al. (eds)
© 2014 Taylor & Francis Group, London, ISBN 978-1-138-00144-2

Prevention through Design (PtD) concept at university. Engineering & Architecture students' perspective

J.C. Rubio-Romero & A. López-Arquillos
Universidad de Málaga, Málaga, Spain

M.D. Martínez-Aires
Universidad de Granada, Granada, Spain

J.A. Carrillo-Castrillo
Universidad de Sevilla, Sevilla, Spain

ABSTRACT: Although the importance of the Prevention through Design concept, many accidents still happen because cited concept is not implemented properly in construction projects. Aim of current research is to quantify the integration of Prevention through Design concept at university courses about construction or design of concrete structures at engineering and architecture. Opinions from 246 students were collected using a specific questionnaire. Students were selected from engineering and architecture under/degrees at university of Granada. Results showed different opinions between students from Old degrees and Bologna degrees. Prevention through Design concept had better integration in courses from old degrees. Absence of education and training in Prevention through Design was found in the construction courses studied. Future engineering and architecture syllabi should improve the presence of Prevention through Design concept.

1 INTRODUCTION

Construction sector has the highest incidents rates in many countries. (Eurostat, 2013). Due that, many authors have investigated the problem in countries as different as Taiwan (Cheng et al., 2010), Scotland (Cameron et al., 2008) Turkey (Etiler et al., 2004) (Müngen & Gürcanli, 2005) Portugal (Macedo & Silva, 2005), South Korea (Im et al., 2009). In Spain, accident rates are not better (Camino et al., 2008, López-Arquillos et al., 2012, Martinez-Aires et al., 2010).

Prevention through Design concept (PtD) is especially relevant when accidents in civil construction are studied. The concept definition is available at the web site of the National Institute for Safety and Health (NIOSH, 2013) as:

> "Addressing occupational safety and health needs in the design process to prevent or minimize the work-related hazards and risks associated with the construction, manufacture, use, maintenance, and disposal of facilities, materials, and equipment".

Cited concept has been addressed in many researches by several authors (Schulte., et al., 2008; Gambatese et al., 2008; Toole & Gambatese, 2008; Gangolells et al., 2010).

Other similar approaches about the importance of design in the construction project's safety, consider the improvement of OHS through design (Design for Safer, Safety by Design, or Systematic Design Approach).

According to existent literature many accidents could be eliminated or reduced with a correct implementation of PtD during the design phase and the development of the project (Haslam, 2005; Gibb, 2006; Gambatese et al., 2008). In Europe, architects and design engineers are required to implement design for construction safety (ILO, 1985) but unfortunately Prevention through Design concept is often not clearly integrated in engineering and architecture university courses. In consequence many professionals are required to implement a concept little-known or unknown along their academic trajectories. In general, safety contents are not integrated on under/graduate curricula as it could be desirable (HSE, 2009) but this integration is difficult due to the already crowed under/graduate curricula (Culvenor & Else, 1997).

Aim of present research is to quantify the integration of Prevention through Design concept at university courses about construction or design of concrete structures at engineering and architecture.

2 METHODOLOGY

With the aim to collect the opinion from the university students a sample of them were chosen from engineering and architecture courses at Granada University (Spain) focused on the design or construction of concrete structures. It is important to consider that university students can be divided in two groups depending on their academic itinerary. Students from degrees approved before the Bologna process, (Old degrees (OD)) and students from degrees created during the Bologna process in order to achieve the European Higher Education Area (Bologna degrees (BD)). Although now number of students is similar in both, evolution of the distribution of the student is quite different. Old degrees are in an extinction process and the number of students is decreasing on them, at the same time that Bologna degrees are in an implementation process in order to substitute old degrees, and the number of students are increasing.

Criteria of selection were that the course was part of a civil or building construction under/degree and included in its academic programs contents about design or construction of structures. A total of 246 students, from four different under/degrees completed the questionnaire. (62.2% from Old degrees and 37.8% were from Bologna degrees).

A Likert-scale questionnaire (Likert, 1932), was designed in order to collect the opinion from the sample selected. Likert scale questionnaire has been demonstrated as a useful tool in others researches about occupational health and safety in construction (Melia et al., 2008; Gittleman et al., 2010; Ismail et al., 2012). This is the reason of the application of this methodology in the questionnaire form designed. Questionnaire was designed to be simple and brief. It contained 11 items grouped in two categories of questions:

- In the first category, questions were related with their general education and training at Occupational Health and Safety issues.
- In the second category, specific questions about the influence of Prevention through Design concept in the course were included.

Respondent were asked to evaluate each item in a Likert-Scale from 1 [Strongly disagree] to 5 [Strongly agree].

3 RESULTS

Results obtained from questionnaires were divided in two different subsections attending to the category of the item.

3.1 *Occupational Health and Safety issues*

Items related with Occupational Health and Safety issues included in the current subsection are the following:

Q1—Occupational Health and Safety of workers at vertical construction is integrated with the rest of the technical concepts at every lecture of the course

Q2—Occupational Health and Safety of workers at vertical construction is only considered at some lectures of the course

Q3—Workers Occupational Health and safety is often confused with structural safety of the construction

Q4—Workers Occupational Health and safety topic is addressed in other specifics courses of the under/degree

Q5—Knowledge on Occupational Health and safety for workers are evaluated in the course

Results from Old degrees' students, and from Bologna degrees' students are showed in figure 1 and figure 2.

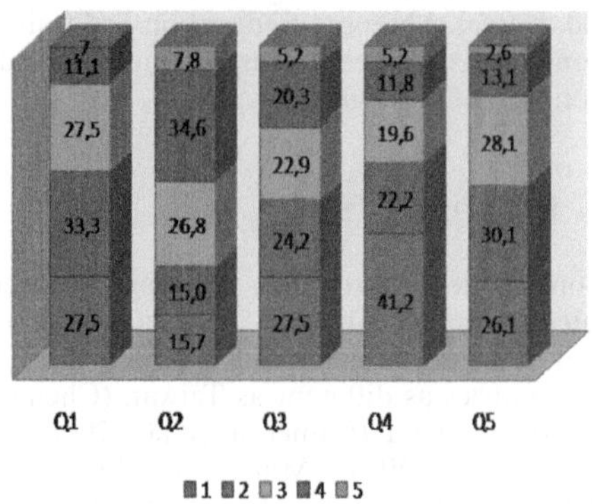

Figure 1. Old degrees respondents. Distribution of results from items Q1–Q5.

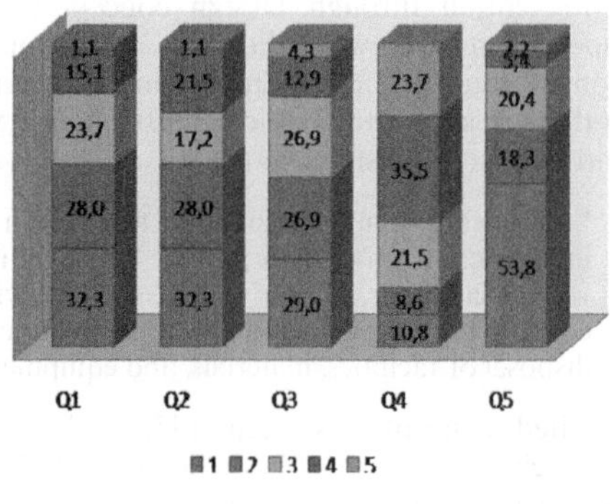

Figure 2. Bologna degrees respondents. Distribution of results from items Q1–Q5.

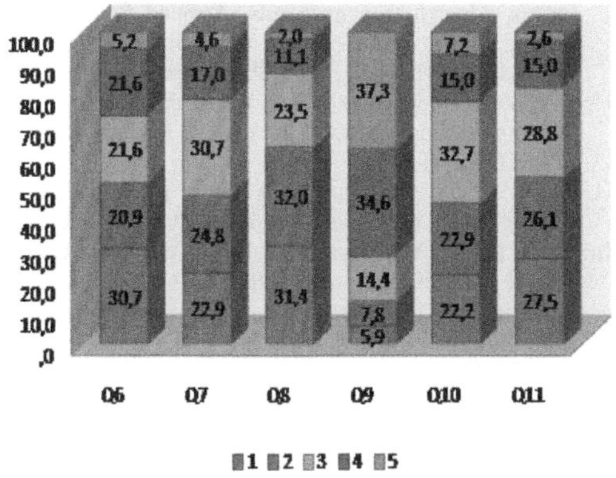

Figure 3. Old degrees respondents. Distribution of results from items Q6–Q11.

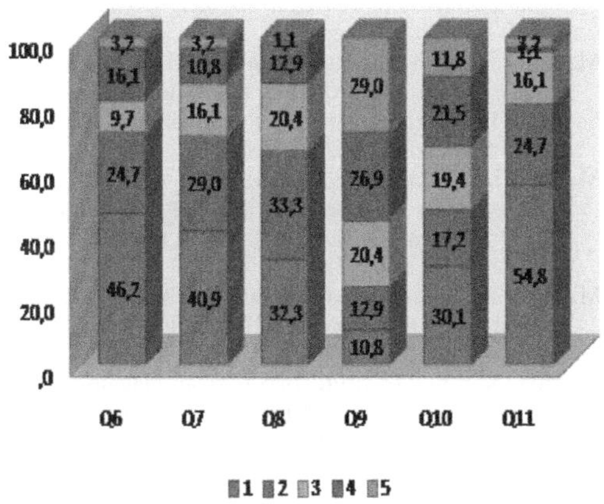

Figure 4. Bologna degrees respondents. Distribution of results from items Q6–Q11.

It is remarkable that integration of the OHS of workers obtained worse results for Bologna degrees, but in contrast OHS issues are addressed in more specific courses in Bologna degrees if we compare them with Old degrees. Values obtained for the item about confusion between.

3.2 *Prevention through Design influence*

In current subsection items included are related with the influence of PtD concept in courses studied. Questions at this category are:

Q6—I know the meaning of Prevention through Design (PtD) concept

Q7—Worker safety on the design of the construction project is considered along the course

Q8—Prevention through Design (PtD) concept is addressed in many courses like this one

Q9—Prevention through Design (PtD) concept is important to improve workers occupational health and safety in vertical constructions

Q10—Prevention through Design (PtD) concept is not important if compare it with the rest of the course contents.

Q11—Knowledge about Prevention through Design (PtD) concept is evaluated in this course

It is important to note that the knowledge about the meaning of PtD in Old degrees is higher than Bologna degrees. Similarly, the importance of the PtD concept is lower between students from Bologna degrees, than students from Old degrees.

4 CONCLUSIONS

Results obtained in current research from students selected, pointed that level of education and training at Prevention through Design between engineers and architecture students is not as good as could be desirable and a big improvement is required. Comparative study between Old degrees and Bologna degrees are not very positives because the evolution of the contents of construction courses about OHS and PtD has been poor. A lower level of knowledge of PtD concept was found linked to a lower consideration of the importance of PtD concept in construction, in consequence better knowledge of PtD concept would improve the importance of the concept between future professionals from construction sector.

4.1 *Future research*

Current analysis should be extended to other universities in order to compare the similarities and differences between the performance and integration of PtD in university construction courses.

ACKNOWLEDGEMENTS

This study was financed by Spanish Government [Ministry of Science and Technology] through the project named Safety from Design in performance of formwork activities at civil works, referenced like BIA2011-27338 into the list of approved project in the National Plan of Research and Development.

The authors would like to thank the students from all construction courses studied for their opinion and suggestions. Without their collaboration this study would not had been possible.

REFERENCES

Cameron, I., Hare, B., Davies, R. (2008). Fatal and major construction accidents: A comparison between Scotland and the rest of Great Britain. *Safety Science*, 46, 692–708.

Camino, L.M., Ritzel, D.O., Fontaneda, I., & González, A.O. (2008). Construction industry accidents in Spain. *Journal of Safety Research*, 39(5), 497.

Cheng, C.W., Leu, S.S., Lin, C.C., & Fan, C. (2010). Characteristic analysis of occupational accidents at small construction enterprises. *Safety Science*, 48(6), 698–707.

Culvenor, J., & Else, D. (1997). Finding occupational injury solutions: The impact of training in creative thinking. *Safety Science*, 25(1), 187–205.

Etiler, N., Colak, B., Bicer, M., Barut, N.(2004). Fatal Occupational Injuries among Workers in Kocaeli, Turkey, 1990–1999. *International Journal of Occupational and Environmental Health*, 10(1), 55–62.

Eurostat, 2013. European Statistics Official; Available at: http://epp.eurostat.ec.europa.eu. Accessed March, 2013.

Gambatese, J.A., Behm, M., & Rajendran, S. (2008). Design's role in construction accident causality and prevention: Perspectives from an expert panel. *Safety Science*, 46(4), 675–691.

Gangolells, M., Casals, M., Forcada, N., Roca, X., & Fuertes, A. (2010). Mitigating construction safety risks using prevention through design. *Journal of Safety Research*, 41(2), 107–122.

Gibb, A.G.F. Haslam, R.A. Hide, S. Gyi, D.E. Duff, A.R. (2006). What causes accidents, Civil Engineering. *Proceedings of the Institution of Civil Engineers*, 159 (Special Issue 2) pp. 46–50.

Gittleman, J.L., Gardner, P.C., Haile, E., Sampson, J.M., Cigularov, K.P., Ermann, E.D.,& Chen, P.Y. (2010). [Case Study] CityCenter and Cosmopolitan Construction Projects, Las Vegas, Nevada: Lessons learned from the use of multiple sources and mixed methods in a safety needs assessment. *Journal of Safety Research*, 41(3), 263–281.

Haslam, R.A. Hide, S.A. Gibb, A.G.F. Gyi, D.E. Pavitt, T. Atkinson S. et al. (2005). Contributing factors in construction accidents. Applied Ergonomics, Invited paper, special edition on ergonomics in building and construction, 36(4) (2005), pp. 401–416.

HSE, (2009). Integrating risk concepts into undergraduate engineering courses. Research Report RR702, HMSO Norwich.

ILO (1985). Safety and health in building and civil engineering work. *International Labour Office*, Geneva.

Im, H.J., Kwon, Y.J., Kim, S.G., Kim, Y.K., Ju, Y.S., & Lee, H.P. (2009). The characteristics of fatal occupational injuries in Korea's construction industry, 1997–2004. *Safety Science*, 47(8), 1159–1162.

Ismail, Z., Doostdar, S., & Harun, Z. (2012). Factors influencing the implementation of a safety management system for construction sites. *Safety Science*, 50(3), 418–423.

Likert, R. (1932). A technique for the measurement of attitudes. *Archives of psychology*.

López Arquillos, A., Rubio Romero, J.C., & Gibb, A. (2012). Analysis of construction accidents in Spain, 2003–2008. *Journal of Safety Research*.

Macedo, A.C., Silva, I.L. (2005). Analysis of occupational accidents in Portugal between 1992 and 2001. *Safety Science*, 43, 269–286.

Martínez Aires, M.D., Rubio Gámez, M.C., & Gibb, A. (2010). Prevention through design: The effect of European Directives on construction workplace accidents. *Safety Science*, 48(2), 248–258.

Meliá, J.L., Mearns, K., Silva, S.A., & Lima, M.L. (2008). Safety climate responses and the perceived risk of accidents in the construction industry. *Safety Science*, 46(6), 949–958.

Mungen, U., Gürcanli, G.E.,(2005). Fatal traffic accidents in the Turkish construction industry. *Safety Science*, 43(5–6), 299–322.

NIOSH. National Institute for Safety and Health (2013). http://www.cdc.gov/niosh/topics/ptd/#face

Schulte, P.A., Rinehart, R., Okun, A., Geraci, C.L., & Heidel, D.S. (2008). National prevention through design (PtD) initiative. *Journal of Safety Research*, 39(2), 115–121.

Toole, T.M., & Gambatese, J. (2008). The trajectories of prevention through design in construction. *Journal of Safety Research*, 39(2), 225–230.

Occupational Safety and Hygiene II – Arezes et al. (eds)
© 2014 Taylor & Francis Group, London, ISBN 978-1-138-00144-2

Prevalence of Work-related Musculoskeletal Disorders (WMSD) symptoms among milkers in the State of Paraná, Brazil

L. Ulbricht, E.F.R. Romaneli, A.M.W. Stadnik, M. Maldaner & E.B. Neves
Federal Technological University of Paraná, Curitiba, Paraná, Brazil

ABSTRACT: Available information regarding the countryside activity, put it as one of the three major occupational risk activities (along with mining and civil construction). The aim of this research was to identify the prevalence of WMSD symptoms among milkers in the state of Paraná, Brazil. This exploratory descriptive survey adopted the Standardized Nordic Questionnaire, modified by Ulbricht (2003), with 1.103 milkers. Milkers work was studied using the ergonomic analysis in order to identify hazard factors and the risk of developing WMSD. It was observed a prevalence of 83% (915) of WMSD in the last 12 months before the interviews. It was also found that female milkers were more affected 90.4% (283) compared to male milkers 80% (632). With this distribution of WMSD by gender of milkers, there has been an Odds Ratio (OR) for developing WMSD among females of 2.35 (confidence interval = 1.03 to 5.38) compared to males.

1 INTRODUCTION

Available information regarding the agricultural activity, put it as one of the three major occupational risk activities (along with mining and civil construction), this situation is exacerbated by the fact that it uses a great part of the total labor force (59% in developing or underdeveloped countries) and uses workers who are not normally involved in other dangerous occupations (under 16 and over 60 years old) (Ulbricht, 2003).

While in urban activities, there is a number of government supervision agencies, whose aim are to maintain the health of the work environment, in the Brazilian agricultural areas this kind of work is still rare and usually focuses on analyzing the relationship of labor and avoid slave labor. Although mortality rates related to work have presented a decrease in the last decade, they have increased in agriculture (Popija & Ulbricht, 2005).

Studies have shown that countryside workers have several risk factors in their activities for the development of Work-related Musculoskeletal Disorders (WMSD) (Patsis & Ulbricht, 2006; Singh & Arora, 2010).

The health of countryside workers was reported as the fifth category (in the worker's health area) further studied (Bezerra & Neves, 2010). The same study has showed that musculoskeletal disorders are the third most studied category. However, there are few studies related to the musculoskeletal disorders in countryside workers, specifically involving milkers.

A survey among milkers in the state of Santa Catarina showed an extremely concerning situation, where 85.16% of milkers had complaints of pain and discomfort due to WMSD (Ulbricht, 2003). Furthermore, it was found that there were several inadequacies in the workplace. It leaded milkers to adopt constraining postures for your musculoskeletal system, there was no adjustment of the milking equipment and that they presented musculoskeletal overload.

In order to expand the knowledge of this occupational health problem, the aim of this research was to identify the prevalence of WMSD among milkers in the state of Paraná, Brazil. The choice of this target was due to the fact that the state of Paraná is the fifth largest producer of milk in Brazil, despite representing only 2.3% of the country area (Brazil, 2009). Its agricultural structure is formed predominantly of small and medium properties with 87% of them having less than 50 ha, basically composed by family properties.

2 MATERIALS AND METHOD

This exploratory descriptive survey adopted the Standardized Nordic Questionnaire modified by Ulbricht (2003) to determine the prevalence of WMSD symptoms in milkers in the state of Paraná, Brazil. A number of 1103 milkers from all regions of the state of Paraná where they were interviewed using a quota sample method (3% sampling error and confidence level of 95%).

Milkers work was studied using the Work Ergonomic Analysis in order to identify risk factors and to develop recommendations that could minimize them. The analysis of complaints of pain was focused in two main regions: wrist and hand and lumbar spine (Ulbricht, 2003).

The risk of developing WMSD among the milkers has been determined using Jonsson's methodology. His methodology studied a population of 84,643 males and female workers of different occupations (Pinzke, 1999).

It was performed descriptive statistics with measurements of location (mean, median and mode) to find the frequency distribution on the axis of variation and scatter (standard deviation and amplitude). Data were analyzed by SPSS (Statistical Package for Social Science) to perform the statistical treatment which used the frequency distribution, the definition of quartiles to be used in exploratory data analysis, the Odds Ratio (OR) was calculated to evaluate the association between musculoskeletal pain with: milkers age and milking system. The OR allows testing the force of association between two variables (Neves et al., 2009).

3 RESULTS

The sample can be characterized by having mean age of 45 years (where the youngest had 15 years and the oldest 74 years). The median was 44 years (showing a roughly symmetric distribution of this sample in terms of age). Separating the sample by gender, respondents were 72% (790) males and 28% (313) were females. In these 1103 milkers, it was observed a prevalence of 83% (915) of WMSD symptoms in the last 12 months before the interview. It was also found that females were more affected 90.4% (283) compared to male milkers 80% (632). With this distribution of WMSD symptoms by gender of milkers, there has been an Odds Ratio (OR) for developing WMSD among females of 2.35 (confidence interval = 1.03 to 5.38) compared to males.

Regarding the organization of work, the work was performed exclusively by the members of the family in the visited properties and most of the milking was performed with manual system. Also there was no hierarchy among the workers (there is no distinction between administrators and executors of tasks) or task division by specialization of labor, as usually happens in any urban profession. All family members learnt to perform all tasks since childhood, as it was observed in several properties where the minor children assisted in the execution of the work.

The activity of milking is seen by many as only the extraction of milk, however, this activity is composed of many subtasks that require much physical overload at the extraction of milk. Some of these subtasks can be described as:

a. ensure the feeding of animals in all periods of the year (requiring planting and harvesting);
b. clean and disinfect the milking area and tools to be used (e.g. buckets, hoses and milking clusters);
c. gather the animals in the pasture and lead them to dairy farm in order to start milking;
d. calm the animal before milking distributing food;
e. immobilize the animal, tying the posterior limbs and tail;
f. clean and lightly massage the udder of the animal and proceed the manual or machine milking,
g. weigh the milk and note the quantity produced by each animal;
h. filter the milk by passing it in suitable devices and store it in cooling tanks;
i. clean the equipment and facilities again;
j. perform the procedures with the cattle such as dehorning, vaccinations, treatments, insemination, among other.

In addition, the ergonomic risks directly related with milking are: constraining postures, static muscle work, repetitive lifting heavy loads, workstations not properly sized. It was also observed accident hazard (risk of falling and slipping); physical hazard (farmers are exposed to radiation, heat or cold depending on the season and region of the state where the property is located and vibration); chemical hazard (e.g. dust, fertilizers and pesticides); and biological hazard (e.g. zoonosis as tuberculosis, brucellosis and leptospirosis, anthrax).

With respect to the distribution of musculoskeletal pain, according to age, four categories were created using the measurement planning by percentiles (each of which has 25% of the sample) and this information was crossed with information about the presence of pain or not. Tables 1 and 2 show the distribution, in the male and female categories, respectively, as well as the calculation of the odds ratio for each category, considering the first category as the reference.

For odds ratio values presented in Tables 1–5, the confidence interval that not contain the value 1.00 indicate statistical significance for this indicator.

According to Tables 1 and 2, it can be observed a tendency to increased risk of musculoskeletal pain with increasing age of the milkers.

Surveyed workers had their workload investigated and it was found that all performed activities on the property resulted in an average of 65 hours

Table 1. Prevalence and the values of Odds Ratio (OR) for musculoskeletal pain, in the male category, with respect to age of milkers, Paraná, Brazil.

	With pain (%)	Without pain (%)	OR
First quartile (15–36 years)	67	33	1.00*
Second quartile (36–44 years)	78	22	1.75
Third quartile (44–52 years)	86	14	3.03**
Fourth quartile (52–74 years)	88	12	3.61**

*Reference; **Interval of Confidence not include the value 1.00.

Table 2. Prevalence and the values of Odds Ratio (OR) for musculoskeletal pain, in the female category, with respect to age of milkers, Paraná, Brazil.

	With pain (%)	Without pain (%)	OR
First quartile (15–36 years)	81	19	1.00*
Second quartile (36–44 years)	92	8	2.70**
Third quartile (44–52 years)	91	9	2.37**
Fourth quartile (52–74 years)	97	3	7.58**

*Reference; **Interval of Confidence not include the value 1.00.

(median of 65 hours) of week labor, indicating a higher workload than urban workers (44h per week).

Analyzing the work of milking in the subtask, extraction of milk, it was identified that the vast ma-jority (71%) of milkers performed their activity twice a day, spending on average 2 hours and 15 minutes (2 hours the median and mode), to milk 14 animals on average (median 10, with an amplitude between 1 and 180 animals per time). There was no significant association between the time spent milking and the presence of musculoskeletal pain.

The mechanical and manual systems of milking were compared with respect to the musculoskeletal pain.

It was found that 83% (911) of the milkers complained of pain or discomfort. When analyzing with respect to the production system, it is observed that the mechanical milking was more damaging where 87% (389) of these milkers were affected if compared with 80% (522) of those who used manual milking. This association was statistically significant.

Table 3. Prevalence and the values of Odds Ratio (OR) for musculoskeletal pain, in all milkers, with respect to milking system, Paraná, Brazil.

	With pain (%)	Without pain (%)	OR
Manual milking	80	20	1.00*
Mechanical milking	87	13	1.67

*Reference.

Table 4. Prevalence and the values of Odds Ratio (OR) for occurrence of musculoskeletal disorders in wrists and hands, in all milkers, with respect to milking system, Paraná, Brazil.

	With pain (%)	Without pain (%)	OR
Manual milking	53	47	1.84**
Mechanical milking	38	62	1.00*

*Reference; **Interval of Confidence not include the value 1.00.

Table 5. Prevalence and the values of Odds Ratio (OR) for occurrence of musculoskeletal disorders in lumbar spine, in all milkers, with respect to milking system, Paraná, Brazil.

	With pain (%)	Without pain (%)	OR
Manual milking	46	54	1.00*
Mechanical milking	59	41	1.69

*Reference.

The analysis of complaints of pain was focused in two main regions: lumbar spine; wrist and hand. It was observed that 47% (514) of the milkers complained of pain or discomfort in the wrists/hands and 51% (562) of the milkers complained of pain or discomfort in the lumbar spine.

The Tables 4 and 5 show the prevalence and the values of odds ratio for occurrence of musculoskeletal disorders in each milking system.

It was observed that the manual milking caused more injury (53%) than the mechanical milking (38%).

Regarding the impact of the milking system in the lumbar spine was observed (Table 5) that mechanical milking is more damaging the lumbar spine (59%) than the manual milking (46%).

4 DISCUSSION

In this study, 83% of milkers reported complaints of musculoskeletal pain or discomfort in the last 12 months. Several authors mention that there is a higher incidence of musculoskeletal disorders in females (Ulbricht, 2003; Singh & Arora, 2010). In this study, 90% of females reported complaints of musculoskeletal pain or discomfort in the last 12 months, versus 80% of males. This result is in agreement with the study of Ulbricht (2003) that assessed the prevalence of musculoskeletal pain in milkers of Santa Catarina (Brazil) and has found a prevalence of 89% among women. Singh & Arora (2010) highlight the need of development of new technologies for females for critical issues in the countryside, such as the adequacy of milking place, lifting and transporting heavy materials.

In this study, it was found a significant association between increasing age and the risk of developing musculoskeletal pain. It was demonstrated by the increased risk associated with increasing chronological age. This result is more significant when only the male subjects are considered. This result was different from that found in a study conducted in Santa Catarina, Brazil, in which the authors haven't found any statistical significance between musculoskeletal pain in the last 12 months and the age of the milkers (Ulbricht, 2003).

The main risk factors found was the early activity in milking (often during childhood and adolescence), assisting the family, combined with a grueling routine (working time load average of 65 hours per week), with no gaps, thereby increasing effort to skeletal muscle that these workers are subjected.

In this study, there was a higher incidence of WMSD symptoms in wrist/hand that performed the workers when manually milking compared with those who performed the milking machine. This result seems obvious when the excessive use of hands in the system without mechanization is considered. Kolstrup (2008) highlights that milking in a rotary system was demanding with respect to high values of velocities and repetitiveness and almost no time for rest for hands/wrists, which might be contributing factors for the development of symptoms and injuries in the hands. This result agrees with one founded in research of Codo & Almeida (1995) involving 620 people and showed that the wrist (20%) and hand (12.3%) were the anatomic regions most affected.

In the mechanized system, on the other hand, one possible explanation for the higher prevalence of musculoskeletal pain in the lumbar spine (59%) between the subjects who perform the milking machine may be the adoption of trunk flexion postures for long periods (representing a static force) to monitoring the process. This corresponds to an odds ratio of 1.69, which means that the mechanical milking increase by 69% the chances of your operator develops lumbar pain when compared with manual milking, in which the posture is usually seated.

5 CONCLUSION

This article demonstrates that the prevalence of WMSD symptoms was 83% in the last 12 months, in the 1.103 milkers studied. It can be considered a high prevalence when compared to others researches founded in literature. In addition, these milkers are exposed to all sorts of hazards (physical, accidents, chemical, biological and ergonomic). This fact places them as a highly vulnerable population to the incidence of WMSD. The results showed that the female is more affected by WMSD symptoms than male and reinforce the need for development of new technologies for improve the conditions of female work in countryside regions such as the adequacy of milking place, lifting and transporting heavy materials.

ACKNOWLEDGMENT

We would like to thank CNPq, CAPES and Araucaria Foundation for financial support to our participation in this event.

REFERENCES

Bezerra, M.L.S. & Neves, E.B. 2010. Perfil da produção científica em saúde do trabalhador; Profile of the scientific production in workers' health. Saudesoc, 19, 384–394.

Brasil. 2009. Instituto Brasileiro de Geografia e Estatística—IBGE. Censo Agrário. Disponível em:<http://www.ibge.gov.br/>. Acesso em: 10 outubro 2009.

Codo, W.; Almeida, M.C.C.G (Org.). 1995. L.E.R.—Diagnóstico, Tratamento e Prevenção. Petrópolis, RJ: Vozes.

Kolstrup C. 2008. Work environment and health among Swedish livestock workers. SLU Printshop, Alnarp Sverigeslantbruksuniv., Acta Universitatis agriculturae Sueciae.

Neves, E.B., De Souza, M.N. & De Almeida, R.M.V.R. 2009. Military parachuting injuries in Brazil. Injury, 40, 897–900.

Patsis, K.S., Ulbricht, L. Desmistificando as LER/DORT In: XXI Jornada Paranaense de Saúde Ocupacional. Londrina: APAMT, 2006.

Pinzke, S. 1999. Towards the Good Work: Methods for Studying Working Postures to Prevent Musculoskeletal Disorders with Farming as Reference Work. Doctoral Thesis from Swedish University of Agricultural Sciences, Alnarp, 1999.

Popija, M., Ulbricht, L. 2005. Detecção de Riscos no Trabalho Agrícola: Um Estudo de Caso. In: XII Seminário Sul Brasileiro da ANAMT e XX Jornada Paranaense de Saúde Ocupacional. Curitiba: APAMT, 2005.

Ulbricht, L. 2003. Fatores de risco associados à incidência de DORT entre ordenhadores em Santa Catarina. Doutorado em Engenharia de Produção, Universidade Federal de Santa Catarina.

Occupational Safety and Hygiene II – Arezes et al. (eds)
© 2014 Taylor & Francis Group, London, ISBN 978-1-138-00144-2

Real determination of metabolic rate to improve the PMV model

E.E. Broday, A.A.P. Xavier & R. de Oliveira
UTFPR—Federal University of Technology of Paraná, Ponta Grossa, Brazil

ABSTRACT: The incorrect determination of metabolic rate can be linked to discrepancies between the model of the PMV and real thermal sensation collected in field studies. Aiming at improving adherence to the PMV and the real thermal sensation, this work has established a new value for the metabolic rate for welder's activities using a metabolic analyzer. The value found was 145.46 W/m², different from the range provided by ISO 8996 (2004) for this activity. It was made a linear regression between PMV and the real thermal sensation in two ways: S × $PMV_{tabulated}$ ($R^2 = 0.1749$) and S × PMV_{real} ($R^2 = 0, 7854$). Values obtained in "real" were used for the correction of the table. The corrected table provides a $M_{predicted}$ and tabulated values can be multiplied by the correction coefficient 1.4546 for welder's activities. The relation between $PMV_{predicted}$ and thermal sensation provides a coefficient of determination of 0.7338.

1 INTRODUCTION

This study will emphasize the thermal comfort that, according to Frontczak and Wargocki (2011), has six different variables: air temperature, mean radiant temperature, air velocity, relative humidity, clothing insulation and the metabolic rate, subject of this research.

The metabolic rate (M) is produced by the body, through the transformation from chemical into mechanical and heat energy. Humphreys (1996), De Dear et al. (1998), Xavier (2000), Vergara (2001), Havenith et al. (2002) and Antonelli (2012) claim that the results of the PMV (Predicted Mean Vote) model, when using the real thermal sensation of people collected in field research, show significant discrepancies. One of the possible factors to this disparity is the determination of the metabolic rate.

PMV is an index that shows the average thermal sensation of a large group of people exposed to the same environment (ISO 7730 [2005]). The PMV is the thermal comfort index used to assess the thermal sensation in a moderated environment.

According to Havenith et al. (2002) and Katavoutas et al. (2009), the estimated metabolic rate for people, developing the same activity, should not be used in generalized way, due to individual differences and possible psychological or sociological influences on the work environment. The tabulated values presented in ISO 8996 (2004) are widespread, as in tables is not considered the way in which the activity is performed, the time of exposure to the environment, the form of execution of the task, and much less to the person's adaptation to the activity, which can lead to distortions (Antonelli, 2012).

This research determined the metabolic rate using an equipment called "metabolic analyzer". This metabolic rate, provided by the equipment, was named as "real". As the industry analyzed in this research is the metal-mechanical, workers who participated in the study are welders.

In this way, the tabulated and real values are compared and the result of the comparison provides a $M_{predicted}$, with the correction equation for tabulated values.

Through the $M_{predicted}$ it is determined a $PMV_{predicted}$. After this determination will be carried a linear regression between the Real the Thermal Sensation and the $PMV_{predicted}$ to, finally, be able to verify that PMV model improvement occurs through the minimization of inaccuracies of the tabulated metabolic rate values.

2 MATERIALS AND METHODS

2.1 *Data collection procedures*

The company chosen for the development of this research is located in the Industrial District in the South of Brazil and the choice was in reason for its accessibility. The company is considered a metal-mechanic industry, offering a surface treatment service based on special coatings for industrial line, having as its main focus the bakery line. The factory plant can be seen in Figure 1.

The survey took place in two points of the company's activities. The point 1 was on the met-

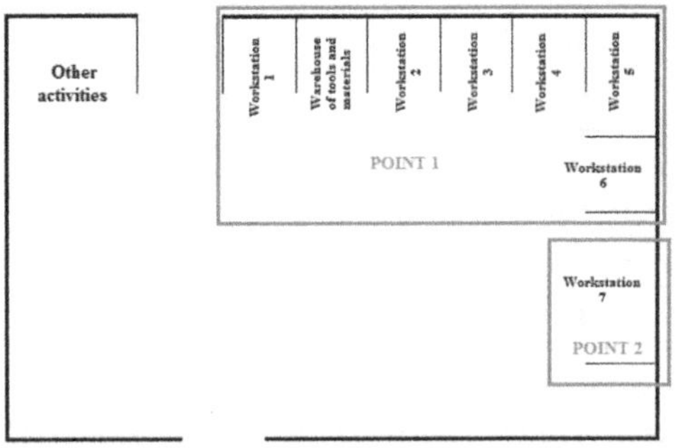

Figure 1. Workstations of the company;
Source: The author.

allurgy, manufacturing of metal boxes to traction batteries. The point 2 is the place where bread pan for food industry is handled. All substrates used in the making of pans or trays can be coated: aluminized steel, carbon steel, stainless steel, tinplate as well as aluminium and its alloys. This research, therefore, reviewed the welders' work of this industry, collecting data on all workstation and all those involved in the environment.

The measurements were carried out from September 2012 to January 2013, on the morning and the afternoon shifts. During the full development of professionals' activities, environmental and personal data were collected. Environmental data, including air temperature, air velocity, temperature of the globe and relative humidity, necessary to determinate metabolic rate calculation and also for the calculation of the PMV, they were recorded by the appliance called *Confortímetro Sensu®* at intervals of 1 minute to 1 minute. The equipment was placed within each workstation, next to the worker, taking turns by the seven workstations used in this research.

Before starting the measurements, the device was assembled so that it would be in balance with the environment, for 20 minutes. The measurement was held for 40 minutes, totalizing one hour for each workstation. After one hour, the unit had recorded 40 data of environmental variables. For the purposes of use of the data, were made the average of these 40 data.

A total of 31 measurements were carried out. According to Triola (2005) with large sample sizes, generally meaning $n > 30$, from a continuous population, the distribution of data is well approximated by a normal distribution. The distribution of data will increasingly approximate a normal distribution as the sample increases. This is described by the Central Limit Theorem. Thus, the Central Limit Theorem allows that approximation and makes it possible to use the normal curve for data evaluation (TRIOLA, 2005).

Table 1. Environmental data.

	Average	Standard deviation
AT	23.97	1.11
AV	0.19	0.16
GT	24.27	1.04
RH	85.00	3.37

where: AT = air temperature (°C); AV = air velocity (m/s); RH = relative humidity (%); GT = globe temperature (°C).

Figure 2. Metabolic analyzer;
Source: The author.

Despite the fact of $n > 30$ not guarantee that the model is stable, this paper considered that in this particular case it is justifiable because there are only two variables in the regression model that will be presented. Data for the environment can be view in table 1.

When it comes to obtaining the real metabolic rate, it was used the metabolic Analyzer VO2000®. However, before using, it was necessary to keep the equipment connected for 30 minutes for calibration. The appliance can be seen in Figure 2:

The metabolic analyzer consists of a mask with a tube through the person must breathe. The wire coming out of the pipe and is connected to the analyzer is very short, making the equipment being placed close to the person, not allowing the worker to have much mobility.

As this work also ran the PMV calculation it was necessary to collect environmental variables through the *Confortímetro* and personal variables through a questionnaire, which during the measurements, the workers were invited to fill it in. In this questionnaire, the employee was asked to fill in your personal information, to inform the clothes he/she was wearing, to choose how he/she was feeling following or according to the index on the seven-point thermal sensation scale (+3 = hot, warm = +2, +1 = slightly warm, 0 = neutral, −1 = slightly cool, −2 = cool, −3 = cold) and to choose how he/she would like to be thermally feeling (+3 = Much warmer, warmer = +2, +1 = A little warmer, 0 = Neither warmer or cooler, −1 = Slightly cooler, cooler = −2, −3 = much cooler). The measurements were performed for an hour in each workstation, and thirty minutes after starting the measurements in each sector, questionnaires were handed out to be filled by employees.

2.2 *Analysis of the values from the obtained metabolic rate*

This work determined a real value for the metabolic rate using a metabolic analyzer VO2000®. Each of these values will generate a value of PMV. Since the goal of this research is to improve the PMV model by decreasing its metabolic rate inaccuracies, it was also calculated a PMV using a tabulated metabolic rate in ISO 8996 (2004), as shown in Figure 3.

In order to verify which of the values of metabolic rate was closer to the model of PMV with the real thermal sensation, it was performed a simple linear regression analysis in three ways:

a. Real thermal sensation as dependent variable and $PMV_{tabulated}$ as independent variable, obtained through the tabulated values of metabolic rate provided in ISO 8996 (2004);
b. Real thermal sensation as dependent variable and PMV_{real} as independent variable, obtained through the equipment VO2000®.

2.3 *Steps to correct the table of ISO 8996 (2004)*

The values determined for the metabolic rate, as arranged in ISO 8996 (2004), are generic and as a result, they do not represent the reality. The fact is that the values of thermal sensation obtained in field study are not consistent with the values of PMV presented in the standard. According to Xavier (2000), this difference may be related to incorrect values of metabolic rate used.

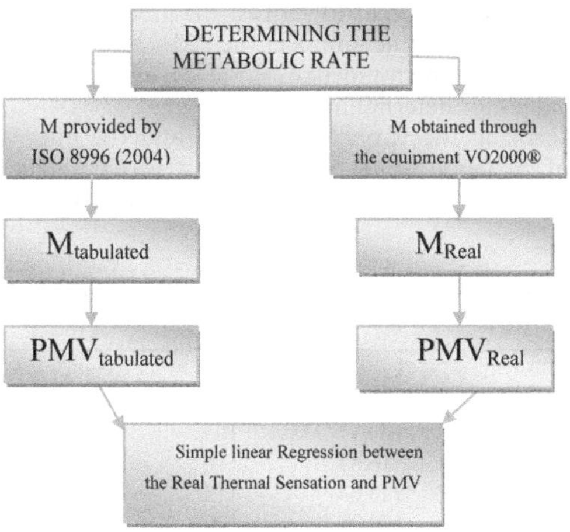

Figure 3. Determination of M; *Source*: The author.

In order to correct the table of ISO 8996 standard (2004) some steps must be followed. The first step is to make a comparison between the values obtained by the metabolic analyzer and the tabulated values of the standard, thus, obtaining the $M_{predicted}$, with the correction equation to the tabulated values. Figure 4 shows the steps that were carried out to obtain the correction function.

After finding the function $M_{predicted}$, it can be determine new values of metabolic rate based on the tabulated values. Obtaining the new values of metabolic rate and using the environmental variables we can determine the $PMV_{predicted}$. This PMV will be related to the real thermal sensation of workers through simple linear regression.

3 RESULTS AND DISCUSSION

The great advantage observed by using the metabolic analyzer is that it provides immediately a value of metabolic rate (in met), without the need to perform any mathematical operation. Figure 5 shows the metabolic analyzer being used in field studies.

As the values of the metabolic rate are given by the equipment in met, it was used the conversion

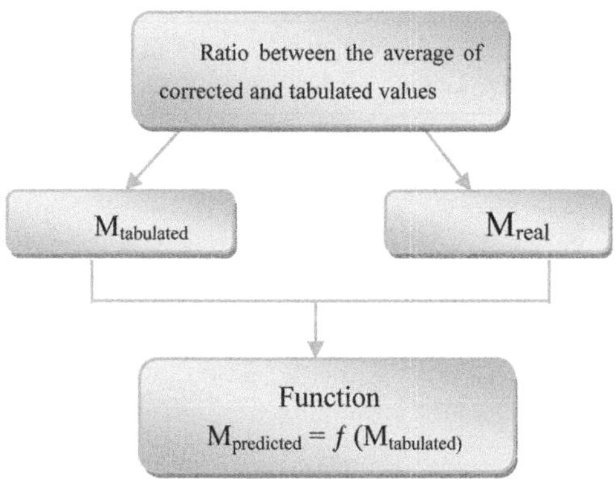

Figure 4. Steps taken to find the correction function; *Source*: The author.

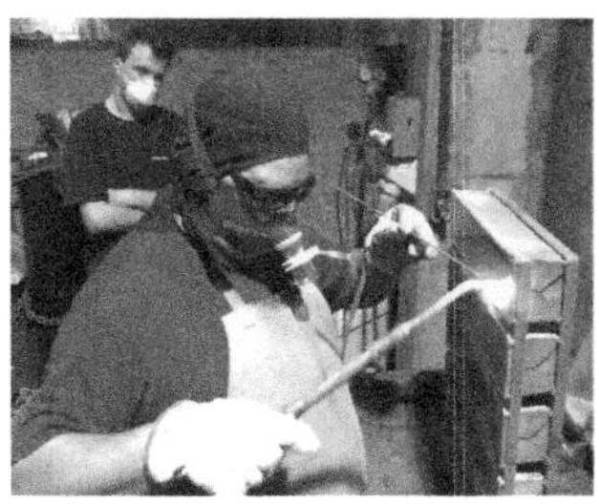

Figure 5. Metabolic analyzer; *Source*: The author.

factor 58.2 W/m² to transform the unit. The values can be seen in table 2 below, which also shows the values of PMV.

It can also be seen in table 2 above, that the average obtained of the real rate by metabolic analyzer for the welder was of 145.46 W/m², different from the interval of values that the ISO 8996 (2004) suggests: 75 to 125 W/m².

3.1 *Linear regression between real thermal sensation and PMV*

To obtain the PMV, besides metabolic rate, other variables are also necessary. These were also collected and used for implementation of the calculation. The graph 1 below shows the relation between real thermal sensation and the real $PMV_{tabulated}$, with a coefficient of determination of 0.1749.

Graph 2 below shows the relation between real thermal sensation and the PMV_{real}, with a coefficient of determination of 0.7854.

The coefficient of determination between the real thermal sensation and $PMV_{tabulated}$ was of 0.1749, which shows a low adhesion between the data. As it can be seen above, the other function has presented a higher coefficient of determination being 0.7854 the largest, obtained by the relationship between the real thermal sensation and PMV_{real}.

Table 2. Values of Real Metabolic Rate with respective PMV.

	M (met)	M (W/m²)	PMV
Average	2,50	145,46	1,30
Standard Deviation	0,39	22,61	0,39

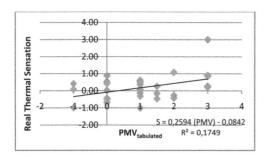

Graph 1. $S \times PMV_{tabulated}$.

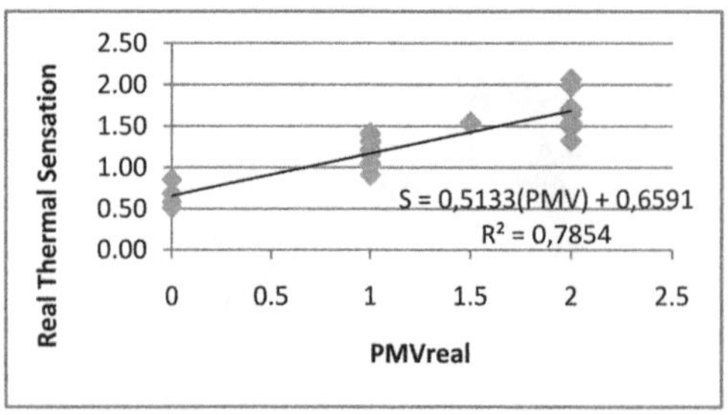

Graph 2. $S \times PMV_{real}$.

3.2 *Obtaining the function of correction ($M_{predicted}$)*

After having the corrected values, it was, then, proceeded table's correction. According to the table of ISO 8996 (2004), the metabolic rate of welders varies between 75 and 125 W/m². The average between the tabulated values of metabolic rate is 100 W/m² for the welder and the average between the corrected values was 145,46 W/m². Dividing the corrected value's average by the average of tabulated values, it is obtained the $M_{predicted}$ in the form of correction equation of the table ISO 8996 (2004) as shown in equation (1):

$$M_{predicted} = 1.4546 \times M_{tabulated} \quad (1)$$

According to equation (1), the values to the activities of a welder, presented in ISO 8996 (2004) should be multiplied by 1.4546 to make the correction. Table 3 below shows the predicted values for the activity of the welder.

As it can be seen in table 3, the activities' interval for the welder varies between 109.1 and 181.83 W/m², therefore, differing from the interval tabulated by ISO 8996 (2004), which varies from 75 to 125 W/m².

Thus, it turns to be recommended to apply a correction coefficient of 1.4648 over tabulated values, so doing this, the probability that the values are more consistent with the reality is significantly increased. In order to verify if the predicted values of metabolic rate improves the PMV model, it was used the environmental variables and predicted values for determining the $PMV_{predicted}$.

Using the variables collected and the interval of metabolic rate of Table 3, it was calculated the $PMV_{predicted}$. With the $PMV_{predicted}$ and the real

Table 3. Values of predicted metabolic rate for welder's activity.

$M_{tabulated}$	$M_{predicted}$	$M_{tabulated}$	$M_{predicted}$
75	109,1	105	152,73
80	116,37	110	160,01
85	123,64	115	167,28
90	130,91	120	174,55
95	138,19	125	181,83
100	145,46	–	–

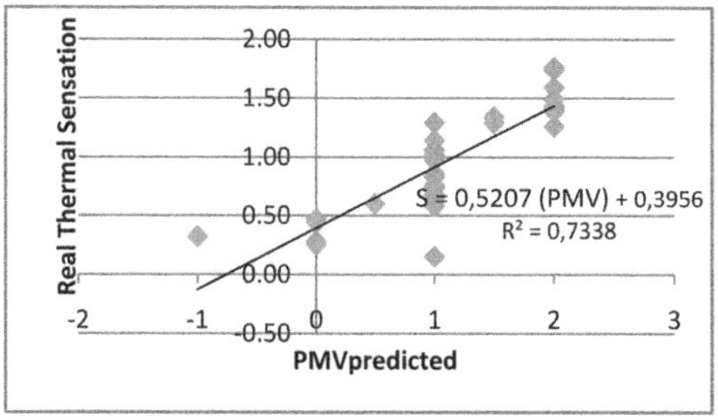

Graph 3. $S \times PMV_{predicted}$.

thermal sensation reported by employees was possible to perform the linear regression, as shown in Graph 3.

The coefficient of determination between the Real Thermal Sensation and $PMV_{predicted}$ was 0.7338, which represents a good adhesion between data. As it can be seen above, there is a significant increase in adhesion of data when comparing the real thermal sensation with $PMV_{predicted}$, compared to tabulated values ($S \times PMV_{tabulated}$).

4 CONCLUSIONS

Through field studies, analyzing the results of the real thermal sensation and their PMV, it's observed that data have low adherence. This can be linked with the metabolic rate. To calculate the PMV, a value for the metabolic rate is required. The tabulated metabolic rate, provided by ISO 8996 (2004), is a possible indicator for the low coefficient of determination found.

When the metabolic rate is measured by the metabolic analyzer and it's realized the confrontation of data between the real thermal sensation and PMV, there is a significant increase in adhesion of these data, thus, demonstrating that a correct determination of values for the metabolic rate is essential for the calculation of the PMV, in order to achieve an improvement of the model.

The use of metabolic analyzer for determining the real metabolic rate was the great difference of this work, because this equipment provides, immediately, the value of metabolic rate. Due to the collaboration of people from the factory, it was possible to use this machine, because at first it was thought that this equipment could bring some embarrassing situation for workers.

The average of the extremes corrected values found was 145.46 W/m². As the average between the extremes values of the activities of welders is 100 W/m², this value was used to find the $M_{predicted}$, which corrects the table of ISO 8996 (2004). Thus, it is recommended that for the activities of a welder, the standard values should be multiplied by the correction coeficient, worthing 1.4546.

This research worked with the metal-mechanic welders. Therefore, the findings can be applied to industries in the same segment. Regarding to variables of thermal comfort, it is known that there is a likelihood of inaccuracies in other variables with a subjective nature, as the clothing insulation. Thus, there may be improvements in the PMV model if it is not using the tables of ISO 9920 (2007) for determining the coefficients of clothing insulation.

It was also considered that the metabolic analyzer, being calibrated, provides the correct values of metabolic rate, which is not a source of error, although it is known that the worker making use of the metabolic analyzer is subject to variations.

This research uses the minimum number of sample recommended in the literature for that data can be considered distributed according to the normal distribution curve. If the sample is significantly higher, it may be that the results will display variability.

The general objective of this research was achieved, to the extent that, through the relationship between the real thermal sensation and $PMV_{predicted}$, it was obtained a determination coefficient of 0.7338, a value much higher when it is related the thermal sensation with $PMV_{tabulate}$ (0.1749). It is verified, then, an improvement in PMV model.

REFERENCES

Antonelli, B.A. (2012). Verificação da adequabilidade do modelo normalizado de conforto térmico utilizando a taxa metabólica determinada pelas sensações térmicas reais de usuários em ambientes industriais. *Dissertação (Mestrado em Engenharia da Produção)*—Universidade Tecnológica Federal do Paraná, Ponta Grossa.

De Dear, R.J., Brager, G.S. (1998). Developing an adaptive model of thermal comfort and preference. *ASHRAE Transactions*. Atlanta: 104, 145–167.

Frontczak, M.; Wargocki, P. (2011). Literature survey on how different factors influence human comfort in indoor environments. *Building and Environment*, 46, 922–937.

Havenith, G., Holmér, I., Parsons, K. (2002). Personal factors in thermal comfort assessment: clothing properties and metabolic heat production. *Energy and Buildings*, 34, 581–591.

Humphreys, M.A; Nicol, J.F. (1996). Conflicting criteria for thermal sensation withinthe Fanger predicted mean Vote Equation. In: CIBSE/ASHRAE JOINT NATIONAL CONFERENCE. *Proceeding*, 153–158.

ISO—International Organization for Standardization. (2004). Ergonomics of the thermal environment—Determination of metabolic heat production. *ISO 8996*, Genebra.

ISO—International Organization for Standardization. (2005). Ergonomics of the thermal environment—Analytical determination and interpretation of thermal comfort using calculation of the PMV and PPD indices and local thermal comfort criteria. *ISO 7730*, Genebra.

Katavoutas, G; Theoharatos, G; Flocas, H.A et al. (2009). Measuring the effects of heat wave episodes on the human body's thermal balance. *Int J Biometeorol,*. 53, 177–187.

Triola, M.F. 2005. *Introdução à Estatística* (10th ed.). Rio de Janeiro: LTC.

Vergara, L.G.L. (2001). Análise das condições de conforto térmico de trabalhadores da unidade de terapia intensiva do hospital universitário de Florianópolis. *Dissertação (Mestrado em Engenharia de Produção)*—Departamento de Engenharia de Produção, Universidade Federal de Santa Catarina, Florianópolis.

Xavier, A.A.P. (2000). Predição de Conforto Térmico em ambientes internos com atividades sedentárias—teoria física aliada a estudos de campo. *Tese (Doutorado em Engenharia de Produção e Sistemas)*—Programa de Pós Graduação em Engenharia de Produção e Sistemas, Universidade Federal de Santa Catarina, Florianópolis.

Occupational Safety and Hygiene II – Arezes et al. (eds)
© 2014 Taylor & Francis Group, London, ISBN 978-1-138-00144-2

Analysis of the role of worker characteristics in three of the most frequent manufacturing accidents

J.A. Carrillo-Castrillo & L. Onieva
University of Seville, Seville, Spain

J.C. Rubio-Romero & A. López-Arquillos
University of Malaga, Malaga, Spain

ABSTRACT: Accident causation involves multiple factors. There are some worker characteristics considered risk factors. Accident investigation can be used to research the mechanisms that explain associations between worker characteristics and accident causation. This study focuses on the association between contributing causes and the worker characteristics, analyzing official investigation reports of three frequent accident types in manufacturing. The high proportion of contributing causes is a serious concern for safety practitioners. The findings of differential causation depending on the worker characteristics indicate that preventive programs should be oriented to specific worker groups.

1 INTRODUCTION

1.1 *The analysis of accident causation*

Accident causes can be classified (International Labour Organization, 1998) as immediate causes (unsafe acts, unsafe conditions) and contributing causes (safety management performance, mental and physical conditions of workers). Contributing causes are part of the latent causes (Reason, 2000).

Within the same company, activity, task and job, there are personal risk factors that affect the likelihood of accidents. The identification of those factors and the understanding of the mechanisms that explain their influence in accident causation can help in a better design of preventive programs.

In the manufacturing sector of Andalusia, as previous studies have identified, a number of individual worker characteristics can affect likelihood of occupational safety injuries (Carrillo *et al.*, 2012) and in case of injury they also affect the severity of injuries (Carrillo & Onieva, 2012). Workers at risk in the manufacturing sector are male workers and young workers.

Although from an epidemiological point of view, identification of groups of workers at risk is an objective; from the preventive point of view, it is more important to identify the underlying mechanisms in order to design specific actions to fight the accident causes. The analysis of accident investigations is one of the most used tools for that purpose.

Analysis of accident causation is not a simple issue. Each accident can have multiple causes and the relationship among different factors is complex (Hale *et al.*, 2007).

Although there are a few studies of causes of accidents in specific industries (Jacinto *et al.*, 2009) or types of accidents (Aneiziris *et al.*, 2013), accident causation in the manufacturing sector is still a relevant field of research (Khanzode *et al.*, 2012). In the literature review performed, we found no analysis of the role of worker characteristics in accident causation in the manufacturing sector.

There is evidence of the association between the type of causes identified and the accident scenarios (Carrillo-Castrillo *et al.*, 2013). This fact indicates the need of study separately each accident type. For this reason this study is based on three of the most frequent accident types in the manufacturing sector.

If there are certain workers with higher risk of accident, it is reasonable to expect that the accident causation is different depending on the worker characteristics.

Although some of the differences can be attributed to active causes and possible unequal distribution of the risk exposure depending on the worker characteristics, we decided to concentrate on the worker influence on contributing causes of the accidents.

However, it will only be possible to test the influence of each of the worker characteristics in the causation pattern of the accidents if enough number of investigated accidents is available.

1.2 *Scope of this research*

The main purpose of this paper is to analyze the role of some human factors in the accident causation in the accidents officially investigated in the Andalusia manufacturing sector.

Manufacturing is the sector with highest number of annual days of absence because of accidents in Europe (European Commission, 2012). For accidents with more than three days of absence in the European Union and Norway in 2007, the latest incidence rate of the manufacturing sector published by Eurostat is 3,097 accidents per 100,000 workers.

It is important to add that Andalusia is one of the biggest regions of Europe, and represents approximately 12% of the Spanish manufacturing sector and employs on average more than two hundred thousand workers.

2 MATERIALS AND METHOD

2.1 *Accident investigation reports*

In order to deeply research in the different causation of accidents in terms of the workers at risk, we have identified three of the most prevalent types of accidents in the manufacturing sector: "loss of control of machinery", "being trapped" and "fall from a height". Our identification of the type of accident is based in the deviation variable included in the European Statistics of Accident at Work methodology (ESAW hereinafter) described in European Commission (2002).

We have analyzed all fatal and severe of those types of accidents investigated by safety officers in the manufacturing sector of Andalucía occurred from 2004–2011. The manufacturing sector is defined in this study according to the European Statistical Classification of Economical Activities (NACE).

Sometimes, when an accident occurs, safety officers of Andalusia Labor Administration conduct an official accident investigation, after the accident notification is received, especially when the accident is severe or fatal. Medical criteria are applied to classify the accident as slight or severe, depending on the severity of injuries and the expected period or recovery. Traffic accidents and non-traumatic deaths such as strokes or heart attacks are not usually investigated.

In this study, all of the accidents are severe or fatal and have been investigated by safety officers. The accidents included occurred from 2004 to 2011.

In 2004, the Labour Authority in Andalusia adopted a common codification system for the causes identified in official accident investigation. Data is also available for accident circumstances following ESAW methodology.

In the coding system, approved by the National Institute for Safety and Health at Work in Spain, there are 255 possible causes (Fraile, 2011). Codes are defined with four digits, the first digit identifies the group of causes, the first two digits identify the subgroup of causes and the other last two digits are the different causes in each subgroup. For example, cause 7206 is "inadequate training / information about risks of preventive measures", classified as group 7 (safety management, latent) and the subgroup 72 (preventive activities).

The number of causes identified per accident investigated is variable, reaching a maximum of nine, but in most cases the number of causes is between three and five.

Active causes are usually classified in the groups with first digit 1 to 5 and most of the causes classified in the groups 6 to 8 are contributing causes. The analysis is focused on the latent causes because the hypothesis is that they might be related to the worker characteristics.

2.2 *Difference of proportions test*

For each group of workers, we compute the proportion of each cause among all the causes identified in the same accident mechanism, thus the proportion of accidents in each group of workers with that cause identified is calculated. One of the groups is considered as reference and the confidence interval for difference of proportions is calculated respect to that reference. The difference of proportion of cases with a specific causes identified is considered significant if the confidence interval for that difference do not include 0.0.

In order to have enough cases for the use of difference of proportions test, only causes with at least 10 cases in each of the accident type are analyzed (Kines, 2002).

3 RESULTS

3.1 *Cross-sectional analysis*

All identified causes are classified into group of causes. Contributing causes are classified in the groups from 6 to 8 (see Table 1).

3.2 *Causation pattern differences*

Only characteristics of workers that show significant differences such as age and type of contract are presented. Besides age, there are other important demographic variables such as nationality and gender. Unfortunately the number of accidents investigated was not enough to analyze them.

Table 1. Number of causes identified for each subgroup of causes in each of the selected accident types.

Subgroup of causes	Loss of control of machinery	Being trapped	Falls from a height
Work organization (6)	84	61	5
Safety management (7)	77	45	81
Personal factors (8)	78	36	77
Total number of causes	311	206	291
Number of cases	83	57	86

Table 2. Contributing causes for "Loss of control of machinery" in each group of workers age.

Contributing Cause (No)	Young (16–24)	Senior (>55)	Normal (25–54)
Inadequate work methods	8 (11%)	2 (11%)	24 (11%)
Inadequate training or information regarding risks or the preventive measures	8 (11%)	1 (5%)	14 (6%)
Other causes related to personal factors	**1 (1%)**[a]	3 (16%)	16 (7%)
Removing guards or protections	4 (6%)	1 (5%)	12 (5%)
Failure to comply with work procedures	1 (1%)	**0 (0%)**[a]	12 (5%)
Failure to use personal protective equipment	**6 (8%)**[a]	1 (5%)	4 (2%)
Total number of causes	71	19	221

[a] Differences significant with normal age ($p < 0.05$)

Using the differences of proportions test, there are significant differences in terms of the causation pattern for each group of workers age (see Tables 2, 3 and 4).

In the "Falls from a height accidents", the cause "Other causes related personal factors" that include the physical and psychological conditions of worker is very unlikely identified in accidents of young workers but frequently identified in accidents of older workers (see Table 4).

For older workers is very unlikely the cause "Not complying with safety rules" in "Loss of control of machinery" accidents, the cause "Fail to identify hazards" in the "Falls from a height" accidents and the cause "Lack of training" in "Being trapped" accidents (see Tables 2, 3 and 4).

For young workers is very likely the cause "Not using personal protection equipment" in "Loss of control of machinery" accidents (see Table 2).

Table 3. Contributing causes for "Being trapped" in each group of workers age.

Contributing Cause (No)	Young (16–24)	Senior (>55)	Normal (25–54)
Inadequate work methods	4 (12%)	1 (17%)	15 (15%)
Inadequate training or information regarding risks or the preventive measures	**7 (21%)**[a]	**0 (0%)**[a]	7 (7%)
Not complying safety rules	**0 (0%)**[a]	1 (17%)	9 (9%)
Total number of causes	34	6	102

[a] Differences significant with normal age ($p < 0.05$)

Table 4. Contributing causes for "fall from a height" in each group of workers age.

Contributing Cause (No)	Young (16–24)	Senior (>55)	Normal (25–54)
Inadequate work methods	2 (13%)	5 (31%)	20 (11%)
Hazards not identified during the risk assessment	1 (7%)	**0 (0%)**[a]	14 (8%)
Failure to use personal protective equipment	2 (13%)	3 (19%)	9 (5%)
Failure to comply with work procedures	1 (7%)	1 (6%)	8 (4%)
Other causes related to personal factors	**0 (0%)**[a]	1 (6%)	9 (5%)
Total number of causes	15	16	180

[a] Differences significant with normal age ($p < 0.05$)

Table 5. Contributing causes for "Being trapped" in each group of workers type of contract.

Contributing Cause	Non permanent	Permanent
Inadequate work methods	13 (16%)	7 (11%)
Inadequate training or information regarding risks or the preventive measures	**12 (15%)**[a]	2 (3%)
Failure to comply with work procedures	**2 (3%)**[a]	8 (13%)
Total number of causes	79	63

[a] Differences significant with permanent contract ($p < 0.05$)

In relation to the contract type of the workers, it was only significant for "Being trapped" accidents (see Table 5). For non-permanent contract workers, it is likely the causes "Lack of training" and "Not complying safety rules".

4 DISCUSSION

4.1 *Causation pattern and worker collectives*

Comparing causes attributed in accidents of workers of the different groups analyzed, we found important differences that evidence how personal characteristics affect the accident causation. Differences in proportions in a specific cause means for a group of workers are a signal of the interest of designing specific programs to prevent them.

The aptitude of senior workers needs to be checked when working in tasks with risk of "Being trapped" or "Loss of control of machinery" as their personal factors are likely cause of those accidents. Specific campaigns advising young workers about the use of personal protection equipment in tasks with risk of "Loss of control of machinery" should be effective.

The more frequent causes found confirm the importance of knowledge, behavioral and management issues in manufacturing accidents (Gardner *et al.*, 1999).

All of the more frequent causes are classified as latent causes. Although technical condition of equipment is important, based on these accident investigations, the more serious concerns are the inappropriate safety management, the lack of proper training and the poor development of preventive activities. Moreover, the link between organizational failures and inappropriate training has been identified as one of the key issues in manufacturing safety (Katsakiori *et al.*, 2010).

Another serious concern is the inadequate training of non-permanent workers (Benavides *et al.*, 2006). With a high proportion of non-permanent workers, increased in the last years, specific measures are needed to assure to enforce the proper training.

Same problem have been identified in young workers for "Being trapped" accidents. Although age influence in accident causation is a complex issue (Salminen, 2004), the inadequate training is a serious threat for new workers (Verjans *et al.*, 2007).

These young workers show a high proportion of failures to use personal protective equipment in "Loss of control of machinery" accidents. It is an evidence of the lack of proper training and the lack of adequate motivation.

Unfortunately, the coding system allowed the use of miscellaneous codes such as "Other causes related to personal factors". The deeper research on this issue can provide useful information about the role of worker characteristics in "Falling from a height" accidents (Bentley, 2009).

Positive findings are the lower proportion of young workers not complying with safety rules.

Although the number of cases for each specific mechanism of accident is limited within the accident investigations available, the conclusions of analyzing the causation pattern for each accident mechanism is a powerful tool to propose effective preventive actions oriented to fight the most likely causes for each group of workers in terms of their individual characteristics.

At the same time, reasonable explanations of the differences of severity and proneness of accidents in terms of the individual characteristics of the worker can be suggested. Further research can confirm these initial results and the possible differences in other regions or sectors.

4.2 *Preventive action identification*

One of the first applications of the results presented is the identification of preventive actions.

Comparing causes attributed in accidents of workers of the different collectives analyzed, we found important differences that demonstrate how personal characteristics are affecting the accident causation.

Differences in proportions for a group of workers in a specific cause mean that specific programs are needed for those causes in those workers.

Preventive actions should be oriented to the most likely causes of severe and fatal accidents depending on the personal characteristics of the worker. According to the significant differences highlighted, programs of training and motivating to comply safety rules are needed for non permanent workers to reduce the risk of "Being trapped" accidents.

4.3 *Future research on this topic*

These results apply to Andalusian manufacturing sector. In other regions or sectors other areas of improvement based on the accident causation can be identified with the methodology used in this paper in order to compare the results.

In further research it would be very interesting to analyze the workers not injured but exposed to the same risks. This would help in the understanding of the safety mechanisms instead of concentrating only on the causes where safety fails (Holcroft & Punnet, 2009).

For that purpose, additional information about workers, such as level of training, previous accident experience or physical and psychological aptitude, can provide additional insight into the role of individual worker characteristics within accident causation, given the same environment and risk exposure.

Future accident investigations need to include that pieces of information in order to not only consider the contributing causes but the whole latent failure environment.

REFERENCES

Aneiziris, O.N., Papazoglou, I.A., Konstandinifou, M., Baksteen, H., Mud, M., Damen, M., Bellamy, L.J. & Oh, J. 2013. Quantification of occupational risk owing to contact with moving parts of machines. *Safety Science* 51(1): 382–396.

Benavides, F.G., Benach, J., Muntaner, C., Delclos, G.L., Catot, N., & Amable, M. 2006. Associations between temporary employment and occupational injury: what are the mechanisms? *Occupational Environmental Medicine* 63(6): 416–421.

Bentley, T. 2009. The role of latent and active failures in workplace slips, trips and falls: An information processing approach. *Applied Ergonomics* (40): 175–180.

Carrillo, J.A., Gómez, M.A. & Onieva, L. 2012. Safety at work and worker profile: analysis of the manufacturing sector in Andalusia in 2008. *Occupational Safety and Hygiene—SHO*. Guimaraes: Portuguese Society of Occupational Safety and Hygiene.

Carrillo, J.A. & Onieva, L. 2012. Severity Factors of Accidents: Analysis of the Manufacturing Sector in Andalusia. *Occupational Safety and Hygiene—SHO*. Guimaraes: Portuguese Society of Occupational Safety and Hygiene.

Carrillo, J.A., Rubio-Romero, J.C., López-Arquillos, A. and Onieva, L. 2013. The role of worker individual characteristics in the severe injuries causation in the Andalusian manufacturing sector. *Occupational Safety and Hygiene—SHO*, Guimaraes: Portuguese Society of Occupational Safety and Hygiene.

Carrillo-Castrillo, J.A., Rubio-Romero, J.C. & Onieva, L. 2013. Causation of Severe and Fatal Accidents in the Manufacturing Sector. *International Journal of Occupational Safety and Ergonomics* 19(3): 423–434.

European Commission. 2002. *European Statistics on Accidents at Work (ESAW) - Methodology (ed. 2001)*. Luxembourg: Office for Official Publications of the European Communities.

European Commission. 2012. Eurostat: Your Key to European Statistics. http://epp.eurostat.ec.europa.eu/.

Fraile, A. 2011. NTP924: Causas de accidentes: clasificación y codificación. Madrid: Instituto Nacional de Seguridad e Higiene en el Trabajo.

Gardner, D., Carlopio, J., Fonteyn, P.N. & Cross, J.A. 1999. Mechanical Equipment Injuries in Small Manufacturing Businesses. Knowledge, Behavioral, and Management Issues. *International Journal of Occupational Safety and Ergonomics* 5(1): 59–71.

Hale, A.R., Ale, B.J.M., Goossens, L.H.J., Heijer, T., Bellamy, L.J, Mud, M.L., Roelen, A., Baksteen, H., Post, J., Papazoglou, I.A., Bloemhoff, A. & Oh, J.I.H. 2007. Modeling accidents for prioritizing prevention. *Reliability Engineering and System Safety* 92(12): 1701–1715.

Holcroft, C.A. & Punnet, L. 2009. Work environment risk factors for injuries in wood processing. *Journal of Safety Research* 40(4): 247–255.

International Labour Organization. *Encyclopaedia of Occupational Health and Safety: Part VIII, Chapter 56*. SW: International Labour Organization.

Jacinto, C., Canoa, M. & Guedes, C. 2009. Workplace and organizational factors in accident analysis within the Food Industry. *Safety Science* 47(5): 626–635.

Katsakiori, P., Kavvathas, A., Athanassiou, G., Goutsos, S. & Manatakis, E. 2010. Workplace and Organizational Accident Causation Factors in the Manufacturing Industry. *Human Factors and Ergonomics in Manufacturing and Service Industries* 20(1): 2–9.

Khanzode, V.V., Maiti, J. & Ray, P. 2012. Occupational injury and accident research: A comprehensive review. *Safety Science* 50(5): 1355–1367.

Kines, P. 2002. Construction workers' falls through roofs: fatal versus serious injuries. *Journal of Safety Research* 33(2): 195–208

Reason, J. 2000. Human errors: models and management. *British Medicine Journal* 320: 768–770.

Salminen, S. 2004. Have young workers more injuries than older ones? An international literature review. *Journal of Safety Research* 35 (5): 513–521

Verjans M., de Broeck, V. & Eeckelaert, L. 2007. *OSH in figures: Young workers—Facts and figures*. Luxembourg: Office for Official Publications of the European Communities.

Occupational Safety and Hygiene II – Arezes et al. (eds)
© 2014 Taylor & Francis Group, London, ISBN 978-1-138-00144-2

A holistic framework for safety performance evaluation

E. Sgourou, P. Katsakiori, I. Papaioannou, S. Goutsos & Em. Adamides
University of Patras, Rion, Patras, Greece

ABSTRACT: This paper presents a conceptual framework based on systems thinking, which aims at a thorough and holistic understanding of safety performance evaluation. The starting point is a measurement matrix that includes prospective and reactive measures of technical, organisational, human and external factors. Interrelations of these factors are explored, through causal loop diagrams. Measurement data are then evaluated through the criteria of efficacy, efficiency and effectiveness. Soft Systems Methodology is proposed to be used for defining these criteria by taking into consideration different perspectives among various stakeholders. By implementing this holistic approach, those in position to take decisions for safety issues could gain systematic knowledge of the existing level of safety in their organisation and thus support and improve their decision making processes.

1 INTRODUCTION

The evaluation of safety performance is an essential process of any safety management system, with three main purposes: (i) to provide information on the current status of a safety management system, (ii) to support decisions regarding improvements, and (iii) to motivate those in position to take the above decisions (Health and Safety Executive, 2001; Hale, 2009).

Safety performance refers to the accomplishment of specific tasks measured against preset goals. These goals can be set inside an organisation (i.e. during the development or the review phase of a safety management system) and outside it (i.e. from legislation or contract obligations). Safety performance does not refer to personal performance, but to the performance of a work-team, a section of an organisation or the organisation as a whole.

In theoretical terms, safety performance is a dependent variable and all direct and indirect factors that influence safety performance are the independent variables. The multitude and complexity of these factors, even in a common workplace, make almost impossible the mathematical calculation of safety performance. For this reason, safety researchers and professionals were led to the use of safety performance indicators, categorised to proactive (leading) and result (lagging) indicators.

In recent years, emphasis is given to the use of more than one safety performance evaluation methods and to the combination of their results for a more complete assessment (Petersen, 2001; Sherif, 2003; Mengolini & Debarberis, 2008). However, the complementary or combining use of safety performance indicators does not satisfy the need for a holistic approach in safety performance evaluation, which is increasing in the present dynamic society (Sgourou et al., 2010).

In this paper, a holistic framework for safety performance evaluation is proposed. The purpose of this framework is twofold: first, to organise and communicate existing knowledge on safety performance evaluation, and second to serve as a theoretical basis for the development of a holistic safety performance evaluation methodology.

2 BUILDING THE BASE: A MEASUREMENT MATRIX

Safety performance has already been described as a dependent variable, influenced by a complex system of direct and indirect factors. Direct factors refer to the result of activities related to safety (i.e. incidents, costs), while indirect factors either facilitate safety management (i.e. resources) or have been found related to direct factors (i.e. safety climate) (Mohaghegh, 2007).

As a first step towards a holistic approach, we suggest that all factors should be divided in four categories: technical, organisational, human and external factors (Sgourou et al., 2010). Technical factors refer to the hardware used (structures, equipment, material) and the physical workplace, including relevant risk control systems and outcomes from the application of such systems. Organisational factors may be connected with the general organisation system in place (structure, responsibilities, communication, adaptation to change etc.), and also with the safety management system in place (safety policy, safety responsibilities, management commitment etc.). Human factors refer to individual characteristics which influence behaviour at work in a way which can affect health and safety (Health and Safety Executive, 1999). Finally, external factors refer to the physical, social and economical environment, which relates with and consequently affects all previous categories. All above factors are inter-connected and cannot be considered in isolation, but in a macroergonomics framework, which incorporates a systems perspective to understanding the organisation (Dray, 1985; Hendrick, 1986).

It is proposed that the factors of all categories are further characterized as either determining factors or results. Determining factors are those, which influence the development and implementation of any formal or informal safety management system, and are further divided into general or safety-related factors. Results are the outcomes resulting from the implementation of the safety management system. Such outcomes include, among others, safety related incidents (injuries, damages, near-misses), occupational illnesses, and related direct and indirect costs.

Thus, we develop a generic measurement matrix (Table 1), which, when completed, may vary according the work and the organisational settings, the company's and employees' characteristics and the given external environment. Moreover, the contents of the proposed matrix may differ according the stage of life cycle a given organization is at (Quinn, & Cameron, 1983), however it is important for a holistic approach that all categories are taken into consideration and assessed. The matrix remains useful even when a safety management system enters maturity because as it happens to any organisation no one of the mentioned factors (technical, organisational, human, external) remain stable in time. The process and knowledge gained may be routinised, but not the essence of the factors.

A variety of methods can be implemented in order to develop a company-specific measurement matrix. One approach is to conduct a literature review or an experts' survey for the identification of factors with major influence to safety performance (Sgourou, 2013). Alternatively, a participatory approach, such as focus groups, surveys and interviews, can offer a more context-specific result, together with an organisational learning possibility.

The factors included in the measurement matrix can be measured with a variety of methods, subjective or objective. For the collection of as much information as possible on the state or value of certain factors, a combination of methods may be selected. This combination can be used for the measurement of different dimensions with different methods (complementary approach) or the measurement of all dimensions with different methods (supplementary approach).

Table 1. A generic measurement matrix.

Factors	Determining		Results
	General	Safety related	
Technical			
Organisational			
Human			
External			

3 EXPLORING THE INTERRELATIONS: A SYSTEMS APPROACH

For the purposes of a holistic approach to safety performance evaluation, we suggest that the interrelations of the identified factors must be explored, through different perspectives and within defined boundaries (Siemeiniuch & Sinclair, 2006). In that way, evaluation teams and, more importantly, decision makers will acknowledge the systemic nature of factors influencing safety performance and the possible delays in certain interrelations.

System dynamics is the method proposed for exploring intra-relations of direct and indirect factors influencing safety performance, as well as of their interrelations with the general organisational and external environment. It was introduced by Forrester (1961) as a method for modeling and analysing the behaviour of complex social systems, especially in industrial settings. System dynamics modeling has the capability to capture the interactions among a range of system variables and to predict the implication of each other over a period of time (Sterman, 2000; Irwin, 2000).

To conceptualize a real world system under investigation, system dynamics focuses on the structure and behaviour (over time) of the system using multiple feedback loops, in the form of causal loop diagrams (CLDs) (Richardson, 1986; Sterman, 2000). A CLD is a graphic representation of how interrelated variables affect one another. The relationships between these variables, represented by arrows, can be labeled as positive or negative. Positive causal link means that the two variables change in the same direction, i.e. if the variable in which the link starts decreases, the second variable also decreases. Negative causal link means that the two variables change in opposite directions. Causal loops can be reinforcing, if they have even (or zero) number of negative links, or balancing if they have odd number of negative links. Identifying reinforcing and balancing loops is an important step for identifying the possible dynamic behaviours of the system, since reinforcing loops are associated with exponential increases/decreases and balancing loops are associated with reaching a plateau. In CLDs, delays are also depicted in both types of loops, reinforcing and balancing. Delays refer to the time intervals between the implementation of an action and the consequences of this action (Goh et al., 2010).

CLDs do not have to include all identified factors, since this will lead to one diagram too complex and difficult to understand. Selected factors can provide a frame with specific boundaries, which will lead to a manageable and useful causal analysis. The set of factors selected and the final form of the respective CLD, depend on the perspectives of different stakeholders, but also on the different perspectives of the same stakeholder (Williams, 2010).

Figure 1 presents a CLD, which investigates the interrelations of management commitment to safety with selected determining factors and results. Management commitment is considered one of the most influential factors in safety performance (Thompson et al., 1997; Vredenburgh, 2002; Michael et al., 2005). The determining factors selected to be explored in relation to management commitment were: (a) technical: equipment suitability & risk controls systems in place; (b) organisational: resources (for safety); (c) human: workforce commitment (to safety); and (d) external: financial crisis. The results selected were: failures, incidents, incident costs, safety costs, conformity with legislation, safe behaviours. Risk level was used as an intermediate variable, serving in clarifying the interrelations between factors (Sgourou, 2013).

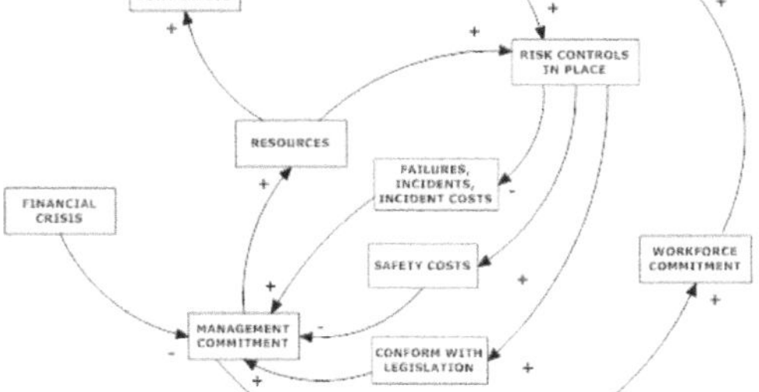

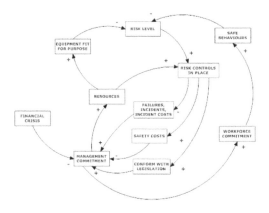

Figure 1. A causal loop diagram exploring the interrelations between factors affecting safety performance.

4 EVALUATING MEASUREMENT DATA: MULTIPLE EVALUATION CRITERIA AND PERSPECTIVES

One of the central ideas of the proposed approach is that measurement produces data; however, this data must be evaluated in order to gain knowledge. Evaluation involves some method of judgment about the measurement data, based on criteria. Judgment requires a viewpoint, since different viewpoints or perspectives may lead to different results (Nicholson, 2004; Williams, 2010). Therefore, the implementation of multiple evaluation criteria and perspectives is critical for the holistic safety performance evaluation.

A safety performance evaluation scheme will facilitate decision on safety issues, if the evaluation criteria have the following three characteristics:

a. Evaluation criteria must reflect the organisation's safety policy and goals. For the development of effective evaluation criteria, safety goals must be clearly set in advance. A well-integrated evaluation scheme in an organisation's goal setting process provides consistency, rationality and motivation to those taking decisions on safety issues.
b. Evaluation criteria must provide for an accurate evaluation of the organisation's overall safety performance. One of the reasons that the traditional approach to evaluate safety performance is still through measurement and statistical analysis of incident-related data, despite the negative critique of this type of evaluation, is that incident data have greater face validity. Any innovative and, possibly, more complex evaluation system must provide specific, comprehensible and, preferably, quantified results. In that way, it will better satisfy the three purposes of safety performance evaluation, as presented in the introduction.
c. Evaluation criteria must represent key areas of importance. This will ensure that the evaluation scheme is comprehensive but not too complex and exhaustive, which again may diminish its face validity and also its flexibility and usefulness.

One methodology that can be used for selecting evaluation criteria and respective performance indicators, and which is consistent with the holistic approach proposed in this paper, is Soft Systems Methodology (SSM) (Checkland, 1981; Checkland & Scholes, 1999). SSM provides a development framework, which is systemic, participative and enables organisational learning. Moreover, it takes into account the current situation in an organisation in terms of resources, culture and power relations, different views regarding the ideal desired situation (safety performance evaluation in our case), and the feasibility of proposed changes. SSM aims at accommodating different perspectives through conceptual models of human activity systems. These models are used to decide on interventions for the resolution, or improvement, of the situation.

To support the monitoring and control activities of the conceptual systems, SSM implements three core criteria: efficacy, efficiency, effectiveness. The same criteria are proposed to be used for the evaluation of an organisation's safety management activities (Sgourou et al., 2012; Sgourou, 2013).

5 CONCLUSIONS

One of the challenges, that decision makers face, is the broad context under which safety performance evaluation takes place. It is more convenient, and usually occurs, to perform an evaluation in a specific problematic area, which requires immediate management involvement. However, according to systems theory, technical, organisational, human and external factors are all connected; a problem in one area may have its origin in another area of the system. A single evaluation of a group of factors will leave out crucial information for decision making.

The current paper proposes a series of steps towards a holistic approach to safety performance evaluation. The initial step of developing a measurement matrix may vary to its content and extend, since it is dependent on the type and size of the organisation as well as to the availability of resources. Following factors' identification, proper measurement methods for each factor have to be selected. The organisation's ability, to measure their identified set of determining factors and results, is related to the available expertise and resources. Criteria for the selection of a method should be set in order to facilitate this process (Sgourou et al., 2010).

After measurements have been collected, they must be evaluated. Evaluation involves some method of judgment about the measurement data. A holistic evaluation must be based on an understanding of interrelations, perspectives and boundaries. Systems thinking provide the theoretical framework for this approach. System dynamics is one of the methods that can support exploring of interrelations. Moreover, the need for different perspectives in safety performance evaluation can be supported by the implementation of the Soft Systems Methodology.

Implementing a holistic approach, in safety performance evaluation, will serve all purposes of safety performance evaluation: (a) it will provide more accurate, realistic and valid information, (b) it will better support decisions, by identifying not only a wider range of problematic factors, but also their interrelations with other parts of the organisational system, and (c) it will motivate those taking decisions, not only through (a) and (b), but also through the learning process of evaluating safety performance by implementing a participatory, systemic methodology.

REFERENCES

Checkland, P. 1981. *Systems Thinking, Systems Practice*. New York: Wiley.

Checkland, P. & Scholes, J. 1999. *Soft Systems Methodology in Action*. Chichester: John Wiley.

Dray, S. 1985. Macroergonomics in organizations: An introduction. In I.D. Brown, R. Goldsmith, K. Combes, & M. Sinclair (Eds.), *Ergonomics International* 520–522.

Forrester, J.W. 1961. *Industrial Dynamics*. Cambridge MA: MIT Press.

Goh, Y.M., Brown, H. & Spickett, J. 2010. Applying systems thinking concepts in the analysis of major incidents and safety culture. *Safety Science* 48(3): 302–309.

Hale, A. 2009. Why safety performance indicators? *Safety Science* 47(4): 479–480.

Health and Safety Executive 1999. *Reducing Error and Influencing Behaviour (HSG48)*. Sudbury, Suffolk: HSE Books.

Health and Safety Executive 2001. *A Guide to Measuring Health & Safety Performance*. Sudbury, Suffolk: HSE Books.

Hendrick, H.W. 1986. Macroergonomics: A conceptual model for integrating human factors with organizational design. In O. Brown, Jr., & H.W. Hendrick (Eds.) *Human factors in organizational design and management II*; 467–478. Amsterdam: North-Holland.

Irwin, J.S. 2000. *Business Dynamics*. New York: McGraw-Hill.

Mengolini, A. & Debarberis, L. 2008. Effectiveness evaluation methodology for safety processes to enhance organisational culture in hazardous installations. *Journal of Hazardous Materials* 155(1–2): 243–252.

Michael, J.H., Evans, D.D., Jansen, K.J. & Haight, J.M. 2005. Management commitment to safety as organisational support: Relationships with non-safety outcomes in wood manufacturing employees. *Journal of Safety Research* 36(2): 171–179.

Mohaghegh, Z. 2007. On the Theoretical Foundation and Principles of Organisational Safety Risk Analysis. Ph.D. Thesis, University of Maryland, College Park.

Nicholson, S. 2004. A Conceptual Framework for the Holistic Measurement and Cumulative Evaluation of Library Services. *Journal of Documentation* 60(2): 164–182.

Petersen, D. 2001. The Safety Scorecard: Using Multiple Measures to Judge Safety System Effectiveness. *Occupational Hazards*, January 2001.

Quinn, R.E. & Cameron, K. 1983. Organizational Life Cycles and Some Shifting Criteria of Effectiveness. *Management Science* 29(1):31–51.

Richardson, G.P. 1986. Problems with causal-loop diagrams. *System Dynamics Review* 2(2): 158–170.
Sherif, M. 2003. Scorecard Approach to Benchmarking Organisational Safety Culture in Construction. *Journal of Construction Engineering and Management*, January/February 2003: 80–88.
Sgourou, E., Katsakiori, P., Goutsos, S. & Manatakis, Em. 2010. Assessment of selected safety performance evaluation methods in regards to their conceptual, methodological and practical characteristics. *Safety Science* 48(8): 1019–1025.
Sgourou, E., Katsakiori, P., Papaioannou, I., Goutsos, S. & Adamides, Em. 2012. Using Soft Systems Methodology as a systemic approach to safety performance evaluation. *Procedia Engineering* 45: 185–193.
Sgourou, E. 2013. A systemic approach to safety performance evaluation of an organisation. Phd thesis, University of Patras, Rion, Patras.
Siemieniuch, C.E. & Sinclair, M.A. 2006. Systems integration. *Applied Ergonomics* 37(1):91–110.
Sterman, J.D. 2000. Business dynamics: systems thinking and modeling for a complex world. New York: McGraw-Hill.
Thompson, R.C., Hilton, T.F. & Witt, L.A. 1997. Where the safety rubber meets the shop floor: A confirmatory model of management influence on workplace safety. *Journal of Safety Research* 29 (1): 15–24.
Vredenburgh, A.G. 2002. Organisational safety: Which management practices are most effective in reducing employee injury rates? *Journal of Safety Research* 33(2): 259–276.

Occupational Safety and Hygiene II – Arezes et al. (eds)

Association of job-related psychosocial factors with environmental conditions in a mixed plan office: The basis for a screening proxy?

D.A. Coelho
Universidade da Beira Interior, Covilhã, Portugal

C.S.D. Tavares & M.L. Lourenço
Instituto Politécnico da Guarda, Guarda, Portugal

ABSTRACT: Based on a cross-sectional study of 32 office workers, this paper reports on association between environmental conditions in the domains of lighting, noise and the thermal environment and psychosocial factors. The six COPSOQ subscales are modeled on environmental and local demographic data. Insecurity towards the future, conflicting demands and workers control over work were significantly associated to environmental and demographic factors, including illuminance level, air temperature, sound level, as well as the state of window blinds and air conditioning in the room. Behavioral signs linked to lighting were significantly associated to psychosocial assessment (social support and leadership quality, esteem as well as insecurity towards the future). Gender biases for conflicting demands, for social support and leadership quality and for esteem favored men. Further research is needed to assess using environmental and demographic data as a rapid assessment proxy for selected psychosocial factors in practitioner screening of large office laborer populations.

1 INTRODUCTION

Several methods have been proposed and widely used for the assessment of the risk of developing work related musculoskeletal disorders (WRMSDs) focusing on work-related physical exposure. These include, but are not limited to, OWAS (Karhu et al. 1977), RULA (McAtamney & Corlett 1993), PLIBEL (Kemmlert 1995), OCRA (Occhipinti 1998), REBA (Hignett & McAtamney 2000), LUBA (Kee & Karwowski 2001), and MAC (Monnington et al. 2003). Additionally, the role of mental stress and work organization on reinforcing and aggravating the musculoskeletal outcomes caused by physical demands and loads has been recognized (Sauter and Swanson 1996, Wahlström 2005). Psychosocial work characteristics have been recognized as important risk factors for the etiology of WRMSDs (Bongers et al. 2006). As a consequence, interventions conducted by ergonomics and occupational health practitioners should focus on more than one factor. Interventions should comprise optimization of the workplace layout in combination with a feedback survey of the psychosocial work environment and individual training focusing on working technique (Wahlström 2005).

This paper aims assessing the level of association between environmental conditions and psychosocial factors. While the associations reported are based on cross-sectional data, and hence can not provide inference into causal relationships, the unveiling of associations may prove beneficial in the screening phase of practitioner diagnosis and intervention. Association between factors that may be more easily and quickly, as well as, objectively assessed and other less tangible factors (e.g. psychosocial and, or, symptoms, complaints and performance problems) may assist in establishing the basis for screening instruments. These kinds of instruments can assist practitioners in resource-constrained environments to quickly identify the people in the work envi-ronment that should be prioritized for an in-depth analysis of their working conditions (Winter et al. 2006). Because this in-depth analysis should occur in multiple dimensions of exposure (including individual and organizational, as well as physical and psychosocial) and health and well-being outcomes, it is bound to be resource consuming, albeit necessary. Screening tools that enable prioritizing represent a pragmatic alternative to universal assessment in resource constrained environments. These necessarily entail trade-offs, as in any less than ideal solution.

1.1 *Proxy measures*

The use of proxies has been reported in the data collection phase of ergonomics and occupational health studies. Martin et al. (2008) reporting on medical device development, consider that developers may have to use proxies in the place of real users particularly early on in the design process (e.g. using healthy participants to test a device for usability instead of patients, or asking clinicians to provide the views of their patients). In a study by Krause et al. (2004) focusing on transit operators, years of professional driving, obtained from a medical examination dataset, was used as a proxy measure of past (cumulative) physical work. In the same study, weekly hours of professional driving during the previous 12 months, assessed by questionnaire, was used as a proxy measure for current physical workload.

These examples represent quite straightforward substitution or approximation of subject views in user requirements elicitation and usability testing as well as of simple variables in the data analysis stage of a long prospective study. This is not the use that is hypothesized in this paper, hypothesizing the feasibility of a regression expression including several environmental and demographic variables as an approximation, hence a proxy measure, for a psychosocial scale. This is bound to clash with validity and reliability considerations, and is hence cautiously proposed as a potential means of screening and not as a definite substitute for actual application of psychosocial questionnaires.

1.2 *Method*

This paper reports on measures of association found between environmental conditions as well as demographic variables and psychosocial factors. It is based on data from a cross-sectional study of 32 administrative workers laboring in a mixed plan computerized office located in a Portuguese city (Tavares et al. 2013, Coelho et al. forthcoming). Environmental data were obtained in the domains of lighting, noise and the thermal environment, using calibrated instruments and standard procedures. Ergonomic and psychosocial data were collected in the Spring of 2012. Noise, illumination and acoustic data were additionally collected in the month of May 2013, as the initial dataset was not encompassing of all workstations (2012 and 2013 acoustic data is presented in the following section for comparison), and no changes in personnel or personnel workplace assignment had taken place in the meantime.

Lighting assessment followed European standard EN 12464-1:2011: Light and lighting. Lighting of work places: Indoor work places". This standard specifies the requirements for indoor lighting in order to provide an adequate work environment. A matrix with measurement points for the work area in each individual workstation in the office was defined according to EN-12464-1:2011, and illuminance (Lux) value for each workstation was calculated as the average of all the values measured in the points of the matrix for the workstation. The state of the window blinds in each office room (open or closed) was also recorded.

In order to assess exposure of workers to noise with an objective measuring method, the generally recognized standard IEC-61672–1:2002 was used as a basis for the procedures. Due to rather stable noise level, and for convenience of measurement, the L_{eq} (equivalent continuous sound level) parameter was substituted for L_{AT}, a time averaged noise level (stable), as workers were not at risk of occupational deafness, but from loss of mental focus. The time period for assessment of L_{AT} was obtained automatically, by using the "stop when stable" mode for an L_{AT} sampling comparison rate over 10 seconds. Hence the duration of time period to attain stability of L_{AT} and the respective L_{AT} were considered for the association analysis. The former provides an indication of the degree of variation and instability of the sound pressure level in the workstation.

Air temperature and relative air humidity were measured in each workstation. Additionally the state of the air conditioning unit of the office room where each workstation was located was recorded. All AC units were set to 25°C and workers could chose to have their office room unit on or off. Outside air temperature during the time of measurements (two consecutive Spring days) ranged from 9 to 22°C and from 12 to 22°C.

The musculoskeletal outcomes were assessed using the DASH—Disabilities of the Arms, Shoulders and Hands—questionnaire (Hudak et al. 1996) and a body diagram (Corlett & Manenica 1980). The assessment resulted in individual DASH scores and individual musculoskeletal complaints, which were subsequently grouped into three major body areas: upper extremities, upper body (torso) and lower extremities. The cross-sectional study also collected physical ergonomics data through deployment of Lima and Coelho's (2011) checklist. It is composed of eighty-eight items, grouped into four domains: posture, seating, workplace equipment and workplace layout. Although part of the cross-sectional study dataset, this is not the paper's focus.

Psychosocial factors were assessed using the Spanish short version of the COPSOQ—Copenhagen Psycho-social Questionnaire (Moncada et al., 2005), translated in to Portuguese by the first two authors, as at the time of instrument selection, a Portuguese version was not finalized and was not publicly and freely available. This instrument surveys the following six major groups of psychosocial factors: the psychological demands of work (concerned with the volume of work in relation to the time available and management of emotions), workers control over work (concerning the opportunities that the job provides for active, meaningful work, that contributes to develop the individual's skills), insecurity towards the future (concerning worries about the future in terms of job loss or unwanted changes in work conditions), social support and leadership quality (concerning relationships with co-workers and superiors), conflicting demands (in relation to the need to compromise between time and tasks for work, family and socialization), and, finally, esteem (concerning respect, rewards and justice experienced in exchange for the work effort). Psychological demands of workers were considered as a dissatisfaction measure, while the workers control over work was considered a satisfaction subscale. Insecurity towards the future was considered a dissatisfaction subscale, while social support and leadership quality were considered a satisfaction subscale. Finally, conflicting demands were considered as a dissatisfaction subscale, and the esteem group was considered a satisfaction subscale.

The paper reports on salient Spearman rank correlation coefficients using the approach reported by Coelho et al. (2013) across the board as well as on the standardized coefficients resulting from significant linear regression (using the backward entry method—Hastie et al. 2009) for the COPSOQ scales modeled on the environmental settings and local demographic data. R squared values and significance of the final models obtained using SPSS™ 19 statistical package for Windows were the criteria for identification of candidate psychosocial scales for the future development of compound expressions hypothesized as proxies for those psychosocial scales, to be used in screening done by practitioners.

2 RESULTS

2.1 *Description of data*

The data extracted from the cross-sectional study is summarily presented in Tables 1, 2 and 3. Table 1 presents the descriptive summary of ordinal categorical data, while Table 2 presents descriptive statistics for all numerical data. Due to incomplete coverage of data collected for the (stable) time averaged sound level (L_{AT}) in 2012, 2013 environmental data are used in the study of association. Comparison between the mean and standard deviation of L_{AT} data collected in 2012 and 2013 (Table 3) suggests a noticeable match, and also suggesting that time and day of measurement did not greatly affect the results, with uniformity and continuity of noise.

Table 1. Case summaries of ordinal categorical data.

Variable	Category 0	Category 1	Total (n)
Gender	Male 15	Female 17	32
WS+ in open-office section	No 25	Yes 7	32
Majority of women in room	No 15	Yes 17	32
Office room lights (state)	Off 2	On 30	32
Window blinds (state)	Closed 19	Open 13	32
Air Conditioning unit (state)	Off 10	On 22	32

+ Workstation.

Table 2. Descriptive statistics summary of numerical environmental and local demographic variables and psychosocial scales.

Variable/Scale	Mean	Stand. Dev.	n
Time averaged sound level (stable)*	49.7	10.17	32
Duration of L_{AT} measurement (s)	161.9	38.74	32
Illuminance level (Lux)	704.2	584.74	32
Air temperature (°C)	22.0	1.20	32
Relative air humidity (percentage)	42.8	2.09	32
No. additional co-workers in room	3.1	2.12	32
Percentage of women in room	53.2	38.62	32
Age category**	37.5	6.72	32
Psychological demands***	12.4	2.77	25
Insecurity towards the future***	8.1	2.99	25
Conflicting demands***	5.6	2.80	21
Workers control over work***	25.9	4.75	25
Social sup. & leadership quality***	27.3	6.96	25
Esteem***	7.8	4.01	25

* L_{AT} measured in dB(A).
** Age categories defined by rounding age to tens of years.
*** COPSOQ short version subscales.

Table 3. Comparison of time averaged sound level (stable) results from 2012 and from 2013.

	May 2012		May 2013	
L_{AT} [db(A)]	Mean	St. Dev.	Mean	St. Dev.
WS 10–16*	65	1.8	64	2.6
All WS (1–32)	–	–	50	10.2

* Workstations (WS) located in one open-plan office room.

The office had several rooms with more than one workstation. The biggest room in the office had 7 workstations, and was considered an open-plan office. Additionally, the office had one room housing 5 workstations and three rooms housing 4 workstations each and 8 rooms with only one workstation each. Hence, a secondary variable was computed, concerning the number of coworkers in the room, besides the individual case. Additionally, and given the importance of gender (for references see Lima and Coelho (forthcoming)) the percentage of women and verification if women were the majority (ties between men and women were treated as null) in each office room were also computed as secondary variables.

The sample population was balanced between women and men, and overall represents a relatively young workforce. Environmental data, viewed in summarized form fall within recommendations for office work, except for a very large standard deviation in the illuminance level. Standard deviation of L_{AT} is large as 10dB actually represents three time the sound pressure level. The average score for the psychological demands and the esteem dimensions fall, according to the reference scores for the instrument (Moncada et al. 2005), in the unfavorable region. The overall psychosocial assessment for the entire sample shows many opportunities for improvement. Workers control over work is the only psychosocial dimension that taken overall falls in the favorable level. Insecurity towards the future, conflicting demands and social support and leadership quality fall over the intermediate reference scores. However, standard deviations for the psychosocial dimensions give an indication of great variability in individual scores. A reduced number of valid cases across the psychosocial scales was due to the voluntary nature of the questionnaire at the time it was applied. Moreover, for some workers who lived alone, the conflicting demands scale was not applicable.

2.2 *Analysis of correlation*

Spearman rank order correlation coefficients were computed for the individual data presented in Tables 1 and 2 in summarized form. Due to paper length restrictions only significant results ($p < 0.05$) are reported. No associations were found between the environmental and local demographic variables and the psychological demands scale. Table 4 presents the results for the insecurity towards the future psychosocial scale.

Table 4. Significant correlation coefficients for association with insecurity towards the future (n = 25).

Variable	rho	p
Time averaged sound level (sta.) [db(A)]	+0.414	0.040
Office room lights (state)	–0.425	0.034
Age category	–0.438	0.028

Conflicting demands only correlated significantly with gender (rho = +0.501, p = 0.006, n = 21). Workers control over work only correlated significantly with percentage of women in room (rho = –0.458, p = 0.021, n = 25). Table 5 presents the results for the social support and leadership quality scale, showing one behavioral/environmental and four local demographic variables.

Esteem correlated significantly (n = 25) with gender (rho = –0.458, p = 0.021) as well as with the state of the office room window blinds (rho = +0.645, p = 0.000).

2.3 *Linear regression analysis*

Due to confounding across multiple correlations with every one single variable, and low sensibility for less than moderate to strong associations,

Table 5. Significant correlation coefficients for social support and leadership quality (n = 25).

Variable	rho	p
Window blinds (state)	+0.409	0.042
Gender	–0.454	0.023
Workstation in open-office section	–0.445	0.026
Majority of women in room	–0.431	0.031
Percentage of women in room	–0.521	0.008

Table 6. Standardized coefficients (B) and respective significance (p) of linear regression models.

Variable / Model	B	p
Insecurity towards the future	–	–
Age category	–0.683	0.001
Time averaged sound level (sta.) [db(A)]	+0.630	0.001
Air Conditioning unit (state)	–0.528	0.013
Gender	–0.504	0.013
Duration of L_{AT} stable measurement (s)	+0.378	0.023
Window blinds (state)	–0.321	0.083
Conflicting demands	–	–
Number of addit. co-workers in room	–0.956	0.002
Air temperature (°C)	+0.749	0.002
Illuminance level (Lux)	–0.470	0.009
Workstation in open-office section	+0.402	0.071
Gender	+0.307	0.058
Workers control over work	–	–
Air Conditioning unit (state)	+0.609	0.027
Age category	+0.570	0.017
Window blinds (state)	+0.560	0.018
Time averaged sound level (sta.) [db(A)]	–0.546	0.012
Duration of L_{AT} stable measurement (s)	–0.476	0.022
Office room lights (state)	–0.377	0.070
Relative air humidity (percentage)	+0.362	0.077

a regression analysis paired with analysis of covariance provides a more reliable indication of the level of association of one variable with multiple variables. Regression is however, a much more complex analysis. Correlation quantifies the strength of the linear relationship between a pair of variables, whereas regression expresses the relationship in the form of an equation (Bewick et al. 2003).

Linear regression results with higher than 50% determination coefficients (R^2) and statistical significance at $p < 0.05$ were obtained for selected models (using the Backward variable entry method) of insecurity towards the future ($R^2 = 66\%$, $p = 0.002$, $n = 25$), conflicting demands ($R^2 = 69.9\%$, $p = 0.002$, $n = 21$) and workers control over work ($R^2 = 52.3\%$, $p = 0.047$, $n = 25$). The standardized coefficients of the three models and their respective significance p-values are presented in Table 6.

3 CONCLUSION

Analysis of correlation suggests that in the sample, there are gender biases for some of the psychosocial scale results, with men obtaining more favorable results than women for conflicting demands, for social support and leadership quality and for esteem. Workers laboring in office rooms where most people are women are bound to experience less control over their work than others.

Association levels and significance thereof differ according to the technique used, e.g. while social support and leadership quality correlated significantly to 5 environmental and demographic variables, a significant and determined linear regression model could not be established. Insecurity towards the future, conflicting demands and workers control over work are significantly associated to, and more than 50% determined by, a weighted combination of environmental and local demographic factors. The latter include the illuminance level, air temperature, the time averaged sound level (stable), as well as the state of the window blinds and of the air conditioning unit in the office room. Due to the small sample size, further research is needed to assess the possibility of using environmental and demographic data as a proxy for selected psychosocial factor in practitioner's screening of large office populations.

The association study reported in this paper, based on a subset of a cross-sectional study dataset, enabled identifying insecurity towards the future, conflicting demands and workers control over work as candidate psychosocial scales for the future development of compound expressions hypothesized as proxies, to be used in the screening stages done by practitioners. Testing the hypothesis embedded in the regression analysis results presented on larger datasets, should enable concluding if environmental and local demographic data may be used as a compound proxy for the aforementioned subset of two dissatisfaction subscales and one satisfaction subscale that are part of the short version of the COPSOQ questionnaire.

The results also show how behavioral signs are significantly associated to more complex psychosocial assessment. Working with the office room window blinds open is associated to greater satisfaction with social support and leadership quality as well as with greater esteem. Working with the office room light turned off correlates with greater insecurity towards the future. Testing these findings on other populations is needed in order to ascertain their validity.

REFERENCES

Bewick, V., Cheek, L., & Ball, J. (2003). Statistics review 7: Correlation and regression. *Critical Care*, 7(6), 451.

Bongers, P.M., Ijmker, S., Van den Heuvel, S., & Blatter, B.M. (2006). Epidemiology of work related neck and upper limb problems: psychosocial and personal risk factors (part I) and effective interventions from a bio behavioural perspective (part II). *Journal of Occupational Rehabilitation*, 16(3), 272–295.

Coelho, D.A., Harris-Adamson, C., Lima, T.M., Janowitz, I., & Rempel, D.M. (2013). Correlation between Different Hand Force Assessment Methods from an Epidemiological Study. *Human Factors and Ergonomics in Manufacturing & Service Industries*, 23(2), 128–139.

Coelho, D.A., Tavares, C.S.D., Lourenço, M.L., & Lima, T.M. (forthcoming) Working conditions under multiple exposures: A case study of private sector administrative workers. *Work*.

Corlett, E.N., & Manenica, I. (1980). The effects and measurement of working postures. *Appl Ergon*, 11(1), 7–16.

Hastie, T., Tibshirani, R., & Friedman, J. (2009). *Linear Methods for Regression* (pp. 43–99). Springer New York.

Hignett, S., & McAtamney, L. (2000). Rapid entire body assessment (REBA). *Appl Ergon*, 31(2), 201–205.

Hudak, P.L., Amadio, P.C., Bombardier, C., Beaton, D., Cole, D., Davis, A., Hawker, G., Makela, M., Marx, R.G., Punnett, L. & Wright, J. (1996). Development of an upper extremity outcome measure: The DASH (Disabilities of the Arm, Shoulder, and Head). *American Journal of Industrial Medicine*, 29(6), 602–608.

Karhu O, Kansi P, Kuorinka I. Correcting working postures in industry: a practical method for analysis. *Appl Ergon* 1977; 8:199–201.

Kee, D., & Karwowski, W. (2001). LUBA: an assessment technique for postural loading on the upper body based on joint motion discomfort and maximum holding time. *Appl Ergon*, 32(4), 357–366.

Kemmlert, K. (1995). A method assigned for the identification of ergonomic hazards—PLIBEL. *Appl Ergon*, 26(3), 199–211.

Krause, N., Rugulies, R., Ragland, D.R., & Syme, S.L. (2004). Physical workload, ergonomic problems, and incidence of low back injury: A 7.5-year prospective study of San Francisco transit operators. *American Journal of Industrial Medicine*, 46(6), 570–585.

Lima, T.M., & Coelho, D.A. (2011). Prevention of musculoskeletal disorders (MSDs) in office work: A case study. *Work*, 39(4), 397–408.

Lima, T.M., Coelho, D.A. (forthcoming) Ergonomic and Psychosocial Factors in the Prevalence of Musculoskeletal Complaints in Public Sector Administration Staff. *Work*.

Martin, J.L., Norris, B.J., Murphy, E., & Crowe, J.A. (2008). Medical device development: The challenge for ergonomics. *Appl Ergon*, 39(3), 271–283.

McAtamney, L., & Nigel Corlett, E. (1993). RULA: a survey method for the investigation of work-related upper limb disorders. *Appl Ergon*, 24(2), 91–99.

Moncada, S., Llorens, C., Navarro, A., & Kristensen, T.S. (2005). ISTAS21: Versión en lengua castellana del cuestionario psicosocial de Copenhague (COPSOQ). *Arch Prev Riesgos Labor*, 8(1), 18–29.

Monnington, S.C., Quarrie, C.J., Pinder, A.D., & Morris, L.A. (2003). Development of Manual handling Assessment Charts (MAC) for health and safety inspectors. *Contemporary Ergonomics*, 3–8.

Occhipinti, E. (1998). OCRA: a concise index for the assessment of exposure to repetitive movements of the upper limbs. *Ergonomics*, 41(9), 1290–1311.

Sauter, S.L., & Swanson, N.G. (1996). An ecological model of musculoskeletal disorders in office work. *Beyond biomechanics: Psychosocial aspects of musculoskeletal disorders in office work*, 3–18.

Tavares, C.S.D., Lima, T.M., & Coelho, D.A. (2013). Analysis of ergonomics in office Work: A case study leading to an intervention in office acoustics. *Occupational Safety and Hygiene*, 307–311.

Wahlström, J. (2005). Ergonomics, musculoskeletal disorders and computer work. *Occupational Medicine*, 55(3), 168–176.

Winter, G., Schaub, K., Landau, K., & Bruder, R. (2006, July). Development and validation of ergonomic screening-tools for assembly lines. *In Proceedings IEA 2006 congress, 16th world congress on ergonomics* (pp. 10–14).

Occupational Safety and Hygiene II – Arezes et al. (eds)
© 2014 Taylor & Francis Group, London, ISBN 978-1-138-00144-2

Incidence of musculoskeletal symptoms in industry workers

J. Trotta & L. Ulbricht
Federal University of Technology—Paraná, Curitiba, Brazil

J.L.H. Silva
Technical Nursing Work, Curitiba, Brazil

SUMMARY: Due to the presence of musculoskeletal injuries in the health service in an electromechanical company, an exploratory study has been conducted with the purpose of establishing the incidence and prevalence of musculoskeletal pain syndromes and their relations. The sample consisted of 1,253 employees and the result was that 33% of the workers reported musculoskeletal pain in an often and daily basis. The most affected body segments were the shoulders (12.69%), followed by the lower limbs and spine. The majority of employees related the pain syndromes to the performance of their labor function (65.94%), classifying the discomfort degree as significant (70.3%) and it was found that the clinical condition tended to become chronic since the symptoms had more than six months of duration (39.86%). We conclude that despite being a preliminary study, there are a several number of company's employees with frequent musculoskeletal pain of significant intensity and subjectively work-related, who delay seeking the occupational health medical service, with impacts on their daily life personal and organizational activities.

1 INTRODUCTION

According to Rodrigues et al. (2013) enhancement of production methods associated with repetitive work developed in long hours without breaks, use of vibrating tools, constant pressure for production and work overload has contributed to increase the prevalence of work-related diseases. Moreover, these factors increase the possibility of accidents causing temporary or permanent disability, revealing the causal link between health and work. Among the work diseases with important prevalence in Brazil are MSDs, which had their first record in the eighties of the last century but were only recognized as occupational disease at the beginning of the next decade. Therefore, according to De Lima & Lima (2013), the disease does not have all its aspects and triggering causes properly discussed (such as biomechanical factors, pathophysiology, psychological, sociological, among others).

The presence of these precipitating agents and aspects can be monitored and controlled before the development a pathological condition (Mendes, 2003 cited Souza, 1992, Merlo, 1999, p. 48).

Historically, there is the understanding that the work, in certain circumstances, can cause injury, illness, lower life expectancy, or even death. Today it is known that this is a broad and generic concept, but there are many nuances in the middle of this climb to become a disease, often imperceptible, but no less serious (Mendes, 2003 apud Souza, 1992; Merlo, 1999, p. 48).

According to Couto (2007), before the onset of an injury other signs appear, especially in the early stages, which is very important as early detection of an imminent problem and it is this phase the ideal to establish the actions to eliminate or at least minimize future consequences. The author further refers that around "65 to 75% of workers complaining of pain do not show other characteristic clinical signs of injury".

Pain is one of the earliest symptoms and serves as a warning that there may be some commitment to health, one of the first signs of impending disease.

According to Kennedy et al. (2010) reported that the symptoms of pain, numbness, tingling in various parts of the body may be warning signs of developing musculoskeletal injuries, enthesopathies and peripheral nerves or other less specific.

However, even before the pain, specifically when referring to musculoskeletal diseases, we can see other symptoms such as tingling, loss of strength, among others that if are diagnosed early and properly, can prevent the onset of the disease. Currently, musculoskeletal diseases are a major problem to be prevented within organizations, since beyond the personal consequences, they cause a high rate of absenteeism, legal liabilities, effects on production, in the organizational environment and the even in the social community to which the employee is part. Therefore, the need for understanding and mapping of the conditions and correlations of these symptoms are extremely important to focus the prevention efforts.

In the midst of this macro scenario and due to the high rate of musculoskeletal complaints daily arriving to the company health service department, the high rate of absenteeism due to diseases and the lack of an adequate mapping of this scenario, we have identified the need for a better understanding of the situation in order to adopt future control actions for this problem.

The purpose of this study was to establish the incidence and prevalence of musculoskeletal pain syndromes and their relationships in an electromechanical industry in the metropolitan region of Curitiba—PR, Brazil.

2 MATERIALS AND METHODS

We performed an exploratory research where the data were collected through a survey instrument in the form of a formatted questionnary that was applied to a population of 1,253 employees (70.40% of the total universe) of an electromechanical multinational company, degree of risk three, with workstations on foot and with a high requirement for repetitive members superiors during the regular review between the period January 2013 to December 2013.

Developed a self-administered questionnaire with 13 questions (Table 1) designed to explore aspects of musculoskeletal symptoms and signs and their relationship with the various body segments, etiology, start time etc. This questionnaire has objective answers true or false. The last question of giving descriptive ability of the employee to express perceptions about the job from him.

The form was developed to understand the needs of employees and to serve as an indicator for future actions in the workplace company. It was approved by the medical department of the company and was conducted a pilot project with ten volunteers before the final application for appropriate adjustments.

The survey instrument first performed an ergonomic census, with questions regarding the workplace, section, equipment, overtime performing and body segments in which the respondent felt pain.

After the identification of pain, it was characterized by the form of discomfort reported (e.g. fatigue, glitch, tingling, etc) and later classified as to the degree as very strong, strong, moderate, mild or very mild.

Table 1. Ergonomic census questionnaire.

ERGONOMIC QUESTIONS

It is not necessary identification. Read carefully and respond as realistic as possible.
This information will be used for scientific work and improvements to the workplace.
I agree to participate :
....

WORKPLACE : ______________________________

SECTOR NAME : ______________________ Nº: __________

WORK EQUIPMENT : ....

Age : ____________ Sex : ________ Date of Admission : ____________

WHAT IS THE HAND THAT YOU WRITE? ( ) Right ( ) Left

HOW MANY OVERTIME HOURS YOU WORK ON AVERAGE PER MONTH? ....

1 - DO YOU FEEL ANY PAIN OR DISCOMFORT IN MUSCLES OR JOINTS FREQUENTLY?
( ) IF NOT, GO DIRECT TO QUESTION 9 . (Go to question 9)
( ) IF YES , ANSWER ALL QUESTIONS BELOW

A- SHOULDERS? ( ) YES ( ) NO	B- NECK? ( ) YES ( ) NO
C- COLUMN? ( ) YES ( ) NO	D- ARMS? ( ) YES ( ) NO
E- ELBOW? ( ) YES ( ) NO	F- FOREARMS? ( ) YES ( ) NO
G- HANDS? ( ) YES ( ) NO	H- FISTS? ( ) YES ( ) NO
I- LEGS? ( ) YES ( ) NO	J - HIP? ( ) YES ( ) NO
k- KNEES? ( ) YES ( ) NO	L- LEGS? ( ) YES ( ) NO
M- ANKLES? ( ) YES ( ) NO	N- FEET? ( ) YES ( ) NO

2 - IS THE PAIN YOU FEEL RELATED TO THE WORK?
( ) YES ( ) NO

3 - FOR HOW LONG ?
( ) UNTIL 1 MONTH ( ) 1st 3 MONTHS ( ) 3 to 6 MONTHS ( ) OVER 6 MONTHS

>>>

4 - WHAT IS THE DISCOMFORT THAT YOU FEEL?
( ) FATIGUE; ( ) SHOCK; ( ) ESTRALOS; ( ) PAIN;
( ) LIMITATION OF MOVEMENTS; ( ) POWER LOSS; ( ) WEIGHT .

5- HOW STRONG IS YOUR DISCONFORT?
( ) VERY STRONG; ( ) STRONG; ( ) MODERATE; ( ) LIGHT; ( ) VERY LIGHT

6- DOES THE PAIN YOU FEEL INCREASE WITH THE WORK?
( ) DURING NORMAL WORKING DAY; ( ) DURING OVERTIME
( ) AT HOME; ( ) NOT RELATE TO WORK

7 - DOES THE PAIN YOU FEEL GET BETTER IN REST ?
( ) NIGHT; ( ) ON WEEKENDS; ( ) ON VACATION;
( ) DURING TURNS IN OTHER TASK: ( ) NO BETTER NEVER

8- DO YOU TAKE SOME MEDICATIONS TO WORK ?
( ) YES; ( ) NO; ( ) SOMETIMES

9 - HAVE YOU ALREADY TAKEN ANY TREATMENT FOR ARTSANY OR MUSCLE DISTURBS IN THE PAST ? ( ) YES; ( ) NO
DESCRIBE THE MEDICATION:
____

10 - WHAT IS THE ACTIVITY WHICH IS CAUSING YOUR DISCOMFORT?
( ) NO; IF YES , DESCRIBE WHAT ARE :
....

11 - ARE YOUFEELING GOOD WITH YOUR ACTIVITY IN THE COMPANY?
( ) Yes , fully satisfied , ( ) Like it, partially motivated ;
( ) I do what I am told to do, I think I hit my goals ;
( ) expected more from the Company
( ) I'm not attracted by the activity that I perform

2 - WHAT IS YOUR SUGGESTION TO IMPROVE THE WORKPLACE?
DESCRIBE: ....
....
....

13 - DO YOU MAKE LABOUR GIMN? ( ) Yes, ( ) No, because:
....

DATE : ______________________ .

The survey instrument investigated if the pain increased with the work performance, if it improved with rest and if a treatment was necessary.

Finally, we investigated the work satisfaction and the performance or not of labor gymnastics.

The data were processed using descriptive statistics, with position and dispersion measures.

3 RESULTS

We have evaluated 1,253 employees (70.40% of all employees of the company), the majority of which are male (75.6%).

Although the average age points to a mass of young workers (30.8 years), 33.04% of them reported musculoskeletal pain in a frequent and daily manner.

The body segments most affected by pain were the shoulders (12.69%), the legs (11.65%), the spine (11.09%), the neck (7.58%), the feet (7.18%), the hands (7.18%), the arms (6.78%), the knees (6.68%), the wrists (6.23%), the forearms (3.35%), the ankle (3.27%), the elbow (2.87%), the hip (2%) and the thighs (2%).

The vast majority of employees (65.94%) related the pain syndromes to the performance of their work tasks. The intensity of discomfort perception was generally regarded by 54.35% of the sample as moderate (mild 23.43%; strong 14.98%; very light 2.90% and very strong 0.94%).

The clinical condition tended to show a chronical aspect since 38.86% of the employees reported feeling pain for more than six months (between 3 and 6 months, 19.08%; from 1 to 3 months, 18.60%; and less than one month, 15.22%).

From table 2 we can see that, as to the lost days, most were related to musculoskeletal or conjunctive tissue diseases.

Table 2. Prevalence of diseases by work days loss in 2012.

Group of Illness	Cumulative lost work days in 2012
CID M—Musculoskeletal diseases or conjunctive tissue	877.12
CID J—Respiratory diseases	599.59
CID S—Injuries and other consequences of external causes	546.7
Cumulative annual	6073,01

Table 3. Prevalence of pain complaints by sectors. Sectors with the highest prevalence of pain complaints.

Sectors with the highest prevalence of pain complaints	%
Assembly A/C	61.11
Adm. Produçtion CM	50.01
Assembly KIT	46.67
Hoses	43.75
CHE Welding	43.75
Logistics	41.24
Evaporator RS	39.18
Information Technology	39.13
Condenser	38.95
Evaporator	38.64

Corporate sectors that were most often appointed as the cause of pain complaints were assembly A/C (61.12%); followed by adm production (50%); welding (43.75%); assembly kit (46.67%); hoses (43.75%) and logistics (41.23%) (table 3).

Most of the employees (50.40%) declared to be very satisfied with the work (24.58% satisfied, 18.83% regularly satisfied, 3.67% dissatisfied and 1,52% very dissatisfied).

4 DISCUSSION

From 2004 to 2008, workplace accidents in Brazil increased significantly while occupational diseases decreased from 2004 to 2008, and the age group of 40 to 59 years were the most affected by occupational diseases (Graup, 2013). In our research, the average age of workers was 30.8 years, a youngest age group that, however, already presented frequent and daily musculoskeletal pain (33.04% of the cases).

Research conducted by Trindade et al. (2013) among workers with an average age of 31.7 years, similar to our study, found a higher prevalence of pain complaints (of 92.7%) in the textile industry. In such study, the majority of the labor force was formed by males (61.5%).

In Cunha & Person (2013) reasearch in a steel industry, musculoskeletal and neuropathic disorders also appeared as the major cause of pain in workers survey participants (11.7%), yet with a prevalence lower than we found in our study.

As we can see in the various studies, musculoskeletal diseases represent a prevalence showing the companies need to develop epidemiological and ergonomic studies so that risk factors can be intercepted (Ulbricht & Stadnik, 2010).

The body segments most affected by pain in our research were the shoulders (12.69%). According to CID-10, occupational diseases of musculoskeletal origin most prevalent in Brazil in 2004–2008 (Graup, 2013) were M65 (synovitis and tenosynovitis) and M75 (shoulder injuries).

The body segments most affected were the shoulders, the back and the fist (41.23%).

The highest prevalence of occupational diseases in Brazil in 2004–2008 were in the service sectors of financial activity and the metallurgical industry, in which the clerk profession presented the largest number of patients throughout the analysis period (Graup, 2013). In our research, company section that caused more painful complaints was the air conditioning assembly.

It is essential to identify the work places with higher risk for MSDs development so that preventive actions can be taken to avoid pathological conditions that can bring many losses. Study by Gontijo et al. (2012) in a steel industry found 17 licenses for health-musculoskeletal problems recorded in a period of 267 days. Those licenses were responsible for 111 days of lost work, corresponding to 41.57% of total lost days at work. The authors emphasize labor productivity losses, financial burden to the social security service and the damages to the worker, generated by these pathologies.

Although the majority of employees (50.40%) were very satisfied with the work, there were pain complaints. Some authors affirm that the greater the job dissatisfaction, the higher is the triggering pain, suggesting that there are biopsychosocial factors involved in disease development (De Lima & Lima, 2013), yet others claim that these factors can modify the degree of sensibility to pain (Trinidad et al., 2013).

Finally, although these symptoms are found in workplaces this company, it can not necessarily be considered as musculoskeletal disorders of occupational etiology.

5 CONCLUSION

The results shows that despite being a preliminary study, there is a significant number of company employees (33.04%) with frequent musculoskeletal pain, which mostly relate the pain symptoms to the labor activities (75.3%), and even though the discomfort is of significant intensity (70.3%), such employees belatedly seek the occupational health medical service, since they report the pain onset for over more than 6 months of evolution (39.86%).

Company sector with the highest prevalence of musculoskeletal pain syndromes were exactly those with more ergonomically unfavorable labor activities (repeatability, inappropriate posture, cargo handling etc).

Therefore, we found that the employees of this company are assuming a defensive and less participative posture with the occupational health service, which leads to a late diagnosis, raising the risk of increasing injuries aggravation and sequelae. This approach also precludes the ergonomic improvements actions often required in their workplaces that could reduce the incidence of work—related musculoskeletal disorders and the labor liabilities for the company.

REFERENCES

Brasil. 1994. *Manual de legislação, segurança e medicina do trabalho.* 27ª edição. São Paulo: Ed. Atlas.

CID-10. 1994. *Organização Mundial da Saúde.* São Paulo: Ed. Universidade de São Paulo.

Couto H.A., Nicoletti,S.J.& Lech, O. 2007. *Gerenciando a LER e os DORT nos tempos Atuais.* Belo Horizonte: Ergo Editora.

Couto, H.A. 1978. *Fisiologia do trabalho Aplicado.* Belo Horizonte: Ed. Ibérica.

Couto, H.A. 1995/6. *Ergonomia aplicada ao trabalho: manual técnico da máquina humana.* Vol. I e II. Belo Horizonte: Ergo Editora.

Cunha, E.M.T. d., & Pessoa, Y.S.R.Q. 2013. O perfil de morbidade dos trabalhadores de uma metalúrgica paraibana/Profile of the morbidity of workers of a metallurgical from Paraiba state, in Brazil. *Trabalho & Educação,* 2013, *21*(3), 61–77.

De Lima, F.G.S.B., De Lima, E.V.; & Da Silva, A.P. 2013. Perícia médica em lesões por esforços repetitivos/distúrbios osteomusculares relacionados ao trabalho: diagnóstico. *Cognitio Unilins,1,* 1–5, from: http://revista.unilins.edu.br/index.php/cognitio/article/view/72.

Dul, J.; Weerdmeester, B. 1995. *Ergonomia prática.* Tradução Itiro Iida. São Paulo, Editora Edgard Blücher.

Gontijo, R.S., Antunes, D.E.V., de Oliveira, V.C., de Oliveira, R.C., & Guimarães, E.A. d. A. 2012. Análise dos distúrbios osteomusculares relacionados ergonomia em aciaria de uma empresa siderúrgica. *Revista de Enfermagem do Centro-Oeste Mineiro.* 2(2):203–210.

Graup, S. 2013. *Cenário epidemiológico de morbidade no ambiente de trabalho no Brasil.* Tese (doutorado) - Universidade Federal de Santa Catarina, Centro Tecnológico, Programa de Pós-Graduação em Engenharia de Produção, Florianópolis.

Kennedy, C.A., Amick III, B.C., Dennerlein, J.T., Brewer, S., Catli, S., Williams, R., Serra, C.; Gerr, F.; Irvin, E.; Mahood, Q.; Franzblau, A.; Eerd, V.; Evanoff, B.; & Rempel, D. 2010. Systematic review of the role of occupational health and safety interventions in the prevention of upper extremity musculoskeletal symptoms, signs, disorders, injuries, claims and lost time. *Journal of occupational rehabilitation,* 20(2), 127–162.

Mendes, R. 2003. *Patologia do Trabalho.* São Paulo: Atheneu.

Oliveira, J.E.G. 2012. *Gestão das condições de trabalho e saúde dos trabalhadores de saúde.* São Paulo: UFMG.

Rodrigues, B.C., Moreira, C.C.C., Triana, T.A., Rabelo, J.F., & Higarashi, I.H. 2013. Limitações e consequências na vida do trabalhador ocasionadas por doenças relacionadas ao trabalho. *Revista da Rede de Enfermagem do Nordeste-Rev Rene,* 14(2), 448–57.

Santos, N. & Fialho, F.A.P. 1997. *Manual de Análise Ergonômica no Trabalho.* 2 ed. Curitiba: Gênesis.

Trindade, L.D.L., Krein, C., Schuh, M.C.C., Ferraz, L., Amestoy, S.C., & Asamy, É.K. (2013). Trabalhadores da indústria têxtil: o labor e suas dores osteomusculares. *Journal of Nursing and Health,* 2(2).

Ulbricht, L.; Stadnik, A.M.W. 2010. Identificação dos Fatores de Risco Presentes na Ordenha: aplicação no estudo dos distúrbios osteomusculares relacionados ao trabalho dos ordenhadores no Paraná. *Anais do XXX ENEGEP* (pp. 1–14). São Carlos: ABEPRO.

Occupational Safety and Hygiene II – Arezes et al. (eds)
 ISBN 978-1-138-00144-2

Influence of hot thermal environment in practice of Aerostep

J. Nogueira
DFIS—University of Aveiro, Aveiro, Portugal

M. Talaia
DFIS—CIDTFF—University of Aveiro, Aveiro, Portugal

ABSTRACT: Nowadays, the improvement of the thermal environment that influences the well-being of occupants in closed spaces has became a topic to be explored. It is known that the human over the years became increasingly sedentary, therefore more user of closed spaces, including for sporting, in detriment of the sport "outdoors". One of the authors of this work is a practitioner of physical activity in closed areas (sports halls and gyms), and his experience has shown that there are difficulties at metabolic and respiratory level during sport activity. Some practitioners arrived to a condition of collapse of the body. In this work it's evaluate the thermal sensation felt by practitioners of the modality of AeroStep, in a multipurpose room of a gym. In order to evaluate the thermal pattern and the thermal indexes *ITH* and *EsConTer*, 16 observations posts were identified. For the studied days the obtained results showed that the thermal environment of the gym is influenced by the thermal environment outside and that the initial conditions of the thermal environment of the gym are determinants for an enjoyable sports practice.

1 INTRODUCTION

The human body is a homoeothermic living. As such, it is natural that the existence of thermal comfort is one of the fundamental factors of human everyday. However, perhaps never in another phase of human existence such factor had been placed in so much risk and concern as currently, due largely to climate change that have occurred gradually. It is recognized by the global scientific community and political, that climate change should be the result of some human activities, particularly in the last one hundred years (IPCC 2007). The human body, either directly or indirectly, changed the environment that surrounds him and is in this context that the intervention for the improvement of the conditions of comfortable thermal environment is important. The literature shows that there is available a diverse set of thermal indexes capable of evaluating the thermal sensation of an individual. According to Markov (2002) satisfy all individuals inserted in a given thermal environment is a task almost impossible.

It is known that in the last years has witnessed a sedentarization of the human being. A thermal environment can be indoor or outdoor. Practice shows that a thermal environment inside is influenced by external thermal environment. However, if at these changes in thermal comfort offered by climate change, add physical activity inherent of the practice of sport, particularly in indoors spaces, it may be facing one of the greater situations of exposure to thermal stress. Kruger *et al.* (2001) showed that in a thermal environment is important to evaluate meteorological parameters (air temperature, relative humidity of air and air velocity, for example), physical aspects (clothing, activity) and subjective aspects (state of mind of individuals).

In the present study is shown as an inside thermal environment is influenced by the thermo-higrometric conditions of the outside environment confirmed by thermal sensation maps. The thermal sensation of the practitioners at the beginning of a class was evaluated through a color scale. Intervention strategies are considered.

2 HUMAN BIOMETEOROLOGY AND THERMAL SENSATION

The Human Biometeorology is the science that studies the relationship between human being and atmospheric weather around them. Studied since antiquity, this science is becoming increasingly important in everyday life, once that the concern to public health, comfort and well-being of humans, increased over the time, with the emergence of new data indicators of atmospheric pollution that man produces since the Industrial Age (Tromp, 1980).

In order to know what improvements can be made in a thermal environment, to offer maximum comfort both physically and psychologically, it became important to consider the sensation of each individual exposed to a certain environment and what their maximum capacity so as the human body does not come in collapse. In a thermal environment the thermal balance of the human body is important because of the exchange of energy in the form of heat that occurs with the environment around the individual. The interaction between the human being and the environment involves exchanges of energy by conduction, convection, radiation and evaporation, all related to meteorological factors. It was established by Fanger (1972), through a research conducted in climate chambers, the thermal balance equation. According Lamberts (2002), can be written (1) where M is the metabolic rate (W.m^2), W is the external work (W.m^2), C the rate of convective heat transfer in the skin (W.m^2), R is the total rate of heat transfer by radiation in the skin (W.m^2), E_{SK} the rate of heat loss by evaporation in the skin (W.m^2), C_{RES} the rate of convective heat transfer through respiration (W.m^2) and E_{RES} the rate of heat transfer by evaporation in the breathing (W.m^2).

$$M - W = (C + R + E_{SK}) + (C_{RES} + E_{RES}) \qquad (1)$$

It is verified through the expression (1) the importance of the individual analysis of each one of the energy exchange processes performed by the human body.

The thermal sensation that humans has before certain environment and that leads to a search for thermal comfort is subjective and depends on various factors, such as age, sex, weight, state of mind, activity that is exercising, clothing and food. It is essential that a study includes bearable limits by the human being, through their thermal sensations, and the more detailed development of indexes that characterize increasingly better the environment that surrounds the human being, in order to have maximum comfort in the future.

3 MATERIALS AND METHODS

In order to perform this study, a formal contact with the manager of the gym, was established.

After the convergence of interests between the two sides and approval was selected a closed space (physically) which occurs the practice of sports activities. Was selected the type of physical activity (Step classes, specifically AeroStep), based on an activity whose practitioners were in a very small intervention area of the gym but that they produced an intense physical effort. Each class had beginning at 6.30 pm and finished at 7.30 pm, having been this the time to collect all data. Then it's proceeded to the elaboration of a grid of 12 points of the gym. In these observation points were recorded: air temperature T (°C) and relative humidity U (%) using the device *"CENTER 317 Temperature, Humidity, Meter"*. The data were recorded before the beginning and at the end of class. Were also recorded data of air temperature T (°C) and relative humidity U (%) next the four walls of the gym and the environment outside the gym before and after each class. The clothing of each of practitioners was considered and the direct observation of the behavior of each participant (identified by a number) was recorded during the class. Was used the seventh scale of ASHRAE (ANSI/ASHRAE 55, 2004) in a corresponding color scale (Talaia & Simões, 2009) to know the real thermal sensation of each practitioner. In the scale of colors, individually, was placed through the use of a "cross" the thermal sensation that each practitioner felt at the beginning and end of the session, as shown in Figure 1. The information "0" is considered comfort. The information of the seventh scale of −3 corresponds to the dark blue color of a very cold thermal environment and +3 correspond to the dark red color of a very hot environment. The position of the "cross" allowed determining the thermal sensation through a linear scale in terms of calculations.

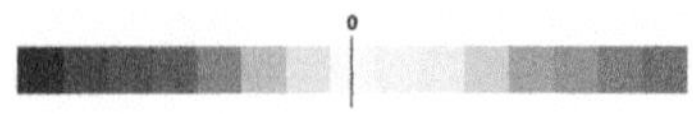

Figure 1. Scale of thermal sensation.

It was applied after data collection, two indexes *ITH* and *EsConTer*, which allowed knowing in the positions of the gym the thermal sensations. The Index of Temperature and Humidity, *ITH*, was in a first phase developed by Thom (1959), which combined the temperature of the wet thermometer T_{wn} (°C) with the air temperature T (°C). However, Nieuwolt (1977) replaces the temperature of the wet thermometer T_{wn} (°C) by the relative humidity U (%). With this modification, Nieuwolt (1977) was intended to facilitate the implementation and evaluation of the index. According Nieuwolt (1977), the *ITH* is calculated from the expression (2), where T (°C) represents the air temperature and U (%) relative humidity of the air.

$$ITH = 0.7T + (U/500) \qquad (2)$$

ITH values for thermal sensation were adapted by Talaia *et al.* (2013), as shown in Table 1. The *EsConTer* index (Talaia & Simões, 2009) is based on a color scale (*Es*), the feeling of comfort (*Con*) and is thermal (*Ter*). The index, based on the information of thermal sensation which is recorded by a color scale, limited by the ends of −3 to +3 depends of air temperature and the temperature of the wet

Table 1. Thermal sensation based on *ITH* (adapted by Talaia *et al.* (2013).

ITH	Thermal sensation
$ITH < 8$	Too cold
$8 \le ITH < 21$	Sun needed for comfort
$21 \le ITH < 24$	COMFORT
$24 \le ITH < 26$	Wind needed for comfort
$26 \le ITH$	Too hot

Table 2. Values recorded for the day May 30.

	Beginning of class	End of class
T (°C)	21.6	21.6
U (%)	62.4	64.4
T_w (°C)	17.2	17.4
ITH	20.0	20.0
EsConTer	0.25	0.26

thermometer. It is determined using the expression (3) where T is the air temperature (°C) and T_w the temperature of the wet thermometer (°C).

$$EsConTer = -3.75 + 0.103(T + T_w) \qquad (3)$$

The calculated values of *ITH* and *EsConTer* were introduced in a program developed in MATLAB who allowed knowing the patterns for air temperature, relative humidity of air, *ITH* index and *EsConTer* index. The location and dynamic of lines, at different hours and days, allowed knowing the points where the practitioners were under higher thermal vulnerability.

Patterns were traced to the beginning and end of each class. The values recorded at the beginning and end of each class allowed evaluating the amount of water that practitioners lost to the environment. This result took in consideration the difference of partial pressure of saturation of water vapor at the temperature of dew point at the beginning and end of each class. The clothing of each practitioner was registered for an evaluation of energy balance in terms of alteration of the metabolism during the practice of AeroStep. The information of each practitioner about their thermal sensation, in the scale of colors, at the beginning and end of each class, allowed recognizing that the stress of the practitioner was due to metabolism. The results were interpreted and allowed to suggest intervention strategies.

4 RESULTS AND DISCUSSION

The gym where the study was developed and the classes of AeroStep present a square format with side of 14 m. Each face of the gym has a window and one of the faces has an access door, which was open during the classes. Another face has four glazed doors (closed) that allowed the entry of external luminosity.

The first recorded data allowed improving the way to collect data and guiding strategies for a better understanding of the thermal pattern of the gym and the most vulnerable places for the practitioners. Numerous information maps were constructed to define intervention strategies.

For option are presented calculated data for the day May 30, 2013. The Table 2 shows the mean values obtained of the air temperature T (°C), relative humidity of air U (%), temperature of the wet thermometer T_w (°C), *ITH* index and *EsConTer* index, for the beginning and the end of the class.

The Table 2 shows that the values of the air temperature are equal. The relative humidity of air show a slight increase as expected. In practice the breathing and transpiration of practitioners during sporting activity make increase the temperature of the dew point (there is humidification of humid air that is confined to the activity space). The indexes values show that for the surrounding air to the practitioners does not register variation, the *ITH* maintains the value of 20.0 and the *EsConTer* passed from 0.25 to 0.26.

The amount of water in the vapor state between the beginning and end of class, in the humid air, was 0.30 g/m^3, which is equivalent to say that each practitioner has lost about 40 cL of water to the environment. Each practitioner with an identifying number, allowed to know their thermal sensation based on a color scale (Talaia & Simões, 2009). The Table 3 shows the values of the thermal sensation of practitioners for the day May 30.

The observation of Table 3 shows unequivocally that the thermal sensation *EsCor* indicated by the practitioners (in number of 8), in the beginning of class, is in agreement with the calculated value by the *EsConTer* index (0.25) and shows to be a comfortable thermal environment. In the Table 3 the difference that exists between the value of thermal sensation *EsCor* of the practitioners at the beginning of class (0.24) and the end of class (1.88), and the value of the *EsConTer* index at the beginning of class (0.25) and the end of class (0.26) is justified by the change of metabolism of the practitioners. In practice, sporting activity has made increase the internal body temperature and the practitioner at the end of class registers a thermal sensation that does not depend of the environment that surrounds her. However, is important to mention that the conditions of the initial thermal environment can

Table 3. Thermal sensation of the practitioners in a scale of colors *EsCor*, for the class of the day May 30.

	Beginning of class	End of class
EsCor (mean)	0.24 ± 0.23	1.88 ± 0.39
EsCor (maximum)	0.47	2.27
EsCor (minimum)	0.01	1.50

determine the collapse of the body. The indicated value of 1.88 is a value that according to the ISO and ASHRAE Standards causes a hot environment, in the seventh scale of ASHRAE (2001, 2004).

By option, some figures are presented for the hottest day of the class of AeroStep registered in this study. The Figures 2 and 3 show the patterns obtained for the beginning and the end of the class of the day June 26, 2013, for the *ITH* index. The Figures 4 and 5 show the patterns obtained for the beginning and the end of the class of the day June 26, 2013, for the *EsConTer* index. The observation of the Figures 2 and 3 and the Figures 4 and 5 shows that the patterns obtained in the gym for the two indexes show an excellent agreement for the beginning and end of class. The values obtained for the air temperature T (°C), relative humidity of air U (%), temperature of the wet thermometer T_w (°C), *ITH* index and *EsConTer* index, for the day June 26, are indicated in Table 4.

The observation of the values indicated in the Table 4 show that the air temperature remained almost constant during the class (increased 0.1°C) and the relative humidity of air increased about 12%. This result indicates that the amount of water in the vapor state increased in the humid air that surrounds the practitioners. The thermal indexes show variation to despise (*ITH* from 23.8 to 24.2 and *EsConTer* from 1.0 to 1.1). For the day June 26 the thermal environment shows a situation slightly hot at the beginning of class. The Table 5 shows the values of the thermal sensation of the practitioners to the day June 26.

The observation of the Table 5 shows, unequivocally, that the practitioners presents at the beginning of class a mean thermal sensation of 1.64 with a standard error of 0.35. It is authors believe that these values can be related to the fact that the practitioners comes into the gym room after coming from external environmental conditions that cause different thermal sensation. The acclimatization of the practitioners should be considered in this type of study. The Figure 6 shows the positioning of the observation points for the registration of values and location of the practitioners in number of 14, for the day June 26, 2013.

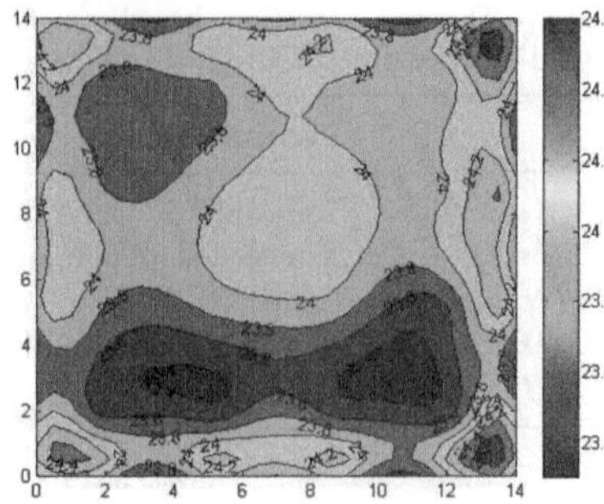

Figure 2. *ITH* at the beginning of the class of June 26, 2013.

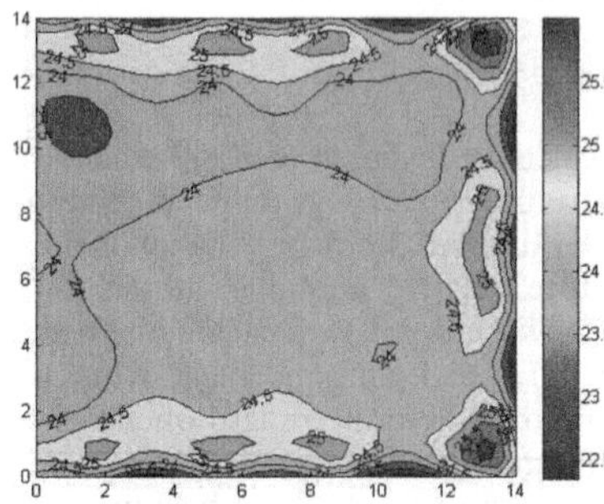

Figure 3. *ITH* at the end of the class of June 26, 2013.

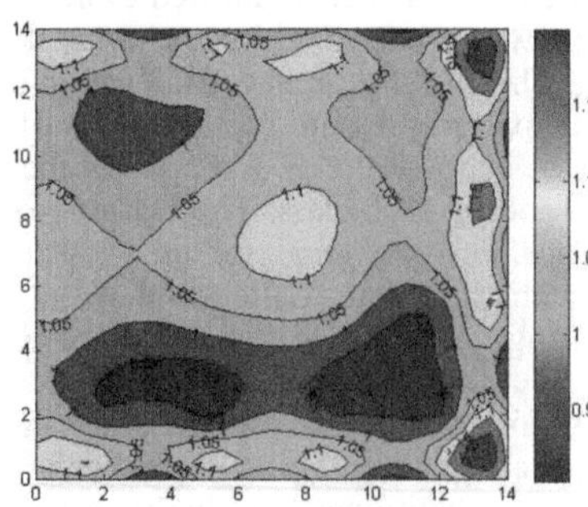

Figure 4. *EsConTer* at the beginning of the class of June 26, 2013.

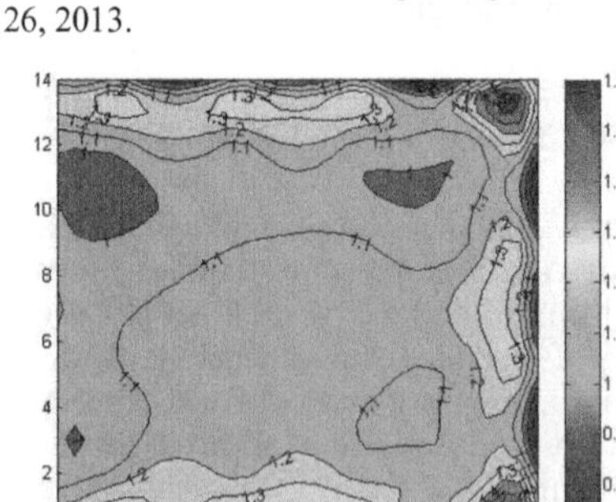

Figure 5. *EsConTer* at the end of the class of June 26, 2013.

Table 4. Values recorded for the day June 26.

	Beginning of class	End of class
T (°C)	26.1	26.2
U (%)	56.5	63.0
T_w (°C)	20.3	21.3
ITH	23.8	24.2
EsConTer	1.0	1.1

Table 5. Thermal sensation of the practitioners in a scale of colors *EsCor*, for class of the day June 26.

	Beginning of class	End of class
EsCor (mean)	1.64 ± 0.35	2.57 ± 0.30
EsCor (maximum)	1.99	2.87
EsCor (minimum)	1.29	2.27

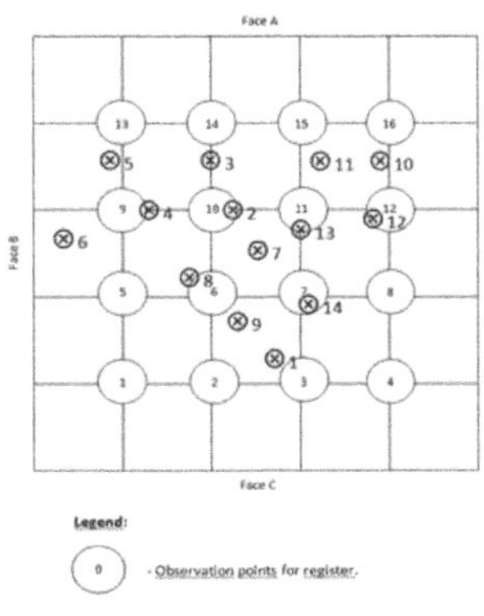

Figure 6. Positioning of the practitioners in June 26, 2013.

At the end of the class, the calculated thermal indexes have continued to indicate a location slightly hot, however the practitioners showed a mean thermal sensation of 2.57 with a standard error of 0.30. The thermal sensation of the practitioners at the end of the class shows that they were in an environment of almost thermal stress, which means very hot environment. In practice, the visual observation made during the class, showed that some practitioners entered in collapse of organism, having stopped the activity three practitioners. The calculated value of water in the vapor state in humid air, increased 1.1 g/m^3 which is equivalent to say that each practitioner has lost about 80cL of water for an hour of activity.

5 CONCLUSION

The applicability of the methods used in this work was only viable, because there was interest and availability of the administration of the gym to the improvement of the thermal conditions for sports practice. It was possible to know the pattern of thermal sensation by application of thermal indexes at the beginning and end of a class of AeroStep.

The practitioners through a thermal scale of colors indicated their thermal sensation, which was compared with determined values through a thermal index. The thermal sensation of the practitioners at the end of each class was significantly higher than the sensation offered by the thermal environment. This situation is justified by the alteration of the metabolism during the class. The practitioners have significantly increased the internal body temperature. The identification of each practitioner, before each class, through a label attached with a number, allowed an observer register the behavior and verifies the existence of collapse of the body when the thermal environment at the beginning of the class already presents slightly hot environmental conditions.

In future the recorded values of clothing in terms of measurement of clo, will be used to determine which metabolism is identified to create the same thermal sensation at the end of the class (ISO 8996, 2004; ISO 7730, 2005).

The results obtained and conclusions were prized by the administration of the gym for making decisions, namely the installation of adequate ventilation and acclimatization systems. Additional studies were required by the administration of the gym in regard to the practice rooms of sauna and gym that uses different equipment to develop sporting performance.

REFERENCES

American Society of Heating, Refrigerating and Air-Conditioning Engineers—ASHRAE (2001). *Handbook of Fundamentals—Physiological Principles for Comfort and Health*, Chapter 8, Atlanta, 1–32.

American Society of Heating, Refrigerating and Air-Conditioning Engineers—ASHRAE 55 (2004). Standards ASHRAE 55, *Thermal Environmental Conditions for Human Occupancy*, ISBN/ISSN: 1041–2336, pp. 55.

Fanger, P. (1972). *Thermal Confort*. 2ª Edição. McGraw-Hill, New-York.

IPCC (2007). *Climate Change 2007: The Physical Science Basis Summary for Policymakers*. Intergovernmental Panel on Climate Change. Paris: IPCC, Working Group I.

ISO 7730 (2005). *Ergonomics of the thermal environment—Analytical determination and interpretation of the thermal comfort using calculation of the PMV and PPD indices and local thermal comfort criteria.*

International Organization for Standardization, Genève, Switzerland.
ISO 8996 (2004). *Ergonomics of the Thermal Environment—Determination of Metabolic Rate*. International Organization for Standardization, Genève, Switzerland.
Kruger, E., Dumke, E., e Michaloski, A. (2001). *Sensação de Conforto Térmico: respostas dos Moradores da Vila Tecnológica de Curitiba*, VI Encontro Nacional de Conforto no Ambiente Construído, Anais do VI ENCAC (CD-ROM), São Pedro—São Paulo, Brasil, UNICAMP/UFSCar/Associação Nacional de tecnologia do Ambiente Construído (ANTAC), Volume 1, 1–7.
Lamberts, R. (2002). *Conforto Térmico e Stress Térmico*, LabEEE Laboratório de Eficiência Energética em Edificações—Universidade Federal de Santa Catarina, Florianópolis, Brasil.
Markov, D. (2002). *Practical Evaluation of the Thermal Comfort Parameters*. Annual International Course: Ventilation and Indoor climate, Sofia, Bulgaria, P. Stankov (Ed), ISBN: 954-9782-27-1, 158–170.
Nieuwolt, S. (1977). *Tropical Climatology*. Wiley, London: Wiley.
Talaia, M., Meles, B. & Teixeira, L. (2013). *Evaluation of the Thermal Comfort in Workplaces—a Study in the Metalworking Industry*. Occupational Safety and Hygiene. Editors Arezes et al. Taylor & Francis Group, London, 473–477.
Talaia, M. e Simões, H. (2009). *Índices PMV e PPD na Definição da "performance" de um Ambiente*. Livro de atas do V Encontro Nacional de Riscos e I Congresso Internacional de Riscos, Coimbra, de 29 a 31 de maio, 83.
Thom, E.C. (1959). The Discomfort Index. *Weatherwise*, 12(1), 57–60.
Tromp, S.W., (1980). *Biometeorology—The Impact of the Weather and the Climate on Humans and Their Environment*. Heyden International Topics in Science, London.

Occupational Safety and Hygiene II – Arezes et al. (eds)
© 2014 Taylor & Francis Group, London, ISBN 978-1-138-00144-2

The thermal feeling of a human being can contribute to the definition of a comfortable thermal environment to the animals

S. Gonçalves
DFIS—University of Aveiro, Aveiro, Portugal

M. Talaia
DFIS—CIDTFF—University of Aveiro, Aveiro, Portugal

ABSTRACT: Experience shows that avoiding milk production breakdowns is very important in a dairy farm, in a way that is important to know the time of the year in which the animals are more vulnerable in terms of warm and cold ambient feelings. In these terms, it is important to the sustainability of milk production in a country to know that dairy farms can be optimized in terms of organization and thermal environment. In this paper it is shown that the thermal environment of a dairy farm can affect the thermal feeling of workers and animals.

Thermal indexes are used with the same characteristics for the human being and animals. A contribution of this paper is to present a mathematical expression, based on human thermal sensation data, which makes possible to know the animal's thermal feeling, allowing the adoption of prevention strategies and improvement of the human's and animal's well-being.

1 INTRODUCTION

The study of thermal comfort has a great economic relevance. The control of meteorological variables allows the optimization of the thermal environment and consequently an increase in production levels (Krüger *et al.*, 2001).

In animal Biometeorology are considered intervention areas to health and animal productivity, namely the thermal well-being which is determined from temperature registration, ventilation, relative air humidity and partial pressure of water steam (Osteras & Leslie, 1997).

The specialized literature shows that the animal is a biological machine that can express all of its potential when working in optimal ambient conditions. When those conditions are not suitable, the well-being and productivity indexes are affected. Note that climatic differences have an impact on the animal condition, whereby in different regions or countries the systems and technology must be adjusted. (Silva, 2000).

The cattle are warm-blooded, capable of maintaining the body temperature independently of the ambient temperature. The surrounding environment of the animal influences its performance, because it triggers the mechanisms of energy transference. By that reason, the thermal environment is a restriction factor to the maximum efficiency of milk production (Huber, 1990).

The body temperature of the cattle may vary between 37°C and 39°C. The optimal temperature of milk production depends on the species, breed and the animal tolerance degree to heat and cold. For instance, to the studied dairy farm and to Holstein Friesian animals (more representative breed), in lactation, the air temperature should be around 24°C and the relative air humidity should near 38% (Baccari Júnior, 1998).

The well-being of an animal is evaluated using physiological and behavioral indicators. The physiological measurements associated to stress are conditioned by the decrease of the well-being (Borell, 1995). Fraser *et al.* (1975) and Perissinotto *et al.* (2006, 2007), exposed that the animal acclimatization improves its well-being and avoids infections and breaks in milk production.

The animal's ability to withstand thermic stress conditions has been physiologically evaluated by changes in rectal temperature and respiratory frequency. The rectal temperature allows to know the animal's physiological adaptability to the thermal environment (Martello, 2002).

According to Baccari Júnior (2001) the cattle raises his breathing frequency and perspiration (in the skin surface) thereby regulating, by evaporation processes, the internal body temperature.

The thermal stress reduces the animal's immunity performance. This situation is limited by the combination of temperature and relative air humidity when it destroys the animal's thermoregulation mechanism.

In practice, the thermal comfort occurs when is verified a situation in which the thermal balance is zero. When the animal is in a thermal comfort zone the required energy to maintain constant the body temperature is minimal.

The cattle's mastitis (from Greek mastos) is one of the most frequent pathologies in milk cattle. This pathology is described by an inflammatory process of the mammary gland mainly caused by bacteria conditioned by multiple factors in which environmental characteristics are influent factors for transmission and proliferation.

Mastitis can damage the economic bond of a dairy farm, causing unexpected liquid losses, due to milk production break.

The most common bacteria involved and more difficult to treat are, for instance, Staphylococcus Aureus, Streptococcus Agalactiae and Corynebacterium Bovis (contagious mastitis) and Escherichia coli (environmental mastitis).

In this paper is studied, to the indicated bacteria, the weather circulation type that increases the epidemiology applying weather maps.

The animal's and human being's thermal sensation are studied and the obtained results allowed to develop a mathematical expression capable of predicting the animal's thermal sensation when the thermal sensation of the human being is known.

2 MATERIALS AND METHODS

In all data colleting phases, was used a CENTER 317 to record temperature, dew point temperature and air relative humidity. The workers registered in a color scale, seventh scale of ASHRAE, their own thermal feeling. This information was very important to verify if the predicted data by the thermal indexes and the color scale presented a significant correlation. The presented study was done in summer days and the worker's outfit was carefully planned to be the same throughout the measuring period.

In the first phase and for seven years of activity were investigated the infections that indicate milk production breaks. Weather maps were used for a period of three days previous to the detection of the infection, in order to investigate the connection between episodes and weather circulation type. It is known that an infection episode occurs about 24 hours before its detection. The synoptic characteristics of the different atmospheric weather circulation types were based in information from Trigo & DaCamara (2000).

The identified infection records in the dairy farm allowed to know which atmospheric circulation types influenced the occurrence of certain bacteria.

In a second phase it was evaluated the thermal feeling of the workers and the animals in two different locations: in the stable (Figure 1) and in the milking parlor (Figure 2). Six control animals were identified.

The cows were milked twice a day (09h00 and 20h45) and its production, registered daily.

The animal's monitored physiological data included the animal's rectal temperature (as an indicator of the rising of the body temperature) and the visual observation of the breathless thermal feeling.

Figure 1. Stable of the dairy farm.

Figure 2. Milking parlor of the dairy farm.

Inside the stable and the milking parlor were registered the air temperature and relative humidity of the air for three different times (09h00, 14h30 and 20h45).

The thermal feeling of the workers was also registered in the same schedules. These records were based on a thermal color scale (Talaia & Simões, 2009) as shown in Figure 3. This scale, seventh of the ASHRAE scale, allows to know the thermal feeling of each worker, knowing that the "dark blue" color is -3 and the "dark red" color is +3. Once this is a linear scale, it was easily known the worker's thermal feeling.

In the specialized literature are presented some thermal indexes whose purpose is to describe a determined thermal environment for the human being. It is convenient to investigate the *HTI* for the human being. According to Corleto (1998), a thermal stress index is a number contains the effect of various indicators in the human thermal environment, whose value describes the thermal stress that an individual is subjected in a hot environment. The *HTI* (Humidity and Temperature Index) was developed by Thom (1959), was later on improved by Nieuwolt (1977), starting to include the air temperature T (°C) and the air relative humidity U (%) for its calculation.

The HTI_{hu} for human beings is calculated through the expression

$$HTI_{human} = 0.7T + T(U/500) \tag{1}$$

The HTI_{human} values for the thermal feeling were adapted by Talaia *et al.* (2013), as shown in Table 1.

The thermal comfort/discomfort of each animal was evaluated from the use of the humidity and temperature index, animal *HTI*.

According to Johnson (1987), the *HTI* was initially developed to human beings, but according to practice it has been shown that the same psychometric variables that cause thermal discomfort in humans, also cause discomfort in milk cows, which can cause a breakdown in milk production (Buffington et al., 1981).

Figure 3. Thermal feeling scale, 0 indicates comfort.

Table 1. Thermal feeling based on HTI_{hu}.

HTI_{hu}	Thermal feeling
$HTI_{hu} < 8$	Very cold area
$8 \leq HTI_{hu} < 21$	Area which needs to be heated
$21 \leq HTI_{hu} < 24$	COMFORT
$24 \leq HTI_{hu} < 26$	Area with ventilation needed
$26 \leq HTI_{hu}$	Too hot area

Table 2. HTI_{animal} for cattle.

ITH	Thermal sensation
$HTI_{an} < 70$	Normal
$70 \leq HTI_{an} < 72$	ALERT
$72 \leq HTI_{an} < 78$	Alert and milk production break
$78 \leq HTI_{an} < 82$	Danger
$82 \leq HTI_{an}$	Emergency

The HTI_{animal} for animals is calculated through the expression

$$HTI_{an} = T + 0{,}36T_0 + 41{,}2 \tag{2}$$

where T = air temperature (°C); and T_0 = dew point (°C).

According to DuPreez (2000) there are reference values (depend on the animal breed, lactation phase and animal's age) to the interpretation of the determined value by the HTI_{an} as shown in Table 2.

The analysis of the shown data in tables 1 and 2 suggests that for the same registered data of air temperature and air relative humidity (the dew point temperature can easily be obtained from air temperature records—saturation steam partial pressure and from air relative humidity) there are common information that can be important to predict the animal's well-being from the human's well-being.

The thermal feeling values of the human being, by using the *HTI* shown excellent correlation with the registered data in the color scale, which allowed to develop a mathematical expression of animal *HTI* based on the information of the human's color scale.

3 RESULTS AND DISCUSSION

The Figure 4 shows, for seven years of the dairy farm, the percentage of registered infections.

The analysis of the surface weather charts when was investigated the class of circulation that improves the mastitis shows that the WCT (Weather Circulation Type) is from type N and NE.

For the studied thermal environments, the air temperature in the milking parlor was always higher than the one recorded in the stable. This result is justified by the construction type of the milking parlor which has less ventilation. However, the time that the cattle remains in this area is low.

In table 3, are shown some registered and calculated values for the stable and milking parlor. In this table T represent the air temperature, U the air relative humidity, HTI_{animal} the animal thermal index, HTI_{animal_m} the minimum thermal index, the

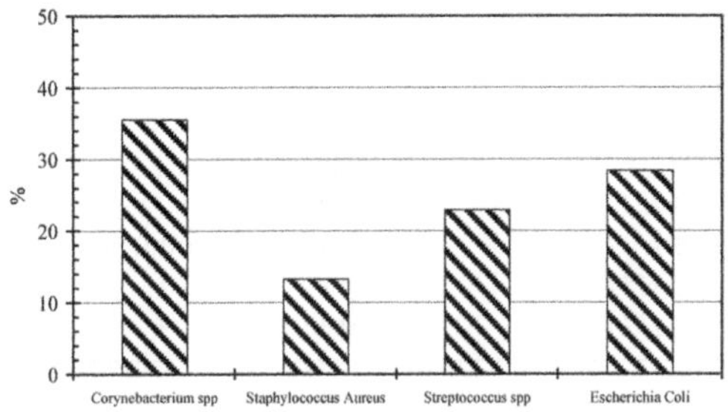

Figure 4. Registered infections: 2006–2012.

Table 3. Registered and determined parameters e in the milking parlor and in the stable.

	Stable	Milking parlor
T (°C)	19.0 ± 2.5	19.2 ± 2.4
U (%)	71.6 ± 10.5	76.5 ± 9.0
HTI_{an}	65.6 ± 3.1	65.7 ± 2.9
HTI_{an_M}	71.3	71.5
HTI_{an_m}	58.4	60.9
HTI_{hu}	16.0 ± 2.0	16.3 ± 1.9
$EsCor$	−0.25 ± 0.30	−0.19 ± 0.18

HTI_{human} the human's thermal index and the thermal feeling of the human being in a color scale.

The table 4 shows the registered data for six animals tagged from A to F. In table 4, T_r represents the animal's rectal temperature, T_{r_M} the maximum rectal temperature, T_{r_m} the minimum rectal temperature and P_m the minimum milk production. The gathered data show that milk production decreases when the animal is in a breathless condition.

The observation of the data presented in table 5 shows that the air relative humidity for the studied location is way above of the pointed value by Baccari (1998). An explanation for the registered data in the dairy farm is its location which is very near the Atlantic Ocean.

The Figure 5 shows the thermal feeling felt by the animals during fifteen days, in the stable and in the milking parlor and unequivocally shows (see diamonds) there are situations that caused alert situations for the animal (breathless). The thermal index *HTI* seems to be a good indicator in order to take intervention strategies (mist blowers).

The Table 3 shows values for the human's thermal index that raises an environment that needs to be heated. However, it was possible, with the gathered data for the human being, in a color scale, *EsCor* (see Figure 3), to find and expression capable of predicting the animal's thermal feeling HTI_{animal}.

$$HTI_{an} = 8.0EzCor + 66.8 \qquad (3)$$

Table 4. Registered data by the tagged animals.

	A	B	C
T_r	38.6 ± 0.6	38.6 ± 0.4	38.4 ± 0.4
T_{r_M}	39.7	39.6	39.3
T_{r_m}	37.0	37.7	37.7
P_M	36	53	25
P_m	29	44	18
	D	**E**	**F**
T_r	38.4 ± 0.4	38.2 ± 0.5	38.3 ± 0.3
T_{r_M}	39.2	39.1	38.9
T_{r_m}	37.6	37.1	37.0
P_M	25	29	33
P_m	20	19	28

Table 5. Calculated parameters in the stable and in the milking parlor.

	Stable	Milking parlor
Air temperature	19.6 ± 2.2	20.0 ± 2.3
Relative humidity	71.6 ± 11.7	73.4 ± 8.3
Animal HTI	66.4 ± 2.5	66.6 ± 2.6
T_minimum	15.0	16.0
T_maximum	25.0	24.8
U_maximum	97.0	84.0
U_minimum	45.0	55.0

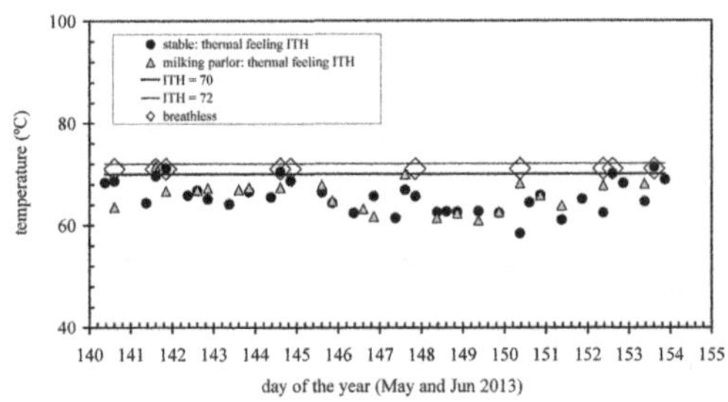

Figure 5. ITH_{an} for the stable and milking parlor, May and June 2013.

This equation allows to know the thermal feeling of the animal when the human's thermal feeling is known through a color scale, ASHRAE seventh scale.

A very important contribution developed from gathered data is presented in table 6 where it is shown values for the human's thermal feeling in the color scale and the corresponding thermal feelings for the animal.

Table 6. Animal's thermal feeling based on *EsCor*.

Human's thermal feeling	Animal's thermal feeling
$EsCor < 0.40$	Normal
$0.40 \leq EsCor < 0.65$	ALERT
$0.65 \leq EsCor < 1.40$	Alert and milk production losses
$1.40 \leq EsCor < 1.90$	Danger
$1.90 \leq EsCor$	Emergency

4 CONCLUSION

It was possible to understand the weather circulation types (WCT) that make worse the disease conditions (*Staphylococcus Aureus*, *Streptococcus Agalactiae*, *Escherichia coli* and *Corynebacterium Bovis*) and decrease the milk production.

For both of the thermal environments (milking parlor and stable) were known human's thermal feelings by applying a thermal index and a color scale and the obtained results showed an excellent correlation.

It was evaluated the animal's thermal feeling using a thermal index and the evaluation of the rectal temperature was considered important in animal's observation in case of "breathless" symptoms.

A very important contribution of this study was the development of a mathematical expression that allowed to know the thermal feeling of an animal from a human's thermal feeling.

Studies that give importance to this new vision of evaluating the animal's thermal feeling from the knowledge of the human's thermal feeling are very convenient.

At this point new data gathering is in course in cooler months.

Values for the human's thermal feeling in the same environments were processed (stable and milking parlor).

It was developed an expression connecting the human's thermal feeling (in a color scale) and the animal's thermal feeling (through HTI_{an} information).

The gathered results in this investigation were appreciated by the veterinarian making possible to act preventively.

In conclusion, a thermal environment for the human being does not indicate the same thermal feeling to the animal. However, this study shows that knowing the human's thermal feeling in a color scale is possible to predict the animal's thermal feeling.

REFERENCES

Baccari Júnior, F. (1998). *Adaptação de sistemas de manejo na produção de leite em clima quente*. In I Simpósio Brasileiro de Ambiência na Produção de Leite 1., pp. 24–67. [Anais] Piracicaba: FEALQ, 1998, Piracicaba.

Baccari Junior, F. (2001). *Manejo ambiental da vaca leiteira em climas quentes*. Londrina: Universidade Estadual de Londrina, 142 p.

Borell, E.V. (1995). Neuroendocrine integration of stress and significance of stress for the performance off arm animals. *Applied Animal Behaviour Science*, 44: 219–227.

Buffington, D.E., Collasso-Arocho, A., Canton, G.H. & Pitt, D. (1981). Black globe humidity index (BGHI) as comfort equation for dairy cows. *Transaction of the ASAE*, 24 (3): 711–14.

DuPreez, J.H. (2000). Parameters for the determination and evaluation of heat stress in dairy cattle in South Africa. *Onderstepoort J. Vet. Res.*, 67: 263–271.

Fraser, D., Ritchie, J.S.D. & Fraser, A.F. (1975). The term "stress" in a veterinary context. *British Veterinary Journal*, 13(1): 653–662.

Hubber, J.T. (1990). *Alimentação de vacas de alta produção sob condições de estresse térmico*. In: Peixoto *et al.* Bovinocultura leiteira. Piracicaba: FEALQ, 33–48.

Johnson, H.D. (1987). *Bioclimatology and adaptation of livestock*. Amsterdam: Elsevier, 279 p.

Krüger, E., Dumke, E. & Michaloski, A. Sensação de Conforto Térmico: respostas dos Moradores da Vila Tecnológica de Curitiba. VI Encontro Nacional de Conforto no Ambiente Construído, Anais do VI ENCAC (CD-ROM), São Pedro – São, (2001).

Martello, L.S. (2002). Diferentes recursos de climatização e sua influência na produção de leite, na termorregulação dos animais e no investimento das instalações. Dissertação em Qualidade e Produtividade Animal, Faculdade de Zootecnia e Engenharia de Alimentos, Universidade de São Paulo, Pirassununga.

Nieuwolt, S. (1977). *Tropical climatology*. London: Wiley.

Osteras, O. & Leslie, K. (1997). *Animal housing and management—prevention of bovine diseases*. Proceedings 9th International Congress in Animal Hygiene. Editor: Saloniem, H., Vol. 1., Helsinki, Finland.

Perissinotto, M., Moura, D.J. & Cruz, V.F. (2007). Avaliação da produção de leite em bovinos utilizando diferentes sistemas de climatização. *Rev. De Ciências Agrárias*, 30(1): 135–142.

Perissinotto, M., Moura, D.J., Matarazzo, S.V., Silva, I.J.O. & Lima, K.A.O. (2006). Efeito da utilização de sistemas de climatização nos parâmetros fisiológicos do gado leiteiro. *Eng. Agríc. Jaboticabal*, 26(3): 663–671.

Silva, R.G. (2000). *Introdução à Bioclimatologia animal*. São Paulo: Nobel, 286 p.

Talaia, M., Meles, B. & Teixeira, L. (2013). *Evaluation of the Thermal Comfort in Workplaces—a Study in the Metalworking Industry*. Occupational Safety and Hygiene. Editors Arezes et al. Taylor & Francis Group, London, 473–477.

Talaia, M. e Simões, H. (2009). *Índices PMV e PPD na Definição da "performance" de um Ambiente*. Livro de atas do V Encontro Nacional de Riscos e I Congresso Internacional de Riscos, Coimbra, de 29 a 31 de maio, 83.

Trigo, R.M. & DaCamara, C.C. (2000). Circulation Weather Types and Their Influence on the Precipitation Regime in Portugal. *Int. J. Climatol.* 20: 1559–1581.

Occupational Safety and Hygiene II – Arezes et al. (eds)
© 2014 Taylor & Francis Group, London, ISBN 978-1-138-00144-2

Methodology of chemical hazard management—a tool for occupational health promotion

M.B.F.V. Melo, M.S.M.L. Souto & D.S.C. Vasconcelos
Universidade Federal da Paraíba, João Pessoa, Paraíba, Brasil

ABSTRACT: Given the complexity of managing chemical risks, it is important to guide their analysis and subsequent control, and the adoption of a systemic and integrated approach. In this context, this article proposes, through bibliographical research based on the existing literature on the subject and Brazilian and international standards of occupational health and safety. Through a case study in a large company in the food industry, it can be verified that the correct application of methodology for chemical risk management is a tool to promote occupational health. This proposed model aims to facilitate the implementation and real efficiency of risk management and especially to enter into business culture focused on continuous improvement.

1 INTRODUCTION

The harmfulness of work environments caused by agents contained therein, mostly resulting from the production process, is also due to an intensely artificial industrial environment where operations are performed with dangerous machinery, potentially toxic chemicals, excessive noise, high temperatures, radiation sources, microorganisms etc. These agents, known for occupational hazards capable of producing damage to the health of workers, are classified by labor legislation as physical, chemical and biological.

Chemical Risks are substances, compounds or materials that have the ability to modify the chemical composition of the working environment and can be absorbed into the body of the worker by ingestion, inhalation or direct contact; the most common forms are cutaneous and respiratory absorption. The body of the worker quickly eliminates some of these risks, but others may concentrate on specific organs or tissues, causing from dizziness to cancer or genetic mutations.

The International Labor Organization (ILO) believes that nearly 2 out of every 3 workers worldwide are exposed to chemicals, it is estimated that 1.5 to 2 billion people are affected. These numbers are indicators that the issues of occupational safety and health are related to public health.

The exposure to unhealthy work environments makes the worker vulnerable to developing the disease which will disable him to work, removing the worker for treatment. Upon returning to work after treatment, if the same unsanitary conditions persist, this worker will probably get sick again and so forth until he is completely unable to work. It is noticed that only providing the treatment of the disease means to act on consequence and not on the root cause, which is the contamination of the work environment and risk exposure.

In this context, this article proposes, through bibliographic research based on the existing literature on the subject and Brazilian and international standards of occupational health and safety, a Methodology for Chemical Hazard Management.

2 METHODOLOGY FOR CHEMICAL HAZARD MANAGEMENT

Given the complexity of managing chemical risks, it is important to guide their analysis and subsequent control, the adoption of a systemic and integrated approach.

Occupational hygiene as "the science and art devoted to the study and management of occupational exposures to physical, chemical and biological agents, through actions of anticipation, recognition, evaluation and control of conditions and workplaces, in order to preserve the health and well being of workers, and considering the environment and the community" (ABHO, 2013). Thus, by definition, the occupational hygiene works in the management of occupational risks using, in an integrated manner, the methodological steps: anticipation, recognition, evaluation and control.

Anticipation involves the analysis of new projects, facilities, products, methods or procedures of work or modification of existing ones, the goal of which is to identify potential risks and the introduction of control measures needed, anticipating the exposure to risk environment.

The Recognition is very important, because if a toxic agent is not recognized, it will not be evaluated or controlled (Brevigliero et al. 2006). This step is based on the recognition of environmental agents that affect the health of workers, which involves deep knowledge of the products involved in the process, working methods, process flow, layout of facilities, number of exposed workers, etc. (Saliba & Corrêa, 2000; Saliba et al. 2002).

Evaluation is an essential step that is compared to a tool for disease prevention of work. It is performing the qualitative and/or quantitative analysis of physical, chemical and biological agents in the workplace. Expertise of evaluation is required, based on calibration of equipment, collection time, type of chemical analysis to be made, etc. (Saliba & Corrêa, 2000; Saliba et al. 2002). Not performing this step or its improper realization will certainly result in diseases among the workers.

The control of these agents should be done preferably through engineering measures, protecting the environment. In general, for all agents, the control measures should be adopted, prioritizing its efficiency, that is, firstly those that refer to the source, followed by referring to the course and finally those relating to workers (Brevigliero et al. 2006). The purpose of this step is to adopt measures aimed at eliminating or minimizing the risk in the environment (Saliba & Corrêa, 2000; Saliba et al. 2002).

3 PROPOSED METHODOLOGICAL MODEL FOR CHEMICAL HAZARD MANAGEMENT

In Brazil, the approaches to health and safety at work are focused on risk control and do not relate to the management aspects, such as the allocation of resources, responsibilities, commitment of decision makers and, especially, lack of sight of the structural aspects of the work process. Thus, the performance of most Brazilian companies is limited to compliance with regulatory standards through specific actions, translating into interventions: the control of occupational health in corrective action on the materials, equipment and workplaces, and the level of personal protective equipment.

This means that the safety function is not perceived as an activity as relevant as the others developed in the company and in that case it should be considered as an integral part of its overall management.

To manage risks at work generally speaking, and chemical hazards in particular, it is necessary that the direction of the organization takes the OSH as a priority issue, hire and/or form boards of employees responsible for the various themed levels, involving workers in discussions about the risks in their workplace and seek to define the objectives, strategies, resources and actions that will make this management.

In summary, the chemical risk management must evolve into a systemic approach, in which the various stages of the management process are viewed as a continuous cycle of activities that feed back the system, enabling a continuous improvement process. It is suggested to use the PDCA cycle, also known as the Deming Cycle, a management tool widely used by companies worldwide, which is the main focus of continuous improvement.

The commencement of this work is a theoretical methodology for chemical risk management that applies the PDCA cycle, to the conceptual and methodological base adopted by the Occupational Hygiene. This proposed model aims to facilitate the implementation and real efficiency of risk management and especially to enter into business culture focused on continuous improvement.

Thus, the proposed model consists of six stages inserted in the four phases of the PDCA cycle, namely: anticipating recognizing and evaluating risks, elaborating an action plan, training, implementing control measures, monitoring the results and acting in the process depending on the results. (Fig. 1)

3.1 *Anticipating, recognizing and evaluating risks*

Anticipation can be accomplished by working with project teams in order to identify the generation of new sources of risk and under which conditions these innovations may be accepted and produced (regulatory risk). In fact, this study occurring during the project means an anticipation of future problems, since at that stage the chemical risks

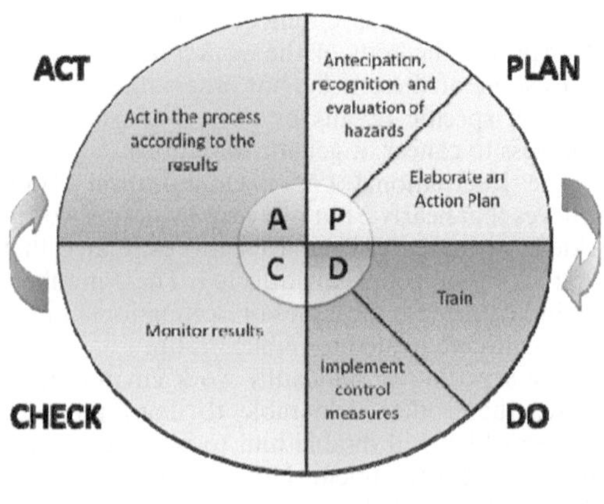

Figure 1. Proposed model (Source: Adapted from Demming, 2002).

only exist potentially. It is about, therefore, preventing diseases and injuries caused by chemicals in its outset.

This step is complemented by dictating standards for preventive orientation of buyers, designers and contractors of services in order to avoid inadvertent exposure to chemicals caused by poor selection of products, materials and equipment.

Recognition involves making a survey of the various types of risks found in a particular workplace. In the case of chemicals, they can be produced or used in the production as raw materials, components, cleaning supplies, among others. The vision at this stage should be of knowing and analyzing the production process and forms of work organization in order to identify the chemical risks that exist there, control measures used and possible repercussions on the health of workers.

To fully realize this survey it is necessary to transit and watch incessantly the workplace and listen to employees who experience different situations in their daily lives and that hold a lot of information. Some techniques and documents can also be used, such as: organization chart, layout, process flow graphic or flowchart, flowchart map, description of the organization of work, description of major equipment and facilities; listing of raw materials, products in process, finished products and waste.

Also during this survey process, when appropriate, it should be noted: the handling, transportation and storage of chemicals, maintenance, repair and cleaning of equipment and containers for chemicals, the release of chemicals resulting from work, disposal and treatment of waste chemicals.

It is recommended to study the process, not only with existing data in the enterprise, but also using specific occupational literature and instruments already set standards in force at the place where the company operates.

In order to make a judgment on the tolerability of exposure to a chemical agent identified in the previous step, proceed to step qualitative and quantitative evaluation.

The qualitative assessment is fast, viable and cost effective, since it is performed through analysis and inspections at workplaces. There are two forms of qualitative assessment: the risk map (described above) and indices of chemical (NR 15, ACGIH, AIHA, etc.) of products used in the company (Brevigliero et al. 2006). However, while recognizing the importance of the second form, on the methodological model exposed here it is proposed to use only the risk map because it is a tool developed with the involvement of workers and their visibility in the workplace and it is easy to understand by all people working in the company.

The quantitative evaluation is aimed at proving or not, the effectiveness of control measures previously adopted, as well as subsidizing the planning of control measures to be adopted. This evaluation may be either instantaneous (with direct reading instruments) or continuous (with sampling equipment).

This type of evaluation quantifies the concentration of the chemical in the workplace, using standardized methods, using appropriate equipment and that is as representative as possible of the real exposure that afflicts workers. In the sequel, it proceeds to the comparison between the exposure information obtained with a suitable criterion—limit of tolerance (Fig. 3). Therefore, it is recommended to use the values of occupational exposure limits adopted by the American Conference of Governmental Industrial Hygienists—ACGIH, or those established in national legislation or collective bargaining negotiations (if more stringent).

In some cases, when it is noticeable the presence of chemical agents in large concentrations, the quantitative evaluation is waived, and the qualitative evaluation proceeds to the adoption of control measures. Actions can be taken quickly in the workplace and substantially reduce the cost of evaluation. After that, at the stage of monitoring it will be essential to the quantitative evaluation.

3.2 *Elaborating an action plan*

After the stage of anticipation, recognition and chemical risk assessment, it should be identified the critical points, with the participation of workers, set priorities, goals and schedule, thus composing the annual action plan. Such a plan should include all activities identified in the stages of recognition, evaluation and control measures as defined.

3.3 *Training*

It precedes the implementation of control measures, raises the awareness and trains employees about the importance of the subject security and worker health, the types of occupational risks, controls measures adopted by the company, including as to the procedure to ensure the efficiency and still provides clarifying information on the possible limitations of protective measures deployed of a collective and/or individual character.

Besides this more general content, the following topics, among others, should be part of this training:

- Handling, transporting and storage of chemicals;
- Maintenance and cleaning equipment;
- Cleaning of containers used for chemicals;
- Correct use of collective and individual protections adopted by the company;
- Utilization of safe operating procedures for routine operations and emergency.

3.4 *Implementing control measures*

According to the Regulatory Standard n. 9, Prevention Program for Environmental Risk the control measures of risks must be come in the following order of priority: 1st—Collective measurements; 2nd—Administrative measures of work organization; and 3rd—Personal Protective Equipment (PPE).

The study, development and implementation of collective measures should obey the following hierarchy (Fig. 2) (Brevigliero et al. 2006):

- Measures to eliminate or reduce the use and training of environmental agents that are harmful to health (in source control): replacement of chemicals; modification methods and work processes; modification projects, equipment maintenance, etc.;
- Measures to prevent the release or dissemination of these agents in the workplace (in the trajectory control): creation of ventilation/exhaustion; cloistered machines and substances; etc.;
- Measures to reduce the concentration levels of these agents in the workplace (control receiver): use of personal protective equipment; conducting periodic training on safe development activities; periodic medical examinations for early detection of diseases.

The measures of work organization are taken with the aim of reducing the exposure of workers through various measures. For the case of chemical risks, the reduction of working hours and/or use of pauses on repetitive tasks are examples of measures of organizational control.

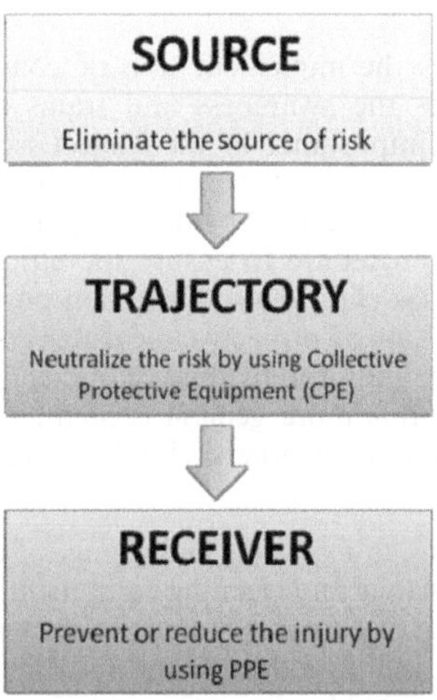

Figure 2. Priority of control measures (Source: Adapted from Brevigliero et al. 2006).

3.5 *Monitoring results*

The monitoring of workers' exposure and control measures should consist of evaluation carried out systematically and repetitively, regardless of the outcome of the comparison of the results of the quantitative analysis with the tolerance limits established (Fig. 3).

When the value of the concentration of the chemical is above the tolerance limits established, control measures should be taken, and monitoring should be performed to verify the effectiveness of the measures adopted.

If the concentration of the chemical agent is below the tolerance limits established and over half of these values, it is understood that this situation is in "level of action", that is, the risk is under control, but monitoring should be performed and preventive actions should be initiated in order to minimize the likelihood that environmental exposures exceed the exposure limits, similarly to what advocates the Regulatory Standard N. 9 of the Ministry of Labor and Employment of Brazil (Brasil, 1978).

For cases in which the concentration values of the chemical are below half of the value of the tolerance limits established, the risk is under control and monitoring should be done to ensure such a situation.

All data from the monitoring stage should be stored so as to constitute a history of technical and administrative workers exposure to chemicals.

3.6 *Acting in the process according to the results*

In the context of continuous improvement, it should close the loop in the process acting as the results obtained in the monitoring, followed by an

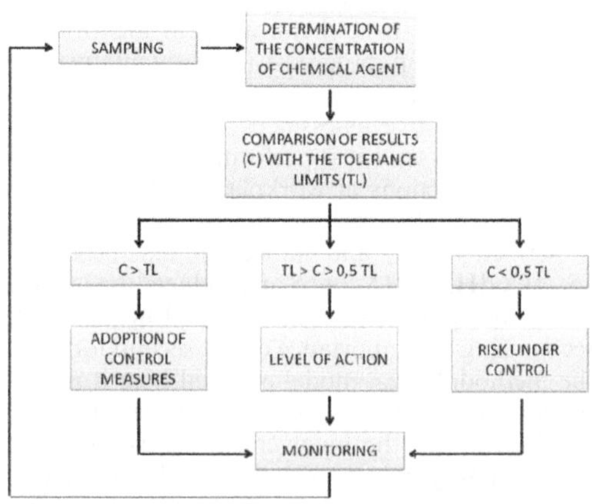

Figure 3. Quantitative evaluation (Source: Adapted from Brevigliero et al. 2006).

analysis and evaluation of the overall risk management of chemicals, in order to seek improvement to restart the next cycle.

4 CASE STUDY

In this paper it is presented a proposal for a Methodology for Chemical Risk Management in the Occupational Health perspective and it is analyzed its validity through a case study of a large company in the food industry. This study consisted of a survey in order to know the model of chemical risk management implemented by this company, with a comparison with the theoretical and methodological model proposed, showing the nonconforming points in the view of the methodology in analysis and unfavorable efficiency risk management of the company.

The existence of repetitive annual action plans, training and control measures focusing only on individual protection, absence of measures that treat the environment and the complete absence of monitoring are the major non-conformities detected that constitute barriers to continuous improvement.

All non-compliances observed culminate for corrective actions focused only on the worker, that take place in an isolated way, thus generating a misunderstanding of the indicators proposed by national and international standards.

The number of cases of work-related diseases is one of the indicators of occupational health in existing Brazilian legislation (Regulatory Standard. 4 of the Ministry of Labor and Employment). However, the non-existence of occupational diseases on workers does not mean the work environment is healthy. The health of the work environment should be guaranteed by the observance of the tolerance limits established.

The company has not studied cases of occupational diseases related to the chemical agents, but concentrations of some agents in the analyzed workstations are above the tolerance limits established in national and international standards. Control measures are focused only on protecting workers individually ensuring compliance with labor legislation, however, the work environment remains unhealthy and the worker's health is slowly being compromised by the action of these harmful chemicals.

5 CONCLUSION

In this paper it is presented a proposal for a Methodology for Chemical Risk Management in Occupational Health perspective.

Finally, this methodology presents the following main advantages:

- Adopts a systemic approach, in which the various stages of the management process are integrated into a continuous cycle of activities;
- Leads to efficient management of chemical risks to the extent that, after monitoring, closes the loop acting in the process;
- Inserts in business culture the idea of continuous improvement;
- Is easy to implement;
- Involves employees' participation;
- Enables service to national and international regulatory standards, since it does not deviate from the conceptual basis of Occupational Hygiene.

In summary, the methodological model for the management of chemical risks proposed presents a systemic view, contributing to the integration of the stages of anticipation, recognition, evaluation and control and their corrective actions to ensure the health of the work environment and of the workers continuously, reflecting the business and environmental sustainability.

REFERENCES

Associação Brasileira de Higiene Ocupacional (ABHO). http://www.abho.org.br. (accessed on 20/01/2013).

BRASIL. Ministério do Trabalho. Portaria nº 3.214, de 08 de junho de 1978—Norma Regulamentadora n.º 9—Programa de Prevenção de Riscos Ambientais. Diário Oficial da República Federativa do Brasil: Brasília, Brasil.

BREVIGLIERO, E.; POSSEBON, J.; SPINELLI, R. (6th ed.) 2006. *Higiene Ocupacional: Agentes Biológicos, Químicos e Físicos.* São Paulo: SENAC.

DEMING, W.E. (1st rd.) 1990. *Qualidade: A revolução da Administração.* Rio de Janeiro: Saraiva.

SALIBA, T.M.; CORRÊA, M.A.C. (5th ed.) 2000. *Insalubridade e Periculosidade: Aspectos Técnicos e Práticos.* São Paulo: LTR.

SALIBA, T.M.; CORRÊA, M.A.C.; AMARAL, L.S. (3rd ed.) 2002. *Higiene do Trabalho e Programa de Prevenção de Riscos Ambientais.* São Paulo: LTR.

Occupational Safety and Hygiene II – Arezes et al. (eds)
© 2014 Taylor & Francis Group, London, ISBN 978-1-138-00144-2

The attitude of washing hands among the health staff

H. Boudrifa & A. Amiar
Laboratory of Prevention and Ergonomics, University of Algiers 2, Algeria

ABSTRACT: The aim of this study is to assess the attitude of the health staff towards washing hands and how much they are prepared to obey to the rules as well as to find out factors that affect this type of preventive behaviour, which can be a threat to their health as well to the patients' health.

The analysis of the results showed that the individuals of the study sample are not fully aware of the different cases that need washing hands before or after these cases. They were not fully aware of the different techniques and conditions for washing hands, nor do they conform to them. The results also indicated that there was a lack in providing the necessary tools and material for washing hands. It is therefore concluded that continuous campaigns for raising awareness of washing hands should be introduced as an absolute necessity.

1 INTRODUCTION

Despite the fact that health care workers are supposed to be well aware of the importance of washing hands, as this can be considered as one of the actions to prevent the spread out of contamination in hospitals, simply because hand-washing is recognized as the leading measure and the most effective way of preventing the spread of infectious diseases (Anderson et al., 2008). Both patients and employees in the health sectors can suffer from illnesses caught inside hospitals because of lack of hygiene, beside the fact that it might raise the gravity of injury by some illnesses among patients who are receiving care at the health services. This might not only have bad effects on the health of patients and staff but can also have so many drawbacks on the economical and organisational levels.

Despite the fact that there are many studies which emphasis on the importance of washing hands in preventing the propagation of infection inside hospitals, yet this might well depend on how the health care workers are committed to wash their hands according to the principle set out.

Moreover, It has been found that compliance to hand washing was in general well below 50% (Pittet D, et al., 2000; Boyce JM, & Pittet D. 2002; Albert RK, & Condie F. 1981; Pittet D, et al., 1999). For example: It has been found that only 40% of people working in the sanitarians structures are committed to policy of washing hands (Khaled Abdaziz Abousaad; 2005).

While khaled et al., (2008) found that the overall hand hygiene compliance among health care workers is 34%, Muhammad Ali Anwar et al., (2009) found that the Overall compliance of washing hand was 38.8%, but it widely varied as a function of patient care activity. Furthermore, it has been found that most subjects considered "lack of sinks, soap, water and disposable towel" as a major barrier towards hand washing compliance (Muhammad Ali Anwar, 2009).

Hence, the aim of this study is to assess the attitude of the health staff towards washing hands and how much they are prepared to obey to the rules as well as to find out factors that affect this type of preventive behaviour, which can be a threat to their health to the patients' health.

2 THE AIMS OF THE STUDY

The aims of this study are:

1. To determine the attitude of hand washing among health care workers towards.
2. To assess the compliance of the health care workers with techniques and the right rules of hand washing.
3. To assess the availability of hand-washing facilities.
4. To determine barriers to hand washing compliance among health care workers.
5. To find out the best ways to overcome these barriers and improve the level of hand washing compliance.
6. Incorporate health care workers into the process of developing hand hygiene campaigns and protocols by gathering information regarding their opinions on hand washing.
7. The results of the present study can be used to determine what issues need to be addressed to increase hand hygiene compliance rates.

3 METHOD

After carrying many interviews and observations in hospitals, a questionnaire based on the third person principle was developed and applied on a final sample of 160 individuals (41,2% & 58,8%) among health care workers in order to study their attitudes towards hands washing. Their age distribution was as is shown in table 1. Their ages were Subjects were asked to mark the frequency of each behaviour (item) on a scale of five points (never, rarely, sometimes, most times, always), as it is shown below:

Do you think that doctors and nurses wash their hands:	Level of frequency of behavior				
	Never	Rarely	Some times	Most times	Always
1. Before every contact with the patient					✘
9. After each examination	✘				

Table 1. Age distribution of the study sample.

Age	20–25	26–30	31–35	36–40	41–45	46–50	51–55	>55	Total
freq	27	39	40	22	18	10	2	2	160
%	16,9	24,4	25,0	13,8	11,3	6,3	1,3	1,3	100

4 RESULTS

4.1 *Attitude of doctors and nurses towards hand washing*

The results of the present study showed that the health care workers have in general low compliance for hand washing in most cases and especially before each case as can be shown in table 2. Perhaps what can be concluded most from the variety of the most frequent items shows that there is high percentage of health care workers who do not comply with hand washing according to what is required in their job. Therefore Hand washing practices among health care workers were not in line with WHO recommendations. Despite the fact that these doctors and nurses are supposed to be well aware of the danger of these attitudes, yet they seem to be affected by daily social habits and cultural characteristics of the population rather than the rules and principals laid down for excursing this job. This situation might be related to the ambiguity of the meaning of prevention to them and the lack of awareness about hand hygiene.

The most common type of HW practiced among most health care workers were limited specifically to the real medical interventions like:

- Before doing a small surgery
- After playing with an animal
- After doing a small surgery
- After inserting tools into the patient body
- After a contact with polluted products (blood, chemical products)
- Before wearing gloves
- After wearing gloves
- After placing a sonde
- Before inserting tools into the patient body
- After giving an injection to the patient
- Before eating
- After each examination
- After going to the toilet
- After every contact with the patient
- After giving an injection.

However, The most common type of the least practiced HW among most health care workers were mostly cases of general behaviors, which might be interpreted as a lack of awareness about the importance of HW in these cases like:

- Before giving an injection to the patient
- Before each examination
- Before every contact with the patient
- After eating
- Before giving an injection
- After throwing rubbish in the bin
- After shaking hands with someone
- Before putting chew tobacco in the mouth
- Before going to the toilet
- After reading a news paper
- After putting on a blouse
- Before getting into the workplace
- After getting out of the workplace
- After taking out the chew tobacco from the mouth
- Before putting chew tobacco in the mouth
- After touching the hair of the head
- Before taking out the chew tobacco from the mouth
- After using the phone
- Before smoking a cigarette
- After smoking a cigarette
- After sneezing or coughing on hands
- Before using the phone
- After handling money.

4.2 *Compliance of health care workers with techniques and right rules of hand washing*

The results of the present study showed that the health care workers are not in compliance in following techniques and rules for Hand washing as is shown in table 3.

Table 2. Decreasing order of attitude of doctors and nurses towards hand washing.

Do you think that doctors and nurses wash their hands:		Answers: Never	Rarely	Some times	Most times	Always	Mean	SD	Mean rank
1. Before doing a small surgery	%	2,5	1,3	6,3	13,8	76,3	4,63	0,833	34,62
2. After playing with an animal	%	3,1	4,4	6,9	10,6	75,0	4,54	0,942	33,99
3. After doing a small surgery	%	0,6	1,9	6,3	36,9	53,1	4,42	0,751	33,45
4. After inserting tools into the patient body	%	0,6	1,9	6,3	36,9	53,1	4,42	0,751	33,45
5. After a contact with polluted products (blood, chemical products)	%	0	1,3	7,5	64,4	26,9	4,18	0,596	32,09
6. Before wearing gloves	%	1,9	5,0	10,6	74,4	8,1	3,83	0,724	29,57
7. After wearing gloves	%	1,3	3,8	16,9	71,9	6,3	3,77	0,660	29,05
8. After placing a sonde	%	1,3	3,8	16,9	71,9	6,3	3,77	0,660	29,05
9. Before inserting tools into the patient body	%	3,8	3,8	20,0	68,1	4,4	3,68	0,749	28,43
10. After giving an injection to the patient	%	5,0	8,1	25,0	55,0	6,9	3,53	0,915	27,03
11. Before eating	%	5,6	8,8	31,3	41,9	12,5	3,46	1,014	26,61
12. After each examination	%	3,1	5,6	51,3	26,9	13,1	3,41	0,903	25,98
13. After going to the toilet	%	11,3	14,4	16,3	50,0	8,1	3,31	1,156	25,53
14. After every contact with the patient	%	7,5	11,3	35,6	36,9	8,8	3,30	1,014	25,35
15. After giving an injection	%	8,8	11,3	43,1	19,4	16,3	3,23	1,130	24,77
16. Before givining an injection to the patient	%	34,4	21,3	18,8	20,0	5,6	2,42	1,303	23,24
17. Before each examination	%	10,6	16,9	56,9	10,0	5,6	2,85	0,930	21,81
18. Before every contact with the patient	%	15,6	23,1	50,6	10,6	0	2,58	0,869	19,62
19. After eating	%	7,5	61,3	16,9	9,4	5,0	2,42	0,946	19,62
20. Before giving an injection	%	2,5	13,1	67,5	10,0	6,9	2,24	0,777	18,20
21. After throwing rubbish in the bin	%	29,4	32,5	33,8	2,5	1,9	2,16	0,937	16,57
22. After shaking hands with someone	%	31,3	28,8	33,8	4,4	1,9	2,18	0,983	16,42
23. Before putting chew tobacco in the mouth	%	67,5	22,5	8,1	1,3	0,6	1,44	0,753	16,02
24. Before going to the toilet	%	53,1	18,8	18,8	6,3	3,1	1,87	1,121	13,71
25. After reading a news paper	%	53,8	17,5	20,6	4,4	3,8	1,84	1,097	13,47
26. After putting on a blouse	%	61,3	15,6	16,3	3,8	3,1	1,72	1,071	12,45
27. Before getting into the workplace	%	64,4	14,4	11,3	5,6	4,4	1,71	1,147	12,39
28. After getting out of the workplace	%	65,6	15,6	11,3	6,9	0,6	1,61	0,983	11,67
29. After taking out the chew tobacco from the mouth	%	20,0	51,9	22,5	2,5	3,1	1,56	0,885	11,67
30. Before putting chew tobacco in the mouth	%	67,5	22,5	8,1	1,3	0,6	1,44	0,753	10,35
31. After touching the hair of the head	%	70,6	15,0	14,4	0	0	1,43	0,735	10,16
32. Before taking out the chew tobacco from the mouth	%	75,0	11,9	9,4	2,5	1,3	1,41	0,846	10,09
33. After using the phone	%	75,0	11,9	9,4	2,5	1,3	1,41	0,846	10,09
34. Before smoking a cigarette	%	70,6	20,6	6,3	1,3	1,3	1,41	0,775	10,08
35. After smoking a cigarette	%	71,3	20,0	6,3	1,3	1,3	1,42	0,776	10,08
36. After sneezing or coughing on hands	%	78,1	7,5	10,6	2,5	1,3	1,40	0,867	9,84
37. Before using the phone	%	81,3	8,1	7,5	1,9	1,3	1,34	0,804	9,40
38. After handling money	%	87,5	5,6	5,0	1,3	0,6	1,20	0,623	8,22

Table 3. Compliance of health care workers with techniques and right rules of hand washing.

Do you think that doctors and nurses follow the following techniques and conditions when washing their hands:		Never	Rarely	Some times	Most times	Always	Mean	SD	Mean rank
1. Take off jewelries	%	31,3	51,3	12,5	3,1	1,9	1,92	0,852	141.813
2. Cut off nails	%	14,4	10,0	65,0	6,9	3,8	2,74	0,911	207.438
3. Rubbing fingers	%	1,9	3,8	8,8	76,9	8,8	3,91	0,616	326.438
4. Taking off artificial nails	%	77,5	11,9	7,5	1,9	1,3	1,37	0,810	336.688
5. Taking off nail varnish	%	68,1	14,4	11,9	3,8	1,9	1,55	0,955	240.500
6. Taking off gloves	%	0	0	1,9	92,5	5,6	4,03	0,263	252.388
7. Taking off esthetic creams	%	63,1	22,5	9,4	3,1	1,9	1,54	0,864	207.375
8. Taking off the watch	%	6,9	12,5	70,6	6,9	3,1	2,89	0,737	459.875
9. Use water	%	0	0	0	20,6	79,4	4,80	0,398	55.225
10. Use water + soap	%	1,3	1,3	6,3	23,1	68,1	4,56	0,777	257.438
11. Use antiseptics (alcohol; Bétadine etc.)	%	16,9	64,4	9,4	5,6	3,8	2,15	0,897	205.000
12. Hygienic wash	%	9,4	71,9	12,5	3,1	3,1	2,17	0,742	343.875
13. Surgical wash	%	26,9	56,3	8,8	2,5	5,6	2,05	0,983	274.375
14. Respect the time of washing which is from 15 to 20 seconds	%	15,0	66,3	9,4	5,0	4,4	2,19	0,897	160.063
15. Respect the steps of each methods	%	78,8	5,6	8,8	3,1	3,8	1,46	1,026	219.688

Table 4. Means for washing hands.

Are the following means for washing hands provided?		Never	Rarely	Some times	Most times	Always	Mean	SD	2χ
1. Water in quantity and quality	%	1,9	3,1	6,2	10,0	78,8	4,62	0,879	348.313
2. Conform lavatories	%	47,5	40,0	8,8	3,1	0,6	1,70	0,810	155.438
3. The necessary disinfectants (alcohol; Betadine etc.)	%	68,1	17,5	8,1	3,1	3,1	1,54	0,975	242.625
4. Tools for drying hands	%	80,0	8,8	5,6	3,8	1,9	1,36	0,869	362.063
5. Water in quantity and quality	%	1,9	3,1	6,2	10,0	78,8	4,62	0,879	348.313
6. Conform lavatories	%	47,5	40,0	8,8	3,1	0,6	1,70	0,810	155.438
7. The necessary disinfectants (alcohol; Betadine etc.)	%	68,1	17,5	8,1	3,1	3,1	1,54	0,975	242.625
8. Tools for drying hands	%	80,0	8,8	5,6	3,8	1,9	1,36	0,869	362.63

4.3 *Means for washing hands*

The results of the present study showed that apart from water hand washing facilities are unavailable as is shown in table 4. The series of questions on the domain of obstacles in adhering to hand hygiene principles revealed that unavailability of hand-hygiene facilities like shortage of soap, disposable towels and gloves was the most frequently reported barriers. Most of the respondents believed that hand irritation associated with hand washing barriers to hand washing compliance among health care workers.

4.4 *Barriers to hand washing compliance among health care workers*

The results of the present study showed that the health care workers are not in compliance in following techniques and rules for Hand washing as is shown in table 5. Moreover, when Friedman ranking test was applied on the results it was found that unsafe behaviours of health stuff seem to be the main factor as Barriers to hand washing compliance among health care workers.

Table 5. Order of the main barriers to hand washing compliance among health care workers.

In your opinion do you think the non-respect of washing hands is due to:		Never	Rarely	Some times	Most times	Always	Mean	Std.	Mean rank
1. Lack of awareness raising	%	2,5	3,1	5,6	22,5	65,6	4,47	0,926	18,67
2. Negligence	%	2,5	5,0	11,3	19,4	61,9	4,34	1,024	18,00
3. Lack and non-conformity of lavatories	%	0	1,9	3,8	80,6	13,1	4,06	0,512	16,00
4. Lack of tools for drying hands	%	1,9	6,9	75,0	15,6	0,6	4,06	0,565	15,90
5. Unconsciousness	%	3,1	2,5	3,8	73,1	17,5	3,99	0,775	15,90
6. Habit	%	1,9	6,9	75,0	15,6	0,6	4,06	0,565	15,90
7. Believing that wearing gloves do not requires hand washing	%	1,9	4,4	6,9	71,9	15,0	3,93	0,747	15,43
8. The lavatories designated for washing hands are dirty	%	1,9	3,8	5,6	81,9	6,9	3,89	0,642	15,15
9. Lack of water	%	2,5	6,3	10,0	70,0	11,3	3,82	0,800	14,81
10. The non-respect of hygiene rules	%	2,5	3,1	10,6	76,9	6,9	3,82	0,701	14,72
11. Lack of a model person (who gives the example)	%	6,3	6,3	15,6	52,5	19,4	3,73	1,048	14,36
12. Lack of The necessaries disinfectants (alcohol; Betadine etc.)	%	5,0	6,9	11,3	71,3	5,6	3,67	0,876	13,93
13. Imitating what is going on	%	5,0	6,9	11,3	71,3	5,6	3,67	0,876	13,93
14. Ignorance of risks	%	6,9	6,9	57,5	16,9	11,9	3,19	0,969	10,55
15. Lack of warm water	%	9,4	13,1	50,0	18,1	9,4	3,04	1,024	9,88
16. The work load	%	10,0	11,9	58,8	12,5	6,9	2,96	0,950	9,25
17. The forgetfulness	%	8,1	6,3	75,6	5,0	5,0	2,92	0,795	8,70
18. The cold	%	12,5	58,8	11,9	10,6	6,3	2,75	0,934	8,21
19. The lack of time	%	13,8	18,1	59,4	4,4	4,4	2,69	0,915	7,80
20. Skin Inflammation because of repeated washes	%	13,8	65,0	14,4	4,4	2,5	2,17	0,813	5,70
21. The fear of skin inflammation	%	1,9	4,4	6,9	71,9	15,0	1,51	1,055	3,89
22. The disinfectants used cause generate allergies	%	83,1	13,8	2,5	0,6	0	1,19	0,452	2,44

5 CONCLUSION

It is therefore concluded that continuous campaigns for raising awareness of washing hands should be introduced as an absolute necessity not only for the hospital stuff but also for all the individuals of the society to make it as cultural behaviour in order to prevent the spread of contagious illnesses and by consequences economise the bill of their treatments especially for under developed countries.

REFERENCES

1. Albert RK, Condie F. Hand-washing patterns in medical intensive-care units. N Engl J Med. 1981;304:1465–6. [PMID: 7248048].
2. Anderson JL, Warren CA, Perez E, Louis RI, Phillips S, Wheeler J, Cole M, Misra R. Gender and ethnic differences in hand hygiene practices among college students. Am J Infect Control. 2008 Jun;36(5):361–8.
3. Boyce JM, Pittet D. Guideline for Hand Hygiene in Health-Care Settings. Recommendations of the Healthcare Infection Control Practices Advisory Committee and the HICPAC/SHEA/APIC/IDSA Hand Hygiene Task Force. Society for Healthcare Epidemiology of America/Association for Professionals in Infection Control/Infectious Diseases Society of America. MMWR Recomm Rep. 2002;51:1–45, quiz CE1–4. [PMID: 12418624].
4. Didier Pittet, Anne Simon, Stéphane Hugonnet, Carmen Lúcia Pessoa-Silva, Valérie Sauvan, and Thomas V. Perneger. (2004), Hand Hygiene among Physicians: Performance, Beliefs. Annals of Internal Medicine Volume 141, Number 1.
5. Elizabeth Pyne, Lois Sater, Timothy McAvoy, (2010), Physician Attitudes toward Hand Hygiene in the Acute Care Setting, Wisconsin Medical Society Fellowship, June–August 2010.
6. Khaled M. Abd E laziz and Iman M. Bakr (2008), Assessment of knowledge, attitude and practice of

hand washing among health care workers in Ain Shams University hospitals in Cairo. The Egyptian Journal of Community Medicine Vol. 26, No. 2.
7. Muhammad Ali Anwar, Sana Rabbi, 2 Muhammad Masroor, Fouad Majeed, Marie Andrades, Shehla Baqi6, (2009) Self-reported practices of hand hygiene among the trainees of a teaching hospital in a resource limited country, J Pak Med Assoc, Vol. 59, No. 9.
8. Pittet D, Hugonnet S, Harbarth S, Mourouga P, Sauvan V, Touveneau S, et al. Effectiveness of a hospital-wide programme to improve compliance with hand hygiene. Infection Control Programme. Lancet. 2000;356:1307–12. [PMID: 11073019].
9. Pittet D, Mourouga P, Perneger TV. Compliance with handwashing in a teaching hospital. Infection Control Program. Ann Intern Med. 1999;130:126–30. [PMID: 10068358].

Occupational Safety and Hygiene II – Arezes et al. (eds)

Scenarios and implications of ambivalence experienced by hotel managers

C.M.G. Leite
Extension and Research Group in Ergonomics, GREPE, Rio Grande do Norte, Brazil

R.J.M. Carvalho
Extension and Research Group in Ergonomics, GREPE, Rio Grande do Norte, Brazil
Production Engineering Department, Federal University of Rio Grande do Norte—UFRN, Brazil

ABSTRACT: This article deals with the situations of ambivalence of power experienced by managers of a luxury hotel in the city of Natal, Rio Grande do Norte, Brazil. We adopted the method of Ergonomic Work Analysis (EWA), in which searches were conducted literature and documents, questionnaires and semi-structured interviews with managers and made observations of their activities. The objective is understand the relationship between the situations of ambivalence power, occupational health and organizational performance. It can be concluded that the ambivalence of power are part of everyday activity working hotel managers, causing them physical and mental, which can compromise the occupational health and organizational performance.

1 INTRODUCTION

There are few studies related to occupational health managers. This article intends to identify the ambivalence in the work of managers of a luxury hotel located in the city of Natal, Brazil, and know the impact of these ambivalences about health and work performance of these managers.

The quest for understanding the work activity of hotel managers justified by the scenario of intense growth of the tourism sector in Brazil and therefore the activity of hospitality, characterized by technological upgrading and market strategies, which have resulted in a progressive movement for change and improvements in hotel management systems, seeking to achieve higher productivity and competitiveness.

Managers assume an intermediate position in the organizational hierarchy, so they need to deal with the demands arising from higher and lower hierarchical equivalents, in addition to external customers and suppliers. Thus, managers are caught by ambivalent and, usually, conflicting demands between those involved in the organization of work, so that these ambiguities can have a negative effect on the health and performance of the work activity.

Assuming that the ambivalence at work may become critical, problematic and lead to sicknesses due to organizational and management models (that do not respect the work characteristics, expectations, limits and needs of managers), strict production goals and extended working hours are imposed, thus being responsible for compromising the free time that managers have to perform everyday tasks, spend time with family, leisure, rest and take care of their own health.

2 ERGONOMICS: UNDERSTANDING TO TRANSFORM

Ergonomics aims to understand the job so it can intervene and transform work as Guerin et al (2001) stated. You must contemplate the work activity in all its dimensions to understand it and understanding goes "*besides recording the positions and dimensions of the furniture, which are also very important, but should be evaluated with other more complex aspects of work activity*" (Bouyer 2011).

According to Bouyer (2011), suffering is part of the work experiences lived at work. Ergonomics furthers our understanding of the origins, consequences and ways of transforming suffering, considering that when the worker cannot change it; this latter is transformed into illness. In this matter, Bendassolli & Soboll (2011) add that suffering at work is signaled by the work individualization and the loss of collective referential, manifesting itself in feeling unease at work, physical diseases, disorders, mental and psychosocial changes.

However, it is assumed that labor is not only a source of suffering, but also of pleasure. The work can also be a device in which the worker builds its identity and the experience of pleasure by subjective recognizing and mobilizing, allowing the individual to rearrange test procedures and, thus, find mechanisms to make the work more pleasurable (Dejours 1994). Ergonomics can boost labor organization so it may be considered that the professional reorganization can be favorable to the worker's skills, his or her needs of his psychosomatic economy with full employment psychomotor skills, and psychical psycho-sensory—thus promoting the scenario of pleasure at work (Dejours 2006).

According to Dejours (2006), the work is also a source of balance for the individual in the act of "work *is not only having an activity, but also to live: to experience pressure, live together, face resistance from the real, build the sense of work, the situation and the suffering*".

Ergonomics can through EWA—Ergonomic Work Analysis (Wisner 1994, Vidal 2003), examine the complexity of the work giving rise to reflection about the activity, since it allows the involved individuals permanent learning of work rules and values, as well as "*make the activity communicable and submit it to the confrontation of knowledge in a process of knowledge-processing activity*" (Bendassolli & Soboll 2011).

3 THE AMBIVALENCE IN THE WORK OF MANAGERS

The concept of ambivalence arises from the Psychology and is inspired from Freud, meaning the possession of opposite feelings for the same person or object. Berrebi-Hoffman et al (2009) add that the ambivalence in Sociology refers to is incompatible normative expectations regarding attitudes, whether in relation to beliefs and behaviors that are assigned a status (i.e., a social position) or a set of a company status.

According to Guivante & Tomiello (2008), "*the continuous movement of meaning and significance can re-generate the discomfort of doubt, the contradictory feelings, the sense of ongoing duel between the appearance and the pursuit of truth*". In this matter, the ambivalence becomes unpleasant when the positive and negative aspects of a subject are present in a person's mind at the same time. These are authors add that this condition can lead to a psychic leak or a deliberate attempt to solve the ambivalence. When the situation does not require a decision to be made, people have less discomfort even when affected by the ambivalent feeling.

Guivante & Tomiello (2008) also emphasize that idea of that the individual feels discomfort in having to choose between alternatives that present themselves is a confusing result of an ambivalent situation, thereby causing a feeling of indecision, insecurity, or loss of control of the situation experienced.

The activity of managers involves aspects not always reconcilable. To Lacomblez & Teiger (2007), it is important to identify the integration and the separation of multiple and heterogeneous elements that are established through a combination of technical criteria such as efficiency and trustworthiness of the production and quality, and also the criteria as human health, skills, safety, and of the criteria that involve social workload.

Davel & Melo (2005) considered that the manager is an "agent" that operates in systems of representation of reality. So he does not plays only a role—derived from the growing bureaucratization of organizations—but part of a process in which, increasingly, are involved aspects to influence in the interaction of symbolization schemes (management of organizational culture, identity), dominance (the issue of control and persuasion) and of assimilation (the ability to continuously learn).

According to these authors, the manager in organizations is also a mediator between "who thinks" and "who is" in an attempt to approach them and to manage the constant changes occurring in the world of work, with the ability to cope the conflicting pressures between production time and quality of service. While, the manager has to control the work of employees and speak on behalf of the internal power structures of organizations, he also need to reconcile personal conflicts and each individual, in addition to dealing with the own emotions and expectations, while subordinated to the upper management of the company.

4 MATERIALS AND METHODS

This is a qualitative research and descriptive, developed in the molds of a field research, a case study (Yin 2010) held in a luxury hotel located in the city of Natal, Rio Grande do Norte, Brazil. We adopted the method of Ergonomic Work Analysis—EWA (Wisner 1987, Guerin 2001, Vidal 2003). The EWA is a method that combines participatory and observational methods to understand the activity of work and, while promote the positive changes of work situations. Literature searches were performed and documented also were adopted methods and techniques interactional (questionnaire, semi-structured, dynamic scripts for conversational action, listening to spontaneous reports) and methods and techniques observational (open and systematic observations aided by videos and photos) (Vidal 2003).

5 RESULTS

The case study began with the statement of demand and global analysis which indicated that the management staff of the hotel under study consists of six (6) managers, (five (5) are male and one (1) is a woman) whose age ranges from 25 to 35. Managers represent 7% (seven percent) of the total employees of the company and 75% of them has a graduation degree.

During the analysis of the activity of managers, it was identified that the working conditions of the hotel managers vary according to high season (from December to February) and low season (from March to November) of tourism the city of Natal and also vary according to the high occupancy rate of the hotel (85 to 100% occupancy) and low (less than 85%) occupancy rate of the hotel.

There are several situations of work in which the manager of the hotel carry out multitasking and ambivalence deal with situations such as in the case of general manager, depicted in Figure 1.

Figure 1 illustrates the group of interactions made by the general manager with internal and external environment to the hotel organization in focus. This manager takes the top hierarchical position in the organization chart of the hotel business, like actions of coordination and supervision of the work of other managers of the hotel, in addition to direct contact with the governing board and external customers such as suppliers, hotels (partners), agencies and operators (partners), guests, government agencies etc.

This work revealed, in addition to the varied interactions, a space for intercultural strife and values, which is expressed on certain occasions as cognitive overload and emotional in the work activity of managers.

Another ambivalent situation could be observed at the time the manager was negotiating with guests about their transfer to another hotel due to the occurrence of overbooking. Therefore, in this

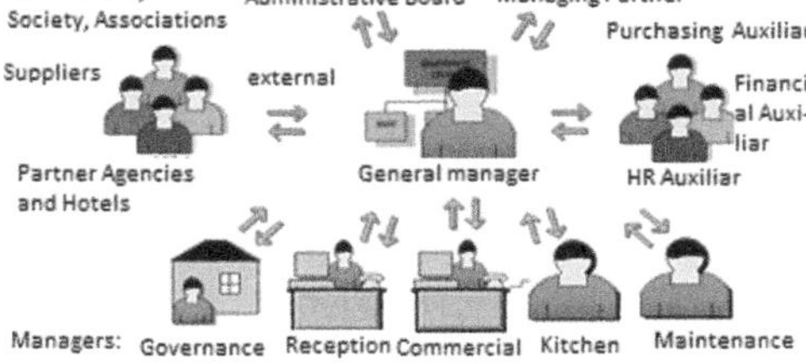

Figure 1. Interactions and ambivalences in the activity of the general manager. *Source*: Field survey, 2011 to 2012.

situation, he was performing the duties typical desk manager, who had resigned the day before and, therefore, the general manager was performing the duties of this suffering, so the pressure of the demands of top management so he decided to overbooking at the same time that he was being pressured by guests who wanted to stay in that hotel.

With regard to the ambivalence of power suffered by the manager of governance, one can understand that the general manager requires responsibility for maintaining the entire hotel clean, in other words, all the private area and common of hotel guests, including the sidewalk around the hotel. This requirement promote an organizational aspect to the manager of governance an since the team of assistants that provides general services is reduced to the structure of the hotel. The same goes for the amount of chambermaids, leading to the occurrence of organizational conflict by the lack of communication between sectors, standardization of operating procedures and reducing staff.

Among the ambivalence experienced by the commercial manager is the need to maintain high occupancy in the hotel, through the orders received by the general management and the board (superior) Hotel, bringing you into contact with partner companies to promote increased sales and disclosures hotel nationally and internationally, while he must be aware of the work of the sales team (less hierarchical) and the return of the guest in the service provided by the hotel on completion of these sales so that thus the hotel guest can indicate other potential new customers.

It can be observed that the manager of the maintenance sector is under pressure from his immediate superior (general manager) for all maintenance services performed at the hotel are made quickly so release the areas of use of hotel guests. Meanwhile, the manager of the maintenance sector is confronted with small permanent staff (less hierarchical) and the lack of adequate tools for the work of the maintenance team.

The ambivalences of the manager of the kitchen were identified in tasks that involving temporal demands required by the waiters of the restaurant hall and of the bar. Were also identified ambivalences in the demands of the general manager to the manager of the kitchen to for quality in providing the services of food and drinks, beyond the demands of customers for good attendance and quality of the products.

It is noticed that managers are faced with situations ambivalent in several times different of work. According to Bouyer (2011), people who have a lot of responsibility and many interpersonal conflicts are more prone to physical and mental illnesses when it exceeds their ability to maintain self control.

6 ERGONOMIC RECOMMENDATIONS

As a means to reduce the harmful impact of ambivalence experienced by managers of this hotel, it is recommended that there be cooperation work between managers, directors and employees among subordinates according to Morgan (2007). Thus, it is recommended that, in the weekly meetings held between all managers and senior management, each manager can pass the demands, problems and suggestions for each sector of work for the board and the other managers, so that they can jointly find solutions to the difficulties raised by sectors. And, in the sectoral meetings, managers can relay to the sector team work of each, the solutions of the board and the other hotel managers to remedy the difficulties presented by them.

7 CONCLUSIONS

The situations of ambivalence lived by the managers in the hotel occurred frequently and at various times distinct seasonality of tourism and hotel occupancy, being understood as situations inherent in the performance of the activity of manager.

By comparing the data with the readings taken can be concluded that the ambivalence of power suffered by managers can generate suffering and illness at work. Therefore, it is recommended the use of class actions work where collaborative work is adopted at the expense of individualized work, as well as managers can take advantage of autonomy and thus adopt the regulations necessary to carry out the work activities.

REFERENCES

Bendassolli, P.F. & Soboll, L.A.P. 2011. Clínicas do trabalho. São Paulo: Atlas, p.12.

Berrebi-Hoffman, I. & Lallement, M. & Nicole-Drancourt, C. & Sarfati, F. 2009. Ambivalência, gênero e modernidade capitalista: a França na era da flexibilidade. Dossiê: economias de gênero: cadernos pagu 32: 9–42.

Bouyer, G.C. 2011. Entre o número e a vivência: qual ergonomia praticar? In Sznelwar, L.I. *Saúde dos bancários*. Publisher Brasil: Editora Gráfica Atitude Ltda, São Paulo.

Daniellou, F. 2004. A ergonomia em busca de seus princípios. São Paulo: Blucher, p. 9.

Davel. E. & Melo, M.C.O.L. 2005. Gerência em ação. Rio de Janeiro, FGV Editora.

Dejours, C. 1994. In Dejours, Christophe. Abdoucheli, Elizabeth. Jayet. Christian. *Psicodinâmica do trabalho: contribuições da Escola Dejouriana à análise da relação prazer, sofrimento e trabalho*. São Paulo: Atlas.

Dejours, C. 2006. A banalização da injustiça social. Rio de Janeiro: editora FGV.

Guérin, F. & Laville, A. & Daniellou, F.& Duraffourg, J. & Kerguelen, A. 2001. Compreender o trabalho para transformálo: a prática da ergonomia. São Paulo: Ed. Edgard Blucher.

Guivante, J.S. & Tomiello, N. 2008. Ambivalência na comunicação das estratégias de sustentabilidade: uma análise da cadeia de valor da Wal-Mart na perspectiva global, nacional e local. Cadernos de Pesquisa interdisciplinar em ciências Humanas. UFSC.

Lacomblez, M. & Teiger, C. 2007. *Ergonomia, formações e transformações*. In Falzon, Pierre. *Ergonomia*. Editor [tradução Ingratta, G.M.J. *et al*]. Ed. Blucher, São Paulo, p. 615.

Maggi, B. & Tersaac, G. 2004. O trabalho e a abordagem ergonômica. In Daniellou, F. *A ergonomia em busca de seus princípios: debates epistemológicos*. São Paulo: Ed. Edgard Blucher, p. 92.

Vidal, M.C. 2003. Guia para Análise Ergonômica do Trabalho na empresa: uma metodologia realista, ordenada e sistematizada. Rio de Janeiro, Editora Virtual Científica.

Wisner, A. 1987. Por dentro do Trabalho: ergonomia: método e técnica. São Paulo. Oboré.

Yin, R.K. 2010. Estudo de caso: planejamento e métodos. Editora Bookman: Porto Alegre.

Occupational Safety and Hygiene II – Arezes et al. (eds)
© 2014 Taylor & Francis Group, London, ISBN 978-1-138-00144-2

Bioaerosols in hospital environment: A short review

Cláudia Vieira
Occupational Health Service, São João Hospital Center, E.P.E, Alameda Prof. Hernâni Monteiro, Porto, Portugal

J. Santos Baptista
Research Laboratory on Prevention of Occupational and Environmental Risks (LABIOMEP/CIGAR), Faculty of Engineering, University of Porto, Porto, Portugal

ABSTRACT: Biological hazards are an important occupational health problem. The intention in this article is to present a short review on the exposure of health workers to biological agents transmitted by aerosols. 637 articles were collected, and 41 were selected for analysis. It is recognized that bioaerosols are the cause of several diseases. However, there are difficulties to establish their prevalence and incidence within occupational range. Also there are no official records on occupational infectious diseases or airway that caused the death of health professionals.

1 INTRODUCTION

Biological agents include a variety of microorganisms, toxins and allergens that can transmit infectious or respiratory diseases (Haagsma *et al.* 2012). Hospital environment requires resources to protect their professional during their activities against occupational diseases (Santos 2008), particularly, biological hazards (Gutiérrez *et al.* 2009). The presence of these agents is particularly severe due to the lack of information on exposure limits (Corrao *et al.* 2012).

1.1 *Bioaerosols and occupational diseases*

The bioaerosols are defined as aerosols or particles of microbial origin including endotoxins, allergens, protozoa, viruses, bacteria and fungus. (Douwes *et al.* 2003). They are an important vector of microorganism's transmission (Corrao *et al.* 2012). The exposure of health professionals to these risk factors makes them susceptible to contract and transmit infections (Moreira, 2010), presenting a higher risk compared with the general population (Corrao *et al.* 2012). Although exposure to bioaerosols be a risk factor (Corrao *et al.* 2012), there is no tracking system with the number of illnesses and deaths (Sepkowitz *et al.* 2005).

The aerosol infectivity depends, among others, of the particle size, the density of pathogens inside it, and the volume inhaled. Exposure to the bioaerosols in working environment is associated, among others, with infectious, inflammatory and allergic diseases (Corrao *et al.* 2012) and cancer, with first two being the most common. Despite the diseases resulting from exposure to bioaerosols be documented, there is a lack of validated data related to the incidence and prevalence from the majority of this diseases (Douwes *et al.* 2003).

1.1.1 *Bacteria*

a) Mycobacterium tuberculosis—Tuberculosis (TB) TB is recognized because of its occupational risk for professionals health (Severo *et al.* 2011; Tafuri *et al.* 2009a; Costa *et al.* 2009; Menzies *et al.* 2007), particularly to the ones that provide direct care to infected patients or those who perform functions in laboratories researching the bacillus.

The bacterium is transmitted from person to person through droplets of 1 μm to 5 μm of diameter, during the acts of sneezing, coughing and conversation. These particles can remain in suspension for a long period of time (Santos 2008). TB can cause illness and death among professionals (Chai *et al.* 2013). Therefore it is necessary to implement control measures and effective screening programs in health facilities for their own Professionals (Salt *et al.* 2007). Training and prevention are equally important to reduce the risk of transmission of the bacillus (Naidoo *et al.* 2013; Cuhadaroglu *et al.* 2002). Workers' immunization, triage and patients treatment, are examples of other measures to develop. (Laraqui *et al.* 2004).

b) Staphylococcus aureus—SA

Staphylococcus aureus is a bacterium frequently located in the skin and nasal mucosa of human beings. It is responsible for a wide variety of infectious diseases (pneumonia, meningitis, septicemia) (Santos *et al.* 2007). Its virulence and ability to acquire resistance to anti-microbial agents, results in a serious global issue for healthcare facilities (Carvalho *et al.* 2009).

Healthcare professionals are a vulnerable group-level to infection by this bacterium, due to occupational exposure. The prevention of occupational infection by *SA* must be accomplished by better surveillance of infected patients (Haagsma *et al.* 2012), rational and controlled use of antibiotics, (Moura *et al.* 2010; Peres *et al.* 2011) by training (Peres *et al.* 2011) and adoption of standard precautions to prevent transmission (Carvalho *et al.* 2009).

1.1.2 *Virus*

Infections caused by viruses are common diseases and one of the causes of morbidity that most affects the quality of life and productivity (Quadros 2008). The droplets are projected by respiration and may contain viruses. Many infection can be transmitted by this route. In most cases, infective dose rarely moves more than a few meters from the source (Santos 2008).

a) Measles, Rubella, Varicella

Healthcare professionals have a higher risk of acquiring and transmitting measles, rubella and varicella (Fedeli *et al.* 2002).

Measles is a highly contagious disease (Simone *et al.*, 2012). As prevention, is recommended vaccination and surveillance for the most vulnerable professionals (Tafuri *et al.* 2009b; Thierry *et al.* 2008); adoption of appropriate work practices for prevention and control of transmission in hospital environment (Thierry *et al.* 2008); training activities for professionals (Tafuri *et al.* 2009b).

The rubella virus is transmitted by air or by contact with infected patients. There is the need to introduce rubella vaccination in healthcare workers (Singh *et al.* 2010).

The varicella virus is transmitted from person to person by droplets inhalation or by direct contact (Holmes *et al.* 2004; Façanha *et al.* 2006). Preventive measures and protection that should include the serological triage of healthcare workers it is essential to determine the state of immunization, without regarding age, professional activity or individual history of infection (Tablei—Taher *et al.* 2010; Façanha *et al.* 2006). It also must ensure the isolation of patients, improve the information about the transmission of the disease and adequate protection of employees (Façanha *et al.* 2006).

b) Coronavírus—Severe acute respiratory syndrome

Healthcare professionals are particularly vulnerable to coronavirus (Dutkiewicz 2004; Hui *et al.* 2010) when in contact with infected patients (Yen, *et al.* 2006). They are, therefore, workers with high risk of exposure to this virus (Chan-Yeung 2004). The coronavirus is spread mainly by droplets and by direct and indirect contact with secretions from infect and, in some situations, by aerosols, for example, intubation or resuscitation (DGS 2013). Currently there is no proven treatment. The control of this highly contagious disease is made by an early recognition, isolation of patients in wards with negative pressure and control of severe infections (Hui *et al.* 2010). The appropriate and permanent use of protective equipment, including mask, is crucial in the prevention of severe acute respiratory syndrome (Nishiyama *et al.* 2008).

c) Influenza—Flu

The healthcare professionals are an important target group for flu vaccination. Vaccination against influenza is recommended to reduce the spread of disease (Akker *et al.* 2011).

However, the professionals vaccination rate is low in the world, due to errors relating safety and efficacy of the vaccine (Couto *et al.* 2012; Kok *et al.* 2011; Boudma *et al.* 2012). In case of disease outbreaks, it is recommended that these professionals wear a face mask with breathing filtration system that minimizes exposure to influenza virus (Danyluk *et al.* 2011).

1.1.3 *Fungi*

Fungi are omnipresent in the environment, once in indoor air can be inhaled (Herr *et al.* 2012). The degree of fungal contamination in a hospital environment can increase dramatically in combination with various factors such as the materials of construction and favorable microclimate (Perdelli *et al.* 2012).

Aspergillus spp—Aspergillosis

Aspergillus spp. is acquired especially by inhalation of spores (Tablan *et al.* 2003). The exposure can cause serious hazards to human health (Reijula *et al.* 2003).

Exposure to this fungus could be caused by many invasive procedures, such as bronchoscopy, the induction of secretion (Moreira 2010), and tracheal suctioning body fluids, treatment of nebulization and the insertion of nasogastric tubes (Tran *et al.* 2012).

1.2 *Statistics*

Despite occupational infectious diseases have compulsory notification, the data is incomplete and is difficult its relationship with the work environment (Pietrangeli 2008). According to the statistics published under the Health and Safety at Work by Eurostat (1999–2007), infectious diseases caused by biological agents are in 8th place in occupational diseases ranking. Still according to the same source, on the degree of disability caused by occupational infectious diseases, it was found that the most causes have some or significant limitation on the ability of working or earning professionals.

1.3 *Need to develop*

Despite its complexity, it's necessary to establish, prevalence and incidence rates of the diseases related to exposure to biological agents, particularly aerosols. Despite all the difficulties, it is necessary to know all the epidemiological logical connections present, as well as the total number of workers exposed (Pietrangeli 2008). Another important field of research is the determination of limits of exposure to biological agents. It is necessary overcome the difficulties resulting from the effects that may influence the interaction with other factors and mechanisms acting on the same target organs (Corrao *et al.* 2012) and individual susceptibility.

The risk assessment is seriously compromised by the absence of methods (Gehanno *et al.* 2009) validated for quantitative evaluation of exposure (Douwes *et al.* 2003) and also by the possible influence of the interaction of microorganisms with non-biological agents in working environment (Corrao *et al.* 2012). There are several recommendations, but there is no sampling or analysis method that has been implemented to establish the relationship between the levels of microorganisms in the atmosphere and the rate of infections (Soares, 2009), which complicates the characterization of exposure to biological agents (Klug *et al.* 2012). The number and type of microorganisms present in the air can be used to determine the source of discomfort for professionals from health institutions (Ekhaise *et al.* 2008). The measurement methods that exist for the collection of air samples may include: sedimentation, filtration, electrostatic precipitation, centrifugal impaction, impaction in liquid and in solid impaction (Santos 2008).

To minimize exposure of health workers to bioaerosols can be implemented appropriate administrative and/or engineering measures and personal protection (Trajman *et al.* 2010). Some of these measures include: Isolation of patients with disease, use of ventilation systems with negative pressure, promotion of adequate insulation of windows and doors, use of appropriate personal protective equipment (Tang *et al.* 2006), control of moisture, maintenance of the equipment, by promoting natural ventilation and placing air filtration systems/exhaust (HEPA), use of disinfectants to decontaminate air and surfaces, maintenance of the doors of wards/rooms properly closed (Vaquero *et al.* 2003; Srikanth *et al.* 2008) and by restricting access (Vaquero *et al.* 2003).

In regard to personal protective equipment it is recommended to use protective masks type P2. These masks retain the particles through a filter and are indicated in the care of patients with tuberculosis, measles, varicella and other diseases (Moreira 2010).

Despite the importance of prevention and protection, it is essential to early identify by all health professionals and patients suffering from infectious diseases (Wong *et al.* 2004).

This work aims to identify and characterize the health effects resulting from exposure of health workers to biological agents transmitted by aerosols in the hospital context. To achieve this objective a literature review was conducted in order to determine the knowledge gaps and research conducted on the issue.

2 MATERIAL AND METHOD

Was developed through a systematic review using as keywords: *biological agents; aerosols; healthcare professionals, bioaerosols, occupational diseases, infectious diseases, tuberculosis, staphylococcus aureus, measles, varicella, aspergillus*. Was used the search engine "MetaLib of exlibris", Google scholar, as well as scientific journals, theses and dissertations. The research has been carried out by subject. Emphasis was given to articles with publication date later than or equal to 2002. The criteria used for exclusion of articles were the absence of biological agents transmitted by aerosols in health care facilities.

The application of the above keywords, provide a total of 637 articles. From these, 596 were excluded, 73 because of publication date before 2002, repeated 46 and 477 by the fact that do not address exposure to bioaerosols in hospital environment.

3 RESULTS

Most publications are unanimous by characterizing health professionals as a professional group with higher risk of exposure to biological agents. The development of symptoms or diseases may be caused by viruses, bacteria or fungi.

Many articles on the subject deal with the fact that currently there are no occupational exposure limits for biological agents, as well as the absence of methodologies for detection and estimation of the risk of exposure to agents transmitted through the air. These gaps makes it hard to have a decision and for the implementation of strategies for risk management in hospitals. Another shortcoming is related to the legislation on the subject which is still basic, with defined parameters, not specific, as defined concentrations are acceptable for certain bacteria, but unacceptable to other bacteria, including pathogenic bacteria and resistant ones.

There are gaps in the knowledge of the atmospheric concentration and diversity of microorganisms in health care and on its interface with exposure to occupational diseases. There is also difficulty in demonstrating the role of microbiological contamination of the air and the etiology of certain diseases. However, the authors are unanimous about the economic impact of these diseases in occupational health professionals, particularly associated with high rates of absenteeism and the resulting decrease in productivity and quality of patients care.

4 CONCLUSION

Sources of pollutants, exposure to biological agents is considered a risk evolving and therefore an emerging risk (Corrao *et al.* 2012). The literature review highlighted the importance of knowledge on biological agents as risk factors for the health of the individual and the community, capable of causing serious diseases and epidemics (Srikanth *et al.* 2008). The potential effects of exposure to bioaerosols on human health are diverse (Douwes *et al.* 2003). The role of these microorganisms in air in health care settings are poorly studied (Srikanth *et al.* 2008). Therefore, it is important to evaluate the quality of the air that we breathe, especially in hospital environments (Ekhaise *et al.* 2008). The increased incidence of these diseases indicates the need for further studies (Srikanth *et al.* 2008).

Few studies have quantified the spread of potentially infectious bioaerosols produced by patients in institutions providing health care and health care worker exposure to these particles (Lindsley *et al.* 2012). The absence of specific methodologies validated risk assessment associated with the lack of exposure limit values, hinder the adequate estimation of occupational risk and their effects (Corrao *et al.* 2012). In this sense, it is urgent to conduct further research in order to minimize the gaps, establish better exposure assessment tools, to validate newly developed methods (Douwes *et al.* 2003) and develop more specific regulation on microbiological monitoring in hospital environment (Santos 2008).

The biological risk at work requires a complex approach in the management and evaluation of the wide variety of biological agents, environments and work techniques that can determine the exposures (Corrao *et al.* 2012).

REFERENCES

Akker, I.L.; Hulscher, M.E.; Verheij, T.J; Dalhuisen, J.R.; Delden, J.J.V.; Hak, E. (2011). How to develop a program to increase influenza vaccine uptake among workers in health care settings? *Implement Science*; 6:47.

Boudma, L.; Barbier, F.; Biard, L.; Farèse, M.E.; Le Corre, B.; Macrez, A.; Salomon, L.; Bonnal, C.; Zanker, C.; Najem, C.; Mourvillier, B.; Lucet, J.C.; Régnier, B.; Wolff, M.; Tubach, F. (2012). Personal Decision-Making Criteria Related to seasonal and pandemic A (H1N1) Influenza-vaccinatio acceptance among French healthcare workers. PloS One; 7(7).

Carvalho, M.J.; Pimenta, F.C.; Hayashida, M.; Gir, E.; Silva, A.M.; Barbosa, C.P.; Canini, S.R.M.S.; Santiago, S. (2009). Prevalence of methicilin- resistant and methicilin-susceptible s. aureus in the saliva of health professionals. Clinical Science; 64(4): 295–302.

Chai, S.J.; Mattingly, D.C.; Varma, J.K. (2013). Protecting heatlh care workers from tuberculosis in China: A review of policy and practice in China and the United States. Health Policy and Planning: 28(1): 100–109.

Chan-Yeung, M. (2004). Severe acute respiratory syndrome (SARS) and healthcare workers. International Journal of occupational and environmental health; 10(4): 421–427.

Corrao, C.R.N.; Mazzotta, A.; La Torre, G.; De Giusti, M. (2012). Biological risk and occupational health. Industrial Health; 50(4): 326–337.

Costa, P.A.; Trajman, A.; Carvalho de Queiroz Mello, F.; Goudinho, S.; Silva, M.V.M.A.; Garret, D.; Ruffino-Netto, A.; Kritski, A.L. (2009). Administrative measures for preventing Mycobacterium tuberculosis infection among healthcare workers in a teaching hospital in Rio de Janeiro, Brazil. Journal of Hospital Infection; 72(1): 57–64.

Couto, C.R.; Pannutti, C.S.; Paz, J.P.; Fink, M.C.D.; Machado, A.A.; Marchi, M.; Machado, C.M. (2012). Fighting misconceptions to improve compliance with influenza vaccionation among health care workers: an educational project. PloS One; 7(2).

Cuhadaroglu, C.; Erelel, M.; Tabak, L.; Kilicaslan, Z. (2002). Increased risk of tuberculosis in health care workers: a retrospective survey at a teaching hospital in Istanbul, Turkey. BMC Infectious Diseases; 2(14).

Danyluk, Q.; Hon, C.Y.; Neudorf, M.; Yassi, A.; Bryce, E.; Janssen, B.; Astrakianakis, G. (2011). Health care workers and respiratory protection: is the user seal check a surrogate for respirator fit-testing? J. Occ. Env. Hyg.; 8(5): 267–270.

DGS. Comunicado sobre novo coronavírus. Despacho nº C52.02.v1. Direção Geral da Saúde Acedido em 20 de Maio de 2013, em http://www.dgs.pt/?cr = 24391.

Douwes, J.; Thorne, P.; Pearce, N.; Heederik, D. (2003). Bioaerosol Health Effects and Exposure Assessment: Progress and Prospects. Ann. Occup. Hygiene: 47(3): 187–200.
Dutkiewicz, J. (2004). Occupational biohazard: current issues. Medycyna Pracy: 55(1): 31–40.
Ekhaise, F.O.; Ighosewe, O.U.; Ajakpovi O.D. (2008). Hospital indoor airborne microflora in private and government owned hospitals in Benin, Nigeria. J. Med. Sc; 3(1): 19–23.
Façanha, M.C.; Monroe, A.C.P. (2006). Occupational and nosocomial transmission of varicela. The Brazilian Journal of Infectious Diseases; 10(2): 156–158.
Fedeli, U.; Zanetti, C.; Saia, B. (2002). Susceptibility of healthcare workers to measles, mumps rubella and varicela. Jornaul Hospital Infection; 51(2): 133–135.
Gehanno, J.F.; Louvel, A.; Rysanek, E.; Pestel-Caron, M.; Nouvellon, M.; Kornabis, N.; Touche, S.; Ripault, B.; Buisson-Valles, I.; Sobaszek, A. (2009). Biological risk assessment among healthcare workers. Journal Archives des Maladies Professionnelles et de l'Environnement: 70; 36–42.
Gutiérrez, M.N.; Sáenz González M.C. (2009). Vaccination and postexposure prophylaxis in health-care workers. Revista española de quimioterapia; 22(4): 190–200.
Haagsma, J.A.; Tariq, L.; Heederik, D.J.; Havelaar, A.H. (2012). Infectious Disease Risks Associated With Occupational Exposure. Occup. Envir. Medicine; 69(2): 140–146.
Herr, C.E.W.; Eikmann, T.; Heinzow, B.; Wiesmuller, G.A. (2012). Environmental medical relevance of molds in living environment. Allergologie: 35(12): 611–619.
Holmes, C.N.; Iglar, K.T.; Mcdowell, B.J.; Glazier, R.H. (2004). Predictive value of a self-reported history of varicela infection in determining immunity in adults. Canadian Medical Association or its Licensors; 171.
Hui, D.S.C.; Chan, P.K.S. (2010). Severe acute respiratory syndrome and coronavirus. Infectious Diseases Clin. North American; 24(3): 619–638.
Klug, K., Jackel, U. (2012). Detection of exposure to biological agentes over the working shift by total cell count analyses. Gefahrstoffe Reinhaltung der Luft; 72 (9): 373–378.
Kok, G.; Essen, G.A.V.; Wicker, S.; Llupià, A.; Mena, G.; Correia, R.; Ruiter, R.A.C. (2011). Planing for influenza vaccinationa in health care workers: na intervention mapping approach. Vaccine; 39(47): 8512–8519.
Laraqui, C.E.H.; Laraqui, O.; El Kabouss, Y.; Moulin, M.; Mahmal, A.; Faucon, D.; Verger, C.; Laraqui, S.; Caubet, A. (2004). Assessement of the risk of tuberculosis among health care personnel. Cahiers Sante; 14 (3): 167–171.
Lindsley, W.G.; King, W.P.; Thewlis, R.E.; Reynolds, J.S.; Panday, K.; Cao, G.; Szalajda, J.V. (2012). Dispersion and exposure to a cough-generated aerosol in a simulated medical examination room. J. Occ. Env. Hyg.; 9(12): 681–690.
Menzies, D.; Joshi, R.; Pai, M. (2007). Risk of tuberculosis infection and disease associated with work in health care settings. The intern. J. Tuberc. and lung Dis. 11(6): 593–605.
Moreira, M.O. (2010). Medidas de precaução padrão no ambiente hospitalar adotadas por alunos do curso de fisioterapia. Curso Fisioterapia. UFPB. Brasil.
Moura, J.; Gir, E.; Rosa, J.; Rodrigues, F.; Cruz, E.; Oliveira, A. Pimenta, F. (2010). Resistência à mupirocina entre isolados de Staphylococcus aureus de profissionais de enfermagem. Acta Paul Enfermagem; 23(3): 399–403.
Naidoo, A.; Naidoo, S.; Gathiram, P.; Lallo, U. (2013). Tuberculosis in medical doctors – A study of personal experiences and attitudes. South Afr. Med. Journal; 103(3): 176–180.
Nishiyama, A.; Wakasugi, N.; Kirikae, T.; Quy, T.; Ha le D.; Ban V.V.; Long, H.T.; Keicho, N.; Sasazuki, T.; Kuratsuji, T. (2008). Risk factors for SARS infection within hospitals in Hanoi, Vietnam. Jap. J. infect diseases; 61(5): 388–390.
Perdelli, F.; Cristina, M.L.; Sartini, M.M; Spagnolo, A.M; Dallera, M.; Lombardi, R.; Grimaldi, M.; Orlando, P. (2012). Fungal contamination in hospital environments. Infection control and hospital epidemiology; 27(1): 44–47.
Peres, D.; Pina, E.; Cardoso, M.F. (2011). Methicillin-Resistant Staphylococcus aureus (MRSA) in a Portuguese hospital and its risk perception by health care professionals. Revista Portuguesa de Saúde Pública; 29(2): 132–139.
Pietrangeli, B. (2008). Biological risk in workplaces: research priorities for risk assessment. Prevention Today. Janary-March.
Quadros, M.E. (2008). Qualidade do Ar em Ambientes Internos Hospitalares: Parâmetros Físico-Químicos e Microbiológicos. Dissertação de Pós Graduação em Engenharia Ambiental. UFSC. Brasil.
Reijula, K.; Tuomi, T. (2003). Mycotoxins of aspergilla; exposure and health effects. Frontiers in Bioscience: 232–235.
Santos, A.C.M. (2008). Microbiologia do ar: monitorização do ar em ambiente hospitalar. Diss MSc Byol. UAv Portugal.
Santos, A.L.; Santos, O.D.; Freitas, C.C.; Ferreira, B.L.A.; Afonso, I.F.; Rodrigues, C.R.; Castro, H.C. (2007). Staphylococcus aureus: visitando uma cepa de importância hospitalar. J. Br. Patol. Med. Laboratorial: 43(6): 413–423.
Sepkowitz, K.A.; Eisenberg, L. (2005). Occupational Deaths among Healthcare Workers. Journal List. Emerging Infectious Diseases: 11(7);1003–1008.
Severo, K.G:P; Oliveira, J.S.; Carneiro, M.; Valim, A.R.M.; Krummenauer, E.C.; Possuelo, L.G. (2011). Latent tuberculosis in nursing professionals of a Brazilian hospital. Journal of Occupational Medicine and Toxicology; 6–15.
Simone, B.; Carrillo, P.; Lopalco, P.L. (2012). Healthcare workers' role in keeping MMR vaccination uptake high in Europe: a review of evidence. Eurosurveillance; 17(26).
Singh, M.P.; Diddi, K.; Dogra, S.; Suri, V.; Varma, S.; Ratho, R.K. (2010). Institutional outbreak of rubella in healthcare center in Chandigarh, India. J. Med. Virol.; 82(2): 341–344.
Soares, I.C.M (2009). Aeromicologia Hospitalar. Dissertação de Mestrado de Biologia Molecular e Celular. UAv.

Srikanth, P.; Sudharsanam, S.; Steinberg, R. (2008). Bioaerols in indoor environment: composition, health effects and analysis. Indian Journal of Medical Microbiology; 26(4): 302–312.

Tablan, O.C.; Anderson, L.J.; Besser, R.; Bridges, C.; Hajjeh, R. (2003). Guidelines for preventing health-care-associated pneumonia, 2003. Acedido a 19 de Maio de 2013 em http://www.vap.kchealthcare.com/media/138482/cdc%20 guidelnes%20for%20preventing%20healthcare%20associated%20pneumonia.pdf.

Tafuri, S.; Martinelli, D.; Caputi, G.; Germinario, C.; Prato, R. (2009a). Na audito f TB prevention on Italian health care workers. J. Prev. Medicine and Hygiene; 50(2): 127–128.

Tafuri, S.; Germinario, C.; Rollo, M.; Prato, R. (2009b). Occupational Risk from measles in healthcare personnel: a case report. Jornaul Occupational Health: 97–99.

Tang, J.W.; Li, Y.; Eames, I.; Chan, P.K.S.; Ridgwas, G.L. (2006). Factors involved in the aerosol transmission of infection and control of ventilation in healthcare premises. Journal of Hospital Infection: 64: 100–114.

Tablei-Taher, M.; Noori, M.; Shamshiri, A.R; Barati, M. (2010). Varicella zoster antibodies among health care workers in a university hospital., Teheran, Iran.

Thierry, S.; Alsibai, S. (2008). Na outbreak of measles in Reims, Eastern France, January-March 2008—A preliminar report. Euroisurbeillance: 13: 1–3.

Tran, K.; Cimon, K.; Severn, M.; Pessoa-Silva, C.L.; Conly, J. (2012). Aerosol generating procedures and risk of transmission of acute respiratory infections to healthcare workers: a systematic review. PloS One; 7(4): e 35797.

Trajman, A.; Menzies, D. (2010). Occupational respiratory infections. Current opinion in pulmonary medicine; 16(3).

Vaquero, M.; Gómez, P.; Romero, M.; Casal, M.J. (2003). Investigation of biological risk in mycobacteriology laboratories: a multicentre study. The international journal of tuberculosis and lung disease; 7(9): 879–885.

Wong K.C.; Leung K.S. (2004). Transmission and prevention of occupational infections in orthopaedic surgeons. The J. of Bone and Joint Surgery Amer. 2004;86-A:1065–1076.

Yen, M.Y.; Lin, Y.E.; Su, I.J.; Huang, F.Y.; Ho, M.S.; Chang, S.C.; Tan, K.H.; Chen, K.T.; Chang, H.; Liu, Y.C.; Loh, C.H.; Wang, L.S.; Lee, C.H. (2006). Using an integrated infection control strategy during outbreak control to minimize infection of severe acute respiratory syndrome among healthcare workers.

Occupational Safety and Hygiene II – Arezes et al. (eds)
 ISBN 978-1-138-00144-2

Analysis of the hearing risk and implantation of environmental improvements in workers of the industry of paper and cellulose and introduction of acoustic improvements

D.L. Mattos, E.A.D. Merino, A.C.C.S. Pinto, K.C.G. Orzatto & V.M. Camargo
Federal University of Santa Catarina, Florianópolis, Santa Catarina, Brazil

C.A. Maciel
Contestado University, Brazil

V. Federle & A.R.P. Moro
Federal University of Santa Catarina, Florianópolis, Santa Catarina, Brazil

ABSTRACT: The noise pollution is today, after the pollution of air and water, the environmental problem that affects the largest number of people. In this context, this study aims to analyze the level of noise coming from the sectors of wood yard (chipper) and Utilities (turbo pump and turbo generator) of a Brazilian company of pulp and paper, located in the southern region, as well as deploy improvements to control the risks, and analyze the situation after the measures deployed, in order to provide acoustic comfort to workers and suit the NR 15, legislation that deals with the limits of exposure prevailing in the country. The noise to which workers are exposed in the pulp and paper industry which was studied are important environmental risk factors. It is known the losses entailed by exposure, as well as its consequences on health in the long and short term.

1 INTRODUCTION

The health of the industry worker is subject in vogue nowadays. The risk exposure can cause occupational diseases and accidents. Noise pollution is one of the most recurrent risks and has great impact to human health.

The noise pollution is today, after the pollution of air and water, the environmental problem that affects the largest number of people. It is perceptible increase of annoyance due to noise and the damage that this has caused to man in his working 8environment and/or environmental. The speed of the manifestation of the damage depends, in addition to the level of noise emissions, factors such as: the time of exposure; the general health conditions; the age, etc. (LACERDA et al. 2005a).

Noise exposure occurs not only in the workplace. In everyday life we are exposed to various noise sources such as cars, television, telephones and others.

The noise is a harmful agent and permanent in workplaces, as much as in our lives, it is even present in our leisure and in urban environments. This propagation of noise in our environment generates a great concern being that the auditory damages are irreversible, and the exposure cause other damage such as: organic; physiological and psycho emotional aspects which results in the reduction of health and quality of life (GONÇALVES & ADISSI, 2008).

According to Granziera (2009) "the production of noise and its impact on health and environmental balance are not easy standardization, much less regarding the measures needed to curb abuses and prevent damage".

Any change in the physical, chemical and biological properties of the environment, caused by any form of matter or energy resulting from human activities that, directly or indirectly, affect the health, safety and well-being of the population; the social and economic activities; the biota, the aesthetic conditions and health of the environment and the quality of environmental resources.

The definition of pollution is linked, in Law 6938/81, which regulates the national policy for the environment in its Article 3 (2) subparagraph III:

III—pollution, the degradation of environmental quality resulting from activities that directly or indirectly:

a. Affect the health, safety and well-being of the population.

b. Create adverse conditions the social and economic activities;
c. Adversely affect the biota;
d. Affect the aesthetic conditions or health of the environment;
e. Launch materials or energy in disagreement with the established environmental standards.

The consequences of noise exposure in workers' health were widely studied and documented (Braga et al. 2002a), for example, listed the main effects of noise on human health are:

a. Hearing Loss: temporary when it is exposed to excessive and permanent noise, when there is a loss of hearing usually caused by prolonged exposure and high-frequency sounds, this last one is irreversible.
b. Interference in speech: this one is affected by hearing loss and sounds that compete for the attention of the listener.
c. sleep disturbance: occurs in environments with noise above 35 dB.
d. Stress and Hypertension: the noise can dilate pupils, cause muscle tension, and increase of heart beats and blood pressure.
e. Stress and Hypertension: the noise can dilate pupils, cause muscle tension, and increase of heart beats and blood pressure.
f. Other symptoms are discomfort, disturbances in the work environment with loss of income.

In recent decades, scientific researches alert to the fact that the man seems to be more accustomed with the noise. In the survey conducted by Yorg & Zannin (2003), for example, when individuals were asked if they felt uncomfortable or harassed by the levels of existing noise in their work environment and/or in their urban environment, the frequent response was: "...we're already accustomed to these noises, with the time everyone gets used" (LACERDA et al, 2005b).

According to Bistafa (2011a, p. 137) "the hearing loss entails significant changes in person that interfere with their quality of life".

The author also reports: "Outside of the work environment, the noise can also cause disorders that interfere in activities such as sleep, conversation, relaxation, concentration, which cause psychological impacts and may harm the mental health" (BISTAFA, 2011b).

1.1 *Solutions to reduce the spread of this noise*

For Brevigliero & Possebon & Spinelli (2006), the measures of control of high noise levels present in certain work environments should take into consideration first interventions in source, subsequently in the trajectory, and therefore personal interventions.

According to Braga et al. (2002b) it's possible make a noise control straight from the source, the path or the receiver, the first involves activities such as acoustic insulation, flame arresters, main among others. Already the route control is done through barriers between the receiver and the source. In the case of the receiver, the control can be done through the limitation of the source, thus reducing their exposure to the noise, and he may also use protective gear.

2 METHODOLOGY

This is a case study carried out in the company Celulose Irani S. A, in a paper unit that has an installed capacity of approximately 17,000 tons per month and produces several lines of roles for flexible and rigid packaging in four paper machines (MP), being MP01, MP02, MP04 and MP05.

In the internal market CELULOSE IRANI operates mainly in the states of São Paulo, Santa Catarina, Paraná, Rio Grande do Sul, Minas Gerais, Goiás, Mato Grosso, Mato Grosso do Sul, Distrito Federal and Rio de Janeiro. In the foreign market it is present in North and South America, Africa, Europe, Middle East and Asia.

The apparatus used for the completion of this stage was a sound pressure level measurer, dosimeter of the brand INSTRUTHERM, Model DOS-500, operating with compensation circuit (A) and slow response (SLOW), at the ear level of the worker and in multiple jobs, facing greater source of noise, for measurements of noise levels continuous or intermittent. It was also used a calibrator of sound pressure level of the brand Instrutherm DOS-500 sound level meter Model OF-500, operating at a frequency of 1000 Hz and level of calibration at 114 dB (A) calibrated in 01/08/13. The measurement was made twice, one before and one after the improvement. Data was analyzed after all dosages made.

The research of the potential risk of an area is done by lifting the sound spectrum of the site. The audible aspect is a curve which provides the variation of sound level with the frequency.

In this context, this study aims to analyze the level of noise coming from the sectors of wood yard (chipper) and Utilities (turbo pump and turbo generator) of a Brazilian company of pulp and paper, located in the southern region, as well as deploy improvements to control the risks, and analyze the situation after the measures deployed, in order to provide acoustic comfort to workers and suit the NR 15, legislation that deals with the limits of exposure prevailing in the country.

Table 1. Tolerance limits for continuous or intermittent noise according to the NR 15.

Noise level dB (A)	Maximum daily allowable exposure
85	8 hours
86	7 hours
87	6 hours
88	5 hours
89	4 hours and 30 minutes
90	4 hours
91	3 hours and 30 minutes
92	3 hours
93	2 hours and 40 minutes
94	2 hours and 15 minutes
95	2 hours
96	1 hour and 45 minutes
98	1 hour and 15 minutes
100	1 hour
102	45 minutes
104	35 minutes
105	30 minutes
106	25 minutes
108	20 minutos
110	15 minutes
112	10 minutes
114	8 minutes
115	7 minutes

2.1 *Stages of the research*

- Identify machines and equipment with high index of noise.
- Make the measurement of this noise with a dosimeter for monitoring and controlling of the noise in these machines and equipment.
- Presentation of the improvement plans for the reduction of the level of the sound in these equipments.
- Monitoring in the installation and control measures of these noises.

The NR 15 that is who legislates on the tolerance of noise exposure in Brazil, was taken as reference for the study. The table of NR 15 was used not only as a reference in the measurement of the noise, but also supported the decision-making in the implementation of improvements. The table defines the tolerance limits for continuous or intermittent noise. It follows in table 1.

3 RESULTS

3.1 *Chipper Bruno*

This is the operation room of the chipper, functioning 24 hours. In this room the noise is high because large part of this structure is closed with glass of low thickness and with the simple aluminum door.

Figure 1. Room without acoustic insulation in Bruno chipper. *Source*: Survey data, 2013.

Figure 2. Acoustic insolation in the room of chipper Bruno. *Source*: Survey data, 2013.

The proposed improvement was to replace the window with a greater thickness and installation of a door with double layer on the aluminum. With this improvement the noise was reduced from 88 decibels to 82 decibels thus improving the work environment in this location.

3.2 *Utilities sector*

The Boiler HPB is used for power generation through the burning of biomass. At this point it is observed in figure 3, that there existed a piping where it was released the high-pressure steam causing a noise level reaching 90 decibels.

In figure 4 it is observed an installed attenuator of noise at the output of the feeding turbo pump of the Boiler HPB, which consequently reduced the noise up to 82 decibels.

Figure 3. Vent piping from turbo pump HPB. *Source*: Survey data, 2013.

Figure 4. Noise attenuator installed. *Source*: Survey data, 2013.

Figure 5. Piping steam output (before the improvement). *Source*: Survey data, 2013.

Figure 6. Noise attenuator—installed. *Source*: Survey data, 2013.

Figure 7. Dual port operation room. *Source*: Survey data, 2013.

Table 2. Record levels of noise before and after the improvements.

Local/Sector	Noise before	Noise after
Chipper Bruno	88 dB (A)	82 dB (A)
Turbo pump	90 dB (A)	82 dB (A)
Turbo generator	98,6 dB (A)	84,3 dB (A)

Also in the Boiler HPB, in the turbo-generator, an attenuator of noise was installed at the steam output, because when it increases the area of fluids (steam) in the output, decreases the speed and therefore the noise, where before there was a noise of 98.6 dB, with the installation of the equipment it was reduced to 84.3 decibels.

4 DISCUSSION

Of the seven locations within the two sectors where the initial measurement was held, it was found high noise levels, including above 85 decibels.

In the chipper Bruno, the noise coming from the wood passing by saws and impact noise are important source of discomfort. After the improvement, 88 decibels dropped to 82 decibels, staying within the established standard.

At the turbo pump sector HPB, in the same way, was registered 90 decibels of noise. After the intervention, the values dropped to 82 decibels, softening the risks to which workers are exposed in the workplace. Yet at the turbo generator, the measurement before the intervention registered 98.6 decibels. After the improvement, it registered 84.3 decibels.

The results pose all sectors in accordance with the NR 15, which establishes the maximum limits in the work environment, in Brazil, and provide acoustic comfort, improving the health and well-being of those in the realization of labor activities daily.

The table 2 shows the results measured and recorded before and after the improvements.

5 CONCLUSION

The noise to which workers are exposed in the pulp and paper industry which was studied, is an important environmental risk factor. It is known the losses entailed by exposure, as well as its consequences on health in the long and short term.

The adequacy of the sectors is an important measure to mitigate such risks. The productivity and health of the worker suffer direct impact by these measures, which improve the day-to-day life in the work environment, improving both aspects, and generating satisfaction for the users of the jobs in question.

Actions as replacing glass windows, implanting more robust structures and implementation of attenuators are ways to improve the quality of life of workers and reduce their exposure to risk.

REFERENCES

Andrade, D.R., Finkler, C., Closs, M., Marini, A.L., Capp, E. 1998. *Efeitos do ruído industrial no organismo.* Revista Pró-Fono, 10(1), p. 17–20.

Braga, B. et al. 2002. *Introdução à engenharia ambiental.* Prentice Hall. São Paulo.

Bistafa, S.R. 2011. *Acústica Aplicada ao Controle de Ruído.* Blucher, 2ª edição. São Paulo.

Breviglieiro, E, Possebon, J, Spinelli, R. 2006. *Higiene Ocupacional – Agentes Biológicos, Químicos e Físicos.* Editora Senac. 3ª edição. São Paulo.

Gonçalves, V.S.B, Adissi, P.J. 2008. *Identificação dos níveis de pressão sonora em Shopping Center na cidade de João Pessoa. Universidade tecnológica Federal do Paraná* – UTFPR Campus Ponta Grossa—Paraná – Brasil ISSN 1808–0448/04(03): p. 146–159. Available in: <http://revistas.utfpr.edu.br/pg/index.php/revistagi/article/view/13/10>. Access in: 30 May 2013.

Granziera, M.L. 2009. *Direito ambiental.* Atlas. São Paulo.

Lacerda, A.B.M. et al. 2005. *Ambiente Urbano e Percepção da Poluição Sonora.* Ambiente & Sociedade–Vol. VIII nº. 2 jul./dez. Available in: <http://www.scielo.br/pdf/asoc/v8n2/28606.pdf>. Access in: 13 may 2013.

Lei Nº 6938 de 31 de dezembro de 1981. *Política nacional de meio ambiente.* Available in: <http://www.planalto.gov.br/Ccivil_03/Leis/L6938.htm>

Norma regulamentadora. NR Nº 15. Atividade e operações insalubres. Available in: <http://portal.mte.gov.br/legislacao/norma-regulamentadora-n-15-1.htm>. Accses in: May, 30, 2013. em: <http://www.mma.gov.br/port/conama/>. Access in: June, 10, 2013.

Yorg, C.M, Zannin, P.H.T. 2003. *Noise evaluation in the Itaipu Binacional Hydroelectric Power.* 27º International Congress on Occupational Health. Iguassu Falls, Brazil.

Occupational Safety and Hygiene II – Arezes et al. (eds)
© 2014 Taylor & Francis Group, London, ISBN 978-1-138-00144-2

Ergonomic evaluation of the cluster during teat cup attachment

C.C. Oliveira
Federal University of Santa Catarina, Florianópolis, Santa Catarina, Brazil

A.A. Bazán
State University of Londrina, Londrina, Paraná, Brazil

L.A. Gontijo & A.R.P. Moro
Federal University of Santa Catarina, Florianópolis, Santa Catarina, Brazil

L. Ulbricht
Federal Technological University of Paraná, Curitiba, Paraná, Brazil

ABSTRACT: This article aims to present the functional features and evaluate ergonomic aspects of the teat cups used in mechanical milking of the Co-Op of Municipalities of the Region of Campo Mourão Region (COMCAM), Brazil. The methodological procedure adopted for this evaluation consisted of two stages. The first identified the socio-demographic profile of the sample (31 female milkers) using a questionnaire and then sought data concerning problems with pain and/or discomfort by applying the Standard Nordic Questionnaire. In the second stage, activity analysis identified ergonomic risks in the execution of the activity and their correlation to the set of teat cups. From the activity analysis and the ergonomic assessment results, recommendations are given that will contribute to establishing parameters towards designing a better set of mechanical milking teat cups.

1 INTRODUCTION

The design of equipment for use in the production process, i.e., industry, should consider parameters relating to the health and safety of its operator, in addition to technical performance requirements. Such health and safety parameters must be defined through ergonomic evaluation (Estivalet et al., 2004).

Concern for milker safety and comfort is increasingly important (Amaro, 2012). This fact can be seen in the increasing demand for special milking parlor floors (soft and non-skid), shoes with specific anatomical insoles, and disinfectants that do not harm the milker health.

However, in general the conception surrounding material handling, equipment, and the tools used during milking has adequately adjusted to meet the demands of worker comfort in their work, but meeting production needs instead. In fact these factors are closely associated with milking speed, which in turn affects total production and the health of the cows' mammary glands, and principally affects the quality of milk composition (Amaro, 2012).

In milking mechanically, one of the tools frequently used involves a set of liners. Studies have shown that inadequacies in the collection cup volume, weight of the unit, and set design expose their female employees, called milkmaids in this study, to various constraints, such as poor posture.

Work load assessment studies regarding the milking parlor operator have been carried out on farms (Pinzke et al., 2001; Kauke et al., 2009). This paper presents a consecutive experimental study regarding the effects of work activity height and manipulated weights while attaching a milking cluster.

The cluster as seen in Swedish studies like those by Stål et al. (2003) and Stål (2000) reported a significant load reduction for the biceps and flexor muscles when using a support arm. In the region of Florianópolis, Brazil, the cluster does not adapt well to female hands. Such research pointed out difficulties in handling the milking equipment because it is very long, making it necessary to use their fingertips to be able to keep this equipment safe and even turn it to connect properly (Ulbricht, 2003).

Given this context, this study aims to present the functional characteristics and assess ergonomic aspects of the set of mechanical milking liners in the Co-Op of Municipalities of the Campo Mourão Region (COMCAM).

2 METHOD

The proposed methodology for this study of the set of mechanical milking liners is characterized as qualitative and analytical descriptive. It counts upon survey data from theoretical references, papers, magazines, and world wide web searches. This research involves milking photographs during connection of the set of teat cups to the animals' teats, seeking to identify risk factors in performing the milk extraction activity, specifically how the set of liners are handled to engage the teat cups to dairy cattle in order to evaluate the equipment's ergonomic design. Thus, an authorized term of informed consent was provided and signed by users.

The ergonomic evaluation consisted of Activity Analysis. Activity Analysis was based on interviews and observations during local visits, as well as video recording and photograph analysis.

The research herein is characterized as quantitative and exploratory through its multiple case studies.

The study was conducted involving 31 milkmaids associated with the Regional Milk Producers Cooperative—COPROLEITE, located in Campo Mourão, PR, Brazil, outlining the area of the Community of Municipalities of the Campo Mourão Region—COMCAM.

The data was collected through the application of questionnaires divided into two parts:

a. General and Occupational Data
This questionnaire took basic information, in the form of 22 questions to evaluate the milker profiles.

b. Identification of Musculoskeletal Symptoms
In order to evaluate musculoskeletal symptoms, the Nordic Standard Questionnaire (Kuorinka et al., 1987) was used.

Ethical care was taken concerning the questionnaire respondents and their personal information kept confidential. The milkmaids were identified by numbered codes (1 to 31).

For better data accuracy extracted from each questionnaire, the structured interview method in the form of a questionnaire was selected. This was due to the typically low levels of education in the population sample, evident from the researchers themselves needing to fill out the questionnaires for those researched.

3 RESULTS AND DISCUSSION

3.1 *Milker profile and milking activity*

Thirty-one (31) milkers were interviewed (sample). The average age of the sample was 45 years old, spanning from 25 years to 78 years (median 45 years). Their average weight and height was 66 pounds and 1.59 meters, respectively.

The education level of most respondents was low, being that no milkmaid had a university degree and only 13.95% had completed high school. Regarding the type of employment, most with 88.37% (28) owned properties.

Referring to area property, most of the properties were small, because 46.63% of the respondents had 1–5 acres. The greater proportion of the sample at 21.47% have 6–10 acre properties.

The interviewees were asked about job satisfaction. Most milkmaids 76.74% reported to like much their work.

The interviewees started their activity with a mean age of 22 years (median 20 years), and the youngest started at six years of age, while the oldest began at 45 years.

On average, the milkmaids have been in this occupation for 19.88 years (median 14 years). The minimum time of work found was one year and the maximum 70 years.

The daily workload was observed on average to be 11 hours. Of this time, an average of 3.40 hours are exclusively dedicated to milking (subtask of extracting milk) and 7.60 hours are meant for activities involving dairy cattle management (feeding, vaccination, insemination, among others) and other activities. The weekly workload including all activities on the property corresponded to an average of 82.74 hours per week, proving to be higher than urban occupations, since milkers have no weekly day off.

Of the 31 interviewed milkmaids, 83.87% only performed dairy cattle activities and 16.12% (five milkmaids) performed other activities, e.g., agriculture and raising silkworms, chicken, fruit, vegetables, among others.

On average, 17 animals are milked at each milking period, with a minimum of four animals milked and at most 44.

3.2 *Existence of pain and/or musculoskeletal discomfort*

Of the 31 women interviewed, 25 had pain and discomfort (80.64%).

A total of 80.64% (25) of the workers were identified with complaints of pain and/or musculoskeletal discomfort. In most cases, such symptoms presented themselves as multifocal. In other words, 17 (54.84%) workers indicated the presence of pain in different body segments simultaneously.

With respect to the temporal aspect, Table 1 presents the body regions with their respective percentages of pain (over the previous 12 months and in the previous seven days) plus departures resulting from pain, as interviews performed through the implementation of Nordic Standard Questionnaire, in which 25 milkmaids complained of pain and/or discomfort.

Table 1. Symptoms of musculoskeletal pain and/or discomfort reported by the COMCAM milkers over different body regions.

Body regions	N° of answers (12 months)	12 months (%)	Removals (%)	7 days (%)
Shoulders	17	68	16	32
Arms	09	36	8	24
Elbows	03	12	4	8
Hands/ Wrists	07	28	12	12
Neck	13	52	8	20
Thoracic	05	20	4	16
Lumbar	18	72	0	24
Hips	02	8	4	4
Legs	07	28	12	8
Knees	08	32	8	16
Ankles/ feet	06	24	8	8

Table 1 aggregates the segments in body regions over the previous 12 months, Upper Limbs (UL), Spine (cervical, thoracic and lumbar region), Lower Limbs (LL) including hip, and multifocal pain including the three segments (upper limbs, lower limbs and the spine).

The highest percentage found in the responses of pain occurs in the spinal region with 67.74% (21 respondents), followed by 64.52% (20 responses) complaints in the region of the upper limbs, and 35.48% in the lower limbs. Over half 54.84% had multifocal pain, meaning pain in three body segments.

This higher prevalence of pain in the spinal region is probably due to surveys, cargo shipments, execution of work with an embarrassing posture (trunk inclination) that occurred during the course of the activity. This requires a larger effort from the spinal region and may consequently generate musculoskeletal overloads.

To place the set of liners, the milkmaid is required to assume a static posture in the shoulder and arm region. This conduct overburdens the shoulder joint that provides the stability necessary to execute the work, resulting in the prevalence of pain in the upper limbs.

Regarding the multifocal aspect of pain among the milkmaids over the 12 months prior to the survey, this study sought to establish data specifically on upper extremities (UE), being that pain from these regions may radiate to other segments of the upper limbs. This radiation can present itself as pain from the cervical (neck) to the upper limbs (Borges & Ximenes, 1997).

Figure 1 shows the frequency and segments most affected by multifocal pain.

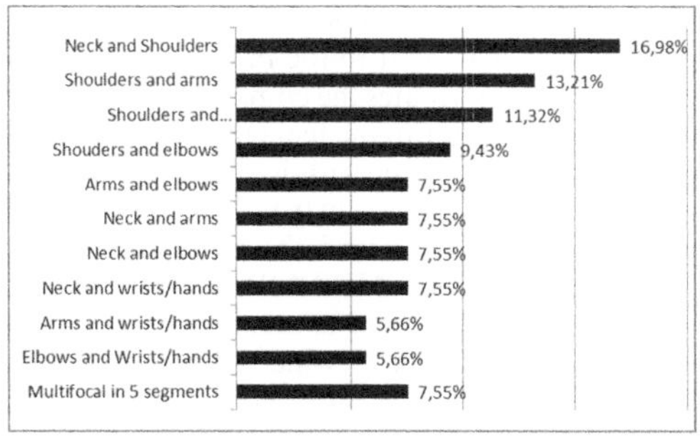

Figure 1. Frequency distribution of multifocal aspect of pain and/or discomfort in upper extremities (UE).

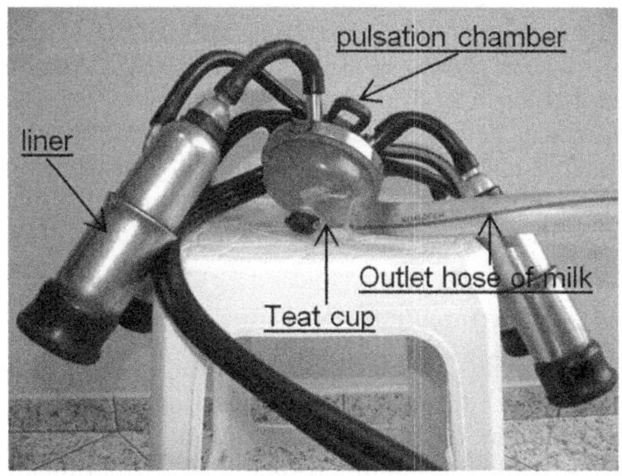

Figure 2. Set of teat cups.

It could be observed in the answers, 16.98%, that the prevalence of pain among the milkmaids showed symptoms in the cervical (neck) segments with irradiation to the shoulders, as well as 13.21% (seven responses) with respect to the shoulders and arms, and 11.32% (six responses) concerning the shoulders and wrists/hands, respectively.

3.3 *Functional analysis of the set of liners*

The term functional refers to the function or performance of the product. In the product, the function characterizes use and utility features. Functional analysis is the recognition and understanding of the characteristics of product use, including ergonomics (Bertoncello, 2001).

The part of milking that is deeply linked to cattle is the teat cups, so they should be anatomically designed and constructed of suitable material, not only to provide comfort to the animal, but also to the workers who handle the equipment.

The set of liners comprises: a cup having a pulsation chamber, four teat cups, a hose connected to the vacuum pump, and an outlet hose to the milk. These can be seen in detail in Figure 2.

The teat cup has the function of collecting milk from the teat, leading it to the milk outlet hose, in other words it distributes the command of vacuum/air from the pulsator to the teat cups. The teat cups are composed of the pulsation chamber, which is the space limited by the inner wall of the teat cup made of stainless steel and the outer side of the rubber from the cup (EMBRAPA, 2007).

Inside the liner there is a vacuum, because it is connected to the collector which acts as a vacuum distributor, through the hose connected between the liner set and vacuum pump controlled by the pulsator. The air entering the chamber must occur quickly, and will depend on the proper functioning of the pulsator (EMBRAPA, 2007).

The teat cups consist of an outside body and an adjusted inner tube which acts as a lining, called a squeezer (Netto et al., 2006). The teat cups are made of rubber and have the function to keep the liner attached to the teats, so the vacuum can open the sphincters, performing the milking, in addition to promoting the massage of the udder and transport the milk to the collector (EMBRAPA, 2007).

3.4 *Activity analysis*

The activity analysis presented here was conducted with respect to the connecting step of the set of liners during milk extraction. The milking machine is based on the principle of extracting air from the teat cups over the teat of the cow through the vacuum pump through which the milk will be extracted via suction (Netto et al., 2006).

The process of attaching a conventional milking cluster includes the worker grabbing the milking system with one hand, holding it beneath the udder while all short liners are bent to avoid letting air in. The other hand then attaches each cup to successive teats. The operation consists of opening the vacuum to milk and engaging the teats of the animal.

In milking, during teat engagement, posture is static and the milkmaid holds a weight of approximately 3.5 kg (weight of all the liners) with arms high above 90°. In the dairy property analyzed, the height of the parapet of the gap is higher than recommended, it is observed in Figure 3 that the milkmaid posture presents arms elevated above 90°.

Another situation that causes erroneous posture involves the protection fixed bars, which are the "herringbone" type at the property researched, which lies approximately in the region of the eyes of the milkers, hindering visualization of the udder by the milkmaid, forcing her to adopt a tilt and neck flexion and anterior flexion of the shoulder to see the animals' tits, i.e., the work field (Figure 3).

While it is waited the decline and withdrawal of the milk, the milkmaid starts to prepare the next animal to be milked (tie and sanitize the animal) and palpating the udders of the cattle to see if it is empty, or hold the liners, to then, disconnect them.

Regarding the dimensioning and structure of the set of teat cups is noted that the size of the cup liners and the fact that the milk outlet hose being connected at the bottom of this cup hinder the handle. These factors, combined to the weight of the product, cause difficulties to the task of connecting the liners to the animals. Figure 4 demonstrate the handling of the set of teat cups and their respective movements to link this device to the animals' udders.

Figure 4 shows that even women of the 5% percentile have difficulty in finding a comfortable and safe position to hold the teat cups. As this part of the liner has a diameter of 11.5 cm (at the top of which fits the pulse camera), they become greater than the length of the palm of 95% of women, which is 10.7 cm (Tanabe et al., 2010). This indicates that the product requires that the milkers,

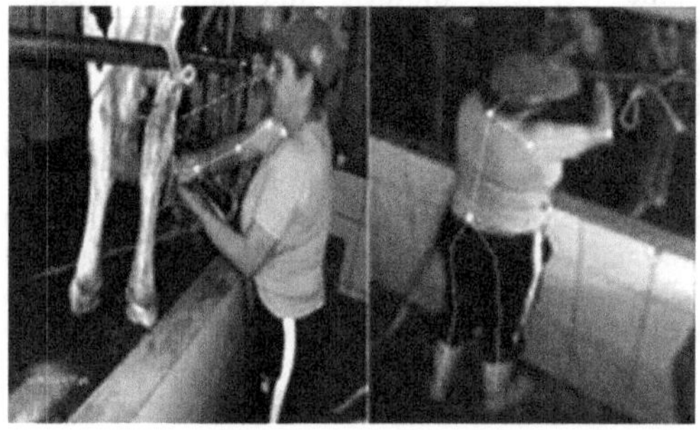

Figure 3. Postures adopted in performing of the activity in dairy property.

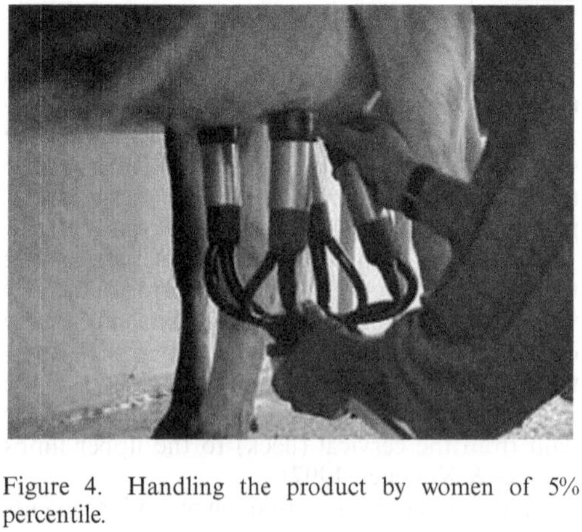

Figure 4. Handling the product by women of 5% percentile.

even those from the higher percentile, use the force of their fingers to hold the set, since support from the palm alone is not enough.

The negative aspects of the current set of liners are as follows:

- Due to the weight of the set of teat cups combining with repeating the same movement (connecting the teat cups to the animal's udders) for an extended period of time can result in muscle damage and wear of the joints and nerves, causing pain and inflammation and even right upper extremity;

Most milkers experience difficulty in holding the equipment because of its long circumference, making use of their fingertips to be able to keep this equipment safe and also turn it up for connection, which can result in injuries resulting from occupational diseases.

4 RECOMMENDATIONS

The analysis of the negative aspects of the set of liners allows for the following recommendations to be considered in establishing design parameters related to postural aspects of work, such as:

- Reduce the circumference of the cup according to the demographic data for women's hands, requiring further study and investigation concerning the ideal circumference of the cup;
- Study other possibilities of lighter materials to reduce the weight of the set;
- Examine the possibility of a new position for the outlet hose of the milk, aiming to facilitate the grip and handling of the product.

5 CONCLUSIONS

According to literature review, the set of liners is a resource widely used by milkmaids who own property with milking machine, but there are no published scientific studies that evaluate this working tool.

This article has shown that upper extremities (UE), can also affect rural workers while outlining a worrysome scenario with respect to the incidence of UE in the area of milk production in COMCAM. This is because 80.64% of milkmaids have this disease that affects workers in their most productive phase, is irreversible in its advanced cases, and is difficult to diagnose.

As noted, there is also the prevalence of pain in upper limbs, because 64.52% of milkmaid complained of pain and/or discomfort, which proves the chances of this working class chain the prevalence of musculoskeletal symptons probably because of the design of the set of liners associated with the required awkward postures during milking. Critical factors related to milkmaid postures arising from the design of the set of liners and the demands from the activity were identified.

This study pointed out the negative aspects of the set of liners from the activity analysis. The combination of these results with the demands of users and its recommendations will enable the establishment of design requirements for the set of liners. These requirements should be validated through testing with prototypes to be performed in real use conditions. Thus, it is essential to study this type of equipment to make it more suitable in handling, in order to preserve the health of milkmaids, allowing them an improvement in their quality of life.

REFERENCES

Amaro, F.R. 2012. Sistemas de ordenha: funcionamento, dimensionamento, manutenção e avaliação. Capítulo 10. Disponível em: <http://marcosveiga.net/ biblioteca/livros/10%20parte1.pdf>. Acesso em: 04 jun. de 2012.

Bertoncello, I. 2001. O papel do fisioterapeuta no desenvolvimento de produtos hospitalares—Análise da cadeira de rodas. Dissertação de Mestrado—Santa Maria/RS/Brazil.

Borges, C.A.; Ximenes, A.C. 1997. Coluna vertebral: semiologia médica. 3. ed. Rio de Janeiro: Guanabara Koogan.

Empresa Brasileira de Pesquisa Agropecuária - EMBRAPA. 2007. Ordenha Mecânica. Disponível em: <http://www.agencia.cnptia.embrapa.br/Agencia 8/AG01/arvore/AG01_63_217200392359.html>. Acesso em: 05 jun. de 2012.

Estivalet, P.S.; Linden, J.C.S.V.D.; Amaral, E.; Paula, S.; Borelli, F.; Fontoura, C. 2004. Avaliação ergonômica de máquina de costura. In: XXIV Encontro Nacional de Engenharia de Produção, Florianópolis: ABEPRO.

Kauke, M.; Hermanns, I.; Höhne-Hückstädt, U.; Schick, M.; Ellegast, R. 2009. Analysis and assessment of workloads via CUELA using the example of the milking procedure. *Bornimer Agrartechnische Berichte* 66: 22–30.

Kuorinka, I.; Jonsson, B.; Kilbom, A.; Vinterberg, H.; Biering-Sorensen, F; Andersson, G.; Jorgensen, K. 1987. Standardised Nordic questionnaires for the analysis of musculoskeletal symptoms. *Applied Ergonomics* 18(3): 233–237.

Netto, F.G.S; Brito, L.G.; Figueiró, M.R. 2006. A ordenha da vaca leiteira. Comunicado Técnico, 319, Embrapa. Porto Velho. Disponível em: <http://www.cpafro.embrapa.br/media/arquivos/publicacoes/cot319_ordenhadavacaleiteira_.pdf>. Acesso em: 04 jun. de 2012.

Pinzke, S.; Stål, M.; Hansson, G.-Å. 2001. Physical workload on upper extremities in various operations during machine milking. *Ann Agric Environ Med.* 8: 63–70.

STÅL, M. 2000. Upper Extremity Disorders in Professional Female Milkers. *Proceedings of the Human Factors and Ergonomics Society Annual Meeting* 44: 608–611.

Stål, M.; Pinzke, S.; Hansson, G.-Å. 2003. The effect on workload by using a support arm in parlour milking. *Int J Ind Ergon.* 32: 121–132.

Tanabe, A.S. et al., 2010. Pesquisa antropométrica aplicada a confecção de EPIs. In: 10 ERGODESIGN, Rio de Janeiro. Congresso Internacional de Ergonomia e Usabilidade de Interfaces Humano-Tecnologia.

Ulbricht, L. 2003. Fatores de risco associados à incidência de DORT entre ordenhadores em Santa Catarina. 239 f. Tese (Doutorado em Engenharia de Produção)—PPGEP/UFSC. Florianópolis/SC/Brazil.

Occupational Safety and Hygiene II – Arezes et al. (eds)
© 2014 Taylor & Francis Group, London, ISBN 978-1-138-00144-2

Risk factors of the most frequent accident types in manufacturing

J.A. Carrillo-Castrillo & L. Onieva
University of Seville, Seville, Spain

J.C. Rubio-Romero & M. Suárez-Cebador
University of Malaga, Malaga, Spain

ABSTRACT: Workers in the manufacturing sector are not equally employed and the differences in job assignment depending on their characteristics. Injury rates for the most frequent accident types can be calculated to provide an estimation of the relative risk. The main strength of this paper compared to previous studies is the quality of worker data from the Continuous Sample of Working Lives obtained from Social Security Office. This data set allows the calculation of relative risks adjusting for some confounding variables. Adjusted relative risk confirms that female workers, foreign, non-manual and older workers have lower injury rates. The methodology presented is useful in the analysis of the risk factors of specific type of accidents and allows a more deep analysis of injury rates and the multivariate effects of each variable.

1 INTRODUCTION

1.1 *The Continuous Sample of Working Lives*

The Continuous Sample of Working Lives (hereinafter CSWL) contains a simple random sampling without stratification of an average 4% among all the affiliated workers, working or not, and pensioners. The database is described elsewhere (Jiménez-Martín & Sánchez-Martín 2007) and offers information about the personal characteristics of the worker and also about all their labor history.

This database has been used previously for several purposes but not in the analysis of injury rates. Microdata are available from the Spanish Social Security Agency upon request.

Although other tools such as Working Conditions Surveys (Carrillo *et al.* 2012) can provide some insight, the sampling error compared with CSWL is much higher and the quality of data gathered for social security purposes is higher. The use of CSWL in injury rate estimation has already been explored (Carrillo-Castrillo *et al.* 2013).

The main purpose of this study is to explore the potential use of CSWL for analyzing the relative risk of the most frequent accident types in manufacturing. At the same time appropriate methodology is proposed to estimate injury rates for the worker categories controlling for possible confounding effects.

Most injury rates are calculated using the number of workers. However the real exposure is related to the time at work. In CSWL the information gathered includes the number of days that each individual worker is active and the number of working hours per day in their contract.

1.2 *Worker characteristics and the risk of accident*

This study analyzes the influence of worker characteristics in injury rates. Each of them has been analyzed previously in the literature.

It is important to find out what worker collectives have higher relative risk in order to prioritize public enforcement and promotion programs. Estimating injury rates for those categories is difficult as there is no reliable information for the population at risk. Analysis of accident notifications usually lacks information about the worker population characteristics and their exposure.

According to previous studies the individual worker risk factors that should be taken into account are: age, gender, nationality and type of contract. Some specific works and tasks are more likely assigned to male, foreign or non-permanent workers.

The majority of studies have reported that young workers had a higher injury rate, especially if they are men (Salminen 2004). This behavior has also been studied at the company level (Pollack *et al.* 2007).

There are important differences in terms of job assignment and foreign workers are expected to be employed in more dangerous tasks (Ahonen *et al.* 2007).

Female workers show lower injury rates as a general trend (Islam *et al.* 2001) but not for every industry or occupation. In some cases, when analyzing at the company level, female workers seem to have more accidents than their male counterparts (Taiwo *et al.* 2008). Some studies conclude that men have higher rates for acute injuries and women for musculo-skeletal ones.

Working in small enterprises is considered a risk factor (Sørensen *et al.* 2007). Small and micro enterprises have more severe and fatal accidents in comparison with medium and big enterprises. However, some manufacturing activities concentrate more hazardous activities in enterprises with a certain minimum size.

Another important issue is the effect of contract type on safety. Non-permanent contracted workers have more accidents (Benavides *et al.* 2006).

All of these worker categories need to be studied with multivariate techniques in order to control the confounding phenomena. CSWL provides sampling of real individual workers including data for each worker in all categories of interest.

2 MATERIALS AND METHODS

2.1 *Data*

The cross-sectional CSWL dataset contains employment variables such as occupational levels, sector of activity, contract type and duration of employment. It also includes relevant variables related to demographics such as location, age and gender and nationality. The duration of the employment comes from the dates of beginning and ending the contract.

Data was gathered from CSWL in 2008. Worker data selected from CSWL are those contracted in the Andalusian Manufacturing Sector. Andalusia is one of the biggest regions in Europe, producing 12% of the Spanish manufacturing sector and employing on average more than two hundred thousand workers. Self-employed worker data are not considered.

The manufacturing sector is defined according to the Statistical Classification of Economic Activities in the European Community Council Regulation EEC N°3037/90 (section D, subsections 15 to 37).

2.2 *Estimation of the proportion of equivalent work days for each category*

The number of equivalent days for each worker is calculated based on the standard work day of eight hours in Spain. The number of effective working days of all workers of the same category in CSWL is added; therefore the equivalent standard days worked for each category can be calculated.

As the number of equivalent work days calculated is a sampling of the real population, the proportion in each of the categories can be calculated.

The data for 13,533 workers with 2,353,401 equivalent work days are available for analysis from CSWL in the Andalusian manufacturing sector in 2008. This represents 6.79% of the workers in the manufacturing sector for 2008.

2.3 *Estimation of the relative risk for the worker categories (not adjusted)*

The relative risk is calculated as the ratio of the injury rates of two groups of workers, one of them is being considered the reference (Benavides *et al.* 2006).

Accident notifications in Andalusia are electronically collected in "Official Workplace Incident Notification Forms" (Jacinto & Aspinwall 2004). All accidents that result in an absence from work of one or more days must be reported. Relapses, travelling to work accidents ("in itinere") or accidents of self-employed workers are not included.

The identification of the type of accident is based on the variables included in ESAW. As previous authors have proposed (Jacinto *et al.* 2008; Rajala & Väyrynen, 2010) and according to the European Commission (2009) itself, there are two important concepts implicit in ESAW codification of an accident. The concept of *Accident Mechanism* is used for this. The accident mechanism is used to identify each central event of the bow-tie and is defined by the combination of deviation and mode of injury (Jacinto *et al.* 2008; Hale *et al.* 2010).

2.4 *Worker categories*

The categories in each worker variable are defined in Table 1. They are the same used in the National Employment Survey in Spain.

Job and qualification is classified into four groups according to the current version of the International Standard Classification of Occupations, ISCO-88, approved by the International Labour Organization.

ISCO-88 classifies groups on the basis of the similarity of skills required to fulfill the tasks and duties of the jobs. Two dimensions of the skill concept are used, skill level and skill-specialization. Those groups are the following: High qualification/non-manual (hereinafter HQNM), low qualification/non-manual (hereinafter LQNM), high qualification/manual (hereinafter HQM) and low qualification/manual (hereinafter LQM).

Table 1. Estimated proportion of effective working days as indicator of population at risk in each category of workers. Data from CSWL of Andalusian manufacturing sector in 2008.

Variable	Category (identifier)	%
Contract type	Permanent (1)	64,7%
	Non-permanent (2)	35,3%
Nationality of worker	Spanish (1)	96,1%
	Foreign (2)	3,9%
Sex of worker	Male (1)	78,9%
	Female (2)	21,0%
Age of worker (years)	Less than 30 years old (1)	27,7%
	Between 30 and 44 years old (2)	43,3%
	More than 44 years old (3)	29,0%
Company size (number of workers)	Micro, less than 9 workers (1)	30,7%
	Small, from 10 to 49 workers (2)	37,4%
	Medium, from 50 to 249 workers (3)	21,9%
	Big, more than 249 workers (4)	10,1%
Type of job and qualification	High qualification/ non-manual (1)	8,7%
	Low qualification/ non-manual (2)	15,1%
	High qualification/ manual (3)	53,0%
	Low qualification/ manual (4)	23,1%

2.5 *Panel data set*

A panel data set is created with the sum of all working days in CSWL for the workers sampled with the same combination of the categories of the variables (see Table 1 for the identifiers of each of the categories). That combination represents all the workers with the same *Affiliation Pattern*. For example, combination "20431311" consists of the number of working days in CSWL of workers with an affiliation pattern in activity 20 (NACE code for manufacturing of wood and wood products), working in big enterprises (identifier 4), with high qualification and manual jobs (identifier 3), with a permanent contract (identifier 1), older than 44 years old (identifier 3), male (identifier 1) and Spanish workers (identifier 1).

There are 1,197 possible combinations identified in the panel for 2008 with affiliation data available. Not all of them had affiliation data available for injury rate estimation. An error of 10% in the affiliation estimation was considered acceptable (see Table 2).

There are 849 combinations with enough population to estimate injury rates. They represent the 96.6% of the working days gathered in CSWL and covered 90.7% of all accidents reported in the period.

Table 2. Affiliation patterns available for injury rate estimation.

Affiliation data	N° combinations	N° accidents	N° equivalent working days
With error less than 10%	849	20,126	2,274,519
With error higher than 10%	348	1,413	78,882
Without affiliation data	222	647	–

Injury rate is estimated for each affiliation pattern as the ratio of the real number of accidents reported and the estimated number of workers using the proportion of equivalent working days for the combination of categories in CSWL.

2.6 *Regression models*

A regression model is adjusted for dependent variables, the natural logarithm of each accident mechanism injury rate. Logarithms are used because most authors considerer that accidents are better modeled with exponential variables (Haviland 2012).

All independent variables are the categories of the qualitative variables available, coded with dummy dichotomous variables. Each qualitative variable with "n" categories is represented by "n-1" dummy variables. Activities identified with the two first digits of NACE code (from 15 to 37) were also included in the model as controls because injury rates of each activity are very different.

The relative risk of one category to their reference is calculated with the expression e^b, where b is the regression coefficient, because dependent variable is logarithmic. Thus, the relative risk is adjusted with the regression model and calculated considering the effect of the other explanatory variables considered.

SPPS v.18 was used. Data and models are available upon request.

3 RESULTS

3.1 *Accident mechanisms and task identification*

Accident mechanisms are identified using the first digit of the ESAW variables *Deviation* and *Mode of Contact*. All of the accident mechanisms identified accounted for more than 1,000 accidents in 2008 (see Table 3).

Table 3. Accident mechanisms identification.
Source: Accident reported in the manufacturing sector of Andalusia in 2008.

Accident mechanism	Deviation (first digit of ESAW code)	Mode of Contact (first digit of ESAW code)	N° accidents in 2008
M1	Body movement under or with physical stress (7)	Physical or mental stress (7)	4,713
M2	Slipping—Stumbling and falling—Fall of persons (5)	Impact with or against a stationary object (3)	1,722
M3	Body movement without any physical stress (6)	Physical or mental stress (7)	1,443
M4	Loss of control (4)	Contact with sharp, pointed, rough, coarse (5)	1,393
M5	Loss of control (4)	Struck by object in motion (4)	1,342

3.2 *Regression models*

The regression model has been adjusted for the dependent variables, natural logarithm of the accident mechanisms. Relative risk estimated with the regression models is presented (Table 4). All coefficients are significant ($p < 0.001$).

4 DISCUSSION

A close examination of the results in Tables 4 shows that young workers have higher injury rates for accident mechanisms M3, M4 and M5 that are related to *Loss of Control* accidents. Although age is a risk factor, there accident mechanisms such as M1 and M2 where the adjusted relative risk is not conclusive.

It must be considered that accident mechanism M1 is basically related to *Musculoskeletal disorders*, and previous analysis showed that for this type of accidents, the injury rates were not clearly higher for younger workers (Carrillo-Castrillo *et al.* 2013).

At the same time, accident mechanism M2 include *Falls* accidents and previous analysis shown that for this particular type of accidents, younger workers have lower injury rates (Salminen 2004).

Female workers have lower relative risk of injury in all of the accident mechanisms analyzed.

Table 4. Adjusted relative risk: regression model for natural logarithm of traumatic injury rate.

Variable	Category	M1	M2	M3	M4	M5
		Accident Mechanisms				
Contract type	Permanent	*	*	1.06	*	*
	Temporal	Reference				
Nationality of worker	Spanish	1.18	1.31	1.24	1.16	1.24
	Foreign	Reference				
Sex of worker	Male	1.24	1.13	1.26	1.28	1.19
	Female	Reference				
Age of worker (years)	[<30]	*	*	1.09	1.08	1.14
	[30–44]	1.08	*	1.09	1.08	1.17
	[>44]	Reference				
Estab. Size (number of workers)	Micro	1.15	1.12	*	1.27	*
	Small	*	*	0.92	1.20	*
	Medium	1.13	1.11	1.10	1.20	*
	Big	Reference				
Type of job and qualification	HQNM	0.68	0.75	0.75	0.79	0.79
	LQNM	0.70	0.78	0.78	0.79	0.77
	HQM	1.13	1.22	1.20	1.24	1.29
	LQM	Reference				
R^2 of the model		0.49	0.39	0.39	0.40	0.36

* Not significant $p > 0.05$

Nevertheless, the adjusted relative risk is not so high, showing that for the same task and exposure, the sex influence is not so important (Taiwo *et al.* 2008).

At the same time, as previously have been noted (Carrillo-Castrillo *et al.* 2013), when adjusted in the regression model, the relative risk of foreign workers is lower. This is consistent with other studies (Ahonen *et al.* 2007) and could be a sign of underreporting.

As other studies have highlighted (Benavides *et al.* 2006), the relation between non-permanent workers and injury rates is not a straight forward issue. In our analysis, the relative risk is not significant for most of accident mechanisms.

In terms of company size, although big companies have lower injury rates, in most of cases the adjusted relative risk not significant. This non-straight forward relationship of injury rates with company size is difficult to analyze (Mendeloff *et al.* 2006).

Of all employment groups, the higher rates have been found for high qualified workers employed in manual tasks. These results are consistent with other studies in manufacturing using Working Conditions Surveys. In Carrillo *et al.* (2012) it was found that being employed as technicians were collectives with higher injury rates.

In spite of the lower relative risk for non-manual workers, their adjusted injury rates are two high considering the different exposure.

5 CONCLUSIONS

5.1 *Advantage of the method proposed*

CSWL provide a sampling error for the worker categories lower than other sources of information. The panel data set designed allows estimating injury rates for specific accident mechanisms and the use of multivariate regression models to control of possible confounders.

The use of the equivalent working days instead of affiliation is a more precise indicator of exposure. Partial-contracts and temporal contracts are included with the real equivalent working days. In those contracts the differences between affiliation and equivalent working days is very important.

5.2 *Implications of the results in safety policies*

The use of CSWL for estimation of the actual exposure of each category of workers for injury rates calculation provides a useful tool in Public Policy design and this paper shows some of the potential uses.

According to the European Safety Framework Directive 89/391/EEC "particularly sensitive risk groups must be protected against the dangers which specifically affect them".

Some of those sensitive risk groups have been identified according to the personal characteristics of workers (European Agency for Safety and Health at Work 2009). In the Andalusian manufacturing sector, specific safety promotion programs are needed for young workers, but they can be specifically designed for those accidents were they have higher relative risk of accident.

Prevention programs should be designed considering that intervention in each group of workers needs to be specific. As this study finding suggest, the influence of worker characteristics in accident causation is different for each accident mechanisms and should be analyzed separately.

REFERENCES

Ahonen, E.Q., Benavides, F.G. & Benach, J. 2007. Immigrant Populations, work and health—a systematic literature review. *Scandinavian Journal of Work, Environmental and Health* 33(2): 96–104.

Benavides, F.G., Benach, J., Muntaner, C., Delclos, G.L., Catot, N. & Amable, M. 2006. Associations between temporary employment and occupational injury: what are the mechanisms? *Occupational and Environmental Medicine* 63(6): 416–421.

Carrillo, J.A., Gómez, M.A. & Onieva, L. 2012. Safety at work and worker profile: analysis of the manufacturing sector in Andalusia in 2008. In Arezes, P. et al. *Occupational Safety and Hygiene – SHO2012.*

Carrillo-Castrillo, J.A., Onieva, L., Rubio-Romero, J.C. & Suarez-Cebador, M. 2013. Relative risk of accident: Worker collectives in the manufacturing sector. In *Occupational Safety and Hygiene*: 181–186. London: CRC Press.

European Agency for Safety and Health at Work. 2009. *Workforce Diversity and Risk Assessment: Ensuring Everyone is Covered.* DOI: 10.2802/11532.

Hale, A.R., Ale, B.J.M., Goossens, L.H.J., Heijer, T., Bellamy, L.J, Mud, M.L., Roelen, A., Baksteen, H., Post, J., Papazoglou, I.A., Bloemhoff, A. & Oh, J.I.H. 2007. Modeling accidents for prioritizing prevention. *Reliability Engineering and System Safety* 92(12): 1701–1715.

Haviland, A.M., Burns, R.M., Gray, W.B. & Mendeloff, J. 2012. A new estimate of the impact of OSHA inspections on manufacturing injury rates, 1998–2005. *American Journal of Industrial Medicine* 55(11): 964–975.

Islam, S.S., Velilla, A.M., Doyle, E.J. & Ducatman, A.M. 2001. Gender Differences in Work-Related Injury/Illness: Analysis of Workers Compensation Claims. *American Journal of Industrial Medicine* 39(1): 84–91.

Jacinto, M.C. & Aspinwall, C. 2004. A survey on occupational accidents' reporting and registration systems in the European Union. *Safety Science* 42(10): 933–960.

Jacinto, C. & Soares, C.G. 2008. The added value of the new ESAW/Eurostat variables in accident analysis in the mining and quarrying industry. *Safety Science* 39(6): 632–644.

Jiménez-Martín, S & Sánchez-Martín, A.R. 2007. An evaluation of the life cycle effects of minimum pensions on retirement behavior. *Journal of Applied Econometrics* 22(5): 923–950.

Mendeloff, J., Nelson, C., Ko, K, & Haviland, A. 2006. *Small Businesses and Workplace Fatality Risk: An Exploratory Analysis*, USA: RAND Corporation.

Pollack, K.M., Agnew, J., Slade, M.D., Cantley, L., Taiwo, O., Vegso, S., Sircar K., Cullen M.R. 2007. Use of employer administrative databases to identify systematic causes of injury in aluminum manufacturing. *American Journal of Industrial Medicine* 50(9): 676–686.

Salminen, S. 2004. Have young workers more injuries than older ones? An international literature review. *Journal of Safety Research* 35(5): 513–521.

Sørensen, O.H., Hasle, P. & Bach, E. 2007. Working in small enterprises - Is there a special risk? *Safety Science* 45(10): 1044–1059.

Taiwo, O.A., Cantley, L.F., Slade, M.D., Pollack, K.M., Vegso, S. & Fiellin, M.G. 2008. Sex Differences in Injury Patterns Among Workers in Heavy Manufacturing. *American Journal of Epidemiology* 169(2): 161–166.

Occupational Safety and Hygiene II – Arezes et al. (eds)

Determination of characteristics predicting the ignition of organic dusts

Z. Szabová, M. Pastier, J. Harangozó & T. Chrebet
Slovak University of Technology in Bratislava, Faculty of Materials Science and Technology in Trnava, Institute of Safety and Environmental Engineering, Trnava, Slovak Republic

ABSTRACT: The minimum ignition temperature of dust clouds is one of important factors required for determination of preventive measures against dust explosion. Determination of these characteristics by calculating is not sufficiently accurate, and therefore the most reliable results are determined experimentally. The article deals with determination of minimum ignition temperatures of flour and starch dusts and use of thermal analysis to predict the risk of explosion.

1 INTRODUCTION

Minimum temperature of hot surface to ignite the dust cloud is set by the standardized procedure of European standard EN 50281-2-1 Electrical apparatus for use in the presence of combustible dust. Part 2-1: Test methods. Methods for determining the minimum ignition temperatures of dust. It is the lowest temperature of the hot inner side of the oven, at which the ignition of dust cloud is occurring in the air inside the furnace (Slabá & Tureková 2012, VST Engineering 2012).

Because the dust clouds are not homogenous in time and/or place, the 1/3 safety factor is used. This indicates that the maximum permitted temperature of the hot inside surface of technological equipment is 2/3 of the ignition temperature of dust clouds (VST Engineering 2012, Tureková et al. 2013, Mračková 2006).

The ignition temperatures of most dusts range from 350°C to 700°C, but exceptionally can also be lower than 300°C (Industrial Explosion Protection Institute 2010, Marková et al. 2011).

2 MATERIAL AND METHODS

For determination of properties of dusts the following methods were used:

- sieve analysis of dust samples,
- determining the minimum ignition temperatures of dust clouds,
- thermal analysis (thermogravimetry (TG) and differential scanning calorimetry (DSC)).

Two types of organic dust samples were used for testing.

- smooth wheat flour,
- finecorn starch.

Chemical composition of dust significantly affects the fire and technical characteristics.

Wheat flour is the crushed inner part of cereal grain that contains the following ingredients: water, proteins, fat, carbohydrates, potassium, phosphorus, iron, sodium, magnesium, calcium, copper, zinc, proteins, tassel and vitamins B1, B2, B3, B6, B12, E. Wheat flour is also known as wheat starch (Chemical Plus sal. 2008).

Corn starch is a polysaccharide consisting of amylase (25%) and amylopectin (75%). These two components are formed of long rows of glucose molecules, including also lipids, proteins and water.

3 RESULTS AND DISCUSSION

Sieve analysis was performed on analytical RETSCH sieve shaker type AS 200 which is used for separation, fractioning and particle size determination and the results of analysis are shown in Figure 1.

Samples of flour and corn starch are mostly consisting of the particles with size over 71 microns. In terms of particle size these samples are relatively similar.

3.1 *Thermal analysis of dust samples*

Thermal analysis of flour and starch samples took place in three stages at a constant heating rate of $10°C.min^{-1}$ and the shape of thermal curves of tested samples is shown in Figures 2 and 3.

The first stage of sample decomposition consisted of the final drying stage. The mass loss rate of flour at this stage of decomposition was 8.72%. The mass loss rate of starch was 10.21%.

No significant differences in the behaviour of dust samples were observed. This does not apply

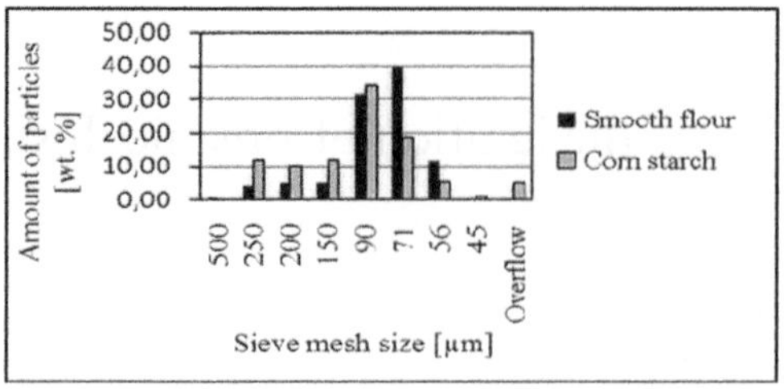

Figure 1. Comparison of representative particles in a sample of flour and starch (sample weight: smooth flour = 30.124 g, corn starch = 30.018 g).

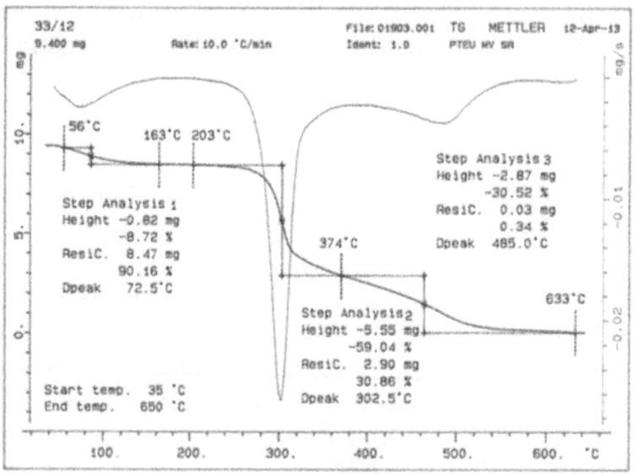

Figure 2. TG analysis of wheat flour.

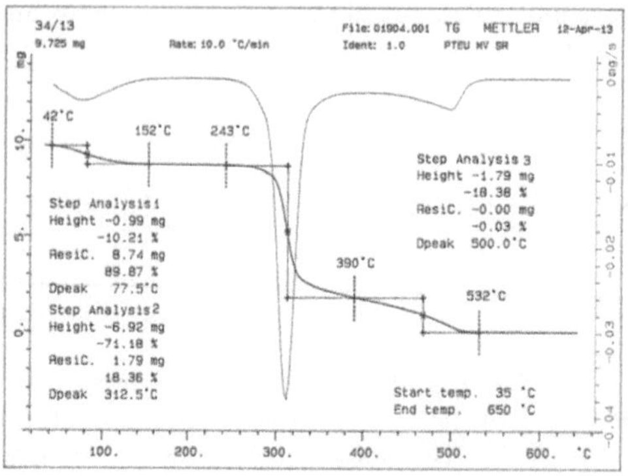

Figure 3. TG analysis of corn starch.

to the second and third stage of thermal decomposition. In the second stage a significant mass loss rate of both samples (flour – 59.04%, starch – 71.18%) was observed. Corn starch thus exerts a lower resistance against thermal degradation at this stage of decomposition.

The third stage of decomposition in case of wheat flour samples took place at temperature 633°C and in case of starch sample at temperature 532°C. Active phase of decomposition of corn starch ended at a lower temperature than in case of wheat flour, which is related to the different composition of dusts. In the third stage of decomposition, the mass loss rate was 30.52% in case of flour and 18.38% in case of starch. From the TG analysis it can be concluded that the dust sample of cornstarch degraded at lower temperatures than the wheat flour.

3.2 *Differential scanning calorimetry*

Differential scanning calorimetry provides information on the endothermic and exothermic processes in the controlled thermal decomposition (Figures 4 and 5).

For both test samples of dust, the temperature intervals with significant endothermic and exothermic changes were observed. In case of flour, the endothermic process in the temperature range 35–129°C and the exothermic process in the temperature range 240–600°C was observed. These endothermic reactions are the result of dehydration processes. The exothermic process of both dust samples is characterized by a rapid increase of reaction enthalpy changes as a result of oxidative transformations. The value of reaction enthalpy for

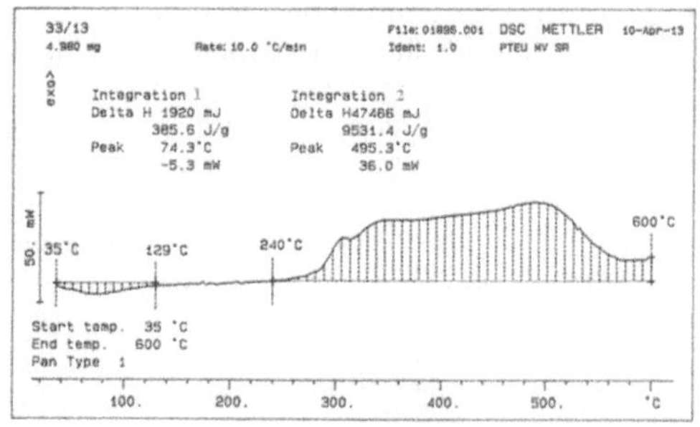

Figure 4. DSC analysis of wheat flour.

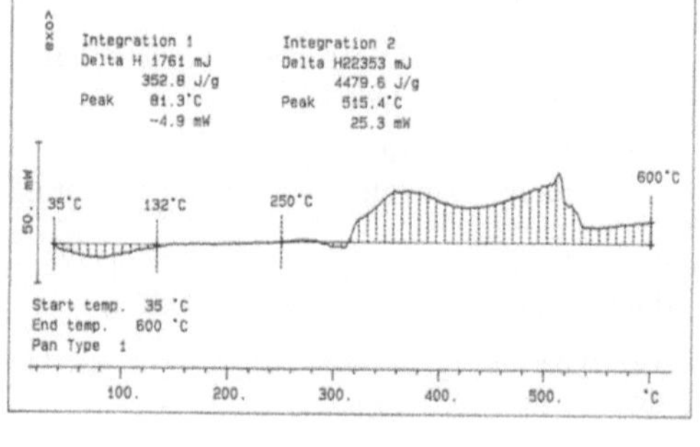

Figure 5. DSC analysis of corn starch.

Table 1. Determination of the minimum ignition temperature of turbid flour (Balluch 2013).

Mass [g]	Pressure [kPa]	Temperature [°C]		Ignition [Y–yes/N–no]	
		Flour	Starch	Flour	Starch
0.1	10	450	450	N	N
	20	450	450	N	N
	50	450	450	Y	N
		440	–	N	–
		430	–	N	–
0.3	10	450	450	N	N
	20	450	450	N	N
	50	450	450	Y	Y
		440	440	N	Y
		430	430	N	N
0.5	10	450	450	N	N
	20	450	450	N	Y
		–	430	–	N
	50	450	450	Y	Y
		440	440	N	Y
		430	430	N	N

exothermic processes, which took place in wheat flour, was higher than the value in case of corn starch. For a sample of wheat flour a more intensive oxidation process has occurred than in case of samples of corn starch, which was conditioned by the chemical composition of samples.

3.3 *Minimum ignition temperature of turbid dust*

Determination of the minimum ignition temperature of this parameter was performed on the standardized equipment according to EN 50281-2-1: 2002 with the results shown in Table 1.

Minimum ignition temperature of turbid flour was 440°C and that of corn starch 430°C.

4 CONCLUSIONS

The achieved results of this study correspond with the published results (Slabá & Tureková 2012, Coneva 2009). It can be concluded that the tested organic dusts are capable of forming an explosive atmosphere and therefore are subject to categorization of areas with probability of an explosive atmosphere.

In this respect, based on the attained results, implementation of corrective measures is necessary. These are classified into three groups in dependence on the controlled factor. This concerns namely the control of dust, ignition sources and damage.

ACKNOWLEDGEMENTS

The submitted work was supported by the project APVV-0057-12.

REFERENCES

Balluch, R. 2013. *Vplyv vlastností organických rozvírených prachov na ich zápalné charakteristiky (The effect of properties of organic turbid dusts on their ignition characteristics)*. [Thesis] – STU Bratislava. The Faculty of Materials and Technology in Trnava. Trnava: MTF STU, p. 77.

Chemical Plus sal. 2008. *Wheat Starch*. [online]. [cit. 2013-09-10]. Available on the net: http://www.chemicalplus.com/PDF/Wheat%20Starch.pdf

Coneva, I. 2009. Problematika horenia a požiaru tuhých, polymérnych materiálov so zameraním sa na vybraté celulózové materiály (Issues of burning and fire of the solid, polymeric materials oriented to selected cellulosic materials): Part 1, *In: Crisis Management*. ISSN 1336-0019. Vol. 8, No. 1, pp. 8–18.

Health and Safety Executive. *ATEX and explosive atmospheres*. [online]. [cit. 2013-03-07]. Available on the net: http://www.hse.gov.uk/fireandexplosion/atex.htm

Industrial Explosion Protection Institute. 2010. *Explosion Characteristics Measurement of Combustible Dusts*. North-eastern University. [online]. [cit. 2012-11-26]. Available on the net: http://www.iepi.com.cn/Service/PDF/Explosion%20Characteristics%20Measurement.pdf

Marková, I., Réh, R., Orémusová, E., Mračková, E. 2011. Evaluation of the combustion heat of selected types of biomass by calorimetric method under adiabatic conditions. *In: Book of abstracts: 1st Central and Eastern European Conference on thermal analysis and calorimetry*. ISBN 978-606-11-1893-9.

Mračková, E. 2006. *Výbušnosť drevného prachu: (smrek, buk, dub, topoľ, drevotriesková doska) (Explosivity of wood dust: (spruce, beech, oak, toplar, chipboard))*. 1. Issue. Zvolen: Technical University, p. 91. ISBN 80-228-1698-1.

Slabá, I., Tureková, I. 2012 *Smouldering and Flaming Combustionof Dust Layer on Hot Surface*. 1st ed. Dresden: IFW, p. 88. (Scientific monographs). ISBN 978-3-9808314-5-1.

Tureková, I., Turňová, Z., Harangozó, J., Kasalová, I., Chrebet, T. 2013. Determination of Ignition Temperature of Organic Dust Layers. *In: Advanced Materials Research*. ISSN 1022-6680 (P). ISSN 1662–8985 (E). Vol. 690–693, pp. 1469–1472.

VST Engineering. 2012. *Výbuchové a požárně-technické charakteristiky látek (Explosion and fire-technical characteristics of matters)*. [online]. [cit. 2012-11-26]. Available on the net: http://www.vst.cz/CZE/informace/vybuchove-charakteristiky-latek.htm

Occupational Safety and Hygiene II – Arezes et al. (eds)

Bibliometric Mapping to analyze the evolution of research on *Ergonomics* using the SciMAT tool

M.D. Martínez-Aires, M. Martínez-Rojas, M. López-Alonso & E.J. Gago
University of Granada, Spain

ABSTRACT: This study presents an analysis of the evolution of ergonomics research as a technological discipline over the last thirty years. For this analyze has been used a software tool called SciMAT (Science Mapping Analysis software Tool). The results show that research has been on the rise, both in the number of publications and in the emergence of topics closely related to ergonomics.

1 INTRODUCTION

Analyzing scientific maps is a technique used in Library Sciences to graphically represent the relationships between documents published by the different disciplines or specific scientific fields. This highlights the specific subareas where research has placed its focus in order to identify, analyze and visualize the intellectual, social and conceptual structure of the field as well as its evolution over time (Cobo *et al.* 2011a).

The construction industry is one of the largest as well as one of the most dangerous. Specifically, musculoskeletal disorders (MSDs) are a major problem in this field. Construction workers may also be exposed to different psychosocial factors specific to their job environment. It has been shown that psychosocial factors are associated with MSDs either independently or in combination with physical factors (Rinder *et al.* 2008). The etiological factors of MSDs were recognized as being work-related as early as the beginning of the 18th century. Since then the literature has grown dramatically; according NIOSH (2001) approximately 4000 articles that focused on occupationally-related musculoskeletal disorders were published in the last decade of the twentieth century.

A large number of tools have been specifically designed to analyze scientific maps (Cobo 2012a). They include:

Bibexcel (Persson *et al.* 2009), CiteSpace II (Chen 2004), CoPalRed (Bailon *et al.* 2005, 2006), INSPIRE (Wise1999), Loet Leydesdorff's Software, Net-work Workbench Tool (Börner *et al.* 2010, Herr *et al.* 2010), Science of Science (Sci2) Tool (Sci2 Team 2009), VantagePoint (Porter & Cunningham 2004) y VOSViewer (Cobo *et al.* 2011b).

A comparative analysis of each tool reveals that each of them has a different set of basic characteristics. As with any tool, cognitive mapping has limitations (Village *et al.* 2013). For example, many of the packages include powerful processing tools (CoPalRed and VantagePoint), others can generate a large number of bibliometric networks, and others are only able to extract one type of network (CoPalRed). From this comparative analysis it thus follows that not all of the steps of the analysis can be performed by each tool (Cobo *et al.* 2012b).

Consequently, a detailed and comprehensive analysis of a scientific discipline based on bibliometric maps should be done using several different software tools to extract the greatest possible amount of information and different perspectives on the same field of study (Cobo 2012a).

SciMAT (developed by the Department of Computer Science and Artificial Intelligence, University of Granada) includes most of the advantages of previous applications, while at the same time limiting the reliance on external tools.

Specifically, SciMAT includes everything needed for a complete analysis of scientific maps in a longitudinal study based on bibliometric impact measures. It also facilitates analysis of the social, intellectual and conceptual evolution of a scientific field (Cobo 2012b).

SciMAT has two key features that are either lacking or poorly implemented in other tools:

- A powerful pre-processing module of the data under study, and
- The use of impact and quality as bibliometric indicators.

SciMAT builds a knowledge base from a set of scientific publications where it stores the relationships between each publication (document) and the different items (authors, keywords, journals, references, etc). This database helps edit and pre-process the information, which improves data quality and the resulting analysis of the scientific maps (Cobo 2012b).

The aim of this paper is to analyze the evolution of ergonomics as a technological discipline over the last few decades. The authors use SciMAT tool to analyze this evolution.

2 METHODOLOGY

In this paper a bibliometric co-word analysis tool is used to identify the ergonomics topics published during the period 1984–2013.

The bibliometric application (reference) used, combines both performance analysis and scientific mapping tools to analyze a research field and to detect and visualize its conceptual subdomains (particular topics/themes or general thematic areas) and its thematic evolution. Additionally, three stages were defined to analyze the themes and thematic evolution of the research field.

2.1 *Detection of research themes*

The data sources used were the ISI Web of Science (ISI Web of Science 2012) and Scopus (Scopus 2013). Database searches for "Construction and Health and Safety" or "Construction and Ergonomics" were performed. During the search process, constraints were established for periods (1984–2013), subject areas (construction, engineering, etc., with several subjects areas such as medicine, earth and planetary science being excluded), or source title, in order to obtain the articles containing the keywords to beanalyzed.

After the results were imported into SciMAT, a significant amount of effort was dedicated to cleaning up the data. First, duplicate documents and documents that did not belong to the study area were eliminated. Second, the data was subject to preprocessing. This is perhaps one of the most crucial steps for improving the quality of the units of analysis (mainly words) and is key to obtaining better results from the scientific mapping analysis. In this process the data was normalized by combining singular and plural forms, as well as grouping together different terms relating to the same concept. Misspelled words were also detected and combined with their corresponding representative. Once pre-processing was completed, 2882 articles and 16756 keywords were available for analysis.

The period analyzed (1984–2013) was divided into six five-year periods. In Figure 1, the distribution of documents per period is shown. Figure 2 shows the overlapping-items graph across the two consecutive periods. The circles represent the periods and their number of associated items

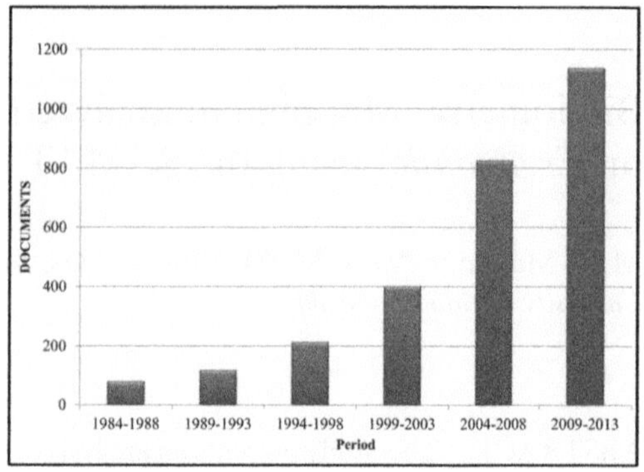

Figure 1. Distribution of documents per year.

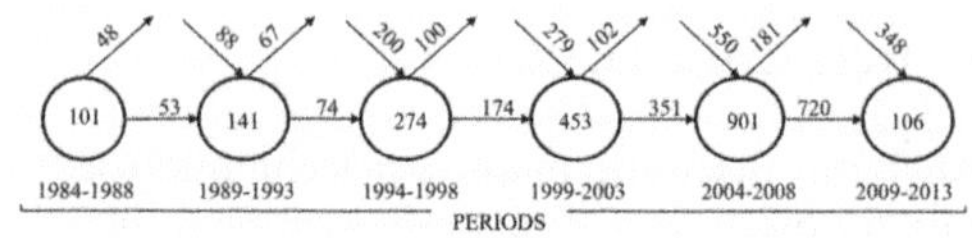

Figure 2. Overlapping-items graph.

(unit of analysis). The horizontal arrow represents the number of items shared by both periods. The diagonal incoming arrow represents the number of new items in, for example, Period 2, and the diagonal outgoing arrow. represents the items that are presented in Period 1, but not in Period 2 (Cobo 2012b).

SciMAT performs various processes to locate keyword networks that are strongly linked to each other and that correspond to centers of interest or to research problems that are the object of significant interest among researchers (Cobo 2012b).

2.2 *Building strategic diagrams*

In which each keyword network or theme can be characterized by two parameters (Callon *et al.* 1991):

- *Centrality*: Measures the strength of external ties to other themes. This value can be understood as a measure of the importance of a theme in the development of the entire research field analyzed.
- *Density*: Measures the strength of internal ties among all the keywords describing the research theme. This value can be understood as a measure of the theme's development.

Themes in the upper-right quadrant (I) (Fig. 3) are both well developed and important for the structuring of a research field. They are known as the drivers or principal themes of the specialty

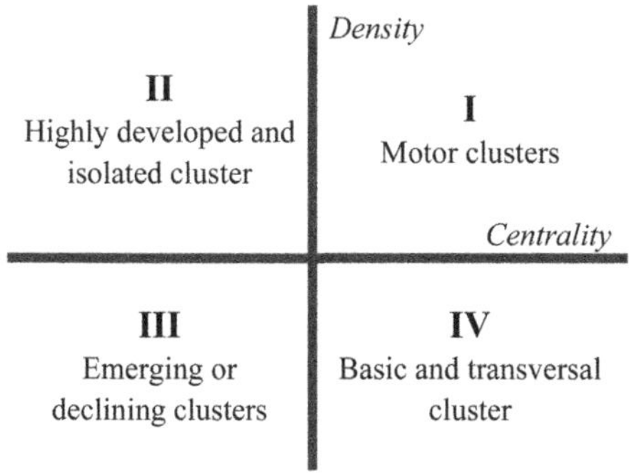

Figure 3. The strategic diagram.

given that they exhibit strong centrality and high density.

Themes in the upper-left quadrant (II) have well-developed internal ties but unimportant external ties and so are of only marginal importance for the field. These themes are very specialized and peripheral in character. Those in the lower-left quadrant (III) are both weakly developed and marginal. They have low density and low centrality and mainly represent either emerging or declining themes. The themes in the lower-right quadrant (IV) are important for a research field, but are not developed. Hence, this quadrant groups transversal and general or basic themes.

Once the knowledge base is ready, the scientific mapping analysiscan begin. To build the maps, the tool has an eleven-step process that must be completed.

The parameters to be analyzed are selected in this stage, such as period, unit of analysis, data reduction (the data are filtered using a minimum frequency threshold), network building, selection of the performance and bibliometricquality measures, etc. (Cobo, 2011a, b).

2.3 *Conducting a performance analysis*

In this phase the most prominent, productive, and highest impact subfields can be detected by measuring (quantitatively and qualitatively) the relative contribution of themes and thematic areas to the whole research field.

3 RESULTS AND DISCUSSION

To analyze the most prominent themes in the ergonomics field for each subperiod, strategic diagrams were built using SciMAT. In Figure 4b the volume of the spheres is proportional to the number of documents associated with each theme.

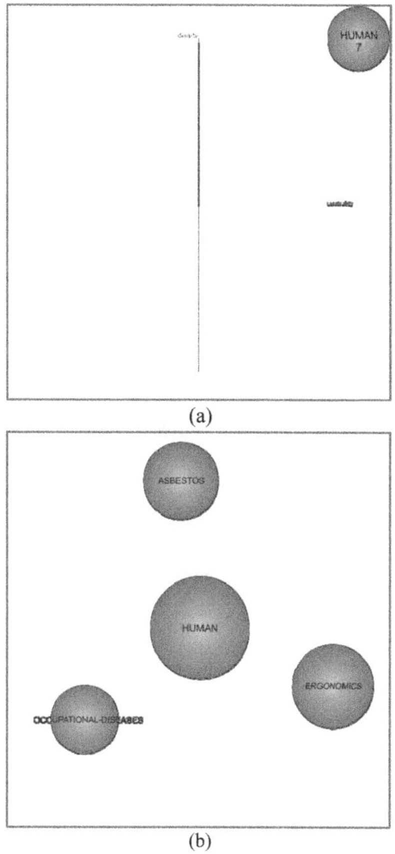

Figure 4. (a) Strategic diagrams for subperiod 1984–1988 (b) Thematic networks related with *human*.

For each of the five-year periods from 1984 to 2013 (the last 30 years) strategic diagrams were generated as well as their corresponding thematic networks. As shown in the Figure 4a, the first period from 1984 to 1988 focuses on issues related to the concept *human*, but in the analysis of the network of related themes *ergonomics* and *occupational diseases* emerge.

In the next two stages, 1989–1993 and 1994–1998, *human* continues to be the focus of publications and other keywords such as *risk* or *regulations* appear. However, it reveals a qualitative and quantitative leap in terms of those themes related to *human*. Table 1 below shows the network for *human* for the 1994–1998 period, where words related to *human* such as *occupational diseases* and *ergonomics* continue to appear, *asbestos* falls of the charts, and new subjects come into view such as *occupational exposure*, *occupational safety* or *workplace*, which underline the emerging concerns in the field of ergonomics.

In the next stage corresponding to the 1999–2003 period (Fig. 5a), *human* is not the only focal point. The term *ergonomics* appears, along with others such as *regulations* or *costs* (Fig.5b). It also shows that in the thematic network for the word *ergonomics* other words emerge from fields such as *design*, *construction method*, *behavior* or *working conditions*.

Finally, Figures 6a, b shows the publications corresponding to the 2008–2013 period, where *ergonomics* and *human* continue, but in the strategic diagram terms such as *management*, *organizations*, *innovation*, or *process* cannow be seen, which reflect

Table 1. Network for *human* for the 1994–1998 period.

Name	Document Count	Document Index	Document Citations
Human	36	11	395
Regulations	36	1	6
Risk	6	6	165

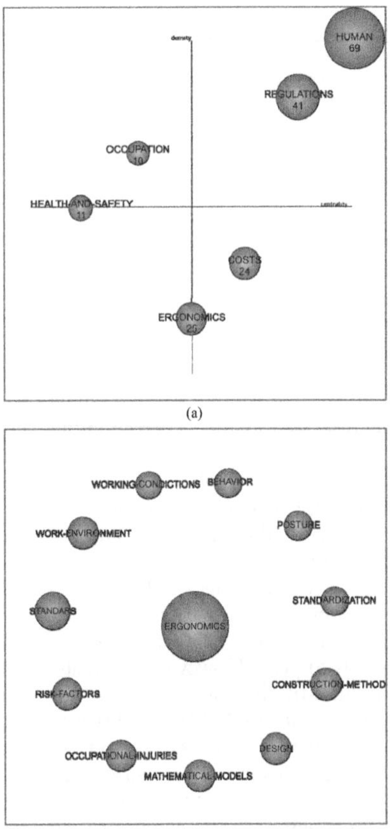

Figure 5. (a) Strategic diagrams for subperiod 1999–2003 (b) Thematic networks related with *human*.

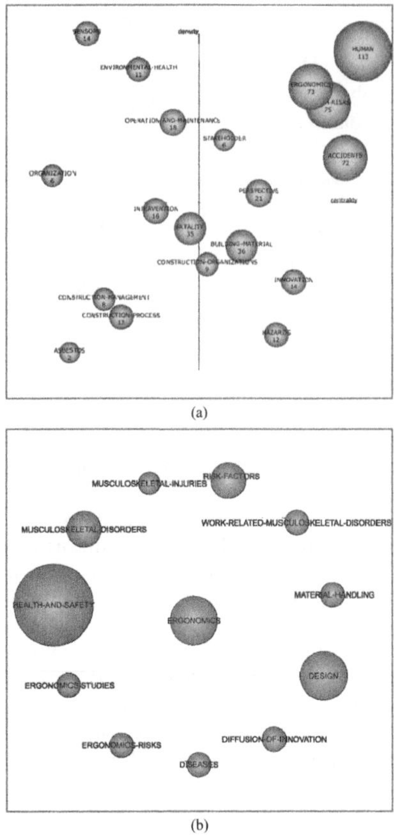

Figure 6. (a) Strategic diagrams for subperiod 2003–2013 (b) Thematic networks related with *ergonomic*.

the trends in research dedicated to understanding the interactions between humans and the elements of production systems.

In the ergonomics network,terms such as *health and safety* or *design*now appear in the highest number of publications, and with regard to the most important relationships *risk factor* and *MSD* or the triangle formed by *musculoskeletal-injuries*, *work-related MSD* and *diseases* can be highlighted.

4 CONCLUSIONS

In this paper a study on the evolution of the subject of ergonomics has been conducted and the conceptual ergonomics themes and thematic areas have been visually presented.

The analysis of the number of papers in each thematic area shows an increasing interest in ergonomics and in related thematic areas.

During the period 1999–2003, the term *ergonomic* appears and displaces the term *human*. Until this moment *human* was a focal point.

From 2008 to nowadays, the term *ergonomic* has been growing up and gaining a more prominent role. It has been strongly introduced into the business management and innovation. Finallu, it has been assumed the close relation between *ergonomic* and *human*.

The results shown here can be used by both experts and novices to better understand the evolution of ergonomics in the papers published from 1984 to 2013. Furthermore, these results could be used to predict new future research trends, given that it is logical to suppose that papers associated with the most productive and highest impact themes and thematic areas will be successfully published.

REFERENCES

Bailón-Moreno, R., Jurado-Alameda, E., Ruíz-Baños, R. & Courtial, J.P.2005. Analysis of the scientific field of physical chemistry of surfac- tants with the unified sci-entometric model: Fit of relational and activity indicators. *Scientometrics* 63(2):259–276.

Bailón-Moreno, R., Jurado-Alameda, E.& Ruíz-Baños, R. 2006. The scientific network of surfactants: Structural analysis. *Journal of the American Society for Information Science and Technology* 57(7):949–960.

Börner, K., Huang, W., Linnemeier, M., Duhon, R., Phillips, P., Ma, N., Zoss, A., Guo, H. & Price, M.A. 2010. Rete-netzwerk-red: Analyzing and visualizing scholarly networks using the network work-bench tool. *Scientometrics* 83(3):863–876.

Callon, M., Courtial, J.P., Turner, W.A. & Bauin, S. 1983. From translations to problematic networks–an introduction to co-word analysis. *Social Science Information Sur Les Sciences Socials* 22(2):191–235.

Chen, C. 2004. Searching for intellectual turning points: Progressive knowledge domain visualization. *Proceedings of the National Academy of Sciences, USA* 101(1):5303–5310.

Cobo, M.J., López-Herrera, A.G., Herrera-Viedma, E. & He-rrera, F. 2011a. An approach for detecting, quantifying, and visualizing the evolution of a research field: A practical application to the fuzzy sets theory field. *Journal of Informetrics* 5(1):146–166.

Cobo, M.J., López-Herrera, A.G., Herrera-Viedma, E. & He-rrera, F. 2011b. Science mapping software tools: Review, analysis and cooperative study among tools. *Journal of the American Society for Information Science and Technology* 62(7):1382–1402.

Cobo, M.J. 2012a. *SciMAT: Herramienta software para el análisis de la evolución del conocimiento científico.* Granada: Universidad de Granada.

Cobo, M.J., Lopez-Herrera, A.G., Herrera, F. & Herrera-Viedma, E. 2012b. SciMAT: A new science mapping analysis software tool. *Journal of the American Society for Information Science and Technology* 63:1609–1630.

Herr, B., Huang, W., Penumarthy, S. & Börner, K. 2007. Designing highly flexible and usable cyberinfrastructures for convergence. In W.S. Bainbridge & M.C. Roco (eds.), *Progress in convergence: Technologies for human wellbeing*: 161–179. Boston: Annals of the New York Academy of Sciences.

ISI Web of Science 2013. Thomson Reuters / Web of Science. Retrieved January 5, 2013, from http://scientific.thomson.com/products/wos/.

NIOSH, 2001. National Institute for Occupational Safety and Health US. *National Occupational Research Agenda for Musculoskeletal Disorders: Research Topics for the Next Decade.* Report by the NORA Musculoskeletal Disorders Team. DHHS (NIOSH) Publication No. 2001–117.

Persson, O., Danell, R. & Wiborg Schneider, J. 2009. How to use Bibexcel for various types of bibliometric analysis. In F. Åström, R. Danell, B. Larsen, & J. Wiborg Schneider (eds.), *Celebrating scholarly communication studies: A Festschrift for OllePersson at his 60th birthday*: 9–24. Leuven, Belgium: International Society for Scientometrics and Informetrics.

Porter, A.L. & Cunningham, S.W.2004. *Tech mining: Exploiting new technologies for competitive advantage.* Hoboken, NJ: John Wiley.

Rinder, M.M., Genaidy, A., Salem, S., Shell, R. & Karwowski, W. 2008. Interventions in the construction industry: A systematic review and critical appraisal. *Human Factors and Ergonomics in Manufacturing* 28:212–229.

Sci^2 Team. 2009. Science of Science (Sci^2) Tool. Indiana University and SciTech Strategies. Available at: http://sci.slis.indiana.edu.

Scopus 2013. Elsevier B.V. Database. Retrieved January 5, 2013, from www.scopus.com.

Village, J., Salustri, F.A. & Neumann, W.P.2013. Cognitive mapping: Revealing the links between human factors and strategic goals in organizations. International Journal of Industrial Ergonomics 43:304–313.

Wise, J.A. 1999. The ecological approach to text visualization. *Journal of the American Society for Information Science* 50(13):1224–1233.

Occupational Safety and Hygiene II – Arezes et al. (eds)

Psychosocial risks factors at work: A study of health professionals

C. Gonçalves
Instituto Politécnico de Coimbra, ESTESC—Coimbra Health School, Saúde Ambiental, Coimbra, Portugal

L. Simões Costa
Instituto Politécnico de Coimbra, ESTESC—Coimbra Health School, Ciências Complementares, Coimbra, Portugal and Centre of Psychology of the University of Porto, Porto, Portugal

A. Ferreira
Instituto Politécnico de Coimbra, ESTESC—Coimbra Health School, Saúde Ambiental, Coimbra, Portugal

ABSTRACT: Psychosocial risks are risks to mental, physical and social health, originated by work conditions and by organizational and relational factors. This study aimed to identify and evaluate the psychosocial risk factors among workers of three services in the health sector who carry out their duties in a Portuguese hospital. Data were compared between services according to the different professional categories and gender. Considering the workers information, we verify that they are exposed to psychosocial factors related to Work Intensity and Time, Emotional Demands, Social Relations at Work and Insecurity at Work. Differences in psychosocial factors where found between the three services, contrary to the other variables.

1 INTRODUCTION

For the human being, work is fundamental in life as part of his personal identity. The prevention of risks associated to work is a right that every worker should have (Decree-Law 102/2009, 2009), since, on one hand, it can be a synonym of something healthy, on the other, may be a place where the exposure to different constraints causes problems to their health and safety. Changes in the organization of work underlie the rise of emerging professional risks, including psychosocial risks, which may cause serious problems of deterioration in social, physical and psychological health due to its multiple consequences (European Agency for Safety and Health at Work-EU-OSHA, 2010).

For some authors, the origin of psychosocial risks focuses on four main sources of tension existing in the workplace: changes in workplaces, work pressures, expectations of the employee and relationships and behaviors (Association Régionale pour l'Amélioration des Conditions de Travail [ARACT], 2011). In fact, in any organization or company, regardless its activity or sizes, working situations leading to tensions are a reality. The fulfillment of goals and objectives of the company and the commitment from employees in achieving them, the relationship between colleagues and hierarchical superiors and the recognition for work done, are a group of components that are a source of balance at work (ARACT, 2011).

The notion of risk implies exposure to danger and the probability of occurring damage. Therefore at work what makes a risk to health to be psychosocial, is not its manifestation, but its origin, so psychosocial risks can be defined as the risk to the mental and physical and social health originated by the working conditions and organizational and relational factors (Gollac & Bodier, 2011; AESST, 2010).

According to Gollac and Bodier (2011) psychosocial risks factors at work can be grouped into six dimensions: The work intensity and time; the emotional demands; the lack/insufficiency of autonomy; the bad quality in social relations at work; conflicts of values and the insecurity at work.

The adoption of different perspectives in occupational health seems to be an appropriate approach, in order to identify and assess more extensively the factors affecting the various dimensions of health, trying to cover other problems and diseases that may exist in a hospital environment (Gollac & Volkoff, 2000). The hospital environment is one of working places with higher exposure to these factors. Working in healthcare institutions expose workers to unhealthy surroundings related to the organization of work, and even to circumstances where workers are exposed to emotions of loss, grief, physical pain and fear of failure (Rios, 2008). There are also requirements related to medical acts and to the speed and responsibility needed to save lives (Cabete, 2000). Yet, some studies confirm the existence of psychosocial factors like pressure and the lack of solidarity among colleagues and emotional exhaustion of health professionals, causing them some health effects (Carvalho, 2010; Ribeiro et al., 2010; Gomes et al., 2009).

Although the relevance of research on psychosocial risks is evident in several emerging studies, in those same studies the consequences of such risks is often emphasized to the detriment of their origin. So, this study is aimed to address the upstream of these risks and what can contribute to their appearance. The identification of these factors in different professions and services in the hospital may be an opportunity on the reflection upon psychosocial risks and open the pave the way to the integration in risk assessments done in organizations. Thus, we intended to verify the existence and origin of psychosocial risks in three professional services of a hospital.

2 MATERIAL AND METHODS

The present research was performed in the Physical Medicine and Rehabilitation Service, Emergency and Clinical Pathology Service, of a hospital in Portugal. The target population was composed by professionals working in the mentioned services. The instrument used was a hetero—applied questionnaire.

The questionnaire was developed by the authors based on the indicators suggested by Gollac e Bodier (2011) for each one of the psychosocial risk dimensions and undergone to a comprehensive test. The final version comprises51 questions, the first part to sample characterization, and the other ones divided in six sections, according to the dimensions mentioned above. The structure of the questionnaire allows three types of answers, such as yes or no, Likert scale and open response. The statistical processing of the data was conducted using the software SPSS Statistics version 19 to Windows.

3 RESULTS

58 individuals participated in this questionnaire; 34 female and 24 male, aged 26 up to 63 years old. Twenty work in the Physical Medicine and Rehabilitation Service, 24 in the Emergency and 14 in the Clinical Pathology Service. Professionally there are 6 medical doctors, 11 nurses, 12 operational assistants, 13 therapists, 8 technical assistants, 6 clinical analysis technicians, 1 biochemist e 1 pharmacist.

The overall results in relation to each category of psychosocial factors are shown below.

The work intensity and time of work: Most of the professionals (81.0%) work in team and half are shift workers (51.7%), or work during the night (44.8%). From the respondents, 51.7% state the impossibility of stopping the work. As to work fast and have to perform different tasks from the normal, 82.8% and 79.3%, respectively, respond positively Most of them have to replace colleagues in their absence, about 82.7%. The interruption of that task to start an unplanned one happens to 79.3% of respondents, and 34.5% of these say that the interruption disrupts work. Relating to consequences of possible errors that may happen in the workplace, it is shown that the majority considers that errors have consequences both for the quality of service and for their and for the patients' safety.

Emotional Demands: 77.5% of individuals experience moments of tension in the relationship with patients. On the other hand 91.4% replied to have the capacity to address these tensions. The fear of accidents or occupational diseases and contagion, does not exist for, respectively, 55.2% and 63.8% of the respondents.

Autonomy at Work: About 86.2% of professionals state to have autonomy to change their work.

Social Relations at Work: The possibility of discussing difficulties at work with colleagues and superiors is possible for more than 90% of the respondents. In the workplace 39.6% have already suffered inconvenience behavior from colleagues, 29.3% from superiors, 55.2% from patients and 58.6% from patients' family. 91.4% acknowledged their work recognized, being higher from the patients.

Ethical Conflicts: It is noteworthy that 20.7% have to go against their professional values, social values and 27.6% perform their work according to personal values.

Job and Work Insecurity: The fear of unemployment is real for 48.3%, lack of enough income for the future for 62.1% and not improving in career for 67.2%.

Most of the professionals (83.3% and 79.2%) in the Emergency Service and just over half (57.1% and 50%) of professionals from Clinical Pathology have shift work and night work. The vast majority of the professionals must work fast and experience moments of tension. The professionals of Physical Medicine and Rehabilitation Service refer to experience more fear of accidents and professional diseases and those in the Emergency are the ones who note their work recognized. In all services there are professionals who have already experienced inconvenient behaviors. The insecurity at work is present in the majority of professionals at the Clinical Pathology Service.

As regards to the results depending on the variable profession, only therapists do night work. The latter and the doctors are those who claim to do overtime and take home extra tasks. Only physicians report to have difficulty in achieving the daily goals. The existence of moment of tension is referred by 100% of nurses and 92% of therapists, and so the ability to manage these moments is bigger for the former than for the letter (91% e 77%). The therapists and the clinical analysis technicians are those who refer higher autonomy to change their work, and the doctor, the operational assistants and the senior technician state to have less (67%, 50% and 50%). Nurses suffer more inconvenient behavior mainly from patients and their relatives. Operational assistants and technical assistants are the professionals who most believe that their work is recognized and less than 50% of nurses and therapists consider the same. From the clinical analysis technicians, only 17% find their work recognized and these professionals are those who claim to work more in disagreement with their professional, personal and social values. Operational assistants mostly reveal job insecurity.

The majority of professionals, who performs shift work, 66.7%, and night work, 62.5%, are male. Women work more hours and take extra tasks home and have to work more quickly. Men report greater perception of consequences of errors in their own and patients' health, show to have more autonomy and to be more afraid of not improving their careers. Both men and women have already been subjected to inconvenient behaviors, women from colleagues and superiors and men from patients and their families. Women are those who state to experience more recognition for their work and men more fear of not improving their career.

4 DISCUSSION

The *work intensity and time of work* dimension includes subjecting to constraints of rhythm and to unrealistic and vague goals, interruption of activity, duration and organization of working time, such as the number of hours, night work and shift work and reconciling work/life outside work (Gollac & Bodier, 2011).

From all workers, many do not have the ability to stop their activity and the most need to work fast. They fell able to accomplish their goals, though they have to replace absent colleagues and perform other unplanned tasks. They are aware of the consequences of the errors they make, especially for their safety. These results indicate that professionals are more subject to constraints of rhythm rather than the problems in terms of duration and work organization. These constraints may contribute to the increased probability of making errors, and the consequences thereof. This is a referred consequence and such consequences are more valued for the institution or for patients rather than for their own safety. Relevant is, as well, the fact that professionals of Physical Medicine and Rehabilitation Service are the least likely of having atypical hours, the ones who have to work fast and the least who value the consequences of errors into their own health. The professionals who work longer shifts or perform night work are from the Emergency. Associated to this fact is the need, for many of them, to work fast. The latter may be exposed to problems related to the lack of a routine, which is meant to be the healthiest for human beings, and to other health problems where some studies revealed nerve problems, fatigue and sleep disturbances associated with achieving this type of shift (Cabete, 2000). But, also, those who do not make these working hours can be at risk of suffering, due to the intensity of their work. In relation to work overtime or at home, therapists and doctors are the most referred since these situations may also have negative consequences and the relationship between work and family, in which they cease to have a social life and time to spend with other people outside the workplace (Camelo & Angerami, 2008). The latter are the ones who claim to fell difficulties in reaching the daily goals. Men are the ones who have more atypical hours, but are women who perform extra hours and although the former have more difficulty in achieving goals and work more quickly, are also men who are more aware of the consequences of the errors in their own safety. There are no gender differences with respect to the reconciliation work family.

The *Emotional Demands* dimension include, among others, the relation with the public, the contact with suffering and fear (Gollac & Bodier, 2011). In a hospital, the relationships between professionals and patients can generate some tensions, relating not only to complaints and requests but also because of the suffering, anguish and anxiety. It is noted that professionals experience moments of tension regularly which are mostly related to the pathology, needs and claims of the patients. Although it is known that hospitals are dangerous and unhealthy environments (Elias & Navarro, 2006), just over half of the respondents claimed to have concerns in relation to accidents, occupational diseases and to illnesses associated with infection by patients. This may possibly be explained by a tendency to not enhance a condition that is part of the work and of routine activity, which itself constitutes a risk but also, in case of infection, there is a rooted professional logic saying that repeated contact with the disease brings immunization. Nursing professionals, associated to the Emergency are those who experience more moments of tension with patients and those who are mostly refer to have the capacity to cope with these tensions. Studies show that health professionals tend to choose this area because they feel more vulnerable to disease and pain, or because, in life circumstances, they have dealt with the loss of someone close (Rios, 2008) (Leite, 2005). Given the atrocities that are happening in hospitals, health professionals may tend to feel as caregivers although they can not cease to have the capacity to deal with these problems (Rios, 2008).

Autonomy includes, among others, autonomy in the task, predictability of work, the monotony and boredom (Gollac & Bodier, 2011). The importance of this factor lies mainly in a certain degree of autonomy in work activity, promoting the worker's ability to make decisions contributing to feel useful, responsible, giving as well meaning to the work they perform (Hennington, 2008; Rodrigues et al., 2006). The results show that, generally, professionals consider having autonomy to change their work but mostly the impossibility to stop working or make small breaks. This seems to show that they feel able to decide on what they do, but in reality, they face situations under their control, such as interruptions or pauses. In fact, it is all about dealing with people, often in situations where "waiting" is not allowed. To testify this, the clinical analysis technicians are the ones who reveal power interrupt work more easily.

The *social relations at work* dimension covers relations with colleagues, with the hierarchy, the support, recognition and internal violence (Gollac & Bodier, 2011). The communication among professionals proves to be very positive since they state they can discuss difficulties. The importance of interaction and communication at work is documented (Camelo & Angerami, 2008) and its absence may give rise to workers feel distressed and frustrated at work (Spindola & Santos, 2004). If communication and positive interpersonal relations are a reality for these professionals, on the other hand, a considerable number states to have already suffered inconvenient behaviors from colleagues, superiors, patients and families. These behaviors occur mostly in the Emergency and with nurses, mainly by patients and their families. In fact, in this service there are conditions that may explain these behaviors, since the number of people, their characteristics and pathologies are very different, and involve anxiety and anguish. This may lead to a greater difficulty in managing time and information and give rise to tensions that can be translated into the revolt by patients and families. Men are the ones who suffer the most from inconvenient behaviors from patients and their families, but in relation to inconvenient behaviors from superiors are the women who affirm to suffer the most. Another worrying reality is that just over half of the professionals feel recognized for their work. This indicates that many of them feel little rewarded by the utility they have at the workplace, which can lead to disinterest and to the feeling of useless work. This reality is most visible in Clinical Pathology Service and among clinical analysis technicians, although nurses and therapists also feel little recognized. Generally, professionals in Emergency are the ones who feel their work recognized and women, more than men, acknowledge this recognition, despite the high emotional demands. This can be justified by the perception of saving lives and the gratitude towards this result. On the other hand, the technical assistants and the operational assistants consider their work more recognized. Particularly the last one, who deal with patients in terms of hygiene, nutrition, have less responsibility about their health and may be seen as professionals who help in difficult situations. On the other hand, therapists, for example, affirm to be poorly recognized, often because the response to the disease is slow and requires long treatments.

The *ethical conflicts* are seen in conflicts of values, that is, when professionals have to act because they are imposed or asked against their professional, social or personal values (Gollac & Bodier, 2011). To do their job, there are professionals who have to act in disagreement with their values. Once again we face a situation that should be theme of analysis, since it is not assumed working involves actions against the workers beliefs. Are the reasons in the fact that someone must work fast, with such autonomy, and not corresponding to what actually happens, suffering from inconvenient behaviors? Analyzing these issues, we find that it is in the Clinical Pathology Service and specifically with the clinical analysis technicians, where most conflicts of values take place and these are the professionals who claim to have not a recognized work.

The *job and work insecurity* concerns to safety at work, salary and career (Gollac & Bodier, 2011). This study reveals that most professionals are afraid of unemployment, lack of income in his personal life and of not improving their career. The operational assistants are particularly afraid of lack of income in the personal lives and career progress. As there are problems of employability and economic stability in the country, it is natural that these professionals feel these insecurities regarding the workplace and particularly if the work contract is unfavorable, which can happen in the less specialized professional categories.

5 CONCLUSION

Psychosocial risk factors are well present in the workplace, and most of them are hardly detected. Thus, both the employers and professionals of occupational health need to be alert to such risks. The results of this study reveal the presence of some psychosocial risk factors for those who work in a hospital setting, as Work Intensity and Time of Work and Emotional Demands. As they are factors related to circumstances that may not be easy to modify, solutions have to be found among professionals and supervisors so that the pressure carried out by daily activity and the impossibility of take breaks can be mitigated. Social Relations at Work should also be subjects of attention, either in relation to inconvenient behaviors or to the recognition of the work performed. The issues concerning the insecurity at work, as they are disturbing, are nowadays a very transversal reality and, therefore, have also to be object of reflection in relation to the current context. Regarding to the variables under study, there were differences in risk factors, especially among services, since they have distinct characteristics which imply realities that can not be matched in terms of risk factors.

We think it is important to emphasize that the questionnaire used in this study is an original instrument, needs and deserves further development and to be applied to other workers. Nevertheless we cannot fail to highlight, as an important result of this study, the innovative potential of the instrument as well as the kind of results that one can get with it.

This study had some limitations due to the fact that some of these professionals work in shifts, and consequently it is not always easy to find different professionals in a given shift and due to the difficulty of not getting much time for breaks, as already stated. Despite these limitations, we understand the importance of undertaking these questionnaires, being present to inquire the professionals. It is important to include these psychosocial factors in the regular risk assessments. Workplaces should be evaluated and these issues should be addressed in the worker' consultation so that these risks can be prevented like all others.

REFERENCES

European Agency for Safety and Health at Work. (2010). *Inquérito Europeu das Empresas de riscos novos e emergentes (ESENER), Gestão da segurança e saúde no trabalho.*

Association Régionale pour l'Amélioration des Conditions de Travail. (2011). Prévenir les risques psychosociaux, Fiche Pratique n°10.

Cabete, D.G. (2000). *Risco, Penosidade e Insalubridade. Uma realidade na profissão de enfermagem.* Lisbon.

Camelo, S.H., & Angerami, E.L. (2008). *Riscos Psicossociais do Trabalho que podem levar ao Estresse: Uma revisão da Literatura.* Cienc Cuid Saúde.

Carvalho, G.D. (2010). *Mobbing: Assédio Moral em Contexto de Enfermagem.* Journal of Research in Nursing.

Decree-Law 102/2009 (Assembly of the Republic 2009).

Elias, M.A., & Navarro, V.L. (2006). *Relação entre o Trabalho, a Saúde e as Condições de Vida: Negatividade e Positividade no Trabalho das Profissionais de Enfermagem num Hospital Escola.* Latin American Journal of Nursing.

Gollac, M. & Bodier, M. (2011). *Mesurer les facteurs psychosociaux de risque au travail pour les maîtriser (Relatório do Collège d'Expertise sur le Suivi des Risques Psychosociaux au Travail).*

Gollac, M. & Volkoff, S. (2000). *Les Conditions de Travail.* Paris: Éditions La Découverte.

Gomes, A., Cruz, J. & Cabanelas, S. (2009). *Estresse ocupacional em profissionais de saúde: um estudo com enfermeiros portugueses.* Psychology: Theory and Research.

Hennington, É.A. (2008). *Gestão dos processos de trabalho e humanização da saúde: reflexões a partir da ergologia.* Rio de Janeiro: Journal of Public Health.

Leite, G.M. (2005). *Uma relação delicada: o médico e o seu paciente.* Scientific E-Journal of Psychology.

Ribeiro, L., Gomes, A. & Silva, M. (2010). *Stresse ocupacional em profissionais de saúde: Um estudo comparativo entre médicos e enfermeiros a exercerem em contexto hospitalar.* Portuguese Association of Psychology.

Rios, I.C. (2008). *Humanização e Ambiente de Trabalho na Visão de Profissionais da Saúde.* Health Soc. São Paulo.

Rodrigues, P.F., Alvaro, A.L., & Rondina, R. (2006). *Sofrimento no Trabalho na Visão de Dejours.* Scientific E-Journal of Psychology.

Spindola, T. & Santos, R. D. (2004). *Trabalho versus Vida em Família. Conflito e Culpa no Cotidiano das Trabalhadoras de Enfermagem.* Science and Infirmary.

Occupational Safety and Hygiene II – Arezes et al. (eds)
© 2014 Taylor & Francis Group, London, ISBN 978-1-138-00144-2

Safety at construction site: Case study at the university hospital of the Federal University of Juiz de Fora

W.C. Silva & M.A.S. Hippert
Universidade Federal de Juiz de Fora—PROAC, Juiz de Fora, Brasil

ABSTRACT: The issue of health and safety gains larger dimensions when applied to larger and more complex constructions, such as hospitals. The objective of this paper is present the model of health and safety management used for the construction of a university hospital at UFJF in Juiz de Fora—MG, in order to create and update the resources for health and safety management in large construction sites. The methodology is action-research as the first author acts as the supervisor of the studied construction. In this construction, a period of just over a year, no serious accident with fatality or removal of any production employees was observed. This study shows the way how a construction company, together with supervision, acts in relation to occupational safety in the performance of a public work. The results indicate the existence of a carefully established management process, followed by constant monitoring, which are the determinant factors for the safety of workers be achieved throughout the performance of the work.

1 INTRODUCTION

In order for the State to fulfill its role in ensuring the basic rights of citizenship is necessary to have the formulation and implementation of policies and government actions which must be guided by cross-cutting and intersectoral approaches. In this perspective, the actions of the worker safety and health require a multidisciplinary, interdisciplinary and intersectoral activity capable to contemplate the complexity of the production-consumption-environment and health relationship (Brazil, 2012).

In the international literature, the view that improvements in safety management are investments has grown in the construction industry, mainly due to increased costs arising from accidents (Gyi, 1999; Rowlinson, 2000; Bridi, 2013).

In this sense, in an articulated and cooperative way the ministries of Labor, Social Welfare and Health, in order to ensure the work, the basis of social organization and fundamental human right, they established that the work must be carried out in conditions which contribute to improving the worker's quality of life, the personal and social fulfillment of workers, and with no harm to their physical and mental health. So, they developed a National Policy on Health and Safety of the Worker (Brazil, 2012).

This policy aims to promote an improved quality of the worker's life and health, through the articulation and integration of ongoing governmental actions in the field of production consumption, environment and health relations (Brazil, 2012).

In Brazil, the legislation that regulates the Occupational Safety and Health work is supported by the Law 6.514/1977 along with the Ordinance 3.214/1978, which introduces the Regulatory Standards (in Portuguese, Normas Regulamentadoras, hereinafter, NR's). At the present time there are 36 NR's governing various activities that put the workers in danger; as an example we can mention: construction, working on heights, working in confined spaces, etc. (Brazil, 2013). The NR's are mandatory for private and public companies and for public institutions of direct and indirect administration, as well as bodies of the Legislative and Judiciary, which have some employee under the Consolidated Labor Laws—CLT. The provisions contained in the NR's are applied, as appropriate, to independent workers, entities or companies which use the service and to the unions representing their respective professional categories (Brazil, 2009).

The identification of good or best practices in occupational safety and health has been a quite common focus in many academic studies in different countries, especially the United States (Abudayyech, 2006).

The issue of health and safety gains larger dimensions when it is applied to larger and more complex constructions, such as hospitals, for instance. Starting from these considerations, the aim of this paper is to present a model for managing health and safety which could be applied in the implementation of the university hospital at UFJF, in Juiz de Fora—MG—Brazil, in order to create and update the resources for health and safety management in large construction sites.

2 METHODOLOGY

The methodology is action-research since the author first acts as the supervisor of the studied construction. Action-research is a link between research and action in a process in which the actors involved participate, along with the researchers, to interactively reach the elucidation of the reality in which they live, identifying collective problems, seeking solutions and experimenting solutions in real situation (Thiollent, 2009).

3 SAFETY AT CONSTRUCTION SITE

Considering the constitutional rights to health, social security and labor and the need of structuring the intra-governmental coordination in relation to safety and health of workers, the Interministerial Working Group was established with the following duties: a) reassess the role, the composition and duration of the Interministerial Executive Occupational Health Group—GEISAT; b) examine the measures and propose integrated and synergistic actions to contribute to enhance the actions for the of workers' safety and health; c) prepare a proposal of National Worker's Occupational Safety and Health Policy, noting the existence of interfaces and common actions between different governmental sectors; d) analyze and propose inter-sector actions to ensure the right to occupational safety and health, as well as the specific actions area, requiring immediate implementation by the respective Ministries, individually or jointly and e) share the information systems regarding to workers' safety and health which exist in each Ministry (Brazil, 2012).

The NR 18 establishes guidelines of an administrative, planning and organization order, seeking to implement control measures and preventive safety system in the processes, conditions and working environment in the Construction Industry. According to the NRs', the following are considered activities of the Construction Industry: activities and services of demolition, repair, painting, cleaning and maintenance of buildings, construction of any number of decks or type of building, including maintenance of infrastructure works and landscaping (Brazil, 2013).

It is forbidden to allow the entrance and the permanence of workers at the construction site, without being assured by the measures contained in the NR 18 and compatible with the stage of the work. The compliance with what is set forth in this NR does not relieve employers from the provisions relating to the conditions and environment of work as determined by federal, state and/or municipal legislation, and in other ones established in collective labor bargaining (Brazil, 2009).

Also, according to Saurin (2002) the NR-18 (Conditions and Work Environment in the Construction Industry), is the only standard directed specifically to the construction industry, becoming the main Brazilian legislation for regulating the safety and working conditions at the construction site.

However, in a large portion of enterprises, a secondary role is given to occupational safety in the management of companies, which stems from the lack of norms, with a too prescriptive nature in certain requirements, which makes the nonconformity easier and discourages the adoption of alternative solutions. This situation tends to change, as states Saurim (2002), when the companies realize the potential benefits of investments in safety and are aware of its interface with all other management processes, such as production planning, projects and budgets and then the gain in productivity is substantial.

Although there is no consensus as to the name, type and number of categories of management practices in occupational health and safety, some categories of practices are investigated further by the academic community, such as the categories "Management Commitment to occupational safety and health", "Training", "Occupational safety and health planning", "Incentive programs", "Involvement of employees" and "Subcontractor management". Some of these practices are also explicitly mentioned by management safety standards, such as the OHSAS 18001 (British, 2007; Bridi, 2013).

4 CASE STUDY: CONSTRUCTION WORK OF THE UNIVERSITY HOSPITAL OF THE FEDERAL UNIVERSITY OF JUIZ DE FORA—BRAZIL

Hospital projects differ from those of residential buildings since they need to be pre-approved by the National Sanitary Surveillance Agency (ANVISA), according to the Board Resolution (RDC) No. 50. According to this resolution, in its article 1, in which it is quoted that is the responsibility of ANVISA "To approve the Technical Regulation for the planning, programming, design, evaluation and approval of physical projects of health care buildings, attached to this Resolution to be observed throughout national territory, the private and public sectors, including:

a. new construction of health care facilities across the country;
b. the areas of the already existing health care facilities to be extended;
c. the reform of existing health care facilities and of those previously not intended as health care facilities" (ANVISA, 2002).

4.1 *Characterization of the construction site*

The Don Bosco University Hospital of the Federal University of Juiz de Fora, consists of 2 existing blocks, Block A and B, and 7 other blocks and a sewage of treatment plant (ETE) which are under implementation, as described below:

Block E: It is the largest building in the complex; its highest part is composed by 13 floors, with two basements. The building is a steel structure with three rigid concrete cores, which is anchored to the metal structure, providing rigidity to the building. The slabs are steel deck, giving greater flexibility in the construction process, i.e., with this construction method, the time spent with removal of the wooden form was saved. Internal partitions are in dry wall, also giving greater agility in performing the building; consequently, as the dry wall is lighter than traditional masonry, the dead load of the building was reduced, leaving the metal structure lighter and cheaper. The building will house infirmary wards, pre-delivery, delivery and post-delivery birthing rooms, nutrition and dieting service, clinical engineering, waste shelter, substation, intensive care unit, transfusion agency, hemodynamics, transplant and burn centers, pharmacy, emergency, pathological anatomy, maternity, neonatal intensive care unit, obstetric center and warehouse.

Block E7: The building consists of a ground floor plus 1 concrete structure floor, the internal partitions are in dry wall, but the dressing rooms and bathrooms are in traditional masonry wall. The building will house changing rooms, study center, library, blood bank and a place to offer services to patients with high blood pressure and diabetes.

Block E8: The building consists of a ground floor plus 4 concrete structure floors, with internal partitions in dry wall. This building will house pharmacy, endoscopy, outpatient, dental and administration.

Block E9: The building consists of a ground floor and 1 concrete structure floor, with internal partitions in dry wall. The block will be occupied in its two floors by outpatient wards.

Delivery Block: a building with 2 floors in reinforced concrete structure.

Living center: a building structure with a concrete floor, which will house a restaurant and café, plus space for contemplation of the landscape.

Block G: Building with 2 floors in reinforced concrete structure, which will house medical psychiatric service.

Sewage treatment plant (ETE): Its base made in concrete and it will be fitted with prefabricated sewage tanks.

The total area to be built is approximately 44,000 m^2 and the contract value of approximately 160 million Brazilian reais. In this construction there is a supervisory team consisting of 7 employees of Federal University of Juiz de Fora who follow the perfomance of services.

4.2 *The safety practiced at the construction site*

The construction site is extensive, with a lot of equipment such as bulldozers, trucks, freight and passenger elevators and a crane, among others.

Photo 1. Overview of the construction site—*Source*: Author, May 2013.

Photo 2. Overview of the construction site—*Source*: Author, May 2013.

Safety is an item to be permanently monitored. The company engaged in the construction has a team of about 6 persons, between safety engineer and occupational safety technicians, who are responsible only for safety.

There is a range of tools for managing and controlling safety at the construction site. Next, the documentation and the procedures which must be performed to ensure safety will be described:

Kick-Off Meeting: held between Employer and Contractor representatives. At this meeting, the Contractor reports to the Employer safety-related requirements and procedures.

Integration: Every employee, including subcontractors, to become active in the construction site will undergo training with Contractor safety personnel. In this, all employees are introduced to the construction site, the risks and the precautions to be taken, the obligation to wear the personal protection equipment (PPE) and collective protection equipment (CPE), which are the procedures that must be adopted in case of accidents, the escape routes and other relevant information to ensure safety.

Daily Safety Dialog (DSD): Daily, before the start of activities, one conversation is performed with employees lasting between 10 and 15 minutes. In this conversation, some issues related to health, safety, environment and quality procedures are discussed.

Preliminary Risk Analysis (PRA): The PRA is held before starting any new activity, in this moment the activity is described, the risks and mitigation measures stating what procedures should be followed, indicating the correct use of personal and collective protective equipment.

Meetings with foremen and supervisors: They are specific meetings for foremen and supervisors with the purpose of multiplying the knowledge and culture about safety at work, because these professionals are the ones responsible for driving the performance of services and who must understand that if is safe to perform the tasks, the risk of accidents is greatly reduced, avoiding loss of time taking proper care with casualties and the actual loss of the injured professional.

Training for working on heights: All employees of the HU construction must undergo this training since a fall in excess of 2 meters in height can generate major damage. In this training, the employees are informed about the risks, how to use the personal and collective protection equipment correctly, and how to act if an accident occurs to him or to a coworker.

Training for working in confined environments: The risks that exist in the implementation of services in closed and confined places are addressed in this training. In this construction, there is no actual need for this training, because every service excavation in confined area is done by mechanical means, i.e., the hand-dug caisson foundations were replaced by mechanically bored piles.

Training for working in power grid: this training is performed by professionals, or specialized individuals or companies. The risks of working with electricity are informed in this training told; moreover they are informed about the working procedures and the proper use of personal and collective protection equipment.

Internal Accident Prevention Commission (CIPA): These meetings take place on the first Friday of each month with the participation of committee members assigned by Tratenge and the subcontractors, plus a representative of Federal University of Juiz de Fora. The whole process of communication and election of the CIPA members is filed at the construction Health and Safety Executive (HSE) office, as well as the training of committee members and nominees.

Specific trainings are also performed in the construction site for operators of circular bench saw, walsiva pistol, cement mixers and hand tools. Moreover, there is a control of operators through the identification badges and credentials displayed in the equipment.

An accredited company was hired for training the rack elevator operators, well as for the training on securing cargo.

Risk map: it is posted throughout the construction site and a copy is filed in the CIPA folder.

Crane Weight Plan: this procedure is taught to the crane operator and the signalers. They are informed about the crane's maximum weights and its possible rotation directions.

Monthly assessment: At the beginning of each month, a operational meeting is held with the Oversight of the Federal University of Juiz de Fora and Engineers and Supervisors of the Contractor. In this meeting, items for the proper conduct of the work are discussed and also five items are evaluated: personnel, equipment/tools, contractor's performance, safety and environment:

Personnel: This item assesses the quality and technical level, quantity, performance and behavior in service, behavior towards others, relationship with the Audit of the Federal University of Juiz de Fora and support to the Engineer's work.

Equipment/tools: Evaluates the quality (condition) and quantity of equipment available at the construction site.

Contractor Performance: This item encompasses the compliance with the schedule, collaboration with the Supervision of the Federal University of Juiz de Fora, coordination with the office/construction, quality of services performed.

Safety: In this part of the evaluation, attention is given to the employee's working method, CIPA performance, jobsite safety and documentation update.

Environment: This item assesses how the Construction Waste Project Management is being implemented.

The above items are evaluated with grades 1–10, corresponding from poor, fair, good and great. The items which are evaluated as 9–10 (best) must be explained by the Supervision of the Federal University of Juiz de Fora as well as the items valued between 1–4 (poor) and 5–6 (regular). If there is a repetition of poor and regular assessments, the Contractor is called for a meeting, in which they discuss the items with value 1–6 and propose some necessary corrections.

Photo 3. Installation of steel structure—*Source*: author, May 2013.

Table 1. Number of accidents in the construction site—author, September 2013.

Period	Number of workers	Number of accidents: Without loss of time	Lost time
August-12	25	0	0
September -12	31	0	0
October -12	53	0	0
November -12	61	0	0
December -12	83	0	0
January -13	60	0	0
February -13	69	0	1
March -13	67	0	1
April -13	114	0	0
May -13	91	0	0
June -13	125	3	1
July -13	149	1	0
August -13	163	1	0
September -13	177	0	1
TOTAL		5	4

The construction began in August 2012 and in just over a year no any serious accident with fatality or removal of any production employees was observed. There were some incidents which led only to property damage. As an example, we can mention the lifting of materials, but it hasn't caused any physical damage because there were some restriction measures on the access and movement of persons in the hazardous areas. This fact is due to the great importance that is given in Safety in Construction site with the autonomy of the Safety sector to suspend or abort any activity that poses a risk to the employees.

Below, the summary table on the rate of minor accidents which occurred on the construction site.

5 FINAL CONSIDERATIONS

In order for a good safety management to take place in a construction site, the compliance with the standards is only the first step for the drastic reduction of the high economic and human losses resulting from accidents.

This work showed how a construction company, together with supervision of the Contractor, acted in relation to the occupational safety in the performance of a public construction. The study showed the existence of a carefully established process, followed by constant monitoring in which some factors were established for the worker's safety to be achieved over time until the construction is completed.

BIBLIOGRAPHY

Abudayyech, O.; Fredericks, T.; But, S.; Shaar, A. An Investigation of management's Commitment to Construction Safety. In: International Jounal of Project Management, v.22, p.167–174, 2006.

ANVISA/Agência Nacional de Vigilância Sanitária/ Resolução—RDC nº 50, de 21 de fevereiro de 2002.—Dispõe sobre o Regulamento Técnico para planejamento, programação, elaboração e avaliação de projetos físicos de estabelecimentos assistenciais de saúde. available in: <http://portal.anvisa.gov.br/wps/wcm/connect/ca36b200474597459fc8df3fbc4c6735/RDC+N%C2%BA.+50,+DE+21+DE+FEVEREIRO+DE+2002.pdf?MOD=AJPERES>. Accessed August 30, 2013.

Brasil/Ministério do Trabalho/Portaria 3214 de 08 de junho de 1978). Aprova as Normas Regulamentadoras-NR–do Capítulo V do Título II da Consolidação das Leis do Trabalho, relativas à Segurança e Medicina do Trabalho. in: ATLAS Manual de Legislação. 62ª ed. São Paulo, 2009.

Brasil/Normas regulamentadoras. Available at: http://portal.mte.gov.br/legislacao/normas-regulamentadoras–1.htm. Access em 26 outubro 2013.
Brasil/Política Nacional de Segurança e Saúde do Trabalhador–Brasília: Comissão Tripartite de Saúde e Segurança do Trabalho, Abril de 2012.
Bridi, M.E.; Formoso, C.T; Pellicer, E.; Fabro, F.; Castello, M.E.V.; Echeveste, M.E.S. Identificação de práticas de gestão da segurança e saúde no trabalho em obras de construção civil. In:Revista Ambiente Construído, Porto Alegre, v. 13, n. 3, p.43–48, jul./set. 2013.
British Standards Instituition. OHSAS 18001: occupational health and safety management system. Londres: BSI, 2007.
Gyi, D.; Gibb, A.; Haslam, R. The Quality of Accident and Health Data in the Construction Industry: interviews with sênior managers. In: Construction Management and Economies, v17, p.197–204, 1999.
Rowlinson, S. Human Factors in Construction Safety Management Issues. In: Coble. R.: Hinze, J., Haupt, T. (Eds.). In: Construction Safety and Health Management. Prentice-Hall: Upper Saddle River, p.59–83, 2000.
Saurin, T.A.; Lantelme, E; Formoso, C.T. Contribuições para Aperfeiçoamento da NR-18: condições e meio ambiente de trabalho na indústria da construção. In: Normalização e Certificação na Construção Habitacional/Editores Humberto [e] Luis Carlos Bonin.—Porto Alegre: ANTAC.—(Coleção Habitare, v. 3) p.174–207, 2003.
Thiollent, Michel. Pesquisa-ação nas organizações. 2.ed. São Paulo: Atlas, 2009.

Occupational Safety and Hygiene II – Arezes et al. (eds)
© 2014 Taylor & Francis Group, London, ISBN 978-1-138-00144-2

Integrated management system in civil construction based on PMBOK guide extension

B.F. Rocha & M.A.S. Hippert
Universidade Federal de Juiz de Fora—PROAC, Juiz de Fora, Brasil

ABSTRACT: Nowadays there is a growing concern about the quality, the environment and worker safety in the construction market. Management systems are being used more often in this segment in order to perform an assistance in the management of these areas. The use of these systems in isolation can cause incompatibility issues, particularly as a way to reduce these problems using an integrated management system. There may be difficulties in developing an IMS using international standards because these are not specific to the construction industry. Based on this concept, it is highlighted the PMBOK guide to extension construction, published by PMI as a guide for developing an integrated management system directed to civil construction. This paper aims to do a literature review in order to show the possibility of developing an integrated management system specific to construction by the vision of PMI.

1 INTRODUCTION

Nowadays there is a growing concern about the quality, the environment and worker safety in the construction market. By seeking a gap in the market, companies are using management strategies that meet the costumers' requirements, employees and society (Mendes, 2007). To satisfy these requirements, Quality Management Systems (QMS), Environmental Management Systems (EMS) and Safety Management Systems are being used (OHSMS).

These three management systems can be understood as a set of guidelines and procedures assembled and aligned to enable the planning and control of a company to increase the quality of the product offered, adequate it to the environmental constraints, minimizing the impacts of its activity and control risks and hazards for existing workers in the workplace (França, 2009).

In isolation, management systems may not get the most efficiency, so it is important to implement an Integrated Management System (IMS). IMS can involve various types of management systems, and so all the subsystems involved in the IMS must be identified and monitored (França, 2009). In this work it will be used the IMS formed by the Quality Management System, Environmental Management System and Safety Management System.

This study aiming to show the possibility of implementing IMS in the construction companies by the vision of Project Management Institute (PMI) instead of ISO standards commonly used in companies in others segments. This possibility will be shown through an analysis the management systems of IMS according to the chapters of Construction Extension to a Guide to the Project Management Body of Knowledge (PMBOK), third edition.

2 INTEGRATED MANAGEMENT SYSTEM

A management system can be defined as a set of procedures, resources, materials and people whose associated components interact in an organized manner in order to accomplish a task and achieve a specific goal (Mendes, 2007).

The Integrated Management System (IMS) encompasses all management systems deployed in the enterprise so that they all work harmoniously obtaining a greater efficiency. The most used IMS by the civil construction companies is formed by the integration of quality, environmental and safety (Mendes, 2007).

The most common IMS model uses the ISO 9000 series of standards as a basis for the quality management system, ISO 14000 series of standards related to the environmental management system and OHSAS 18000 management system for the safety and health at work (França, 2009). But this model is not specific to any area of knowledge. In order to assemble an IMS oriented to the construction area, we can use the PMBOK Guide in its extension to civil construction called Construction Extension to a Guide to the Project Management Body of Knowledge as the basis for a specific IMS that can be complemented with the international standards in subjects that it is not addressed.

The third edition of Construction Extension to a Guide to the Project Management Body of Knowledge is a companion guide to the PMI PMBOK launched by PMI aiming to cater specifically the construction industry. In this new guide, some processes were removed and adapted as they are, in general, used for management in an industry, and it is not an applicable concept in the construction industry. Moreover, it presents four areas of knowledge: Security Management Project, Environmental Management Project, Financial Management Project and Project Claims, which together, account from 13 to 44 existing processes in the PMBOK® Guide Third Edition (Arbache, 2011).

By deploying the three systems, some building companies do it in isolation, putting different responsibilities for each management system dissociating its focuses. This reflects in organizational conflicts, may causing incompatibilities within the focuses. The IMS model by PMI performs all processes in an integrated way, with responsibility by a team of processes managers. The definition of a single coordination is essential in the implementation of the IMS, because the leader is personally involved in the communication of goals, plans and to motivate and reward employees. (Zeng *et al*, 2006).

In civil construction, the deployment of IMS is a key factor in achieving productivity, competitiveness, quality, and respect for the environment, cost and run time jointly. Furthermore, the implementation and maintenance of IMS facilitates the relationship and group work in construction companies (Dias, 2003). Many other advantages can be obtained with the use of IMS and among them we can highlight (Degani *et al*, 2002):

- Reduction in costs (with internal audits, training, insurance, etc);
- Benefits to the Company in meeting the requirements of customers and other stakeholders;
- A more effective law fulfillment;
- Single system facilitates the understanding and involvement of the employees;
- Scratches, damages and waste prevention (environmental accidents, fines and other penalties, etc.);
- Simplifies documentation;
- Avoids conflict between procedures;
- Improves management processes due to standardization.

Before deploying an IMS it is interesting that the construction company performs an analysis of its situation, structure and performance requirements in order to achieve greater efficiency in the implementation of the IMS. Moreover, it is necessary the commitment of everyone involved within the organization, firmness of purpose and the adaptation of models of norms to the reality of the company (Degani, 2003).

According to Sjoholt (2003), for the IMS to be effective and there is a commitment of those involved, knowledge must be used about how to involve people to adopt new knowledge. In order to make it happen, it is necessary that the leader raises workers' awareness of the importance of procedures and convinces managers that they will be awarded with improvements if they apply it and control it correctly.

3 QUALITY MANAGEMENT SYSTEM

The quality management system can be developed according to the recommendations of Chapter 8 of Construction Extension to a Guide to the Project Management Body of Knowledge, third edition. It must be prepared to ensure that the project meets the needs for which it was undertaken. In the case of civil construction, it is important to ensure that the project specifications are met within the agreed schedule and budget, as well as to maintain the standard of quality of materials used and workmanship. We can divide the quality management in three processes: quality planning, quality assurance and quality control.

Quality planning aims to identify which are the quality standards relevant to the project and what will be done to meet them. For this, it must be established a quality policy, or in other words, what are the intentions and principles related to the quality of the organization. This policy is a policy of the company, a specific policy for the project, or the client may require a specific quality policy. However, it is important that there is a consensus and integration between the terms of the company's quality policy executor, the client and other stakeholders, creating a single policy for that project.

Besides the quality policy, still in the planning process, the scope statement should be done, which documents the main project deliverables and objectives thereof, the declaration of the product containing details of technical issues and any other concerns that may affect the planning and the rules and regulations that the project should follow as local building codes, engineering standards, both at the national and state and municipal.

Putting all the above information together, we can perform an analysis with the aid of flowcharts, experiments projects, costs related to quality, value engineering to identify the best alternative in terms of methodologies and materials in order optimize the construction and thus, prepare the management plan quality, operational definitions and checklistis that will be used throughout the project to ensure quality.

With this plan in hand, the process of quality assurance begins, which is up to the manager to evaluate the performance of the project, ensuring that it is meeting the placed requirements and being verified through audits of quality. Audits are performed to evaluate the effectiveness of the processes and to verify compliance with all procedures, identifying non-compliances and proposing corrective and preventive actions, thus ensuring improvements in the project's quality.

The last proposed process is quality control, which monitors the results of the project in order to verify if they meet the established quality standards and to identify ways to eliminate causes of dissatisfaction. This process should be performed during all phases of the project. In order to control the quality, tools are used as usual inspections (technical or administrative) control charts (curve S, Gantt chart, Pareto diagram) for materials and analysis of trends that are used in specific areas to predict the effectiveness of program quality control.

After performing all these checks, reports that verify if the items inspected were accepted or not, are prepared, and for those which were not accepted, there must contain a solution or an alternative rework. These unaccepted items generate a separate report of non-compliance, demonstrating the deficiencies found and the necessary steps to resolve them. In addition, there is a catalog of rework, which consists of a list of unaccepted items that must be executed again. The rework should always be avoided in order to reduce the waste of materials which generate an increase in costs and schedule changes.

When deploying the system of quality management using PMBOK, questions may arise on how to perform the processes suggested, as the PMBOK is a guide which discusses what should be done, but not how it should be done. In case of doubt, the guide can be used in conjunction with ISO 9001:2008 standard which provides a list the requirements to be fulfilled by a quality system.

4 SAFETY MANAGEMENT SYSTEM

Security management is basically a subset of risk management also addressed in the PMBOK. Good security practices can reduce or eliminate workers' accidents and injuries, improving performance and reducing costs. The safety management system can be developed according to the recommendations in Chapter 13 of Construction Extension to a Guide to the Project Management Body of Knowledge, third edition. It must be prepared to ensure the project is executed with all necessary care to avoid accidents that may cause injury or property damage. The PMBOK proposes three processes to establish a project safety management: safety planning, the execution of a security and management and safety records.

The security planning consists in developing a plan to manage the risks for the security of the project. First of all, the security policy to be adopted must be set, since many contractors and engaged people have their own security policy, as well as in quality, a common security policy for everybody involved in the project should also be adopted. The contracts with other companies should also have specific requirements in order to ensure safe execution.

The location of the project should also be considered, as it can have a major influence on its safety. In addition, one should always follow the rules, regulations and national, state and local laws related to safety. After these initial data have been collected, an analysis about hazards should be made, i.e., a review of the construction process in order to identify all risks to the workers involved in construction as well as for other employees.

It is very important to attend to the choice of the contractors. When choosing a subcontractor company, one should always choose those that have a good record of safety performance. Furthermore, it is important to educate all employees. Currently, one of the most effective awareness is rewarding those who develop more effectively work with insurance benefits as jackets, dinners, and even financial compensation.

Gathering all these data, we can draw up a security plan of the project, which is a key document in order to provide guidelines for reducing accidents and property protection. This plan should contain a list of safety individual equipment and collective items that entrepreneurship and employees must possess in order to avoid accidents, first aid supplies, telephones and addresses useful in the event of an emergency. Above all things, it should include procedures to be followed in order to work in the construction area in a safe way, both for possible risks and for specific risks as working at height, electricity, etc. This plan may also contain a program to inform new employees about the security policy and ensure that they comply with the safety standards of the company. Still in the planning phase, one should be responsible for determining the safety of the project, to work with employees on awareness, prevention and audit. It should also establish the costs that will be spent for this security management and add them to the project's budget.

The second process is the implementation of the security plan, which involves the application and implementation of the plan developed in the previous process. To make it happen, tools are used as personal protective equipment and collective extinguishers, warnings, periodic checking of equipment, training, regular health checks, safety inspections and accident investigations. When this plan is executed with success, a small number of lesions are achieved, reduced spending on insurance employees, an improvement in productivity and the in the company's reputation in the consumer market.

The third process is the administration and safety records, which consists in reporting safety activities through reports containing all security-related information on the project. These reports may follow specific standards of the company, or standards established by regulations, rules or contracts. These reports may contain records of inspections, meetings and training record, record of accidents and their investigation. These reports should be reported to governmental agencies in the event of accidents, as well as serve to establish and monitor safety performance goals established by the company.

As in quality management, the PMBOK provides what must be done, but not how. As there is no international standard ISO for safety, the OHSAS 18000 standard is used as the basis for security procedures. Moreover, it is very important to check the laws in force in the project site, as these provide more detailed procedures of security that must be deployed.

5 ENVIRONMENTAL MANAGEMENT SYSTEM

The environmental management system can be designed by following the recommendations of Chapter 14 of Construction Extension to a Guide to the Project Management Body of Knowledge, third edition. It must be prepared to ensure that the impacts generated by the execution of the project are within the limits of the law. Three processes are proposed by PMBOK for the preparation of an environmental management system: The environmental planning, environmental guarantee and environmental control.

The first process is the environmental project which consists in identifying the environmental impacts which the project can generate and develop solutions. It should be done along with the planning of the enterprise where there is a greater possibility of changes without generating costs. First of all, one must have in hands the scope of work, in which should be included which ones permit the development needs and environmental laws and regulations that the project must follow. Furthermore, an analysis should be made of where the project will be built with the environmental characteristics of the site, information on similar project, and construction features of the project that might generate some impact. It is also necessary to know the company's environmental policy, and as in previous management systems, ensure that this is in accordance with the policy of the contractors and also with the project.

After gathering this information, one should perform an analysis of the impacts found and seek a solution for it. This analysis should be done by seeking to reconcile the interest of stakeholders, with environmental interests through flowcharts and benchmarking with companies that have achieved good results in similar environmental situations. Once the analysis is completed, it is prepared the environmental management plan which basically describes how the management team will implement its environmental policy. Basically, this plan must contain features of the site before and after the project, the anticipated impacts of construction thereof, as well as the emergency measures in case of any unplanned impact. It should also be prepared a checklist to assist in controlling the impacts during the implementation of the project.

The second process proposed by PMBOK is the environmental guarantee. In this process, all proposed activities and implemented systems related to environmental policy are checked in order to ensure that the project will satisfy the environmental standards proposed. This process must be performed throughout the project. The verification is done through audits and analysis results of the environmental control. Moreover, one should make the recycling of the materials and encourage environmental awareness of all stakeholders through training. Performing this process correctly can increase the effectiveness and efficiency of the project, reducing environmental impacts.

The third process is the environment control, which involves monitoring specific results to determine whether the project complies with existing environmental standards, identifying ways to eliminate causes and effects of unsatisfactory results. This control is done through quality control tools and risks, analyzing the audits, reports, checklists and other environmental guarantee tools. From this control, one must do improvements in environmental policy and accept or reject the work performed, determine rework on unaccepted items, make adjustments in processes and complete checklists that have been started in the earlier process.

The environmental management system by PMBOK was developed based on ISO 14001:2004 standard, this way, it already provides some information on how to make environmental management, different from other management systems presented. While providing information to run, they are still very small when compared to ISO 14001:2004 standard. Thus, in case of doubt, the manager should use the standard for detailed information on how to prepare the environmental plan and how to execute it efficiently.

6 CONCLUSION

The Integrated Management System—Quality, Safety, Environment—is already something quite widespread in the industry overall, but still not so used in civil construction, mainly small companies. The use of international standards for building management systems provides a general theoretical basis, which can be used in any industry, having the necessity of being adapted by the adaptation responsible for the construction market.

This approach highlights the Construction Extension to a Guide to the Project Management Body of Knowledge which is intended as a guide to managing projects for construction. As a basis on this tab, it is possible to set up an IMS designed specifically for this segment, using international standards as a complementary bibliography in case or doubts.

This study conducted a literature review in order to show the possibility of developing a IMS according to the recommendations of the PMBOK. The implementation of this integrated system on a construction company is the theme of the work that will be done in sequence.

REFERENCES

Arbache, A.P. 2010. Projetos sustentáveis estudos e práticas brsileiras. Ed. Editorama. Chapter 3. In: http://www.arbache.com. Accessed in August 10, 2013.

Degani, C.M.; Cardoso, F.F.; Melhado, S.B. 2002. Análise ISO 14001:1996 X ISO 9001:2000 integrando sistemas. In: Proceedings. IX Encontro Nacional de Tecnologia do Ambiente Construido. Foz do Iguaçu, Brazil.

Degani, C.M. 2003. Sistemas de gestão ambiental em empresas construtoras de edifícios. Master Theses of Escola Politécnica, Universidade de São Paulo. In http://www.teses.usp.br. Accessed in August 12, 2013.

Dias, L.A. 2003. Integrated management systems in constructions (IMSinCONS). *In:* Proceedings. CIB W99—Safety and Health on Construction Sites International Conference on Construction Project Management Systems: the Challenge of Integration. EPUSP, São Paulo, Brazil.

França, N.P. 2009. Sistema integrado de gestão—qualidade, meio ambiente, segurança e saúde: recomendações para implementação em empresas construtoras de edifícios. Master Theses of UNICAMP. In: http://www.bibliotecadigital.unicamp.br. Accessed in August 05, 2013.

Mendes, M.F.R. 2007. O impacto dos sistemas QAS nas PME portuguesas. Master Theses of Universidade do Minho. In: http://repositorium.sdum.uminho.pt. Accessed in August 10, 2013.

PMBOK extension—Construction Extension to a Guide to the Project Management Body of Knowledge. 2007. Project Management Institude. Third Edition.

Sjoholt, O. The evolution of management systems in construction. 2003. *In*: Proceedings. CIB W99—Safety and Health on Construction Sites International Conference on Construction Project Management Systems: the Challenge of Integration. EPUSP. São Paulo, Brazil.

Zeng, S.X.; Lou, G.X.; Tam, V.W.Y. 2006. Integration of manegement systems: the views of contractors. Arquitectural Science Review. v.49, n. 3, p. 229–235.

Occupational Safety and Hygiene II – Arezes et al. (eds)
© 2014 Taylor & Francis Group, London, ISBN 978-1-138-00144-2

Application of the HSC_PEI2012 model in construction work

M. López-Alonso, P. Ibarrondo-Dávila & M. Rubio-Gámez
University of Granada, Granada, Spain

ABSTRACT: In this study it is developed a new model for the calculation and the management of the health and safety cost for the construction companies HSC_PEI2012. We implement these model in a case study, focusing on a construction project carried out in 2008 in Andalusia (Spain). This study reveals that health and safety costs are substantial and remain invisible to the company to a very large degree (92.42%), because the items that make up this cost are dispersed within other accounting entries, thus remaining unidentified on the income statement.

1 INTRODUCTION

The construction sector is one of the most productive sectors in Spain. It represents a significant percentage of the Gross National Product (GNP) 10.5%, (INE, 2011), and generates jobs directly and indirectly. However, many authors as Loosemore (2007) highlight the fact that the construction sector is the source of a large number of occupational accidents in comparison to other sectors, and also state that the efforts made to reduce the workplace accident rate in construction have so far been unsuccessful.

The human and economic repercussions of workplace accidents and occupational illnesses are currently a cause of growing social concern. Moreover, if the accident rate is analyzed from another perspective, the costs of occupational accidents and illnesses, as calculated by the ILO (2008), amount to 4% of the GNP. Surprisingly, according to the *Encuesta Nacional de Gestión de la Seguridad y Salud en lasEmpresasEspañolas* (ENGE) [National Survey on Health and Safety Management in Spanish Companies] (2009), there seems to be little or no entrepreneurial awareness of the high economic cost of workplace risks for construction workers.

Generally speaking, large construction firms with more than 250 workers have management tools at their disposal for the monitoring of work activity. Such tools facilitate access to information and provide guidance in regards to the implementation of organizational innovations. However, only 27% of these companies possess data of pertaining to the economic repercussions of workplace accidents. Consequently, there is a lack of information concerning their actual cost. It is thus impossible to generate economic indicators that can orient decision-making in matters of occupational safety and health, and which can be used to estimate the cost-effectiveness of investing in risk prevention at the workplace.

Although knowledge of safety costs is crucial for risk management in construction companies, this requires a strict accounting of the safety-related activities. Generally speaking, the accounting practice of construction firms has no explicit heading for safety and health costs. Most of these items are consigned to different categories and are thus scattered throughout the budget. This makes it extremely difficult to calculate their financial impact on the company's economic results (Oxenburgh and Marlow, 2005).

As a solution for this problem, this paper presents a model designed for the systematic inclusion of information concerning health and safety costs in businesses, with a special focus on construction companies. The HSC_PEI2012 model is based on the results obtained in a research study performed in 2008, which analyzed a sample of 40 construction worksites in Andalusia (Spain), in which the building construction was in different operational phases.

2 MODEL FOR THE MANAGEMENT OF HEALTH AND SAFETY COSTS IN CONSTRUCTION COMPANIES: HSC_PEI2012

The design and implementation of a health and safety management system for quality costs should at least include the following phases (Morse *et al.*, 1987):

- Raising awareness of the problem of safety costs among company directors and obtaining company support.
- Specification of health and safety costs.
- Development of implementation procedures.

- Implantation of the model throughout the organization.
- Evaluation and monitoring.

In the case of construction companies, senior management first needs to make a serious commitment to occupational health and safety by implanting an effective cost management system for risk prevention in the company. Moreover, company directors should also be willing to foster and develop a monitoring and evaluation system that specifically focuses on health and safety costs.

The second phase delineates the type and number health and safety costs that will guide the management system and identify the different concepts to be included. This is crucial since these items will be the focus of the calculation process and the basis for the elaboration of indicators and reports in consonance with the information needs of the personnel responsible for risk prevention in the company.

In line with Andreoni (1986) and Brody *et al.* (1990), the classification in the HSC_PEI2012 model differentiates the costs derived from the implementation of prevention measures in the company from the costs stemming derived from the absence of such measures and the number of accidents produced as a result. In other words, there is one budget category for risk prevention costs and another category for accident costs. However, a more in-depth analysis of the items included in each of these budget concepts led us to propose a different and more innovative way of specifying and classifying safety costs in construction companies (Barsky, 1997).

In our proposal, health and safety costs for construction projects are regarded as the value of the consumption of production factors, goods, and services, which is the result of all those activities performed in the company to improve work conditions and reduce the occupational accident rate in construction work as well as that derived from the occurrence of incidents and/or accidents.

Costs related to health and safety in the workplace can be classified as follows:

- *Safety Costs*: The cost of guaranteeing health and safety in the company includes the risk prevention costs or the prevention measures implemented by the company as differentiated from the evaluation and monitoring costs, stemming from the actions taken by the company to verify, assure, and adequately maintain the safety and health measures implanted.
- *Non-safety Cost*: The cost of not guaranteeing safety and health in the company includes those costs assumed by the company as a result of work-related accidents as well as any other costs that may arise because of non-compliance with safety regulations. Accordingly, in this respect, there are tangible costs and intangible costs. Tangible costs are those identified with the accident that caused them and which can be quantified by conventional calculation methods. In contrast, intangible costs, which stem from lack of safety, cannot be economically quantified and/or do not have operational indexes that can be used to measure their repercussion on the organization. These include loss of image, low employee morale, labor conflicts, and market loss (Gosellin, 2004).
- *Extraordinary costs*: This category includes all losses generated by events that cannot be prevented by human or technical management measures and which are irremediable (e.g. catastrophes). In our opinion, this category is for all the cost items that the company and its directors are unable to control. As uncontrollable costs, they are not included in a structured model for safety and health costs in the company.

Each of these health and safety cost categories can have direct and indirect costs, depending on the nature of the agent that causes them, on the type of accounting system used in the company, and on the constraints of the data collection process. Furthermore, the basic unit of analysis, taken as a reference for calculating the health and safety cost impact on construction companies, is each *construction project planned and/or executed.*

The third phase is the design of the procedures required for the smooth and seamless operation of the management system. In other words, it is for the creation and development of the Risk Prevention Plan.

The fourth phase focuses on implementing the procedures designed in the third phase. In all likelihood, it will be necessary to make adjustments in the plan when the procedures actually begin to be used. In this phase, managers will obtain the indicators and reports created, all of which will be systematically distributed among the set of previously defined internal users.

The fifth and final phase establishes the schedule for the periodic evaluation and monitoring of the safety and health management system. These sessions will bring to light possible incidents and analyze the statistics as assessed by the company directors. This has the purpose of assessing the implementation and operation of the management system for safety and health costs. Based on the results obtained, the system will be revised to include those risk prevention changes required to adjust it to established policies.

The HSC_PEI2012 model supports company decision-making such that it provides a steady supply of information regarding health and safety-costs to company managers.

3 CASE STUDY

A case study was performed to validate the HSC_PEI2012 model for the calculation, analysis, and monitoring of safety costs in construction companies. The construction project selected was MA/1/08/30-07-2007(1st), which belonged to the sample from the research study previously carried out in Andalusia (Spain). The descriptors of this construction project were the following:

- Private construction of 122 single houses.
- Completion time: 22 months.
- Contract award discount: 0.0%.
- Budget implementation: 35.844.000, 00 €.
- Number of accidents: 8.

The following considerations have a linear production with the same amount of money for each month, 1,629,272.73 €, as well as an average monthly number of workers (company employees, subcontractors, and companies) for the calculation of the different costs.

A comparative analysis of health andsafety costs was performed of the information provided by the management system used in the company and the safety costs obtained with the HSC_PEI2012 model for the calculation, analysis, and monitoring of these costs, based on the survey data collected in previous research and based on the basic parameters of our model. The tangible cost of accidents was calculated with the model proposed by the Navarre Occupational Health Institute en el año 2003 (INSL, 2003), based on the Safety Technical Note 273 of the Spanish National Institute for Health and Safety at Work (INSHT, 1991).

4 ANALYSIS OF RESULTS

In relation to the tangible costs of non-safety, the cost management system used by the company did not contemplate the identification of costs derived from the occurrence of an accident. In contrast, the HSC_PEI2012 model provided detailed information regarding accident-related cost items. When this model was applied to the construction project in the case study, based on the survey data, this cost was found to be 3.676,39 €.

Table 1 shows the estimated amounts for health and safety costs in the construction project. As can be observed, there is a significant difference between the total health and safety cost as calculated by the company system and the cost estimated by the HSC_PEI2012 model. It should be pointed out that when our model was applied to the construction project, it was impossible to calculate many of the health and safety costs since sufficient data were not available (Lewis, *et al.*, 2002). However, this issue could easily be resolved if the construction company decided to use the HSC_PEI2012 model for the analysis, calculation, and monitoring of health and safety costs.

Table 1. Safety and non-safety costs.

	Current system	Model proposed	Difference
Prevention costs (1)	105.902,73	191.388,51	85.485,78
Evaluation and monitoring costs (2)	15.728,88	1.408.578,49	1.392.849,61
Tangible costs of the lack of safety (3)	–	3.676,39	3.676,39
Total safety cost	121.631,61	1.603.643,39	1.482.011,78
Difference	7,58%	100,00%	92,42%

Application of the proposed model to Construction Project MA/1/08/30-07-2007.

i. Summatory of the costs of the measures taken to implement a risk prevention.
ii. Summatory of the costs of the actions taken to verify and maintain measures on health and safety issues in order to reduce or minimize accident risks.
iii. Summatory of those costs which can be identified with a specific accident.

1. The Prevention Costs in construction project obtained in the construction site MA/1/08/30-07-2007(1st) amounted to 105.902,73 € until the moment the construction site was visited, and the costs of evaluation and monitoring came to 15.728,88 €. Both items represent a total of 0.003% of the project budget. However, when the HSC_PEI2012 model was used to calculate the costs that were previously not considered, this percentage soared to 0,04% of the project budget. Since the construction had not finished, this means that the percentage could even increase. If one takes into account that the amounts associated with any construction project are always very high, the estimated differences in this case, which could be higher, which constitutes a large budgetary slippage.
2. The analysis based on data collected during a visit to the construction site of Project MA/1/08/30-07-2007(1st) showed that:

- There are safety and health costs that are generally not considered and which should be monitored because this would improve work conditions and make the construction worksite safer for employees.
- The HSC_PEI2012 model for the calculation of safety and health costs in construction companies

was able to accurately and exhaustively describe the wide range of different cases that could arise in the construction sector.

3. This analytical method could be applied by the company each month to each of their construction projects in the execution phase. This would help to estimate the costs for the following month and thus optimize decision-making in the area of health and safety risk prevention, based on indicators.

5 CONCLUSIONS

The HSC_PEI2012 model, designed to calculate and control costs related to health and safety at work, determines first the cost objects that are useful for decision making in this field. Accordingly, we propose the following cost classification: Safety cost, with two components, prevention cost and monitoring and evaluation costs; and Non-safety costs, consisting of two components, tangible and intangible costs.

The application of the safety cost model proposed in this paper, to a case study of a construction project, has shown that companies face substantial health and safety costs and that to a very large extent these remain invisible to company managers. In the construction project examined here, these hidden costs represent at least 92.42% of the total safety costs.

Model HSC_PEI2012 provides quantitative information that is of great value for decision-making in the company in relation to the management of health and safety, and can contribute to improving working conditions on construction sites.

REFERENCES

Andreoni, D. 1986. *The Cost of Occupational Accidents and Diseases. Occupational Safety and Health Diseases* Edn. International Labour Office, Geneva.

Barsky, I., & Dutta, S.P. 1997. Cost assessment for ergonomics risk (CAFER). *International Journal of Industrial Ergonomics* 20: 307–315.

Brody, B., Létourneau, Y., & Poirier, A. 1990. An indirect cost theory of work accident prevention. *Journal of Occupational Accidents* 13(4): 255–270.

Gosselin, M. 2004. *Analyse des avantages et des coûts de la santé et de la sécurité au travail en entreprise: Développement de l'outil d'analyse*. Rapport de Recherche R–375.

Instituto Nacional de Estadística.Retrieved August, 8 2012, 2011, from http:/www.ine.es/

Instituto Nacional de Seguridad e Higiene en el Trabajo 1991. *NTP-273: Costes no asegurados de los accidentes: método simplificado de cálculo*. Madrid.

Instituto Nacional de Seguridad e Higiene en el Trabajo 2009. Encuesta Nacional de Gestión de la Seguridad y Salud en las Empresas. *Instituto Nacional de Seguridad e Higiene en el Trabajo.* Madrid.

Instituto Navarro de Seguridad Laboral 2003. *Análisis de costes de accidentes de trabajo*. Pamplona.

International Labour Office 2008. ILO Standards on occupational safe and healthy jobs. Ginebre.

Lewis, R.J., Krawiec, M.; Confer, E., Agopsowicz, D., & Crandall, E. 2002. Musculoskeletal disorder worker compensation costs and injuries before and after an office ergonomics program. *International Journal of Industrial Ergonomics*29: 95–99.

Loosemore, M., & Andonakis, N. 2007. Barriers to implementing OHS reforms—The experiences of small subcontractors in the Australian Construction Industry. International Journal of Project Management 25(6): 579–588.

Morse, W.J., Roth, H.P. & Poston, K.M. 1987.Measuring, Planning and Controlling Quality Costs. *National Association of Accountants*: 2.

Oxenburgh, M. & Marlow, P. 2005, The Productivity Asessment Tool: Computer-based cost benefit analysis model for the economic assessment of occupational health and safety interventions in the workplace.*Journal of Safety Research*, 36(3):209–214.

Occupational Safety and Hygiene II – Arezes et al. (eds)

Scholar ergonomics—primary schools in Tartu (Estonia) study case

F. Lima, J. Almeida, J.P. Figueiredo & A. Ferreira
IPC, ESTeSC, Coimbra Health School, Portugal

ABSTRACT: A good part of posture problems that most people get, start up as a child, not paying attention to the positions taken and bringing long-term adverse effects to their health. The present study aimed to conduct an evaluation of ergonomic desk and chair sets school in the city and outskirts of Tartu, Estonia, to children from 1st to 4th classes, being the sample of the study constituted by four Schools of the Estonian Basic Education. The main objective of this paper is to realize if in fact the school furniture is compatible with the study population. Regarding the collection methodology, two types of instruments were defined: surveys previously adapted, self-administered to 155 parents of the students in study (132 of whom delivered the questionnaire after filling); observation of postures, furniture and general conditions in the classroom (non-participant observation). The data were statistically analyzed using the SPSS software version 21. The data analysis showed that the school furniture used in the schools under study was not ergonomic, since it mostly did not respect the existing recommendations, not attending to the needs of students as a whole, showing a clear mismatch between the anthropometric characteristics and the dimensions of the furniture.

1 INTRODUCTION

The first systematic work that man carries in his life is the student. Even as a child, he comes into contact with the first furniture designed for a specific task that will possibly accompany him for many years (Oliveira & Lucia, 2006). One of the essential elements for a good education of children is the school furniture, this being a necessary and paramount. To enable it to fulfil its role, the furniture must meet the characteristics of the target population (Prado-Léon et al., 2001) (Paraizo et al., 2009). If the furniture is produced not taking into account these characteristics, accidents may occur due to incorrect size of the product that does not meet the dimensional requirements of children and health problems can arise at the visual level, circulatory and skeletal muscle (Prado-Léon et al., 2001). The use of furniture, with a design that meets the specific requirements of the activities, is a key factor for the adoption of proper posture and consecutively to a greater productivity of the individual. On the other hand, furniture that requires to the individual a postural pattern that employs continuous efforts and constraints skeletal muscles, will present a hardship to the individual character of the conduct of its activities. The "poor" design of chairs and desks, for example, imposes constant physical arrangements, improvisations and accommodations for students, to compensate deficiencies of the project or industrial design (Mouro & Reis, 2010). Designing suitable furniture posture is one of the actions of Ergonomics, which transform technology to adapt it to the man, as well as assess and improve the environment and working conditions (Rocha & Sousa, 2008). Ergonomics is very important for the work to be a source of health and productivity and its objectives include safety, satisfaction and well-being. The efficiency will come as a result of this union of factors (Ferreira et al.,2009). Postural problems begin in childhood, as the child enters school, considering that activities in sitting posture can lead to the onset of pain in the lumbar region (Paraizo et al., 2009).The common pathologies caused by improper sitting posture can lead to postural deviations and later functional incapacity in adulthood (Marcolino et al., 2009).

2 MATERIAL AND METHODS

This study was conducted between January and April and the data collection was carried out in four primary schools located in the city and periphery of Tartu (Estonia). The samples were encoded in the X, W, Y and Z to ensure data confidentiality.

The study was a Level II, descriptive-co relational and transversal. From the universe of students in four Estonian Primary Schools, a sample of 132 children has been studied, aged between 7 and 11 years, of which 132 returned the questionnaire after filing it. The conception of the sampling design was established on the type of sample

Table 1. Colour code and its dimensions (cm) of students (European Standard BS EN 1729-1:2006).

Colour code	White	Orange	Violet	Yellow	Red	Green	Blue	Brown
Stature	80 95	93 116	108 121	119 142	133 159	146 176,5	159 188	174 207
Height of popliteal	20 25	25 28	28 31,5	31,5 35,5	35,5 40,5	40,5 43,5	43,5 48,5	+48,5
Seat height	21 ± 1	26 ± 1	31 ± 1	35 ± 1	38 ± 1	43 ± 1	46 ± 1	51 ± 1
Seat depth	22,5 ± 1	25 ± 1	27 ± 1	30 ± 2	34 ± 2	38 ± 2	42 ± 2	46 ± 2
Height of desk	40 ± 1	46 ± 1	53 ± 1	59 ± 1	64 ± 1	71 ± 1	76 ± 1	82 ± 1

as non-probabilistic and as to the technique, as accidental.

For collection of data, two types of instruments have been defined: assessment of conditions in the classroom (non-participant observation) and previously adapted surveys, self-administered to 155 parents of students, of whom 132 returned the questionnaire after filling. These instruments for data collection were composed of three parts: the posture of students, furniture and general conditions; personal data, health condition and conditions of the classroom, respectively.

The recommendations of the European Standard BS EN 1729-1:2006, which specifies the dimensions of school furniture depending on the height of the children, were taken as reference. According to the height and popliteal height, they are grouped by colour, as can be seen in Table 1.

Considering the anthropometric assumptions, it was used the following recommendations for the treatment and analysis of data: seat height adapted to the height of the popliteal in order to allow the feet are perfectly flat on the floor; seat that allows the child to sit with legs flexed at 90 degrees (or higher), but without compressing the area of the popliteal and, simultaneously, be able to support the back in the lumbar support; height of the work surface (from the seat) between the distance elbow-seat and 3 to 5 cm below this measure Anthropometric (Gonçalves & Arezes 2012).

The data collected were processed using the SPSS software version 21. For the evaluation of the as-sumptions regarding the type of statistic to use (parametric or nonparametric) were calculated, for the study variables (of the research hypotheses), the values for the symmetry (skewness / the standard error), flatness (kurtosis / the error standard) and normal distribution (Shapiro-Wilk). For symmetry and flatness, the observed values should be between −2 and +2 and the distribution of significance should be $\alpha < 0{,}05$, so it can be normal. Simple descriptive statistics were used: measures of location (mean and median) and measures of dispersion (standard deviation and variance). Statistical tests were applied χ^2 Independence, Fisher's Exact Test and t-Student for independent samples. The interpretation of statistical tests were based on a significance level of $\alpha = 0.05$ with confidence inter-val (CI) of 95%. As in hypothesis testing criteria were defined: for α significant (≤ 0.05) the H0 is rejected, namely, were observed differences/associations between groups. For $\alpha > 0.05$, H0 is not rejected, namely, there were no significant associations or differences between groups. About the applicability of χ^2 of independence were taken into account the following assumptions: N > 20; all expected frequencies (Eij) were > 1; at least 80% of the expected value (Eij) are ≤5 (18).

3 RESULTS

The sample of 132 respondents who answered the questionnaire was characterized by having 48,5% female and 51,5% males. The W School is the one with the greatest number of students surveyed (34,1%), followed by X School with 28,8%, Y School with 19,7% and finally the Z School with 17,4% of all respondents.

It was also found that the majority of individuals frequented the 1st grade (41,7%), the 2nd grade was attended by 28,8% of children, the 3rd year represented a percentage of 21,2% and the 4th grade represented 8,3% of students.

When evaluating the variable "Average hours a student is sitting in class," it was found that the X School showed a higher average on this variable ($\bar{x} = 6.53$; s = 1.31), and that the Z School had the lowest average ($\bar{x} = 4.2$; s = 0.43.). It can be said that the relationship between the school year and the average number of hours a student is sitting in class is directly proportional.

In the analysis of the variable "Complains of poor visibility to the frame", it was found that of the 132 respondents, only 19 answered "Yes" to this question, representing 14,4% of the total, and of these 19 students, 15 were in the first two rows of the classroom.

Environmental conditions in the classroom are very important. Thus, it has been found, in relation to temperature and lighting, that 51,9% of children classified the lighting in the classroom with Good.

Regarding temperature, 46,1% of students considered that it was good, and the total of 132 respondents, 6 children classified the temperature in the classroom as not adequate (too hot), representing 4,6% of the total.

Of the total respondents, for the variables relating to symptoms/diseases, 12,9% had muscle pain, 5,3% had back pain, 27,3% manifested headache, 24,2% accused irritability, 9% accused stress 37,9%, accused fatigue, 21,2% had a lack of concentration in classroom, 6,8% admitted feeling anxiety and sleep disorders and only one student felt revealed another type of symptom/disease.

In the evaluation of the variables relating to body areas where students felt pain, it was observed that 12,9% of students had pain in the neck, 9,8% had shoulder pain, 3.8% of students admitted pain in the upper limbs, 9% had lower back pain and 41,7% had pain in the lower limbs. However, no student was submitted to surgery, or in the lumbar region or cervical, although 12,1% had lesions of the musculoskeletal forum and 3% of respondents admitted any anatomic anomaly.

Regarding the height of respondents, we found that, according to the European Standard BS EN 1729-1:2006, 101 pupils fit into the yellow colour code, 18 in the red colour code and only 1 in the code green. According to the recommendations in this European standard, it was found that the measures recommended for the height of the chair seat to colour code yellow is 35 ± 1 cm, for colour code red is 38 ± 1 cm and for colour code green is 43 ± 1 cm. For measures of seat height, only 3 were with the recommended height. However, the three children in question had a difference between the height of the seat and popliteal height of 5 cm, 5 cm and 3 cm (being ± 2.5 cm the recommended).

Thus, it was proposed to obtain an average estimate that resulted from the difference between the height of the chair seat and popliteal height, and the result was as follows: of the 121 students of whom had the necessary measures for the calculation in question, 43 students (35,1%) had a proper seat height, 60 children (49,6%) had a seat too high for their popliteal height and 18 students (14,9%) had a seat too low.

Decided to analyze the height of the chair seat in relation to gender, noting that there isn´t any association between exposure to inadequate seat heights by gender (p-value = 0.471). However, it is noted that 60 children were too high chair seats, of which 55,0% were male and 45,0% female.

Still in relation to the chair, we attempted to evaluate if the depth of this was within the recommended actions and it was found that of the 120 students, only 3 had a chair proper depth, 1 student had a too short chair in depth and the others children had chairs too long. By analyzing the relationship of the ergonomic evaluation of the depth of the chair according to the use thereof, for male children and female, it was found that there was not an association between exposure to chair ergonomically unsuitable as a function of gender (p-value = 0.513). However, it should be noted that only 3 children (2,5%) had a chair ergonomically suited to the depth and 96,7% were sitting in chairs "too long", being mostly male children (50,9%).

Making a comparison between the "grade" with "ergonomic evaluation of depth" and the "height of the chair seat", it was shown that there were statistically significant differences (p-value ≤ 0.05). It was found that, as to the depth of the chairs, the 1st, 2nd and 4th grade do not have a suitable chair in this dimension. In total, the students who attend these years of schooling have long chairs in depth. Of children attending the 3rd year, 3 (12%) had a suitable chair in depth, while the majority of students (84%) had long chairs regarding this dimension.

According to the recommendations in the European Standard BS EN 1729-1:2006, it was found that the measures recommended for the height of the desk to colour code yellow is 59 ± 1 cm, for colour code red is 64 ± 1 cm and to the green colour code is 71 ± 1 cm. It was found that in relation to the height of the desk, 19 students (15.8%) had a desk suitable to their stature, 5 children (4.2%) had a desk too low in relation to their height and consequently, 96 students (80%) of a total of 120, had too high desks.

Regarding the actual height of the desk in the classroom by gender, it was found that there is no statistically significant difference between the ergonomic evaluation of the actual height of the desk in the classroom due to the gender (p-value ≥ 0.05). It is important to note that of the 120 children, 96 (80%) had a desk too low for your height.

In respect of the suitability of the height of the desk according to the European Standard BS EN 1729-1:2006, depending on the grade, it was found that respondents who attended the 1st, 2nd, 3rd and 4th year, 84.9%, 69.3%, 76% and 100%, had a desk too high, respectively. We can also add that the 1st grade had 15.1% of secretaries with appropriate height, the 2nd year had 24.2% and 3rd year showed 2.5% of desks suitable for this dimension.

With regard to the elbows, they should be located at approximately the height of the work plan. It was found that of the 19 students who had to their disposal a desk suitable for their height, according to the European standards, the difference between the elbow (90°) and the height of the work plan, varied in a range from 0 cm to ± 14 cm. There are also situations in which the desks would not be appropriate (according to the standard), but the difference between the elbow (90°) and the height of the desk were approximately in the same plane.

From the 121 students, 13 had a gap between the elbow (90°) and work plan with a maximum of 2 cm, 10 students had a difference of 3 cm to 5 cm, and all other students had greater than or equal to 6 cm.

Comparing the "ergonomic evaluation of the height of the table (90° elbow)" to the "pain in the cervical area" we found that the two students (which had a gap between −3 and −5 cm), had no pain complained zone cervical. Of the remaining 119 students with a range greater than or less than −3 cm, 15 (12.4%) reported pain in this area. There was also a comparison between the "Evaluation of ergonomic desk height (90° elbow)" and "shoulder pain", where it was observed that one of two students (which differed between −3 and −5 cm) complained with pain in the area, the remaining 12 students who also complained with shoulder pain possessed entirely a desk too high. However, it was realized from this analysis that this was not decisive to explain the pain of the students want in the cervical or shoulder (p-value > 0.05), and it cannot be said that particular pain was due to the influence height of the work plan. Comparing the results obtained, based on the recommendations of the European Standard BS EN 1729-1:2006, as "pain in the cervical area" and the "shoulder pain", it was found that there were no statistically significant differences (p-value > 0.05,) or for the pain in the shoulders, or pain in the cervical area, when compared to the height of the desk.

As regards the comparison between the height and depth of the chair with "lower limbs" students, as well as the depth of the seat to the "lower back pain", it was found that the pain in lower limbs and lower back pain are not determined by the furniture, so we cannot affirm that it is influenced by it (p-value > .05).

4 DISCUSSION

From the results obtained through observations and questionnaires, it was observed that there are several mismatches in the sample, noting that the school furniture does not meet the needs of all students.

In all the schools visited, the teachers revealed that a relatively short period of time, approximately 30 minutes, children stood up to stretch for 5 minutes, as if their positions were not correct they were warned. By the observation made, it can be seen that it was true and that teachers have a constant preoccupation with the posture of their students.

Children observed mostly had become less restless and more responsive activities after performing these stretches in the classroom. The factor restlessness and lack of attention in the activities can be an important factor when it comes to ergonomic aspects of tables and chairs.

It was found that children with less stature move their legs wider, possibly due to the height of the chair seat is too high for their height. It was also observed that in children with greater stature these movements are decreased, implying that the chairs are more comfortable for taller students. The activities in low chairs cause the lateral displacement of the arm, increasing the burden on the column. Likewise, if the chairs were too high, the head and torso will tend to tilt forward increasing the load on the spine (Gasparini & Rozestraten, 2008).

The analysis of chairs and tables used in the four schools in study showed that they do not meet, in most cases, the recommendations of the European standard BS EN 1729. But it must be added that even if the school furniture followed the recommendations of normative, would not be mostly appropriate for the students in the study. This is not only because this standard, for each colour code has a difference between statures in most cases more than 20 cm (for the same height of desk, chair seat height and depth of the chair seat) but also because it does not presents different body segments (which do not grow proportionally to the same population).

Regarding the pain felt by the students, the legs are the area that has the highest prevalence, 41.7%. The pains experienced in this area may be influenced by both the height and the depth of the seat.

Given the popliteal height of children, it is known that 66.3% did not have a proper chair height and 97.5% had no proper seating depth of the chair. The use of very high seating surface results in compression of the posterior muscles of the thigh, hampers blood circulation in the lower limbs and, if the feet are not supported, body balance may be decreased (Gonçalves & Arezes, 2012) (Pheasant & Haslegrave, 2006). If, conversely, the seat height is below the popliteal height, the user will tend to bend over the backbone (Gonçalves & Arezes, 2012) (Murphy & Buckle, 2007). The legs tend to remain extended forward, leaving the feet without stability (Gonçalves & Arezes, 2012).

In addition to the seat height, we still have to consider the depth of it. The seats too deeply can cause compression in the tissues and cause discomfort to the user. Seats shallow can cause lack of support at the bottom of the thighs, giving the feeling that the user is falling from the seat (Souza et al., 2010) (Gili, 2002).

Followed by the leg pain, the pain felt by the students with the highest prevalence were pains in the cervical (12.9%), shoulder (9.8%) followed by lower back pain (9%). As noted in the results, the elbows should be located at approximately the height of the work plan. Following this logic, it was found that of the 121 students, 13 children had a gap between the elbow (90°) and work plan with a maximum of 2 cm, 10 students had a difference of 3 cm to 5 cm, and all other students had greater than or equal to 6 cm.

The height of school tables lines up of utmost importance since it infers the health of users. A low table force children to a forward tilt, thus overloads the structures of the spine, causing back pain. On the other hand, a desk too high demands an exaggerated abduction of the upper limbs, increasing the load on the spine and leading to the onset of pain in the neck and shoulders (Carvalho et al., 2012).

The fact that the school furniture does not meet the needs of comfort and functionality, translates into a compensatory behaviour, adjustment to the environment by children in an attempt to improve the comfort and relieve body areas affected.

In school furniture, fixed height, some children are forced to adapt to the furnishings, acquiring awkward postures. Thus, the heights of tables and chairs are mostly unsuitable for use by children in study, which can cause postural problems of relative gravity.

5 CONCLUSIONS

As a final product and conclusive of this study, it proved the incompatibility current ergonomic school furniture and anthropometric measurements of the children in the study. It is necessary to have mechanisms to enforce existing ergonomic knowledge, transforming it into beneficial actions in school activities.

Following the statistical analysis made of the results obtained, it was stressed that in these schools, particularly in the classes considered in the sample, there are high levels of conflict between the body size of children and the existing furniture in primary schools in question.

It was also found the variability of anthropometric dimensions either among the children of the four academic years, or between the children who attend the same year or are in the same age group. It should also be noted that this demonstrates that the furniture fixed dimensions will not adequately accommodate the children that will use that, taking this a direct impact on postures adopted by students.

For a future study, it would be interesting and necessary to raise awareness for the importance of school furniture tailored to the needs of the population, since it is proven that can be reflected in the general condition of the student and, consequently, their health.

REFERENCES

Carvalho, G. Ferreira A. & Tracana R. 2012. International Conference of Phsysical Education, Leisure and Health. In Proceedings of VIII SIEFLAS. *Prevention of bad postures in children from 7 to 12 years, in the coast and the interior of Portugal.*

Ferreira, S. Stadler, R. & Pilatti, L. 2009. *Recommendations applicable to the ergonomic school furniture to prevent postural problems in students with low vision.* Magazine I National Symposium of Science e Technology Teaching.- Gasparini, G. & Rozestraten, R. 2008. *The importance of postural reeducation of children from 1st grade of elementary school for the learning process.*

Gili, G. 2002. *Dimensionamento humano para espaços interiores.* Barcelona. 320p.

Gonçalves, M. & Arezes, P. 2012. *Analysis of ergonomic conditions of the classrooms of the 1st cycle of basic education.*

Marcolino, M. Meneguetti, M. & Mendes, G. 2009. *Anthropometric measurements to design school sets of chair and table.* VI International Meeting of Cesumar Scientific Production.

Moro, A & Reis, P. 2010. *Human costs of the sitting posture: an approach to ergonomic school furniture.*

Murphy, S. Buckle, P & Stubbs, P. 2007. *Cross-sectional study of self-reported back and neck pain among English schoolchildren and associated physical and psychological risk factors.* Applied Ergonomics 38: 797–80.

Oliveira, J. & Lucia, R. 2006. Ergonomic Analysis of school furniture for the definition of criteria.

Paraizo, C. Moraes A. & Gomes, V. 2009. *Biomechanical costs of posture of the child in school computer.*

Pheasant, S. & Haslegrave, C. 2006. Bodyspace: anthropometry, ergonomics and the design of work, 3rd Edition, CRC Press, 332 pgs.

Prado-León, L. Avila-Chaurand, R & González-Muñoz, E. 2001. *Anthropometric study of Mexican primary school children.* Applied Ergonomics 32: 339–345.

Rocha, A. & Souza, J. 2008. *School effects on postural and ergonomic work about postural behavior.*

Souza, A. Fialho, P. Minette, L. & Silva, J. 2010. *Evaluation of ergonomic wood chairs and wood products.* Tree Magazine 34; 1; 157–164.

Occupational Safety and Hygiene II – Arezes et al. (eds)
© 2014 Taylor & Francis Group, London, ISBN 978-1-138-00144-2

Damages derived from mobbing and professional contingencies—state of the art on the Portuguese legal framework and case law

Ana Cristina Ribeiro Costa
Doctoral student and invited teacher at Universidade Católica Portuguesa—Escola de Direito do Porto
Lawyer at Gama Lobo Xavier, Luis Teixeira e Melo e Associados, Sociedade de Advogados, R.L., Portugal

ABSTRACT: As happens with other jurisdictions, the Portuguese law sanctions moral harassment in the context of the labour law and provides the worker a right to be compensated in the civil terms, for the pecuniary and moral damages suffered. However, the injuries that generate a disability for work due to a phenomenon of mobbing in the workplace are not yet considered as pathologies compensated in the same terms as occupational contingencies (workplace accidents or occupational diseases). In fact, although the Portuguese doctrine has been studying the phenomenon of mobbing for the past years, the case law is still unwilling to accept the compensation and sanction of these conducts in the same terms as the other EU jurisdictions have been allowing.

1 INTRODUCTION

Mobbing is a topic that has gained an increasing importance, due to escalating numbers of cases that are reported and go back to courts. This phenomenon may be characterised as a set of intimidating, humiliating or embarrassing, harmful or unwanted conducts of various kinds, occurring in the context of an employment relationship, that objectively violate the fundamental rights of the worker, namely, his dignity and physical and moral integrity (Pereira, 2009, p. 72). Moreover, a systematization of these behaviours is demanded: the repetition of the offensive conduct, which should be extended in time, but also the combination of several acts that complement each other in achieving the purpose it was intended[1]. Therefore, mobbing may cause damage to the physical and psychological health of workers[2]. Indeed, it is responsible for cognitive, psychological, psychosomatic and hormonal changes, as regards the nervous system, muscle tension and sleep, and may ultimately lead to suicide. In September 2012, the Portuguese Health authorities changed the Mental Health Programme for 2007–2016, as they figured that the number of mental diseases and suicides will increase with the financial crises that the country is facing. But the truth is that, despite the recent media coverage, the phenomenon tends to hide in the walls of work units, confining itself to the space frequented by the suicidal colleagues, unrevealing, thus, potentially traumatic events.

The theme that will be analysed requires the assessment of the possibility of a phenomenon of mobbing and the suicide of a worker to be considered as work-related accidents within the Portuguese legal framework. Thus, the possibility of the injuries that generate a disability for work or the death subsequent to a suicide due to a phenomenon of mobbing to be considered as eligible for compensation under the same conditions as a work-related accident will be evaluated. The possibility of such injuries to be qualified as occupational diseases and, as such, eligible for compensation under the same conditions that these contingencies are will also be considered (for further developments, vd. Costa, 2010 and Costa, 2012).

1. *Vd.* the decision from the second instance Court *Tribunal da Relação de Coimbra*, dated 07-03-2013, reported by Jorge Manuel Loureiro, demanding the requirement of intentionality.
2. The existence of damages is not a requirement of the existence of mobbing, but only a possible consequence. In this sense, the decision from the second instance Court *Tribunal da Relação do Porto* (TRP), dated 08-04-2013, reported by Maria José Costa Pinto.

2 PRACTICAL RELEVANCE OF THE PROFESSIONAL CONTINGENCIES SYSTEM

The relevance of this study is gauged by the response to the following question: is the provision of civil liability for the practice of mobbing on the article (art.) 28 of the Portuguese Labour Code (Law nr. 7/2009, 12-02—hereinafter only CT) not adequate and sufficient? The convenience of this research lies not only with the advantages that the institutes of professional contingencies[3] may adduce for the analysis of this matter, but also with the need to give adequate treatment to the figure of mobbing, which has worryingly gained an increasing dimension.

In fact, there are many advantages of the schemes of work-related accidents and occupational diseases. Firstly, these systems allow a more efficient access to a pension or compensation from the worker, to the extent that they include less expensive and quicker mechanisms than the access to the courts through the civil jurisdiction[4]. Secondly, these schemes have alternatives that the regime of civil liability does not provide, such as the provision of para. 8 of art. 283 CT: the obligation of the employer to ensure an occupation for the injured worker in well-suited functions. The Law nr. 98/2009, 04-09, that regulates the regime for the compensation of the work-related accidents and occupational diseases (hereinafter only referred to as LAT), also provides for other mechanisms that the mobbed cannot access through civil liability, without forcing the scope of this institute, such as the possibility of professional rehabilitation and adaptation of the workplace (arts. 44 and 154'LAT), as well as the allowance for frequency of education for professional rehabilitation (arts. 69 and 108 LAT).

It should be noted, furthermore, that these schemes consent constant updating of allowances and their revision in the light of improvement or worsening of the health status of the victim, which is an advantage over the regime of civil liability. Moreover, the claims under the right to compensation established in the LAT are inalienable, unattachable and unrenounceable, enjoying the guarantees provided for in CT (see art. 78 LAT). As such, they're preferential credits, which constitutes another difference with respect to claims arising from civil liability.

Finally, the consideration of certain situations as professional contingencies can bring the advantage of exempting from the hard evidence that there was a phenomenon of mobbing, because it is not necessary to prove the elements of this concept to conclude for the existence of an accident or an occupational disease (although the proof of its constituent elements might be necessary in order to meet the requirements of causality).

Subsequently, various benefits of the submission of the particular regime of work-related accidents to cases of mobbing may be indicated, with regard to safeguarding the State treasury, as the State won't have to pay the pensions due to the victims of that phenomenon[5], who currently seek the pension by the social security's scheme for common diseases. Anyway, in the Portuguese system, if the event integrated into the phenomenon of mobbing is considered as a work-related accident, it makes sense for insurers to appeal for the classification of that event as guilty, so they can get rid of their responsibility, in the cases where the accident is due to a third or a co-worker (art. 17 LAT) or to the employer (in the cases of art. 18 LAT). As regards the advantages of the system of work-related accidents for workers, it should be noted that there is a presumption of irrelevance of the pathological predisposition of the workers, which has consequences in terms of evidence of causation, because the existence of such predisposition could affect the compensation for damage in the scheme of civil liability, to the extent that it may exclude the causation. In addition, the employee's claims arising from work-related accidents benefit from a stronger guarantee of payment, to the extent that the Workers' Compensation Fund is responsible for the payment of pensions established in the LAT, when the responsible entity is not able to comply its obligation (art. 1, para. 1, subpara. a) of Decree Law nr. 142/99, of 30-04, and art. 82 LAT).

Finally, the benefits of the qualification of damages caused by mobbing as occupational diseases shall be described, emphasising the preventive and, consequently, economic dimensions (as the decline on the number of situations verified will lead to lower costs for companies and for the society). Moreover, it may contribute to a broader discussion about the various psychopathological phenomena connected with work.

In common with the occupational diseases in the broadest sense, it shall be pointed, also, the guarantee of solvency of the entity that is

3. "Professional contingencies" is a concept that comprehends both work-related accidents and occupational diseases.

4. In Portugal, judicial proceedings due to professional contingencies are considered urgent and, as such, they run even on court holidays. Cfr. art. 26, para. 1, subpara. e) in the Portuguese Labour Procedural Code (Decree-law nr. 480/99, 09-11).

5. It should be noted that, in what concerns work-related accidents, in Portugal, the employers are obliged to assign an insurance for each worker, which is a particularity of our professional contingencies system, that does not exist in many other European legal systems. Respecting occupational diseases, the compensation is guaranteed by the Social Security System.

responsible for the compensation (the social security system), which constitutes an important benefit for the worker[6]. Moreover, the regime of civil liability will hardly protect a situation of disease as a cause of the damage suffered by the worker. Finally, it shall be noted, as regards the benefits of the system of typical occupational diseases, that there is no need for concrete proof of causation.

As such, we advocate an alternative application of the civil liability and the professional contingencies regimes, to the extent that they are different responses to the same phenomenon. In fact, whenever there is a work-related accident there may be, in parallel, a civil liability proceeding against the real guilty, be it the employer, a co-worker or a third. In addition to this, there is the misdemeanour responsibility of the employer, also provided by law, and the possible criminal liability of the harasser, when the conduct is an illicit typified act (as the para. 2 of art. 18 LAT consents). Thus, the legislature did not intend to exclude the possibility of different solutions for the same event, so we only propose another possibility of reaction to the phenomenon of mobbing. The two paths are, in fact, autonomous and independent. Therefore, one does not adversely affect the success of the other. However, it may be argued that when the victim chooses not only the rules of professional contingencies, but also the criminal and civil liability, they must be "coordinated" in order to avoid accumulation of compensations that generates unjust enrichment.

On the contrary, it may be affirmed that the application of civil liability to cases of mobbing, under para. 3 of art. 29 CT, excludes the possibility of applying the professional contingencies regime, insofar as this is, itself, a specific regime of civil liability. Anyway, disregarding the discussion concerning its nature, which does not fall within this study, we shall state that the regimes respond to different situations, as mentioned above. Another argument against our theory is that even though the victim is compensated, the system of professional contingencies reparation lacks the protection of a key aspect for his recovery: the effective sanction of the harasser, which has a therapeutic effect for the injured. However, this disadvantage arises only when these regimes are separately considered. Indeed, if applied in combination with the regime of civil liability, as already advocated, there will be no gap in protection.

6. Even though, in the present state of the art, one can no longer affirm that the solvency of the Portuguese social security system is a guarantee...

3 MOBBING AS A WORK-RELATED ACCIDENT

It is essential to analyse the concept of work-related accident in the Portuguese legal system, in order to conclude whether the conducts qualified as mobbing can be therein considered. The condition that has been identified by the scholars as fundamental to the definition of a work-related accident, as well as to the distinction between this and the occupational disease, is the suddenness (Domingos, 2007, pp. 41 e 42)[7]. As such, the work-related accident should be datable, determinable in time, or at least "*of short and limited duration*" (Franco, 1979, p. 62). Therefore, to some scholars and case law, this requirement will be a strong deterrent to the consideration of a situation of mobbing as a work-related accident (Parreira, 2003, p. 238)[8]. However, the most recent case law, as well as some doctrine, has questioned this, as one must acknowledge the existence of "(...) *grey areas in which the suddenness fades away towards a slow evolution, as for example, resulting from the continuous action of a working tool or the aggravation of a pathological disposition or of pathogenic diseases contracted by reason of the work*" (Alegre, 2000, p. 37, Ravisy, 2000, p. 81). In fact, there are several instances in national case law that confirms the mitigation of the requirement of suddenness. These findings emphasize the tenuous boundary between work-related accidents and occupational diseases, although the STJ states that there is a work-related accident when the cause of the injury is not immediate, but is limited to a short and finite period of time, even though the effects suffer a gradual evolution (JOURDAN, 2006, p. 29)[9].

7. *Vd.* also the decision from the second instance Court *Tribunal da Relação de Lisboa*, dated 10-11-2005, reported by Manuel Gonçalves.

8. In the same sense, *vd.* the decision of the TRP, dated 10-03-2008, reported by Ferreira da Costa, and, after the appeal, the decision of the Supreme Court of Justice (STJ), dated 13-01-2010, reported by Sousa Grandão. Anyway, the decision from TRP reported by Domingos Morais, dated 10-09-2007, should also be mentioned, as it states that "*(...) the concept of accident at work is constantly being updated due to changes in social, behavioural (watch out, for example in certain cases of mobbing) and geographical mobility of workers, new situations that enhance multiple and complex causes of accidents at work (...)*" (*Colectânea de Jurisprudência – CJ* -, nr. 201, XXXII, IV, 2007, pp. 236 and 237). It should be noted that, in this case, the event suffered by the worker wasn't qualified as a work-related accident, as there was no evidence of a causal link between the event and the suffered damage.

9. Cfr., for instance, the decision of the Supreme Court of Justice, dated 21-11-2001, reported by Mário Torres.

M.A. Domingos focuses on the analysis of the problem of the suddenness (or not) of the action, breaking down the elements of that concept: the unpredictability[10] and the limitation in time. The author also states that the "*grey areas*" arise from circumstances in which the action has a duration that is variable but continued in time, without verification of the elements that characterise the existence of occupational disease, as a particularly dangerous work environment or products used (Domingos, 2007, p. 43).

Moreover, the law does not characterise the event that constitutes a work-related accident, so no one can say that the suddenness corresponds to a legal requirement.

As regards the requirement of repetition for the existence of mobbing, there are also authors who understand that the concept does not always require the practice of several conducts, and that isolated facts should not be excluded from the definition if they're of such gravity that produce the same result that several minor incidents, or entail serious consequences for the worker. Moreover, national legislation does not seem to require that characteristic, referring to harassment in the singular, as an "unwanted behaviour" (art. 29, para. 1, CT).

Note also that M. A. Domingos accepts that a hypothesis of "overwork" triggering an injury on the worker's health can be considered as a work-related accident (Domingos, 2007, p. 44). Hence, one can conclude that the author admits that a situation which is not sudden, but that corresponds to an event that may be considered as an unique cause, can be considered a work-related accident, in case it causes damage to the health of the victim. As such, for some authors, behaviours that may be grouped, yielding a unique phenomenon which may be called mobbing, will fit the concept of suddenness for the purpose of consideration as a work-related accident.

Thus, to reach the result of consideration of mobbing as a work-related accident, two paths may be followed: either extending the concept of work-related accident, or conceiving mobbing as an unique behaviour. However, in our opinion, this second reasoning proves to be unsuccessful, as mobbing is a whole range of conducts, not an isolated one, even when it causes very serious damage, since it isn't the severity or the consequences that characterise this phenomenon. Therefore, even considering the entire process as a single event causing damage, mobbing will never have the characteristics of suddenness and certainty (yet) inherent in the concept of a work-related accident. On the contrary, conceptually, the first path does not shock us, given the scale of the situations that the legislature has been predicting. Think of the protection of the work-related accidents that do not occur in time or place of work, such as worker protection during the credit hours for seeking a new job, where there isn't any proper link between the employer and the employee. Thus, insofar as it is considered that the scheme of work-related accidents is based on the risk of the availability of the work provision, *ie*, that derived from the fact that the worker "*offers others the availability of his work*" (Gomes, 2000, pp. 208 e 211, Leitão, 2001, pp. 560 e 579), it is possible to say that it makes sense to consider such as work-related accidents. In conclusion, due to the expansion of the concept of a work-related accident, it does not surprise us that it may be extended to cases of mobbing.

4 MOBBING AS AN OCCUPATIONAL DISEASE

Our welfare system is mixed: the diseases listed in the legally prescribed and periodically reviewed table (at this point, the *Regulamentar* Decree nr. 6/2001, 05-05, republished by the *Regulamentar* Decree nr. 76/2007, 17-07), where a presumption of causation works (between catching the disease and the nature of the work), shall be referred to as typical occupational diseases, while those not listed shall be entitled work diseases or atypical occupational diseases (para. 2 of art. 94 LAT and para. 3 of art. 283 CT).

It is common to the various European jurisdictions to exhaustively stipulate the illnesses that constitute occupational diseases, and not to consider the consequences of mobbing as such. Therefore, the legal framework prevents a possible classification of mental disorders caused by mobbing in the workplace as occupational diseases, although it seems there is no impediment as regards other diseases (e.g., physical) that may arise as a result of mobbing, as long as they are under the legal framework. In this sense, R. G. Pereira seems to accept that diseases such as "*reactive depression or heart problems*" listed in the table of typical occupational diseases, are compensated as such, when they arise from mobbing (Pereira, p. 210). However, the legislative instrument that typifies the occupational diseases connects them to specific risks which, in our opinion, prevents that judgement.

10. In what concerns the requirement of unpredictability, cfr. the decision from the STJ, dated 30-05-2012, reported by Gonçalves Rocha, *CJ*, nr. 248, II, 2012, 261–264. This is an important decision, as it states that an event in which a drunk passenger of a flight causes trouble, causing stress and depressive anxiety in a hostess, is deemed to be a work-related accident.

Another Portuguese author supports the need for a "*reformulation of the concept of occupational disease*" so that it can include any injury arising from mobbing (Pacheco, 2007, p. 249). We disagree with the mentioned position, as the changes should not be regarding the concept of occupational disease, but the cast of legally determined occupational diseases (notably by extending this list, allowing the inclusion of certain psychic and psychological pathologies). We do not advocate, however, a qualification of injuries arising from mobbing as occupational diseases, because we understand that the worker must always prove the existence of a causal link between the event and the damage, *ie*, he must demonstrate that there was a situation (that, *in casu*, constitutes mobbing) which originated a specific lesion in his health, which should be protected under the schemes of professional contingencies.

Moreover, the national law already provides a concept under which those damages could be compensated: the above mentioned work diseases or atypical occupational diseases. These are intermediate species between work-related accidents and occupational diseases, category which is necessary, given the unpredictability of the origin and sources of diseases. In fact, this provision sets a special rule for determining the causal link, to the extent that it is necessary to demonstrate the relationship between the performed activity and the pathology or functional disorder, proving that this is not derived from normal wear and tear of the body. While in the typical occupational diseases the characteristic element is the concrete pathological process, in atypical occupational diseases the essential element is the origin of that process. Work shall be the triggering factor and sole cause of the pathology, and not just an occasionality.

However, R. Redinha believes that this hypothesis, although desirable, is not possible, because of the narrow limits of the concept of atypical occupational diseases. The author believes that the law requires a causal link between the disease and the concerned activity, and that in mobbing the disease is not caused by the activity itself, but by a deliberately painful exercise of it (Redinha, 2003, p. 846). Thus, R. Redinha seems to argue that mobbing, as a phenomenon which is strange to the normal labour relations (Álvarez Sacristán, 2003, p. 4), cannot be considered as a normal exercise of a particular activity.

Nevertheless, in our opinion, the connection of the illness to the activity takes place by the mere exercise of the activity, *i.e.*, it's enough that this condition is a consequence of the worker's availability for work, regardless of the conditions of its practice (Redinha, 2003, p. 846), and not necessarily within an activity exempt from any risk. Otherwise, injuries suffered by the worker with the employer's fault, for instance, because of the violation of certain rules of safety and health at work, wouldn't be reimbursed as professional contingencies. Therefore, if it is proven that the injury, functional disorder or disease that the worker suffers (due to a situation of mobbing) is caused by the labour activity, and it does not represent the normal body's wear and tear, one may considerer it will be an atypical occupational disease. Moreover, due to the wording of the law, the cause shall be unique, as the pathology will only be necessary and direct consequence of the carried out activity if no any other factor contributes to enhance the appearance of the disease. Although the requirement of a unique cause seriously restricts the scope of this thesis, it may be stated that the narrow normative forecast does not allow another reading.

5 CONCLUSION

The National Mental Health Plan (2007–2016) envisages the need for an intersectorial coordination in prevention and promotion activities, having emphasized this area of "*employment policies and promotion of mental health in the workplace, reducing and management of the stressors related to work and unemployment, reduction in absence due to psychic diseases*" and the "*awareness and information in various sectors such as (...) the workplace*"[11].

Thus, in our opinion, the Portuguese case law should accept the possibility to consider the consequences of mobbing and worker's suicide as eligible for compensation under the professional contingencies legal framework, similarly to what has been happening in other jurisdictions of the EU countries, such as Spain, France and Belgium (Lerouge, 2010).

REFERENCES

Alegre, C., 2000, Regime Jurídico dos Acidentes de Trabalho e das Doenças Profissionais, Coimbra, Almedina.

Álvarez Sacristán, D.I., 2003, «Tratamiento jurídico de los daños morales causados por los accidentes de trabajo», *Actualidad Jurídica Aranzadi*, XIII, nr. 567, 2–7.

Costa, A.C.R., 2010, «O ressarcimento dos danos decorrentes do Assédio Moral ao abrigo dos Regimes das Contingências Profissionais», *Questões Laborais*, XVII, nrs. 35–36, january-december, 143–146.

11. Resolution of the Ministers Counsel nr. 49/2008, 06-03.

Costa, A.C.R., 2012, «O Acto Suicida do Trabalhador—A Tutela ao abrigo dos Regimes das Contingências Profissionais», *Questões Laborais*, XIX, nr. 40, july-december, 203–251.

Domingos, M.A., 2007, «Algumas Questões Relacionadas com o Conceito de Acidente de Trabalho», *Prontuário de Direito do Trabalho*, nrs. 76–77–78, 37–61.

Franco, J.A.P.M., 1979, «Acidentes de Trabalho e Doenças Profissionais», Direito do Trabalho, BMJ, Suplemento, Lisboa, 1979.

Gomes, J., 2000, «Breves Reflexões sobre a Noção de Acidente de Trabalho no Novo (mas Não Muito) Regime dos Acidentes de Trabalho», in I Congresso Nacional do Direito dos Seguros, coord. António Moreira e M. Costa Martins, Coimbra, Almedina, 205–218.

Jourdan, M., 2006, L'Accident (sur le Chemin) du Travail: Notion et Preuve, Études Pratiques de Droit Social, Waterloo, Kluwer.

Leitão, M., 2001, «A reparação de danos emergentes de acidentes de trabalho», in Estudos do Instituto de Direito do Trabalho – vol. I, coord. Pedro Romano Martinez, Coimbra, Almedina, 537–579.

Lerouge, L., 2010, «Moral Harassment in the Workplace: French Law and European Perspectives», *Comparative Labor Law & Policy Journal*, vol. 32, n.° 1, 109–152.

Pacheco, M, 2007, Assédio Moral em Portugal. «O Elo mais Fraco», Coimbra, Almedina.

Parreira, I.R., 2003, «O Assédio Moral no Trabalho», in V Congresso Nacional do Direito do Trabalho, Memórias, Coimbra, Almedina, 209–247.

Pereira, R.G., 2009, Mobbing ou Assédio Moral no Trabalho. Contributo para a sua Conceptualização, Coimbra, Coimbra Editora.

Ravisy, P., 2000, Le Harcèlement Moral au Travail, Delmas Express, Paris.

Redinha, R., 2003, «Assédio Moral ou Mobbing no Trabalho», in AAVV, Estudos em Homenagem ao Professor Doutor Raúl Ventura, vol. II, Coimbra, Coimbra Editora, 833–847.

Occupational Safety and Hygiene II – Arezes et al. (eds)
© 2014 Taylor & Francis Group, London, ISBN 978-1-138-00144-2

Evolution of skin temperature at different temperature and humidity conditions

E. Quelhas Costa, J. Santos Baptista & Jorge Carvalho
Research Laboratory on Prevention of Occupational and Environmental Risks (PROA/LABIOMEP), Faculty of Engineering, University of Porto, Portugal

ABSTRACT: The skin temperature changes according to external environmental conditions and can be an important physiological parameter to characterize the thermal sensation of the human body. The aim of this paper is to present results concerning the evolution of the skin temperature over three days in different environmental conditions. For that purpose, five healthy young volunteers were recruited to participate in the experiments. The subjects were fully informed about the trials before giving their written consent to participate. Trials were conducted with three days of acclimation in a climatic chamber at 20°C/30% relative humidity (HR); 20°C/60% HR and 32°C/60% HR. In spite of the different individual characteristics, three of the subjects reached a very similar skin temperature level on the third day at 20°C and 30% HR. But it was also found different behavior among individuals at the same temperature of 20°C with different relative humidities (30% HR and 60% HR).

1 INTRODUCTION

Thermal environment is a key factor to control in occupational settings. In response to different thermal environments, humans have distinct physiological and behavioral responses to both, heat and cold (Parsons, 2003). On the other hand, the skin temperature changes due to the thermal environment, and for this reason it is an important criterion for characterizing the thermal sensation of the human body. According to Kataoka et al. (1998) skin temperature is an effective indicator to evaluate human sensations. Under extreme conditions of heat and humidity, fatigue settles quickly and can cause problems. So, not only the risk of failures and accidents is a constant threat to those who are working in these conditions (Tanabe et al. 2006) but also the productivity.

2 MATERIALS AND METHOD

2.1 *Subjects*

Five healthy young volunteers, aged between twenty and twenty seven years old, were recruited to participate in the experiments. All the subjects were healthy and were examined to exclude any significant disease. The protocol was approved by the Ethics Committee of the University of Porto (n°04/CEUP/2012). The subjects were fully informed of any risk and discomfort associated with the experiment and verbal and written informed consents were obtained from each subject prior to the participation in the study. Subjects were asked to avoid coffee, alcohol, smoking and any significant physical activity at least 12 hours prior to each test.

2.2 *Procedure*

After an initial familiarization session, the subjects made three consecutive days of acclimation for each condition of temperature and relative humidity.

All the tests were performed in a climatic chamber (Figs. 1, 2). The first one was carried on at 20°C of temperature and 30% HR and tests were performed at10 am; 2 pm and 4 pm, with three different subjects. The second one was carried on at 20°C, 60% HR and the third trial at 32°C, 60% HR and were performed by two different subjects at 4 pm. All subjects were wearing shorts and t-shirts.

A sedentary activity was simulated in a climatic chamber properly tested for the effect (Guedes, et al. 2012) and parameters and equipment's suited to the study were selected (Costa and Baptista, 2013), as well as the procedure for measuring the internal temperature of the body (Costa et al. 2012). In this situation, besides to skin temperature (ISO 9886, 2004), were monitored the internal temperature (intra-abdominal) using ingestible temperature sensors (TIS), the brain activity using EEG and heart rate using ECG.

The other variables inside the climatic chamber, including the air velocity, were kept constant through-

Figure 1. Outside of the climatic chamber.

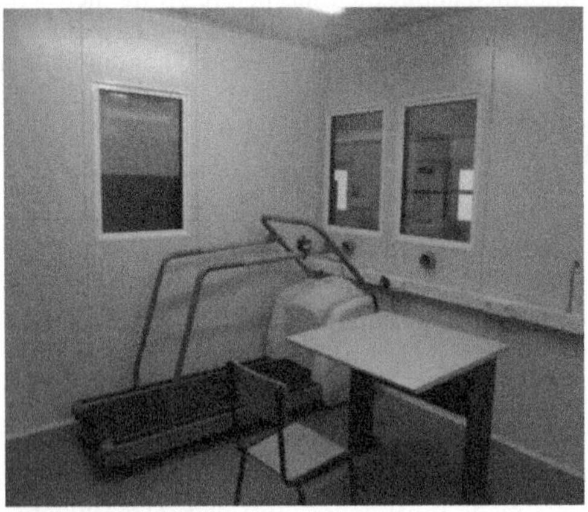

Figure 2. Inside of the climatic chamber.

out the experiments. Each experiment, performed during January, February and March of 2013, lasted one hour, during which the subjects performed a set of tests to simulate a sedentary activity.

The trials were done according to predetermined criteria, aiming to confirm if they are appropriate for the intended use. In this sense, and once selected the equipment for assessing the physiological response, the necessary tests were performed to confirm the experimental results regarding the equipment. All experiments involved non-invasive safe procedures.

3 SKIN TEMPERATURE MEASUREMENTS

The skin temperature was monitored at 14 points of the body as shown Figure 3. This monitoring is conducted in accordance with standard ISO8996 (International standard—Ergonomics Evaluation of Thermal Strain by Physiological Measurements)

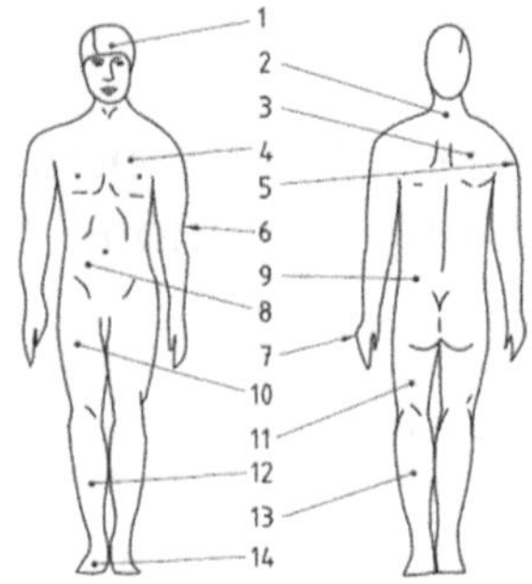

Figure 3. Temperature measuring points on the skin according ISO9886.

Table 1. Places of skin temperature measurements (adapted from ISO9886).

Numbers of points measured
1- Forehead
2- Neck
3- Right scapula
4- Left upper chest
5- Right arm in upper location
6- Left arm in lower location
7- Left hand
8- Right abdomen
9- Left paravertebral
10- Right anterior thigh
11- Left posterior thigh
12- Right skin
13- Left calf
14- Right instep

and some studies like (Ely et al. 2009; Yao et al. 2008 Costa and Baptista 2013). Table nº 1 shows the points where the sensors were placed.

After data collection the mean skin temperature is obtained by Equation 1, ISO9886 (1992).

$$tsk = \frac{1}{14}\sum_{i=1}^{14} tski \tag{1}$$

The skin temperature was measure with "Plux temperature sensors" by *bioPLUXreserach.* This sensors were designed for applications involving continuous or intermittent temperature readings. This NTC (negative temperature coefficient) thermistor, have been developed for biomedical applications with the possibility of being used in a temperature range from 0 to 50°C. The Plux temperature sensors present robust, stable, accurate and low tolerance values. Its geometry and rapid response also provides an increase in the confidence of the obtained results.

The *bioPLUXresearch* is a transmission device via Bluetooth of information received by sensors. The information sent via Bluetooth, comes in an encrypted format so, later, the values have to be converted using a mathematical treatment.

The temperature can be obtained from an equation relating the resistance of the thermistors with temperature. The equation in question is the Steinhart-Hart equation,

$$T(^{\circ}K) = \frac{1}{a_0 + a_1 \ln(R_{NTC}) + a_2[\ln(R_{NTC})]^3} \qquad (2)$$

(Source: Technical data sheet *termpPLUX Temperature Sensor*)

4 DATA AND OBTAINED RESULTS

In this section is presented the data from the subjects and results about the behaviour of the skin temperature over three days under different environmental conditions: 20°C_30%; 20°C_60% and 32°C_60%.

4.1 *Skin temperature at 20°C 30% HR*

The first trials run at the conditions shown in the Table 2, and the outdoor weather are shown in the Table 3.

The meteorological data were collected at *Laboratory of Physics of Constructions*—CFL (2009) Meteorological Station LFC / FEUP. http://experimenta.fe.up.pt/estacaometeorologica.

Table 4 represents the anthropometric characteristics of the subjects in each trial in all conditions.

Concerning the local skin temperature, Figure 4 shows its variation at 20°C/30% HR, on the 1st day of exposure, between different parts of the body between the three subjects.

The local skin temperature on the third day was plotted in Figure 5. There were no significant differences between the results obtained at different times in the first and third trial, for this reason the folowing trials were carried out at 4 pm.

In this first trial, it was shown the evolution of the behaviour of the temperature between the

Table 2. Conditions of the first trials.

Days	*	Climatic chamber	Laboratory conditions
22 January2013	GA1;PC2;RS3	20°C	20–22°C
23 January2013	GA4;PC5;RS6	30% HR	48–70%
24 January2013	GA7;PC8;RS9		

*GA, PC, RS are the subject code and the number is the order of test.

Table 3. Outdoor weather values during the first trials performed in different days.

*	Time	t_{ar} (°C)	HR %	Rad Atm (W/m²)	Atmospheric pressure hPa	Win speed m/s
GA1	9 am	7,70	76,70	286,30	994,80	1,90
PC2	2 pm	8,75	72,40	286,30	994,00	3,62
RS3	4 pm	8,27	76,23	287,24	994,00	2,01
GA4	9 pm	6,50	92,05	315,46	1005,36	1,02
PC5	2 pm	7,90	100,00	317,80	1003,20	2,20
RS6	4 pm	9,70	100,00	317,50	1001,70	1,60
GA7	9 am	9,71	90,85	258,14	1002,00	0,85
PC8	2 pm	14,50	60,20	229,70	1004,10	1,30
RS9	4 pm	13,70	64,40	231,90	1004,00	1,00

*GA, PC, RS are the code of each subject.

Table 4. Individuals Anthropometric characteristics.

	Height (m)	Initial weight (kg)	Final weight (kg)	BM*
GA1	1,77	74,00	73,90	23,62
PC2	1,75	133,35	132,70	43,79
RS3	1,73	78,00	77,80	25,97
GA4	1,77	73,40	73,25	23,43
PC5	1,75	132,70	132,60	43,54
RS6	1,73	77,00	76,80	25,72
GA7	1,77	73,40	73,30	23,40
PC8	1,75	130,25	130,20	42,77
RS9	1,73	76,80	76,80	25,66
TR1	1,71	66,90	66,65	22,88
TR2	1,71	67,71	67,71	23,55
TR3	1,71	67,72	67,71	23,55
FJ1	1,78	84,65	84,60	26,54
FJ2	1,78	85,20	85,20	26,71
FJ3	1,78	84,65	84,40	26,53

*Index of muscle mass ($Kg.m^{-2}$)

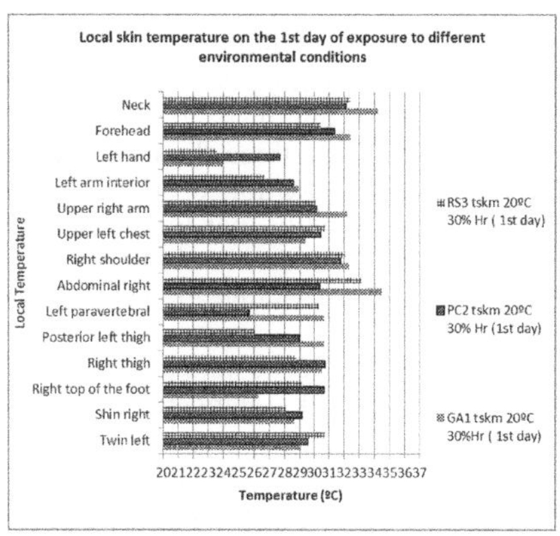

Figure 4. Variation of local skin temperature at 20°C_30% on the first day.

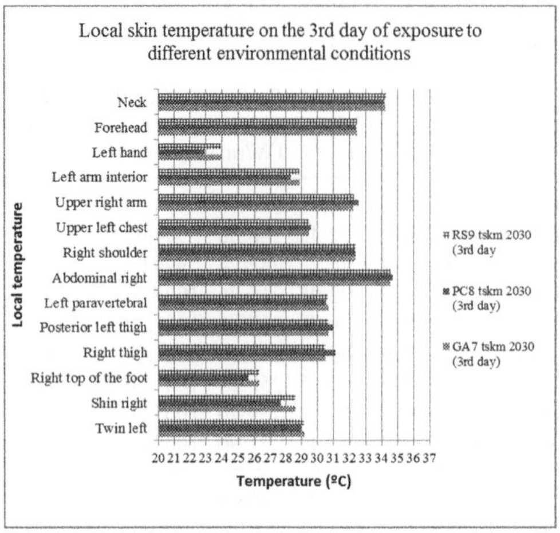

Figure 5. Variation of the local skin temperature at 20°C 30% HR on the third day.

first and third day and is visible the stability of the temperature achieved in the third day, mainly in head temperature in three volunteers as can be seen in the Figure 5. Despite individual differences between the subjects the skin temperatures have a tendency to stabilize between the three subjects, after three days of acclimation at the same temperature and relative humidity.

4.2 *Skin temperature at 20°C and 32°C 60% HR comparing with 20°C 30% HR*

The second trials took place in two different conditions and it was compared with the previous one (20°C 30% HR). These trials run respectively on February 19, 20 and 21 at 4 pm and on March 6, 7 and 8 at 4 pm. The conditions of the climatic chambers are shown on Table 5.

The outdoor weather conditions of each trial are shown in Table 6.

In the second trials Figure 6 and Figure 7, it can be seen that at the same temperature of 20°C, but with different humidities, 30% and 60%, the skin temperature, after three days, not stabilized between subjects as mentioned in the first trial. This raises the possibility that for the same temperature, an increase in the humidity (30% HR to 60% HR) may lead to a change in the physiological response of the body.

In the third trial at 32°C, were not observed significant differences in skin temperature after three days of acclimation.

The mean skin temperature of the body shown in Table 7, in the second trials, increase slightly from the first to the third day only in the first two conditions (20°C_30% HR and 20°C_60% HR) and in the third condition (32°C_60% HR) remains

Table 5. Conditions of the second trials.

Days	*	Climatic chamber	Laboratory conditions
22 January 2013	RS3	20°C30% HR	20–22°C
23 Januart 2013	RS6		48–70% HR
24 January 2013	RS9		
19 February 2013	TR1	20°C60% HR	
20 February 2013	TR2		
21 February 2013	TR3		
6 March 2013	FJ1	32°C60% HR	
7 March 2013	FJ2		
8 March 2013	FJ3		

*Code of the subjects

Table 6. Outdoor weather values during the second trials performed in diferent days.

	Time	t_{ar} (°C)	% HR	Rad Atm ($W.m^{-1}$)	Atmospheric pressure hPa	Win speed $m.s^{-1}$
RS3	4 pm	8,27	76,23	287,24	994,00	2,01
RS6	4 pm	9,70	100,00	317,50	1001,70	1,60
RS9	4 pm	13,70	64,40	231,90	1004,00	1,00
TR1	4 pm	12,75	77,28	265,13	994,00	1,45
TR2	4 pm	12,28	84,72	298,96	999,00	2,09
TR3	4 pm	14,29	94,61	301,45	993,60	3,51
FJ1	4 pm	13,89	99,66	308,72	983,68	3,61
FJ2	4 pm	12,13	95,15	310,17	985,20	2,97
FJ3	4 pm	15,00	70,46	253,28	991,00	3,08

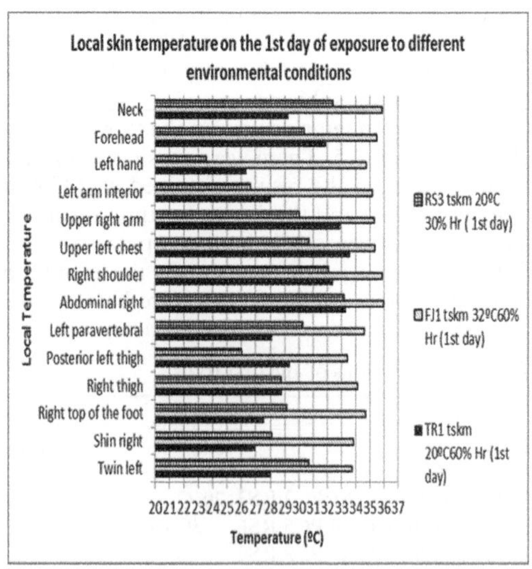

Figure 6. Variation of the local skin temperature at three different conditions on the first day.

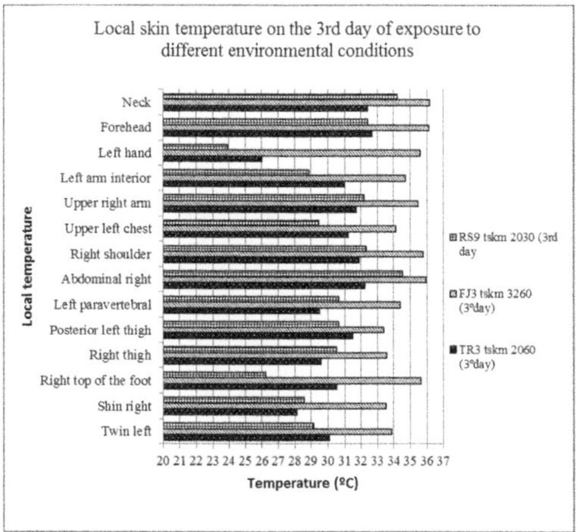

Figure 7. Variation of the local skin temperature at three different conditions.

Table 7. Mean skin temperature of the body and local maximum and minimum temperature between the second trial.

Conditions at the climatic Chamber	Tskm (°C) of the body		Diference between maximum and minimum local skin temperature (°C)*	
	1° day	3° day	1° day	3° day
20°C_30%	29,38	30,27	10	10
20°C_60%	29,77	30,61	7	6
32°C_60%	34,87	34,90	3	2,5

*Differences between the maximum and minimum temperatures of the sites of the body

practically unchanged, while the difference between the maximum and minimum local skin temperatures of different parts of the body decrease with the increasing of relative humidity (when is changes from 30% HR to 60% HR) and also with increasing of environmental temperature (when it changes from 20 to 32°C). This situation indicates a general greater adaptation of the body to the environment between the first to the third day of aclimation.

5 CONCLUSIONS

This study mainly investigates the response of the skin temperature (local and mean) of five subjects to different environmental conditions of temperature and humidity. Three assays were performed, the first at 20°C and 30% HR, the second at 20°C/60% HR and the third one at 32°C/60% HR.

Regarding the thermal sensation, two questionnaires were performed in each test: one to report the thermal sensation at the beginning of the tests and another to report the thermal sensation at the end of the trials. The subjective evaluation indicated a temperature of 20°C and 60% HR as the neutral thermal sensation.

In spite of the different individual anthropometrics characteristics, in the first trials, the three subjects reached a very similar local skin temperature level after three days. It was also found that changing only the humidity from 30% HR to 60% HR and maintaining the temperature at 20°C the subject submitted to these conditions did not reach a similar skin temperature level as the one recorded among individuals of the first test.

Finally, at 32°C and 60% HR it was found greater stability between different parts of the body between the first day and the third day.

REFERENCES

Costa, E.Q., Guedes, J.C. & Baptista, J.S. (2012). 'Core Body Temperature Evaluation: Suitability of Measurement Procedures'. *Thermology International,* Vol. 22 n° Appendix 1 to Number 3, pp. 33–41.

Costa, E. Quelhas & Baptista J. Santos (2013). 'Thermal Environment and Cognitive Performance: Parameters and Equipment.' Proceedings n° 69 of SHO2013, Guimarães pp. 267–272.

Ely, B.R., Ely, M.R., Cheuvront, S.N., Kenefick, R.W., DeGroot, D.W., & Montain, S.J. (2009). 'Evidence against a 40°C core temperature threshold for fatigue in humans'. *J Appl Physiol* 107, 1519–1525. doi: 10.1152/japplphysiol. 00577.2009.

Guedes J.C., Costa, E.Q., & Baptista, J.S. (2012). 'Using a Climatic Chamber to Measure The Human Psychophysiological Response Under Different Combinations Of Temperature And Humidity'. *Thermology International,* Vol. 22 n° Appendix 1 to Number 3, pp. 49–54.

Kataoka, H., Kano, H., Yoshida, H., Saijo, A., Yasuda, M., & Osumi, M. (1998). 'Development of a skin temperature measuring system for non-contact stress evaluation'. Paper presented at the Engineering in Medicine and Biology Society, 1998. *Proceedings of the 20th Annual International Conference of the IEEE.*

Parsons, K.C. (2003) *Human thermal environments: the effects of hot, moderate, and cold environments on human health, comfort, and performance*, 2nd ed., London. Taylor & Francis.

Shin-ichi Tanabe, Naoe Nishihara, and Haneda, M. (2006). 'Indoor Temperature, Productivity, and Fatigue in Office Tasks'. *HVAC&R Research.* 2006, Vol. 13.

Yao, Y., Lian, Z., Liu, W., & Shen, Q. (2008). Experimental study on physiological responses and thermal comfort under various ambient temperatures. [Comparative StudyResearch Support, Non-U.S. Gov't]. *Physiology Behavior,* 93(1–2), 310–321. doi: 10.1016/j. physbeh.2007.09.012.

Occupational Safety and Hygiene II – Arezes et al. (eds)
© 2014 Taylor & Francis Group, London, ISBN 978-1-138-00144-2

Ergonomic assessment of a workstation in a paints production line

J.M. Monteiro & J.C.A. Calvão
Tintas CIN, S.A., Maia, Portugal

J.M. Monteiro, M.E. Pinho & M.A.P. Vaz
Faculty of Engineering, University of Porto, Porto, Portugal

ABSTRACT: Despite the increasing automation of industrial processes, the manual handling remains a major cause of work related MSDs. This paper presents an ergonomic study of a workstation in a filling line of a paints and varnishes production plant which involves manual handling. It aimed to assess the risk of MSDs that workers are exposed to. Four methods were used: EEPP™ (Energy Expenditure Prediction Program), revised NIOSH Lifting Equation (National Institute for Occupational Safety and Health), REBA (Rapid Entire Body Assessment) and QEC (Quick Exposure Check). The risk of work-related MSD was identified and some corrective measures for their mitigation were proposed.

1 INTRODUCTION

Musculoskeletal Disorders (MSD) are common and well known in many countries, and represent substantial costs either financially, or on the quality of life (Baldwin, 2004, David, 2005). Although not exclusively caused by the professional activity, they constitute an important part of the occupational diseases recorded.

MSD are the leading cause of disability in the European Union (EU) (David et al., 2008), representing more than a third of the occupational diseases recorded (Mirmohamadi et al., 2004), either in the United States, in the Nordic countries and also in Japan (Punnett and Wegman, 2004). They represent a significant cost in health systems and are the cause of significant losses in productivity (Baldwin, 2004), and therefore are considered a significant and growing problem in the EU (Schneider et al., 2010). Although there is no concrete data for the EU, with regard to the Nordic countries their estimated costs range between 0.5% and 2% of the Gross National Product, and it is believed that this value can increase over time (Buckle et al., 1999, Schneider et al., 2010, Uva et al., 2008).

MSD are also the subject of great debate in the scientific literature, since the physical and ergonomic work characteristics often cited as risk factors of musculoskeletal disorders, include fast working pace and repetitive movements, vigorous efforts, non neutral postures and exposure to mechanical vibration, often neglecting nonprofessional factors (Punnett and Wegman, 2004).

Several factors may contribute to the onset of MSD. These factors may exert an individual or a combined action and may be related with three components, namely work activity, individual factors and organizational and psychosocial factors (Serranheira, 2007, Uva et al., 2008).

The manual handling of loads is one of the major causes of work-related MSD, and despite the automation of many industrial processes, it is still often used to perform several tasks (Perista, 2007).

The main focus of this paper is therefore the ergonomic analysis of a semiautomatic filling line in a paint and varnishes industry. It is intended, therefore, to identify situations that can trigger the development of work-related MSD and, thereby, contribute to the risk reduction. Overall this study aimed to: (1) Evaluate MSD risk associated with handling of loads in the workstation selected, (2) Perform a comparative analysis of the results obtained by the application of the selected methods, (3) Enumerate some of the preventive strategies that can be taken to mitigate the risk.

2 MATERIALS AND METHODS

2.1 *Workstation*

Figure 1 shows the assessed workstation (filling line).

In this station the worker supplies cans to the filling line. The complete station comprises a filling machine and an automated conveyor.

Figure 1. Image of the workstation evaluated.

Figure 2. Worker 1 performing the palletizing task at level 0.

Table 1. Characteristics of the tasks performed.

	Tasks characteristics			
Subtask	Initial Height (m)	Final Height (m)	Load (g)	Duration (s)
Fill the table	1.20	0.78	455	612
Supply covers	0.70	1.75	900	612
Box	0.70	0.91	5500	5760
Palletize—level 0	0.70	0.15	11000	477
Palletize—level 1	0.70	0.36	11000	477
Palletize—level 2	0.70	0.57	11000	477
Palletize—level 3	0.70	0.78	11000	477

Figure 3. Worker 2 performing the palletizing task at level 0.

2.2 *Subjects*

Two male employees participated in the study. They were 26 and 28 years old, and weighted 91 and 84 kg, respectively. Their heights were 1.81 and 1.60 meters.

2.3 *Tasks features*

The tasks performed by both workers are essentially manual handling of loads whose characteristics are compiled in Table 1.

Both employees perform the same subtasks and, therefore, there is no differentiation. Figure 2 and Figure 3 show both workers performing the palletizing task at level zero.

2.4 *Used methods*

Numerous methods have been developed either to assess the risk of developing MSD, or to monitorize the effects of corrections implemented (Takala et al., 2009).

Generally, the use of risk assessment methods demands an analysis of the job performed, through its decomposition in distinct and successive events in time, which allows the observation of important aspects such as the application of force, frequency and posture.

According to David (2005), available methods for MSD risk evaluation may briefly be classified according to three categories: worker's reports, observational methods and direct measurements using monitoring tools.

In the present study four methods were used: EEPP™—which estimates the rates of metabolic consumption (University of Michigan, 2010), revised NIOSH Lifting Equation—which determines the Recommended weight limit and the Lifting Index (Dempsey, 2002, Waters, 2004), REBA—which is a postural analysis tool (Hignett and McAtamney, 2000) and QEC—which is an assessment tool that takes into account the evaluation of the worker. (David et al., 2008).

These methods were selected based on two conditions. First, they did not interfere in the performance of the workers activity and, second, they had low associated logistics (in particular concerning to financial charges).

In a first step, the use of EEPP™ and revised NIOSH Lifting Equation allowed the identification of the most penalizing task for the workers and then, this was reassessed through REBA and QEC.

3 RESULTS AND DISCUSSION

Regarding the overall assessment of the task, the methodology EEPP™ pointed as metabolic consumption an overall value of 3.33 kcal/min and 3.10 kcal/min for workers 1 and 2, respectively.

Taking as reference the NIOSH guidelines 1981, that establishes 3.5 kcal/min as the Limit action for an average 8 hour work day (which means that for an energy expenditure lower or equal to 3.5 kcal/min, 75% of the female population and 99% of the male population—or 90% of the population global—can develop the activity without being subjected to increased risk of fatigue or MDS (University of Michigan, 2010)) it was found that none of the employees presents a metabolic rate higher than this value, and therefore they do not seem to be exposed to an increased risk of developing MSD. However, if the analysis is made using as reference the NIOSH 1991 guidelines (University of Michigan, 2010) which considers as reference the value of 3.2 kcal/min, then worker 1, once he exceeds this value, is at the risk of developing MSD.

EEPP™ takes into account an individual characteristic of the workers—their body weight—which differentiates both workers. In fact, while all other variables involved in the calculation of the metabolic rates have the same value for the two workers, worker 1 has a higher body weight than worker 2. In their study, Garg et al. (1978) reported that light changes in physical parameters, commonly used to describe the activity may result in significant changes of metabolic rates. Our study seems to support the findings of Garg et al.

On the other hand, the multi-task analysis performed through NIOSH Lifting Equation led to a Lifting Index of 1.23. As this value is between 1 and 3 (Waters et al., 1994), it means that the global demand of the job may exceed the workers capabilities, and these may therefore be exposed to an increased risk of MSD.

Contrary to what happens with the EEPP™, NIOSH Lifting Equation does not take into account the biometric data of employees. In Table 2 are summarized the results obtained for each of the various subtasks according to methodologies EEPP™ and revised NIOSH Lifting Equation.

Table 2. Results obtained with the methods EEPP™ and revised NIOSH Lifting Equation.

	EEPP™		NIOSH Lifting Equation	
	Metabolic Rate (kcal/min)		FIRWL	STRWL
Subtask	Worker 1	Worker 2	Worker 1&2	Worker 1&2
Fill the table	2.90	2.70	0.03	0.039
Supply covers	3.34	3.10	0.06	0.073
Box	2.90	2.70	0.29	0.405
Palletize—level 0	6.56	6.14	0.71	0.857
Palletize—level 1	5.26	4.91	0.68	0.812
Palletize—level 2	3.91	3.67	0.64	0.773
Palletize—level 3	3.21	3.03	0.64	0.773

Frequency—Independent Recommended Weight Limit (FIRWL) identifies subtasks that can present potential problems of strength associated with low frequency liftings, while Single-Task Recommended Weight Limit (STRWL) is generally used to identify tasks with excessive physical demands that would result in fatigue. As shown by the analysis of Table 2, both methodologies point the palletizing tasks, particularly at level 0 and level 1, as the ones having the highest values of metabolic rates and FIRWL and STRWL.

Once identified the most damaging subtask, this one was focused on a new study, performed by the application of methodologies REBA and QEC. Regarding the application of REBA, through the analysis of Table 3, can be realized that like previously pointed by EEPP™ and by the NIOSH Lifting Equation, it is at the lower levels (where the vertical load manipulation distance is greater) that the risk intensifies. This situation occurs in both workers, regardless of their different body heights.

Contrary to expectation, the application of REBA shows that worker 2 (whose height is lower) seems to be the most exposed to the risk of MSD. This may happen because he adopts a working position with a larger trunk flexion.

However, this analysis can not only be done based in the difference between workers heights since the positioning of the pallets for loading also seems to play an important role. The positioning of the pallet is a free choice of the employees at the beginning of the work journey and it could be noticed that worker 2 always places the pallet at a greater distance from the conveyor, which causes him to bend over.

With respect to methodology QEC and a since it also includes the evaluation of workers, it was decided not to split the palletizing task. Table 4 shows that shoulder and hand/wrist areas

Table 3. Results obtained through REBA.

	REBA			
	Worker 1 (1.81 m)		Worker 2 (1.60 m)	
	Score	Risk	Score	Risk
level 0	3	High	4	Very High
level 1	3	High	4	Very High
level 2	2	Moderated	3	High
level 3	2	Moderated	3	High

Table 4. Results obtained through QEC.

	QEC			
	Worker 1 (1.81 m height)		Worker 2 (1.60 m height)	
	Score	Exposition level	Score	Exposition level
Back	30	Moderate	34	High
Shoulder	34	High	36	High
Hand/Wrist	36	High	42	Very High
Neck	10	Moderate	14	High
Driving	1	Low	1	Low
Vibration	1	Low	1	Low
Work Pace	4	Moderate	4	Moderate
Stress	9	High	4	Moderate

present a High and/or Very High risk in both workers and that, globally, the task seems to inspire concerns in the development of MSD, since there are different zones where the risk is classified as High or Very High.

In a study performed by Chiasson et al. (2012) where the results of the implementation of eight distinct methods used in the evaluation of the risk associated with manual handling of loads (including REBA and QEC) were compared, the authors found as the less rigorous in the evaluation of the overall risk. However, a study conducted by Motamedzade et al. (2011) showed a strong correlation between the two methods either in the identification of the harmful tasks, or in the determination of the potential risk to the onset of MSD.

Further, Motamedzade et al. (2011) demonstrated that regardless of the type of task, the results obtained by the methodologies QEC and REBA were similar. In fact, and although it cannot be made a partial comparison (since in the application of the QEC it was decided not to split the task into subtasks), both methods point to the existence of Very High, High and Moderate risk in both workers thus stressing the need for an ergonomic intervention. This agreement between methods seems to support the conclusions made by Motamedzade et al. (2011) and Chiasson et al. (2012). Summarily, all the methodologies identified palletizing task (and their associated subtasks) as a risk situation, pointing towards the need of an intervention which may lead to ergonomic risk mitigation.

Due to logistic constrains it has been impossible to validate the results obtained by using direct methods. However, although the used methods have been considered the most proper, taking into account the available means, authors are aware of their strengths and weaknesses. In fact, all the methods have their restrictions and therefore several methods were used to overcome the limitations of each of the others and thereby contribute to a better assessment of the workstation. The revised NIOSH Lifting Equation does not take into account worker's biometrical characteristics. This could be a strength once it is not discriminatory, but it is also a weakness since some valuable individual information is lost. EEPP™ takes into account the variability among workers and allows to distinguish between them.

However, none of these methods analyses posture or takes into account worker's perceptions. That was the main reason why it was decided to apply REBA, which is a powerful postural analysis, and QEC which includes worker's assessment.

Conversely, the successful appliance of the methods selected is also dependent of the correct division of the task into subtasks and the ability of the evaluator to apply them evenly. This is one of the major limitations of QEC. Finally, although extremely valuable, worker's assessment is often partial and can lead to incorrect results.

4 CONCLUSIONS

Through the methodologies EEPP™, revised NIOSH Lifting Equation and REBA it was able to determine that are the subtasks combining the handling of the heaviest load with the largest vertical distance to travel that most penalizes workers.

Through the application of REBA it was possible to conclude that the comparison between workers can not only be made attending to the biometric data since it was found that the positioning of the pallet load seems to have a significant influence on the posture adopted by the workers.

Among all the methodologies applied, only the revised NIOSH Lifting equation does not take into account the distinction between workers. This can be seen as an advantage, since it is not discriminatory, but also as a constraint, once that workers characteristics (individual factors) are not taken into account.

As the methods EEPP™, revised NIOSH Lifting Equation and REBA show that, in the performance of the palletizing task, the effort/risk level is higher in the level 0 than in the level 3. The results of this study thus seem to point towards the need for defining a strategy to prevent the risk of MSD associated with this task. Given the impossibility to change the load handled (as it is the final product), the intervention could start by acting at the level of both loading and unloading heights.

An example of a possible intervention could consist in keeping the pallet at a constant height (equivalent to the level 3) and to use an automatic adjustment system with springs or, otherwise, to use a pallet carrier to support the pallet at the level considered optimal. In view of REBA results, to adjust the positioning of the pallet (with markers on the ground) and check the postural reaction of worker 2 could be an effective intervention. It should, however, be noted that the herein proposed risk mitigation interventions should be integrated in the company's general prevention plan of occupational hazards and to include the participation of workers, as recommended by the principles of participatory ergonomics. Finally, it should be noted that given the reduced sample size it was not possible to make any kind of extrapolation of results.

ACKNOWLEDGMENT

The authors would like to thank Master in Occupational Safety and Hygiene Engineering (MESHO), of the Faculty of Engineering of the University of Porto (FEUP), all the support in the development and international dissemination of this work.

REFERENCES

Baldwin, M.L. 2004. Reducing the costs of work-related musculoskeletal disorders: targeting strategies to chronic disability cases. *Journal of Electromyography and Kinesiology 14*(1): 33–41.

Buckle, P., Devereux, J., Safety, E.A.F., & Work, H.A. 1999. *Work-Related Neck and Upper Limb Musculoskeletal Disorders*: Office for Official Publications of the European Communities Luxembourg.

Chiasson, M.-È., Imbeau, D., Aubry, K., & Delisle, A. 2012. Comparing the results of eight methods used to evaluate risk factors associated with musculoskeletal disorders. *International Journal of Industrial Ergonomics 42*(5): 478–488.

David, G. 2005. Ergonomic methods for assessing exposure to risk factors for work-related musculoskeletal disorders. *Occupational Medicine 55*(3): 190–199.

David, G., Woods, V., Li, G., & Buckle, P. 2008. The development of the Quick Exposure Check (QEC) for assessing exposure to risk factors for work-related musculoskeletal disorders. *Appl Ergon 39*(1): 57–69.

Dempsey, P.G. 2002. Usability of the revised NIOSH lifting equation. *Ergonomics 45*(12): 817–828.

Garg, A., Chaffin, D.B., & Herrin, G.D. 1978. Prediction of metabolic rates for manual materials handling jobs. *The American Industrial Hygiene Association Journal 39*(8): 661–674.

Hignett, S., & McAtamney, L. 2000. Rapid entire body assessment (REBA). *Appl Ergon 31*(2): 201–206.

Motamedzade, M., Ashuri, M.R., Golmohammadi, R., & Mahjub, H. 2011. Comparison of ergonomic risk assessment outputs from rapid entire body assessment and quick exposure check in an engine oil company. *Journal of Research in Health Sciences 11*(1): 26–32.

Perista, H.C., J. 2007. *Managing musculoskeletal disorders-Portugal*. EWCO: Eurofond.

Punnett, L., & Wegman, D.H. 2004. Work-related musculoskeletal disorders: the epidemiologic evidence and the debate. *Journal of Electromyography and Kinesiology 14*(1): 13–23.

Schneider, E., Irastorza, X., Copsey, S., Verjans, M., Eeckelaert, L., & Broeck, V. 2010. OSH in figures: Work-related musculoskeletal disorders in the EU—Facts and figures. *Luxembourg: European Agency for Safety and Health at Work*.

Serranheira, F.M.S. 2007. *Lesões músculo-esqueléticas ligadas ao trabalho: que métodos de avaliação do risco?* Universidade Nova de Lisboa.

Takala, E.P., Pehkonen, I., Forsman, M., Hansson, G.A., Mathiassen, S.E., Neumann, W.P., et al. (2009, Aug 9–14). *Systematic evaluation of observational methods assessing biomechanical exposures at work*. Paper presented at the 17th World Congress on Ergonomics, IEA2009, China.

University of Michigan. (2010, October 25, 2010). EEPP. Retrieved April 4, 2012, from http://www.engin.umich.edu/dept/ioe/ENGEXP/

Uva, A.S., Carnide, F., Serranheira, F., Miranda, L.C., & Lopes, M.F. (2008). *Lesões Musculoesqueléticas Relacionadas com o Trabalho - Guia de Orientação para a Prevenção*.

Waters, T.R. 2004. National efforts to identify research issues related to prevention of work-related musculoskeletal disorders. *Journal of Electromyography and Kinesiology 14*(1): 7–12.

Waters, T.R., Putz-Anderson, V., & Garg, A. 1994. *Applications Manual for the Revised NIOSH Lifting Equation*: US Department of Health and Human Services, Public Health Service, Centers for Disease Control and Prevention, National Institute for Occupational Safety and Health, Division of Biomedical and Behavioral Science.

Occupational Safety and Hygiene II – Arezes et al. (eds)
ISBN 978-1-138-00144-2

Formworks—maximizing the prefabrication of each set on the ground

A. Reis
Brisa Engenharia e Gestão, S.A., Quinta da Torre da Aguilha Edifício Brisa 2785-599, São Domingos de Rana, Portugal

J. Santos Baptista
Research Laboratory on Prevention of Occupational and Environmental Risks (PROA/LABIOMEP), Faculty of Engineering, University of Porto, Portugal

ABSTRACT: There is no *in situ* concrete without formwork. The penetration of *in situ* concrete in construction is highly significant. The formwork is the main stage of its implementation. However, this activity combines high risks of falling from heights and of crushing, in the processes of assembling, use and disassembling of the formworks. Formwork projects are often designed considering its structural stability, but without providing the necessary safety conditions for its implementation. This article intends to present and justify rules of good practice, associated with maximization of the prefabrication in the ground. Such procedures must be always on the basis of the design of formwork projects. Was conclude that the outcome of this exercise will make equipment become more ergonomic, more effective, and hence, more secure.

1 INTRODUCTION

The Prevention Chapter is the main aspect in the role of Occupational Health and Safety Technicians (OHST) and Construction (Design and Management) co-ordinators (CDM co-ordinators). These technicians are responsible for safety planning during construction. It is at this stage where, in the construction industry can be open, effectively, a window of opportunity for their greatest contribution to the improvement of safety in the construction sites. This contribution can be made mandatory for all the players to create effective production cycles, embodied on safety planning.

For Safety planning fulfill its role it is necessary to clarify with validated practices, operation by operation, all the kinematics of the production processes and their interaction with workers. Thus it is possible to design efficient production cycles. To act otherwise is an illusion, not so unusual as we might think, and that as contributed to feed the precariousness and improvisation that have been made during formwork use in construction sites and civil engineering.

The risks can only be defined and prevented, in an activity such as assembling and disassembling of formwork, not only in possession of information to forecast and anticipate all kinematics associated to operations, but also its interaction with workers, throughout the different stages of the construction process. No formwork is built as a result by a magical step.

The performance of traditional formwork, which this study focuses, is associated to a scale problem and structures considerably larger than human. The size difference between the workers, the formwork and/or the construction element in concrete is huge. And also the time between the beginning of formwork execution and its final state. In the building process are present the risks of falling from height, falling of materials and smashing, often and with high magnitude. Depending on the equipment effectiveness, shown on an appropriate formwork design, these risks may or may not be located at reasonable intervals, duly and previously controlled.

All formwork elementary parts are constructed with a specific and desired degree of prefabrication. However, only after all of its components get together, according to the project (and assembling instructions), is that the equipment is effectively stable, complete and ready to receive concrete.

Since the 3 m pillar to the one with 100 m height, is the formwork equipment that dictates the dignity with which the work is performed. Whole operation happens and lives around the formwork, even in the most adverse circumstances. It's there that the cycles are completed and this is the ground on which workers trust to support of their bodies and their lives.

The shortcomings in activities planning come from the gross disconnection with ergonomic and anthropometric knowledge. These gaps are associated, mainly, to processes of assembling, use and disassembling of formwork. These processes should have on the essence of its foundation its main actor, the: Man. With about 1,7 m high in the Portuguese case (Arezes et al., 2006), the man effectively allows the formwork materialization.

From this scenario results, necessarily, the impossibility of seeing all necessary *movie images* for risk assessment procedures and to implement the preventive measures. This gap implies, in turn, unpredictability of the risks, associated with the inevitable improvisation. This fact impairs the effectiveness of the production cycle and, necessarily, the safety of operations, which often turn out to be held precariously in construction sites.

The formwork design is the core element, in the development of all operations related to the assembling, use and disassembling. It is the formwork project that defines the safety level of activities implementation related to its mounting and dismounting. Therein is concretized the dignity, or lack thereof, that the work actually will happen later in the construction site.

The way formwork projects are designed is of increasing importance, since the simple calculation of its stability to the construction sequence. The information included in formwork project should address risk assessment and avoid the need to equate any preventive measures and/or mitigating the risks involved. It should also show a cycle of activities considered safe and in accordance with the defined safety policy in contract, by the contracting authority. The production cycle can't be hurt of functional deficiencies that have to be resolved after by the responsible for safety planning, which are not inherently Civil Engineers or Formwork Designers.

In this context, this work proposes safety rules in assembling and disassembling formwork, maximizing its prefabrication on ground and seeks to define the respective basic procedures.

2 MATERIALS AND METHOD

2.1 *As a result of professional experience*

The matter in discussion results of the analysis and follow-up for eleven years of major public works projects. This labor was complemented with a systematic survey about construction methods, based on an ergonomic perspective, where the *in situ* concrete assumes a prominent place, for their inseparable co-existence with the formwork and the risks of high magnitude, involved, that are associated with. The monitoring was conducted daily in duties of responsibility for safety in the construction, of contractual origin.

The three main areas of development, in the task of ensuring the preservation of human life (employees and contractors) during the execution of work, were ranked as follows:

1. Technical validation of works safety planning;
2. Ensure that the training to the different stakeholders is properly done;
3. Motorisation of the conditions in which the activities take place in the construction site, keeping the contracting authority properly informed.

The OHST and the CDM co-ordinators, for carrying out the integration of the activity of assembling and disassembling of formwork, of a particular building element in a Safety Plan, must perform a detailed and critique analysis of the formwork project.

2.2 *Bibliographic review*

In parallel to the field work developed, a survey was conducted by scientific studies that addressed best practices in project activity, assembling and disassembling of formwork, as well as international legal and regulatory frameworks on the same subject.

The most relevant elements found were from institutions from U.S.A (SSFI. 2006), Australia (Territory, 2011) and Spain (Pino, 2010).

3 PREFABRICATION ON THE GROUND

3.1 *Vertical elements: Pillars*

Among other rules of good practice, to consider, maximizing prefabrication on the ground plays an important role in the effectiveness of the production cycle and, of course, safety operations. The higher the number of operations performed at ground level, the lower the number of operations performed in height, which in turn implies lower risk of falling. Everything is possible. Mounting piece by piece all formwork *in situ* (Fig. 1) or fully manufacture it at ground level, for subsequent application of site, where it will actually be molded the concrete element (Fig. 2).

The first option often incorporates high improvisation, which degenerates into high risks of falling from height and falling of materials during operations of assembling and disassembling. It often results in remaining operations, such as concreting, with identical high improvisation scenarios incorporating high risk of falling from height (Fig. 3).

Figure 1. Workers assemble pillars' formwork, of a building, piece by piece.

Figure 2. Transportable formwork with vertical access and work integrated platform, pillars of a building. (Source—Formwork Pillars Quattro. PERI).

Figure 3. Concreting a pillar of a building, held in poor conditions, associated with *in situ* formwork assembled piece by piece.

The inside formwork of a coffin pillar, a special work of art of a road or rail infrastructure, can often reach a height of 4.5 m. Its implementation can be done in several ways. Among them, spliting the interior mold in two or four parts. This option implies assembling and disassembling each one of the four sets in all the construction cycles. For this it is necessary to make the reception, binding and unbinding of the lift system, fixing the links between each one, mounting and unmounting the required working platforms for each operation (either intermediate or concreting). Apply the integrated vertical access in the connection between the bottom and their respective platform. And finally, raise the bottom and the support of the internal formwork in a integrated way with its suspension platform (Fig. 4).

Alternatively, this coffin pillar can be molded inside with a single inner formwork that only shrinks a few *cm* to be removed as a single piece. At the same time, all its features, including accesses and working platforms would remain intact, if they were specially designed to adapt to this process of formwork compression and decompression (Fig. 5, 6 and 7).

In this second option the inner formwork promotes a substantial reduction of operations and

Figure 4. Withdrawal of the 4 sets that are inside the formwork, to prepare a new rise of the coffin pillar, of a bridge in a road infrastructure.

Figure 5. Single interior set that is mold inside the formwork with working platforms and access that keeps intact during compression, a coffin pillar of a bridge of a road infrastructure.

Figure 6. Single interior mold set, inside the formwork with working platforms and access that keeps intact during compression set, a coffin pillar of a bridge of a road infrastructure.

Figure 7. Withdrawal of the whole interior set which is mold inside the formwork, to prepare a new elevation of a coffin pillar of a bridge of a road infrastructure.

a proportional decrease to the magnitude of risk during operation, in particular falling from height, falling of materials and smashing.

Another example that involves the increment of operations on height is related to the design of work platforms. Operations as referred on Figure 8 may be repeated 32 times in each cycle. Alternatively, a simple change in their design, not only could eliminate these operations as the equipment would provide protected access and horizontal continuity between sets, during its process of assembling and disassembling (Fig. 9).

3.2 *Horizontal elements*

It can be performed a deck of a formwork (including the slab and beams of a span of a work of art from a road or rail infrastructure) with sets fully implemented at the ground level (Figs. 10 and 11), or only partially executed at this level (Fig. 12).

Figure 8. Worker, using personal protective equipment (safety harness) to remove the floor of continuity between platforms of adjacent sets, a coffin pillar of an overpass of a road infrastructure.

Figure 9. Design of platforms keeps horizontal access protected and continuity between sets, immediately after application, a coffin pillar of an overpass of a road infrastructure.

Figure 10. Sets of formwork previously prefabricated on the ground, a tray of a flyover of a road infrastructure.

Figure 11. Workers on sets of formwork, previously prefabricated on the ground, and still in position, board of a flyover of a road infrastructure.

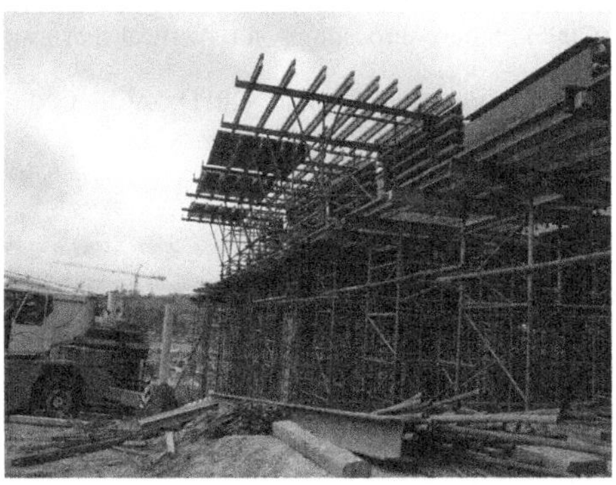

Figure 12. Sets of formwork of a board, flyover of a road infrastructure, applied *in situ* partially prefabricated.

In the latter case, all the lining is executed (usually formwork is performed in phenolic panels or marine plywood), incrementing, largely, the number of operations to perform in height, and the carpenters will necessarily to build these structures to the working process.

In both cases there must be the reception of the sets in height for its positioning and untying from the lifting equipment.

Additionally, in the first case, after the application of all sets, remains only to perform the catches (openings between sets, transversely of the board, usually with a width of approximately 20 to 30 cm, which need to be filled with the lining of the formwork) where the risk of falling from height, comparatively, is greatly minimized (Fig. 11). In the second case, must be executed around the ceiling with the carpenters positioned on the sets, where the risk of falling from height reaches other proportions and openings has another dimension (Fig. 12).

In these cases protection against the risk of falling from height is difficult to implement, usually is used only the individual protection (horizontal lifelines, on and along the bottom of the beam in its entire length and fixed at the end of the pillars). This solution has weaknesses/limitations related to position of lifeline at the bottom of the beam, which coincides with the upper bound of the pillars. This requires the positioning of workers on the sets, to an upper point (proportional to the height of the rib) and eccentric, in the zone which shapes the bottom of the central slab and lateral of the sets that are to be applied.

The fact that the formwork is not modeled previously with a specific geometry, by sections associated with the sets integrally prefabricated at ground level, it necessarily implies a disassembling performed randomly without any criteria, in difficult conditions. With workers positioned between the upper part of the falsework and the bottom of the formwork structure, rare are the cases where there is no risk of falling from height with strong magnitude. In these cases it is inevitable disorder and unpredictability of the risks associated with falling materials, smashing, falling from height, among others.

It is one of the most risky and difficult to control existing works in the construction of road infrastructure, that few watch and the majority avoids be witness, which explains in part the longevity of this precariousness, far from modern and socially evolved laws concerning safety at work. (Fig. 13).

Alternatively, the modeling of the predefined sets allows a planned demoulding prior to the implementation of holes in the tray. These holes aim to create four fixing points, on each one of the sets for previous attachment to the fresh concrete casting segment and subsequent return to the ground. It is thus eliminating the risk of falling from height and minimized, to negligible, levels the risk of falling materials/smashing, during operations. Activity that begins after withdrawal of all fastenings of the formwork, made the takeoff of the sets, and the dismantling of all underlying falsework.

Figure 13. Dismantling of formwork, of a flyover of a road infrastructure, after dismantled centering.

4 CONCLUSIONS

The formwork is the main stage of the implementation of the *in situ* concrete, associated with high risk of falling from height and smashing during activities for assembling, use and disassembling.

The maximization of prefabrication on the ground plays an important role in the effectiveness of the production cycle and minimizing the risk of falling from height and falling materials.

The design and development of the formwork project should include all elementary components and respective movement sequences, as well as other operations that require human interaction involving risks. This analysis should be carried out step by step in the project, until the withdrawal of the last piece of equipment likely to generate damage to the health of workers or others.

The result of this exercise makes the equipment, and their assembling and disassembling processes, more ergonomic, more efficient and therefore safer and often more effective, in direct proportion to the relatively higher number of cycles that each device has to meet.

ACKNOWLEDGEMENTS

The authors thank to Masters Degree in Safety Engineering and Occupational Hygiene (MESHO), Faculty of Engineering, University of Porto (FEUP), full support and international dissemination of their work.

REFERENCES

ACI & ASCC. American Concrete Institute and the American Society of Concrete Contractors. (2008). The Contractor's Guide to Quality Concrete Construction.

Arezes, Pedro M. (2006). ISHST. Instituto para a Segurança, Higiene e Saúde no Trabalho. Estudo Antropométrico da População Portuguesa.

Oberlender, Robert L. Peurifoy and Garold D. (2011). Form-work-for-Concrete-Structures-4th-Edition.

Pino, José Mª Tamborero. (2010). Encofrado vertical. Muros a dos caras, pilares, muros a una cara e sistemas trepantes. INSHT. NTP 834/835/836/837.

Shapira, Aviad. (1999). Contemporary trends in formwork standards—a case study.

SSFI. Scaffolding, Shoring and Forming Institute, INC. (2006). Safety Procedures for Vertical Formwork. (*www.ssfi.org*).

Territory, Australian Capital. (2011). ACT Code of Practice for Formwork.

Occupational Safety and Hygiene II – Arezes et al. (eds)
 ISBN 978-1-138-00144-2

Formworks—provisionally stability of the sets

A. Reis
Brisa Engenharia e Gestão, S.A., Quinta da Torre da Aguilha Edifício Brisa 2785-599, São Domingos de Rana, Portugal

J. Santos Baptista
Research Laboratory on Prevention of Occupational and Environmental Risks (PROA/LABIOMEP), Faculty of Engineering, University of Porto, Portugal

ABSTRACT: The *in situ* concreted elements are only possible with the use of formwork. Formwork has a decisive importance in safety conditions during work. However, the formwork projects are often designed just to his state of formwork ready and filled with concrete, thinking only of final structural stability. The safety conditions required for its realization are not taken into account. In this work, it is intended to present rules of good practice for the preparation of formwork projects associated with the provisional stability of its elementary parts, inherent in its processes of assembly and disassembly. An adequate perspective to minimize the risk of falling from height and smashing arising from these operations.

1 INTRODUCTION

Formwork is a mold which receives concrete in its fluid state and enables its molding to the desired shape until it is be self-supporting. There is no *in situ* concrete without a mold, whatever its nature.

The formwork systems addressed in this article are those of a dimension considerably taller than a man. They require, during the process of assembly and disassembly, the direct involvement of labor under and over its components. This often involves harsh activities and difficult positions, in height, and is associated with the risk of structural collapse and/or objects falling on workers and other persons.

The phases of assembly and disassembly of the formworks, during the execution of a work, can be the most dangerous task on a construction site. If the economic and quality issues are important, safety should be a primary concern. The formwork should be able to withstand loads to which it is subjected without the risk of collapse, or cause danger to workers and others (Contractors 2008).

Traditionally, on construction sites, the highest importance is given to projects focused on the permanent structures at the expense of interim structures (Shapira 1999). Likewise, during construction, temporary structures, such as the formwork, are treated as a simple means to an end. Its construction is often neglected. Despite its importance throughout the construction process formworks have not been treated as structures as of right (Shapira 1999). However, in addition of being forms that shape the construction, they also are workplaces, and they also deserve dignity.

The formwork should be designed ergonomically to interact with employees and with the dynamic associated to the processes for their assembly and disassembly. There are not simple innocuous devices associated with the achievement of concrete.

When designing a formwork system its sequence of assembly and disassembly methodology should be considered (Oberlender 2011).

It should take into account the safety implications for workers involved in the assembly, use and dismantling of the system (Territory 2011). The formwork project is the nuclear document in preparation of safety planning. Among other aspects to consider, it should study the provisional stability of all elementary parts and equipment components. Each of them will interact with workers and must prevent all physical damage or even affect their health throughout assembly, use and disassembly processes. Ensuring stability is an engineering problem, often of high complexity, which cannot be neglected.

Thus, the main objective of this work is to point out the main aspects to consider in formwork projects such as safety, hygiene and health at work perspectives.

2 MATERIALS AND METHOD

The results achieved primarily arise from eleven years of continuous monitoring and progressively systematic construction methods, guided by an ergonomic perspective, where the *in situ* concrete assumes a prominent place, due to its inseparable coexistence with the formwork and inherent risks of high magnitude associate with.

Monitoring was conducted daily, in result of contractual responsibility for safety in construction, in major public works contracts.

The integration of assembly and disassembly activities of formwork system in safety planning is only basic step and, therefore, will be in a position to be accepted, allowing to explicit whole kinematics of production cycles, leading to the achievement of the planned work. Thus, it must be evident, operation after operation, all interaction sequences associated with employees working in the production cycle.

It is required that safety planners undertake detailed analysis of the formwork project, with an ergonomic and anthropometric view, underlying the kinematics of the production cycle. Among other aspects to consider, is the verification of provisional stability of sets that constitute the equipment. The involvement of these sets with workers is associated with serious risks, including the drop in height and crushing.

The work was supplemented by a survey of scientific studies focused on the best practices in project activity, assembly and dismantling of formwork, as well as legal and regulatory international frameworks on the same subject.

3 CONDITIONS OF PROVISIONAL STABILITY OF SETS

The formwork is inherently a high-risk activity. Here is an example, of an accident which occurred in 2008 (Fig. 1). This configured a serious accident involving a fall of workers, during formwork assembly from a board of an overpass in a highway infrastructure.

The operations associated with the assembly and disassembly of formworks, generally coincide with more significant magnitude risks, involved in carrying on construction of *in situ* concrete, which includes the fall of materials and crushing, as previously mentioned. Currently the formwork equipment is considered to be completed and stable only after application of all assemblies and devices that constitute the respective equipment. It turns out that almost all formworks are built in stages, by increasing the elementary parts with a certain degree of prefabrication. (Fig. 2).

Regardless of the type of building element, depending on the difference in size between workers and formwork and/or concrete construction element to perform, using each set of formwork usually involves:

1. Connecting the lifting system to the formwork set still on the ground, for subsequent handling;
2. The formwork set reception by workers positioned at height, usually in the area of fall risk, to then proceeding to its positioning and subsequently untying the load lifting equipment. Needless to say that they are all using individual anti-fall protection and are positioned over the

Figure 1. Fall of formwork set in an overpass, during assembly, in a highway infrastructure, 2008.

Figure 2. Start-up set from a "I" pillar of an overpass, in a highway infrastructure.

Figure 3. Formwork set application in a tray of an overpass, in a highway infrastructure, with an associated overturning moment after its placement.

fresh concrete construction element, or over the formwork system itself, which assembling is in progress.

These are problematic cases, for example, some sets of peripheral formwork in artwork trays, in highways or rail infrastructures, where the application of each set can be associated with an overturning moment, that requires special treatment in the design phase and, that should be considered for provisional stability, required before untying the loads lifting equipment. It can thereby, be prevented and minimized, the risk of falling from height and crushing, of workers, which occur with higher frequency throughout the assembly and disassembly process of the sets (Fig. 3).

Depending on the lifting accessories type and the lift system design, the worker can position himself only in two possible locations: either under the set, or on it, to proceed with the disengagement of the lifting device. These workplaces require, therefore, distinct protection measures to prevent the risk of falling from height, not forgetting the overturning moment associated with the application set plus men.

Other activities that require workers positioning on the sets throughout assembly process are usually, guardrails placement on the outskirts of the formwork system, and the implementation of the lining of the sets (usually held in phenolic panels or marine plywood). In both activities, depending on the degree of assemblies' prefabrication, the work dimension and the risk magnitude can be very different.

In the first case, if the prefabrication of the assembly in the soil includes peripheral guard bodies, it is only necessary after the application of sets, to implement continuity guardrails. These will only fill the apertures located in the interface sets zone. For the second activity, the prefabrication of assemblies includes ground level coating of the formwork. After its application, it is only necessary to generate closure of the gap between sets, transversely of the board, typically with 20 to 30 *cm* in width. However, in any circumstances, both activities will involve a risk of overturning of the set, and therefore, require prior study of their stability. Therefore, it should be addressed in the formwork design appropriately to its setting, even before each set being disconnected from the load lifting equipment (Figs. 4, 5).

In case of vertical formwork, on walls of one face, two faces, or pillars, the use of props and kicker braces, forming isostatic figures that ensure the temporary stabilization of the whole, are a solution. All manufacturers hold them and have proved essential in most applications. However, they are often forgotten in projects (when they exist), and consequently during the execution of the work *in situ*. (Figs. 6, 7).

These sets are prefabricated at ground level and are associated with vertical formwork systems with different heights. They often have about 4.5 m in height. They have no vertical dimension limits defined. To be placed in use they must be attached to the load lifting equipment. They are received and positioned with workers on the soil, in case it is the first stage of startup concreting or in case of high following elevations. In these last situations workers are placed on the platform of the formwork base (main platform), which coincides with the upper bound of the last concrete leg of the building element or the run-level upper bound.

In all these situations, workers have to access the top of this set, for loosening the lifting system. This operation can be performed only after the temporary stabilization of the set and before it is released from the lifting equipment. This

Figure 4. Attachment points and lift system of formwork set of a board of a flyover, in a road infrastructure, with associated overturning moment after its placement in situ.

Figure 5. Applying a formwork set in a flyover board of a road infrastructure, with associated overturning moment after its placement in situ.

makes it mandatory to design the applications so that the formwork is assembled on the ground before being elevated. With no provision in the project of the temporary stabilization system of assemblies, the work will take place with a high risk of crushing and falling from heights, uncontrolled and unjustified in case of accident. Sets without provisionally stabilizing elements, agreed and mentioned in the project, are associated with poor operations, and high risk of instability of the assemblies (Figs. 8, 9).

In the project, the units shall be designed to be temporarily stable, dimensioning and designing appropriate devices for that purpose. This would avoid the need for detection and correction request of this fault, through the development of a responsible security plan. This would also avoid the risk of being improvised by these stabilizing devices

Figure 6. Sets of outer formwork of a coffin pillar, of a flyover in a road infrastructure, applied with previous incorporation of pull props to allow a temporary stabilization, before proceeding to the untying of lifting loads equipment.

Figure 7. Set of circular pillar formwork in an overpass, of a highway infrastructure, busy with anticipated application of pull props for temporary help to stabilize, before proceeding to the untying of lifting loads equipment.

agents, in an attempt to resolve the problem in detriment of safety.

Often working platforms that integrate formwork, including its standing tablets (which are their floor), as well as its railings, are not ade-

Figure 8. Sets prefabricated on soil without provisional stability elements. (Source—Contractors, A. C. I. A. T. A. S. O. C. The Contractor's Guide to Quality Concrete Construction. In., 2008).

Figure 9. The Worker moves to the top of the set to unlink the lifting loading means, with precarious shoring assembly.This is done after its receipt in situ.

quately characterized in the design of formwork. In this, the elements of protection are often just referred to. This lack of detail often results in work carried out in floors with standing planks, simply supported. This results in unstable platforms that increase the magnitude of falling from heights, risks inherent to their own work.

This type of solutions is even more problematic when combined with prefabricated sets in soil, with integrated platforms. That is incongruous, if such omissions are in project (Fig. 10).

In this context, work platforms of prefabricated sets on ground, must have standing planks properly fixed to the formwork system consoles, and scaled for the purpose they are intended (Fig. 11).

Similarly the guard rails should be featured in formwork design, including their bobs, footers, intermediate guards and upper guards, as well as the establishment of the first elements for the structure of the formwork.

Often, by shortages of elementary nature, the production cycle is hopelessly endowed with an

Figure 10. Work platforms at the base of a wall formwork with planks standing simply supported.

Figure 11. Foot planks properly screwed and fixed to the consoles of a formwork system.

undesirable level of security, which will be repeated in all the production cycles of the equipment.

4 CONCLUSIONS

The aim of this study was to compare the stability of temporary formwork sets, with the risk of falling from height and crushing during their assembly and disassembly.

It was demonstrated that the provisional stability of the assembly and disassembly system assumes an important role in the effectiveness of the production cycle, minimizing the risk of falling from heights and falling material during the operations of mounting and dismounting of the sets.

These devices should not only be regarded in the design phase as load bearing elements and calculated only for the final state, to achieve concreting, in its state of concrete formwork and ready to share (formwork completed and the fluid mass concrete in its maximum height and maximum thrust).

The formwork project is the nuclear document in the development of safety planning. No formwork appears built without the application of elementary parts that require the involvement of workers and equipment, with associated risks of high magnitude. The equipment supervisor should, in formwork design, integrate all the information necessary for the protection of workers during their assembly and disassembly processes, necessary for their realization.

The production cycle cannot be affected by functional deficiencies that must be resolved by the safety planning responsible, who are not inherently Civil Engineers or Formwork Designers.

ACKNOWLEDGEMENTS

The authors thank to Masters in Occupational Safety and Hygiene Engineering (MESHO), Faculty of Engineering, University of Porto (FEUP), their full support and international dissemination of this work.

REFERENCES

ACI & ASCC. American Concrete Institute and the American Society of Concrete Contractors. (2008). The Contractor's Guide to Quality Concrete Construction.

Arezes, Pedro M. (2006). ISHST. Instituto para a Segurança, Higiene e Saúde no Trabalho. Estudo Antropométrico da População Portuguesa.

Oberlender, Robert L. Peurifoy and Garold D. (2011). Formwork-for-Concrete-Structures-4th-Edition.

Pino, José Mª Tamborero. (2010). Encofrado vertical. Muros a dos caras, pilares, muros a una cara e sistemas trepantes. INSHT. NTP 834/835/836/837.

Shapira, Aviad. (1999). Contemporary trends in formwork standards - a case study.

SSFI. Scaffolding, Shoring and Forming Institute, INC. (2006). Safety Procedures for Vertical Formwork. (*www.ssfi.org*).

Territory, Australian Capital. (2011). ACT Code of Practice for Formwork.

Occupational Safety and Hygiene II – Arezes et al. (eds)
© 2014 Taylor & Francis Group, London, ISBN 978-1-138-00144-2

Exposure assessment of health ionizing radiation in nuclear medicine at Coimbra

C. Sousa, J.P. Figueiredo, N.L. Sá & A. Ferreira
IPC, ESTeSC, Coimbra Health School, Portugal

S. Paixão
IPC, ESTeSC, Coimbra Health School, Portugal
Centro de Estudos de Geografia e Ordenamento do Território, Portugal

F. Alves
IPC, ESTeSC, Coimbra Health School, Portugal
Instituto de Ciências Nucleares Aplicadas à Saúde, Coimbra, Portugal

ABSTRACT: The risk of ionizing radiation in the area of Nuclear Medicine can damage the health of health professionals and the patient. Thus, the main goal of this study focused on the evaluation of the dosimetric data of workers exposed and its relation to the behaviours and practices of professionals. The dosimetric data have been collected by means of annual records as also the evaluation of behaviours and practices have been performed through a questionnaire administered to 54 respondents from three different institutions. Data collection occurred between January and May 2013. The analysis of the results indicated that the professionals who do the administration and handling of radiopharmaceuticals, have higher doses of whole body and extremities. However, the dosimetric data sometimes do not dictate the reality of work, insofar as professionals do not apply good behaviours and good practices in the execution of their tasks.

1 INTRODUCTION

Recently, the application of ionizing radiation has been growing in the field of Medicine. Despite the knowledge of its adverse effects on health, and to be considered an environmental hazard and occupational benefits inherent to its application, both in diagnosis and therapy are immense and certainly exceed the risks that may be associated with them (Feinendegen, 2004).

Nuclear medicine is a conventional art which employs small amounts of radionuclides as diagnostic markers for the disease, or for larger amounts therapy. Although this increase is a positive trend for the benefit of patients, the risk of radiation exposure has to be assessed (Gauri, 2006).

Given the great technological innovation that currently exists, the equipment in the area of diagnosis are equipped with multiple detectors and may even be some equipment hybrid (Michael, 2008). The medical specialty of Nuclear Medicine now living generation hybrid equipment, able to provide anatomical and functional information, such as SPECT/CT (Computed Tomography Single Photon Emission) and PET/CT (Positron Emission Tomography) which may have simultaneously and radiation detectors and emitters/detectors radiation X (Cherry, 2003).

As reference value for the evaluation of workers exposed to ionizing radiation, one was used in the legislation in force. Regarding dose limits for exposed workers, the legislation used was the decree law nr ° 222/2008 of 17 November (DR, 2002).

The radiological protection adopts three criteria (distance, time and shielding) to reduce radiation exposure:

- Minimizing the exposure time—the workers professionally exposed (TPE) should reduce the residence time in proximity to the radiation source because the radiation exposure is directly proportional to the exposure time);
- Maximizing the distance from the source (most effective)—the variation in intensity of exposure (dose received by TPE) is conversely proportional to the square of the distance from the source primary or secondary) (ICRP, 2008).

Ionizing radiation can cause two types of effects on the body: deterministic and stochastic.

Deterministic effects (early) are observed shortly after exposure. Are related to the malfunction/loss of function of tissues and organs, mainly due to the death of a significant number of cells. These effects associated with exposure to high doses of radiation. The stochastic effects (late) result from changes at the cellular level, namely chains of deoxyribonucleic acid (DNA) and the resulting chromosomal aberrations involve increased risk of onset of cancer and inherited mutations and generally checks the existence latency period years (BEIR, 2005).

This study aims to evaluate and compare the levels of ionizing radiation to which health professionals are exposed in three institutions, and also to evaluate the behavior of professionals in relation to radiological risk in Nuclear Medicine.

2 MATERIAL AND METHODS

This research study and data collection took place from February to April 2013.

A literature review has been carried out in order clarify and define the main concepts and ideas of all research. The type of study was descriptive-correlational level II sampling was not probabilistic and convenience. In a second phase of the study used the retrospective cohort and subsequently the transversal cohort.

Initially, a dosimetry annual collection of each professional has been made in the unit of Nuclear Medicine from 2010 until 2012. The sample was composed of three institutions in the city of Coimbra and 174 workers exposed to ionizing radiation (24 professionals A, 102 from institution B and 48 of C).

The questionnaire included 18 questions, organized under the following set of variables: Personal Data; Professional Data; Individual monitoring.

Criteria for assessing the behaviour and practices were adopted in accordance with the duties of professionals each institution. For the classification of behaviours considered: good (meet "always" good behaviour); enough (if mostly admit, "almost always" or "sometimes"), insufficient (if mostly admit, "rarely" or "never") and practices adopted are: good (meet all practices) enough (but fails to use a dosimeter from other practices) and insufficient (not wear dosimeter).

For the statistical description descriptive measures were used, respectively, central tendency and dispersion and absolute and relative frequencies. For the verification of statistical hypotheses Chi square and ANOVA Independence I factor were applied designs.

The interpretation of statistical tests were based on a significance level $\alpha = 0.05$ with a confidence interval of 95%. For a significant α ($\alpha \leq 0.05$), differences were observed between groups or associations, however, for $\alpha > 0.05$, associations or differences were not considered statistically significant.

Throughout the study it was necessary to take into account some ethical issues, in particular in relation to the behaviours and practices of professionals in dealing with the patient, sometimes it was difficult to create a barrier between the professional and the patient.

3 RESULTS

3.1 *Doses of whole body*

The average annual dose of the professionals exposed to ionising radiation in the unit of Nuclear Medicine has been evaluated since 2010 until 2012.

It has been found that there are statistically significant differences ($p \leq 0.05$), according to the Bonferroni multiple comparison test between the average doses of whole body and the professional classes, particularly among the professional class of nurses compared to the Doctor one.

Over the three years it was found that there was a difference of average whole-body doses for the class of nurses reaching the highest average dose with $x = 4.6$ mSv in 2012, followed by Physical with $x = 3.1$ mSv in 2011. In the remaining classes have not reached significant differences of the average dose over the three years.

We evaluated the annual average dose of whole body of workers exposed to ionizing radiation in the nuclear medicine unit of the institution B from 2010 until 2012.

It has been found that there are statistically significant differences ($p \leq 0.05$), according to the Bonferroni multiple comparison test between the average doses of whole body and the professional classes, particularly among the class of the OA and NMT; Nurse and NMT, NMT and AMA; NMT and Nurse; NMT and Medical; NMT and Physical; TDT and radiopharmaceutical Physician and

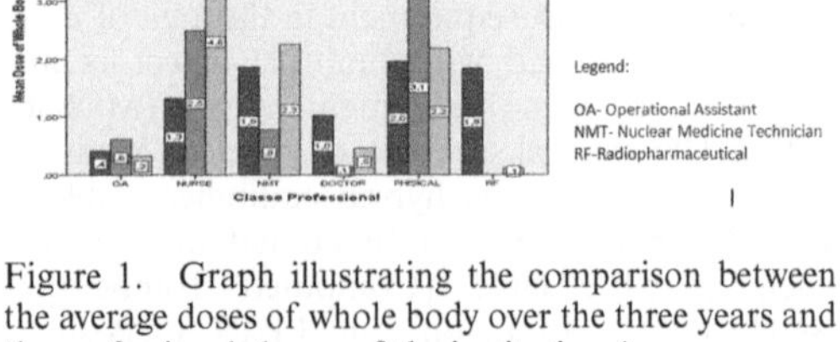

Figure 1. Graph illustrating the comparison between the average doses of whole body over the three years and the professional classes of the institution A.

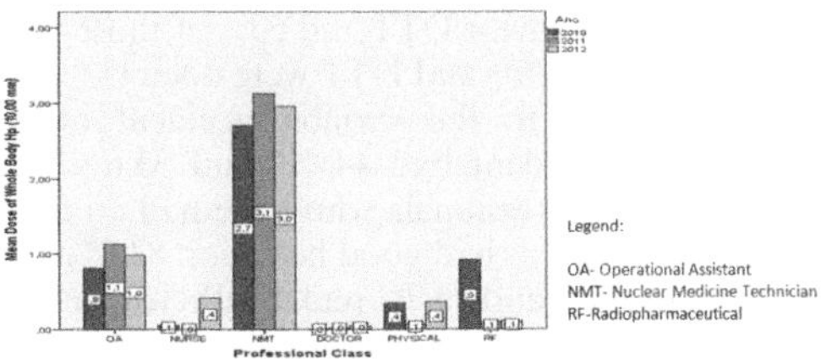

Figure 2. Graph illustrating the comparison between the average doses of whole body over the three years and the professional classes of the institution B.

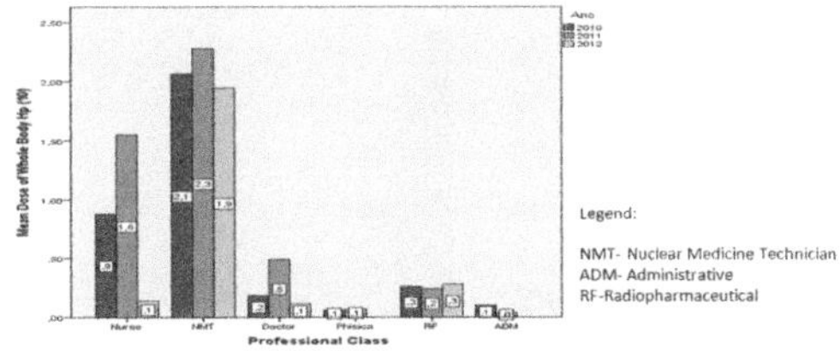

Figure 3. Graph illustrating the comparison between the average doses of whole body over the three years and the professional classes of the institution C.

NMT; Physical and NMT; radiopharmaceutical and TDT.

Over the three years it was found that there was a difference of average whole-body doses for the Class of DTT dose reaching the highest average with $x = 3.1$ mSv in 2011, followed by the AO with $x = 1.1$ mSv in 2011. In the remaining classes have not reached significant differences of the average dose over the three years.

We evaluated the whole body average annual dose of workers exposed to ionizing radiation in nuclear medicine unit of the institution C from 2010 until 2012.

It was found that there are statistically significant differences ($p \leq 0.05$), according to the Bonferroni multiple comparison test between the average doses of whole body and the professional classes, particularly among the class of NMT and Medical Physician and NMT, Administrative and NMT. These mean differences are statistically significant ($p \leq 5$).

Over the three years it was found that there was a difference of average whole-body doses for the Class of NMT dose reaching the highest average with $x = 2.3$ mSv in 2011, followed by the nurse with $x = 1.6$ mSv in 2011. In the remaining classes have not reached significant differences of the average dose over the three years.

3.2 *Extremity doses*

We evaluated the dose average annual end of occupationally exposed to ionizing radiation in nuclear medicine unit of the institution A from 2010 until 2012.

It was found that there are no statistically significant differences ($p \geq 0.05$), according to the Bonferroni multiple comparison test between the average doses of whole body and the professional classes.

Over the three years it was found that there was a difference in average doses end for the Class of RF dose reaching the highest average with $x = 14$ mSv in 2011, followed by the nurse with $x = 8$ mSv in 2011. In the remaining classes have not reached

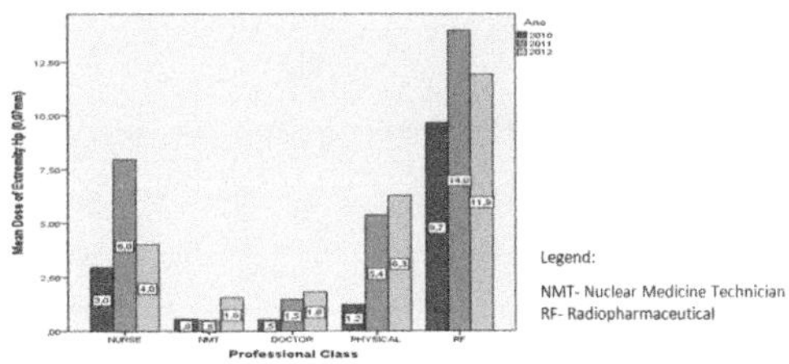

Figure 4. A graph illustrating the comparison between the average dose over the end of three and the professional classes institution A.

significant differences of the average dose over the three years.

We evaluated the dose average annual end of occupationally exposed to ionizing radiation in nuclear medicine unit of the institution B from 2010 until 2012.

It was found that there are no statistically significant differences ($p \geq 0.05$), according to the Bonferroni multiple comparison test between the average doses of whole body and the professional classes (Table 4).

Over the three years it was found that there was a difference in average doses end for the Class of RF dose reaching the highest average with $x = 39.8$ mSv in 2010, followed by DTT with $x = 25.1$ mSv in the year 2010. In the remaining classes have not reached significant differences of the average dose over the three years.

We evaluated the dose average annual end of occupationally exposed to ionizing radiation in nuclear medicine unit of the institution C from 2010 until 2012.

It was found that there are no statistically significant differences ($p \geq 0.05$), according to the Bonferroni multiple comparison test between the average doses of whole body and the professional classes.

Over the three years it was found that there was a difference in average doses end for the Class of NMT dose reaching the highest average with $x = 48.6$ mSv in 2011. In the remaining classes have not reached significant differences of the average

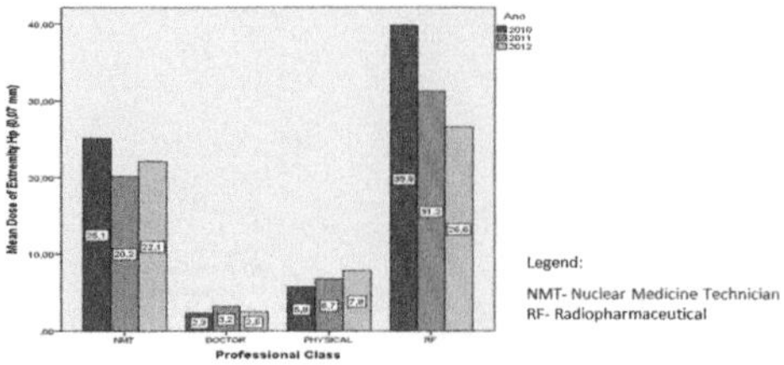

Figure 5. Graph illustrating the comparison between the average dose over the end of three and the professional classes institution B.

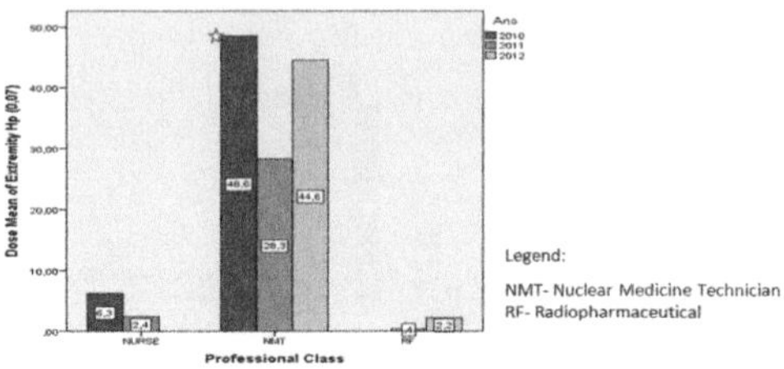

Figure 6. Graph illustrating the comparison between the average dose over the end of three years, and the establishment of the professional class C.

dose over the three years, and only records for two years.

3.3 *Behaviors and practices*

A total of 54 respondents from the three institutions responding to the questionnaire conducted, 59.3% reported having a good practice, a practice enough 22.2% and 18.5% an insufficient practice. At The institution stood out with 33.3% of those who had good practices MEAs and 80% of those who had insufficient practice nurses. In institution B 22.2% of professionals had good practices were AAM and DTT and 100% of professionals had been malpractices DTT. The institution C 37.5% of professionals had good practices were administrative and 75% of those who had been malpractices DTT.

Regarding the behavior it was found that a total of 54 professionals, 3.7% had good behavior, with 85.2% and 11.1% enough behavior with insufficient behavior. In The institution stood out with 50% of good behavior physicists and physicians, 27.3% with enough behaviors physicians and 100% of the professionals who had been misbehaving nurses. In institution B 0% of professionals had good behavior, 27.3% of professionals had enough behavior were physicians, and 50% of professionals who have been misbehaving DTT and physicians. C in the institution did not exist with no professional good behavior, 53.8% of those who had been enough behavior DTT and 50% of those with insufficient behavior and DTT were doctors.

With respect to the variable accident of 54 professional accident had 44.4% and 55.6% had no accident. Professionals who have had an accident, none of them had good behavior, 83.3% had enough behavior and 16.7% had insufficient behavior. Professionals who have not had an accident only 6.7% showed good behavior, 86.7% and 6.7% enough behavior insufficient behavior. Professionals who have had an accident 45.8% admitted to having good practice, 25% and 29.2% enough practice insufficient practice. Professionals who have not had accidents 70% had good practice, 20% fair and 10% poor. During the three institutions that suffered no professional accident had good behavior, institution B stood out from the other institutions because 80% of professionals who have had an accident enough behavior. Regarding the practices of most professional accident that took the three institutions had a insufficient practice except institution B only 7.1% had poor practice.

The variable year career was also statistically significant in relation to the behaviors and practices of professionals. Professionals with good behavior all had 11–20 year career, with enough behavior vast majority (47.8%) had 1–10 years of career and misbehavior large part (66.7%) had> 20 years career. Professionals with good practices only 15.6% had> 20 years of career, with enough practice 50% had 1–10 years of career and malpractice vast majority (70%) had> 20 year career Questionnaire.

4 CONCLUSION

We conclude that the dosimetric data sometimes do not dictate the reality of work, to the extent that professionals, despite the knowledge and specific training does not show good behaviour and good practices in implementing its tasks. The results show that most professionals have training but in practice it does not apply, then this training is not sufficient neither effective. Only care about the same as it has direct consequences to their own health and not when it is a mere formality.

We also conclude that, in addition to the behaviours and practices of professionals, specific policies and work realities of the three institutions directly influence the display of the professional to ionizing radiation. In this case, it was found that the distribution and rotation of tasks by different professionals were instrumental in minimizing the risk.

According to some studies, on dosimetric extremes of the professionals who handle radiopharmaceuticals, it has been found that doses of radiopharmaceuticals and physicians involved in intravenous administration were higher than in technicians. However, the doses to the edges do not exceed the prescribed maximum annual, except for the tips of the fingers of a physician responsible for the iodine treatment with 131, which supervises all treatments throughout the year. For other professionals the exposure limit was not exceeded because the tasks are rotated and do not deal with constant radioactivity throughout the year (Gauri, 2006).

Similarly, a more recent study, it was found that nurses and technicians are those at higher doses. This is due to the fact that these are professionals who prepare radiopharmaceuticals and even inject into patients to treat or diagnose. However, it was shown that even with these results in techniques regarded as safer, events have occurred that justify the need for a permanent monitoring by dosimetry (Pinto, 2009). Generally, professionals are exposed routinely monitored through whole body dosimeter (insufficient on the chest) and extremity dosimeter (ring) to the fingers (Abrantes, 1989).

This study also has shown that experience influence the behaviours and practices. It was found that respondents with more years of professional career longer fails at the level of behaviours and practices. It can be concluded that the mechanization of the following conditions change work habits.

Given this reality only through regular training can improve behaviours and practices of professionals, sensitize them and alerting them to issues of radiological protection.

Currently, there are authors who advocate the non-linear model which assumes a linear relationship of nature and admits no safe dosage for low levels, i.e., any level of radiation can cause harmful effects (cancer) (Bilska, 2011). This model has been subject of much controversy and discussion about the effects on health. However, according to legislation, the employer is obliged to ensure the radiological protection at a level which will eliminate occupational exposure (Falzon, 2007).

This study had some limitations:

- Cost/benefit against certain criteria of radiological protection, in particular with regard to the distance and the time of exposure. It was found that many professionals had difficulty in adapting these criteria in their daily tasks, to the extent that the patient himself is a source of radiation, which may raise questions of professional ethics.
- Situations that involve the use of some protective equipment, in particular, lead apron, is quite heavy which affects the physical condition of the professional, the use of gloves is another practice that often fails because it is faster to do administration without gloves, gloves sometimes complicate the procedure.
- Difficulties in adapting the questionnaire to different professionals, to the extent that each class had professional different tasks in three institutions in which the study was applied.

Research areas for future research:

- Based on the questionnaire used in this work, it would be interesting and pertinent to conduct a study at national level, in order to obtain concrete and recent data on the types of behaviours and practices experienced professionals in different institutions in mainland Portugal and the Autonomous Regions.

REFERENCES

Abrantes, A., Tavares, A., Godinho, J. Manual de métodos de investigação em saúde. 1ª Edição. Lisboa: APMCG, 1989.

Bilska, Hanna et al., Eibi, Bogumita; Bankowska, Katarzyna. "Occupational exposure at the Department of Nuclear Medicine as a work environment: A 19-year follow-up". POL J Radiol, 2011; 76(2): 18–21.

Cherry, S.R., Sorenson, J.A., Phelps, M.E.; 2003, Physics in Nuclear Medicine, 3ed. Estados Unidos, Elsevier Science.

Decreto-Lei n.° 180/2002 de 8 de Agosto. DR-I Série-A, n°182. Ministério da Saúde.

Executive summary. In: Board on Radiation effects research ed. Health risks form exposure to low levels of ionizing Radiation; BEIR VII- Phase 2. Washington, DC: The National Academy, 2005.

Feinendegen, L.E., Pollycove, M. Responses to Low Doses of Ionizing Radiation in Biological Systems. Nonlinearity Biol Toxicol Med. 2004, Vols. 2(3): 143/171.

Falzon P. Natureza, objetivos e conhecimentos da ergonomia. Elementos de uma análise cognitiva da prática. In: Falzon P(org.), Ergonomia. 2007; 3–19.

Gauri S. Pant, Sanjay K. Sharma, and Gaura K. Rath. "Finger Doses for Staff Handling Radiopharmaceuticals in Nuclear Medicine", department of radiotherapy, all India Institute of Medical Sciences, New Delhi, India, 2006.

GB Saha. Fundamentals of nuclear pharmacy. Michigan, USA: Springer, 1998.

ICRP. Publication 105: Radiological protection in Medicine. Annals of the ICRP, 2008.

Michael G. Stabin. "The Importance of Patient-Specific Dose Calculations in Nuclear Medicine". Nuclear Engineering and Technology. Vol. 40, Nr. 7, 2008.

Pinto, A., Schiappa, J. et al. "Health Professionals Radiological Exposure Evaluation". Faculty of Medical Sciences, New University of Lisbon, 2009.

Occupational Safety and Hygiene II – Arezes et al. (eds)
© 2014 Taylor & Francis Group, London, ISBN 978-1-138-00144-2

Risk assessment: Information for qualitative estimation of probability and severity in high risk work context

A.A. Ribeiro
Instituto Superior de Línguas e Administração de Leiria, Leiria, Portugal

J.P.C. Fernandes Thomaz
Centro de Estudos de Gestão do Instituto Superior Técnico, Universidade de Lisboa, Lisboa, Portugal
Instituto Superior de Línguas e Administração de Leiria, Leiria, Portugal

R.A.C. Veiga
Instituto Superior de Línguas e Administração de Santarém, Santarém, Portugal

ABSTRACT: The process of risk management is a structured study which consists of three basic steps; risk analysis, risk assessment and risk control. Risk analysis is the first step of the risk assessment process and comprises the identification of all hazards present in the activities, and the estimation of probability and severity according to the method of risk assessment chosen.

This article intends to present the results of a study conducted in order to identify the information considered essential for a qualitative estimation of probability and severity to facilitate the technician's job during risk assessment when they use the Semi-quantitative Risk Assessment Methods (SqtRAM).

1 INTRODUCTION

The occurrence of accidents allowed in most cases to establish a causal link leading to the notion of risk and the understanding of the need for its management. In today's legal framework special emphasis is given to duties and responsibilities of the participants involved in Risk Assessment (RA) as a crucial element of the prevention process, providing knowledge of the existence of hazards (nature-magnitude) and contributing with information for risk control and development of preventive actions.

Although the freedom to select the methodology, the method chosen must be able to discriminate the different Levels of Risk (LR) in each situation, taking into account the parameters of the Probability (P) and Severity (S), and its weight to achieve this goal.

The Semi-quantitative Risk Assessment Methods (SqtRAM) are used in most cases because they are generalist methods, easy to apply and a tool available to carry out the obligations imposed by law, in opposition to quantitative methods, which are more complex and therefore, involve higher costs.

1.1 *Definition of the research issues*

In the definition of the objectives, we attempted to find answers for the matter of qualitative risk estimation. To achieve the objectives of the study the following central research question was drawn up:

CRQ—Which are the markers of essential information for qualitative estimation of the variables Probability and Severity when using SqtRAM?

This question guided to the following Derivative Questions (DQ) and hypotheses (H):

DQ1—Is the risk level obtained with SqtRAM significantly influenced by the information chosen for estimation of Probability and Severity?

H0—There are significant differences in risk level obtained when we use distinct information for estimation.

H1—There are no significant differences in risk level obtained when we use distinct information for estimation.

DQ2—Does the type of assessed risk influence the choice of information for qualitative estimation of Probability and Severity, in SqtRAM?

H0—The type of assessed risk influences the choice of information for estimation.

H1—Type of assessed risk does not influence the choice of information for estimation.

1.2 *Estimation of probability and severity variables*

The process of risk assessment and management consists in a structured study of all aspects inherent to the work and is composed of risk analysis, risk assessment and risk control. Risk analysis is performed to identify all hazards in the organizations activities and the estimation of probability and severity according to the method chosen—$R = P \times S$ (Nunes 2010 & Freitas 2008).

Estimating the risk, means measuring its magnitude, which is the product of the probability of the damage (estimated probability) by the severity (estimate of the damage), as objectively as possible. The risk estimation, should consider the systems and control measures already implemented, as well as information that may influence the results (Lluna 2003 & Romero 2004).

It is extremely helpful to understand the estimation of these two variables to illustrate the preventive measures that have the main objective of reducing the likelihood of exposure (actions taken before exposure to the risk) and collective or individual protective measures aiming to reduce the consequences due to exposure (actions required after exposure to reduce risk impact).

Lluna (2003) refers that the estimation highlights the sensitivity of the results due to its influence on preventive measures, as an important aspect when deciding on the measures to be implemented.

1.3 *Estimation of the probability and severity*

Estimating means forming an opinion based on an approximate judgment (based on predefined reference parameters). The question is, estimating what?

The probability is evaluated by different approaches. It is related to the possible occurrence of damage loss or consequence due to the conditions of use, exposure or interaction with the material component of work that presents a danger. In particular, when the working conditions determine the type of worker exposure, considering the number of times and/or duration of exposure to risk.

Estimating the severity, according to the approach taken, is interpreted as the consequences caused by the occurrence of exposure to risk (the potential severity of the damage).

The exposure to a particular occupational risk in the workplace which can generate different consequences, estimation of severity should meet the following points: (i) the damage perceived as more serious for the person or group with the likelihood of exposure, (ii) attend the exposure of workers considered most vulnerable to risk (pregnant women, children, disabled, elderly, illiterate) and (iii) take into account cases where the risk being assessed has a high level of risk resulting from the combination with the variable probability.

2 RESEARCH METHODOLOGY

This study used a qualitative research approach, with the main objective of identifying the information needed for qualitative estimation of probability and severity in risk analysis, the first step of risk management.

To reach this goal, the content analysis of data (interviews) was done through the interpretation of the concepts expressed in a qualitative and cognitive analysis following two complementary techniques:

i. Triangulation technique as a way to avoid individual analysis on the basis of personal opinion of the researcher, so that appealed to three independent analysts (Northcutt & McCoy 2004);
ii. Cognitive mapping technique for aggregation of concepts and their clarification, according to the Strategic Options Development and Analysis approach (Eden & Ackermann 1998).

This method was selected because it allows us to describe and interpret qualitative estimation of risk, without any control or interference in; existing work conditions, instruments for data collection and data analysis.

2.1 *Steps of the research*

The research took place in five key steps:

Step 1—Definition of indicators that characterize the utilization of explosives for rock dismantling in extractive industry (workplace case study);

Step 2—Observation, systematically directed to the collection of data in the workplace;

Step 3—Definition of the criteria for selection the sample of interviewees;

Step 4—Execution of interviews to the sample and presentation of workplace/activity observed;

Step 5—Transcription of interviews and data analysis with qualitative analysis techniques;

Step 6—Analysis of estimation information identified in "counterpoint" to the variables of three semi-quantitative risk assessment methods.

2.2 *Data collection techniques*

In data collection were applied the following two instruments:

i. A planned and participated observation of the workplace where explosives are used for rock dismantling in a quarry (high risk job)—Step 2;
ii. Individual, semi-structured interviews carried out with risk assessment teachers and risk managers of medium and large companies—Step 4.

2.3 *Observation of workplace—case study*

The observation of the workplace went through the following phases: definition of objectives, site and work selection, identification of indicators to characterize the workplace and implementation of observation plan (timing of visits and data collection).

During the observation of rock dismantling with explosives the following indicators were assessed:

i. Organization, processes and work procedures,
ii. Technical, technological and products used,
iii. Physical environment, and
iv. The human factors (step 1 and 2).

2.4 *Individual semi-structured interview*

The selection procedure of respondents was made by a non-probabilistic rational or typical cases sampling (Fortin 2006)—Step 3.

This option was due to two key factors, the specific subject of study, and the difficulty of defining and accessing the entire population of teachers of risk assessment and risk managers of medium and large companies. To form the sample the following criteria and purpose were taken into account:

i. The technical and scientific character of the data and avoid the saturation of information (data repetition);
ii. Be risk assessment teachers or responsible for risk management in medium-sized (50–249 workers) and large companies (250 or more workers);
iii. Elimination of the sample, all participants at any stage of the study, which do not comply with the criteria described above.

The interview consists of two parts:

Part 1—aimed to realize if the way the estimation develops influences the results of risk assessment and to identify the essential information required for qualitative estimation of risk probability and severity.

Part 2—intended to identify the required information for qualitative estimation of probability and severity from the perspective of the risk type, by analyzing the observation in real workplace situation (the use of explosives for rock dismantling).

Sixteen face interviews with risk assessment experts were carried out: two were test interviews, other two were excluded because respondents have not confirmed the requirements during the interview and twelve were analyzed (with same number of teachers and risk managers).

In the interview, open questions were used to allow respondents to express their opinion, what they think and know without being influenced by suggestions made by the researcher and avoid the effects of fixed format due to semi-structured model.

2.5 *Processing and analyses of interview data*

To perform the analysis of the interviews, a team of three analysts, all teachers and risk managers in large companies, with the same technical-scientific requirements of the sample of respondents, and which did not participate in any stage of the study was formed. They performed a qualitative analysis of the content of the interviews by applying the complementary techniques of data analysis to avoid the influence of the researcher-interviewer and ensure an independent analysis (triangulation).

The interactive analysis of data with the team of analysts took place in four phases according to the following thematic areas:

i. Influence of the form of risk estimation in risk assessment;
ii. Identification of essential information for estimation of probability and severity of occupational risk in general;
iii. Influence of the type of risk on the information required (markers of information);
iv. Importance of information available at the time of risk estimation and its influence on risk assessment.

The analysis of interviews was done through the interpretation of the concepts expressed in the contents in a qualitative and cognitive approach verifying in:

Phase 1—the placement of importance-ordered responses per question (1 to 12);

Phase 2—the elaboration of individual cognitive maps and respective structure of concepts;

Phase 3—the construction of the aggregate map, based on the individual structures obtained in the interactions with each analyst, carried out in the previous stage (ordered responses by area and subject);

Phase 4—the meeting with the three analysts for discussion and looking for a consensus about the final structure of concepts obtained from the group aggregate map according to the purpose of the interview (without researcher intervention).

2.6 *Participants in real workplace situation*

The explosives team composed of six workers: production coordinator, team leader, two fire loaders and two charger fire apprentices.

3 RESULTS AND DISCUSSION

3.1 *Results of the qualitative analysis*

In the initial Question (Q1) of the interview concerning the form adopted to perform the estimation of probability and severity in a given work situation, and its possible influence on the Level of

Risk (LR), all respondents answered affirmatively and state that the way the technician chooses to proceed with the risk analysis estimation of the two variables have a direct influence on the results of risk assessment (level of risk), when using the SqtRAM.

In the questions relating to the influence of the type of risk (Q3 and Q4), respondents answered largely yes (10 in 12) that the kind of risk influences the information needed to estimate the variables probability and severity in presence of risks related to safety, to occupational hygiene and in high-risk activities. Two respondents reported that the information required is independent of the type of risk analysis, because it should always be gathered according to each situation in the following areas: (i) work organization and procedures; (ii) technical, technological means; (iii) workplace environment; and (iv) human factors.

In the question concerning the influence of the information provided through observation of real work environment (Q5c), workplace case study, all respondents answered affirmatively, that the information available about the workplace is essential to carry out the estimation. Respondents indicated for instance, that the lack of information to estimate the two variables is a factor that leads the technician assigning higher values influencing the results (level of risk) and consequently, the risk control measures proposed.

Regarding the possible influence of developing risk estimation on the level of risk and on risk assessment results (Q1, Q2b, Q3, Q5c), the answers were always affirmative. The following key points contribute to respondents conviction: the chosen form to estimate (by sector, workplace, function, installation, type of risk); the information gathered prior to the application of the assessment method (tendency to increase value of estimation when the information available is limited) and the type of risk (areas of high risk work, safety and occupational hygiene).

3.2 *Qualitative estimation of probability (markers)*

Several questions were asked to identify markers of information (Q2a, Q5a), understand its influence on the magnitude of risk obtained (Q2b, Q5c) and realize if the type of risk affects markers of information necessary to estimate the probability (Q3).

After applying the techniques used for content analysis of the interviews, the need for the following information to estimate probability was identified:

i. Information on the organization and work processes—the type and sequence of tasks, products used, workload (type and duration of tasks, schedules and breaks), frequency of the task execution or duration of exposure to risk, results of previous risk assessments and controls implemented, outcomes of inspections, compliance status, mobility and rotation of workers and information about accidents (frequency rate);
ii. Technical, technological and material—adequacy of the technical and technological means to demands, operational conditions of machinery, equipment and tools, existence and standing of fulfillment of maintenance plan (safety manuals) and type of products used;
iii. Workplace environment—information relating to the physical environment involving the workplace (occupational hygiene and safety conditions);
iv. Human factors—training, information, experience and the characteristics of each worker with special relevance in behaviors related to safety (individual safety culture).

3.3 *Qualitative estimation of severity (markers)*

Regarding the estimation of the possible consequences caused by the occurrence of exposure to risk, the potential severity of the damage, several questions were raised to identify markers of information needed to estimate this variable (Q2c, Q5b), understand the influence of information available on risk assessment results (Q2d, Q5c) and understand if the type of risk affects markers of information required for the qualitative estimation of severity (Q4).

After analysis of the interviews content, the following information considered essential for qualitative estimation of severity was identified:

i. Collective and individual protections are identified as crucial concepts for the estimation;
ii. Registration and characterization of accidents (severity rate and absenteeism);
iii. Indications of potential result due to improper use of the means (manuals, safety data sheets);
iv. Level of emergency planning/training (to reduce damage in the event of an accident).

3.4 *Other information (contributing items)*

Participants also highlighted a set of concepts that are not considered as markers of information for qualitative estimation of the probability and severity, but that should be taken into account as contributing factors and be considered in the risk analysis step, namely: (i) time available to do the risk assessment, (ii) access to the workplaces, (iii) the experience of the evaluator, (iv) methodology and schedule risk assessment, (v) method of estimation, (vi) interaction with employees, (vii) the observation of workload trip and avoid the analysis of each individual task in itself, (viii) unpredictable factors (changes in expected conditions) and (ix) correction of subjectivity in successive risk assessing according to performance indicators and management of OHS.

3.5 *Information available versus variables of semi-quantitative risk assessment methods chosen*

The methods of risk assessment were selected in the semi-quantitative group, based on the following criteria: (i) be considered a semi-quantitative method, (ii) use a simple matrix with two variables or a matrix composed of more than two variables and (iii) be accepted as the most widely used because they are free and available in the literature.

After the identification of the essential information, in order to understand its influence while using semi-quantitative risk assessment methods (mandatory information and technical responsibility) it was performed a juxtaposition of the information identified by the analysts versus the variables of the following chosen methods: (i) simplified method for evaluating accident risks—NTP 330 (National Institute of Safety and Hygiene at Work of Spain), (ii) method of William T. Fine (Cabral & Veiga 2006) and (iii) the simple matrix method—SMM (Pinto 2008).

The method NTP 330 determines the need for information that makes possible the estimation of the Level of Exposure—LE (often performing the task and/or exposure time). To estimate the Level of Disability—LD, information should be gathered on the laws and regulations applicable to the work under analysis to detect compliance/non-compliance through the application of verification techniques in terms of organization and work processes, training, means used and the physical environment of the place. The Level of Probability—LP, will be found only after gathering the information to estimate the Level of Exposure—LE and the Level of Disability—LD, is the product of the two variables and depends on how they were estimated ($LP = LE \times LD$).

The Level of Consequence (LC), is the possibility of damage or injury. This method gives more importance to this variable than to the Level of Probability (LP), as seen through the values of reference established in the scales of valuation, but leaves to the evaluator the responsibility to collect the information considered necessary to estimate the potential damage and reduce subjectivity.

In WT Fine, the second method chosen, the juxtaposition of information is focused only on the degree of hazard, which is the product of three variables ($R = Fp \times Fe \times Fc$), the exposure factor (Fe), the probability of the accident to occur (Fp) and the consequences if the accident happens (Fc). The risk exposure (Fe) determines that the evaluator should gather information on worker exposure, as he does in method NTP 330 (often performing the task and/or exposure time). To estimate Fp the evaluator is left with the decision of the necessary information, without any of this being determined by the method. The same happens with the variable Fc, which corresponds to the damage/injury possible, once again, it is up to the evaluator to gather the information necessary to estimate the potential damage and thereby reduce the subjectivity of the estimation.

When using the third chosen method, the Simple Matrix Method (SMM), the level of risk is calculated as a function of the independent variables Probability (P) and Severity (S). This method leaves the decision on the information deemed necessary to estimate the two variables entirely to the evaluator, becoming even more relevant the definition of information considered essential to perform the estimation of risk and the contributing factors presented in the paragraph "other information" (e.g. experience of the evaluator and the methodology used), to estimate realistically the two variables.

4 FINAL REMARKS

4.1 *Results of research hypotheses*

Regarding the derivative questions raised in this study (DQ1 and DQ2), the validation of the hypotheses points to the importance of the available information in qualitative risk analysis, taking into consideration that:

i. There are significant differences in the level of risk when we use different information (amount and type of information available) to estimate the variables probability and severity of accidents. Lack of information leads the evaluator in technical terms to assign higher values in accordance with the reference scales, also when the type of information is inadequate leads to a higher subjectivity of risk assessment and makes it difficult to fix the deviations in successive evaluations;
ii. The type of risk influences the choice of information needed to estimate these two variables. The criteria to gather information must meet the risk typology of the activities and workplace analyzed, in accordance with work safety, occupational hygiene and high risk activities contexts.

4.2 *Contributions to the risk assessment*

In this work information considered essential was identified for the qualitative estimation of risk when semi-quantitative risk assessment methods are used, concerning the organization and work procedures, technical and technological means employed, physical environment and human factors which characterize the working conditions.

In the qualitative analysis of interview contents, a set of concepts identified as contributing factors were also raised and should be taken into account by the person responsible for risk assessment.

The information identified substantiates the descriptors of the chosen methods, the technical responsibility and the reduction of subjectivity in the estimation of risk.

REFERENCES

Cabral, F. & Veiga, R. 2006. *Higiene, segurança e saúde e prevenção de AT* (vol. 1, 20ª ed.). Lisboa: Verlag Dashöfer.

Eden, C. & Ackermann, F. 1998. *Making strategy: The journey of strategic management*. London: SAGE Publications.

Freitas, L.C. 2008. *Manual de SST*. Lisboa: Edições Sílabo.

Fortin, M. 2006. O processo de investigação: *Da conceção à realização* (3ª ed.). Loures: Lusociência.

Lluna, G. 2003. *Sistema de gestión de riesgos laborales e industriales*. Madrid: Edições Fundación Mapfre.

Northcutt, N. & McCoy, D. 2004. *Interactive qualitative analysis: A systems method for qualitative research*. Thousand Oaks, CA: SAGE Publications.

Nunes, F. 2010. *Segurança e saúde no trabalho: Manual técnico* (3ª. ed.). Amadora: Edições Gustave Eiffel.

Pinto, P. 2008. *Manual segurança: Construção, conservação e restauro de edifícios* (3ª. ed.). Lisboa: Edições Sílabo.

Romero, J.C. 2004. *Métodos de evaluación de riesgos laborales*. Madrid: Ediciones Diaz de Santos.

Occupational Safety and Hygiene II – Arezes et al. (eds)
© 2014 Taylor & Francis Group, London, ISBN 978-1-138-00144-2

Occupational risks behind teaching activity

Carla Barros-Duarte
Faculdade de Ciências Humanas e Sociais da Universidade Fernando Pessoa, Porto, Portugal

Teresa Lajinha
Faculdade de Ciência e Tecnologia da Universidade Fernando Pessoa, Porto, Portugal

Maria Reina, Fátima Rocha, Tiago Santos & Inês Soares da Costa
Faculdade de Ciências Humanas e Sociais da Universidade Fernando Pessoa, Porto, Portugal

ABSTRACT: A broad definition of health at work, in which physical, psychological and social dimensions interact, has been gaining increasing attention. The education sector, where physical and psychosocial risks are both present in teaching activity, affecting their health, and wellbeing, reflects this holistic conception of occupational health. The main objective of this study was to search the main factors in teacher's working conditions that influence teacher's general state of health, as they perceived it. The study was carried with a sample of teachers of an elementary school at Porto, Portugal, using the survey instrument INSAT (Barros-Duarte & Cunha, 2010). The survey scores were object of a statistical correlation analysis. It was found that the statistical significant work characteristics that teachers perceive as directly affecting their health, are related with both environment and physical constraints, as well as organizational and relational constraints.

1 INTRODUCTION

Accordingly to the World Health Organization, "Health is a state of complete physical, mental and social well-being and not merely the absence of disease or infirmity". In this definition of health as the well-being of an individual, there is a subjective dimension that represents more than being in good health (absence of disease), but also feeling with good health (Canguilhem, 1999). In this sense, the study of health raises a reflection focused on the perspective of the person and a guidance addressed to well-being (Maggi, 2003). This favours a more comprehensive approach to health (Volkoff, 2002) in order to give more visibility to less obvious relationships between health and work.

The study of health at work reflects this approach. Work can be beneficial to (mental) health, increasing the feeling of social inclusion, status and identity, but it can also be a source of anxiety, depression and burnout, leading to a decrease in the level of professional performance and an increase in levels of absenteeism, accidents and turnover (Harnois & Gabriel, 2000). In fact, the analysis in question is characterized by a great complexity. In addition to its character differed in time, the relationship between health and work are not simple nor unique (Gollac & Volkoff, 2000). Not only a common work feature can have multiple consequences on health, but also a particular health problem can have several causes, in which non-professional factors are connected. On the other hand, the health state itself influences the worker ability in carrying out his professional activity and, as a result, interferes with the potential conditions under which he does his work.

Health at work calls for a broad definition, in which physical, psychological and social dimensions interact. These dimensions have been for a long time object of attention by ILO/WHO Joint Committee to "promote and maintain the highest degree of physical, mental and social well-being of workers" (ILO, 1985). Occupational well-being, like general well-being, may be understood as a multi-dimensional phenomenon. It comprises more than affect, it manifests itself in employee cognitions, motivations, behaviors, and self-reported physical health (Van Horn, Taris, Schaufeli & Schreurs, 2004). This approach takes a comprehensive, holistic and integrative perspective of occupational health focusing on well-being at work (Barros-Duarte & Lacomblez, 2006) in the sense that all work should provide workers the possibility of having an active role in its conduction, reflecting the ability of acting on themselves and their work (Clot, 2008).

These concerns, associated with psychological complains, have been gaining increasing attention, conducting to the development of scientific studies about psychosocial risks, understood as risks for mental, physical and social health, caused by working conditions and organizational and relational factors which might interact with mental functioning (Gollac & Bodier, 2011). The education sector does not escape this reality, where physical and psychosocial risks come together and accompany teaching activity, exerting a negative impact on health and teachers' welfare. New practices of organization and innovation in schools, multiple tasks, intensification of work and employment and job insecurity, put teachers in a more vulnerable situation, leading to changes in physical and psychological health (Carlotto & Palazzo, 2006).

Psychosocial problems tend to be more severe in education, health and social work sectors, when compared to manual occupations (European Commission, 2012), which Travers & Cooper (1993) had already concluded in the 90's that, as compared with other highly stressed occupational groups, teachers experienced lower job satisfaction and poorer mental health. As major causes of psychological strain among teachers there have been identified three job demands: disruptive pupil behaviors, work overload and poor physical work environment (Hakanen, Bakker & Schaufeli, 2006). Time pressure and workload, students' lack of motivation, inadequate behavior, indiscipline. coping with change, being evaluated by others, conflicts with colleagues, problems dealing with administration/management. role conflict and ambiguity, and poor working conditions were identified as important sources of stress in Portuguese schools (Ferreira & Martinez, 2012).

When answering to the European Working Conditions Survey Question: Does your work affect your health? a relationship between work and health is not obvious, as it is influenced by the level of available scientific evidence, cultural and country norms and stereotypes, gender, occupational and other differences (Eurofound, 2012). Answering this question in the education sector was the propose of this study, which aims to explore the factors involved in teachers working conditions that seem to explain their perception of how work can affect their health.

2 DATA AND METHODS

The study was carried out on a population of 84 teachers from the Elementary School E.B. 2, 3 Gomes Teixeira, in Porto, in December 2012, with a response rate of 73%. The sample (60 teachers) is composed of 15 males (24.2%) and 45 female (72.6%), where 11 individuals are single (17.7%), 42 are married (67.7%), 4 (6.5%) widowed, 2 (3.2%) divorced and one person in consensual union (1.6%). 72.6% of the individuals have a bachelor degree and 19.4% have a master degree. The average age of the individuals is 49 years old, with a standard deviation of 8.5 years. The mean time of teaching experience is 22.15 years, with a standard deviation of 10.6 years.

The survey was conducted with the permission of Directorate-General for Innovation and Curriculum Development and, after preliminary meetings with the direction of the school, the survey instruments were delivered and an awareness action regarding the theme of the study was carried out among the teachers. Along with the survey instrument, it was placed a leaflet so that all teachers could read and have access to more complete information regarding the purpose of the study and the intended analysis. In order to ensure the confidentiality of data, teachers would put the answered surveys in sealed envelopes and place them into an urn located in teachers´ lounge.

The poll instrument used in the study was INSAT—Health at Work Survey (Barros-Duarte & Cunha, 2010). INSAT is a self-applied survey instrument, centered in the person, where the main objective is to analyze the past and present working conditions and their consequences in worker health and well-being. It begins with a set of questions designed to collect general personal information (gender, age, sector of activity, etc.) and is subsequently composed of queries grouped into seven categories: (1) work; (2) conditions and characteristics of work; (3) conditions of life outside work; (4) training and work; (5) work and health; (6) my health and my work; (7) my health and my well-being.

The data analyzed in this study concerns the second and sixth categories of information obtained via INSAT, namely, "conditions and characteristics of work" and "my health and my work". It was intended to study the behavior of the variable: "I believe that my health is or has been affected by the work that I do", which is an item obtained in the sixth category of the survey. This variable was measured in a Likert scale, from 1 to 5, where 1 represents the maximum accordance and 5 represents total disagreement, with 4% of the responses for 1; 18% for 2; 40% for 3; 28% for 4 and 10% for 5. To do this study it was necessary to evaluate the relation between this variable and factors included in the category of INSAT concerning "conditions and characteristics of the work". This category, per se, is grouped into information that aims to describe the "environment and physical constraints", "organizational and relational constraints" and the "characteristics of the work".

For the selection of factors in the category "conditions and characteristics of the work" that could explain the influence of work on health, a first selection criterion was the relative frequencies of the variables measured in this category. Namely, it was established that the factors selected would have to present a relative frequency equal or greater than 50%, that is, the majority of respondents recognize the existence of these work characteristics in their present job, in their past job or in both (Rudow, 1999; Gaillard, 2001; Hakanen, Bakker, & Schaufeli, 2006; Gollac, & Bodier, 2011).

Although the factors were evaluated in terms of degree of discomfort perceived in an ordinal scale from 1 to 5, where 1 represents the maximum discomfort and 5 represents the absence of discomfort, for the application of the criterion based on relative frequency, this scale was transformed in a binary scale with two results: *been affected* (1 to 4 in Likert scale) and *not been affected* (5 in Likert scale).

To estimate a mathematical model that could describe the relationship between health perception, as dependent variable, and the factors included in the category "conditions and characteristics of work" (and with a relative frequency of 50% or more for the result *been affected*), as independent variables, it was assessed the degree of linear correlation between the dependent variable (evaluated in terms of degree of discomfort perceived in the 1 to 5 Likert scale) and each of those factors. The degree of linear correlation was evaluated by calculating the Pearson linear correlation coefficient.

The last step was to estimate, by the ordinary least squares method, a multiple linear regression model containing independent explanatory variables (factors belonging to the category "conditions and characteristics of work") that have statistically significant linear correlation with the dependent variable: "I believe that my health is or has been affected by the work that I do".

3 RESULTS AND DISCUSSION

Based on the relative frequency equal or greater than 50% criterion, it was selected a total of 17 explanatory variables in the category "conditions and characteristics of the work" (Table 1). Among these variables, 14 belong to the groups "environment and physical constraints" and "organizational and relational constraints" and 3 belong to the group "characteristics of work".

Considering the number of sample elements and the number of explanatory variables selected, it was decided, before the estimation of the mathematical model, that to get a reasonable number of degrees of freedom, it would be necessary to reduce the number of explanatory variables in the model.

Table 1. Potentially explanatory factors of the category "conditions and characteristics of the work".

Variable	(%)
Working Conditions	
Environmental (I'm exposed to...)	
Noise	57
Intense heat or cold	55
Physical (I'm forced to...)	
Remain standing with displacement for a long time	50
Work Rhythm (I'm exposed to...)	
Solve sudden situations/problems without help	72
Work beyond my assigned timetable	70
Frequent interruptions	68
Have to do several things at the same time	63
Work relationships (I'm exposed to...)	
Verbal aggression	67
Intimidation	50
Dealing with the public (There is...or, I have to...)	
Endure the demands of the public	63
Deal with situations of tension in the relationship	70
Be exposed to verbal aggression from the public	73
Be exposed to physical aggression from the public	50
Have to deal with suffering of others	57
Work Characteristics (My work is a type of work...)	
With moments of work overload	63
Lack of suitable equipment	53
With inadequate facilities	50

Among the 17 variables pre-selected only two of them where maintained as potentially explanatory. The condition imposed for the variables to maintain their explanatory character was based in the assessment of the degree of linear correlation, by calculating the Pearson linear correlation coefficient between the dependent variable (evaluated in terms of degree of discomfort perceived by the respondents) and each of the 17 explanatory variables pre-selected. Through the computation of Pearson linear correlation coefficient it was found that only two of the 17 pre-selected variables presented a statistically significant linear correlation with the dependent variable (Table 2). The two variables are "remain standing, with displacement, for a long time" (with a significance level of 5%, in the category "environment and physical constraints") and "my job is or was a job with moments of work overload" (with a significance level of 1%, in the category "organizational and relational constraints").

Table 2. Pearson linear correlation coefficient values for the 17 possible explanatory variables.

Variable	Pearson coefficient
Working Conditions	
Environmental (I'm exposed to...)	
Noise	0.063
Intense heat or cold	0.152
Physical (I'm forced to...)	
Remain standing with displacement for a long time	0.352
Work Rhythm (I'm exposed to...)	
Solve sudden situations/problems without help	0.185
Work beyond my assigned timetable	0.279
Frequent interruptions	0.133
Have to do several things at the same time	0.271
Work relationships (I'm exposed to...)	
Verbal aggression	0.014
Physical aggression	0.014
Intimidation	0.133
Dealing with the public (There is...or, I have to...)	
Endure the demands of the public	0.037
Deal with situations of tension in the relationship	0.136
Be exposed to verbal aggression from the public	0.011
Be exposed to physical aggression from the public	0.041
Have to deal with suffering of others	0.171
Work Characteristics (My work is a type of work...)	
With moments of work overload	0.513
Lack of suitable equipment	0.261
With inadequate facilities	0.235

Using those two significant explanatory variables, it was possible to estimate a multiple linear regression model by the ordinary least squares method. The estimated regression model, with an adjusted determination coefficient of 26.2%, is given by the following equation:

$$\hat{Y} = 0.611 + 0.292X_1 + 0{,}415X_2$$

where: $\hat{Y}$ = degree of discomfort for the variable "I believe that my health is or has been affected by the work I do and/or I did"; X_1 = degree of discomfort for the variable "remain standing, with displacement, for a long time"; and X_2 = degree of discomfort for the variable "my job is or was a job with moments of work overload".

The estimated regression model has a globally significance level of 5% and the coefficients of the explanatory variables of the model are also statistically significant (with a significance level of 5% for the coefficient of X_1, and a significance level of 10% for the coefficient of X_2).

From the obtained results, it can be state that, overall, the statistical model explains 26.2% of total variations of the dependent variable around its sample mean, and that each of the explanatory variables, taken individually, has a positive influence in the explained variable, with a higher intensity for the second variable (approximately twice of the first). Indeed, *ceteris paribus*, when the degree of discomfort for the variable "remain standing, with displacement, for a long time" increases a point on its scale, the effect on the scale of the degree of discomfort associated with the explained variable "I believe that my health is or has been affected by the work I do and/or did" will be increased in 0.292 points. Additionally, when the degree of discomfort on the variable "my job is or was a job with moments of work overload" raises a point on its scale, *ceteris paribus*, will lead to a positive effect of 0.415 points on the degree of discomfort scale of the explained variable.

In accordance with earlier studies, it was found that teachers' job demands refer to physical, psychosocial or organizational aspects of a job that requires sustained physical or mental effort, and are associated with physiological and psychological costs (Gasparini, Barreto & Assunção, 2005; Hakanen, Bakker & Schaufeli, 2006; Ferreira & Martinez, 2012). Work content and context can be experienced as hindrance, leading to poor health conditions. More precisely, workload and work pace seem to have a decisive role (European Comission, 2012) in the individual perception stated by: "I believe that my health is or has been affected by the work that I do". In fact, physical workload and time pressures were pointed out as possibly having a significant influence in health and work context (Ferreira & Martinez, 2012). Even a mental disease can be predicted by a variety of job pressures, but it is predominantly linked to job pressures that arise from 'ambiguity of the teacher's role' (Travers & Cooper, 1993; Hakanen, Bakker & Schaufeli, 2006).

4 CONCLUSIONS

From the set of 17 variables with a percentage of respondents equal or greater than 50% that compose the factors associated with "conditions and characteristics of the work", only two of them have a relation with statistical significance with the variable "I believe that my health is or has been affected by the work I do and/or did". From these two variables, one belongs to the category "environment and physical constraints" and the other to "organizational and relational constraints". The explanatory variable concerning "organizational and relational constraints" is connected with work overload, and the explanatory variable related to environment and physical constraints is linked with "remain standing, with displacement, for a long time". The estimated multiple linear regression model constructed to explain the behavior of the variable "I believe that my health is or has been affected by the work I do and/or did" allowed to conclude that the single influence of the variable representing "organizational and relational constraints" has nearly twice the influence of the variable linked with "work environment and physical constrains", though, both variables have a less than proportional positive effect on the variable "I believe that my health is or has been affected by the work I do and/or did" (almost half a point for the first variable and nearly a quarter of a point for the second variable). The results obtained in this school and with this sample reinforce studies that have already been developed giving clues for more integrative statistical analysis involving other characteristics of the work.

In fact, exposure to risk may have a direct impact on health. The education sector, where physical and psychosocial risks are both present in teaching activity, affecting their health, and wellbeing, reflects the holistic conception of occupational health: work characteristics that teachers perceive as directly affecting their health, are related with both environment and physical constraints, as well as organizational and relational constraints.

REFERENCES

Aristides I., Ferreira & Luis F. Martinez (2012). Presenteeism and burnout among teachers in public and private Portuguese elementary schools, The International Journal of Human Resource Management, 23:20, 4380–4390, DOI: 10.1080/09585192.2012.667435.

Barros-Duarte, C. & Cunha, L. 2010. INSAT2010—Inquérito Saúde e Trabalho: outras questões, novas relações. Laboreal 6 (2): 19–26.

Barros-Duarte C. & Lacomblez M. 2006. Santé au travail et discrétion des rapports sociaux. *Pistes* 8 (2): 1–17. Retrived from http://www.pistes.uqam.ca/v8n2/articles/v8n2a2.htm

Canguilhem, G. (8 ed.) 1999. *Le normal et le pathologique.* Paris: PUF.

Carlotto, M.S. & Palazzo, L.S. 2006. Síndrome de burnout e fatores associados: um estudo epistemológico com professores. *Cad. Saúde Publica* 22 (5): 1017–1026.

Clot Y. 2008. *Travail et pouvoir d'agir*. Paris: PUF.

Eurofound 2012. *Fifth European Working Conditions Survey*. Luxembourg: Publications Office of the European Union.

European Commission 2012. *Management of psychosocial risks at work: An analysis of the findings of the European Survey of Enterprises on New and Emerging Risks*. Luxembourg: Publications Office of the European Union.

Ferreira A.I. & Martinez L.F. 2012. Presenteeism and burnout among teachers in public and private Portuguese elementary schools. International Journal of Human Resource Management 23 (20): 4380–4390.

Gaillard, A.W.K. (2001). Stress, workload, and fatigue as three biobehavioral states: A general overview. In P.A. Hancock, & P.A. Desmond (Eds.), Stress, workload, and fatigue. Human factors in transportation (pp. 623–639). Mahwah7 Lawrence Erlbaum Associates.

Gasparini, S.M., Barreto, S.M. & Assunção, A.A. 2005. O professor, as condições de trabalho e os efeitos sobre sua saúde. *Educação e Pesquisa*. 31 (2): 189–199.

Gollac, M. & Volkoff, S. 2000. *Les conditions de travail.* Paris: Éditions La Découverte.

Gollac, M. & Bodier, M. 2011. *Mesurer les facteurs psychosociaux de risque au travail pour les maîtriser* Collège d'Expertise sur le Suivi des Risques Psychosociaux au Travail. Retrived from http://www.college-risquespsychosociaux-travail.fr/rapport-final,fr,8,59.cfm.pdf

Hakanen J.J., Bakker A.B. & Schaufeli W.B. 2006. Burnout and work engagement among teachers. *Journal of School Psychology* 43: 495–513.

Harnois, G. & Gabriel, P. 2000. *Mental Health Mental health and work: Impact, issues and good practices.* Geneva: WHO, ILO.

ILO 1985. ILO *Convention No. 161 concerning occupational health cervices.* Geneva: International Labour Organization.

Maggi, B. 2003. De l'agir organisationnel: un point de vue sur le travail, le bien-être, l'apprentissage. Toulouse: Octarès.

Rudow, B. 1999. Stress and burnout in the teaching profession: European studies, issues, and research perspectives. In A.M. Huberman (Ed.), Understanding and preventing teacher burnout: A sourcebook of international research and practice (pp. 38–58). New York7 Cambridge University Press.

Travers, C.J. & Cooper, C.L. 1993. Mental health, job satisfaction and occupational stress among UK teachers. Work & Stress: An International *Journal of Work, Health & Organisations* 7 (3): 203–219.

Van Horn, J.E., Taris, T.W., Schaufeli, W.B. & Schreurs, P.J.G. 2004. The structure of occupational well-being: A study among Dutch teachers. *Journal of Occupational and Organizational Psychology* 77: 365–375.

Volkoff, S. 2002. Des comptes à rendre: usages des analyses quantitatives en santé au travail pour l'ergonomie. [Internet]. Centre d'Etudes de l'Emploi [cited 2008 Jul 02]. Available from:www.eurofound.europa.eu/ewco/surveys/EwCS2005/index.htm

Occupational Safety and Hygiene II – Arezes et al. (eds)
© 2014 Taylor & Francis Group, London, ISBN 978-1-138-00144-2

Noise exposure and military police: A review

K.C.S. de Lima & L.B. da Silva
UFPB, João Pessoa, Brazil

ABSTRACT: The policeman's work is guided by military noise exposure; this variable is responsible for signs of attention deficit, lack of concentration and irritability when making decisions, making these factors causing the increase in the number of accidents. The purpose of this article is to conduct a systematic review in order to identify the state of the art as well as the possible paths that can lead to a more thorough search within the theme of police and noise. The methodology used was a systematic review. The research was conducted in the databases of Brazilian national and international journals, with publications available in the last ten years. As a result, it was found 32 articles selected for the composition of the prior art. Among these items was possible to identify the predominant research about noise produced by shooting firearms, demonstrating that the theme 'police and noise' has gaps that can lead to new research into the subject.

1 INTRODUCTION

Policing is ruled by dynamism required for the nature of action, and includes a varied repertoire of means (monitoring, regulating, patrolling, guarding and others) and ways to achieve an end which aims to meet the expectations of social groups as regarding safety (Muniz & Machado, 2010).

The military police activity comprises a set of situations and actions that can cause health problems and affect occupational safety at work. With the increasing of violence, this activity can bring health complications, such as stress, mental problems, disturbances in the body and many others (Nogueira, Rosa 2012).

The exposure of police officers in situations involving decision-making under stress with emotional control demand is added to the most diverse environmental conditions present in the places where activities are carried out, which are unpredictable and variable (Nogueira, 2012). Working conditions are responsible for frames of occupational diseases and inefficiency of the task, which are addressed by Ergonomics, the science that studies the interactions of the system man-machine-environment (Wilson, 2000).

According to Parsons (2000), Environmental ergonomics should elucidate how environmental conditions (acoustic, thermal, luminous and related to air quality) interfere with working activities. When inappropriate, those are responsible for most problems among workers, followed by factors related to the organization and cognition task (Costa, Campos & Villarouco, 2012). It is common consensus among some authors (Parsons 2000, Arezes 2002, Bistafa 2011, Frontezak & Wargocki 2011) that noise is a risk agent, harmful to health and the cause of many illnesses.

The work of the police officer is guided by exposure to noise, since it is present in the shooting of a firearm, radio communicator in city traffic, sirens of vehicles in urban noise, among others (Heupa, Gonçalves & Coifman 2011, Guida et al. 2010). These variables are responsible for signs of attention deficit, lack of concentration and irritability in decision-making, factors which contribute to the increase in the number of accidents (Rosa, 2012).

According to Celli, Ribas & Zannin (2008), one of the main problems caused by noise exposure without the regular use of personal protective equipment is the Noise Induced Hearing Loss - NIHL. This issue has been reported in the literature by some authors that relate hearing loss to high levels of occupational noise (Chang et al. 2012, Metidieri et al., Yankaskas 2013).

Studies involving the activity of the military police and noise exposure have significant national and international production (Guida et al., 2010). In order to identify and understand the focus of these studies, the aim of this article is to conduct a systematic review in order to raise a significant state of the art, and identify possible gaps that may lead to a research of greater contribution to the theme.

2 METHODOLOGY

The theme "Military Police and Noise" has publications with varied research focuses, and in order to identify this production was performed a

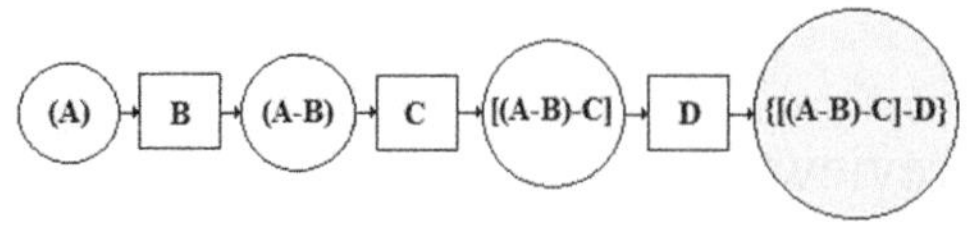

Figure 1. Scheme of systematic review. Font: Authorship.
(A) Selected publications; (B) Reading of the titles for identification of repeated publications; (A-B) Number of publications not repeated; (C) Inspection of publications with incomplete content; [(A-B)-C] Publications complete and available; (D) Complete reading to review publications which do not contemplate the theme; {[(A-B)-C]-D} Selected publications.

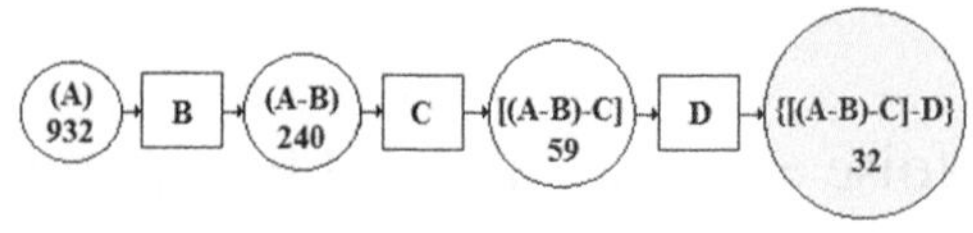

Figure 2. Results of systematic review. Font: Authorship.

Table 1. Combinations of keywords.

Keywords	Noise	Police	Noise exposure	Military noise
Firearms	53	350	11	13
Guns	40	96	7	2
Military	253		31	
Cops	17			
Police	30		6	
Noise Risk		4		
Hearing Loss		15		
Headsets				4
Radio				

systematic search of the literature published during the last ten years in constant journals located on databases CAPES (Coordination for the Improvement of Higher Education of the Ministry of Education and Culture of Brazil), SciELO and Science Direct.

The methodology adopted for this research was based on De-la-Torre-Ugarte-Guanilo, Bertolozzi & Takanashi (2011), according to the following procedures: 1. Select the databases which must follow a criterion of greater coverage of publications; 2. Search with combinations of keywords; 3. Full and partial reading of the articles, to select only those that are representative.

Combinations of keywords used during the selection of articles were composed by the following terms: noise, police, noise exposure, military noise, firearms, guns, military, cops, risk noise, headsets, radio, hearing loss, whose selection procedures are systematized in Figure 1:

3 RESULTS

Table 1 shows results obtained with the combinations of keywords. The blank spaces demonstrate the researches on the databases with the keywords did not originate publications. Figure 2 presents the results of the selection of the representative articles synthesized according to the procedures articulated in Figure 1.

The main focuses of the 32 selected papers presented in Table 2 were the noise produced by firearms (37.5%), exposure to the military band (6.25%), urban traffic police (12.5%) and various other approaches clustered in a general category (43.75%).

4 STATE OF THE ART

According to Guida et al. (2010), the literature regarding the exposure to noise in the police is large and representative, especially in the last ten years. Many of these studies were aimed at measuring the negative potential of the noise on performance and health of these professionals, as this is a risk agent present in many of the activities.

The systematic review about this topic reports to the existent diversity and indicates the main trends of studies that comprise a significant state of the art.

Silva et al. (2004) evaluated a group of military officers divided by sectors within the organization—administrative, artillery and support—with an average of 12.75 years in service, through audiometric tests, interviews and questionnaires. The results stated that 38.1% of the military have hearing loss and 64.59% do not use protection in shooting drills.

Job et al. (2004) studied the association between exposure to noise of firearms with alterations in mood and cases of tinnitus in young officers. The study concludes that even with the use of hearing protection at the end of a shooting practice, the exposure caused an increase in the incidence of tinnitus and changes in mood states.

Lehtomäki & Pakkonen (2005) and Neves & Soalheiro (2010) studied the efficiency and the degree of attenuation of ear protectors used in shot activities by soldiers, carrying out measurements inside the ear canal protected with protective plug and headphones. In both publications, the conclusion is that for full effectiveness of protective equipment is necessary to choose the correct and suitable protector for each individual depending

Table 2. State of the art.

Authors	Urban traffic	General	Military band	Fire-arms
Silva et al., 2004	□	□	□	■
Reid, Dick & Semple, 2004	□	■	□	□
Job et al., 2004	□	□	□	■
Magann et al., 2005	□	■	□	□
Lehtomaki & Paakkonen, 2005	□	□	□	■
Ingle et al., 2005	■	□	□	□
Republic of Korea Air Force, 2007	□	□	□	■
Gonçalves, Tochetto & Gambini, 2007	□	□	■	□
Chiovenda et al., 2007	■	□	□	□
Santos, Juchem & Rossi, 2008	□	□	□	■
Jokitulppo et al., 2008	□	□	□	■
Mrena et al., 2008	□	■	□	□
Celli, Ribas & Zannin, 2008	□	□	□	■
Gonçalves et al., 2009	□	□	■	□
Lesage et al., 2009	□	■	□	□
Guida et al., 2010	□	■	□	□
Rocha, Atherino & Frota, 2010	□	■	□	□
Neves & Soalheiro, 2010	□	□	□	■
Nageris, Attias & Raveh, 2010	□	■	□	□
Helfer et al., 2010	□	■	□	□
Guida, Diniz & Kinoshita, 2011	□	□	□	■
Vaillancourt et al., 2011	□	■	□	□
Heupa, Gonçalves & Coifman, 2011	□	□	□	■
Colle et al., 2011	□	■	□	□
Muhr & Rosenhall, 2011	□	■	□	□
Guida, Souza & Cardoso, 2012	□	□	□	■
Abel, Nakashima & Smith, 2012	□	■	□	□
Cason, 2012	□	■	□	□
Yuan et al., 2012	■	□	□	□
Yankaskas, 2013	□	■	□	□
Barkokebas Jr. et al., 2013	□	□	□	■
Caciari et al., 2013	■	□	□	□

■ Explored in research.
□ Not explored in research.

on the degree of exposure and in accordance with the current standards.

In the article published by the Republic of Korea (2007), as well as publications of other authors (Celli, Ribas & Zannin 2008, Guida, Diniz & Kinoshita 2011, Guida, Souza & Cardoso 2012), there is an investigation of how many soldiers have lost induced hearing because of the noise produced by firearms, being protected or not by hearing protectors.

At both of these publications, the results lead to a considerable percentage of cases of the disease in early or advanced stages, in addition to other symptoms: tinnitus, modifications of thresholds acoustic irritability, as well as psychological and behavioral changes.

In Barkókebas Jr. et al. (2013) a study was conducted with the special operations group of the military police, especially with snipers, whose results showed that exposure suffered by these professionals possess limits within the Brazilian normative.

Santos, Juchem & Rossi (2008) tested the auditory processing assessment to identify officers with changes in the conditions of speech due to noise exposure. The results demonstrated that changes in speech are manifested even before hearing loss.

In Heupa, Gonçalves & Coifman (2011), the research was conducted based on the comparison between a controlled group (belonged to the administrative staff of the military) and another group exposed to noise. The results indicated that all the military have some auditory dysfunction, even with regular use of hearing protection.

Other articles explore issues related to the noise generated by urban traffic (Ingle et al. 2005, Chovenda et al. 2007, Caciari et al. 2013). This measured noise is intense; it is above the allowed values and the exposure occurs for more than eight hours. The results suggest that the emotional stress framework of the military is considered high and is associated with exposure to noise, which over the years can also characterize frames of hearing loss in several stages.

Officers of the Military Band also have exposure to noise and can suffer from hearing loss, according to Gonçalves, Tochetto & Gambini (2007) and Gonçalves et al. (2009). What is common in these articles is that prolonged exposure to sound pressure levels produced by music may cause hearing problems, regardless of his military career.

Some researches with different approach were found during the systematic review, these study other aspects related to the policeman subject and noise. At Reid, Semple & Dick (2004), for example, there has been an investigation of the exposure to the noise of dogs used by the police and special operations during training of animals.

In a research conducted by Magann et al. (2005), the investigation of the harmful effects of noise was held in military pregnant women who are exposed to low-frequency spectra found in their jobs in the administrative sector. The results of this study show negative correlations between the variables noise and problems in pregnancy by reinforcing the relationship of noise and other labor problems.

In Mrena et al. (2008), Muhr & Rosenhall (2011) and Cason (2012), the research interest was to attest to the effectiveness of the control program in the auditory military organizations. Abel, Nakashima & Smith (2012) on the other hand, investigated the communication efficiency in terms of noise produced by the Canadian military vehicles.

Yuan et al. (2012) researched the association between certain human genes and reduced hearing ability in military with great noise exposure, in order to identify if there is a predisposition in some of them. Yankaskas (2013) made a revision about the publications related to noise-induced hearing loss and tinnitus in the military police. As a result of this study, this author could verify that in the effects found, tinnitus is more frequent than hearing loss.

In general, most of the researches reported deals with the investigation of hearing problems found in the military throughout their career from performing audiometric and oto-acoustics exams through questionnaires, interviews, statistical analysis and composition of an audiological profile related to the risk of hearing loss. These results demonstrate a clear association between hearing loss and noise exposure (Jokitulppo et al. 2008, Lesage et al. 2008, Guida et al. 2010, Rocha, Atherino & Frota 2010, Nageris et al. 2010, Helfer et al. 2010, Vaillancourt et al. 2011, Colle et al. 2011).

5 CONCLUSION

This systematic review enabled the selection of only the items with representativeness on the international scenario, concerning research with military police. In which, it was noted the predominance of researches on the effects produced by exposure to noise of firearms in these professionals, regarding their hearing health.

However, according to some authors (Heupa, Gonçalves & Coifman 2011, Guida et al. 2010) due to the variety of situations encountered during the workday, this can not be the only source of acoustic discomfort. Considering that the activity of the police is governed by exposure to urban traffic noise, sirens of vehicles, environmental noises and the sounds coming from the radio communicator.

According to Rosa (2012), Muniz & Machado (2010) there is a very important role in the work of the military police for the society, in which communication is an essential tool for a good performance. These authors also mention that the radio equipment, which supports these professionals and represents one of the most effective means of communication during events, is considered a major risk factor in causing exposure to high noise levels during use, being variable and influenced by the number of calls received.

With the increasing levels of violence, murders, thefts and accidents, this number of calls becomes very significant and therefore responsible for the increased noise level. Taking into account the territorial extension of Brazil and the problems accentuated under the social point of view that permeate Brazilian cities, this equipment becomes an essential tool for the success of the police activity (Guida, Muniz & Machado 2010).

Through such findings and also according to Rosa (2012) it is possible to recognize that the gaps in scientific findings on the topic are significant, and research on the effects of noise from the radio communicator and other sources inherent to patrol are necessary to ensure a good performance of these professionals.

REFERENCES

Abel, S.M.; Nakashima, A. & Smith, I. 2012. Divided Listening in Noise in a Mock-up of a Military Command Post. *Military Medicine* 177(4): 436.

Arezes, P.M.F.M. 2002. *Percepção do Risco de Exposição Ocupacional ao Ruído.* Tese. Guimarães: Escola de Engenharia da Universidade do Minho.

Barkókebas JR., B. et al. 2013. Study on the impact of exposure to noise in Professional snipers. *Work* 41: 3269–3276.

Bistafa, S.R. 2011. *Acústica aplicada ao controle do ruído.* 2 ed. São Paulo: Blutcher.

Caciari, T. et al. 2013. Noise-induced hearing loss in workers exposed to urban stressors. *Science of the total environment* 463(464): 302–308.

Cason, E.M. 2012. Air Force Hearing Conservation Program Data 1998–2008: A Cross-Sectional Analysis of Positive Threshold Shifts. *Military Medicine* 177(5): 589.

Celli, A.; Ribas, A. & Zannin, P.H.T. 2008. Effect of impulsive noise on military personnel – A case study. *Bras. J. Otorhinolaryngol.* 77(6): 747–753.

Chang, T.Y. et al. 2012. Noise frequency components and the prevalence of hypertension in Workers. *Science of the Total Environment* 416: 89–96.

Chiovenda, P. et al. 2007. Environmental noise-exposed workers: Event-related potentials, neuropsychological and mood assessment. *International Journal of Psychophysiology* 65: 228–237.

Colle, A. et al. 2011. Occupational exposure to noise and the prevalence of hearing loss in a Belgian military

population: A cross-sectional study. *Noise and Health* 13(50): 64.
Costa, A.P.L.; Campos, F & Villarouco, V. Overview of ergonomics built environment. *Work* 41: 4142–4148.
De-La-Torre-Ugarte-Guanilo, M.C.; Takanashi, R.F. & Bertolozzi, M.R. 2011. Revisão sistemática: noções gerais. *Rev. Esc. Enferm. USP* 45(5): 1260–1266.
Frontezak, M. & Wargocki, P. 2011. Literature survey on how different factors influence human comfort indoor environments. *Building and Environment* 46: 922–937.
Gonçalves, C.G.O. et al. 2009. Percepção e o impacto da música na audição de integrantes da banda militar. *Rev Soc Bras Fonoaudiol.* 14(3): 515–20.
Gonçalves, M.S.; Tochetto, T.M. & Gambini, C. 2007. Hiperacusia em músicos de banda militar. *Rev Soc Bras Fonoaudiol.* 12(4): 298–303.
Guida, H.L.; Diniz, T.H. & Kinoshita, S.K. 2011. Acoustic and psychoacoustic analysis of the noise produced by the Police force firearms. *Braz J Otorrinolaryngol.* 77(2): 163–70.
Guida, H. et al. 2010. Perfil audiológico dos Policiais Militares do Estado de São Paulo. *Arq. Int. Otorrinolaringol.* 14(4): 426–432.
Guida, H.L.; Souza, A.L. & Cardoso, A.C.V. 2012. Relação entre os achados da avaliação audiométrica e das emissões otoacústicas em policiais militares. *Arq. Int. Otorrinolaringol.* 16(1): 67–73.
Helfer, T.M. et al. 2010. Epidemiology of hearing impairment and noise-induced hearing injury among U.S. military personnel, 2003–2005. *Am. J. Prev. Med.* 38(1): 71–77.
Heupa, A.B.; Gonçalves, C.G.O. & Coifman, H. 2011. Effects of impact noise on the hearing of military personnel. *Brazilian Journal of Otorhinolaryngology* 77(6): 747–53.
Ingle, S.T. et al. 2005. Noise exposure and hearing loss among the traffic policemen working at busy streets of Jalgaon urban centre. *Transp. Research Part D* 10: 69–75.
Job, A. et al. 2004. Moderate variations of mood/emotional states related to alterations in cochlear otoacoustic emissions and tinnitus onset in young normal hearing subjects exposed to gun impulse noise. *Hearing Research* 193: 31–38.
Jokitulppo, J. et al. 2008. Military and Leisure-Time Noise Exposure and Hearing Thresholds of Finnish Conscripts. *Military Medicine* 9: 906.
Lehtomaki, K. & Paakkonen, R. 2005. Protection efficiency of hearing protectors against military noise from handheld weapons and vehicles. *Noise and Health* 7(26).
Lesage, F.X. et al. 2009. Noise-induced hearing loss in French Police officers. *Occupational Medicine* 59: 483–486.
Magann, E.F. et al. 2005. The effects of standing, lifting and noise exposure on preterm birth, growth restriction, and perinatal death in healthy low-risk working military women. *The Journal of Maternal-Fetal and Neonatal Medicine* 18(3): 155–162.
Metidieri, M.M. et al. 2013. Noise-Induced Hearing Loss: literature review with a focus occupational medicine. *International Archives of Otorhinolaryngology* 17(2): 208–212.
Mrena, R. et al. 2008. The effect of improved hearing protection regulations of military noise-induced hearing loss. *Acta Oto-Laryngologica* 128: 997–1003.
Muhr, P. & Rosenhall, U. 2011. The influence od military service on auditory health and the efficacy of a hearing conservation program. *Noise and Health* 13(53): 320.
Muniz, J.O. & Machado, E.P. 2010. Polícia para quem precisa de polícia: contribuições aos estudos sobre policiamento. *Caderno CRH* 23(60): 437–447.
Nageris, B.I.; Attias, J. & Raveh, E. 2010. Test-retest tinnitus characteristics in pacients with noise-induced hearing loss. *American Journal of Otorhinolaryngology-Head and Neck Medicine and Surgery* 31: 181–184.
Neves, E.B. & Soalheiro, M. 2010. A proteção auditiva utilizada pelos militares do Exército Brasileiro: há efetividade. *Ciência & Saúde Coletiva* 15(3): 889–898.
Nogueria, G.E.G. 2012. Condições de trabalho e saúde mental do trabalhador da segurança pública. *Revista de Psicologia, Saúde Mental e Segurança Pública* 1(4): 53–58.
Parsons, K.C. 2000. Environmental ergonomics: a review of principles, methods and models. *Applied Ergonomics* 31: 581–94.
Reid, A.; Dick, F. & Semple, S. 2004. Dog noise as a risk factor for hearing loss among Police dog handlers. *Occupational Medicine* 54: 535–539.
REPUBLIC OF KOREA AIR FORCE. 2007. Noise-Induced Hearing Loss Caused by Gunshot in South Korean Military Service. *Military Medicine* 4: 421.
Rocha, R.L.O.; Atherino, C.C.T.; Frota, S.M.M.C.F. 2010. High-frequency audiometry in normal hearing military firemen exposed to noise. *Braz J Otorhinolryngol* 76(6): 687–94.
Rosa, J.G. 2012. *Trabalho e Qualidade de vida dos policiais militares que atuam na modalidade de policiamento da Rádio Patrulha do 9º Batalhão de Policia Militar de Criciúma/SC.* Monografia. Criciúma: Universidade do Extremo Sul Catarinense - UNESC.
Santos, C.C.S.; Juchem, L.S. & Rossi, A.G. 2008. Processamento auditivo de militares expostos a ruído ocupacional. *REV CEFAC* 10(1): 92–103.
Silva, A.P. et al. 2004. Avaliação do perfil auditivo de militares de um quartel do Exército Brasileiro. *Rev Bras Otorrinolaringol.* 70(3): 344–50.
Wilson, J.R. 2000. Fundamentals of ergonomics in theory and practice. *Applied Ergonomics* 31: 36–39.
Vaillancourt, V. et al. 2011. Evaluation of Auditory for Royal Canadian Mounted Police Officers. *Journal of the American Academy of Audiology* 22(2): 313–331.
Yankaskas, K. 2013. Prelude: Noise-induced tinnitus and hearing loss in the military. *Hearing Research* 295: 3–8.
Yuan, B.-C. et al. 2012. A predictive model f the association between gene polymorphism and the risk of noise-induced hearing loss caused by gunfire noise. *Journal of the Chinese Medical Association* 75: 36–39.

Occupational Safety and Hygiene II – Arezes et al. (eds)
© 2014 Taylor & Francis Group, London, ISBN 978-1-138-00144-2

Occupational exposure to particulate matter and fungi in a composting plant—case study in Portugal

S. Viegas
Environmental Health RG, Lisbon School of Health Technology, Polytechnique Institute of Lisbon, Portugal
Center for Malaria & Tropical Diseases (CMDT), Public Health and Policy, Escola Nacional de Saúde Pública, Universidade Nova de Lisboa, Portugal

M. Almeida-Silva
Environmental Health RG, Lisbon School of Health Technology, Polytechnique Institute of Lisbon, Portugal
C2TN, Instituto Superior Técnico, Universidade de Lisboa, Loures, Portugal

R. Sabino
Environmental Health RG, Lisbon School of Health Technology, Polytechnique Institute of Lisbon, Portugal
Mycology Laboratory, National Institute of Health Dr. Ricardo Jorge, Lisbon, Portugal

C. Viegas
Environmental Health RG, Lisbon School of Health Technology, Polytechnique Institute of Lisbon, Portugal

ABSTRACT: The handling of waste can be responsible for occupational exposure to particles and fungi. The aim of this study was to characterize exposure to particles and fungi in a composting plant. Measurements of particulate matter were performed using portable direct-reading equipment. Air samples of 50 L were collected through an impaction method with a flow rate of 140 L/min onto malt extract agar supplemented with chloramphenicol (0.05%). Surfaces samples were also collected. All the samples were incubated at 27°C for 5 to 7 days. Particulate matter data showed higher contamination for PM_5 and PM_{10} sizes. *Aspergillus* genus presents the highest air prevalence (90.6%). *Aspergillus niger* (32.6%), *A. fumigatus* (26.5%) and *A. flavus* (16.3%) were the most prevalent fungi in air sampling, and *Mucor* sp. (39.2%), *Aspergillus niger* (30.9%) and *A. fumigatus* (28.7%) were the most found in surfaces. The results obtained claim the attention to the need of further research.

1 INTRODUCTION

Composting is an important process of solid waste management and it can be used for treatment of a variety of different wastes (green waste, household waste, sewage sludge and more) (Duquenne et al., 2012). This is a natural self-heating process involving the biological degradation of organic matter under aerobic conditions. Installations for composting vary greatly in size (from domestic to large-scale facilities), degree of enclosure (open, partially enclosed, enclosed facilities), design (static windrow systems, aerated static piles, bioreactors, etc.) and the type of wastes composted (Swan et al., 2003).

The handling of waste and compost that occurs frequently in the process (compost turning, shredding, and screening) has been shown to be responsible for the release of dust and airborne microorganisms and their compounds in the air of the composting facilities. Therefore, dust, mesophilic and thermophilic microorganisms as well as volatile organic compounds, endotoxins and glucans compose the bioaerosols in those settings and have been found at high levels in numerous composting facilities and may present an exposure hazard especially for workers (Marchand et al., 1995; Duquenne et al., 2012). Consequently, several microorganisms and thermophilic fungi, such as *Aspergillus fumigatus* have been reported (Swan et al., 2003; Duquenne et al., 2012). Furthermore, this kind of contamination in composting facilities has been associated with increased respiratory and dermal pathologies among compost workers (Bünger et al., 2000, 2007).

In spite of the several published studies, exposure to bioaerosols and dust in composting facilities located in Portugal is still insufficiently characterized. Taking this in consideration the aim of the present study was to characterize and assess the exposure to particulate matter and fungi in a totally indoor composting plant located in Portugal.

2 MATERIALS AND METHODS

2.1 *Description of the composting plant*

The capacity of the studied plant is of 40 thousand tons of biowaste per year and can go until 60 thousand tons per year. The facility consists in a building in which the different composting operations are performed. In this building there is no sorting because all the waste sorting process is done in a different plant. Waste is unloaded in a reception area which is confined and prepared to receive trucks with the waste and to avoid the emission of odors for the outdoor environment. First phase of the process is a waste pretreatment that intends to remove undesirable materials from the process (glass, rocks, plastics, metals…). The next phase is anaerobic digestion and after dehydration there is closet composting action, followed by an open composting with forced aeration. All the process takes thirteen weeks.

2.2 *Particulate matter*

One measurement of Particulate Matter (PM) was performed in 7 workplaces (Maintenance workshop, Centrifuges, Maturation Park, Pre-treatment, Control Room, Waste screw and Cabinet of the Forklift) using a portable direct-reading equipment (Lighthouse, model 3016 IAQ) to measure 5 different sizes ($PM_{0.5}$; PM_1; $PM_{2.5}$; PM_5; PM_{10}). This option was considered because the differentiation between particle`s size fractions is important in order to estimate with more detail the possible penetration of dust into and within the respiratory system (WHO, 1999; Brunekreef and Forsberg, 2005). Measurements were conducted near the workers nose and during tasks performance. The criteria follow for chosen the measurement places were the ones where the workers spend more time. All measurements were conducted continuously with duration of 5 min and were considered the mean, maximum and minimum values obtained for each particle size.

2.3 *Fungal contamination*

Air samples of 50 L were collected from 6 indoor sampling sites (Maintenance workshop, Centrifuges, Maturation Park, Pre-treatment, Control room, Waste screw) were collected through an impaction method with a flow rate of 140 L/min onto malt extract agar (MEA) supplemented with chloramphenicol (0.05%), using the Millipore air Tester (Millipore). An outdoor sample was also collected since this was the place regarded as reference.

Surfaces samples were collected by swabbing the surfaces of the same indoor sites, using a 10 by 10 cm square stencil disinfected with 70% alcohol solution between samples according to the International Standard ISO 18593 (2004). The obtained swabs were then plated onto MEA.

All the collected samples were incubated at 27ºC for 5 to 7 days. After laboratory processing and incubation of the collected samples, quantitative (colony-forming units—CFU/m^3 and CFU/m^2) and qualitative results were obtained with identification of the isolated fungal species. For species identification, microscopic mounts were performed using tease mount or Scotch tape mount and lactophenol cotton blue mount procedures. Morphological Identification was achieved through macro and microscopic characteristics as noted by Hoog et al. (2002).

3 RESULTS

3.1 *Particulate matter*

Particulate matter data showed that the workplace "Room Process Control" presented the lowest values of contamination. In opposite, the workplace with higher contamination was Forklift Cabinet for $PM_{2.5}$, PM_5 and PM_{10} sizes. In the case of $PM_{0.5}$ and PM_1 the Maintenance Workshop presented the higher values (Figure 1). The particles concentration obtained in all workplaces were significantly different ($p < 0.05$) with exception for results obtained in: Room Process Control and

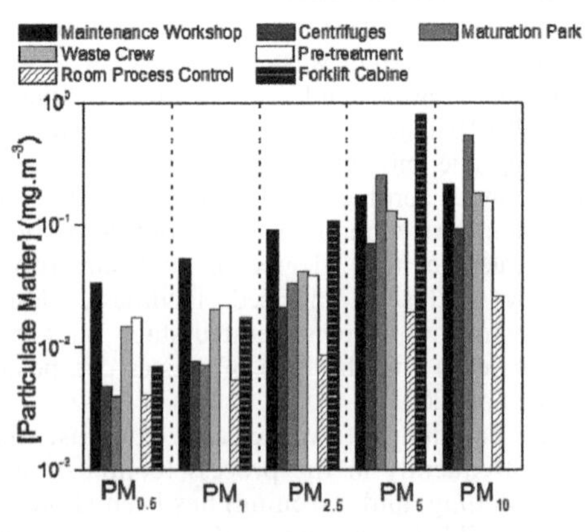

Figure 1. PM mean values distribution by workplace. The results are presented in $mg \cdot m^{-3}$.

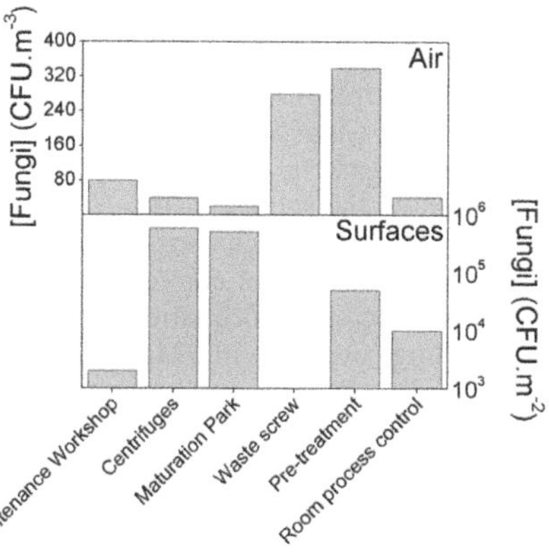

Figure 2. Fungal load distribution in the different workplaces.

Maturation Park in $PM_{0.5}$ (p = 0.07); Pre-Treatment and Maturation Park in PM_5 (p = 0.35); and Maintenance Workshop and Maturation Park in PM_{10} (p = 0.12).

3.2 *Fungal contamination*

Pre-treatment and Waste screw were the sampling sites with the highest fungal load in air and Centrifuges and Maturation park has the highest fungal load in surfaces samples (Figure 2).

Besides both sampling sites with higher air fungal load, also Maintenance Workshop presented more $CFU.m^{-3}$ than the outdoor sample.

Nine different species of filamentous fungi were identified in air samples with a total of 982 isolates. *Aspergillus* genus presents the highest prevalence (90.6%). *Aspergillus niger* (32.6%), *A. fumigatus* (26.5%) and *A. flavus* (16.3%) were the most prevalent fungi in air sampling, but A. *sydowii*, *A. versicolor* and *A. glaucus* were also isolated. *Neosartorya fumigata* was identified in air sampling. *Mucor* sp. and *Penicillium* sp. were also identified.

Four different species were isolated in surfaces samples with a total of 1810000 isolates. *Aspergillus* genus also presents the highest prevalence (60.8%). *Mucor* sp. (39.2%), *Aspergillus niger* (30.9%) and *A. fumigatus* (28.7%) were the most found, but *A. flavus* was also isolated.

4 DISCUSSION

Depending on the waste treatment technology employed, aerosol formation capable of transporting breathable particles with microorganisms has already been published. Regarding composting plants several hazardous agents may be released during the waste composting process (Kummer and Thiel, 2008; Nadal et al., 2009).

Concerning PM data it was possible to measure 5 different sizes that can be distinguished either in the inhalation fraction (PM_5 and PM_{10}) and respiratory fraction ($PM_{0.5}$; PM_1 and $PM_{2.5}$) (WHO, 1999; Brunekreef and Forsberg, 2005). Moreover, studies developed in waste management setting consider normally only the inhalable dust or total dust and do not distinguish or provide data related with the respiratory fraction (Wouters et al., 2006; Park et al., 2011).

In this study, high values of contamination were observed with particular emphases in Forklift Cabinet and Maturation Park. In this last case, the contamination found is probably due to the constant movement of forklifts in this area promoting re-suspension of PM. There are negative health effects related with this type of exposure demonstrated in literature, namely penetration into the gas exchange region of the lung (PM_5) and also the possibility to produce disease by impacting in the upper and larger airways below the vocal cords (PM_{10}) (Vincent and Mark, 1981; Brunekreef and Forsberg, 2005). But $PM_{2.5}$ is also a concern in the Forklift Cabinet and in the Maintenance Workshop, being already in the respiratory fraction this particle size can penetrate in the alveolar region and be involved in systemic effects (Brunekreef and Forsberg, 2005).

In the Maintenance Workshop $PM_{0.5}$ and PM_1 have also some expression probably due to the type of work develop that can involve activities such as sanding, drilling and cutting and not so much waste management activities. However, these workers spend a great part of their time in the places where waste management operations are done, and therefore, the exposure is very diverse during the day and it is dependent of many factors, such as the task developed, materials and equipment used, environmental conditions and also work practice (Eduard and Bakke, 1999).

The same tendency in particle's size distribution was found in other research work developed in different occupational settings that involves exposure to organic dust, namely in swine and poultry production (Basinas et al., 2013; Viegas et al., 2013a,b). In those settings, and with similar values of contamination by PM, it was possible to observe respiratory health effects related with exposure (Donham et al., 1989; Viegas et al., 2013a).

It is important to refer that only one measurement of 5 minutes was taken in each workplace, and variations in PM results can be expected along the day, related with the waste characteristics and also with the tasks being developed. Nevertheless, the measurements were performed during a "normal" day in what concern with the process and the amount of waste that was being handled.

Besides mass concentration, there are other aspects that must be contemplated when considering PM health effects: chemical characteristics and number of particles. Additionally, particles may act as a carrier and a source of nutrients for fungi (such as *Aspergillus*, *Penicillium* and *Mucor* genus) (Seedorf et al., 1998), and bacteria (Halstensen et al., 2013). PM is also rich in endotoxins from the cell wall of gram-negative bacteria and is also associated with mycotoxins produced by several fungi (Mayeux, 1997). These biologically active compounds adhered to PM, along with coexisting toxicant gases that can be carcinogenic, allergic and irritant, promote concern regarding exposure to mixtures and possible additive and synergistic health effects (Von Essen and Donham, 1999).

As stated by Spencer and Alix in 2006: "Composting operations are industrial facilities, like factories. As such, the same reasoning and concepts of industrial hygiene that apply to any other type of factory also apply to compost factories." Some of the preventive and protective measures that can me mentioned are for instance: pave roads; keep roads, areas and equipment clean; dampen loads; enclose and ventilate potentially dusty process areas; maintain closed cabin door or windows from forklifts; provide masks for workers; consider spraying and much others (Spencer and Alix, 2006).

In this kind of industry it is impossible to eliminate all fungi, not only because the permanent bioaerosol generating operations carried out inside the facilities (Marchand et al., 1995), but also due to the high amounts of waste, that is the perfect subtract for fungal growth and development in suitable levels of temperature and humidity (Wouters et al., 2000). When waste is handling in indoor environments, the bioaerossol's exposure is higher (Wouters et al., 2006), increasing the potential health effects due to fungi exposure.

With respect to the health risks derived from exposure to fungi, nowadays, there are no guidelines set by the National Institute of Occupational Safety and Health (NIOSH) concerning the allowable load at the workplace (Vilavert et al., 2009). The World Health Organization (WHO) considers the value of 150 $CFU.m^{-3}$ as a reason for concern, especially when potentially pathogenic species of fungi are present (Goyer et al., 2001) as the ones most prevalent in our study. Pretreatment and Waste Screw sampling sites surpass the WHO value. Moreover, 3 from the 6 sampling sites presented higher fungal load than outdoor sample meaning that there are sources of indoor fungal contamination, such as waste (Wouters et al., 2000).

Besides the quantitative assessment, is crucial to analyze the fungal species present, since adverse health effects dependend on fungal species (Rao et al., 1996; Hoog et al., 2000). According to the American Industrial Hygiene Association (AIHA 1996) in the *Field Guide for the Determination of Biological Contaminants in Environmental Samples*, the identification of the species *A. flavus* and *A. fumigatus*, both of them identified in the analyzed plant, requires implementation of corrective measures.

Mycotoxins occur in occupational environments whenever fungi are present (Mayer et al., 2008) and the results obtained were useful to confirm the relevance to assess toxinogenic strains, from *A. flavus* and *A. fumigatus* in future research. Additionally, knowledge regarding the presence of mycotoxins in bioaerosols from composting facilities is still widely uncertain and is very wide the possibilities of toxins and metabolites that can be present needing to be consider for additional research work (Fischer et al., 1999). In agreement with this, Brera and colleagues (2002) found aflatoxin in airborne dust samples from different occupational settings. Previously, also Autrup and colleagues (1991) measured aflatoxin levels in dust samples from animal-feed production. More recently, some published work developed in Portugal (Viegas et al., 2012, 2013c) has shown that there is occupational exposure to aflatoxin B1 in environments where there is exposure to organic dust concomitantly to high fungal contamination, like in the studied composting plant.

5 CONCLUSIONS

The results obtained in this study claim the attention to the need of further research regarding occupational exposure in composting plants. Additionally, these results allowed the definition of more detailed sampling strategies to better characterize exposure to the risk factors.

Despite the need of more detailed data, the information obtained claims attention to the need for applying adequate preventive and protective measures.

REFERENCES

Albrecht, A., Fischer, G., Brunnemann-Stubbe, G., Jäckel, U. & Kampfer, P., 2008. Recommendations for study design and sampling strategies for airborne microorganisms, MVOC and odours in the surrounding of composting facilities. Int. J. Hyg. Environ. Health 211: 121–131.

American Industrial Hygiene Association, 1996. Field Guide for the Determination of Biological Contaminants in Environmental Samples. AIHA.

Basinas, I, Schlunssen, V., Takai, H., Heederik, D., Omland, O., Wouters, I.M., Sigagaard, T. &

Kromhout, H. 2013. Exposure to inhalable dust and endotoxin among Danish pig farmers affected by work tasks and stable characteristics. Ann. Occup. Hyg. (In press).

Brera, C., Caputi, R., Miraglia, M., Iavicoli, I., Salerno, A., & Carelli, G. 2002. Exposure assessment to mycotoxins in workplaces: Aflatoxins and ochratoxin A occurrence in airborne dusts and human sera. Microchem. J. 73:167–173.

Brunekreef & Forsberg, B. 2005. Epidemiological evidence of effects of coarse airborne particles on health. Eur Respir J 26:309–318.

Bünger, J, Schappler-Scheele, B., Hilgers, R. & Hallier, E. 2007. Int. Arch. Occup. Environ. Health 80: 306–312.

Bünger, J., Antlauf-Lammers, M., Schulz, T.G., Westphal, G.A., Muller, M.M., Ruhnau, P. & Hallier, E. 2000. Occup. Environ. Med. 57: 458–464.

Déportes, I., Benoit-Guyod, J.L. & Zmirou, D. 1995. Hazard to man and the environment posed by the use of urban waste compost: a review. Sci. Total Environ 172: 197–222.

Donham, K. Haglind, P. Petersen, Y. Rylander, R. & Belin L. 1989. Environmental and health studies of farm workers in Swedish confinement buildings. Br J Ind Med. 46: 31–37.

Douwes, J., Thorne, P., Pearce, N. & Heederik, D. 2003. Ann. Occup. Hyg 47: 187–200.

Duquenne, P., Simon, X., Koehler, V., Goncalves-Machado, S., Greff, G., Nicota, T. & Poirota, P. 2012. Documentation of bioaerosol concentrations in an indoor composting facility in France. J. Environ. Monit. 14(2): 409–419.

Eduard, W. & Bakke, B. 1999. Experiences with task-based exposure assessment in studies of farmers and tunnel workers. Norsk Epidemiologi 9(1): 65–70.

Eitzer, B.D., 1995. Emissions of volatile organic chemicals from municipal solid waste composting facilities. Environ. Sci. Technol. 29: 896–902.

Fung, F. & Hughson, W.G. 2003. Health effects of indoor fungal bioaerosol exposure. Appl. Occup. Environ. Hyg. 18: 535–544.

Goyer, N., Lavoie, J., Lazure, L. & Marchand, G. 2001. Bioaerosols in the Workplace: Evaluation, Control and Prevention Guide. Institut de Recherche en Santé et en Sécurité du Travail du Québec.

Halstensen, A.S., Heldal, K.K., Wouters, I.M., Skogstad, M., Ellingsen, D.G. & Eduard, W. 2013. Exposure to Grain Dust and Microbial Components in the Norwegian Grain and Compound Feed Industry. Ann Occup Hyg (In Press).

Herr, C.E.W., Nieden, A., Bodeker, R.H., Gieler, U. & Eikmann, T.F. 2003. Ranking and frequency of somatic symptoms in residents near composting sites with odor annoyance. Int. J. Hyg. Environ. Health 206: 61–64.

Hoog, C., Guarro, J., Gené, G. & Figueiras, M., (2th ed), 2000. Atlas of Clinical Fungi. Centraalbureau voor Schimmelcultures.

Komilis, D.P., Ham, R.K. & Park, J.K. 2004. Emission of volatile organic compounds during composting of municipal solid wastes. Water Res. 38 (7): 1707–1714.

Kummer, V. & Thiel, W.R., 2008. Bioaerosols–sources and control measures. Int. J. Hyg. Environ. Health 211: 299–307.

Marchand, G., Lavoie, J. & Lazure, L. 1995. J. Air Waste Manage. Assoc. 45: 778–781.

Nadal, M., Inza, I., Schuhmacher, M., Figueras, M.J. & Domingo, J.L. 2009. Health risks of the occupational exposure to microbiological and chemical pollutants in a municipal waste organic fraction treatment plant Int. J. Hyg. Environ. Health 212(6): 661–669.

Park, D., Ryu, S., Kim, S. & Yoon, C. 2011. An Assessment of Dust, Endotoxin, and Microorganism Exposure during Waste Collection and Sorting. J. Air & Waste Manage. Assoc. 61:461–468.

Rao, C., Burge, H. & Chang, J. 1996. Review of quantitative standards and guidelines for fungi in indoor air. J Air Waste Manage Assoc. 46: 899–908.

Schlosser, O., Huyard, A., Cartnick, K., Yaez, A., Catan, V. & Do Quang, Z. 2009. Water Environ. Res. 81: 866–877.

Seedorf, J., Hartung, J., Schröder, M., Linkert, K.H., Phillips, V.R., Holden, M.R., Sneath, R.W., Short J.L, White, R.P., Pedersen, S., Takai, T., Johnsen, J.O., Metz, J.H.M., Groot Koerkamp, P.W.G., Uenk, G.H. & Wathes, C.M. 1998. Concentrations and Emissions of airborne Endotoxins and Microorganisms in Livestock buildings in Northern Europe. In: Journal of Agricultural Engineering Research 70: 97–109.

Spencer, R. & C.M. Alix. 2006. Dust Management, Mitigation at Composting Facilities. BioCycle. 47(3):55.

Swan, J.R.M., Kesley, A., Crook, B. & Gilbert, E.J. 2003. Occupational and Environmental Exposure to Bioaerosols from Composts and Potential Health Effects—A Critical Review of Published Data, Norwich.

Viegas, S., Faísca, V.M., Dias, H., Clérigo, A., Carolino, E. & Viegas, C. 2013a. Occupational Exposure to Poultry Dust and Effects on the Respiratory System in Workers Journal of Toxicology and Environmental Health, Part A: Current Issues, 76(4–5): 230–239.

Viegas, C., Viegas, S., Almeida-Silva, M., Veríssimo, C. & Sabino, R. 2013b. Environmental impact caused by fungal and particles contamination of Portuguese swine. WIT Transactions on Biomedicine and Health. 16: 11–24.

Viegas, S., Veiga, L., Verissimo, C., Sabino, R., Figueredo P., Almeida, A., Carolino, E. & Viegas, C. 2013c. Occupational exposure to aflatoxin B1: the case of poultry and swine production. World Mycotoxin Journal. 6(3):309–315.

Villavert, L., Nadal, M., Figueras, I. & Domingo, M. 2009. Baseline levels of bioaerosols and VOC's around a municipal waste incinerator prior to the construction of a mechanical – biological treatment plant. Waste Manag, 29(9):2454–61.

Vincent, J. & Mark, D. 1981.The basis of dust sampling in occupational hygiene: A critical review. Ann Occup Hyg. 24: 375–390.

Wijnand, E. & Bakke, B. 1999. Experiences with task-based exposure assessment in studies of farmers and tunnel workers. Norsk Epidemiologi 9 (1): 65–70.

World Health Organization 1999. Hazard prevention and control in the work environment: Airborne dust. WHO/SDE/OEH/ 99.14.

Wouters, IM., Spaan S., Douwes J., Doekes G. & Heederik, D. 2006. Overview of personal occupational exposure levels to inhalable dust, endotoxin, beta(1–N3) glucan and fungal extracellular polysaccharides in the waste management chain. Ann Occup Hyg 50: 39–53.

Occupational Safety and Hygiene II – Arezes et al. (eds)
© 2014 Taylor & Francis Group, London, ISBN 978-1-138-00144-2

Safety culture and labor accidents in the town councils in the District of Leiria

D. Chagas
Universidade de Léon, Departamento de Ciências Biomédicas, León, Espanhã

M. Dias-Teixeira
REQUIMTE, Instituto Superior de Engenharia do Politécnico do Porto, Porto, Portugal
CIEG—Centro de Investigação em Economia e Gestão, Universidade Lusófona de Humanidades e Tecnologias, Lisboa, Portugal
ULP—Universidade Lusófona do Porto, Porto, Portugal
ISLA—Instituto Superior de Línguas e Administração de Santarém, Santarém, Portugal

ABSTRACT: This study of safety's culture and work accidents intend to evaluate the procedures for the safety and health of the workers in the prevention of work accidents, concerning the local administration.

1 SAFETY CULTURE

The quality of life in the workplace, in particularly the facts that favored by safety and health conditions at work, contributes for the man's personal and professional accomplishment. These conditions are the basis for an efficient work performance and, therefore, they influence directly and positively the productivity, with consequent economic gain.

In the same way, they contribute in a determining reason, not only for the increase of the competitiveness, but also for the decrease of the local administration accident rate, constituting indispensable matter in any program of prevention of occupational risks.

It's this qualitative improvement of life at work environment that is designated "worthy work" (OIT, 2007).

Safety's culture is without a doubt an important approach for the safety's improvement in the organizations and prevention of accidents. There is a strong consensus that safety's culture plays a decisive role in the death rate and specifically for the existence of a safe atmosphere (Silva, 2008). The involvement of the employer and the workers' participation are essential to promote a safety culture in the workplace (OIT, 2010).

For Turner et al., safety's culture is a group of beliefs, norms, attitudes, social practices and techniques involved in the minimization of the workers' exposure to conditions considered dangerous or with the potential to cause lesions (Freitas, 2008).

In Singapore, the Department of Labor approved a new law of safety and health in the workplace (March 2006) that forces the employers to create a safety and health management system in the workplace, to identify and to manage the risks and to promote a prevention culture regarding safety and health (OIT, 2007).

It is important to create incentives and stimulus to motivate the workers' participation, in order to allow them to identify with the employed safety to the several risk situations that they are facing in the execution of the work (Rolo, 1999).

The safety in the workplace is a comprehensive way of prevention, in other words, it is a group of measures and applied actions to prevent the workers, of accidents and occupational diseases in the course of the daily routine. It integrates a group of appropriate methodologies into the prevention of work accidents, the recognition and the control of risks associated to the workplace and to the productive process (Fonseca, 2002).

For some authors the safety should be more than the use of personal protective equipment (PPE), verify possible risks with the equipment and to maintain the work area in order.

Although the promotion and the safety's prevention and health at work are contemplated in our labor legislation, many are the town councils that neglect it.

The local administration has to be conscience that, in addition to the need to comply with the legislation, the absence of a policy for safety and health at work and of means of prevention of accidents. These facts involve added costs regarding insurance aggravation, equipment damage and increase of the absenteeism taxes, among others.

The work accidents affect the worker in a temporary or permanent way, on a personal and professional level. When the lesion causes death, the consequences are even more serious. The typification of the accidents, according to the causes, has demonstrating excessively that the main cause is the human factor.

Off course the enforcement of the law can never be neglected but, to limit to that, is also a relative passive attitude.

Greater attention should be paid to the work conditions and to the workers' satisfaction levels, recognizing that, a Local Administration carries out not only an economical function but also an important social role.

2 OBJECTIVE

The aim is to evaluate the safety's culture and the work accidents in the Town Councils of the District of Leiria, regarding the procedures for the safety and health of workers, in the prevention of work accidents.

3 METHODOLOGY

To develop this work, a descriptive investigation was conducted, using the survey by survey method to all town councils in the district of Leiria. The survey is constituted with closed answers and it varies in a scale of "ns/nr", "yes" and "no", it is constituted by 11 questions and divided in 2 groups, with regard to safety's culture and the work accidents.

The questions of the survey that intend to evaluate the safety's culture in the town councils were contemplated to characterize the organization aspects.

Procedures for the safety and health of the workers: we intend to evaluate in what way and if the town councils adopt measures for the safety and health of the workers, specially the way that they prevent the risks. Here we evaluate the existence of personal protective equipment, information regarding the tasks, where they are used, education in the usage of the equipments and if the safety and health measures are achieved.

Knowledge of the risks: we intend to evaluate the knowledge that the workers have about the risks in their workplace, prevention and safety measures on first aid, in case of an accident.

Information about risks and dangerous substances: we intend to evaluate the information that the workers receive regarding dangerous substances that they work with and all the risks that are associated with that task.

To evaluate the work accidents and how to characterize them better, we divided this section in: accidents in the workplace, professional categories, legal work relations, nature of the lesions, area of the body that was affected, consequences of the accident and the number of workdays of absence.

The information received will permit the characterization of the type of worker as well as the consequences regarding the accident.

Later an evaluation of work accidents per injured worker was performed. To characterize the accidents a form was elaborated. The intention of this form was to obtain, not only information of individual character, but also to collect additional data that allowed a detailed characterization of the accident, as well as the costs that were associated to it.

The form is divided in three groups.

Identification of the worker that suffered the accident: in this group we ask about the gender, age, literary qualifications, salary and if that was his routine job during the accident.

Data of the accident: in this group of questions we intend to evaluate the consequences of the accident, the participant of the accident, the place where the accident took place, as well as the time shift of the worker.

Costs concerning the accident: in this group of questions we intend to quantify the direct costs (salaries, health and medical costs and compensations) and indirect cost, like the time wasted by the workers and his fellow colleagues, administrative costs (including the analysis of the accident), costs regarding the repairs of the equipment, costs with substitute workers and extra hours, products and lawyers. The general cost are quantified immediately with the cost of the accident, like first aid equipment, transportations of the worker involved in the accident, penalties and fines, compensations regarding a third party, and other costs.

The study was conducted from information regarding work accidents from 2005 to 2009.

The workers' social-demographic data, such as the gender, age and literary qualifications were obtained through the Town Councils by a form per worker. The collecting of these data was made in the sense of characterizing the sample as individual factors in the study of the human resources.

The cross sectional descriptive study was accomplished in August of 2010, to the Town Councils, with a sampling of 148 inquired.

The adopted research methodology intends to clarify safety's culture and its practice, in the Town Councils, as well as, the work accidents.

4 RESULTS

The results demonstrate that, in medium terms, the Town Councils dispose of personal protective equipment (PPE) for their workers.

Most of the Town Councils (76.1%), also have equipment and protection clothing.

It was proven that only 34.8% of the Town Councils give appropriate education to the workers for them to execute their activity in safety and all the information in the use of personal protective equipment (PPE) and in the tasks that the equipment is used.

In the sample used (62.0%), it was verified that only 26.1% of the Town Councils implement the safety, hygiene and health services at work.

Another pertinent aspect is the year of implementation of these same services. It was verified that it was in the years of 2008 and 2009 that was registered a larger number, 6.5% in both in witch the Town Councils implement the safety, hygiene and health services at work

Of the obtained results it was proven that 60.9% of the workers have knowledge of their health and safety risks in his/her workplace. 65.2% have notion of the necessary measures regarding prevention and protection in his/her workplace and 55.4% have knowledge of first aid measures in case of accident.

As for the information given to the workers on the dangerous substances that they handle or with which they can enter in contact, only 45.7% of the Town Councils do it, on the other side 63.0% give information on the risks that are associated with dangerous substances.

The obtained data indicate that the work accidents happened in larger number in the year of 2009 (9.9%) and they are centered, largely, in the operational assistant professional category (caretakers) (8.8%).

In concern of the worker's juridical relationship, it was verified that 9.9% are effective/permanent workers, 7,7% have fixed-term contract and 1.1% have a contract of indefinite period.

As for the most frequent accident, it was verified that injuries and lesions were most comon (3.3%), as well as dislocations and sprains (3.3%) and the ones regarding tasks that demand the application of excessive effort, overload and over-exertion (2.2%).

It was verified that the anatomic area most affected, after the work accident, was the thorax/ abdomen 6.6%, followed by hands (4.4%) and arms (2.2%).

Of the results obtained relatively to the severity of the accidents, it was verified that 16.4% are quick injuries, that had as origin in cuts, crushes, fractures, lesions, on-load and over-exertion and dislocations/sprains.

1.1% had permanent injuries having as its caus lesions. The serious accidents, but not mortal, (1.1%), resulted of overload and over-exertion. These lesions were located mainly in the superior members.

It was in the year of 2009 that the workers (10.9%) missed more days at work, due to accidents.

The largest number of work accidents happened with men workers (64.7%) comparatively with the one of the opposite gender (35.3%), in spite of the number of workers studied were women (58.8%).

It was equally verified that most of the accidents happened in the 40 year-old age group. Most of the accidents took place in workers with the 4th and the 9th grade, in a total of 35.3% pointing that the low level of qualifications can contribute to the increase the gravity of the accident, fact, that also can be associate to the type of activity that the workers carry out.

The average of the monthly wage of the injured worker is 600 € and 88.2% of them were doing a routine work at the moment of the accident.

Of the total of the accidents, 94.1% resulted in temporary incapacity and 5.9% in permanent incapacity. In our study, the workers were absent a total of 801 days, what gives an average of 2.1 day of work lost by accident.

Of all the accidents the participant was the employer and it took place during the normal schedule period. Most of the accidents happened in the exterior (82.3%).

The global amount of the indirect costs assumed by the Town Councils, in the period in study, from 2005 to 2009, are 1 935.94 €, while, the amount of the direct costs are 34 330.33 €.

The amount of the general costs, or immediate costs, are 1 608.50 €.

As for the human resources, it was verified that the largest number of workers are women, 58.8% against the 41.2% of men.

The workers' medium age is 31 to 50 years.

As for the level of qualifications, it was verified that the majority possessed the 4th grade of education (33.9%).

5 DISCUSSING THE RESULTS

The analysis of the results obtained points that the Town Councils give personal protective equipment to their employees. In relation to the information/ education that the workers receive regarding the use of the equipments, so that they exercise their activity safely, points to the answer "no". The educational factor is fundamental, because all the workers should know the potentialities, the limitations and the correct method of use (Freitas, 2008).

The execution of the services of safety and health at workplace in the Town Councils is centered in the answer "no". *Regarding the Law Lei nº 102/2009 de 10 de Setembro*, all workers are obliged to organize services of safety, hygiene and health services at work, in order to involve all the workers that are responsible for those services (República A, 2009). In reference to the knowledge that works have about the risks in their workplace and their health the answers point to a "yes".

As for the work accidents, the majority is centered in the minor injuries and had consequent reversible lesions.

The shortage of information associated with the work accidents that happened in the period between 2005 and 2009, that constitute the sample, didn't allow to acknowledge some of the pertinent information, in the extent of the proposed objectives. This was due, on one side, to the fact that not all the Town Councils possessed records of those occurrences for the whole established period, on the other hand, it was verified that of the accident victims the only approach made by the Town Councils was related with a mere formal execution of the law, through the fulfillment of the participation of the work accident to the insurance company, *art.º 15 of DL n.º 143/99, of 30/04* (Ministers C, 1999).

The factors that contributed significantly to accidents with sick leave, influencing the average of lost days, were: the ages between 40 and 45 years; the low level of qualifications; carrying out task in fixed schedule; the accidents concerning injuries and lesions; dislocations/sprains; falls and cuts. Additionally, this contributed to the statistics of work accidents that happened in Portugal in the year of 2009, in which the injured worked with hand tools (27.0%) or exercised manual transport (25.1%) (GEP, 2012).

The costs with the accidents ascended 37 874.77 €. The consequences of the work accidents regarding costs for the company are just a minor factor. The social impact and the loss of human lives overcome anything else (Carnero, 2010).

The costs of the work accidents have a very significant weight, either relatively to the case in study, either for inference, to the national reality. Such fact has economic repercussions, affecting not only the national wealth, as well as the society in general.

Through the observation of the work activities it was verified that the workers, carrying out the same functions, have the same behavior, associated with the risks that they are exposed to, if it is not solved, could be, hereafter, the cause of serious accidents at work.

6 CONCLUSION

The main problem that was established for this work was to discover if the way the safety's culture is implemented in the Town Councils corresponds to the workers' needs.

After the analysis of the results it can be concluded that:

1. In general, the Town Councils have at the workers disposal personal protective equipment. However, the employer should strengthen the consciousness of the worker, informing him on his/her of the purpose and the correct way of use of the equipment.
2. The Town Councils don't promote education/information actions adapted to the activities developed by their employees.
3. It is necessary to create of rules and procedures regarding safety measures, in attempt to avoid work accidents and, consequently, to increase the worker's motivation.
4. The work accidents in the Town Council have a significant contribute, given the gravity of their consequences. Most of the accidents happened with men, with low academic qualifications and in the 40 year-old age group. The variable "professional service/experience years" seems to indicate that it is difficult for to accept the security rules, because the concepts learned in their basic education are today obsolete and the change of behavior doesn't happen easily.
5. The incidence of the lesion per anatomic area is: crush of the hands, dislocations/sprains of the superior and inferior members and of the thorax/abdomen including thorax/abdomen fractures.
6. The cost of the work accidents, calculated regarding the daily salary of each professional group, ascended the total of 37 874.77 €.

In general, the Town Councils present an absence of favorable practices that characterize a culture of positive safety. This verification occurs of the fact of the existing consistency between the observed critical aspects and the measures identified by the inquiries. However, and in spite of other imperatives that speak higher, the implementation of these "good practices" are neglected and postponed on a daily basis.

In a word, safety's culture understands behavior, compromising, cooperation, capacity, investments, maintenance, inspection, in short, a series of factors that depend on continuous actions.

BIBLIOGRAPHY

AEP. 2004. *Manual formação pme – Higiene e Segurança no Trabalho*. Associação Empresarial de Portugal.

Associação NDST. 1992. *Acção de Formação. Higiene, Segurança e Saúde no Trabalho*. Associação Nacional dos Deficientes Sinistrados no Trabalho.
Carnero, MC, et al. 2010. *Modelling and forecasting occupational accidents of different severity levels in Spain*. Reliability Engineering & Sistem Safety, 95, 1134–41.
Fonseca, A. 2002. *Higiene e Segurança no Trabalho*. AEP – Associação Empresarial de Portugal.
Freitas, LC. 2008. *Segurança e Saúde do Trabalho*. Lisboa: Sílabo.
GEPP, MSSS. 2012. *Colecção estatística – Acidentes de Trabalho*. Gabinete de Estratégia e Planeamento e Ministério da Solidariedade e da Segurança Social.
Ministros, C. 1999. *Decreto-Lei n.º 143 – Reparação dos danos emergentes de Trabalho*. Diário da República, 101, 2323–32.
OIT. 2007. *Locais de Trabalho Seguros e Saudáveis. Tornar o Trabalho Digno uma Realidade*. Bureau Internacional do Trabalho (BIT).
OIT. 2010. *Riscos emergentes e novas formas de prevenção num mundo de trabalho em mudança*. Autoridade para as Condições do Trabalho.
República, A. 2009. *Lei n.º 102 – Regime jurídico da promoção e prevenção da segurança e saúde no trabalho*. Diário da República, 176, 6167–92.
Rolo, JC. 1999. *Sociologia da Saúde e da Segurança no Trabalho*. SLE – Electricidade do Sul.
Silva, SCA. 2008. *Culturas de segurança e Prevenção de Acidentes numa Aborgagem Psicossocial: Valores Organizacionais Declarados e em uso*. Fundação Calouste Gulbenkian – Fundação para a Ciência e a Tecnologia.

Occupational Safety and Hygiene II – Arezes et al. (eds)
© 2014 Taylor & Francis Group, London, ISBN 978-1-138-00144-2

A study regarding the ergonomic conditions in an area of winding transformers

G.N. Bolzan, G.S. Freitas & L.A.S. Franz
Federal University of Pelotas, Rio Grande do Sul, Brazil

ABSTRACT: The search of the improvement in health and safety of the worker gains more and more space in the last years, although it is still a challenging for employers and safety experts. This point of view repeat itself in different areas in which can be cited the case of the production and maintenance of electrical equipment. This work aims to investigate which are the main ergonomic demands in responsible work stations by winding in a firm that manufactures and reconditions the transformers, located in the city of Lagoa Vermelha, in the state of Rio Grande do Sul, Brazil. For this, it was used techniques of ergonomic analysis including application of a semi structured questionnaire and postural analysis by the method RULA. It was observed that the environmental factors do not influence in the operator performance, being more critical those factors associated to posture and rhythm of work. The method RULA showed the urgent need of intervention with aims to improvement in the work stations covered by this study.

1 INTRODUCTION

Some parameters are considered essential to a better quality of life as an access to potable water, to electrical energy and to the basic sanitation (BRASIL, 1989). In the particular case of electrical energy, it has whole a transmission system to that arrives at establishments and houses with the best efficiently as possible. In this distribution system, in several points, found equipment called transformer, important in the conversion of electrical potential, what imply directly in the obtainment of different levels of electrical tension and electrical current. Smaller currents and bigger tensions imply in cables of smaller diameter and vice versa. In this sense, and with an adequate designing, the transformers can provide expressive financial gains by means of the adequate use of material resources. The transformers can be considered electrical equipment of low complexity which works basically through the use of coils (a lot of times isolated by specific oil), connections and iron core.

During the manufacture of transformers, there is an operation known as winding, frequently performed using places and procedures with high potential for occurrence of ergonomic demands of different types. According to Iida (2005), one of the critical and challenge aspects in terms of ergonomic demands refers to necessary postures to the performing activities in the work stations, when observed the interaction man versus machine.

A lot of occupational diseases can be related to the worker posture in the work station, which belong a group of diseases known as WMSD's (Work-Related Musculoskeletal Disorders) and that are a great concern among OHS (Occupational Health and Safety) experts. This group of diseases has strong association to wrong postures, excessive and inadequate forces and other factors of risk that can trigger these disorders (CARDOSO, 2006). Beyond the postures, there are other factors that can affect the worker health, as environmental factors of physical and chemical nature, as noise, vibration, lighting, thermal comfort and aggressive substances (COUTINHO *et al.*, 2010).

Therefore, the present work aims to investigate which are the main ergonomic demands in work station responsible by winding in a firm that manufacture and reconditions transformers, located in the city of Lagoa Vermelha, in the state of Rio Grande do Sul, Brazil.

2 ERGONOMICS AND THE PRODUCTION OF TRANSFORMERS

The specific literature points several factors that can bring ergonomic risks to worker and specially the occurrence of WMSD's. Among the main present factors in the activities of workers which trigger the lesions or sensations of discomfort, there is the inadequate posture, need of force application, rhythm of movement, repeatability, duration, recovery time, heavy dynamic effort and localized vibration. These conditions associated to characteristics as heat, cold, noise and luminance and additional factors as stress, cognitive demand, organization of work and work load potentiate the occurrence of WMSD's.

2.1 *Ergonomics and the work stations*

Observing a work station in the scope of ergonomic conditions, we can decrease it to a basic analysis made up by man, by machine and by environmental (IIDA, 2005). From the viewpoint of the man, it is observed among other aspects, the postures and the manner of movement, in a way that the maintenance of these by an extended time can bring losses to health of the worker and also the loss of his productivity. According to Oliveira *et al.*, (2011), during the diary journey, a worker can assume hundreds of postures, even existing basically three postures that the human body takes, working as much as resting: lying, sitting down and standing up.

In the event of the same posture be kept for an extended period, or in the case of overly repeated movement, in both can occur damage to worker. Some important damage in these cases is the tendinitis, the bursitis, and the fatigue (OLIVEIRA *et al.* 2011). Tendinitis result of inflammations from micro lesions of tendinous tissue, in the same way bursitis is inflammations in the bursa, which is tiny fluid-filled sac that wrap the articulations, performing the function of shock absorber between bones, tendons and muscle tissues (OLIVEIRA *et al.* 2011; VARELLA, 2010).

The fatigue, in turn, demonstrates itself according to three factors: physiological factors, psychological and environmental and social factors (IIDA, 2005). Physiological factors are related to intensity and duration of physical and mental work. To psychological, the relation exist with a monotony, lack of motivation and finally the environmental and social factors that are related to lighting, noise, temperature and behavior with superiors and colleagues.

Other damage that deserves attention by its incidence is the low back pain that is characterized as a pain which vary between sudden, intense and extended in the region of low back of vertebral column. Factors that can be the generator of low back pain gain more highlight as the repetitive work, actions of push and pull, falls, maintaining of postures for extended periods sitting down or standing up (PERES, 2005; ULIAN *et al.*, 2005).

In the context of the environment and machine, elements that also forms the basic unit of analysis, several aspects can step in the work station, among them, the physical arrangement or layout, which treats how the various tools and equipment are positioned and the ones that are used by the worker. Factors like lighting, noise and temperature can also be embraced, depending on the context in which the analysis is performed. According Iida (2005), a correctly design of lighting and colors helps to increase the satisfaction in the work and improve the productivity, while it reduces the fatigue and accidents.

The thermal comfort, by its part, if disregarded, can produce in the individual fatigue, physical fatigue, decrease of performance, increased risk of accidents and errors at work and even exposes the worker's body to various diseases. According to Iida (2005), in temperatures above 24°C, workers tend to feel torpor, when below 18°C individuals involved with sedentary jobs or with low physical activity begin to feel tremors.

The noise also configures itself as an important element to be considered in the respect to environment, seeing that the discomfort resulting of undesirable exposition levels can generate hearing loss, reaching the point of deafening the individual, as well as change the mental performance negatively, it can also cause physiological damaging reactions to the body, having some consequences as the increase of stress and fatigue (KROEMER & GRANDJEAN, 2005). In Brazil, the norm NR15 (BRASIL, 2005) establish the maximum levels of noise that allow the minimum comfort to occupants of a place. Above the limits of noise recommended; more than disturb any human activity, noise can, during the not extended period of time, cause serious damage to hearing (ROSSI *et al.* 2011).

2.2 *Distribution of power and the importance of the using of transformers*

The Brazilian electricity sector actually is divided in segments such as generation, transmission, distribution and marketing. A large part of the power generated in Brazil occur in hydroelectric power plants (74,9%), and the rest is divided between wind energy (0,4%), natural gas (5,8%), biomass (5,5)%, importation (6,3%), coal and derived (1,3%), nuclear (2,6%) and petroleum based fuels (3,1%). Immediately after its generation, the energy must to be provided to consumers through special electrical system, involving high and low voltages with direct (DC) or alternates (AC) current, in order to achieve the using of electricity-conducting materials in the most economically manner (LEÃO, 2011).

The distribution grid is fundamental to make the connection between the generation plants and unities of high potential. After going through this initial stage, the energy follows to intermediary transmission grids with lowest electric potential; it arrives finally to the residential consumers, commercial establishments, services and industrial consumers of medium and small sized (LEÃO, 2011). Across many points of this system, near the final user and where there is changes of electrical power, there are the transformer model that has part of its process of manufacturing covered in the present study.

The transformer is basically static electrical equipment that has the function of transport electric power using electromagnetic induction. In it, the value of tension and the current are alternated, but the power and frequency remain unchanged, considering the case of an ideal transformer (NEVES & MÜNCHOW, 2010).

3 MATERIALS AND METHODS

This study was developed in the period of January to March of 2013. Previously to this period, the necessary contacts were performed to scheduling of field activities and initial understanding of the possible challenges to be overcome in achieving them. In this period, visits to the object local of study were routed and performed the necessary data collection.

3.1 *Object of study*

The firm object of the present study manufactures and performs the maintenance in transformers genus 15 or 25 used typically to transformation of potentials of tension in the range between 127 and 23100 Volts, contemplating equipment to capacities until 1500 kVA.

The firm had, in the epoch of the study, 50 workers allocated in some separate departments according to the performed tasks. Physically, the firm is divided in several sectors, having an area built in 4000 m^2 where deliveries in medium 500 transformers per month, from manufacture or maintenance. In the sector of winding, where it is found the work station embraced in the present study, it is performed the making of coils. This sector is responsible by the production of coils which will be used in the transformers, acting in the station six workers, all of them female, being one responsible for manage the same, and the others five for the manufacture of coils.

It was analyzed just the productive process of coils HT (high tension), because this was a process that could be followed. This process had its steps divided in phases, being them separated by similar movements, in this way is possible have a better result in the application of the method RULA.

3.2 *Instruments of surveying and analyzing*

To the performing of surveying in fields was used equipment to the registration of images, audio and videos, as well as a semi structure questionnaire having 19 questions. The register of images and videos was used to analysis of postures and rhythm of work. In turn, the questionnaire and register of audio aimed to bring support to the comprehension the perception of the workers and the managers face the demands of the work imposed in their routine work.

The method RULA (Rapid Upper Limber Assessment) (McAtamney & Corlett, 1993) was used by the characteristic of the analyzed work station and by its low complexity of application. This method considers the postures according to the relative angles between the limbs and trunk, obtaining scores that define the level of action to be followed. The evaluation that follows the method RULA supplies a score ranging incrementally from the value 1 and that distribute itself in four levels of criticality, being the levels with lower scores which one has acceptable postures and the level with more scores that one which demands urgent alterations. To get these scores, the observed angles in the limbs are divided in groups and classified in charts. The Chart A is considered the superior limbs and the Chart B the neck, trunk, legs and feet. Data obtained from these charts are used in a third chart (Chart C), where it is obtained the final score, which determinates the criticality of the posture by analysis.

3.3 *Conduction of analysis*

The study of the obtained data in field was accomplished through analysis of contents, in the case of obtained answers by questionnaire and, through the statistical analysis in the case of numeral data. More specifically, in the case of method RULA, it was used the insertion and primary treatment of data in full electronic worksheet with specially fields developed for the referred analysis tool.

4 RESULTS

The process of winding covered by this study is characterized for to be predominantly manually, and the firm manager related that it is an important font of worries associated at the health problems and the wellbeing of workers. During the early dialogues with the manager, it verifies that the firm acknowledged that the sector has critical aspects in relation to discomfort, mainly involving inadequate postures. It was found the occurrence of pain reports by workers, mainly in lumbar region, shoulders and wrists. Also there was the relocation of a collaborator who had problems involving pain and discomfort. These facts signaled that the sector is a priority place to actions aiming ergonomic improvements.

When analyzing the task, it was verified that winding is performed in a controlled temperature room using air conditioner system and with artificial lighting provided by fluorescent lamps. In this place there are the whole equipment and materials used daily by workers from that place. The raw material used is delivered to the transformers setting up station, being produced two types of high tension (HT) coils and low tension (LT) coils.

To the production of the coils, eleven steps must to be followed and also the step of preparation of materials and forms, done before or during the process of winding and that involves preparation and separation of roles, forms, threads and specifically accessories to each coil, which will be confectioned.

The coils HT production, the focused process of this study, is synthetically described steps in the Chart 1. However, during the analysis of the activity, it was verified that the previewed phases in the task could be divided in four steps (useful in applying the method RULA), taking into account the similar movements occurred during the execution of work and as showed in the Chart 1.

From the records of images and videos, it was filled up the electronic worksheet, taking into account the left and right side of the operator, considering the four existent phases in the work station and its frequency. After the fulfilling, it was possible establish some inferences about the under review activity.

It was found, for example, that the operator dedicated a large part of his time of effective execution of labor activities to the phases 2 and 3 (approximately 78%). This aspect can reveals itself as critical, because, if the operator stays in activity in his work station for approximately 6 hours, thereby, more than 4 hours of his workday will be dedicated exclusively to the execution of the coils. In the Figure 1, it is showed an image with two photographic registers of activities apprehended in these phases.

As aggravating, it was found that is exactly in the phases 2 and 3 that the higher scores are obtained, during the application of the method RULA, as considering the left side as its right side, as can be observed by the graph of Figure 2.

When comparing the analysis contemplating the left and right sides of the operator, perceives a slight difference in terms of levels of criticality identified. At the same time it is considered the left side, in which 75% of the time, the operator finds itself under scores 5, 6 and 7, to the left side this percentage increase to 79%. However, the difference became irrelevant if we consider that a score 5 already points to the need of immediate changes in the workplace.

Chart 1. Steps of winding process.

Phase	Steps	Name of the step	Steps's description
1	1	Cylinder	Placing the cylinder role.
	2	Winding Direction	Checking the direction of the winding, RIGHT or LEFT
	3	Production of the Starting Terminal	Preparing the yarn, isolate and verify the terminal, trapping to the intern cylinder
2 e 3	4	Production of the First Spire	Placement and gluing of the first isolation layer
	5	First Layer Winding	Reeling all the first layer checking the number of coils per layer with the counter machine assistance
	6	Winding	Continuing the winding placing the isolation between layers
	7	Derivations Production	In case of derivations, stop in the determined coil and perform the procedure of derivation
4	8	Gap's Production	In case of having a gap (a spacer elements of coils), stop the winding and product the gap as specified
	9	Last Layer Bracing	At the last layer, place the lace to bracing
	10	Making the Final of the Coil	Placing the insulation tube of the last spire, cut the yarn, passing the terminal by the lace and tightening secure the terminal
	11	Inspect the Coils	To verify specific control items for all produced coils and to make the register

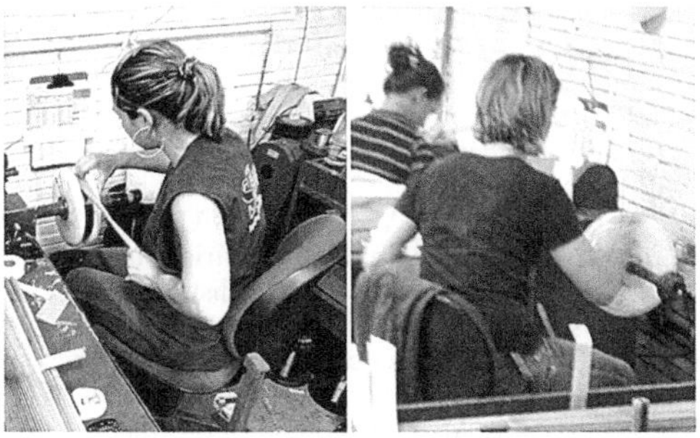

Figure 1. Workers executing actions at Phases 2 and 3, under analysis.

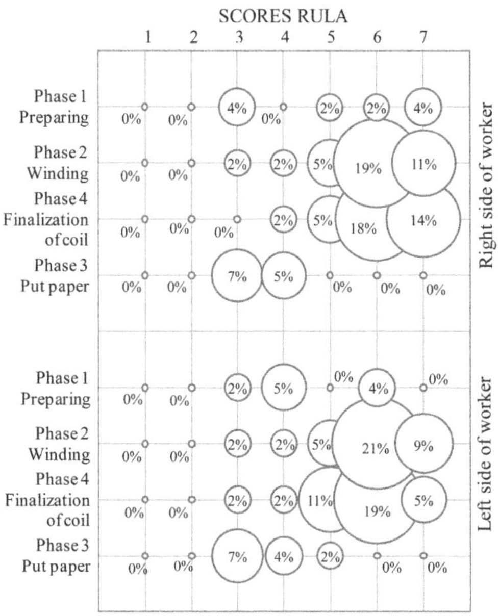

Figure 2. Distribution of scores according to analysis by method RULA.

A interesting fact and relevant of being highlighted is that there wasn't a occurrence during the analysis of postures using the method RULA, where the score was between 1 and 2 (posture is acceptable if do not kept or repeated for extended periods), which serves as other indicative of need of an action aiming immediate improves at the place.

From the semi structured questionnaire, it was observed that environmental factors as lighting and thermal comfort are not cited as problems. However, aspects related to noise and organization of the layout of the work station, emerge how a factor that could receive improves, as well as the needed of development of solutions to transport the finished coils. As the operator remains sitting on the most part of his workday, it was questioned about his perception in relation to the chair and the equipment used in the work station. Relatively to this issue, workers pointed that the chair is not uncomfortable. Their complaints refer more specifically to the required position in the activity, which demands a special seatback. Also, because of the kind of processes, the chairs can't to be have armrests, resulting in muscular fatigue and pains in the end of the workday. Concerning at the distribution and manipulation of material in the workstation, it was not verified problems under the perception of the workers. They reported that the distances are comfortable and there is space for distributed the material and tools according the manner that can judge it convenient. The vibration on the equipment and of the workstation during the winding of the coils is not perceived as relevant by the operators in terms of discomfort. The relevant physical discomforts realized by operators are located in the region of cervical and lumbar spinal, occurring also the existence of discomforts in some of the operators in their shoulders, legs, arms and forearms.

5 CONCLUSIONS

The present study revealed that winding (phase 2) and a phase of finalization of coils (phase 3) are the most critical activities in terms of ergonomic demands during the fabrication of coils of transformers. It was found from that the workers who act in these stations are subjected to high risks of musculoskeletal lesions, not just because of the postures, but also by the higher extended period dedicated to the activity in the work stations during the workday. In the context of environmental, it was found that these are not the main cause of workers' discomfort or dissatisfaction. However, the reports points to the need of studies of improvement in the layout of stations, including furniture.

Combining the results obtained with the method RULA and the semi structured questionnaire is possible to infer that improvements in the workstation are urgent. The results of the two survey tools (method RULA and questionnaire) are complementary and suggest that is necessary to consider rhythm of work, physical characteristics of workstation and postures.

REFERENCES

BRASIL. (1989). Lei n. 7.783, de 28 de junho de 1989. Dispõe sobre o exercício do direito de greve, define as atividades essenciais, regula o atendimento das necessidades inadiáveis da comunidade, e dá outras

providências. Retrieved: July, 17 2013 from <http://www.planalto.gov.br/ccivil_03/leis/l7783.htm>
BRIGANÓ, J.U.; MACEDO, C.S.G. 2005. Análise da mobilidade lombar e influência da terapia manual e cinesioterapia na lombalgia. Semana: Ciências Biológicas e da Saúde, Londrina, 26(2).
CARDOSO, M.M. (2006). Avaliação Ergonômica: Revisão dos Métodos para Avaliação Postural. Produção on line, 6(3).
COUTINHO, J.P.; SILVA, A.T.C.; ARAÚJO, I.F. 2010. Análise das demandas ergonômicas em uma microempresa têxtil localizada na cidade de Campina Grande/PB. Bauru: XVII Simpósio de Engenharia de Produção.
IIDA, I. 2005. Ergonomia: Projeto e Produção (2th ed.). São Paulo: Edgard Blücher Pubisher.
KROEMER, K.; GRANDJEAN, E. 2005. Manual de Ergonomia: Adaptando o Trabalho ao Homem (5th ed.). Porto Alegre: Bookman Publisher.
LEAO, R.P.S. 2009. GTD – Geração, Transmissão e Distribuição de Energia Elétrica (apostila). Fortaleza: Federal University of Ceará;
McATAMNEY, L.; CORLETT, E. 1993. RULA: A Survey Method For Investigation Of Work-Related Upper Limb Disorders. Applied Ergonomics, 24(2), 91–99.
NEVES, E.G.C.; MÜNCHOW, R. 2010. Eletrotécnica (Cadernos didáticos). Pelotas: Universidade Federeal de Pelotas Publisher.
OLIVEIRA, W.A.; SILVA, J.M.N.; BATISTA, D.M.B. 2011. Análise ergonômica de operários em um posto de trabalho de uma indústria calçadista. Bauru: XVIII Simpósio de Engenharia de Produção.
PERES, C.C. 2005. Ações coletivas para a prevenção de LER/DORT. Porto Alegre: Boletim da Saúde. 19(1).
REIS, E.S. 2001. Análise ergonômica do trabalho associada à cinesioterapia de pausa como medidas preventivas às LER/DORT em um abatedouro de aves. Dissertation, Production Engineering Department. Florianópolis: Federal University of Santa Catarina, Brazil.
ROSSI, M.A.; FRANCO, D.; MANTELATO, B. 2011. Ação ergonômica aplicada em posto de trabalho na fabricação de roupas íntimas. Bauru: XVIII Simpósio de Engenharia de Produção.

Occupational Safety and Hygiene II – Arezes et al. (eds)
© 2014 Taylor & Francis Group, London, ISBN 978-1-138-00144-2

Respiratory function study in a group of car repair workers

S. Correia, A. Ferreira, J.P. Figueiredo & J. Conde
Instituto Politécnico de Coimbra, ESTeSC, Coimbra Health School, Saúde Ambiental, Portugal

ABSTRACT: Concern about air pollution is not new, but recently the issue about quality of indoor air has been emerged. It is important to approach this subject, since humans usually spend more time inside buildings than in exterior environments. Therefore, this research intends to study the respiratory function of the mechanics workers in the district of Coimbra. It was used different methodologies for data collection: equipment to measure the various parameters of indoor air quality, the pulmonary physiology examination was made by Spirometry and Oscillometry Impulse, a checklist and a questionnaire. Environmental monitoring, there was no presence of alarming values, except O_3 which showed to be no risk to the health workers. However, there was a clinic trend, that most of the individuals with ventilatory disorders (80.0%) were exposed to concentrations of the pollutant above the legislated.

1 INTRODUCTION

In developing countries smoking habits, occupational exposure to dusts or chemicals and air pollution indoors, are the most significant risk factors for progression of chronic obstructive pulmonary disease (COPD) (Hart et al, 2006). The World Health Organization considered indoor pollution as among the biggest threats to the public health since studies have revealed that the levels of pollutants in the indoor air can exceeded 2 to 5 times, and occasionally more than 100 times the levels of pollutants in the atmospheric air, this occurrence could be explained by the higher probability of microorganism development, as well as the absence of air exchanger ventilation system. Taking into account that people spend most of their days in indoor environments, it is fundamental to control the building projects with ventilation systems and other equipment to protect public health (Filho, 2008).

Despite the knowledge about the atmospheric pollution and its association with increased mortality and hospitalizations of patients with COPD, there is not much accurate information about the impact of pollution indoors, particularly in the auto repair industry (Park et al, 2002). Considering that fossil fuels are leading source of pollutants and toxic chemicals, it is very relevant to study the relation of the indoor air quality in auto repair workshops and what adverse effects it could bring to the health of workers. Reports state that chemicals are the most extensive and complex group of risk factors from professional nature, there are over 3000 compounds, among the around 100 000 substances used, which have an allergenic action and/or irritation to the eyes, skin, and upper airways. Specifically in industrialized countries there is a prevalence increase of the allergic disorders, attributed both to the new lifestyle adopted, as to the existence of new environmental and occupational risk factors (Gioda & Neto, 2003).

The study subjects are characterized by carshop workers subjected to a large amount of pollutants that have the main origin in fossil fuels. Their respiratory system is very exposed to the pollutants harmful effects, but also highly vulnerable to the chemicals used into the automobile maintenance and repair. Therefore, it is of great academic interest to study and analyse the respiratory function of those workers.

To sum up, the aim of this research is to demonstrate that the pollutant concentrations measured above legal values, are responsible for changes in respiratory physiology, evaluated through the Spirometry and Oscillometry Impulse (IOS).

2 MATERIAL AND METHODS

This was a descriptive-correlational and cross-sectional study using a convenience sampling procedure that involved the selection the most accessible workshops of approximately 700 workshops existing in the district of Coimbra. A total of 10 car workshops were selected studying the respiratory function of 97 workers. These workshops performed maintenance and repair works of motor vehicles. Although some workshops also have a heavy vehicles repair section, which were not taken into account in the study.

Data collection was obtained through four assessment instruments: the Respiratory Function Study

(RFS); the air quality monitoring; an inquiry to the workers; and a checklist to the workshop facilities.

The RFS was executed by Spirometry and IOS. Both exams were made to each employee, throughout a normal working day, through the equipment *Jaeger* Brand, *MasterScreen IOS* Model. Analytical data obtained were subjected to treatment using Software *PET—IOS version 4.24*, later being treated statistically by the *IBM-SPSS version 20*.

The variables studied by spirometry used: Forced Vital Capacity (FVC), Forced Expiratory Volume expired in the first second (FEV_1), Tiffenau Index (TI), which is defined as the FEV_1 over the FVC (FEV_1/FVC), Peak Expiratory Flow reached during a loop (PEF), Forced Expiratory Flow at 50% of the FVC ($FEF_{50\%}$), the mean flow between the points $FEF_{25\%}$ and $FEF_{75\%}$ is also a very important parameter called the $FEF_{25-75\%}$. The IOS parameters used: the resistance measured at 5 Hz (R_5), the resistance measured at 20 Hz (R_{20}), the reactance measured at 5 Hz (X_5), and resonant frequency (Fres), which is the frequency where the reactance is zero. The reference values for normative respiratory parameters mentioned could be visualized on Table 1.

Each worker filled out an inquiry with variables such as, smoking habits (or pre-existing diseases that could divert the investigation), years of service, symptoms, and the use of PPE (Personal Protective Equipment). Comparisons and associations were determined using Chi-square independence test; Odd Ration; One sample T-test; and Independent-samples T-test.

The respiratory function study data of workers were crossed with the results from air quality monitoring of the workshops, which was conducted in a parallel investigation.

At each workshop, the environmental measurements were repeated four times in each sector (concessionaire, mechanical, collision and painting) for the period of a normal working day, which correspond the following periods: early morning (8:30 am–10:30 am), late morning (10:30 am–12:00 pm), early afternoon (1:30 pm–3:30 pm) and late afternoon (3:30 pm–5:00 pm). Accordingly 64 measurements were made per day, in ten distinct days, one in each car workshop studied.

Table 1. Normative reference values for the assessment of respiratory function of workers.

Spirometry[1]		IOS[2]	
Parameters	Lower limits	Parameters	Lower limits
FVC	80%	R_5	150%
FEV_1	80%	R_{20}	150%
TI	80%	X_5	$X_{5Teórico} - 0{,}2$
PEF	80%	FRES	>10 Hz e <14 Hz
$FEF_{25-75\%}$	70%		

[1]Adapted from *Rodrigues et al, (2002)*;
[2]Adapted from *Valle, (2002)*.

In the collection of environmental data was used a monitoring portable Enviromental Monitoring Instrument (EVM), *Quest Technologies 3M Company* Brand, *R.10V* Model with a laser photometer that performs in real-time, measures and help to identify the pollutants present in the air. The equipment was used to monitoring the gases and the dust. Environmental analytical data was processed using the Software *QuestSuite Professional II*, and subsequently treated statistically by the *IBM-SPSS program version 21*.

It was considered as a reference for the statistical analysis the Threshold Limit Value—Time Weighted Average (TLV-TWA) - average exposure on the basis of a 8h/day, 40h/week work schedule, which practically all workers could be exposed day after day without adverse health effects. Portuguese Standards (NP 1796:2007) published constants for Ozone (O_3), Carbon Monoxide (CO), Nitrogen Dioxide (NO_2), Sulfur Dioxide (SO_2), Carbon dioxide (CO_2) and Total Dust (Table 3). About this last pollutant, despite not being establish a TLV-TWA, the American Conference of Governmental Industrial Hygienists (ACGIH) believes that even particles biologically inert, insoluble or poorly soluble, can lead to adverse effects, therefore, it recommends, that their concentrations must be kept below 10 mg/m^3 (10000 µg/m^3) for inhalable particles in the air.

Regarding the physical and structural facilities, it was applied to each workshop a checklist that included the following issues: workshop location, number of workers in the sectors assessed, general installation conditions (lining materials, heating systems, air renewal systems, type of ventilation, presence of no smoking sign, use of PPE and the existence of safety and hygiene services.

3 RESULTS AND DISCUSSION

Characterizing the sample: 97 male individuals with an average of 44 years, between 21 and 62 years. Distributed in the different sectors as follows: 56,7% work in the mechanical repair field, 17,5% in the painting field, 14.4% in the collision field, and 11.3% belongs to the concessionaire or management post jobs. Most individuals reported spending about eight hours a day, five days a week at the workplace, but there was a minority who worked more than five days a week. Furthermore, it was established that 82.6% (excluding administrative/concessionaire staff) used at least one PPE. Concerning about the use of the face mask, 69.8% said that they did not use it. It had also been checked that, overall, workers performed medical examinations every year through the Department of Health and Safety at Work.

The RFS was classified according to the results of the two exams: Spirometry and IOS, which had a statistically significant association (p-value < 0,05) (Table 2). The workers were divided only into two groups, the group of subjects with normal RFS (89,7%) and another group composed by individuals who had RFS with ventilatory disorders (10,3%). Not giving importance to different disorders types (8,3% of V. D. Peripheric Obstruction; 1,0% of V. D. Obstruction; and 1,0% of V. D. Mixed), since each of them presented very small percentages.

It was studied the relation between the smoking habits and pre-existing diseases with respiratory function of workers, in order to exclude any possibility of respiratory disorders caused by smoking or disease acquired outside of work, that could tamper with the rest of the investigation. The asthma, allergic rhinitis, sinusitis, and sleep apnea syndrome were the pre-existing diseases identified by the workers.

The ventilatory disorders are not related by pre-existing diseases, since all the individuals with respiratory disorders were found in the group of healthy subjects. As well, there is no significant association between the occurrence of respiratory disorders and smoking habits (p > 0,05), and furthermore, it can be observed that the prevalence of having disorders and being a smoker was 3,1% and ex-smokers was 4,1% revealing to be similar to the prevalence between the presence of disorders and be non-smoker (3,1%). This supports that the smoking habits were not the main origin of these lung diseases, and, in another hand, it could be also explained by continuous occupational exposure to pollutants, as it was found in the Guanabara et al, (2011) research.

In environmental context, it was evaluated the average concentration of the different pollutants at the car workshops, considering the TLV-TWA

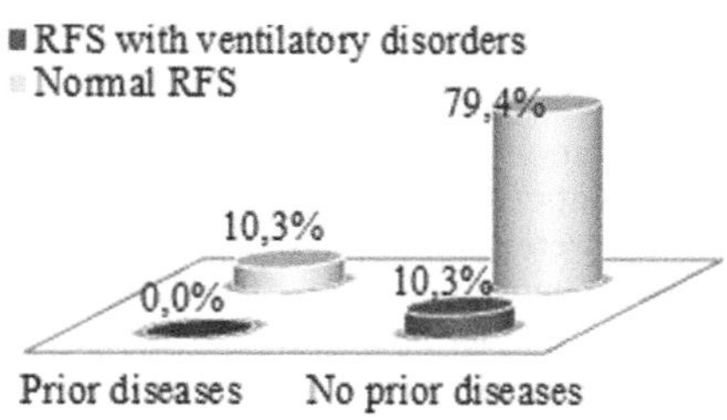

Figure 1. Relation between prior diseases and RFS of the workers.

Table 2. Respiratory function study.

			Spirometry				
			V. D. Peripheric Obstruction	V. D. Obstruction	V. D. Mixed	Normal	Total
IOS	Peripheric Obstruction	N	3	1	0	0	4
		% Total	3,1%	1,0%	0,0%	0,0%	4,1%
	Central Obstruction	N	5	0	1	0	6
		% Total	5,2%	0,0%	1,0%	0,0%	6,2%
	Normal	N	0	0	0	87	87
		% Total	0,0%	0,0%	0,0%	89,7%	89,7%
	Total	N	8	1	1	87	97
		% Total	8,3%	1,0%	1,0%	89,7%	100,0%

Statistical test: Chi-square independence $\chi^2 = 118{,}22$; gl = 6; p-value = 0,000.

Table 3. Concentration average values of pollutants.

Pollutants	N	Mean	Std. deviation	Limit values for exposure	Mean difference	Test; gl; p
Ozone	160	0,205	0,218	0,05 ppm	0,155	9,017;159; <0,0001
Carbon Monoxide	160	0,500	0,925	25,0 ppm	–24,500	–335,085;159; <0,0001
Nitrogen Dioxide	160	0,016	0,054	3,0 ppm	–2,946	–5889,820;159; <0,0001
Sulfur Dioxide	160	0,001	0,008	2,0 ppm	–1,999	–3199,0;159; <0,0001
Carbon Dioxide	160	594,48	1321,560	5000 ppm	–4405,516	–84,333;639; <0,0001
Total Dust	160	91,23	54,675	10000 $\mu g/m^3$	–9908,772	–4584,768;639; <0,0001

Statistical test: One sample T-test.

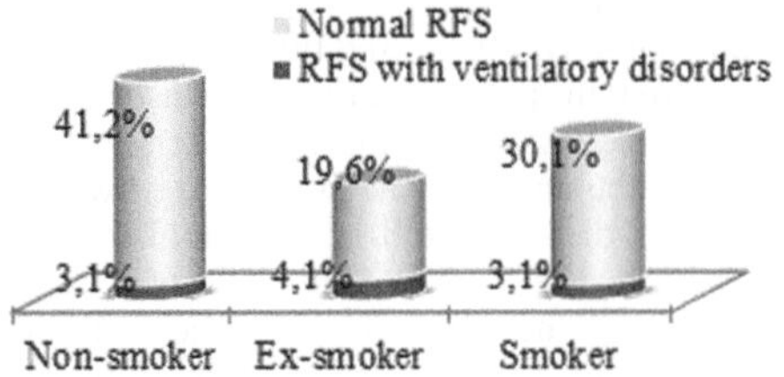

Figure 2. Relation between smoking habits and RFS of the workers.

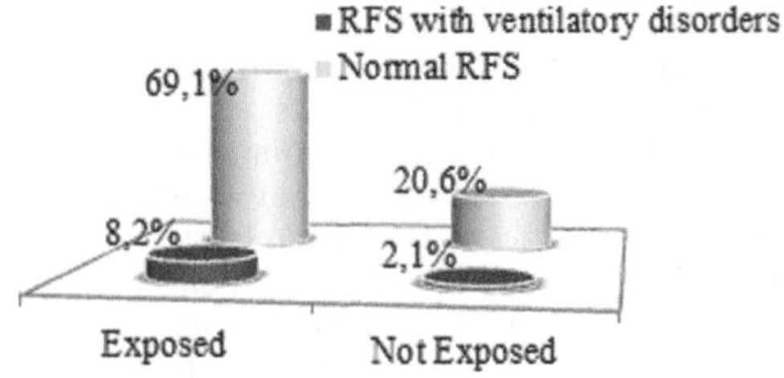

Figure 3. Relation between the type of exposure to O_3 and RFS of the workers.

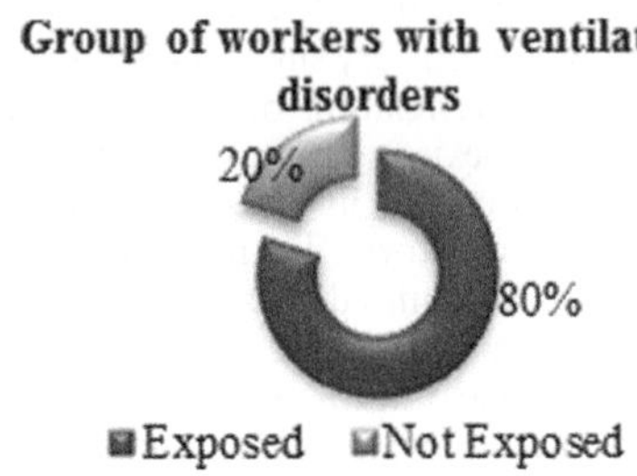

Figure 4. Relation between exposure to O_3 and RFS of the workers.

for each of them, in order to determine which pollutants have exceeded the legislating values, which could be harmful for the health of workers. Consider the Table 3.

In a general way, the levels of indoor environmental pollution did not prove to be alarming. The pollutants CO, NO_2, SO_2 and CO_2, such as Total Dust were statistically significant below the permissible exposure limit ($p < 0.0001$), being SO_2 concentrations practically absent.

These results could be explained as Claro et al, (2010) reported. The existence of extracting gas and dust systems, air renewal systems and the using of PPE play a crucial role in reducing the risk of occupational exposure to chemical pollutants. These results could be also justified by the fact that sector employers and the physical and structural facilities at the workshops are subject to more stringent standards to protect the health of workers, leading most companies to resort to management Environmental Quality, Hygiene and Safety systems and, therefore, ensuring better working conditions. However, the level of the O_3 concentration average value estimated at workshops exceeded the legislated reference value by 0,155 ppm.

Thus, the relationship between the lung function study and the CO, SO_2, NO_2, CO_2 and Total Dust was not evaluated, since it was found below than the TLV-TWA—average exposure on the basis of a 8h/day, 40h/week work schedule, which practically all workers could be exposed day after day without adverse health effects. The same did not happen to the O_3 concentration values that exceeded the TLV-TWA at some workshops.

As a result, the exposure to O_3 concentrations above the TLV-TWA values was not considered to be a risk to the health workers ($p > 0,05$). However, from a total of individuals with respiratory disorders ($n = 10$), most of them (80,0%) had been exposed to concentrations of the pollutant above the TLV-TWA (Figure 4). In another words, there is a tendency for who had respiratory disorders to be ex-posed to higher concentrations. In terms of the prevalence estimated of lung disorders due to O_3 exposure, it was found that the prevalence of having disorders was 8,2% being exposed and 2,1% being unexposed (Figure 3).

Ferreira et al. (2009) was more consistent in his study in São Paulo, which compared the lung function among three groups: bikers, taxi drivers and administrative staff (control group), and showed that the impairment of pulmonary function was proportional to the exposure intensity to O_3, with values significantly worse for the group most exposed (the bikers). Such as Sendão (2008) demonstrated a statistically significant negative association between exposure to O_3 and lung physiology.

Furthermore there was no statistically significant relation between the number of years of service and the respiratory disorder prevalence. Nevertheless, workers with lung disorders had worked on average 30 years in the workshops (automotive) branch, while the healthy group had only an average years of 26 years of professional activity in this same branch. Gomes, (2002) reported that prolonged exposure to O_3 provides an annual fall steeper of the FEV_1 and FVC parameters. And the study by Sendão (2008) that was performed over a five years in Lisbon showed that rising level of O_3 immediately increased the number of admissions to the emergency services for asthma exacerbation.

The O_3 is a very insoluble gas that penetrates deeply into the lung affecting the bronchioles and alveoli, where it could lead to toxic reactions. The 69.1% of workers who were exposed to high concentrations of O_3 obtained the normal RFS one of the possible explanations may be, by the presence of defensive mechanisms, which remove particles deposited in the respiratory tract. The first mechanism to act is the sneezing, triggered by large particles that are unable to go beyond the nostrils. Other important mechanisms are the cough and the mucociliary system. The particles that reach the most distal portions airways are phagocytosed by alveolar macrophages, and then it is removed by the mucociliary system or lymphatic system. However, when theses defensive mechanisms are no longer acting effectively, these particles cause irritation leading to a narrowing or an airways inflammation, which consequently increase the risk of bronchial asthma, mortality from lung cancer or the exacerbations pre-existing diseases. Studies on exposure of animals indicate that O_3 induce alveolar epithelial injury along with an inflammatory cell response, characterized by an influx of neutrophils in the airways (Park et al, 2002).

Due to the factors mentioned, it is pertinent to apply control systems and fault detection systems for extracting gas and dust in workshops where values of O_3 were detected above the TLV-TWA, in order to reach acceptable values. The use of PPE is another aspect studied in this research, because it plays a crucial role in reducing risk of occupational exposure to chemical pollutants. In this case, the focus goes to the facial mask, since it is the most important equipment for respiratory protection from contact with such substances. Even if there is an exhaust air system, it is indispensable using a suitable respiratory protective appliance. There should be greater awareness among employees about the importance of the use of PPE, especially the mask, because 69,8% of the workers with dangerous activities to the respiratory system (excluding administrative/concessionaire sectors) did not use the facemask.

4 CONCLUSIONS

In conclusion, the respiratory function of workers did not change significantly face the single pollutant O_3, which had high illegal concentrations at some workshops. However, there is a relevant clinically prevalence for individuals with ventilatory disorders, which could not be explained by smoking or previous diseases, were mostly exposed to O_3 concentration values above the TLV-TWA. In addition, the individuals group with ventilatory disorders worked (being exposed) more time than the group with RFS standard. However, despite this exposure scenario occurs we cannot assure that there is a relationship between exposure and disease, due to the small sample number. The promotion of safety culture, especially at the level of adoption of safe behaviors, as well as shared responsibility in achieving good safety should be improved, particularly a regard to the correct use of PPE.

As a suggestion, the study should continue evaluating the smaller workshops and the uncertified ones, with the addition of the concentrations of Volatile Organic Compounds (VOC's) assessment, since there were technical problems in the measuring instruments.

REFERENCES

Claro, L., Andrade, I., Figueiredo, J.P., Ferreira, A., Almeida, J., Paixão, S., Sá, N., Santos, C., Simões, H. 2010. Exposição ocupacional ao monóxido de carbono. Colôquio internacional de segurança e higiene ocupacionais, Sho2010.Guimarães, Feb. 11–12.

Ferreira, C.A.S., Pereira, L.V., Santos, F.S., Rego, P.L., Costa, I.P. 2009. Análise da função pulmonar de motoboys: o efeito da poluição atmosférica. Prevenção, a melhor forma de cuidar. O Mundo da Saúde, 33(2): 170–174.

Filho, A.V. 2008. Avaliação da Qualidade do Ar de Cabines de Veículos Automotores Recém-manufacturados. Pós-Graduação em Química, Universidade de São Paulo.

Fong, K.K., Mui, K.W., Chan, W.Y., Wong, L.T. 2010. Air quality influence on chronic obstructive pulmonary disease (COPD) patients' quality of life. Indoor Air Journal, 20(5): 434–441.

Gioda, A., Neto, R.F. 2003. Considerações sobre estudos de ambientes industriais e não industriais no Brasil: uma abordagem comparativa. Caderno Saúde Pública, 19(5): 1389–1397.

Gomes, M.J.M. 2002. Ambiente e pulmão. Jornal Pneumologia, 28(5).

Guanabara, L.C.R. 2011. Associação da função pulmonar de indivíduos fumantes e não fumantes com a qualidade do ar, nas cidades de Cubatão e Bertioga. Dissertação Pós-Graduação em Saúde Pública da Faculdade de Saúde Pública. Universidade de São Paulo.

Hart, J.E., Laden, F., Schenker, M.B., Garshick, E. 2006. Chronic Obstructive Pulmonary Disease Mortality in Diesel-Exposed Railroad Workers. Environmental Health Perspectives, 114(7): 1013–1017.

Park, C.P., Lee, J.S., Jang, A.S., Choi, I.S., Koh, Y.I. 2002. The relationship between alveolar epithelial proliferation and airway obstruction after ozone exposure. Allergy, 57: 737–740.

Parker, D.L., Waller, K., Himrich, B., Martinez, A., Martin, F.A. 1991. Cross-sectional study of pulmonary function in autobody repair workers. American Journal of Public Health, 81(6).

Portugal, Instituto Português da Qualidade (IPQ). Norma Portuguesa, NP 1796:2007.

Rodrigues, J.C., Cardieri, J.A., Bussamra, H.F., Nakaie, C.M.A., Almeida, M.B., Filho, L.V.S., Adde, F.V. 2002. Provas de função pulmonar em crianças e adolescentes. Jornal Pneumologia, 28(3).

Sendão, M.A.F. 2008. Impacto da poluição atmosférica na saúde da população residente em Lisboa. Dissertação de Mestrado em Engenharia do Ambiente. Universidade de Aveiro.

Valle, E.L.T. 2002. Resistência das vias aéreas: técnica da oscilação forçada. Jornal Pneumologia, 28(3).

Occupational Safety and Hygiene II – Arezes et al. (eds)

Simultaneity of risk in the civil site construction: An analysis of accumulated risk

Nuno Ferreira
Joaquim Peixoto Azevedo & Filhos, Lda, Vila Verde, Braga, Portugal

Gilberto Santos
Escola Superior de Tecnologia, Campus do IPCA, Barcelos, Portugal

ABSTRACT: Nowadays a lot has been improved concerning the evaluation of risk, being common to have it done task by task, function by function and machine by machine. However the risk associated to a task, function or machine isn't isolated from the most variable conditions inherent to a building civil site. Therefore a study was made to the simultaneity of risks associated to the tasks happening at the same time in the building for an Aged People Hostel, placed in the district of Braga, Portugal. This was aimed at determining the associated risk to each task attaching it to the chronogram of work. When determining the periods of work with the possibility of higher risk, some organizational modifications were introduced, relative to work chronogram in order to diminish the peaks of the previous obtained risks. We have stated that after the implementation of organizational measures the highest peaks were significantly reduced.

1 LITERATURE REVIEW

According to Nunes (2009) mentioned by Carneiro (2011) the evaluation of risks in the working place has begun after half a century following the beginning of the Industrial Revolution in England due to a concern connected with working accidents and other risk factors, common in the sphere of the first industries. Precisely on that occasion the first laws were issued in the branch of social security. Thus, risk analysis is a core process of safety management (Ying Lu, 2013). On other hand, Safety is a prominent feature in complex systems, and there is an abundance of different traditions how to deal with this (Lars Harms-Ringdahl, 2004). By definition, risk analysis deals with uncertain situations, that is, with situations in which we do not have complete and accurate knowledge about the state of the system. (Gurcanli, 2009).

However it was in the United States of America that a movement of prevention was created and developed, due to added efforts done by the government, entrepreneurs and specialists.

In 1928 *The American Engineering Council* had already mentioned the connection existing between the direct and indirect costs of accidents and assigned to the indirect costs the payment of non-productive salaries, financial loses and break in the outwork of production.

In 1941, H.W. Heinrich published a study relative to the direct and indirect costs of accidents of work where he displayed a method to study the causes of accidents which was known as the domino theory. This was based in a causality effect which determined an accident as a sequential group of five factors: ascendancy and social ambience, human failure, insecure act or dangerous condition, accident and personal damage (Nunes, 2006, sic by Carneiro, 2011).

In 1947 R.H. Simonds has presented a method for the calculation of costs connected to four types of accidents which might cause disabled lesions, cases of medical help, cases of first aid help and accidents without injuries (Carneiro, 2011).

In 1953 *The International Conference of Work* has defined in Recommendation nº 97 two basic methods for the protection of heath concerning the workers: medical care of each worker and technical measures to prevent, reduce and eliminate risks in work ambience (Carneiro 2011).

In 1966 Frank E. Bird Jr. has published the results of a study where he analyzed 90 thousand accidents happening in a siderurgical enterprise during 7 years.

According to the chronological previous description we can notice that the studies already done and the measures taken thereafter had as motivation the costs that such accidents had caused. Thus, proactive hazard identification and elimination is always safer and more cost effective than reactive hazard management (Gangolells et al, 2010).

However with the industrial and social development already registered, the needs had obtained another goal. At the end of the 20th century the prevention to the exposure to a risk factor, originator of a lesion or professional disease has become the goal to attain in what concerns safety and health at work (Meneghetti, 2010, sic. by Carneiro 2011). On other hand, research and practice have demonstrated that decisions made prior to work at construction sites can influence construction worker safety (Gangolells et al, 2010). Effective risk management of construction projects requires a reliable risk assessment and risk treatment plan (Yao-Chen Kuo, 2013).

2 METHODOLOGY USED

During all the construction of the Aged People Hostel, the necessary working days to accomplish each task were registered, istum est, the date of the beginning, the date of conclusion and if applicable the total number of periods of work. At the end we have obtained a chronogram of real work.

Simultaneously to the register of works, using the method of composed matrix (NTP330, 1995), one of the most widespread in the area of construction, several evaluations of risk were done for each of the tasks developed. We highlight the risks of falls from higher level, fall at the same level, ergonomic postures incorrect and electrical contacts as the most observed in majority tasks developed. However, due to the high number of information collected at this stage it is not possible to present all data.

The calculated risk was distributed along the chronogram of the work to get a real chronogram, with the quantification of associated risks to each task, that is to say, a risk map.

With the map of the obtained risk, from the adding of the working chronogram with the risk evaluations it is possible to quantify daily the level of accumulated risk, in functions of the works taking place on the spot.

Having determined the phases of the work with the possibility of a higher risk, several organizational measures were taken in order to diminish the peak risks. Without interfering in the critical activities, some alterations were introduced to limit the number of tasks developed simultaneously. These modifications were done exclusively on the days of the beginning and ending of tasks, having always preserved the working days necessary to the conclusion of works.

After having introduced the alterations to the chronogram of the initial work, we have obtained a new histogram of accumulated risk and we have done a thorough comparison.

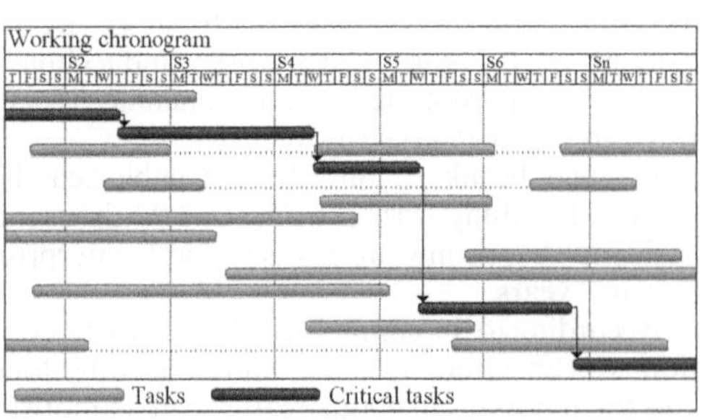

Figure 1. Schematic representation of the chronogram in the initial works.

3 RESULTS

3.1 *Initial histogram*

The obtained histogram (Fig. 2) characterizes the way of construction of these types of buildings. This type is characterized by two distinctive phases normally separated by a determined period of time. The first period includes works of formwork and stripping of formwork of every element of concrete which compose the building. The second includes the execution of all necessary works to finish the work. Generally between these phases there is a small appeasement or even a total stop of every kind of work. This appeasement or pause is due to the time of hardening of the elements of the building concrete, which attain their project resistance after 28 days.

The first month of work displays a level of residual risk due to the settlement of the dock and the cleaning of the ground for the implementation of the building. These works include the installation and preparation of every type of support equipments, technical nets, sanitary basis and social fittings.

The four future months of the project are characterized by distinctive and delineated levels of the

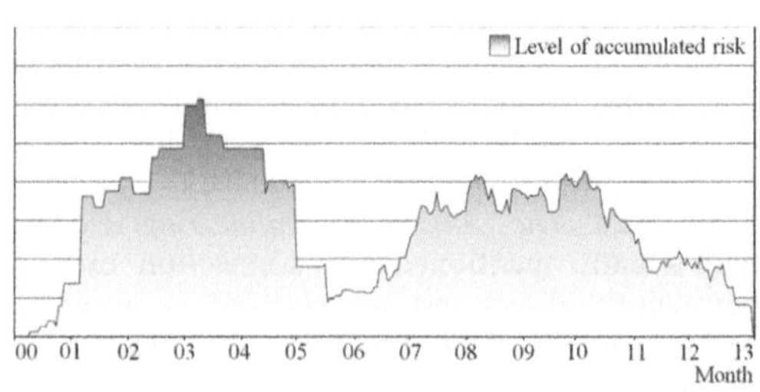

Figure 2. Histogram of level of accumulated risk.

histogram. This is due to the reduced number of different tasks developed simultaneously namely the execution of elements in concrete. Generally the execution of tasks in this period displays the highest risk index of all the enterprise, because it involves works in height, digging works and movement of hanging cargoes.

Afterwards there is a reduction in the number of tasks executed simultaneously, as the execution works of concrete come to an end. This appeasement has the approximate period of 28 days, the required period for the concrete elements to attain the necessary resistance to take the propping out. Through the graphic we can conclude that the works will never completely cease as the accumulated risk is not at all null.

The following four months are characterized by irregular and broad peaks. This is due to the high simultaneity of developed tasks. Normally the risks associated to each tasks in this moment are lower, however the high simultaneity of tasks and people working in this building causes a rather high accumulated risk.

The last two months of work are still characterized by high simultaneity, but the task associated to the finishing, tests and cleaning softens deeply the accumulated level of risk.

Summarizing, from the histogram of accumulated risk we can conclude that the construction is characterized by two phases with different points of action. Although in the first case we have included the tasks with the level of individual risk much more dangerous we have also stated that besides the similitude of the developed tasks, these ones are duly intermingled in the building process. That is to say that the number of developed tasks taken simultaneously is highly reduced.

The second phase characterized by the high number of developed activities taken simultaneously, due to its different specific points and also to the several fronts of work states a level of risk accumulated rather meaningful. The activities with a low risk connected to the high simultaneity of these same tasks show a level of risk accumulated that can´t be undervalued.

3.2 *Implementation of established measures*

Having determined the phases of the work with a possibility of a higher risk, we can create established rules with the aim of diminishing the risk peaks.

So without limiting the correct development of work, i.e. without interfering in the critic activities, we must establish organizational measures to diminish the number of works done simultaneously. In the first phase, period during which the tasks have the highest risk level, the simple reduction of one or two tasks simultaneously may cause meaningful reductions in the level of accumulated risk. On the other side in the second phase of the work, period during which the tasks have a lower level of risk, the reduction of tasks done simultaneously may create a graphic with peaks in more defined levels.

With the intention of diminishing the accumulated risk level, several small organizational alterations were proposed, in what concerns the chronogram of works with the aim of diminishing the peaks obtained in each of the phases.

Some modifications were exclusively inserted on the beginning and ending dates, preserving always the necessary working days considered in the chronogram of the initial project. Therefore as a clear and schematic title we will describe some of the modifications done on the initial chronogram of work:

- The work of setting the concrete in pillars is divided through three periods of work. The second period was anticipated eight days of work;
- The two periods of work which constitute the task of setting the concrete in stairs were anticipated eight days of work;
- The first phase concerning the proofing of walls was anticipated one day, while the second phase was delayed three days;
- The work of putting the first layer of filling up in pavements, on the very beginning subdivided in three periods of work, was transformed in two periods anticipating the second period in eleven days.

Through the graphic of figure 3 and 4 we show the schematic representations of the chronogram before and after the establishment of organizational measures, respectively.

After the introduction of these changes in the chronogram of the initial works, we have obtained a new histogram of accumulated risk (Fig. 5).

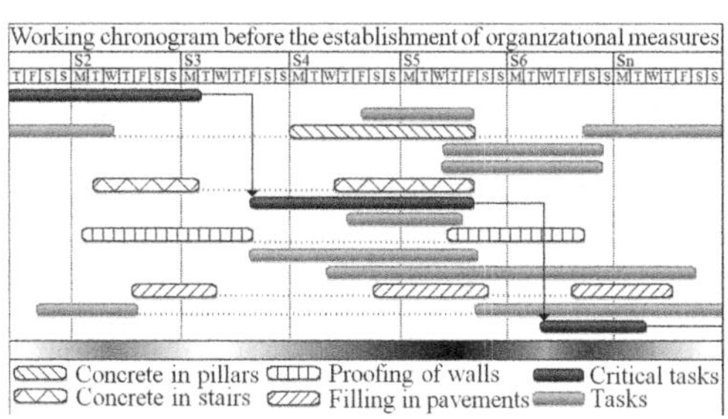

Figure 3. Schematic representation of the working chronogram before the establishment of organizational measures.

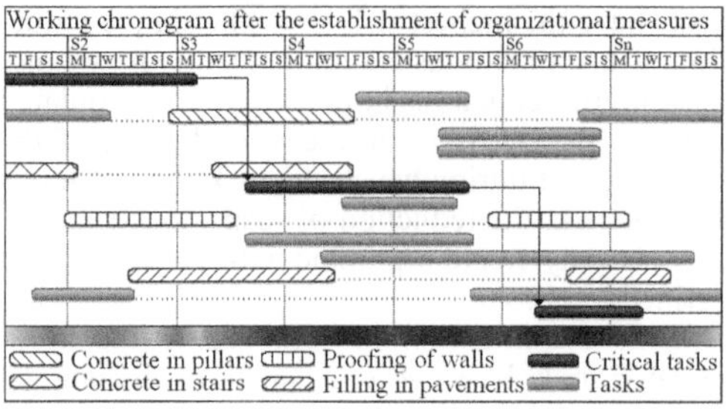

Figure 4. Schematic representation of the working chronogram after the establishment of organizational measures.

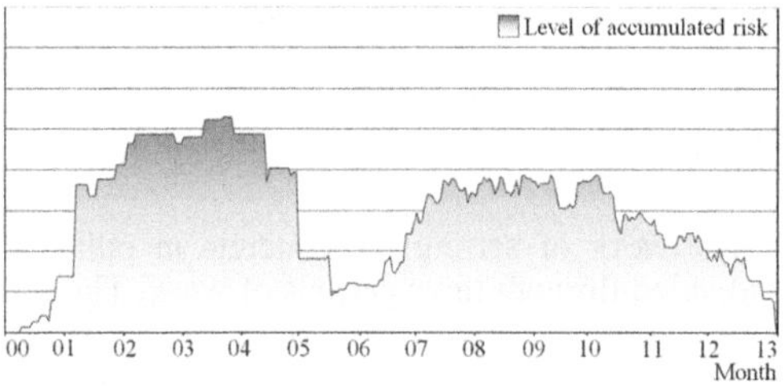

Figure 5. Level of risk accumulated after the introduction of organizational measures.

3.3 *Comparison of results*

We have stated that after the implementation of organizational measures of some non-critical tasks we have obtained a histogram with a level of accumulated risk much simpler and more homogeneous. Although in some points the level of accumulated risk has increased, we have verified that the highest peaks were doubtlessly more reduced.

We have also verified that the potential of risk was not diminished. This was simply redistributed along time, in order to reduce the peaks of accumulated risk existing on the very beginning.

For a better visibility and comparison of the histogram concerning the initial accumulated risk and after the settlement of organizational measures we have made the juxtaposition of the different histograms. We can perfectly see the redistribution of the level of accumulated risk through the graphic of Figure 6.

In order to turn this view excellent, we have isolated the areas of decrease and increase of accumulated risk using different colors. The areas of decrease of accumulated risk are displayed in light gray and the areas of increase are in dark gray (Fig. 7).

Through the graphic of Figure 8 we can perfectly verify the redistribution of the level of accumulated risk. We have verified that the peak of the accumulated risk on the beginning of the fourth

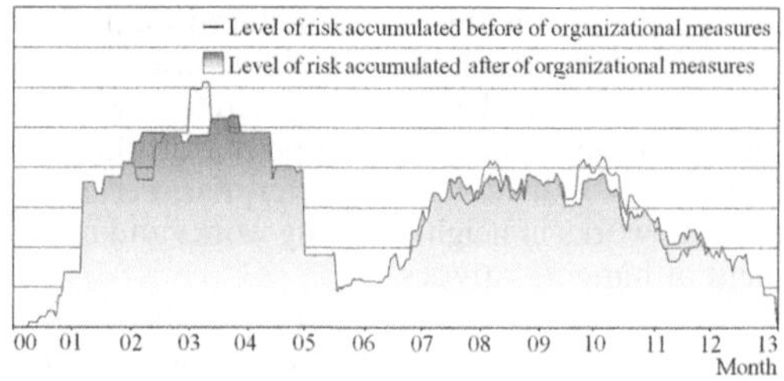

Figure 6. Comparison of the levels of accumulated risk before and after the implementation of organizational measures.

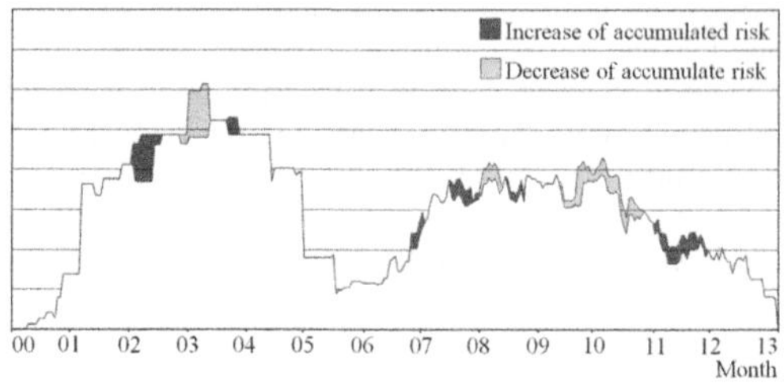

Figure 7. Areas of increase and decrease of accumulated risk.

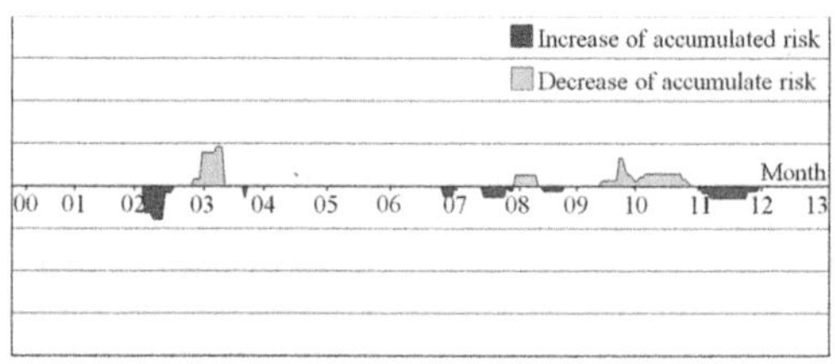

Figure 8. Areas of increase and decrease of accumulated risk.

month of work was redistributed along the beginning of the third month and the remaining increase during the fourth month of work. On the other side the irregular peaks identified on the beginning of the ninth and tenth month were redistributed along the eighth, ninth and mainly twelfth month of work.

Knowing that the potential of risk was just redistributed along time we can also conclude that the adding of the areas of development in the level of accumulated risk (dark gray areas) is equal to the adding of the decreasing areas in the level of accumulated risk (light gray areas).

Really the obtained results would be much more interesting if the potential of verified risk after the implementation of the organized measures was truly inferior to the potential of risk existing on the very beginning. This fact isn´t verified as long as the results weren´t affected by an act of pondering according to the number of tasks developed simultaneously. This ratio would have as a main goal to raise the accumulated risk in gear of the highest number of simultaneous developed tasks and soften the accumulated risk in gear of diminishing of the developed functions.

On the other way the rate of pondering should also be influenced by the number of executed tasks together with the same level of individual risk. That is to say the number of developed tasks done simultaneously should be grouped in different steps of tasks with a similar individual risk. In this way we could avoid that a certain number of works with a more serious level of individual risk would get the same degree as a certain number of works with a less serious level of individual risk.

In synthesis the ratio of a real pondering would be a relationship among the different levels of pondering according to the individual risk of each task.

4 CONCLUSION

With a critical analysis done to the displayed results we have to consider the importance of strictness in the risks evaluation. The systematic changing of workforce, the continuous change of conditions and sceneries of work, the exposure to the weather conditions and all the other variable factors render difficult the task of identification, observation, evaluation and gathering of data.

From the interpretation of the histogram of accumulated risk we can conclude that this construction is mainly characterized by two distinctive phases. Although in the first phase we have included the works with a most serious level of individual risk, we have stated that besides the similitude of the works developed, these same are duly mixed in the constructive process. That is to say the number of works executed simultaneously becomes much reduced. The second phase characterized by the high number of activities developed simultaneously, due to their different characteristics and also to the front points of work represents a level of accumulated risk rather meaning full. That is, the activities of a lower risk connected to the high simultaneity of tasks show a level of accumulated risk which can´t be underestimated.

With the implementation of organizational measures in the chronological register of tasks, we were able to reduce the peaks and transform the histogram of accumulated risk into a simpler and predictable item. We have also verified that the potential of existing risk was not diminished. This was simply redistributed along the period of time, in order to reduce the peaks of accumulated risk existing on the very beginning.

In summary without interfering in the critical activities, some organizational processes had to be taken in order to diminish the number of tasks fulfilled simultaneously. During the first phase of the construction, period during which the tasks possess a higher level of risk, the simple reduction of one or two works done at the same time may create rather meaningful reductions in the level of accumulated risk. On the other side during the second phase of the construction, period where the works have a level of lower risk, the reduction of works done at the same time may create a graphic with peaks in better defined grades.

In future studies it would be very important to define the pondering rate to apply to the total amount of individual risks. Besides being directly related with the simultaneity of works it would allow to get a more realistic quantification about the level of accumulated risk. This pondering would also allow, through the implementation of organizational measures a real reduction of the risk ratio accumulated in the construction.

Besides the implementation of organizational measures this work has helped to determine the curve of accumulated risk so characteristic in this type of building construction. In further studies it would be important to determine the curve of accumulated risk characteristic of other types of works in the civil construction and proceed with a thorough comparison.

On limit it would be of the highest importance for the security manager to have at his disposal a wide range of risk curves characteristic of each type of construction. Knowing the type of characteristic curve expected for a special construction, the responsible for the control, monitoring and evaluation of risks could beforehand take measures to reduce the level of risk existing in the construction.

REFERENCES

Bird, Frank E. 1975. Admnistración Moderna del Control Total de Perdidas. Madrid: Consejo Interamericano de Seguridad.

Carneiro, F.C. 2011. Avaliação de riscos: Aplicação a um processo de construção. Tese de mestrado. Universidade de Aveiro.

Gangolells M., Casals M, Forcada N., Roca X., Fuertes A. 2010. Mitigating construction safety risks using prevention through design. Journal of Safety Research 41, pp. 107–122.

Gurcanli, G.E., Mungen U., 2009. An occupational safety risk analysis method at construction sites using fuzzy sets. International Journal of Industrial Ergonomics 39, pp. 371–387.

Heinrich, H.W. 1941. Industrial Accident Prevention, A Scientific Approach. New York and London: McGraw W—Hil Book Company, Incl.

Lars Harms-Ringdahl 2004. Relationships between accident investigations, risk analysis, and safety management. Journal of Hazardous Materials 111, pp. 13–19.
Meneghetti, A.A. 2010. A Importância da Auditoria Comportamental para a Prevenção de Acidentes na Indústria Petroquímica. Niterói: Universidade Federal Fluminense.
NTP330, 1995. Sistema simplificado de evaluación de riesgos de accidente. Espanã: Ministerio de Trabajo y Asuntos Sociales—Instituto Nacional de Seguridad e Higiene en el Trabajo.
Nunes, F. 2009. Segurança e Higiene do Trabalho. Escola Profissional Gustave Eiffel.
Simonds, R.H. 1947. Safety Management Accident Cost And Control. R.D. Irwin.
Yao-Chen Kuo, Shih-Tong Lu. 2013. Using fuzzy multiple criteria decision making approach to enhance risk assessment for metropolitan construction projects. International Journal of Project Management 31, pp. 602–614.
Ying Lu, Qiming Li, Wenjuan Xiao 2013. Case-based reasoning for automated safety risk analysis on subway operation: Case representation and retrieval 2. Safety Science 57, pp. 75–81.

Occupational Safety and Hygiene II – Arezes et al. (eds)
© 2014 Taylor & Francis Group, London, ISBN 978-1-138-00144-2

High-Density surface Electromyography applications & reliability vs. muscle fatigue—a short review

T. Sa-Ngiamsak, J. Castela Torres Costa & J. Santos Baptista
Research Laboratory on Prevention of Occupational and Environmental Risks (PROA/LABIOMEP), University of Porto, Portugal

ABSTRACT: Muscle fatigue has been documented in various occupations. The aim of this study was to evaluate the applications of High Density surface Electromyography (HD-sEMG), in context of its variables/factors correlating with muscle fatigue, beside with its feasibility and reliability in muscle fatigue assessment. The search was performed over 33 electronic databases through integrated and metasearch search methods. Seven studies were included in the review. Four of them associated with HD-sEMG applications in muscle fatigue assessment and three other studies involving with reliability in muscle fatigue assessment. Evaluation by HD-sEMG is feasible and reliable in muscle fatigue assessment. There are many variables/factors correlated with muscle fatigue. Its reliability in terms of repetition and reproducibility of a diagnosis were also proved. This review indicates that applications of HD-sEMG are feasible and reliable in order to assess muscle fatigue whether static or dynamic contraction.

1 INTRODUCTION

Muscle fatigue has been documented in different occupations (Troiano et al., 2008). Muscle fatigue results in the declination of force generate by the muscle. It can be caused by repetitive or sustained work, short work cycles, and localized muscle loadings (Troiano et al., 2008). Muscle fatigue is developed as result of a chain of metabolic, structural and energetic changes in the muscle, due to the insufficient oxygen and nutritive substances supplied through blood circulation, as well as a result of changes in the nervous system efficiency (Cifrek et al., 2009).

Intramuscular electromyography signal recording so-called invasive technique was first introduced by Adrian and Bronk in 1929 (Henneberg, 2000). It is necessary to use the insertion of needles into the muscle and it has become a classic tool to investigate individual motor properties, particularly in clinical examination. This use is due to an important feature of the needles type electrode, that by limiting the diffusion effect can detect electrical potentials near the active muscle fibers in term of conductive capacity (Merletti et al., 2008). However in some cases the use of needles insertion is not either desirable or comfortable, such as in the children clinical examination, sport or ergonomics (Merletti et al., 2008). Surface electromyography (sEMG) is a noninvasive technique that can be used as an alternative to such limitation, although its properties couldn't be replaceable to the invasive technique in clinical examination (Merletti et al., 2008). Surface EMG is widely used to measure muscle action potentials, through the placement of surface electrodes on the skin overlying a muscle or group of muscles (Drost et al., 2006). The basic sEMG configuration has a single channel, that can only provide the examining information only over the small area of muscle, and it is still unable to investigate the spatial distribution of muscle activity, as well as the application of intramuscular technique (Madeleine et al., 2006). In the last decade, the evolution of sEMG technology has been developing rapidly. The new technique of two-dimensional array systems, so-called high-density surface electromyography (HD-sEMG) was presented and has been increasingly applied in this study area. This technique provides the possibility to investigate the spatial distribution of single muscle activity, moreover it also allows the visual investigation through the topographical mapping of muscle activity (Madeleine et al., 2006). Nevertheless there are some questionable outcomes of such HD-sEMG applications and their reliability. Can they actually be used as a major tool in an attempt to investigate muscle fatigue?

This work was conducted to present the current situation of HD-sEMG applications, in context of its variables/factors correlating with muscle fatigue, beside with its feasibility and reliability in muscle fatigue assessment. The objective focuses on getting an actual vision, clear and comprehensive of the applications and reliability of HD-SEMG.

2 METHODS

2.1 *Search strategy*

The literature search was conducted on March 29, 2013 and April 8, 2013. The search was performed over 33 electronic databases through integrated and metasearch search methods. The database type was E-journals. The key search terms focus on words such as "surface EMG", "fatigue", "musculoskeletal disorders", "muscle", "ergonomics" and were used in all the database with the appropriate Boolean operators (such as And and Or). In addition another literature searches were also performed through the google search engine and reference list of those relevant articles. Only full papers were considered and insufficient information formats, such as abstracts published in term of conference or workshop proceedings were not included.

2.2 *Screening and eligibility criterias*

After duplication removal, all the articles found were considered as excluded after being screened following these criterias:

- No related subject
- Publication before 01-01-2005
- Language not accessible
- Text not available.

Articles were considered eligible if they meet the following criterias:

- Studies which whose objectives are involved with HD-sEMG
- Studies which testify the effectiveness, application or reliability of HD-sEMG assessment
- Studies which provide the information about sEMG based muscle fatigue evaluation in biomechanics
- All the above considered studies are performed in humans but not in animals.

3 RESULTS

3.1 *Study selection*

The details of the selection of all relevant articles both, excluded articles and included articles were performed through several criterias, based on the PRISMA statement for reporting systematic reviews and meta-analyses of studies (PLoS_Medicine, 2009). All those details can be shown in Figure 1.

3.2 *HD-sEMG applications in muscle fatigue assessment*

Four of the seven included studies examined muscle fatigue by using HD-sEMG (table 1). Two of them were conducted with static contraction, one study was conducted with dynamic contraction, and another one with both static and dynamic contraction. All of these studies were assessed on Biceps brachii, upper trapezius and lumbar muscle.

The first study from table 1. (Troiano et al., 2008), Amedeo Troiano and his team conducted the experiment to assess fatigue in isometric contractions of the upper trapezius muscle. The results presented an EMG surface amplitude increased during the contraction meanwhile mean frequency of the power spectrum (MNF), fractal dimension (FD), and muscle fiber conduction velocity (CV) decreased, This provided indications of fatigue and could be predictive of endurance time (ET). The second study from table 1 (Falla and Farina, 2005), Deborah Falla and Dario Farina presented an experiment to compare average muscle fiber conduction velocity (CV) between patients with chronic neck pain and an healthy control group during dynamic contraction. It appeared that CV was decreased by fatigue during dynamic contraction, but CV of the upper trapezius muscle was higher in people with chronic neck pain than in the control group. It was concluded that, this may be associated with the histological and morphological changes which had previously been identified in people with pain over the trapezius muscle.

The third study from table 1 (Kallenberg and Hermens, 2009), L.A.C Kallenberg and H.J. Hermens performed an experiment to assess motor properties between a chronic stroke patients group and healthy control group, by using High-Density Surface EMG during both dynamic and static contraction. For dynamic contraction larger motor unit action potentials (MUAPs) reflected in a higher value of the root mean square of motor unit action potentials (RMS_{MUAP}) were found in stoke group when compared with the control group. Furthermore there was a correlation between a clinical

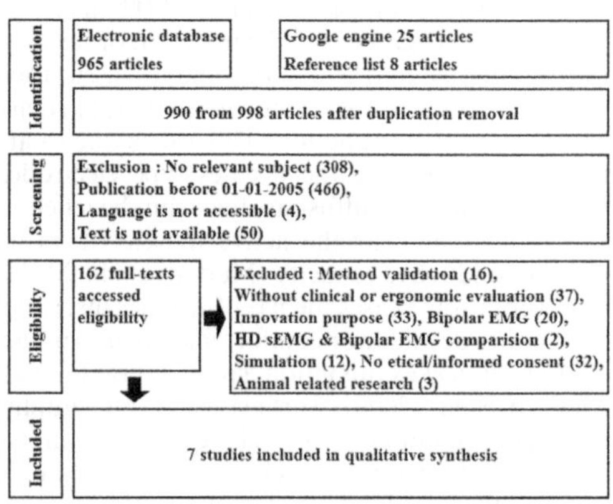

Figure 1. Selection criterias of the relevant articles.

Table 1. HD-sEMG applications in muscle fatigue assessment.

Authors	Year	Country	Number of electrodes	Number of subjects	Contraction types	Muscle types
Amedeo Troiano et al.	2008	Italy	64	14	Static	Upper trapezius
Deborah Falla & Dario Farina	2005	Australia	4	19 patients 9 controls	Dynamic	Upper trapezius
L.A.C Kallenberg & H.J. Hermens	2009	The Netherlands	16	18 patients 20 controls	Static & Dynamic	Biceps brachii
Yong Hu et al.	2007	Hong Kong	20	5 patients 30 controls	Static	Lumbar

Table 2. HD-sEMG reliability in muscle fatigue assessment.

Authors	Year	Country	Number of electrodes	Number of subjects	Comparative reliability	Muscle types
GyuTau kim et al.	2007	USA	N/A	5	Intramuscular	Rectus femoris
Marco Barbero et al.	2011	UK	64	10	Inter & Intra-rater	Upper trapezius
Laura A.C. Kallenberg et al.	2009	The Netherlands	8	12	Reproducibility	Upper trapezius & Sternocleidomastoid

scale, the Fugl-Meyer score (Physical performance assessment tool), and the ratio of RMS_{MUAP} at the affected side divided by unaffected one.

In the fourth study of table 1 (Hu et al., 2007), Yong Hu and his team performed an experiment by using the sEMG topography pattern identified Low Back Pain (LBP) rehabilitation. They found that in all healthy subjects in the control group, sEMG topography patterns presented symmetric muscle activities. Meanwhile sEMG topography patterns obtained from 5 LBP patients group presented obviously asymmetric and different from the normal pattern. This was able to conclude that sEMG topography illustrates the distribution of muscle activities, which provides a visible result of lumbar muscle coordination.

3.3 *HD-sEMG reliability in muscle fatigue assessment*

The other three of the seven included studies dealed with reliability in muscle fatigue assessment. Their details are presented in table 2. There were three different methodologies performed to assess reliability of HD-sEMG applications in muscle fatigue assessment. The first study from table 2 (Kim et al., 2007), GyuTau Kim and his team performed an experiment to assess reliability of muscle fatigue indices correlation. Measurements were done by surface EMG and compared with intramuscular EMG or invasive technique. It appeared that muscle fatigue indicts including with root mean square (RMS), average rectified value (ARV) and mean frequency of the power spectrum (MNF) typically changed in accordance with the results obtained from intramuscular EMG or invasive technique. It appeared that MNF decreased meanwhile RMS and ARV increased as muscular fatigue progressed. A linear correlation was also observed and it was concluded that the change rates in RMS and ARV values, of intramuscular EMG are almost the same rate as surface EMG. In the second study presented in table 2 (Barbero et al., 2011) Marco Barbero and his team performed the experiment to assess reliability of HD-sEMG applications by evaluating intra-rater and inter-rater reliability and the suitability of surface EMG. It appeared that both intra-rater and inter-rater reliability analysis showed an almost perfect agreement. This study provided a strong evidence that visual estimation of innervation zone location is a reliable procedure. The third study from table 2 (Kallenberg et al., 2009), Laura A.C. Kallenberg and her team performed an experiment to assess the within-day and between-days reproducibility of variables obtained from linear array sEMG during three functional tasks: a shoulder abduction, ironing and head turn task. The ICC (Intra-class correlation coefficients) value for investigation of EMG variables were higher than 0.7 during the three tasks, meanwhile ICC values > 0.6 are commonly accepted as a good reliability. This high reliability level implies that, they could be used to distinguish between-subjects, what may prove to be important for clinical practice.

4 DISCUSSION

From the seven included studies, four were found associated with the applications of HD-sEMG, and three of them were found to be associated with the reliability of HD-sEMG assessment in muscle fatigue. For the applications of HD-sEMG both static and dynamic contraction demonstrated that, there are many variables/factors correlated with muscle fatigue. Namely during the contraction surface EMG amplitude will be increased, meanwhile MNF, FD and CV decreased. This provides the sign of fatigue and could be productive of ET (Troiano et al., 2008). While surface EMG amplitude increased during the contraction as fatigue progressed, it also caused the root mean square (RMS) value of motor unit action potential (MUAP) to be increased as well. This fact was remarkably related with one study that compared the RMS value of motor unit action potential (MUAP) between healthy people (Control group) and the stroke group. As the result, it was perceived that RMS of stroke group was higher than the one which obtained from healthy group. This implies that the muscle activity in stoke group is higher than the healthy group (Kallenberg and Hermens, 2009). In addition the comparison between healthy people (Control group) and patients group with chronic neck pain during the dynamic contraction of one study, demonstrated that the CV of the upper trapezius muscle was higher in people with chronic neck pain than in the control group. It was concluded that, this may be associated with the histological and morphological changes which had previously been identified in people with pain over the trapezius muscle (Falla and Farina, 2005). Moreover another study was performed by using the sEMG topography pattern to identify LBP rehabilitation, Its result showed that, sEMG topography patterns obtained from the LBP patients group presented obvious asymmetry and different from normal pattern. This leads to the conclusion that sEMG topography illustrates the distribution of muscular activities, which provides a visible result of lumbar muscle coordination (Hu et al., 2007)

For the reliability of HD-sEMG assessment, three from the seven included studies were found with three different systematic comparative methodologies. All of those assessments had demonstrated the reliability of HD-sEMG applications. One study suggested the reliability of fatigue indices measured by surface EMG compared with intramuscular EMG or invasive technique. It appeared that muscle fatigue indices obtained from surface EMG including RMS, ARV and MNF typically changed in accordance with the results obtained from intramuscular EMG or invasive technique, namely MNF decreased as muscle fatigue progressed and particularly RMS and AVR increased almost at the same rate obtained from the intramuscular EMG or invasive technique. This implies that the results of surface EMG are in accordance with the intramuscular EMG or invasive technique, which is generally used in clinical diagnosis (Kim et al., 2007). Another two studies proved the reliability of HD-sEMG applications in terms of visual estimation of the EMG signal and reproducibility of variables obtained from HD-sEMG. They were performed through different evaluation methods, the intra-rater and inter-rater reliability methods and the reproducibility method. These methodologies presented a high reliability in the applications of high density surface EMG, according to both intra-rater and inter-rater reliability analysis, indicating an almost perfect agreement (Barbero et al., 2011). And for the ICC indicator in reproducibility analysis, it presented a score higher than 0.7 which is considered acceptable as a good reliability (Kallenberg et al., 2009). These evidences implied that the applications of HD-sEMG are reliable in statistics, in term of multiple repetition of diagnosis. Namely the visual estimation of EMG signal over innervation zone and reproducibility of variables/factors obtained from HD-sEMG recordings are a reliable procedure.

5 CONCLUSION

The result of this review indicates that, the applications of HD-sEMG are feasible and reliable in order to assess muscle fatigue. The results of the various studies lead to the conclusion that there are many variables/factors correlated with muscle fatigue and also accordant with intramuscular EMG or invasive technique either static or dynamic contraction. And its reliability in term of multiple repetition of diagnosis and reproducibility was also proved. In addition its remarkable applications to evaluate muscle activity through the sEMG topography pattern over particular muscle, might show some musculoskeletal disorders (MSDs) and symptoms. From all these evidences have revealed that HD-sEMG is increasingly more useful in applying as a muscle fatigue investigation tool.

REFERENCES

BARBERO, M., GATTI, R., CONTE, L. L., MACMILLAN, F., COUTTS, F. & MERLETTI, R. 2011. Reliability of surface EMG matrix in locating the innervation zone of upper trapezius muscle. Journal of Electromyography and Kinesiology, 21, 827–833.

CIFREK, M., MEDVED, V., TONKOVIC, S. & OSTOJIC, S. 2009. Surface EMG based muscle fatigue evaluation in biomechanics. Clinical Biomechanics, 24, 327–340.

DROST, G., STEGEMAN, D.F., ENGELEN, B.G.M.V. & ZWARTS, M.J. 2006. Clinical applications of high-density surface EMG: A systematic review. Journal of Electromyography and Kinesiology, 16, 586–602.

FALLA, D. & FARINA, D. 2005. Muscle fiber conduction velocity of the upper trapezius muscle during dynamic contraction of the upper limb in patients with chronic neck pain. Pain, 116, 138–145.

HENNEBERG, K. 2000. Principles of Electromyography. In: BRONZINO, J.D. (ed.) The Biomedical Engineering Handbook. Second Edition ed. Boca Raton: CRC Press.

KALLENBERG, L.A.C. & HERMENS, H.J. Motor Unit Properties in the Biceps Brachii of Chronic Stroke Patients Assessed with High-Density Surface EMG. Proceedings of the 4th International ThB1.3 IEEE EMBS Conference on Neural Engineering, 2009 Antalya, Turkey.

KALLENBERG, L.A.C., PREECE, S., NESTER, C. & HERMENS, H.J. 2009. Reproducibility of MUAP properties in array surface EMG recordings of the upper trapezius and sternocleidomastoid muscle. Journal of Electromyography and Kinesiology, 19, e536–e542.

KIM, G., AHAD, M.A., FERDJALLAH, M. & HARRIS, G.F. 2007. Correlation of Muscle Fatigue Indices between Intramuscular and Surface EMG Signals. IEEE, 378–382.

LLC. HU, Y., MAK, J.N.F. & LUK, K. Application of Surface EMG Topography in Low Back Pain Rehabilitation Assessment. Proceedings of the 3rd International IEEE EMBS Conference on Neural Engineering, 2007 Kohala Coast, Hawaii, USA.

MADELEINE, P., LECLERC, F.R., ARENDT-NIELSEN, L., RAVIER, P. & FARINA, D. 2006. Experimental muscle pain changes the spatial distribution of upper trapezius muscle activity during sustained contraction. Clinical Neurophysiology, 117, 2436–2445.

MERLETTI, R., HOLOBAR, A. & FARINA, D. 2008. Analysis of motor units with high-density surface electromyography. Journal of Electromyography and Kinesiology, 18, 879–890.

PLOS_MEDICINE 2009. The PRISMA Statement for Reporting Systematic Reviews and Meta-Analyses of Studies That Evaluate Health Care Interventions: Explanation and Elaboration. Guidelines and Guidance.

TROIANO, A., NADDEO, F., SOSSO, E., CAMAROTA, G., MERLETTI, R. & MESIN, L. 2008. Assessment of force and fatigue in isometric contractions of the upper trapezius muscle by surface EMG signal and perceived exertion scale. Gait & Posture, 28, 179–186.

Occupational Safety and Hygiene II – Arezes et al. (eds)
© 2014 Taylor & Francis Group, London, ISBN 978-1-138-00144-2

Dispersion of quarry's dust—pilot study

A. Campos
Research Laboratory on Prevention of Occupational and Environmental Risks (PROA/LABIOMEP), Faculty of Engineering, University of Porto, Porto, Portugal

M.L. Matos
LNEG—Laboratório Nacional de Energia e Geologia, S. Mamede de Infesta, Portugal

J. Góis, M.C. Vila, M.L. Dinis & J. Santos Baptista
Research Laboratory on Prevention of Occupational and Environmental Risks (PROA/LABIOMEP), Faculty of Engineering, University of Porto, Porto, Portugal

ABSTRACT: The mining industry involved, among other factors, exposure to dust, which affects workers, and also the surrounding populations. In this article, the dispersion of dust in the air environment was studied and the dust sampling was based on the French standard NFX 43 007:2008. The dust flux deposition was determined in twenty-four sampling points located inside and around the quarry. It was verified that the points with higher level of dust correspond to the operating area access, crushing plant, and a national road, with the average of 369.49 $mg.m^{-2}.day^{-1}$, 337.70 $mg.m^{-2}.day^{-1}$ and 152.65 $mg.m^{-2}.day^{-1}$, respectively. The lowest value of dust flux deposition corresponded to a point located in a limit of the quarry, with the value of 14.37 $mg.m^{-2}.day^{-1}$. Through this study it was concluded that the surrounding population could be affected by the dust quarry, although, the dispersion will greatly depends on the meteorological conditions, among other factors.

1 INTRODUCTION

1.1 *Mining industry*

The dust exposure, and in particular, the exposure to silica crystalline affect the human health and can originate silicosis, one of the oldest industrial diseases. In 2001, the European Occupational Disease Statistics (EODS) has classified the silicosis as the sixth occupational respiratory disease (Karjalainen & Niederlaender, 2004). In fact, in mining industry there are the added risks associated with exposure to chemical agents such as dust, which could be dangerous for human health. The exposure to this agent could affect not only the workers of the quarry but also the surrounding population.

In mining industry there are several processes that originate fine particles, such as the site preparation, drilling, blasting, crushing, loading, stockpiling, transportation, among others (Petavratzi et al. 2005, NEPSI 2006). Taking into account all these sources it is almost inevitable that the surrounding areas are not affected by fugitive dust (Cattle et al. 2012).

Dust may cause potential risks to human health (workers and surrounding populations), to environment, working conditions, and workers' productivity. Its effects may change depending on the particles concentration, size, shape, sharp edges and their chemical composition (Petavratzi et al. 2005, Polichetti et al. 2009).

1.2 *Pilot study*

The pilot study was carried out in a company located in the North of Portugal. The activity is characterized by quarry exploitation (gneiss), and it is responsible for the production of crushed aggregates and riprap to coastal protection. This company is constituted by 180 workers, 30 of them are associated to the quarry, and the operators' number is around 10. For this, the work procedure includes different phases: rock removal, selection of the material for riprap, transporting the remaining material to the processing plant for the production of crushed aggregates, storage the material by granulometry in silos with large capacity or in open air stocks and, at the end, the loading system.

In the process unit there are two crushing circuits:

- a crushing plant with large capacity—primary crusher, secondary crusher, vibrating screens, and the conveyors belts which carry the material

directly to the silos. It is important to highlight that the conveyors belts are encapsulated and the silos are closed.
- a mobile crushing circuit which receives the smaller materials. On this one there are belts that are not covered and the produced stock is stored in the open air.

Through the above mentioned characteristics it was verified that by the fragmentation process, transport and deposition, or by the exposure in the prevailing wind direction, mainly in what concerns the storage in the open air, these circuits present a strong potential to produce dust.

Another element of this system is the sand washing. However, due to the use of water, this unit is not responsible for the dust production. In this quarry there is also a decantation unit for sludge and a filter press.

For exploitation activities, in a general view, there is some equipment operating on a defined place and with appropriate functions, such as:

- *Rock*/mining area/drilling and blasting
- *Dumper*/mining area—primary crushing plant/ transportation of material to the crushing plant
- *Wheel Loader*/mining area and stock zones/ transportation of material with different dimensions to the stock location or trucks loading
- *Track excavator*/mining area/excavation and, trucks and dumpers loading

The main objective of this work was focused on the analysis of the dispersion of dust, which affect the workers and the surrounding population, by environmental air, in the neighborhood of a quarry, and the variation of dust flux deposition over time.

2 MATERIALS AND METHODS

The general procedure used to achieve the main objective was based in the following tasks: to measure the total dust flux deposition in different sampling points within the quarry and its surroundings; to record the meteorological conditions during the several sampling periods in order to evaluate their influence on dust dispersion; to assess how local residents are affected by dust coming from the quarry; to study the effect of tree cover in the quarry surroundings; to compare the levels found within the quarry and its surroundings with the values obtained for the closer communication infrastructures (national road and municipal roads).

2.1 *Sampling procedure*

Through the analysis of dust flux deposition in the quarry's surroundings, this study tried to understand the dispersion plume in the study area and thus, to assess if the workers/population are affected by particles coming from the quarry. The methodology for collecting dust samples was based on the French standard NFX 43 007:2008—Air quality—Ambient air—Determination of the mass of dry atmospheric depositions—Sampling on deposit plates—Preparation and treatment, which details the placement of plates, coated with resin, the locals to analyze and subsequent treatment of the samples collected. This standard seemed to be the most suitable for the results that this study was seeking.

In a first step the twenty-four sites were selected in order to evaluate the total dust flux deposition (Fig. 1). This selection was made aiming to cover four types of areas: surrounding the quarry, quarry zone, stocks zone and national road.

The four groups of sites correspond to the following points:

- Surrounding the quarry: 1, 2, 3, 4, 5, 6, 7, 23, 24;
- Quarry itself: 8, 9, 10, 11, 12, 13, 17, 18, 19, 20, 21;
- Stock zone: 14, 15, 16;
- National road: 22.

The second step was to clean the plates with alcohol and then with dichloromethane to remove all the impurities and then, put them in an oven at 105°C for one hour. When the cleaning phase is finished, the plates should be adequately numbered and stored in separate containers for subsequent transport to the sampling site. Although the cleaning procedure is not described in the standard NFX 43 007:2008, the previous version of it, in the year 1973, it is referred that the plates should be dried in the oven at 105°C. The plates are made of stainless steel, with the dimensions of 5×10 cm.

In the selected sites, the plates used for sampling were covered with resin (polydimethylsiloxane)

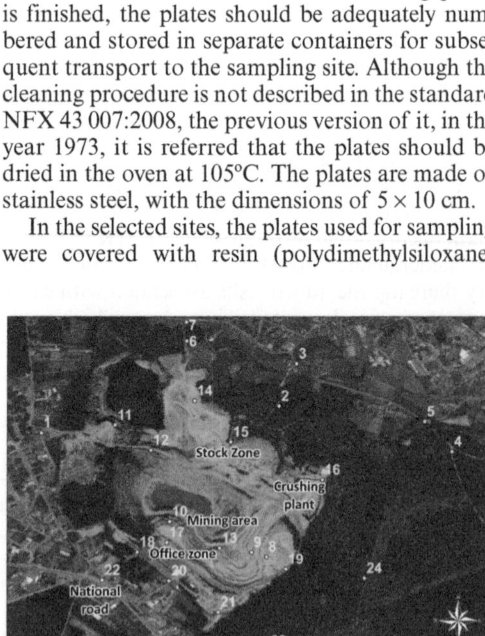

Figure 1. Location of sampling points within and surrounding the quarry.

using a glass sprayer with a pump. This procedure should be performed in an appropriate site to minimize possible interference in the results. The resin in the plates aims to allow the adhesion of dust after deposition.

Regarding the plates position and according to NFX 43 007:2008, they should be placed at a height exceeding 1.5 meters above the ground. However, this was not always possible. In the specific case of the plates bordering on certain benches, iron stakes were located to allow placement of the plates in horizontal position. In these cases, the plates were placed approximately at 0.5 meters above the ground, since a greater height would require a support system more resistant, otherwise they would be more susceptible to fall due to the vibrations caused by explosions or the projection of small fragments.

After the placement of the plates, the GPS coordinates of all sampling points were registered for the following spatial representation. In this study three (each one for a period of approximately 20 days) samplings campaigns were done in order to ensure the representation of the values obtained and to allow the study of the influence of meteorological conditions on dust deposition. The first sampling was performed between April and May, the second one during May and June and the third in June and July.

2.2 *Laboratory procedure*

After the sampling period, the plates were collected and treated in the laboratory according to the following procedure:

a. Connect the Erlemeyer flask with screw to the pump;
b. Weigh the virgin filter in a precision balance (three weightings);
c. Place the filter on top of the Buchner funnel, place the top part and the spring that allows the join of the both parts;
d. Using a pipette, place a small amount of dichloromethane in the surface of the plate containing dust;
e. Scrape with a spatula;
f. When the plate is completely cleared, proceed to the filtration connecting the pump;
g. Remove the filter from the Buchner funnel and weigh the sample with a precision balance (three weightings).

The dust flux deposition is calculated through the mass of the sample on each plate, the surface area, and the sampling time:

Dust flux deposition ($mg.m^{-2}.day^{-1}$) = $m/(a \times t)$ (1)

where m = mass of dust (mg); a = surface area that collect dust (m^2); and t = exposure time (days).

3 RESULTS

3.1 *Dust flux deposition*

After the laboratory treatment the dust flux deposition was calculated taking into account the mass of the sample on each plate, the timing sampling and the area of the plate, Table 1.

The same results are presented in the Figure 2, to better understand the behavior of three samplings, and it is possible to observe that the data are consistent for all of them.

3.2 *Spatial representation*

For the spatial representation the GPS coordinates and dust flux deposition were introduced in ArcGIS and using the interpolation method of "Radial Basis Function", available in the software used, the Figure 3–5 were obtained.

Table 1. Dust flux deposition for each sampling point, for the three samplings and correspondent average ($mg.m^{-2}.day^{-1}$).

	1st sampling	2nd sampling	3rd sampling	Average
No	Dust flux deposition	Dust flux deposition	Dust flux deposition	Dust flux deposition
1	100.95	4.44	49.84	51.75
2	53.81	3.59	27.30	28.23
3	48.25	–	3841	43.33
4	58.57	10.43	57.46	42.15
5	111.11	9.91	43.17	54.73
6	**20.95**	3.42	**18.73**	14.37
7	61.59	2.22	25.40	29.74
8	119.05	4.44	79.68	67.72
9	133.33	8.55	95.24	79.04
10	134.29	14.53	32.38	60.40
11	89.84	3.59	33.06	42.16
12	78.10	11.11	96.83	62.01
13	218.41	4.10	58.10	93.54
14	106.67	7.01	31.75	48.47
15	246.98	36.24	105.40	129.54
16	**598.41**	**198.80**	**215.87**	**337.70**
17	130.48	4.10	40.95	58.51
18	266.03	13.50	67.30	115.61
19	**591.67**	**48.55**	**468.25**	**369.49**
20	218.33	2.74	56.83	92.63
21	155.33	**1.71**	64.76	73.93
22	**304.00**	**46.32**	**107.62**	**152.65**
23	124.33	19.49	68.89	70.90
24	37.14	3.93	69.52	36.87

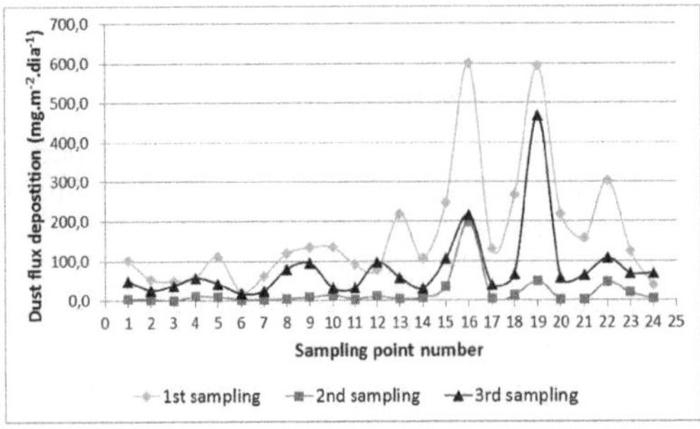

Figure 2. Dust flux deposition for each sampling point.

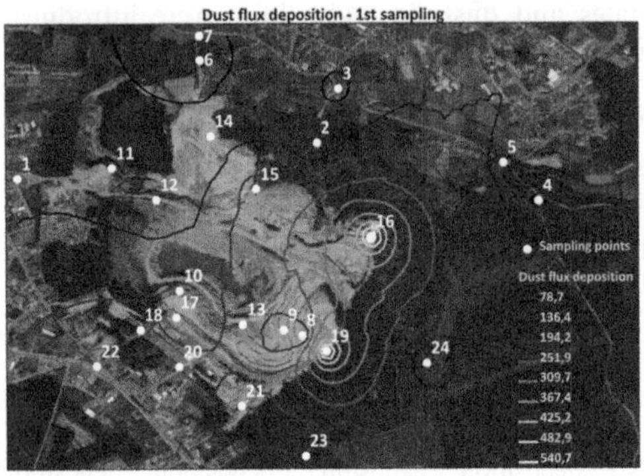

Figure 3. Dust flux deposition for the first sampling.

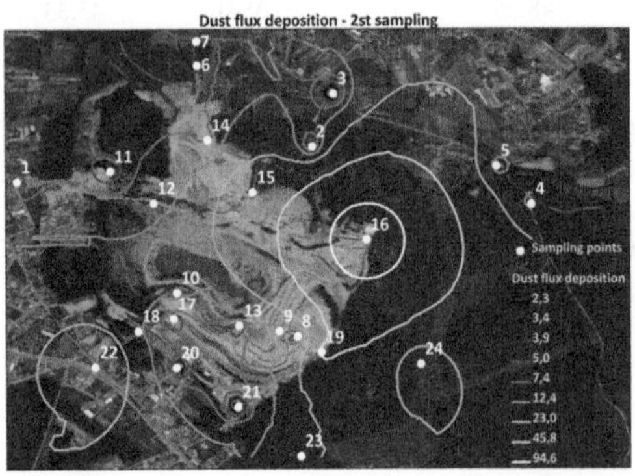

Figure 4. Dust flux deposition for the second sampling.

4 DISCUSSION

In 1st and 2nd samplings, the greatest dust flux deposition is presented, in decreasing order, at the points 16, 19 and 22. They are located close to the crushing plant (first crusher), on the quarry's internal road (with circulation of trucks and other

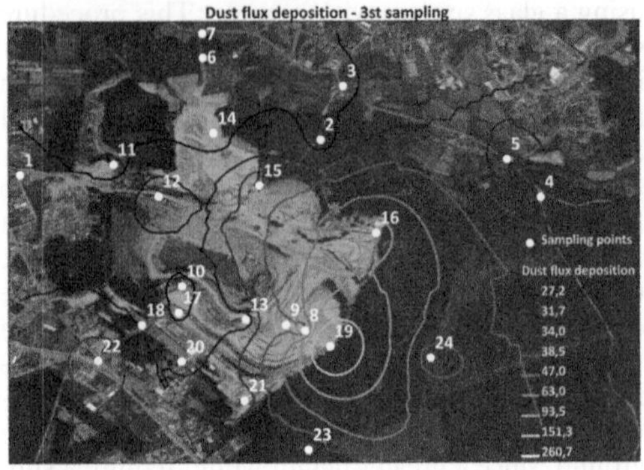

Figure 5. Dust flux deposition for the third sampling.

equipment), and national road, respectively. These results were expected, since the crushing plant (16) originates a lot of dust which falls and deposits on the plate. In point 19, the high value is justified by the fact that it is a local of frequent circulation of trucks and other equipment and is located downwind from the quarry. The point 22 reflects the influence of national road because this point is located upwind from the quarry, and points number 17 and 18 present lower values, so, only the traffic may influence this value.

Regarding the 3rd sampling, one of the possible explanations for the fact that point 19 has more dust than point 16, may be the lower use of the crushing plant. The lower workload can be an influential factor for the observed behavior.

The lowest dust flux deposition corresponds to the sampling point no 6 for both 1st and 3rd sampling periods. Despite being on limit of the quarry, this point is not affected by dust coming from it, since it is located upwind from the quarry. The existing of a tree cover (mainly composed by evergreen trees) also contributes to the observed value. In the 2nd sampling the point correspondent to the lowest dust flux deposition was point 21, located in a construction site, in a low traffic area.

Regarding point no 24, although it is located downwind from the quarry, the low dust flux deposition in the 1st sample allows to confirm the barrier effect of the tree cover existing around, preventing the dust dispersion. In the 2nd and 3rd samplings this issue was not verified due to the beginning of an external work to the quarry which influenced the dust flux deposition.

Comparing the three samplings, it was observed that the 1st sampling presents the highest values, with an average of 166.98 $mg.m^{-2}.day^{-1}$. This could be related to the meteorological conditions, mainly due to the lower precipitation levels, since the wind velocity didn't vary greatly over the three sampling periods. On the other hand, the lowest average of dust flow deposition, 20.12 $mg.m^{-2}.day^{-1}$, was obtained in the 2nd sampling as well as the lowest values for all sampling points, when compared with the first one. In this period the precipitation reached the highest level, 0.12 $mm.h^{-1}$. This reason may have originated the reduction of dust flux deposition and also have influenced the adhesion of dust to the resin/plate.

The intermediate values of wind velocity and precipitation correspond to the intermediate value of dust flux deposition, registering an average for the twenty-four sampling points of 81.36 $mg.m^{-2}.day^{-1}$. Regarding the last samplings it is important to note that the average precipitation between the 2nd and 3rd sampling period is not very distinct, however in the 2nd there were 5 periods of precipitation during all the sampling period, and in the 3rd sampling there was only one initial period of precipitation.

The values obtained e.g. lower in the 2nd sampling are in accordance with the study of Dinis et al. (2012), where the lowest values of precipitation cause major dust flux deposition.

On the other hand, the results obtained may be related to intensity of work occurring in the quarry that are related with more or less detonations/explosions, trucks traffic, work level in the crushing plant, among others. However, the data used in this study didn't take into account this type of information. Nevertheless, in the Figure 2 is possible to observe that the results are coherent during the three sampling campaigns.

5 CONCLUSION

Under the scope of the study of dust dispersion in the air environment, some meteorological parameters were registered, such as wind direction and velocity, temperature and precipitation. Thus, it was concluded that the daily average of wind velocity presented the highest values in the 2nd sampling, 14.3 $km.h^{-1}$, and the lowest result in the 1st sampling, 13.4 $km.h^{-1}$. Regarding the wind direction, applying a circular method of circular statistics, the prevailing winds direction was classified as NNW. The temperature increased during all the sampling periods, ranging from 10 to 27°C. In what concerns precipitation, the daily average was 0.03 $mm.h^{-1}$ in the 1st period and 0.12 $mm.h^{-1}$ in the 2nd one.

The collection of samples to the study of dust dispersion was based on the standard NFX 43 007:2008—Air quality—Ambient air—Determination of the mass of dry atmospheric depositions—Sampling on deposit plates—Preparation and treatment. The results showed that the differences obtained between dust flux deposition in each sampling could be related with the meteorological conditions. The highest values were registered in the 1st sampling, which corresponds to the period with lowest precipitation and wind levels. On the other hand, the 2nd sampling corresponded to the period with lowest dust flux deposition, and in opposite, the highest precipitation and wind levels.

In general, the three samplings allow the conclusion that the critical points, of the 24 in analysis, are the local of access to the quarry, the crushing plant and the point located in the national road. In these sites, the average of three samplings for dust flux deposition was 369.49, 337.70 and 152.65 $mg.m^{-2}.day^{-1}$, respectively. The site with lowest dust flux deposition is located on the quarry limit, close to the stock zone. This site is not affected by the prevailing wind direction and the tree cover contributes to the reduced value, with the average of 14.37 $mg.m^{-2}.day^{-1}$.

Through this study it was possible to conclude that the surrounding population could be affected by the dust quarry, however, the dispersion will greatly depends on the meteorological conditions, such as the wind and precipitation, the season of the year and the workload. It was also verified that the national road is also a significant source of dust. Therefore, in many places, mainly northwest of the quarry, the recorded levels are due to the movement of vehicles or other activities that produce dust, and not exactly the quarry activity. It was also observed that, in the point located on the national road, the dust flux deposition is well above the values recorded in the municipal roads, such as the points 1, 3, 5 and 7. According to the results, the tree cover in the quarry surrounding fulfills its functions; however, the presence of a new work outside the quarry influenced the results desired for this site, especially in the 2nd and 3rd samplings.

ACKNOWLEDGMENT

The authors would like to thank Master in Occupational Safety and Hygiene Engineering (MESHO), of the Faculty of Engineering of the University of Porto (FEUP), all the support in the development and international dissemination of this work.

REFERENCES

Aneja, V.P., Isherwood, A., & Morgan, P. 2012. *Characterization of particulate matter (PM10) related to surface coal mining operations in Appalachia.* Atmospheric Environment, 54: 496–501.

Cattle, S.R., Hemi, K., Pearson, G.L., & Sanderson, T. 2012. *Distinguishing and characterizing point-source mining dust and diffuse-source dust deposits in a semi-arid district of eastern Australia*. Aeolian Research, 6: 21–29.

Dinis, M.L., Góis, J., Fiúza, A., Carvalho, J.S., & Castro, A. 2012. *Particulate matter flux deposition in the vicinity of a coal-fired power plant*. Environmental Science and Technology, 488–494.

Karjalainen, A. & Niederlaender, E. (2004). *Occupational Diseases in Europe in 2001*. Statistics in focus—Population and social conditions. Eurostat.

NEPSI. 2006. European Network for Silica. *Guia de Melhores Práticas para a protecção da saúde dos trabalhadores através do correcto manuseamento e utilização da sílica cristalina e produtos relacionados*.

Petavratzi, E., Kingman, S., & Lowndes, I. 2005. *Particulates from mining operations: A review of sources, effects and regulations*. Minerals Engineering, 18:1183–1199.

Polichetti, G., Cocco, S., Spinali, A., Trimarco, V., & Nunziata, A. 2009. *Effects of particulate matter (PM10, PM2.5 and PM1) on the cardiovascular system*. Toxicology, 261:1–8.

Occupational Safety and Hygiene II – Arezes et al. (eds)
 ISBN 978-1-138-00144-2

Study of Radon in dwellings—case study in Moimenta da Beira

C. Carvalho, A. Ferreira & J.P. Figueiredo
Instituto Politécnico de Coimbra, ESTeSC, Coimbra Health School, Saúde Ambiental, Portugal

ABSTRACT: Exposure to ionizing radiation causes changes in the cells of living organisms causing lung cancer. Radon is descended from uranium in rocks and soils and has a tendency to concentrate in enclosed spaces such as homes. This study aimed the evaluation of the radon concentration inside houses of the Moimenta da Beira. This was a study of the described-correlational level II, and the type of cohort is transverse. The sample included 19 dwellings. The collection of data used is the passive type detectors LR-115 and also a questionnaire on the characteristics of the dwelling. It was concluded that the average radon concentration inside houses of Moimenta da Beira (687,63 Bq/m^3) shows values well above the legislated (400 Bq/m^3). Indicators as "building materials", "old", and "ventilation options", among others do not influence the levels of concentration, concluding that the type of existing soil is the main source of radon.

1 INTRODUCTION

Exposure of human beings to ionizing radiation from natural sources is a lasting and inevitable aspect of life on Earth (UNSCEAR, 2000) and, despite this, its perception by society is unknown, since its main worries are relative to anthropogenic radiation, such as medical activity and nuclear power plants. Thus, natural radiation includes the radiation produced by cosmic rays and by the disintegration of radioactive isotopes of U, Th and K existent on rocks (Neves et al. 2009). Natural radioactivity is responsible for approximately 85% of all ionizing radiation that people are exposed to, and radon is responsible for approximately half of this value (Rêgo, 2004).

Radon is a radioactive gas generated naturally by the decay of uranium (occurring in soils, rocks and water), which may generate other radioactive elements upon its disintegration (such as polonium, bismuth and lead), all of them causing hazardous effect on people's health condition due to ionizing radiation (Dias et al. 2009). Granitic rocks usually have higher radon concentrations, whilst limestone rocks have lower concentrations. Radon's release to the atmosphere is also conditioned by the permeability and porosity of the soils and rocks. Atmospheric pressure, humidity and temperature may also influence the release of the said gas (Ferreira et al. 2006). For these reasons, radon's concentration in the atmosphere is not constant, changing from one region to another accordingly to the exposure time. In this way, the regions in Portugal with higher concentration of granitic rocks are more exposed to radon than other regions where granite is scarce, since the exposure results not only from the soil itself but also of the type of rock used in the construction of the dwelling or structures (Coelho, 2006).

On the exterior, radon tends to scatter, exhibiting lower concentrations, usually below 10 Bq/m^3, whereas inside the dwellings, radon's concentration may reach levels well above 400 Bq/m^3 or even 1000 Bq/m^3, for this gas has the ability to infiltrate through the soil to the inside of dwellings through cracks, poorly sealed joints and drainage infrastructures. Recent constructions are also better sealed to assure better thermal stability and energetic savings, resulting in a higher exposure. Since radon is denser than air, it has the ability to build up on the lower floors of the dwellings (Dias et al. 2009, Ferreira et al. 2006). Changes on radon's concentration levels along the year also occur. On the winter, concentration levels are frequently higher due to less airing on the dwellings. The opposite situation is evidenced outside the dwellings, that is, on the summer concentrations on the exterior are higher than on the winter due to the rise of temperature and fall of humidity (Rêgo, 2004).

According to the European Environment Agency, the exposure to radon is one of the greatest proven causing agents of cancer linked to environmental causes. The emission of α particles, on its own and by some of its descendants, is an enhancing factor of pathologies on the cellular level, specially of the respiratory tract (Coelho et al. 2006). In confined spaces, radon is an important risk factor to public health, being estimated that 10% of all cases of lung cancer are caused by exposure to this gas. However, the exposure to radon inside dwellings hasn't been considered a major public health issue in Portugal, contrarily to what happens in many developed countries such as Ireland (Veloso et al. 2011). Moimenta da Beira is located amongst the Douro valley and is based above a granitic spot, being also observed relevant shale areas. Taking into account that some of the studies carried in Portugal indicated radon levels above the advised levels on national decrees (Decree—law nº 79/2006 of April 4th), and also indicated the higher levels in regions of granitic nature, an appreciation of the natural radioactivity in the region is considered suitable.

The main goal of this study was to evaluate the radon levels in 19 dwellings in Moimenta da Beira, relating these values to the presence of granite.

2 MATERIAL AND METHODS

The research had as target population, dwelling in Moimenta da Beira. The sample included 19 dwellings, meaning this was a nonprobability sample in what concerns the type of the sample, and a convenience sampling in what concerns its technique. It was a level II descriptive–correlational research design and a cross-sectional study. For data harvesting purposes, a survey provided by IST (Institute Superior Technical) of the Technical University of Lisbon was used. This survey aims to study the technical features of the dwellings and passive detectors of LR-115 type (a special film provided by the same institution). Radon's emitted radiation pressures this film, causing microscopic holes on it. After special treatment, radon's concentration on the air where the film was exposed was determined by the whole count on the film. A detector was placed inside each dwelling, where it remained for approximately 80 days, thus representing the exposure during the winter season. To standardize the evaluation amongst the different sampled dwellings, all the detectors where applied 1,5–2 meters above the floor, in floors mostly directly above the ground (ground floors). These where directly applied on the wall of the division where its inhabitants spent most of their time. The geologic map of the county was used for a better choice of the dwellings to be evaluated, to guarantee that all of them where laid above the same type of soil.

The dwellings' sample choice was based on the percentage of granite contained on the building material of the infrastructures with specialized aid from an architect for obtaining the said value.

IBM SPSS version 21 was the chosen software for data processing. As descriptive measurements, the following where applied: Frequency (Absolute and Relative), Location and Dispersion Measurements. The statistical tests applied in the study where: Pearson's Linear Correlation Test; *t*-Student for one sample; ANOVA with fixed factor; *t*-Student for independent samples and Spearman's Ordinal Correlation Coefficient. The interpretation of these statistical tests was performed with a significance $\alpha = 0{,}05$ a confidence interval of 95%.

3 RESULTS

The study was conducted in Moimenta da Beira, in 19 dwellings which had granite in its constitution (in different percentages). Every dwelling was laid above a similar type of soil, as seen in the geological map of the county. Concerning the evaluation of the radon's concentration values inside the dwellings, the study aimed to verify if the latter respected the advisable limits proposed in the effective national decree (Decree—law nº 79/2006).

The estimated radon's average concentration was significantly different (*p-value* < 0,05) when compared to the average value defined in the legislation (400 Bq/m^3) concerning the safety standards related to dwellings. The observed mean differences of the estimated values are statistically significant in view of the criteria legally defined ($p < 0{,}05$). It was evidenced that the dwellings presented a mean difference of 287,6 Bq/m^3, which implies a significantly higher average radon concentration level comparatively to the reference value. Figure 1 shows that only 3 out of the 19 studied dwellings evidenced average radon's concentrations below 400 Bq/m^3. The remaining dwellings showed values above this legally recommend maximum value.

The dwellings 1, 11 and 19 where the ones that evidenced higher average concentrations of radon, placing them the farthest from the legal recommended value. After this preliminary analysis, it was evaluated if de radon's concentration levels inside the dwellings was related to the presence of granite in the building materials, but a significant correlation between the percentage of radon in the dwelling and the average radon concentration levels wasn't evidenced ($r_{(19)} = 0{,}192$; *p-value* = 0,431).

Table 1. Observed average differences of radon relatively to the criteria of 400 Bq/m^3.

	N	$\overline{x} \pm s$ (Bq/m^3)	Mean difference	Reference value (400 Bq/m^3)
Value of radon's concentration inside the dwellings	19	687,6 ± 363,9	287,6	$t(18) = 3{,}445$ *p-value* = 0,003

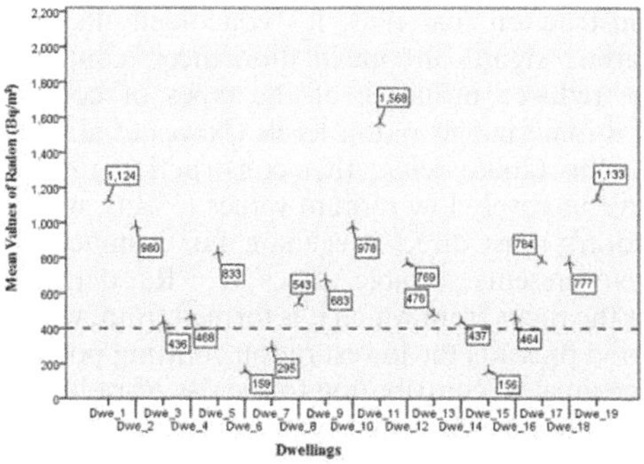

Figure 1. Average values of radon in each room and cool compared to the recommended value.

Table 2. Observed mean radon levels as a function of the construction's materials.

Materials used in the construction	N	$\bar{x} \pm s$ (Bq/m³)	IC95% $\bar{x}$	F(gl); *p-value*
Granite	5	850,0 ± 506,5	[221,2; 1478,8]	0,317(2,16) 0,733
Brick and concrete	7	671,43 ± 292,8	[400,6; 942,3]	
Granite, brick and concrete	7	587,9 ± 326,7	[285,7; 863,1]	
Total	19	687,3 ± 363,9	[512,2; 863,1]	

The possibility of changing of radon's concentration levels inside the dwellings depending on their age was also studied. These are within the interval of 3 to 173 years, with an average of 35,47 years. However there is no connection amongst the analyzed values ($r_{(19)} = -0{,}341$; *p-value* = 0,153) for a significance level of 0,05. However, it can be stressed that more recent dwellings present higher concentration values than older ones. According table 2, it was evidenced that radon's average concentration levels might change with the type of materials present in the dwellings.

As can be observed, the average analytical values of radon inside the dwellings where higher in dwellings built mostly using granite ($\bar{x}$ = 850,00 ± 506,47 Bq/m³), comparing to the remaining dwellings with other types of construction materials. However, it was found that there aren't significant differences amongst the average analyzed radon's concentration levels in dwellings with different building materials (*p-value* > 0,05). It was aimed to find if the floor's coating would influence radon's concentration levels inside the dwellings, i.e., radon's infiltration through the floor. Radon's analytical average values inside the dwellings where higher in dwellings with tiled floors ($\bar{x}$ = 784,67 ± 337,76 Bq/m³), relative to wood (323,75 ± 196,75). In spite of this fact, there aren't statistically significant mean differences between radon levels and the material that coats the floor (*p-value* > 0,05). It was then aimed to find if the amount of ventilation options would influence the variability of radon's average concentration inside the dwellings.

Table 3. Description of radon's average observed values as function of the number of ventilation options.

Number of ventilation options	N	$x \pm s$ (Bq/m³)	IC95% $\bar{x}$	F(gl); *p-value*
One ventilation option	8	534,8 ± 273,5	[306,1;763,4]	3,42(2,15) 0,06
Two ventilation options	2	473,0 ± 7,1	[409,5;536,5]	
More than two ventilation options	8	857,6 ± 425,0	[502,3;1212,9]	
Total	18	657,6 ± 367,4	[488,58;854,1]	

Taking the previously presented values in account, radon's concentration inside the dwellings was higher in dwelling which had more than two ventilation options ($\bar{x}$ = 857,63 ± 425,04 Bq/m³), comparatively to the remaining dwelling with one or two ventilation options. It was concluded that there aren't statistically significant mean differences between radon's concentration levels and the number of different option types on the dwellings (*p-value* > 0,05).

Lastly, it was studied if radon's concentration levels were related to the number of ventilation options (doors and windows) existent in the dwellings' divisions. It can be observed that there isn't significant statistical evidence to assert that radon's concentration levels inside the dwellings and the number existent of doors and windows are related ($\rho_{Doors} = 0{,}166$; *p-value* = 0,497 e $\rho_{Windows} = -0{,}214$; *p-value* = 0,379), whereby it wasn't verified any change in radon's concentration with the number of doors and windows.

4 DISCUSSION

Since people spend about 65 to 80% of their time in closed environments, the interior air quality assumes an important role in studies which aim to protect and promote public health (Sanguessuga, 2012).

Radon is considered a chemical contaminant of the interior air and, since high concentrations represents risks for public health, it is necessary to be aware and act on this issue. Thus, this leads to a need to conduct studies related to this problem in order to adopt the necessary radon mitigation measures based on the obtained results (Oliveira, 2006). Due to its lithological conditions, mainly the presence of granitic rocks, Moimenta da Beira constitutes a great potential relatively to the existence of high radon levels. In this investigation, it was concluded that the average concentration of radon inside the dwellings in Moimenta da Beira was of 687,6 ± 363,9 Bq/m^3, a value which significantly exceeds the national and European maximum recommended value (400 Bq/m^3), reaching values as high as 1568 Bq/m^3. This reference value (400 Bq/m^3) doesn't represent a stark barrier but a maximum advisable limit, and from this value on it is necessary to apply mitigation measures (Radiological Protection Institute of Ireland). Thus, the analyzed radon levels can be justified by the high amount of granitic rocks existent in the soil, being that many of the dwellings are laid directly above granitic blocks (Duarte et al. 2004).

With the analysis of the results, significant differences between radon's concentration levels and the percentage of granite existent in the dwellings were not found, but there are dwellings with a high percentage of granite which evidence high radon concentration levels. This rock has got U and Th which, being chemical elements with high radioactive potential and since radon belongs to the radioactive decay chain of uranium, justifies the fact that higher radon concentration levels can be found inside dwellings built from granite (Ferreira et al. 2006).

In what concerns the dwellings' age, it was verified that most recent dwellings have higher radon concentration levels than older ones. There might be a relation between these values and the frequent refurbishments to which the dwellings were subjected along the years (Dias et al. 2009). Thus, it's concluded that the dwellings age is a weak decisive factor in the mean differences of radon's concentration levels inside the dwellings, also concluding that the recent dwellings exceed the recommended limit defined by the European Union of 200 Bq/m^3 for dwellings under construction or recent dwellings (Simões et al. 2006). On the other hand, Ribeiro (2009), verified that older dwellings tend to have higher radon concentrations, which can be explained by the fact that they were built from stone, possessed few windows, were of smaller dimensions and presented more structural defects (Ribeiro, 2009). Also in the analysis of the change in radon's concentration levels with the used construction materials, it's concluded that there weren't significant mean differences, confirming the reduced influence of the types of construction materials in radon levels (Neves et al. 2009). Coelho (2006) refers that construction materials have relatively low radium values (^{226}Ra), which is radon's most direct precursor. For instance, concrete presents variable values of ^{226}Ra, depending on the rocks from which it is formed from, whereas wood presents the lowest radon emitting potential. The greatest contribution to the rise of radon concentration levels is that of the radon coming from the soils (Coelho et al. 2006). In spite of this, for its risk to potentiate radon concentrations, the materials applied in the construction of dwellings must be selected in order to minimize the said risk, mainly in higher risk geological zones such as Moimenta da Beira. It may also me stressed that the type of floor's coating also presents a low contribution to the change of radon's concentration levels inside the dwellings (Dias et al. 2009).

In what comes to the variability of radon average levels with the number of ventilation options, it was evidenced that there aren't statistically significant differences between these variables. Higher average radon concentrations were registered in dwellings with more than two ventilation options comparing with dwellings with only one or two ventilation options. This values may come as a consequence of house airing habits, ie, although a dwelling may have more ventilation options, the practice of the airing/ventilation of the house may be less frequent comparatively to dwellings with only one or two ventilation options, but where there may exist more frequent and effective airing/ventilation habits (Sanguessuga, 2012, Dias et al. 2009). To complement this, there wasn't evidence of any relation between the amount of doors and windows of the division and its average radon level. This fact can be justified by the same reasons presented previously. An adequate ventilation of the dwelling is one of the most important measures to control the interior radon concentration levels (Ferreira et al. 2006).

Thus, it can be concluded that the change on the average radon concentrations may depend mainly on the type of soil on which the dwellings were constructed. Some studies refer that 95% of all radon contained within dwellings comes from the subsoil, 5% from the construction material and less than 1% is released from the consumed water (Neves et al. 2009, Deco Proteste, 2012, Ferreira et al. 2006).

With the high radon concentration levels found inside the dwellings, public health becomes relevant in the identification and implementation of mitigation measures of high radon levels inside dwellings, thus promoting the inhabitants health and lowering the high radon concentration levels. It also seeks the creation of environmental conditions in the dwellings, and public spaces which favor the health and well-being of the population and, subsequently, of its social and personal success, favoring the binomial health/environment (Deco Proteste, 2012, Simões et al. 2006). It must be stressed that the fact that radon is detected inside dwellings doesn't imply that they have to be demolished. Rather than this, it's necessary to apply mitigation measures, but one should also keep in mind that the most effective and less costly measure to minimize radon levels is prevention. One of the main measures to be applied is an adequate and usual airing/ventilation of the dwelling, by simply opening the doors and windows in order to allow air circulation within the dwelling. This ventilation must be regulated since it lasts for few hours, being registered a rapid reinstatement of radon's concentrations (Ferreira et al. 2006).

Another measure which must be used in the case of existing dwellings is the identification and sealing of cracks, fractures or joint of pipes and electric cables or of poorly sealed plumbing, since they enhance the entrance of radon gas in the dwelling (Martins, 2012). Through adequate repairs, it is possible to reduce radon below the reference value in the vast majority of buildings. The presented mitigation measures are feasible not only in existent dwellings but also in new constructions, in which prevention is of utmost importance (Zeeb et al. 2009).

Thus, it is necessary to conduct field studies in granitic regions, which allow the evaluation if the said region is prone to reveal high radon concentration values, which is common practice in other countries such as Finland, Sweden, Ireland, United Kingdom, amongst others (Zeeb et al. 2009). Another measure, in order to prevent lung cancer, is the minimization of indoors smoking where radon's concentration levels are high, since smokers have a significantly increased risk of contracting lung cancer, thus becoming hazardous to public health (Oliveira, 2006, Zhang et al. 2010).

The results of measurements in dwellings are not automatically applicable to workplaces, due to different forms of utilisation, constructional conditions, time exposure, heating and ventilation conditions. But the special conditions in the workplaces have considerable influence on the radiation dose of the workers (Reichet et al. 1999). Workers who may be exposed to high levels of radon can be sites with particular characteristics, such as spas, radon in the atmosphere of these establishments comes from degassing of water used in hydrotherapy, in which contains high concentrations of dissolved radon. Also occurs in non-radioactive ores mines and may also occur in buildings constructed in areas with high radon exhalation in soils, such as hospitals, schools, offices, etc. So, can present radon concentrations above the recommended and cause continued exposure of workers by inhalation of gas (Carvalho et al. 2006).

5 CONCLUSIONS

The results of this study allowed understanding that the average radon concentration in Moimenta da Beira presents values way above the ones mentioned in national and European legislation. It was verified that indicators such as "construction materials", "age", "ventilation options" and "floor coating" didn't influence radon's concentration levels inside the dwellings, concluding that the soil composition, mainly granitic, was the main source of radon.

Thus, it is essential that risk charts are prepared by the autarchies to be included in the territorial planning in the Municipal Master Plan, Urbanization Plans and Detail Plans, allowing the marking of high risk areas, so that the necessary mitigation measures in the high radon risk potential areas can be applied. It must be stressed that the national decree (Decree—law nº 79/2006, of April 4th) refers that a survey must be conducted in buildings laid above granitic zones, namely in the district of Viseu. It is important that the civil parishes or the county become more aware about this issue, and thus providing anti radon measures to reduce the exposure of the inhabitants to this type of radiation, lowering the chance of occurrence of lung cancer. In Portugal, the architects and civil engineers themselves don't receive any sort of training in what concerns this particular issue, which reveals the scarce importance that this issue currently has institutionally.

It is thus necessary to raise awareness and consciousness in Moimenta da Beira, both in relation to the autarchy, the population and the builders, so that mitigation measures are imposed, enabling a reduction in radon's concentration levels inside the dwellings. As future work, it would be interesting to further develop this study, using a more representative amount of samples so that a greater variability of the values might be found. Concerning this theme, it is still necessary to perform various studies and surveys in order to achieve a clearer vision of the several factors which influence radon's concentration inside dwellings. Another study may focus individuals who work directly with granite, in order to understand if they are more likely to be exposed to higher radon concentrations.

REFERENCES

Carvalho, F. Radão e edifícios—Qualidade do ar interior. Revista Industria e Ambiente, nº52, p.22–26, 2008.

Carvalho, F.P. and Reis, M.C. Radon in Portuguese Houses and Workplaces. In Proceed Internacional Conference Healthy Buildings HB 2006.

Coelho, F. M e Neves L.J.P.F e Gomes. M.E.P. Distribuição de radão em habitações da região de Vila Real: condicionantes geológicas. [Dissertação] - Universidade de Trás-os-Montes e Alto Douro, 2006.

Deco Proteste. Dossiês Qualidade do ar interior. [Online] [Citação:10 de Dezembro de 2012]. http://www.deco.proteste.pt/saude/nc/dossie/qualidade-do-ar-interior/1.

Dias, A.I, Ferreira. A e Figueiredo, J.P. Influência do Granito na variação dos níveis de radão nas Habitações Estudo de caso Vila Pouca de Aguiar [dissertação]. Coimbra: Escola Superior de Tecnologia e Saúde de Coimbra, 2009.

Duarte, P., Silva e J.J.C. e Reis, M.J. Radão no Interior de Habitações do concelho de Nisa [Dissertação]. Instituto Tecnológico e Nuclear, Sacavém e Universidade de Évora, 2004.

Ferreira, MJMM e Coelho, MJP. O Radão nos edifícios—minimização da perigosidade [dissertação]. Porto: Universidade Fernando Pessoa, 2006.

Martins, Unidade de Saúde Pública. Radão um gás radioativo de origem natural. [Online] [Citação:13 de Dezembro de 2012]. http://usp-be.blogspot.pt/2012/11/radao-um-gas-radioativo-de-origem.html.

Neves, F; Pereira. A. Radiatividade natural e ordenamento do território: o contributo das Ciências da Terra [artigo cientifico] 2004.

Oliveira, C.A.C. A Radioatividade e o ambiente no ensino secundário. Universidade de Lisboa. Faculdade de Ciências, Departamento de Física: s.n., Outubro 2006.

Radiological Protection Institute of Ireland. Understanding Radon Remediation—A householder's Guide.

Rêgo. F; As Radiações no ensino. Universidade de Lisboa, Faculdade de Ciências Departamento de Física. Lisboa: s.n., Maio 2004.

Reichelt, A. Reineking, A. Lehmann, K.-H. Porstend Örfer, J., Sclawedt, J. Streil T. and Radon in Workplaces, Germany 1997.

Ribeiro, C. Emissão de radão em materiais de construção. Escola Superior de Tecnologia e Gestão. Instituto Politécnico da Guarda. Guarda: s.n., Dezembro 2009.

Sanguessuga, M.S.G. Síndroma dos Edifícios Doentes. Escola Superior de Tecnologia e Saúde de Lisboa. Instituto Politécnico de Lisboa. Lisboa: s.n., Abril 2012.

Simões, L., et al. Concentração de Radão em Espaços Interiores da Área de Viseu [dissertação]. Viseu: Departamento de Ambiente da Escola Superior de Tecnologia do Instituto Politécnico de Viseu, 2006.

UNSCEAR (2000) Sources and Effects of Ionizing Radition. United Nations Scientific Committee on Effects of Atomic Radiation (UNSCEAR). United Nations, New Yorkm vol I, p.84.

Veloso, B, Nogueira J.R. e Cardoso, M.F. Lung cancer and indoor radon exposure in the north of Portugal—An ecological study [Dissertação]. Porto, 2011.

Zeeb, H; Shannoun F, Who handbook on Indoor Radon, a public health perspective, France 2009.

Zhang, W.; Chow, Y.; Meara, J. and Green, M. Evaluation and equity audit of the domestic radon programme in England. Centre for Radiation, Chemical and Environmental Hazards, HPA, United Kingdom, 2010.

Occupational Safety and Hygiene II – Arezes et al. (eds)
 ISBN 978-1-138-00144-2

Occupational exposure to car workshops air pollutants

Élia Oliveira, Ana Ferreira & João Paulo Figueiredo
Instituto Politécnico de Coimbra, ESTeSC—Coimbra Health School, Saúde Ambiental, Portugal

ABSTRACT: Since fossil fuels are the main source of pollutants and toxic chemicals that pollute the air, it is pertinent to consider how is the auto body shop air quality and what harm it can bring to workers healthcare. Was used the portable reading device in real time for monitoring occupational and the application of a checklist. Was conducted a descriptive-correlational study, level II, cross-sectional cohort. The workshops have undergone a positive development with the adoption of for protecting the worker's health systems and standards, were not registered air pollutants levels in alarming concentrations. Only the ozone concentrations average values exceeded the ELV (0.05 ppm) of the reference standard. Were recorded higher concentrations of air pollutants in the "Painting" and "Plate" sectors. Thus, the importance of environmental workplace monitoring should be relive, also the workers health surveillance and especially highlight the training and information program adoption importance.

1 INTRODUCTION

The car workshops are, par excellence, a place with a strong potential for hazards and consequent risks (Binder *et al.*, 2001). In addition to the potentially dangerous substances existing in these workplaces, we must consider the target, the exposed worker, the route of entry into the human body, the exhibition site (position/work equipment), the exposure time or duration, as well as the exposure frequency to the substance may or may not exercise their adverse effects that can range from simple irritation to a fatal disease (Matos *et al.*, 2012; Crepaldi, 2012). The combustion that often occurs in car workshops is generally inefficient and the poor ventilation can result in high exposure, which is associated with respiratory symptoms and infections (Binder *et al.*, 2001). The air quality issue in terms of occupational health is enshrined in the Portuguese Standard 1796:2007 and more recently in Decree-Law no.° 24/2012 of 6 February (Ministério da Economia e do Emprego, 2012). Most of the exposure limit values are developed for a single chemical. However, the working environment is often composed of various substances causing multiple exposures, either simultaneously or sequentially. Thus, if the sum: C1/VLE1+C2/VLE2+ … +Cn/VLEn>1. Then the mixture exposure limit value is considered exceeded.

The absence till now of air renovation minimum values, the weak control and lack of proper maintenance of the facilities during operation, are the reason for the emergence of problems in terms of air quality. The occupational risks prevention also depends on a high degree of workers adopt appropriate behaviors depending on the security requirements imposed by chemical agents (Fernades *et al.*, 2012).

Thus, the aim of this study was to evaluate the workers occupational exposure, compare the results obtained in order to perceive what physical and structural characteristics influence the values obtained. We also intend to investigate which differences of values recorded between the different workshops sectors and to establish possible complex relationships between the characteristics of the sectors and the values obtained. Thus we intend to sensitize and inform employers to implement preventive mechanisms or practical, simple and effective to protect somehow the workers concerned.

2 MATERIAL AND METHODS

Was conducted a descriptive-correlational study, level II, cross-sectional cohort (observational). The type of sampling used was not probabilistic, and the sampling technique for convenience. In the 700 workshops existing in Coimbra district universe, were studied ten workshops as sample. The workshops recorded in the study perform maintenance and repair of motor vehicles. For occupational monitoring were conducted 64 measurements per day in each workshops studied. Were evaluated different sectors, including the dealer, mechanical painting and plate, this cycle was repeated four times during the day corresponding to the measurement periods designated Early Morning

(8:30–10:30 a.m.), Morning Late (10:30–12 a.m.), Early Afternoon (13:30–15:30 p.m.) and Late Afternoon (15:30–17 p.m.). The measurements were taken between December 2012 and February 2013 months with similar climatic conditions during this period.

In the data collection was used portable monitoring Environmental Monitoring Instrument (EVM), 3M Company Brand Quest Technologies, model R.10V with a laser photometer that performs the real-time air pollutants measurement. That equipment was used to take measurements of particulate matter (dust), gases, temperature, relative humidity and dew point. Analytical data obtained were subjected to treatment using the Software QuestSuite Professional II and were processed statistically by use of the IBM-SPSS version 21.

Was considered as a reference for comparison in statistical inference the national Exposure Limit Value—Weighted Average (ELV), meaning weighted average concentration for a day's work of eight hours and a 40-hour week, which is considered practically all workers may be exposed day after day without adverse health, constant in NP 1796:2007 for Ozone (O_3), Carbon Monoxide (CO), Nitrogen Dioxide (NO_2), Sulfur Dioxide (SO_2) Carbon Dioxide (CO_2) and "Particulate Matter" (IPQ, 2007). This last pollutant despite not being established a ELV, the American Conference of Governmental Industrial Hygienists (ACGIH) is convinced that even particles biologically inert, insoluble or poorly soluble can cause adverse effects and recommends that their concentrations in the air must be kept below 10 mg/m^3 (10000 µg/m^3) for inhalable particles. For comparison of relative humidity and temperature were used as reference the ELV stipulated by the General Regulation of Health and Safety at Work in Commercial Offices and Services approved by Decree-Law No. 243/86 of 20 August (Ministério do Trabalho e Segurança Social, 1986).

On the assessment of the physical and structural facilities, was applied to each workshop a checklist that had as main fields of analysis: workshop's spatial location, number of workers, installation conditions, use of Personal Protective Equipment and Hygiene, Health and Safety Service existence. The analysis of the data was ensured using descriptive statistics.

3 RESULTS

Was performed a workshop characterization. It was concluded that the parishes types distribution is averagely urban, the sampled workshops area average is 3509.972 m^2 and feature an 7m-ceilinged average, the oldest workshop is 40 years and the most recent has 6 operation years (+ - 17). Regarding the total number of workers, it appears that two workshops have fewer than 10 employees, 3 of them have between 10 and 20 workers (inclusive), 4 have between 20 and 30 and 1 workshop had more than 30 workers. It can be seen that, in general, the sectors that included more workers were mechanics and reception/administration. Half of workshops held air conditioning system as heating system. Regarding the fume and dust extraction systems, all the workshops were provided with this system type, however one of the workshops was at the

Table 1. Normative reference values.

PN 1796:2007	O_3	CO	NO_2	SO_2	CO_2	Particulate matter
ELV-WA (ppm)	0,05	25	3	2	5000	10000
DL nº 243/86	Temperature			Humidity Relative		
Reference Values	Between 18°C and 22°C			Between 50% and 70%		

Table 2. Pollutants concentration mean values measured by probe (ppm).

Pollutants types (n = 640)	N	Mean	Standard deviation	Exposure limit value	Value mean difference	Test; df; p
Ozone	160	0,205	0,218	0,05	0,155	9,017;159; <0,0001
Carbon Monoxide	160	0,500	0,925	25,0	24,500	−335,085;159; <0,0001
Nitrogen Dioxide	160	0,16	0,054	3,0	−2,946	−5889,820;159; <0,0001
Sulfur Dioxide	160	0,001	0,008	2,0	−1,999	−3199,0;159; <0,0001

Dependent Variable: Pollutants measured by the average probe Statistical tests: t-Student test for one sample.

Table 3. Ozone concentration mean values measured by employment sector.

Pollutant: Ozone (ELV 0,05 ppm)	N	Mean	Standard deviation	95% confidence interval	
				Lower value	Upper value
Mechanical	40	0,194	0,216	0,125	0,263
Plate	40	0,224	0,248	0,145	0,303
Painting	40	0,229	0,227	0,156	0,301
Concessionaire	40	0,174	0,177	0,118	0,231
Total	160	0,205	0,218	0,171	0,239

Statistical test: ANOVA F: 0.557, df: 3,156, p-value = 0.644.

Table 4. Carbon monoxide concentration mean values measured by sector.

Pollutant: Carbon monoxide (ELV 25 ppm)	N	Mean	Standard deviation	95% confidence interval	
				Lower value	Upper value
Mechanical	40	0,550	0,986	0,235	0,865
Plate	40	0,700	1,159	0,329	1,071
Painting	40	0,475	0,784	0,224	0,726
Concessionaire	40	0,275	0,679	0,058	0,492
Total	160	0,500	0,925	0,356	0,644

Statistical test: Brown-Forsythe F: 1.474, df: 3; 134.272, p-value = 0.224.

Table 5. Nitrogen dioxide concentration mean values by employment sector.

Pollutant: Nitrogen (ELV 3 ppm)	N	Mean	Standard deviation	95% confidence interval	
				Lower value	Upper value
Mechanical	40	0,015	0,066	–0,006	0,036
Plate	40	0,020	0,052	0,004	0,037
Painting	40	0,020	0,056	0,002	0,038
Concessionaire	40	0,010	0,038	–0,002	0,022
Total	160	0,016	0,054	0,008	0,025

Statistical test: ANOVA F: 0.314, df: 3,156, p-value = 0.815.

moment the system non-functional. This system was in all the workshops evaluated in the painting area concretely, especially in paint box and some workshops had also a piping systems suction gas and dust in the mechanics area. These tubes were connected to the exhaust pipes to eliminate the existence of the air pollutants.

Was proposed to evaluate the various pollutants concentrations average variation present in maintenance and repair motor vehicles shops taking into account the ELV for each one.

Were made ratings using different probes for each of the pollutants which were made 160 measurements for each pollutant in different workshop. Consider the following table:

We found that workshops O_3 concentration level value exceed the legislated value, this difference ratio was 0,155 ppm. Regarding the estimated CO, NO_2 and SO_2 average variation find that these pollutants were significantly below over the permissible exposure limit ($p < 0.0001$) and the SO_2 concentrations are practically nonexistent.

Was suggested to evaluate the O_3 concentrations average variation in the different maintenance shops and repair of motor vehicles working sectors. Consider the following table:

It was found that there were no significant differences in O_3 concentration mean for different working sectors ($p > 0.05$), however the mean values were higher in "Plate" and "Painting" sectors comparing to "Mechanical" and "Dealer" sectors.

It was also evaluated the CO concentrations variation average in the different labor sectors. Consider the following table:

Was notary no significant differences in recorded CO concentration mean among different work sectors ($p > 0.05$), however the sectors "Plate", "Mechanical" and "Painting" showed higher values compared to values recorded in the "Dealer" sector.

It was suggested to assess NO_2 variability concentrations mean values compared to job sectors. Consider the following table:

It was found that there were no significant differences in NO_2 concentration's mean between different work sectors ($p > 0.05$), however the mean values were higher in "Plate" and "Painting" sectors.

Similarly, it was felt appropriate to analyze the CO_2 concentrations average variation between different work sectors. Observe the following table:

Gauged that there were no significant CO_2 mean concentration differences between different work sectors ($p > 0.05$), however the mean values were higher in "Plate" and "Painting" sectors.

Were study "Particulate Matter" concentration's variations between different work sectors. Consider the following table:

We found that "Particulate Matter" average concentrations are different depending on the job sector assessed. Mean differences occurred, according to the *Games-Howell* multiple comparisons test among the "Mechanical" sector compared to the "Painting" sector ($p < 0.0001$) and compared to "Dealer" ($p = 0.005$) sector. We also found differences between the pollutant studied in the "Plate" sector compared to "Dealer" sector ($p < 0.0001$). Finally, also recorded differences between

Table 6. Carbon dioxide concentration mean values by sector.

Pollutant: Carbon dioxide (ELV 5000 ppm)	N	Mean	Standard deviation	95% confidence interval	
				Lower value	Upper value
Mechanical	40	542,190	1698,067	277,060	807,330
Plate	40	661,260	1325,206	454,350	868,180
Painting	40	634,370	1165,129	452,450	816,290
Concessionaire	40	540,110	1005,036	383,190	697,040
Total	160	594,480	1321,560	491,900	697,070

Statistical test: ANOVA F: 0.357, df: 3.636, p-value = 0.784.

Table 7. "Particulate matter" concentration mean values measured by sector.

Pollutant: particulate matter (ELV 10000 μg/m^3)	N	Mean	Standard deviation	95% confidence interval	
				Lower value	Upper value
Mechanical	40	85,090	45,143	78,040	92,140
Plate	40	96,930	44,504	89,980	103,870
Painting	40	113,040	75,520	101,250	124,830
Concessionaire	40	69,860	35,828	64,270	75,460
Total	160	91,230	54,675	86,980	95,470

Statistical test: Brown-Forsythe F: 19.429, df: 3; 455.801, p-value < 0.0001.

Table 8. Temperature mean values measured by employment sector.

Temperature (reference values: between 18 and 22°C)	N	Mean	Standard deviation	95% confidence interval	
				Lower value	Upper value
Mechanical	40	17,516	1,833	17,230	17,802
Plate	40	17,171	1,680	16,908	17,433
Painting	40	17,322	1,636	17,067	17,578
Concessionaire	40	17,889	1,970	17,581	18,196
Total	160	17,475	1,8005	17,335	17,614

Statistical test: Brown-Forsythe F: 4.837, df: 3; 622.123, p-value = 0.002.

the "Painting" compared to "Dealer" sector ($p < 0.0001$).

Similar mode was pertinent to reviewed the temperature average changes between the work sectors. Consider the following table:

The temperature values were different depending on the evaluated sector. Mean temperature differences occurred, according to the multiple comparison *Games-Howell* test, among "Plate" compared to "Dealer" sector ($p = 0.003$). We also found temperature differences between "Painting" compared to "Dealer" sector ($p = 0.028$). Were no significant mean relative humidity differences between different work sectors ($p > 0.05$), however the mean values were higher in "Plate" and "Painting" sectors.

Then we evaluated "Particulate Matter" concentrations average values evolution during the day considering defined measurement periods. Consider the following graph:

We found that "Particulate Matter" average concentration were different according to measurement periods (Statistical test: Brown-Forsythe F: 24.625, df: 3; 536.882, p-value < 0.0001). Mean differences occurred, according to the *Games-Howell* multiple comparisons test, between "Early Morning" compared to "Early Afternoon" measurement period ($p < 0.0001$) and compared to "Afternoon Late" measurement period ($p < 0.0001$).

We also found differences between "Morning Late" compared to "Early Afternoon" measurement period ($p < 0.0001$) and compared to "Afternoon Late" measurement period ($p < 0.0001$).

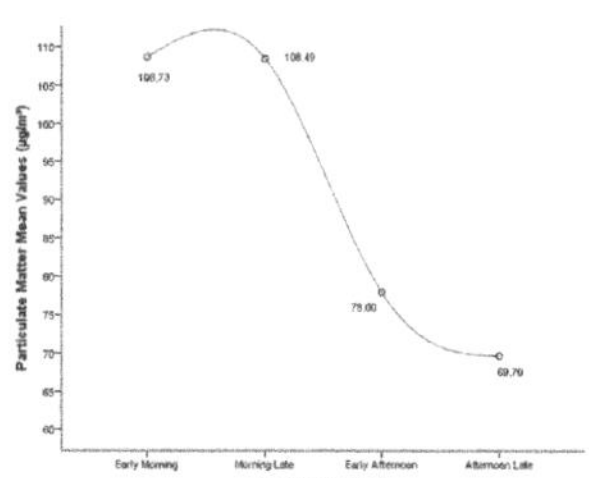

Graph 1. "Particulate Matter" average concentration measured by measuring period.

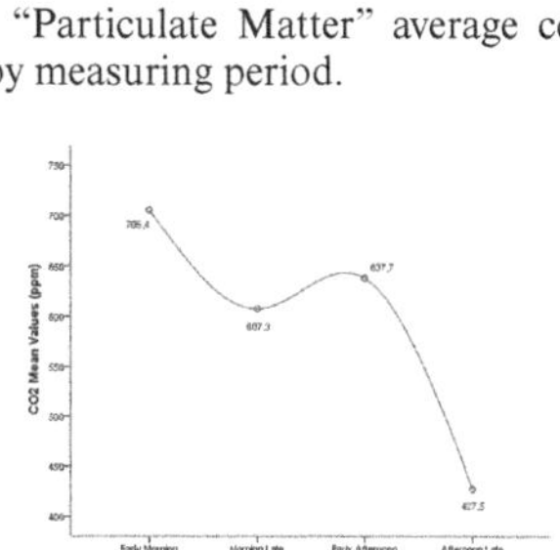

Graph 2. Carbon dioxide mean concentrations by measuring period.

The following graph shows CO_2 concentrations average variation during the day given the measurement periods defined. Consider:

We found that there were no significant CO_2 concentrations mean differences values in different measurement periods recorded (Statistical test: Brown-Forsythe F: 1.291, df: 3; 521.445, p-value = 0.277), however it was notorious that mean values decrease between "Early Morning" and "Morning Late" measurement period and also a decrease, more pronounced, between "Early Afternoon" and "End of the Afternoon" measurement period occurs, so there were a CO_2 concentrations decrease throughout the day.

It was also appropriate to study variation of the O_3, CO, and NO_2 average concentrations during the day given the measurement periods defined. Consider the following graphs:

We found that this pollutants average concentration were not significantly different according to the period in which the measurement was made. (The statistical O test: Brown-Forsythe F: 5.336, df: 3; 137.597, p-value = 0.002; Statistical test CO: Brown-Forsythe F: 2.8, df: 3; 117.521, p-value = 0.043; Testing Statistical NO: Brown-Forsythe F: 2.221, df: 3; 94.656, p-value = 0.91) It can be seen however that O values are increased throughout the day evidenced primarily between "Early Morning" and "Late Afternoon" measurement periods (+0.186 ppm) relative to CO values we can observe a reduction in mean concentration between "Early Morning" and "Morning Late" measurement periods (–0.375 ppm) and the same occurred between "Early Afternoon" and "Afternoon Late" (–0.325 ppm), although less pronounced. As regards NO_2 average concentration was evidenced also a decrease throughout the day.

According to VLE O_3 documentation this pollutant cause adverse effects on lung function, the CO is responsible for causing carboxyhemoglobinmy, NO_2 causes upper respiratory tract and lower respiratory tract irritation and CO causes asphyxiation. Once pollutants evaluated cause a similar toxicological on the same target organ, namely the respiratory system, and since they are simultaneously in workplace air is important to examine if those air mixture presents alarming values

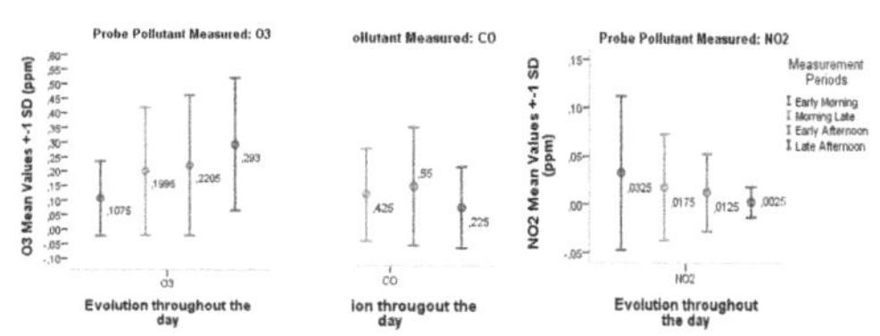

Graph 3. O3, CO and NO2 mean concentration by measuring period.

Table 9. Mixing agents exposure limit values.

Laboral sectores	Mixing sgents ELV: O_3, CO, NO_2, CO_2 and particulate matter	
Mechanical	4,22	>1
Plate	4,66	
Painting	4,74	
Concessionaire	3,61	

for workers health in different work sectors. Was proceeded therefore the VLE analysis for mixing agents. Consider the following table:

We found that the result is greater than unity so the mixing agents ELV is exceeded.

4 DISCUSSION

Considering the physical and structural workshop characteristics evaluated and according to pollutants occupational monitoring results, it is concluded, in agreement with "Occupational Exposure to carbon monoxide" (Claro, *et al.*, 2010) study that the existence of extraction systems gas and dust, air renewal systems and the use of PPE play a key role in reducing the risk of occupational exposure to chemical pollutants. Generally speaking, workshops evaluated did not show air pollution alarming levels. This is due to the fact that in recent years the vehicles workshop concept in Portugal has evolved positively. Currently the physical and structural workshops characteristics and employers are subject to more stringent standards regarding the workers' health protection. This fact can also be explained by the increase of the number of companies that relied on the implementation of a Work Safety and Health Management System, the majority of the workshops sampled were certified. The Portuguese Standard for Work Safety and Health Management Systems is enshrined in OHSAS 18001.

Despite one workshop with the system fume extraction and dust occasionally damaged in paint sector, this fact is not reflected higher levels of pollutants. This may be due to the fact that said sector is not much used during the measurement time and also due to the existence of a large gate nearby.

Although only the O_3 values have exceeded the pollutants evaluated ELV are considered mixing pollutants additive effects and was found values above the unit and should be subject to watch and must be ensured renewal of air continues throughout the day.

In the workshops which were detected values of O_3 above the ELV is pertinent the application of a extraction gas and dust monitoring and detecting failures system mainly in painting and plate in order to reach acceptable O_3 values.

Despite the CO, CO_2, NO_2, SO_2 and Particulate Matter" concordance with normative values in all workshop evaluated, the temperature and relative humidity recorded values were in disagreement with the reference range in some workshops so these parameters should be targeted lookout for companies that provide Work Health and Safety Services to in these workshops. Similarly, it is important to take into account aspects such as floor maintenance, was notorious bad floor conditions in some workshops. Despite not having been recorded "Particulate Matter" worrying values, the floor should be regular, waterproof and washable so as not potentiate the particles accumulation.

5 CONCLUSIONS

We conclude, therefore, that the adoption of chemical agents workers protection, especially in terms of prevention, present yet a significant challenge when it comes to changing habits by employers and also by the workers themselves. In this sense it is extremely important to implement an effective program of training and information, according to the professional category and specific risks, to provide workers with the necessary tools for this activity development, minimizing occupational exposure to chemical agents. The promotion of safety culture, including adoption of safe behaviors, as well as shared responsibility in achieving good safety, should be strengthened. We also conclude that there is need to continue the study of occupational exposure to car workshops air pollutants, so we propose to extend the study to uncertified and smaller workshops, the assessment of Volatile Organic Compounds (VOC's) concentrations and air velocity measurements since that, for measurement equipment technical reasons was not possible to use in this study. It would be equally relevant reactivate the partnership initially considered in this study with the Cardiology course, which due to both parties impossibilities was not possible to implement effectively, in order to relate the results of occupational health monitoring with workers respiratoy function results.

REFERENCES

Binder, Maria Cecília Pereira, *et al.* Condiições de Traballho em Offiiciinas de Reparação de Veíicullos Automotores de Botucatu (São Paullo): Nota Prévia. 2001.

Claro, Liliana, *et al.* Exposição ocupacional ao monóxido de carbono. 2010.

Conferencia Internacional del Trabajo. 1959.

Crepaldi, Luciane. Levantamento de riscos ambientais na atividade de manutenção de veículos automotores. 2012.

Fernandes, Maura, Borges, Vítor e Gomes, Alexandre E. A exposição a produtos químicos num serviço anatomia patológica. 2012.

Gioda, Adriana e Neto, Francisco Radler de Aquino. Considerações sobre estudos de ambientes industriais e não industriais no Brasil:uma abordagem comparativa. 2003, pp. 1389–1397.

Hoppe, L.F. Qualidade dos ambientes interiores e o papel da saúde ocupacional. 1999, pp. 43–51.

Husman, T., *et al.* Respiratory infections among children in moisture damaged schools. 2002, pp. 485–490.

M.S., Reisch. Paints & coatings. 1999, pp. 77:22–33.

Matos, M. Luísa, Baptista, J. dos Santos e Diogo, M. Tato. Exposição ocupacional a poeiras em lavra a céu aberto. Breve Revisão. 2012.

Mehtal, S., Smith, K.R. e Balakrishnan, k. Using household characteristics to predict respirable particulate levels in rural households of Andhra Pradesh, India.5. 2002, pp. 506–601.

Portugal, Ministétio da Economia e do Emprego. Decreto-Lei nº. 24/2012 de 6 de Fevereiro.

Portugal, Instituto Português da Qualidade (IPQ). Norma Portuguêsa NP 1796:2007.

Portugal, Ministério do Trabalho e Segurança Social. Decreto-Lei nº 243/86 de 20 de Agosto.

Rodrigues, C., Guedes, J.C. e Baptista, J. Santos. Qualidade do ar em bolocos operatórios—breve revisão. 2012.

TLV's and BEI's—Based on the Documentation of the Threshold Limit Values for Chemical Substances and Physical Agents & Biological Exposure Indices. 2006. American Conference of Governmental Industrial Hygienists.

Udonwa, N.E. Uko, E.K. Ikpeme, B.M. Ibanga, I.A. and Okon, B.O. Exposure of Petrol station Attendants and Auto Machanics to Preminum Motor Spirit Fumes in Calabar, Nigeria. 2009.

Uva, António de Sousa e Faria, Mário. Exposição profissional a substâncias químicas: diagnóstico das situações de risco. Revista Portuguesa de Saúde Pública. Janeiro/Junho de 2000.

WHO. Indoor Air Pollutants: Exposure and Health Effects. 1982.

Occupational Safety and Hygiene II – Arezes et al. (eds)
© 2014 Taylor & Francis Group, London, ISBN 978-1-138-00144-2

Food safety of transported meals in primary schools and kindergartens

B. Alves, A. Carvalho, L. Tavares, A. Ferreira, N.L. Sá, S. Paixão, H. Simões & J. Almeida
Instituto Politécnico de Coimbra, ESTeS-C, Coimbra Health School, Saúde Ambiental, Portugal

ABSTRACT: Food safety is currently an alarming topic, being directly related to the lifestyle adopted by modern society. It is estimated that about 90% of foodborne illnesses are caused by microorganisms. The aim of this study was to evaluate the impact caused by the transportation of food, in what concerns the storage temperatures and microbiological contamination of food. The sample consisted of 51 meals in Primary Schools and Kindergartens in the centre of Continental Portugal, served by two different catering companies (A and B). In relation to the average temperature of the main dish, one observes that there is a statistically significant average difference ($p < 0.000$) when compared to the reference value of 65 °C, having been registered an average of 58.45 ± 7.57 (°C). It was observed that there was a statistically significant negative correlation between the distance from the kitchen and dining storage temperature of the main dish ($p < 0.05$), as well as soup ($p < 0.000$). Microbiological contamination has been observed in various food samples (microorganisms at 30 °C, *Escherichia coli*, coagulase-positive staphylococci and *Listeria monocytogenes*). It can be concluded that transported food remained stored at risk temperatures, so it is recommended to decrease the shipping time and the revision of the equipment used for that purpose.

1 INTRODUCTION

The concepts of food safety and food quality are not static and are deeply linked to the evolution of societies and to the change of behaviour and eating habits of consumers (Margarida, 2011).

Some of the measures to avoid or reduce cross contamination can be related to heat treatment. This should successfully eliminate vegetative cells of potentially pathogenic microorganisms such as *Listeria monocytogenes*, *Escherichia coli*, *Salmonella spp.*, coagulase positive staphylococci (group to which the *staphylococcal aureus* belong to) and reduce levels of Mesophiles (Fig. 1) (Silva, 2005).

Thus, perishable foods stored at temperatures above 5 °C and below 65 °C can allow the development of microorganisms, making it a major critical control point in transported food.

Therefore, there must be a further strengthening of food safety and quality in kindergartens and primary schools, since in recent years there has been a progressive increase in outbreaks of Foodborne Diseases. This population is characterized by an increased susceptibility to infections, because their immune system is still developing, and because very often the meal given by the school is the only daily meal some children have (Pistore, 2006).

The aim of this study was to evaluate the impact caused by the transportation of food, at the level of storage temperatures and microbiological contamination of food.

2 METHODOLOGY

The sample consisted of 51 refectories in Primary Schools and Kindergartens in central Continental Portugal, served by two separate catering companies (A and B). The samples were collected between October 2012 and February 2013. The food sampling was based on the Portuguese Standard 1828:1982.

The analytical methods used are described in Table 1.

The limits given in Table 2 and 3 are conformed to the reference values of the National Institute of Health Dr. Ricardo Jorge as well as in Regulation

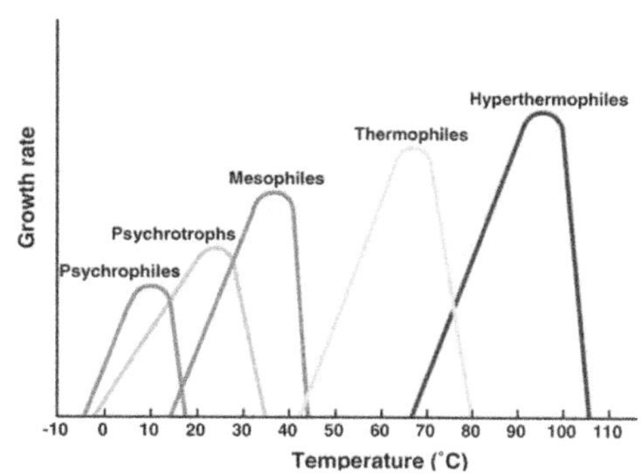

Figure 1. Effect of temperature on microbiological growth.

Table 1. Laboratory method used.

Parameter	Method
Microorganisms at 30 °C	ISO 4833:2003
Escherichia coli	ISO 16649–2:2001
Coagulase positive staphylococci	ISO 6888–1:1999
Salmonella spp.	ISO 6579:2002
Listeria monocytogenes	ISO 1129–1:1996/ Amd 1:2004

Table 2. Limit values of the different microorganisms evaluated (Food not subjected to thermal treatment).

Parameter	Limit
Microorganisms at 30 °C	$<5.0 \times 10^7$
Escherichia coli	$<1.0 \times 10^3$
Coagulase positive staphylococci	$<1.0 \times 10^2$
Salmonella spp.	Negative/25 g
Listeria monocytogenes	Negative/25 g

Table 3. Limit values of the different microorganisms evaluated (Food subjected to thermal treatment).

Parameter	Limit
Microorganisms at 30 °C	$<1.0 \times 10^4$
Escherichia coli	$<1.0 \times 10^1$
Coagulase positive staphylococci	$<1.0 \times 10^2$
Salmonella spp.	Negative/25 g
Listeria monocytogenes	Negative/25 g

EC No. 2073/2005, amended by Regulation EC No. 1441/2007 of December 5th 2007.

The temperature measurement and the sampling collection were performed as soon as the trays were removed from the isothermal containers. We used a HANNA instruments, Model HI 4060 thermometer.

To calculate distances between different canteens and their respective refectories we used "Google Earth" (Version 7.0.3.8542).

The type of study is *Descriptive-Correlational*, the sample type is not probabilistic, for convenience on the technique. For the analysis and processing of data, we used the software SPSS 19.0 for *Windows Evaluation Version*. The interpretation of statistical tests was based on a significance level of $p \leq 0.05$, for a confidence interval of 95%.

3 RESULTS

After statistical analysis, it was obtained an average temperature for the soup of 74.67 °C, with a standard deviation of 9.41, observing the existence of a statistically significant difference ($p < 0.000$) compared to the reference value of 65 °C, observing an average difference of 9.67 °C in relation to the reference value.

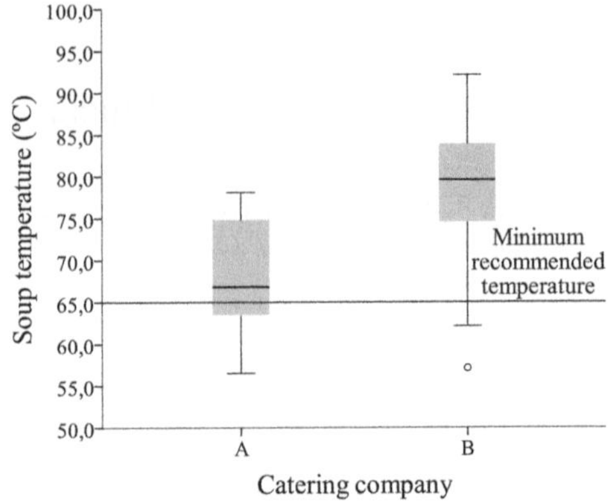

Figure 2. Soup temperature by catering company.

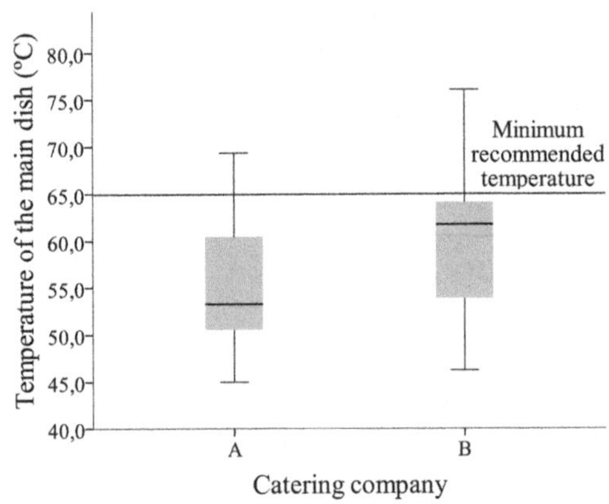

Figure 3. Temperature of the main dish by catering company.

In what concerns the average temperature of the main dish, it was observed that there was a statistically significant difference ($p < 0.000$) in relation to the reference value of 65 °C, having been registered an average of 58.45 ± 7.57 (°C).

In the 51 samples of soup it was obtained a median of 75.6 °C, with a maximum recorded of 92.1 °C and minimum of 56.5 °C (Fig. 2). In the 53 samples of the main dish it was obtained a median of 59.4 °C, with a maximum recorded of 76.1 °C and a minimum of 45 °C. It was found that over 75% of the samples were below the recommended temperature (Fig. 3).

It was also found that, on average, the meals that are served by the company A are maintained

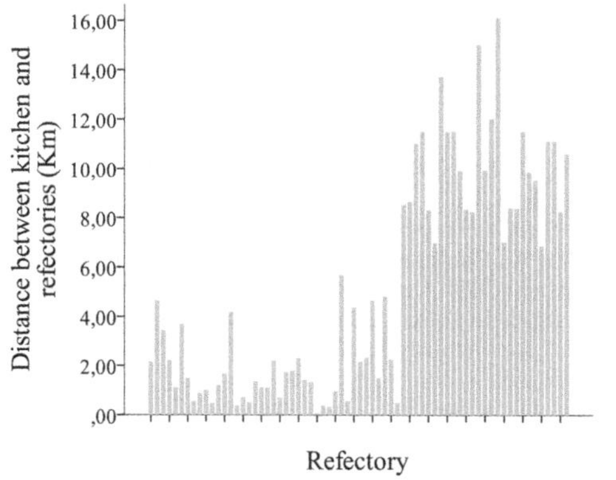

Figure 4. Distance between kitchen and refectories (Km).

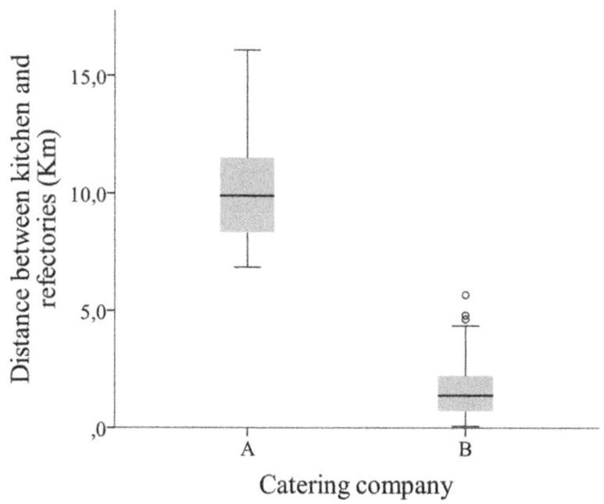

Figure 5. Distance between kitchen and refectories (Km).

at temperatures lower than those of company B (Fig. 5).

On average, the distance between the canteen and the refectories was 5.07 ± 4.47 km. It was found that there was a significant correlation ($p < 0.05$) and negative distance between the kitchen to the refectory and the storage temperature of the main course, as well as the soup (Fig. 4).

3.1 *Food not subject to thermal treatment*

In the case of *Escherichia coli*, coagulase positive Staphylococcus and *Salmonella spp.* we did not obtain samples with values above the threshold value. However, for the microorganisms at 30 °C, 14.3% of the samples had results above the limit values, and we also obtained a sample with worst results than the threshold value for *Listeria monocytogenes*.

Table 4. Non-compliant samples in food not subject to thermal treatment.

Parameter evaluated		Non-compliant samples n	Non-compliant samples %
Microorganisms at 30 °C	n = 21	3	14,3
Escherichia coli	n = 20	0	0
Coagulase positive staphylococci	n = 20	0	0
Salmonella spp.	n = 20	0	0
Listeria monocytogenes	n = 20	1	5,0

Table 5. Non-compliant samples in food subject to thermal treatment.

Parameter evaluated		Non-compliant samples n	Non-compliant samples %
Microorganisms at 30 °C	n = 34	5	14,7
Escherichia coli	n = 34	4	11,8
Coagulase positive staphylococci	n = 34	1	2,9
Salmonella spp.	n = 34	0	0
Listeria monocytogenes	n = 34	0	0

3.2 *Food subjected to thermal treatment*

In the case of coagulase-positive Staphylococci, we detected a sample with results above the threshold value, as well as 11.8% of the samples with values greater than the threshold value of *Escherichia coli*. In 14.7% of the samples the values of microorganisms at 30 °C were higher than the limit value (Table 5).

4 DISCUSSION

It was found that as the distance from the canteen to the refectory increases, the food preservation temperature decreases, along with the fact that there were incorrect storage temperatures. This fact indicates the existence of a risk of microbiological growth during the process of transporting meals, since the temperature is a limiting factor for microbial growth, as demonstrated in several studies (Lacasse, 1995; Pinto & Neves, 2008).

In what concerns the food not subject to heat treatment, the results are indicative of food exposure to inadequate treatment. It is noted that the detection of *Listeria Monocytogenes* is extremely serious since it has a high pathogenicity, being its pathological symptoms similar to meningitis. The risk groups are pregnant women, the elderly, children and immunocompromised (Pinto & Neves, 2008; Vidal-Martins et al. 2005; Greig & Ravel, 2009).

In the case of food subjected to heat treatment, the detected presence of microorganisms at 30 °C higher than the limit value indicated the presence of microbial contamination and the possible presence of pathogenic microorganisms or potential spoilage of food, indicating inadequate sanitary conditions of handling, processing, production or storage (Vidal-Martins et al. 2005; Greig & Ravel, 2009). The presence of E. coli is an indicator of faecal contamination and / or the presence of enteropathogens in food (Sand, 2010). The contamination by coagulase positive staphylococci represented a situation with some gravity due to the formation of toxins by this microorganism, indicating possible incorrect procedures by food handlers (Cortez et al. 2004).

5 CONCLUSIONS

It is concluded that the food "soup" was stored at appropriate temperatures. However, it was found that the main dish was often packaged at risk temperatures. It is recommended to review the procedures for the preparation and distribution of food, in order to decrease the shipping time to a minimum. It is also recommended to review the equipment used for transporting food in order to use equipment with a high thermal insulation in order to preserve foods at suitable temperatures, avoiding risk temperatures to the growth of microorganisms.

Despite what one would expect, the foods that were subjected to heat treatment showed a higher frequency of samples with results above the limit values, results which may reflect inadequate storage conditions in terms of temperatures. From all the evaluated parameters the Microorganisms at 30 °C presented the highest values. which, together with other results, reflect the need for health and hygiene improvements.

BIBLIOGRAPHY

Comissão das Comunidades Europeias. 2005. Regulamento (CE) N° 2073/2005 da Comissão. *Jornal Oficial da União Europeia* 338: 1–26.

Comissão das Comunidades Europeias. 2007. Regulamento (CE) N° 1441/2007 da Comissão. *Jornal Oficial da União Europeia* 322: 12–29.

Cortez, A., Carvalho, Â., Amaral, L., Salotti, B., & Martins, A. 2004. Coliformes fecais, Estafilococos coagulase positiva (ECP), Salmonella spp. e Campylobacter spp. em linguiça fresca. *Alimentação e Nutrição* 15(3): 215–220.

Greig, J.D. & Ravel, A. 2009. Analysis of foodborne outbreak data reported internationally for source atribution. *Internatinal journal of food Microbiology* 130(2): 77–87.

Lacasse, D. 1995. Introdução à Microbiologia Alimentar. *Ciência e Técnica*. Lisboa: Instituto Piaget.

Lorenzetti, D. 2006. *Influência do tempo e da temperatura no desenvolvimento de microrganismos psicrotróficos no leite cru de dois estados da região sul*. Curitiba: Universidade Federal do Paraná.

Margarida, S. 2011. *Segurança Alimentar nos processos confecção a quente em restauração tradicional*. Dissertação de Mestrado. Lisboa: Universidade Técnica de Lisboa, Faculdade de Medicina Veterinária.

Pinto, J. & Neves, R. 2008. HACCP - *Análise de riscos no processamento alimentar*. Porto: Publindústria.

Pistore, A. 2006. *Evaluation of the hygienical sanitary knowledge of the manipulators school feeding: bedding for continuous and adjusted training* 20(146):17–20.

Sand, S. 2010. *Considerações sobre o uso de Escherichia coli como indicador de contaminação—vantagens e desvantagens. VIII Jornada de Oficinas Analíticas, Rio Grande do Sul, 8–10 Junho 2010*. Rio Grande do Sul: Instituto de Ciências Básicas da Saúde.

Silva, E. 2005. *Manual de controle Higiénico-Sanitáriio em serviços de alimentação*. São Paulo: Varela.

Vidal-Martins, A., Rossi, O. & Lago, N. 2005. Mesophilic heterotrophic microorganisms and spore forming bacteria from Bacillus cereus group in ultra high temperature milk. *Arq. Bras. Med. Vet. Zootec.* 57(3). Belo Horizonte: Faculdade de Ciências Agrárias e Veterinárias.

Occupational Safety and Hygiene II – Arezes et al. (eds)
© 2014 Taylor & Francis Group, London, ISBN 978-1-138-00144-2

Eco-construction—contribution to occupational risk occurrence

F. Rodrigues & A. Ribeiro
Department of Civil Engineering, Geobiotec, University of Aveiro, Portugal

ABSTRACT: Eco-construction concept is currently being implemented to face the environmental concerns. This paper analyzes Eco-construction regarding occupational hazards prevention through the study of the main areas underlying an assessment system for sustainable buildings that may have greater influence on the risks prevention. Then it was performed a risk assessment having regard to these criteria. The methodology used in this paper was based on a research and a review of contents in the area of sustainable development, risk assessment methodologies, the hazards and risks associated with Eco-construction, as well as measures to mitigate these risks. It was also conducted an exploratory study related to the application of preventive measures for the Eco-construction of three buildings. From the risk assessment performed through a qualitative method, it was concluded by its similarity with the conventional construction, however, this one have higher incidence of some risks according with the specificity of some activities performed.

1 INTRODUCTION

Eco-construction or sustainable construction is a concept that is growing and getting stronger all over the World. It has as principal aim to create and manage a healthy built environment, taking into account the ecological principles and using the resources efficiently. However, to achieve these goals it is also essential to know which is the Eco-construction contribution (positively and negatively) for the occupational risks prevention. So, applying a risk assessment at the areas and criteria used by sustainable assessment systems for the construction it is possible to evaluate that contribution.

LiderA, the acronym for Leadership for the Environment in Sustainable Building, is an assessment and voluntary acknowledgement system for sustainable building and built environment. Considering the assessment areas and criteria of this system, that ones with higher occupational risk were identified and its risk assessment was done through a qualitative method. From the risk assessment results it was possible to identify the common risks to the conventional building construction as well that ones that have greater incidence in Eco-construction. The aim of this paper is to present these results.

2 RISK PREVENTION AT ECO-CONSTRUCTION

2.1 *Eco or sustainable construction*

Eco-Construction had its origin in the green building concept. These buildings present several positive characteristics and attributes, especially regarding to energy efficiency, resources depletion, impact on the environment, users health protection and users comfort. The sustainable construction contributes through its features to the sustainable development minimizing the life-cycle-costs, the environmental impact, the resources depletion, and contributing to the protection of the workers and all the players at the health and safety level.

So, it turns out that there is an overlap of concepts between green and sustainable buildings. In spite of green buildings can comply with all the sustainable concepts, to avoid confusions, many authors use sustainable construction designation for simplicity reasons (Lützkendorf & Lorenz, 2007).

2.2 *Risk prevention at LEED buildings*

The LEED system (Leadership in Energy and Environmental Design) has been developed in the United States of America, in 1999, as an environmental performance classification system, having as main goal to develop and implement design and construction practices, environmentally responsibles (LEED, 2009).

Recent research studies point out that building construction accordingly LEED requirements lead to higher occupational risks and consequently to a higher rate of workers' injuries comparatively with the traditional construction of buildings (not LEED) (Dewlaney & Hallowell, 2012). The results of this study were obtained through 26 interviews with professionals in the sector, indicating that the activities presenting the highest risk are: interventions in degraded industrial areas, rainwater reuse systems, the heat island effect in roofing, innovative technologies for the treatment of wastewater, energy performance optimization, renewable energy systems, the coexistence at site of a large number of people who are not familiar with the construction processes, construction waste management, air quality monitoring and management systems, control of the lighting system and the use of 75% of the natural light in the spaces. In this study designers and manufacturers identified the prefabrication and the use of alternative materials, as methods to minimize and/or avoid injuries directly related to the dangers resulting from the sustainable construction. These results can be used to increase the construction workers safety since this is an aspect of sustainability that was not addressed in the LEED program (Dewlaney & Hallowell, 2012).

Therefore, it is necessary to identify safety strategies for a reduction of occupational risks associated to the characteristics of the design and the means and methods to achieve the sustainable construction.

2.3 *Sustainability assessment*

The term "sustainable construction" was firstly proposed by Kibert (1994) to describe the responsibilities of the construction concerning the concept and objectives of sustainability. According to this author the existing knowledge and the diagnosis of the construction industry, in terms of environmental impacts, show that there is a need for changes to achieve sustainability objectives. The United States Green Building Council (USGBC, 2013), recognize that a good environmental performance is characterized by achieving reduced negative environmental impacts. So, building sustainability assessment can be carried out according to sustainable management of built area: water and energy efficiency, and renewable energy use, conservation of materials and resources and indoor environmental quality (Degani & Cardoso, 2002). The sustainability assessment purpose is to gather data and report information to serve as a basis for decision-making, during the different phases of a building lifecycle. Nowadays, there are several tools that can be applied in the construction sustainability assessment during the design, execution and operation phase of buildings (Degani & Cardoso, 2002; Fowler & Rauch, 2006; Gu et al., 2006; Raymond, 2005): SimaPro, BEES, ATHENA, LISA, NABERS, LEED, BREEAM, CEEQUAL, SBTool. In Portugal was developed a tool named LiderA—Sustainability Assessment System that uses a set of 43 criteria in 6 categories with 22 areas to measure the sustainability level. So, the method evaluates 6 main areas (categories) that include specific areas of intervention based on 43 criteria which assess sustainable measures (see Fig. 1). These areas have different weight and are site integration (14%), the resources consumption (32%), socio-economic adaptability (19%), environmental load impacts and comfort (15%), environment management and innovation (8%). The comfort most significant in the assessment is the consumption of resources due to its higher relative importance (Pinheiro, 2013).

To comply with the economic and social sustainability demands, the LiderA system also has to avoid the occupational hazards in the fulfillment of each one of its criteria. So, it is essential to do a risk assessment during the design and the execution phase for all the system areas (Fung et al., 2010). According to each one it is possible to know the main activities targeting the building sustainability, namely at the level of site integration (volume, materials, architecture), the resources consumption (renewable energies—photovoltaic and solar panels, wind energy, biomass, geothermic, water reuse systems, and materials), the environmental comfort (the constructive systems to obtain the indoor thermal and acoustic comfort, high air quality, natural lighting). For each one is than possible to identify the hazards and the preventive measures to be adopted.

It is also important to highlight the benefits for safety that can be achieved through the construction sustainability measures.

2.4 *Risk assessment*

The base of risk prevention is the risk assessment during which all the dangerous factors have to be identified and eliminated or controlled through preventive or protective measures. The

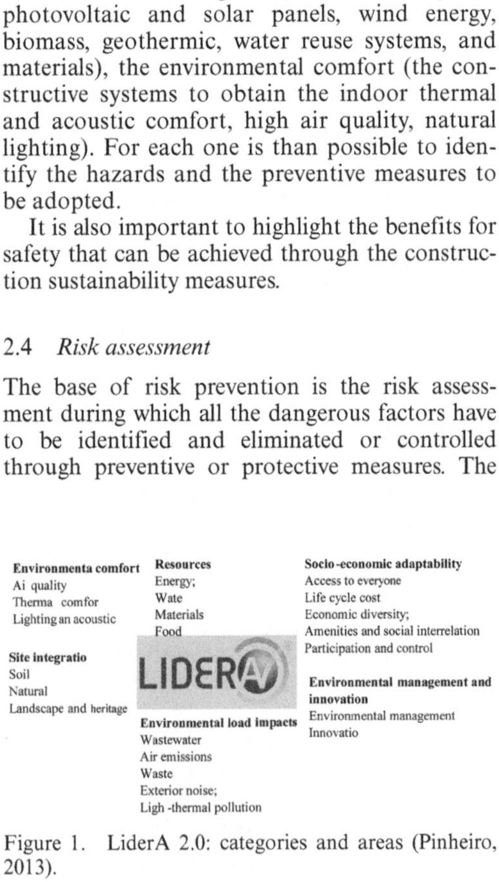

Figure 1. LiderA 2.0: categories and areas (Pinheiro, 2013).

risk assessment methods include processes, tools and reference values. According to the methodology the assessment techniques are classified in three main categories: qualitative, quantitative and semi-quantitative. The qualitative ones lead to a qualitative risk evaluation by qualitative analyses (Marhavilas et al., 2011). Several qualitative methods can be used as: checklists (Ayyub, 2003; Harms-Ringdahl, 2001; Marhavilas et al., 2009; Reniers et al., 2005 cit. Marhavilas et al., 2011), what-if analyses (Ayyub, 2003; Reniers et al., 2005 cit. Marhavilas et al., 2011), safety audits (Ayyub, 2003; Harms-Ringdahl, 2001; Reniers et al., 2005 cit. Marhavilas et al., 2011), task analyses (Doytchev & Szwillus, 2009; Kontogiannis, 2003), Sequentially Timed Event Plotting—STEP technique (Kontogiannis et al., 2000 cit. Marhavilas et al., 2011) and the Hazard and Operability Study—HAZOP (Ayyub, 2003; Baysari et al., 2008; Harms-Ringdahl, 2001; Labovsky et al., 2007; Yang & Yang, 2005 cit. Marhavilas et al., 2011). In this work it was used a qualitative assessment method named simple matrix 4×4 considering its simple application. It use an integrated matrix taking into account the events probability (Table 1) and the severity level (Table 2), in a 4 levels scale (Carvalho, 2007; Pires 2013).

Table 1. Probability scale.

Probability	
Low	Rare or low frequency and/or exposition duration combined with low probability of a hazardous event occurrence.
Average	Rare or low frequency and/or duration of exposure combined with high probability of a hazardous event occurrence.
High	Frequent or long duration of exposure combined with low probability of a hazardous event occurrence.
Very High	Frequent or long duration of exposure combined with high probability of a hazardous event occurrence.

Table 2. Severity scale.

Severity	
Low	Accident/occupational disease without work stoppage.
Average	Accident/occupational disease with work stoppage.
High	Accident/occupational disease with permanent partial incapacity.
Very High	Fatal accident/occupational disease.

Table 3. Risk levels gradation.

	Severity			
Probability	Low	Average	High	Very High
Low	Low	Low	Low	Low
Average	Low	Average	Average	Average
High	Low	Average	Very High	Very High
Very High	Average	Average	Very High	Very High

To obtain the risk gradation the method applies a three risk levels scale (Table 3) defining the intervention priority (Pires, 2013).

3 CASE STUDY

3.1 *Methodology*

To each one of the categories of the LiderA system (see Fig. 1), considering the design and the execution phase of a building, the risk assessment through the method described above was made. To support this assessment for each category, tables with the identification of the hazards, the corresponding risks and the preventive and protective measures, were done (Ribeiro, 2013). Three project models constructed under the LiderA system requirements were studied to identify the differences between the traditional constructions targeting the contribution for occupational risks occurrence.

3.2 *Results and discussion*

From the six studied categories (site integration, resources consumption, socio-economic adaptability, environmental load impacts, environmental comfort, environment management and innovation) in this paper it will be depicted the risk assessment of the site integration and of the resources consumption. To the first one the main hazards were identified in the following operations: a. demolitions that permit to reduce the soil use and to promote the territorial valorization; b. soil excavation that can be associated to j. polluted sites. In the sequence of these hazards there are risks of exposition to: f. noise, g. dust inhalation, h. vibrations, i. particle projection. The existence of k. adjacent structures and infrastructures lead to the existence of risks inherent to the maintenance of its stability. The risks can still occur as the result of c. inadequate professional training, d. lack of protective or preventive means, e. inadequate tools and equipments selection. Applying the qualitative risk assessment method indicated in 2.4 it was obtained the risk levels depicted in Table 4.

To the resources consumption (energy, water, materials) the majority of the hazards are related with the equipments and materials that have to be applied. It was concluded that higher probability and severity occur from (Table 5): a. works at height, f. electricity, g. inadequate professional training, h. fatigue, j. falling objects, m. trenching, k. inadequate tools and equipments selection, l. lack of protective or preventive means, p. falling materials, s. noise.

The remaining risks range from very low (following the occurrence of, d. individual stress, e. organizational stress) to average (in the sequence of b. various dangers to health, c. fragility of materials, i. weather conditions, n. hit-and-run by vehicles, o. machines collision, q. particles projection, r. dust inhalation, t. vibrations).

It was found that the main hazards and risks identified are similar to those of traditional construction. However, despite the similarity between the activities inherent of the two types of construction and consequently between the respective hazards and risks, the sustainable construction projects analyzed, develop activities that are not normally performed in conventional construction. Some of these works are related to the execution of green roofs, placement of shading systems, rainwater and wastewater capture, storage and treatment systems.

Accordingly with Dewlaney & Hallowell (2012) the installation of solar collectors or panels is associated with the sustainable construction, but currently the placement of these renewable energy production systems is compulsory in all the new and retrofitted buildings (RCCTE, 2006; EPBD, 2010). The installation of these equipments is not a short duration activity and involves hazards, such as work at height, exposure to hazards of electrical and chemical origin, several health hazards, which may lead to falling objects and professionals' fatigue (HSE, 2013a). So, it is essential implement preventive and protective measures as in the conventional as in the sustainable construction.

Table 4. Risk levels to the site integration category (Pires, 2013).

	Severity			
Probability	Low	Average	High	Very High
Low				
Average				
High			b.	a. k.
Very High		g. h. i.	c. f.	d. e. j.

Table 5. Risk levels to the resources consumption (Pires, 2013).

Probability	Severity			
	Low	Average	High	Very High
Low		d. e.		
Average			b. c.	n. o.
High		i.	f. h.	a. j. m. p.
Very High		q. r. t.	g. s.	k. l.

It was verified that the work carried out in the construction of the studied projects (sustainable construction according to the LiderA system) feature as the main hazard the work at height, as the conventional construction. It should be noted that more than 60% of the deaths that occur during the execution of the work at height involve essentially falls from ladders, scaffolds, working platforms and the surround of the covers, and may also occur falls in fragile areas from roofs or skylights (HSE, 2013b).

On the other hand it is important to highlight that Eco-construction demand the application of non-toxic materials being many of them natural, involving a lower level of occupational risks during their handling and application.

4 CONCLUSION

Taking into account the different criteria of the LiderA system, particularly the local integration and resources consumption, it was possible to perform the occupational risk assessment during the different phases of design and the execution phase of a building. It was concluded that the main existing hazards are related to demolition activities, land movements and work at height enhanced by inadequate training, lack of protection/prevention means, inadequate selection of equipment, soil contamination, and adjacent location to other buildings or infrastructures, fragility of materials, physical and organizational stress. It was found that these hazards are common to both, the traditional construction and Eco-construction, but this one have higher probability of exposure to some of them, given its specific activities. Regarding materials applied in this type of construction being non-toxic contribute to eliminating/minimizing the hazards of chemical origin, both in handling and application operations, during the exploration of the building as well as in the maintenance, repair and demolition. If they are high durability materials will also have a reduced need for maintenance, repair and replacement, contributing to the probability decrease of exposure to operations that contain risk of falling materials and people, and of musculoskeletal injuries. As in traditional construction is indispensable in the design phase identify hazards and implement specific measures aiming at their elimination/prevention. During the execution phase these measures have to be effectively put into practice together with the information and training of workers, because it is through them that are possible to prevent, reduce and mitigate the hazards.

As the construction industry contributes to the higher rate of fatal accidents in Portugal, it is crucial that these emerging branches of the construction sector implement an effective prevention of occupational hazards, to eliminate these fatalities.

REFERENCES

Ayyub, B.M. (2003). *Risk analysis in engineering and economics*. Florida: Chapman & Hall/CRC.

Baysari, M.T., Mcintosh, A.S. & Wilson, J.R. (2008). Understanding the human factors contribution to railway accidents and incidents in Australia. *Accident Analysis and Prevention* 40 (5): 1750–1757.

Carvalho, F. (2007). Risk Assessment Comparison of Different Methods of Risk Assessment in Real Work Situation. Master thesis (in Portuguese). Technical University of Lisbon, Lisbon.

Degani, C.M. & Cardoso F.F. (2002). Sustainability during the building life cycle: the importance of the design phase (in Portuguese). Polytechnic University of São Paulo, São Paulo. Retrieved April 14, 2013, from http://www.pcc.usp.br/files/text/personal_files/francisco_cardoso/Nutau%202002%20Degani%20Cardoso.pdf.

Dewlaney, K.S. & Hallowell, M. (2012). Prevention Through Design and Construction Safety Management Strategies for High Performance Sustainable Building Construction. *Construction Management and Economics* 30: 165–177.

Doytchev, D.E. & Szwillus, G. (2009). Combining task analysis and fault tree analysis for accident and incident analysis: a case study from Bulgaria. *Accident Analysis and Prevention* 41(6): 1172–1179.

EPBD (2010). Directive 2010/31/EU of the European Parliament and of the Council of 19 May 2010 on the energy performance of buildings (recast). Official Journal of the European Union. 18.6.2010. Retrieved February 10, 2013, from http://eur-lex.europa.eu/LexUriServ/LexUriServ.do?uri = OJ:L:2010:153:0013:0035:EN:PDF.

Fowler, K.M. & Rauch, E.M. (2006). Sustainable Building Rating Systems Summary, Pacific Northwest National Laboratory, US Department of Energy. Retrieved March 14, 2013, from http://wbdg.org/ccb/GSAMAN/sustainable_bldg_rating_systems.pdf.

Fung, I.W.H.; Tam, V.W.Y.; Lo, T.Y. & Lu, L.L.H. (2010). Developing a Risk Assessment Model for construction safety. *International Journal of Project Management* 28 (6): 593–600.

Gu, Z.; Wennersten, R. & Assefa, G. (2006). Analysis of the most widely used Building Environmental Assessment methods. *Environmental Sciences* 3: 175–192.

Harms-Ringdahl, L. (2001). *Safety analysis, principles and practice in occupational safety* (2nd ed.). London:Taylor & Francis.

HSE (2013a). Solar panel installation. Retrieved May 31, 2013, from http://www.cskills.org/uploads/GS001_Safe%20solar%20panel%20installation_tcm17–33755.pdf.

HSE (2013b). Assessing all work at height. Retrieved May 31, 2013, from http://www.hse.gov.uk/construction/safetytopics/assess.htm?ebul = gd-cons/may13 = 14.

Kibert, C.J. (1994). Establishing Principles and a Model for Sustainable Construction. Proceedings of the 1st International Conference on Sustainable Construction, Tampa, University of Florida, CIB Publication TG 16, Rotterdam.

Kontogiannis, T. (2003). A Petri net-based approach for ergonomic task analysis and modelling with emphasis on adaptation to system changes. *Safety Science* 41(10): 803–835.

Kontogiannis, T.; Leopoulos, V. & Marmaras, N. (2000). A comparison of accident analysis techniques for safety-critical man-machine systems. *International Journal of Industrial Ergonomics* 25: 327–347.

Labovský, J.; Svandová, Z.; Marko_S, J. & Jelemenský, L. (2007). Model-based HAZOP study of a real MTBE plant. *Journal of Loss Prevention in the Process Industries* 20: 230–237.

LEED (2009). Leadership in Energy & Environmental Design. LEED for New Construction and Major Renovations v.3. US Green Building Council. Retrieved March 29, 2013, from http://www.usgbc.org/ShowFile.aspx?DocumentID = 5546.

Lützkendorf, T. & Lorenz, D. (2007). Integrating Sustainability into Property Risk Assessments for Market Transformation. *Building Research & Information* 35(6): 644–66.

Marhavilas, P.K.; Koulouriotis, D.E. & Gemeni, V. (2011). Risk analysis and assessment methodologies in the work sites: On a review, classification and comparative study of the scientific literature of the period 2000–2009. *Journal of Loss Prevention in the Process Industries* 24: 477–523.

Marhavilas, P.K.; Koulouriotis, D.E. & Voulgaridou, K. (2009). Development of a quantitative risk assessment technique and application on an industry's worksite using real accidents' data. *Scientific Journal of Hellenic Association of Mechanical & Electrical Engineers* 416: 14–20.

Pinheiro, M.D. (2013). Voluntary system for assessment of sustainable construction. Retrieved April 24, 2013, from www.lidera.info/?p = index&RegionId = 3&Culture = en.

Pires, F.C.M. (2013). Prevention of Occupational Risks in the replacement of the press rolls in a paper pulp mil. Master thesis (in Portuguese). Instituto Superior de Línguas e Administração, Vila Nova de Gaia (in Portuguese).

Raymond, J.C. (2005). Building environmental assessment methods: redefining intentions and roles. *Building Research & Information* 33: 455–467.

RCCTE (2006). Decree-Law n° 80/2006. D.R. I Série-A. n° 67 (4 of April), 2468–2513 (in Portuguese).

Reniers, G.L.L.; Dullaert, W.; Ale, B.J.M. & Soudan, K. (2005). Developing an external domino prevention framework: Hazwim. *Journal of Loss Prevention in the Process Industries* 18: 127–138.

Ribeiro, A.F. (2013). Contribution of Eco-construction for the prevention of occupational hazards. Master thesis (in Portuguese). University of Aveiro, Aveiro.

USGBC (2013). United States Green Building Council. Retrieved March 14, 2013, from www.usgbc.org.

Yang, S.-H. & Yang, L. (2005). Automatic safety analysis of control systems. *Journal of Loss Prevention in the Process Industries* 18: 178–185.

Occupational Safety and Hygiene II – Arezes et al. (eds)
 ISBN 978-1-138-00144-2

Risk exposure prevention tool to health care workers

J. Sá, Miguel Tato Diogo & R. Cruz
Faculdade Engenharia da Universidade do Porto, Porto, Portugal

ABSTRACT: Health professionals are exposed to a high presence of risk factors at work. However, this exposure can be controlled and minimized when control measures to security of workers are implemented. This study's main objective is to develop a proposal for a simulator model that supports the risk management in institutions, as regards on occupational exposure of workers, based on the risks in the workplace and the vulnerability of exposed workers. The field work consisted in determining occupational hazards and assess the risks of the sectors, resulting in the definition of risk levels to sectors and the registration of workers' characteristics, in order to determine their vulnerabilities. From the obtained data, were established three criterions levels to local access, namely: access allowed, conditioned access and access to authorized personnel. The simulation model proposal developed can be an important resource that allows endow the organizations of tools for a more effective prevention.

According to the European Agency for Safety and Health at Work, every year dies about 5580 people in the European Union as a result of accidents and 159 000 due to other diseases (European Agency for Safety and Health at Work, 2012).

The activities carried out in the health area, such as the handling of hazardous substances and technical equipment used, involve exposure to different professional risk factors, such as chemical, physical, biological and psychosocial. Being workers exposed of several professional groups, as long the nurses up to the workers of cleaning are exposed to the contact with agents causing accident (Galego, 2009).

The applied methodology to a risk management system, has as first step at the hazards identification in the work environment, the inherent risks in the performed tasks and the people involved. The second step consists in the evaluation and acceptability of risks, which according to his greatness, are established priorities of action. Once evaluated, it is necessary to implement measures to control the risk, and the wager lies in health prevention through control of exposure of the health professionals, i.e., avoiding the risk.

1 OBJECTIVE

The main objective of this work is to develop a simulator model proposal that monitoring and control the risk exposure of health care workers. A system that answers the question on how to protect workers from occupational risk, i.e., a system that supports organizations in managing of risk exposure of their employees, thereby reducing accidents at work and occupational diseases. This study aims to contribute to the definition of measures that focus on the prevention and control of risks in the work environment, i.e., a program of occupational risk management who will be based on the identification, assessment and monitoring of risk situations likely to happen in order to ensure the health and safety of employees.

2 METHODS

The study was applied to a Private Health Unit, with 50 employees, distributed by diverse work sectors. The study was applied to a Private Health Unit, with 50 employees, distributed by diverse work sectors. In the developing of the simulator model proposal, it only was considered the following sectors: laboratory, nurse room, medical offices, ambulatory surgery and receptions.

The first was to collect data, with methodological based on research in the field, by method observational and descriptive, using tools, such as checklist and grille risk assessment, constituents manual Technical Guideline No. 1, the Regulatory Authority Health (2010). The field research was divided into four areas of data collection: i) areas and facilities, which were identified in the floor plants, all existing sectors and assigned an identification code to each, ii) human entity, having been made a characterization of human resources, according to age, gender, education and sector work in the health unit; iii) existing risks in the health unit, made initially through a risk mapping and posterior a sector risk assessment, which allowed the determination of the acceptability of the involved risk, iv) risk exposure, with a record of the paths of movement for workers, patients and hazardous materials, and a workers characterization on the basis of training, experience and skills in order to determine their level of susceptibility to the risk that they are exposed. In the second part we proceeded to the elaboration of a proposed model simulator that simulates a system of monitoring and access control to sectors, to professionals, the health unit. To design the system, were treated the data gathered in relation to the risk presented, starting on the determination of the acceptability of risk in sectors previously defined, establishing a criterion level of risk present in each sector. The data of the characteristics of human resources has allowed perform the record of workers' vulnerability to risk. With these support elements, was elaborated a table that lists the risk levels of the sector with the workers vulnerability, resulting in a permissions data base to be used by the simulator. The data collection relatively to the paths of the workers, allowed the study of the location of the equipment to be used for the access control system. This allows realize the monitoring of worker exposure by restricting entries on sectors that present risk situations, defined in the data base permissions. The access control system to be applied for simulation was radiofrequency active, using as equipment modules reading, identification cards and information modules.

3 RESULTS

From the treatment of the obtained data of the risk mapping and sector risk assessments it was determined the level of risk acceptability in each sector that results in ascription of risk levels. The established criterion has three levels, the Low level that corresponds to areas where the risk is almost inexistent, the Medium level associated to places with the presence of biohazard, controlled and with measures to protection and the High level to places with a raised presence biohazard, controlled but due to the executed tasks cannot be avoid it. The data collected, regarding the risk exposure of workers, results in a new criterion to take into account, the vulnerability of the worker. It was determined two groups of vulnerability. The Vulnerable group receives workers with no experience, no qualifications in area of biological agents, pregnant workers and/or breastfeeding women and external persons to the health unit. The Non Vulnerable group includes workers with qualifications in the area and who have proven professional experience in the biological agents' area. The fact of chose the focus on the biological risk, was solely because it's the type of risk more relevant in the studied sectors. Based on the established criteria, the sector risk level and workers vulnerability, it is possible to develop the simulator model proposal to control worker exposure.

The developed simulator model allows control the workers access to the existing sectors in the health unit, with biological risk exposure. Through the relation the risk level of a sector with the workers vulnerability to the risk exposure, it is essential to manage the access or the time of permanence by a worker in a sector. This simulator model proposal allows an effective management, acting in prevention and reducing the likelihood of risk.

4 DISCUSSION

In the developing of this study the question that ruled was "How to prevent occupational risk exposure of vulnerable workers?" The manual of Portuguese Health Regulatory Authority refers that "The risk of a health worker contracting diseases related to work is about 1.5 times greater than any other worker" (Ministério da Saúde, 2010). Some accomplished studies, (Araújo, 2012) (Van Wijk, 2009), confirm that the biological risk factors are the main causing agents of accidents and those that most concern health professionals. The health professional's susceptibility comes from its exposure to risk situations such as needdlestick and sharp injuries arising from their work activities, refers the European Agency for Safety and Health at Work (2008). In this circumstance adds to the reality of the results of daily exposure to risks, involving health does not always manifest them immediately and may take days, weeks and even years (HSE, 2011). It is essential to develop prevention strategies that identify and control the risks to which workers can be exposed, emerging the simulator model proposal, which allows the prevention of workers risk exposure. This model simulates the access control to sectors in accordance with the level of risk that they present. Like a normal access control system requires a permissions criterion, it was established three criteria for access, according to Table 1. The definition of these criteria, relates the sector risk levels with the workers vulnerable groups.

Determined the access criteria, combining the classification of risk exposure levels of sectors with the vulnerable groups of workers, have been guaranteed the necessary data for the development of the data base on which the proposed simulation model will work. The simulator model proposal will have as operating base the analysis of the data bases that contains the registration of workers with the respective code, the work sector and the

Table 1. Criteria access model proposed simulator.

Access criterion	Description
Access allowed	Allows access to all workers, patients and visits to areas where the level of risk is low. In sectors with risk level Medium and High, only have access workers of the Non Vulnerable group.
Conditioned access	This criterion implies access conditions to sectors, the conditions are access with escort and access with a time limit. The access restriction applies to workers of the Vulnerable group, once the sectors to accede have a Medium risk level.
Authorized personnel access	This criterion determines access to sectors where only for authorized workers. In High risk sectors, only workers that belong to the group Non Vulnerable are allowed.

Table 2. Database to the proposed simulation model.

Database	Description
Workers	Workers identification code.
Worker vulnerability	Each worker is associated to a vulnerability group.
Risk	Each sector is associated with a risk level.
Local	Each access readers are associated to a sector.

vulnerability group, the assigned risk level of each sector and the data base of each system reader associated to the sectors, as presented in Table 2.

The performance of the simulator model is based on the flowchart shown in Figure1.

The working method of the simulator model starts with the analysis of the worker vulnerability. Starts with the worker identification through the data base "Worker", it makes the decision test—"Worker vulnerable?"—through the data table "Worker—Vulnerability". If returns a negative answer—Non-vulnerable worker—it results in allowed access to the sector, opening the door. If is a positive answer–Vulnerable worker—the flowchart reads the code sector associated with the access reader through the data table "Local", and then perform the hypotheses test, related to the risk level attributed to sector detected, using the table data "Risk". According to the type of risk detected in the hypothesis testing, it takes place the opening of the door for low and medium risk and the non-opening door to the high risk.

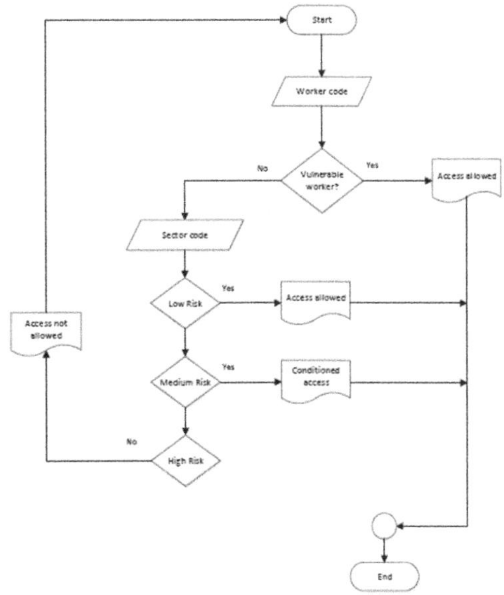

Figure 1. Flowchart of the proposed model simulator.

Table 3. Sector access scenarios—simulator model proposal.

Scenarios	Description
Scenario I	Access allowed—Non Vulnerable worker
Scenario II	Access allowed—Vulnerable worker
Scenario III	Conditioned Access—Vulnerable Worker
Scenario IV	Non authorized Access—Vulnerable Worker

With this running base, the criterion of permission set based on the level of risk exposure of workers, results in the existence of four situations types of sectors access, as the risk level shown in Table 3.

The simulator model proposal presented is adaptable to changes in the existing structure of human resources by the update of the databases, whether by entry of a new worker or when there is a change in the workers vulnerabilities, such as in situations of workers pregnant and/or breastfeeding.

5 CONCLUSIONS

The proposed simulated model as a tool for risk management, objective of this study, has a high interest and application, particularly in situations where it isn't possible to eliminate the exposure to the main source of risk, or change work procedures. However, any entity engaged in an economic activity and that has a risk management system, will have in this simulator proposal an excellent option of a system to control risk exposure of their workers. With the application of this simulator, it is ensured that all workers are covered by the preventive measure applied, both in the study to developing of the same as the fact that the exposure control does not depend on the workers themselves. With the establishment of data bases, from the conducted registrations, for the operation of the simulator, it is possible to make a perfect access management, taking into account the workers characteristics and the risk that they are exposed. This study also allowed perceiving the importance of integrating the workers characteristics, when carrying out the risk assessments, integrating in the process of risk management the perception that professionals have of their jobs, the functions performed and the inherent risks that expose them.

ACKNOWLEDGMENT

The authors wold like to thank Master in Occupational Safety and Hygiene Engineering (MESHO), of the Faculty of Engineering of the University of Porto (FEUP), all the support in the development and international dissemination of this work.

REFERENCES

Araújo, T.M.; Caetano, J.A.; [et al] (2012). Acidentes de trabalho com exposição a material biológico entre os profissionais de Enfermagem. Revista Enfermagem, III Série, nº7. 2012. Available in: http://www.scielo.gpeari.mctes.pt/scielo.php?pid=S0874-02832012000200001&script=sci_arttext.

European Agency for Safety and Health at Work. (2008). Avaliação de riscos e ferimentos por picada de agulha. Factsheet 40. 2008. Available in: https://osha.europa.eu/pt/publications/e-facts/efact40/view.

European Agency for Safety and Health at Work. (2012). Participação dos Trabalhadores na Segurança e Saúde no trabalho—Guia Pratico. Available in: https://osha.europa.eu/pt/publications/reports/workers-participation-in-OSH_guide.

Galego, M. (2009). A Saúde dos profissionais dos cuidados de saúde primários. Sociedade Portuguesa Medicina no Trabalho. Revista Saúde e Trabalho, nº 7: 47–67.

Health and Safety Executive. (2011). Successful health and safety management. HSE Books. England 2011.

Ministério da Saúde, Administração Regional da Saúde de Lisboa e Vale do Tejo. (2010). Orientação Técnica N.º 1 - DSP Gestão dos Riscos Profissionais em Estabelecimentos de Saúde. Available in: http://www.arslvt.min-saude.pt/Documents/ARS_Manual%20Gest%C3%A3o%20Risco%20Profissional_pag_a_pag.pdf.

Van Wijk, P. Th. L.; Schneeberger, P.M.; Heimerisk, K. Boland, G.J.; Karagiannis, I.; Geraedts, J. and Ruijs, W.L.M. (2009). Occupational blood exposure accidents in the Netherlands. European Journal of Public Health, Vol. 20, No. 3, 281–287. Available in: http://eurpub.oxfordjournals.org/content/20/3/281.full.pdf+html.

Occupational Safety and Hygiene II – Arezes et al. (eds)

The construction of the Brazil colony: The role played by the carpenters

J.A.B. Castañon & M.T.G. Barbosa
Universidade Federal de Juiz de Fora—PROAC, Juiz de Fora, Brazil

C.M. Ferreira
IFET—Campus Barbacena, Universidade Federal de Juiz de Fora—PROAC, Juiz de Fora, Brasil

ABSTRACT: In 1549, the Jesuits arrived in the colony to build the administrative structure. The shortage of skilled labor challenged Jesuits to educate and train officers in various areas, and to request the support of Portugal to send carpenters and other officers. Craft Corporations acted in qualifying and had a mutual character by that time. The carpenters participated in civil construction using various techniques, most of them currently present in Brazil. This article aims to report on the formation and participation of carpenters in civil construction in colonial Brazil, contributing through a comparative analysis of three aspects of the colonial era to the present day, namely, society's relationship with the mechanical officers, similarity between the goals of craft corporations and labor unions, and finally, an analysis of the strengths of the construction techniques and materials used, considering aspects implemented today, regarding the concept of sustainable development for the construction industry.

1 INTRODUCTION

When they landed in the colony, the Jesuits found a fertile land with plenty of raw material, wood of excellent quality, combined with promising vegetation, many mineral resources and water, but which contrasted with the shortage of skilled labor, composed primarily of slaves brought from Africa and native Indians. The latter knew some primitive building techniques and the various possibilities for the use of wood (Pellegrini and Hoffmann, 2009).

Thus, the mission of the Jesuits sent to Brazil in 1549 for territorial occupation and the Portuguese settlement, involved several challenges for the catechetical project to meet with success, the main one being the training of manpower for the erection of buildings. Once their work started, the Jesuits learned the mechanical arts to perfection to teach and build the facilities (Bittar and Ferreira, 2012).

The period from 1549 to 1570, known as the heroic age, was devoted to the training of these professionals (Fonseca, 2007, p. 1 and 2).

The Jesuits requested the support of Portugal to send priests to be teachers and evangelists in Brazil, and with the same eagerness, "brothers who were skilled in the Arts and Crafts" (Leite, 1953, p.20 quoted Fonseca, 2007, p.4).

With the support from Portugal, the work of the Jesuits focused on popular classes, i.e., the *mestizos*, the poor Portuguese, the slaves and the Indians, who should take care of the 'know-how' and on whom on normality of everyday life depended (Müller, 2005).

Another challenge was the productive complex itself involving the cutting, transport and assembly of materials (wood, earth, lime) used in the construction of schools, churches, buildings of the stables, sugar mills and dwellings of the slaves who would be the set of buildings of the Society of Jesus.

Craft Corporations played an important role in social-economic colonial life, in the relations between crafts and the religious links with lay brotherhoods in their mutual characteristics that ensured protection to their workers, the economic role they played in urban life and the political role with the Senate Chamber. In addition, other features were related by their function as centers of protection, credit, and social security, and also for establishing the rules and criteria for carrying out the craft and to legitimate the product acceptance on the city streets (Martins, 2012).

This work aims to study the participation of carpenters in construction at the time of colonial Brazil, considering the training mechanical of officers by the Society of Jesus, the role played by the Craft Corporations and the construction systems which were used, the latter relating to the current sustainability concept.

2 THE SOCIETY OF JESUS

Calainho (2005) reports that, in 1549, the first Governor-General of Brazil, Tomé de Souza landed in Brazil, accompanied by Ignatians, led by Manuel da Nobrega, who were involved in the organization of villages for converting the natives, facing many challenges in the missionary project of the Society of Jesus. Prominent among the challenges were the conditions found in a land of fickle weather, dangerous wildlife, aggressive natives, harsh conditions of subsistence and the mission of converting the Gentiles into the Christian faith, as well as the establishment of an administrative structure with a progressive territorial occupation.

The arts and crafts of construction were needed for developing Colonial Brazil, where the immense participation, contribution, dedication and efforts of the Jesuits to evangelize and educate the Gentiles can be highlighted, as per Leite (1953, p.19) apud Fonseca (2007, p. 4).

Bittar and Ferreira (2012), claim that Manuel da Nóbrega, in his colonization plan, devised the teaching of the first letters to indigenous children at the a-b-c houses, origin of Jesuit colleges and radiators *locus* of Christian culture, based on two supporting pillars: the first one related to the pedagogy of the teaching and learning process of the language and Christian principles, and the second relates to the material support to ensure operation through allotments, planting, farming and cattle raising and the slave labor.

Worthy of note was the participation of the priests Juan Azpilcueta Navarro (1st Expedition—1549) and José de Anchieta (3rd Expedition—1553), engaged in the task of grammatization of the Tupi language, allowing it to be taught at home.

According to Bittar and Ferreira (2012, p. 5), two key documents were developed to regulate the operation of the Society of Jesus: *Constitutions* (1558) and the *Ratio Studiorum* (1599) the latter related to the pedagogy of teaching and learning to be adopted the colleges. The Constitutions brought in draft form in 1553, stated that: "The Company will receive the property of the colleges with temporal goods belonging to them, and appoint them a Rector who has the talent most appropriate to the craft. This person will undertake the responsibility for the conservation and management of temporal goods [...]" (Society of Jesus, 1997, p. 122).

The authors claim that the Jesuit colleges of the sixteenth century were born inside the cane sugar (mill) and cattle farms, which architectural structure was described as such by Lúcio Costa (1945, p. 13).

Bittar and Ferreira (2012) argue that the Jesuits set up workshops and became masters of crafts. These were linked to the farms and were attached to schools. In the deployment of the first Jesuit college at Bahia, called "Todos os Santos" in 1554, the teaching of trades was envisaged in its educational system structure, which were the mechanical arts.

3 MECHANICAL OFFICERS

Conditions encountered by the Jesuits when they arrived in Brazil caused the creation of mechanical arts workshops, which were a matter of survival to produce manufactured goods consumed in the properties of the Society of Jesus (Bittar and Ferreira, 2012).

Cunha (2005) reports the existence of prejudice related to the officers of the mechanical arts, characteristic of Western culture, inherited from the ancient Greeks, who would have started the belief that contemplation was proper to the wise while action was the function of fools.

Woodwork then becomes one of the mechanical trades essential in the development of colonial Brazil. Currently, the reintroduction of building techniques used in the colonial period takes place due to the economy of resources and contribution to sustainability.

4 CRAFT CORPORATIONS

In the colonial period, a first formal initiative was teaching professions by means of the Craft Corporations aiming to regulate the handicraft production process. This association was composed of workers of the same trade, who performed tasks considered craft, requiring physical strength and skill, as those developed in carpentry, construction and potteries. To survive, they collected contribution quotas among its members to organize a common heritage and provide relief to members in times of financial need, illness or funerals. The more organized ones had a protective or "patron" saint, of devotion by the workers (Müller, 2005).

Within this context, it appears that Craft Corporations played an important role in social-economic life, the relationship between crafts and religious links with lay brotherhoods in their characteristics that ensured mutual protection of workers, the economic role they played in the urban life, as well as the political profile with the senate chamber. In addition, other features were related through their function as centers of protection, credit, and social security, and also for the establishment of rules and criteria for the performance of craft and legitimizing the products' acceptance on the city streets (Martins, 2012).

Müller (2005) reports that the Craft Corporations acted as "service providers" and its goal was the transmission of knowledge of a particular craft. Besides teaching, the teachers offered and hired work, judged the friction arising between members and applying tests to less qualified members. In 1824, the Imperial Constitution extinguished the Corporations due to treaties signed between the

Portuguese Crown and England, prohibiting the exercise of certain functions and, consequently, making the association of workers impossible.

Interesting to note the similarity between the role of labor unions in the present day and the role played by the Craft Corporations, in the colonial times, seeking association of mechanical officers in defense of matters of common interest and benefits for the good of the class.

5 CONSTRUCTIVE SYSTEMS

5.1 *The activities of carpenters*

The variety of wood parts confectioned by carpenters for use in construction did not always allow the employment of safe tools and easy to handle. Large saws are used for sawing large parts, as shown in Figure 1. In this operation the carpenters, in spite of working in pairs, used a lot of physical strength, ergonomically unsuitable positions and exposure to accident hazards.

Figure 2 shows the transport of boxes of tools (hammer, jigs, bench, saw, etc.) made by the carpenters. Looking at it, it appears that the carpenter has central leg bandages featuring, possibly, treatment due to accident in their work activities or protection, currently known as personal protection equipment.

Figure 1. Jean-Baptiste Debret. *Serradores*. 1822. Watercolor, 17.3 cm × 24 cm. MEA 0266 Collection of Castro Maya Museums/IBRAM/Minc, Rio de Janeiro. Playing Horst Merkel.

Figure 2. Jean-Baptiste Debret. *Carpinteiro indo para o trabalho*. 1821. Watercolor, 18.7 cm × 25.1 cm. MEA 0212 Collection of Castro Maya Museums/IBRAM/Minc, Rio de Janeiro. Playing Horst Merkel.

5.2 *The role of carpenters in various sectors*

Carpenters trained by the Jesuits contributed in various sectors with their work force and knowledge acquired, using wood in different ways.

5.2.1 *Mills*

Figure 3 shows the structure of a sugar cane wooden mill of that was of major importance in the colonial economy. Most of the parts were made of wood as well as the structure of pillars and roof of the shed behind the mill, demonstrating a great work provided by carpenters.

5.2.2 *Building*

Fonseca (2007) writes of the Jesuit style in the construction of the first temples that had a few Renaissance traccs as architectural features, regular lines, uncomplicated and unpretentious. Since this is a work of great size in Colonial Brazil, Bueno (2012) describes fragments of a common reality of construction as shown in Figure 4.

5.3 *Construction techniques*

The presence of carpentry services is remarkable in many types of construction techniques used in Colonial Brazil. Among them are the *taipa de*

Figure 3. Sugar mill, Lithograph by Johann Moritz Rugendas, 1835, In: Picturesque Voyage through Brazil. São Paulo: Martins, 1954. Library of the University of Brasilia.

Figure 4. João Francisco Muzzi. *Feliz e pronta reedificação da Igreja do Antigo Recolhimento de N. S. do Parto*. 1789. Oil on canvas, 100.5 × 124.5 cm. MEA 4031. Collection of Castro Maya Museums/IBRAM/Minc, Rio de Janeiro. Playing Jaime Acioli.

pilão (rammed earth) and its mold, the wattle and daub and the tangle of wood which were used, the *enxaméis* and their wooden structural parts and paneling.

5.3.1 *Taipa de pilão*

Colin (2011) describes that the *taipa de pilão* (rammed earth), presented in Figure 5, was a technique of Moorish origins practiced by Portuguese, Spaniards, and Black Africans commonly used in Europe until the mid-nineteenth century. Because of the abundance of red clay, the relative ease of implementation, satisfactory durability and the excellent conditions of protection it offers when given proper maintenance, it was the most employed material in the constructions in Colonial Brazil.

5.3.2 *Pau a pique, taipa de sebe, taipa de mão, barro armado or taipa de sopapo*

Widely used in Colonial Brazil this technique, shown in Figure 6, consists of a master structure of wooden parts, consisting of pillars—vertical pieces buried in the ground, *baldrames*—lower horizontal parts, and *frechais*—upper horizontal parts, with sections between 50 × 50 cm 40 × 40 cm or 20 × 20 cm.

The struts have lengths up to 15 m, of which 2–4 m are buried (Colin, 2011). Among its qualities the strength, durability and low cost are highlighted, for all the materials are natural, renewable and of low environmental impact.

5.3.3 *Enxaimel*

According to Colin (2011), this ancient technique (Figure 7) is very similar to the main structure of

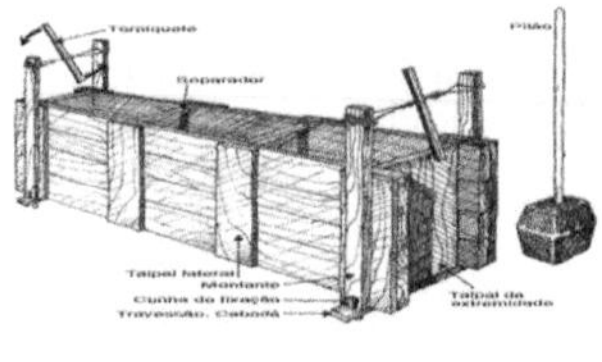

Figure 5. *Taipal* and *pilão*.

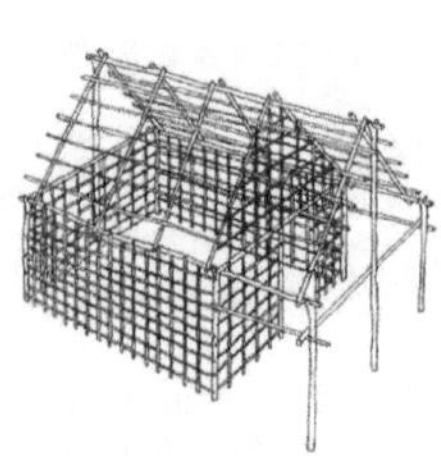

Figure 6. Construction in cleansed wattle and daub cottage. Bardou Picture 1981.

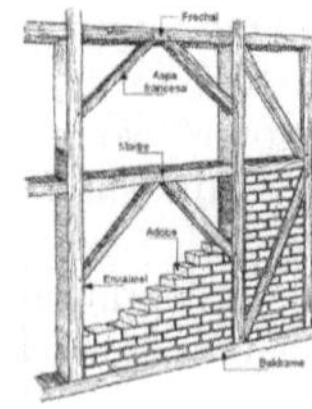

Figure 7. Muro de enxaimel. Imagem BARDOU, 1981.

Figure 8. *Tabique* picture. www.masisa.com.

the wattle-and-daub, differing in relation to the sealing that is filled with bricks or adobe. The gaps between the vertical parts (*enxaiméis* or struts) and the horizontal parts (*baldrames*, purlins and *frechais*) received the reinforcement of sloping parts in the corners and in diagonal in the frames, called St. Andrew's cross or French quotation marks.

5.3.4 *Tabique*

Colin (2011), states that "*tabique* is a partition made with wooden beams frame and lining with board (Figure 8), being a service of great simplicity and ease of implementation, especially for internal partitions". Despite the great simplicity of the *tabique*, its function is of responsibility whose striking example is the Church of Our Lady of Pilar in Ouro Preto, where the outer walls are of stone masonry and the wall of the aisle in timber, giving it its polygonal shape (Colin, 2011).

6 THE CONSTRUCTIVE SYSTEMS AND THE CARPENTER MASTERS IN THE PRESENT DAY

Currently, environmental sustainability is a term commonly used to describe the proper use of Earth's natural resources. It is considered a systemic concept, which relates to the continuity of the economic, social, cultural and environmental aspects of human society.

It is proposed to be a means of configuring civilization, so that society, its members and its economies can satisfy their needs in the present, while preserving biodiversity and natural ecosystems. Coupled to this debate, the Rio +20 was held recently with the theme: "The Future We Want". The conference came in order to help define a

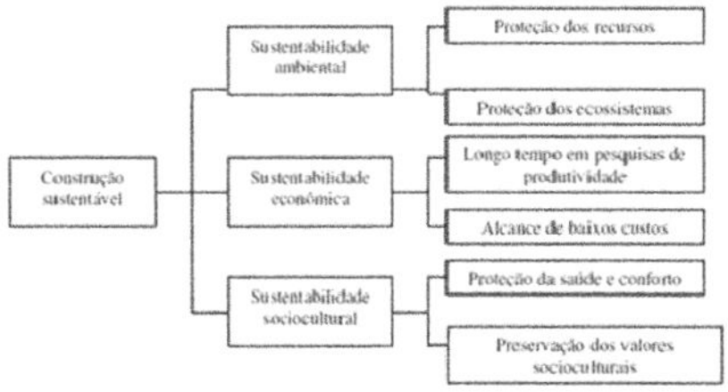

Figure 9. The initial dimensions of sustainable construction (Carvalho, 2009).

sustainable development agenda for the next few decades.

Discuss measures to induce social equity and environmental protection to ensure a habitable planet was the main focus. During the conference topics were discussed about the significant changes in environmental and geopolitical terms, the influence of economic and population growth in the welfare of humanity, the environment and ecosystems; awareness that growth combined with the dangers of environmental degradation, the loss of biodiversity and desertification, among others (Sousa, 2012).

In this context, we highlight the current concept of sustainable construction which is characterized by considering environmental, social (social-cultural) and economic (cost-benefit) factors, as illustrated in Figure 9.

The importance of environmental, social, technological and economic aspects for the choice of materials and techniques to be used in construction has adopted a new trend allied to the concept of sustainability of buildings. The materials derived from renewable and degradable resources, such as wood, earth present as the ones most suitable. Therefore, the techniques and building systems developed in the period of colonial Brazil were rescued, resurfacing the officers-carpenters, as well as their technical expertise ensuring sustainability in construction.

7 CONCLUSIONS

The work force made up of officers-mechanics labeled as fools, non-wise, since they belonged to the lower classes should be considered. This thought remains nowadays and, as stated Cunha (2005), it is a feature of the Western culture, a heritage of the ancient Greeks, they would have started the belief that "contemplation was proper to the wise while action was the function of fools". Currently the scarcity of this skilled labor is perceived, which by the consideration laid out above, has low pay, not being an attraction for the training of professionals with such expertise who want to pursue their studies in more promising areas. It is a problem to be faced, since it runs the risk of returning to the shortage of carpenters, just as in colonial times.

With regard to Craft Corporations (a form of organization of professionals), whose livelihood depended on the contribution of shares of associates, it had a mutual profile, acting in the interests of its members benefiting them with protection, credit, and social security, the establishment of rules and criteria for the carrying out the trades and legitimizing the acceptance of the products in the streets. Currently, with a similar role, the Labor Unions stand out, representing the workers in collective bargaining and agreements and seeking benefits for their categories.

And finally, the architecture at the time of colonial Brazil consists mainly of material derived from renewable and biodegradable resources, such as earth and timber. Thus, in relation to the discussion on the selection of traditional and innovative materials to be employed in construction, it is the environmental and social criteria which are considered. Therefore, this constructive system is reasserted, since this architecture complies in full with the needs of sustainability which have been entering the construction decisions, opting for the maintenance of its constituent materials.

REFERENCES

Bittar, M. & Ferreira, A.J. 2012. Artes liberais e ofícios mecânicos nos colégios jesuíticos do Brasil colonial. Disponível em: <http://www.scielo.br/pdf/rbedu/v17n51/12.pdf>. Acesso em: 4 abr. 2013.

Bueno, B.P.S. Sistema de produção da arquitetura da cidade colonial Mestres de ofício, riscos e traças. Disponível em: <http://www.revistas.usp.br/anaismp/article/view/39816/42680>. Acesso em: 5 abr. 2013.

Calainho, D.B. 2005. Jesuítas e Medicina no Brasil Colonial. *Tempo 10: 61–75. Jesuítas e Medicina no Brasil Colonial.* Disponível em: <http://www.scielo.br/pdf/tem/v10n19/v10n19a05.pdf>. Acesso em: 10 abr. 2013.

Carvalho, M.T.M. 2009. Metodologia para avaliação da sustentabilidade de edificações de interesse social com foco no projeto. *Tese (Doutorado em estruturas e construção civil)* Faculdade de Tecnologia, Universidade de Brasília.

Colin, S.V. 2011. Técnicas construtivas do período colonial—I. Disponível em: <http://phonteboa.blog.terra.com.br/tecnicas-construtivas-do-periodo-colonial.pdf>. Acesso em: 10 abr. 2013.

Costa, L. 1978. Arquitetura Jesuítica no Brasil. In: Arquitetura Religiosa, São Paulo, MEC/SPHAN/USP/FAU.

Cunha, L.A. 2005. 2ª ed. O ensino de ofícios artesanais e manufatureiros no Brasil escravocrata. São Paulo: Editora UNESP; Brasília, DF: FLACSO.
——. 2000. O ensino industrial manufatureiro no Brasil. Disponível em: <http://www.scielo.br/pdf/rbedu/n14/n14a06.pdf>. Acesso em: 8 abr. 2013.
Fonseca, S.M. 2007. A formação para o trabalho na Companhia de Jesus no Brasil Colônia (1549–1759). Disponível em: <http://anpuh.org/anais/wpcontent/uploads/mp/pdf/ANPUH.S24.1271.pdf>. Acesso em: 11 abr. 2013.
Leite, S. Arte e Ofícios dos Jesuítas no Brasil (1549–1760). Lisboa/Rio de Janeiro, Edições Brotéria/Livros de Portugal, 1953.
Martins, M. de S.N. 2012. A arte das corporações de ofícios: as irmandades e o trabalho no Rio de Janeiro colonial. Disponível em:<http://www.revista.ufpe.br/revistaclio/index.php/revista/article/viewFile/225/>. Acesso em: 08 abr. 2013.
Müller, M.T. 2005. A lousa e o torno—o SENAI e a Educação Profissionalizante no Brasil. In: *V JORNADA DO HISTEDBR* "História, Sociedade e Educação no Brasil", Sorocaba. Disponível em: <http://www.histedbr.fae.unicamp.br/acer_histedbr/jornada/jornada6/trabalhos/586/586.pdf>. Acesso em: 30 mar. 2013.
Pelegrini, S. de C.A. & Hoffmann, A.C. 2009. A técnica de construir em madeira: um legado do patrimônio cultural para a cidade de Maringá. Disponível em: <http://www.pph.uem.br/cih/anais/trabalhos/732.pdf>. [Acesso em: 2 abr. 2013.
Souza, P.M. da S. Construção Sustentável–Contributo para a construção de sistema de certificação. *Dissertação (Mestrado em Engenharia Civil) Faculdade de Ciências e Tecnologia*. Universidade Nova de Lisboa. Lisboa, 2012.

Occupational Safety and Hygiene II – Arezes et al. (eds)
 ISBN 978-1-138-00144-2

Safety analysis in the building construction carpentry work

C.M. Ferreira
IFET—Campus Barbacena, Universidade Federal de Juiz de Fora—PROAC, Brazil

M.A.S. Hippert & M.M. Borges
Universidade Federal de Juiz de Fora—PROAC, Juiz de Fora, Brazil

ABSTRACT: In 2013, construction industry was one of the segments that most boosted Brazil's economic development. The 2011 occupational accidents recorded statistics increased by a rate of 6.9%. This was specifically relevant in the circular saw operation, where carpenters had the upper limbs exposed to potential injuries and fractures. The Brazilian legislation regulates collective and individual protection measures and also establishes management and training programs for employee's accident prevention. However these are not sufficient to reduce significantly statistics of workplace accidents, thus creating a social cost for the country. From this context, this work stresses needs of investments in a broad sense in order to establish advances in health, wellness, life quality and social aspects of construction industry employees, more specifically in carpentry field. For that, the research is focused in identify specific needs of carpentry work in order to establish parameters for machine, protection devices, methods and work processes innovations.

1 INTRODUCTION

The construction industry in Brazil is experiencing a period of intense growth requiring a large quantity of construction workers. It is a sector that has many environmental risks and impacts, requiring many investments for the establishment of social advances, especially in the aspects of health, wellness and life quality of its employees. Among the diverse tasks in the process of building construction, one can highlight the carpenter's work that uses the circular saw mainly in the tasks of cutting wood for the concrete molds of the building structure.

Despite the fact that legislation on safety and occupational health had instruments to collective and individual protection and also regulate workers training programs for the accident prevention in construction management, one can verify that it is still not enough to eliminate workplace accidents, contributing in that way to create a social cost to Brazilian society.

There are many risks in the circular saw operation, where the carpenter who operates this equipment endangers mainly their hands. Statistics reporting high accident rates in the construction demonstrate that the body part most affected is the hand, indicating that despite the effort to comply with the law, by employers, carpenters and also with the supervision of the government, these accidents are not really being avoided. It also enhances that the sector requires innovations in methods and working processes.

From this scenario, this paper reports an ongoing research that aims to analyze the role of carpenter work in construction sector. The research is justified by the need for modernization in wood cutting machines, in order to make them safer and consequently reduce the labor woes that affect the carpenters. For that it was performed a statistical and literature review on the subject that addresses the current Brazil's construction scenario. This review looked at the number of occupational accidents, safety legislation; occupational medicine applied to the carpentry industry, the carpenter's training functions and activities, the risks in operating the circular saw and the actions of individual and collective protection. From this theoretical basis, field analyzes and interviews are being conducted with the aim of identifying the specific needs of the carpenter in order to establish the design parameters for machine, protection devices and methods.

2 THE BUILDING SECTOR IN THE BRAZILIAN CONTEXT

2.1 *Building sector*

The Brazilian economy has started the year of 2013 at an intense growth rate continuing the trend of acceleration occurred since the second half of 2012, offsetting the remaining difficulties in the international economy. Regarding the labor market, the year 2013 continues with a very robust dynamic performance, where the 4.6% rate of unemployment recorded in December 2012, was the smallest of the historical series and the January 2013 (5.4%) rate was the lowest for the month since the series beginning. It can be noted that the generation of formal employment continued with considerable numbers, reaching 1.3 million of new jobs in 2012 and that there was also an improvement in the job quality due to the increased scholarity level of the working population. The improvement of job quality and the minimum wage increase had produced an increased income, which, associated with to government social programs, led to a reduction of inequality in the country (BRASIL, 2013).

Teixeira (2013) states that in recent years the construction industry was one of the Brazil's economic sectors with highest growth. Programs geared specifically for housing and major infrastructure needs for the country development, contributed to the market heating, creating job vacancies in various areas.

In 2012 it was launched the National Commitment to improve the working conditions in the construction industry, aiming to ensure adequate conditions for construction workers in diverse aspects such as the recruitment and selection, training and qualifications, health and safety, union representation in the workplace and working conditions and community relations. Their guidelines are applicable to all activities of the construction industry through membership, and can cover a company, a single work, sets of works and work fronts, public or private (BRAZIL, 2013b).

2.2 *Statistics of occupational accidents*

To positively impact the way they are raised statistics of occupational accidents in Brazil, from April 2007, the INSS has instituted a new system of granting accident benefits (BRAZIL, 2013c).

According to the Protection magazine (2013), construction accidents totaled 59,808 in 2011; from this total, 39,301 were typical accidents. Comparing with 2010 data, the sector had the most significant increase of accident records, i.e. an increase of 6.9% in the incidents recorded in the area since last year the sector generated 55,920 accidents. One can highlight that from this total, 21,700 accidents where in the building construction sector inside the broad range of civil construction. It corresponds to a rate of 36.3% of the accidents occurred in the 2011 year.

Table 1 contains data on human body's parts most affected in accidents recorded in the year 2011, classified according to the code of the International Classification of Diseases—ICD. These data recorded injuries of the wrist and hand (67,465) and fractures at the wrist and hand (33,339) which together account for 17.2% of reported accidents (REVISED PROTECTION, 2013).

Table 1. Statistics of occupational accidents registered in Brazil.

Statistics	2011	2010	2009
Contributing employees	66.026.612	61.307.072	55.785.995
Acidentes típicos registrados	423.167	417.295	424.498
Typical accidents recorded	6.944.791	6.236.298	5.112.004
Typical accidents in construction	39.301	36.611	35.265
Injury of wrist and hand	66.259	70.797	75.428
Fracture at wrist and hand	33.504	30.722	32.679
Traumatic amputation at the wrist and hand	4.841	5.067	6.181
Forearm injury	5.459	5.975	6.632
Trauma to the eye and eye socket	2.373	2.774	2.990

Source: Statistical Yearbook of Social Security—AEPS (2013).

According to the AEPS data (2011), from 2007 to 2010 the parts of the body most affected in the records of typical occupational accidents were basically the same as 2011. This confirms that there is a need for more effective protection actions. These should be taken to protect employees who expose their hands in the danger zone of the machine, including the carpenters operating circular saws.

3 WOODWORK SECTOR

Nazar (2007) justifies the importance of detailed studies on the concrete forms dimensioning and materials choice and in the design, implementation and structure costs of a building. This kind of study will reflect in the labor and other items, including items that are not directly connected to the concrete structure.

Assahi (2011) states that the wood form is one of the many subsystems that makes up the building system. These multiple subsystems are interdependent and contribute to the outcome of the whole. However, the mold has a unique feature in this context, it is the subsystem that starts the whole process, referencing the others, establishing and standardizing the level of excellence required for the entire building.

3.1 *Carpenter*

The Labor and Employment Ministry—MTE, according to the Brazilian Occupations Classification briefly describes the form carpenter activities as: plan carpentry work, assembly wood forms, prepare construction site and assemble metallic molds, cook up wooden forms for slab panels, build scaffolding and wood protection structures and wooden roof, anchor bridges slabs of large spans, assembly of doors and window frames, Finalize services such as dismantling of scaffolding, cleaning and lubrication of metal forms, selection of reusable materials, storage of parts and equipment (BRAZIL, 2013b).

According to Gomes (2011), the Brazilian Occupations Classification (CBO), the carpenter has the 7155-35 code number and their activities are de-scribed as: cut pieces of wood to build forms for beams, columns for bracing slab and closing openings peripheries and also installing hinged doors, including the separation of reusable timber and nails.

In Brazil (2013b), it is regulated for these occupations exercise the necessity of schooling between fourth and seventh grades of high fundamental and basic course of professional qualification. The required workload is variable and is described below. Up to two hundred hours for scenario, roofs, scaffolding and for the assembler carpenters (buildings), two hundred to four hundred hours to the specialized carpenter (mining, carpentry, formwork for concrete and civil works of art), more than four hundred hours for the building carpenter. After one to two years of professional experience, the carpenter can be considered in the full profession exercise.

In the implementation of a building structure, the forms are made and assembled by carpenters, where after placement of the steel frame, the concrete is poured. After cured and demolded, the structural parts such as blocks, beams, columns and slabs are finished. In this process several tools are used in carpentry services, such as the hand tools (saws, torques's, feet-crowbars and hammers) and power tools as the circular saw and drills. Among them the circular saw is the one that offers greater risk of accidents (SAMPAIO, 1998).

3.2 *Risks on circular saw operation*

The circular saw bench is a cutting machine consisting of a circular disc provided with cutting edges on the periphery, mounted on a shaft. The drive is made by an electrical engine, transmitting rotary motion and cutting power through pulleys and belt (MORAES, 2009).

According to Sampaio (1998), the wood is not a homogeneous material and may have irregularities causing serious accidents. These irregularities came from loose or rotten knots, dried or resinous areas, stretch marks, parts of sapwood color contrasted and warping due to improper storage. Among the main causes of accidents with the circular saw one can highlight the follow: body contact with the active part of the disc or the descending part, the rejection of the work piece or one of its parts, contact with moving parts of the machine, exposure to electricity, inappropriate working methods and machine manufacturing defects.

The NR 9 classifies environmental risks in physical, chemical, ergonomic, biological and accidents. In the area of carpentry there are the following. Physical hazards: noise (issued by the circular saw), heat, cold, moisture and UV radiation due to climatic factors; chemicals from dust generated from cutting wood; ergonomic agents due to improper posture, lifting and manual transport weight, intense work pace, work standing for long periods, repetition of movements and time pressure Brazil (2013e). The risks of accidents are falling on the same level or height, upper limb injuries, electric shock, projection of fragments of the face and eyes (SESI, 2008).

4 SAFETY WORK LAW APPLIED TO CARPENTRY

The Brazil's Federal Constitution of 1988 confirms the fundamental right of workers related to reducing the inherent working risks through standards of health, hygiene and safety (BRAZIL, 1988). In fact, Brazil has, since 1978, a set of Regulatory Standards (NR) regarding safety and occupational medicine. However, despite the existence of more than 30 years, the number registered accidents are still high (BRAZIL, 2013d).

Brazil (2013e), 6.514/77 Law and 3.214/78 Decree, regulates the Safety and Health at Work, where the Regulatory Standards (NR) are obligatory for private and public companies and for public institutions of direct and indirect administration, as well by the organs of the Legislative and Judiciary, who have employees under the Consolidated Labor Laws (CLT). The provisions contained in NRs apply, as appropriate, to independent workers, entities or companies and the unions representing their respective professional categories.

Compliance with Regulatory Standards (NR's) does not exempt companies from compliance with other provisions included in building codes and health regulations of the states and municipalities, and others, arising from conventions and collective bargaining agreements (BRAZIL, 2013e).

As Saurin (2002) states, compliance with mandatory working standards of safety and health is the minimal effort to prevent accidents and occupational diseases, especially by the lack of studies showing that their full compliance is, alone, sufficient for a significant and lasting reduction of accident rates.

4.1 *Collective Protection Equipment—EPC*

According to Moraes (2009), the NR 18 (Conditions and Working Environment in the Construction Industry Program—PCMAT), establishes the guidelines for administrative, planning and organization, that aim to implement control measures and preventive security systems on the processes and on environmental working conditions in the Construction Industry.

The sub-item 18.7.1 states that the machines and equipment operations required to conduct the carpentry activity can only be performed by skilled worker under this NR.

The sub-item 18.7.2, establishes the provisions that a circular saw must meet:

a. Be provided with stable table, with closure of their lower faces, front and rear, built of sturdy wood and first quality, metallic or similar equivalent resistance, without irregularities, with sizing enough to perform the tasks;
b. Have the motor housing electrically grounded
c. The disc should be kept sharp and locked and must be replaced when present cracks, bends or broken teeth;
d. Mechanical force transmissions must necessarily be protected by fixed and resistant shields, and cannot be removed under any circumstances, during the execution of the work,
e. Be provided with protective hood and disc cleaver splitter, with identification of the manufacturer and sawdust collector (MORAES, 2009).

The NR 12 (Work Safety in Machinery and Equipment and their Attachments) defines technical references, principles and protective measures to ensure the health and physical integrity of workers and establishes minimum requirements for the accidents prevention and occupational diseases in the phases of design and use of machinery and equipment of all kinds, and also the manufacture, importation, marketing, exhibition and sale for any reason, in all economic activities, without prejudice to compliance with the other provisions of the Regulatory Standards—NR approved Ordinance n.° 3214 of 8 June 1978, the official technical standards and in the absence or failure of these, the international standards (BRAZIL, 2013e).

4.2 *Individual Protection Equipment—EPI*

The NR 6 considers Individual Protective Equipment (EPI), any device or product suited for individual use by the employee, for the protection of risks likely to threaten the worker safety and health. Basic carpenter's EPI are: helmet, gloves, leather boots and safety glasses. In the case of circular saw operators, other EPI are necessary as the apron shaves, face shield, ear shell and disposable mask (BRAZIL, 2013e).

4.3 *Training*

Santos and Rosenberg (2006) argue that the education process could be the instrument of adherence to safety standards for workers, i.e., one resort to the educational methodologies to reproduce the hegemonic power of rationality, thus educating to instruct, prevent and produce.

Employers have a duty to train their workers in accident prevention and it is set in various NRs. Workers have a duty to participate in the training, assimilating the guidelines of the employer.

According to Brazil (2013e), NR 1—General Provisions, states that in the training the workers should be informed about occupational hazards that may arise in the workplace, the means and measures adopted by the company for preventing and mitigating those risks; and the results of environmental assessments conducted in the workplace.

Likewise, NR 18 states that all employees should receive training on admission with minimum duration of six (6) hours and shall be applied within the working hours, before the employee starts its activities. These admission training should include information about the conditions and work environment; risks inherent to its function; proper use of Individual Protective Equipment—EPI; information about Collective Protection Equipment—EPC existing in the construction site. Periodic training should also be made whenever it becomes necessary and at the beginning of each phase of the work. In these training, workers must receive copies of procedures and operations to be performed safely (MORAES, 2009).

5 CONCLUSION

For these reasons, efforts to prevent circular saw workplace accidents should always be implemented, both the actions of collective protection (EPC), as the related to the use of Individual Protective Equipment (EPI), in addition to the training provided in the Regulatory Standards. It also can be stressed that accidents at work prevention will be enhanced with the effectively participation of employers, workers and government, each fulfilling their responsibilities.

The statistics draw attention by the number of accidents recorded in typical construction, highlighting the injuries of the wrist and hand and fractures at the wrist and hand as the body parts most affected, in the last three years analyzed. Those are the parts stated as the carpenter function risks. Moreover, the numbers show that there is yet much to be done to improve construction work environments. The methods and processes must be safer, the machinery and equipment must be modernized and employers and employees working together on prevention to provide safer environments and consequently a better quality of life for all.

This work is not exhaustive, but points to the need of deeper discussions on solutions that will contribute to the prevention of accidents in the activities of carpentry construction forms. Specifically the ongoing research is toward the direction of the design of new equipments and protection devices based on a bibliography and statistical review and also on a field survey among workers and employees in the context of construction sites at Juiz de Fora city.

BIBLIOGRAPHY

Assahi, P.N. 2005. *Sistema de Forma para Estruturas de Concreto.* Disponível em: <http://www.deecc.ufc.br>. Acesso em: 26 ago. 2013.

BRASIL. 2013a. Ministério da Fazenda. *Relatório Economia Brasileira em perspectiva.* Distrito Federal. Disponível em:<http://www.fazenda.gov.br>. Acesso em: 5 ago. 2013.

______. 2013b. *Novo pacto pelo trabalho na construção civil.* Disponível em: <http://www.brasil.gov.br/noticias/arquivos/2012/02/29/novo-pacto-pelo-trabalho-na-construcao-civil>. Acesso em: 2 ago. 2013.

______. 2013c. Ministério da Previdência Social. *Anuário Estatístico de Acidentes do Trabalho 2011.* Brasília, DF. Disponível em: <http://www.previdenciasocial.gov.br/anuarios/aeat>. Acesso em: 9 ago. 2013.

______. 2013d. *Constituição da República Federativa do Brasil de 1988.* Brasília, DF. Disponível em: < http://www.planalto.gov.br/ccivil_03/constituicao/constituiçao.htm>. Acesso em 8 jul. 2013.

______. 2013e. Ministério do Trabalho e Emprego. *Normas Regulamentadoras.* Brasília, DF. Disponível em: < http://portal.mte.gov.br/legislacao/normas-regulamentadoras-1.htm>. Acesso em: 12 ago. 2013.

Gomes. H.P. 2011. *Construção civil e saúde do trabalhador: um olhar sobre as pequenas obras.* 191f. Tese (Doutorado) – Escola Nacional de Saúde Pública Sergio Arouca, Rio de Janeiro, 2011.

Moraes, G.A. 2009. *Norma regulamentadoras comentadas e ilustradas. Vols. 1, 2 e 3.* 7a ed. Rio de Janeiro. Editora: Gerenciamento Verde Editora e livraria virtual. RJ.

Nazar, N. 2007. *Fôrmas e escoramentos para edifícios: critérios para dimensionamento e escolha do sistema.* 1a ed. São Paulo. Editora Pini. SP.

Revista Proteção. 2013. *Matérias/Anuário Brasileiro de Proteção 2013/Estatísticas de Acidentes Brasil.* Disponível em:< http://www.protecao.com.br/materias/anuario brasileiro de protecao2013/brasil/J9y4Jj>. Acesso em: 26 jul. 2013.

Sampaio, J.C.D.E.A. 1a ed. Manual de aplicação da NR-18. São Paulo: Editora Pini-Sinduscon: SP, 1998.

Santos, A.K. & Rozemberg, B. *Estudo de recepção de impressos por trabalhadores da construção civil: um debate das relações entre saúde e trabalho.* Caderno de Saúde Pública, Rio de Janeiro, vol. 22, n°5, p.975–985, maio, 2006.

Saurin, T.A. & Lantelme, E. & Formoso, C.T. 2000. *Contribuições para Aperfeiçoamento da NR-18: condições e meio ambiente de trabalho na indústria da construção.* Porto Alegre: Universidade federal do Rio Grande do Sul. 140 p. Relatório de Pesquisa.

Serviço Social da Indústria (SESI). 2008. Divisão de Saúde e Segurança—DSST. *Manual de segurança e saúde no trabalho: Indústria da construção civil - Edificações.* São Paulo: SESI. 212p.

Sousa, P.M. da S. *Construção Sustentável – Contributo para a construção de sistema de certificação.* Dissertação (Mestrado em Engenharia Civil) Faculdade de Ciências e Tecnologia. Universidade Nova de Lisboa. Lisboa, 2012.

Teixeira, S. 2013. Série Áreas Promissoras 2013: Construção Civil. Disponível em: <http://www.catho.com.br/carreirasucesso/noticias/serie-areas-promissoras-2013-construcao-civil>. Acesso em: 2 ago. 2013.

Occupational Safety and Hygiene II – Arezes et al. (eds)

Elementals ergonomic constituents for administratives environments

D.A. Cassano, M.C. Vidal & M.V.C. Aguilera
Federal University of Rio de Janeiro, Rio de Janeiro, RJ, Brazil

ABSTRACT: This paper present elementals ergonomic constituent for administrative environments. As a tool for optimizing the process of workplaces project, this study contain ergonomics concepts in the respective workplaces. According to the data collected and the analysis (qualitatively with the identification of most recurrent problems and their specificities) of consultancies carried out during 15 years through the Laboratory GENTE (Group of Ergonomics and New Technologies) from the Federal University of Rio de Janeiro, it is clear that the environment of administrative workplaces are not constituted from a wide architectural project, which should be considered issues related to ergonomics. The aim of this work is to present some possibilities of architectural project solutions based on a repertory of best practices. The ergonomic patterns of administrative workplaces environment developed here will initiated the creation of a database, aiming to articulate essential information about the workspace created for professionals involved in the design process.

1 THE ADMINISTRATIVE WORK ENVIRONMENTS

1.1 *The concept to create administrative workplaces environments*

Attempting to support project professionals to create workplaces environments in the sense of architectural conception (under analysis, evaluation or project) we developed the concept of elemental constituents of project to work with the notion of ergonomic constituent. The notion of elementals ergonomic constituents is materialized through the ergonomic patterns (Alexander et al, 2013) which are conformed in a structure, combine according to the relational characteristics and inserted in its context. This provides to the designed environment its aspect of effectiveness and humanization of the space, besides, it helps with the sustainability and maintenance of the environment in the everyday practice and the probable future.

The formulated ergonomic patterns provide characteristics minimally satisfactory to work environments (Santos, 2010) and aid to the architectonic propagation once it is define by the logic current production. The underlying idea is the development of project reference of work environments, providing the conceptual construction of a reference situation for the practice of conception ergonomics. This development is composed potentially of a ergonomics pattern catalog, follow by a set of tables, leading to a database where the professionals of project will be able to articulate the programing elements that they have with ergonomics presets according to a gathered collection of good practices.

1.2 *The importance of administrative workplaces environments*

Production Engineering deals with issues related to planning, programming and controlling of purchase, production and distribution of products. Whereas the productive operations differ according to the nature of their specific inputs and outputs having as a final result a product or a service. The technical production of any organization involves in an orderly way: materials, equipment, people and information. Information produced and data processed to give support to the several functions in the process of all of the stages of the operation. This information is created, treated and storage in an information system that which is managed and operated by people who produce document through computerized and automated means that we call administrative work activities. Indeed, administrative work activities are at the core of occupations and concerns of Production Engineering field.

According to Slack (2009), "all incoming resources (input) have transformers resources, which are divided in "facilities" and "people". These two elements are considered the structural base of all operation. The facilities play a very important role in a productive organization in the sense of here is where it can be visualized both the workplace and the workers as being the one in charge of the products and services produced. Considering that the administrative work dominates most of the current labor scenarios, this study focuses in issues related to the project of these facilities.

2 THE ARCHITECTURE OF ADMINISTRATIVE WORKPLACES

2.1 *The backstage: Reality projetual practiced*

After 15 year of consultancy activity with the Laboratory GENTE, several subjects were studied, ranging from factories, refineries and even control rooms. It is important to emphasize that the analysis of administrative environment emerged in a significant number among the consultants conducted. Due to the fact that they are present in most of the organizations studied whether they were units of refineries, plants or commercial.

The demands concerning to administrative environments were mostly corrective action. These environments that were in full operation, presented dysfunctions and were distant from the criteria of efficiency established by the organizations. At this moment is when ergonomics becomes necessary and put it into practice in order to fix malfunctions aiming to make some progress in work environments to a state that is favorable to work activities, and consequently, to workers.

As a result of the ergonomics corrective action, there is a disconnection in the process of conception and execution of the project. Thus, appear generic workplaces: where different activities are perform in each place with questionable typologies that are not positively according with the activities. Supporting the idea of the possible origin of malfunctions, Vidal (2010) contends that beyond "a simplistic design, there are variabilities that were not taken into account at the time of projecting the work environment". The unawareness of the variabilities, their sources and their impacts, lead to a mismatch between the designer and the application of the knowledge coming from the performed everyday activities, considering primarily prescribed work activities at the time to design the work environment. We assume that at the time of designing work environments, in most of the cases, there is lack, from the professionals involved in projects, of domain about the ergonomic precepts (Cassano, 2008). It is important to comprehend and integrate ergonomic concepts in projects of work environments to create a comfortable, efficient and safe workplace (Vidal, 2010).

2.2 *Ergonomic patterns of work environments*

The ergonomic patterns of work environments provide meaningful recommendation to deal with certain conception problem of work environments. It is a guideline for people involved in the project (Fonseca, 2012), opening possibilities and inspirations for the architecture of administrative work environments. We emphasize that this study aims to supplement a gap for the bibliographic material that has been produced so far as to design with ergonomics. Bibliographic production directed to the area of ergonomics when dealing with the subject in question, pointing specifically jobs with computers and shows up quite incomplete as to address the architectural space in its entirety. With what is presented you can not feed the design process of administrative work environments because cares about demonstrating the postures in the workplace and their anthropometric issues. The aim of ergonomic patterns is to indicate adequate architectonic typologies based on the production line and based on the collection of good practices for this type of work. Furthermore, is to be used as a support for project professionals to minimize the time required to develop projects and to avoid the use of wrong concepts usually applied in projects. It is important to note that the major stakeholders to use the ergonomic patterns are those who may be called the ergonomic agents—person in charge of the integration among project professionals, manager and workers involved in the study (Vidal, 2010).

The priority of this proposal is to create architectonic concepts for administrative environments in a structured chart providing to those involved in the project meaningful data to discuss necessary issues and facilitate the decisions of the project.

The ergonomic patterns are in accordance with the of the organization, in the sense of the organization development level to respond to ergonomic issues according to major decisions related to the functioning of their internal processes (Vidal 2009). Meaning, how organizations should incorporate ergonomic issues in a strategic way, going beyond the safety and health of workers (Dul & Neumann, 2008).

This approach reduces costs (avoiding to adjust workplaces in the future) and increases productivity with adequate workplaces; providing greater satisfaction among workers and less risk coming from the absent of ergonomic (Masculo et al., 2011).

3 METHOD

The analysis of collected data of several consultancies carried out by the Laboratory GENTE has shown the incidence of similar problems in different administrative environments and from different backgrounds. In this paper we present empirical contents that are characterized as the cause of negative effects whether they are: physique, cognitive, organizational or environmental. Then, a group of useful characteristics has been selected from the identified problems for a deep analysis.

This study has three stages of analysis:

a. Statement of relevance: the patterns effectiveness is demonstrated by testing illustrative
   - Establish a database of patterns for project (structuring)
   - Use the Database in concrete situations (implementation)
b. Statement of purpose: existence of working environments with general characteristics and typologies both questionable
   - Characterize environments
   - Demonstrate impacts questionable
c. Statement of merit: tool's application in redesign
   - Elicit premises (method of correction)
   - Re-design (design methodology)

3.1 *Detailing the method*

Below, we detail the method used, explaining each step of the investigation.

A. To get a demonstration of relevance:

1. Output modeling cross-elemental characteristics resulting in a matrix;
   - classify information for the most common environments with purpose statistician;
   - cataloging situations in disagreement and procedures performed as a function of the architecture imposed;
2. The transmutation of matrix types of environments vs. types of problems in elemental contents structured;
   - illustrate the problems from reality-type work;
3. Application for an essay illustrating conceptional analysis and conceptual design.
   - provide conceptual solutions to problems;
   - store data in order for research and future projects.

B. To get a demonstration of purpose:

1. Election of thirty (30) consulting works performed by the Laboratory GENTE;
   - Identification of the negative impacts.
2. The collection and analysis of data showing the incidence of some class of similar problems in different environments and completely different backgrounds;
   - Meeting the features that should be considered in its architectural programming fundamentals with ergonomics.
3. Explain what impacts these groups situations presents.

C. To get a demonstration of merit:

1. Method of correction
   - Select a group of situations with relevant characteristics for analysis;
   - List most relevant impacts;
   - Establishing causal aspects;
   - Affirm improvement opportunities.
2. Re-design (design methodology)
   - Define characteristics to use the ergonomics standards;
   - Apply standards in reshaping the environment, such as an illustrative essay.
3. Practical demonstration
   - Perform the above process in a number of relevant office's cases.

4 RESULT

4.1 *Taxonomy of the absence of ergonomics problems in an environment of administrative work: Empirical analysis*

A taxonomy of problems lack of ergonomics in an environment of administrative work presented here aims to meet the characteristics of the environments that should be considered in its architectural programming with ergonomics fundamentals.

In practice, ergonomics aims to provide appropriate conditions for the worker to develop their activities as well. And that at the end of your workday complications do not arise which adversely affect your comfort or that may affect your health. Including also seek to depart from the reality of work problems that may cause injuries or accidents. And this applies to any kind of work, without distinction.

Due to the nature of the work in an environment of administrative work are not registered high rates of accidents. Therefore, for most managers and organizations, administrative workers are affected only by the effects from maintaining awkward postures and repetitive movements which are in front of computers for long periods of time. Mistakenly fail to give due attention to environmental characteristics, forgetting to analyze the working method as well as the entire system of work where the employee is inserted. The generalization of the administrative work makes another important question is also not treated as a priority, which refers to the examination of the cognitive aspect of the activity.

By analyzing the consulting work performed by the Laboratory GENTE in administrative working environments were checked several aspects that point discomfort and disturbance of activity, and discontinuities (in the case of existence of local ergonomic standards) or undesired results (of various kinds). Appreciated the ergonomic aspects were classified into 15 classes of problems:

A. Physical effects
   - Explanations for awkward postures (what strength);
   - Interference and uncomfortable.

a. Absences of several objects;
b. Footrest;
c. Place for accommodation of material and equipment;
d. Inadequacies of furniture and other things;
e. Order (exposed wiring, carpet loose, slippery etc.)

B. Problems of Ambience
   - Explanations for wrong organized physical space architecturally.

f. Acoustic environments
g. temperature
h. Ambient lighting

C. Cognitive Requirements
   - Explanations for problems relating to the process of development of mental processes, which are affected: a way of thinking, attention, memory, perception, language and reasoning.

i. Interference neighborhood and proximity;
j. Difficulty communication;
k. Dispersion of spatial information;

According to Vidal (2010) "A human characteristic is to seek quick fixes, but this trend should not be taken for Ergonomics." One should not believe that the solutions for offices and for the problems of the binomial "human-computer" restrict themselves to correcting postures and offering the best market securities. Although the location of administrative work is seemingly quiet and away from issues of great impact, all the positive and negative aspects of work situations should be investigated, evaluated, designed, discussed with the main stakeholders involved in the situation, and finally put into practice.

4.2 *A catalog of ergonomic patterns*

Fifteen years of data collection provided us the structuring of a pilot catalog of ergonomic patterns, where we classified architectonic problems that could be solved. The suggested solutions are represented by a floor plan or a perspective drawing followed by tables that contain work situations with certain aspects referenced. The tables will be supplemented with more data according to their usage, providing a meaningful database about the architecture of administrative work environments, which consists on: problem-type, description, illustration and solution concepts (table 1).

Starting from a registration of the work situation, the table will be presented with two design solutions: an innovative set and the other based on a reference situation searched the literature. Thus, we show an example of a real work situation (figure 1) and propose a solution to the problem projetual concerned (figure 2) and the point of view of a reference situation (figure 3).

Table 1. Pilot catalogue of ergonomic patterns (part of the exemple).

Problems-type		Description	Ilustration	Solution concepts	
Type of work: administrative assistants Function: contract analysis					
Category Class	Physique Scope	The worker can not easily access the folders with documents	Registration of the work situation	I	II

Figure 1. Registration of the work situation.

Figure 2. Proposal solution for activities that require the handling of multiple files.

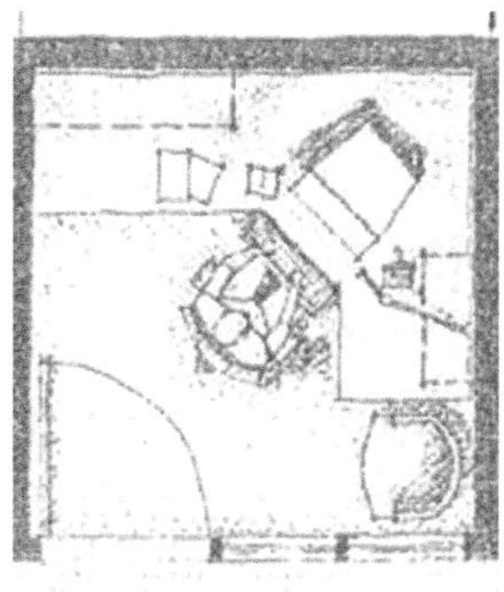

Figure 3. Reference situation baseline for administrative work environments that require more space for files and interactivity (Brill et al. 2001).

5 DISCUSSION

This study's main proposition is that the use of ergonomics standards for administrative environments could be designed to ensure their effectiveness and humanizing aspects of spaces, as well as assist the sustainability and maintenance of the environment in everyday practice future. The creation of ergonomic standards for administrative environments presents itself here, contrary to the thinking that leads to a possible immobilization of ideas or lack of creative propositions. As the placement of Saunders (2002), in the last century was established as common sense for intellectual efforts that should not be attributed ergonomic standards or rules, because they block the designers freedom. However, there are situations that require specific care projetuals and techniques that are providers of such care, which goes beyond the limits of creativity of designers. We present the elementals ergonomic constituent as a tool that will allow designers to review their concepts through examination of current situations.

It is important to pointed out that the ergonomic patterns do not attempt to solve all of the existing problems of work environment conception. The use of the ergonomic patterns support the goal of the ergonomic maturity of an organization through the accomplishment of its application. Since the people involved in the project will have advanced knowledge about the work activity in study. Moreover, they will be able to use effectively the elementals ergonomic constituent, articulating the production organization with the ergonomic solutions for facilities of professional environments that form the artificial environment in which we work.

REFERENCES

Alexander, C., Ishikawa, S., Silverstein, M. 2013. *Uma Linguagem de Padrões*; tradução: Alexander Salvaterra; Porto Alegre: Bookman.

Brill, M., Margulis, S., Konar, E., & Bosti. 2001. *Disproving widespread myths about workplace design*. Jasper, IN: Kimball International.

Cassano, D.A. 2008. *Arquitetura de ambientes de escritórios e ergonomia. Estudo de casos múltiplos no setor de serviços de uma mesma empresa*. Dissertação de mestrado, Rio de Janeiro: UFRJ.

Dul, J.; Neumann, W.P. 2008. Ergonomics contributions to company strategies. *Applied Ergonomics*, Elsevier.

Fonseca, B. B.; Aguilera, M. V. C.; Vidal, M. C. R.2012. *Conceptual design pattern for ergonomic workplaces*. Work (Reading, MA), v. 41, p. 797–803.

Masculo, F., Vidal, M.C. 2011. *Ergonomia: Trabalho adequado e eficiente*. Rio de Janeiro: Elsevier/ABEPRO.

Saunders, W. S., 2002. Pattern Language Reviewed, in *Harvard Design Magazine*, Winter/Spring, Number 16.

Santos, Marcelo S; Carvalho, Paulo Victor; Vidal, Mario Cesar R. 2010. Ergonomic Pattern Mapping A new method for participatory design processes in the workplace. In: 3rd AHFE, Miami, FL. 3rd International Conference on Applied Human Factors and Ergonomics.

Slack, N.; Chambers, S.; Johnston, R. 2009. *Administração da Produção* – 3ª Ed São Paulo: Atlas.

Vidal, M.C.R, 2010. *Principios para un abordaje macroergonómico: útil, practico y aplicado*. 1. ed. Bogota: Editora Guadalupe.

Vidal, M.C., Guizze, C.L., Mafra, J.R., Bonfatti, R.J., Santos, M.S., Pacheco, R., Moreira, L.R.; 2009. The ergonomic maturity of a company enhancing the effectiveness of ergonomic process. *Proceedings of the XIV Triennial Congress of the IEA*.

Occupational Safety and Hygiene II – Arezes et al. (eds)
© 2014 Taylor & Francis Group, London, ISBN 978-1-138-00144-2

Research on occupational exposure to radon in Portuguese thermal spas

A.S. Silva, M.L. Dinis & A. Fiúza
Research Laboratory on Prevention of Occupational and Environmental Risks (PROA/LABIOMEP), Geo-Environment and Resources Research Centre (CIGAR), Faculty of Engineering, Universityy of Porto, Portugal

ABSTRACT: Radon exposure in thermal spas have been the target of increasing and recent studies in some countries such as Italy, Brazil, Spain, Greece, China, Turkey, but without any development in Portugal. Moreover, the findings in terms of health impacts have not been unanimous and conclusive, once radon and its short-living decay products are the main contributors to natural radiation exposure for general population. The purpose of this work is to contribute to the understanding of the effect on workers' health from the exposure to radon in thermal spas, taking into account the contribution from outside of the occupational environment.

The adopted methodology proposes a continuous literature review; the selection of the sites and a group of workers to monitor; the assessment of radon exposure; the acceptability and effectiveness of the developed research; the management and analysis of data and finally, the conclusions and recommendations. The study should provide conclusions on whether the exposure to radon in thermal spas translates into an occupational hazard, its magnitude and the need to implement a radiological protection program reflecting the measures to prevent and minimize the occupational exposure.

1 INTRODUCTION

Spa derives from the name of the Belgian town of Spa, known in Roman times as *Aquae Spadanae*, meaning "health through water". The concept has become popular in the late twentieth century, being the spas known as places for relaxation and socialization, and more than therapeutic spaces, currently, are preferential sites of social life (Köteles, 2007).

In Portugal, the concept was originated in 2004 through the publication of the legal regime of thermal activity, allowing the use of thermal waters for purposes not directly therapeutic. In this way, thermal spas are defined as units of health-care where the properties of a natural mineral water are used for cosmetic or beauty treatments as well for spiritual or psychological activities (Portuguese Legislation, 2004).

Portuguese natural mineral waters have become of a great interest mainly due to the quality, diversity and favourable effects on health being classified as one of the most important of the European waters (APIAM, 2010).

Mineral and thermal waters are a particular kind of ground-water, distinguished by specific chemical or physical properties such as higher mineralization, concentration of certain constituents, dissolved gas, radioactivity or temperature. A radioactive gas, radon, may be present in most of these environments as mineral or thermal waters are usually connected with specific and unique geological and tectonic structures.

Radon results from the natural decay of radium-226 which in turn is a decay product of uranium-238. Since radon is a gas it may easily escape into air from the material in which is formed and since uranium and radium occur widely in rocks, soils and water, radon is naturally present in the air that we inhale, outdoors and indoors, as well as dissolved in groundwater (IAEA, 2003; Robertson et al., 2013). In addition, radon has a half-life of 3.82 days which allows a large period of time for migration before it decays into another nuclide (Dinis and Fiúza, 2005).

Airborne radon disintegrates to a series of short-lived radioactive decay products (progeny: ^{218}Po, ^{214}Bi, ^{214}Pb and ^{214}Po) that are hazardous if inhaled (Dinis and Fiúza, 2005). Radon's progeny are not gaseous, but rather particulate and they will attach themselves to dust particles or other particulates suspended in the air. Once inhaled, they will permanence inside the body (following radon decay) and irradiate lung tissue based on their decay and associated respective half-lives (varying from a fraction of a second to less than 30 minutes) (IAEA, 2003).

Radon and progeny are recognized as the most significant natural source of human exposure to natural radiation (UNSCEAR, 2000) and the most important cause of lung cancer incidence except for smoking (US-EPA, 2003; WHO, 2009).

It is estimated that about 8% to 25% of the current deaths from lung cancer are thought to be due, in part, to radon exposure in indoor air, from past situations (Puskin and Yang, 1988; in Mose and Mushrush, 2008).

Radon can present a hazard in a wide range of workplaces. In fact, high levels of radon have been found in workplaces from several countries with concentrations exceeding 1000 Bq/m^3 (IAEA, 2003). In particular, for thermal spas workplaces, considerable high concentrations of radon and progeny have been observed (Steinhäusler, 1988; Lettner et al., 1996; Szerbin, 1996; Trabidou et al., 1996; Datye et al., 1997; Vogiannis et al., 2004a,b,c; Radolic et al., 2005; Song et al., 2005; Manic et al., 2006; Bonotto and Santos, 2007; Somlai et al., 2007; Gnoni et al., 2008; in Nikolopoulos et al., 2010), causing significant additional radiation exposure to patients and workers. The EU has mentioned this fact in the directive 96/29/EURATOM, proposing spa therapy as a professional activity of enhanced natural radiation exposure (CEC, 1996; in Nikolopoulos et al., 2010).

Numerous studies have been conducted in several countries to determine the occupational exposure to radon in thermal spas. Recently, a study carried out in thermal spas from Ischia Island showed that the average annual effective dose to workers due to radon exposure was higher (0.01 to 7.03 mSv/y) than the limit (3mSv) imposed by the Italian law.

The health effects of exposure to radon in indoor environment (indoor air), and in particular, in thermal spas, depend mainly on the concentration of inhaled radon, the ventilation rate of the place, frequency and duration of the exposure (Labidi et al., 2006). In Portugal, the monitoring of exposure to radon is compulsory only in buildings constructed in granite areas, particularly in the districts of Braga, Vila Real, Porto, Guarda, Viseu and Castelo Branco, in particular due to the geological nature of these regions. This obligation came into force in 2006 within the new regulations for indoor air quality (Portuguese Legislation, 2006).

This work describes the methodology to assess the exposure to radon taking into account the contribution from outside of the workplace (workers homes) in different Portuguese thermal spas contributing to the understanding of the effect on workers' health from this sector. The current investigation is predicted to be undergoing until 2014.

The main research issue relies on whether radon exposure in Portuguese thermal spas reflects an occupational hazard, and if so, the need to implement a radiological protection program reflecting the measures to eliminate or decrease the exposure until acceptable and safe levels.

2 MATERIALS AND METHOD

The adopted methodology proposes to include in this study all thermal spas facilities existent in Portugal. However, a first selection is being carried out based on their location and geological features. In a second stage a group of spas workers, as well as their houses, are to be selected.

Radon concentration measurements were established to be carried out seasonally (January and July) in thermal spas water, spas indoor air and workers houses indoor air, for a maximum period of 3 weeks. Workers are also predicted to be monitored at the workplace. The main stages of the adopted methodology are briefly described below.

As a first step of the proposed methodology, a continuous literature review is being carried out updating the methodology.

The selection of the thermal spas to include in this study is undergoing and as mentioned before, it will depend both on the local geology and hydrogeology as well as the acceptability of the thermal spa facility to participate in this research. So far about 7 facilities were already selected and identified. The first set of measurements is to be carried out in January 2014.

A group of workers from each spa will be selected and identified; a control group will also be considered. A structured questionnaire and an observation checklist are to be used to collect information from the selected population: workers from the selected thermal spas and control group. The questionnaire explores specific characteristics such as age, gender, medical history (illness and declared pathologies) and life style. In particular for spas workers the questionnaire will also include a characterization of the professional tasks in order to identify the places from the thermal spas where the exposure theoretically will be higher, its duration and frequency.

A few houses from the selected group are also to be included in the study in order to consider the contribution from outside of the workplaces. The selection will be based on local geology and construction materials.

Similar tools (questionnaire and check list) are to be used to collect information on both selected thermal spas and workers housings. The data considers the location, the type of construction, the ventilation rate, the number of rooms and usage.

A few measurements of radon concentration in air will also be conducted outside of the granite areas to be used as control areas considering the dependence of radon concentration with the local geology.

Radon measurements are to be carried out in thermal spas indoor air and workers housings twice a year for a period of 3 weeks and in spas water (captures and places of treatment) only once as it is expected that radon concentration in water will not vary with time. Radon and daughters in indoor air are to be measured with CR-39 detectors (Pershore Mouldings, LTD) while radon concentrations in the thermal spas water are to be measured using the radon monitor AlphaGUARD 2000 Pro (Saphymo) (Pereira, 2001). This equipment may also be used, eventually, to measure radon concentration in air.

Radiation measurements will be carried out with a gamma ray scintillometer, SPP2 from Saphymo (building materials and water). Gamma Scout dosimeters, GS3, will be used for dose measurements (γ, $\beta + \gamma$, $\alpha + \beta + \gamma$) in indoor air of both thermal spas and worker houses.

For workplaces with many rooms or work areas, it is important that more detailed measurements of radon concentrations are made in a sufficient number of these locations in order to make an appropriate assessment of worker exposures to radon. Individual monitoring may be useful, in particular, in working environments with significant spatial and temporal variations in radon concentrations and in situations with significant differences in staff working patterns. In addition, when passive area monitoring indicates high radon levels and the staff's work patterns make a dose assessment based on area monitoring difficult to conduct, then individual dosimeters will be used.

For dose assessment the adopted methodology follows the International Basic Safety Standards for Protection against Ionizing Radiation and for the Safety of Radiation Sources (IAEA, 2003) where its stated that in cases where individual monitoring is inappropriate, inadequate or not feasible, the occupational exposure of the worker shall be assessed on the basis of the results of monitoring of the workplace an information on the locations and durations of exposure of the workers.

The average annual effective dose due to occupational exposure to radon, regarding the time by which the workers stayed in a single type of room, are to be calculated. The incremental risk resulting from this occupational exposure will also be assessed.

The analysis of the dose values and risk interpretation are to be combined with existing illness or pathologies among the monitored workers.

The complete study will be compiled in a final report expressing conclusions and recommendations directed to each thermal spa facility.

3 RESULTS AND DISCUSSION

There are 37 thermal spas in Portugal: 14 in the northern region, 20 in the central region and 3 in the region of Alentejo and Algarve. In 2011, this sector was composed by 910 professionals corresponding approximately to 60% of the spas workforce (Silva et al, 2013) (figure 1).

Although the wide range of thermal spas in Portugal and the significant number of workers that may be exposed to radon, there are no previous research studies focusing on this issue and up to day there is no data regarding the exposure of workers to radon and its daughters.

On the contrary, many countries had already carried out several research studies on this concern:

i. In a study conducted in the thermal baths of Riccione (Italy) radon concentration in air ranged from 6 to 70 Bq/m^3 and these values are in the same range of values measured in other thermal baths from Italy (Emila-Romagna region) (Desideri et al., 2004);
ii. Radon concentrations in air were measured in 10 spas from Croatia and the values ranged from 11 to 109 Bq/m^3;
iii. Two surveys carried out in Spain showed concentrations of radon varying from 3560 to 6650 Bq/m^3 and from 800 to 5200 Bq/m^3 (Rodolic et al., 2005);
iv. In thermal hotels of Guangdong (China) it was observed that radon released into the air was 110–401% higher in the bedrooms and 510–1200% higher in bathrooms compared to the average values when there was no use of thermal water (Song et al., 2011).

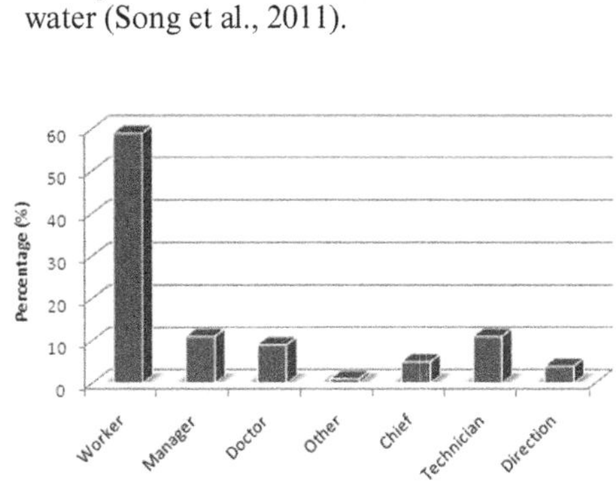

Figure 1. Staff in Portuguese thermal spas, 2011.

v. In thermal spas from Ischia Island (Italy), a wide range of radon concentrations was registered: 30–3983 Bq/m^3 (Pugliese et al., 2013).
vi. Radon concentration measured in Ikaria spas (Greece) showed that mineral water is the principal source of high concentrations of radon in indoor air of thermal spas, in addition to the soil, subsoil and construction materials (Nikolopoulos and Vogiannis, 2007).
vii. In Ischia Island thermal spas the radon concentration measured in water varied among plants in the range 7–98 Bq/L, which are comparable with the results of several studies in other countries (Greece and Venezuela). Higher values were reported by other authors from several countries (Slovenia, Spain, Greece and Poland) (Pugliese et al., 2013).

As projected results, this research study should give similar results to the ones mentioned above allowing concluding whether there is an occupational exposure to radon in Portuguese thermal spas. The occupational exposure should be quantified through dose and risk assessments. The final achievable outcome should be a radiological protection program adjusted to each thermal spa monitored.

As expected results from this study, it is predictable to obtain:

i. Radon concentration measured in the water (Bq/L);
ii. Ranges of radon concentration in air for a representative number of thermal spas and its variability: seasonal, from facility to facility regarding their geological location and relevant rooms from each thermal spa;
iii. It is expected that radon concentration will be higher in the rooms with poorer ventilation rate and with more water volume usage;
iv. In workers housings it is expected that radon concentration will be higher, in particular in locations with a granite background and a poor ventilation system;
v. Dose resulting from the occupational exposure to radon concentration reflecting the basic parameters of exposure: inhalation rate, frequency and duration of exposure, in accordance with recommendations of international radiation protection (UNSCEAR, IAEA, ICRP and USEPA);
vi. The average annual effective dose is likely to be higher than the recommended legal limit, in particular for the spas with deficient or inexistent ventilation system;
vii. The risk for workers resulting from the occupational exposure is likely to result in a value higher than the internationally recommended limit: 10^{-5} on occupational exposure (EPA, 1997);
viii. Radiological registration system of each thermal spa with effective occupational exposure.

The acceptability and effectiveness of this research study will be given by the workers motivation to change behaviors and adopt new protection measures. In addition, the employer's awareness and acceptance of the results should allow an intervention and implementation of a radiological protection program.

4 CONCLUSIONS

This study presents a methodology to evaluate the occupational exposure to radon in Portuguese thermal spas waters and it should provide conclusions whether exposure to radon in thermal spas represents an occupational health risk, in which magnitude and the need to implement a radiological protection program reflecting the measures to prevent and minimize the occupational exposure to safe levels. Dose limits, intervention and action levels will also be taken in consideration according to official values.

This work will also attempt to investigate the receptivity of workers and their motivation to change behaviors and adopt new safety measures. Likewise, it is important to understand the interest and acceptance from thermal spas employers, to allow a possible intervention. The main conclusion will derive from:

i. Radon concentrations measurements in spas indoor air and workers houses carried out seasonally, twice a year (January and July). Radon concentrations measurements in thermal spas water are also included;
ii. The annual effective dose for workers taking into account the working time and the measured radon concentration in air, compared with legal limits from the radiological protection;
iii. An incremental risk for workers due to occupational exposure in Portuguese thermal spas will be the outcome from the obtained results.

The results from this research study will provide specific data for occupational exposure in this sector. The results should be compiled in a registering system, as complete as possible, for the assessment of radon exposure in this particular occupational environment.

REFERENCES

APIAM (2010). Águas Minerais Naturais e Águas de Nascente, Livro Branco, APIAM, Associação Portuguesa dos Industriais de Águas Minerais Naturais e de Nascente.

Bonotto, D.M., Santos, T.O. (2007). In: Brebbia CA, editor. Health risk related to radon in a thermal spa utilizing waters from Guarani aquifer. Environmental Health Risk. vol. IV. Southampton: WIT Press; 2007. p. 137–46.

CEC (Council of the European Union) (1996). Council Directive 96/29/Euratom of 13 May 1996. Off J Eur Communities 1996 L-159 of 29.6.

Datye, V.K., Hopke, K., Fitzerland, B., Raunemaa, A. (1997). Dynamic model for assessing 222-Rn and progeny exposure from showering with radon-bearing water. Environ Sci Technol 1997;31(6): 1589–96.

Desideri D, Bruno M.R. and Roselli C. (2004). 222Rn determination in some thermal baths of a central eastern Italian area. J. Radianalyt Nucl Chem 2004; 261(1): 37–41.

Dinis, M.L., Fiúza A. (2005). Simulation of Liberation and Dispersion of Radon from Uranium Mill Tailings, in: "Environmental Contamination from Uranium Facilities and Remediation Measures" pp 63–70. Proceedings of the International Workshop on Environmental Contamination from Uranium Facilities and Remediation Measures, 262 p., Lisbon, Portugal, 11–13 February 2004. IAEA Scientific & Technical Publications 2005—Austria. Hardcover STI/PUB/1228, ISBN 92-0-104305-8.

EPA (1997). United States Environmental Protection Agency, Health Effects Assessment 14 Summary Tables, Radionuclide Carcinogenicity Slope Factors: HEAST. http://www.epa.gov/radiation/heast/docs/heast_ug_0401.pdf, 01 April 2011.

Gnoni, G.; Czerniczyniec, M.; Canoba, A. and Palacios, M. (2008). Natural radionuclide activity concentrations in spas of Argentina. AIP conference proceedings, vol. 1034, p. 242–245.

IAEA (2003). Radiation Protection against Radon in Workplaces other than Mines, jointly sponsored by the International Atomic Energy Agency and the International Labour Office. Safety Reports Series No. 33, IAEA, Vienna.

Köteles, G.J. (2007). Radon Risk in Spas? vol. 5, no. 41, p. 1.

Labidi, S., Essafi, F. and Mahjoubi, H. (2006). Estimation of the radiological risk related to the presence of radon 222 in a hydrotherapy centre in Tunisia. Journal of Radiological Protection, vol. 26, no. 3, p. 309.

Lettner, H, Hubmer, A.K., Rolle, R., Steinhäusler, F. (1996). Occupational exposure to radon in treatment facilities of the radon-spa Badgastein, Austria. Env ron Int 1996;22:S399–S407.

Manic, G., Petrovic, S., Vesna, M., Popovic, D., Todorovic, D. (2006). Radon concentrations in a spa in Serbia. Environ Int 2006;32:533–7.

Mose, D.G. and Mushrush, G.W. (2008). Prediction of indoor radon based on soil radon and soil permeability. Journal of Environmental Science and Health, Part A: Toxic/Hazardous Substances and Environmental Engineering, 34:6, 1253–1266, doi:10.1080/10934529909376894.

Nikolopoulos, D. and Vogiannis, E. (2007). Modelling radon progeny concentration variations in thermal spas. Science of the total environment, vol. 373, no. 1, 82–93.

Nikolopoulos, D., Vogiannis, E., Petraki, E., Zisos, A., Louizi, A. (2010). Investigation of the exposure to radon and progeny in the thermal spas of Loutraki (Attica-Greece): Results from measurements and modeling. Science of the Total Environment 408 (2010) 495–504.

Pereira, A.J.S.C., Dias, J.M.M., Neves, L.J.P.S. e Godinho, M.M. (2001). O Gás Radão em Águas Minerais Naturais: Avaliação do Risco de Radiação no Balneário das Caldas de Felgueira (Portugal Central). Memórias e Notícias, Publicações do Departamento de Ciências da Terra e do Museu Mineralógico e Geológico da Universidade de Coimbra, n.° 1, Coimbra.

Portuguese Legislation (2004). Decreto-Lei n.° 142/2004, D.R. I Série 136 (3632–3640), Ministério da Saúde, http://dre.pt/pdf1sdip/2004/06/136A00/36323640.pdf (13-09-01).

Portuguese Legislation (2006). Decreto-Lei n° 79/2006, D.R. I Série 67 (2416–2468), Ministério das Obras Públicas, Transportes e Comunicações, http://dre.pt/pdf1sdip/2006/04/067A00/24162468.pdf (13-09-01).

Pugliese, M., Quarto, M. & Roca, V. (2013). Radon concentrations in air and water in the thermal spas of Ischia Island. Indoor and Built Environment, doi: 10.1177/1420326X13480053 ibe.sagepub.com.

Puskin, J.S., and Yang, Y. (1988). A retrospective look at Rn-induced lung cancer mortality from the view point of a rela (figutive risk model: Health Physics, v. 54, p. 635–643.

Radolic, V., Vukovic, B., Smit, G., Stanic, D., Planinic, J. (2005). Radon in the spas of Croatia. J Environ Radioactivity 2005; 83(2): 191–198.

Robertson, A., Allen, J., Laney, R. and Curnow (2013). The Cellular and Molecular Carcinogenic Effects of Radon Exposure: A Review. International Journal of Molecular Sciences (14), 14024–1406. doi:10.3390/ijms140714024.

Silva, A.S., Dinis, M.L., Diogo, M.T. (2013), Occupational Exposition to Radon in Thermal Spas, published in the proceedings book: "Occupational Safety and Hygiene-SHO 2013".

Somlai, J., Torma, A., Dombovári, P., Kávási, N0, Nagy, K., Kovács, T. (2007). Contribution of 222Rn, 226Ra, 234U and 238U radionuclides to the occupational and patient exposure in Heviz-spas in Hungary. J Radioanal Nucl Chem 2007;272(1):101–6.

Song, G., Zhang, B., Wang, X., Gong, J., Chan, D., Bernett, J. and Lee, S.C. (2005). Indoor radon levels in selected hot spring hotels in Guangdong, China. Science of The Total Environment, p. 63–70.

Song, G., Wang, X., Chen, D. & Chen, Y. (2011). Contribution of (222) Rn-bearing water to indoor radon and indoor air quality assessment in hot spring hotels of Guangdong, China. J Environ Radioact, 102(4), 400–406. doi: 10.1016/j.jenvrad.2011.02.010.

Steinhäusler, F. (1988). Radon spas: source term, doses and risk assessment. Radiat Prot Dosim 1988;24:257–9.

Szerbin, P. (1996). Natural radioactivity of certain spas and caves in Hungary. Environ Int 1996;22: S389–98.

Trabidou, G., Florou, H., Angelopoulos, A., Sakelliou, L. (1996). Environmental study of the radioactivity of the spas in Ikaria Island. Radiat Prot Dosim 1996;63(1):63–7.

UNSCEAR (2000). United Nations Scientific Committee on Effects of Atomic Radiation, Sources and Effects of Ionizing Radiation. United Nations, New York, United Sales publication E.00.IX.3.
US-EPA (2003).United States Environmental Protection Agency, Air and Radiation, Assessment of Risks from Radon in Homes. http://www.epa.gov/radiation/docs/assessment/402-r-03-003.pdf, June 2003.
Vogiannis, E., Niaounakis, M., Halvadakis, C.P. (2004a). Contribution of 222Rn-bearing water to the occupational exposure in thermal baths. Environ Int;30:621–9.
Vogiannis, E., Nikolopoulos, D., Louizi, A., Halvadakis, C.P. (2004b). Radon variations during treatment in thermal spas of Lesvos Island (Greece). J Environ Radioact;75:159–70.
Vogiannis, E., Nikolopoulos, D., Louizi, A., Halvadakis, C.P. (2004c). Radon exposure in the thermal spas of Lesvos Island-Greece. Radiat Prot Dosim;111(1):121–7.
WHO (2009). WHO handbook on Indoor Radon, A Public Health Perspective, ISBN 978 92 4 154767 3. http://whqlibdoc.who.int/publications/2009/9789241547673_eng.pdf.

Occupational Safety and Hygiene II – Arezes et al. (eds)

ISBN 978-1-138-00144-2

Vibration and tipping of agricultural machinery: A reality to be adjusted

G.A. Gomes de Moraes & J.E.G. Santos
Universidade Estadual Paulista "Júlio de Mesquita Filho", Botucatu, São Paulo, Brazil

A.L. Andreoli, J.A. Cagnon & V.G. Meneghetti
Universidade Estadual Paulista "Júlio de Mesquita Filho", Bauru, São Paulo, Brazil

ABSTRACT: The present study aimed to develop a methodology for the collection, transfer, storage and processing of vibration levels emitted in jobs occupied in agricultural machinery. The reason of this work is the study the vibration dose applied to operators of heavy vehicles and its relation to occupational health, linking the still high number of accidents involving farm machinery in relation to overturning (tipping). There is a need for the development and improvement of efficient tools in measuring vibration and tilt machine work, which minimize damage to health and accident risks for operators.

1 INTRODUCTION

Accidents and incidents involving farm machinery although it's not well known and few statistics on the subject, are common in rural areas. Among the potential risk factors can be cited: lack of training available for the operator who will operate the machine, lack of knowledge regarding the operational limits of the equipment and lack of attention of the operator in the execution of a given task.

Other mechanisms include the use of machines that do not conform to the principles of ergonomics an comfort provided by law, work performed under insalubrious conditions, (heat/cold, sun, dust, noise and vibration machines and physical effort too) and the lack of proportionality between the model used tractor and agricultural implement used in execution of a given task.

All these elements together or even alone can generate a series of conditions that may contribute to the occurrence of an accidents at work, contributing to the occurrence of accidents, operations on slopes, high speed during operations, the operator unpreparedness, and the use of alcoholic beverages described by Santos (2012).

Studies by Kahil and Gamero (1997) show that the tractor operator is exposed to problems of noise, vibration, heat, physical efforts (exertion) among others, making it more susceptible to accidents, making necessary to correct these deficiencies, existing station operation therefore can significantly reduce accidents occurring in the rural areas.

In the specific case of risk overturning (tipping) side by using a portable inclinometer, the operator has the opportunity to check the potential risk of overturning the tractor during operation, it's caused by improper use of the equipment or the condition in which the work environment imposes.

The vibration is a harmful physical agent present in various work activities of our daily life, as it puts labor to localized vibrations (also called vibration hand-arm vibration) and whole-body vibrations.

This agent is intrinsic to heavy vehicles and agricultural machinery such as tractors and backhoes, and its transmission to the body and members of the operator is made by contact with the vehicle seats and even by your steering wheel, pedals and levers of manipulations.

During your workday, the tractor operator is exposed to low-frequency vibrations as described by Alonço (2004), Anflor (2003), Balbinot (2001) and Berasategui (2000).

There are several studies to absorb and alleviate these low frequency vibrations. Even so, there is a growing need for further research for understanding how the vibration is transmitted to the body and its influence on human comfort, performance, health and safety of farm tractors operators, described by Alonço (2004).

Given the legal requirements increasingly present in rural environment, is mandatory in the implementation of the present determinations of the Regulatory Norm 31 in Brasil (2012) or even in other Regulatory Standards that will bolster the rural worker, it is essential to develop instruments that provide subsidies for the evaluation of vibration dose to the operator that will be submitted throughout the workday.

Still exist, the constant need to develop equipment or improve those already found in the market, which allow greater security for operators of agricultural machinery.

It is necessary also to create reliable methodologies that can give a meaning of the data collected by these devices, that are subsequently compared with the limits established by current labor laws, allowing to verify the actual working conditions that the worker is exposed in that job post.

The ability to collect vibration levels in real time, or allowing the attainment of conditions that the worker is exposed at the moment at which the activity is being performed is very representative, since it allows faithful characterization of the mentioned job post.

In order to obtain the collection of that data and its subsequent interpretation to be reliable, it is also essential that the appliance is installed in a location that allows this reading.

Besides counting for it, with desirable characteristics in an electronic system such as sensitivity, efficiency in the collection and transmission of data, storage capacity of your memory, robustness and low cost.

These devices will be installed in agricultural machinery such as tractors, harvesters, self-propelled sprayer, without the final sale prices become significantly burdened, and with that providing safety and comfort to the machine operator.

2 OBJECTIVE

The aim of this work is to develop a methodology that allows, with the use of a device that has the desirable characteristics described, collecting reliable data of exposure to vibration levels that operators of agricultural machinery may be exposed.

Also, understand them efficiently so as to obtain the vibration dose emitted throughout a workday.

After done this interpretation, we want to compare them with the limits established by Brazilian labor law, so that the workstation is resized to decrease this exposure, or the exposure time is not higher than stipulated by law.

The main goal is that all these data are obtained through a single device that in addition to providing this information, can also inform situations of imminent risk of side overturning (tipping) the machine, provided due to the excessive slope of the land topography.

3 MATERIALS AND METHODS

3.1 *Materials*

Such research will use a 3-axis accelerometer (electromechanical device capable of measuring dynamic acceleration, or caused by movement or vibration) of the capacitive type, which has an accelerometer with desirable characteristics, such as high sensitivity, stability, range, resolution, reliability and low cost, that will include a system of radio communication frequency, also incorporated a system of storage and data transfer. Allied to this equipment will be coupled a graphical indicator of longitudinal and lateral slope (inclinometer) so that the operator can view his situation during the operation of farm machinery.

The set will be installed within a robust metal housing, compact, with support for fixing the agricultural machine.

For the interpretation of vibration levels and calculation of doses will be used a personal microcomputer staff which will compare with the levels established by the Brazilian legislation, which is standardized and regulated by Norm 15, establishing the maximum levels of exposure to doses of vibration.

3.2 *Methods*

The methodology of the present study will consist primarily in choosing a suitable location for the installation of the device on the machine; the installation site should capture the vibration levels that truly will be issued to the operator. This location should also represent the balance of the machine, as well as capture vibration data, the equipment will also indicate the limits of land slope on which the machine should be operated, without the occurrence of accident risk by rolling over or side tipping.

After proper installation of the equipment on the machine, it must be operated in working conditions to its everyday routine, the conditions that the operator will actually be exposed.

Although ISO 5008 standard specifies methods for measuring and reporting the whole body vibration that the operator of a tractor wheel or other field equipment is exposed, and also the operating conditions of the machine and the ordinate of the artificial test tracks, this methodology holds that the data readings should be performed in actual working conditions in the field, that is, the exposures are truly representative, so that it described the conditions of the position occupied by the operator.

This methodology is innovative, since it can be employed in any job, whether in relation to the

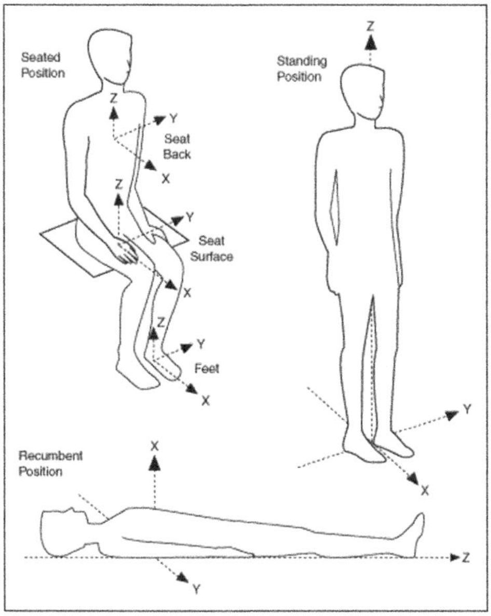

Figure 1. Directions of the coordinate system for whole-body mechanical vibration in humans.

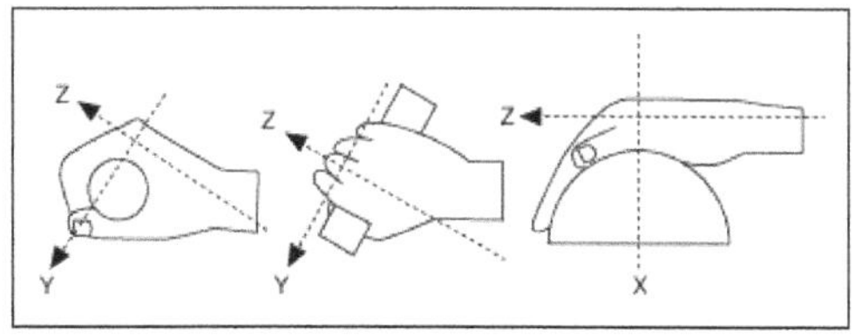

Figure 2. Mechanical vibration system directions specifically for hands.

type of machine used, the intended activity, condition of the slope topography, in other words, any agricultural activity, regardless motor vehicle type.

As already described, after the correct installation of the equipment in the vehicle, the employee will perform his normal activities.

Throughout the workday, the equipment will collect vibration levels emitted, which will be stored in a computer memory (storage system). After the workday, these data will be downloaded on a personal microcomputer which, by means of specific software, will be possible to calculate the doses of vibration that were transmitted to the operator.

Once calculated doses vibrations emitted, these will be compared to the dose limits established by Brazilian legislation, contained in Regulatory Norm number 15, which characterizes insalubrious activities and operations. This standard is based on the tolerance limits defined by the International Standardization Organization—ISO, in its standard ISO 2631 which establishes guidelines for exposure assessments of whole body vibration and ISO/DIS 5349 laying down guidelines for assessments vibration exposure specifically for hands.

After the interpretation of the collected data it will be possible to know if the working conditions are insalubrious or not.

The user of this equipment will have the option to collect vibration data later, after the end of the workday, or in real time via radio frequency communication system, allowing the interpretation of the work station conditions at the time of task execution. If it is detected that the working conditions are characterized as being insalubrious, an intervention may take place by relocating the operator work station without him suffering as a consequence of excess exposure, considered insalubrious.

Also it is part of this methodology, to create a protocol for the prevention of accident by side tipping or rolling over of the vehicle, since the equipment used will have an electronic device that also indicate the limits of land slope in which the machine can be operated without risk of an accident occurring.

When the operator is working on a slope land susceptible to tipping, a signal indicator light and a warning signal will be issued, alerting the operator to the imminent risk of side tipping of the vehicle.

The methodology proposes to make: the collection, interpretation, calculation of vibration doses compared with the doses permitted by law and warning of imminent risk of accident by the use of only one equipment, which is efficient, safe and reliable.

It will allow correlating if the task performed with a particular machine can be harmful to worker health and also that the work station can be occupied so as not to be regarded as unhealthy, since adjustments may be made in the machine structure aiming to reduce vibration or so that exposures during work hours are scaled according to Brazilian legislation.

As these data can be performed in real time, the dynamism of the process will provide a clear idea of the conditions of the work position occupied. When doses are close to the limits, an alarm will sound which will enable the operator to leave his occupation in that job.

For the data obtained to be reliable and the results representative and close to expected, it is indispensable that the installation of such equipment in the machine is in a suitable location, enabling the measurement of the correct data and retracting the conditions found in that job.

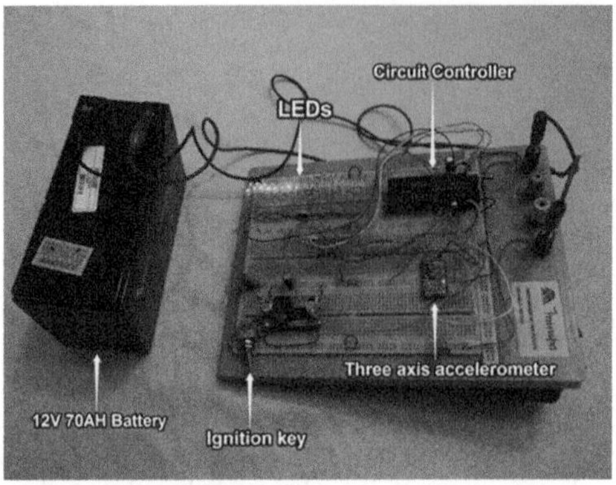

Figure 3. Top view of the inclinometer with warning lights on imminence of tipping.

The tests to be performed using the proposed methodology should follow a standardized application and enforcement, so that no external influences will interfere the data obtained. Standardization regarding test conditions, such as type of machine to be evaluated, model, year of manufacture or usage time is essential because it can directly influence the results.

The systematic compilation of data to be released on the microcomputer also should follow a pattern, so that the dose calculation will not be influenced.

4 RESULTS AND DISCUSSION

It is hoped with this methodology, to obtain an overview of the working conditions of jobs held by operators in agricultural machinery.

Above all, from the point of view of exposure levels to which the operators may be exposed, as some models of agricultural machinery can deliver a level of excessive vibration, which may cause serious damage to workers' health.

Among the expected results is the possibility of measuring working conditions in real time, which is extremely positive, as it allows the visualization of the work station and if it is providing some damage to the health of the operator, he can be removed from the execution of this task.

Provides knowledge of these levels with dynamism, optimizing the use of labor employed safely and efficiently, without the worker has damaged his health.

The way technology scouting, transfer and storage of data is inserted into the device, provides a simplified equipment for operator use and also low cost, which will not influence the final cost of agricultural machinery, allowing for installation in machines as an optional security device, or may be used in the safety of the operator even on machines that do not count on this item from the factory.

Only with the use of this methodology and using the same equipment, you can get data from vibration levels and land slope in which the machine is operating, preventing the operator from having accidents by the risk of tipping sideways and that he suffers the consequences of the damage caused by vibration to the worker body.

It is worth mentioning that this research is in the development stage still, having no further applications or results obtained from tests yet, getting even restricted to simulations of what is expected with the use of this methodology.

5 CONCLUSIONS

The proposed methodology as well as being innovative, is promising to its users.

Regarding its applicability by the employer, which will have an efficient way to evaluate the conditions of employment of his collaborators, allowing his employees not to operate the machines in ideal test conditions that the law determines, avoiding absences due to accident and labor lawsuits by workers.

Also with regard to employees, which will have an efficient way to protect their health, working under conditions that will not damage their health.

Unfortunately in Brazil, news on accidents in the agricultural environment are usually common still, going against the trend of an increasingly mechanized agricultural larger fleet.

The use of methodologies such as these are trying to change this reality, making the field work to become safer and providing life quality for rural workers.

REFERENCES

Alonço, A.S. *Metodologia de projeto para a concepção de máquinas agrícolas seguras*. 2004. Tese (Doutorado em Engenharia Mecânica) - Universidade Federal de Santa Catarina, Florianópolis, 2004. 221 p.

Anflor, C.T.M. *Estudo da transmissibilidade da vibração no corpo humano na direção vertical e desenvolvimento de um modelo biomecânico de quatro graus de liberdade*. Dissertação (Mestrado em Engenharia) - Universidade Federal do Rio Grande do Sul, Porto Alegre, 2003. 105 p.

Balbinot, A. *Caracterização dos níveis de vibração em motoristas de ônibus: Um enfoque no conforto e na saúde.* Tese (Doutorado em Biomecânica) - Universidade Federal do Rio Grande do Sul, Porto Alegre, 2001. 281 p.

Berasategui, M.R. *Modelización y simulación del comportamiento de um sistema mecánico con suspensión aplicado a los asientos de los tractores agrícolas*. (Doctorado En Ingenieria Mecánica) – Departamento de Ingenieria Mecánica, Universidad Politécnica de Madri, Madri, 2000. 259 p.

Brasil. Ministério do Trabalho e Emprego. Portaria 3.214/78 do Ministério do Trabalho e Emprego – *Normas Regulamentadoras*. Brasília, 2012. 658 p.

International Organization for Standardzation. ISO 2631: Mechanical Vibration and Shock - Evaluation of Human Exposure of Whole- Body Vibration: General requirements. Geneva, 1978. 31 p.

International Organization for Standardzation. ISO 5008: Agricultural wheeled tractors and field machinery—Measurement of whole-body vibration of the operator Mechanical Vibration. Geneva, 2002. 12 p.

International Organization for Standardzation. ISO 5349: Mechanical Vibration - Measurement and Evaluation of Human Exposure to Hand-Transmitted Vibration. Geneva, 1979. 18 p.

Kahil, M.A., Gamero, C.A. Níveis de ruído: Avaliação ergonômica de alguns tratores e equipamentos agrícolas. *Energia na agricultura*, Botucatu, v. 12, n.3, p. 46–53. 1997.

Santos, J.E.G. *Curso de especialização em engenharia de segurança do trabalho, Módulo: Segurança Agropecuária.* Bauru-SP, Brazil: Faculdade de Engenharia Mecânica, Universidade Estadual Paulista "Júlio de Mesquita Filho", 2012. 104 p.

Occupational Safety and Hygiene II – Arezes et al. (eds)

Noise emission levels during spinning classes and the effects on workers' health

P.H. Bahniuk, L. Ulbricht, W.L. Ripka & A.M.W. Stadnik
Federal University of Technology, Paraná, Curitiba, Brazil

ABSTRACT: The noise appears as the most frequent physical agent in the workplace, being highly harmful and considered a public health and ergonomics problem. From this finding, we chose to analyze level of sound transmission that professionals are exposed during spinning classes and possible consequences of this agent on health. This is a descriptive study that used as a research tool a dosimeter to measure the level of noise that 33 spinning teachers were exposed, in Curitiba city. Furthermore, it was made an interview in order to identify the possible symptoms presented by those professionals. It was concluded that the sound transmission on the professional's working environment is high, and on this subject, several of the analyzed gyms are in disagreement with the Ministry of Labour and Employment legislation.

1 INTRODUCTION

One of the motivational components in the gyms is the music (Silva et al. 2009). The purpose of music in fitness classes is to mark the rhythm of each phase of the class taking into account the number of beat-per-minute of a song (Domingues Filho 2005). It also determines the intensity of the class (Silva et al. 2009) and distracts the discomforts caused by physical activity (Miranda & Godeli 2003), whereas when in high intensity there is the increase in production of adrenaline by the organism (Grandjean 1998).

One of the types of gymnastics offered by the gyms is spinning, also known as indoor cycling, which was created in 1991 by Jonathan Goldberg (Spinning 2011). The classes are aimed at the improvement of conditioning and physical performance of their students, and bring well-being to the participants (Silva & Oliveira 2002). Music plays key role in the methodology employed in the modality (Domingues Filho 2005).

However, the music when excessively high can bring risks to the health of the professional who teaches the class, and for the people who practice the activity in search of health. So music can be considered a noise, which can be defined as annoying sounds that have the potentiality to cause harm to the human body. Factors such as intensity, frequency and duration of this noise emission caused by exercise classes can be extremely harmful to people exposed.

Iida (2005) classifies noise as an auditory stimulus that does not contain useful information for the running task. The World Health Organization handles noise as a public health problem since the year 1980 (Rios 2003).

The Ministry of Labour and Employment (MTE) proposes a scale containing the tolerance limits to noise in the workplace, and for eight hours of daily exposure the maximum value should not exceed 85 decibels (dB), and that each additional five decibels to this value, the contact time should be decreased by half (Brasil 2011).

Based on these relationships, we consider relevant to verify if there are relations between the noise levels of spinning classes and health complaints presented by the professionals who work to this modality.

2 MATERIALS AND METHODS

The research is a descriptive study, with the description of the characteristics of a given population and the establishment of relationships between variables (Gil 2007).

Observational approach was used armed-as a way to collect qualitative and quantitative data because they are considered reliable for analyzing accurately the individual or group in their environment innate (Thomas et al. 2007).

As tools have been used: a) personal noise dosimeter Instrutherm Portable Digital—DOS 500, which was in the possession of the teacher all class (with the microphone near the ear canal of the same), b) a directed interview with the objective of gather information as time of career, work hours, sound perception at classes and inherent aspects of health.

In the interview were discussed 18 health symptoms highlighted in the literature as: auditory symptoms (tinnitus, sensitivity or irritation face of loud sounds, clicks, pains in hearing device, volume increase in stereos and television and difficulty with speech comprehension) and systemic (frequent headaches, throat problems, sleep, gastric disorders, vomiting, loss of appetite, nausea, fainting, high stress, anxiety, depression and high blood pressure).

The sample consisted of 33 spinning teachers from 33 gyms in the city of Curitiba, in a nondeterministic choice. As inclusion criteria we adopted to sign a Term of Free and Clarified Consent and having a minimum experience of one year in the sport. We analyzed a class of each teacher by checking the average noise levels during class and found in it the maximum value. Through the equipment software DOS 500 was calculated Equivalent Continuous Sound Level (LEQ) for eight hours of work and the real LEQ that individuals are exposed on the workload of spinning.

Data were compared with the levels prescribed in literature and the regulatory standards of the area.

For statistical analysis it was resorted descriptive statistics to organize and summarize the data set with measures of position and dispersion (average, mode and standard deviation). It was resorted inferential statistics in order to explain the occurrence of the findings, it was considered a probability error of type I of 0.05 in all tests (Maroco 2007).

We used the chi-square test of independence in association with Fisher's exact test, in this statistical analysis software SPSS to see if the analyzed groups differ in certain characteristics. It was also used the F-test of variance analysis—ANOVA (Maroco 2007).

3 RESULTS

Able to verify that the noise emission of spinning classes for analyzed professionals in Curitiba was in accordance to Brazilian regulations established by the NR-15 (a standard) from MTE (Brasil 2011). There was obtained high noise level in such classes, with an average of 82.5 dB and 13 of these individuals (39.4%) possessed values above 85 dB and therefore it is at risk and in violation of the law. The maximum value found in classes reached 110.6 dB. However, only 12 individuals (36.36%) reported feeling uncomfortable with the noise emission during lessons.

As the symptoms may indicate problems related to high noise emissions, it was found that many symptoms are present among the studied which only five (15.15%) have not admitted any health complaint. On average, there were 2.79 different complaints by individuals, and seven (21.21%) had at least five different symptoms.

The vast majority of respondents (84.85%) were found to have some type of symptom related to the possible effects of noise on human organism. It was resorted to the chi-square test to verify the relationship between the symptoms found in evaluated with LEQ gauged during spinning classes.

It has been verified the relative risk which they are exposed, taking into account the equivalent sound level projected to the daily workload. The calculation examined the odds ratio for the appearance of symptoms in individuals with LEQ above 85 dB and below this value. The results showed that individuals who are in the group above the permitted have 8.62 times greater chance of the appearance of one of 18 health symptoms addressed.

When using the LEQ designed for eight hours you will find again relationship between noise and appearance of health symptoms ($p < 0.00$).

Comparing an average of health complaints of individuals who find themselves with LEQ above 85 dB to workload with spinning classes with those who have LEQ below this value is obtained, respectively, 3.31 and 2.25 complaints per individual but there is no significant difference ($p = 0.291$) between these figures.

The study also attempted to elucidate the possible correlation between time of career and symptoms presented, indicating that the range of greatest risk was among individuals between 11 and 15 years of career (with an average of 3.5 complaints per individual). These averages did not obtain statistically significant differences in the test F of ANOVA variance analysis.

By linking the time of career with the symptom of tinnitus perceived by the individuals it is noted the existence of a relationship between the variables ($\chi^2 = 5.241$ with associated probability of $p = 0.022$), this being one of the most frequent symptom among individuals. Analyzing Cramer's V we note that approximately 22% of the variation in counting the frequencies of the evaluated having tinnitus can be explained by time of career. Similar results were found in comparison with the need to increase the volume in leisure activities ($\chi^2 = 8.183$ with associated probability of $p = 0.037$, Cramer's V 40%) and the symptom of stress ($\chi^2 = 7.792$ with associated probability of $p = 0007$, Cramer's V 49%). The other symptoms were not statistically significant.

With respect to auditory effects, table 1 shows that eight teachers have difficulty in speech com-

Table 1. Related auditory complaints.

Symptoms	Number of individuals	
Increase the volume in activities	18	54.55%
Irritation face of loud sounds	10	30.30%
Speech comprehension	8	24.24%
Tinnitus	6	18.18%
Clicks	1	3.03%
Frequent pain at hearing	1	3.03%

Table 2. Related systemic complaints.

Symptoms	Number of individuals	
Sleep disorders	9	27.27%
High anxiety	9	27.27%
Frequent headache	8	24.24%
Reporting stress	6	18.18%
Disorders in blood pressure	6	18.18%
Gastric problems	4	12.12%
Nausea	2	6.06%

prehension, one of the main symptoms of Noise Induced Hearing Loss (NIHL). A large number (54.54%) mentioned having to increase the volume in daily leisure activities such as watching television and listening to radio, presenting itself as the symptom most commonplace in the study.

And table 2, among the systemic complaints were found nine individuals with sleep disorders, nine presenting high anxiety, eight with frequent headache, six reporting stress, six with disorders in blood pressure, four with gastric problems and two with nausea. Other symptoms of systemic order were not reported.

4 DISCUSSION

The average age in this study was 31.6 years (78.79% aged between 20 and 35 years). In the study by Souza (2011) with 66 weight trainers also the focused age was between 20 and 35 years (42.42% of total). Antunes (2003) in a study of 130 teachers of gymnastics and weight of São Paulo the higher prevalence oh the group was up to 30 years, showing similarity. The youth of the teachers can be a reflection of a socio-cultural issue in the workplace where it is expected a professional with good physical appearance and which spreads the idea of body worship.

Regarding gender, the data published in the literature are also similar to this study. Souza (2011) found the prevalence of male teachers (78.79%), as well as Milano (2007) who investigate the 72 teachers spinning in the city of Rio de Janeiro and also found a higher prevalence of males (58.3%).

The average time of career of the participants was 8.54 years with the spinning, whereas 51.5% had six years of career or less. Antunes (2003) when analyzing the professional profile of bodybuilding and fitness instructors in the State of São Paulo got 60% of their sample having five years or less experience in their area of expertise. Andrade & Russo (2010) researching teachers in the city of Niterói found professional with 20 years of experience in the area, and of these, 65.63% were less than six years of career. Milano et al. (2007), studying professionals working with spinning obtained an average of 3.91 years of work (standard deviation of 2.71 years). Palma et al. (2009) in a study with 15 teachers of spinning in the city of Rio de Janeiro found the average time of 10.7 hours/week working with the modality, a value very similar to professionals in this research that showed an average of 10.84 hours per week.

As the average noise emission measured at one hour, Silva et al. (2009) in a research in the Federal District with spinning classes, had a lower average of 85.91 dB, and also a lower minimum value (52 dB), than those reported in this study. However, they found a higher maximum value that reached 112 dB. Palma et al. (2009), in a study conducted in Rio de Janeiro with the same method, found an average value of 89.81 dB, with minimum of 74.4 dB and maximum of 101.6 dB, all values lower than those identified in this investigation.

Taking into account the assessed average noise emission of 91.25 dB and the limits of tolerance provided by the MTE, professionals could be exposed only during three hours and 30 minutes to such values. While analyzing the maximum value of 110.6 dB, it would take only 15 minutes of exposure to these individuals find themselves in disagreement with the standard and with greater propensity to deleterious effects that exposure to noise can cause (Brasil 2011).

The standard 10152 of the Brazilian Association of Technical Standards (ABNT 1997) states that the level of acoustic comfort in enclosed pavilions that perform shows or sports activities would be 45 dB with an acceptable amount of up to 60 dB, immissions much lower than those measured in this study.

These immission levels of noise described by other studies reinforce the prerogative presented in this study, demonstrating that the professionals who work with spinning (specifically) and aerobic exercise (in general) are exposed to high levels of noise in their local job.

The relationship between noise and vocal overload was identified in this study, as observed among the 33 professionals, 11 (33.33%) complained of throat problems such as pain, hoarseness and/or frequent inflammations. Milano et al. (2007) found a value of 6.94% and, at the other extreme. Palma et al. (2009) found results well above identifying 53.3% of individuals with this complaint. The high incidence of this problem in this study and also in Palma et al. (2009) deserve worth mentioning.

Auditory complaints were trivial whereas 72.72% of the sample admitted having at least one of the symptoms frequently in the past three months. Palma et al. (2009) in their study found 26.7% of patients reporting hearing problems.

A study analyzing the medical records of a referral center for occupational health in the city of Campinas found 80% of tinnitus among patients with NIHL, however, found no statistical association between tinnitus and age and between tinnitus and time of exposure to noise (Ogido & Costa 2009), suggesting that tinnitus would already be connected to the prognosis of hearing loss.

Through the interview, it was possible to found that the demand for the use of high sound level in the classes of these teachers do not come from customers who perform the activity, mostly, since only 12 of the interviewees (36.36%) reported that their students ask to increase the loudness of the music during lessons and on the other hand, there are 16 teachers reporting that most of their students ask to attenuate the noise. It is worth to highlight the proposition made by Lacerda et al. (2001) that are not uncommon professionals who believe that sound very intense increases the performance of students, keeping them motivated but the most affected are the professionals themselves who spend much of their work in front of excessive noise.

5 CONCLUSION

Through statistical analysis it was demonstrated a direct relationship between the LEQ, both designed to eight hours of work such as specific for the modality under study, and the symptoms present in the past three months of these individuals. By analyzing the relative risk of individuals with LEQ above 85 dB compared to those who were below this threshold it was found 8.62 times higher risk of symptoms onset between individuals of the first group compared to the second.

It can be concluded that the analyzed group is at risk for exposure to noise and their employers have not taken adequate measures established by MTE for prevention and mitigation of these effects such as conducting periodic exams.

A concerning data raised in this study was that only four individuals (12.12%) had admission exams in their local labor and only one (3%) had audiometric testing imposed by their employers.

Besides the ergonomic measures of protection, the use of microphones during spinning classes should also be encouraged for this population since 11 individuals had frequent complaints of sore throat, hoarseness and/or inflammation in this region. Only 39.3% of the group uses such equipment during your lessons and vocal overload during the classes is high since these individuals need to pass information to their students in an environment with high levels of noise.

Regarding the possible influence of long career on the health symptoms, there was obtained direct relationship between time of career and perceived tinnitus by the participants, this being one of the most apparent symptoms in people diagnosed with NIHL.

It is worth noting the dichotomy between the professionals in this area since they have a purpose to disseminate the idea that the practice of regular exercise would lead to improved the physical fitness components related to health and, controversially, perform this practice in environments that have the potential to cause serious damage to their own organism and of your students due to the high level of noise in their classes.

REFERENCES

ABNT–Associação Brasileira de Normas Técnicas 1997. *NBR 10152/2000: Níveis de ruído para conforto acústico*. Rio de Janeiro.

Andrade, I.F.C. & Russo, I.C.P. 2010. Relação entre os achados audiométricos e as queixas auditivas e extra-auditivas dos professores de uma academia de ginástica. *Revista da Sociedade Brasileira de Fonoaudiologia* 15(2): 167–173.

Antunes, A.C. 2003. Perfil profissional de instrutores de academias de ginástica e musculação. *Revista Digital Efdeportes* 9(60).

Brasil-Ministério do Trabalho e Emprego 2011. *Normas regulamentadoras*.

Domingues Filho, L.A. 2005. *Ciclismo indoor: Guia teórico prático*. Jundiaí: Fontoura.

Gil, A.C. (4 ed.) 2007. *Como elaborar projetos de pesquisa*. São Paulo: Atlas.

Grandjean, E. 1998. *Manual de ergonomia: adaptando o trabalho ao homem*. Porto Alegre: Artmed.

Iida, I. (2 ed.) 2005. *Ergonomia: projeto e produção*. São Paulo: Edgard Blucher.

Maroco, J. (3 ed.) 2007. *Análise Estatística*. Lisboa: Silabo.

Milano, F. et al. 2007. Saúde e trabalho dos professores que atuam com ciclismo indoor. *Revista Digital Efdeportes* 12(109).

Miranda, M.L. & Godeli, M.R.C.S. 2003. Música, atividade física e bem-estar psicológico em idosos. *Revista Brasileira Ciência e Movimento* 11(4): 87–94.

Ogido, R. & Costa, E.A. 2009. Prevalência de sintomas auditivos e vestibulares em trabalhadores expostos a ruído ocupacional. *Revista Saúde Pública* 43(2): 377–380.

Palma, A. et al. 2009. Nível de ruído no ambiente de trabalho do professor de educação física em aulas de ciclismo indoor. *Revista Saúde Pública* 43(2): 345–351, abr., 2009.

Rios, A.L. 2003. Efeito tardio do ruído na audição e na qualidade do sono em indivíduos expostos a níveis elevados 194p. Master's Dissertation. Faculdade de Medicina de Ribeirão Preto da Universidade de São Paulo.

Silva, P.S.B. et al. 2009. Nível de ruído sonoro em aulas de ciclismo indoor em academias do distrito federal. *Educação Fisica em Revista* 3(3).

Silva, R.A.S. & Oliveira, H.B. 2002. Prevenção de lesões no ciclismo indoor–uma proposta metodológica. *Revista Brasileira Ciência e Movimento* 10(4): 07–18.

Souza, D.F. 2011. *Perfil dos instrutores de musculação: um estudo sobre as estratégias utilizadas na formação professsional 59p.* Trabalho de Conclusão de Curso. Universidade Federal do Rio Grande do Sul.

Spinning 2011. *About Madd Dogg Athletics, Inc.* In: <www.spinning.com>. Accessed on: 23 April In 2011.

Thomas, J.R. et al. (5 ed.) 2007. *Métodos de pesquisa em atividade física.* Porto Alegre: Artmed.

Occupational Safety and Hygiene II – Arezes et al. (eds)
© 2014 Taylor & Francis Group, London, ISBN 978-1-138-00144-2

A model for assessing maturity of Integrated Management Systems

J.P.T. Domingues, P. Sampaio & P.M. Arezes
Production and Systems Department, University of Minho, Guimarães, Portugal

ABSTRACT: Maturity models have been adopted in several fields aiming at people, objects, software and social systems assessment. The first reported maturity model (Maslow motivation theory) assesses human needs according five levels. Integrated management systems (IMS) are spreading among organizations focusing resources optimization, processes efficiency and documentation reduction. This article intends to report a two-component model that assesses the IMS maturity through a five level path. The back-office component is a structural statistical based model and the front-office component is a Capability Maturity Model integrated (CMMi) based model. Additionally, this paper aims at reporting the methodologies adopted in developing such model.

1 INTRODUCTION

1.1 *Integrated management systems*

The most commonly reported subsystems combined into a single Integrated Management System (IMS) are the Quality Management System (QMS) implemented according ISO 9001 standard, the Environmental Management System (EMS) implemented according ISO 14001 standard and the Occupational Health and Safety Management System (OHSMS) implemented according OHSAS 18001 standard.

The integration phenomenon analysis should consider the assessment of several variables: motivation pro integration, benefits collected, obstacles transposed, integration level achieved, integration strategy, audits and IMS typology, integration model adopted are among those variables.

European Foundation for Quality Management (EFQM) model and ISO 9004 standard propose integration paths aiming the fulfilment of several stakeholders' requirements. These integration models assume that a pre-existing QMS is implemented. An implemented QMS may be advantageous or disadvantageous. On one hand, a pre-existing QMS may reduce the required documentation, on the other hand, it may protect its own nature by preventing an appropriate resources dissemination among other management systems to be implemented.

1.2 *Maturity models*

Maturity models may have a practical application or be just a conceptual abstraction being sustained on maturity and capability concepts. Maturity models may be classified as comparative, descriptive or prescriptive according to their application. Some maturity model definitions found in the literature are presented in Table 1.

Some requirements must be attained in order to develop a successful maturity model. Some of these requirements relates to model design while other assures broader model dissemination. Cost, adoption easiness and applicability are among some requirements that relates to a broader dissemination.

2 METHODOLOGY

Two surveys have been held online: the first one focusing organizations (Appendix-Table A.1) and the second one an academic and industrial experts group (Appendix-Table A.2).

A 30 Question/Statement (Q/St) online survey was held focusing on Portuguese organizations with more than one certified management subsystem according to the following standards: ISO 9001, ISO 14001 and OHSAS 18001/NP 4397. The survey was conceptually supported on a Likert type scale, categorical and multiple option answers being it structure reported elsewhere (Domingues *et al.*, 2012). A pre-test performed on three companies was used to validate the survey due to the reported limitations of using online surveys including sampling, representativeness, and selection bias and response rate issues. A 15% response rate supports the following results and may be classified as a limitation on the current research.

The final model is sustained on two components. The back-office component is a statistical structural model while the front-office component is a CMMi based model. Figure 1 presents the conceptual approach between these components.

Table 1. Some maturity model definitions.

Source	Definition
OPM3 (2003)	... a structured elements array describing process or product characteristics.
SEI (2006)	... it contains essential elements from effective processes from one or more disciplines and describes an improvement evolutionary path from low maturity levels to high maturity levels improving quality and efficiency.
Korbel & Benedict (2007)	... it is an evaluation framework enabling an organization to compare its projects with the best practices from competitor, defining, an improvement structured path.
Becker *et al.* (2009)	... it consists on a maturity levels sequence for an objects class, representing an anticipated, desired or typical evolutionary path for those objects in a discrete series stages.
Kohlegger *et al.* (2009)	... instruments used to evaluate the maturity capabilities from certain elements and select the appropriate actions to achieve a high maturity by the elements.
Lee *et al.* (2010)	... model the temporal development of an organizational function, person, initiative or technology.
Demir & Kocabaş (2010)	... reflect certain reality issues, commonly described as capabilities, and defined qualitative attributes that are used to classify the competences from the targeted object upon well defined areas...
Bing *et al.* (2010)	... it is a conceptual framework and a systemic model aiming at continuously controlling the development and improvement processes from a product in order to achieve a well defined state.
Röglinger & Pöppelbuß (2011)	... a sequential levels array that, altogether, constitutes an anticipated, desired or logical path from an initial state till maturity.
Jia *et al.* (2011)	... a critical tool to evaluate the organizations capability and help them implementing structured modifications and improvement actions.

The front-office component is accessible to the organizations aiming to assess their IMS maturity. Organizations must populate this model with information required by each key process agent (KPA).

The back-office component (not accessible to users) processes the inputs and generates outputs to be presented by the front-office component.

Figure 2 presents the final model development path.

The results from the first survey held online sustained the experts' survey development. Additionally, after string to numerical variable transformation, the variance from each answer set result defines the KPAs allocation among the 5 maturity levels constituting the CMMi based model component. Statistical analysis assessment identified the variables influencing the latent variable-IMS maturity. A multiple regression linear model sustains the structural statistical based component, that is, the back-office component.

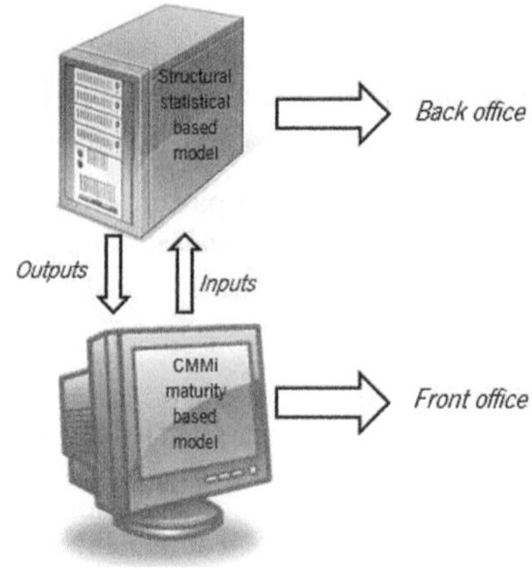

Figure 1. Conceptual IMS maturity assessment model.

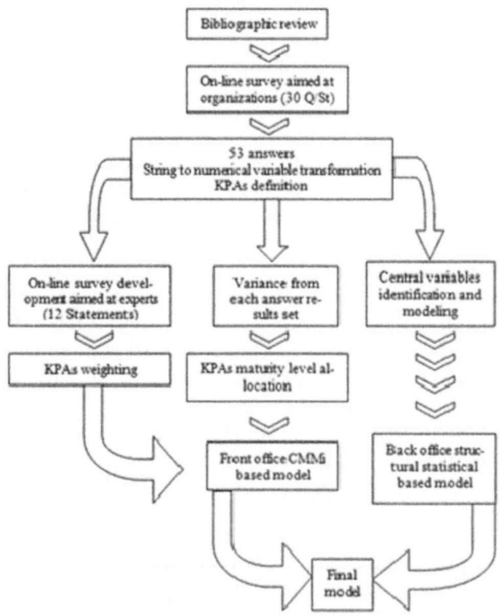

Figure 2. Final model development path.

2.1 *Back-office component development*

Statistical results analyses were performed by IBM SPSS version 20. Multiple and simple linear regression models were adopted as the development path to achieve the final structural statistical component (back-office). Data normality was assessed through Kolmogorov-Smirnov and Shapiro-Wilk non-parametric statistical tests. Non-parametric Kruskal-Wallis statistical test was adopted to allow the comparison of more than two independent groups. It is used when we wish to compare three or more sets of scores that come from different groups.

A simple linear regression model was tested in order to assess data quality and coherence provided by respondents. A multiple linear regression model was adopted in order to identify which variables (central variables) were statistically influencing the latent variable. Pearson correlation was assessed between statistical non-influential variables and the central variables.

2.2 *Front-office component development*

Front-office component is a CMMi based model achieved by iterative development and through KPAs identification. After the initial literature review a parameters set (KPAs) was identified as influencing the integration process contributing for high integration levels achieved by the companies IMS. After string to numerical transformation the results variance, collected from the companies' survey, defined the level allocated to each KPA. The results collected from the experts survey defined the weight attributed to each KPA.

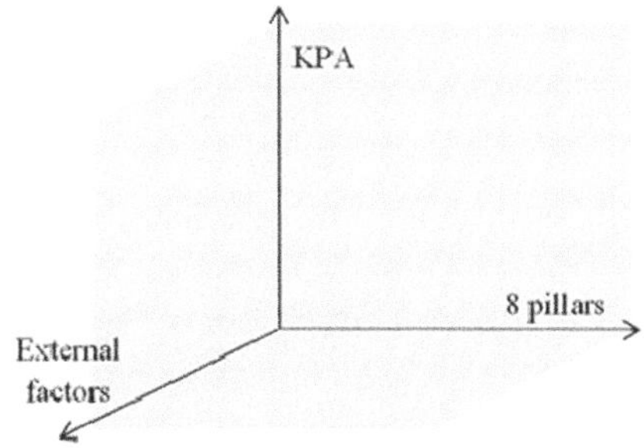

Figure 3. Three dimensional front-office nature.

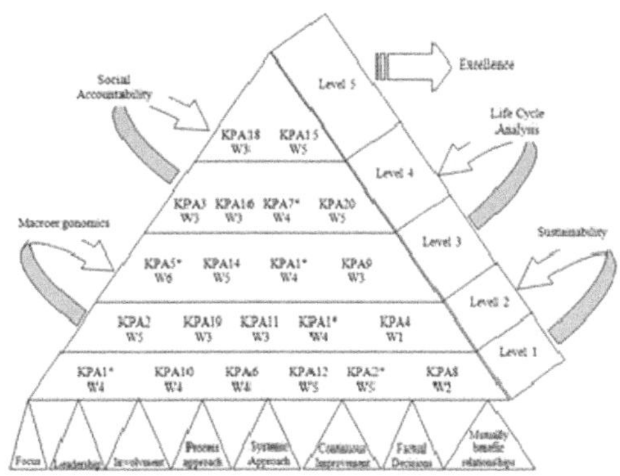

Figure 4. Final front-office component.

3 RESULTS AND DISCUSSION

3.1 *Front-office component*

Front-office component has a three dimensional nature as one may see in Figure 3. One dimension relates to the KPAs assessment, other to the eight excellence management pillars and the final to the external factors assessment (macroergonomics, life cycle analysis, social accountability and sustainability).

The front-office final version component is displayed in Figure 4 sustained on the eight pillars of excellence management: focus, leadership, involvement, process and systemic approach, continuous improvement, factual decision and mutually beneficial relationships.

Table 2 presents the score and the requirements to be met in order to assess IMS maturity according to Figure 4. Both the score and requirements should be met. In order to assess the score we should multiply the weight from each KPA that assessed IMS comply with. If the conditions for each level are met, action to be taken is to proceed to the upper level. The levels may be classified according to Crosby denomination (Crosby, 1979).

Table 2. Conditions to assess IMS maturity.

Level	Score	Requirement	Action	Crosby
5	–	KPA.18; KPA.15✓	Excellence	Certainty
4	≥ 60	KPA.7* ✓	↑ level 5	Wisdom
3	≥ 72	KPA.17* ✓	↑ level 4	Enlightment
2	≥ 60	KPA.13* ✓	↑ level 3	Awakening
1	≥ 160	KPA.21*KPA.1*✓	↑ level 2	Uncertainty
Base	Eight excellence management pillars assessment		↑ level 1	–

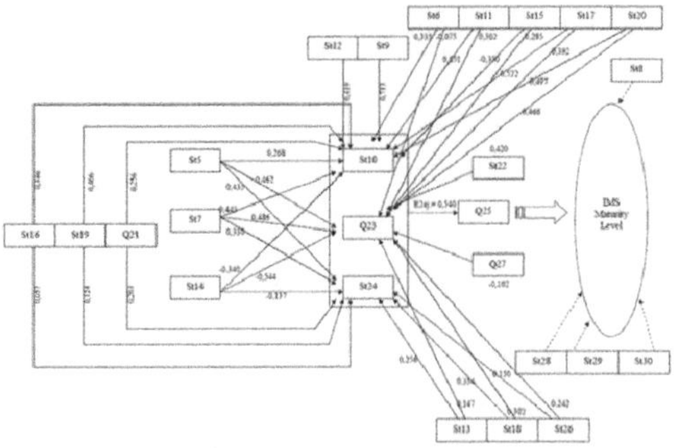

Figure 5. Final back-office component.

The eight excellence management pillars act as a zero level or a pre requirement to access the maturity assessment pyramid.

Access to a superior maturity level is enabled only if each external factor is considered and complied with.

3.2 *Back-office component*

Figure 5 presents the final back-office model version.

Three variables variation (Q/St10, Q/St23 and Q/St24) are statistically responsible for 54% of Q/St25 variation. Top management integrated vision (Q/St10), organizational integration level classification (Q/St23) and audit typology (Q/St24) explains mostly the Q/St25 (Integration level characterization) variation.

The other variables are related to these central variables and the Pearson correlation index is presented.

4 CONCLUSIONS

The developed final maturity model allows the conclusion that three variables contribute the mostly to the "IMS Maturity" latent variable. Those variables are the top management integrated vision (Q/St10), the integration level classification (Q/St23) and the audit typology (Q/St25).

Concepts like sustainability, macroergonomics, life cycle assessment and social accountability and the eight excellence management pillars are related also to the IMS maturity.

It is possible to conclude that management systems integration phenomenon is characterized by a high number of variables involved. The adopted strategy, the implementation process, the audit typology, the IMS typology, the motivation, the benefits, the obstacles and the achieved integration level are among those variables. Additionally, literature review identified external concepts that relates to IMS maturity. Topics like life cycle assessment, macroergonomics, sustainability and social accountability are concepts that should be taken into account, enabling a deeper integration, and a higher maturity IMS.

A back-office and front-office based maturity model development assessing IMS was the ultimate contribution from this paper. This contribution, on its own, and sustained on multimethodological tasks, allowed some other contributions, namely, the management systems integration phenomenon worldwide macroanalysis, macro-indicators development and their evaluation were reported elsewhere (Domingues *et al.*, 2011).

REFERENCES

Becker, J., Knackstedt, R. & Pöppelbuβ, J. 2009. Developing maturity models for IT management- A procedure model and its application. *Business and Information Systems Engineering*. 3: 213–222.

Crosby, P. 1979. *Quality is free*. New York: McGraw-Hill.

Domingues, J.P.T., Sampaio, P. & Arezes, P.M. 2011. Management Systems Integration: should "Quality" be redefined?. *In proc. of 55th EOQ Congress*, June, Budapest, Hungary.

Domingues, J.P.T., Sampaio, P. & Arezes, P.M. 2012. Integrated management systems: On the path to maturity and efficiency assessment. *In proc. of SHO 2012,* ISBN: 978-972-99504-8-3, 9-10 Feb., Guimarães, Portugal, pp. 177-179. *(Full paper published on CD).*

Jia, G., Chen, Y., Xue, X., Chen, J., Cao, J. & Tang, K. 2011. Program management organization maturity integrated model for mega construction programs in China. *International Journal of Project Management*, 29: 834–845.

Kohlegger, M., Maier, R. & Thalmann, S. 2009. Understanding maturity models: Results of a structured content analysis. *In proc. of I-KNOW '09 and I-SEMANTICS '09*, 2–4 Sep., Graz, Austria.

Korbel, A. & Benedict, R. 2007. Application of the project management maturity model to drive organisational improvement in a state owned corporation. *In proc. of AIPM Conference*, 7–10 Oct., Tasmania, Australia.

Lee, J.L., Suh, E.-H. & Hong, J. 2010. A maturity model based CoP evaluation framework: A case study of strategic CoP's in a Korean company. *Expert Systems with Applications*, 37: 2670–2681.

OPM3, 2003. Organizational Project Management Maturity Model. Newtown Square, Pennsylvania, USA, Project Management Institute.

Röglinger, M. & Pöppelbuß, J. 2011. What makes a useful maturity model? A framework for general design principles for maturity models and its demonstration in business process management. *In proc. of 19th European Conference on Information Systems*, June, Helsinki, Finland.

SEI 2006. Software Engineering Institute.

APPENDIX

Table A1. Companies survey questions/statements.

Q/St	Possible answers
St1- The company main activity is:	Unstructured
Q2- How many workers employ the company?	Unstructured
Q3- Where is geographically located the company?	North; Centre; Lisbon; Alentejo; Algarve; Madeira; Azores
St4- The management system is certified according the following standards:	ISO 9001 + ISO 14001; ISO 9001 + OHSAS 18001; ISO 14001 + OHSAS 18001; ISO 9001 + ISO14001 + OHSAS 18001; Other
St5- Quality, Environmental and Occupational Health and Safety policies are integrated.	Totally disagree; Disagree; Nor agree or disagree; Agree; Totally agree
St6- Training related to management systems integration had been provided to top management.	“ ”
St7- Integration concept had been taken into account during IMS implementation.	“ ”
St8- Management system is bureaucratized.	“ ”
St9- The tools, methodologies and goals from each management sub-system are harmonized/aligned:	“ ”
St10- Top management reveals integrated vision.	“ ”
St11- Management procedures are integrated.	“ ”
St12- Organizational interactions derived from IMS implementation are perceived by responsible and top management.	“ ”
Q13- The implementation process was supported on a guideline or in a framework.	“ ”
St14- Integration occurs at a documental level.	“ ”
St15- Authority from Environmental and/or OHS responsible is residual.	“ ”
St16- IMS is an add-value.	“ ”
St17- Integrated objectives are defined.	“ ”
St18- On the company organizational structure there is a clear responsible by the IMS.	“ ”
St19- The company monitors their processes based on KPI's, MPI's and OPI's.	“ ”
St20- The company promoted the implementation of integrated indicators.	“ ”
Q21- How do you classify the integration level of sub-systems standards?	Very easy; Easy; Reasonable; Difficult; Very difficult
Q22- If the company did not had implemented an IMS the overall performance comparing with the actual reality would be:	Lower than the present status; Equal to the present status; Higher than the present status
Q23- How do you classify the management system integration level?	1- Documental/ 2- Management tools plus 1)/ 3- Policies and objectives plus 1) and 2)/Common organizational structure plus 1), 2) and 3)
Q24- Audits performed to management sub-systems are?	Integrated; Simultaneous; Overlapped; Sequential
Q25- In a 1 to 5 scale how do you characterize the IMS?	1- Minimum integration level/2- Low i.l./3- Medium i.l./4- High i.l./5- Total/Maximum i.l.
Q26- The strategy followed during integration process was:	• Sequential • “All In”
Q27- Organizational items not susceptible of being integrated are identified?	• Yes • No
Q28- The main motivations to implement the IMS were:	Internal/Mainly internal/External/ Mainly external
Q29- The main benefits resulting from the integration of the management system were:	“ ”
Q30- The main obstacles found during IMS implementation were:	“ ”

Table A2. Experts survey questions/statements.

ID	Statement
S1	The predominance of internal origin motivations, obstacles and benefits before, during and after the integration process.
S2	Environmental manager and/or OHS manager responsibility is not residual and formally there's a clear responsible by the IMS on the company organizational structure.
S3	The company monitors their processes based on integrated indicators (KPI's, MPI's and OPI's).
S4a	Workers have the perception that the management system overall performance is superior in an integrated context and that top management reveal integrated vision.
S4b	Workers have the perception the integrated system is an add value and the company performance would be lower in a non-integrated context.
S5	The identification of organizational items not susceptible of integration.
S6	Integrated audits performed on the management system.
S7	An "all in" sequence integration *versus* a sequential process.
S8	Same organizational tools and methodologies between sub-systems and objectives alignment.
S9	Implementation process supported on a guideline or in a framework.
S10	Implementation responsible has the opinion that sub-system standards are easy or relatively easy to integrate.
S11	The company has an integrated policy of Quality, Environment and Occupational, Health and Safety and management procedures are integrated as well.
S12	Integration does exist at a documental level and workers have the perception that the system is bureaucratized.
S13	Integration level perception by the workers matches with the integration level achieved by the company.

Occupational Safety and Hygiene II – Arezes et al. (eds)
© 2014 Taylor & Francis Group, London, ISBN 978-1-138-00144-2

Effects of indoor air quality on respiratory function of children in the 1st cycle of basic education of Coimbra, Portugal

A. Ferreira
Instituto Politécnico de Coimbra, ESTeSC-Coimbra Health School, Saúde Ambiental, Portugal

S.M. Cardoso
Faculdade de Medicina, Universidade de Coimbra, Portugal

ABSTRACT: Air pollution is a health problem with long term consequences, which are responsible for respiratory diseases. This study sought to evaluate the effect of exposure to air pollutants in terms of respiratory function in children in the 1st cycle of primary education (CEB) of Coimbra, Portugal. 1019 children were assessed by spirometry and air quality in 81 classrooms and outdoor air quality were further evaluated. Of the various pollutants evaluated, carbon dioxide (CO_2), formaldehyde (HCHO), volatile organic compounds (VOCs) and particles of diameter smaller than 10µm (PM_{10}) were those that were sometimes above the maximum concentration of reference. In multivariate analysis, some pollutants were associated with deterioration of lung function. This study reinforces the interest in the study of children's exposure to air pollutants, particularly because these influence the airways.

1 INTRODUCTION

The good quality of the air we breathe is considered a basic requirement for health and human well-being. In addition to the natural emission sources, in the last century, with industrialization and technological development, there has been a significant increase in anthropogenic air pollution, with the amount of pollutants released into the atmosphere to reach the highest levels, with concentrations that can vary from hundreds to millions of tons per year (WHO, 2000).

Currently, the majority of the population spends more than 90% of their time in buildings where pollution levels can be up to 100 times higher than outdoor air, thus contributing to the development and aggravation of a significant number of diseases, including respiratory ones (Graudenz et al., 2006, Platts-Mills, et al. 2000; Koistinen et al., 2008; Kim et al., 2009; Bernstein et al., 2008).

The aim of the study was to investigate the association between exposure to indoor air pollution and lung function of children in two assessment moments, in autumn/winter season and in spring/summer season.

2 MATERIAL AND METHODS

The study consisted of the evaluation of air quality and the health of children in the 1st and 4th grade (November 2010 to June 2011), from primary schools of Coimbra, Portugal.

To characterize air quality we evaluated temperature (T°) Relative Humidity (RH) concentrations of carbon monoxide (CO), CO_2, ozone (O_3), nitrogen dioxide (NO_2), Sulphur dioxide (SO_2), VOCs, HCHO and particulate matter ($PM_{2,5}$ and PM_{10}) within schools, as well as on the surrounding outer area, in the autumn/winter season and in the spring/summer season.

Measurements were performed in accordance with the Technical Note NT-SCE-02.

We considered the maximum concentration of reference for the concentration of air pollutants that is referred in Decree-Law no. 79/2006, 4th April.

For NO_2 and SO_2 we used as reference the average values recorded in the analytical measurements of the outside air.

According to the Decree-Law no. 80/2006, 4th April, the environmental conditions of comfort reference are 25°C and 50% RH for the cooling season. ISO 7730 discloses a range of relative humidity from 30% to 70%.

For measuring the concentration of O_3 we used AEROQUAL series 500; to NO_2 and SO_2 the device QRAE; to VOCs the device PHOTOVAC; to $PM_{2,5}$ and PM_{10} the device TSI DUSTTRACK; and to CO and CO_2 we used the appliance TSI 9555-P (also used for the measurement of meteorological variables, T° and RH).

In what concerns the assessment of respiratory function, this was evaluated by spirometry.

The examination was conducted in accordance with the recommendations of the American Thoracic Society and the European Respiratory Society (Miller et al., 2005). We used a Jaeger spirometer and the software for spirometry was PFT-IOS 4.24.

The results were expressed as a percentage of the predicted value according to reference equations of Polgar (Quanjer et al. 1995).

For the final analysis, we considered as variables the FVC (forced vital capacity), that is, the maximum volume vented with the maximum effort from the position of maximum inspiration to the end of a maximum expiration; the FEV_1 (forced expiratory volume in first second of the FVC); the IT, the resulting index of the ratio of these parameters (FEV_1/FVC); the peak of the expiratory flow (PEF) and debits (FEF_25; FEF_50; FEF_75 FEF_75_25).

For the statistical treatment of the data we used the data analysis software IBM SPSS Statistics v. 19.

Several tests were performed to study the statistical hypotheses, including the t-Student test for independent samples, the t-Student test for matched samples, Wilcoxon T, Independence χ^2, McNemar and ANOVA for a single factor.

The interpretation of the statistical results was developed based on a significance level of $p \leq 0.05$ with a confidence interval of 95%.

3 RESULTS AND DISCUSSION

It was found that the average age of students attending the 1st grade was 6.20 ± 0.42 years and that of the students of the 4th grade was 9.25 ± 0.48 years. Most children were male (51.63%) and 98.63% of the children were Caucasian. It was observed in the 493 female children a relatively balanced distribution attending either the 1st or the 4th grade. A similar trend was observed in male students.

Of all environmental parameters analysed the one which presented the most significant results and with high potential risk was CO_2.

The average concentrations of CO_2 inside the classrooms were in general well above the CMR (≤984 ppm), reaching values of 1942 ppm. It was found that the average concentration of VOCs in two schools exceeded the CMR in both seasons.

The CMR of PM_{10} within the classrooms from four schools was exceeded. In what concerns the HCHO, it was found that only in one classroom, in the spring/summer season, the average concentration was above the CMR (0.08 ppm).

In Table 1, we can analyse the variation of the average spirometric parameters measured in the first phase (autumn/winter) and the spirometric parameters measured in the second (spring/summer), by comparing the respective estimated averages.

We noted that all parameters showed statistically significant differences except for FEF_75_25. It was found in all spirometric parameters an increase of the respective averages between the different seasons.

We found that, in autumn/winter, there were 549 children with a normal spirometric standard, 13 children with obstructive respiratory disorder and 457 children with restrictive respiratory disease. In spring/summer, we now have 612 children with normal spirometric standard, 10 with obstructive respiratory disorder and 397 with restrictive respiratory disease.

There was a significant improvement of the values obtained in the spring/summer compared to the figures in the autumn/winter season.

In order to understand how exposure to pollutants that at some point had concentrations above the CMR was associated with respiratory problems in children, we classified the classrooms as "presence of risk" (when the average concentrations obtained exceeded the CMR), and "no risk" (values below CMR).

Thus we established a relationship between the presence and absence of exposure and spirometric parameters that children had during the autumn/winter season and spring/summer season.

It was found that relatively to CO_2, there were significant differences in terms of the IT, PEF FEF_25, and FEF_50 FEF_75 spirometric parameters. We found that children who are not exposed to CO_2 levels above the CMR have higher spirometric values in relation to children who are exposed.

Although there are no statistically significant differences at the level of FVC, FEV1 and FEF_75_25 we found the same trend in other parameters except FVC which decreased in children who were in a situation of absence of risk. In what concerns PM_{10}, it was found that there were statistically significant differences at the level of FEF_75_25.

We found that its value decreased when children are not exposed to PM_{10} values above the CMR.

We also found that, although there are not statistically significant differences, all other spirometric parameters exhibit lower values when children are in the presence of risk. With respect to VOC, we found that there were statistically significant differences at the level of IT and FEF_75.

These two values are higher when children are not exposed to amounts of VOCs higher than the CMR.

For other spirometric parameters and despite the fact that there are no statistically significant differences, we found that in the case of FEV_1, and FEF_25 FEF_50 these are higher when children

Table 1. Comparison of the averages of the spirometric parameters from phase 1 to phase 2.

Spirometric parameter	Average	N	Standard Deviation	df	Phase 2 < Phase 1	Phase 2 > Phase 1	Phase 2 = Phase 1	t	gl	p-value
FVC–autumn/winter	80,7479	1019	10,42045	–1,57	423	589	7	–7,575	1018	< 0,0001
FVC–spring/summer	82,3147	1019	9,80597							
IT–autumn/winter	96,1503	1019	5,01745	–0,56	301	438	280	–4,526	1018	< 0,0001
IT–spring/summer	96,7050	1019	4,73573							
FEV1–autumn/winter	91,1531	1019	11,48488	–2,35	360	635	24	–10,678	1018	< 0,0001
FEV1–spring/summer	93,5072	1019	10,94368							
PEF–autumn/winter	81,6057	1019	14,30103	–5,33	277	735	7	–16,612	1018	< 0,0001
PEF–spring/summer	86,9382	1019	14,07759							
FEV_25–autumn/winter	87,0789	1019	16,05417	–5,20	313	700	6	–15,138	1018	< 0,0001
FEV_25–spring/summer	92,2832	1019	15,87417							
FEV_50–autumn/winter	93,4251	1019	19,81111	–3,33	415	587	17	–8,130	1018	< 0,0001
FEV_50–spring/summer	96,7581	1019	20,26186							
FEV_75–autumn/winter	101,9728	1018	28,93930	–3,94	422	586	10	–5,250	1017	< 0,0001
FEV_75–spring/summer	105,9112	1018	28,70347							
FEV_75_25–autumn/winter	80,3000	4	5,62968	–12,30	2	2	0	–1,262	1018	< 0,296
FEV_75_25–spring/summer	92,6000	4	23,73689							

T-Student test for matched samples and Wilcoxon T test;
<—When the values of phase 2 were below those of phase 1;
>—When the values of phase 2 were higher than those of phase 1;
=—When the values of phase 1 and phase 2 remain the same.

are at no risk, contrary to the values of FVC and PEF that, given the same situation (absence of risk), are relatively lower than those of the children who were in the presence of risk. In the spring/summer season, it was found that in what concerns CO_2, there are statistically significant differences at the level of IT, FEV_1, PEF and FEF_25 spirometric parameters.

We found that children who are not exposed to CO_2 levels above the CMR have higher spirometric values in relation to children who are exposed. Although there are no statistically significant differences at the level of other spirometric parameters, we found the same trend, ie, when children are exposed to CO_2 levels above the CMR the spirometric values are always lower than in children in the absence of risk.

With regard to PM_{10}, we found that there were statistically significant differences in terms of IT, and FEF_50 FEF_75. We found that its value is higher when children are not exposed to PM_{10} values above the CMR.

It was noted that even though there are no statistically significant differences all other spirometric parameters have lower values when children are in the presence of risk, ie, when they are exposed to PM_{10} values above the CMR, except that FVC is slightly higher in children not exposed to values above the CMR.

With regard to VOCs, it was found that in all spirometric parameters there are no statistically significant differences, however, it was found that all but IT are higher in children in the absence of risk.

As in this study, in some parameters, other studies have assessed the impact of pollutants on the health of children and observed changes in respiratory function in children exposed to pollutants.

In Austria, a study that followed 975 children during three years found that there is an association between reduced lung function and the increase of PM_{10}, SO_2, NO_2 and O_3 (Horak et al. 2002).

Another study in rural Holland found decreased pulmonary respiratory function two weeks after the increase of SO_2 and particulate matter involving 1,000 children from six to twelve years old (Brunekreef et al., 1989).

In Rio de Janeiro, Brazil, a study in which the authors sought to determine which was the association between daily exposure to air pollution and respiratory function in children from six to fifteen, found that even within acceptable levels in most period, air pollution, especially PM_{10} and NO_2, was associated with the decrease of respiratory function (Castro et al., 2009).

4 CONCLUSIONS

In conclusion, there are schools that have poor IAQ, long associated with the non-renewal of air for keeping buildings closed. A thermal comfort deficient promotes the lack of ventilation, and thereby prevent the entry of cold air remain windows and doors.

The results obtained in this research show that the schools studied, mostly correspond to buildings that long ago are not the object of rehabilitation, having many years of existence. Care presented in their constructive aspects, lack of air conditioning system and lack of ventilation or mixed and opening windows or doors provided ventilation. As a result of this situation, especially in the fall/winter were measured especially high levels of CO_2 in the indoor environment, clearly revealing the IAQ deficiencies due to insufficient ventilation. They also found high concentrations of PM_{10}, VOCs and HCHO in some classrooms. The exposure to these pollutants affects their health, especially the respiratory system. It was found that the concentrations of pollutants in the indoor air are higher than those schools outside, indicating the importance of emission sources interiors.

It is essential that schools perform continuous monitoring, so as not to expose children to hazardous situations. It is also important to improve systems for air renewal in order to make this more effectively and efficiently renewed. However, you have to modify and change attitudes and behaviors on the part of the occupiers of buildings, performing the simple habit of opening the windows often. It is urgent to promote and support education initiatives for environmental health related to the quality of indoor air.

BIBLIOGRAPHY

Air Quality Guidelines for Europe. *WHO Regional Publications. European Series*. 2nd Edition 2000, nº91.

Bernstein JA, Alexis N, Bacchus H, Bernstein IL, Fritz P, Horner E, et al. The health effects of non-industrial indoor air pollution. *J Allergy Clin Immunol* 2008;121:585–91.

Brunekreef B, Lumens M, Hoek G, Hofschreuder P, Fischer P, Biersteker K. Pulmonary function changes associated with an air pollution episode in January 1987. *JAPCA*.1989;39(11):1444–7.

Castro, Hermano; Cunha, Márcia; Mendonça, Gulnar; Junger, Washington; Cunha, Joana; Leon, António. Effect of air pollution on lung function in schoolchildren in Rio de Janeiro, Brazil. *Revista de Saúde Pública*. 2009;43(1):26–34.

Graudenz GS, Latorre MR, Tribess A, Oliveira CH, Kalil J. Persistent allergic rhinitis and indoor air quality perception – an experimental approach. Indoor Air 2006;16:313–9.

Horak F, Studnicka M, Gartner C, Spengler JD, Tauber E, Urbanek R, et al. Particulate matter and lung function growth in children: a 3-yr follow-up study in Austrian school children. *Eur Respir J.* 2002;19(5):838–45. DOI: 10.1183/09031936.02.00512001.

Kim H, Bernstein JA. Air pollution and allergic disease. Curr Allergy Asthma Rep 2009;9:128–33.

Koistinen K, Kotzias D, Kephalopoulos S, Schlitt C, Carrer P, Jantunen M, et al. EU Forum: The INDEX project: executive summary of a European Union project on indoor air pollutants. Allergy 2008;63:810–9.

Miller MR, Hankinson J, Brusasco V, Burgos F, Casaburi R, Coates A, et al. Standardisation of spirometry. *Eur Respir J* 2005;26:319–38.

Platts-Mills TA, Vaughan JW, Carter MC, Woodfolk JA. The role of intervention in established allergy: avoidance of indoor allergens in the treatment of chronic allergic disases. *J Allergy Clin Immunol* 2000;106:787–804.

Quanjer PH, Borsboom GJ, Brunekreef B, Zach M, Forche G, Cotes JE, et al. Spirometric reference values for white European children and adolescentes: Polgar revisited. Pediatr Pulmonol 1995;19:135–42.

Occupational Safety and Hygiene II – Arezes et al. (eds)
© 2014 Taylor & Francis Group, London, ISBN 978-1-138-00144-2

Risk analysis and reliability assessment of energy facilities

Juraj Sinay, Adrián Tompoš & Slavomíra Vargová
Department of Safety and Quality of Production, Faculty of Mechanical Engineering, Technical University of Košice, Slovakia

ABSTRACT: Current trends indicate a growing need for paying attention to occupational health and safety, whose appropriate management contributes to the optimization of production processes, brings economic effects, benefits to cultural and social status of the company and the country in which it operates. In order to consider particular work safe, all operations must be performed in a safe environment, using safe machinery and safe working procedures. In other words, everything the worker/employee encounters and performs must be safe.

1 INTRODUCTION

All existing industrial plants have some elements in common, yet each possesses some unique characteristics. Therefore, it is necessary to deal with both types, in order to timely identify objects, systems and situations leading to faults or damage to humans, environment or material property, and prevent such damage by well selected and applied preventive measures.

One of risk assessment methods usable in the phase of threat/danger identification and estimation of their consequences is the Failure mode and effects analysis (FMEA). It uses detailed quantitative determination of occurrence of potential failure causes and severity of their possible effects, and reveals their causes, in order to identify which failures fall into the intolerable region, i.e. either require immediate repair or will require improvement in future, unless preventive measures are taken.

2 ENERGY SYSTEMS AND MOTORS

Energy systems and motors include diesel engines present in numerous machines, locomotives, ships and power generators. Examples include generator sets for production of electricity in places without connection to the power grid, available as canopied or container modules. The generators can be used as portable or stationary for permanent operation, or as emergency power-supply. They can operate independently, be interconnected in parallel operation or in parallel with the electric grid. They are also used to power construction machines, crushers, special purpose vehicles, towing units, agricultural machines, compressors, pumping devices and power generators for building sites and industries (Karwowski & Marras 1999).

3 RISK MANAGEMENT

Risk management is a systematized sequence of algorithm solving steps, which are useful for defining particular hidden attributes, potentials and abilities of an object to cause damage, and for describing how they can lead to damage or injury. The solution gradually proceeds from risk analysis and evaluation to final adoption and implementation of measures. Risk management also includes monitoring of the efficiency of adopted measures, since it is an occupational safety and health (OSH) management system tool. The principles of OHS management system, similarly to other quality and environment management systems, involve constant improvement efforts (Cox & Tait 2003).

To sum up, risk management involves describing and determining boundaries of the given system, identifying the probability and extent of potential effects of the negative incident related to the work performed or other activity. This identification serves as basis for the assessment of the risk severity (Sinay, Tompos & Sviderova 2011).

Risk assessment (in relation to OSH) means the process of observation and examination of potential causes of damage in a workplace in the form of an injury and/or health damage and considering the appropriateness of adopted measures and their efficiency. On the basis of risk analysis, the estimated risk extent and its severity is evaluated (risk evaluation) and the necessity of its minimization is considered (risk acceptability) (Brauer & Brauer 2005).

The most commonly used is the Health and Safety Executive (HSE) approach (Fig. 1), consisting of three risk management levels (Pacaiova et al., 2008).

4 FAILURE MODE AND EFFECTS ANALYSIS

Failure mode and effects analysis (FMEA) uses the inductive reasoning (bottom-up approach) and reviews systems and subsystems (components) to identify failure modes that influence the system functions. Concurrently, their causes and effects are determined. FMEA is predominantly used for qualitative evaluation of potential failures of relatively uncomplicated components, defines which modes or causes of failures result in dangers with severe effects (intolerable level), and those that fall into the acceptable or negligible region. In order to specify potential effects of failure, the analysis is complemented by qualitative assessment of the following: safety—S, environment—E, quality—Q and downtime—D (Stamatis 2003).

Qualitative evaluation is expressed in a triplet of values, namely failure severity of the observed system—Sev (the extent of the effect severity in the given system), failure occurrence—O (how frequently the given failure or its cause occurs in a time unit) and the value of the failure detection—D (what the probability is/how long it will take/how easy it is to detect the failure cause).

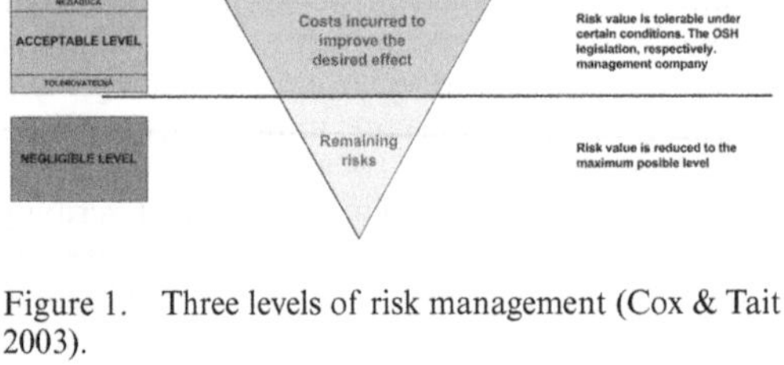

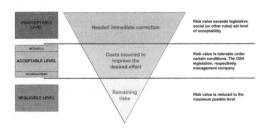

Figure 1. Three levels of risk management (Cox & Tait 2003).

Also are there answers for impact for each failure mode focused on Safety—S, Environment—E, Quality—Q and Production—P.

All three parameters are evaluated using grades 1 to 10, while the qualitative evaluation, influence interval and their values depend on evaluator's subjective opinion which requires and utilizes professional knowledge, skills and experience in the given area (Reason, J. T., 1997).

The selected system analysis is suitable for achieving solutions that prevent (detect) or effectively tackle (minimize negative effects of) failures or their causes, which will significantly influence the quality of the observed system. The application of method principles proposed by German Association of the Automotive Industry (standard VDA 4.2), has tightened the conditions for evaluating Risk priority number (RPN) values, as can be seen in the following table (Tab.1) (Carlson 2013).

FMEA analysis was followed by Weighted Pareto Analysis, in which downtimes due to maintenance were observed and weight coefficients were allocated according to Table 2.

5 RESULTS

The failure mode and effect analysis focused on mechanical and electrical parts (Fig. 2) of the engine, in order to accurately identify the most frequent failures and their causes.

The analysis results identified the 20 per cent of failure causes, whose elimination/minimization removes/reduces 80 per cent of undesired effects (failures and their consequences). The failure causes related to mechanical part of the motor include the following: machine overload, shavings form piston, gummed piston ring, mechanical damage of material, fatigue of material, high pressure, insufficient lubrication, unsuitable state of the cooling fluid and broken valve. Failure causes related to the electrical part of the motor include high current, fatigue of material, mechanical damage, short circuit and mechanical wear of material (Fig. 3).

Table 1. System FMEA of electric part of vehicle.

Company/division:				System - FMEA of Produkt								→	Team: AdrianTompos, Slavomira Vargova			FMEA Number:
System: Electric part of vehicle													Prepared by: Adrian Tompos			6
Subsystem: Glowing													Date: 10.02.2012			Page: 1 of 1
Function of subsystem: Temporary heating up to operating temperature													Actions			
Item	Potential Failure Mode	Potential Cause	Potential Consequences	S	E	Q	P	SEV	OC	DET	RPN	>	Recomended action(s)	Resp. Target	RPN*	Action taken
Coil	Overburn of coil	High current flow (power supply)	Diesel does not get warm up to operating temperature - Cold Start	YES	NO	YES	YES	3	10	5	150	1	Daily check incasdescent by operator		80	
Cabel	Cracked cabel	Mechanical damage	Intermittent supply of electric power	YES	NO	YES	YES	5	8	2	80	2	Checking of correctnes of mounting and attaching of cables		40	
		Vibrations						3		2	48	3			24	
		Material fatigue						5		2	80	4			40	
	Burned cabel	High current flow (power supply)	Complete interruption of electric power	YES	NO	YES	YES	3	10	1	30	5	Use thermal protection for cables		20	

Table 2. Allocating weight to individual failure causes in relation to downtimes.

Weight	Downtime up to	Time unit
1	2	Hour
2	4	Hour
3	6	Hour
4	12	Hour
5	24	Hour
6	2	Day
7	5	Day
8	10	Day
9	15	Day
10	20	Day

6 CONCLUSIONS

The research focused on defining, selecting and reducing statistically most significant causes of failures in diesel engines and their parts, and achieved provably positive results. This does not imply that further selection or reduction of causes is undesirable or unwelcome. Quite contrary, any improvement effort has a meaning and place in the endless process of improvement.

The direct effect of risk reduction is demonstrated in the reduction of costs related to decreasing downtime; however, it also depends on managerial decisions.

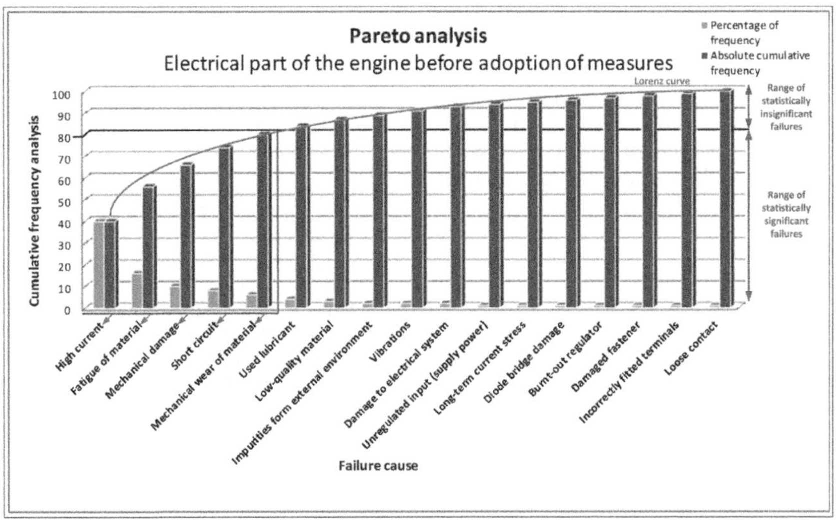

Figure 2. Pareto analysis chart for electrical parts of the engine.

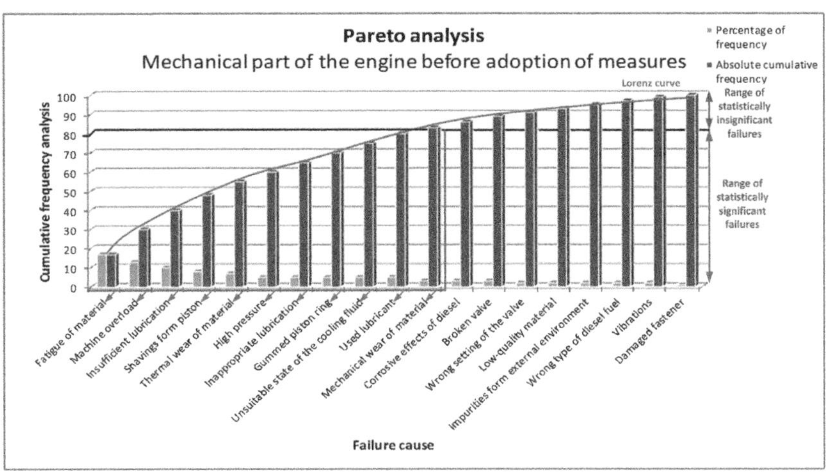

Figure 3. Pareto analysis chart for mechanical parts of the engine.

Its secondary effect, yet not lesser in importance, is related to the process causality ranging from mechanical safety to safe behaviour. It is an attitude that needs to be adopted: "the more reliable the system is, the fewer interventions are necessary; the fewer interventions there are, the lower is the exposure to the undesirable, hazardous or threatening, i.e. risky environment". We are thus approaching the range of safe behaviour, which emphasizes the "willingness to think safe and act safe".

Both effects form a starting point or a primary impulse towards new solutions for other parts of the system in relation to risk management.

ACKNOWLEDGMENTS

The research reported in this paper was conducted within project APVV-0337-11 Research into new and newly emerging risks related to industrial technologies within integrated safety and security as a precondition for sustainable development management.

REFERENCES

BRAUER, Roger L. & BRAUER, Roger (2005). *Safety and Health for Engineers, Wiley-Interscience*, USA, 758 p. ISBN 978-0471291893.

CARLSON, Carl: Effective FMEAs: *Achieving Safe, Reliable, and Economical Products and Processes using Failure Mode and Effects Analysis* (*Quality and Reliability Engineering Series*), Wiley, USA, 2013. 464 p. ISBN 978-1118007433.

COX, Sue & TAIT, Robin (2003). *Safety, Reliability & Risk management: An Integrated Approach*, Butterworth-Heinemann, Lightning Source UK TD., Milton Keynes UK, 320 p. ISBN 0-7506-4016-2.

KARWOWSKI, Waldemar & MARRAS, William S. (1999). *The Occupational Ergonomics Handbook*, CRC Press, USA, 2088 p. ISBN 978-0849326417.

PAČAIOVÁ, Hana & SINAY, Juraj & GLATZ, Juraj (2009). *The safety and risks of technical systems*, Košice: TU, SjF, 246 p. ISBN 978-80-553-0180-8.

PAČAIOVÁ, Hana et al. (2008). *Prevention of Major Industrial Accidents: A Handbook for Management of the Law requirements for U.S. Steel Košice, s.r.o.*, Košice: Technická univerzita v Košiciach, SjF, KBaKP, 44 p. ISBN 978-80-553-0024-5.

Quality production: *FMEA principles and benefits*: [online]. [s. a.]. [accessed: 2012-02-20]. available at: <ttp://www.kvalitaprodukcie.info/fmea-%E2%80%93-principy-a-prinosy/>.

REASON, James T. (1997). *Managing the Risks of Organizational Accidents*, Ashgate Publishing Company, USA, 252 p. ISBN 978-1840141054.

SINAY, Juraj et al. (2007). *Quality improvement tools*, ManaCon Prešov, Prešov, pp. 65–82, 181–192, ISBN 978-80-89040-32-2.

SINAY, Juraj & TOMPOŠ, Adrián & ŠVIDEROVÁ, Katarína (2011). *Theory and practice of health and safety at work,* Edícia študijnej literatúry, Košice, ISBN 978-80-553-0791-6.

STAMATIS, D.H. (2003). *Failure Mode and Effect Analysis: FMEA From Theory to Execution*, Second Edition, Amer Society for Quality, USA, 2003. ISBN 978-0873895989.

Occupational Safety and Hygiene II – Arezes et al. (eds)
 ISBN 978-1-138-00144-2

Microbiological parameters of surfaces in school canteens

C. Laranjeiro, C. Santos, B. Alves, A. Ferreira & J.P. Figueiredo
Instituto Politécnico de Coimbra, ESTeSC-Coimbra Health School, Saúde Ambiental, Portugal

ABSTRACT: Microbiological surveillance of cooked meals in school canteens is an important barrier to contamination associated with improper handling practices. With a view to improving the quality of services provided is essential to quantify the presence of microorganisms associated with surfaces (boards, dishes, utensils) as well as at the hands of manipulators. Worked checklists were applied in 62 schools, as well as microbiological tests affected by the presence of microorganisms in a environment with 30°C, *Escherichia coli* and coagulase positive staphylococci in handlers and surfaces.

It is concluded that the presence of microorganisms in environment with 30°C was statistically significant or in the hands as the surfaces (showing the board on average higher contamination. The presence of *Escherichia coli* and coagulase positive staphylococci were found on 2,9% of boards and were absent in the other areas. Being that 4.3% of schools that had a domestic machine, present contamination by Escherichia coli and coagulase positive staphylococci. Still preserving the handlers that do not comply with the assumptions on hand washing, showed greater amounts of microorganisms to 30°C in their hands.

1 INTRODUCTION

In recent years, it has been experiencing an increase excessive occurrence of Foodborne Diseases (FBD) worldwide (Havelaar, 2009). Contamination risks are higher in schools due to food preparation well in advance, which favors prolonged exposure to possible contaminants (Lagaggio, 2002). In addition, inadequate hygiene on-site preparation and distribution also contribute to this. The FBD results are regarded as unsuitable for food preparations, as the microbiological contamination can occur during all stages of the preparation of these (Piggot, 2008). Storage at room temperature favors the proliferation of pathogenic bacteria; the microorganisms can remain on the surface for hours or days after contamination (Declan, 2007; Kusumaningrum, 2002).

The FBD are mainly caused by ingestion of viable pathogens (infection) or toxins they produce (toxicity) in amounts sufficient to develop the pathology commonly referred to as foodborne infections (Afifi, 2012). Among these diseases, there is a gastrointestinal disease that can be caused by foodborne infection. Among the microorganisms important for the occurrence of the foodborne infection, stresses the importance of *Escherichia coli*, the main indicator of fecal contamination and coagulase positive staphylococci, presence indicator material nasal or oropharyngeal (which can also be colonizer of the human microbiota). Possible forms of food contamination by coagulase positive staphylococci are through nasal secretions and *Escherichia coli* in the faeces and contaminated water (Murray, 2009).

Hand sanitizing should therefore ensure the elimination of visible and invisible dirt and destruction of pathogenic microorganisms and decay to levels that not undermine the health of consumers and product quality (Wirtanen, 2003).Therefore, despite the washing procedures being a control point to prevent cross-contamination in kitchens, if not performed properly may be responsible for contamination of utensils or surfaces (Mattic, 2003; Gent, 2010).

The goal of this research resided in establishing a direct relationship between the presence of microorganisms, when significant, and hygiene conditions in existing establishments under study, as well as the type of treatment used in the washing of surfaces, including utensils and tableware thin.

2 MATERIAL AND METHODS

For this study, we used data already collected in the period from 1 October 2012 to 18 March 2013. Therefore, had to be based on a number of schools in the Central Region, a total of 62 schools (kindergartens and primary schools), located mostly in urban).

These data result from a check sheet designed and implemented by the organization to provide services for each school, taking into account the legislation, which supports the evaluation the hygienic conditions of the establishments, as well as microbiological smears resulting in utensils, hands and other surfaces handlers belonging to each school, resulting from the work performed by the authority aforesaid.

Regarding the form of check, it was divided into two main parts: the first relating to the characterization of the physical facilities and environment and the second proposed to obtain information relating to food handlers.

The person responsible for taking the sample proceeded to collect it, effecting a proper handwashing and using subsequently the appropriate personal protective equipment, including sterile gloves, mask, cap and gown.

For the realization of smears, where the samples collection were over flat surfaces, are dishes or trays, resorted to the use of molds sterilized with a total area of 25 cm^2.

When analyzing the surface was irregular, such as utensils or hands handler, had been based on a movement comprising approximately 1 cm^2 and the same was repeated 25 times, which allowed to quantify the sampling area, which was important at the time of analysis. Thus, for the harvest, we used a dry swab and solution "Letheen Broth Base Modified ISO 21149," to preserve and transport swab after the smear. Emphasize that harvests carried out in establishments placed in cooler chilled and delivered in accredited Laboratory, within three hours.

Regarding the quantification of microorganisms in the samples collected were used as reference standards ISO 18593:2004 point 8 and 9 and ISO 4833:2003, ISO 18593:2004 point 8 and 9 ISO 16649-2:2001 and ISO 18593:2004 point 8 and 9 ISO 6888-1:1999, referring to the methods adopted and accredited by the Laboratory for each test.

Microbiological parameters, according to the microbiological criteria based on national and international standards, the maximum acceptable due to the presence of microorganisms at 30°C was <1.0 e+2 Colony Forming Units (CFU)/cm^2, the maximum acceptable due to the presence of *Escherichia coli* was <1 CFU/cm^2, the maximum acceptable limit due to the presence of coagulase positive staphylococci was <1 CFU/cm^2, which are benchmarks for the National Institute of Health Dr. Ricardo Jorge (INSA).

Given also the type of washing away the utensils and tableware thin (industrial machine, domestic or manual washing), intended to evaluate the data obtained concerning microbiological contamination, in order to see if there was an association between the presence of microorganisms and different types of washing.

The statistical treatment of the data was performed with the software IBM® SPSS® Statistics version 21.0 for Windows.

The interpretation of statistical tests were based on a significance level of p-value ≤ 0.05 with confidence interval (C.I.) of 95%.

3 RESULTS

3.1 *Analysis of the results*

Given the checklists applied in each school, it was observed that in the "*Physical Facilities and Environment*", in the "windows", the parameter "windows" kept did meet up in 55 of the 56 schools analyzed (98.2%), verifying a significant absence of windows fitted with mosquito nets, 45 of the 55 schools did not meet with this item (81.8%). For the class "Wash Hands," it is found that the presence of sink areas manipulation was accomplished in 52 of 62 analyzed schools (83.9%), with no specific procedure for handwashing observed in 47 schools (75.8%).

As for the "Food Handlers", in the "hand washing" it was found that it was performed with appropriate product in 83.6% of cases, and washing hands often the parameter that did meet in only 24 of the schools studied (39.3%). Regarding the category "hygienic habits", in only 13.1% of establishments existed fly killer.

After application of tools for collecting data, we compared the mean values estimated for analytical microorganisms to 30°C, both on surfaces and in the hands of the manipulator (Table 1).

The average of hands of the handler CFU/cm^2 for the 40 observations (388.2 ± 465.9) was greater than average of CFU/cm^2 for 36 observations presented surfaces (293.9 ± 529.5).With the application of the Student t test verified the existence of a statistically significant difference between the average CFU/cm^2 microorganisms at 30°C and the reference value (1.0 e + 2 CFU/cm^2) for both surfaces and for hands of the manipulators (p-value ≤ 0.05), and these higher values than the maximum permissible value.

Table 1. Differences in the amount of 30°C microorganisms present on surfaces and hands.

	n	$\bar{x} \pm s$	t	p-value
Surface	36	293.9 ± 529.5	2.127	0.035
Hands Handler	40	388.2 ± 465.9	3.912	0.000

Reference value: 1,0 e + 2 CFU/cm^2; n = Sample size; $\bar{x}$ = Average; s = Standard deviation; t: Student t test for one sample.

It was intended to subsequently certify that the surface on which the cleaning have been carried out more effectively and consequently, the larger amount of which had colonies of microorganisms to 30°C (Table 2).

There were statistically significant differences relative to the amount of microorganisms in different areas on which crops were conducted. The results suggest that was on the surface of the trays ($\bar{x} \pm s = 711.6 \pm 491.3$) have found that higher quantities of colonies of microorganisms at 30°C.

Followed by the analysis of the data collected for the presence of colonies of *Escherichia coli* and positive coagulase positive staphylococci, and the reference value <1 CFU/cm^2. Regarding the quantification of *Escherichia coli* has been studied this parameter in 35 schools. In 2.9% of the sample surfaces were detected amounts of colonies greater than the reference value (<1 CFU/cm^2). The results were analyzed with the test X^2 of independence revealed that there was no statistically significant difference for the presence of colonies of the bacterium *Escherichia coli* in trays compared to their presence in the remaining surfaces (plates and utensils) for a significance level of 5% (p-value > 0.05).

Regarding the presence of coagulase positive staphylococci to surfaces analyzed, it was found that the situation was similar to that mentioned above, namely 2.9% of the samples there were colonies of coagulase positive staphylococci. It was considered that the results were analyzed with the test X^2 of independence did not reveal a statistically significant presence of colonies of coagulase positive staphylococci in relation to other surfaces observed for a significance level of 5% (p-value > 0.05). These organisms were not detected in samples collected from trays obtained.

We sought to determine the presence of microorganisms at 30°C according to the type of washing, in order to assess whether there are differences, given the type of wash. There were no statistically significant differences for the presence of microorganisms at 30°C associated with the type of wash used for cleaning surfaces (Industrial Machine, Domestic Machine, Hand Washing, Machine Domestic and Industrial), for a significance level of 5% (p-value >0.05). And the results obtained by the Kruskal Wallis test. The results achieved suggest that the use of different types of washing machine for kitchen equipment, in general, all allow to obtain similar results in terms of cleaning.

Table 2. Differences in the amount of 30°C microorganisms present on the surfaces analyzed.

	n	$\bar{x} \pm s$
Utensil	19	103.9 ± 434.9
Trays	7	711.6 ± 491.3
Dishes	9	402.8 ± 597.9
Total	35	302.3 ± 534.8

F(B-F): 3.689; gl = 2;19,240; p-value = 0.044; n = Sample size; $\bar{x}$ = Average; s = Standard deviation.

Even if inquired into, given the type of wash, which was presented evidence that colonies of bacteria (*Escherichia coli* and coagulase positive staphylococci), with greater frequency. Analyzing the results and the type of washing, it was found that in no case was detected the presence of *Escherichia coli* in surface plates and utensils. Of the 22 schools that used the machine home, 95.7% showed no significant microbiological contamination by this bacterium. Observing the results, obtained using the test X^2 of independence, it was shown that there was no statistically significant difference between the microbial load of wash type used (p-value > 0.05).

There was evidence of association between the type of washing away the presence of coagulase positive staphylococci, and only schools that were using domestic machines showed contamination by this microorganism (4.3%) (X^2 of independence, p-value ≤ 0.05). These results reveal non-compliance with the proper cleaning of domestic machine.

Also examined whether the handlers met with key assumptions for hand hygiene, and the consequent burden of microorganisms at 30°C.

In view of Table 3, it was found that there were no statistically significant differences between the presence of microorganisms at the hands of the manipulators and performed the same way as the cleaning of the same, and these average values higher than the maximum permissible value in both cases. It was found that on average, of 40 observations on 18 cases was performed correctly handwashing.

4 DISCUSSION

Based on the results it was concluded that in general the conditions of "physical facilities and environment" school and the "food handlers" were found to be adequate. However, there were some

Table 3. Differences in the amount of 30°C microorganisms associated with the type of wash.

Hand handwashing	n	$\bar{x} \pm s$	t
Complies	18	260.1 ± 458.4	−1.605
Does not comply	22	493.1 ± 455.4	−1.604

p-value = 0.732; n = Sample size; $\bar{x}$ = Average; s = Standard deviation; t: Student t test for one sample.

shortcomings, particularly the lack of mosquito nets, lamps with protective clothing appropriate hand hygiene frequently and the use of ornaments. But it is important to place lamps with protection, the existence of devices to prevent entry of insects or other pests, such as bed nets that must be cleaned frequently (Declan, 2007).

Hands should be cleaned frequently to avoid cross-contamination and the use of ornaments must be removed because they are a risk factor, may inadvertently fall and incorporate foods or favor bacterial growth in food with which they come in contact (Borges, 2007).

This study highlights the presence of microorganisms at 30°C, above the reference value (1,0 e+2 CFU/cm^2), whether in the hands of the handlers as the surfaces analyzed. Situation corroborated by other studies, including one in which 68 underwent food handlers in 12 power units microbiological evaluation of the hands, checking up on 88% of results microorganism count to 30°C higher than the reference value (Andrade, 2003). In another study, carried out an assessment on the terms of microbiological food utensils used in the power unit, it was observed that none of the utensils analyzed yielded values of microorganisms at 30°C within the range considered acceptable. Being stressed that proper maintenance and a good state of preservation utensils are essential to prevent wear allows the multiplication of microbial population (Kochanski, 2009).

Given the number of microorganisms present in the surface analyzed, it was found that the tray is made larger than the surface microbial contamination, following the dishes and utensils. This may be related to the nature of the surface, which must be smooth and impervious material, easy to wash and sanitize non-toxic. The frequent use of the trays causes your wear and allows multiplication of microbial population and should be properly maintained and are always in good condition (Murmann, 2008). According to a study, the plastic material is contaminated with a higher microbial load, followed by ceramics and glass. This has been associated with rough topography of the plastic ware, caused by wear of this type of material due to characteristics such as low hardness and scratch facility (Sigua, 2011).

The presence of *Escherichia coli* as well as coagulase positive staphylococci on surfaces analyzed (2.9%), denotes a poor implementation of good practices in hygiene and conservation areas, and indicators of health risk to consumers (Borges, 2011). The presence of *Escherichia coli* is an indicator of fecal contamination that may occur during the processing of raw materials of animal origin, mainly due to an inefficient personal hygiene of food handlers, equipment and surface environments confection food (Ray 2004). The presence of coagulase positive staphylococci occurs mainly through handlers (Jerônimo, 2011).

However, these results associated with the presence of microorganisms on surfaces may also be indicators ineffective cleaning process. The mechanical washing involves controlling an automatic way to certain specifications relating to its operation, including the temperatures of washing, rinsing and disinfection operations, pressure, speed and time of the wash cycle (VDACS, 2010). Recommended temperatures for washing must be kept above 43°C, or in the case of hot water immersion be used as a process for disinfecting water temperature above 77°C for at least 30 seconds (Schmidt, 2007).

The recurrent use of such equipment requires a simultaneous maintenance and periodic verification to ensure the effectiveness of the cleaning process, should be done regularly monitored particularly the verification of filters, temperature, conservation and hygienic condition of the equipment and accessories washing, and if these are not fulfilled, it is possible recontamination between wash cycles.

For the quantity of microorganisms at 30°C as well as the presence of *Escherichia coli* and coagulase positive staphylococci present in 4.3% of the surfaces and for the type washing found that the differences between each wash type were not significant, because microbial load was similar in different types of washing.

Some studies on the cleaning process of household food only reveal that the mechanical washing has proved more effective than manual cleaning, which corroborates with another previous study that concluded that the mechanical washing systems better control the level of microbiological utensils used in food service establishments, than the most basic level of handwashing (Sigua, 2011; Wernersson, 2003). Other studies show some causes to explain these differences as the discomfort that may be caused in employees to perform a manual wash at temperatures above 40°C, unlike the mechanical washing normally uses higher temperatures and is not limited by this constraint human (Pfund, 2004). Other factors contributing to the variation in efficacy between these two washing systems are the number and size of vessels which are to be cleaned, the microbial load initially present, the type of dirt adheres to the tools, as well as the duration of the cleaning process (Sigua, 2011). The data relating to handwashing suggested that handlers who do not comply on average with these assumptions expressed greater amounts of microorganisms at 30°C in their hands.

Thus, hand hygiene plays an important role in controlling the spread of infectious diseases (Afifi,

2012). The handlers have great importance in all stages of quilting meals; they may facilitate the spread of pathogens in the workplace (Green, 2006). It is also crucial to the adoption of good practices in the workplace and the implementation of appropriate training programs to bridge some of these gaps and hence provide healthy meals for their children.

5 CONCLUSIONS

School canteens must ensure the safety, health and adequate conservation that provide meals, but may be gaps function as the non-adoption of good practices by food handlers, for reasons of neglect or ignorance, and consequent occurrence FBD. So, good hygiene practices when handling food presented as an important means to reduce cross-contamination between surfaces and handlers. Since it is essential to ensure the hygiene of both the physical facilities and working environment, either of their own handlers.

With the assessment made in the process of cleaning the surfaces, together with the data obtained by microbiological evaluation showed up faults with possible impact on the quality of meals served in canteens of schools evaluated.

The presence of microorganisms at 30°C, *Escherichia coli* and the coagulase positive staphylococci, reveals the need for a targeted investment to meet the hygiene and safety by food handlers, through a plan of training that includes not only the component as well as theoretical, practical component in the workplace. But also checks the mechanical washing equipment in order to maintain your condition and set temperature thus allowing an adequate score after the cleaning process, eliminating the microbial load present.

Stresses the importance of proper hand hygiene and the training of handlers in order to make them aware of the repercussions of their role and responsibilities in the prevention of contamination and have knowledge and skills necessary for the performance of their duties.

The present study has limitations and with regard to the presence of *Escherichia coli* and coagulase positive staphylococci to surfaces, it was not possible to quantify the number of CFU having been researched only their presence or absence, or using only qualitative criteria. It is considered that the results were not as impressive as what was expected, so it would be appropriate to continue the same in order to see if it maintains the same pattern of results.

The present study aimed to highlight the importance of hand hygiene and surfaces in order to ensure a food supply to microbiological quality, safe and without risk to the users of school canteens.

REFERENCES

Afifi, H.S. & Abushelaibi, A.A. 2012. Assessment of personal hygiene knowledge, and practices in Al Ain, United Arab Emirates. Food Control 25: 249–253.

Andrade, N.J. *et al.* 2003. Avaliação das condições microbiológicas em unidades de alimentação e nutrição. Cienc. Agrotec. Lavras 27(3): 590–596.

Borges, J.G. *et al. Qualidade higiénico-sanitária de alimentos oferecidos em escolas públicas do município de Lavras.* Universidade Federal de Lavras. [Serial online] 2011 dez. Disponível em: URL: http://www.proec.ufla.br/conex/ivconex/arquivos/trabalhos/a126.pdf.

Borges, F.; Silva, B.L.; Gontijo-Filho P.P. 2007. Hand Washing: changes in the skin flora. *American Journal of Infection Control* 35: 417–420.

Declan, J.B. & Maunsell, B. 2007. *Guia para Controlo da Segurança Alimentar em Restaurantes Europeus.* Instituto Nacional de Saúde Dr. Ricardo Jorge: Lisboa.

Gent, C.L.; Sylla, Y.; Faille, C. 2010. Bacterial re-contamination of surfaces of food processing lines during cleaning in place procedures. *Journal of food Engineering* 96: 37–42.

Green LR, *et al.* 2006. Food worker hand washing practices: an observation study. *J Food Prot* 10(69): 2417–2423.

Havelaar, A.H. *et al.* 2009. Future challenges to microbial food safety. *International Journal of Food Microbiology*.

Jerônimo, H.M.A. *et al.* 2011. Ocorrência de Staphylococcus spp. e S. aureus em superfícies de preparo de alimentos em unidades de alimentação e nutrição. *Nutrire: Rev Soc Bras Alim.* 1(36): 37–48.

Kochanski S, *et al.* 2009. Avaliação das condições microbiológicas de uma unidade de alimentação e nutrição. *Alimentação e Nutrição* 20: 663–668.

Kusumaningrum, H.D. *et al.* 2002. Effects of dishwashing liquid on foodborne pathogens and competitive microorganisms in kitchen sponges. *J. Food Prot* 65: 61–65.

Lagaggio, V.R.A. *et al.* 2002. Avaliação microbiológica da superfície de mãos dos funcionários do restaurante universitário, da Universidade Federal de Santa Maria, *RS. Hig Aliment* 16(100): 107–110.

Mattic, k. *et al.* 2003. The survival of foodborne pathogens during domestic washing-up and subsequent transfer onto washing-up sponges, kitchen surfaces and food. *International Journal of Food Microbiology* 3(25): 213–226.

Murmann, L. *et al.* 2008.Quantification and molecular characterization of Salmonella isolated from food samples involved in salmonellosis outbreaks in Rio Grande do Sul, Brazil. *Braz. J. Microbiol* 3(39): 529–534.

Murray, P.R. *et al.* 2009. *Microbiologia médica.* 5th ed. Elsevier: Rio de Janeiro.

Pfund, R. 2004. Avoiding the hidden handwashing hazard. *Plumbing Engineer* 41–42.

Piggot, D.C. 2008. Foodborne ilness Emergency. *Medicine Clinical of North América* (26): 475–497.
Ray, B. 2004. Fundamental food microbiology. 3rd ed. Boca Raton: CRC Press.
Schmidt, R.H. 2007. Basic elements of equipment cleaning and sanitizing in food processing and handling operations. In Information Sheet FS14 University of Florida Extension: Institute of Food and Agricultural Sciences; 1–12.
Sigua, G. *et al.* 2011. Comparative efficacies of various chemical sanitizers for warewashing operations in restaurants. *Food Control* 22: 13–19.
Virginia Department of Agriculture and Consumer Services [VDACS], 2010.
Wernersson, E.S, *et al.* 2003. Hygiene in warewashers utilizing blasting granules that foodservice establishments use. *Food Protection Trends* 23: 797–807.
Wirtanen, G. & Salo, S. 2003. Dinsinfection in food processing – efficacy testing of disinfectants. *Environmental Science and Biotechnology* 2: 2293–2306.

Occupational Safety and Hygiene II – Arezes et al. (eds)
© 2014 Taylor & Francis Group, London, ISBN 978-1-138-00144-2

Good practices of hygienic processes on NEMI's professionals

C. Rodrigues, C. Santos, H. Simões, A. Ferreira & J.P. Figueiredo
Escola Superior de Tecnologia da Saúde de Coimbra, Coimbra, Portugal

ABSTRACT: The prevention and control of health care associated the infections assumes a preventive behaviour of healthcare professionals. This study to intend to evaluate the adhesion of professionals to Standard Precautions, namely hand Hygiene during emergency assistance. This study was forwarded to teams of professionals of three ambulances of the National Emergency Medical Institute Center Region.

The majority of professionals don't do hygiene their hands before contact with the patient (78.8). The adhesion to hand hygiene practice after the contact with bodily fluids and after the contact with the patients was low. The professionals after contact with patients's proximities they proceeded to hygiene the hands in 97.7 of attendances. It was verified that 66.7 professionals they used accessories. The technicians of the Emergency Ambulance which even though not attending training workshops within the last 2 years, they reveal the best results in comparison to the after contact with the patient and after the risk of exposure of bodily fluids in comparison with nurses.

Considering these results there should be a greater investment on training, heading towards the increase of knowledge and motivation of professionals for the need to adopt Standard Precautions, having as a goal to establish the control of infections.

1 INTRODUCTION

The Infection Associated to Health Care (IAHC) is an infection acquired by the patient as a consequence from the health care which could, as well, affect health care professionals during the course of their activity (Cristino, Correia, Carvoeiro, Costa, Silva & Silva, 2007).

The epidemiological surveillance is an essential tool in the performance evaluation of the health institutions regarding the prevention of IAHC's and patient's safety. According to the results of the Enquiry of the Infection Prevalence, within the National Prevention Program and Health Care Associated Infection Control, the prevalence of IAHC, in 2010, was of 11.7% ((Pina, Silva & Ferreira, 2010).

In this sense, the General Health Board defines, in its normative circular regarding the practice of Hand Hygiene (HH), five moments essential to hand hygiene: *before contact with the patient, before procedures, clean/aseptic; after risk of exposure to bodily fluids; after contact with patient; after contact with the environment involving the patient* (Direcção-Geral da Saúde, 2010).

In order to prevent the transmission of infectious agents, and to a better control of microorganism transmission, the technical training and technical basis of emergency and pre-hospital professionals when it comes to acknowledging the importance of all bodily fluids, secretions and excretions and the adoptions of standard precautions for all patients is essential, forcing all health care professionals to take responsibility (Viana, 2010).

In Portugal, the National Campaign for Hand Hygiene in 2011 has revealed a global adhesion rate on behalf of health care professionals to Hand Health Hygienic process of 66%, having had a record of 2% in comparison to 2010 (Noriega, Costa & Gaspar, 2012).

Studies reveal that the awareness on behalf of healthcare professionals, concerning basic infectious agent transmittal mechanisms and the need to adopt the SP presents low adhesion.

Due to the high number of risk procedures and to the constant presence of bodily fluids in its interior, the ambulances are classified as critical locations (European Union, 2011).

This study had for a main purpose to evaluate the adhesion of good practices when it comes to the adoption of SP namely the HH of 23 health care professionals, throughout 132 attendances in 3 ambulances of emergency of the National Emergency Medical Institute (NEMI) for the period of seven days.

2 MATERIAL AND METHODS

The study developed had as a sample 3 ambulances, 23 professionals (4 nurses and 19 emergency ambulance crew members) and 132 NEMI's emergency assistance of the center area. The sample was attained in a non probable manner as for its type and, for convenience for its technique.

The data collection was done for a period of seven days throughout the month of May 2013, through the administration of a questionnaire to professionals who were notified by an authorization issued by the precedence of NEMI.

A Check-list was applied, concerning the structural conditions and good practices existing in the ambulances, adapted through guidelines defined by the World Health Organization (WHO).

The questionnaire, adapted from the normative circular concerning the good practices guidelines for the healthcare units hands hygiene according the WHO's guidelines, was done to evaluate the good practices applied in each moment of the emergency assistance. The moments of the hygienic process in study were: *before contact with the patient*; this moment reported to the hygiene made through the patient transport, before healthcare by healthcare professionals; *before processing aseptic procedures*, when there's need to proceed to these; *after the risk to exposure to bodily fluids*; after the contact to bodily fluids; *after the contact to the patient*; this moment represented the end of the interventions on the patient by ambulance professionals; and *after the contact with patient proximities*; in this moment the professionals have finalized the healthcare giving, not finding themselves in contact with the patient.

The answer possibilities presented on the check-list applied were "*yes*" and "*no*". The answers to the questionnaires were "bacterial solution", "not realized" and "soap and water", and the remaining evaluation questions of good general practices were "*yes*" and "*no*".

The data analysis was made by the statistical software IBM SPSS version 2.1. When we checked the investigation hypotheses verification we used the statistic chi-square of independence, the Fisher Exact Test and the Adherence Chi-Squared Test. The interpretation of statistical tests was realised based on a $\alpha = 0.05$ significance with a confidence interval of 95%. For a α statistically significant ($\alpha \leq 0.05$), associations were observed within groups, however, for $\alpha = 0.05$ the associations observed were not considered statistically significant, for a Type I error.

3 RESULTS

As for the check-list applied to the three ambulances in study it was possible to verify that all of them had material suited for the proper hand hygiene. The bacterial solution for the HH and the procedural gloves were available within the unit. The pocket vial, containing the bacterial solution, for individual transport was not available in the ambulances. The three ambulances had a person responsible for restocking the bacterial solution vials and the gloves. The adhesion to the SP, concerning the appropriated use of the uniform, was not totally practiced in the ambulances, once the units had only long sleeves uniforms.

In two ambulances, the professionals did not attend specific training on HH, within the last two years, being that these two units gather a total of 19 professionals.

Out of 132 assistances made by professionals in study, it was verified in a significant manner (p-value = 0,000) in which 78,8% of assistances, professionals did not fulfill an accurate practice of HH before contacting the patient.

In need of realizing aseptic procedures (48,5%), professionals have not gone through with HH in 78,1% of assistances, while in 21,9% it was proceeded with the use of a bacterial solution. Having in account that for a total of 132 assistances (100%) in 51.5% did not apply to this condition, once did not aseptic procedures was taken (Table 1).

The presented data in the table 1 for HH after bodily fluid exposure, having occurred this condition in 90.0% of assistances, has presented a non adhesion in 42.5% of assistances; however these values are not statistically significant (p-value = 0.100). As for HH after contact with the patient it has revealed a higher adhesion (53.0%) on behalf of professionals.

It possible verify *After contact with the patient proximities* the professionals have proceeded with HH in 97.7% of assistances, being that in 41.6% of assistances the technique used was antiseptic Friction, in 34.8% it was the wash with soap and water and in 21.2% both techniques were used, initially wash with soap and water and after antiseptic friction with bacterial solution (Table 1).

The adhesion found in the table 1 to SP, concerning the use of ornaments, it was verified that in just 33.3% of 100% assistances. As for the use of short hair or held up during assistance to the patient, most professionals has joined this SP (93.2%).

From the 23 healthcare professionals 4 was nurses and 19 Emergency Ambulance Technicians (EAT). Of these 13 belong to the male gender and 10 to the female gender.

It is possible to observe that in 100% of attendances (n = 132) the adhesion to HH with bacterial solution *before contact with the patient* by the male gender was of 39.3% (n = 11) and of 60.7% (n = 17) by the professionals of the female gender.

There were significant differences registered between the male and female genders (p-value

Table 1. Adhesion of Hand Hygienic process on different moments.

	Hygienec process	Observed frequency	Expected frequency	Residuals	X^2; gl; p-value
Before patient contact	Bacterial solution	28	66	−38	43,758 1
	Not realized	104	66	38	0
	Total	132			
Before doing aseptic procedures	Bacterial solution	14	44	−30	20,25 1
	Not realized	50	44	6	0
	Total	132			
After risk of exposure to bodily fluids	Bacterial solution	69	44	25	2,7 1
	Not realized	51	44	7	0,1
	Total	132			
After contact with the patients	Bacterial solution	70	66	4	0,485 1
	Not realized	62	66	−4	0,486
	Total	132			
After contact with the patient	Bacterial solution	55	33	22	47,818
	Water and Soap	46	33	−30	3
	Water and Soap/ Bacterial solution	28	33	13	0
	Not realized	3	33	−5	
	Total	132			

≤ 0,05). The female gender presented a higher adhesion to HH in different opportunities in comparison to the male gender. (Table 2)

In the adoption of SP the male gender demonstrated an adhesion significantly superior to the female gender. In 88 (100%) of attendances in which we verified the use of ornaments, 69.3% belongs to the female gender and 30.7% to the male gender. Analyzing the SP, regarding the appropriate use of their hair, the female gender presented a higher adhesion (60.2%) in proportion to the male gender (39.8%) (Table 2).

According to table 3, HHP regarding to professional category has manifested a significant association *after the risk of exposure to bodily fluids* and *after the contact to the patient*. The EAT stood out showing a 100% HH at these moments in comparison to nurses (0%) in the mentioned moments. *Before the contact with the patient and after the contact with the patient's proximities* it was not verified a statistically very significant association (p-value >0, 05).

The adhesion by both professional categories to PP showed no significant differences, verifying the use of ornaments during assistances from nurses (11.4%) as the TAE (88.6%) (Table 3).

In the results of Table 4, were verified that were statistically significant as to the initial behaviour adopted by the professionals that is, before contact with the patient and the behaviour adopted after, regarding the need to proceed to the HHP once again.

It was established that in 132 attendances (100%), in 47.1% of them HH was not complied to *before patient contact* and in 70% it was not done at all *after contact with the patient.*

In 64 of attendances (100%) HH was not proceeded to *before patient contact* in 11.5% of them, or *before doing aseptic procedures* in a 42.9%.

Of the 120 attendances (100%) the hygienic process was not made at all in 53.1% of attendances *before contact with the patient* and in 75.4% *after the risk of exposure to bodily fluids.*

In this sense, it is possible to observe a prevalence of the initially adopted behaviour in face of the one adopted after. (Table 4).

Table 2. Relationship between gender amongst professionals and the practice of Hand Hygienic process.

Moment of Hygiene		Gender			
		Male	Female	Total	X²; gl; p-value
Before contact with patient	n	11	17	28	0,312
	% Line	39,30%	60,70%	100%	1
	% Column	19,00%	23,00%	21,20%	0,367
	% Total	8,30%	12,90%	21,20%	
Before doing Aseptic procedures	n	3	11	14	8,829
	% Line	21,40%	78,60%	100%	1
	% Column	8,30%	39,30%	21,90%	0,005
	% Total	4,70%	17,20%	21,90%	
After risk with exposure to bodily fluids	n	21	48	69	8,427
	% Line	30,40%	69,60%	100%	1
	% Column	42,00%	68,60%	57,50%	0,003
	% Total	17,50%	40,00%	57,50%	
After contact with patient	n	21	49	70	11,756
	% Line	30,00%	70,00%	100%	1
	% Column	36,20%	66,20%	53,00%	0,001
	% Total	15,90%	37,10%	53,00%	
After contact with patient and proximities	n	56	73	129	0,644
	% Line	43,40%	56,60%	100%	1
	% Column	96,60%	98,60%	97,70%	0,409
	% Total	42,40%	55,30%	97,70%	
Are ornaments used during patient assistance	n	27	61	88	18,837
	% Line	30,70%	69,30%	100%	1
	% Column	46,60%	82,40%	66,70%	0
	% Total	20,50%	46,20%	66,70%	
In patient assistance is hair, when not short, held up	n	49	74	123	12,323
	% Line	39,80%	60,20%	100%	1
	% Column	84,50%	100%	93,20%	0
	% Total	37,10%	56,10%	93,20%	

Table 3. Relationship between professional category and Hand Hygiene practice.

Moment of hygienic process		Professional category			
		Nurses	Emergency ambulance technician	Total	X²; gl; p-value
Before contact with the patient	n	4	24	28	0,021
	% Line	14,30%	85,70%	100%	1
	% Column	20,00%	21,40%	21,10%	
	% Total	3,00%	18,20%	21,20%	0,576

(Continued)

Table 3. (Continued).

Moment of hygienic process		Professional category: Nurses	Emergency ambulance technician	Total	X²; gl; p-value
After risk of exposure to bodily fluids	n	0	69	69	32,471
	% Line	0,00%	100%	100%	1
	% Column	0,00%	69,00%	57,50%	0
	% Total	0,00%	57,50%	57,50%	
After contact with the patient	n	0	70	70	26,613
	% Line	0,00%	100%	100%	1
	% Column	0,00%	62,50%	53,00%	0
	% Total	0,00%	53,00%	53,00%	
After contact with the patient and its proximities	n	20	109	129	0,548
	% Line	15,50%	84,50%	100%	1
	% Column	100%	97,30%	97,70%	0,608
	% Total	15,20%	82,60%	97,70%	
Are ornaments used during patient attendances	n	10	78	88	2,946
	% Line	11,40%	88,60%	100%	1
	% Column	50,00%	69,60%	66,70%	0,074
	% Total	7,60%	59,10%	66,70%	
In patient assistance the hair, when not short, is held up	n	20	103	123	1,725
	% Line	16,30%	83,70%	100%	1
	% Column	100%	92,00%	93,20%	0,217
	% Total	15,20%	78,00%	93,20%	

Table 4. Healthcare professional's behavioural analyses: Relationship between the adopted behaviour before contact with the patient and the previous moments of hygienic process.

Moment of hygienic process		Before contact with the patient: Bacterial solution	Not realized	Total	X²; gl; p-value
Before contact with the patient	n	21	49	70	6,887
	% Line	30,00%	70,00%	100%	1
	% Column	75,00%	47,10%	53,00%	0,007
	% Total	15,90%	37,10%	53,00%	
Before doing aseptic procedures	n	8	6	14	17,339
	% Line	57,10%	42,90%	100%	1
	% Column	66,70%	11,50%	21,90%	0
	% Total	12,50%	9,40%	21,90%	
After risk of exposure to bodily fluids	n	17	52	69	4,31
	% Line	24,60%	75,40%	100%	1
	% Column	77,30%	53,10%	57,50%	0,031
	% Total	14,20%	43,30%	57,50%	

4 DISCUSSION

This study has revealed satisfactory results as to the available material inside the ambulances.

However the units did not have at their disposal the portable vials with bacterial solution not allowing, in case of need, the HH during patient assistance on the outside of the ambulance. This condition may have influenced the adhesion variation by the professionals to hand hygiene, once a higher accessibility to the bacterial solution might have contributed to higher values of adhesion (Sjöberg & Eriksson, 2010).

Having in account the HH moments in analyses, *after contact with the proximities of the patient* was the moment of hygienic process in which professionals demonstrated most adhesion (97.7% of attendances). This result coincides with the one in a similar study, executed in Valmland in Sweden, in which HH *after contact with patient proximities* presents the highest adhesion percentage (72%). Such as it was established in that study, HH *before contact with the patient* exhibits the lowest adhesion (34%) by professionals in comparison to this study in which 21.2% of attendances HH was not proceeded to (Emanuelssona, 2012).

The values obtained regarding the *use of ornaments* are similar in both studies, being that in this study in 66.7% of attendances it was observed the use of ornaments in comparison to 74% of attendances in the comparison study (Emanuelssona, 2012). The *use of short hair* is a practice that revealed constant in both studies, presenting an adhesion in more than 90% of attendances.

In this sense the low adhesion to HH stood out in the moment *before doing aseptic procedures* (21.9%) and *after the risk of exposure to bodily fluids* (42.5%). In this manner the satisfactory adhesion to HHP praises the permanence of attitudes adequate throughout the assistance, requiring motivation from the professional as well as technical training (Direcção-Geral da Saúde, 2010).

The National Campaign for Hand Hygiene has revealed an increase in the HH adhesion of 2% in 2011, in comparison to 2010, having a global adhesion rate of 66%, however these were not the expected values, previewing a higher adhesion. The campaign also has evaluated the HH in different moments where an adhesion of 56% of professionals was verified before contact with the patient, 83% *after exposure to blood and bodily fluids, 77% after contact with the patient, 70% before clean/aseptic procedures and 64% after contact with the patient's surrounding environment*. (Noriega, Costa & Gaspar, 2012).

In the present study, the moment *after contact with the patient*, HH has revealed a higher adhesion (53%) in relation to the non adhesion by professionals.

Attending to the relationship between female and male genders facing HH in the different moments, the female gender has manifested a preventive attitude which was higher than the one of the male gender. These results may reflect the cultural behaviors. Since over time, the female gender has shown more care with their appearance and they are therefore more aware about hygiene habits, such as HH. It is crucial sensitizing the male gender to adopt these behaviors.

After the analyses of the relationship between the practice of HH and the professional category, it was established that, in fact, this relationship has prevailed in a significant manner in two moments of the hygienic process: *after risk of exposure to bodily fluids and after contact with the patient.* It was the EAT those who proceeded to the hand hygiene in 100% of attendances in comparison to the null adhesion by the nurses (0%). The EAT presented a higher adhesion to HH, however these professionals, in the last two years, did not attend specific HH training in proportion to nurses who have. As to SP adhesion, concerning the *use of ornaments* during the process of health care giving, significant differences were not observed, having both categories stood out for the non compliance of the referring SP.

Analysing the initial behaviour adopted by professionals that is, *before contact with the patient*, and the behaviour adopted after, in face with the need to do HP again in a moment or another, it was obtained a statistically satisfactory relationship. These results have shown that professionals who have not initially adhered to HHP have also not done it afterwards (Trampuz & Wildmer, 2004).

The Infection prevalence questionnaire of 2010, on a national level, presented a prevalence of HCAI of 9.8% (Pina, Silva & Ferreira, 2010). The Infection prevalence questionnaire acquired in Portuguese Hospitals in 2012 has revealed a rate of infection of 10.6% values superior to the European average. (Pina, Paiva, Nogueira & Silva, 2012).

5 CONCLUSIONS

The HCAI's constitute a problem with consequences in the security and health of professionals and patients all over the World. This is a relevant data if we consider the goal of continuously improving the care giving, as well as all costs associated with HIAC's treatment (Direcção-Geral da Saúde, 2010).

The environment inside the ambulances is characterized by the existence of high risk areas of microorganism transmission, direct transmission and/or cross-transmission, by the contact with the environment of emergency, and the hands of the health care professionals are its main vector (Noriega, Costa & Gaspar, 2012). HH is considered the most important isolated action in the control of infections in health care services.

The results shown in this study, in comparison with ones obtained in previous studies, demonstrated that the sensitization of the healthcare professionals relatively the SP adhesion, as well as HH, is low.

Understanding the importance of HP and the complexity that its incorrect procedure, or non procedure, brings to the institutions, professionals and patients, healthcare professionals need to be aware on the need hygienic process the hands, to establish the control of infections direct and/or crossed in health care giving environments.

REFERENCES

Cristino, J.A., Correia, M., Carvoeiro, M., Costa, C., Silva, E., & Silva, M. (2007). Programa Nacional de Prevenção e Controlo, Lisboa.

Lena Emanuelssona, L.K. (2012). Ambulance personnel adherence to hygiene routines: still protecting ourselves but not the patient. p. 5.

Noriega, E., Costa, A.C., & Gaspar, M.J. (2011–2012). Camapanha Nacional de Higiene das Mãos.

Occupational health and safety risks in the healthcare sector: Guide to prevention and good practice. (2011). European Union.

Pina, E., Paiva, J.A., Nogueira, P., & Silva, M.G. (2012). Prevalência de Infeção Adquirida no Hospital e do uso de Antimicrobianos nos Hospitais Portugueses.

Pina, E., Silva, G., & Ferreira, E. (2010). Inquérito de Prevalência, Portugal.

Portugal. Direcção-Geral da Saúde, Circular Normativa N°13/DQS/DSD, de 14 de Agosto de 2010. Orientação de Boa Prática para a Higiene das Mãos nas Unidades de Saúde.

Sjöberg, M., & Eriksson, M. (2010). Hand Disinfectant Practice: The Impact of an Education Intervention. The Open Nursing Journal.

Trampuz, A., & Widmer, A. (2004). Hand hygiene: a frequently missed lifesaving opportunity during patient care.

Viana, R.D. (2010). Manual de Segurança e Boas Práticas para o Profissional TAE do INEM.

Occupational Safety and Hygiene II – Arezes et al. (eds)
© 2014 Taylor & Francis Group, London, ISBN 978-1-138-00144-2

Conceptual model of competency profile for work equipment inspections

Joana Branco & Miguel Tato Diogo
Faculty of Engineering of the University of Porto, Porto, Portugal

ABSTRACT: The Work Equipment Directive provides work equipment inspections, by a competent person. However, the specific requirements for this competency are, still, without definition. In this sense, this article aims to build a competency's conceptual model for these cases, based in criteria established by law or by official authorities for inspection professionals. 32 already published profiles were analyzed, cataloged according to knowledge, skills and experience for its exercise. Through this inventory, it was possible to propose a three-dimensional space (limited by the resultant values of each mentioned variable) where it can be considered to exist the necessary competency for the person who performs work equipment inspections.

1 INTRODUCTION

1.1 *Work equipment inspections*

The work equipment inspections are expected in the Directive 89/655/CEE (with the amending given by Directives 95/63/CE e 2001/43/CE), of Council, 30th November, concerning the minimum safety and health requirements for the use of work equipment by workers at work, transposed to the Portuguese law by the Decree-Law 50/2005, through its 6th article. In this article can be read that if the work equipment safety depends on its installation conditions, the employee should guarantee periodic inspections and, if necessary, periodic tests by a competent person. In the definitions' section of the same document is referred that the competent person is the one that has theoretical and practical knowledge and experience in the work equipment's type. The designation of work equipment takes broad outlines, which can be from a simple portable tool to a lifting machine carrying large loads, for example. According to Health and Safety Executive (HSE), work equipment includes machines (circular saws, drilling machines, photocopiers, cutting machines, tractors, dump trucks and presses), hand tools (screwdrivers, knives, saws and cleavers), lifting equipment (forklifts, work platforms, cranes and lifting vehicles) and other equipment (ladders and pressure cleaning systems).

Work Safety & Prevention Services (WSPS) refers that inspections should be performed by a competent person in regular intervals to assure equipment is at safety conditions for use.

In Europe, according to Eurostat, more recent records of accidents (mortal and with 3 or more days of absence) with work equipment remounts to 2005. This records fall into manufacturing where 394 fatal accidents and 197424 accidents with 3 days of absence were recorded, representing more than 10% of all accidents at work. One factor that HSE points to be the cause for that is the lack of regular maintenance and inspections in work equipment.

To WSPS, for the inspections to be performed in the right way, the person responsible for them should have the necessary training. It's crucial that she knows the work equipment installation's layout, the hazards and processes associated to it, the safety controls and access the history of accidents.

The inspection can be useful to diagnose causes of accidents, as well as to detect the improper functioning origin. The equipment operator's incorrect behavior is critic in the sequence of events that leads to the accident. This kind of behavior can be minimized through the increase of supervision and training actions. (Dzwiarek, 2004) In other hand, perform inspections positively influences the reduction with accidents costs. (Knapp, Bijwaard, & Heij, 2011).

1.2 *"Competency" concepts*

The theme "competency" has been the target of a raising number of studies, especially during the first decade of this century. The tendency is clearly to raise the interest about the subject. (Silva & Amorim, 2012) A review to several individual competency's concepts conclude that the most authors refers the words "abilities", "attitudes", "knowledge", "skills", "experience" and other qualities, as

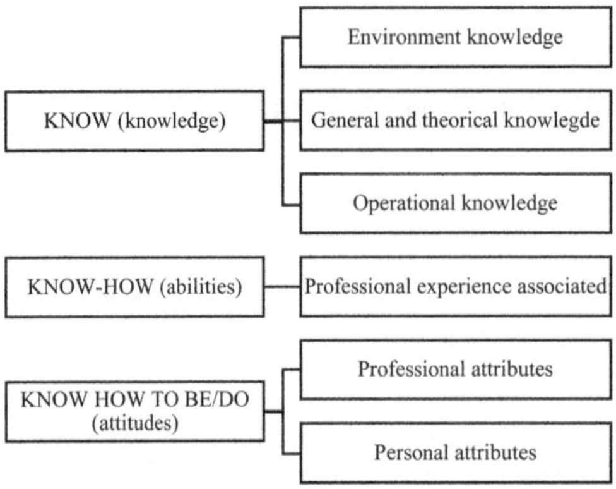

Figure 1. Competency's resources and possible unfolding.

"values". (Soares & Andrade, 2010) (Assumpção, 2011) (Fernandez, et al., 2012) (Hughes & Ferrett, 2011) Another similar analysis defends the structured composition shown in Figure 1, as an outcome from the conjunction of the underlying variables to the word "competency" defended by several authors. (Assumpção, 2011).

Regarding to the competency related to safety and hygiene at work tasks, the Institution of Occupational Safety and Health (IOSH) refers that competency is associated to the capability of someone doing something effective and efficiently, being composed by three essential pillars: knowledge, skills and experience, which should always be maintained in the appropriate portion. Regardless the way competency is establish or developed, all levels of employees need to keep their competency up to date.

1.3 *Published inspector profiles*

The Portuguese Acreditation Insititute (IPAC, I.P.) provides a directory of accredited inspection bodies (confer www.ipac.pt/pesquisa/acredita.asp), being organized by sectorial and vehicle inspection. Regarding to pressure equipment, the American Petroleum Institute (API) has certification programs for pressure vessel inspector—API 510, piping inspector—API 570 and aboveground storage tank inspector—API 653.

For gas installations, the 6th article of Decree-Law 263/89 predicts the existence of a gas technician, whom should assure, with rigor, the project enforcement, follow up and control its material execution, as well as checking the used materials, accordingly with regulation standards.

The Ordinance 1211/2003 presents the fuel oil products installation inspector's profile.

In case of electrical installations, the Regulatory Decree 31/83 (with the reading given by the Decree-Law 229/2006) approves the statute for its responsible technician, whom should perform inspections with a frequency specified by installation's features.

The Decree-Law 123/2009 (rectified by Rectification Declaration 43/2009) establishes the applicable rules regarding the construction of infrastructures suitable for the accommodation of electronic communications networks. This document refers that the accomplishment of evaluation procedures are an installers' obligation, and the requirements for this career are also introduced on it. The National Authority for Communications (ICP-ANACOM) presents the evaluation procedures for telecommunication infrastructures, which tasks include the inspection of its elements, the record of those inspections and testing.

The inspector's profile of lifts, hoists, escalators and moving walkways is founded in Decree-Law 320/2002. This kind of inspectors should perform maintenances and inspections.

Regarding to the transport of dangerous goods, the Decree-Law 41-A/2010 (transposes the Directives 2006/90/CE and 2008/68/CE) presents the safety adviser profile, whose duties pass to guarantee all the activities in optimal conditions of safety.

The inspection of vehicles is regulated by Decree Law 258/2003, which one establishes the conditions for the inspector's license. This professional activity aims to inspect motor vehicles and its trailers.

Although not belonging to the list of accredited bodies presented by IPAC, I.P, there is a profile of inspection for security conditions against buildings' fires, whose requirements are founded in Ordinance 64/2009.

2 MATERIALS AND METHODS

Using the requirements established in the documents listed above, a database was created, which allowed having the perception of extension and patterns of those requirements for the different inspection profiles found. For this sense, an inventory of these profiles was prepared, which are identified in Table 1, having one or more valid ways to access them.

The mentioned inventory was made following the adopted concept of "competency", which includes three different variables: knowledge, skills and experience. The variable "knowledge" was standardized through the equivalences between the different scholar's levels and the levels of National Qualification Framework. For "skills", the analysis was more complex, since its nature is qualitative. Initially, the existent keywords in each profile were identified and counted. Then, by combining the synonymous keywords, families of keywords were created in order

Table 1. Collected profiles and number of its access ways.

Profile's name	Number of ways of access
Pressure vessel inspector	4
Piping inspector	4
Aboveground storage tank inspector	4
Gas technician	1
Fuel oil products installation inspector	1
Technical responsible for the operation of electrical installations	3
Installer of accommodation of electronic communications net-works	3
Inspector of lifts, hoists, escalators and moving walkways	2
Safety Advisor	2
Technical inspector of motor vehicles and their trailers	4
Inspector of security conditions against buildings' fires	4

to make the analysis as objective as possible. The variable "experience", as it has always been found in whole numbers of years, was analyzed directly.

3 RESULTS

It was found that 41% (13 out of 32) of the analyzed profiles required qualification level 6 (corresponding to degree or BSc), as can be seen from the graphic of Figure 2. Through extremes and quartiles diagram it was found that the 1st quartile was 3, 4 the median and 6 the 3rd quartile. This means that 50% of qualification levels required by the profiles were comprised between 3 and 6. It was found that there is a bias to the right.

127 occurrences of keywords were identified. The average number of keywords per profile, rounded off by excess units, is 4.

According to the graphic in Figure 3, more than 29% required skills in published profiles were related with task of inspection and/or supervision. On the other hand, 6 analyzed profiles did not require any skill.

The 5 families with the 4 highest incidence rate were "inspection, inspect, supervision, supervise", "build, installer, operator, repair", "training", "design" and "professional, techniques", having the last 2 the same percentage.

Regarding the variable "experience", values between 0 and 5 years were collected, with an average of 2 years. The graphic in Figure 4 ascertains that 11 (34%) of the profiles found didn't require any experience for the performance of the function in question.

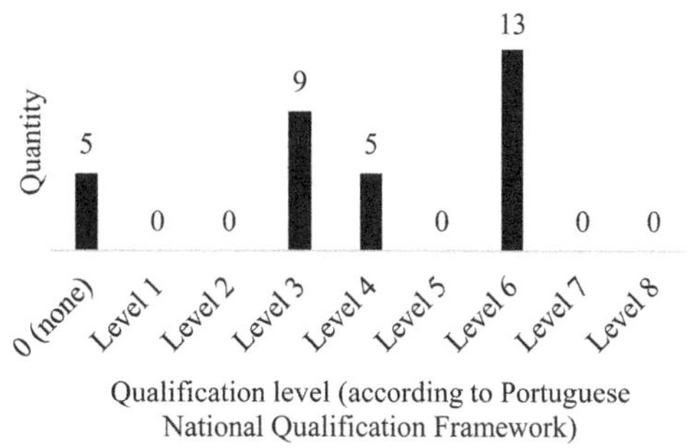

Figure 2. Qualification levels of the collected profiles.

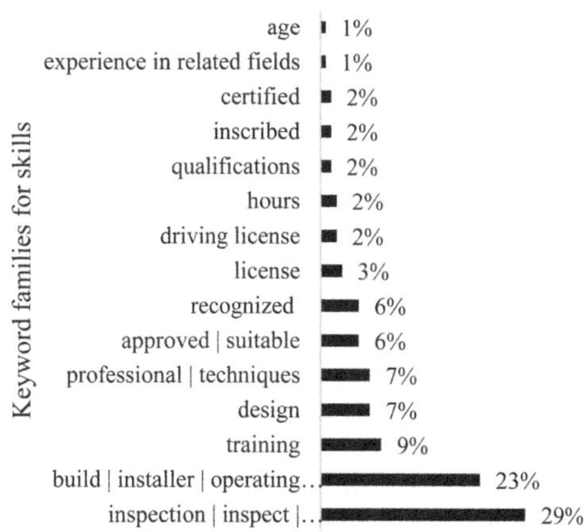

Figure 3. Distribution of skills collected profiles.

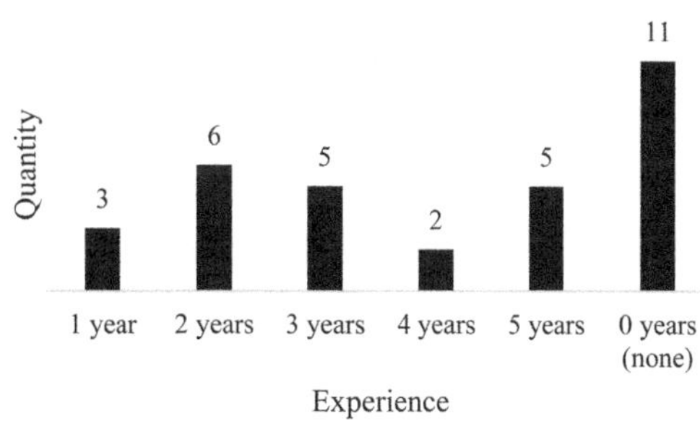

Figure 4. Experience required in collected profiles.

An extremes and quartiles diagram for experience was also built; it revealed that 50% of the data was concentrated between 0 (none) and 3 years of experience. There was a bias to the left.

4 DISCUSSION

With reference to the results of the statistical analysis of 32 collected competency profiles, boundaries were defined for each of the three variables adopted for the three-dimensional definition of "competence": "knowledge", "skills" and "experience". With regard to the variable "knowledge", harmonized according to the Qualification Levels cited above, it was found that it aggregated between levels 3 and 6, thus having a minimum of 3 and maximum of 6. The variable "skill", since it is a qualitative variable, required a more delicate treatment. Since the average of keywords per profile was 4, the 5 most rated families of keywords were chosen. Finally, it was found that the variable "experience" concentrated between 0 and 3 years, being these the lower and upper bounds, respectively, used for defining the axis.

However, it is necessary to note that the definition of competence does not include the mandatory framework in this range; it is just a conceptual analysis according to the profiles collected. The limits set for each variable were placed on the axes and thus proceeded to build a three-dimensional space, an area where it was acknowledged that the competency profile exists. This construction is shown in Figure 5.

Regarding the variable "experience", there are some considerations to make the upper and lower limits established for the delimitation of the area of competence.

Statistical analysis of the competency profiles collected concluded that the first and third quartiles of the variable "experience" were 0 (zero) and 3, respectively, thus considering these values to be limit of the scope of competence. However, taking 0 (zero) as the lower limit of the "experience" nullifies the meaning and purpose of the existence of a third axis, so there is no reason to be constructed

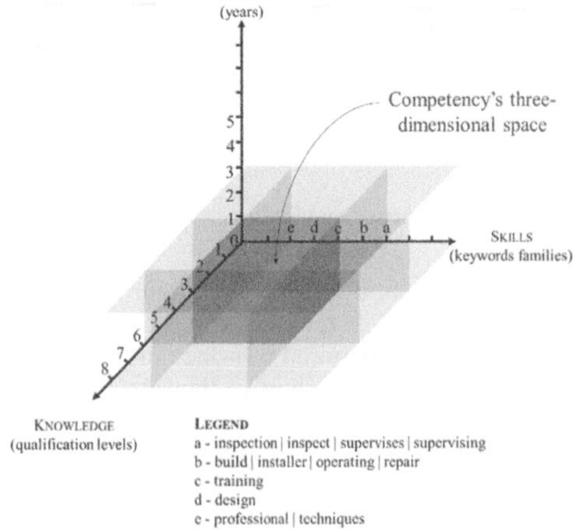

Figure 5. Conceptual model of competency profile.

a three dimensional model, eventually not meeting the definition of "responsibility" advanced by Institution of Occupational Safety and Health. For the upper limit, although in the representation space of competence the value chosen is 3, this is a merely illustrative sample obtained from the inspector profiles collected. It is easily understood that the fact of having a minimum experience greater than 3 cannot be a ground for exclusion of a potential competent person.

5 CONCLUSIONS

According to several authors, competent person is the one that brings together knowledge, skill and experience in the field—in this case, in verification and testing of work equipment. However, it also has to have personal characteristics and attitudes that are more complex to measure. In this sense, the analysis of competency profiles for professional inspectors, which characteristics of competence were already published, allowed to create one generic model profile: the Conceptual Model for the definition of "competent person" for verification of work equipment, as required in Decree-Law 50/2005.

The Conceptual Model of competency profile is an open prototype; if more profiles are found published in official sources, their requirements could be treated using the proposed methodology. The introduction of new profiles needs the statistical analysis to be reformulated, which will allow to readjust the boundaries established for this model. More profiles obtained mean a more robust Conceptual Model of competence for verification of work equipment.

ACKNOWLEDGMENT

The authors would like to thank Master in Occupational Safety and Hygiene Engineering (MESHO), of the Faculty of Engineering of the University of Porto (FEUP), all the support in the development and international dissemination of this work.

REFERENCES

Assumpção, L.C. 2011. Uma visão sobre formação das competências individuais, profissionais e or-ganizacionais. RICI: R.Ibero-amer: 1–21.

Dzwiarek, M. 2004. An Analysis of Accidents Caused by Improper Functioning of Machine Control Systems. *International Journal of Occupational Safety and Ergonomics*: 129–136.

Fernandez, N., Dory, V., Ste-Marie, L.-G., Chaput, M., Charlin, B., & Boucher, A. 2012. Varying conceptions of competence: an analysis of how health sciences

educators define competence. *Medical Education*: 357–365.
Hoffmann, T. 1999. The meanings of competency. *Journal of European Industrial*: 275–285.
Hughes, P., & Ferrett, E. (5th Edition) 2011. *Introduction to Health and Safety at Work*. Oxford, UK: Elsevier Ltd.
Knapp, S., Bijwaard, G., & Heij, C. 2011. Estimated incident cost savings in shipping due to inspections. *Accident Analysis and Prevention*: 1532–1539.
Pieper, R. 2012. How to evaluate the risks of work equipment and installations for health and safety? Research and activities of the German Committee for Plant Safety and consequences for regulation. *Work* 41: 3321–3324.
Silva, L.D., & Amorim, T.N. 2012. Estudos sobre competências: Uma análise dos artigos publicados nos EnANPADs.
Soares, A.V., & Andrade, G.A. 2010. Gestão por Competências-Uma Questão de Sobrevivência em um Ambiente Empresarial Incerto: 484–491.
Workplace Safety & Prevention Services (WSPS) 2011. How to Conduct Workplace Inspections.

Occupational Safety and Hygiene II – Arezes et al. (eds)
© 2014 Taylor & Francis Group, London, ISBN 978-1-138-00144-2

Assessment levels of dustiness—case study applied to technicians of gel nails

T. Margalho, A. Ferreira, J.P. Figueiredo & F. Silva
Instituto Politécnico de Coimbra, ESTeSC, Coimbra Health School, Saúde Ambiental, Portugal

ABSTRACT: Due to the complexity of the work environment in establishments of gel nails, that covers the use of the gel nail technician, the approach for evaluating the exposition tend to mischaracterize potential dangers as disturbances. The main objective of this study is to evaluate the overall dustiness from the filing gel, verifying the effectiveness of ventilation systems and identify what constitutes good practice for the future. Of the 30 workers, only 12 agreed to participate, where these were selected to a second type of non-probability sampling, where the technician was a matter of choice. An analytical measurement of the parameters mentioned above and applied checklist for verification of working conditions and worker was realized. It is concluded that the Technicians of gel nails are not sensitized on this subject, having inadequate ventilation control and minimizing dusting, there is still plenty of cases that do not use this protective equipment.

1 INTRODUCTION

Artificial fingernails, also known as sculptured or acrylic nails, have become increasingly popular since the 1970's, endangering nail technicians exposed to dozens of potentially dangerous chemicals, including acrylic acids, solvents and biocides in powder or vapors (Connecticut Department of Public Health, 1997). The exposure to chemicals is one of the factors that contributed to put in risk the health of workers in the performance in their functions. Thus, it is important to quantify the concentrations of the workplace whether from the point of view of the exposure assessment and health diagnosis (Oliveira, 2011).

The exposure to these chemicals at work, as well as the total dustiness produced by hand filing or electric drills, were examined by a small but growing number of studies that have found possible links between the work of the nail technicians and adverse results for health, including respiratory problems, such as occupational asthma, neurological and musculoskeletal and other health problems, including cancer. Since polymer powders containing silica may also be used, the dusts generated when filing the gel nail, presents an additional health risk (Connecticut Department of Public Health, 1997) (Habia, 2007). A survey of occupational hygienists by the U.S. National Institute of Occupational Safety and Health (NIOSH) (Health and Safety Executive, 2008) reported that nail salons were considered to be in great need of an evaluation of perceived safety and health risks. Currently, the assessment of occupational exposure of workers to chemicals in the workplace is one of the concerns of the technical of hygiene and safety, whether by legal obligation either the risk that the chemicals may present to health workers. This assessment consists in determining the concentration of these agents in the workplace air through methodologies and equipment indicated in rules and their subsequent comparison with reference values, which represent acceptable levels of exposure. It is recommended that reference values to use if the exposure limits (VLE) defined in Portuguese Standard 1796 of 2007 (NP1976, 2007) (Prista, et al). The first artificial nails were made by applying methyl methacrylate (MMA) dental acrylate to the nails. In 1974, the U.S. Food and Drug Administration outlawed a Yesilar chemical, methyl methacrylate (MMA), used in fingernail products (Healthy Air, 2007). It should also be noted, that occupational diseases of workers in nail salons, are generally not tracked or monitored. Doctors may not recognize that certain symptoms or diseases are related to occupational hazards such as exposure to harmful chemicals (Massachusetts Department of Public Health, 1997). The aim of this research resided in the assessment of total dust from the filing gel, determining if the technicians are exposed to high concentrations of particles, if they exceed the limits set forth, the efficacy of ventilation systems and the use of means of protection (use of mask).

2 MATERIAL AND METHODS

The study lasted for one year between October 2012 until June 2013, having been held a data collection in January of 2013. This study was developed from an observational character, level II, from a descriptive-correlational type and cross-sectional nature. This research work had as work related places with manicure in the council of Coimbra. The target population consisted in the technicians of gel nails, that it was composed of 30 workers, that only 12 agreed to participate, where these were selected to a second type of non-probability sampling, where the technician was a matter of choice. This study was divided in 4 phases, recruitment, checklist administration, the measure of dustiness level in the exposed Technicians and lastly the analysis of the statistical data collected, using IBM SPSS version 21. Checklist: Related to the working conditions of the work and worker. This checklist was focused in the presence (or absence) of ventilation (forced or natural), as well as the use of a vacuum cleaner as a way to counteract the particles from the filling gel and the type of used filling and protective equipment, as masks.

Measurement of dustiness: This essay addresses the measurement of dust in nail salons (total dust, breathable dust and inhalable), and compare the results obtained with the Exposure Limit Values (ELVs) as defined by law and/or normalization applicable. On the day that was applied the checklist, these measurements were carried out for the purpose was using the portable meter Lighthouse, Model 3016 IAQ, the mode of mass concentration approaching the density in $\mu g/m^3$, providing up to six channels at the same time of particle counting, this device displays the data counted particle in the dimensions of 0.3–10.0 μm, and measuring the temperature and relative humidity being used to collect quantitative data $PM_{2,5}$, PM_{10} and total suspended particle\s. The device was placed in the work area near of each technical worker during the filling process which could vary from 5 to 15 minutes, realized only at a fraction of the time taken to finish this process, which may vary by customer 1–2 hours, where the gel removal is performed. The obtained data was recorded by the equipment used in the checklist for subsequent analysis. The tests used were the t-Student test for one sample, t-Student test for independent samples and Fisher's exact test. The interpretations of statistical tests were based on a significance level of *p-value* = 0,05 for a Confidence Interval (CI) of 95%. As reference values to the determination of the maximum concentration of exposure of the evaluation parameters, values determinated by the Portuguese legislation, norm 1796 of 2007, which imposes the limit of professional exposure to chemical agents, referred in the Attachment B for Particles (insoluble or poorly soluble) without classification [PSOC], which recommends that air concentrations should be kept below 10 mg/m^3 for inhalable particles (Prista, et al.). The reference values used in the verification of the conditions for temperature and humidity were the Decree Law No. 243 of 20 August 1986, concerning the Hygiene and Safety in Commercial Establishments, Offices and Services, defined in section II, the temperature of the workplace should vary between 18°C and 22°C and humidity between 50% and 70% (Dec-Lei n°243, 1986). This investigation has exclusively academic interest, subtracting any financial or economic interest.

3 RESULTS

This study included 12 technicians of gel nails, female gender with age between 20 to 35 years. Regarding the exposure limit value, it was tried to verify whether on average occurred exceedances to reference values stipulated in the current legislation. Consider table 1.

As it can be seen, we verified that the estimated average values of total suspended particulate evaluated in the study sites have revealed to be below the reference level (dx = −5,41 mg/m^3) in a significant way (*p-value* <0,05). Already the estimated average values to the relative humidity expressed values higher than to the reference value in a significant way (*p-value* <0,05). It is noted that the temperature was not presented in Table 1 is indicated as the mean values were estimated in the optimum range according to the legislation in force, i.e. between 18°C and 22°C. In the following tests, it was used the value of particulate material supplied by the apparatus in the unit $\mu g/m^3$. We proposed to evaluate the average distribution of particulate material resulting of filing gel nail by using two distinct methods of work. Consider the following table 2.

In fact, we came to realize that there were no significant differences of $PM_{2,5}$ (*p-value* = 0,274),

Table 1. Comparison of average values of total suspended particles and relative humidity with the values stipulated in the legislation.

	N	$\bar{x}$	S	Reference value
Total suspended particles (mg/m^3)	12	4,59	7,49	10 mg/m^3
Relative humidity (%)	12	82,85	13,12	50–70%

Test: One-sample t-test.

Table 2. Comparing the amount of particulate material in relation to filing methods used.

	Filing methods	N	$\bar{x}$	S
Particulate matter 2,5	Electric drill	9	79,92	82,56
	Hand filing	3	22,85	7,64
Particulate matter 10	Electric drill	9	712,95	556,03
	Hand filing	3	241,92	154,04
Total suspended particles	Electric drill	9	5585,27	8513,29
	Hand filing	3	1621,08	1212,82

Test: Independent-samples t-test.

Table 3. Efficiency of general ventilation in relation to particulate matter.

	General ventilation	N	$\bar{x}$	S
Particulate matter 2,5	Yes	10	74,12	80,02
	No	2	23,31	6,03
Particulate matter 10	Yes	10	660,30	550,71
	No	2	269,62	190,02
Total suspended particles	Yes	10	5216,67	8112,81
	No	2	1481,96	1582,17

Test: Independent-samples t-test.

PM_{10} (*p-value* = 0,190) in total suspension (*p-value* = 0,454) produced between the electric drill method compared to the manual method (hand filing). However, we can say that the method for electric drill was that which produced, on average, a greater amount of particulate material. In order to reduce the amount of particulate material, two ventilation systems are used—General and located. In table 3 we can verify the relation to the effectiveness of general ventilation means.

We have come to realize that in fact there were no significant mean differences of particulate matter $PM_{2,5}$ (*p-value*=0,546), PM_{10} (*p-value*=0,408) in total suspension (*p-value* = 0,360) compared to the method of general ventilation. However, we have noted to those that have general ventilation are exposed to greater amounts of particulate material. In addition to analyzing the effectiveness of the means of general ventilation it is necessary to understand the effect of the means localized ventilation compared to particulate matter, which can be seen in the following table 4.

We can observe that there is a tendency for those who have any kind of localized ventilation, presented a greater efficiency in the absorption PM_{10} and for total suspended particles. The results analyzed with the test *t-Student*, in fact, we came to realize that there were no significant mean differences of $PM_{2,5}$ (*p-value* = 0,199), PM_{10} (*p-value* = 0,156) in total suspension (*p-value* = 0,095). Within the localized ventilation systems we can find different types, were analyzed for this study found two distinct types—Nail Dust Suction and Drill with Vacuum, we can see the results in the Table 5.

In relation to local system ventilation used, in this case study are portable or included with the electric drill, we can see a trend of the, Nail Dust Suction Collector, aspires the best type of $PM_{2,5}$. Already with the use of the Drill with Vacuum the opposite happens, performs a higher suction on larger particles, as well as total suspended particles. These results were analyzed with the *Independent-samples t-test*, there were no significant mean differences, for a significance level of $PM_{2,5}$ (*p-value* = 0,461), PM_{10} (*p-value* = 0,510) total suspension (*p-value* = 0,090).

Table 4. Efficiency of local ventilation systems used to compare to the amount of particulate matter.

	Local ventilation	N	$\bar{x}$	S
Particulate matter 2,5	Yes	7	89,87	91,38
	No	5	31,75	22,49
Particulate matter 10	Yes	7	516,76	271,73
	No	5	704,99	786,39
Total suspended particles	Yes	7	1537,10	907,09
	No	5	8874,19	10678,95

Test: Independent-samples t-test.

Table 5. Efficiency of each local ventilation systems used to compared to the amount of particulate matter.

	Types of local ventilation	N	$\bar{x}$	S
Particulate matter 2,5	Nail dust suction collector	5	71,88	73,20
	Drill with vacuum	2	134,83	151,67
Particulate matter 10	Nail dust suction collector	5	564,85	257,88
	Drill with vacuum	2	396,54	369,51
Total suspended particles	Nail dust suction collector	5	1900,33	788,47
	Drill with vacuum	2	629,03	375,95

Test: Independent-samples t-test.

In addition to the local systems of ventilation, it is necessary to check that filing method is used more efficiently, which can be seen in the following table 6.

Since only one case was found for those with localized ventilation and uses a hand file during filling gel, it is important to focus only on those who do not have any kind of localized ventilation, which reinforces there is a tendency to be

Table 6. Amount of particulate matter compared to filing methods used, when there is not any kind of local ventilation.

Local ventilation		Filing methods	N	$\bar{x}$	S
No	Particulate matter 2,5	Electric drill	3	36,41	29,71
		Hand filing	2	24,75	9,75
	Particulate matter 10	Electric drill	3	1067,73	859,87
		Hand filing	2	160,88	89,76
	Total suspended particles	Electric drill	3	14036,15	11287,48
		Hand filing	2	1131,25	1225,69

Test: Independent-samples t-test.

Table 7. Efficiency of general ventilation in conjunction with the local ventilation to compare to the amount particulate matter.

General ventilation		Local ventilation	N	$\bar{x}$	S
Yes	Particulate matter 2,5	Yes	5	116,49	97,03
		No	5	31,75	22,49
	Particulate matter 10	Yes	5	615,62	242,84
		No	5	704,99	786,39
	Total suspended particles	Yes	5	1559,16	778,64
		No	5	8874,19	10678,95

Test: Independent-samples t-test.

more exposed to all kinds of particulate material, in particular if using the electric drill compared to the other filling method discussed. The results analyzed with the test *t-Student* there were no significant mean differences, for a significance level of $PM_{2,5}$ (*p-value* = 0,253), PM_{10} (*p-value* = 0,634), in total suspension (*p-value* = 0,224). Relation to the two ventilation systems—General and located, the same comparison is made to verify the effectiveness of these together, we can see the results in the table 7.

To finalize the types of ventilation, it was found in table 7, which uses the localized ventilation in addition to the general ventilation, there is a tendency to be more effective for larger particulate matter. But who does not have localized ventilation, found themselves more exposed to a greater number $PM_{2,5}$ and total suspended particulates. The results analyzed with the test *t-Student* there were no significant mean differences, for a significance level of $PM_{2,5}$ (*p-value* = 0,094), PM_{10} (*p-value* = 0,814), in total suspension (*p-value* = 0,165).

After a review of the types used in response vent to particulate, the personal protective equipment was analyzed (such as using mask) related to respiratory problems or existing allergies. There was no pattern of association between the use of protective (mask) during the work activity and the presence of respiratory problems or allergies (*p-value* = 0,50). However, the 9 people who wear masks indicated 55.6% had respiratory problems or allergies. It was also found that the 3 people who do not used mask, 33.3% (n = 1) also suffered from respiratory. Then it analyzed the relationship between those who use personal protective equipment to the fact that it has no symptoms or condition that makes this profession. There was no pattern of association between the use of protection (mask) during the work activity and the presence of symptoms (*p-value* = 0,159). However, the 9 people who wear masks indicated 55.6% had symptoms. It was also found that the 3 persons who didn't use mask, any of them had suffered from symptoms. Finally, we tried also check the relationship between those who have symptoms compared to amount of exposure of particulate matter. From the evaluated gel nail workers, who claims to have some kind of symptom from practicing this profession, have a tendency to be exposed to larger quantities of $PM_{2,5}$ and PM_{10}. The results analyzed with the test t-Student, there were no significant mean differences, for $PM_{2,5}$ (*p-value* = 0,767), PM_{10} (*p-value* = 0,564) in total suspension (*p-value* = 0,935).

4 DISCUSSION

After analyzing the results, it can be seen that the total concentration of total suspended particulate, on average, didn't exceed the reference value, despite of the exposure to a chemical agent to low concentrations than the considered limits does not invalid that some of the individuals exposed may exhibit results at an higher intensity, not expected adverse effects or the worsening of pre-existing situations (NIOSH, 1999).

As defined in the NP1796, a American Conference of Governmental Industrial Hygienists (ACGIH) also defines as a limit value to this kind of place of 10 mg/m^3, where in realized studies by Cora Roelofs e Tuan Do, confronts results realized by other two studies. For the medium time of 8 hours, medium concentration of 1,4 mg/m^3, for any kind of filling despite the ventilation used being that in another analised study, it shows values of 0,15 mg/m^3 for those who do not possess ventilation and uses the manual method, 0,28 mg/m^3 for those who do not possess ventilation and uses an electrical drill and 0,24 mg/m^3 for those that possess ventilation and an electrical drill (Health and Safety Executive, 2008). Analyzing the relative humidity values, it was found that on average the beauty salons analyzed it had exceeded the reference value, which may be related to inefficient ventilation. In contrast the temperature values remained within the proposed limits of the legislation. In the present study, we observed based on the assessment of levels of dustiness found during filling gel, there is a tendency for the electric drill to create more dust compared to hand filing. In the video performed by APFANails, explains that use of electric drill creates very fine dust, which can increase the potential for overexposure. As these particles are thinner, they will easily float more in the air, as opposed to larger particles that fall more easily, what happens to the method of sharpening—hand filing, which will create fewer particles and particulate matter such as this is larger, falls easily on the table and is less likely to be inhaled (Schoon Doug, 2011). As a way to combat this exposure to this particulate system are used for general ventilation (natural or forced) and localized. Several studies, like those conducted by the National Institute of Occupational Safety and Health (NIOSH), states that despite not finding hazardous exposures, recommends that salons must ensure adequate ventilation, both general and localized, to provide fresh air and absorb the contaminated air (Habia, 2007) (Health and Safety Executive, 2008). In fact who has general ventilation remains exposed to all sorts of particulate material that may be caused by lack of maintenance or aeration compartment. It can also be the effect of the season in which measurements were taken, in winter, where there may be a greater accumulation of dust in the salon. The most effective type of ventilation in a area is what captures fumes and dust at its source before it reaches the air we breathe. These are "local ventilation systems" which adequately protect the gel nail Technicians and customers when used in conjunction with general ventilation, if properly installed and operated alone can significantly reduce chemical exposure and improve the environment of the salon (Healthy Air, 2007). As we can observe in this study, who did not have localized ventilation, even in addition to the general will be more exposed to PM_{10} and total suspended particulate. In addition to the advice of a good ventilation in the salon, the electric drills create a volume of dustiness near where they are used, (United States Environmental Protection Agency, 2007) especially those of a small size ($PM_{2,5}$), being difficult to capture by these ventilation systems, even integrated into these drills. Types of localized ventilation systems suggest a portable ventilation source that removes contaminants before crossing the breathing zone. Ventilation from underneath or from the side, pulls the contaminated air from near the area where the product is placed or used to prevent reaching the face and being inhaled (Marlow, et al., 2012).

From the analysis of the relative humidity values, it was found that on average the salons analyzed exceeded the reference value, which may be related to inefficient ventilation. In contrast the temperature values remained within the proposed limits of the law. In order to control and minimize inhalation of these powders, it is also recommended by NIOSH, the correct use of disposable dust masks, with rankings "N95", whereas surgical masks, and unique Technicians used by this case study, well as a study of Prevention and Control held in Boston, in which a sample of 71 nails gel workers, 64 (90%) reported using a mask with 89% of people 64 who wears, claimed to be paper or cloth as a surgical mask, which will not provide sufficient protection and should not be used. Such protection is especially useful in cases of pre-existing asthma, allergies or other respiratory problems and should never be used in an adequate ventilated salon (Healthy Air, 2007) (Roelofs, et al., 2012), although there are studies that claim located with good ventilation can reduce the need of technicians they use personal protective equipment (Massachusetts Department of Public Health, 1997), since others have suggested that the protective equipment will be much more effective if the amount of particles is greatly reduced (United States Environmental Protection Agency, 2007).

5 CONCLUSIONS

It is concluded from this study, that the gel nails workers are exposed to large amounts of particulate material and that increasingly do not opt for the best ways to reduce and control such exposure, not using the most appropriate type of mask, well as the methods of localized ventilation, showing a general ventilation inefficiency in combating the dustiness. After the elaboration of the research, it was found that there is a greater need to focus in this area, sensitizing the gel nail workers to practice Hygiene and Safety in relation to exposure to chemicals, by performing a thorough research and specifically including the larger number of workers possible, in a way that risks may be identified and minimize to those who are exposed during the filling process of the gel, relative to the dustiness that this practice implies. It is necessary to sensitize these workers for recurrent use of the appropriate mask to the type of dust you are exposed to, as well as good ventilation either locally or at the general level.

REFERENCES

Air, Healthy. 2007. Reducing Air Pollution from: Nail Salons. Owner/Operator Information Sheet. 2007.

Barbosa, Fernando, Matos, Luísa and Santos, Paula. As diferentes metodologias de recolha e análise de Poeiras Ocupacionais: Equipamentos e Técnicas.

Connecticut Department of Public Health. 1997. Occupational Airways. Occupational Health & Special Projects Program. 2, 1997, Vol. 3.

Dec-Lei n°243, de 20 de Agosto de 1986. Portugal. Ministério do Trabalho e Segurança Social. Regulamento Geral de Higiene e Segurança do Trabalho nos Estabelecimentos Comerciais, de Escritório e Serviços.

Doug, Schoon. 2011. Dangers of electric filing (aka drilling) the natural nail. APFANails.

Habia. 2007. Code of Practice for Nail Services. Code of Practice for Nail Services The standards setting body for Hair, Beauty, Barbering, African Caribbean Hairdressing, Nails and Spa. 2007.

Health and Safety Executive. 2008. Health and safety in nail bars. 2008, Vol. RR627.

Healthy Air. 2007. Reducing Air Pollution from: Nail Salons. 2007.

—. 2007. Reducing Air Pollution from: Nail Salons. 2007.

Marlow, David A., Looney, Timothy and Reutman, Susan. 2012. An Evaluation of Local Exhaust Ventilation Systems for Controlling Hazardous Exposures in Nail Salons. NIOSH Alice Hamilton Laboratory (Cincinnati, OH): CDC—Workplace Safety and Health, 2012. 812113.

Massachusetts Department of Labor Standards. Nail Salon Ventilation. Workplace Safety and Health Program.

Massachusetts Department of Public Health. 1997. Worker Exposures to Dusts and Vapors in Nail Salons. SENSOR Occupational Lung Disease Bulletin. 1997.

Nail Manufacturers Council. Guidelines for Controlling and Minimizing Inhalation Exposure to Nail Products.

NIOSH. 1999. Controlling Chemical Hazards During the Application of Artificial Fingernails. Hazard Controls. 1999.

NP 1796. Segurança e Saúde no Trabalho. Valores limites de exposição profissional a agentes químicos. 2007.

Oliveira, António José Pereira de. 2011. Avaliação dos níveis de empoeiramento—Estudo de caso na indústria têxtil. 2011.

Porter, Catherine A. and California Healthy Nail Salon Collaborative. 2009. Dismantling Barriers to Health and Safety in California Nail Salons. Overexposed & Underinformed. 2009.

Prista, João e Sousa, António Uva. Exposição Profissional a Agentes Químicos: os indicadores biológicos na vigilância de saúde dos trabalhadores. s.l.: OPSS—Observatório Português dos Sistemas de Saúde.

Roelofs, Cora and Do, Tuan. 2012. Exposure Assessment in Nail Salons: An Indoor Air Approach. International Scholarly Research Network. ID 962014, 7 pages, 2012, Vol. 2012.

United States Environmental Protection Agency. 2007. Protecting the Nail Salon Workers. 2007. 744-F-07-001.

Occupational Safety and Hygiene II – Arezes et al. (eds)
© 2014 Taylor & Francis Group, London, ISBN 978-1-138-00144-2

Assessment of electromagnetic fields in arc and resistance welding

A.C.M. Garrido & M.L. Dinis
Research Laboratory on Prevention of Occupational and Environmental Risks (PROA/LABIOMEP), Geo-Environment and Resources Research Centre (CIGAR), Faculty of Engineering, University of Porto, Portugal

ABSTRACT: Welding is used in a wide range of heavy and light industry, construction and maintenance, including vehicle construction. Workers using welding processes are exposed to electromagnetic fields (EMF). The International Commission on Non-Ionizing Radiation Protection (ICNIRP) has published guidelines on limits of the exposure to static magnetic fields and on guidelines limiting the exposure to time-varying electric, magnetic and electromagnetic fields. Moreover, the European legislation is based on these guidelines being the most recent and updated limits published in the Directive 2013/35/EU. The ICNIRP reference levels for magnetic fields may be exceeded in both arc and resistance welding. This study aims to assess the workers exposure to EMF in arc and resistance welding and compare the results with legal limits. Several measurements of EMF were carried out in different industries at workplaces using welding processes. The maximum measured field was near the cables with 1065 μT for single sided spot welding; 761,7 μT for arc welding (MIG) and 646,7 μT with resistance welding using a pneumatic "C" clamp. In some cases, the results were above the Action Levels (ALs) of the Directive 2013/35/EU.

1 INTRODUCTION

Electromagnetic fields (EMF) are present everywhere but they are invisible to the human eye. The electromagnetic field can be understood as a combination of an electric field and a magnetic field that can be emitted as waves by many natural and artificial sources such as electricity.

With the industrialization, the increasing of the mobility and communications, the human exposure to electromagnetic radiations has also increased.

Since the publication of the study of Wertheimer and Leeper (1979) "Electrical wiring configurations and childhood cancer" alleging that childhood leukemia was higher in households located near electric power lines, the international scientific community and general population began to look at this issue and to the associated adverse effects. Since then, several studies have been carried out, and in 2002 the International Agency for Research on Cancer (IARC, 2002) classified the magnetic fields with Extremely Low Frequency (ELF: 1 Hz–100 kHz) as possibly carcinogenic to humans (Group 2B), the static fields (electric and magnetic) and the electric fields with frequency extremely low as not classifiable regarding its carcinogenicity to humans (Group 3). In 2011, IARC (2011) classified the electromagnetic field in the range of radiofrequency (RF: 100 kHz–300 GHz) as possibly carcinogenic to humans (Group 2B).

Welding is widely used in industry, construction, maintenance and in particular in vehicle construction. Workers performing welding tasks are exposed to magnetic fields from the welding processes. There are many different welding processes, however the most used are arc welding (MMA: Manual Metal Arc; MIG/MAG: Metal Inert Gas/Metal Active Gas; TIG: Tungsten Inert Gas; Plasma; SMAW: Shielded metal arc welding) and resistance welding.

Depending on the specific process and technology used, the fundamental frequency ranges from 0 Hz to some hundreds of kHz (Grassi et al., 2012).

There are several internationally accepted guidelines concerning the exposure to electromagnetic fields. The most known are those from the ICNIRP (International Commission on Non Ionizing Radiation Protection), in which the European legislation is based.

A review of the literature in this field (Melton, 2005) shows that the ICNIRP reference levels for magnetic fields may be exceeded in both arc and resistance welding.

Recently, the Directive 2013/35/EU (EU, 2013) was published concerning the minimum health and safety requirements regarding the exposure of workers to risks arising from electromagnetic fields and repeal the Directive 2004/40/EC.

The aim of the present study is to provide satisfactory understanding of the magnetic fields which are present in the immediate vicinity of some arc and resistance welding processes, measuring the exposure to magnetic fields and comparing these values with the limits established in the Directive 2013/35/EU.

2 MATERIAL AND METHODS

This study was performed in three different companies from different sectors of activity.

Company A is a small enterprise in the light metalworking, which works with two similar MIG welding machines: ESAB MIG C280. This company has 2 welders; one does welding tasks during approximately 4 h/day and the other during 1 h/day. The first one has about 24 years of welding experience, and the second about 10 years.

Company B is a medium-sized company with 70 employees, of which 31 are welders. They work in three shifts of 8 hours, with 11 similar MIG welding machines: ESAB MIG 5004i. All do welding tasks during 6 h/day. Concerning the 31 welders, 20 perform welding processes since 10 years ago in this company, 3 started their professional work as welders 4 years ago and 4 were already welders when they joined the company.

Company C is from the sector of vehicle repair and maintenance. In this sector of activity, welders usually perform welding tasks for a short period of time. They used to use arc welding (MIG/MAG, SMAW, plasma and oxyfuel) and resistance welding. Nowadays, the most used are MIG/MAG and spot welding. In this company, some measurements of the EMF were carried out in a resistance welding machine using a single sided spot weld and a "C" clamp: Blackhawk/CompuSpot WEL 750. This company has 8 welders who weld on average 1 hour per day, distributed approximately as follows: 10 minutes with resistance spot welding, 10 minutes with MIG/MAG, 10 minutes with oxyfuel and 30 minutes with small sop welder.

The measurements were performed in each company for 5 shifts of 8 hours each one. The welders were also observed during one shift without disturbing their jobs to analyze the welding process, the type of pieces welded and to check the periods in which the welding tasks were performed.

The magnetic field data was acquired using a isotropic and triaxis magnetic sensor (Aaronia NF5035). This instrument measures magnetic fields in the range of 0,1 nT to 2 mT, from 0 Hz to 30 MHz and with an accuracy of 3%.

The measurements were conducted at several positions, close to the welders head, chest, waist and hand/electrodes and also near the welding cables and machine. In addition, a few measurements were carried out at distances of 20 cm, 50 cm, 1 m and 2 m from the welding cables and machine.

Figure 1. Spectrum analyzer of the isotropic magnetic field (Aaronia).

Each measurement was carried out with a minimum period of 1 minute. The electric field was also measured, but the registered values were very low compared to the action levels defined in the Directive 2013/35/EU.

3 RESULTS

All measurements were performed in a real work environment, during typical welding processes with usual welded parts and using habitual specifications.

The measurements were done in the range 0 Hz–30 MHz, according the manufacture recommendations concerning band width.

For each machine, the magnetic fields were measured at several positions (Meas. Posit.): near welding cables (1); welding machine (laterals without cables-2, lateral with cables-3); electrodes (4); welder head (5); and welder chest/waist (6). Concerning the welding cables and the machine, some measurements were done at distances of 20 cm, 50 cm, 1 m and 2 m from these 2 positions (Table 1, Table 2, Table 3 and Table 4).

The results of the measurements conducted in company A with ESAB MIG C280 welding machine are presented in Table 1. The measurements were registered with a welding current of 140–175 A, a wire diameter of 0,8 mm, a wire speed of 9–10 $m.min^{-1}$ and a shielding gas with 98% Argon + 2% CO_2. The welding cables were 3 m long.

The results of the measurements carried out in company B with ESAB MIG 5004i welding machine are presented in Table 2. The measurements were taken with a welding current of 265–285 A, a welding voltage of 26,2–29 V, a wire diameter of 1,2 mm, a wire speed of 10–12 $m.min^{-1}$ and a shielding gas with 75% Argon + 25% CO_2. The welding cables were 5 m long.

Table 1. Measurements of the magnetic field [µT] for ESAB MIG C280 welding machine.

Meas. Posit.	Freq. [Hz]	Distance [cm] 1	20	50	100	200
1	0	760,3	463,9	141,0	79,2	75,6
	53	225,7	–	–	–	–
	98	95,0	–	–	–	–
	204,4	48,6	–	–	–	–
2	0	108,5	–	–	–	–
3	0	496,0	477,9	71,8	62,3	–
4	0	480,0	–	–	–	–
5	0	322,3	–	–	–	–
6	0	231,0	–	–	–	–

Table 2. Measurements of the magnetic field [µT] for ESAB MIG 5004i welding machine.

Meas. Posit.	Freq. [Hz]	Distance [cm] 1	20	50	100	200
1	0	761,7	185,0	105,0	76,5	66,1
	45	133,0	–	70,9	–	–
	50,2	111,1	–	–	–	–
	53,4	116,3	77,85	–	–	–
	56,2	104,4	–	–	–	–
	80	1,6	–	–	–	–
	98	6,4	–	–	–	–
	132	60,0	–	–	–	–
	154	49,3	–	–	–	–
	281,8	26,2	–	–	–	–
2	0	214,9	–	–	–	–
3	0	485,0	477,9	71,8	62,3	–
4	0	635,5	–	–	–	–
	53,4	82,2	–	–	–	–
5	0	105,0	–	–	–	–
6	0	495,0	–	–	–	–

Table 3. Measurements of the magnetic field [µT] for a resistance welding WEL 750 with "C" clamp.

Meas. Posit.	Freq. [Hz]	Distance [cm] 1	20	50	100	200
1	0	646,8	–	–	–	–
	53,8	95,5	–	–	–	–
2	0	190,0	–	–	–	–
	53,8	115,7	–	–	–	–
3	0	591,0	–	–	–	–
4	0	590,4	–	–	–	–
	53,8	111,2	–	–	–	–
5	0	95,0	–	–	–	–
6	53,8	108,0	–	–	–	–

Table 4. Measurements of the magnetic field [µT] for a single point spot welding Blackhawk/CompuSpot WEL 750.

Meas. Posit.	Freq. [Hz]	Distance [cm] 1	20	50	100	200
1	0	1064,5	510,0	494,5	–	–
3	0	237,5	–	–	–	–
4	0	375,9	247,0	–	–	–

The results of the measurements performed in company C with Blackhawk/CompuSpot WEL 750 resistance welding machine, using a "C" clamp are given in Table 3. The measurements were taken with a welding current of 8100 A, a welding time of 300 ms and a welding pneumatic force of 400 daN. The cables and the torch were cooled with a fluid consisting of 50% ethyl glycol and 50% water. The welding cables were 2,5 m long.

The results of the measurements carried out in company C with Blackhawk/CompuSpot WEL 750 resistance welding machine, using a single point spot welding are given in Table 4. The measurements were performed with a welding current of 8600 A and a welding time of 70 ms. The cables were air cooled and were 2,5 m long.

4 DISCUSSION

From the obtained results it is possible to observe that the magnetic field is higher near the cables, for all cases studied. The results also show that the magnetic field is predominantly at one frequency with harmonics at other frequencies.

The highest magnetic field is measured near the cables for a single point spot welding, 1064,5 µT. As the higher the current intensity, the higher the magnetic field, it is expectable that the magnetic field obtained with this spot welding machine will be greater if we use higher welding current (maximum 12 000 A). However, in this kind of applications, this is not usual the case.

Concerning the resistance welding machine with "C" clamp, the magnetic field is 646,8 µT near the cables, 591,0 µT near the machine in the side with cables and 590,4 µT near the electrodes.

For MIG welding machines (C280 and 5004i) the results were very similar. The highest values are obtained near the cables, 761,7 µT and 760 µT, for the ESAB MIG 5004i and ESAB MIG C280, respectively.

Although the measurements were carried out at various positions, from the point of view of occupational safety and health, the most important

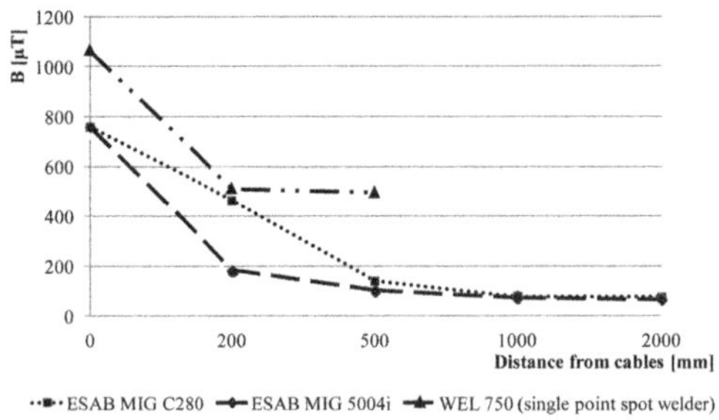

Figure 2. Variation of the magnetic field (B) with the distance to the cables.

measurement is the one that usually presents higher values (near welding cables). In fact, in practice and for all manual welding processes the contact between the welder and the cables is inevitable; therefore the compliance with the Action Levels (ALs) near the cables will ensure worker protection.

According to the Directive 2013/35/EU, the Exposure Limit Values (ELV's) applied to the sensory effects and health effects for external magnetic flux density (B_0) were not exceeded. However, the action levels for magnetic flux density of static magnetic fields (ALs(B_0)), concerning interference with active implanted devices, e.g. cardiac pacemakers, were exceeded near the cables, for all cases. In particular, for the case of a single point spot welder WEL 750, the ALs(B_0) were exceeded up to 50 cm away from the welding cables. Although the magnetic field is predominant at 0 Hz, we could verify that the ALs were not exceeded for the harmonics.

For each welding equipment and process, the reduced data set has been tabulated showing the individual significant frequency of the magnetic field. These values have been compared with the Directive 2013/35/EU Action Levels (ALs). According to the ICNIRP (1998a; 2010b), the ratio of the measured value (B_n) and the reference level (R_n, which in this case is the Action Level AL_n) will be calculated and the sum of these individual contributions should be less than or equal to 1 (Equation 1).

$$\sum_n \frac{B_n}{AL_n} \leq 1 \tag{1}$$

In spite of the results obtained in this study, it should be noted that in the considered exposure conditions, the reference levels set by the Council Recommendation (EC, 1999) concerning the exposure of the general public to electromagnetic fields (0 Hz to 300 GHZ) were exceeded, meaning that the access to these areas should be limited to welders.

The risk assessment concerning the exposure to magnetic fields by the welders should consider the Lifelong Exposure (LE) for each individual, as some welders work (almost) the whole day (like in company B), other welders may perform the welding task for certain periods in a day (like in company A), and others weld only just a few hours in some days of the week (like company C). According to Man et al. (2007), the LE to magnetic fields (B) is calculated by LE = B[μT] × the average usage in hours per day × the number of days in a year where welding tasks were performed × the number of years in which welding tasks were performed.

According to the data collected during the measurements it is possible to conclude that different LE can be found for the workers of the same company, with a similar welding machine. So the cumulative exposure can be very different for each welder. This is a very important aspect to be considered in the risk assessment to EMF exposure and most probably has influence in the potential adverse health effects.

5 CONCLUSION

From the measurements of the magnetic fields carried out in the vicinity of an arc and resistance spot welding equipment, in addition to an exhaustive literature review, the following conclusions may be drawn:

- Different processes produced different magnetic fields strengths;
- The highest values measured were close to the welding cables for all cases;
- In manual welding processes the permanent contact between the welder and cables is inevitable, whereby it is acceptable to use these exposure values as the welder exposure;
- In some conditions the Action Levels (ALs) of the Directive 2013/35/EU (EU, 2013) are likely to be exceeded;
- The access to welding areas should be limited to welders;
- People with active implanted devices (e.g. pacemakers) should be kept away from the welding areas;
- In some cases, it may be important to consider the contribution of the fundamental frequency and the most important harmonics, concerning the exposure to EMF;
- The lifelong exposure should be considered in the analysis of cumulative and potential adverse health effects;
- Further investigation is required to assess the exposure if the Directive 2013/35/EU (EU, 2013) Action Levels (ALs) are exceeded.

ACKNOWLEDGEMENT

The authors would like to thank Master in Occupational Safety and Hygiene Engineering (MESHO), of the Faculty of Engineering of the University of Porto (FEUP), all the support in the development and international dissemination of this work.

REFERENCES

EC (1999), Council Recommendation n.° 1999/519/EC on the limitation of exposure of the general public to electromagnetic fields (0 Hz–300 GHz).

EU (2013), Directive 2013/35/UE of the European Parliament and the Council of 26 June, concerning the minimum health and safety requirements regarding the exposure of workers to risks arising from electromagnetic fields.

Grassi F., Spadacini G., Pignari S.A. (2012), Human exposure in Arc-Welding Processes: Current versus Previous ICNIRP Basic Restrictions, IEEE.

IARC (2002), Working Group on the Evaluation of Carcinogenic Risks to Humans. Non-ionizing radiation, Part 1: static and extremely low-frequency (ELF) electric and magnetic fields. IARC Monographs on the Evaluation of Carcinogenic Risks to Humans; Vol. 80: 1–395.

IARC (2011), IARC classifies radiofrequency electromagnetic fields as possibly carcinogenic to humans. Press release n.° 208, 6p.

ICNIRP (1998a), Guidelines for limiting exposure to time varying electric, magnetic and electromagnetic fields (up to 300 GHz), Health Physics, vol. 74, n.° 4, p. 494–522.

ICNIRP (2010b), Guidelines for limiting exposure to time varying electric, magnetic and electromagnetic fields (1 Hz to 100 kHz), Health Physics, vol. 74, n.° 4, p. 494–522

Man A.K., Shahidan R. (2007), Variations in Occupacional Exposure to Magnetic Fields Among Welders in Malaysia. Radiation Protection Dosimetry, Vol. 128, No. 4, p. 444–448.

Melton G.B. (2005), Health AND Safety Executive Research Report 338, HSE books, Suffolk.

Wertheimer N., Leeper E. (1979), Electrical wiring configurations and childhood cancer. American Journal of Epidemiology 109(3), p. 273–284.

Occupational Safety and Hygiene II – Arezes et al. (eds)

Influence of pedal use in developing of chronic low back pain and work disability

S. Barata & S. Gagulic
Physical Therapy and Ergonomics, Escola Superior de Saúde Jean Piaget, Viseu, Portugal

P.M. Arezes
Human Engineering, University of Minho, Portugal

ABSTRACT: Low back pain is a common dysfunction in workers, contributing to decreased quality of life and productivity. The aim of this study is to evaluate the influence of the pedal in industrial machines for developing chronic back pain, as well as the impact of back pain on quality of life and ability to work. The sample consists of 84 sewers of an automotive industry, divided into two groups: with and without using the pedal. Was applied to the scale of pain and disability of Quebec, the quality of life questionnaire (SF-36) and the Beliefs, Fear and Avoidance questionnaire (QCME). After descriptive statistics and inferential analysis, it was found that back pain does not suffer an influence statistically significant, use the pedal. It is concluded that the workers back pain is not influenced by using the pedal, while causing a decrease in quality of life and increased fears concerning the failure.

1 INTRODUCTION

Is estimated that nearly two million Portugueses suffering from chronic pain. Only with the back pain, the costs with health care, absenteeism and loss in productivity, stand at around two billion euros per year (Figueiredo, 2008). Back pain has been defined as the activity intolerance due to lumbar symptoms or symptomatology of the lower limbs (Kendall, 1997). In addition to the time factor, the classification of low back pain should consider the function (Airaksinem et al., 2004) and the psychological and social aspects of pain (Van Tulder et al., 2000).

Currently, we can say that the pain is associated with chronic low back pain and functional disability, not directly resulting from the pain, but the existence of cognitive-behavioral change and environmental translated, into each individual as beliefs and attitudes, sometimes more disabling than pain itself (Waddell, 2004). According to Benard et al (1997), psychosocial factors related to employment and the workplace, can develop musculoskeletal disorders of the upper limbs and spine, particularly with regard to the perception of the performance of an intensive, monotonous with limited control, unclear and low social support work. With all technological advances, the incidence of occupational low back pain caused by overload on the lumbar spine has increased, leading to a number of physical disorders and financial losses caused by absenteeism, decreased productivity and therefore corporate earnings (Goumoens et al., 2006). The use of the pedal of the sewing machine may be a factor aggravating injury, or by the same driving force exercised repeatedly, or by the asymmetry of movement that might cause its use (Dellaman, 2002). The design of the proper pedal, should also take into account not only the force needed to operate the foot, but also by the output biomechanics and kinematics of the foot pedal (Bullinger, 1991).

2 OBJECTIVES

The main objective of this study is to evaluate the contribution of the frequent use of the pedal in industrial machinery, to the development of chronic low back pain as well as determining the impact of low back pain in limiting the activities of daily living and work disability. Specifically we intend to know the relationship between low back pain and inability to perform activities of daily living and functionality; verify the influence of the use of pedal, bilateral or unilateral, for the development of limiting the activities of daily living and occupational health; assess the physical and mental component operators and finally check the beliefs of fear and avoidance, work-related.

3 METHODS

The study took place in a branch company of auto-mobile in the district of Aveiro, specifically in the field of sewing, work stations operated in the standing position. This place was chosen for the job requirement and multi-task perfor-nied, as well as the representativeness of the population. The field study was conducted between April and May 2010.

For this study, it was considered a target population of workers of the company, who met the inclusion criteria: present some form of back pain symptoms, aged over 18 years who can read and write without cognitive and no pregnancy.

The sample was randomly chosen stratified each. The study population was divided into two groups: one group A, which does not use the pedal and to serve as control group) and sample B using the pedal. The last group was further divided into two subgroups: sample B1, which uses an alternating pedal and the sample B2 using the pedal in a unilateral fashion.

The study is descriptive-correlational and cross, in order to find relationships between variables at the same time you want to collect information regarding the frequency of health problems at the time of the study (Fortin, 1999).

After the process of contact and authorization of the company, the employees were informed about the purpose of the study, to collect informed consent duly signed. Participants were instructed as to the filling of the instruments chosen.

In total 100 questionnaires were delivered, only 90 have been collected and used in the study, 84 fully completed questionnaires. The confidentiality and anonymity were assured by assigning a code to each worker to protect your identity and your computer where they found the database was protected by a password.

The variable limitation of daily activities and functional, will be assessed using the scale of Low Back Pain and Disability Quebec, consisting of a measuring instrument composed 20 items, developed based on the conceptual model of disability ICIDH (Jette, 1994.) The answers may be given on a numerical scale of 6 points, where 0 corresponds to no difficulties and 5 to the total inability to perform the activity. The score is calculated by simply proceed to the sum of each item. Values can range from 0 to 100, 0 being no failure no failure and 100 total perform these activities (Kopec, et al, 1996; Schoppink et al, 1996).

We also used the reduced health questionnaire (SF-36) to assess the perception of health status. Contains 36 items, grouped into measures mental and physical. The SF-36 is split into eight dimensions such as physical functioning, physical performance, bodily pain, general health perception, vitality, social functioning, role emotional, mental health (Pais 2005).

Finally we applied the Questionnaire Fear-Avoidance Beliefs (QCME), self-administered, with 15 items on a single page, which are divided into two sub-scales: Fear-avoidance and physical activity and fear-avoidance and work. This instrument is used for screening and quantification of fear-avoidance beliefs, and work-related physical activity in individuals with low back pain (Gonçalves, 2004).

Statistical analysis of the collected data was performed in Statistical Package for Social Sciences (SPSS) version 17.

4 RESULTS

The sample was composed of 84 workers females with a mean age of 36 years to a maximum of 57 and a minimum of 19 years presenting a normal distribution. Regarding working hours, all working were fixed shifts with 8 hours, with an average of 10.4 years of work.

The body mass index showed a higher percentage in the normal range, indicating a trend towards pre-obese, according to data from the World Health Organization (Table 1).

This value representative of a slight excess weight, can be considered as a possible risk factor for the development of musculoskeletal injury association work and thus as limiting the generic health status of workers. (Heliovaara 1991).

With regard to the use of pedal study (Table 2), 15 workers (17.9%) do not use the pedal work, using pedal 30 to the right (35.7%), 10 workers (11.9%) using the left pedal and 29 individuals (34.5%) use bilaterally pedal.

The scale of low back pain and disability Quebec, presents an index of Cronbach's alpha of 0.95, indicating good internal consistency.

By comparing the workers who use with the pedal foot that do not use it has been found that workers without the use of pedal have a lower average full scale in Quebec than using pedal (Table 3).

Table 1. Body Mass Index (BMI).

BMI	N	(%)
<18,5—Underweight	3	3,6
18,5–24,9—Normal Weight	46	54,8
25–29,9—Pré-Obesity	28	33,3
30–34,9—Obesity I	5	6,0
35–39,9—Obesity II	1	1,2
>40—Obesity Obesidade III	1	1,2

Table 2. Distribution of the sample as a function of the pedal variable percentage.

Pedal	N	Percent (%)
Without pedal	15	17,9
Right Pedal	30	35,7
Left Pedal	10	11,9
Bilaterally	29	34,5

Table 3. Analysis of variance between the total Quebec depending on the use or non-use foot pedal.

Pedal	N	$\bar{x}$	SD	F	P
Without	15	11,7	12,1	1,5	0,2
Right	30	22,7	21,8		
Left	10	16,7	12,4		
Both	29	21,2	16,0		

Table 4. Statistics and scoring reference in sub-scales QCME.

	N	Reference	Min	Máx	$\bar{x}$	SD
FA	84	15	0	35	12,8	8,1
WR	84	34	0	70	21,3	13,4

Table 5. Analyses of variance between sub-scale work/employment questionnaire fear and avoidance beliefs related to the use or non-use foot pedal.

Pedal	N	$\bar{x}$	SD	F	P
Without	15	20,6	13,6	1,6	0,1
Right	30	22,4	13,8		
Left	10	13,1	4,5		
Both	29	23,5	14,3		

There is greater use of the right pedal and pedal bilaterally pedal than the left.

Given the possible limitation of functional daily activities and there are no significant effect on the use of the pedal in the analysis of variance. Since the number of observations in each group is the same, there was a robust test for Brown-Forsythe, with the result ap 1.8 to 0.1 and there were no significant differences. Conducting Post-Hoc Tukey Test, to find the differences between the different uses of pedal, also revealed no significant differences for a result of 0.26.

The questionnaire applied fear and avoidance, comprises two sub-scales: fear-avoidance and physical activity (FA) and fear-avoidance and work (WR). A score above 15 on the subscale of physical activity QCME was proposed as an indicator of elevated fear-avoidance beliefs for users with low back pain seek-ram primary care or osteopathic treatment (Fritz, 2002). A score of sub-scale work/employment QCME than 34, is associated with an increased risk of these users do not return to work, while scores of less than 29 of the same sub-scale were associated with a decreased risk of their users do not return to work. Participants in the study have an average of 21.3.

The questionnaire presented QCME index of Cronbach's alpha of 0.93, indicating good internal consistency (table 4).

Considering the effect of the variable limitation of occupational activities, the generality of the results is not statistically significant verification purposes in relation to the use of pedal (Table 5). It appears however that the average worker using the right foot on the sub physical activity level is too close to the score 15, and hence presents a greater predisposition to developing chronic back pain.

In relation to SF36, the best health profiles were obtained in the Mental Health dimension, with values of 69.7, above the average used for the construction of the questionnaire with a sample of 1434 healthy individuals (Pais, 2005). The average values referred to below, are located in physical performance dimension (difference 27) followed by bodily pain scale (difference 25.9) (Table 6). The significance of lower scores for physical performance are described as problems with work or other daily activities in con-sequence of physical health (Pais, 2005).

The mean of the study population are below the average reference for all health questionnaire items except the item mental health in the study population has a higher average (69.7).

This survey showed an index of Cronbach's alpha of 0.91 in item 1 and 0.8 in the remaining items. This result demonstrates good internal consistency (Ribeiro, 1999).

Looking at the values of Pearson's correlation verified the existence of statistically significant associations ($p < 0.05$) between the overall score and sub-scale of the Quebec labor/employment ($r = 0.34$) with a moderate positive correlation.

In the analysis of the association between the overall score of esca her pain and inability to Quebec and dimensions of the health questionnaire, there is the existence of statistically significant associations ($p < 0.01$) between the amount of pain scale and inability to Quebec and all dimensions of the SF36.

The fact that there are negative associations indicates that the strength of the association between these variables is done in reverse and not in the same sense. Ie, the higher the score on the pain scale and inability to Quebec (greater limitation in

Table 6. Statistics of the study population in reduced health questionnaire SF36.

	Reference	Min.	Máx.	$\bar{x}$	SD
FF	89,2	15	100	70,1	22,9
DF	86,2	0	100	59,2	41,9
DC	72,7	0	84	46,8	21,9
SG	70	0	100	60	18,5
VT	60,6	5	80	39,5	16,4
SM	66,6	12,5	100	69,7	22,6
FS	76,6	0	100	65	41
DE	73,2	8	80	47,8	17,8

PF—Physical Functioning, PP—Performance Physical, BP—Bodily Pain, GH—General Health, VT—Vitality, MH—Mental Health, SF—Social Functioning, RE—Role Emotional. *Average standard for female participants (age 25–44 years).

the activities of daily living and functional) lesstion scores in the dimensions of the SF36 (lower quality of life). When comparing the influence of the use of pedal with the scale of pain and inability to Quebec we can infer that our sample presents detectable changes in limitation of daily activities and functional for a confidence interval of 90%. Despite these results, the limitations do not appear to be related to the different use of the pedal, but above the level of beliefs. Although the company has bet on preventive measures in the workplace, such as the adaptation of workstations and ergonomic measures darias secondary control stations as rotation of the labor and gymnastics program, the population holds the beliefs of fear and avoidance relates to the work, so it is therefore important to foster an approach that emphasizes the component with cognitive-behavioral, directed to various aspects.

The worker education about the multidimensional pain associated with the integration of physical activity programs may result in behavioral changes that reduce significantly the self-reported disability.

5 CONCLUSIONS

The exponential increase of occupational low back pain is a problem that has been worrying different organisms, with an increased incidence and prevalence, emerging solutions and therefore more effective in terms of clinical practice.

Ergonomics conditions may be related to the occurrence of the first episode of pain, however, there is no evidence that control of these conditions will influence the risk of recurrence or progression to a chronic condition. The results from the Quebec scale, seem to demonstrate that the use of pedal represents no effect on disability and functionality of the workers, not directly linking low back pain with the attitude during work hours, but a multifactorial etiology. However, lower back pain and consequent disability and functionality limitations, influences the quality of life of workers and increases fears due to the inability to work. Results show the importance of fear and avoidance beliefs about work for the development of a chronic disability. It will therefore be noted that, beliefs, related to pain, are modified to become the target for intervention in working with back pain, to prevent the chronic situation and absenteeism caused by this clinical condition.

Considered as limitations of the study, the absence of a socio-demographic characteristics of the sample depth, as the educational level, the extra labor activities, history of smoking, previous professional activities. The lack of characterization of low back pain such as previous medical history, physical state (tests or evaluation musculoskeletal) and complaints of pain time was considered a limitation of the results presented.

The fact that the sample is composed only of females did not allow for comparisons between genders and understand the influence of this variable on the differences of the limitations of physical activity, health status and beliefs of fear and avoidance related to work in the Automotive Industry.

The study also revealed the need to develop new lines of research directed to this group of people, particularly in terms of intervention and monitoring clinical perspective based bio-psycho-social and educational.

REFERENCES

Airaksinen, O., & Hildebrandt, J., et al (2004). European guidelines for the management of chronic nonspecific low back pain Working Group on Guidelines for Chronic Low back pain. *Management of low back pain.* Palma de Maiorca: http://www.backpaineurope.org/web/files/.

Andersson, GB., & Frymoyer, JW., et. al. (1999). Epidemiology and cost. In POPE—Occupational low back pain, assessment,. St. Louis: Mosby.

Bekkering, G., & Hendrikset HJM et al (2003). Dutch physiotherapy guidelines for low back pain. *Physiotherapy. 89*, 82–96.

Bernard, B. (1997). *Musculoskeletal dissorders and workplace factors A critical review of epidemiologic evidence for work related musculoskeletal disorders of the neck, upper extremity and low back.* USA: Public Health Service Center for Disease Control an Prevention National Institute for Occupacional Safety and Health.

Cruz, Eduardo & Gonçalves, Eurico et al. (2005) Adaptação e validação para a População Portuguesa

da escala de dor e incapacidade de Quebec. Setúbal: Instituto Politécnico de Setúbal-Escola Superior de Saúde
Delleman, N.D. (2002). Sewing machine operation: workstation adjustment, working posture, and workers' perceptions. *International Journal of Industrial Ergonomics*, 341–353.
Figueiredo. (2008). *Nós somos o que repetidamente fazemos*. Obtido de http://estudo geral.sib.uc.pt/jspui/bitstream/10316
Fortin, M.-F., (1999). *O processo de investigação, da concepção à realização.* Loures: Lusociência. Edições técnicas e científicas, Lda
Fritz, J. (2008). Investigation of elevated fear-avoidance beliefs for patients with low back pain: a secondary analysis involving patients enrolled in physical therapy clinical trials. *Journal Orthopaedic and Sports Physical Therapy*, 50–58.
Gonçalves, Eurico.(2004). *Adaptação e validação para a população Portuguesa da escala FABQ por E Área Disciplinar da Fisioterapia. Escola Superior de Saúde—Instituto Politécnico de Setúbal. Dezembro de 2004.* Setúbal: Escola Superior de Saúde—Instituto Politécnico de Setúbal.
Goumoens, P.S. & Fritsch, C. et al (2006). Low back pain in 2006: back to the root. *Rev Med Suisse*, 1268–1270, 1272–1264.
Hass, M. & Muench J., et al. (2005). Chronic disease self management program for low back pain in the elderly. *Manipulative Physical Therapy 28(4)*, 228–238.
Heliovaara. (1991). Relation between functional characteristics of the trunk and the occurrence of low back pain. Associated risk factors. *Spine*, 23(3):359–65.
Kendall, O. (1997). *Guide to assessing psychosocial yellow flags in acute low back: risk factors.* New Zealand,: National Health Committee, Ministry of Health.
Kopec J. & Esdaile, J.M. et al (1996). The Quebec back pain disability scale: Conceptualization and development. *J Clin Epidemiol. 1996; 49: 151–161*, 49: 151–161.
Pais, J. Ribeiro, (2005). *O importante é a saúde: Estudo de adaptação de uma técnica de avaliação do Estado de Saúde—SF-36*. Lisboa: Fundação Merck Sharp & Dohme.
Schibye, B.S. & Skov, T. et al (1995). Musculoskeletal symptoms among sewing machine operators. Scand J Work Environ Health. *Scand J Work Environ Health.*, 427–434.
Schoppink EM et al (1996). Reliability and validity of the Dutch adaptation of the Quebec. *Physical Therapy*, 268–275.1.
Serranheira, F. & Pereira, Mário et al (2003). Auto referência de sintomas de Lesões Musculo-Esqueléticas Ligadas ao Trabalho numa grande empresa em Portugal. *Revista Portuguesa de saúde Pública.*
Van Tulder, M. & Ostelo R., et al. (25 de 2000). Behavioural treatment for chronic low back pain: a systematic review within the framework of the Cochrane Back review group. *Spine, Vol 25 nº20*, 2688–2699.
Waddell, G. (2004). *The back pain revolution.* Australia: Churchill Livingstone.
Waddell, G. (2009). *A Fear-Avoidance Beliefs Questionnaire (FABQ) and the role of fear-avoidance bealifs in chronic low back pain and disability*. Glasgow, UK: Orthopaedic Department Western Infirmary.

Occupational Safety and Hygiene II – Arezes et al. (eds)
 ISBN 978-1-138-00144-2

3D anthropometric data collection for occupational ergonomics purposes: A review

S. Bragança & P.M. Arezes
DPS, University of Minho, Guimarães, Portugal

M.A. Carvalho
DET, University of Minho, Guimarães, Portugal

S. Ashdown
HED, Cornell University, Ithaca, NY, USA

ABSTRACT: This paper presents a literature review of anthropometric data collection to address occupational ergonomics issues. One of the uses of anthropometry is to assess the negative effects associated with working postures. Using new techniques, such as 3D body scanners, it is possible to have very reliable data to use in the enhancement of workstation design or other ergonomic interventions, in order to prevent work-related musculoskeletal disorders.

1 INTRODUCTION

People spend most of their lives at work, therefore it is extremely important that the work environment is healthy, safe and comfortable in order to avoid occupational injuries and/or diseases. Optimising workplace design is also a central factor in insuring workers' efficiency and safety on the job. User-centered design approaches, where ergonomic principles and anthropometrics are considered, should be preferred (Wichansky, 2000). This type of design aims to minimize the stress imposed on the users and to eliminate harmful postures. When their workplaces are inadequate users may experience Work-Related Musculoskeletal Disorders (WRMSD). Workers' WRMSD are very prejudicial for companies since they are one of the major causes of reduced work capacity, absenteeism or productivity losses (Escorpizo, 2008). Hence, user-friendly workplaces are decisive in workers' welfare. In order to design this type of work environment, it is essential to have a full understanding of the human body and to create workplaces which are suitable for users taking into consideration different body dimensions and different activity requirements.

Anthropometry is the branch of the human sciences that deals with body measurements, such as size, shape, strength and working capacity (Pheasant, 2006). When applied in occupational studies the data acquired (anthropometric measurements) can be used to assess the interaction of workers with their tasks, tools, machines, vehicles, and Personal Protective Equipment (PPE). This last issue, PPE, is very important especially in regard to determining the degree of protection afforded against hazardous exposures. An inadequate fit of the personal protective equipment does not provide workers with sufficient protection from health and injury exposures, such in the case of facemasks or hearing protection devices (Hsiao & Halperin, 1998). Thus, it is extremely important that the designs are compatible with normal anthropometric measurements of a workforce, since misfit could result in undesired incidents.

However, currently the amount of data on the size and shape of industrial workers is limited. Most of the data used by safety and ergonomics researchers are based on data drawn from studies of military personnel that are quite different from the average workforce populations. As anthropometric characteristics vary according to several factors (e.g. gender, age and race), creating anthropometric databases that reflect the full variation of the population typically requires considerable resources (time, know-how, funds, equipment and workforce, etc.). Nevertheless, nowadays, there are a growing number of anthropometric databases attempting to represent the characteristics of entire populations (Barroso et al., 2005). However, as the study of Hsiao et al. (2002) concluded, there are even significant anthropometric differences among occupational groups, meaning that, for example, a truck driver and a firefighter are, or can be, anthropometrically different from each other and from the average civilian population.

One of the many applications of anthropometry for occupational ergonomics is the assessment of the body modifications that are associated with different working postures (sitting and standing). This paper aims to present a literature review with the identification of the negative effects underlying each working posture, as well as investigating the literature on studies designed to determine if they contribute to the appearance of WRMSD, and also present some ergonomic interventions that are designed to reduce their magnitude.

2 COLLECTION OF ANTHROPOMETRIC DATA

The variance in body dimensions is frequently reported by calculating means, standard deviations, and percentiles (Roebuck et al., 1975). Despite being useful to create general and broad parameters for the design of workplaces and products, detailed fit information was missing for use in cases such as personal protective equipment. Until the development of 3D body scan technology anthropometric studies were conducted by manually measuring each study participant using tools such as anthropometers, calipers, and tape measures. 3D body scanners have revolutionized anthropometric data acquisition, being more practical, accurate, fast and, comparably, less expensive. There are several types of imaging techniques to create full body images. These imaging technologies, include 2D video silhouette images converted to 3D models, white light phase based image capture, laser-based image capture, and radio-wave linear array image capture (Treleaven & Wells, 2007; Istook & Hwang, 2001). Body scanning systems normally consist of one or more light sources, one or more vision or capturing devices, software, computer systems and monitor screens to visualize the data capture process (Daanen & Water, 1998). In most cases, the 3D body scanner captures the outside surface of the human body by using optical techniques. This means that there is no longer the need for physical contact with the subject's body, but the image based data collection introduces the question of privacy. There are different opinions regarding the privacy of the body scanner. If in the one hand it provides more privacy, since it avoids the need to actually touch the body, on the other hand the highly accurate more personal images produced by the scanners are potentially more invasive since they can be stored insecurely and transferred directly from the scanner over local networks or the internet. Nevertheless, with these advances in anthropometric science and computer-based human-form modeling it is now possible to give a different perspective to the collection of anthropometric measurements.

2.1 *Body variations due to working posture*

As can be imagined, the shape and size of the human body can be affected by repetitive physical activities performed during a working period. Moreover, the body may also be influenced by the working posture adopted during a workday, i.e. when people spend most of their time sitting or standing. The following considerations reflect the effects on the human body of excessive sitting and excessive standing.

2.2 *Excessive sitting*

There are many people who spend approximately 8 to 9h of their day in a sedentary behavior and a large part of this sedentary time is spent at work (Healy et al., 2011). Some studies demonstrated that most working adults spend 1/2 to 2/3 of their time at work in a sitting position (Tigbe et al., 2011). In some jobs the time spent on sedentary behavior can reach 90%, such as the case of call centers, reported in Toomingas et al. (2012). Undoubtedly, sedentary behavior is directly related to obesity (Pi-Sunyer, 1999). Moreover, sedentary behavior has been shown to be an independent risk factor for obesity, diabetes, some cancers and death from any cause (Katzmarzyk et al., 2009). An effect of prolonged sitting that has been very much analyzed is leg swelling (Table 1).

However, sitting may be less energy consuming than standing and less stressful on the lower extremity joints (Grandjean, 1988). Nevertheless, several authors refer the increased risk of low back

Table 1. Effects of prolonged sitting.

Author	Identified effects
Pottier et al., 1969	Volume increase causes: hydrostatic pressure, thermal increase and obstruction of blood circulation
Shvartz et al., 1982	Chair's seat compresses the veins in the thigh and hip areas, causing poor blood circulation to the legs
Seo et al., 1996	Higher lower leg swelling due to the activity level required to the leg muscles to sustain the body
Winkel & Jorgensen, 1998	Swelling and discomfort of the lower extremities
Carpentier et al., 2004	Venous disorders and vascular effects

pain in seated jobs (Kroemer & Robinette, 1969) and the greater disc pressure for a seated posture than for a standing posture (Andersson et al., 1979). Lehman et al. (2001) conclude that working in a seated position can also require greater shoulder abduction, which causes more stress on the shoulder joints and shoulder/neck. Many health specialists, such as orthopedists and physical therapists assume that de-conditioning of the trunk and lumbar spine structures occur due to long-term sitting without longer active periods of standing, walking or running (Mörl & Bradl, 2013). The same authors affirm that this de-conditioning may be a reason for low back pain and for the accelerated degeneration of lumbar spine structures. Due to all these adverse effects, some authors suggest that it is important to combat occupational sedentariness, by rethinking, and redesigning the way people work (McCrady & Levine, 2009). Mörl and Bradl (2013) stated that to reduce the high prevalence of low back pain in sedentary work, reasonable prevention is necessary. While seated, the lumber muscles have low activation rates, so it is possible to conclude that the use of special office chairs to protect the spine or to train the paravertebral muscles will fail since the muscle activation depends more on the task than on the office chair used (van Dieën et al., 2001). Thus, some researchers conclude that the only way to prevent adverse effects is to increase physical activity in the workplace, promoting postural changes. Still, it is difficult to define with precision the amount of time that should be spent on each working posture since the optimal proportion of standing to sitting is unknown (Messing et al., 2008).

2.3 *Excessive standing*

There are still many professions, such as retail workers, cleaners, security guards, supermarket checkout employees, quality control and assembly workers, and health care staff that require workers to adopt a standing posture during the whole work day (Bahk et al., 2012). According to Balasubramanian et al. (2009), the standing posture can be divided in dynamic standing (in which a worker intermittently walks while on the job) and stationary standing (in which a worker does not walk, but stands still, while on duty). Most industrial jobs are characterized by a stationary standing posture, however, the dynamic ergonomic posture is not universally employed (Messing & Kilbom, 2001). Depending on the job, a standing posture provides a more stable condition for the low back by preserving the natural lordosis of the lumbar spine (Andersson, 1979). Standing also allows for dynamic use of the arms and trunk, which is better for handling loads, and enables workers to cover larger workspace areas because of the ability to move (Lehman et al., 2001). According to Bridger (1995), standing work is better than sitting work since the reach is greater, the body weight can be used to apply forces, requires less space for the legs, the lumbar disk pressure is lower and it can be maintained with little muscular activity. The investigation carried out by Balasubramanian et al. (2009) showed that during 1h of mechanical assembling operations, the subjects demonstrated the appearance of fatigue in lower extremity muscles at a much faster rate in stationary standing than in dynamic standing. The same authors indicated that along with the fatigue the perceived pain and discomfort in the lower extremity muscles was also relatively high during the stationary standing. Other authors refer different effects of the prolonged stationary standing posture, such as the ones presented in Table 2. One of the most studied effects of prolonged standing postures is the appearance of varicose veins and leg cramps. Several studies demonstrated that the risk for varicose veins is associated with different aspects, e.g. age, female gender, family history, pregnancy, obesity and prolonged standing or sitting (Beebe et al., 2005; Ahti et al., 2010). In the work of Bahk et al. (2012), it has been shown that women had a higher prevalence of varicose veins and nocturnal leg cramps than men. However, the occupational characteristics of the job could be more predictive of the prevalence of varicose veins than gender itself. Concerning nocturnal leg cramps, the same study showed that women had higher prevalence of leg cramps than men, regardless of their work posture.

Table 2. Effects of prolonged stationary standing.

Author	Identified effects
Hansen et al., 1998	Chronic venous insufficiency, leg swelling, discomfort and tiredness
Dempsey, 1998	Muscle fatigue aggravation, neck and shoulder stiffness
Krause et al., 2000	Progression of carotid atherosclerosis
Cham & Redfern, 2001	Lower extremity discomfort
Messing et al., 2008	Lower extremity discomfort fatigue and swelling; low back pain, and entire body fatigue
Ngomo et al., 2008	Orthostatic intolerance
Tissot et al., 2009	Musculoskeletal disorders in the back
Bahk et al., 2012	Higher prevalence of varicose veins and nocturnal leg cramps

3 DISCUSSION

Many of the effects mentioned before, such as foot and leg swelling, reduced circulation, varicose veins, and lower extremity discomfort are associated with both prolonged sitting and prolonged standing (Sadick, 1992). As such, it is important to note that a posture that causes pain or discomfort is generally harmful for workers since it can lead to WRMSDS. These disorders will reduce the working capacity and consequently cause productivity losses and can lead to work disability (Escorpizo, 2008). King (2002) stated that the effects of musculoskeletal disorders were associated with absenteeism, lack of productivity and decreased well-being. It is possible to conduct a study of the variance in anthropometric measurements to assess if the workplace, as well as the working postures adopted, are contributing to the development of WRMSD. In these cases the use of 3D anthropometric data allows the study to be more effective since, when compared to the traditional anthropometry methods it has many advantages. Traditional anthropometry uses devices such as calipers and tapes to determine the dimensions of the human body. According to Pargas et al. (1997), apart from being tedious, inconsistent and inaccurate, when the manual measurement procedure is made by different people it might have several variations: compaction of flesh during measurement, inconsistent land-marking (palpating for specific points generally located at bone prominences) and tension of the measuring tape. Even if measurements are taken by the same person it is possible to have lack of consistency throughout the day when that person gets tired (Pargas et al., 1997). With the 3D body scanners capturing the body dimensions is fast and can be reproduced almost exactly the same way all the time. One of the most important benefits of this type of measuring procedure is that the data of the subjects can be stored and accessed when necessary (Daanen & Water, 1998). Furthermore, the number of anthropometric variables that can be derived from a scanned human body is almost without limits. Nevertheless, Daanen and Water (1998) pointed out some disadvantages of 3D-scanning compared to traditional anthropometry, e.g., the initial investment on a body scanner, camera blocking effects in arm pits and crotch, light absorption by the hair and skin, moving artifacts and reliable data processing and handling.

Once problems are identified it is very important to act in order to prevent poor working conditions. This can be accomplished by implementing ergonomic interventions. There are many cases where the ergonomic interventions used during prolonged standing tasks reduced, but did not eliminate completely, the risk of WRMSD (Hasegawa et al., 2001; Messing & Kilbom, 2001; Chiu & Wang, 2007). Using sit/stand chairs or workstations, wearing soft shoes, using shoe insoles, wearing compression stockings, using foot rests, standing on soft surfaces or standing on floor mats are examples of improvements to the work environment that can be made to reduce leg swelling, discomfort and fatigue in the lower extremities (Hansen et al., 1998; Madeleine et al., 1998; Chester et al., 2002; King, 2002). Even though sit/stand chairs are a popular solution, their use might not be very effective, as shown in the work of Chester et al. (2002) demonstrated where using sit/stand chairs caused the most swelling (when compared to standing and sitting postures). Also, some publications showed that implementing sit/stand workstations in an office environment leads to lower levels of whole body discomfort without resulting in a significant increase in performance (Karakolis & Callaghan, 2013; Davis et al., 2009). Several publications showed that the floor type likely plays an important role in discomfort while standing. Many workplaces have installed soft floors or floor mats in order to reduce the leg muscle discomfort during prolonged standing (Madeleine et al., 1998). Lin et al. (2012) discovered that subjective discomfort ratings were related to floor type, shoe condition, and standing time. Making various proscribed leg movements (Lin et al., 2012), having frequent sitting breaks and including an optional seat or a footrest increases the variety of body positions available for a worker and encourages frequent changes between them, resulting in less discomfort and swelling in their lower extremities (Sartika & Dawal, 2010). The work from Winkel and Jorgensen (1998) demonstrated that leg and foot activity reduces swelling and increases the blood circulation. Thus, a static work posture, whether it is standing or sitting is discouraged since changes in work posture are important in reducing fatigue (Kroemer & Robinette, 1969). The study of Hansen et al. (1998) showed that standing work without any motion or walking caused greater musculoskeletal discomfort than a combination of standing or walking tasks. Considering the differences between and similarities of the two postures, the choice must reflect the requirements of the tasks to be performed. At the same time ergonomic design should be considered since it might reduce the risk of acquiring lower extremity disorders and may have a positive impact on productivity enhancement (Balasubramanian et al., 2009).

4 CONCLUSIONS

As workers' productivity and well-being relies on working conditions, evaluating the negative effects caused by the work postures assumes a very important role. Working postures that are 'wrong' (or extreme) or that are adopted for long periods of time may result in WRMSD. These disorders may put at risk companies' competitiveness since they are and absenteeism. Accordingly, knowledge about the adverse effects of different working postures is essential. As such, being able to determine the anthropometric changes related to each work posture is one of the new concerns of anthropometry applied to occupational ergonomics. To do so, it is now possible to use new measuring techniques, such as the case of 3D anthropometry data by using 3D scannersas a more efficient and rapid data collection method than traditional anthropometry.

The authors of this paper are already involved in a research project, as part of a PhD project that is based on this identified need, i.e. the need to understand the implications of the working posture for the workers' anthropometrics. In this project, aspects that are being considered are: the determination of the modifications in the human body that occur with each posture (as well as understanding how quickly they happen) or identifying the anthropometric changes that can be more harmful for the workers. And last but not the least, determining the percentage of time that should be spent in each posture for the greatest health and productivity outcomes.

REFERENCES

Ahti, T., Makivaara, L., Luukkaala, T., Hakama, M., & Laurikka, J. (2010). Lifestyle factors and varicose veins: does cross-sectional design result in underestimate of the risk? *Phlebology*, 25(4), 201–206.

Andersson, G., Murphy, R., Ortengren, R., & Nachemson, A. (1979). The influence of backrest inclination and lumbar support on lumbar lordosis. *Spine*, 4(1), 52–58.

Bahk, J., Kim, H., Jung-Choi, K., Jung, M.-C., & Lee, I. (2012). Relationship between prolonged standing and symptoms of varicose veins and nocturnal leg cramps among women and men. *Ergonomics*, 55(2), 133–139.

Balasubramanian, V., Adalarasu, K., & Regulapati, R. (2009). Comparing dynamic and stationary standing postures in assembly task. *Int. J. Industrial Ergonomics*, 39, 649–654.

Barroso, M., Arezes, P., da Costa, L., & Sérgio Miguel, A. (2005). Anthropometric study of Portuguese workers. *Int. J. Industrial Ergonomics*, 35(5), 401–410.

Beebe-Dimmer, J., Pfeifer, J., Engle, J., & Schottenfeld, D. (2005). The epidemiology of chronic venous insufficiency and varicose veins. *Annals Epidemiology*, 15(3), 175–184.

Bridger, R.S. (1995). Introduction to ergonomics.

Cham, R., & Redfern, M. (2001). Effect of flooring on standing comfort and fatigue. *Human Factors*, 43(3), 381–391.

Chester, M. Rys, M., & Konz, S. (2002). Leg swelling, comfort and fatigue when sitting, standing, and sit/standing. *Int. J. Industrial Ergonomics*, 29(5), 289–296.

Chiu, M., & Wang M. (2007). Professional footwear evaluation for clinical nurses. *Applied Ergonomics*, 38, 133–141.

Daanen, H., & van de Water, G. (1998). Whole body scanners. *Displays*, 19(3), 111–120.

Davis, K., Kotowski, S., Sharma, B., Herrmann, D., Krishnan, A. (2009) Combating effects of sedentary work: postural variability reduces musculoskeletal discomfort. *Proc. Human Factors and Ergonomics Society*.

Dempsey, P. (1998). A critical review of biomechanical, epidemiological, physiological and psychophysical criteria for designing manual materials handling tasks. *Ergonomics*, 41(1), 73–88.

Escorpizo, R. (2008). Understanding work productivity and its application to work-related musculoskeletal disorders. *Int. J. Industrial Ergonomics*, 38(3–4), 291–297.

Grandjean, E. (1988). Fitting the Task to the Man (4th ed.). London: Taylor & Francis.

Hansen, L., Winkel, J., & Jorgensen, K. (1998). Significance of mat and shoe softness during prolonged work in upright position: based on measurements of low back muscle EMG, foot volume changes, discomfort and ground force reactions. *Applied Ergonomics*, 29(3), 217–224.

Hasegawa, T., Inoue, K., Tsutsue, O., & Kumashiro, M. (2001). Effects of a sit-stand schedule on a light repetitive task. *Int. J. Industrial Ergonomics*, 28(3–4), 219–224.

Healy, G., Clark, B., Winkler, E., Gardiner, P., Brown, W., & Matthews, C. (2011). Measurement of Adults' Sedentary Time in Population-Based Studies. *American Journal of Preventive Medicine*, 41(2), 216–227.

Hsiao, H., Long, D., & Snyder, K. (2002). Anthropometric differences among occupational groups. *Ergonomics*, 45(2), 136–152.

Hsiao, H., & Halperin, W. (1998). Occupational safety and human factors. *Proceeding of the Environmental and Occupational Medicine*, Philadelphia, PA: Lippincott—Raven.

Istook, C., & Hwang, S. (2001). 3D body scanning systems with application to the apparel industry. *J. Fashion Marketing and Management*, 5(2), 120–132.

Karakolis, T., & Callaghan, J. (2013) The impact of sit-stand office workstations on worker discomfort and productivity: A review. *Ap. Ergonomics*.

Katzmarzyk, P., Church, T., Craig, C., & Bouchard, C. (2009). Sitting Time and Mortality from All Causes, Cardiovascular Disease, and Cancer. *Medicine and Science in Sports and Exercise*, 41(5), 998–1005.

King, P. (2002). A comparison of the effects of floor mats and shoe in-soles on standing fatigue. *Applied Ergonomics*, 33(5), 477–484.

Krause, N., Lynch, J., Kaplan, G., Cohen, R., Salonen, R., & Salonen, J. (2000). Standing at work and progression of carotid atherosclerosis. Scandinavian *Journal of Work Environment & Health*, 26(3), 227–236.

Kroemer, K., & Robinette, J. (1969). Ergonomics in the design of office furniture. *Industrial medicine and surgery*, 38(4).

Lehman, K., Psihogios, J., & Meulenbroek, R. (2001). Effects of sitting versus standing and scanner type on cashiers. *Ergonomics*, 44(7), 719–738.
Madeleine, P., Voigt, M., & Arendt-Nielsen, L. (1998). Subjective, physiological and biomechanical responses to prolonged manual work performed standing on hard and soft surfaces. *Europ. Journal of Applied Physicology*, 77, 1–9.
McCrady, S., & Levine, J. (2009). Sedentariness at Work: How Much Do We Really Sit? *Obesity*, 17(11), 2103–2105.
Messing, K., & Kilbom, A. (2001). Standing and very slow walking: foot pain-pressure threshold, subjective pain experience work activity. *Applied Ergonomics*, 32(1), 81–90.
Messing, K., Tissot, F., & Stock, S. (2008). Distal lower-extremity pain and work postures in the Quebec population. *American Journal of Public Health*, 98(4), 705–713.
Morl, F., & Bradl, I. (2013). Lumbar posture and muscular activity while sitting during office work. *Journal of electromyography and kinesiology*, 23(2), 362–368.
Ngomo, S., Messing, K., Perrault, H., & Comtois, A. (2008). Orthostatic symptoms, blood pressure and working postures of factory and service workers over an observed workday. *Applied Ergonomics*, 39(6), 729–736.
Pargas, R., Staples, N., & Davis, J. (1997). Automatic measurement extraction for apparel from 3D body scan. *Optics and Lasers in Engineering*, 28(2), 157–172.
Pheasant, S., & Haslegrave, C. (2006). Bodyspace: anthropometry, ergonomics, and the design of work: CRC Press.
Pi-Sunyer, F.X. (1999). Comorbidities of overweight and obesity: current evidence and research issues. *Medicine and Science in Sports and Exercise*, 31(11), S602-S608.
Pottier, M., Dubreuil, A., & Monod, H. (1969). Effects of sitting posture on volume of foot. *Ergonomics*, 12(5), 753.
Roebuck, J., Kroemer, K., & Thomson, W. (1975). Engineering anthropometry methods: Wiley-Interscience, NY.
Sadick, N. (1992). Predisposing factors of varicose and telangiectatic leg veins. *Journal of Dermatologic Surgery and Oncology*, 18(10), 883–886.
Seo, A., Kakehashi, M., Tsuru, S., & Yoshinaga, F. (1996). Leg swelling during continuous standing and sitting work without restricting leg movement. *Journal of occupational health*, 38(4), 186–189.
Shvartz, E., Gaume, J., Reibold, R., Glassford, E., & White, R. (1982). Effect of the circutone seat on hemodynamic, subjective and thermal responses to prolonged sitting. *Aviation Space and Environmental Medicine*, 53(8), 795–802.
Straker, L., Abbott, R., Heiden, M., Mathiassen, S., & Toomingas, A. (2013). Sit-stand desks in call centres: Associations of use and ergonomics awareness with sedentary behavior. *Applied ergonomics*, 44(4), 517–522.
Tigbe, W., Lean, M., & Granat, M. (2011). A physically active occupation doesn't result in compensatory inactivity during out-of-work hours. *Preventive Medicine*, 53(1–2), 48–52.
Tissot, F., Messing, K., & Stock, S. (2009). Studying the relationship between low back pain and working postures among those who stand and those who sit most of the working day. *Ergonomics*, 52(11), 1402–1418.
Treleaven, P., & Wells, J. (2007). 3D body scanning and healthcare applications. *Computer*, 40(7), 28–34.
Toomingas, A., Forsman, M., Mathiassen, S., Heiden, M., & Nilsson, T. (2012). Variation between seated and standing/walking postures among male and female call centre operators. *Bmc Public Health*, 12.
Wichansky, A. (2000). Usability testing in 2000 and beyond. *Ergonomics*, 43(7), 998–1006.
Winkel, J., & Jørgensen, K. (1986). Swelling of the foot, its vascular volume and systemic hemoconcentration during long-term constrained sitting. *Europ. Journal of applied physiology and occupational physiology*, 55(2), 162–166.

Occupational Safety and Hygiene II – Arezes et al. (eds)

Occupational hygiene and work safety in a shoe company in Campina Grande

L.O. Rocha, M.B.G. Santos, I.F. Araújo & E.S.P. Sales
Federal University of Campina Grande, Campina Grande, Paraíba, Brazil

ABSTRACT: Damage to health and accidents can be devastating and disabling, resulting in illnesses caused by agents of physical hazards, ergonomic and chemicals present in the stage production of footwear, as well as accidents, leading to amputation of limbs and crushing. This study aims to identify environmental risks and recommend control measures upon propositions for improvements in a footwear industry in the city of Campina Grande—PB. The methodology covers literature, case study site visits questionnaires, checklists and interviews with the managers of the organization with a focus on environmental hazards in the machinery and usability of personal protective equipment, PPE. It was concluded that the company needs to create an Internal Commission for Accident Prevention, CIPA, introducing signaling areas of risk, as well as conduct trainings and capacity building for employees to implement measures of work safety in their jobs.

1 INTRODUCTION

The accident at work is considered the most important health problem around the world, to be potentially disabling and fatal and affect mainly young people or economically productive, which leads to social and economic consequences of high relevance to society.

Despite the severity of their negative consequences, workplace accidents continue to occur in high numbers and victimizing thousands workers. The work done unsafely, with risk of death or injury to the physical integrity of the worker, there is a decent job, because, according to the International Labour Organization, to be considered decent must be exercised safely.

In Brazil, the Work safety Law consists in 36 Regulatory Standards, complementary instruments, such as regulations and decrees as well as well as the consideration of the International Labour Organization, ratified by the country Companies must comply with extensive laws Brazilian work safety and maintain on its staff a team responsible for this sector. This concern is justified, since damage to health and accidents can be devastating and incapacitating the worker linked to any productive sector, causing personal distress, social problems and economic losses for the country.Brazilian footwear industry consists of about four thousand companies, thus generating 260,000 jobs. It is estimated installed capacity of 560 million pairs per year, which 70% is allocated to the domestic market and 30% for export, with a total turnover of $ 8 billion per year.

Brazil stands as the third largest producer of shoes, according to the MTE (2010) have been reported 6,831 accidents in a population of 377,330 registered workers in the shoe industry, generating an incidence rate (18.1/1000). This rate, although higher than that considers all economic sectors of the country (15.0/1000), is much lower than the incidence rate for the industry in general (29.7/1000). These accidents occur as a result of unsafe conditions, which employees are exposed in the workplace and/or unsafe acts, which are acts committed by employees, resulting from ignorance of the risks, or attitudes that neglect the risk and danger. Measures to prevent these agents causing accidents deserve to be studied, since only the use of individual protection equipment (PPE), is not sufficient to minimize this accident scenario.

One can find in this industry, agents chemical, which causes damage through the use of toxic substances such as, glues, solvents, which can cause abnormal liver function, nervous system, blood and skin, as well as agents physical risks, such as exposure to excessive noise, causing changes in hearing, poor lighting and exposure to excessive heat, which can trigger vision problems and heat stress, respectively; agents ergonomic hazards such as awkward postures and repetitive movements, may cause RSI/WMSD.

However, most of the injuries are from accidents typical of work by officers from mechanical injury, found mainly in the sectors of machinery cutting and assembly companies, which can reach in hands, fingers and arms, and may result in amputations located and even total loss of members as a result of crushing, pressings and deep wounds.

This study is based on regulatory standards—NR, specifically NR-4 (specialized services in engineering safety and occupational medicine), NR-5 (commission of internal accident prevention), NR-12 (safety in machine and equipment) and NR-15 (activities and operations unhealthy) to evaluate and propose solutions in order to control and minimize occupational hazards in the working environment in the shoe company.

Given the above, this research aims to identify environmental risks and control measures by proposing suggestions for improvements in a shoe company in the city of Campina Grande, Paraíba State, Brazil.

2 CHARACTERIZATION OF THE COMPANY

The Karmélia Shoes was founded in May 1999, is a small business that produces an average of 1800 pairs of shoes per month. The organizational structure is composed by a president, manager, supervisor, secretary and over twenty (20) employees, a total of 24 (twenty four) employees. The steps of the production process of shoes cover the following sectors: cutting, shaping, sewing, assembly, finishing and shipping. The company operates from Monday to Friday in the morning and afternoon shifts, 8 hours per day.

3 MATERIALS AND METHODS

To achieve the objective of this research, we used three methodological aspects:

- Literature review, through research in primary and secondary sources, in books, technical articles and scientific journals, as well as consulting to various sites related to the topic discussed.
- Identification of environmental agents by applying sheets of check or inspection check list of work safety in areas of the company, which addressed the following aspects: machinery and equipment, fire protection, infrastructure of buildings, the facilities and services in electricity, transport and handling, storage and material handling and work safety signs.
- Identification of the profile of the importance of design and usability of PPE, using semi structured interviews and direct personal questionnaires with both the management, the employees of the production sector. The questions were related to the time of use, comfort level, motivation, knowledge of the importance and symptoms related to the consequences of non—use of PPE's. The instruments of data collection techniques were implemented during regular visits to the organization, aiming to observe, in situ and in situ, the procedures adopted under the aegis of safety and health at work.

4 RESULTS AND DISCUSSION

4.1 *Profile of company employees*

Regarding the profile of the employees who were interviewed, the average time of service in the organization was seven (7) years, which the extremes were found (1) year for the newest employee and thirteen (13) years for the older employee in the company.

According to the questionnaire, it was found that all employees have had experience in the footwear industry, and in relation to the variable age, the age of the hand labor was among 22 (twenty two) years to 54 (fifty-four) years with its range of 32 (thirty two) years.

It was found that 66.67% of the employees are male and 33.33% female. There is a higher concentration of men at work, because it is considered heavy and traditional local culture, where women take over the functions related to finishing and packaging of footwear.

4.2 *Sizing CIPA*

According to Norm 05 of the Ordinance 3.124/78, the Ministry of Labour and Employment—MTE, the Commission for the Prevention of Accidents—CIPA—aims at the prevention of accidents and illnesses resulting from work in order to make it compatible permanently working with the preservation of life and health promotion worker.

The NR-5 also provides us who should be the CIPA, its organization, duties, operation, training participants CIPA will the electoral process, as well as its design according to the number of employees and the National Classification Economic Activities—NCEA.

CIPA is mandatory in the private, public, semipublic companies, government agencies directly and indirectly, charities, cooperatives, and other institutions that admit workers as employees. Its composition (number of representatives) depends on the number of employees in the establishment and activity branch (SALIBA, 2010).

The company does not have a CIPA, but according to the National Classification of Economic

Activities NCEA, the organization is classified with the NCEA 15.31-9 (production leather shoes), and according to Table II of NR-5, the company belongs to grouping of the economic sector C-5 (Footwear and similar), moreover, the company has 24 employees. Based on this information, the NR in Table I-5 is scaled size of CIPA, which is one member and one alternate member, and the organization required to comply with legislation.

4.3 *Sizing SESMT*

By Table I, NR-4, the company has 3 degree of risk and by matching the number of employees and the degree of risk in Table II of NR-4, the company is not legally obliged to provide a SESMT. But the organization has a SESMT in partnership with other public and private institutions in the city which is responsible for the clinical examinations dimensionless and monitoring the health of workers.

4.4 *Signaling and fire risk*

The company works with flammable chemicals such as adhesive glue, and easily combustible materials such as leather, PVC and rubber. In relation to fire prevention, the organization has a mobile system of combat that are four fire extinguishers distributed by factory plant, the equipment conform to the classification of types of fires. The emergency signaling is deficient. Lack signaling in some missing fire extinguishers and exit signs to the factory and hazardous areas such as energized air.

4.5 *Machines and equipment*

The machines presented are semi-new, following the specifications of the NR12 with work safety devices (lock accidental activation and emergency stop), in which live parts are properly insulated operator, as well as belts and gear motors are protected by components that prevent access to dangerous parts of the machine.

Air compressors are arranged in an isolated room to reduce the noise produced by the operation, which is a measure of collective protection. They receive preventive maintenance with the use of ultrasound techniques to detect cracks and corrosion. Complying with NR12, because they have a protection on the mobile belt force distribution in the engine.

The rocker cutting machine is already equipped with work safety device drive (preventing the machine from accidentally triggered), starting and stopping (if any accident happens to the equipment operator). Furthermore, the operation control of the machine is two hands, to maintain the operator's hands out of the danger area, or cutting the material to be worked. Complying with NR12, which deals with the work safety equipment and mechanical operations.

The manual cutting is performed when the minimum lot size shall not compensate financially for ordering knives Rocker mechanical cutting the shape of shoes being produced. This activity has risks of accidents at work, because the employee uses sharp tools (knives, box cutters, industrial scissors) to cut the material, there is no use of PPE in this activity.

In the area of notch, the part of the footwear undergoes little wear at the edges to form the product, there was identified the sanding dust from the leather piece and the lack of use of PPE (mask and goggles) is ergonomic hazards because the machinery and accent work are not adapted anthropometric measurements of the contributor, thus causing a greater physical wear during work.

In this sector of folds, the leather undergoes deformation to fit the format type of footwear to be produced. It was identified that employees use protective masks (for dust and waste from the leather) and there accent adaptation measures physical operator.

In industrial sewing seams are made from different parts which will make up the footwear, according to NR 12, the sewing machines have work safety devices which prevent the needle in the event of breakage, does not reach the face of the operator.

In the sector of sanding, the sole of the footwear is crafted to receive adhesive glue and nails to be placed. Dust particles were observed and the materials that are thrown toward the operator. The machine is coupled to an extractor where it is a part responsible for retaining the dust. Although this job was observed negligible risk of an accident if the operator touches the fingers on the disc of sandpaper, because of existent protection.

The machine to conform, which is the equipment responsible for shaping the heel of the shoe, this job was found crash risk, because the operator's fingers must hold the leather shoes and hit the power button, the physical wear of workday, the employee may forget to hand in the danger area where the machine will conform the sole of the shoe thus causing an accident.

The hydraulic presses are in accordance with the NR 12, and in accordance with the work safety standards the drive is only possible via a front panel in the operator's hands are away from the danger area, where the press will cause the pressure to shape the footwear. In addition, the machines have drives and emergency shutdowns insulation of electrical parts.

The nail machine, which is made the preaching of the sole of the shoe, has work safety system and emergency stop footswitch, reducing the risks of accidents, but the task should be performed by the operator with goggles to prevent any nail is not released in the direction of his face, reaching the eyes and cause serious injury to the employee.

In real bonding is the use of adhesive glue the base of aromatic hydrocarbons, no contact with the glue operator's hands in that continuous exposure to this chemical can cause occupational diseases during the life of the operator. No use of the filter mask to avoid the absorption of chemicals through the airways, and there is no use of other PPE that reduce or mitigate the exposure of employees to chemical adhesive glue.

4.6 *Electrical installations*

As the NR 10, the electrical installations of the company are arranged not offering accident risks for workers. There grounding cables in machines, minimizing the risk of electric shock by leakage current in metal housing. Furthermore, the electrical outlets are arranged so that the fitting with the forces of the cables and folium machines do not disconnect causing injury. In the general framework, the breakers and wiring properly sized shape to withstand the electrical current, minimizing the risk of fire.

4.7 *Physical risks*

As NR 15 in Annex 01, were found agents Physical risks: noise in the room of compressors and heat Annex 06, indicated by all workers in all jobs in the company on the hottest days where ventilation artificial is insufficient.

As questionnaire, Figure 1 shows according to the company's employees the main agents of physical risks found that workers are exposed.

There is a higher concentration in risk agent noise, but the same overall had less impact on the equipment which caused more noise intensity is enclosed in a separate room from the rest of the plant, reducing the intensity of the noisy source. No interference of external factors on the company as the weather. Ventilation is artificial, but according to officials, in warmer weather ventilation is not sufficient to ameliorate the symptoms of heat and the coldest days the plant is influenced by the external environment, cooling the environment.

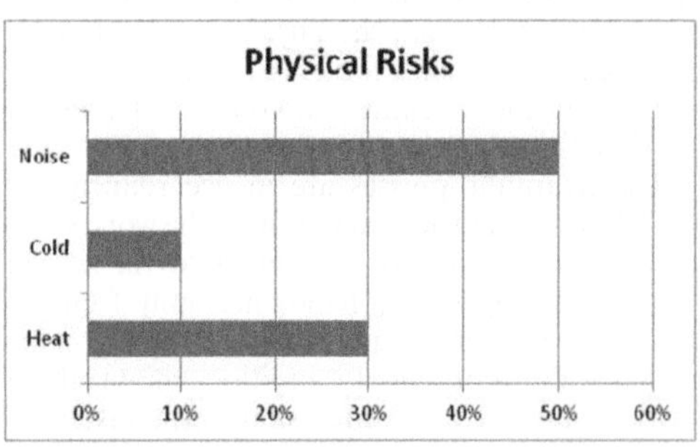

Figure 1. Agents of physical hazards found in the company.

4.8 *Chemical hazards*

The dust in the sectors of chamfer grinding and bending the shoe, the adhesive glue the base of aromatic hydrocarbons were identified as the main agents of chemical hazards that workers are exposed. No use of equipment to avoid skin contact of the operator with the glue and protective mask to avoid contamination of the airways by volatile chemical glue. As this chemical agent is very volatile, spreads easily by the factory and not all the other workers wear protective equipment to mitigate the effects of the strong smell of the product. Prolonged exposure to this chemical may cause the worker to work-related diseases in the future. There dispersion glue throughout the plant due to the volatility of the product causing headaches by strong chemical smell.

4.9 *Ergonomic risk*

The lack of standardization of the height of countertops fitting shoes and lack of accents adjustable anthropometric measures of operators, cause discomfort and fatigue potentiate the workday.

Main ergonomic hazards found in the company

- Lack of height adjustment in the seats of the machines (notching and sew).
- Banks without adjustable height and back support in the industry "stamp" brand in the sole of the shoe.
- Lack of support for the feet in the folding machine.
- Lack of standardization accents and the heights of the countertops assembly of shoes.

4.10 *Accident risks*

The floor irregular gaps, crack in the wall and a large amount of material and tools, in the process, block traffic routes and emergency exit from the company.

4.11 *Signs and fire prevention*

The signaling is deficient in some specific company missing information on the type of extinguisher, the indication of danger spots (energized) and emergency exit. It would be necessary to sign the emergency exit signs with bold colors and contrast, as the NR 26, the dangerous parts of each machine and the locations of extinguishers range in floor plate and indicative of the type of extinguisher on the wall. Adopt training for prevention and fire-fighting explaining classes, handling and types of extinguishing agents.

4.12 *Use of PPE*

Was found in the company that the use of PPE by employees, is not charged or supervised by management of the organization, and its use is almost as optional (employee uses when he wants to). In some jobs it was found the use of PPE by some operators who knew the importance of the equipment in the control of environmental risks and the integrity of their health, as in other jobs employees were not wearing PPE for the discomfort that the equipment caused (even knowing the importance of the use) or lack of education, training and equipment.

4.13 *Sanitary facilities and drinkers*

The company adheres to the scaling of the number of toilets NR-24, which the number of employees and gender are two toilets (one male and one female). The cooler comprises four workers once being supplied to need.

4.14 *Work safety and health at work in general*

Work safety in the company is still not treated with due attention, CIPA has not been established, there is no political training and incentives for safety and health at work are part of the routine activities of the organization. As adaptation brochure SESI/DN (2008), the organization should take the following immediate steps to prevent accidents and occupational diseases:

4.15 *Immediate measures to adapt to workplace safety*

- Use of workbenches in height and position avoiding fatigue proper motion and lumbar unnecessary efforts.
- Ban transportation of loads whose weight could compromise the health or safety of workers.
- Implantation of artificial ventilation dilutive in the workplace.
- Adoption of short breaks to rest especially in activities with a high degree of repeatability.
- Adequate facilities observing the intensity of fixtures, positioning, distribution and height.
- Use quality tools appropriate to the work that will be performed.
- Correct use of tools and always keep them in good condition.
- Save the tools in suitable devices, keeping them avoiding accidents.

5 CONCLUSIONS

It requires the creation of a CIPA in the company as well as the adoption of policies on education, awareness, and providing oversight regarding the use of PPE, particularly regarding the use of filter mask protection sectors beveling, bending and grinding shoes, use of protective eyewear for the employee who is operating the machine nails and gloves, filter mask and apron for the operator responsible for bonding.

- The risk map should be used to train and educate operators about the importance of prevention of environmental risks in the areas of business;
- Should be implemented signaling: types and locations of fire extinguishers; dangerous parts (energized) the company; emergency exit;
- The use of brush and appropriate tool should be required to pass the glue on shoes and avoid contact with the fingers of the operator;
- Hood should be used in the sector of glue to lessen the effects of the dispersion of cola in the enterprise environment, allowing the renewal of the air;
- Should be adopted countertops with accents and adjustable heights standardized measures of operators in the assembly sector, so as to be adopted with accents adaptable backrest and footrest in the areas of sewing, bevel and bend the shoe.

REFERENCES

Abreu, V. 2010 American Industrial Hygiene Association—AIHA, cited in the Encyclopedia of Occupational Work safety and Health, the International LabourOrganisation—ILO.

Barbosa F., Nunes, A. Workplace Work safety & Environmental Management. 3rd edition São Paulo: Editor Atlas, 2010.

Leaflet—Work safety and Occupational Health-SST in the industry. Brasilia: SESI/DN, 2008. Available in: on September 10, 2013.

Macêdo, W.S. Analysis and proposal of a model of the physical arrangement of prevention and fire fighting and panic in a shopping popular. 2011. 92 P.T.C.C.— Course of Production Engineering, Department of Production Engineering, UFCG, Campina Grande, in 2011.

Ministry of Labour and Employment of Brazil. List of regulatory Standards (4, 5, 12 and 15). Available at: <http://portal.mte.gov.br/legislacao/normas-regulamentadoras-1.htm>. Accessed August 10, 2013.

Saliba, T.M. Basic course in occupational health and work safety. Sao Paulo: LTr. 2010.

Occupational Safety and Hygiene II – Arezes et al. (eds)
 ISBN 978-1-138-00144-2

Safety management at small establishments with major hazards: The case of galvanic industry

P. Agnello, S. Ansaldi, P. Bragatto & A. Pirone
INAIL Italian Workers' Compensation Authority, Research Certification and Verification Area, Italy

ABSTRACT: European REACH Regulations and Seveso Directives are driving the Italian Galvanic Industry toward a management system for the major hazards adapted for small and micro sized companies. This is a good chance to experiment the new idea on safety management in small enterprises that has been discussed for a few years in scientific community. To support galvanic industry a simplified model has been developed. In the proposed model, major accident control and occupational safety have been harmonized and integrated. A web based software has been implemented, aimed to support the application of safety procedures at the shop floor. The software is based on a sound digital representation of safety, which connects procedures and equipment. A strength of the software is the mobile reporting of anomalies and near misses. In such a way all workers may be involved in finding the weaknesses of the system and cooperate into continuous improvement. The inspections at a few galvanic plants have provided the opportunity for testing the proposed model and software.

1 INTRODUCTION

Beneficial effects of Safety Management System (SMS) for reducing occupational accident and diseases are widely recognized. The SMS implementation defines in a better way roles and procedures with short term benefits, whilst the continuous improvement process should bring long term benefits. Furthermore it is mandatory for controlling major accident hazard, according to the Directive Seveso II 96/82/CE, as well as the new replacing Directive Seveso III 2012/18/UE. These benefits are confirmed for many industries by empirical studies, including Bottani et al. (2009). SMS has been designed mainly following the requirements and needs coming from major companies, maintaining the compliance with the legislation in force; SMS should also be a powerful system for Small Sized Enterprises (SMEs), its adoption as is may be difficult but with further efforts it can be suitable. The weaknesses of SMS for SMEs have been discussed by many authors in recent years. Reiman and Rollenhagen (2011) say that SMS tends to be too static, whilst in SMEs has to be more dynamic. Kristensen (2011) observes that, under the new economy, operators are urgently needed to deal with constantly changing and novel competitive situations, while SMS activities (e.g. audits) may stick to bureaucratically ordained tasks that are repetitive and easily ignored. According to Grote (2012) there are two basic approaches to managing uncertainty in organizations: (1) minimizing uncertainty, and (2) coping with uncertainty. These two paradigms aim to either maximize stability or flexibility. Flexibility at small, innovative companies is essential for handling high levels of uncertainty caused by frequent external changes and/or variances and disturbances in work processes, for fulfilling high demands on innovation. Following Grote's schema, incidents and near misses reporting have different meaning in larger (assumed stability oriented and uncertainty reducing) and smaller (assumed flexibility oriented and uncertainty coping) companies. At SMEs incidents and near misses reports are expected as a source of mutual support and information sharing. McGuinness et al. (2012) highlight how it is important to learn from incidents, such as near misses, in order to find out how to prevent accidents in the organization. The personnel must participate in incident investigation and analysis of events for learning outcomes. These remarks are even more relevant for the SMEs engaged in activities with major hazards, including the loss of hazardous materials, with consequent fire, explosion or toxic release in air or in water. A typical example of small business with major hazard are the galvanic plants that fall under the Seveso Directives even though enterprises typically have just a dozen of workers.

According to this considerations, two aspects have been focused in the present paper: the integration of Occupational Safety and Health (OSH) with major accident hazard (Seveso) procedures and the potential of web based solution for supporting the actual application of the procedures at the shop floor, involving all workers in safety improvement. The target of this research is the Italian galvanic industry, but the current solution has been implemented by adopting a general model which is also suitable for other specific small sized companies.

2 ITALIAN GALVANIC INDUSTRY

The Italian galvanic industry is still the strongest in Europe and involve some 800 companies. The typical size of the companies is 8–10 workers, with some one million euro annual turnover. The most plants are chrome and nickel plating (55%), the other types include zinc-plating (20%), gold and silver-plating (5%). Most companies work for third parties and supply many industries, including durable consumer goods, automotive, electronic, medical equipment and aerospace. The market is highly competitive and enterprises must be flexible and adaptive to survive. Plants may be specialized for the treatment of metal or non metal surfaces. The core of the plants are the electrolytic bath contained in open tanks where parts to treat are dipped. The major accident hazards include the leakages from the bath and the consequent air or water contamination, the loss of cyanides, as present in a few processes, the fires in pre and post treatment sections, the toxic undesired emission due to the fall of metal items in acid tanks.

Due to the new classification of hazardous substances, according to the REACH Regulation, many galvanic plants in Italy are falling under the Seveso Directives. Actually there are some 150 galvanic plants notified to the competent authorities as National Fire Corps (2013), but this number could increase, as supposed. According to the Seveso II Directive these companies are implementing an SMS, which must be audited by the competent authorities.

As most companies have a very simple organization and poor budget, it is not easy to comply with demanding regulatory requirements. The overlapping of different legislations, Seveso and OSH, is a further issue that makes the management more complicated. Thus in times of economic difficulty from the market, the management system could be seen as a burden rather than as an opportunity of improvement and it is very difficult to answer to the needs of a flexible organization oriented to minimizing uncertainty and to change managing. Therefore, it is the urgency to provide the operators of small galvanic plants with specific and adequate tools for dealing safety issues. The Research, Certification and Verification sector of INAIL is in charge of inspecting a number of galvanic plants, thus the significant experience about SMS in galvanic industry has been exploited to provide this industry with adequate solutions.

3 OBJECTIVES

Aim of the research is to propose and elaborate specific and easy procedures for the implementation of a health and safety management system in galvanic establishments.

It is presented an integrated management system for the prevention of occupational accidents and diseases and control of major accident hazards supported by a specific digital model that helps a dynamic management. This application is useful because adopts an SMS model focalized on near misses management. This point is required for Seveso establishments, and is an opportunity for improving the safety model. The application design and implementation would be able to involve the operators' needs with the aim to remove the gap between SMS and workers, who often feel some safety procedures and documents as a bureaucratic over layer, far from their daily operations. As discussed in § 1, the study of near misses gives the possibility to analyze the "*weak signals*" in order to prevent events with consequences, by overcoming the weaknesses of the SMS. Near misses and anomalies registration and discussion should be strongly encouraged, as they are able to point out which item in SMS was missing, bad applied, or simply misunderstood. These points are essential for a continuous improvement in the safety system.

4 METHODS

The proposed method is based on the strong conviction that safety model should be considered as an alive model, continuously improved during the plant operation, always updated in order to reflect the real situation on the plant, but also to promptly provide workers with the necessary safety documentation available into specific situations. For this purpose a few available methods have been used. They have been improved and customized to achieve the objectives of the research. The two pillars of the present research are the Safety Management Models for SMEs and the Safety Digital Representation.

4.1 *Safety management models for SMEs*

In Italy, in order to support SMEs to implement SMS, many efforts have been done for the last decade by Regulators, Entrepreneurs' Associations and local safety Agencies as UOPSAL AUSL Forlì (2006). On the basis of recognized safety management guidelines and recommended practices, such as BS/OHSAS 18001/2007, have been developed a number of models, which include work procedures and forms for specific industries. These models do not deal with major accident hazard but may be considered as a good basis to elaborate the integrated safety management system for galvanic plants, subjected to Seveso Legislation.

4.2 *Safety digital representation*

In a previous paper Bragatto et al. (2007) introduced a digital model for linking the safety documents (procedures and risk assessment) with the digital representation of the establishment, in terms of components and devices. The capabilities of this model include the support for controlling and updating the changes in the plant. In a more recent paper Bragatto et al. (2010), the digital model was improved so that all items of the safety management model, including operative instructions, procedures, and forms, may be linked each other. This advanced safety digital model may be a backbone, around which everything revolve, as the safety documents should contain the knowledge and experience capital of the establishment, which could improve safety management at every stages.

4.3 *Developed tools*

The methods previously outlined have been improved with two key points: a specific safety model and an adequate tool for the analysis of near misses and anomalies.

4.4 *Safety management model for seveso plant*

The model presented in this study has been obtained integrating procedures of OSH models with specific forms for major accident hazard management and electroplating activities. The integration has been developed taking advantages by the experience of some of our Inspectors involved into SMS inspections at galvanic plants, accordingly to the article 18 of Seveso II Directive. Furthermore, their Safety Reports have been considered for analyzing the risk assessment results.

The safety model is built gradually, usually starting from a collection of documents, and then step by step they are organized through SMS reference model. On the other hand, as previously discussed, the operative instructions are documents already linked to assets, i.e. equipments or devices.

4.4.1 *Near misses and anomalies analysis*

In daily activities are essential the registration and the analysis of near misses. Near misses registration means to record what happens in the plant, above all explicitly linking this event to the equipment or device where it occurred. Thus near miss becomes part of the safety model, and its analysis starts from the equipment involved and moves back along the paths in the safety model, in order to find out which piece of component or safety document failed, in other word to look for a *bug* in SMS. The detected *bug* might be meaningless and insignificant, but it may be a precursor of a more serious event. Indeed, adopting such a safety model, it is possible automatically to point out if the near miss may occur in a pathway running toward a top event. The analysis is carried out looking at the operative instructions and following at the procedures which may be failed. Finally, it is important to find out the solution (*fix the bug*), indicating how to set the document, if it needs to be "*revived*", to improve its learning, or some changes. The method takes also advantages of the bow-tie representation, which connects devices, instructions and procedures as preventive or protective barriers to an accidental event. Bow-tie representation is considered an essential part of the safety digital model.

4.4.2 *Web based tool*

This methodology has been implemented in a web-application named IRIS-Galvanic, in order to be available not only in a desktop version, but also for mobile devices, providing a tool usable directly on the field in daily operation activities.

5 RESULTS

5.1 *SMS integrated model*

To obtain the integrated management system for galvanic industries, specific procedures have been elaborated and/or other instructions, forms and schemes have been added to the existing procedures.

Figure 1 shows some key points of a procedure compilation, i.e. its goal, its instructions, and the forms and other procedures to which is referred.

The final result is shown in Table 1, in which it is indicated the name of the Safety Procedure (SP) and, for each one, only the new or modified forms (F), specified by a progressive number.

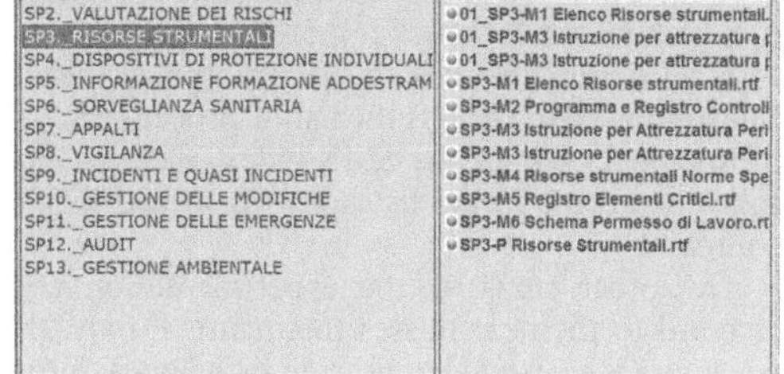

Figure 1. An example of procedure.

Table 1. Procedures and forms of integrated model.

SP1	*Starting and Review* F12 Seveso Information Form
SP2	*Risk Assessment* F11 Risk assessment in the galvanic plant with attachment "Galvanic plant: phases process, health & safety risks and protection measures" F12 Risk analysis (bow-tie)
SP3	*Instrumental Resources* F3 Instruction for dangerous equipment F5 Critical Item Registry F6 Permit To Work
SP4	*Personal Protective Equipment (PPE)* F5 PPE for galvanic activities
SP5	*Information and Training—Periodic meeting* F5 Check list Information and Training
SP6	*Health Surveillance* F3 Specific rules to galvanic plant
SP7	*Contracts*
SP9*	*Accident and Near Miss* F1 Operational Experience Analysis F2 No-compliance and investigation in the event of injury—accident or near injury
SP10*	*Management of Change*
SP11*	*Emergency Plan*
SP12*	*Audit*

* New procedures elaborated for major accident hazard.

5.2 *Test case*

IRIS-Galvanic has been tested in a sample of three galvanic plants. The top events indicated by each plant are of the same type: overflow of material such as chromium or nickel, loss of containment, e.g. due to rupture of bottle, or formation of toxic cyanide, where used. Each plant has registered less than ten near misses into five years, the major parts are related to initial rupture of the pools in particular the protective cover, or some devices, e.g. valves, or level detector.

The first step of the test has been the compilation, for each plant, of SMS model and the definition of bow-ties for every top event analyzed.

The Figure 2 shows the SMS structure at the left side, on the right there is the list of procedures and forms associated to the SMS index selected. The first documents in the list are forms filled by the company.

The other steps of the experimentation correspond to the near misses treatment. Firstly, the near miss is registered capturing its primary information (e.g. description, equipment involved, who

Name of Enterprise	*SP3 – P* *Instrumental Resources*	SMS elaborated by Ed. N. date Rev. N. date Pag.
1. AIM	Managing plants, machineries and work equipment: purchase, use, maintenance, preventive and periodical audits.	
2. OPERATIVE INSTRUCTIONS	- Selection of a new instrumental resource. Risk assessment. - Supply and installation - Use of instrumental - Inspection and Maintenance. Changes. Special Rules. Adjustments.	
3. REFERENCE	SP3-M1 Instrumental Resources SP3-M2 Registry of controls SP3-M4 Special Rules SP3-M5 Critical Item registry SP3-M6 Permit to Work SP2-P Risk Assessment SP4-P PPE SP5-P information & Training SP7-P Contracts	

Figure 2. SMS structure and filled forms associated.

detects it, which working activity was going on). The analysis follows a sequence of steps, that are:

- check if the near miss refers to a chain which may bring toward a top event;
- look at if some operative instructions failed, the reason why, and what solution would be suggested;
- as before for the procedures, too;
- a summary, called simply lesson learnt, which collects all the previous results, showing the eventual comments added to some documents.

All these phases are contextualized by taking advantages of the safety digital model, thus some candidates are proposed to the analyst. For instance, in searching top event reference, the candidates are identified by looking if the item involved in the near miss has some relations with them, e.g. is a preventive o protective device. In operative instructions analysis, the application automatically proposes a set of documents having some relations with the item or with the top event identified. Thus the set of pertinent documents is drastically reduced.

The Figure 3 shows an example of near miss analysis result (lessons learned panel), where the advantages of using such a method and adopting a safety digital model should be clear. Indeed, a direct link between near misses and top events can be pointed out, and the documents which may have failed are identified.

After its brief description (in the example: "dripping in the pool due to protective problems"),

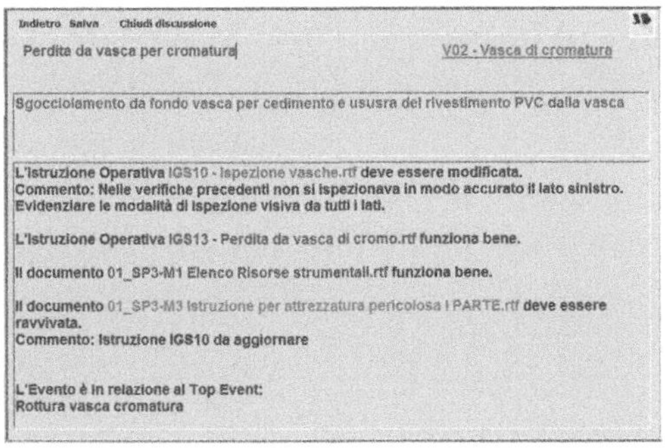

Figure 3. An example of summary of near miss analysis.

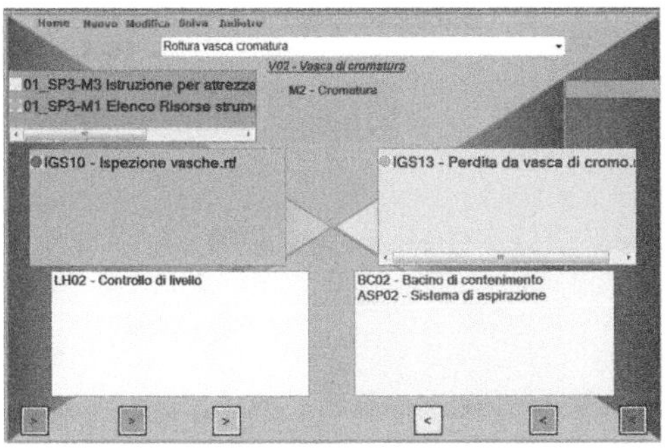

Figure 4. An example of bow-tie representation.

the panel lists the operative instructions and procedures which have been involved in the analysis. The comments, added into the documents during the analysis phase, are shown, the related top event is outlined, too. To describe the documents status, different colors are used: red if changes are required, yellow if they only need to be "*revived*", green if everything works fine. In the example, it has been required a modification in the operative instruction related to the daily inspections of the pool, while a simple tag to remember such an update is associated to the form referring to management of critical components. Furthermore, this panel should point out that adopting such a methodology, the safety documents would become more familiar to the worker, filling them as an opportunity of improving their work rather than a bureaucratic over layer.

Finally, an improvement of bow-tie representation is also done. In Figure 4, an update of bow-tie after the previous analysis is shown. The procedures have been added to the list of preventive documents, and all the documents reflect their current status by using the colors as described above.

6 CONCLUSION

The first testing phase is confirming the model is adequate and valid for small sized galvanic plants, since real test cases are used for managing and analysing near misses. Further tests are planned at some pilot firms for studying the model in a long time, one year and more. The basic ideas could also be transferred, with minor changes, to other industry featuring small (and micro) companies facing large risks, such as major accident hazard. The proposed approach, instead, is not suitable for larger and complex establishments, where generic procedures could not fit the specific problems of complex factories.

AKNOWLEDGEMENTS

The research has partially funded by Italian Ministry of Welfare—Grant PMS/08/50/P3.

REFERENCES

Bottani, E., Monica, L., Vignali, G. 2009. Safety management systems: Performance differences between adopters and non-adopters. *Safety Science* 47: 155–162.

Bragatto, P., Monti, M., Giannini, F., Ansaldi, S. 2007. Exploiting process plant digital representation for risk analysis. *Journal of Loss Prevention in the Process Industries* 20 (1): 69–78.

Bragatto, P., Agnello, P., Ansaldi, S.M., Pittiglio, P. 2010. The digital representation of safety systems at "Seveso" plants and its potential for improving risk management. *Journal of Loss Prevention in the Process Industries* 23 (5): 601–612.

Grote, G. 2012. Safety management in different high-risk domains—All the same?. *Safety Science* 50 (10): 1983–1992.

Kristensen, P.H. 2011. Managing OHS: A route to a new negotiating order in high-performance work organizations?. *Safety Science* 49 (7): 964–973.

McGuinness, E., Utne, I.B., Kelly, M. 2012. Development of a safety management system for Small and Medium Enterprises (SME's). Advances in Safety, Reliability and Risk Management—*Proceedings of the European Safety and Reliability Conference, ESREL 2011*: 1791–1799.

National Fire Corps 2013. Le attività a rischio di incidente rilevante in Italia. http://www.vigilfuoco.it/aspx/download_file.aspx?id=14617

Reiman, T. & Rollenhagen, C. 2011. Human and organizational biases affecting the management of safety. R*eliability Engineering and System Safety* 96 (10): 1263–1274.

UOPSAL AUSL Forlì 2006. Sistema di gestione della sicurezza sul lavoro per imprese fino a 30 addetti. http://www.ausl.fo.it/Portals/0/Documenti/Modulistica/Modello%202006.11.11%20-%20Estratto.pdf

Occupational Safety and Hygiene II – Arezes et al. (eds)
© 2014 Taylor & Francis Group, London, ISBN 978-1-138-00144-2

Qualitative evaluation of occupational exposure to chemical agents in a workplace of a research laboratory

C.S. Castillo
Sponsored by CNPq, Brazil
Research Centre for Industrial and Technology Management, School of Engineering of the University of Minho, Guimarães, Portugal

S. Bragança, A. Fernandes, I.F. Loureiro & N. Costa
Research Centre for Industrial and Technology Management, School of Engineering of the University of Minho, Guimarães, Portugal

ABSTRACT: In order to identify the presence of chemical agents in the work environment, a workplace of a research laboratory was analyzed. For this purpose, colorimetric tubes were used in three of the nine macro activities of the most exposed worker. The impossibility of changing the technical specifications of the linear length of the product X (in process of patenting); and the replacement of the chemical agents, resins and solvents by other more innocuous, was assumed. After being detected the presence of toluene, methanol, n-hexane and benzene, the following sequence of improvements were proposed: (1) move the machine X to the entries C1-086 and C1-087 identified on the layout, (2) remove the division between those areas; install exhaust ventilation to do the correct atmospheric suction of the air/exposed gases and, (3) use of individual protection equipment.

1 INTRODUCTION

In order to identify the exposure to chemical agents in a research laboratory, a qualitative analysis was made. The exposure was related to nine macro activities of a process X (Miranda & Araújo 2013, Pati-nha et al. 2013). This process has shown to be the most critical in terms of occupational exposure to chemical agents in the production process (in the machine X) of a batch of product X (about 9 linear meters in horizontal position for subsequent cutting into predetermined segments).

Regarding the chemicals that were used on the process X, three of the nine macro activities were considered to be the most critical (the macro activities 1, 2 and 9). The criteria used for this identification was the information provided on the chemical sheets for each of the product that was used, namely the composition and the toxicity. To perform these three macro activities the following products are used: (1) mold release agent; (2) air-curing resin; and (9) resin used in the process, added to a cleaning solvent.

Regarding the specifications of the process X, it was assumed the impossibility of changing the technical specifications of the linear length of the product X; and replacing the harmful release agents, resins and solvents by other more innocuous.

The information provided on this paper allowed (i) the study of the contribution of the Dosage (D) model' sensitivity coefficients to support the prioritization on the decision making process to intervention, and (ii) the definition of a proposal to a qualitative evaluation of the occupational exposure on the laboratory research, whose process X/ product X investigated are in process of patenting.

The main objectives of this work were: (i) identify the individuals that were exposed to chemical agents in the area where the process X was developed; (ii) characterize the areas where the process X was performed; and (iii) identify the possibility of existence of chemical agents in the laboratory environment.

2 THEORETICAL BACKGROUND: DOSAGE ABSORBED BY THE HUMAN BODY

The quantities that constitute the Dosage (D) model (Eq. 1) are: D = Dosage absorbed by the human body (the amount of contaminant that an organism absorbs in a given time and that can cause damage); T = Time of exposure to the identified chemical agent; and C = Concentration of the chemical agent in the area of the selected workplace (Miguel 2012, Parsons 2000).

$$D\,[\text{year} \times \text{mg} \times \text{m}^{-3}] = T\,[\text{year}] \times C\,[\text{mg} \times \text{m}^{-3}] \quad (1)$$

Several definitions were taken into consideration. *Response*: effect caused in the body due to exposure to the toxic agent; *Dose/response* as the relationship between the dose intensity and the proportion of individuals who have a particular effect (or intensity of effect), where the dose/response curve reflects the variations between individuals of the same population.

The exposure to chemicals in the research laboratory is made by inhalation and cutaneous contact. Therefore the parameter LC_{50} must be used to evaluate the toxicity of a substance (representing the unique value of a substance capable of causing death in 50 [%] of tested animals). The lower the LC_{50}, the more toxic will be the substance (Miguel 2012, Parsons 2000).

Dimensional analysis consistent with the units of the International System (IS) (ISO/IEC 2008), where the unit 'nd' means non-dimensional, is used.

$$D\,[\text{s} \times \text{kg} \times \text{m}^{-3}] = T\,[\text{s}] \times C\,[\text{kg} \times \text{m}^{-3}] \times 31.54\,[\text{nd}] \quad (2)$$

The sensitivity coefficients of the Dosage (D) model as a function of [s], [kg] and [m] (Eqs 3–5) are considered. These variables can be modified to minimize the sensitivity coefficients of the Dosage (D) quantity in the Equation 2 (ISO/IEC 2008).

$$\frac{\partial D}{\partial [s]} = [\text{kg}] \times [\text{m}^{-3}] \times 31.54\,[\text{nd}] \quad (3)$$

$$\frac{\partial D}{\partial [kg]} = [s] \times [m^{-3}] \times 31.54\,[\text{nd}] \quad (4)$$

$$\frac{\partial D}{\partial [m]} = [s] \times [kg] \times [m^{-4}] \times -94.61\,[\text{nd}] \quad (5)$$

3 METHODOLOGY

The methodology used in this study was divided in two main parts described in the following sections.

3.1 *Identification of people directly/indirectly exposed and characterization of the laboratory area that circumscribes the process X, regarding the occupational (technological, individual/collective and organizational subsystems) exposure to chemical agents during the three most critical macro activities*

To help on the characterization of the working situation, Checklists were developed and applied (Loureiro 2013). Direct observation and site visits were made to assist in the analysis of the characterization. To help the visualization of the area, the layout of the research laboratory is included in Figure 1. The process X is located in the section of the plant identified by the code C1-084.

3.2 *Identification of the possibility of the existence of chemical agents during the activities developed in the area of process X, which uses more critically chemical agents during three specific macro activities*

The Concentration (C) of the chemical agent (previously identified) in [ppm] and Dosage (D) in [ppm × h] were the major quantities to be measured. However as the direct reading through the colorimetric tubes does not enable a very reliable indication of the exact values of (C) and (D), only the identification of the possible existence of chemical agents in the environment was made.

A more conservative measure, aiming to protect people less exposed was done to minimize the effects of absorption of chemical agents. Initially, it was performed an approach of the sociotechnical system allowing the identification of the most exposed worker in the individual/collective subsystem. As reference, it was adopted the most exposed worker. It should be taken into account the toxicological effects related to the occupational exposure to chemical agents, especially those with higher levels of Concentration (C) and Time of exposure (T), when compared with their respective Exposure Limit Values (ELV's) WA (Weighted Average)/ST (Short Term)/MC (Maximum Concentration) (Miguel 2012, NP-1796 2007, Parsons 2000).

The sampling collection was planned as follows (Miguel 2012, Parsons 2000):

Measurement method: Colorimetric indication with active sampling of chemical agents by direct reading (punctual and short duration reactive tubes) and passive sampling (tube by diffusion or long duration), was used. The reaction between the chemical agents present in the environment under

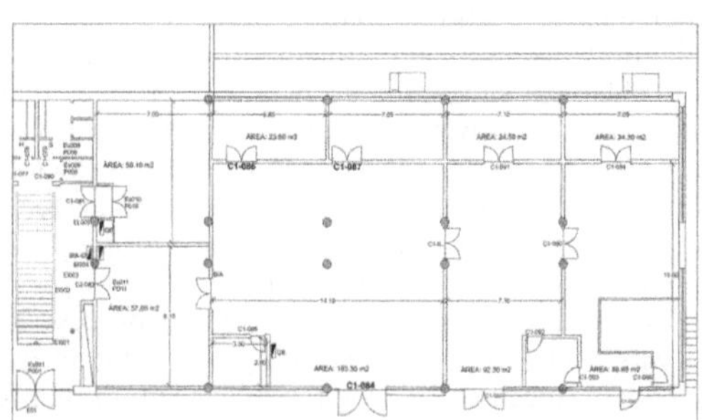

Figure 1. Layout of the research laboratory.

analysis and the chemical product presented inside the tubes was detected by colorimetric change.

Measuring equipment: A Dräger pump was used to collect the air present on the environment, closest to the most exposed worker's airways. After the test, the colorimetric tube was broke rendering it useless for future measurements. In each test a thermo-anemometer was used to measure temperature [°C] and air velocity [m × s^{-1}]. Besides being registered the exposure time (and measurement time) for each of the 3 most critical macro activities, it was also considered 1 pseudo activity without exposure to chemical agents.

Sampling strategies: due to time constraints and despite the identification of several persons presented on the research laboratory, this study was focused solely on the worker that directly dialed with the process X. In this way, it is authors believes that basing the analysis on the complaints of this most exposed worker, the necessary protection to other people who also attend the workplace were also estimated. The information provided on the chemicals sheets were used for the preliminary identification of the chemical products used on the research laboratory (1 external mold release agent, 1 cleaning liquid X, 1 cleaning liquid Y, 1 epoxy resin, 1 polyester resin iso for infusions, 1 vinylester resin X, 1 vinylester resin Y and 1 sealant). Two specifications of the chemical products presented on the sheets contributed to the colorimetric tubes selection: (i) the composition; and (ii) the toxicity. Of the many available colorimetric tubes, only 5 were selected (1 tube for instantaneous measuring/toluene; 1 tube of short duration/benzene; 1 tube of short duration/methanol; 1 tube of short duration/n-hexane; and 1 diffusion colorimetric tube for detection of the ammonia, in which a device for fixation of the respective tube was used to ensure greater proximity to the respiratory system of the most exposed worker during a measurement period of 2 hours). The measurements were made during the complete time-work of the most exposed worker. The air samples were collected, using reactive tubes for punctual measuring and short duration, representing the conditions at the time of collection. It is important to notice that the long-term sampling for the diffusion of the ammonia considered the duration of the measurement lasted in 2 hours. The detection started with the colorimetric tubes at the lower possible concentrations, eliminating unnecessary testing for higher concentrations.

4 RESULTS

In order to characterize the individual/collective sub-system, a general description of the workers presented on the research laboratory was made. Three persons were identified: a male and two females; being the average age of 30 years. Men has 1.75 [m] and 75 [kg]. Women had 1.70 [m]/1.65 [m] and 70 [kg]/73 [kg], respectively. They remain in their jobs daily 8 hours, from Monday to Friday, with the following breaks per day, 15 minutes + 15 minutes + 60 minutes. Despite these workers do not work directly with the process X, all complained of headache, mood, nausea and loss of appetite. Even though it was identified a mechanical ventilation it is a fact that it does not have the capacity to extract the contaminated air. By a careful observation of the layout it was notice that the only available sections for a possible relocation of the machine X were the C1-086 and C1-087 sections.

The Table 1 (whose content is presented in three sequential parts) shows the relationship of the major variables of interest used to identify the possibility of existence of chemical agents in the environment where the three most critical macro activities take place. The data of 'Measurement range' column on the Table 1 (first part) was obtained from the technical manuals of the colorimetric tubes used on this study. The value of the effective atmospheric pressure [hPa] presented on the Table 1 (second part) was obtained on the sampling day (06/14/2013, 12h:00min to 14h:00min) through the IPMA (2013).

5 DISCUSSION

The situations where a colorimetric reaction was identified, as a function of direct atmospheric exposure to the chemical agents, it was recommended to make a new sampling collection and subsequent

Table 1. Key variables for identifying the presence of chemical agents in the environment of the three most critical macro activities of the process X of workplace (first part).

Most critical macro activities	Variables of interest to identify the possibility of existence of chemical agents in the workplace of the most critical macro activities of the process X, considering the most exposed worker in an environment with closed windows / doors (exhaustion and ventilation systems off)				
	Chemical agent	Correction factor of atmospheric pressure* [nd]	Measurement range [ppm]	Colorimetric variation on the reactive tube [presence X absence]	Exposure time [minutes]
(1)	Toluene Methanol n-hexane Benzene	0.84	50 - 400 50 - 3000 50 - 1500 0.5 - 10	absence presence presence presence	10
(2)	Toluene Methanol n-hexane Benzene	0.84	50 - 400 50 - 3000 50 - 1500 0.5 - 10	presence presence presence presence	15
(9)	Toluene Methanol n-hexane Benzene	0.84	50 - 400 50 - 3000 50 - 1500 0.5 - 10	presence presence presence presence	15
(1) + (2) + (9) + (pseudo activity)	Ammoniac	0.84	10 - 750 (considering 2 [h] of measurement)	absence	120

* Correction factor of atmospheric pressure [nd] = 1013 [mbar] ÷ effective atmospheric pressure [mbar].

Table 1. Continued (second part).

Most critical macro activities	Chemical agent	Approximate temperature and within the acceptable range for use of the instrument [°C]	Approximate air velocity $[m \times s^{-1}]$	Effective atmospheric pressure [mbar]	Relative standard deviation of the instrument [± %]
(1)	Toluene	23	< 0.01	1201.90	10 - 15
	Methanol				15 - 20
	n-hexane				15 - 20
	Benzene				30
(2)	Toluene	23	< 0.01	1201.90	10 - 15
	Methanol				15 - 20
	n-hexane				15 - 20
	Benzene				30
(9)	Toluene	23	< 0.01	1201.90	10 - 15
	Methanol				15 - 20
	n-hexane				15 - 20
	Benzene				30
(1) + (2) + (9) + (pseudo activity)	Ammoniac	23	< 0.01	1201.90	—

Table 1. Continued (third part).

Most critical macro activities	Chemical agent	Dosage [ppm × h]	Concentration [ppm]	Corrected dosage** [ppm × h]	Corrected concentration*** [ppm]
(1)	Toluene	—	< 50	—	42.14
	Methanol		3000		2528.50
	n-hexane		100		84.28
	Benzene		10 (equivalent to 2 strokes*)		8.43
(2)	Toluene	—	< 50	—	42.14
	Methanol		500		421.42
	n-hexane		between 50 and 100		84.28
	Benzene		10 (equivalent to 2 strokes)		8.43
(9)	Toluene	—	< 50	—	42.14
	Methanol		500		421.42
	n-hexane		100		84.28
	Benzene		10 (equivalent to 2 strokes)		8.43
(1) + (2) + (9) + (pseudo activity)	Ammoniac	< 10	0 (dosage [ppm × h] ÷ total duration of measurement in 2 [h])	0	—

* Strokes means the amount of compression of the Dräger pump for the admission of a given volume $[m^3]$ of air and / or gas in the investigated environment.
** The corrected dosage [ppm × h] = value of the dosage measured by the tube [ppm × h] × value of the correction factor of atmospheric pressure [nd]. In cases where it was only possible to indicate a range measured dosage (and how there was not detectable colorimetric variation), was considered as the minimum acceptable criteria for calculating the corrected dosage [ppm × h].
*** The corrected concentration [ppm] = value of the concentration measured by the tube [ppm] × value of the correction factor of atmospheric pressure [nd]. In cases where it was only possible to give a range measured concentration, it was considered as a criterion (more conservative protection of persons possibly exposed) the maximum permissible value for the calculation of the corrected concentration [ppm].

analysis in an accredited laboratory. This proposition aims to make a quantitative analysis of the identified chemicals. When no colorimetric change in the reactive tube was identified, it was recommended in one hand the use of other colorimetric tubes for detection of eventual unidentified chemical agents and on the other hand, the use of an alternative measurement range [ppm]. In these cases, the same sampling procedure described above was proposed (Loureiro 2013, Miguel 2012, ISO/IEC 2008).

Regarding the Equations 3 to 5, the length quantity in [m] represents one of the dimensions that constitute the volume quantity in $[m^3]$, where the mass quantity [kg] of the chemical agent (previously identified) is deposited, so [m] is inversely proportional to the Concentration (C), considering [kg] constant (ISO/IEC 2008). The results of quantities in [s], [kg] and [m] presented on those equations contributes to assess the correspondent sensitivity coefficients. This analysis will minimize the effects of the chemicals absorption.

By increasing the dose or accentuating the inclination of the line of the dose/response relationships; considering the X axis as the log dose $[mg \times kg^{-1}]$ and the Y axis as the response [%], the normal thing to happen is that it will also increase the number of affected individuals in the exposed population—being able to predict the danger of toxic substances. Thus, the faster this absorption tends to occur, and more quickly the adverse toxicological effects of the chemical agents begin. Furthermore, to maintain (D) constant it is needed to balance the relationship between (T) and (C). That is, the higher the value of (T), the lower will be the value of (C) or vice versa. Where the scaling of the weighting of values to be assigned to the variables (T) and (C) aims at minimizing the value of (D) below the maximum acceptable levels in the relevant legislation, e.g. the (ELV), by considering the most population with a specific occupational exposure to chemical agents in the atmosphere for 8 hours per day/40 hours per week (Miguel 2012, NP-1796 2007, Parsons 2000).

According to the obtained results, a list of priorities for intervention was developed. It was taking into account that the priority of action aims to minimize the effects of the chemical agents when inhaled; meaning that the airway is more significant than the cutaneous or digestive way. Therefore, to reduce the value of Dosage (D) by the most exposed worker, suggestions on how the main variables (T) and (C) can contribute to the maximization of the response time, were presented (or minimization of sensitivity; regarding the Equation 1, hence minimizing the inherent toxicological effects). In general, the variables that affect the increase of the response time of the Dosage (D) model (Eqs 1–2) are (Miguel 2012, ISO/IEC 2008, Parsons 2000):

- (1st) *Engineering measures*: Chemical agent isolation (which complicates the accessibility to the production system); mechanical exhaustion (higher cost associated); air ventilation (higher cost associated and difficulty of dimensioning if there is need for air conditioning for the productive environment); polymerization speed of the chemical agents to the air and/or to the temperature (higher cost associated and complexity in the dimensioning of the technical specifications/performance of the polymerized rope and of the respective productive parameters).

- (2nd) *Organizational measures*: Working hours adjustment (higher cost associated and difficulty with the training and harmonization of working hours).
- (3rd) *Measures related to the adoption of personal protective equipment*: Use of personal protective equipment. Even though there is a higher cost associated and discomfort, it is recommended: (1st) suitable filter mask and (2nd) special gloves for cutaneous protection. These measures considered the information provided on the chemical sheets.

It is important to notice that there are also other variables not included in the Dosage (D) model (Eqs 1–2) that may have influence on the response time of the Dosage (D) such us; susceptibility of certain human beings who have different settings in terms of metabolism [$W \times m^{-2}$], health status, emotional state/trait, genetic predisposition, respiratory capacity and surface area of their bodies (Miguel 2012, Parsons 2000). It also should be remarked that a greater protection of workers exposed to chemicals when compared to the ELV's, could be achieved by considering the existence of additive effects for the calculation of the weighted average of the concentration of chemical agents (Loureiro 2013).

The fact that a sociotechnical approaching to the organizational system was used, allowed the identification of other persons on the area where the process X was performed. Generally, it is possible to say that there are about 50 students that have classes on the same or adjacent areas where the process X was performed. Even though this study was focused on the most exposure worker, a procedure for chemical exposure evaluation should be planned for that group as they can also be directly and/or indirectly exposed. Due to the number of people it is proposed a definition of homogeneous groups to access the exposure (Loureiro 2013).

If the performance, productivity and technical specifications for the product X is not affect than it is suggested: (i) the replacement of the harmful chemicals identified in the study by other more innocuous (or exhibiting properties which minimize the sensitivity of the Dosage (D) model) (Loureiro 2013, Miguel 2012), and (ii) the modification of some specifications of the product X to allow that the production system X can be repositioned within the sector itself C1-084.

Regarding the symptoms described by the persons directly and/or indirectly exposed to chemicals in the study, it is suggested to systematically carrying out biological monitoring to assess the effects of the dosage directly on the human body (Loureiro 2013, Miguel 2012).

6 CONCLUSION

The greater the synergy between measures of engineering, work organization and those related to the adoption of personal protective equipment, the greater tends to be the complexity in reconciling simultaneity their advantages and disadvantages.

After detecting the atmospheric presence of toluene (except in the first macro activity), methanol, n-hexane and benzene in the workplace; they were primarily proposed three changes for improvement in the following sequence: (1st) displacement of the machine X to plant spaces whose inputs are C1-086 and C1-087, containing about 13.90 meters (greater than the 9 meters in length required for producing the product X), considering the possibility of eliminating the partition between those spaces; (2nd) installation of exhaustion; and (3rd) according to Miguel (2012), Parsons (2000), use of proper respiratory mask and special gloves.

The identified constraints during the development of this work allowed the definition of several steps that should be planned in the future in order to obtain a quantitative analysis of the chemicals identified on the research area under analysis.

REFERENCES

Instituto Português do Mar e da Terra (IPMA) 2013. *Current Atmospheric Pressure (hPa)*. Available in: <http://www.ipma.pt/pt/otempo/obs.superficie.mapa/>. Accessed: 14 jun. 2013.

ISO/IEC 2008. Guide 98-3. *Uncertainty of measurement: Guide to the Expression of Uncertainty in Measurement (GUM: 1995)*.

Loureiro, I.F. 2013. '*Protocolo de higiene industrial: avaliação da exposição a agentes químicos*'. 'Material de Apoio à Unidade Curricular de Higiene Industrial do Mestrado em Engenharia Humana, Universidade do Minho'.

Miguel, A.S. 2012. '*Manual de higiene e segurança do trabalho*'. 12 (ed.). 'Porto Editora'.

Miranda, L. & Araújo, F. 2013. *Fibrenamics*. Available in: <http://www.fibrenamics.com/>. Accessed: 6 jun. 2013.

Montgomery, D.C. 1991. *Introduction to statistical quality control*. 2 (ed.). New York: John Wiley & Sons.

NP-1796 2007. '*Segurança e saúde no trabalho. Valores limite de exposição profissional a agentes químicos*'. Available in: <http://www.ipq.pt/backFiles/prNP001796_2007.pdf>. Accessed: 24 mai. 2013.

Parsons, K. 2000. Environmental ergonomics: a review of principles, methods and models. *Applied Ergonomics*, 31(6): 581–594.

Patinha, S., Vyhnalek, M., Alves, A., Cunha, F., Rana, S. & Fangueiro, R. 2013. Production process of Braided Composite Rods (BCR). Presented in: '*Materiais 2013*', 25–27 June 2013, Coimbra, Portugal.

Occupational Safety and Hygiene II – Arezes et al. (eds)
© 2014 Taylor & Francis Group, London, ISBN 978-1-138-00144-2

Some constraints in the evacuation—a short review

Ana L. Lourenço & J. Santos Baptista
Research Laboratory on Prevention of Occupational and Environmental Risks (PROA/LABIOMEP), Faculty of Engineering, University of Porto, Portugal

Paulo Oliveira
Lusófona University of Porto, Portugal

ABSTRACT: The evacuation strategies have evolved over time. However, there are gaps for vulnerable groups such as children and elderly. In this paper it is intended to understand the approaches and their outcomes, to the problematic of evacuating these particular populations. Through literature review, it was found that the physical and psychological characteristics are crucial to saving lives, so it is essential to understand these limitations, predict behaviors and adjust emergency plans. On the other hand, evacuation is safe and effective when performed timely, but not all people have the same travel speed. Someone considered normal has a travel speed of 1.30 $m.s^{-1}$, while for an elderly and a child, on average, speed decrease, respectively to 0.67 $m.s^{-1}$ and 0.80 $m.s^{-1}$. For these reasons it is essential to carry out emergency drill involving these groups, testing the effectiveness of planning, minimizing any possible impact on occupants.

1 INTRODUCTION

In preventive policy, the main objective is to prevent that certain event happens. However, when this is not possible, the safety of individuals may depend on their ability to move to a safe location (Castle 2008).

With social integration policies, came to coexist in the same space people with different mobility conditions. Nevertheless, in case of evacuation, is necessary answer them all in timely and with the same effectiveness.

Detected this lack, the evacuation is a theme increasingly studied in their different contexts, either at the level of research and at the level of regulation applicable to projects and buildings and responsibility at the level of safety of human life.

In this work it intended to make a synthesis of existing knowledge about the evacuation issue of children and the elderly.

2 MATERIALS AND METHODS

It was performed a literature review about the topic under study through the system "metalib from exlibris" at http://metalib.fe.up.pt and from "Google" at http://scholar.google.com. In the first source was searched the following databases: MetaPress, Sci-enceDirect, IEEE Xplore, Wiley Online Library. Highwire Press and ACM Digital Library. The key-words used were as follows: evacuation, elderly, children, emergency, nursing homes and vulnerabilities.

3 EVACUATION CONSTRAINS—PLANNING AND PREVENTION FOR EVACUATION

For an effective preparation of emergency response is essential to know structures and processes (Claver, Dobalian, Fickel, Ricci, & Mallers 2013). In preparation for the emergency should be established different scenarios taking into account all the needs of different groups, particularly the most vulnerable, such as children and the elderly. The goal is to predict and reduce constraints of those who care for them (Chen, Wilkinson, Richardson, & Waruszynski, 2009).

In that sense, McCann (2011) considers that the main features for planning the response to emergency situations are:

a. Plan the emergency dynamically, i.e. in constant update. An emergency plan outdated is just a paper.
b. Plan and anticipate problems. Reduces doubts in an emergency situation and allows mitigating the potential effects; in other words, certain phenomena are inevitable, but their destructive effects can be reduced through careful planning.

c. Compile all information about the needs of each person or group, allowing a rapid but efficient response.
d. Do not perform plans too elaborated and hard to interpret.
e. Prepare plans simply and concisely in order to be understood by most of its performers.
f. The purpose of planning is to engage stakeholders in a process that raises awareness and promotes resistance to face an emergency by empowering those involved and giving them competences and responsibilities.

Although these considerations seem trivial, there are gaps in its implementation, particularly in groups with special needs. These special needs can be synthesized, for example, in functional, communication difficulty, supervision, medical and transport (Chen et al. 2009). In this way, those who have a daily responsibility to care for the most vulnerable people, should acquire as much knowledge as possible about safety (Kang, Seo, & Yang 2011) and develop an effective emergency management (Ambrose, Cardei, & Cardei 2010; McCann 2011).

In an emergency situation that requires evacuation, is necessary to know the individual characteristics that may compromise the most sensitive groups (Chen et al. 2009; Claver et al. 2013). This recognition requires a careful and organized coordination in order to be successfully accomplished (Ambrose et al. 2010), as well as the tests of efficient planning (Zhu, Jia, & Shao 2012). With these constraints, training becomes very useful by minimizing feelings of stress and panic of the occupants (Taaffe, Kohl, & Kimbler 2005).

4 HUMAN BEHAVIOR IN EVACUATION

4.1 *General considerations*

The evacuation movement must be disciplined and timely, to a safe place (Gwynne, Kuligowski, & Nilsson 2012; Kang et al., 2011).

In the design phase and construction of a building, should be employed technical solutions and measures to protect both the structure and the occupants, of the impact from an emergency situation, because a successful evacuation also depends on such conditions (Kholshchevnikov, Samoshin, Parfyonenko, & Belosokhov 2012). It is essential to address the particularities of groups such as the elderly (>65 years) and children (<7 years) since they are more susceptible to falls and being confused with an event that they are not familiar (Christensen, Brown, & Hyer 2012; Kang et al. 2011).

Factors such as age, physical ability, the situation at the beginning of the emergency, the knowledge about the building, emergency procedures and the level of involvement of the occupant in planning and training evacuation processes, determine his performance (Gwynne et al. 2012; Stein, Dueñas-Osorio, & Subramanian 2010). However, there is a risk associated with this process because the evacuation requires that people engage in non-routine situations (Taaffe, Johnson, & Steinmann 2006). Thus, it is crucial a plan that establishes escape routes, safety zones, rules of conduct and the sequence of actions to be undertaken while the evacuation takes place (Sommerville, Storer, & Lock 2009).

However, it is not always possible to activate the emergency plan. Conduct an evacuation is an intensive and exhausting labor, especially in institutions of care for elderly and/or children. It is essential to carry out drills in order to reduce problems in potential real emergencies (Castro, Persson, Bergstrom & Cron 2008; Taaffe et al. 2006). There are several studies on movement of adults and persons with reduced mobility, in different types of buildings, with and without obstacles, in normal and under emergency situations (Cherniak & Zadorozhny 2010; Christensen, Collins, Holt, & Phillips 2006; Kang et al. 2011; Kholshchevnikov et al. 2012; Lei, Li, Gao, & Wang 2012; Manley & Kim 2012; Tsai et al. 2011; Zhu et al. 2012).

4.2 *Children*

As for adults, there is research on evacuation flows in nurseries and kindergartens, with children between 2 months to 7 years. Probable values of pre-movement time and the dependence between speed and density are determined, considering evacuation time, the sum of pre-movement time and movement time (Kholshchevnikov et al. 2012). The pre-movement time consists on the initial stage of evacuation, in other words, the time it takes a child to interpret the alert, make the decision to evacuate and the time to be ready to evacuate.

In evacuation drills are recorded the times of the actions of the initial phase and movement parameters of children throughout the different sections of the escape routes. To observe the pre-movement time, children are grouped by age in different activities such as, play, sleep, or change a diaper.

In this context it has been found that the teacher/auxiliary, plays a decisive role in the pre movement development due to the characteristics of the children (Hutton 2010). He is who interprets the alert sign, makes a decision, informs and prepares children for evacuation. According to the study of Kholshchevnikov (2012) it was found that the preparation time increases with the decreasing of the atmospheric temperature due to the need of dressing protective clothing. In classes of 10 to 15 children, the total time of preparation were on average 0.6 minutes in the summer; 5 minutes in the spring and autumn and 7.5 minutes in the winter. It was also found that, if replaced the clothing placement by blankets, the preparation time was reduced to 1.1 minutes in the winter. Regarding the travel speed was found that in a child between 3 to 4 years is 0.8 $m.s^{-1}$ and between 5 to 7 years is 0.86 $m.s^{-1}$ (Kholshchevnikov et al. 2012).

To respond successfully to the emergency, it is important an assessment procedures plan of systematic variation, using an approach able to explain unexpected behavior (Manley & Kim 2012).

4.3 *Elderly*

In the preparation of emergency, physical and psychological characteristics related to aging, such as, e.g., the loss of muscle mass, reduction of bone density and decreased hearing and visual acuity, contribute to the loss of independence that influence and hamper the evacuation process. The elderly tend to have more functional restraints than those with less than 65 years (Castro et al. 2008; McCann 2011).

In nursing homes, most users have developed motor limitations, in that the travel speed and the response time is slower compared to healthy adults, highlighting their limitations when it is necessary rapid evacuation (Kang, Seo, & Yang 2011).

One of the key questions for the calculation of time for a safe evacuation is the ability to represent a reliable and credible occupants response to each scenario (Gwynne et al. 2012). The lack of understanding of the subject, may lead to inappropriate behavior during the emergency and, in turn, to inaccurate results in evacuation planning. Castel (2008) examined evacuation plans of 2134 nursing homes and found that only 37% included information about the special needs of residents (Claver et al. 2013).

Some studies have established comparison between RSET (*Required Safe Egress Time*), ie, the sum of the response time and movement time with ASET (*Available Safe Egress Time*), in other words, the time until conditions become unsustainable on escape routes (Gwynne et al. 2012; Kang et al. 2011). The main purpose of these studies was to identify and analyze in a practical way, factors related to the occupants from fire drills. To do this, they have defined plausible scenarios in a real context with, for example, occupants sleeping or performing a daily task. After the evacuation exercise the chosen scenarios have been quantitatively represented in five key components for its explanation. (Gwynne et al. 2012):

1. The response time;
2. The travel speed;
3. The evacuation routes chosen by the evacuees;
4. The availability of evacuation routes;
5. The link between the speed, the flow of people, density and population size.

The comparison between evacuation time of a person that is considered normal, ie, without physical limitations, and an elderly, was carried out by Kang (2011), while in the first, time varies between 2 and 6 minutes, in elderly rises to a value of 3 to 8 minutes. Regarding to the traveling speed, the first is 1.3 ms^{-1}, while for an elderly decreased to 0.67 ms^{-1}.

When questioned the capacity and speed of movement of people who use wheelchairs, was found that, in a horizontal section an electric wheelchair user circulates at 0.89 ms^{-1}, and a manual wheelchair user at 0.69 ms^{-1}.

On the other hand, a person that is in a wheelchair but whose movement is aided by, the moving speed increases to 1.30 ms^{-1}, i.e., is matched to the speed of a person without limitations (Christensen et al. 2006).

5 VULNERABILITIES

5.1 *Preliminary considerations*

The prediction of human behavior during the evacuation is not simple, but it is essential in safeguarding lives (Kholshchevnikov et al. 2012).

The fact that people coexist more or less with a given phenomenon influences their attitude and behavior (Dash & Gladwin 2007).

Fear and panic are psychological states that influence human behavior during an evacuation situation. Although it is difficult to explain clearly, panic can be caused by factors such as the lack of security lighting or signs an incorrect design of evacuation routes and also by the presence of large amounts of smoke. Can result in inappropriate responses and put into question the entire evacuation process. However, by its most delicate particularities, the most vulnerable groups, in an emergency situation as referred to in the previous point, are children, the elderly and people with physical or cognitive limitations (Dash & Gladwin 2007).

5.2 *Children*

Depending on age, children have many particularities and vulnerabilities and, therefore, are more easily at risk (Gribble & Berry 2011; Hutton 2010), making it clear that know well each involved (Claver et al. 2013) is crucial in the process of evacuation. This vulnerability, results from smaller attributes which may be exercised by making them dependent on others for their physical and psychological well-being. Easily are identified infants and children as at-risk populations in an emergency, comparing their characteristics with groups without any limitation (Chen et al. 2009).

Children between two and five years, already have their main physical abilities formed, and the kinetic experience is processed psychologically (Kholshchevnikov et al. 2012). However, their anatomical and psychological development puts them in risk of deprivation and stress (Hutton 2010). For this reason, it was believed that involving children in real context simulations could cause physical or psychological trauma (Kholshchevnikov et al. 2012). Although, it is now known that involving children in evacuation drills has benefits since they become familiar with procedures and behaviors to take as the notice of evacuation routes (Austin, Hannafin, & Nelson 2013). When benefit from advance preparation children acquire a safety culture and assimilate information for future threats (Brown et al. 2012).

The knowledge of the risks, by itself, is not enough to motivate action (Dash & Gladwin 2007). In an institution with children, workers, in addition to the normal concerns with themselves, have to extend these concerns to the people in their care. They must, therefore, know the residents, their needs and their physical and psychological characteristics (Castro et al. 2008), in order to preserve their quality of life (Shiraishi 2012). However, not only the workers but also the residents/users should be aware of emergency exits in order to be prepared for possible evacuations without constraints (J. J. Christensen et al. 2012; Kang et al. 2011; McCann 2011).

In summary, the time spent preparing and forming all possible involved for a possible evacuation, reduces negative consequences and clarifies future adjustments (Brown et al. 2012).

5.3 *Elderly*

In the elderly, the limitations arise primarily from the effects of chronic diseases, biological, psychological and social changes associated with aging that affect their independent functioning (Dobalian 2006).

Based on research of Claver (2013), it is known that almost 80% of nursing homes users, with 65 or more years of age require assistance in, at least, four or more daily activities. In addition to the challenges due to physical health problems or functional limitations, it is estimated that over 50% of nursing homes residents (Dobalian 2006) suffer from some degenerative disease or some kind of cognitive deficit (Brown et al. 2012).

Cognitive decline affects motor coordination and disables some activities that involve more complex tasks requiring, sometimes, permanent vigilance (Claver et al. 2013; Shiraishi 2012). These limitations reduce the ability to respond appropriate and independently in complex and dangerous situations. Therefore, an unplanned and disorganized evacuation can cause stress and trauma (Brown et al. 2012), since people with cognitive impairment cannot understand the reason for the interruption of their daily activities (Shiraishi 2012) while employees prepares the evacuation (J.J. Christensen et al. 2012).

When the movement starts, falls and injuries caused by these (Camilloni et al. 2011) may lead to loss of independence, and/or immobilization (Shiraishi 2012). Every year, between 28% to 35% of people aged 65 or more, falls. The rate rises to 32% to 42% in people over 70 years. For people aged 65 or more, a fall injury is enough to limit normal activities and 47.7 per 1000 inhabitants have diseases like Alzheimer, Parkinson's and Diabetes that increase the rate (Graafmans et al. 1996).

6 CONCLUSIONS

Towards a safe and effective evacuation is important to plan and train evacuation involving vulnerable groups and considering the characteristics associated to them. Knowing that, in case of evacuation, the main objective is to protect human lives, the evacuation time is crucial. On average, children and the elderly reach, half the travel speed of a person considered normal (Kang et al. 2011; Kholshchevnkov et al. 2012), moreover they have several physical and psychological characteristics that make them least protected and dependent on third parties. In this context, it is crucial the previous knowledge of individual circumstances, in order to develop solutions that involve these groups in a safe and efficient evacuation.

ACKNOWLEDGMENT

The authors would like to thank Master in Occupational Safety and Hygiene Engineering (MESHO), from Faculty of Engineering of the University of Porto (FEUP), all the support in the development and international dissemination of this work.

REFERENCES

Ambrose, Arny, Cardei, Mihaela, & Cardei, Ionut. (2010). *Patient-centric hurricane evacuation management system. Paper presented at the Performance Computing and Communications Conference* (IPCCC), 2010 IEEE 29th International.

Austin, E.N., Hannafin, N.M., & Nelson, H.W. (2013). *Pediatric disaster simulation in graduate and undergraduate nursing education.* J Pediatr Nurs, 28(4), 393–399.

Brown, L.M., Dosa, D.M., Thomas, K., Hyer, K., Feng, Z., & Mor, V. (2012). *The effects of evacuation on nursing home residents with dementia.* Am J Alzheimers Dis Other Demen, 27(6), 406–412.

Camilloni, L., Farchi, S., Rossi, P.G., Chini, F., Di Giorgio, M., Molino, N., & Guasticchi, G. (2011). *A case-control study on risk factors of domestic accidents in an elderly population.* Int J Inj Contr Saf Promot, 18(4), 269–276.

Castle, N.G. (2008). *Nursing home evacuation plans.* Am J Public Health, 98(7), 1235–1240.

Castro, Carmen, Persson, Diane, Bergstrom, Nancy, & Cron, Stanley. (2008). *Surviving the storms: Emergency preparedness in Texas nursing facilities and assisted living facilities.* Journal of gerontological nursing, 34(8), 9–16.

Chen, J., Wilkinson, D., Richardson, R.B., & Waruszynski, B. (2009). *Issues, considerations and recommendations on emergency preparedness for vulnerable population groups.* Radiat Prot Dosimetry, 134(3–4), 132–135.

Cherniak, Andrii, & Zadorozhny, Vladimir. (2010). *Towards Adaptive Sensor Data Management for Distributed Fire Evacuation Infrastructure.* 151–156.

Christensen, J.J., Brown, L.M., & Hyer, K. (2012). *A haven of last resort: the consequences of evacuating Florida nursing home residents to nonclinical buildings.* Geriatr Nurs, 33(5), 375–383.

Christensen, Keith M, Collins, Shawnee D, Holt, Judith M, & Phillips, Curtis N. (2006). *The relationship between the design of the built environment and the ability to egress of individuals with disabilities*: National Emergency Training Center.

Claver, M., Dobalian, A., Fickel, J.J., Ricci, K.A., & Mallers, M.H. (2013). *Comprehensive care for vulnerable elderly veterans during disasters.* Arch Gerontol Geriatr, 56(1), 205–213.

Dash, Nicole, & Gladwin, Hugh. (2007). *Evacuation decision making and behavioral responses: Individual and household.* Natural Hazards Review, 8(3), 69–77.

Dobalian, A. (2006). Advance care planning documents in nursing facilities: results from a nationally representative survey. Arch Gerontol Geriatr, 43(2), 193–212. doi: 10.1016/j.archger.2005.10.007.

Graafmans, WC, Ooms, ME, Hofstee, HMA, Bezemer, PD, Bouter, LM, & Lips, PTAM. (1996). *Falls in the elderly: a prospective study of risk factors and risk profiles.* American journal of epidemiology, 143(11), 1129–1136.

Gribble, K.D., & Berry, N.J. (2011). *Emergency preparedness for those who care for infants in developed country contexts.* Int Breastfeed J, 6(1), 16.

Gwynne, S., Kuligowski, E., & Nilsson, D. (2012). Representing evacuation behavior in engineering terms. Journal of Fire Protection Engineering, 22(2), 133–150.

Hutton, David. (2010). *Vulnerability of children: more than a question of age.* Radiation protection dosimetry, 142(1), 54–57.

Kang, Jae-Gyu, Seo, Janghoo, & Yang, Jeong-Hoon. (2011). *Research on the Enhancement of Escape Safety of Small Nursing Homes.* Journal of Asian Architecture and Building Engineering, 10(1), 271–278.

Kholshchevnikov, V.V., Samoshin, D.A., Parfyonenko, A.P., & Belosokhov, I.P. (2012). *Study of children evacuation from pre-school education institutions.* Fire and Materials, 36(5–6), 349–366.

Lei, Wenjun, Li, Angui, Gao, Ran, & Wang, Xiaowei. (2012). *Influences of exit and stair conditions on human evacuation in a dormitory.* Physica A: Statistical Mechanics and its Applications, 391(24), 6279–6286.

Manley, Matthew, & Kim, Yong Seog. (2012). *Modeling emergency evacuation of individuals with disabilities (exitus): An agent-based public decision support system.* Expert Systems with Applications, 39(9), 8300–8311.

McCann, David G.C. (2011). *A Review of Hurricane Disaster Planning for the Elderly.* World Medical & Health Policy, 3(1), 5–30.

Shiraishi, Y. (2012). *Challenges to elderly safety in Safe Community movements in Japan.* Int J Inj Contr Saf Promot, 19(3), 260–266.

Sommerville, Ian, Storer, Tim, & Lock, Russell. (2009). *Responsibility modelling for civil emergency planning.* Risk Management, 11(3–4), 179–207.

Stein, Robert M, Dueñas, Osorio, Leonardo, & Subramanian, Devika. (2010). *Who Evacuates When Hurricanes Approach? The Role of Risk, Information, and Location.* Social science quarterly, 91(3), 816–834.

Taaffe, Kevin, Johnson, Matt, & Steinmann, Desiree. (2006). *Improving hospital evacuation planning using simulation.* Paper presented at the Proceedings of the 38th conference on Winter simulation.

Taaffe, Kevin M, Kohl, Rachel, & Kimbler, Delbert L. (2005). *Hospital evacuation: issues and complexities.* Paper presented at the Simulation Conference, 2005 Proceedings of the Winter.

Tsai, Jason, Fridman, Natalie, Bowring, Emma, Brown, Matthew, Epstein, Shira, Kaminka, Gal, Sheel, & Ankur. (2011). *ESCAPES: evacuation simulation with children, authorities, parents, emotions, and social comparison.* Paper presented at the 10th International Conference on Autonomous Agents and Multiagent Systems-Volume 2.

Zhu, Nuo, Jia, Bin, & Shao, Chun-Fu. (2012). *Pedestrian Evacuation with the Obstacles Based on Cellular Automata.* 448–452.

Occupational Safety and Hygiene II – Arezes et al. (eds)
 ISBN 978-1-138-00144-2

Can the external environment affect the occupational safety conditions and unsafety behaviours?

I.F. Loureiro
Research Centre for Industrial and Technology Management, School of Engineering of the University of Minho, Guimarães, Portugal

M. Rodrigues & C. Vale
Research Centre on Environment and Health, Allied Health Sciences School of Polytechnic of Porto, Portugal

R. Azevedo
Centro de Apoio Técnico à Segurança no Trabalho CATST, ISMAI, Portugal

ABSTRACT: Portugal is undergoing an economic crisis affecting the European Union since 2007. The financial unsustainability of Portugal required foreign aid and consequently austerity and structural measures have been implemented. These measures are causing negative impacts on the economy income and living conditions of the population of Portugal. Studies in this field can stimulate new challenges in Occupational Safety and Health, as well as can be useful to planning strategies that attempt to minimize this situation. The present study adopted a quantitative methodology using a self-completed questionnaire named External Environment Questionnaire. The results indicate that the instability experienced by respondents and the organizations that they work for, may contribute to profound changes in the working conditions.

1 BACKGROUND AND MOTIVATION

Portugal is undergoing an economic crisis affecting the European Union since 2007. Three main factors can be identified as being responsible for the Portuguese crisis: the lack of market confidence and the lack of accountability and integrity regarding the government policies. The financial unsustainability of the country required foreign aid and consequently austerity measures have been implemented. These measures are causing negative impacts on the economy income and living conditions of the Portugal population. It is assumed that work conditions were also affected even though the extension of the effects remains unknown. Papadoupolos et al. (2010) identified some of possible negative consequences related to changes in the work environment. Following these thoughts, it is essential to analyze whether there are significant changes regarding the occupational risks exposure, resulting from the impact of social economic factors in the worker' everyday life. Studies in this area, can stimulate new challenges in the area of Occupational Safety & Health (OSH) at Work, as well as can be useful to planning strategies that attempt to minimize this situation.

2 INTRODUCTION

The European Union is going through an economic crisis since the year 2007 (Ali 2012; Fernando et al. 2009). Due to this situation, some countries were required to take austerity measures and structural reforms to reduce the budget deficits and the public debt. These measures led to negative impacts on income including reductions in employment. Indeed, unemployment and underemployment are increasing namely in the most vulnerable countries, such as Portugal, Spain, Greece, Italy and Ireland (Leahy et al. 2012). In Portugal, employment levels have fallen sharply since 2008. The evolution of employment over the years is presented in Figure 1.

As it is possible to observe, employment in Portugal decreases from the beginning of the financial crisis (2007/2008) resulting in increased unemployment regarding the same period of time. This was expected due to the implementation of austerity measures that may be encouraging workers to subject themselves to more precarious working conditions (Leahy et al. 2012).

Safety culture, among other things, has been related to the companies' work conditions (Fernadez-Muniz 2009; Rodrigues et al. unpubl.). Accordingly,

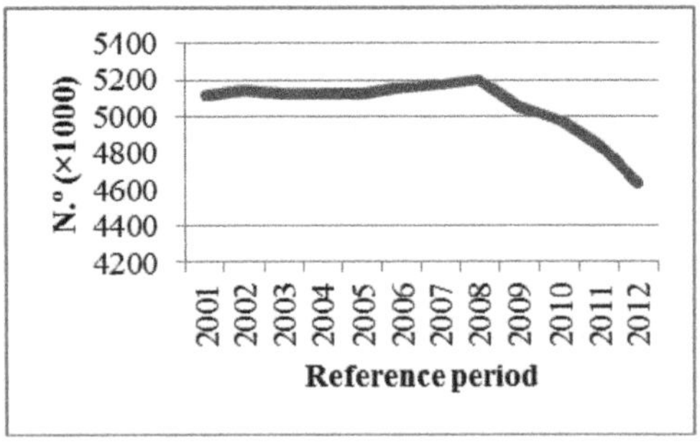

Figure 1. Employment evolution of the Portuguese population, 2001–2012 (INE, 2013).

and considering that it is expected a deterioration of the companies' safety culture with the crisis, unsafe situations can be more prevalent in nowadays. Employers are responsible for ensuring the optimal conditions and thus increasing their profits. Rasmussen (2000) stated that "Any workplace is an integrated part of a complex, dynamic socio-technical system". This is a complex type of system were different levels may be identified, from productive process to normative governmental rules. Each level is related to a research area (engineering, psychology, economics, political science), being study separately by different academic subjects. Several external conditions may have influence on each operation level affecting the system control: environment stressors. Several stressors can be identified: structural changes on the political rules, public awareness, market conditions changes and financial pressure, workers' skills changes and levels of education and technological changes. Each of these stressors may have an effect on a particular system level. Usually, the different levels of the socio-technical system are studied separately by the decomposition of the system into their functional systems. Considering this point of view, the strategies of the managements are often dissociated from the productive process in terms of human factors, that is, the situations where companies give too much importance to political changes. As a consequence, they implement several structural changes in the organizations like less payment or mass redundancies. In this case, workforce is directly affected and it is possible to assist to a reduction in the quality of work (in this case work must be seen from several dimensions: social, cultural and professional, among others). In these kinds of system-approach decisions (managers' decisions or supervisors' decisions) are usually separated from the working context and studied as an isolated phenomenon. It is important to notice that in modern societies, individuals are placed on a system organization not into a workplace per se. According to Loureiro et al. (2012), the understanding of people interactions and related mechanisms of systems regulation is important as they can be responsible for maintaining the system balance. A balanced system means that both social (well-being) and economic goal (total system performance) were achieved and as consequence the competitiveness of the companies is increased (Kogi 2006). Dul et al. (2012) refer that performance and well-being are outcomes of fitting the environment to human. Performance is related to productivity, efficiency, effectiveness, quality, innovativeness, flexibility, safety and security, reliability and sustainability. Well-being is concerned with health and safety, satisfaction, pleasure, learning and personnel development. These two outcomes may have influence on each other and must be understood as strongly connected. In recent years, studies on Kompier (2006) and Papadopoulos et al. (2010) reported that the work environment has undergone significant changes: the time and intensity of work, the schedule and worker flexibility, the extension of working hours for the weekend, the increasing irregular schedules, the precarious work (Benach et al. 2002; Kompier 2006; McNamara et al. 2013; Papadopoulos et al. 2010). Precarious work is the result of the replacement of permanent employment by home-based work, temporary work, and other informal arrangements characterized by decreased job security and lower pay. It was shown by Virtanen et al. (2005) that workers under those working circumstances do not have access to OSH training.

Worker flexibility enables the efficient use of human resources and can bring benefits to the organization (Lepak & Snell 1999). However, according to Guest et al. (2006) and Raquel (2009) there is a strong correlation between the flexibility and the living quality standards. Martin (2012) recognizes that the income that comes from work is the main source of income for most families and that workplaces are unstable places involving social risks such as loss of employment and consequently reduction of income. A review performed on the impact of the austerity measures implemented until 2011 on the daily life, concluded that the poorest people were the most affected by the crisis as in proportion they lost more income than people with higher wealth (Callan et al. 2011). It is therefore important to study how workers feel about the present situation of Portugal and how it can affects their situation on the organization.

In this regard, the main objective of this study was to analyze the employment population feelings about the influence of the crisis regarding some aspects that may have influence on the organizational safety climate.

3 METHODOLOGY

In order to perform this study, an observation tool was developed and data collection was carefully defined. The study adopted a quantitative methodology using a self-completed questionnaire named External Environment Questionnaire (EEQ). The EEQ included several questions to characterize the respondents profile and study their opinion about the current situation of Portugal and how does it affect their perception about security issues.

A pre-test was done to verify if the questions were well formulated and correctly understood by the population. To perform the pre-test, a sample including 10 individuals randomly selected was used. These individuals were asked to fulfill a guide that was specially made for this purpose. The EEQ was divided into six different parts: (1) characterization of the respondent, (2) employment status of the respondent, (3) economic situation of the respondent, (4) perception of the respondent about the economic situation of Portugal, (5) perception of the respondent about security issues, and (6) perception of the respondent regarding the engagement of the organization for which they work in relation to security issues. Briefly, the first part refers to the respondent characterization and gathers information such as age, gender, level of education and characterization of household. Part 2 included questions related to the respondent occupation namely: the tasks, the years of service, the sector of activity, the type of job contract, the hours and days of work per week, and the main source of income. Part (3) included a set of questions to analyze the annual income of the respondents, usual forms of shopping payment, services and credits and identification of the type of loans contracted by the respondent. In Part (4) respondents were asked about the national unemployment rate, the time they would take to get a new job and the main consequences of the austerity measures. Respondents were also asked about the occupational accidents, the number of day's loss and the importance of safety issues in the workplace and their feelings regarding the possibility of occurrence of different scenarios (as example: possibility of losing their job in a short period of time, possibility of having to subject their self to an professional activity with less payment than the current one; possibility of an increase in permanent costs (electricity, water, fuel, food, etc.) (Part 5). Finally, the last part of the EEQ gathers information about: the perception of respondents regarding the current financial situation of the organization, the organization' awareness to security issues and the recovery workers on safety at work.

The target population for this survey was the working population. To be valid and representative a minimum number of five hundred respondents were considered. This survey was conducted in 8 cities of the north of Portugal. The questionnaires were sent by mail and/or email through a list of contacts or by hand from June 14 to July 11, 2013.

After data gathering, a descriptive analysis was performed on all questions and depending on the relevance, correlations between variables were tested. Due to the fact that the variables under analysis are discrete, a study on the chi-square test (χ^2) was accomplished. Conventionally significance value must be less than 0.05 (Field 2009).

4 FINDINGS AND DISCUSSION

Only the results considered to be relevant to the purpose of this study will be presented.

This study surveyed 600 people, have been obtained 510 valid questionnaires. Nearly half of the respondents were male (43.1%). On average, respondents have 38.31 years (SD = 11.40; interval range 17–76 years old). The respondents aged above 65 years old have a par-time job (this age is considered to be the retirement age in Portugal).

The majority of the respondents are from Portugal (98.6%) and the remaining is from Ukraine, Brazil, Venezuela and Angola. Regarding their education background, most of them received primary education (37.7%), 32.7% received senior high school education, 28.6% went to college and 1.0% received an unspecified type of qualification. Results showed that 34.5% of respondents have three people within the household. These respondents have just one minor in charge (χ^2 = 795,937, $p < 0.001$). The respondents developed their professional activity mainly in the industry and services. When asked about the tasks usually performed, most of them marked the enforcement tasks option (45.7%), 18.8% scored the service tasks, 23.7% usually develop tasks related to the decision making process and 11.8% have surveillance tasks. Results showed that 57.3% have a dependent employment; 31.4% are public workers, 6.4% have their own business, 3.3% are self-employed, 1.4% is included on the independent workers group and 0.2% is active members of production cooperatives. On average, respondents work 5.2 days per week and 7.5 hours per day. A vast majority of respondents (94.7%) has no other source of income (work income) and pecuniary are the main reason for respondents having a second job. Respondents assert that the current unemployment rate in Portugal is very high (74.3%). Almost half of respondents (45.9%) reported that they felt their financial condition significantly affected by the policy measures resulting from the Troika (Fig. 2).

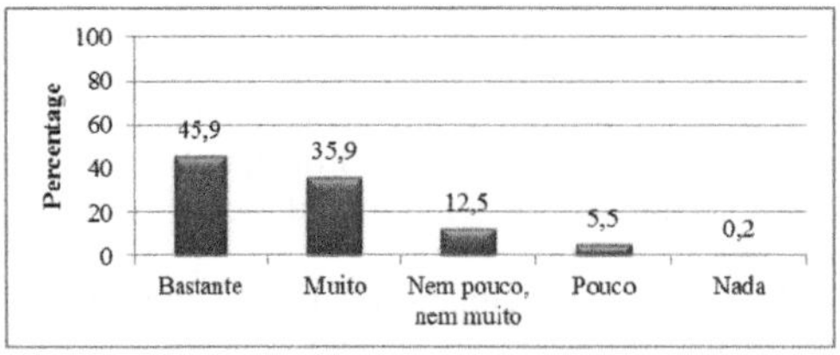

Figure 2. Influence of current policy measures, based on the recommendations of the Troika, on the economic situation of the respondents.

These respondents felt that it would be impossible to get a job if they lost their current position (χ^2 = 64,880, p < 0.001). These results seem to indicate some uncertainty among the respondents compared to the stability of their employment status.

Although most respondents complies the workplace safety rules, 16.5% of respondents have had some kind of occupational accident. These accidents mostly occurred among workers with primary education (χ^2 = 24,685, p < 0.001) and in the industry sector (χ^2 = 24,141, p < 0.002). Given the seriousness of the accident, 73.8% of respondents had injured absent from work through illness being the total working days lost in 3144. When asked about the acceptability of certain scenarios in order not to lose their current position, 34.5% of the respondents reported that they were willing to perform tasks with higher risk, 12.5% were willing to perform higher risk tasks without prior training, 8.5% were willing to withdraw the protections of the machines to increase productivity and 58% were willing to do overtime. 57.8% of respondents feel affected by not having protection equipment for using due to lack of investment by the organization. Even though the companies they work for usually invest in training activities related to safety at work for most of them, the lack of investment in OSH is due to the Portugal current financial status. In this regard, it is important to notice that, 39.4% of respondents said that the companies they work for are being seriously affected by the current crisis. In average, a vast majority of the respondents thinks that there are worst working conditions, increased stress at work, prioritize of production in relation to compliance with the safety rules, increased of the general fatigue, worst peer relationships, increase of the overall demotivation for the work.

5 FINAL CONSIDERATIONS

The increase in working hours, either by accumulation of a second job or by the addition of extra hours, may contribute to the onset of adverse effects, as noted in the study by McNamara et al. (2011). This adverse effects given by Papadoupoulos et al. (2010) and McNamara et al. (2011) can be categorized into psychological disorders and/or behavioral disorders, such as stress, fatigue, decreased concentration overtime, second jobs and increased competition among colleagues. The indebtedness, the increasing of the number of bills to pay and the lower salaries are factors that may contribute to the workers have to ensure their own jobs. Results showed that there are even a small percentage of respondents who claims to make overtime in order to increase the family budget. Studies by Vahtera et al. (1999) showed that fear of loss of employment may lead workers to accept lower or no remuneration for overtime and due to that fact works can be more exposure to physical exertion. Most of the respondents is aware of the high unemployment rate in Portugal and are poorly confident in being able to find a new job in the same area while the crisis prevail. The results indicate that the instability experienced by respondents and the organizations that they work for, may contribute to the decrease in investment in safety training, and OSH. These results are in agreement with studies by Koukoulaki (2010) identifying which companies are forced to increase productivity, giving priority to production in relation to compliance with the safety rules in the workplace, thereby increasing the pressure at work. The perception of the respondents agrees with studies such as those of Papadopoulos et al. (2010) and McNamara et al. (2011) suggesting that the adverse effects (stress, fatigue, lack of motivation) on workers' health can be the result of changes in the working conditions.

This study also points out the importance that workers give to the preservation of their job during an economic crisis scenario. In fact, we can conclude that workers are able to jeopardize their safety in order to maintain their current job. This can be seem trough the performance of tasks with higher risk without prior training, withdraw the protections of the machines to increase productivity and willing to do overtime. This is a subject of major concern once it can jeopardize safety culture in an organization and increase the occurrence of work-related accidents. As a consequence organizations should encourage workers to perform their activities safely specially during major economic crisis scenarios.

REFERENCES

Ali, T.M. 2012. The Impact of the Sovereign Debt Crisis on the Eurozone Countries. Procedia—Social and Behavioral Sciences, 62: 424–430.

Benach, J., Muntaner, C., Benavides, F., Amable, M., & Jodar, P. 2002. A new occupational health agenda for a new work environment, Scandinavian Journal of Work, Environment & Health 28:191–196.
Callan, T., Leventi, C., Levy, H., Matsaganis, M., Paulus, A., & Sutherland, H. 2011. The distributional effects of austerity measures: a comparison of six EU countries.
Dul, J., & Neumann, W.P. 2009. Ergonomics contributions to company strategies. Applied Ergonomics, 40: 745–752.
Fernando, A., Gandra Martins, I., Sousa Andrade, J., Rabello Castro, P., & Bação, P. 2009. A Crise Financeira Internacional (1.a ed.): Imprensa da Universidade de Coimbra.
Fernández-Muñiz, B., Montes-Peón, J.M., & Vázquez-Ordás, C.J. 2009. Relation between occupational safety management and firm performance. Safety Science, 47(7): 980–991.
Field, A. (2009). Discovering statistics using SPSS: (and sex and drugs and rock 'n' roll). (2nd edn) Los Angeles: Sage
Instituto Nacional de Estatística, Statistics Portugal. 2013. (Série 2011). Retrieved from http://www.ine.pt/xportal/xmain?xpid=INE&xpgid=ine_indicadores&indOcorrCod=0006505&contexto=bd&selTab=tab2
Kogi, K. 2006. Participatory methods effective for ergonomic workplace improvement. Applied Ergonomics, 37: 547–554.
Kompier, M. 2006. New systems of work organisation and workers' health. Scandinavian Journal of Work, Environment & Health, 32(6): 421–430.
Koukoulaki, T. 2010. New trends in work environment—New effects on safety. Safety Science, 48(8): 936–942.
Leahy, A., Healy, S., & Murphy, M. 2012. Study of the impact of the crises and austerity on people with special focus on Greece, Ireland, Italy, Portugal and Spain. Caritas Europa Report.
Lepak, D., & Snell, S. 1999. The human resource architecture: Toward a theory of human capital allocation and development, Academy of Management Review, 24: 31–48.
Loureiro, I., Leão, C.P. Arezes, & P.M. (2012). Ergonomic Tridimensional analysis: critical ergonomic factors identification in a commercial environmental. Work: A Journal of Prevention, Assessment and Rehabilitation, 41(1): 636–641.
McNamara, M., Bohle, P., & Quinlan, M. 2011. Precarious employment, working hours, work-life conflict and health in hotel work. Applied Ergonomics, 42(2): 225–232.
Martin, E. 2012. Buffering income loss due to unemployment: Family and welfare state influences on income after job loss in the United States and western Germany. Social Science Research, 41(4): 843–860.
Papadopoulos, G., Georgiadou, P., Papazoglou, C., & Michaliou, K. 2010. Occupational and public health and safety in a changing work environment: An integrated approach for risk assessment and prevention. Safety Science, 48(8): 943–949.
Raquel, M. 2009. Exploring Fragility: Industrial Delocalization, Occupational and Environmental Risks, and Non-Governmental Organizations. International Journal of Environmental Research and Public Health, 6(3): 980–998.
Rasmussen J. 2000. Human factors in a dynamic information society: where are we heading? Ergonomics, 43(7): 869–879.
Rodrigues, M.A., Arezes, P., & Celina P.L. unpublished. Multilevel model of safety climate for furniture industries. Accepted for publication on Work: A Journal of Prevention, Assessment and Rehabilitation.
Virtanen, M., Kivimäki, M., Joensuu, M., Virtanen, P., Elovainio, M., & Vahtera, J. 2005 Temporary employment and health: a review. International Journal of Epidemiology, 34(3): 610–622.

Occupational Safety and Hygiene II – Arezes et al. (eds)
ISBN 978-1-138-00144-2

Screening for psychoactive substances at the workplace—a review

P.H. Marques
Universidade Europeia, Laureate International Universities, Lisboa, Portugal
UNIDEMI—R&D Unit, Faculdade de Ciências e Tecnologia, Universidade Nova de Lisboa, Lisboa, Portugal

C. Jacinto
UNIDEMI—R&D Unit, Faculdade de Ciências e Tecnologia, Universidade Nova de Lisboa, Lisboa, Portugal

ABSTRACT: Although the goal of reducing occupational accidents is often used as an argument to justify workplace screening programs on psychoactive substances, the literature on the topic shows lack of scientific basis for such assumption. Among other reasons, few publications offer quantitative studies, and even fewer are able to provide sound statistical evidence of this cause-to-effect relationship. This paper provides a literature review on the subject and discusses the current knowledge in this field. The review shows that more research is needed to establish causal links between individual submission to psychoactive substances testing and subsequent accident reduction. In the aftermath of this review the authors suggest that such relationship should be evaluated by comparing empirical evidence of accident rates among workers previously tested, with accident rates among workers untested, all other things being equal.

1 INTRODUCTION

The main goal of this paper is to review the relevant literature concerning scientific understanding of the relationship between testing for Psychoactive Substances (PAS) at work, whenever screening is legally admissible, and subsequent change in occupational accident rates. Furthermore, authors intend to discuss adequate ways for studying the alleged effect of accidents prevention as being a result from testing for psychoactive substances at work.

A literature search was performed with automatic search in databases of scientific publications, academic theses, legislation and jurisprudence. Among these electronic sources, Internet more accessed sites were:

- www.euro.who.int—Regional Office Europe of World Heath Organization;
- www.ncbi.nlm.nih.gov/pubmed—USA National Library of Medicine;
- www.elsevier.com—ELSEVIER Websites;
- www.scopus.com—SCOPUS;
- www.scirus.com—SCIRUS;
- www.thecochranelibrary.com—The Cochrane Library;
- www.ewdts.org—European Workplace Drug Testing Society;
- www.ilo.org—International Labour Organization;
- www.act.gov.pt—Portuguese Authority for Working Conditions;
- www.nida.nih.gov—USA National Institute on Drug Abuse;
- www.tribunalconstitucional.pt—Portuguese Constitutional Court;
- www.parlamento.pt—Portuguese Parliament;
- www.dgsi.pt/jstj.nsf—Portuguese Supreme Court of Justice.

The remaining material was physically searched in books and specialty magazines, academic theses, technical catalogs and abstracts of scientific events.

The keywords used for searching, and combinations of them, were:

- Workplace Drug testing;
- Workplace Alcohol testing;
- Occupational accidents;
- Prevention;
- Human risk factor;
- Safety behavior;
- Risk Control.

1.1 *Psychoactive Substances*

According to the International Labour Organisation (ILO), in a broad sense, a psychoactive substance is one that, when ingested (inhaled or injected, or swallowed or aspirated, or smoked) changes the way brain works and it may alter mood, behavior and cognitive processes (ILO, 2003)-or, saying it in another way, is any substance consumed by a person to change how to feel, to think, or how to behave (ILO, 2003). Thus, psychoactive substances are considered to be alcohol, illicit drugs and even legal drugs either consumed with or without medical prescription. In the context of this paper, the term "psychoactive substances" (PAS) shall refer jointly to alcohol and illicit drugs, and only these, because no other substances have been covered by the current review.

The different levels of substance abuse are classified as: intoxication; harmful regular consumption; addiction (ILO, 2003).

PAS abuse can result from a combination of factors such as working conditions, social and family circumstances and intrinsic factors of the individual's personality.

PAS accessibility and social reality are recognized by ILO (2003) when declaring that: the world is witnessing a rising tide of substance abuse; accessibility of psychoactive substances is increasing; consumption and trafficking are growing; alcohol and drugs are everywhere; abuse of these substances is affecting society in ways that were unknown a few decades ago.

1.2 *Risks of Psychoactive Substances at work*

As it appears that PAS can be found at home, on the road, in leisure, and almost everywhere, one cannot assume they do not exist in the workplace, even if they are less visible and/or more difficult to detect. In many organizations, the hidden reality of PAS abuse lasts for too long, as a joint result of consumers avoiding being detected and organization preferring not to confront the problem (Happel, 2008).

As to their effects, all PAS cause more or less long dysfunctional impact on work (Kauert, 2008). Abusing PAS turns the user unfit for work and life.

Workers abusing PAS are risky persons in the employment context, regardless of their visible or not risk behavior—even striving to be preventive, the impact of dysfunctional PAS affects ability to control the risks—, thus it can be concluded that organizations must control work activities under the influence of PAS (Baer and Hess, 2008).

1.3 *Control measures of Psychoactive Substances at work*

According to the Organisation for Economic Cooperation and Development (OECD), in the particular case of the transportation sector, there is a broad international consensus on the need to control risks of working under the influence of PAS, and it is frequent to equate the use of various control measures that are legally viable (OECD International Transport Forum, 2010). For instance, within rail transportation context, the disastrous potential of human error in railway traffic turns this risk unacceptable to passengers, to the carrier company and to society at large. Pondering on the one hand, the individual's right to private life and freedom to consume PAS and, on the other hand, the right to life and physical integrity of workers and third parties (e.g.: passengers), it is obvious that the latter take precedence. Thus, to ensure this, risk control of PAS is justified on the workplace context. Therefore, it was considered important by the Occupational Health and Safety Group of International Union of Railways (UIC), to screen workers for PAS in all workplaces in which activities impact rail safety (UIC, 2008).

Independently of other risk control measures, screening by testing is necessary to protect jobs and workers against the dangers of PAS misuse—as claimed by the National Institute of Drug Abuse, the authority of United States of America (USA) on this matter (Gust and Walsh, 1989; Hanson (1993); Zwerling, 1993).

2 BACKGROUND LITERATURE

By scrutinizing international guidance, namely, American, Australian, European and Portuguese guidelines for programs of prevention, control and rehabilitation of PAS abuse and the practices already implemented in many organizations, one realizes their margin for evolution. As time goes by, more and more organizations are applying tests to job candidates and employees, although the reasons for screening and the circumstances in which tests are performed vary considerably; among the most common purposes are requirements for a work environment free of substance abuse, a job performed safely and improving outcomes of organizations.

In Europe, the shift from general prevention of PAS abuse to programs now including also screening tests—though little is known—is being made by an increasing number of organizations and is increasingly faster. The low perception of this trend may be due to the (rather small) number of organizations that disclose adoption of measures to control risks that are poorly promoted by European states—such as PAS screening tests.

2.1 *Psychoactive Substances testing at work*

In many countries, PAS testing faces restrictions regarding individual rights and data protection law. On the other hand, a growing number of companies and countries are legally coping with PAS risks (Strang *et al.*, 2012) by screening for PAS in employees in the scope of Occupational Safety and Health. This is, for instance, the Portuguese case, where testing programs that comply with certain requirements, exist with legal support, based on the recognition by the major legal authorities, that the collective safety and health outweighs the individual rights to privacy (PCC, 1995; PP, 2009; PSCJ, 1998).

Few publications are known to report test application, even in activity sectors where it would be more likely to expect, as in the chemical process industry or the transportation sector. From the chemical industry, a limited number of contributions are given by DEGUSSA (Breitstadt, 2008), EVONIK INDUSTRIES (Schiffhauer and Breitstadt, 2008) and ROCHE (Seiffert, 2008); in the second case, the few contributors are French *Société Nationale des Chemins de Fer Français* (SNCF) (Wenzek and Ricordel, 2008), British Network Rail (NR) (Network Rail, 2008) and Portuguese *Comboios de Portugal* (CP) (Marques, 2009; Marques *et al.*, 2011).

Economic studies on the prevention of PAS abuse at work, conducted in USA with a large number of studied cases and advanced statistics (Livingston, 1975; Ozminkowski *et al.*, 2003; Rummel *et al.*, 2004; Wickizer *et al.*, 2004, Miller *et al.*, 2007), demonstrated the relevance of balancing the costs of prevention of PAS abuse and the correspondent financial return on investment in prevention.

In the study by Ozminkowski *et al.* (2003)—covering 1791 manufacturing employees—it was concluded that the relationship between frequency of drug tests and injuries' medical expenses was statistically significant, and had a shape of "U". This led to the conclusion that medical expenses resulting from accidents can be minimized if employees are tested for drugs at an average frequency of 1.68 times per year.

The study by Wickizer *et al.* (2004)—about 14,500 workers of 261 companies with drug prevention programs were compared with 650,000 workers in 20,000 companies without such programs—has demonstrated a statistically significant association between drug prevention programs and lower rates of workplace accidents, in construction, manufacturing and services. There was also a small net savings (unspecified) associated with these programs, and considered sharper in the construction sector.

A more recent study, carried out by Miller *et al.* (2007)—which covered employees of a large carrier—showed a statistically significant relationship between the alcohol and drug abuse prevention program and lower rates of work accidents. It was also found a cost-benefit ratio of 1:26, i.e., for each 1 $US spent in the preventive program, 26 $US were saved in accident reduction.

In the twentieth century, quantified evidence of change in accident rates due to tests was summarized by Jess Kraus, in a systematic review of 740 publications on the testing of alcohol or drugs at work, of which only 6 quantified their effects on accidents, whilst the others addressed philosophical, social, moral, legal, management aspects and test protocols (Kraus, 2001). Of these papers, for those with numerical detail, Kraus determined the prevented fraction of accidents as a consequence of having set the PAS testing programs. He concluded to be impossible either refuting or supporting that the introduction or continued application of testing in the workplace has resulted in accidents reduction; according to Kraus, the reason for such inconclusive outcome was due to several methodological shortcomings of the original studies. For the same reason, the author considered that evidence of "random and unannounced" tests preventing more accidents than "non-random" tests, was limited and questionable.

Already in the present century, Cashman *et al.* (2009) carried out a review aiming to determine the effect of drug and alcohol testing in the prevention of injuries resulting from labor accidents of professional drivers; for this new study 6,000 articles and other publications were surveyed, out of which 19 merited further scrutiny, finally leading to only 2 publications with sufficient data and quality for the intended purpose; Both comprised time series of test trials in the USA. Noting that the tests had an effect of reducing accidents in the short term, the authors concluded that there was limited and insufficient evidence to consider that tests, alone, were effectively preventive in the long term, and stressed the need for more evaluation studies.

3 DISCUSSION OF PREVENTIVE EFFECT OF TESTING AND OTHER RESEARCH QUESTIONS

Specialized literature agrees that the number of PAS abusers detected over time tend to decrease with continued application of tests (Taggart, 1989; French *et al.*, 2004; Miller *et al.*, 2007; Wenzek and Ricordel, 2008). In the literature, the deterrent effect of workplace testing is generally attributed to inhibition resulting from personal perception of being held liable in case the state of abuse be analytically confirmed by further tests.

However, it is questionable whether the reduction of abusers detected over time is only due to the deterrent effect of tests. It is worthy noting that, in many organizations, these tests are applied exclusively during Occupational Medicine check-ups (that is, with previous notice of the timing of biological sampling). This is the case, for instance, of the company SNCF (Wenzek and Ricordel, 2008); in such circumstances, many abusers can learn, over time, to attend the tests in a state of abstinence from PAS, or even to deliver a biological sample of an abstinent, pretending to be of its own. If so, the mere existence of the notice of tests may be sufficient, by itself, to reduce detected abuses—not necessarily meaning that tests have an actual reducing effect on PAS abusers. This procedural limitation may bias the results of many organizations that are applying tests—even the few who report their conclusions in scientific publications. This pinpoints the relevance of studying unanticipated (that is, applied by surprise) screening tests in the workplace, in order to better assess their spontaneous effect over time.

As for the aforementioned studies (Ozminkowski *et al.*, 2003; Wickizer *et al.*, 2004, Miller *et al.*, 2007)—about the preventive effect of labor programs to control PAS abuse, made in USA for different professions and sectors of activity, with large number of cases and advanced statistics—one should take their conclusions with some caution, notwithstanding the great importance of these studies and the interest of their findings. This cautiousness is due to the relationship between PAS prevention programs and accidents at work, being investigated trough aggregate results of the organizations studied, regardless of the temporal order concerning individual subjection to preventive measures and the respective effect (expected reduction of occupational accidents). In other words: the relationships assessed in those studies failed to discriminate per worker, if subjection to preventive measures occurred before or after accident occurrence. Given that any measure of risk control applied to someone can only influence events occurred *a posteriori*—since an effect can manifest itself only after the stimulus—some methodological limitations became evident; the mentioned studies investigated the effect that PAS abuse prevention programs may have had (or not) over accidents occurred indiscriminately before and after the measures.

Among the rare publications reporting frequency of testing, none has proved to have the effect of minimizing the occurrence of accidents. It follows that the empirical evidence of a possible frequency level, that could be more preventive (best-frequency, or optimal frequency), remained undiscovered.

4 CONCLUDING REMARKS AND PROSPECTS FOR PROGRESS

This literature review unveiled a number of weaknesses in the scientific understanding of the relationship between PAS tests at work and subsequent occurrence of accidents, regarding statistical evidence. In particular, it was noted the paucity of statistically significant evidence of the alleged preventive effect of PAS tests at work. Furthermore, it was not found quantified evidence of a best-frequency for individual subjection to screening tests, i.e., a specific frequency that could minimize subsequent accidents.

Although the goal of preventing occupational accidents is often invoked to justify PAS screening programs at work, there is little statistically relevant evidence of the presumed causal link between individual subjection to tests and subsequent accident reduction. To overcome this limitation, the presumed causal relationship between individual subjection to tests (the preventive stimulus) and subsequent accident reduction (the desired effect) should be evaluated in a more accurate manner. The authors suggest this be done by comparing empirical evidence of accident rates among workers previously tested, with accident rates among workers untested, all other things being equal.

REFERENCES

Baer, H.P., Hess, M., 2008. Dangerous Behavior in Companies—Individuals at Risk and High-Risk Behavior. In: *The Worker—Risk Factor and Reliability*, Breitstadt, R. and Kauert, G. (Eds.), Shaker Verlag, Aachen: 57–65.

Breitstadt, R., 2008. Drug Screening as Contribution to Safety. In: *The Worker—Risk Factor and Reliability*, Breitstadt, R. and Kauert, G. (Eds.), Shaker Verlag, Aachen: 100–103.

Cashman, C.M., Ruotsalainen, J.H., Greiner, B.A., Beirne, P.V., Verbeek, J.H., 2009. Alcohol and drug screening of occupational drivers for preventing injury. *Cochrane Database of Systematic Reviews 2009*, Issue 2. Art. No.: CD006566. DOI: 10.1002/14651858.CD006566.pub2.

French, M.T., Roebuck, M.C., Alexandre, P.K., 2004. To test or not to test: do workplace drug testing programs discourage employee drug use?. *Social Sciences Research*, 33(1): 45–63.

Gust, S.W., Walsh, J.M., 1989. Drugs in the Workplace: Research and Evaluation Data. *NIDA Research Monograph*, National Institute of Drug Abuse, Rockville, Vol. 91.

Hanson, M., 1993. Overview on Drug and Alcohol Testing in the Workplace. *Bulletin of Narcotics*, Vol. 45(2): 3–44.

Happel, H.-V. 2008. Sociopsychological Aspects of Drug Use. *The Worker—Risk Factor and Reliability*, Breitstadt, R. and Kauert, G. (Eds.), Shaker Verlag, Aachen: 51–56.

ILO, 2003. Drug testing. *Alcohol and drug problems at work—The shift to prevention*. International Labour Organisation, Geneve. 2003: 87.
Kraus, J.F., 2001. The effects of certain drug-testing programs on injury reduction in the workplace: an evidence-based review. *International Journal of Occupational and Environmental Health*, Vol. 7(2): 103–108.
Livingston, W., 1975. Betriebliche Alkoholismus-Programme in U.S.-Firmen: Ein Untersuchungsbericht des Long Range Planning Service, Stanford Research Institute, London.
Marques, P.H., 2009. Occupational Safety Management in the Portuguese Trains, focused on behavioural risk control. In: *Riscos Industriais e Emergentes*, Guedes Soares, C., Jacinto, C., Teixeira, A.P. & Antão, P. (Eds.), Edições Salamandra, Lisboa (ISBN 978-972-689-233-5), Vol. 2: 911–930 (in Portuguese).
Marques, P.H., Jesus, V., Vairinhos, V., Abajo Olea, S., Jacinto, C., 2011. The control of alcohol and drugs and occupational accidents at the "Trains of Portugal": data analysis. In: Arezes *et al.* (Eds), *International Symposium on Occupational Safety and Hygiene—SHO'2011* (ISBN: 978-972-99504-7-6), Guimarães, Portugal, 10–11 Fev 2011: 373–377.
Miller, T.R., Zaloshnja, E., Spicer, R.S., 2007. Effectiveness and benefit-cost of peer-based workplace substance abuse prevention coupled with random testing. *Accident Analysis and Prevention*, 39: 565–573.
Network Rail, 2008. Drugs and Alcohol Policy Testing. In: *Health and Safety Handbook*, NR, London, version 1.0: 4–9.
OECD—International Transport Forum, 2010. Summary Document. In: *Drugs and Driving, Detection and Deterrence*. Organisation for Economic Cooperation and Development, OECD Publishing, Paris: 13
Ozminkowski, R.J., Mark, T.L., Goetzel, R.Z., Blank, D., Walsh, J.M., Cangianelli, L., 2003. Relationships Between Urinalysis Testing for Substance Use, Medical Expenditures, and the Occurrence of Injuries at a Large Manufacturing Firm. *The American Journal of Drug and Alcohol Abuse*, 29(1): 151–167.
PCC—Portuguese Constitutional Court, 1995. Acórdão nº 319/95. *Acórdãos do Tribunal Constitucional*, Vol. 31 (in Portuguese).
PP—Portuguese Parliament, 2009. Lei n.º 7/2009. *Diário da República—Ia Série, nº30, de 12 de fevereiro*, pp. 926–1029 (in Portuguese).
PSCJ—Portuguese Supreme Court of Justice, 1998. *Acórdão do processo 97S243*, de 24/06/1998. www.dgsi.pt/jstj.nsf/954f0ce6ad9dd8b980256b5f003fa814/56e61e8c22fafdd6802568fc003b7ae1?OpenDocument, accessed March 2011 (in Portuguese).
Rummel, M., Rainer, L., Fuchs, R. 2004. *Alkohol im Unternehmen*, Verlag Hogrefe, Göttingen.
Schiffhauer, N., Breitstadt, R., 2008. Keeping Tabs on Drug Abuse. *Dräger Review*, 96, Drägerwerk AG & Co. KGaA, Lübeck: 10–15.
Seiffert, B., 2008. A Field Report of a Drug Screening of Trainees. In: *The Worker—Risk Factor and Reliability, Breitstadt*, R. and Kauert, G. (Eds.), Shaker Verlag, Aachen: 104–108.
Strang, J., Babor, T., Caulkins,J., Fischer, B., Fozcroft, D., Humphreys, K., 2012. Drug policy and public good: evidence for effective interventions. *The Lancet*, 379, pp.71–83.
Taggart, R.W. 1989. Results of the drug testing program at Southern Pacific Railroad. *National Institute of Drug Abuse Research Monograph No. 91*, U.S. Government Printing Office, Washington, DC.
UIC Occupational Health and Safety Group, 2008. *Developing management arrangements for the Control of Safety Risks related to the Influence of Alcohol, Drugs and/or Psychoactive Medication*, International Union of Railways, Paris.
Wenzek, M., Ricordel, I., 2008. Cannabis et sécurité du travail. Photographie de l'evolution de sa detection au sein des contrôles de stupéfiants depuis 2004 à la SNCF. *Annales Pharmaceutiques Françaises*, vol. 66: 255–260 (in French).
Wickizer, T.M., Kopjar, B., Franklin, G., Joesch, J., 2004. Do Drug-Free Worplace Programs Prevent Occupational Injuries? Evidence for Washington State. *Health Services Research*, 39 (1): 91–110.
Zwerling, C., 1993. Current Practice and Experience in Drug and Alcohol Testing in the Workplace. *Bulletin of Narcotics*, 45 (2): 115–196.

Occupational Safety and Hygiene II – Arezes et al. (eds)
© 2014 Taylor & Francis Group, London, ISBN 978-1-138-00144-2

Ergonomic and anthropometric evaluation of a footwear modification: A case study

R. Lima & P.M. Arezes
Center for Industrial and Technology Management, University of Minho, Guimarães, Portugal

M.A. Carvalho
Centre for Textile Science and Technology, University of Minho, Guimarães, Portugal

ABSTRACT: This study aims to evaluate the changes performed on a specific type of shoes, taking into account the ergonomic and anthropometric aspects involved in its use. Due to the need to evaluate the bone structure of the considered subject case study, a conventional X-ray was performed, which has demonstrated a shortening of the left leg of 29 mm and a probable adduct foot. Anthropometric evaluation of the subject confirmed the asymmetry in the left lower limb, located between the knee and the left foot in 27 mm. The corrective changes were implemented in the left boot. Asymmetries and associated pathologies were identified and the footwear chosen by the user was adapted according to the specificities identified. The anthropometric analysis showed that the adopted compensation on the left boot, and according to the observed data, contributed to a postural improvement. The proposed objectives were achieved.

1 INTRODUCTION

The intervention of ergonomics is basically centred in the improvement of the human living and working environments, as well as in the design of means to ensure a maximum of comfort, safety and efficacy (Carneiro, 2010). The contribution of anthropometry for ergonomics allows to clarify the inadequate sizing of the shoes and inadequate postures adopted by its users, which also allows to prevent the onset of injuries, usually musculoskeletal (Barroso & Costa, 2010).

According to Nurse *et al.* (2005) insoles for shoes and orthotics are typically considered mechanical interventions. Interventions in footwear are commonly used to alter gait patterns, improve comfort and treat a number of diseases of the lower limbs (Razeghi, 2000; Nurse, 2005). Use of orthopedic devices to enhance the function of the skeleton (Crabtree, 2009).

As indicated by Barroso e Costa (2010), the characteristics and limitations of physical disabilities are so variable that the help devices have, frequently, to be "custom made" for the user.

The development and adaptability of the footwear must meet the anthropometric aspects of the users, considering the ergonomic features related both to form and comfort.

Considering the previous, the current paper is the description of a case study related to a young man with a congenital deficiency in his left leg, where he has no posterior tibial tendon. He almost has no gastrocnemius muscles, which prevents him from performing the extension of the foot and the flexion of the leg (Seeley *et al.*, 2003). He has also underwent a surgery, in which the short peroneal tendon was transferred to the Achilles tendon and the extensor of the *hallux* to the posterior tibial, tenodesis of the *hallux* extensor with common extensor (Nascimento, 2010). Currently, he presents the appearance of an adduct foot. The left leg is 3 cm shorter, being the asymmetry practically totally located between the knee and the foot. The leg is slimmer from the knee to the foot. The left foot is smaller and has muscle atrophy.

The aim of this study was to evaluate the appropriate changes in a pair of shoes, taking into account the ergonomic aspects, particularly those relating to anthropometric issues of the subject considered in this case study.

2 METHODOLOGY

It was assumed to be essential, in addition to the medical report, to become aware of the real situation of the user, having for this purpose evaluated the bone structure through the prescription of conventional X-rays, as well as with an anthropometric evaluation, using the Protocol of Anthropometric Dimensions and Percentiles (UMinho) using a fix anthropometer and a laptop.

2.1 *Material*

In this study the following anthropometry equipment was used:

- Fix anthropometer (Figure 1), used to collect most of the anthropometric dimensions;
- Portable anthropometer (Figure 2) "*Harpenden*" anthropometer. *Holtain Limited*: *Crosswell*; *Crymych*;
- Dyfed. Used to collect some of the anthropometric dimensions;
- Protocol of Anthropometric Dimensions and Percentiles (UMinho).

2.2 *Procedure*

The anthropometric measurements (with the exception of the biacromial and bideltoid shoulder width) were gathered through a static fix anthropometer (Figure 1), in the standing position and in the sitting position on a chair with a 450 mm height and adjustable to a horizontal surface, the back in the upright position, the legs with a 90° angle and the feet well supported on the ground. During the measurement process, the user was barefoot, wearing shorts and a t-shirt.

The biacromial and bideltoid shoulder width, was measured with the portable static anthropometer (Figure 2). The dimensions used in this study were all the dimensions existing in the UMINHO table (Table 1), namely: standing height, height of the eyes in relation to the ground, shoulder height in relation to the ground, elbow height in relation to the ground, wrist height in relation to the ground, sitting height in relation to the seat, eye distance in relation to the seat, distance shoulder seat, distance elbow seat, thigh thickness, thigh maximum length, distance thigh-popliteal, knee height in relation to the ground, popliteal height in relation to the ground, shoulder width (bideltoid), shoulder width (biacromial) hips width, chest (bust) thickness, abdominal thickness, distance elbow-wrist, functional vertical reach (standing), functional vertical reach (sitting), functional anterior reach, lumbar height in relation to the seat and weight. Taking into consideration all positions and ranges, having as reference the guidelines of the main static anthropometric dimensions (Figure 3). At the end an adjustment was made, removing the seat height (450 mm) in the measurements in which it was associated to. As the user showed asymmetries between the left and the right side, all the performed measurements take these asymmetries into consideration and were duly registered.

All the measurements were performed by the same measurer and registered in mm by a second person. The measurements were performed in the *Ergonomy Laboratory* of the *Department of Production and Systems from the School of Engineering of University of Minho*.

The data collection method used for the gathering and treatment of the data was in accordance with the protocol elaborated by the laboratory. The anthropometric data presented in the protocol constitute the first databank of anthropometric data of the Portuguese adult working population, with relevance to *design* ergonomic implementation (Barroso *et al.*, 2005).

The corresponding percentiles for the registered measurements of the user were calculated. This calculation was performed for each dimension. In the cases where asymmetries were observed, the

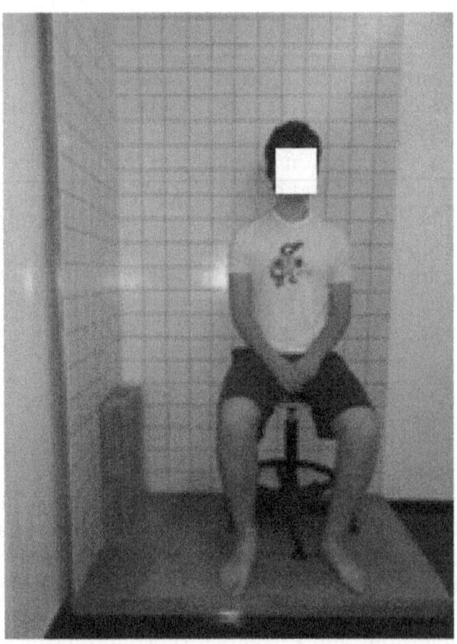

Figure 1. Fix anthropometer with subject.

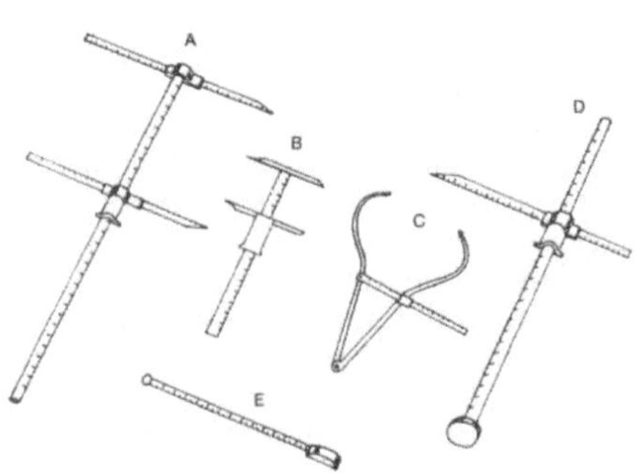

Figure 2. Portable anthropometer.

percentiles related to right and left side were calculated as well.

The weight factor was evaluated by calculating the *Body Mass Index* (BMI).

Table 1. Anthropometric data UMINHO—data from the Portuguese population.

Tabela UMINHO – dados antropométricos população portuguesa adulta								
Dimensão antropométrica	População masculina				População feminina			
	5	50	95	dp	5	50	95	dp
1. Altura de pé	1565	1690	1815	76	1456	1565	1674	66
2. Altura dos olhos (rel. ao solo)	1463	1585	1707	74	1355	1465	1575	67
3. Altura do ombro (rel. ao solo)	1277	1395	1513	72	1181	1290	1399	66
4. Altura do cotovelo (rel. ao solo)	966	1050	1134	51	889	965	1041	46
5. Altura do punho (rel. ao solo)	664	735	806	43	619	685	751	40
6. Altura sentado (rel. ao assento)	818	920	1022	62	799	865	931	40
7. Distância olhos-assento	716	810	904	57	696	760	824	39
8. Distância ombro-assento	576	630	684	33	496	590	684	57
9. Distância cotovelo-assento	206	255	304	30	191	250	309	36
10. Espessura da coxa	134	180	226	28	124	165	206	25
11. Comprimento máximo da coxa	518	590	662	44	517	570	623	32
12. Distância coxa-poplíteo	419	485	551	40	421	470	519	30
13. Altura do joelho (rel. ao solo)	459	525	591	40	434	480	526	28
14. Altura do poplíteo (rel. ao solo)	347	400	453	32	327	365	403	23
15. Largura dos ombros (bideltóide)	426	475	524	30	379	445	511	40
16. Largura dos ombros (biacromial)	299	335	371	22	251	300	349	30
17. Largura das ancas	341	380	419	24	342	400	458	35
18. Espessura do peito (busto)	221	265	309	27	226	275	324	30
19. Espessura abdominal	204	260	316	34	201	260	319	36
20. Distância cotovelo-punho	320	350	380	18	292	320	348	17
21. Alcance funcional vertical (de pé)	1875	2030	2185	94	1719	1860	2001	86
22. Alcance funcional vertical (sentado)	1117	1250	1383	81	1071	1165	1259	57
23. Alcance funcional anterior	628	730	832	62	621	675	729	33
24. Altura lombar (rel. ao assento)	166	215	264	30	174	220	266	28
25. Peso (Kg)	57	75	93	11	49	65	81	10

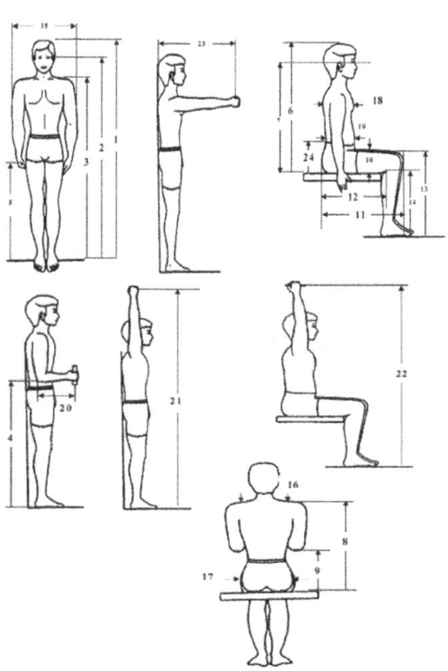

Figure 3. Main static anthropometric dimensions.

2.3 *Changes performed on the footwear*

Due to the asymmetry between the two lower limbs, the necessary changes to the footwear were undertaken. After performing the corrective changes on the boots, the soles and insoles were measured at different points.

The right foot, without alterations, presents a 36 mm heel and a 13 mm half sole. The left foot, subject to the alterations, has a 45 mm heel and a 23 mm half heel, having an increase of approximately 10 mm. The insole has a thickness of 10 mm on the foot sole zone, on the heel zone 15 mm and on the inner part, to correct the adduct foot it has 5 more mm, with a total of 20 mm.

All in all, there is an increase of 20 mm on the foot sole zone and approximately 30 mm on the heel zone.

These changes were confirmed and approved by the user, who tried them out and was pleased with the changes and the level of aesthetics of the boots, mentioning that: "*one can note very little of the alterations on the soles, and on the inside the foot is comfortably settled*".

3 RESULTS DISCUSSION AND ANALYSIS

3.1 *X-ray analysis*

The obtained radiography in extra-long chassis noted a large radius and right concavity back-lumbar scoliosis (Figure 4), centered at L1-L2 (Sarmento, 2012). Scoliosis is an abnormal lateral curvature of the spine, often accompanied by abnormal secondary curvatures (Seeley et al., 2003).

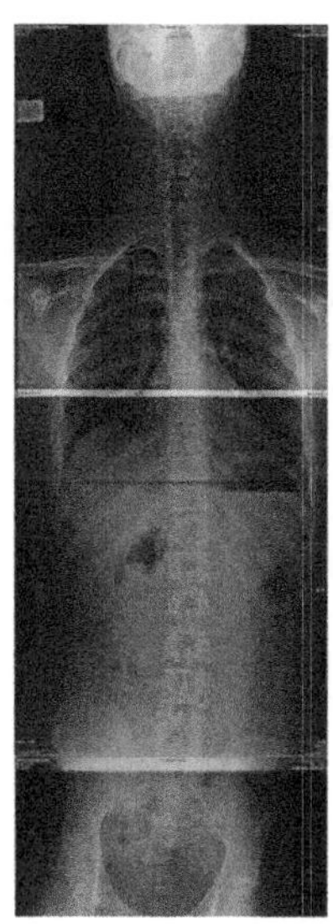

Figure 4. Spine in extra-long chassis (90 cm).

The X-ray (Figure 5) reveals a shortening of the left leg of about 29 mm compared with the right one (Sarmento, 2012).

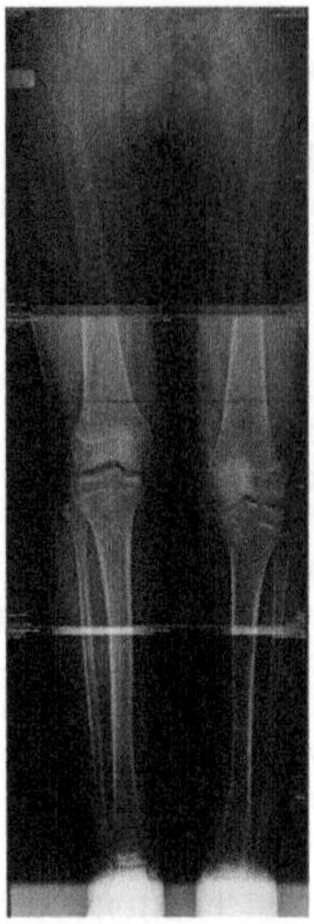

Figure 5. Lower limbs in extra-long chassis (120 cm).

The radiograph of the feet extremities (Figure 6) seem to note a certain degree of hypoplasia of the third and forth metatarsal to the left and possible adduct foot (Sarmento, 2012).

3.2 *Anthropometric analysis prior to footwear alteration*

The user has the height percentile of 97% compared with the Portuguese population (Table 2). The remaining percentiles are around 90%, lying within the normal range for a 16 years old young man.

The physiatry service found that he had showed 30 mm less length on the left leg. The relearning of the gait was immediately started with new instructions about the proper posture and feet position-

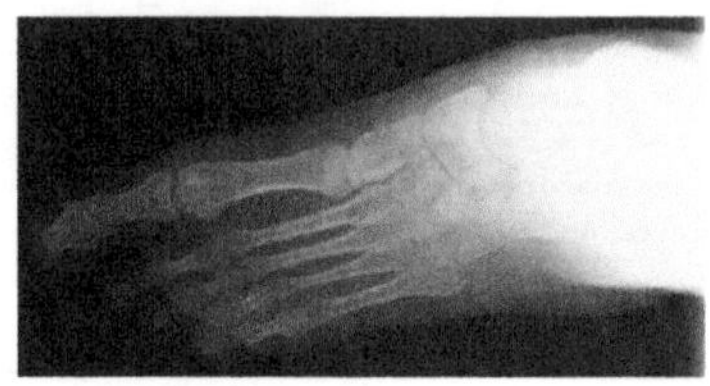

Figure 6. Foot, front view.

Table 2. Anthropometric dimensions and corresponding percentiles.

Anthropometric dimensions (mm)	Left	Right	Mean	SD	Z left	Z right	P left (%)	P right (%)
Stature	1835		1690	76	1,91		97	
Eyes height	1700		1585	74	1,55		94	
Shoulder height	1500	1505	1395	65	1,62	1,69	95	95,5
Wrist height	790		735	43	1,28		90	
Elbow height	1097	1100	1050	51	0,92	0,98	82	84
Distance elbow-wrist	390		350	18	2,22		98,5	
Previous functional range	810		730	62	1,29		90	
Vertical functional range (standing)	2165		2030	94	1,44		92,5	
Seated height	970		920	37	1,35		91	
Eyes height (in relation to the seat)	835		810	34	0,74		77	
Lumbar height (in relation to the seat)	240		215	20	1,25		89	
Maximum thigh thickness	175	185	175	17	0,00	0,59	50	72
Knee height	560	587	525	30	1,17	2,07	88	98
Popliteal height	402	428	400	26	0,08	1,08	53	86
Length thigh-popliteal	500		485	32	0,47		68	
Maximum thigh length	640		590	33	1,52		93,5	
Chest thickness	254		265	23	−0,48		32	
Abdominal thickness	255		265	32	−0,31		38	
Vertical functional range (sitting)	1268		1250	55	0,33		63	
Shoulder-seat height	630		630	33	0,00		50	
Elbow-seat height	270		255	30	0,50		69	
Shoulder width (biacromial)	372		335	22	1,68		95	
Shoulder width (bideltoide)	500		475	30	0,83		80	
Rump width	410		380	24	1,25		89	
Weight (kg)	90,4							

Table 3. Anthropometric data after alteration of the boots.

Anthropometric dimensions (mm)	Left	Right	Mean	SD	Z left	Z right	P left (%)	P right (%)
Shoulder height	1500	1505	1395	65	1,62	1,69	95	95,5
Elbow height	1097	1100	1050	51	0,92	0,98	82	84
Maximum thigh thickness	175	185	175	17	0,00	0,59	50	72
Knee height	582	587	525	30	1,90	2,07	97	98
Popliteal height	433	428	400	26	1,27	1,08	90	86

ing so that the signals reached the nervous system and the new changes were assimilated (Figueiredo, 2006). The left leg shows 27 mm less length between foot and leg in relation to the right one.

The entire left lower limb presents atrophy at a muscular level. While walking, since the right leg carries more load, it is over strengthened, presenting 50% on the left thigh and 72% on the right.

At the shoulder level there is barely no asymmetry. Due to the spine scoliosis it readjusts at each situation, correcting the decompensations, which may exist on the lower limbs through the spinal curvatures.

With a BMI of 26.8, is above the ideal weight, once it exceeds the BMI of 25. This overweight harms his gait and shows another negative factor, which is the fact that his gait does not appear symmetrical.

3.3 *Anthropometric analysis after footwear alteration*

During the anthropometric evaluation it became quite evident that the present asymmetry between left and right leg was located between the foot and the knee. Therefore, a new assessment was done, using only the data in which the asymmetries between left and right side were observed, in order to verify if the alterations showed any improvements of the posture.

The measurements were done in the same way as the previous ones, with the difference that the user was wearing the altered boots. Since the right boot had a height of 35 mm, this difference was deducted from both sides, having then achieved the results showed on Table 3.

As far as the differences of the shoulder and elbow height are concerned, no improvements were observed, since the same differences are shown. This is due to the fact that the spine readapts to new situations. The thigh difference remains the same as well, and due to the same reason of atrophy of the left side, the leg is thinner.

Regarding the knee and popliteal height, these present a difference of 5 mm only, due to the atrophy on the left leg. Globally, the compensation done

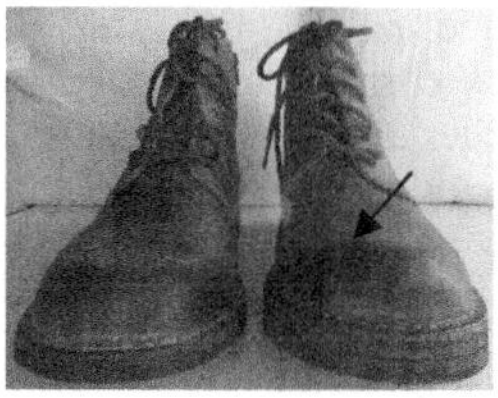

Figure 7. Finalized boots in front profile → changed boot.

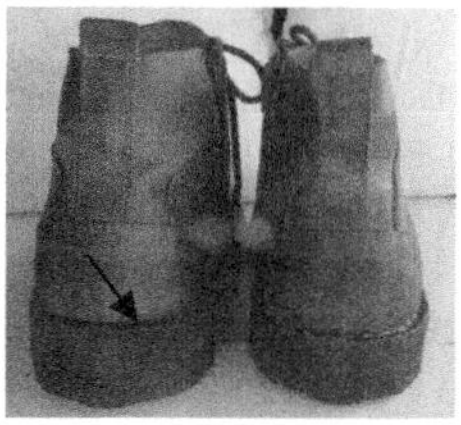

Figure 8. Finalized boots in back profile → changed boot.

on the left boot, and according to the observed data, a posture improvement was confirmed.

4 CONCLUSIONS

The pathologies and asymmetries were identified and confirmed. Through the X-ray an asymmetry of approximately 29 mm between the lower limbs and the adduct foot was identified. The anthropometric evaluation confirmed this deviation. The changes were designed and implemented based on these data.

The corrective changes were only implemented in the left boot, with an increase of 20 mm in the insole (20 mm in the heel zone, 10 mm in the foot sole) and 10 mm in the sole. The total increase was of 20 mm in the foot sole zone and of approximately 30 mm in the heel zone. The corrective changes on the boots proved a pronounced improvement at an anthropometric level, correcting the posture with the elevation of the left foot of approximately 30 mm.

The spine tends to restore the natural position, so that the scoliosis detected by X-ray does not progress and may even vanish.

The anthropometric analysis showed that the adopted compensation on the left boot, and according to the observed data, contributed to relevant postural improvement.

In general, it can be concluded that the main defined aims were achieved. The changes were performed on the boots chosen by the user who was also part of the case study. The asymmetries and associated pathologies were identified, the footwear was corrected according to the identified specificities and without changing its original appearance and, consequently, maintaining the acceptance of the user.

REFERENCES

Barroso, M., Arezes, P., da Costa, L., & Sérgio Miguel, a. (2005). Anthropometric study of Portuguese workers. *International Journal of Industrial Ergonomics*, 35(5), 401–410.

Barroso, P.M., & Costa, L.G. (2010). Antropometria. Guimarães: Dep. Produção e Sistemas Esc. de Engenharia Universidade do Minho.

Carneiro, P. (2010). Ergonomia. Guimarães: Dep. Produção e Sistemas Esc. de Engenharia Universidade do Minho.

Crabtree, P., Dhokia, V.G., Newman, S.T., & Ansell, M.P. (2009). Manufacturing methodology for personalised symptom-specific sports insoles. *Robotics and Computer-Integrated Manufacturing*, 25(6), 972–979.

Figueiredo, C. (2006). *Estabilidade Médio-Lateral vs. Quedas em Idosos*. Tese de Mestrado, Universidade do Porto: Porto, 66 pgs.

Grove, A.T. 1980. Geomorphic evolution of the Sahara and the Nile. In M.A.J. Williams & H. Faure (eds), *The Sahara and the Nile*: 21–35. Rotterdam: Balkema.

Nascimento, A. (2010). Relatório Cirúrgico. Coimbra. Portugal.

Nurse, M., Hulliger, M., Wakeling, J., Nigg, B., & Stefanyshyn, D. (2005). Changing the texture of footwear can alter gait patterns. *Journal of electromyography and kinesiology: official journal of the International Society of Electrophysiological Kinesiology*, 15(5), 496–506.

Razeghi, M., & Batt, M.E. (2000). Biomechanical analysis of the effect of orthotic shoe inserts. A review of the literature, *Sports Med. 29*, 425–438.

Sarmento, D.L.M. (2012). *Radiografias*. Póvoa de Lanhoso.

Seeley, R.R., Stephens, T.D., & Tate, P. (2003). *Anatomia e Fisiologia*. (L. Loures, Ed.) (6a edição).

Occupational Safety and Hygiene II – Arezes et al. (eds)
 ISBN 978-1-138-00144-2

Cultural and linguistic adaptation and validation of the long version of Copenhagen Psychosocial Questionnaire II (COPSOQ II) in Portuguese

S. Rosário
Research Laboratory on Prevention of Occupational and Environmental Risks (LABIOMEP/CIGAR), Faculty of Engineering, University of Porto, Porto, Portugal

J. Fonseca
Faculty of Medicine, University of Porto (FMUP), Porto, Portugal

J. Torres da Costa
Faculty of Medicine, University of Porto (FMUP), Porto, Portugal
Faculty of Engineering, University of Porto (FEUP), Porto, Portugal

ABSTRACT: The aim of this study was the cultural and linguistic adaptation and validation of the long version of *Copenhagen Psychosocial Questionnaire II* (COPSOQ II) in Portuguese. The original version of the instrument has been translated and adapted in a Portuguese sample of 239 workers belonging to various professional sectors. This study adopted a methodology based on the COSMIN (COnsensus-based Standards for the selection of health/Measurement INstruments) recommendations designed to study the psychometric properties of instruments. The questionnaire revealed good internal consistency for the total scale, with values of coefficient alpha test ($\alpha = 0.903$) and retest ($\alpha = 0.917$) reliability. The item-total correlations were estimated, and the items tend to present moderate to strong correlations with the total scale. The two weeks test-retest reliability of the Portuguese long version of COPSOQ II as measured by the intraclass correlation coefficient was ranged from 0.832 e 0.852 for all the dimensions. The Portuguese long version of the COPSOQ II seems to offer the necessary validity for use by researchers and practitioners interested in assessing the psychosocial work environment.

1 INTRODUCTION

In recent decades, we have seen profound changes in regard to the nature of workers, work, working conditions and professional life, due to factors such as globalization, the free market economy, new information technologies, the economic crisis and subsequent recession (EU-OSHA, 2007; Kompier, 2006).

The contemporary work contexts are characterized by a greater emphasis given to knowledge, work based on the information, the greater dependence on new technologies, greater labor market flexibility, with increased use of production and management systems just-in-time/lean, frequent organizational changes including downsizing, restructuring, subcontracting (Dollard, Winefield, & Winefield, 2003; Ferrie, 2008; Schnall, 2009; Sparrow, 2000) that have had significant implications in the way of work organization, and especially in the field of health and safety at work (EU-OSHA, 2012). This global trend has manifestations as increasing job insecurity, corporate pressure on their workers (EU-OSHA, 2002; OCDE, 2012), job insecurity and work intensification (NIOSH, 2002). These social and technological changes led to an increase of mental disorders in the workplace. Moreover, they are already identified relationships between stress and musculoskeletal, cardiac and digestive system, assuming that, if prolonged stress related to work can lead to serious cardiovascular diseases (ILO, 2013).

The psychosocial risks are recognized by European and national instances as one of the greatest challenges for occupational health and safety, as they are able to lead to serious physical and mental health deterioration of workers encompassing social and economic consequences (EU-OSHA, 2007; ILO & WHO, 2000; Leka, Griffiths, & Cox, 2003; WHO, 2001). In literature it is possible to note a direct or indirect role of the psychosocial environment of work in organizational health indices, and in fact, numerous studies have shown that stress is related to a worse performance, increased absenteeism due to accidents at work and occupational diseases, which takes on a human dimension impossible to neglect.

The occupational safety and health legislation confers a central place of the risk assessment in preventive approaches (Directive 89/391/CEE, 1989).

In 2012, the Committee of Senior Labour Inspection (SLIC, 2012) supported the European Campaign of Psychosocial Risk that aimed to promote and maintain good physical, mental and social workers and also improve the quality of assessments risks. The target groups of this campaign were the sectors of health, transport and services to be considered by this Committee the most problematic. The evaluation of occupational hazards requires the use of valid and reliable methods in order to identify risk factors in organizations.

The Copenhagen Psychosocial Questionnaire (COPSOQ) (Kristensen, Hannerz, Hogh, & Borg, 2005) is a tool developed by the Danish entity internationally renowned, recently revised, oriented evaluation of psychosocial occupational risk, considered an instrument directed to prevention. Given the absence of a Portuguese version validated instrument of the psychosocial risk assessment and its necessity in the labour context, the aim of this work was the linguistic and cultural adaptation and validation of the long version of the Copenhagen Psychosocial Questionnaire II (COPSOQ II) (Pejtersen, Kristensen, Borg, & Bjorner, 2010) in Portuguese.

2 MATERIALS AND METHODS

The sample comprised 239 workers from a total of twenty one organizations belonging to various professional sectors[1] participated in this study. Each participant responded to the questionnaire at baseline assessment and after two weeks (7–17 days) to assess reproductability. The assessment took place in the facilities of the organizations.

In this study 153 of the participants were female (65.3%) and 83 male (34.7%) and further the age range of the workers was 20 to 63 years old (mean age = 40.1 ± 10.5 years). The sample size was adopted in this study is based on the recommendations of COSMIN methodology (Mokkink et al., 2010). This methodology classifies the sample size greater than 100 subject as an excellent dimension to the objective proposed. All the participants were informed about the aim of the study and they all gave their consent. All the information collected was kept confidential and was anonymized before statistical analysis.

The COPSOQ has been recently reformulated, in the year 2007 and is available in three distinct versions: the long version for research (128 items), the medium-length version to be used by work environment professionals (87 items) and a short version for workplaces (23 items).

The COPSOQ is based on a set of theories more widely accepted nowadays, such as: (1) the job characteristics model, (2) Michigan organizational stress model, (3) the demand-control (support) model, (4) the socio-technical approach, (5) the Effort-reward Imbalance model, and (6) The Vitamin model, being able to evaluate the most relevant psychosocial dimensions inherent in the work context. All versions dimensions which feature measure indicators of exposure (e.g. psychosocial risks) and indicators of their effect (e.g. health, satisfaction and stress). The long version of COPSOQ II also includes a group of demographic issues. In this study we used the long version of COPSOQ II Questionnaire to assess the psychosocial factors at work, health and well-being of workers. The 41 specific scales of the long version of COPSOQ II instrument are presented in Table 1.

The duration of administration of the questionnaire is 30 minutes. Most responses of 128 items, which are part of 41 scales, are answered on a likert scale. All the scales go from 0 to 100, with high values representing a high level of the concept being measured, and the dimensions are calculated as average of the scores of items included.

For the development of this work it was previously requested permission to the authors of the questionnaire.

The adaptation process of an instrument meets a set of procedures to ensure the linguistic equivalence, conceptual equivalence and psychometric equivalence (SAC, 1995). In order to obtain the linguistic and conceptual equivalence of cross-cultural adaptation of the questionnaire was carried out the proper procedure for translation and back translation.

The long version of COPSOQ II was translated by a bilingual professional and independent reviewed by two health professionals to assess the adequacy of the Portuguese language. The first author [SR] consensualized this versions. The objective were to ensure that the resultant version was equivalent to the original version, conceptually and linguistic. The back-translation was made by a bilingual and independent translator.

The consensus version was applied to a group of professionals (five workers in each professional sector) with similar characteristics to the study

1. An hospital, a food industry, a construction industry, a financial consulting company, human resources and informatics, four commercial establishments, an university, a vertical grouping E.B 2.3 (total includes 11 schools).

population. This procedure permitted testing the operational equivalence, the understanding of the instructions and the items.

The Portuguese version is available at http://paginas.fe.up.pt/~mesho/idex.php/page/view/C.%20C.%20Internacionais.

Table 1. Scales and dimensions of COPSOQ II.

Dimensions	N° of questions
Work environment factors	
Quantitative demands	4
Work pace	3
Cognitive demands	4
Emotional demands	4
Demands for hiding emotions	3
Influence at work	4
Possibilities for development	4
Variation of work	2
Meaning of work	3
Commitment to the workplace	4
Predicability	2
Rewards	3
Role clarity	3
Role conflicts	4
Quality of leadership	4
Social support from colleagues	3
Social support from supervisors	3
Social community at work	3
Person-work interface factors	
Job insecurity	4
Job satisfaction	4
Work-family conflict	4
Family-work conflict	3
Values at the workplace	
Trust regarding management	4
Mutual trust between employees	3
Justice and respect	4
Social inclusiveness	4
Individual Factors	
Self-rated health	1
Sleeping troubles	4
Burnout	4
Stress	4
Depressive symptoms	4
Somatic stress	4
Cognitive stress	4
Self-efficacy	6
Offensive behaviour	
Sexual harassment	1
Threats of violence	1
Physical violence	1
Bullying	1
Unpleasant teasing	1
Conflicts and quarrels	1
Gossip and slander	1
Number of dimensions	41
Number of questions	128

Data analysis included descriptive statistics (mean and standard deviation); the assessment of internal consistency of the total scale and subscales through Cronbach's alpha; the test-retest reliability through intraclass correlation (ICC) and assessment of the item total correlation. Good internal consistency was considered a value greater than 0.60 according to Nunnaly (1978).

With respect to the item-total correlations, Moreira (2004) argues that the item-total values are acceptable when they are above 0.30. The statistical analysis was performed using SPSS version 20.0 for Microsoft Windows.

3 RESULTS AND DISCUSSIONS

The results obtained regarding the Cronbach's alpha coefficient for the overall scale of the sample (n = 239), the test ($\alpha = 0.903$) and retest reliability ($\alpha = 0.917$) revealed a high internal consistency (Table 2) almost all dimensions had Cronbach's alpha coefficients greater than 0.60 (Nunnaly, 1978).

Regarding the subscale that assesses the factors of conflict between work and family, there is internal consistency slightly lower than expected ($\alpha = 0.523$) in the test, and retest phase shows a higher value ($\alpha = 0.835$). Despite showing a lower value in the test, all other dimensions present value greater than 0.60, indicating an acceptable correlation with the total score.

The comparison results of Cronbach's alpha coefficient of the Portuguese version (test and retest) with the original version of Denmark is documented in Table 3. More specifically, the values of Cronbach's alpha for the four subscales of the instrument have to be satisfactory, as they tend to be greater than 0.60 (Nunnally, 1978), except the subscale "variation of work" ($\alpha = 0.22$) of the scale factors in the work environment and the subscale "social inclusiveness" ($\alpha = 0.56$) of the dimension values in the workplace.

Regarding item-total correlations, in general, results relating to the items appear to have moderate or strong correlation with the total, in all of the questionnaire.

Table 2. Internal consistency of the total scale and subscales (Cronbach's alpha).

Dimensions of COPSOQ II	Cronbach's Alfa test	Cronbach's Alfa retest
Work environment factors	0.892	0.901
Person-work interface factors	**0.523**	0.835
Values at the workplace	0.820	0.653
Individual factors	0.875	0.892
Cronbach's alfa (global scale)	**0.903**	**0.917**

Table 3. Cronbach's alpha values of the Portuguese version (test-retest) and original version (DK).

Dimension	Questions N°	Cronbach's α DK	Cronbach's α test	Cronbach's α retest
Quantitative demands	4	0.82	0.75	0.74
Work pace	3	0.84	0.74	0.79
Cognitive demands	4	0.74	0.78	0.78
Emotional demands	4	0.87	0.77	0.77
Demand for hiding emotions	4	0.57	0.64	0.64
Influence at work	4	0.73	0.73	0.73
Possibilities for development	4	0.77	0.71	0.79
Variation of work	2	0.50	**0.22**	0.50
Meaning of work	3	0.74	0.83	0.78
Commitment for the workplace	4	0.77	0.63	0.60
Predictability	2	0.74	0.62	0.64
Rewards	3	0.83	0.67	0.75
Role clarity	3	0.78	0.64	0.73
Role conflicts	4	0.67	0.70	0.65
Quality of Leadership	4	0.89	0.90	0.86
Social support from supervisors	3	0.79	0.83	0.79
Social support from colleagues	3	0.70	0.70	0.74
Social community at work	3	0.86	0.76	0.85
Job insecurity	4	0.77	0.67	0.71
Satisfaction with work	4	0.82	0.67	0.69
Work-family conflict	5	0.80	0.80	0.86
Family-work conflict	2	0.79	0.81	0.92
Trust regarding management	3	0.77	0.65	0.67
Mutual trust between employees	3	0.80	0.63	0.65
Justice and respect	4	0.83	0.75	0.78
Social inclusiveness	4	0.63	**0.56**	0.61
Self-rated health	1	–	–	–
Sleeping troubles	4	0.83	0.91	0.93
Burnout	4	0.81	0.81	0.83
Stress	4	0.86	0.89	0.89
Depressive symptoms	4	0.78	0.73	0.79
Somatic stress	4	0.68	0.70	0.75
Cognitive stress	4	0.83	0.86	0.87
Self-efficacy	6	0.80	0.78	0.86

The values of the item-total correlations ranged from 0.25 to 0.86 on the total scale. Assuming that according to Moreira (2004) the values of the item-total correlations are acceptable when they are above 0.30 in this particular case, it appears that all the items of the subscales are higher than against the full scale but three subscales of the questionnaire pertaining to the dimensions of values in the workplace (e.g. including "trust in leadership", "trust in colleagues" and "social inclusiveness").

In general, these results indicate that the questionnaire has internal consistency quite satisfactory.

The results obtained concerning means and standard deviations of the Portuguese version (test and re test), were very approximate the values of the original version, however the standard deviation values revealed a higher level of variation (Pejtersen et al., 2010). This high level of variation may be related with the imbalance between the series of the professional activities included and the number of participants.

In regard to the overall results obtained for average for subscales, it was found that the values obtained in the study are superior to the original version except in 12 subscales.

The two week test-retest reliability of the Portuguese long version of COPSOQ II as measured by the intraclass correlation coefficient was ranged from 0.832 to 0.852 for total scale.

The intraclass correlation coefficients of the full scale with the different subscales obtained proved quite satisfactory, superior to 0.80.

The results of test-retest reliability, allow to conclude that the COPSOQ II features a good temporal stability and reliability (test-retest), in the time interval considered. To conclude, all results presented allowed to confirm the long version of COPSOQ questionnaire II.

According to the authors, this instrument aims to constitute a valid instrument, capable of application in various professional sectors; improve communication between the various researchers and professionals who assess the work environments; enable national and international comparisons; improve and facilitate the assessment and interventions in the workplace through the operationalization of complex concepts and theories.

4 CONCLUSIONS

Whereas in Portugal we are facing scarcity of instruments adapted to our population, particularly in the area of psychosocial risk assessment, the present work aims to be a contribution to the evaluation and study of the implications of the psychosocial factors of labour contexts in the health and well-being of workers.

Furthermore, the legislation concerning safety and occupational health gives a central place of risk assessment in preventive approaches, being compulsory in every type of organization. Thus, the adaptation and validation of this instrument is intended to be a resource to companies regarding of occupational risk assessment, while meeting the legal plan provided for currently and mainly contribute to the taking of measures of prevention and promotion of health and well-being in the labour context.

The Portuguese version of long COPSOQ II shows to be a robust instrument, seems to offer the necessary validity for use by researchers and practitioners interested in assessing psychosocial work environment.

This version is available in MESHO (FEUP): http://paginas.fe.up.pt/~mesho/index.php/page/view/C.%20C.%20Internacionais.

ACKNOWLEDGMENTS

The authors would like to thank Master in Occupational Safety and Hygiene Engineering (MESHO), of the Faculty of Engineering of the University of Porto (FEUP), all the support in the development and international dissemination of this work.

REFERENCES

Council Directive 89/391/EEC of 12 June 1989 on the introduction of measures to encourage improvements in the safety and health of workers at work (1989).

Dollard, M., Winefield, A., & Winefield, H. (2003). Occupational stress in the service professions. In M. Dollard (Ed.), *Introduction: costs, theorectical approaches, research designs* (pp. 1–43). London: Taylor & Francis.

EU-OSHA. (2002). *New forms of contractual relationships and the implications for occupational safety and health*. Luxembourg: European Agency for Safety and Health at Work.

EU-OSHA. (2007). *Expert forecast on emerging psychosocial risks related to occupational safety and health*. Luxembourg: Office for Official Publications of the European Communities.

EU-OSHA. (2012). *Drivers and barriers for psychosocial risk management: an analysis of the findings of the European Survey of Enterprises on New and Emerging Risks*. Luxembourg: European Agency for Safety and Health at Work.

Ferrie, J., Westerlund, H., Virtanen, M., Vahtera, J., & Kivimäki, M. (2008). Flexible labor markets and employee health. *Scandinavian Journal of Work, Environment and Health, 34*(6), 98–110.

ILO. (2013). *The Prevention of Occupational Diseases. World Day for Safety and Health at Work*. Geneva: International Labour Organization.

ILO, & WHO. (2000). *Mental Health and Work: Impact, Issues and Good Practices*. Geneva World Health Organization.

Kompier, M. (2006). New system of work organization and worker's health. *Scandinavian Journal of Public Health, 32*, 421–430.

Kristensen, T., Hannerz, H., Hogh, A., & Borg, V. (2005). The Copenhagen Psychosocial Questionnaire-a tool for the assessment and improvement of the psychosocial work environment. *Scandinavian Journal of Work Environmental Health, 31*(6), 438–449.

Leka, S., Griffiths, A., & Cox, T. (2003). *Work Organization and Stress*. Geneva: World Health Organization.

Mokkink, L., Terwee, C., Patrick, D., Alonso, J., Stratford, P., Knol, D., & Bouter, L. (2010). The COSMIN checklist for assessing the methodological quality of studies on measurement properties of health status measurement instruments: an international Delphi study. *Quality of Life Research, 19*, 539–549.

Moreira, J. (2004). *Questionários: Teoria e prática*. Coimbra: Livraria Almedina.

NIOSH. (2002). *The changing organization of work and the safety and health of working people: Knowledge gaps and research directions* (Vol. 2002–116). Columbia DHHS: National Institute for Occupational Safety and Health.

Nunnaly, J. (1978). *Psychometric theory*. New York: McGraw-Hill.

OCDE. (2012). *Sick on the Job?*. Paris: OCDE Publishing.

Pejtersen, J., Kristensen, T., Borg, V., & Bjorner, J. (2010). The second version of the Copenhagen Psychosocial Questionnaire. *Scandinavian Journal of Public Health, 38*, 8–24.

SAC. (1995). Instrument review criteria. *Bulletin of Medical Outcomes Trust, 3*(4), I–IV.

Schnall, P., Rosskam, E., Dobson, M., Gordon, D., Landsbergis, P., & Baker, D.. (2009). *Unhealthy Work: Causes, Consequences and Cures*.

SLIC. (2012). The Committee of Senior Labour Inspectors. *European Campaign on Psychosocial Risks at Work*.

Sparrow, P. (2000). New employee behaviours, work designs and forms of work organization. What is in store for the future of work?. *Journal of Managerial Psychology, 15*, 203–218.

WHO. (2001). *The World Health Report. Mental Health: new understanding, new hope*. Geneva: World Health Organization.

Occupational Safety and Hygiene II – Arezes et al. (eds)

Anthropometric study of the student population of a Portuguese faculty

L.S. Sousa & M.E. Pinho
Faculty of Engineering of the University of Porto, Porto, Portugal

P.M. Arezes
Engineering School of the University of Minho, Guimarães, Portugal

ABSTRACT: In recent years, there has been an increasing need of knowledge of the human population's anthropometrics, particularly in Portugal. Changes in the anthropometric characteristics of the Portuguese people have been reported in some studies, leading to an increasing interest on the study of the evolution of these characteristics. This study aimed to collect and analyze the anthropometric dimensions of the student population of FEUP. For that purpose, a stationary anthropometer was built and a portable anthropometer was also used. Two hundred and six students (131 males and 75 females) with ages ranging from 18 to 35 years old, were measured. A set of 14 static anthropometric measures was collected. The results show significant differences in some anthropometric dimensions, when compared with studies performed by other authors, as well as between genders for almost all anthropometric dimensions.

1 INTRODUCTION

Ergonomics aims to fit the environment and the task, whether it is work or leisure, to the user so that they can work easily, effectively and safely (Mokdad & Al-Ansari, 2009). This aim may be summarized by the principle of user-centered design. Pheasant & Haslegrave (2006) defined it as the fit of the design of an object, a system or an environment that are intended to human use to the physical and mental characteristics of its human users, as well as for the demand of the task.

One way of designing the task, based on the physical characteristics of the user, is doing it according to her/his anthropometric characteristics, such as body size (reach, body segment length, and height), shape (segment circumferences, widths), strength and working capacity (Barroso et al., 2005).

Until 2005, there was a lack of anthropometric data from the Portuguese population. Padez (2002) conducted a study which gathered data obtained by the army that measured the stature of the 18 year-old young men that went to military service.

Later Barroso et al. (2005) measured a sample of the Portuguese population, creating the first anthropometric database of this population comprising 25 static anthropometric dimensions. Although the latter provides an accurate database, the student population of a university is, in its majority, much more narrowed to young people (18 to 25 years old). Thereby, the use of this database would probably result in some error due to two main reasons: (1) its age range goes from 18 to 65 years old; (2) the time gap between the two studies. The first one reflects the accentuation of the spinal curvature in people with ages greater than 40 years old (Pheasant & Haslegrave, 2006) which results in a lowering of the people's stature. The second reason may be supported by the expected change in some characteristics due to the secular trend (Arezes et al., 2006) that, accordingly to Padez (2002), estimated an increase of 9.9 mm per decade in Portuguese male stature, while other study suggested that the anthropometric databases should be updated every five years due to effects in the national anthropometric properties caused by regional altitude and climate, as well as sexual ones (Kaya et al., 2003).

The need for the anthropometric study of the Portuguese student population shows a relevant pertinence. Therefore, this paper aims to study the anthropometric dimensions for a sample of the Portuguese students of Faculty of Engineering of the University of Porto (FEUP).

2 METHOLOGY

2.1 *Subjects*

One hundred and thirty one men and 75 women from a universe of 7295 students were measured.

The sample was mostly composed by students from the integrated master programs, but also from undergraduate and master programs. With ages range between 18 and 35 years old, students were selected when passing through the hall of the building where most of the classes are taken.

2.2 *Equipment used*

Measurements were taken on a stationary anthropometer (Figure 1) built for the specific purpose of this study. Wood panels were arranged as a corner and covered with graph paper. To calibrate the anthropometer was used a self-retracting tape measure which insured the match of the lines of the paper with the lines of the measure. As aid to the measurements, grid lines were drawn every 100 mm and the corresponding value was marked.

For the collection of some anthropometric dimensions a portable anthropometer (Holtain's Harpender anthropometer) was used. Those dimensions required postures that could influence the measurement. Therefore was used the portable anthopometer to ensure that the gathered data had the best accuracy and precision.

For the sitting measurements a bench was also used.

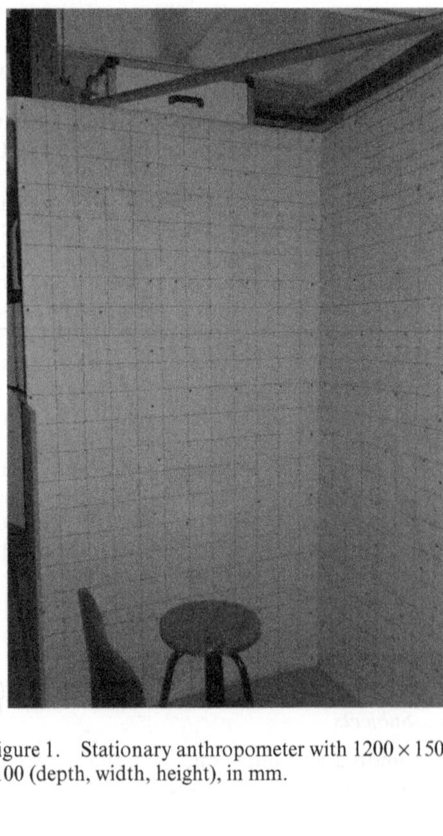

Figure 1. Stationary anthropometer with 1200 × 1500 × 2100 (depth, width, height), in mm.

2.3 *Procedure*

2.3.1 *Data collection*

A total of 14 static anthropometric dimensions were measured for each individual. Six dimensions were measured with the individual standing, while the remaining were obtained while the individual was seated.

The anthropometric measures were taken with the subject in a relaxed and erect posture. Each student was measured in thin cloths (T-shirt, shirt or thin sweatshirt), jeans, skirts or dresses. The standing dimensions were taken with the student standing erect to the anthropometer without shoes. The sitting dimensions were taken with the student seated erect onto the anthropometer, with knees bent 90°, and feet (without shoes) flat on the floor. The body dimensions were measured as described in ISO 7250-1:2008 (ISO, 2008).

The dimensions measured in both standing and sitting positions as well as the apparatus used for the measurements are detailed in Table 1.

2.3.2 *Data treatment*

First, Kolmogorov–Smirnov test was performed to study the normality of the anthropometric data's distribution. Then, mean and standard deviation were calculated for all the measured dimensions. Later, Student's t-test was used to test if the anthropometric dimensions of female and male populations were statistically different.

In order to measure data dispersion, the Coefficient of Variation (CV) was calculated as shown in equation 1:

$$CV = \frac{s}{m} \times 100\% \quad (1)$$

where CV = Coefficient of Variation; s = standard deviation; and m = mean.

Table 1. Dimensions measured in standing and sitting positions.

Dimensions	
Standing	Sitting
Abdominal depth*	Buttock–knee length
Elbow–knuckle length	Buttock–popliteal length
Eye height	Popliteal height
Forward grip reach	Sitting elbow height
Hip breadth*	Sitting eye height
Shoulder breadth (bi-deltoid)*	Sitting stature
Stature	Thigh thickness*

* Measured with the portable anthropometer.

3 RESULTS AND DISCUSSION

A great majority (84%) of the participants are younger than 25 years old (Table 2).

The sample used includes subjects from several lecture courses of FEUP, and several curricular years. That is reflected on the range of ages of the sample, 18–35 years. However, there is a predominance of younger students, 18–24 years. Accordingly to statistics from FEUP, the sample characteristics (age and female percentage) ensure the intended representativeness of the student population.

The coefficient of variation (CV) (Table 3) of this study was compared with the CV obtained in the previous study of Barroso et al. (2005) and with the characteristic value range of Pheasant & Haslegrave (2006).

Comparison between the results obtained and the previous study from Barroso et al. (2005) indicates a significant difference for the CV values in almost 80% of the anthropometric dimensions. It can be explained by the uneven sample size, which could be corrected by increasing the sample size. When compared with the characteristic ranges defined in Pheasant & Haslegrave (2006), lower CV was found in one anthropometric dimension (male and female stature). Nevertheless, about 70% of the dimensions are between the recommended ranges. Exceptions also occur with the values obtained for forward grip reach, hip breadth, popliteal height, thigh thickness and abdominal depth whose values are higher than the reference values by Pheasant & Haslegrave (2006). For the last two, a possible explanation can be the fact that these dimensions are associated with soft body tissue, particularly fat and muscle. Other possible explanation for the differences between them can be the fitness habits of the different people.

With regard to the other dimensions, explanation for the CV values found is possibly associated with the need for a larger sample.

Table 4 displays mean and standard deviation values for the 14 anthropometric dimensions measured in both genders, as well as the p-value obtained by Student's t-test. The results indicate that male have greater anthropometric dimensions, except for the hip breath. In this case, there is no significant difference between both genders. This can be explained by the participant's young age, which bodies are still developing. The p-value for sitting elbow height was very close to the value of 0.05 but it was still below the significance level established. In all the other dimensions, males are bigger than females.

Table 5 and Table 6 compares the mean values, for the male population, obtained in this study with the Portuguese population anthropometric database from Barroso et al. (2005), with British adults aged 19 to 25 (Pheasant & Haslegrave, 2006), with Malaysian adults aged 19 to 25 (Chong & Leong, 2011), and with Iranian adults aged 20 to 30 (Mououdi, 1997).

The comparison between the results of the current study and the Portuguese anthropometric data

Table 2. Distribution of the participants by age group and gender.

	Male		Female	
Age group (years)	N	%	N	%
Less than 20	47	35.9%	18	24.0%
20–24	71	54.2%	48	64.0%
25–29	8	6.1%	6	8.0%
30–35	5	3.8%	3	4.0%

Table 3. Coefficient of variation: Results of the current study, values found by Barroso et al. (2005) and the characteristic value range according to Pheasant & Haslegrave (2006).

	Coefficient of variation (%)				
	Current study		(Barroso et al.)		
Dimensions	Male	Female	Male	Female	(Pheasant & Haslegrave)
Stature	3.7	3.4	4.5	4.2	4–11
Abdominal depth	11.8	14.9	12.1	13.7	5–9
Buttock–knee length	5.1	6.4	5.6	5.6	4–11
Buttock–popliteal length	5.7	6.2	6.7	6.3	4–11
Elbow–knuckle length	5.3	4.9	5.1	5.4	4–11
Eye height	4.0	3.5	4.7	4.5	3–5
Forward grip reach	5.2	4.6	8.6	5.0	3–5
Hip breadth	7.4	9.4	6.4	6.8	5–9
Popliteal height	6.6	5.1	6.3	6.3	3–5
Sitting elbow height	9.9	10.6	11.8	11.8	4–11
Sitting eye height	4.0	3.7	4.2	4.6	4–11
Sitting stature	3.5	3.5	4.1	4.1	3–5
Shoulder breadth (bi-deltoid)	7.0	7.6	6.4	7.0	5–9
Thigh thickness	12.0	16.4	9.9	9.2	5–9

by Barroso et al. (2005) shows a few differences, some of which are significant. For example, the stature has an increment of 61 mm in male and 55 mm in female. This difference may be related to three probable causes: (1) the age range of the sample once the spinal curvature increases above 40 years old (Pheasant & Haslegrave, 2006); (2) the secular trend, which is known to occur in other studied populations (Pheasant & Haslegrave, 2006), and was also observed in Portugal by Padez (2002); and (3) the provenience of the sample (university students that, due to different financial conditions, may adopt distinct diets).

Table 4. Mean (M), Standard Deviation (SD), in mm, of the male population (n = 131) and the female population (n = 75) and the comparison of the means.

	Male		Female		
Dimensions	M	SD	M	SD	*p*-value
Stature	1751	65	1625	56	<0.00001
Abdominal depth	264	31	249	37	0.00460
Buttock–knee length	615	31	580	37	<0.00001
Buttock–popliteal length	490	28	466	29	<0.00001
Elbow–knuckle length	357	19	328	16	<0.00001
Eye height	1635	65	1515	53	<0.00001
Forward grip reach	733	38	676	31	<0.00001
Hip breadth	347	26	346	33	0,76219
Popliteal height	420	28	392	20	<0.00001
Sitting elbow height	245	24	238	25	0.04678
Sitting eye height	808	32	758	28	<0.00001
Sitting stature	925	32	868	30	<0.00001
Shoulder breadth (bi-deltoid)	482	34	424	32	<0.00001
Thigh thickness	190	23	171	28	<0.00001

As for the other populations, it might be observed that some differences exist, some of which are statistically significant. For instance, for stature, the highest differences found are those registered between the Portuguese and the Malaysian population, 38 and 65 mm for female and male populations, respectively. The contrast found in the anthropometric dimensions of the different populations highlights the usefulness of this study and of the presented results.

4 CONCLUSIONS

Fourteen body measurements of Portuguese university students were summarized in this paper and they enabled the anthropometric characterization of the Portuguese student population of FEUP. This can be used for the design of more ergonomic classroom and auditoria furniture.

The contrast between the data of this study and the data of the Portuguese adult anthropometric database emphasizes the need to extend the database, as well as its segmentation in age groups for more accurate use of that resource.

One of the main limitations of this study is the sample size that, due to the lack of enough time,

Table 5. Mean values, in mm, for the anthropometric dimensions from the current study and different young adult male populations.

	Current study		Barroso et al. (2005)		Pheasant & Haslegrave (2006)		Chong & Leong (2011)		Mououdi (1997)	
Dimensions	M	SD	M	SD	M	SD	M	SD	M	SD
Stature	1751	65	1690	76	1760	73	1713	48	1725	58
Abdominal depth	264	31	265	32	240	26	227	27	–	–
Buttock–knee length	615	31	590	33	595	32	561	21	579	27
Buttock–popliteal length	490	28	485	32	500	34	450	23	–	–
Elbow–knuckle length	357	19	350	18	–	–	–	–	–	–
Eye height	1635	65	1585	74	1650	72	1589	52	1616	56
Forward grip reach	733	38	730	62	790	36	764	57	–	–
Hip breadth	347	26	380	24	350	31	354	29	–	–
Popliteal height	420	28	400	26	445	30	431	16	431	21
Sitting elbow height	245	24	255	30	245	32	229	28	260	26
Sitting eye height	808	32	810	34	795	36	786	37	–	–
Sitting stature	925	32	920	37	915	37	874	112	912	26
Shoulder breadth (bi-deltoid)	482	34	475	30	465	29	435	31	455	24
Thigh thickness	190	23	175	17	160	16	–	–	–	–

Table 6. Mean values, in mm, for the anthropometric dimensions from the current study and different young adult female populations.

Dimensions	Current study		Barroso et al. (2005)		Pheasant & Haslegrave (2006)		Chong & Leong (2011)		Mououdi (1997)	
	M	SD	M	SD	M	SD	M	SD	M	SD
Stature	1625	56	1565	66	1620	61	1560	61	1597	101
Abdominal depth	249	37	260	36	220	22	195	21	–	–
Buttock–knee length	580	37	570	32	565	29	529	26	563	29
Buttock–popliteal length	466	29	470	30	475	29	434	28	–	–
Elbow–knuckle length	328	16	320	40	–	–	–	–	–	–
Eye height	1515	53	1465	66	1515	60	–	–	1486	58
Forward grip reach	677	31	675	33	705	31	701	46	–	–
Hip breadth	346	33	400	27	350	29	351	22	–	–
Popliteal height	392	20	365	23	400	27	402	31	365	22
Sitting elbow height	238	25	250	28	230	28	217	30	259	37
Sitting eye height	758	28	760	35	745	33	721	40	–	–
Sitting stature	868	30	865	35	855	35	828	32	861	36
Shoulder breadth (bi-deltoid)	424	32	445	31	395	24	383	19	392	24
Thigh thickness	171	28	165	15	150	16	–	–	–	–

is smaller than would be recommendable. That was reflected in the CV's values observed.

Hopefully this study will become the necessary starting point for new studies in this field, particularly at a university level, to enhance the anthropometric knowledge of the Portuguese young-adult students.

ACKNOWLEDGMENT

The authors would like to thank Master in Occupational Safety and Hygiene Engineering (MESHO), of FEUP, all the support in the development and international dissemination of this work.

We would also like to thank professor Joaquim Góis for sharing his knowledge on statistics which was of greater use for this work.

REFERENCES

Arezes, P.M., Barroso, M.P., Cordeiro, P., Costa, L.G. d., & Miguel, A.S. (2006). *Estudo Antropométrico da População Portuguesa*: ISHST - Institudo para a Segurança, Higiene e Saúde no Trabalho.

Barroso, M.P., Arezes, P.M., da Costa, L.G., & Sérgio Miguel, A. (2005). Anthropometric study of Portuguese workers. *International Journal of Industrial Ergonomics, 35*(5), 401–410.

Chong, Y.Z., & Leong, X.J. (2011). Preliminary Findings on Anthropometric Data of 19–25 Year Old Malaysian University Students. In N. Osman, W. Abas, A. Wahab & H.-N. Ting (Eds.), *5th Kuala Lumpur International Conference on Biomedical Engineering 2011* (Vol. 35, pp. 193–196): Springer Berlin Heidelberg.

ISO. (2008). ISO 7250–1:2008 Basic human body measurements for technological design—Part 1: Body measurement definitions and landmarks. International Organization for Standardization, Geneva, Switzerland.

Kaya, M.D., Hasiloglu, A.S., Bayramoglu, M., Yesilyurt, H., & Ozok, A.F. (2003). A new approach to estimate anthropometric measurements by adaptive neuro-fuzzy inference system. *International Journal of Industrial Ergonomics, 32*(2), 105–114.

Mokdad, M., & Al-Ansari, M. (2009). Anthropometrics for the design of Bahraini school furniture. *International Journal of Industrial Ergonomics, 39*(5), 728–735.

Mououdi, M.A. (1997). Static anthropometric characteristics of Tehran University students age 20–30. *Applied Ergonomics, 28*(2), 149–150.

Padez, C. (2002). Stature and stature distribution in Portuguese male adults 1904–1998: The role of environmental factors. *American Journal of Human Biology, 14*(1), 39–49. doi: 10.1002/ajhb.10017.

Pheasant, S., & Haslegrave, C.M. (2006). *Bodyspace: Anthropometry, Ergonomics, And The Design Of Work*: Taylor & Francis Group.

Occupational Safety and Hygiene II – Arezes et al. (eds)
 ISBN 978-1-138-00144-2

Variability management: A still to be noticed role of workers

A.S.P. Moraes & P.M. Arezes
University of Minho, Guimarães, Portugal

R. Vasconcelos
University of Porto, Porto, Portugal

ABSTRACT: Despite the efforts in reducing or eliminating variability, it is possible to assume that it remains as an inherent aspect in any production system. Activity-centered Ergonomic analyses have been showing the importance of operating strategies to deal with variability found in organisations and to accomplish the production goals. The aim of this paper is to show the relevance of variability situations in the analyzed context. It also aims to demonstrate that the role of workers in dealing with such situations is not recognized inside the organisation, not even by the workers themselves. The research was based on action research strategy, and the main method applied was the Ergonomic Analysis of Wok Activity. It was conducted in a manufacturing company, part of the automotive chain.

1 INTRODUCTION

Variation is an inherent characteristic of any system. According to Fujita (2006, p.19) "no system (i.e., combination of artifact and humans) can avoid changes". He states: "changes occur continuously throughout a system's lifetime ... this should be regarded as a destiny" (Fujita 2006, p.19).

Considering organisations as socio-technical systems, it is possible to argue that changes in organisations can be originated from humans and from its artifacts. Regarding the former, the diversity and variation among individuals is well known—the inter-individual diversity. In addition, it is possible to consider intra-individual variability, due to short-term variation, such as circadian cycles, or for example, variations because of aging. Considering the technical components of organisations, it is possible to identify many sources of variation: differences among products and services, machines and equipment, differences in materials from different suppliers, differences in production goals and their executing conditions.

Guérin et al. (2007, p.78) classify the variability in companies in two major categories: (i) normal variability, that is expected to be found due to the type of work performed, e.g., differences between taxi journeys, the different requests of clients, different products on a same production line, and so on; (ii) incidental variability, such as a poorly finished part or component which cannot be assembled, a tool that breaks, or a file missing data.

Management techniques and technological improvements have been developed to eliminate or minimize variation in organisational settings. They aim to keep the processes stable and consistently operating at the target level of performance with only normal variation. Some examples of such efforts are statistical process control methods, quality tools and automation investments. Obviously, the impacts of such efforts in improving the efficiency and performance of the systems cannot be ignored.

However, some may argue that such efforts are useful when dealing with part of variability that can be foreseen, and therefore controlled. According to Guérin et al. (2007), there is another type of variability that has a random aspect, and therefore, cannot be predicted and completely eliminated. This reinforces the assumption that variability is an intrinsic aspect of any system.

Some pertinent questions emerge from this assumption: If variation cannot be completely eliminated from organisations, i.e., there is always residual variability in a system, who is responsible for dealing with it? And how is this usually done?

Activity-centered ergonomic studies have been demonstrating that usually the workers of the operating staff, also called operators, are those responsible for dealing with variability. Some analyses have highlighted how workers deal with variations in the production processes, developing alternative behaviors, the so-called "operating strategies". Some authors advocate that without the operators' role, the production goals would not be accomplished (Garrigou et al. 1995). However, this is not easily recognized by managers or well integrated within the formal structure of organisations, and it is still common to find managers or other staff members that view operators as mere "procedure followers".

Recent studies in Resilience Engineering (Trotter et al. 2013, Reiman 2011) have demonstrated an increasing interest in such strategies, which have been treated as "improvisations". The central idea of this conceptual framework is that the development of these behaviors is not only a source of risk and violation, but also can have a positive influence in human performance. However, the positive aspects of improvisation have not been tipically investigated.

The goal of this paper is to show the relevance of variability situations encountered in the analyzed context. It also aims to demonstrate that the role of workers in dealing with such situations is not recognized inside the organisation, sometimes not even by the workers themselves.

2 CASE CONTEXT

The variability situations were collected from research conducted in a manufacturing company, part of the automotive chain. The study focused on the calendering process, an intermediate process that produces continuous sheets from rubber compounds incorporated with reinforcing materials such as textile fabrics or wire cords. The calendering process is an important step in the production of tires, because the quality of the sheets is critical to tire performance. In this factory, eleven different products, divided in two types, according to the reinforcing material used are produced in the calendaring process.

The calendering machine, a calender, is a heavy-duty machine equipped with three or more chromeplated steel rolls, which revolve in opposite directions, at specific speeds (Rodgers & Waddell 2005). Beyond the steel rolls, other accessory equipment ensure the production process: let-off stations and creel rooms for unwinding the reinforcing materials; extruders, heating and feeding mills for preparing the rubber compound; accumulators for avoiding machine slowdown; heating and cooling drums; tension controllers and so on. A number of measurement and control systems guarantee the quality of the final sheets.

The overall equipment, including the accessory systems, measures around 84 meters in length, 16 meters wide and 8 meters high. It weighs 150 tons., consumes about 395 kW/h and produces more than 50,000 meters of material/day.

Calendering production follows a 24/7 schedule, requiring 5 work shifts. Each calendering work shift is formed by 6 machine workers. The work in the calendar is divided among the operators, who are allocated in 5 different areas of the equipment. Of all the operators that work on the machine, 25 are company employees, and around 4 are outsourced workers.

3 METHODS

The variability situations were identified in the context of an ergonomic intervention, which was based on an action-research strategy (Stringer 2007). The intervention consisted of two phases, in accordance with the Future Work Activity method (Daniellou & Garrigou 2002). The first phase aimed at characterizing reference situations and the Ergonomic Analysis of Work Activity (Guérin et al. 2007) was the main method applied.

Field studies enabled the research party to familiarize itself with the technical process, progressively gaining the workers trust (De Keyser 1992). They also assisted in mapping the workers activities, including their actions and their visualization and communication needs (Wisner 1987). This phase also included the identification of problems experienced by the workers and the risks involved in the work setting.

Data were collected through systematic observation of the work activities and the machine routines. Open and semi-structured interviews, spontaneous and concurrent verbalizations were also used as data-collection techniques. Data collection also included the analysis of the relevant documents available, such as company specs, work instructions and safety procedures.

When the variability situations were occurring, data were being recorded in real-time: the place and the time of the occurrence, the number of workers involved in the situation and the actions performed by each one, the communications established among them, and so on. Later, the workers were questioned about what they had done, why they had carried out such actions, the frequency of the specific occurrence or similar situations, etc. All data were recorded in a logbook.

The field research lasted for seven months, comprised between November 2010 and June of 2011. It took around 41 data-collecting days, each day consisting of approximately 6 hours of direct contact with the machine operators. As the company works in rotative shifts, all the shift groups were involved in the research. In order to include the weekend shifts, data collecting was also performed during weekends.

4 RESULTS

Although variability situations were expected to be found from the beginning of the research, they

Table 1. Variability situations occurrence.

Variability situations observed	146 situations
Data collection days	41 days
Variability situations/day	Min. 1, Ave. 3.6, Max. 8
Hours of data collection*	200 hours
Frequency of occurrence	43,8 min
Percentage of occurrences during the set-up or start-up	24,7%

* approximately.

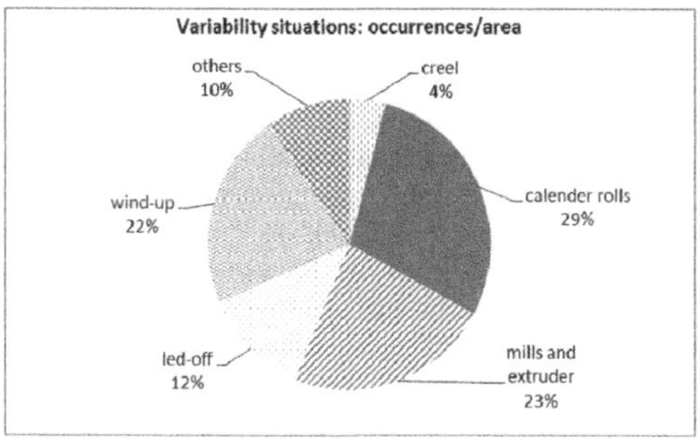

Figure 1. Areas of occurrence.

proved to be of great importance in the analyzed context. They were observed throughout all the data collection days and happened in every work shift. Table 1 presents the overall data from such situations.

The situations were caused by a wide range of sources: changes in raw materials, breakdown of accessory equipment, and unexpected machine errors that resulted in a complete stop of the machine. The results presented in Figure 1 show that variability situations also occurred in all of the different areas of the machine.

5 DISCUSSION

The number of variability situations observed demonstrates their relevance in the analyzed context. However, a deeper analysis of such situations reveals the importance of the role of the workers in dealing with them.

The strategies developed by the operators to deal with the variability situations were of paramount importance to return the calendering process to a controlled condition, and accomplish the production goals. Such strategies aimed, in some cases, to anticipate and avoid future problems in the process. In other cases, they aimed at correcting an incident or a deviation that was not noticed in advance and restoring the production process to a normal condition. It is important to highlight the fact that such strategies were developed individually and also collectively.

5.1 *The recognition of the worker as a variability manager*

The role of the workers in dealing with the variability situations was not noticed in the organisation at the time of this research. This lack of recognition was found in different staff members, from different hierarchical levels, and was also found among the workers.

When describing the calendering process and the work performed on it, workers representations are similar to the stages described in the company procedures. According to them, the calendering process is composed by two stages: first, the set-up of the machine and start-up; and second, the stage of process controlling.

The importance of the set-up between two product types and the machine start-up was clear and well-known. This phase is considered the most critical in the process. However, the phase of process control was considered as valueless. A verbalization demonstrates the difference between the representations of the workers regarding both stages:

"the hardest part is the machine start-up, feeding more or less rubber in, also changing the fabric and rubber ... after, it is all rolling ... you only have to control it, nothing else ..."

The idea about the differences in the work performed in each phase of the calendaring was also highlighted by one supervisor. He stated when passing close to the machine:

"so much is done there, then they just do nothing"

However, as shown in Table 1, less than a quarter of the total number of variability situations observed occurred during the set-up of product type or start-up of the machine. More than 75% of the situations observed occurred during process control phase of the calendering.

5.2 *The process control phase as variability management*

At first sight, the workers discourse seems to undervalue the need of dealing with variations during their work. This is even more evident when referring to the process control stage. However, during the development of the research, this role was observed. When recording the situations and questioning about the occurrence of variability situations, the importance of workers' role in managing the process variables was revealed. One of the workers' statements shows the existence of variations in the process:

"if nothing fails, that is all ... but as things fail ..."

Another worker stated:

"there are a lot of annoying tasks to do because it is not perfect ..."

The evidence of the random aspect of the variations can be observed in two verbalizations from two different workers. One operator stated:

"these kind of things happen"

However, another one stated:

"there are things that don't happen every day ..."

5.3 *Differences among the workers when managing variability*

Another important aspect revealed by the workers, was the difference among their individual strategies to deal with the variability situations. Some of the verbalizations demonstrate this aspect:

"each one works in his own way"
"each one has a specific timing for doing things ... but everything must go well ... there are lots of tricks, I don't even know all of them, and I have been working here for 4 years already ..."

5.4 *Some considerations about the research methods*

Despite the effort in quantifying the number and frequency of the variability situations, it is possible to infer that they are much less than the total number of situations experienced by the calendar operators in their everyday work. This is because of the limitations in the methods used in the research, which was based on direct observation and was conducted by only one researcher.

The dimensions of the machine and the division of the tasks among the operators cannot be ignored as data collection constraints. The more complex variability situations were easier to identify and therefore enter the data collection. This is because in such cases, more workers were involved and a greater need of communication arose among them.

However, it was possible to observe that many situations were solved by one of the workers without even being noticed by others. In such cases, it was a matter of luck for the researcher to have been close to the operator in order to be able to identify and understand why the situation occurred and then discuss it with him.

6 CONCLUSIONS

Variability can be considered as an inherent characteristic of any industrial setting. However, it is even more prevalent than is usually acknowledged (Garrigou et al. 1995).

Activity-centered ergonomic studies have been pointing that the management of variability situations by workers has paramount importance for achieving production goals. However, this role is not always acknowledged in organisations.

This paper reveals that variability situations had great relevance in the analyzed context. It also evidences the fact that there was no recognition of the role of the workers in managing such variations. This lack of recognition was found inside the organisation, including the workers themselves.

The recognition of the workers as variability managers can be a first step towards recognizing that work is not only following procedures, but also dealing with a great number of variables For sure, such acknowledgement can help in putting light on the importance of workers in accomplishing production goals, and in the development of more efficient risk management techniques.

ACKNOWLEDGEMENTS

This work was partially supported by a grant from the Brazilian National Council for Scientific and Technological Development (CNPq/Brazil).

REFERENCES

Daniellou, F. & Garrigou, A. 2002. Human factors in design: sociotechnics or ergonomics? In M. Helander & M. Nagamachi (eds), *Design for Manufacturability: A systems approach to Concurrent Engineering and Ergonomics*: 55–63. London/Bristol: Taylor & Francis.

De Keyser, V. 1992. Why field studies? In M. Helander & M. Nagamachi (eds), *Design for Manufacturability: A systems approach to Concurrent Engineering and Ergonomics*: 305–316. London/Bristol: Taylor & Francis.

Fujita, Y. 2006. Systems are ever-changing. In E. Hollnagel, D.D. Woods & N. Leveson (eds.), *Resilience engineering. Concepts and precepts*. England: Ashgate Publishing Limited.

Garrigou, A., Daniellou, F., Carballeda, G., Ruaud, S. 1995. Activity analysis in participatory design and analysis of participatory design activity. *International Journal of Industrial Ergonomics* 15(5): 311–327.

Guérin, F., Laville, A., Daniellou, F., Duraffourg, J. & Kerguelen, A. 2007. *Understanding and transforming work: the practice of Ergonomics*. Lyon: ANACT—National Agency for the Improvement of Working Conditions.
Reiman, T. 2011. Understanding maintenance work in safety-critical organisations—managing the performance variability. *Theoretical Issues in Ergonomics Science* 12(4): 339–366.
Rodgers, B. & Waddell, W. 2005. Tire Engineering. In James E. Mark, Burak Erman and Frederick R. Eirich (eds), *The Science and Technology of Rubber*. 3rd Edition: 619–662. San Diego: Elsevier Academic Press.
Stringer, E.T. 2007. *Action Research*. 3rd Edition. Thousand Oaks: Sage Publications.
Trotter, M.J., Salmon, P.M., & Lenné, M.G. 2013. Improvisation: theory, measures and known influencing factors. *Theoretical Issues in Ergonomics Science* 14(5): 475–498.
Wisner, A. 1987. Por dentro do trabalho. Ergonomia: método e técnica. São Paulo: FTD/Oboré.

Occupational Safety and Hygiene II – Arezes et al. (eds)
© 2014 Taylor & Francis Group, London, ISBN 978-1-138-00144-2

Whole Body Vibration in open pit mining—a short review

C. Bernardo
Research Laboratory on Prevention of Occupational and Environmental Risks (LABIOMEP/CIGAR), Faculty of Engineering, University of Porto, Porto, Portugal

M.L. Matos
LNEG—Laboratório Nacional de Energia e Geologia, S. Mamede de Infesta, Portugal

J. Santos Baptista
Research Laboratory on Prevention of Occupational and Environmental Risks (LABIOMEP/CIGAR), Faculty of Engineering, University of Porto, Porto, Portugal

ABSTRACT: The main objective of this study was to synthesize scientific knowledge in order to characterize Whole Body Vibration (WBV) in open pit mining: sources, measurements methodologies and main effects on health. A review based on PRISMA Statement was done. Inclusion criteria were research on WBV in mining, published after 1990 and written in English. Thirty-three studies were included in the review. Rock drills, shovels and dumpers are strongly present in this industry. They are potential sources of workers exposure to WBV. All bibliography followed methodological recommendations of ISO 2631-1:1997. For WBV characterization routine works, individual tasks, occurrence records, pavement features, were topics considered. Dumper is the equipment that can be the more dangerous; it has the highest Root Mean Square (RMS) acceleration values. Back pain is the main complaint in workers due to WBV exposure.

1 INTRODUCTION

All risk categories are present in mining environment (Donoghue 2004). Consequently, mining is classified as high risk activity and it is being the target of numerous studies, always aiming to eliminate/reduce accidents and prevent occupational diseases. In open pit mining, production is based on the extraction and processing of rocks. It includes fundamental operations such as drilling, charging and blasting, loading, transport and crushing (Matos & Ramos 2010). These operations are carried out with equipment that generates vibrations which workers are exposed. Rock drills, shovels and dumpers are common equipment in this industry, and potential sources of Whole-Body Vibration (WBV) (Aye & Heyns 2011; Leduc, Eger *et al.* 2011; Kunimatsu & Pathak 2012).

According some authors (Bovenzi 1996; Pope, Magnusson *et al.* 1998; Pope, Wilder *et al.* 1999; Cann, Salmoni *et al.* 2003; McPhee 2004) there are a strong connection between exposure to WBV and low back pain. It is important and interesting to companies and professionals to reduce this exposure and consequently workers complaints.

This article intends to be a synthesis of scientific knowledge on WBV in open pit mining. The main objective is to present the most recent and relevant scientific information available in order to characterize the jobs in this workplace, measurement methodologies and the main known effects on workers' health.

2 RESEARCH METHODOLOGY

A literature review was carried out based on PRISMA® Statement (Preferred Reporting Items for Systematic Reviews and Meta-Analysis).

Searches were conducted in data bases such as Web of Knowledge (which include Current Contents Connect, Derwent Inovation Index, Essential Science Indicators, Journal Citation Reports, Web of Science + Proceedings), Scopus and Academic Search Complete, and scientific journals such as Annual Reviews, EBSCO Electronic Journals Service, Elsevier (ScienceDirect), Springer, Taylor & Francis and Wiley InterScience. Articles from other sources, like references of the articles selected in the first stage of the systematic review, were not excluded.

The keywords used were: occupational vibration, whole-body vibration, hand arm vibration, mining, extractive industry, mining equipment, open cast mining. These words were combined with different Boolean operators, searching on title, abstract and keywords.

The exclusion criteria were:

- Language: The review was restricted to studies published in the English language;
- Publication date: Articles published before 1990 were excluded (this date was set out because it was intended to cover older equipment still in use in the industry);
- Subject (relevance to the purpose of the review): For inclusion, the study should deal with occupational vibration in mining.

All identified studies were screened initially by title and abstract, and then more carefully in full text.

3 RESULTS AND DISCUSSION

3.1 *Research results*

Literature search has produced 607 potentially relevant papers. Following screening process steps shown in Figure 1. Thirty-three articles were included in the review. Most of them were discarded due to violating the inclusion criteria, although many duplicates were also excluded.

3.2 *Sources of vibration in mining*

Classes of equipment and their open pit mining operations (WBV producers) were studied (Table 1).

3.2.1 *Rock drills*

Rock drills are, according to several authors, the equipments that produce lower Root Mean Squared (RMS) acceleration values over the occupational exposure. Howard, Sesek *et al.* (2009) reached an acceleration of 0,30 $m \cdot s^{-2}$ and Van Niekerk, Heyns *et al.* (2000) 0,16 $m \cdot s^{-2}$ on vertical

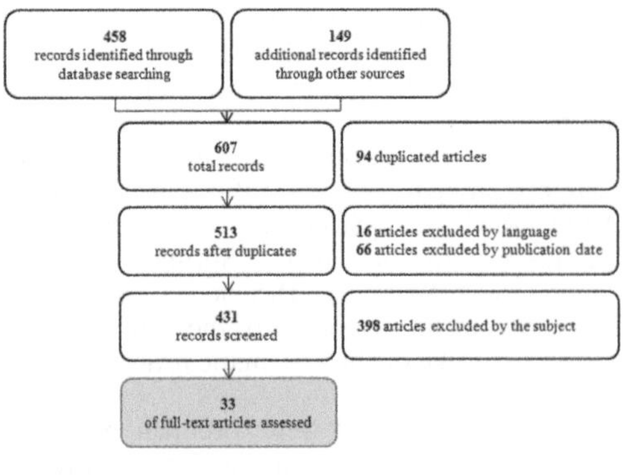

Figure 1. Flow diagram of studies selected for review.

Table 1. Equipment and operations associated with WBV.

Machine	Activity
Rock drill	Positioning Drilling Travelling
Shovel	Trenching/excavating Swinging and loading Travelling
Dumper	Travelling Loaded by shovel Unloading

Kunimatsu and Pathak (2012).

axis (zz). According to these authors, at this workplace, the operator will be safeguarded from the risk of exposure to WBV. These studies were the only ones to analyze WBV in rock drills.

3.2.2 *Shovels*

Due to its flexibility, low-cost operations, maintainability and multifaceted operating capability, shovels are widely used in surface mines (Frimpong, Galecki *et al.* 2011).

The studies aiming the characterization of WBV exposure in shovels, in open pit mining, found very different values for RMS. Some studies point to WBV values below 0.5 $m.s^{-2}$ (Howard, Sesek *et al.* 2009; Aye & Heyns 2011; Dentoni & Massacci 2013). However, authors like Vanerkar, Kulkarni *et al.* (2008) call attention to the risk conditions that may be subject these workers, with WBV values reaching 2.3 $m.s^{-2}$. The same happens with the predominant axis, there is no a clear definition. The vertical (zz) and the longitudinal (xx) are presented as the main axis according Aye & Heyns (2011) and Dentoni & Massacci (2013).

Thus, the discrepant characterization of exposure to WBV in these equipments is notorious. This is probably due to the multiple operations carried by shovel, which lead to different situations.

3.2.3 *Dumpers*

Dumpers are strong vehicles widely used in open pit mining environments to transport large quantities of rock to crushing plants or dump locations. Their capacity range is from 7 to 350 tonnes (Mandal & Srivastava 2010; Eger, Stevenson *et al.* 2011).

The majority of studies analyzed (Van Niekerk, Heyns *et al.* 2000; Kumar 2004; Eger, Stevenson *et al.* 2008a; Howard, Sesek *et al.* 2009; Mandal & Srivastava 2010; Smets, Eger *et al.* 2010) assigns to these machines considerable vibration values, above 0.5 $m.s^{-2}$. The predominant axis is the vertical (zz). The dominant frequencies are set below 4 Hz.

Table 2. Summary of studies characterizing exposure to WBV in rock drilling, shovel and dumper in open pit mining.

Machine	Authors (year)	RMS [m.s^{-2}]	Main axis	Dominant frequencies [Hz]
Rock drill	Howard, Sesek *et al.* (2009)	0,30	–	–
	Van Niekerk, Heyns *et al.* (2000)	0,16	z	–
Shovel	Dentoni & Massacci (2013)	0,52–0,97	x	–
	Kunimatsu & Pathak (2012)	0,5–2,3	–	–
	Aye & Heyns (2011)	0,21–0,51 0,25–0,39	z x	–
	Howard, Sesek *et al.* (2009)	0,45	–	–
Dumper	Smets, Eger *et al.* (2010)	0,80	–	2–4
	Mandal & Srivastava (2010)	1,10	z	–
	Howard, Sesek *et al.* (2009)	0,58	–	–
	Eger, Stevenson *et al.* (2008a)	0,77	z	1,0–1,25 (x) 1–1,25 (y) 3,15–4 (z)
	Kumar (2004)	0,37–11,73	z	2–4
	Van Niekerk, Heyns *et al.* (2000)	0,75	z	–

3.2.4 *Synthesis*

Table 2 shows the analyzed studies towards the characterization of WBV exposure in rock drills, shovels and dumpers in open pit mines. It is focused the RMS acceleration value, main axis and dominant frequencies.

In open pit mines, in terms of WBV exposure were highlighted by its relevance, rock drills, shovels and dumpers (Howard, Sesek *et al.* 2009).

Rock drill workers had the lowest exposition to WBV (only office workers are below). They are followed by the Shovels and Dumpers operators by increasing degree of risk. This ranking is consistent with the values collected in this review. Thus the dumper will be the workplace targeted for priority intervention.

3.3 *Methodologies applied to the measurement of vibrations transmitted to the whole body*

In the equipment under consideration the contact points between the worker and the machine are, par excellence, feet and seat (Kunimatsu and Pathak 2012). The vibration transmitted to the hand-arm through contact with the controls of vehicles or the steering wheel is not relevant (Hill, Langis *et al.* 2001; Donoghue 2004; Gallagher and Mayton 2007; Kunimatsu and Pathak 2012). However, these workers are exposed to impact shocks (Eger, Stevenson *et al.* 2011).

The bibliography, focused on WBV, follows almost consensually, the ISO standard 2631-1:1997-Mechanical vibration and shock—Evaluation of human exposure to whole-body vibration—Part 1: General requirements (ISO 1997). Besides being followed methodological principles for the measurement of vibration, are also followed the calculation methods of required parameters to evaluate vibration exposure. One of the methodological recommendations of ISO 2631-1:1997 that should be emphasized in this work is the measurement and analysis of vibration exposure separately, taking into account the existence of several periods with different characteristics (Chapter 5.5 Duration of measurement).

Typically, a mining environment involves well defined operations with short duration, associated with different vibration amplitudes. Thus, measurements should be made for each operation and the result obtained from the combination of these.

The most striking example is the dumper. In these, worker exposure to vibration changes considerably with the task. Usually a work cycle includes tasks such as unloaded travel, loading, loaded travel and dumping (Smets, Eger *et al.* 2010).

Salmoni, Cann *et al.* (2010) and (Smets, Eger *et al.* 2010) said the expected, that the waiting times (standby) are those where the vibration exposure is lower as opposed to transport tasks (movement of the vehicle loaded and unloaded) where the vibration values are higher. Different authors warn that, for the determination of daily exposure to WBV, all tasks should be considered as well as its duration time.

For this reason, some studies, together with the measurement of vibration levels, followed visually the operations, noting the time period of the tasks and any observations which may have influence in the results (Salmoni, Cann *et al.* 2010; Smets, Eger *et al.* 2010; Leduc, Eger *et al.* 2011). Other authors recorded the operations in video (Eger, Stevenson *et al.* 2008b; Torma-Krajewski, Wiehagen *et al.* 2009; Salmoni, Cann *et al.* 2010).

Another very important aspect of this activity is a strong correlation between the features and conditions in which the work is developed and the resulting exposure to WBV (Kunimatsu and Pathak 2012). The vibration suffered by workers when operating with vehicles depend on, among others, the external excitation force, the mass of the vehicle and environmental factors such as the condition of the pavement, the characteristics of the material handled, operations organization and worker experience (Mandal & Srivastava 2010; Frimpong, Galecki *et al.* 2011). As expected, exposure to WBV does not depend on the type of operation (raw material), but from the working conditions and the type of equipment used as stated by the authors as Vanerkar, Kulkarni *et al.* (2008).

Equipment for determining the vibrations are triaxial accelerometers. The accelerometer is normally placed in a rounded surface rubber which is fixed with tape to the operator's seat. By the bibliography, the vibration measured is often the one that is transmitted through the seat, to the seated body as a whole. The vibration transmitted through the feet and the seat backrest is not normally considered (Leduc, Eger *et al.* 2011).

Other parameters analyzed in WBV studies are:

- machines characteristics (Kumar 2004; Howard, Sesek *et al.* 2009; Eger, Stevenson *et al.* 2011; Dentoni & Massacci 2013);
- operations that the equipment performs (Kumar 2004; Eger, Stevenson *et al.* 2008; Salmoni, Cann *et al.* 2010; Smets, Eger *et al.* 2010; Leduc, Eger *et al.* 2011);
- work routines (Torma-Krajewski, Wiehagen *et al.* 2009; Smets, Eger *et al.* 2010);
- pavement conditions (Mayton, Amirouche *et al.* 2005; Howard, Sesek *et al.* 2009; Salmoni, Cann *et al.* 2010; Dentoni & Massacci 2013);
- position and seat features (Mayton, Amirouche *et al.* 2005);
- vibration type (continuous, intermittent, impact) and its direction and source (Leduc, Eger *et al.* 2011);
- number of times that performs the task or all work cycle per day, average duration of the task or work cycle (Mandal & Srivastava 2010; Smets, Eger *et al.* 2010);

Simultaneously, other important information are collected as e.g. demographic information, medical and occupational historic and lifestyle (Eger, Stevenson *et al.* 2008; Mandal & Srivastava 2010; Smets, Eger *et al.* 2010; Eger, Stevenson *et al.* 2011; Leduc, Eger *et al.* 2011).

3.4 *Main effects of whole-body vibrations*

Establish direct relations between WBV exposure and its effects are not easy. Magnusson, Pope *et al.* (1998) reported difficulties in establishing a clear relationship between exposure and effect, mainly due to the high number of factors influencing the risk associated with exposure to WBV. Mansfield (2004) suggests a holistic and cautious approach. The effect of exposure to WBV most often reported in the literature is the back pain. This relation is established in several epidemiological studies (Teschke, Nicol *et al.* 1999; Lings & Leboeuf-Yde 2000; Gallais & Griffin 2006; Gallagher & Mayton 2007). However, the same type of complaints occurs in most work activities, and is demonstrably influenced by human degenerative process associated with age, so it is not easy to establish a direct causal link. Other reported symptoms are sciatic pain, low back pain and widespread pain in the back, and may be associated illnesses such as herniated discs and early degeneration of the spine (Teschke, Nicol *et al.* 1999; Gallagher & Mayton 2007; Eger, Stevenson *et al.* 2011).

In the particular case of heavy equipment operators, which include mine workers, epidemiological studies try to associate back pain and the operation of vehicles. Truck drivers, shovels and dumpers are at high risk, because are exposed to WBV with values often higher than those recommended. It is also accepted that the risk of musculoskeletal injuries increased with increasing time of exposure (Teschke, Nicol *et al.* 1999; Bovenzi, Rui *et al.* 2006; Gallagher & Mayton 2007).

Teschke, Nicol *et al.* (1999) and Gallais & Griffin (2006), with their analyzes of epidemiological studies on exposure to WBV in drivers, concluded that there were several competing factors for the appearance of back pain. Age, working postures, materials handling and "heavy" work, smoking, falls or other events that cause pain, stress and work pressure, physical condition and morphology of the body, including weight, height are some of the examples reported. In other words, confirmed the large number of confounding factors related to this problem.

In fact, the adoption of postures is a very important aspect regarding to drivers health. This is an additional factor of physical load in the column that, together with exposure to vibration, is a source of back pain reported by professional drivers (Teschke, Nicol *et al.* 1999; Bovenzi, Rui *et al.* 2006).

4 CONCLUSIONS

Rock drills, shovels and dumpers are strongly present in open pit mining. They are potential sources of workers exposure to WBV. The dumper is the one with higher values of WBV, followed by the shovel and finally by the rock drill, which is the least dangerous to workers health. The WBV are typically measured with the operator seated and through the seat. The tasks within a work cycle are detailed with timing and recording occurrences that may influence. The machine characteristics, the operator tasks, work routines, pavement conditions, vibration type, seat characteristics are topics that should be included in an exposure to WBV study. It is not proved a direct causal relationship between the drive a mining vehicle and the appearance of back pain. However there is evidence that the most frequent complaint among workers is back pain.

ACKNOWLEDGMENT

The authors would like to thank Master in Occupational Safety and Hygiene Engineering (MESHO), of the Faculty of Engineering of the University of Porto (FEUP), all the support in the development and international dissemination of this work.

REFERENCES

Aye, S. and Heyns, P. S. (2011). "The evaluation of whole-body vibration in a South African opencast mine." *Journal of the Southern African Institute of Mining and Metallurgy* 111(11): 751–758.

Bovenzi, M. (1996). Low back pain disorders and exposure to whole-body vibration in the workplace. Seminars in perinatology, Elsevier.

Bovenzi, M., Rui, F., Negro, C., D'Agostin, F., Angotzi, G., Bianchi, S., Bramanti, L., Festa, G., Gatti, S. and Pinto, I. (2006). "An epidemiological study of low back pain in professional drivers." *Journal of sound and vibration* 298(3): 514–539.

Cann, A. P., Salmoni, A. W., Vi, P. and Eger, T. R. (2003). "An exploratory study of whole-body vibration exposure and dose while operating heavy equipment in the construction industry." *Applied occupational and environmental hygiene* 18(12): 999–1005.

Dentoni, V. and Massacci, G. (2013). "Occupational exposure to whole-body vibration: unfavourable effects due to the use of old earth-moving machinery in mine reclamation." *International Journal of Mining, Reclamation and Environment* 27(2): 127–142.

Donoghue, A. (2004). "Occupational health hazards in mining: an overview." *Occupational Medicine* 54(5): 283–289.

Eger, T., Stevenson, J., Boileau, P.-É. and Salmoni, A. (2008a). "Predictions of health risks associated with the operation of load-haul-dump mining vehicles: Part 1—Analysis of whole-body vibration exposure using ISO 2631-1 and ISO-2631-5 standards." *International journal of industrial ergonomics* 38(9): 726–738.

Eger, T., Stevenson, J., Callaghan, J. and Grenier, S. (2008b). "Predictions of health risks associated with the operation of load-haul-dump mining vehicles: Part 2—Evaluation of operator driving postures and associated postural loading." *International journal of industrial ergonomics* 38(9): 801–815.

Eger, T., Stevenson, J. M., Grenier, S., Boileau, P.-É. and Smets, M. P. (2011). "Influence of vehicle size, haulage capacity and ride control on vibration exposure and predicted health risks for LHD vehicle operators." *Journal of Low Frequency Noise, Vibration & Active Control* 30(1): 45–62.

Frimpong, S., Galecki, G. and Chang, Z. (2011). "Dump truck operator vibration control in high-impact shovel loading operations." *International Journal of Mining, Reclamation and Environment* 25(3): 213–225.

Gallagher, S. and Mayton, A. (2007). "Back injury control measures for manual lifting and seat design." *Mining Engineering* 59(12): 41–49.

Gallais, L. and Griffin, M. J. (2006). "Low back pain in car drivers: A review of studies published 1975 to 2005." *Journal of sound and vibration* 298(3): 499–513.

Hill, C., Langis, W. J., Petherick, J. E., Campbell, D. M., Haines, T., Andersen, J., Conley, K. K., White, J., Lightfoot, N. E. and Bissett, R. J. (2001). "Assessment of hand-arm vibration syndrome in a northern Ontario base metal mine." *Chronic Dis Can* 22(3–4): 88–92.

Howard, B., Sesek, R. and Bloswick, D. (2009). "Typical whole body vibration exposure magnitudes encountered in the open pit mining industry." *Work* 34(3): 297–303.

ISO (1997). ISO 2631-1:1997 (Mechanical vibration and shock—Evaluation of human exposure to whole-body vibration—Part 1: General requirements), International Organization for Standardization.

Kumar, S. (2004). "Vibration in operating heavy haul trucks in overburden mining." *Applied Ergonomics* 35(6): 509–520.

Kunimatsu, S. and Pathak, K. (2012). "Vibration-Related Disorders Induced by Mining Operations and Standardization of Assessment Process." *MAPAN* 27(4): 241–249.

Leduc, M., Eger, T., Godwin, A., Dickey, J. P. and House, R. (2011). "Examination of vibration characteristics, and reported musculoskeletal discomfort for workers exposed to vibration via the feet." Journal of Low Frequency Noise, *Vibration & Active Control* 30(3): 197–206.

Lings, S. and Leboeuf-Yde, C. (2000). "Whole-body vibration and low back pain: A systematic, critical review of the epidemiological literature 1992–1999." *International archives of occupational and environmental health* 73(5): 290–297.

Magnusson, M., Pope, M., Hulshof, C. and Bovenzi, M. (1998). "Development of a Protocol for Epidemiologal Studies of Whole-Body Vibration and Musculoskeletal Disorders of the Lower Back." *Journal of sound and vibration* 215(4): 643–651.

Mandal, B. and Srivastava, A. (2010). "Musculoskeletal disorders in dumper operators exposed to whole body vibration at Indian mines." *International Journal of Mining, Reclamation and Environment* 24(3): 233–243.

Mansfield, N. J. (2004). Human response to vibration, CRC Press.

Matos, M. L. and Ramos, F. (2010). Indústria extrativa: análise de riscos ocupacionais e doenças profissionais. SHO—Colóquio Internacional sobre Segurança e Higiene Ocupacionais, 10 e 11 de Fevereiro de 2010, Guimarães.

Mayton, A. G., Amirouche, F. and Jobes, C. C. (2005). "Comparison of seat designs for underground mine haulage vehicles using the absorbed power and ISO 2631-1 (1985)-based ACGIH threshold limit methods." *International Journal of Heavy Vehicle Systems* 12(3): 225–238.

McPhee, B. (2004). "Ergonomics in mining." *Occupational Medicine* 54(5): 297–303.

Pope, M., Wilder, D. and Magnusson, M. (1999). "A review of studies on seated whole body vibration and low back pain." Proceedings of the Institution of Mechanical Engineers, Part H: *Journal of Engineering in Medicine* 213(6): 435–446.

Pope, M. H., Magnusson, M. and Wilder, D. G. (1998). "Low back pain and whole body vibration." *Clinical orthopaedics and related research* 354: 241–248.

Salmoni, A., Cann, A. and Gillin, K. (2010). "Exposure to whole-body vibration and seat transmissibility in a large sample of earth scrapers." Work: A Journal of Prevention, *Assessment and Rehabilitation* 35(1): 63–75.

Smets, M. P. H., Eger, T. R. and Grenier, S. G. (2010). "Whole-body vibration experienced by haulage truck operators in surface mining operations: A comparison of various analysis methods utilized in the prediction of health risks." *Applied Ergonomics* 41(6): 763–770.

Teschke, K., Nicol, A.-M., Davies, H. and Ju, S. (1999). "Whole Body Vibration and Back Disorders Among Motor Vehicle Drivers and Heavy Equipment Operators A Review of the Scientific Evidence." *Occupational Hygiene* 6: 1Z3.

Torma-Krajewski, J., Wiehagen, W., Etcheverry, A., Turin, F. and Unger, R. (2009). "Ergonomics: Using Ergonomics to Enhance Safe Production at a Surface Coal Mine—A Case Study with Powder Crews." *Journal of occupational and environmental hygiene* 6(10): D55–D62.

Van Niekerk, J., Heyns, P. and Heyns, M. (2000). "Human vibration levels in the South African mining industry." *Journal-South African Institute of Mining and Metallurgy* 100(4): 235–242.

Vanerkar, A. P., Kulkarni, N. P., Zade, P. D. and Kamavisdar, A. S. (2008). "Whole body vibration exposure in heavy earth moving machinery operators of metalliferrous mines." *Environmental Monitoring and Assessment* 143(1–3): 239–245.

Occupational Safety and Hygiene II – Arezes et al. (eds)
© 2014 Taylor & Francis Group, London, ISBN 978-1-138-00144-2

Factors affecting the safety in Portuguese architectural heritage works

J.P. Couto
Department of Civil Engineering, University of Minho, Campus of Azurém, Guimarães, Portugal

M.L. Tender
Portuguese Association of Coordinators and Safety Managers, Porto, Portugal

ABSTRACT: Portugal continues to present indices of accidents in the construction industry over the desired. This study aims to analyze the principal factors and causes that can origin risk conditions. Based in that analysis it's possible to begin working in preventing measures in order to decrease those indices. The focus on rehabilitation has increased in Portugal and it is in this kind of construction that difficults appear, mainly motivated by the lack of planning, of local knowledge to work and skilled manpower. The architectural heritage is a short group of works that has relevant characteristics that emphasize problems usually existent in traditional rehabilitation sites. Thus, on the basis of accompaniment of several heritage rehabilitation sites, this work puts forward the main difficults of safety management regarding this type of activity.

1 INTRODUCTION

1.1 *Background—construction sector*

The construction sector is considered by many as one of the major pillars of a country's economy, but also one of the most dangerous industries (Enshassi, 1996). The policies adopted to increase productivity in the sector vary from country to country and depend on the strategy of each company. However, occupational safety and health have been widely recognized as two of the most influential aspects in companies' overall performance and have been gradually no longer seen as a luxury and came to be seen as a necessity to avoid losses, injuries or even deaths (Abdul-Rashid et al., 2007). Over time, increasing competencies in project management has enabled greater emphasis on this issue, focusing on workers' safety and health (Wong et al., 2002).

However, over the past few years, the construction scenario in Portugal has undergone changes, as the activity of the construction sector has slowed abruptly and significantly and the new construction has become residual. Thus, the maintenance and rehabilitation of existing buildings has become the resort for those who have chosen not to go international, being also mandatory for the repopulation of the historic centres. This paradigm shift brings about the possibility of an evolution in the concepts of intervention in the area of rehabilitation, together with the use of new methodologies.

The need of rehabilitate is reinforced due to the natural degradation and lack of maintenance over the last decades, constructions begun to show obvious signs of structural degradation consequences that brings consequences on stability and aesthetics of the buildings.

1.2 *Background—architectural heritage sector*

The rehabilitation of architectural heritage is characterized by a direct action on monuments and cultural assets, through qualified interventions regarding the identification, recovery, repair, maintenance, restoration and implementation of various projects, both in the built heritage and respective surrounding, and also in the mobile and embedded heritage (painting, imaginary, furniture, etc., and the so-called assets "by destination", such as carvings, wall paintings or tiles).

At the European level, the history of the framework has several decades. The list that follows presents some of the most important document references:

a. *Athens Charter, October 1931*
"[...] to recommend that, before any consolidation or partial restoration is undertaken, a thorough analysis should be made of the defects of those monuments, recognising that each case has its own specificity."

b. Venice Charter, May 1964

"The conservation and restoration of monuments must have recourse to all the sciences and techniques which can contribute to the study and safeguarding [...]."

"It is essential to the conservation of monuments that they be maintained on a permanent basis."

c. Vila Vigoni Charter (Cultural Assets of the Church), March 1994

"Repairs deemed necessary [...] should be entrusted only to trained personnel, with recognized expertise."

d. Declaration of Principles of the Society for the Preservation of the Built Heritage, 1995

"It is considered that traditional techniques and materials are a wealth of intrinsic value that must be preserved and that often constitute the most appropriate solutions for heritage preservation [...]."

1.3 *Safety in construction*

In order to better understand the state of Rehabilitation in Portugal, a study was carried out within the scope of a master thesis in sustainable construction and rehabilitation developed at the University of Minho, where a questionnaire survey was implemented to 57 actors in the field of rehabilitation, with the objective of identifying the key factors that affect safety in rehabilitation. It was concluded that the main reasons for flaws in the coordination and management of the safety of this type of work (that originate serious breaches) on safety are (Araújo, 2009):

- Lack of technical expertise in organization and planning;
- Lack of communication;
- Skilled labor.

Internationally, there are a substantial number of studies aiming at clarifying the reasons related with safety performance.

Health and Safety Executive, represented by Healey (2011) carried out other study on the construction of Olympic Park site (which included reabilitation works) where it was found a number of elements contributed to the development of an effective safety culture on that site (developed by Olympic Delivery Authority (ODA), the project owner), including:

- The strategic role of project owner across the Park, with safety being set as a priority and integrated into the companies from the outset through standards and requirements;
- The clarity throughout the supply chain of the organizational standards and requirements, including the desire for cultural alignment;
- The empowerment of main contractors to develop their own processes and systems to deliver the ODA's objectives;
- Recognition of the prestige of working on the Olympic Park and striving for excellence in all activities, including health and safety;
- The scale of the project and the length of the construction phase meant that there was sufficient time for initiatives to become embedded, and could be tailored to ensure their efficacy and success;
- Belief by workers in the genuine commitment within organizations, as the message was consistent and reiterated across the Olympic Park over time.

Another study on the topic was carried out by Adbdul-Rashid et al. (2007), with the objective of identifying the main factors affecting the safety in great works in Egypt, using an inquiry as boarding methodology, implemented in some companies of the parents. This study was supported on a bibliographical research that made it possible to identify a set of 72 factors, grouped in 12 categories. The factors indicated in table 1 are part, among others, of this set of factors selected for the implementation of the inquiry.

The authors concluded that the factors considered most relevant by companies were the need for awareness of the company's management and project managers for the implementation of the safety management system, and need for frequent monitoring.

Table 1. Factors affecting safety performance in international literature (Abdul-Rashid et al., 2007).

Literature	Factors affecting safety performance
Ng et al. (2005)	Implementation of safety management system in accordance with legislation;
	Compliance with occupational safety and health legislation, codes and standards;
	Definition of safety responsibility;
	Development of safety policy;
	Provision of safe working environment;
	Development of emergency plans and procedures;
	Development of safety committee;
Fung, et al. (2005)	Effective accident reporting;
	High line management commitment;
	Active supervisor's role;
	Active personal role.

2 FACTORS OF ACCIDENTS IN REHABILITATION WORKS IN PORTUGAL

In Portugal, these types of studies are scarce, being only known some exploratory and preliminary approaches.

Official figures show clearly the preponderance of the construction industry in fatal accidents recorded over the years in Portugal. As documented in Figure 1, although the workpower in construction declined significantly, indices accident still maintain highly unacceptable (namely due falling from heights, burying and crushing).

In order to organize and systematize the relevant information on this matter, the Authority for Working Conditions (ACT) suggested the organization of the main factors contributing to accidents, and the characterization of the construction industry for economical and organizational terms, as follows:

- The proliferation of small and microenterprises;
- The reinforcement of subcontracting;
- The lack of training of decision makers and workers;
- The economic competitiveness.

This latter term leads to serial budget cuts and very demanding deadlines, lacks on planning activities—absence of safety coordination in the design phase—with the intensification of the pace and duration of the work and, more recently assumed, emergent risks.

Beyond these, the ACT presents other causes of construction accidents, namely the absence of safety structures, the poor work planning and timing, the lack of protection and the use of false protections, among others (Paula, 2008 cited by Araújo, 2009).

Confirming this scenario, it has been widely recognized that, in general, the risks associated with workplace accidents are higher in rehabilitation works than in new construction (Egbu, 1999).

Through the study conducted in 2007 by Roberts et al, it was concluded that only 35% of the rehabilitation projects consider safety measures in order to prevent risks in carrying out the work on the ground. A study, conducted through a survey of about 250 of the rehabilitation professionals in Portugal (Rodrigues & Teixeira, 2009), reveled that the reasons given by respondents for the lack of preventive measures are:

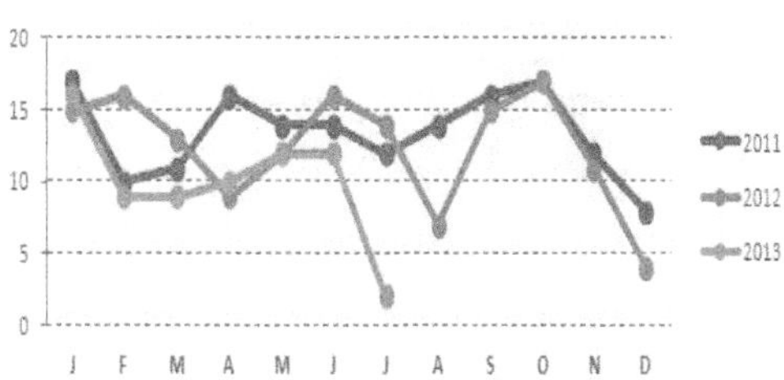

Figure 1. Fatalities in the construction from 2011 to 2013 (ACT, 2013).

- Devaluation of safety issues by project owner and designers;
- Lack of risk assessment by designers;
- Lack of safety coordinator during the design phase;
- Poor interaction between team and project coordinator;
- Lack of technical compilation.

As such, it can be concluded that the existence of flaws compromises the existing legislation right from the project phase.

The diversity of participants in this type of work, in representing various enterprises, is also one of the important factors in accidents because many are unaware of the reality of working conditions. Moreover, turnover of skilled labor affects the correct execution of the works, thus leading to the increase in the risk of accidents. Besides this, the pressure on companies to comply with the times and costs means that the work is carried out under hard conditions, often leading to increased number of people in work, and thus increasing the risk of accidents (Couto, 2008).

3 PARTICULARITIES ON SAFETY IN REHABILITATION OF ARCHITECTURAL HERITAGE

3.1 *Background and general characterization*

The rehabilitation sector presents some very specific characteristics regarding interventions in architectural heritage, and which may be classified according to the following aspects:

a. *Project/Recruitment/Execution*
   - Low execution values;
   - Unpredictability/changes in the project options considering the lack of record and history of the existing structures;
   - Buildings with irregular outlines (creating limitations to the installation of temporary structures);
   - Small number of specialised training actions;
   - Inadequacy of the existing legislation to common practices;
   - Need to use traditional materials and techniques;
   - Subjection to the emergence of archaeological findings (which can have significant effects in terms of cost and deadline);

- Lack of inclusion of security systems that enable subsequent secure interventions (this point will be further detailed in the next subchapter).

b. Stakeholders
- Small number of specialized companies in the market (around 10);
- Small number of skilled technicians for tasks' performance;
- Micro and small companies;
- High expertise;
- Low value budgets/low profit margins;
- Companies' management (in 1st degree) with low educational level;
- Companies with organized safety services, but with reduced practical results: compliance with the minimum standards for safety—workers' training, assembly of Collective Protection Equipment (CPE), use of Individual Protection Equipment (IPE);
- The standard worker involved in these works has particular characteristics: 45 years old; over 20 years of experience in the field and often in the same company; low educational level; reluctant in using protection/over-confidence, especially regarding the risks of falling from heights; lack of habit in the use of CPE and IPE;
- Low quality and poorly maintained equipments.

c. Construction site
- Small or nonexistent spaces for the set up of the construction site, storage and handling of equipment and materials;
- Sanitary facilities with difficult location and maintenance;
- Need for the protection of existing structures/materials (structural pieces or ornaments);
- Operating facilities (implies protection to third parties, namely against materials fall);
- Guarantee of protection against vandalism (fences higher than 4.0 m);
- Difficulty in accessing some sites, particularly in the buildings' roofs;
- Simultaneity of tasks in the same space—e.g. risk of materials fall into the archaeological excavation site.

d. Confined spaces
These spaces usually emerge in interventions between the tile roof and the roof of the building; they are, normally, tight spaces and their interior is not generally known, which could be explained by the existing inaccessibility during the project stage.

This factor is aggravated by the use of materials with special risks in the handling of masonry, wood, steel, etc. (products that are potentially flammable, irritant and toxic).

3.2 *Case study*

Taking the example of an intervention in a national monument, we quote an example of a choice made by the project owner that is of particular importance to prevent this type of work.

The work in question consisted of the refurbishment of the lower roof of an architectural monument with two roofs at two different heights.

Undergoing early intervention was equated the need to ensure access and workplaces with the necessary security conditions for the upkeep of the lower roof. The tasks identified to be developed during maintenance with need of protection were the repair/replacement of roof tiles and guttering repair on the tower in this lower level.

Identified risk of falling from height and valued as high, the project owner, in partnership with other stakeholders in the work, decided to installing fall protection system.

The inclusion of fall protection system against falls from a height is of particular importance given the high number of accidents by falling from heights prevailing in national statistics.

So it was established a procedure for carrying out maintenance work at height: for these operations was determined that the worker would begin his work by accessing, from the interior of the monument and properly equipped with safety harness, to the upper roof.

When it was placed on top of the higher roof the worker would have to approach the top cover and fasten his safety harness, before exceeding the perimeter of the battlements, to fall arrest block belonging to the safety system existing in the outer side of the battlements (Figs. 2–3).

Overcoming the said perimeter worker, already attached to the safety line final descends by stainless steel steps (installed during this project), to access the lower roof (Fig. 4).

We emphasize that, in the course from the moment the worker passes through the battlements till his working stage at the lower coverage, he is protected against falls from a height because

Figure 2. Safety line anchored to stone walls.

Figure 3. Worker with fall arrest block.

Figure 4. Worker descending through the steps.

Figure 5. Worker on lower roof.

he's wearing a safety harness attached to the safety line.

During his stay in lower roof worker meets with the risk of falling from height minimized since the fall arrest block prevents any sudden movement that may occur.

When work is completed, the worker returns to cover higher up the steps in the existing wall and always fixed the system of protection against falls from a height.

This option taken by the project owner, made possible that the risk of falling from height during maintenance and repair was minimized to levels considered acceptable (see Fig. 5).

4 CONCLUSIONS

Rehabilitation of architectural heritage is, naturally, an added advantage for the prevention and preserving of the country's heritage. However, it seems equally obvious that this should not be developed to the detriment of the safety conditions of those involved. The costs with accidents are considerably higher than the costs involved in measures to foster safety. However, even with these arguments, some construction and rehabilitation companies are not sufficiently concerned about this problem or apply preventive measures, thus concurring to the fact that construction is one of the industries that contributes most to accidents in Portugal.

The monitoring of works of architectural heritage rehabilitation by the 2nd author, within the scope of his professional occupation as Health and Safety Construction Coordinator, and with the purpose of studying the reasons for non-compliance with some safety rules, allowed confirming some of aforementioned reasons during the study for the rehabilitation works, namely:

- Depreciation of security issues on the part of the designers;
- Gaps in prevention planning on the part of some designers;
- Gaps in safety coordination in the project stage;
- Lack of technical compilation.

The factors identified in the works of architectural heritage rehabilitation emphasize the importance of ensuring:

- Thorough supervision, from the beginning of the project, by the Construction Owner;
- Integration of preventive measures for the stage of the maintenance of the rehabilitated heritage.

The guarantee of the existence of these assumptions from the beginning of intervention may ensure that the rehabilitation operation and subsequent maintenance operations are carried out within acceptable levels of safety.

REFERENCES

Abdul-Rashid, I., Bassioni, H. & Bawazeer, F. (2007). Factors affecting safety performance in large construction contractors in Egypt. In: Boyed, D. (Ed), Procs *23rd Annual ARCOM Conference,* 3–5, September 2007, Belfast, UK, 661–670.

ACT (2013). Authority for the conditions of the work, Cabinet of strategy and planning. Accessed September 2013, available at: http://www.act.gov.pt.

Araújo, J.D. (2009). Optimization of the management of Rehabilitation projects. *Dissertation of master's degree in Sustainable Construction (in Portuguese).* University of Minho, Guimarães.

Couto, J.P. (2008). Influences in terms of accidents in the PortugueseConstruction. *GESCON2008-International Conference on Construction Management*, December, FEUP, Portugal.

Egbu, C. Skills (1999). Knowledge and Competencies for Managing Construction Refurbishment Works. *Construction Management and Economics*, 17, 29–43.

Enshassi, A. (1996). Factors Affecting Safety on Construction Projects. *Department of Civil Engineering, IUG*, Gaza Strip, Palestine, 14–17.

Fung, Ivan W.H., Tam, C.M., Tung, Karen C.F., & Man, Ads S.K. (2005). Safety Cultural Divergences among Management, Supervisory and Worker Groups in Hong Kong Construction Industry. *International Journal of Project Management*, Vol. 23. No. 7, 504–512.

Healey, N., Sugden, C. (2011). Safety Culture in the Olympic Park. Health and Safety Executive, Research report RR942, iii.

Ng, S. Thomas, Cheng, Kam Pong & Skitmore R. Martin (2005). A Framework for Evaluating the Safety Performance of Construction Contractors. *Building and Environment,* Vol. 40. No. 10, 1347–1355.

Rodrigues, M.F., Teixeira, J.C. (2009). Rehabilitation of Buildings Operations: Coordinating Health and Safety. *International Seminar on Occupational Safety and Hygiene*, Guimarães.

Wong, F.K.H, Chan, A.P.C., Fox, P., Kenny, T.C. & Easther, F.N. (2002). Identification of Critical Factors Affecting the Communication of Safety-Related Information between Main Contractors and Sub-Contractors in Hong Kong. *Research Project funded by occupational Health and Safety Council*, 1–2.

Occupational Safety and Hygiene II – Arezes et al. (eds)
 ISBN 978-1-138-00144-2

Patterns of Whole-Body Vibrations in open pit mining

C. Bernardo
Research Laboratory on Prevention of Occupational and Environmental Risks (PROA/LABIOMEP), Faculty of Engineering, University of Porto, Porto, Portugal

M.L. Matos
LNEG—Laboratório Nacional de Energia e Geologia, S. Mamede de Infesta, Portugal

J. Santos Baptista
Research Laboratory on Prevention of Occupational and Environmental Risks (PROA/LABIOMEP), Faculty of Engineering, University of Porto, Porto, Portugal

ABSTRACT: The main objective is to detect Whole-Body Vibration (WBV) patterns in different mining equipments along their working cycles. Three activities/equipment were studied (rock drill, shovel and dumper) in a north Portugal quarry. The WBV measurement and analysis was conducted in accordance with ISO 2631-1 (1997). It was studied three WBV transmission ways: 1) seat surface, 2) seat backrest and 3) cabin floor (feet). Rock drill shown a WBV pattern, where there was an extensive drilling phase, with residual accelerations values, and a short phase when the vehicle have to move to do another hole. In shovel, was not detected any associated pattern. In dumper, can be distinguished all tasks: loading, loaded and unloaded travel, dumping. Big differences were found when WBV is transmitted through the seat backrest, with the longitudinal x-axis dominating. Rock drill and dumper shown WBV patterns, as opposed to shovel.

1 INTRODUCTION

Mining operations in quarry involve many activities that generate vibrations. These vibrations can be transmitted to workers through their hands, feet or seats. Rocks drills, shovels and dumpers are strongly present in open pit mining. In these ones, the vibration is transmitted mainly to the worker body—Whole-Body Vibration (WBV) (Hill, Langis *et al.* 2001; Donoghue 2004; Gallagher & Mayton 2007; Kunimatsu & Pathak 2012).

Typically, a mining environment involves operations well defined and with short duration which are associated with different vibration amplitudes. Thus, measurements should be made for each operation and the result obtained from the combination of these (ISO 1997).

Some work has been developed in task characterization in the mining equipments such as rock drills and shovels. Nevertheless, dumpers are often studied. On one hand, they have the highest WBV exposition; on the other hand they have well-defined and very different tasks into its working cycle.

Dentoni & Massacci (2013) studied the WBV exposure in shovel, in two different tasks separately: haulage and material shifting. They found that the haulage task leads to greater exposure than the material shifting. In both tasks, the main axis is the longitudinal, x-axis.

In the dumper analysis, the tasks are loading, loaded travel, dumping and unloaded travel. The majority of time was spent during loaded travel and the least amount of time was spent during the dumping phase. The remaining time was almost equally divided between unload travel and the loading phase (Smets, Eger *et al.* 2010). The highest Root Mean Square (RMS) vibration magnitudes were founded in loaded and unloaded travel (Kumar 2004; Smets, Eger *et al.* 2010). The predominant frequencies of vibration generally lay in the band of 2–4 Hz (Kumar 2004).

The main objective is to detect WBV patterns in different mining equipments (rock drill, shovel, and dumper) along their work duty cycles. This allows the definition of a work duty cycle, identifying and comparing working tasks among themselves and between different equipments.

1.1 *Facilities, operators and machinery*

The experimental work was carried out between April and June 2013 in a quarry in the north of Portugal, dedicated to the production of crushed rock aggregates.

According to the objectives set, were selected for the study three jobs, ie three different equipments: rock drill (n = 1), shovel (n = 1) and dumper (n = 1). The characteristics of the equipment are presented in Table 1.

Workers were in a sitting position, controlling the vehicle with joysticks (rock drill and shovel) and a hand wheel (dumper). In the quarry, vehicles moved on a gravel road. Each job is associated with a worker. All of them have a daily vibration exposure, with a work schedule of 8 hours per day, divided by one hour lunch break. The workers were instructed to do their typical work routines.

1.2 *Measurement of Whole-Body Vibration*

WBV exposure measurements were conducted in accordance with the protocol defined by International Standards Organization (ISO) standard 2631-1 (1997). An SV 106 tri-axial accelerometer manufactured by SVANTEK, Poland was used. It measured vibration in three translational axes (longitudinal = x-axis; lateral = y-axis; vertical = z-axis). In order to measure vibration at the operator/seat and operator/floor interface, the accelerometer was mounted in a rubber seat pad and fixed in three different ways: (1) supporting seat surface; (2) seat backrest and (3) cabin floor. This is shown in Figures 1, 2 and 3, respectively. It was possible to study three-way transmission of WBV in the sitting position: input through the buttock, back and feet of the operator. The rubber pad was well fixed with tape trying to stay as close as possible to the surface.

1.3 *Data collection procedure*

For each workplace, typical work duty cycles were monitored. In rock drill, it included (1) Drilling; (2) Vehicle movement and positioning. In shovel, (1) Loading trucks or dumpers; (2) Material shifting and placing. In dumper, (1) Loading the bucket with rock; (2) Driving with loaded bucket to a dumping zone; (3) Dumping the loaded bucket; (4) Driving with an empty bucket back to the development heading to load another bucket.

The time spent performing each task in the cycle was dependant on the skill of the operator, layout of the quarry and environmental factors (e.g. road conditions). Therefore, the total data collection time per workplace varied between 1 and 3 h.

Table 1. Features of the equipment under study.

Equipment	Rock drill	Shovel	Dumper
Brand	Atlas Copco	CAT	Terex
Model	Rock D7	374 D	TR 45
Year of manufacture	2007	2011	2005

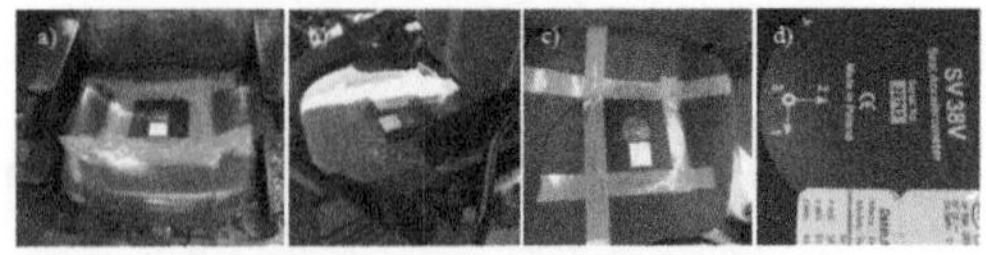

Figure 1. Placement of the whole-body accelerometer in the supporting seat surface a) rock drill, b) shovel and c) dumper.

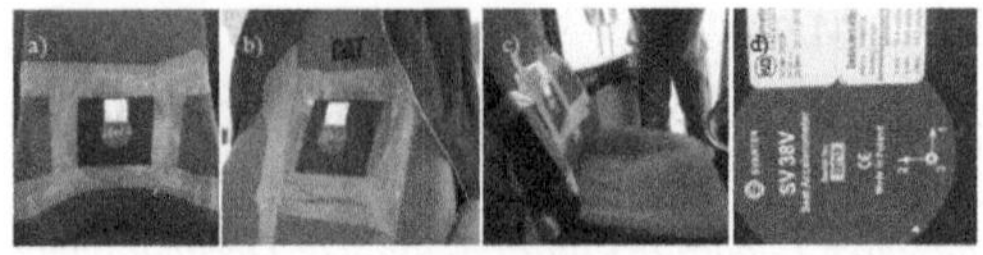

Figure 2. Placement of the whole-body accelerometer in the seat backrest of the a) rock drill, b) shovel and c) dumper.

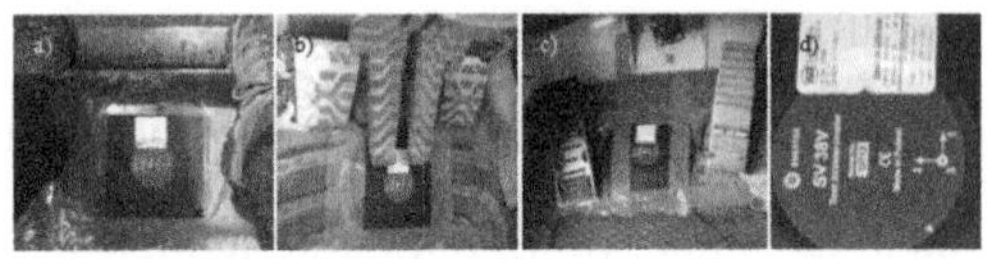

Figure 3. Placement of the whole-body accelerometer in the cabin floor of the a) rock drill, b) shovel and c) dumper, d) Orientation of the coordinated axis (Fig. 1, 2 and 3).

The tasks were timed and the vibration measurements were accompanied by an observation and occurrences registration, sometimes with the aid of video recordings.

1.4 *Analysis of Whole-Body Vibration exposure*

WBV analysis was conducted in accordance with ISO 2631-1 (1997) and carried out with SVAN PC ++, 1.5.10 version, developed by SVANTEK Poland.

The analysis included a graphical monitoring, establishing a pattern of instantaneous acceleration versus time. The work duty cycle and each task were identified.

2 RESULTS

2.1 *Work phases*

The percentage of total daily work time spent in each work phase was calculated for all vehicles (Table 2).

In rock drill, the drilling phase is the dominant (93%). In shovel, almost half of the time (49%) was spent during the material shifting and placing phase. The loading phase also has a considerable

Table 2. Work phases in rock drill, shovel and dumper duty work cycles.

	Average time of the task [min]	% in total daily work time
Rock drill		
(1) Drilling	12,1	93
(2) Vehicle movement and positioning	1,1	7
Shovel		
(1) Loading	6,0	42
(2) Material shifting and placing	7,3	49
Waiting moments	2,6	9
Dumper		
(1) Loading	4,6	34
(2) Loaded travel	3,6	29
(3) Dumping	0,6	5
(4) Unloaded travel	3,2	25
Waiting moments	2,3	7

percentage of the time (42%). In dumper, the majority of the time (34%) was spent during loading and the least amount of time (5%) was spent during the dumping phase, which is very short (less than a minute). The loaded travel takes a little bit more time than the unloaded one. There are some waiting moments (7%) mainly because dumper has to wait until the shovel can load it.

The time spent in each phase of the work duty cycle of the dumper is in accordance with Smets, Eger *et al.* (2010) have found.

2.2 *Whole-Body Vibrations patterns (instantaneous acceleration versus time)*

2.2.1 *Rock drill*

In Figures 4, 5 and 6 is shown an example of a rock drill WBV pattern (instantaneous acceleration, $m{\cdot}s^{-2}$, versus time), transmission through the seat surface, seat backrest and cabin floor, respectively. The tasks are indicated.

2.2.2 *Shovel*

In figures 7, 8 and 9 is shown an example of a shovel WBV monitoring (instantaneous acceleration, $m{\cdot}s^{-2}$, versus time), transmission through the seat surface, seat backrest and cabin floor, respectively. The tasks are indicated.

2.2.3 *Dumper*

In figures 10, 11 and 12 is shown an example of a dumper WBV patterns (instantaneous acceleration, $m{\cdot}s^{-2}$, versus time), transmission through the seat surface, seat backrest and cabin floor, respectively. The tasks are indicated.

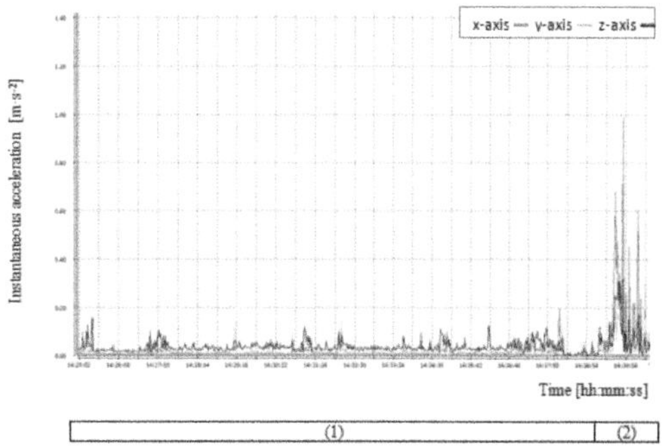

Figure 4. Example of a rock drill WBV pattern with transmission through the seat surface.

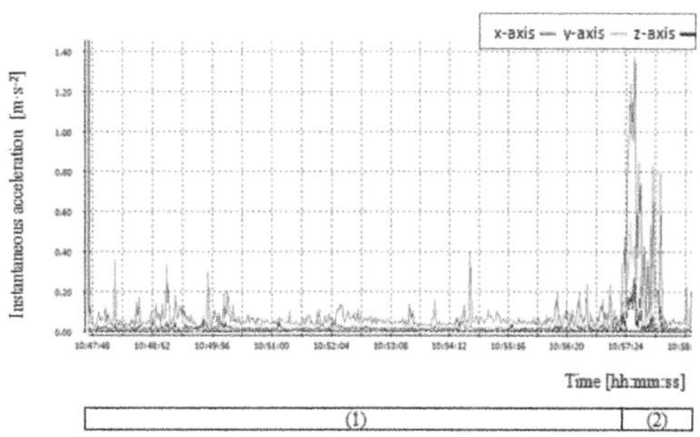

Figure 5. Example of a rock drill WBV pattern with transmission through the seat backrest.

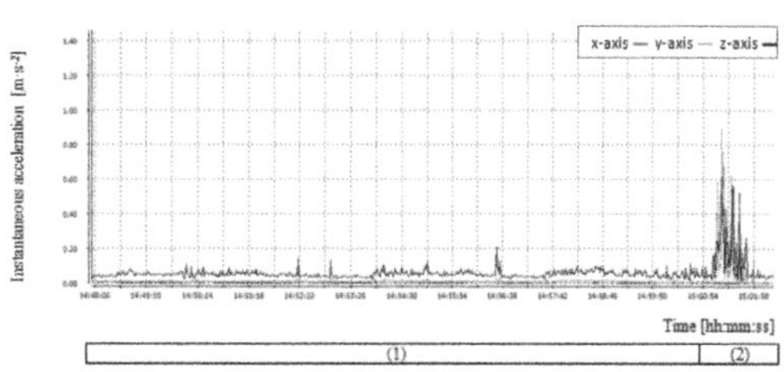

Figure 6. Example of a rock drill WBV pattern with transmission through the cabin floor.

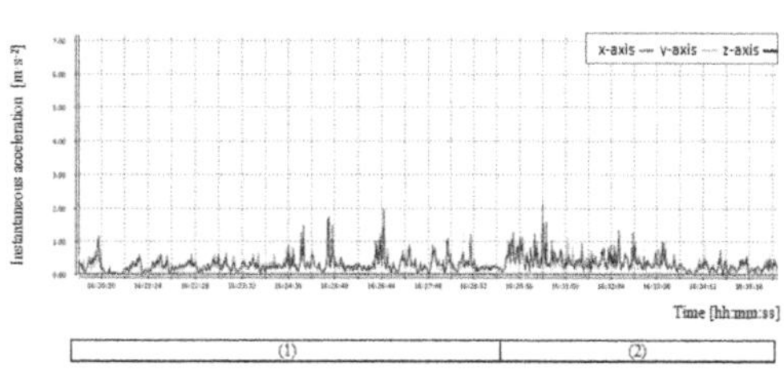

Figure 7. Example of a shovel WBV monitoring with transmission through the seat surface.

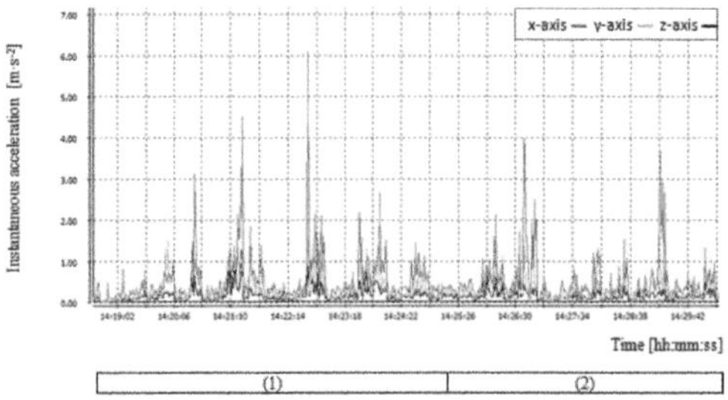

Figure 8. Example of a shovel WBV monitoring with transmission through the seat backrest.

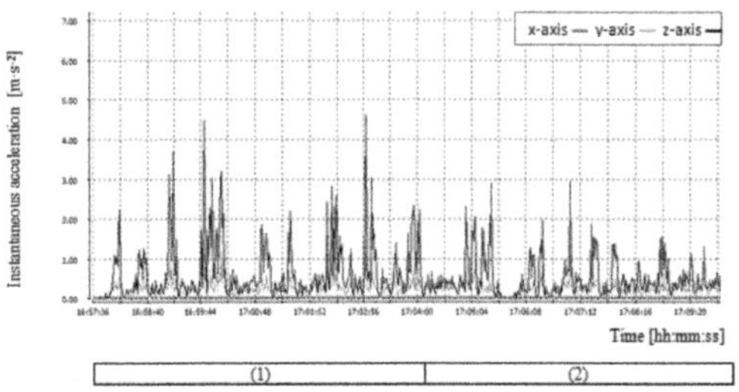

Figure 9. Example of a shovel WBV monitoring with transmission through the cabin floor.

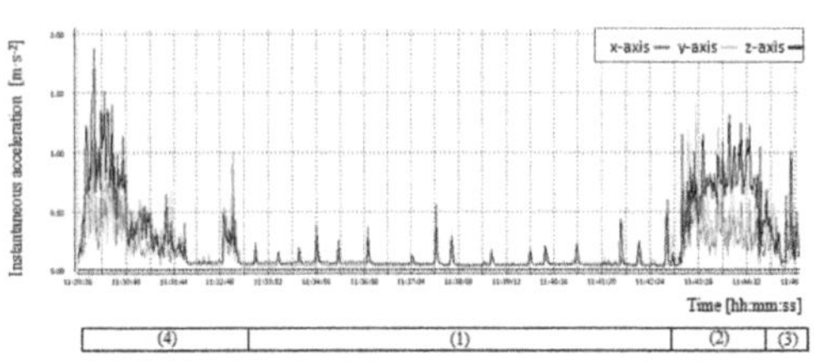

Figure 10. Example of a dumper WBV pattern with transmission through the seat surface.

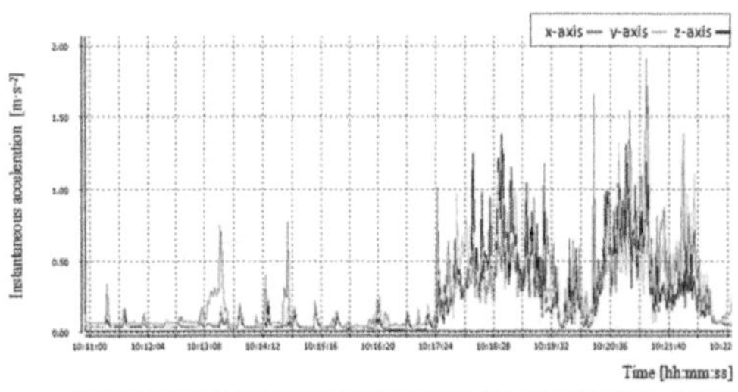

Figure 11. Example of a dumper WBV pattern with transmission through the seat backrest.

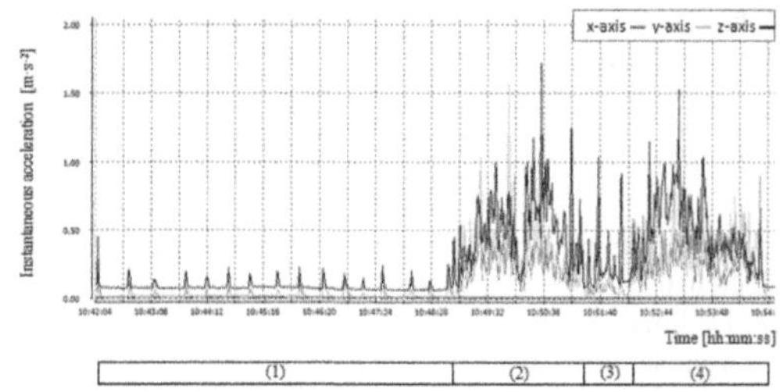

Figure 12. Example of a dumper WBV pattern with transmission through the cabin floor.

3 DISCUSSION

The analysis of Figure 1, 2 and 3 allows the identification of a WBV pattern of the rock drill work duty cycle, which is reproducible along the work day. In this pattern can be distinguished two clear tasks: (1) drilling and (2) vehicle movement and positioning.

The task (1) is the longest and it is characterized by low values of acceleration, below 0.20 $m \cdot s^{-2}$. It is not possible to distinguish the entrance of the different drilling sticks in the execution of the same hole, nor are identifiable behaviours associated with other specific situations as contact stick with a harder rock portion, jamming sticks or other technical problems. The task (2) is very short in the work duty cycle of the rock drill. Nevertheless it implies higher vibration levels, with a characteristic peak.

The shape of graphics (Figure 4, 5 and 6) is similar, but the dominant axis is not. When the WBV transmission is made through the seat surface and cabin floor (feet), the dominant axis in task (1) is the vertical zz and in the task (2) the longitudinal xx. When the seat backrest is analysed, the vibration acceleration is higher (mostly in task 2), and the dominant axis is the longitudinal xx in all the monitoring.

The operator spends more time during drilling phase, which is the one with lowest vibration values. In turn, the most critical phase in terms of exposure to WBV is that one where he spends less time.

Figure 7, 8 and 9 correspond to the work duty cycle of the shovel, including the tasks (1) loading and (2) material shifting and placing. However, these tasks are not distinguishable individually and there is not a visible pattern of vibration. The shape of the graphics is very irregular and rugged. The work made by the shovel is highly dependent on how the disassembled material is selected and the dismount that is necessary, so the work can be extremely variable. The main axis is the vertical zz when the vibration is transmitted through the seat surface and cabin floor (feet) and the longitudinal when the vibration is transmitted through the seat backrest.

In Figures 10, 11 and 12 it is possible to identify a WBV pattern, typical from dumper work duty cycle. It includes (1) loading the bucket with rock; (2) driving with loaded bucket to a dumping zone; (3) dumping the loaded bucket; (4) driving with an empty bucket back to the development heading to load another bucket.

Across all WBV transmission way, loaded and unloaded travel had the highest vibration magnitudes, as checked by Salmoni, Cann *et al.* (2010) and Kumar (2004). It would appear that the heavy load dumper carries has damping effect on the magnitude of vibration.

In loading operation (1) the dumper is parked, while it was loaded by shovel. Thus, this operation is characterized by residual vibration levels interrupted by peaks corresponding to the buckets loaded by shovel. This was also confirmed by the observations '*in loco*' and by the registration of the number of buckets loaded. The rock knocking on the box dumper causes the impact. The magnitude of vibration felt depends on factors like the type of material loaded, the amount of rock already loaded, the practice of workers, among others.

At (3) dumping there is a sudden drop in the level of vibration by vehicle stop and a new vibration peak, by the material discharge, and by downloading the box of the dumper and consequent impact on the operator's cabine.

The dominant axis when the vibration is transmitted through the seat surface and cabin floor (feet) is the vertical zz in all monitoring and the longitudinal xx when the vibration is transmitted through the seat backrest.

The operator spends the majority of time during the loading phase, which has the least vibration values associated. Nevertheless, considerable time was spent during loaded and unloaded travels, and here the vibration levels are the highest.

4 CONCLUSIONS

There is a WBV pattern in rock drill and dumper work duty cycles. In rock drill there is a long phase of drilling with residual vibration and a short phase, while the vehicle is moving, with a peak of vibration in longitudinal x-axis. In dumper all the phases are clearly distinguishable. Loaded and unloaded travels had the higher acceleration values, and loading the least ones. The operator spent more time in loading phase, and the least in dumping. In shovel, there is no WBV pattern identified.

ACKNOWLEDGMENT

The authors would like to thank Master in Occupational Safety and Hygiene Engineering (MESHO), of the Faculty of Engineering of the University of Porto (FEUP), all the support in the development and international dissemination of this work.

REFERENCES

Dentoni, V. and Massacci, G. (2013). "Occupational exposure to whole-body vibration: unfavourable effects due to the use of old earth-moving machinery in mine reclamation." *International Journal of Mining, Reclamation and Environment* 27(2): 127–142.

Donoghue, A. (2004). "Occupational health hazards in mining: an overview." *Occupational Medicine* 54(5): 283–289.

Gallagher, S. and Mayton, A. (2007). "Back injury control measures for manual lifting and seat design." *Mining Engineering* 59(12): 41–49.

Hill, C., Langis, W.J., Petherick, J.E., Campbell, D.M., Haines, T., Andersen, J., Conley, K.K., White, J., Lightfoot, N.E. and Bissett, R.J. (2001). "Assessment of hand-arm vibration syndrome in a northern Ontario base metal mine." *Chronic Dis Can* 22(3–4): 88–92.

ISO (1997). ISO 2631-1:1997 (Mechanical vibration and shock—Evaluation of human exposure to whole-body vibration—Part 1: General requirements), International Organization for Standardization.

Kumar, S. (2004). "Vibration in operating heavy haul trucks in overburden mining." *Applied Ergonomics* 35(6): 509–520.

Kunimatsu, S. and Pathak, K. (2012). "Vibration-Related Disorders Induced by Mining Operations and Standardization of Assessment Process." *MAPAN* 27(4): 241–249.

Salmoni, A., Cann, A. and Gillin, K. (2010). "Exposure to whole-body vibration and seat transmissibility in a large sample of earth scrapers." Work: A Journal of Prevention, *Assessment and Rehabilitation* 35(1): 63–75.

Smets, M.P.H., Eger, T.R. and Grenier, S.G. (2010). "Whole-body vibration experienced by haulage truck operators in surface mining operations: A comparison of various analysis methods utilized in the prediction of health risks." *Applied Ergonomics* 41(6): 763–770.

Occupational Safety and Hygiene II – Arezes et al. (eds)
© 2014 Taylor & Francis Group, London, ISBN 978-1-138-00144-2

Tracking Surgical Instruments: From a management perspective to safety issues

A.M. Sampaio & R. Simoes
School of Technology, Polytechnic Institute of Cávado and Ave, Campus do IPCA, Barcelos, Portugal
Institute for Polymers and Composites IPC/I3N, University of Minho, Campus de Azurém, Guimarães, Portugal

A.J. Pontes
Institute for Polymers and Composites IPC/I3N, University of Minho, Campus de Azurém, Guimarães, Portugal

ABSTRACT: This paper focuses the importance of tracking Surgical Instruments (SI). Since the goal of healthcare is to be able to deliver necessary quality services, while, reducing healthcare costs, it was necessary to understand the problems related to these particular devices. The aim was to search for research that already tackled these issues and what were their suggestions/solutions to these problems. Therefore the problems of tracking SI in healthcare environments are recognized and analyzed, centered in the facilities through where the instruments flow. To validate the literature review, an investigation on the real context was performed. This was conducted in a participant-observer research in three hospitals to correlate the findings and specify the problem. This investigation shows that, if the primary scope of tracking SI was a way to achieve optimization of the devices process flow, problems of patient safety are also issues that could be addressed trough management control.

1 THE MANAGEMENT PERSPECTIVE

Hospitals are under constant pressure to control health costs (Aguado et al. 2007). In order to cut costs, the optimization of asset management within healthcare appears to be an important issue (Bates et al. 2001). In hospitals, high-value mobile assets (e.g. blood, beds, instruments, surgical devices) are often misplaced, lost, or stolen. It is common for a hospital to loose 10% of its inventory annually (Wicks et al. 2006). This is mainly due to the fact that no one in a hospital has real-time information about where movable assets are because these resources are frequently moved due to emergency necessity responses in different locations. As such, it is a "normal" situation for nurses to spend a great deal of time tracking down appropriate assets so that they are delivered on time. In fact, hospitals have difficulty to locate 15–20% of their assets due to inappropriate and ineffective monitoring procedures (Li et al. 2006). In this context, healthcare organizations are searching for answers to reduce operational costs while, at the same time, facilitating healthcare delivery in order to offer patients with highly reliable care service. Bates et al. (2001) conclude that appropriate use of technology in healthcare could result in process simplification and substantial improvement in patient safety. Establishing a traceability system is a key enabler to enhancing patient safety, improve the quality of care and at the same time reduce operational costs (Tu et al. 2009). In the healthcare sector, one of the environments that are being especially targeted for improvement is the Operating Room (OR). This is due to the fact that the OR is the highest revenue generator within a typical hospital, but at the same time, it is also the one that has the largest costs. Another important fact is that the quantity and complexity of SI, the high cost of equipment and staff time, and the inherent risk involved in many surgeries, makes the OR and all the processes that are involved directly or indirectly with surgeries (e.g. sterilization procedures), the main setting for improvements in patient safety and efficient management operations through technology.

From the stated above, it is clear that the healthcare sector is a multifaceted structure that faces a great demand to reduce operational costs, but at the same time, improve patient care and safety. In the case of patient safety, reducing errors is a required task in all the environments but mostly in the OR. Therefore, it is necessary to further develop the knowledge on the main problems in the OR, and relate them to the management perspective. In other words, investigate the importance of implementing asset management in surgery focused on human-errors to improve patient safety. At the same time, procedures directly linked with surgeries (i.e. sterilization procedures) will also need to be investigated in order to analyze the general surgical workflow.

1.1 *Operating room safety problems*

Healthcare professionals work under an omnipresent threat of medical errors. Given the increasing complexity of both the modern healthcare environment and the patient population, reducing medical errors is becoming a high priority task for healthcare policy makers and the medical community alike. While the incidence and outcomes of specific types of surgical errors are relatively well described, the knowledge of why these errors occur is incomplete (Gawande et al. 2003).

One of the most perplexing examples of such a medical error is the occurrence of a retained foreign object, following a surgical procedure, such as sponges or SI (Gawande et al. 2003). Retained Foreign Objects (RFO) may cause harm to the patient and can result in serious professional and medico-legal consequences (Imran & Azman 2005). As such, it is a problem that can seriously compromise healthcare quality and patient safety. Despite being theoretically completely preventable, it has been estimated that a foreign object is retained anywhere between 1 in 9000 operations to 2.4 per 10,000 (Greenberg et al. 2008) with one study involving 47 claims resulting in a average cost of 52,581 US dollars in compensation and legal expenses. The most important preventive measure to ensure that foreign objects are not retained in the surgical cavity is to manually count the SI before and after the surgeries and compare the counts (AORN 2008). Nevertheless, in a study of coronary artery bypass surgery, approximately 0.7% of operations were associated with an incorrect count and the cost of each incorrect count was over 900 US dollars. Additionally, the protocols mentioned can account for up to 14% of operative time (Greenberg et al. 2008). In this way, many perioperative nurses question the efficacy of counting SI for every procedure since most hospitals are now concerned with efficiency and decreasing operating times rather than counting (Goldberg 2010). The literature related to the counting of SI (Halvorson 2010) describes error-prone processes and the major issues of performing routine tasks when pressured or distracted. Characteristics that contribute to the error-prone nature of counting include that counting often is highly automatic and prone to unexpected interruptions and that surgeons tend to move to a different task before completing final validation that the count is correct (Gibbs et al. 2005). In summary, when counting practices were first established the OR environment was much different than it is today. Not only has the instrumentation and technology become much more complex, but procedures have also become more sophisticated and complicated. Despite this, no new technology or strategy has emerged to help surgeons and/or nurses ascertain the accuracy of their counts or perform them with more consistency, accuracy, or efficiency (Halvorson 2010). Therefore, it seems that there is a significant need for electronic tagging and detection systems for SI.

1.2 *General sterilization problematic*

As stated, one of the most important objectives of healthcare facilities is improving quality of care and patient safety, which is to a high extent related to hygiene conditions. At this level, one of the major challenges is to prevent nosocomial infections (Rutala & Weber 2008).

In general, SI are seen as a high-risk item since it penetrates sterile tissues such as body cavities. These items are called critical items because of the risk of infection. As such, they must be sterilized. The use of inadequately sterilized SI represents a high risk of transmitting pathogens (Rutala & Weber 2008). Inadequate destruction or inactivation of pathogens (e.g. bacteria, fungi, viruses, spores and other microorganisms) left on an instrument by one patient can result in serious adverse clinical outcomes in the next patient. Improperly sterilized SI can promote the development of Surgical Site Infections (SSI).

SSI is a dangerous infection that has been associated with mortality and healthcare costs with huge economic impact. As stated by Alexander et al. (2011) the large increases in healthcare costs may be related to the increased incidence of SSI, which may lead to an annual increase of 900 million US dollars in healthcare costs. Although SSI is the most commonly reported nosocomial infection and accounts for 14–16% of all infections among hospital inpatients, published data on SSI may be underestimated because many SSI that are developed after patient discharged are not recorded (Smyth & Emmerson 2000). Therefore, the sterilization of SI is an imperative to prevent SSI.

For the sterilization of instruments there are several methods (e.g. Ethylene Oxide, Hydrogen Peroxide Gas Plasma, Dry Heat, Ozone). The oldest and most commonly used is steam sterilization under pressure—also called autoclaving (Goldberg 2010). Mainly because it is the most reliable in destroying prions, that are associated with the Creutzfeldt-Jacob disease and other degenerative neurologic syndromes, although some cases still exist (Brown et al. 2000). As such, sterilization processes are able to overcome the common microorganisms, but uncertainties about efficacy towards prions destruction subsist. Therefore, the identification of instruments that were involved in patients with degenerative neurologic syndromes would be an important achievement regarding patient safety. Therefore, the optimization of the sterilization process flow is an imperative. As the changing context of healthcare structures encourages hospitals to improve the quality of services while reducing the associated costs, the sterilization process is a potential area for important cost savings. The sterilization activity stands as a major focus to improve efficient management of the instrumentation. SI should be accurately traced in all their movements. From the different OR to the SU, and at the same time it must be possible to identify them individually in the SU.

2 UNDERSTANDING THE PROBLEM IN REAL CONTEXT

Although the main findings of the problem of tracking SI have been already acknowledge, these were at a theoretical level. In order to understand the actual problem that SI pose in healthcare, an investigation of the current situation in the real context was seen as a way to corroborate the previous findings and to acquire a better knowledge of what really was important.

2.1 *Understanding the problem*

As stated this research began with the problem of understanding the importance of tracking the SI. In this scope several questions arose. Namely, if tracking SI were necessary because instruments were stolen? Loose? Mixed? Damaged? Or the main scope was on improving productivity? Faster and simpler management? Patient safety? Therefore, after these general questions, specific questions were developed as a guide to the overall aim of stating and specifying the problem (Tab. 1).

These questions were divided in pre-operative, surgery and post-operative groups. This was done to understand the overall cycle with the specific knowledge of each phase.

2.2 *Participants*

This research was developed in three hospitals (Two public and one private) located in northern Portugal. All the hospitals in this research are general hospitals that offer clinical care, with an emergency room open 24 hours per day and 7 days per week, perform surgeries in a regular basis (every weekday with more than four operating rooms), and have an indoor SU.

2.3 *Procedure*

The research was implemented and conducted in a participant-observation strategy as prescribed by Zenios (2009), in three major hospitals for the same surgical procedure (hernias). All surgeries were performed in the traditional way (no laparoscopes). The research began by sending authorization letters to the clinical directors of the three hospitals, explaining the aim of the research and the need to visualize surgical procedures. The main aim was not in the surgical procedures by themselves but in following the SI through all the steps, and spaces in order to understand and close the cycle. The focus was on the cycle that involved surgical procedures in the OR; therefore, the cycle that involves the emergency room had no part in this research.

In the observations, several records were taken on a notebook, and with a photo camera. Additionally, seven performance measures of routine tasks were also recorded in terms of time.

Table 1. Guide questions for the research.

Pre-operative	Surgery	Post-operative
What steps are required? Who is involved? How were the SI taken to the surgery? From where? How are they prepared? What are the safety measures?	How many people were in the OR? How many people are involved directly with the SI? How are they been used/handled? Was it necessary for more/different SI during the procedure? Why?	What steps are required? Who is involved? Are the SI "clean" just before the procedure? How are they been handle/used? How is the sterilization done?

General
How long did the overall cycle take (SI)?
How many people are involved in the overall cycle? In what steps? What are their various roles?
What are the limitations or difficulties associated with the current cycle?
How is the device used? In all the steps?

2.4 *Main findings*

From the observations done, it was possible to see that the overall cycle (Fig. 1) is not so different between hospitals. All hospitals have the cycle divided in eleven zones (A to K), been those zones identical for all hospitals. Another thing in common is the type of people involved (1 to 4). All observations show that there are four types of participants in the overall cycle. Although operational assistants work on this cycle, the only responsibility that they have is transporting SI from the SU to the OR and vice-versa. The sterilization technicians are responsible for the maintenance of the

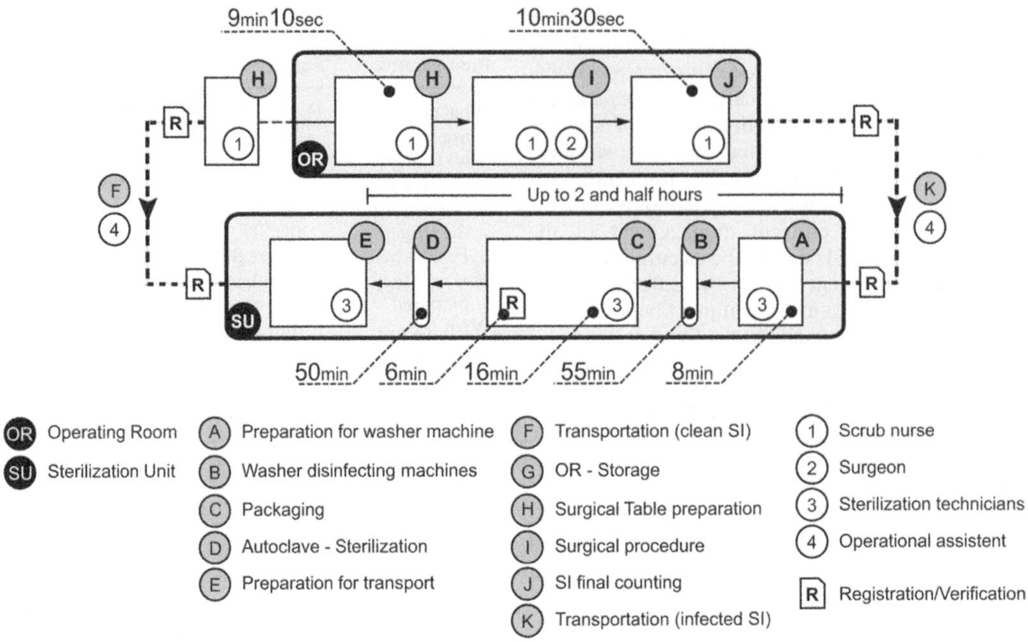

Figure 1. Overall cycle of surgical instrument (people involved, tasks per zone, average task-time measures).

SI. There work is concerned with operational and technical procedures focused on the SI. They clean, wash, verify, pack and sterilize the instruments. Their work relates to the safety of the main procedures of the product (surgeries), from a patient, scrub nurse and surgeons perspective. They are the basis of the SI cycle. Working in the background in order to maintain the proper functions of the instruments. They are responsible for the management of the SI.

Scrub nurses are responsible for the preparation of surgeries. In the OR the SI are their responsibility.

They assist the procedure but the issues of safety are their major concern; this is why they count the SI before (Zone H) and after the surgery (Zone J), assuring that no SI is left behind.

It is possible to say that, as sterilization technicians are responsible for the SI in the SU, in the OR that responsibility is of the scrub nurse. In turn, surgeons are only responsible for the surgical procedure. They are not concerned with any other task other than performing the surgical operation and they just work in zone I. Although the number of zones is the same, the tasks in zones with the same letter are different from hospitals. It is possible to say that, the major differences can be seen in SU area. Divergences are seen in the private hospital, as they do not have the task of cleaning (Zone A) the SI before washing machine process (Zone B), and do not store surgical sets in the SU (Zone E). The absence of cleaning is due to the notion that the washing machine process (Zone B) as already a pre-cleaning, so doing it by hand is not necessary (although the other two hospitals perform this task). The lack of storing the SI in the SU is due to the reduced number of sets that the private hospital has opposed to the other two. In this way, their OR-Storage is sufficient for storing all instruments. The private hospital is also different from the other two in the case of preparing the sets, since they don't mix instruments. In other words, they perform the cycle in the SU for each instruments set. The inner cycle of the SU is done separately—a cycle for each set.

Another major difference can be seen in the number and place of the documentation/verification process. The hospital relies on written documents to manage the SI cycle. There are two types of documents: the documents that follow the SI set and the delivery and pick-up registrations. The first one is a document that lists all the SI contained in that set (by name and number of SI). The main purpose of this document is to make it possible for the scrub nurse to know exactly what SI the set has. The registrations are performed in order to track the sets and assign a worker to the delivery and picking-up tasks.

In the case of the two public hospitals the management of the SI/Sets are done in the same way. The main document is performed in each cycle, by a sterilization technician (Zone C), and the document follows the set through all zones till reaching zone A. This is how they control that a set has completed a cycle. The registrations are done when entering and leaving the OR and SU areas and the OR-Storage. In the case of the private hospital, the main document is the same for several cycles. Outside the OR the registrations are the same as the other two hospitals.

What could be concluded from the observations is that SI requires significant attention inside a healthcare provider organization such as hospitals. Aside from the typically large number of SI that a hospital has, they represent an asset normally associated to the most difficult procedures. These devices, far from their general use, have a very distinct cycle that involves lots of people. If, from a general point of view, it can be said that a major preoccupation is centered in the number of people that interact directly with these instruments and the different places where they are used, other problems such as leaving instruments inside the human body, makes the task of tracking SI a complex but necessary approach. Nowadays there is no consensus regarding the documentation/verifications tasks. Nevertheless what is worth saying is that this constant need to track the SI and the way that is done are very time-consuming. Another important fact is that in order to prevent accidents in the OR, strict protocols are implemented in healthcare providers, since the scrub nurse needs to perform the counting of the SI before and after surgical procedures. This task is, obviously, difficult, error prone, and also very time-consuming. Another aspect is that currently, it is near-impossible to tell how many times a SI has been used. For security reasons, individual management of the SI need to be performed, so the issue of tracking becomes much more than just a question of preventing accidents; it is also a way to reduce task times, optimize nurse efforts, and manage the number of uses of each instrument.

3 CONCLUSION

Asset management in healthcare is an imperative if hospitals hope to reduce costs. These costs have different causes and are of different types. What is seen as an obvious cause to raising costs is the particular structure of hospitals. Given the increasing complexity of modern healthcare environment issues related to whether the assets move and are found and, the number of assets available pose a challenge to improve management control. In this way, one system that healthcare providers are considering is the integration of tracking technologies. Although tracking technologies is seen as an important system in the management of assets that leads to a reduction of costs, several other problems exist where these technologies could be preventive. In this paper it was stated that in hospitals medical errors might lead to adverse events, some with irreversible consequences. This was what it was acknowledged from the analysis of the problems concerning the surgical procedures and the sterilization process. These two main facilities that involve any surgery were investigated for three different, but at the same time cumulative, reasons. First, the OR, as this is the facility with the highest costs in any hospital but at the same time the one that involves the most difficult procedures, second the sterilization unit as this is the facility that assures the integrity of the instruments used in the surgery and, third the instruments as they are the main focus of this problem. As such, in the OR, the major problems identified were the occurrence of RFO after surgery. Although the existence of preventive measures, these procedures were seen as an error-prone practice.

In the SU the facility responsible for the prevention of SSI, the problems were associated with the need to identify instruments that were in contact with patients with degenerative neurologic syndromes, since sterilization processes do not always achieve the sterility assurance level.

The analysis of the problems of these two facilities has demonstrated the need to incorporate electronic tagging in SI, which could permit tracking and identification of the instruments in their working cycle. In the OR it would be much easier to count instruments and to prevent the occurrence of RFO and in the SU the identification of SI would be possible and the efficient management of instruments improved.

REFERENCES

Aguado, F., Alvarez, M. & Barcos, L. 2007. RFID and health management: is it a good tool against system inefficiencies? *International Journal of Healthcare Technology and Management* 8(3/4): 268–297.

Alexander, J., Solomkin, J. & Edwards, M. 2011. Updated recommendations for control of surgical site infections. *Annuals of surgery* 253(6): 1082–1093.

AORN 2008. *Perioperative Standards and Recommended Practices*. AORN Inc.

Bates, D., Cohen, M., Leape, L., Overhage, J., Shabot, M. & Sheridan, T. 2001. Reducing the frequency of errors in medicine using information technology. *Journal of the American Medical informatics Association* 8(4): 299–308.

Brown, P., Preece M., & Brandel J. 2000. Iatrogenic Creutzfeldt-Jacob disease at the millennium, *Neurology* 55: 1075–1081.

Gawande, A., Studdert, D., Orav, E., Brennan, T. & Zinner, M. 2003. Factors for retained instruments and sponges after surgery. *New England Journal of Medicine* 348: 229–235.

Gibbs, V., McGrath, M. & Russel, T. 2005. The prevention of retained foreign bodies after surgery. *Bulletin of the American college of surgeons* 90(56): 12–14.
Goldberg, J. 2010. What the perioperative nurse needs to know about cleaning, disinfection, and sterilization. *Perioperative Nursing Clinics* 5(3): 263–272.
Greenberg, C., Diaz-Flores, R., Lipsitz, S., Regenbogan, S., Mulholland, L., Mearn, F., Rao, S., Toidze, T. & Gawande, A. 2008. Bar-coding surgical sponges to improve safety: A randomized controlled trial. *Annals of surgery* 247(4): 612–616.
Halvorson, C. 2010. Review of best practices and literature on instrument counts. *Perioperative Nursing Clinics* 5: 27–44.
Imran, Y. & Azman, MZ. 2005. Asymptomatic chronically retained gauze in the pelvic cavity. *The Medical Journal of Malaysia* 60: 358–359.
Li, S., Visich, J.K., Khumawala, B.M. & Zhang, C. 2006. Radio Frequency Identification Technology: *Applications, Technical Challenges and Strategies. Sensor Review* 26(3): 193–202.
Rutala, W. & Weber, J. 2008. Guideline for Disinfection and Sterilization in Healthcare Facilities. Retrieved January 10, 2012. From www.cdc.gov
Smyth, E. & Emmerson, A. 2000. Surgical site infection surveillance. *Journal of hospital infection* 45(3): 173–184.
Tu, Y., Zhou, W. & Piramuthu, S. 2009. Identifying RFID-embedded objects in pervasive healthcare applications. *Decision Support Systems* 46: 586–593.
Wicks, A.M., J.K. Visich, & S. Li. 2006. Radio frequency identification applications in hospital environments. *Hospital Topics* 84(3): 3–9.
Zenios, S., Makower, J., Yock, P., Brinton, T., Kumar, U., Denend, L. & Krummel 2009. Biodesign: *The process of Innovating New Medical Technologies.* Cambridge University Press.

Occupational Safety and Hygiene II – Arezes et al. (eds)
© 2014 Taylor & Francis Group, London, ISBN 978-1-138-00144-2

Occupational noise and hearing loss in a metalworking company

A. Beça & A.S. Miguel
Universidade do Porto, Faculdade de Engenharia, Porto, Portugal

J. Góis
Universidade do Porto, Faculdade de Engenharia, Porto, Portugal
CERENA—Centro de Estudos em Recursos Naturais e Ambiente, Portugal

ABSTRACT: In metalworking the noise has been always identified as one of existing hazards, originating the need to evaluate and quantify the associated risk. Records of noise assessments were collected in the company, a questionnaire was applied to all workers and audiometric exams were done. It was performed an additional analysis through surveys workers to try to relate the dependent variables audiometric with the independent variables sound level, duration of exposure to noise at workplace and workers' age. It was intended to study the hearing losses in workers potentially exposed to occupational high noise and check the existence of a correlation between this losses, age and exposure to noise. It is expected to obtain a better perception of the noise control measures and techniques to try to lessen the exposure of workers to excessive noise. This work was focused on a population of 36 individuals, only those ones exposed to noise levels greater than 85 dB (A).

1 INTRODUCTION

The Occupational Health and Safety activities assume, nowadays, an extraordinary importance, either by improving the conditions of life and work or by the development of the productive activity itself and also by promoting the company's image in the market.

Productive activities of metalworking, by virtue of several factors, particularly the nature of their technological processes, present a great diversity of occupational hazards, which must be known, and require appropriate preventive methodologies.

For all this, the increase of the safety and health at work in companies, in the sector, are not only necessary, but also timely and should be undertaken in the context of an occupational health and safety management system. Moreover, these activities are surrounded by a vast and demanding legislative frame, coming from the European institutions.

Exposure to high sound pressure levels, in the workplace, can cause permanent hearing loss, so it deserves a rigorous approach on the part of employers. Portuguese legislation on this matter refers to aspects related to collective protection, personal protective equipment and information and training to workers.

Given the negative influence of noise on workers, it is of paramount importance the development of research aiming to identify and assess the extent of the exposure to high sound pressure levels in the way it can interfere with workers' quality of life.

2 MATERIALS AND METHODS

It was intended to study hearing losses in workers potentially exposed to occupational noise and verify the existence of a correlation between these losses, age and noise exposure. Different evaluations of noise were made in the company and a survey was applied to all workers. Audiometric tests to workers were undertaken as well.

2.1 *Noise study*

Consultation of available noise studies has revealed that the manufacturing process generates quite high sound pressure levels. Studies of noise, based on the calculation of daily personal exposure of workers, were carried out by an external company, in the years 2006, 2010 and 2012.

The Equation 1 allows to calculate the overall value of the daily personal exposure to noise (Miguel, 2012).

$$L_{EX,8h} = \left(\frac{1}{T} \sum_{1}^{n} ti \times 10^{0,1 \bullet LEX,8hi} \right) \quad (1)$$

where $L_{EX,8h}$ = Value of individual daily noise exposure, that the worker would be exposed, during the period T; T = Total exposure time; t_i = Partial exposure time, which corresponds to $L_{EX,8hi}$; $L_{EX,8hi}$ = Value of individual daily noise exposure during the period of time t_i.

2.2 *Survey*

The conducted survey allows knowing the medical, personal and also professional history of the workers with interest to the study. The exposure time to occupational noise is variable for each worker and it is calculated according to the year of entry into the company and the year in which the worker carries out the audiometry. The questionnaire was filled individually in paper.

The first part of the questionnaire seeks information about the employee, the appropriate job and his qualifications.

The second part of the questionnaire refers to the characterisation of possible situations of exposure to excessive noise in the professional prior to entry into the company under study. Then, the questionnaire refers to the history of noise exposure of non-professional context. This part seeks to characterise and record any cases of exposure to excessive noise, off-site and working environment in leisure time. For example, one was victim of blasting or explosion, practices hunting or shooting, usually attends or has frequently activity in nightclubs or musical performances, etc.

The third part of the questionnaire refers to a personal evaluation, family and clinical events that potentiate hearing loss. For example if one has a family member with deafness problems, if one has suffered a head injury, has caught problems in the ears or has received treatment with potentially ototoxic drugs.

The fourth part seeks to question the use of hearing protection equipment, the type of protection that workers use and the individual hearing state, variables that characterise them (Pereira, 2009).

2.3 *Hearing losses*

For the calculation of hearing losses and to make interpretation more intuitive, in terms of quantification and classification of hearing losses, the BIAP criterion was used. The BIAP, International Bureau of Audio Phonology, calculates hearing losses through the average of pure tone frequencies 500, 1000, 2000 and 4000 Hz. In case the losses are asymmetric, it weighs hearing losses in the 2 ears, using the coefficient value 7 for the better ear, 3 for the worse and divides the sum by 10. The result is rounded up (Mendes, 2011).

2.4 *Analysis on major components*

The Principal Components Analysis is a multivariate statistical technique in the field of factor analysis, which allows viewing and reduction of data contained in tables of large dimensions, crossing a number of individuals with quantitative variables that characterise them (Pereira, 2009). The Principal Components Analysis is applicable to tables of input consisting of real numbers and allows finding the factors that best explain the similarities and oppositions between individuals and variables. These factors, classified by descending order of their importance for the explanation of the starting table, constitute a system of orthogonal axes, where is possible to view, in graphical form, the projections of the constituents of the data matrix. The interpretation of graphs is based on a set of clear rules, highlighting the most significant relationships (Pereira, 1990).

3 RESULTS AND DISCUSSION

The consultation of available noise studies has revealed that the manufacturing process is the generator of high sound pressure levels, with values of $L_{EX,8h}$ per sector exceeding 85 dB (A).

From the questionnaire and by analysing the answers, it is possible to provide a more specific characterisation of the sample. It can be seen that most of the workers are male (77.4%), the average age is 50.2 years and the average length of stay in the company is quite high (27.4 years). Regarding the use of hearing protection, it appears that over 90% of workers use "always/often" hearing protection devices. Regarding the perception of individual hearing state, it is noted that about 64.5% of workers say they hear well, while about 25.8% indicate the sense of ringing in the ears.

Audiometric test performed in 2013 is shown in Table 1.

According to the calculation and classification of hearing losses, using the BIAP criterion, the rating distribution of hearing losses, shown in Figure 1, is obtained.

With these results, it can be seen that 26% of workers have a hearing loss that is considered "normal" loss, 52% "light" loss and 22% of workers have a "moderate" loss. No worker is in a state

Table 1. Summary of study variables and hearing losses calculation.

Variable	Mean value	Deviation	Maximum value	Minimum value
Age (years)	50.2	6.2	62.0	37.0
Time noise exposure (years)	27.4	7.4	40.0	12.0
$L_{EX,8h}$ (dB(A))	90.9	4.9	102.6	79.3

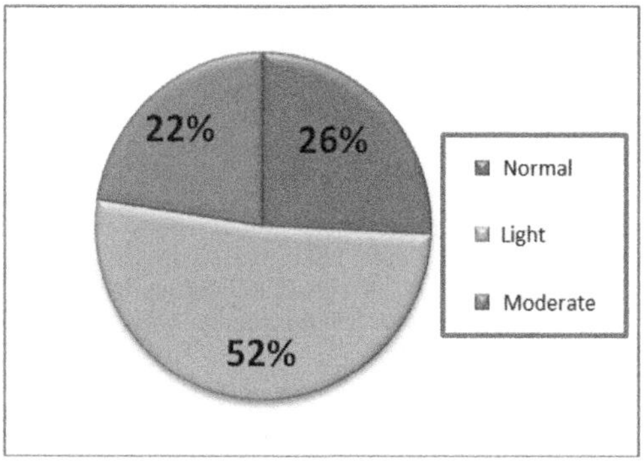

Figure 1. Distribution of hearing losses in 2013, according to the BIAP criterion.

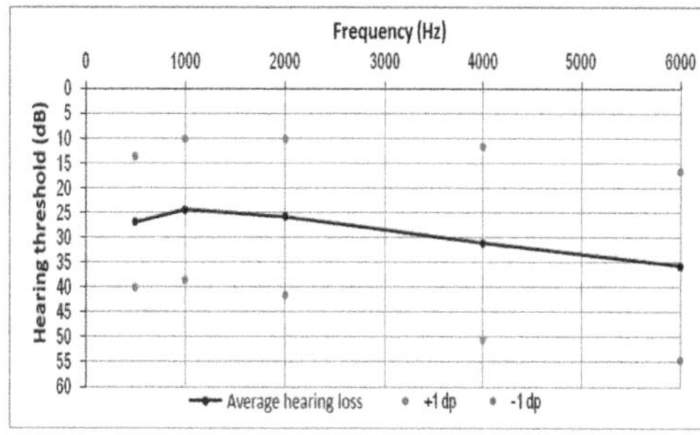

Figure 2. Average hearing losses in the studied population.

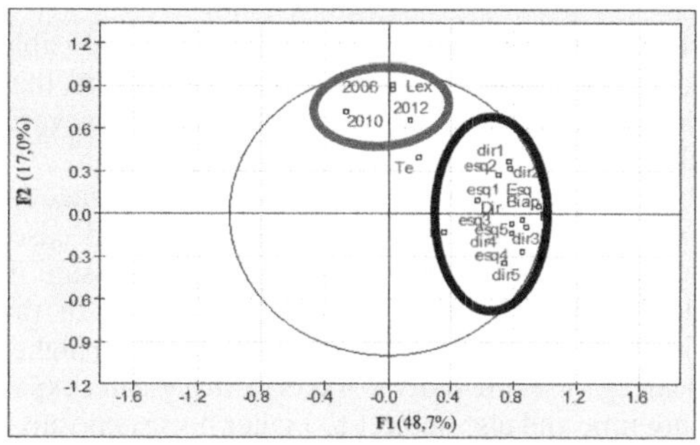

Figure 3. Factorial plan with first two factors, variables projection.

of "severe" loss (above 71 dB) or in a situation of "total" loss. The maximum value of "moderate" loss was 59 dB and the range of this loss was classified between 41 and 70 dB.

The effective use of hearing protection, over the years of exposure, does not mean that in the past the worker has worked unprotected, altering the real $L_{EX,8h}$ of the workers. Workers probably would not have been so careful with the use of personal protection equipment or have faced a possible non-availability of the same.

It is important to know not only the detailed history and antecedents of the study population, but also whether the hearing losses have stemmed from an ear disease.

By averaging the values of audiometric tests by frequency band (Figure 2), it is possible to observe that the most significant values are those corresponding to the frequencies 4000 and 6000 Hz. It was also found that for these frequencies, standard deviation values are higher.

Based on various collected data and compiling all the information on a sheet of EXCEL, it is possible to proceed to a test in an attempt to correlate the different variables with the aid of the ANDAD. This is organised into several modules that allow applying some multivariable techniques for statistical analysis, namely Principal Component Analysis. These methods are used to describe and summarise data frames, usually with higher dimensions, which can be embodied in a matrix. The data were organised in a matrix of 36 individuals (rows) and 19 variables (columns).

Application of Principal Component Analysis allowed evaluating the consistency of the analysed data, being possible to check the similitude of registrations in 2006, 2010 and 2012. The same conclusions are valid for the remaining variables projected very close to each other, Figure 3.

4 CONCLUSIONS

The sample size doesn't allow objective conclusions, mainly due to the applied methodology and the uncertainty associated with the data. Sample size population should be higher so reliable conclusions can be drawn.

The realisation of this work consisted to evaluate the sound pressure levels and study the hearing loss in the company where the author monitors everyday the workers. Thus, with the application of PCA was possible to identify individuals with particular characteristics hearing.

The most significant hearing losses appear for octave central frequencies of 4000 and 6000 Hz.

Considering individual perception, there is not a direct relationship with the hearing losses, since workers with "normal" loss have indicated this problem. But from the 7 workers with "moderate" loss, only 5 have showed the perception of limitations in terms of audition. From 16 workers with "light" hearing loss, only 6 have reported the perception of hearing limitations.

From the PCA, it was possible to reduce the dimensionality of data starting, allowing the crossing of individuals with quantitative variables that

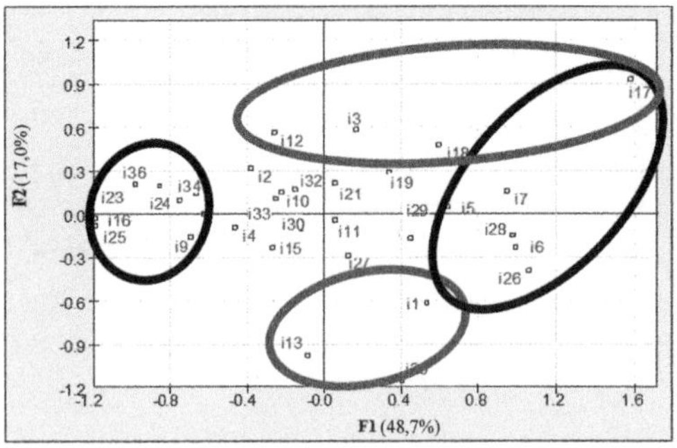

Figure 4. Factorial plan with first two factors. Workers projection.

characterise them, Figure 4. Thus, it was possible to verify, in this case, that the first two factors, that define the factorial plan, explain 65.7% of the variability contained in the data matrix.

The other factors, given the small percentage of explanation, only allow to identify isolated cases (variables or workers), without much expression in the sample. It is possible, from the analysis of the first factorial plan, to verify that workers with higher hearing losses are older workers, with a greater exposure time and also subject to higher noise exposure.

ACKNOWLEDGMENTS

The authors would like to thank Master in Occupational Safety and Hygiene Engineering (MESHO), of the Faculty of Engineering of the University of Porto (FEUP), all the support in the development and international dissemination of this work.

REFERENCES

Arezes, P. (2002). *Percepção do Risco de Exposição Ocupacional ao Ruído*. Universidade do Minho.

Arezes, P., Miguel, A., A exposição ocupacional ao ruído em Portugal. Riscos Ocupacionais, VOL 20, Nº1 – Janeiro/Junho 2002.

Barradas, E. (2011). *Estudo da Exposição Ocupacional ao Ruído em Estações de Tratamento de Águas e Respectiva Avaliação Audiométrica,* Dissertação de Mestrado em Engenharia de Segurança e Higiene Ocupacional, Universidade do Porto.

Mendes, A. (2011). *Ruído Ocupacional em ambiente industrial, Dissertação de Mestrado em Engenharia de Segurança e Higiene Ocupacional*, Universidade do Porto.

Miguel, A.S. (1992). *Proteção Auditiva Individual em Ambientes Industriais*. Universidade do Minho.

Miguel, A.S. (2012). *Manual de Higiene e Segurança do Trabalho. 12a Edição*: Porto Editora.

Pereira, A. (2009). *Avaliação da Exposição dos Trabalhadores ao Ruído (Análise de Casos*), Dissertação apresentada na Faculdade de Ciências e Tecnologia da Universidade Nova de Lisboa para obtenção do grau de Mestre em: Instrumentação Manutenção Industrial e Qualidade.

Pereira, H. (1990). *Análise de dados geológico – Mineiros*. Aplicações e Estudo metodológico. Relatório sobre as disciplinas de matemáticas aplicadas à Engenharia e de Minas e introdução à Investigação Operacional, para as provas de agregação na área de Geomatemática do curso de Minas do IST.

Occupational Safety and Hygiene II – Arezes et al. (eds)
© 2014 Taylor & Francis Group, London, ISBN 978-1-138-00144-2

Analysis of FEUP school furniture's suitability

L.S. Sousa & M.E. Pinho
Faculty of Engineering of the University of Porto, Porto, Portugal

P.M. Arezes
Engineering School of the University of Minho, Guimarães, Portugal

ABSTRACT: Students spend a considerable part of the day sitting down at school, while doing their school work with school furniture that may not match their anthropometric dimensions. This results in some potential anatomical–functional changes and problems. The aim of this study was to assess the level of mismatch between 9 school furniture (5 chairs and 4 tables) used in the Faculty of Engineering of the University of Porto (FEUP) and students' anthropometric dimensions. The sample consisted of 206 volunteer students (131 male, 75 female) with ages ranging from 18 to 35 years from different courses. Six anthropometric measures and 5 dimensions of school furniture were collected. For the evaluation of classroom furniture a match criterion was defined. Classroom furniture dimensions in each match criterion and student's anthropometric characteristics were compared. In this analysis, a significant mismatch between the dimensions of school furniture and students' anthropometric dimensions was found.

1 INTRODUCTION

The classroom is a learning environment in which the furniture is an important physical element. It is expected to facilitate learning and provide a comfortable and stress-free environment. Ergonomically designed classroom furniture provides students with an environment that potentiates their performance and learning process. On the other hand, inadequate school furniture may lead to the adoption of poor sitting postures in the classroom (Geldhof et al., 2007, Koskelo et al., 2007).

It is common that schools and universities choose fixed-type chairs and tables because of the higher price and maintenance involved with adjustable chairs (Straker et al., 2006). However, studying in fixed-type furniture may induce constrained postures (Parcells et al., 1999, Gouvali and Boudolos, 2006), which exposes students' body to static loads, usually seen as a risk factor for the development of musculoskeletal symptoms and discomfort (Cascioli et al., 2011).

Workstations with adjustable seats are preferred, given that people differ in size and postural preference, thus they have a significant positive effect on muscle tension and sitting posture, promoting health and comfort (Koskelo et al., 2007, Thariq et al., 2010). Moreover, there are evidences that the use of adjustable school furniture may have been related to better academic grades (Koskelo et al., 2007).

Students spend a considerable part of the day at school, and the majority of that time is spent in sitting position while doing their school work (Castellucci et al., 2010, Macedo et al., 2013).

Since the use of fixed-type furniture in schools is common and students spend most of their time sitting in school, school furniture should match student's requirements.

However, several studies show that often these requirements aren't taken into account when designing the furniture (Parcells et al., 1999, Corlett, 2006), and therefore mismatch occurs between the dimensions of school furniture and student's anthropometric measures (Castellucci et al., 2010, Gouvali and Boudolos, 2006, Panagiotopoulou et al., 2004, Parcells et al., 1999). The poor sitting posture in the classroom is one of the main negative effects on students of that mismatch (Dianat et al., 2013). Poor postures adopted by students are considered as one of the factors for increased risk of developing musculoskeletal disorders (Cardon et al., 2004).

The design of ergonomically correct side-mounted desktop chairs for university students has been neglected, and Thariq et al. (2010) study show that the side-mounted chairs, often used in university classrooms, aren't capable of satisfying the postural and comfort requirements of university students in their learning environment.

The aim of this study was to assess the level of mismatch between student's anthroppometric dimensions and 5 types of chairs and 3 types of tables available in the Faculty of Engineering of the University of Porto (FEUP).

2 METHODOLOGY

2.1 *Sample*

Anthropometric measurements were taken from students of FEUP in May and June of 2013. From a universe of 7295students, 131 men and 75 women (206 in total) were measured. The sample was mostly composed by students from the integrated master programs, but also from undergraduate and master programs. The ages varied from 18 to 35 years old. Students were selected by convenience sampling. They were invited to participate, when passing through the hall of the building where most of the classes are taken.

2.2 *Type of furniture*

FEUP school furniture is diverse and so 5 types of chairs were measured:

C1. Old design chair with flat perpendicular surfaces (seat and back support);
C2. New design chair with rounded front edge and concavities in the seat and back support. The angle between the seat and the back is higher than 90°;
C3. New design chair with a more eccentric design, a rounded front edge and concavities in the seat and back. The angle between the seat and the back is slightly larger than 90°;
C4. Chair with adjustable height, with wheels and footrest. Its design is alike the C2 chair;
C5. Side-mounted desktop chair with a folding seat. The side-mounted desktop is on the right side of the chair.

The measured tables were:

T1. Old design table with metal frame and wood table top;
T2. New design table with metal frame and wood table top;
T3. Adjustable table with new design metal frame and wood table top;
T4. Table from the chair type 5 with armrest and a small space to lay notebooks.

In FEUP classrooms, the following furniture combinations are used: (1) C1 with T1 (Comb1); (2) C2 with T2 (Comb2); (3) C3 with T2 (Comb3); (4) C4 with T3 (Comb4); (5) C5 with T4 (Comb5).

2.3 *Dimensions of chairs and desks*

The dimensions of the classroom furniture were taken by the same measurer with a metal tape measure.

The criteria, in Figure 1, used for the measurement of each dimension are defined as follows:

Seat Height (SH): vertical distance from the floor to the middle point of the front edge of the seat.

Seat Depth (SD): distance from the back to the front of the seat.

Seat width (SW): horizontal distance between the lateral edges of the seat.

Seat to Desk Clearance (SDC): distance from the top of the front edge of the seat to the lowest structure point below the desk.

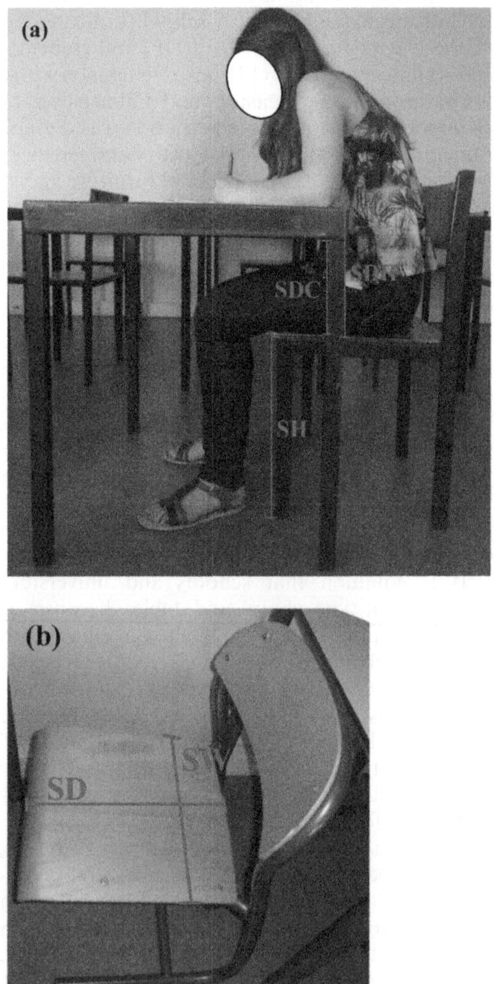

Figure 1. Representation of the classroom furniture measures: (a) lateral view and (b) top view.

Seat to Desk Height (SDH): vertical distance from the top of the middle of the seat to the top of front edge of the desk.

Leg Room at Knee Height (LRKH): horizontal distance from the back support to the opposite chair (this dimension was only considered in the lecture theatre due to chairs fixed-type).

Leg Room at Floor Level (LRFL): horizontal distance from the vertical alignment of the back support to the opposite chair (this dimension was only considered in the lecture theatre due to chairs fixed-type).

2.4 *Anthropometric dimensions*

A total of 14 static anthropometric dimensions were measured for each individual. Six dimensions were measured with the individual standing, while the remaining were obtained while the individual was seated. Although 14 anthropometric dimensions have been measured, in the present study only 6 of them were taken into account: Hip breadth (HB), Buttock–knee length (BKL), Buttock–popliteal length (BPL), Popliteal height (PH), Sitting elbow height (SEH), and Thigh thickness (TT).

The anthropometric measures were taken with the subject in a relaxed and erect posture. Each student was measured in thin cloths (T-shirt, shirt or thin sweatshirt), jeans, skirts or dresses. The standing dimensions were taken with the student standing erect to the anthropometer, and without shoes. The sitting dimensions were taken with the student seated erect onto the anthropometer, with knees bent 90°, and feet (without shoes) flat on the floor. The body dimensions were measured as described in ISO 7250-1:2008 (ISO, 2008).

2.5 *Procedures*

Applied anthropometry and ergonomic principles were considered to evaluate the classroom furniture. Equation(s) enabling the determination of each furniture dimension were used. These equations are based in two types of criteria: "one way" or "two way".

"One way" means that only a minimum or a maximum value is required. Two levels ("Match" and "Mismatch") were defined.

"Two way" criteria require the establishment of two limits, and 3 levels: A minimum limit (High mismatch) and a maximum limit (Low mismatch), and, in between the limits, the dimension is adequate (Match).

2.5.1 *Popliteal height and seat height mismatch*

Based on published literature (Castellucci et al., 2010, Dianat et al., 2013), the mismatch between popliteal height (PH) and seat height (SH) is defined by the equation 1:

$$(PH + 30)\cos 30° \leq SH \leq (PH + 30)\cos 5° \quad (1)$$

2.5.2 *Buttock-popliteal length and seat depth mismatch*

Based on existing studies (Castellucci et al., 2010, Dianat et al., 2013, Parcells et al., 1999), the mismatch between buttock-popliteal length (BPL) and seat depth (SD) is defined by the equation 2:

$$0.80 \times BPL \leq SD \leq 0.95 \times BPL \quad (2)$$

2.5.3 *Hip width and seat width mismatch*

Based on existing studies (Castellucci et al., 2010), the mismatch between hip width (HW) and seat width (SW) is defined by the equation 3:

$$HW < SW \quad (3)$$

2.5.4 *Thigh thickness and seat to desk clearance*

Based on published studies (Castellucci et al., 2010, Parcells et al., 1999), the mismatch between thigh thickness (TT) and seat to desk clearance (SDC) is defined by the equation 4:

$$TT + 20 < SDC \quad (4)$$

2.5.5 *Sitting elbow height and seat to desk height mismatch*

Based on recent studies (Castellucci et al., 2010, Dianat et al., 2013), the mismatch between sitting elbow height (SHE) and seat to desk height (SDH) is defined by the equation 5:

$$SEH \leq SDH \leq SEH + 50 \quad (5)$$

3 RESULTS AND DISCUSSION

3.1 *Anthropometric data*

Table 1 displays, for both genders, mean and standard deviation values for the 7 anthropometric dimensions measured.

Sample includes students from several lecture courses of FEUP, as well as curricular years. That is reflected on its range of ages (18 to 35 years old), although there is a predominance of younger students (18 to 24 years old). According to FEUP's statistics, the sample composition (age and female percentage) ensures the intended representativeness of the student population.

3.2 *Furniture dimensions*

Tables and seats measures are presented in Table 2.

3.3 *Mismatch between FEUP school furniture and students' anthropometric dimensions*

The mismatch between furniture measurements and students' anthropometric dimensions, by gender, is presented in Table 3 (male population) and Table 4 (female population).

For the seat height, a "High mismatch" was found for most students (92% in C3, 72% in C3) while only 1% of the students are included in the "Low mismatch" group. In the case of the "High mismatch" most of the students will not be able to support their feet on the floor, generating increased pressure on the soft tissues of the thighs (Gouvali and Boudolos, 2006). Even though chair C1 has a lower "High mismatch" (47%) it still has a low "Match" with only 51% of the students. Chairs C4 (98%) and C5 (88%) have high levels of Match with the male population. For the female population the case worsens larger "High mismatches" in chairs C1 (91%), C2 (95%) and C3 (92%). Chair C4 remains with a good "Match" for 95% of the female population while C5 "Match" lowers to 51%.

For the seat depth chair C4 has the lower "Match" with only 46% of the male population. All the other chairs have "Match" levels with more 70% of the population. With the female population only chairs C1 (84%), C3 (82%) and C5 (72%) have better "Match" with the hip breadth. C2 chair has a "High mismatch" with 69% and C4 with 61% of that population. The "High mismatch" may cause compression on the thighs and block the blood circulation causing discomfort (Gouvali and Boudolos, 2006) and also restrain the use of the back rest inducing kyphotic postures (Castellucci et al., 2010).

Concerning the seat width, all the chairs have high levels of "Match", once all of them are larger than 90%. For the female population only C1 chair has a lower "Match" with 88% of that population, unlike all the other chairs that have a "Match" with more than 90%.

Table 1. Mean (M) and Standard Deviation (SD) values, in mm, of the anthropometric dimensons by gender.

	Male (n = 131)		Female (n = 75)	
Dimensions	M	SD	M	SD
Buttock–knee length	615	31	580	37
Buttock–popliteal length	490	28	466	29
Hip breadth	347	26	346	33
Popliteal height	420	28	392	20
Sitting elbow height	245	24	238	25
Thigh thickness	190	23	171	28

Table 2. Furniture dimension, in mm.

Furniture	SH	SD	SW	SDC	SDH	LRKH	LRFL
C1	444	409	382	–	–	–	–
C2	473	390	400	–	–	–	–
C3	480	420	396	–	–	–	–
C4	396 to 522*	390	400	–	–	–	–
C5	418	428	410	–	–	–	–
Comb1	–	–	–	229	279	–	–
Comb2	–	–	–	243	294	–	–
Comb3	–	–	–	238	289	–	–
Comb4	–	–	–	<590*	290 to 402*	–	–
Comb5	–	–	–	224	202	787	959

* Values vary due to the adjustability of the furniture.

Table 3. Mismatch between the furniture and the students anthropometric dimensions.

	Male population										
Furniture		C1	C2	C3	C4	C5	Comb1	Comb2	Comb3	Comb4	Comb5
SH	High mismatch	47%	72%	92%	1%	5%	–	–	–	–	–
	Match	51%	27%	7%	98%	88%	–	–	–	–	–
	Low mismatch	2%	1%	1%	1%	7%	–	–	–	–	–
SD	High mismatch	4%	30%	5%	1%	7%	–	–	–	–	–
	Match	79%	70%	85%	46%	87%	–	–	–	–	–
	Low mismatch	17%	0%	10%	53%	6%	–	–	–	–	–
SW	Match	92%	100%	96%	100%	98%	–	–	–	–	–
	Mismatch	8%	0%	4%	0%	2%	–	–	–	–	–
SDC	Match	–	–	–	–	–	86%	94%	91%	100%	71%
	Mismatch	–	–	–	–	–	14%	6%	9%	0%	29%
SDH	High mismatch	–	–	–	–	–	5%	50%	38%	0%	0%
	Match	–	–	–	–	–	66%	48%	57%	100%	4%
	Low mismatch	–	–	–	–	–	29%	2%	5%	0%	96%

Table 4. Mismatch between the furniture and the students anthropometric dimensions.

Furniture	Male population	C1	C2	C3	C4	C5	Comb1	Comb2	Comb3	Comb4	Comb5
SH	High mismatch	91%	95%	99%	5%	48%	–	–	–	–	–
	Match	9%	5%	1%	95%	51%	–	–	–	–	–
	Low mismatch	0%	0%	0%	0%	1%	–	–	–	–	–
SD	High mismatch	11%	69%	16%	61%	27%	–	–	–	–	–
	Match	84%	31%	83%	39%	72%	–	–	–	–	–
	Low mismatch	5%	0%	1%	0%	1%	–	–	–	–	–
SW	Match	88%	100%	93%	100%	99%	–	–	–	–	–
	Mismatch	12%	0%	7%	0%	1%	–	–	–	–	–
SDC	Match	–	–	–	–	–	93%	97%	95%	100%	93%
	Mismatch	–	–	–	–	–	7%	3%	5%	0%	7%
SDH	High mismatch	–	–	–	–	–	29%	65%	53%	0%	0%
	Match	–	–	–	–	–	67%	34%	45%	100%	47%
	Low mismatch	–	–	–	–	–	4%	1%	2%	0%	53%

The seat to desk clearance is compatible with 86% of the population in Comb1, 71% in Comb 5 and with more than 90% in Comb2 to Comb4. All combinations are compatible with more than 90% of the female population. This situation of mismatch produces mobility restriction due to the contact of the thighs with the desk (Parcells et al., 1999).

As for the seat to desk height, Comb2 has a "High mismatch" with 50% of the male population and Comb3 with 38%. This requires them to work with shoulder flexion, causing muscle work load, discomfort and pain in the shoulder region (Parcells et al., 1999). There is a "Low mismatch" in Comb5 with 96% of the population and with 29% in Comb1. Only Comb4 has a perfect "Match" with 100% of the students. In the female population the larger "High mismatch" belongs to Comb2 (65%), Comb3 (53%) and Comb1 (29%). Comb 5 has a "Low mismatch" with 53% of the female population and Comb4 has an 100% "Match".

4 CONCLUSIONS

This study aimed to analyze the relationship between some anthropometric dimensions of 206 students from a Portuguese faculty and the dimensions of the school furniture.

The results show that none of the chairs are adequate for the user population. However, C4 chair, with its adjustability, has characteristics that allow it to better adjust to the students needs. With a seat with a lesser depth would be adequate to most of the students.

Seat Height and Seat to Desk Height, were the furniture dimensions with a higher level of mismatch, which may result in discomfort and pain on the posterior surface of the knee and shoulder.

The results of this study underline the fact that classroom furniture is usually acquired and selected without any previous ergonomics concern which results in its inadequacy.

ACKNOWLEDGMENT

The authors would like to thank Master in Occupational Safety and Hygiene Engineering (MESHO), of the Faculty of Engineering of the University of Porto (FEUP), all the support in the development and international dissemination of this work.

REFERENCES

Cardon, G., Clercq, D.D., Bourdeaudhuij, I.D. & Breithecker, D. 2004. Sitting habits in elementary schoolchildren: a traditional versus a "Moving school". *Patient Education and Counseling,* 54, 133–142.

Cascioli, V., Heusch, A.I. & Mccarthy, P.W. 2011. Does prolonged sitting with limited legroom affect the flexibility of a healthy subject and their perception of discomfort? *International Journal of Industrial Ergonomics,* 41, 471–480.

Castellucci, H.I., Arezes, P.M. & Viviani, C.A. 2010. Mismatch between classroom furniture and anthropometric measures in Chilean schools. *Applied Ergonomics,* 41, 563–568.

Corlett, E.N. 2006. Background to sitting at work: research-based requirements for the design of work seats. *Ergonomics,* 49, 1538–1546.

Dianat, I., Karimi, M.A., Hashemi, A.A. & Bahrampour, S. 2013. Classroom furniture and anthropometric characteristics of Iranian high school students: Proposed dimensions based on anthropometric data. *Applied Ergonomics,* 44, 101–108.

Geldhof, E., Clercq, D.D., Bourdeaudhuij, I.D. & Cardon, G. 2007. Classroom postures of 8–12 year old children. *Ergonomics,* 50, 1571–1581.

Gouvali, M.K. & Boudolos, K. 2006. Match between school furniture dimensions and children's anthropometry. *Applied Ergonomics,* 37, 765–773.

Iso 2008. ISO 7250-1:2008 Basic human body measurements for technological design—Part 1: Body measurement definitions and landmarks. International Organization for Standardization, Geneva, Switzerland.

Koskelo, R., Vuorikari, K. & Hänninen, O. 2007. Sitting and standing postures are corrected by adjustable furniture with lowered muscle tension in high-school students. *Ergonomics,* 50, 1643–1656.

Macedo, A., Martins, H., Santos, J., Morais, A., Brito, A. & Mayan, O. 2013. Conformity between Classroom Furniture and Portuguese Students' Anthropometry: A Case Study. *International Symposium on Occupational Safety and Hygiene.* Guimarães - Portugal.

Panagiotopoulou, G., Christoulas, K., Papanckolaou, A. & Mandroukas, K. 2004. Classroom furniture dimensions and anthropometric measures in primary school. *Applied Ergonomics,* 35, 121–128.

Parcells, C., Stommel, M. & Hubbard, R.P. 1999. Mismatch of classroom furniture and student body dimensions: Empirical findings and health implications. *The Journal of adolescent health: official publication of the Society for Adolescent Medicine,* 24, 265–273.

Straker, L., Pollock, C. & Burgess-Limerick, R. 2006. Excerpts from CybErg 2005 discussion on preliminary guidelines for wise use of computers by children. *International Journal of Industrial Ergonomics,* 36, 1089–1095.

Thariq, M.G.M., Munasinghe, H.P. & Abeysekara, J.D. 2010. Designing chairs with mounted desktop for university students: Ergonomics and comfort. *International Journal of Industrial Ergonomics,* 40, 8–18.

Occupational Safety and Hygiene II – Arezes et al. (eds)
© 2014 Taylor & Francis Group, London, ISBN 978-1-138-00144-2

Modern machines: How to set up the health of the pilot?

E.T. Martins & L.B. Martins
Universidade Federal de Pernambuco Recife, Pernambuco, Brasil

I.T. Martins
Universidade Estácio de Sá, Rio de Janeiro, Brasil

ABSTRACT: The flaws in the commitment of decision-making in emergency situations and the lack of perception related to all elements associated with a given situation in a short space of time indicate, often, lack of situational awareness. Automation always surprises the crews and often prevents them from understanding the extent of this technology that is very common in aircraft units with a high degree of automation. These facts are discussed in a subtle way by pilots who can not do it openly, as it might create an impression of professional self-worthlessness (self-deprecation). This leads to common questions like: What is happening now? What will be the next step of automated systems? This type of doubt would be inadmissible in older aircraft because the pilot of those machines works as an extension of the plane. This scenario contributes to emotional disorders and a growing hidden problem in the aeronautical field.

1 INTRODUCTION

The emotional stability and physical health of workers on board aircraft are faced with the factors and conditions that enable professionals to carry out their activities and develop normally, The emotional stability and physical health of workers on board aircraft are faced with the factors and conditions that enable professionals carrying out their activities and develop normally, despite the fact that these conditions can occur in adverse format (Eugenio, 2011). The modern history of aviation with its great technological complexity is showing pilots as redundant components that integrate embedded controls in modern aircraft. This leads us to say that the value of this class of worker as a permanent social group in society does not receive, currently, the proper priority and the respect need. In research on the health of the pilot, there are three major perspectives that have been investigated that influence his stability, as well as the mental and emotional development of the modern airline pilot (Henriqson, 2010): The previous life of the individual directly tied to experience, age, genetic and physiological vectors, the social/cultural environment and the formal education. This has changed the actions of pilots, especially in relation to emergency procedures. There are few studies that correlate the reduction of accidents with the cognitive and technological changes. The increased cognitive load relates to these changes and requires assessment. The benefits presented by new technologies do not erase the mental models built, with hard work, during times of initial training of the aircraft career pilots in flying schools.

The anguish of the public to receive information in the investigation of an accident in aviation must be resolved. In search of who or what to blame, the pilot is guilty and immediately appointed as one of the underlying factors involved. This is the real evidence of some fact are neglected. The reading of the Black-Boxes of the accidented aircrafts are interpreted like 70% to 80% of accidents happen due to human error, or to a string of failures that were related to the human factor (FAA, 2010). This must be a deep error. Like some of the origins of the real errors we can mention stress and the failure to fully understand the new procedures related to technological innovations linked to automation. Complex automation interfaces always promote a wide difference in philosophy and procedures for implementation of these types of aircraft, including aircraft that has different controls instruments even manufactured by the same manufacturer. In this case, we frequently can identify inadequate training that contributes to the difficulty in understanding procedures by the crews. Accident investigations concluded that the ideal would be to include, in the pilot training, a psychological stage, giving to him the opportunity of self-knowledge, identifying possible "psychological breakdowns" that his biological machine can present that endangers the safety of flight. Would be given, thus, more humane and scientific support to the crew and to everyone else involved with the aerial activity, minimizing factors that cause incidents and accidents. Accident investigators concluded that the ideal situation for pilot training should include a psychological phase (Dekker, 2003), giving him or her, the opportunity of self-knowledge, identifying possible "psychological breakdowns" that biological features can present and can endanger the safety of flight. More human and scientific support should be given to the crew and everyone else involved with the aerial activity, reducing factors that can cause incidents and accidents. Accidents do not just happen. They have complex causes that can take days, weeks or even years to develop (Reason, 1990). However, when lack of attention and/or neglect take place resulting in a crash, we can be most certain there was a series of interactions between the user and the system that created the conditions for that to happen (Rasmussen, 1982). We understand that human variability and system failures are an integral part of the main sources of human error, causing incidents and accidents. The great human effort required to manage and performing actions with the interface as the task of monitoring, the precision in the application of command and maintaining a permanent mental model consistent with the innovations in automation will make it vulnerable to many human situations where errors can occur. The human variability at aviation is a possible component of human error and we can see the consequences of these errors leading to serious damage to aircraft and people. It is not easy, in the new aviation, to convey the ability to read the instruments displays

This can conduct to the deficiency and the misunderstanding in monitoring and performing control tasks: lack of motivation, the fact that it is stressful and tiring, and generate failures in control (scope, format and activation), poor training and instructions that are incomplete, wrong or ambiguous. The mind of the pilot is influenced by cognition and communication components during flight, especially if we observe all information processed. They are very critical considering that the pilot is constantly getting this information through the cockpit instruments. There are a permanent flux of informations about altitude, speed and position of aircraft and the operation of its hydraulic power systems during the flight. If any problem occurs, several lights will light up and warning sounds emerging the increase of the volume and type of man-machine communication which can diminish the perception of detail in information that must be processed and administered by the pilot. All this information must be processed by one's brain at the same time as it decides the necessary action in a context of very limited time. There is a limit of information that the brain can deal with which is part of natural human limitation. It can lead to the unusual situation in which, although the mind is operating normally, the volume of data makes it operate in overload, which may lead to failures and mistakes if we consider this man as a biological machine. All situations in which a planned sequence of mental or physical activities fails to achieve its desired outputs are considered as errors (Reason, 2008). Thus, it is necessary that steps be taken toward reducing the likelihood of occurrence of situations which could cause a problem. The flight safety depends on a significant amount of interpretations made by the pilot in the specific conditions in every moment of the flight. Accidents do not only occur due to pilot error, but also as a result of a poor design of the transmission of information from the external environment, equipment, their instruments, their signs, sounds and different messages. In these considerations, the human agent will always be subject to fatality, which is a factor that can not be neglected. Figure 1 shows the human-machine interaction where difficulties with cognitive and operational perspectives needs and also physical and emotional aspects take place in a human being during the occurrence of system-level of flight. It is difficult to convince, in a generic way, people with merely causal explanations due to human complexity. Further analysis of the problem will always end with the identification of a human error, which was probably originated in the design phase, at the manufacturing stage, or given simply as a result of an "act of God".

Aeronautical activities, designing human-machine systems becomes very necessary to characterize and classify human error. Human activities have always been confronted with the cognitive system.

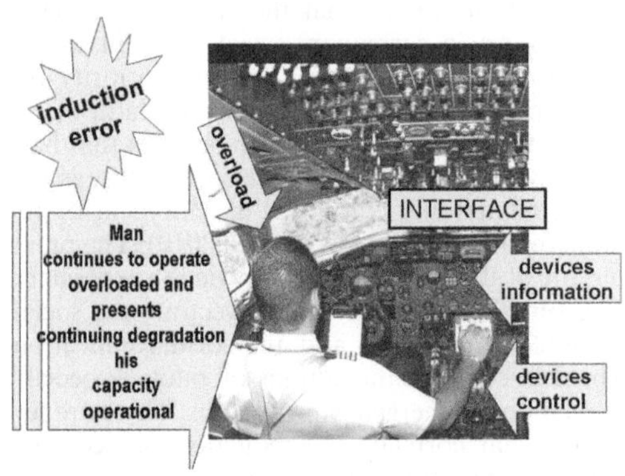

Figure 1. Diagram of the interaction between man and machine.

2 SCENARIO

On the result of the causality of accidents, we must consider the human contribution to accidents, distinguishing between active failures and latent failures due to the immediate adverse effect of the system aspect. The main feature of this component is that it is present within the process of construction of an accident long before declaring the event like an accident, being introduced by higher hierarchical levels as designers, responsible for maintenance and personnel management. We can always guarantee, with respect to organizational accidents, that the layers of defenses, that are the protective barriers, were constructed to prevent the occurrence of natural or man-made disasters. This statement is derived from the design philosophy that treats the defense in depth.

3 ACTIVITIES PURSUED BY THE PILOT AND THE INCREASED COGNITIVE LOAD

The following factors are an integral part of cognitive activity in the pilot: fatigue, body rhythm and rest, sleep and its disorders, the circadian cycle and its changes, the G-force and acceleration of gravity, the physiological demands in high-altitude, night-time take-offs and the problem of false illusion of climbing. But, other physiological demands are placed by the aviators. It is suggested that specific studies must be made for each type of aircraft and workplace, with the aim of contributing to the reduction of incidents arising from causes so predictable, yet so little studied. We must also give priority to airmen scientists that have produced these studies in physiology and occupational medicine, since the literature is scarce about indicating the need for further work in this direction. Human cognition refers to mental processes involved in thinking and their use. It is a multidisciplinary area of interest includes cognitive psychology, psychobiology, philosophy, anthropology, linguistics and artificial intelligence as a means to better understand how people perceive, learn, remember and how people think, because will lead to a much broader understanding of human behavior. Cognition is not presented as an isolated entity, being composed of a number of other components, such as mental imagery, attention, consciousness, perception, memory, language, problem solving, creativity, decision making, reasoning, cognitive changes during development throughout life, human intelligence, artificial intelligence and various other aspects of human thought (Sternberg, 2000).

The procedures of flying an aircraft involve observation and reaction to events that take place inside the cabin of flight and the environment outside the aircraft (Green et al, 1993). The pilot is required to use information that is perceived in order to take decisions and actions to ensure the safe path of the aircraft all the time. Thus, full use of the cognitive processes becomes dominant so that a pilot can achieve full success with the task of flying the "heavier than air." With the advent of automated inclusion of artifacts in the cabin of flight that assist the pilot in charge of controlling the aircraft, provide a great load of information that must be processed in a very short space of time, when we consider the rapidity with which changes occur, an approach that cover the human being as an individual is strongly need. Rather, the approach should include their cognition in relation to all these artifacts and other workers who share that workspace (FAA, 2005).

4 THE DEPLOYMENT OF THE TASKS LEADING TO ACCIDENTS

A strong component that creates stress and fatigue of pilots, referred to the design of protection, detection and effective handling of fire coming from electrical short circuit on board, is sometimes encountered as tragically happened on the Swissair Airlines flight 111, near Nova Scotia on September 2, 1998. The staff of the Federal Aviation Administration (FAA), responsible for human factors research and modern automated interfaces (FAA, 2005), reports a situation exacerbated by the widespread use an electrical product and a potentially dangerous wire on aircrafts, called "Kapton". If a person has to deal with an outbreak of fire, coming from an electrical source at home, the first thing he would do is disconnect the electrical power switch for the fuses. But this option is not available on aircraft like the Boeing B777 and new Airbus. The aviation industry is not adequately addressing the problem of electrical fire in flight and is trying to deal recklessly (FAA, 2005). The high rate of procedural error associated with cognitive errors, in the automation age, suggests that the projects in aviation have ergonomic flaws. In addiction, is has been related that the current generation of jet transport aircraft, used on airlines, like the Airbus A320, A330, A340, Boeing B777, MD11 and the new A380, that are virtually "not flyable" without electricity. We can mention an older generation, such as the Douglas DC9 and the Boeing 737. Another factor in pushing the pilots that causes emotional fatigue and stress is the reduction of the cockpit crew to just two. The next generation of large transport planes four engines (600 passengers) shows a relatively complex operation and has only two humans in the cockpit.

The flight operation is performed by these two pilots, including emergency procedures, which should be monitored or re-checked. This is only possible in a three-crew cockpit or cockpit of a very simple operation. According to the FAA, the only cockpit with two pilots that meets these criteria is the cabin of the old DC9-30 and the MD11 series. The current generation of aircraft from Boeing and Airbus do not fit these criteria, particularly with respect to engine fire during the flight and in-flight electrical fire.

The science of combining humans with machines requires close attention to the interfaces that will put these components (human-machine) working properly. The deep study of humans shows their ability to instinctively assess and treat a situation in a dynamic scenario. A good ergonomic design project recognizes that humans are fallible and not very suitable for monitoring tasks. A properly designed machine (such as a computer) can be excellent in monitoring tasks. This work of monitoring and the increasing the amount of information invariably creates a cognitive and emotional overload and can result in fatigue and stress. According to a group of ergonomic studies from FAA (FAA, 2005) in the United States this scenario is hardly considered by the management of aviation companies and, more seriously the manufacturers, gradually, introduce further informations on the displays of Glass cockpits. These new projects always determine some physiological, emotional and cognitive impact on the pilots.

5 CONCLUSION

The unexpected automation surprises are reflected in a complete misunderstanding or even the misinformation of the users. It also reveals their inability and limitations to overcome these new situations that were not foreseen by the aircraft designers. Our studies showed a different scenario when the accident is correlated with systemic variables. It has identified the problems or errors that contribute to the fact that the crew is unable to act properly. These vectors, when are in conjunction, may generate, eventually, a temporary incompetence of the pilot due to limited capacity or lack of training in the appropriateness of automation in aircraft. We developed a study focusing on the guilt of pilots in accidents when preparing our thesis. In fact, the official records of aircraft accidents blame the participation of the pilots like a large contributive factor in these events. As a result of our research for thesis for Ph.D., found that the work-component presses 50.84% of accidents where there is involvement of the pilot somehow. Also the health vector is present with 49.03% in this type of occurrence. Participation in the pilot accidents and incidents (directly, unjustified and justifiably so) added to the vector health occurs at 88.89%, which reinforces the hypothesis that the pressure work promotes changes in performance of the work. The intensities of the vectors work/health and stress/fatigue go hand in all kinds of involvement in accidents where the pilot conjecture that the pressure vector work has strong influence throughout the system. Using our criteria of contribution of pilots in accidents participation justified and not contribution is 68%. That leaves 32% of accidents with possible guilt or liability of pilots. This scenario is radically different from the traditional world-level where is charged over 80% of direct or indirect participation of pilots in accidents.

Modifying this scenario is very difficult in the short term, but we can see as the results of our study, which the root causes of human participation, the possibility of changing this situation. The cognitive factor has high participation in the origins of the problems (42% of all accidents found on our search). If we consider other factors, such as lack of usability applied to the ergonomics products, choise of inappropriate materials and poor design, for example, this percentage is even higher.

Time is a factor to consider. This generates a substantial change in the statistical findings of contributive factors and culpability on accidents. The last consideration on this process, as relevant and true, somewhat later, must be visible solutions. In aviation, these processes came very slowly, because everything is wildly tested and involves many people and institutions. The criteria adopted by the official organizations responsible for investigation in aviation accidents do not provide alternatives that allow a clearer view of the problems that are consequence of cognitive or other problems that have originate from ergonomic factors. We must also consider that some of these criteria cause the possibility of bringing impotence of the pilot to act on certain circumstances. The immediate result is a streamlining of the culpability in the accident that invariably falls on the human factor as a single cause or a contributing factor. Many errors are classified as only "pilot incapacitation" or "navigational error". Our research shows that there is a misunderstanding and a need to distinguish disability and pilot incapacitation (because of inadequate training) or even navigational error.

Our thesis has produced a comprehensive list of accidents and a database that allows extracting the ergonomic, systemic and emotional factors that contribute to aircraft accidents. These records do not correlate nor fall into stereotypes or patterns. These patterns are structured by the system itself as the accident records are being deployed.

We developed a computer system to build a way for managing a database called the Aviation Accident Database. The data collected for implementing the database were from the main international entities for registration and prevention of aircraft accidents as the NTSB (USA), CAA (Canada), ZAA (New Zealand) and CENIPA (Brazil). This system analyses each accident and determines the direction and the convergence of its group focused, instantly deployed according to their characteristics, assigning it as a default, if the conditions already exist prior to grouping. Otherwise, the system starts formatting a new profile of an accident (MARTINS, 2007).

This feature allows the system to determine a second type of group, reporting details of the accident, which could help point to evidence of origin of the errors. Especially for those accidents that have relation with a cognitive vector. Our study showed different scenarios when the accidents are correlated with multiple variables. This possibility, of course, is due to the ability of Aviation DataBase system, which allows the referred type of analysis. It is necessary to identify accurately the problems or errors that contribute to the pilots making it impossible to act properly. These problems could point, eventually, to an temporary incompetence of the pilot due to limited capacity or lack of training appropriateness of automation in aircraft. We must also consider many other reasons that can alleviate the effective participation or culpability of the pilot. Addressing these problems to a systemic view expands the frontiers of research and prevention of aircraft accidents.

This system has the purpose of correlating a large number of variables. In this case, the data collected converges to the casualties of accidents involving aircraft, and so, can greatly aid the realization of scientific cognitive studies or applications on training aviation schools or even in aviation companies (MARTINS, 2010). This large database could be used in the prevention of aircraft accidents allowing reaching other conclusions that would result in equally important ways to improve air safety and save lives.

REFERENCES

Automação no cockpit das aeronaves: um precioso auxílio à operação aérea ou um fator de aumento de complexidade no ambiente profissional dos pilotos, 1rd ed. vol. 1. São Paulo: Brasil, 2011, pp. 34–35.

Dekker, S. "Illusions of explanation- A critical essay on error classification," The International Journal of Aviation Psychology, New Jersey, vol. 13, pp. 95–106, Sept. 2003.

FAA—Federal Aviation Administration, "DOT/FAA/AM-10/13, Office of Aerospace Medicine, Causes of General Aviation Accidents and Incidents: Analysis Using NASA Aviation," Safety Reporting System Data, Washington DC, press, Sept. 2010, 2011, 2012.

Green, R.G. & Frenbard M. Human Factors for Pilots. Avebury Technical. Aldershot, England, (1993).

Henriqson, E. A coordenação como um fenômeno cognitivo distribuído e situado em *cockpits* de aeronaves- Coordination as a distributed cognitive phenomena situated in aircraft cockpits Aviation in Focus (Porto Alegre), v. 1, n. 1, p. 58–76 (ago/dez. 2010).

Martins, Edgard "Ergonomics in Aviation: A critical study of the causal responsibility of pilots in accidents," "Ergonomia na Aviação: Um estudo crítico da responsabilidade dos pilotos na causalidade dos acidentes," Msc. Monography, Universidade Federal Pernambuco, Pernambuco: Brasil, Mar. 2007, pp. 285–298.

Martins, Edgard "Study of the implications for health and work in the operationalization and the aeronaut embedded in modern aircraft in the man-machines interactive process complex," "Estudo das implicações na saúde e na operacionalização e no trabalho do aeronauta embarcado em modernas aeronaves no processo interativo homem-máquinas complexas," thesis, Centro de Pesquisas Aggeu Magalhães, Fundação Osw Cruz, Perna:Brasil, Aug. 2010, pp. 567–612.

Rasmussen, J. Human errors: A taxonomy for describing human malfunction in industrial istallations. Journal of Occupational Accidents, v. 4, p. 311–333, (1982).

Reason, J. Human Error, Cambridge, Cambridge. University Press, (1990, 2012).

Sternberg, R.J. Cognitive psychology. Porto Alegre: Ed Artmed. (2000).

Occupational Safety and Hygiene II – Arezes et al. (eds)
 ISBN 978-1-138-00144-2

Chemical agents exposure prevention: A training effectiveness approach

Ana P. Soares, António Pinheiro, António Parente & Miguel Tato Diogo
Faculty of Engineering of University of Porto, Porto, Portugal

ABSTRACT: The main objective of this study is to identify the level of improvement of workers knowledge level before and after a training action. To assess this it is necessary to make a study, through questionnaires in two different occasions: before and after the training event, respectively. From the results obtained, it is possible to draw quantitative conclusions regarding the study of the binomial performance/effectiveness of training. From obtained results it can be concluded that workers exposed to chemical agents are now able to identify potential risk situations in the development of their tasks. Furthermore, after the training the workers showed to have more knowledge about the different signage (Warning, Fire, etc.) and the risks associated with chemicals use, allowing them to identify potentially danger situations.

1 INTRODUCTION

The use of chemicals is common in almost any sector of activity. From maintenance activities, cleaning, or integrating parts of the production chain, chemicals are everywhere. Since the use of chemicals is more frequent it presents more risks to the workers' health. The risks associated with chemical uses can be responsible for serious damages to human health like poisoning, suffocation or burns.

According to the third European survey on working conditions (2000), by the European Foundation for the Improvement of Living and Working Conditions, "*16% of Europe's workers are in contact with hazardous substances and 22% being exposed to toxic fumes. The exposure to hazardous substances can occur at any time in the workplace.*" (Trabalho A. E., Introdução às substâncias perigosas no local de trabalho—Facts 33, 2003).

The International Labour Organisation (ILO) estimates 35 million annual worldwide cases of work-related diseases by exposure to chemicals with the occurrence of 439,000 deaths including, among other related causes, 36,000 deaths due to pneumoconiosis, 35,500 deaths from respiratory chronic diseases, 30,700 cardiovascular deaths and 315,000 deaths from cancer (Office, 2004).

Health and Safety at Work is a process of continuous improvement. The training and information thus needs to be constantly updated and made available to employees in order to comply with safety standards and prevent occupational hazards.

The human resources training has an important role in the reduction of accidents, since workers will know the risks to which they are exposed and the precautions they need to have in their duty.

The employee is entitled each year to 35 hours of training, which may be given by the employer, for training organization certified or recognized educational establishment (art. 131., No. 2 and 3 of the current Labour Code, approved by Law no. 7/2009, of February 12, and amended by Law n.° 47/2012 of 29 August, making the fourth amendment to the Portuguese Labour Code).

According to the OECD (Organization for Economic Cooperation and Development), the administration has the responsibility to ensure that all the staff are trained and able to use their competences for the safe performance of their work and are competent to deal with emergencies (OECD, 2003).

Some authors claim that workers trained and with cognitive capacity are better able to respond to certain situations and tensions arising from their tasks, as well as feeling more motivated to perform their job.

According to Quelhas (2009), "*the measures to prevent accidents depend directly on the type of economic activity exercised, work environment, technology and techniques used as well as the adoption of a practical education and information among workers*".

The present study aims to demonstrate the importance of training related to the chemicals used in automotive components industry, by measuring the level of improvement of competence for workers who attend the session. Training arises, in this context, as a need to prevent accidents and to increase awareness for the use of Personal Protective Equipment (PPE) in the factory sections where the exposure to these substances is high and where the study was made.

2 TRAINING AS A PREVENTIVE MEASURE

"*The training refers to planned efforts to facilitate the learning of a specific competence*" (Robson, et al., 2010).

According to a study in Thailand which include small and medium enterprises (Kongtip, 2007), it was concluded that health and safety training was considered essential to prevent accidents and occupational diseases. 85% of these companies have written safety policy and safety rules and 87% of companies gave the workers information about the policy and safety rules. When accidents occurring more than 90% of companies were reporting and investigating them, but only 75.4% used the statistical analysis (Kongtip, 2007).

A study of working accidents in Portugal in 1998, based on surveys in the enterprises, concluded that only 31.5% of the victims reported claimed have been training in Safety and Health at Work. A total of 62.8% of victims claimed not having any type of training within these areas (Departamento de Estatística do Trabalho, 1998).

Soares (2007) suggested that training should be part of the safety culture of the company. "*One of the strategies to be adopted to create a culture of health and safety requires education, training and information to all stakeholders, promoting the acquisition of values, responsibility and commitment in this area, in which the health and safety technician plays a central role and coordination*" (Soares, 2007, p. 1043).

It is important to talk about effectiveness in training, once the present study is focused on the effectiveness of Safety and Health training. Efficacy "[...] *is the institutional criteria that reveal the administrative capacity to achieve the goals and proposed outcomes*" (Sander, 1995).

Adapted to training, efficacy may be associated with competence acquisition, changes in behavior and practice of this new knowledge in the workplace. Some studies (e.g. Becker & Morawetz 2004, Executive, 2003) have shown that training can induce protective behaviors among users of hazardous substances, when the training is effective.

In fact, the workers only learn and training is only effective if the worker has acquired the information transmitted during the training. "*After the conclusion of training, the employee cannot effectively transfer what they have learned, unless he has the ability to retain the content of training*" (Bhatti, Battour, Sundram, & Othman, 2013, p. 281).

There are studies (e.g. Robson, et al., 2010) that demonstrate that attitudes and behaviors are significantly associated with the prevention of accidents.

2.1 *Perspective before and after training*

A study pertinent to this field was made by Becker & Morawetz (2004). This study was made through questionnaires distributed to 55 employees before and 14 to 18 months after the chemical training. It was revealed that after the session, workers shown an increase in the use of resources, the success rates increased for those who tried to change and there was an overall success in making improvements. The conclusion of the study shows that "*workers are more willing and trying to change the conditions of the workplace after training and its effectiveness in achieving change is substantially higher than before they were trained. The study confirms previous work and reinforces these conclusions statistically tested using comparisons of impact measures pre- and post-training*" (Becker & Morawetz, 2004).

Another study involving the comparison of 22 different studies in the literature review concluded that there is a strong evidence in the behavior of workers concerning Occupational Safety and Health (OSH) issues when the training is effective (Robson, et al., 2010). "*The training is widely recognized as an important component of the occupational risk management. However, the expense and effort required to conduct such training requires further research on factors that make training effective*" (Robson, et al., 2010, p. 1).

Lingard (2002), a researcher from Australia has developed a 24-week study that aimed to assess how the first-aid training affects the motivation of employees of a small construction industry, in preventing occupational accidents and diseases; as well the training effect on health and safety behavior by means of interviews made in the workplaces. The motivation of the participants to control the risks of health and safety at work was explored during interviews before and after the action of first-aid training. The measurement of change in the workers behavior was made by observation by a researcher directly at the workplace also before and after the training.

The results suggest that "*first-aid training had a positive effect on the behavior of participants regarding safety and occupational health. Through training, employees became more aware that their own behavior is an important factor in the prevention of working accidents and occupational diseases*" (Lingard, 2002, p. 209).

3 MATERIALS AND METHODS

3.1 *Methodology*

This study was conducted according to the diagram in figure 1: setting goals, literature review, organization and implementation of a training program within the company, questionnaires in the same training, analysis of results and discussion and conclusion.

The present study was based on the collection of information in a manufacturing company, by means of training and surveys.

The fieldwork was carried out in a real context, with the application of questionnaires before and after the training session. This training session was devoted to the chemical risks, within a company that produce automotive components. The training session lasted 60 minutes, divided in 40 minutes for the contents exhibition, 10 minutes to answer the questionnaires and 10 minutes to display a movie concerning the correct use of chemical agents.

The questionnaires were designed to obtain direct and accurate answers related to knowledge of the trainees about the safety and handling of chemicals, including the signaling. With their answers the goal was to make a comparison with the competence level before and after the training session to better asses if there was an improvement with the aforementioned training session.

The presented surveys had 8 groups relating to different themes such as safety signs, symbology of chemical agents, handling and storage of chemicals, emergency: situations of accidents/spills, first-aid, risk for the human body, safety data sheets and personal protective equipment.

3.2 *Company description*

The work was carried in a Portuguese manufacturing enterprise, dedicated to the manufacture of components for the automotive industry, namely upholstery. Aiming for a greater market share, the company has diversified its product range producing complete seats and upholstery (covers) for automobiles, seats for buses, boats and trains, child seats, metal parts for car seats, mechanisms for car seats and molded foam for car seats.

The work has focused on the production of banks, particularly in foams, glues and paint, and the production of metal structures (painting), where the handling and storage of chemicals, which are considered dangerous substances, are present in the production process. Polyurethane foams are widely used in automotive interior components: seats, headrests, armrest, etc.

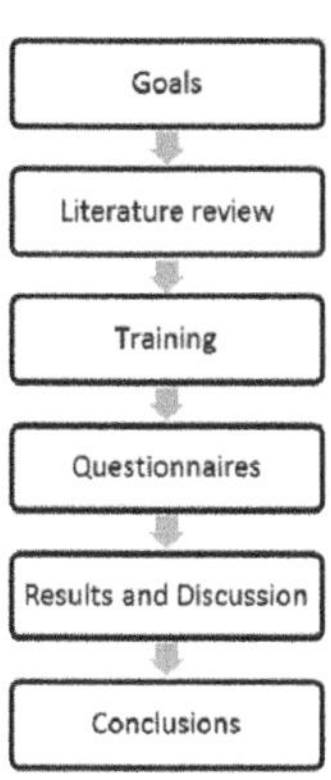

Figure 1. Methodology used in the study.

3.3 *Materials*

3.3.1 *Training "handling and storage of chemicals"*

The contents covered in this training were provided by multimedia support and were developed so the trainees could obtain the largest amount of information concerning chemical hazards due to handling and storage, as well to facilitate their response to the after the training action questionnaire.

There were 33 trainees attending this training. The sessions were distributed over 4 groups of workers in different days and hours, because of the shifts. The workers have two shifts from 8:00 am to 4:00 pm and from 4:00 pm to 12:00 pm so there was a need to include the sessions in these times.

The attending workers are part of the sections Glues, Paint, Foams and Metal, which are the sections where more chemicals are used.

3.2.2 *Questionnaires*

According to the established training plan, surveys with the same questions were proposed and delivered to the trainees before and after the training session. These were exactly the same on both occasions, since the goal was to understand the level of knowledge and training effectiveness.

The questionnaires helped to realize the importance of training in preventing chemical accidents. If workers receive more training they will be more capable to understand a particular risk, prevent accidents or minimize the chemicals effects on the human body through the proper use of personal protective equipment. Also through the questionnaires it was possible to assess what was the workers knowledge level prior to training.

4 RESULTS AND DISCUSSION

The results were based on a sample of 33 workers, aged 28 to 61 years old with a mean age of 42.12 (standard deviation 7.45) and mode 40. The median indicates that at least 50% of respondents have age not exceeding 42 years.

Two trainees are temporary workers, however were also called since they are also exposed to chemical agents and their training is also relevant.

Of the participants surveyed 36% (n = 12) are male and the remaining percentage of 64% (n = 21), are females. These data shows a sample with a higher percentage of women, since the jobs in glue and paint sections are mainly occupied mostly by women. Due to the physical effort required for the placement of molds, foam injection and mold release these tasks are performed by men.

This study has shown a sample of 33 workers, since it represents the number of trainees who attended the training, but it can be extended to larger actions.

The results obtained in the survey's multiple-choice questions are presented in Table 1.

Regarding the correct answers for the 1st group of the questionnaire (Safety Signs), it can be seen in Figure 2, that only 10 trainees answered correctly to the "Fire" signal before the training (31%, n = 10), however, after the training 30 workers answered correctly (91%, n = 30) as shown in Figure 3. This answer was the one that stood out from the rest.

As illustrated in the graph of Figure 2, before the training, the trainees responses were very varied and registering some absenteeism (21%, n = 7). The most frequent response, though wrong, was related with the warning signs (27%, n = 9).

After the training and for the same signal (fire), the difference in the answers is visible and there has been no absenteeism (Figure 3). There were a total of 30 correct responses in 33 (91%). The most frequent response, though wrong, was related to emergency signal (6%, n = 2).

94% of trainees answered correctly about the signal "Obligation" before, while 97% answered correctly after.

Regarding the correct answers for the 2nd group of the questionnaire (Symbology of Chemical Agents—pictograms), only 11 trainees responded correctly to the question about the "Oxidizing" symbol before training (33%, n = 11) while 22 answered correctly after training (67%, n = 22). Comparing now the answers for the two moments (before and after training), regarding the symbol "Corrosive", the graph in Figure 4 depicts the improvement after the training and its respective percentage. To 58% (n = 19) of correct answers before training, there was a total of 82% (n = 27) correct answers after training.

Some results did not shown a significant improvement and had a similar number of responses before and after training. This was the case of "Emergency" signaling: before training 31 correct answers and 33 corrects after the training.

Table 1. Results obtained for the surveys made before and after the training sessions. Frequency represents the amount of correct answers.

	Before		After	
	Frequency	%	Frequency	%
Safety signs				
Warning	27	82	28	85
Prohibition	28	85	30	91
Emergency	31	94	33	100
Fire	10	30	30	91
Obligation	31	94	32	97
Symbology of chemical agents				
Toxic	22	67	24	73
Explosive	24	73	28	85
Highly Flammable	26	79	27	82
Threat to the Environment	22	67	28	85
Corrosive	19	58	27	82
Irritant or Harmful	18	55	27	82
Oxidising	11	33	22	67
Personal protective equipment				
Hand protection	30	91	33	100
Respiratory protection	24	73	27	82
Body protection	31	94	33	100
Face protection	32	97	32	97
Eye protection	32	97	33	100
Feet protection	32	97	33	100

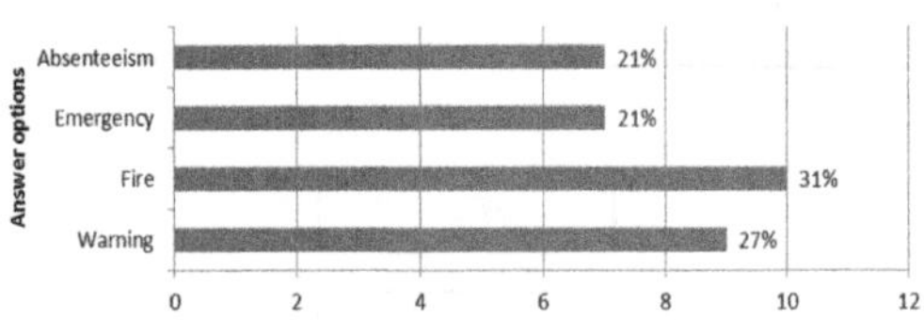

Figure 2. Results of answers about fire signal: before.

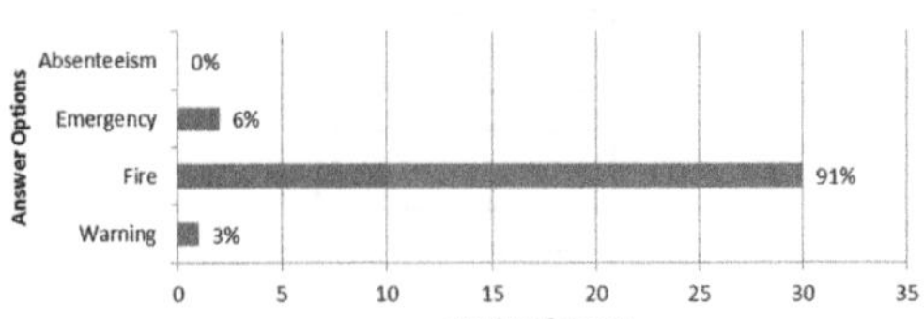

Figure 3. Results of answers about fire signal: after.

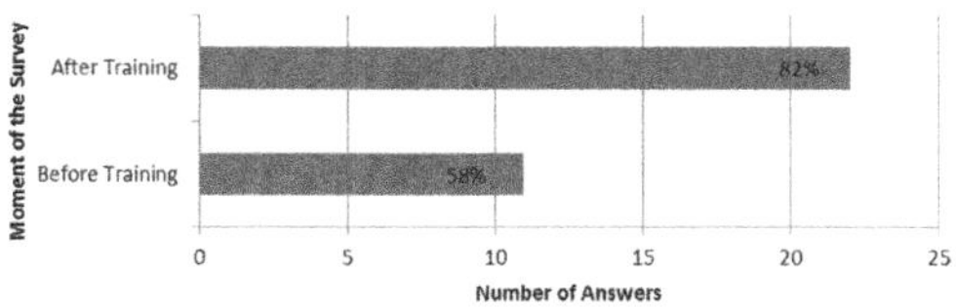

Figure 4. Comparison of responses regarding the corrosive symbol.

Regarding the 3rd group (Handling and storage of chemicals), 76% (n = 25) of the workers said that they know the risks of the chemicals they use before the training and 91% (n = 30) after the action.

In the 6th group (Risk for the human body), the obtained data also revealed that before the training 42% (n = 14) of trainees said they knew what were the ways of contact with chemical agents, whereas after training a percentage of 91% (n = 30) was observed.

5 CONCLUSIONS

From the present study it is possible to conclude that training and information are, as preventive measures, important tools that contribute to the workers knowledge and may help reducing the risk of work related incidents, accidents, and illnesses. Indeed, the proposed training was effective, due to its knowledge improvement.

The results have shown a significant improvement of the acquired competences by the trainees. This competence increase was measured between 30% and 100%, representing, in the second survey (after the training), all of the correct answers.

The obtained data revealed that several workers had some misconceptions regarding a few themes. The results also suggest that the workers are more familiarized with "Emergency" signage than "Fire" signage. Regarding the "Fire" signage it was observed a tendency to associate it with the "Warning" signage. The results also revealed that before the training only 10 trainees knew what "Fire" signage was and after the action 30 trainees responded correctly. They are however acquainted with the "Warning", "Obligation" and "Prohibition" signage. For the presented case, it is suggested that the workers follow a train in fire prevention and fire emergency drills.

The method and the sample can be considered restrictors, however, the goals of the work meet the results, since it is intended to show the improvement of knowledge.

The employees were quite participatory, both in training and in responding to questionnaires, and somehow, it made possible to finish the work.

As future work it would be interesting to continue to contribute to studies in the area of training effectiveness and its impact on reducing occupational injuries.

This study can also be extended to another level by applying the questionnaires after training, with a longer time range, in order to be detected improvement actions implemented in each job.

ACKNOWLEDGMENTS

The authors would like to thank Master in Occupational Safety and Hygiene Engineering (MESHO), of the Faculty of Engineering of the University of Porto (FEUP), all the support in the development and international dissemination of this work.

REFERENCES

Becker, P., & Morawetz, J. (2004). Impacts of Health and Safety Education: Comparison of Worker Activities Before and After Training. *American Journal of Industrial Medicine*, 63–70.

Bhatti, M.A., Battour, M.M., Sundram, V.P., & Othman, A.A. (2013). Transfer of training: does it truly happen? *European Journal of Training and Development, Vol. 37 No. 3*, 273–297.

Departamento de Estatística do Trabalho, E. e. (1998). *Estudo sobre a Sinistralidade em Portugal - Acidentes de trabalho e de trajeto.*

Executive, H. (2003). *Effective communication of chemical hazard and risk information using a multimedia safety data sheet.*

Kongtip, P.Y., Yoosook, W., & Chantanakul, S. (2007). Occupational health and safety management in small and medium-sized enterprises: An overview of the situation in Thailand. *Safety Science*, 1356–1368.

Lingard, H. (2002). The effect of first aid training on Australian construction workers' occupational health and safety motivation and risk control behavior. *Journal of Safety Research*, 209–230.

OECD. (2003). *Guidance on Safety Performance Indicators.* OECD Environment, Health and Safety Publications.

Office, I.L. (2004). *Safe work and safety culture. The ILO report for word day for safety and health.*

Robson, L., Stephenson, C., Schulte, P., Amick, B., Chan, S., & Bielecky, A., et al. (2010). *A systematic review of the effectiveness of training & education for the protection of workers.* Ontario: Institute for Work & Health; National Institute for Occupational Safety and Health.

Sander, B. (1995). *Gestão da educação na América Latina: construção e reconstrução do conhecimento.* Campinas, SP: Autores Associados.

Soares, C.G. (2007). *Riscos Públicos e Industriais* (Vol. 1). Lisboa: Edições Salamandra, Lda.

Trabalho, A.E. (2003). Introdução às substâncias perigosas no local de trabalho - Facts 33.

Occupational Safety and Hygiene II – Arezes et al. (eds)
© 2014 Taylor & Francis Group, London, ISBN 978-1-138-00144-2

Safety management on a refinery maintenance turnaround

A.G. Cardoso
CIGAR, Faculdade de Engenharia da Universidade do Porto, Portugal

M. do Sameiro Queirós
CIGAR, Faculdade de Engenharia da Universidade do Porto, Portugal
e Instituto de Soldadura e Qualidade, Porto Salvo, Portugal

J.M. Meireles
Galp Energia, Refinaria de Matosinhos, Leça da Palmeira, Portugal

Miguel Tato Diogo
CIGAR, Faculdade de Engenharia da Universidade do Porto, Portugal

ABSTRACT: Programmed maintenance is an essential part of safety management in high risk industries such as oil refineries and their life cycle. Maintenance turnaround tasks are diverse and constitute an exceptional universe of risks. In this article we intend to discuss if during maintenance turnarounds there is an increase of undesired events and human factor role. Maintenance activities usually represent an increase in accidents and incidents frequency. Accidents can have origin in a series of causes amongst human factor, being human error the most unpredictable. Considering this, we suggest that effective safety management depends on a comprehensive risk assessment that should integrate an assessment of workers perceived risk. When integrated in a human risk factor management program, it could allow determining measures specifically focused on workers, enhancing their risk perception, hence safe behaviour and aid on diminishing the elevated accident frequency preceded by human failure during maintenance turnarounds.

1 INTRODUCTION

Safety management is becoming more essential to every activity, in particular in high risk industries such as oil refineries where work processes represent a significant risk for people and for the environment (Grote, 2012).

Chemical process plants safety depends on human, technological and organizational factors that turn managing health and safety on these plants a challenging task (Reniers et al., 2009). Therefore maintenance of these installations is vital for keeping acceptable safety levels as it prevents equipment degradation and increases production efficiency as well as reliability (Aissani et al., 2009).

In this paper we will discuss if maintenance turnarounds register a higher frequency of undesired events and human factor role.

2 OIL LIFE-CYCLE

Crude oil, or petroleum, mainly components are hydrocarbons that consist on molecules formed by the linkage between carbon and hydrogen atoms. According to biogenic theory petroleum originates from remains of biological matter in sedimentary rocks (Höök et al., 2010). Transformation processes such as bacteriogenesis and thermogenesis, originate hydrocarbons that are expelled from the compacting source, through permeable rocks until they are trapped in impermeable rocks called reservoir rocks, where oil is protected from oxygenation and microbial degradation (Tsatskin & Balaban, 2008).

Oil extraction is possible using geochemical techniques that allow detecting and marking the reservoir area (Peters & Fowler, 2002). The reservoir may exist on earth or sea and extracted crude oil is usually transported to oil refineries by tankers or pipelines, being then stored in tanks whether of floating or fixed ceilings (Elvers, 2008).

Crude oil molecules complexity requires the submission to a series of physical and chemical processes that allow their separation and treatment. An oil refinery is a highly complex industrial structure constituted by a series of operating units that work in an integrated and complementary way to ensure crude oil and other raw materials originate a range of energetic products (Alkhamis & Yellen, 1995, Babich & Moulijn, 2003).

Generally division between units occurs for its diversity, connection and specificity of the operations performed in each unit provided by a number of production equipment such as, furnaces, distillation towers, heaters, exchangers, pumps, pipes and tanks (Csao, 2007). Hazardous chemicals are transported in and around these units in large volumes through pipeline and road transportation. Oil refineries operate often continuously, in a steady state operation and at very high capacities (Reniers et al., 2009) for this reason, are submitted to turnarounds (maintenance interventions).

3 MAINTENANCE OPERATIONS

Programmed maintenance is contemplated in complex industrial plants (e.g.: oil refineries) being usually plant turnarounds, that take place to execute equipment maintenance, to prepare the plant for new productive cycles and ensure that these happen in a safe and reliable manner (Pokharel & Jiao, 2008, Alkhamis & Yellen, 1995), complying with law requirements. The inherent complexity of this industry's production system and the risk it represents demands to consider safety, reliability and maintenance limitations (Aissani et al., 2009, Alkhamis & Yellen, 1995).

Considering types of work executed during a turnaround maintenance we can identify work on equipment that cannot be taken out of operation unless the whole process is shutdown; work on equipment that does not need to be taken out of operation and work on defects identified during operation but couldn't be repaired during plant operation (Salih O. Duffuaa & Daya, 2004). Maintenance operations may be predictive, preventive or corrective as summarised in table 1.

In order to prolong unit life it is necessary to execute regular preventive activities (Alkhamis & Yellen, 1995). Maintenance period consists on a deviation from routine operations which requires a careful scheduling and planning by a multidisciplinary team as well as highly coordinated and detailed procedures, having the human factor a critical role in tasks coordination (Salih O. Duffuaa & Daya, 2004). Chemical process plants require a strong safety system which is achieved by regulatory procedures and everyone's respect for existing safety culture within the organization enhancing the primary importance of the fact that implemented procedures are followed so human, environment and plant integrity is preserved (Kalantarnia et al., 2010).

Table 1. Types of maintenance operations.

Type of maintenance	Characteristics	Author
Preventive	Defined and programmed	(Wallace & Merritt, 2003)
Predictive	Adapted continuously to equipment conditions	
Corrective	Happens when there is a deviation from normal operation conditions	(Padmanabhan, 2010)

Maintenance activities represent a unique universe of occupational risks often potentiated by simultaneous work, non-routine operations (e.g.: opening of an equipment) from which it may occur electrical discharge, unexpected contact with hazardous chemical substances or with residual product in high quantities or under pressure (Wallace & Merritt, 2003). Staff training and expertise, adequate supervision, management systems and work control by project managers, all contribute to the success of a turnaround (Bevilacqua et al., 2009). The elevated number of works carried out during a turnaround require a high number of workers, with different skills in areas such as metal-mechanical, electricity, industrial cleaning, automation and control, among others (Salih O. Duffuaa & Daya, 2004).

3.1 *Safety management on a turnaround*

One of the main targets of chemical process plants must be to give a safe work environment. Safety is achieved through the implementation of risk management systems, procedures, training and above all, everyone's participation (Summers, 2003). Refining plant safety is a sensitive issue and safety management systems are considered best-practice methods for managing risk (Amyotte et al., 2007). The importance in ensuring safe practices at work can be verified by the fact that all over the world, refineries have been adopting health and safety requirements published by safety management organizations that demand the existence of systems of responsibilities, procedures, adequate resources and technological know-how (Reniers et al., 2009).

Risk management is vital for a processing plant since it assures that the likelihood of normal operation deviations happening is minimal and if such occurs, its effects are controlled and minimized (Sanders, 2004). Risk assessment methods may be used at any moment of the installation, from the design phase until the operational and allow to evaluate human activities impact on hazardous systems (van Duijne et al., 2008). It is common to use in chemical process plants methods such as the Hazard Identification (HAZID), Preliminary Hazard Analysis (PHA), Hazard Operability (HAZOP) (Khan et al., 2003) and risk-based inspection (Bertolini et al., 2009). Another useful method to evaluate workplace risks is the one based on BS 8800:2004 that allows controlling intolerable risks (Lind et al., 2008).

3.2 *Risk function*

Risk is a difficult concept as well as its assessment. Its characterization efficacy depends on a good problem formulation, scientific knowledge and an appropriate assessment method (Meacham, 2008). Risk function considers the probability of an undesired event occurring and the severity of its consequences, being their function used to characterize determined event materialization and allows managing uncertainty (Aven et al., 2007, Ferdous et al., 2013). This reveals to be particularly important in managing tasks and work processes' safety, considering task complexity and tasks interdependencies (Grote, 2012).

Safety level has a dynamic behaviour meaning that it can be influenced by a number of factors resulting from changes in plant activities such as changes from normal operation to shutdown, turnaround and start-up, repairs or procedural changes (Knegtering & Pasman, 2013).

Risk management integrates information of different domains, hazard identification and characterization, assessment of workers exposed and exposure consequences (Ling et al., 2012). It must contemplate all the intervenients of the workplace meaning, workers, physical installations and organizational measures (Makin & Winder, 2008).

3.3 *Risk factors associated with maintenance activities*

Maintenance behind schedule induces risk factors since it may originate an unsafe condition and lead to an emerging threat (Knegtering & Pasman, 2013). When incorrectly executed, in an insufficient or excessive manner or when delayed, maintenance can be a triggering event (Okoh & Haugen, 2013).

Work in refineries constitutes a very particular scenario where work environment often necessitates frequent adaptation which makes working in construction projects a challenging and particular hazardous task, capable of inducing fatigue and stress-related risk factors (Chan, 2011). In 2011 were registered 9 fatalities associated with manufacturing contractors of which 5 resulted from construction and maintenance activities as well as 135 lost work time injuries and in the last 5 years, construction/maintenance/operations activities and road accidents have been the main causes of fatalities in European downstream industry (Concawe, 2011).

Maintenance has been indicated as one of the significant causes of some major accidents related not only with process equipment but also with deficiencies in maintenance management contributing in 30 to 40% of all accidents or precursor events (Okoh & Haugen, 2013). Turnaround accidents can have origin in man, technology or organization (Okoh & Haugen, 2013). Accidents are often caused by human factor (Amyotte et al., 2007, Anderson, 2005, Broadribb, 2013) and of all preceding events, the most unpredictable is human error for which effective safety management depends on workers perceived risk (Chan, 2011).

3.3.1 *Risk perception*

Perception is the mechanism that allows humans to evaluate the external environment stimulus and so is a determinant factor in human behavioural response (Green & Heekeren, 2009). Perceived risk consists on the subjective evaluation of the probability of the materialization of determined risk and on the extent to which each person worries about its consequences. (Cezar-Vaz et al., 2012, Sjöberg et al., 2004). Each individual has the capacity of interacting with the external environment and, consequently, potentiate or diminish the probability of an event happening (Slovic et al., 2007).

Inadequate risk judgments may originate inappropriate decisions which in turn can initiate non safe behaviours and human errors (Rundmo, 1995). Unsafe acts of a refinery employees were responsible for 96% of work injuries (Myers et al., 2010). Human behaviour, particularly unsafe practices, must be an integral part of process safety and should be included so that there is improved risk control (Myers, 2013). In occupational context, risk perception is based on an organization existing risks (Areosa, 2012). It is considered that risk perception has a rational and an emotional component, being the latter an important prognostic of human behaviour (Moen, 2006, Slovic et al., 2004). Individuals consider an activity or technology based on their thoughts and feelings about it which represents a complex relationship between reason and emotion which can be a starting point to improve risk management (Slovic et al., 2004).

4 DISCUSSION

A turnaround should represent an installation-wide effort including workers in general, management, manufacturing and maintenance staff (Bevilacqua et al., 2009).

During a turnaround, work nature and rhythm suffer significant changes. Due to its impact on a refinery production and finance capacity, these are carried out in a short period in which are executed not only normal tasks but also non-routine. These represent periods of intense work, higher work demands, stress and a significant increase in workers number, generally contract workers. Usually greater job demands are positively related to risk perception which can lead to the increase of unsafe practices (Baugher & Roberts, 1999). Turnaround periods register an elevated accident frequency (Souza & Freitas, 2002) and human failure in maintenance operations is a significant cause of major accidents (Anderson, 2005).

Considering risk function, unsafe practices may potentiate probability of undesired events occurring and represent one of the possible factors that justify the fact that maintenance activities are characterized by higher accidents frequency.

It has been shown that major accidents immediate causes often involve human error of operators and maintenance personnel however, these errors occur due to management issues (Anderson, 2005). On table 2 are listed some potential causes of undesired events.

Being human factor a significant cause of occupational and major accidents during maintenance operations, one may consider implementing a human risk factor management program. This should include contract workers since these represent major part of workforce during turnarounds. It would guarantee that maintenance specialized contract workers would enrich the operations carried out. It is also of interest to assess workers risk perception as it is fundamental in workers behaviour. Given turnaround maintenance importance in industrial plants, we suggest that measures focused on worker are implemented in order to enhance risk perception and diminish human factor originated accidents likelihood. This could be accomplished by training in areas identified as needed by workers risk perception assessment, by approaching specific themes during toolbox talks or safety meetings, for instance.

5 CONCLUSION

In high risk industries, safety must be the main characteristic of the organization's culture and integrate the individual, task and organizational characteristics related to health and safety (Cooper, 2000). A careful preparation is imperative in every task taken in a refinery and specially, during a turnaround. Tasks performed in a turnaround, represent high demands in terms of safety and have to be perfectly coordinated so that errors won't happen. This will reduce accidents likelihood and improve plant productivity.

Table 2. Potential accident causes.

Factor	Potential causes	Author
Human	Unsafe practices	(Myers, 2013)
	Human-machine interface	
	Interaction with work environment	
	Individual human factors	
	Stress	
	Fatigue	(MacKenzie et al., 2007)
	Miscommunication	
Task/ Process	High demands	(Myers, 2013)
	Malfunctioning equipment	(MacKenzie et al., 2007)
Management/ Organizational	Inadequate training	(Anderson, 2005)
	Inappropriate procedures	
	Poor supervision	
	Failure to identify hazards	

Technological advances have allowed substituting human activities in most of process operations which improved significantly industrial plants safety performance and reliability. However, human activity is still a system intervenient, with great responsibility that may represent a risk factor. Thus, we suggest that there should be a focus on human activities in a proactive perspective. Simple measures such as specific and focused training and communication based on workers risk assessment may help organizations to diminish human factor accidents likelihood.

ACKNOWLEDGMENTS

The authors would like to thank to Occupational Safety and Hygiene Engineering Master (MESHO) of the Faculty of Engineering of the University of Porto (FEUP), for supporting the development and international dissemination of this work and to Galp Energia—Matosinhos Refinery for allowing to better understand safety role in a turnaround.

REFERENCES

Aissani, N., Beldjilali, B. & Trentesaux, D. 2009. Dynamic scheduling of maintenance tasks in the petroleum industry: A reinforcement approach. *Engineering Applications of Artificial Intelligence,* 22, 1089–1103.

Alkhamis, T.M. & Yellen, J. 1995. Refinery units maintenance scheduling using integer programming. *Applied Mathematical Modelling,* 19, 543–549.

Amyotte, P.R., Goraya, A.U., Hendershot, D.C. & Khan, F.I. 2007. Incorporation of inherent safety principles in process safety management. *Process Safety Progress,* 26, 333–346.
Anderson, M. 2005. Behavioural Safety And Major Accident Hazards Magic Bullet Or Shot In The Dark?: HSE.
Areosa, J. 2012. As perceções de riscos dos trabalhadores: qual a sua importância para a prevenção de acidentes de trabalho? *Impacto Social dos Acidentes de Trabalho*, 66–97.
Aven, T., Vinnem, J.E. & Wiencke, H.S. 2007. A decision framework for risk management, with application to the offshore oil and gas industry. *Reliability Engineering & System Safety,* 92, 433–448.
Babich, I.V. & Moulijn, J.A. 2003. Science and technology of novel processes for deep desulfurization of oil refinery streams: a review☆. *Fuel,* 82, 607–631.
Baugher, J.E. & Roberts, J.T. 1999. Perceptions and Worry about Hazards at Work: Unions, Contract Maintenance, and Job Control in the U.S. Petrochemical Industry. *Industrial Relations: A Journal of Economy and Society,* 38, 522–541.
Bertolini, M., Bevilacqua, M., Ciarapica, F.E. & Giacchetta, G. 2009. Development of Risk-Based Inspection and Maintenance procedures for an oil refinery. *Journal of Loss Prevention in the Process Industries,* 22, 244–253.
Bevilacqua, M., Ciarapica, F.E. & Giacchetta, G. 2009. Critical chain and risk analysis applied to high-risk industry maintenance: A case study. *International Journal of Project Management,* 27, 419–432.
Broadribb, M.P. 2013. Too close for comfort. *Process Safety Progress*, n/a-n/a.
Cezar-Vaz, M.R., Rocha, L.P., Bonow, C.A., Da Silva, M.R.S., Vaz, J.C. & Cardoso, L.S. 2012. Risk Perception and Occupational Accidents: A Study of Gas Station Workers in Southern Brazil. *International Journal of Environmental Research and Public Health,* 9, 2362–2377.
Chan, M. 2011. Fatigue: the most critical accident risk in oil and gas construction. *Construction Management and Economics,* 29, 341–353.
Concawe 2011. European downstream oil industry safety performance, Statistical summary of reported incidents.
Cooper, M.D. 2000. Towards a model of safety culture. *Safety Science,* 36, 111–136.
Csao 2007. *Construction Multi-Trades Health and Safety Manual*, New Edition.
Elvers, B. 2008. *Handbook of Fuels: Energy Sources for Transportation*, Wiley.
Ferdous, R., Khan, F., Sadiq, R., Amyotte, P. & Veitch, B. 2013. Analyzing system safety and risks under uncertainty using a bow-tie diagram: An innovative approach. *Process Safety and Environmental Protection,* 91, 1–18.
Green, N. & Heekeren, H.R. 2009. Perceptual decision making: a bidirectional link between mind and motion. *Progress in Brain Research,* Volume 174, 207–218.
Grote, G. 2012. Safety management in different high-risk domains—All the same? *Safety Science,* 50, 1983–1992.
Höök, M., Bardi, U., Feng, L. & Pang, X. 2010. Development of oil formation theories and their importance for peak oil. *Marine and Petroleum Geology,* 27, 1995–2004.
Kalantarnia, M., Khan, F. & Hawboldt, K. 2010. Modelling of BP Texas City refinery accident using dynamic risk assessment approach. *Process Safety and Environmental Protection,* 88, 191–199.
Khan, F.I., Sadiq, R. & Amyotte, P.R. 2003. Evaluation of available indices for inherently safer design options. *Process Safety Progress,* 22, 83–97.
Knegtering, B. & Pasman, H. 2013. The safety barometer: How safe is my plant today? Is instantaneously measuring safety level utopia or realizable? *Journal of Loss Prevention in the Process Industries,* 26, 821–829.
Lind, S., Nenonen, S. & Kivistö-Rahnasto, J. 2008 Safety risk assessment in industrial maintenance. *Journal of Quality in Maintenance Engineering,* 14, 205–217.
Ling, M.-P., Lin, W.-C., Liu, C.-C., Huang, Y.-S., Chueh, M.-J. & Shih, T.-S. 2012. Risk management strategy to increase the safety of workers in the nanomaterials industry. *Journal of Hazardous Materials,* 229–230, 83–93.
Mackenzie, C., Holmstrom, D. & Kaszniak, M. 2007. Human Factors Analysis of the BP Texas City Refinery Explosion. *Proceedings of the Human Factors and Ergonomics Society Annual Meeting,* 51, 1444–1448.
Makin, A.M. & Winder, C. 2008. A new conceptual framework to improve the application of occupational health and safety management systems. *Safety Science,* 46, 935–948.
Meacham, B.J. 2008. A Risk-Informed Performance-Based Approach to Building Regulation. *7th International Conference on Performance-Based Codes and Fire Safety Design Methods.* Auckland.
Moen, B.E., Rundmo, T. 2006. Perception of transport risk in the Norwegian public. *Risk Management,* 8, 43–60.
Myers, P.M. 2013. Layer of Protection Analysis—Quantifying human performance in initiating events and independent protection layers. *Journal of Loss Prevention in the Process Industries,* 26, 534–546.
Myers, W.V., Mcsween, T.E., Medina, R.E., Rost, K. & Alvero, A.M. 2010. The Implementation and Maintenance of a Behavioral Safety Process in a Petroleum Refinery. *Journal of Organizational Behavior Management,* 30, 285–307.
Okoh, P. & Haugen, S. 2013. Maintenance-related major accidents: Classification of causes and case study. *Journal of Loss Prevention in the Process Industries.*
Padmanabhan, H. 2010. Condition Based Maintenance Of Rotating Equipments On OSI PI Platform – Refineries/Petrochem Plants WIPRO.
Peters, K.E. & Fowler, M.G. 2002. Applications of petroleum geochemistry to exploration and reservoir management. *Organic Geochemistry,* 33, 5–36.
Pokharel, S. & Jiao, J.R. 2008. Turn-around maintenance management in a processing industry: A case study. *Journal of Quality in Maintenance Engineering,* 14, 109–122.
Reniers, G.L.L., Ale, B.J.M., Dullaert, W. & Soudan, K. 2009. Designing continuous safety improvement within chemical industrial areas. *Safety Science,* 47, 578–590.

Rundmo, T. 1995. Perceived risk, safety status, and job stress among injured and noninjured employees on offshore petroleum installations. *Journal of Safety Research,* 26, 87–97.
Salih O. Duffuaa & Daya, M.a.B. 2004. Turnaround maintenance in petrochemical industry: practices and suggested improvements. *Journal of Quality in Maintenance Engineering,* 10, 184–190.
Sanders, R.E. 2004. Practicing chemical process safety: a look at the layers of protection. *Journal of Hazardous Materials,* 115, 141–147.
Sjöberg L., Bjørg-Elin Moen & Rundmo, T. 2004. Explaining risk perception. An evaluation of the psychometric paradigm in risk perception research.
Slovic, P., Finucane, M.L., Peters, E. & Macgregor, D.G. 2004. Risk as Analysis and Risk as Feelings: Some Thoughts about Affect, Reason, Risk, and Rationality. *Risk Analysis,* 24, 311–322.
Slovic, P., Finucane, M.L., Peters, E. & Macgregor, D.G. 2007. The affect heuristic. *European Journal of Operational Research,* 177, 1333–1352.
Souza, C.a.V.D. & Freitas, C.M.D. 2002. Perfil dos acidentes de trabalho em refinaria de petróleo. *Revista de Saúde Pública,* 36, 576–583.
Summers, A.E. 2003. Introduction to layers of protection analysis. *Journal of Hazardous Materials,* 104, 163–168.
Tsatskin, A. & Balaban, O. 2008. Peak oil in the light of oil formation theories. *Energy Policy,* 36, 1826–1828.
Van Duijne, F.H., Van Aken, D. & Schouten, E.G. 2008. Considerations in developing complete and quantified methods for risk assessment. *Safety Science,* 46, 245–254.
Wallace, S.J. & Merritt, C.W. 2003. Know when to say "when": A review of safety incidents involving maintenance issues. *Process Safety Progress,* 22, 212–219.

Occupational Safety and Hygiene II – Arezes et al. (eds)
© 2014 Taylor & Francis Group, London, ISBN 978-1-138-00144-2

Safety footwear and its protective components, a way to promote health and safety at work

C.S. Ferreira & M.J. Abreu
Centro de Ciência e Tecnologia Têxtil (2C2T), Departamento de Engenharia Têxtil, Universidade do Minho, Guimarães, Portugal

Joel Viera da Silva & João P. Mendonça
CT2M, Departamento de Engenharia Mecânica, Universidade do Minho, Campus de Azurém, Guimarães, Portugal

ABSTRACT: The use of Personal Protective Equipment (PPE) has become a necessity in view of the number of accidents at work. The use of adequate protective footwear and other types of PPE's can help reduce the injuries caused by accidents at work and also their severity.

Nowadays, there are three types of protective shoes in the industry: safety shoes, protection shoes and occupational shoes. The use and specifications of this type of PPE respect the European and ISO standards.

In this paper, a brief review of safety footwear and its components will be present focusing anti-penetration inserts. Moreover, the ongoing research of an innovative project is described presenting an illustrative case study. The project is in development at Minho University and aims to improve the existing shoes in terms of comfort using new and innovative materials.

1 INTRODUCTION

1.1 *Main statistics of accidents at work*

According to the European Statistics on Accidents at Work (ESAW), every year in the EU, about 3 million workers were victims of accidents at work leading to more than three days of absence from work; Additionally, about 5000 workers died in accidents at work (Communities, 2004).

The European Union (EU) along with the member states are creating strategies aiming the increase of the use of PPE's and consequently the decrease of the number of accidents at work.

In 2010 there were about 2 634 196 accidents at work in Europe of which 707 292 affected the lower extremities (EUROSTAT, 2013).

During the same period of time, in Portugal, the numbers were equally large; 215 632 accidents of which 52 467 affected lower extremities Table 1 shows the evolution of accidents at work according to the nature of the worker injury between 2002 and 2010. Despite the substantial decrease of the number of injuries in legs and feet after the adoption of the Community Strategy on Health and Safety at work in March 2002, there was a decrease on the rate of serious injuries such as internal injuries and member's amputations. This fact led to an increase in the minor injuries.

In Portugal, these type of accidents leads to an incapacity of the employee to work for normally 15 days. Besides the human suffering, these accidents have a heavy impact on productivity and consequently financial losses.

Although there is a positive trend towards the use of PPE's, in particular safety footwear, presently, a large number of employees still neglect the use of this kind of protection.

Goldcher and Acker in 2005 were able to present some reasons for the fact above mentioned:

- Uncomfortable at work;
- Excessive weight, which lead to muscle fatigue at the end of the day (heel pain, calf pain);
- Lack of flexibility due to the reinforcement of the base;
- Models unsuited to the morphology of some feet;
- Occurrence of foot injuries, in part secondary to the previous two complaints: redness, blistering, hyperkeratosis, among others;
- Lack of aesthetic and ergonomics;
- Inadequate ventilation which promote sweating, maceration and fungal infection.

With this in mind, it is necessary to improve the safety footwear available in order to be accepted by a larger number of employees, thereby reducing the number of accidents at work and the

Table 1. Evolution of accidents at work according to the nature of the worker injuries between 2002 and 2010. Adapted from (INE, 2013).

Year	2002	2004	2006	2008	2010
Total	55 988	47 979	54 399	57 016	52 567
Unknown nature of the lesion	4%	6%	12%	9%	3%
Minor injuries	42%	36%	30%	43%	44%
Bone fractures	10%	7%	6%	6%	5%
Dislocation sprains and distensions	23%	27%	34%	32%	39%
Amputations	3%	0,05%	0,04%	0,05%	1%
Concussions and internal injuries	14%	19%	12%	7%	6%
Shock	1%	1%	2%	0,38%	1%
Multiple injuries	0,36%	0,11%	0,34%	0,21%	0,09%
Others	3%	3%	3%	3%	2%

financial losses that affects both employers and employees.

1.2 *ISO standard*

The normative commitment that guarantees a specific protection to the user constrains the new design proposal. Thus, the introduction of novelties in the design of this components enables the improvement in order to make it more ergonomic, lightweight and even personalized with the customer's foot.

The Table 2 present the standards and specific subject for each norm relevant for safety footwear.

Table 2. European standards for safety footwear.

Standard	Subject
EN ISO 12468	Requirements and test methods for toecaps and penetration resistant inserts.
EN ISO 13287	Test method for slip resistance.
EN ISO 20344	Requirements and test methods for safety, protective and working shoes for professional use.
EN ISO 20345	Specifications for safety shoes for professional use.
EN ISO 20346	Specifications for working shoes for professional use.
EN ISO 20347	Specifications for working shoes for professional use.

1.3 *Safety shoes*

Shoes with suitable protection have improved in order to protect the employee foot of all the environmental risks. There are various types of injuries and the most common are:

- Fall by slipping;
- Injuries in toes due to falling objects;
- Crushing toes by heavy vehicles;
- Piercing the foot by sharp objects resulting in internal injuries and severe infections.

So, penetration resistant inserts and toecaps are the most important components of safety shoes.

The penetration resistant inserts should also prevent injuries in the heel, metatarsal and muscles of the foot.

According to the legislation, there are various types of protective footwear classified according to the protection that they offer: as safety shoes, protection shoes and occupational shoes.

The type of protection conferred by any of the above mentioned types of shoes is shown in Table 3. In this table, the numbers are related to the level of protection offered by the shoes. The letters are related to the type of shoes, S for safety shoes, P for protective shoes and O for occupational shoes. SB[a] and PB[b] refers to all materials that compose the shoe. S[c,] P[c] and O[c] are related to shoes made from all materials, except natural or synthetic polymers and S[d], P[d] and O[d] are related to shoes made from natural and synthetic polymers.

The main difference between safety, protective and occupational shoes is the toecap protection. In the case of safety shoes, toecaps must resist to a mechanical impact of 200 J, while the impact supported by the protective shoes should be 100 J. In the case of the occupational shoes, the toecap protection is not required.

1.4 *Safety footwear evolution*

The safety shoe industry has suffered improvements in order to increase the level of the user's comfort and to decrease the manufacturing costs of the safety shoe and it's components.

This development can be considered to have evolved in three different ways: manufacturing process, materials and design.

Table 3. Basic requirements of personal protective shoes. Adapted from (Goldcher & Acker, 2005).

	EN ISO 20345						EN ISO 20346						EN ISO 20347				
	Safety footwear						Protective footwear						Occupational footwear				
Level of protection	SB[a]	S1[c]	S2[c]	S3[c]	S4[d]	S5[d]	PB[b]	P1[c]	P2[c]	P3[c]	P4[d]	P5[d]	O1[c]	O2[c]	O3[c]	O4[d]	O5[d]
Basic protection	+	+	+	+	+	+	+	+	+	+	+	+	+	+	+	+	+
Closed seat region		+	+	+				+	+	+			+	+	+		
Antistatic properties		+	+	+	+	+		+	+	+	+	+	+	+	+	+	+
Energy absorption of hell		+	+	+	+	+		+	+	+	+	+	+	+	+	+	+
Water penetration resistance			+	+					+						+		
Penetration Resistance				+						+		+			+		
Cleated outsole				+						+		+	+		+		

Once the reliability of a safety shoe not depends only on the safety components, but also in the shoe structure where it's inserted, it becomes important to ensure the production of a steady and comfortable shoe. With advances in the insole and outsole materials and the remaining shoe components as well as the advances in the production processes, it is now possible to produce footwear with high quality and most competitive prices. Materials of toecap and penetration resistant inserts are another way of evolution for safety footwear.

The most common material for these components is the steel since it offers the best price and better mechanical properties, however, in the past, steel has been losing its importance to other materials such as aluminum or composite materials. These materials have the advantage of being lighter, but on the other hand they need to have higher thickness and size to confer the same protection as steel. Flexible materials such as aramide based structures have became very used in protection insoles because they offer a higher comfort level and they have the advantage of being metal free. (Sartor, 2008).

At last but not least, there was an evolution in the safety components design. Nowadays there are a larger number of shapes than few years ago. The design change of this component allowed the development of safety footwear for various situations, making the safety shoes no longer restricted in terms of fashion design and style. Costa, et al. in 2013 shows a redesign of a toe cap component. Currently it is possible to find in the market casual shoes, sports shoes and even lady boots with a high level of protection. The companies in this sector have given a great importance to this subject, allowing the improvement of the ergonomics, comfort, cost and design of the final product. This leads these companies to engage in creating strategies that ensure the reduction of development time for these products.

1.4.1 *Penetration resistant inserts*

As one of the most important components of protection footwear, the principal function of penetration resistant inserts is to prevent the penetration of sharp objects. This occurs in almost every industrial sector, but mainly in the construction sector.

Initially it was only possible to confer protection against penetration using steel plates of great thickness, which made the safety shoe rigid and extremely uncomfortable. In order to optimize the comfort and the protection level at the same time, these components evolved in terms of materials and thicknesses. Nowadays it is possible to find in the market penetration resistante inserts with a wide range of materials and structures.

Of the existing anti-penetration inserts the most used by the industry are metallic or made of aramid fibers (Kevlar® or Twaron®). Metallic inserts have a thickness of $\approx$0.5 mm and are still very used since they have a low cost production. The aramid fiber inserts are very flexible, what makes the shoe more comfortable, but on the other hand, their cost is superior to the metallic inserts. In the market, are also available inserts made from composite materials or made from the combination of materials, but they are patented.

In 1991, Kenji Okayasu, (Okayasu, 1991) developed a new penetration resistant insert aiming the improvement of the flexibility of the existent inserts. Okayasu used small metal plates joined together allowing the insert to bend in the front part, the metal plates are involved in an involcro from plastic or rubber in order to prevent the insert damaging in the sole of the shoe. The main disadvantages of this insert are it complex manufacture process and the fact that it flexibility inflict restrictions in the movement of the foot.

In 1994, Albertus Aleven (Aleven, 1994) developed a insert composed by four layers. This insert provides for the first time stiffness and flexibility in different parts of the insert. The first layer is a olymeric protector layer with a specific protector zone of stainless steel in the front part of the insert. The joint between this two layers is the area where the foot will bend while walking, allowing the caused tension to be absorved by the polymeric layer. The insert has an upper layer made of a fabric with anti-fungal properties and a bottom layer made of polyester. The variation of the thickness throughout the insert area allows the optimization of the flexibility in the required areas. The disadvantage of this type of insert is it complex and expensive manufacture process, additionally, with the use of the shoes, the metallic and the polymeric layers can be separated, leaving an area of the foot vulnerable to sharp objects.

In 1999, Frederick Harrison (Harrison, 1999) developed a polymeric penetration resistant insert that has a toecap and a special area to protect the heel. The main advantage of this insert is the fact that in the manufacture process of the shoe, the steps of the assemble of the toecap is eliminated once this component is already a part of the insert. This insert is made of a polymer capable of absorb a great amount of energy and comfer a good flexibility. A metallic plate is assembled in the polymer in order to improve the protection level of the insert, however, this metallic plate provide to much stiffness to the insert.

In 2003, Luigi Bettaglia (Battaglia, 2003) presented an improvement of the simple metallic inserts by applying longitudinal ribs in the back part of the insert, which form on the upper surface of the sole grooved ribs protruding from the opposite surfaces so as to form the convexities. These channels, stiffens the area between the heel and the arch of the foot, giving it greater resistance to flexion and torsion.

In 2008, Leo Sartor and his collegues (Sartor, 2008) developed a insert with two distinct parts. The front part of the insert is comprised of multiple layers of aramide fiber conferring a good flexibility. The back part of the insert is made of a composite material and acts like a structural element and prevents the heel torsion and therefore heel injuries. This insert combines for the first time the advantages of the metallic inserts and the advantages of the inserts made of aramide fibers. The disadvantage of Leo Sartor insert is the complex manufacture process.

2 DEVELOPMENT OF A NEW PENETRATION RESISTANT INSERT

With the improvement of safety footwear in mind, the main goal of the mentioned research is the development of a new penetration resistant insert. For that, a detailed analysis of the existing inserts was being carried out, whose results are going to be presented below.

Table 4. Relation between the characteristics of the inserts and the foot area where they are required.

Characteristics	Layout
Flexibility	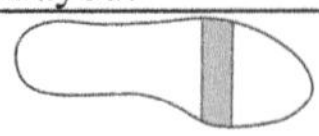
Anti-torsion	
Heel protection	
Confort	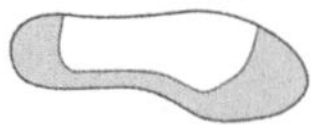
Adhesion to outsole	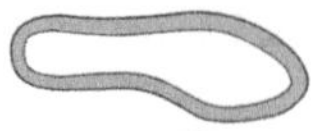
Penetration resistance	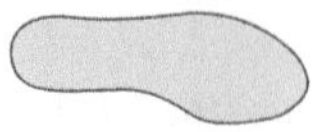
Antifungal	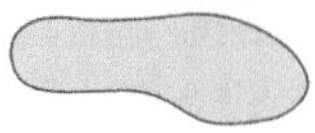
Impact damping	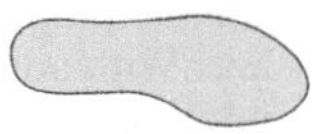
Energy distribution	

The anti-penetration inserts still have several flaws that undermine the comfort and reliability of safety shoes. It is necessary to improve these protection components using the latest technology and innovative materials. These improvements will contribute to a greater acceptance by the workers.

For this purpose, besides selecting the most important characteristics and specifications in a penetration resistant insert, it is also important to realize where these specifications are important in the foot area. Table 4 shows the analysis done by the researchers relating the most important characteristics and the area of the foot where they are required. To accomplish a better improvement, it is necessary to assemble the maximum information possible about the existing penetration resistant inserts. Through the analysis of the safety footwear standards, the penetration resistant inserts patents, and some visits to various safety footwear producers, it was possible for the reseachers to build the next table.

Therefore, chemical, metrological and destructive tests were performed to the normally used materials, Figure 1.

Through these tests it was possible to acquire the necessary information about the tensile strength, thickness and chemical composition. This information was analyzed and will allow the selection of

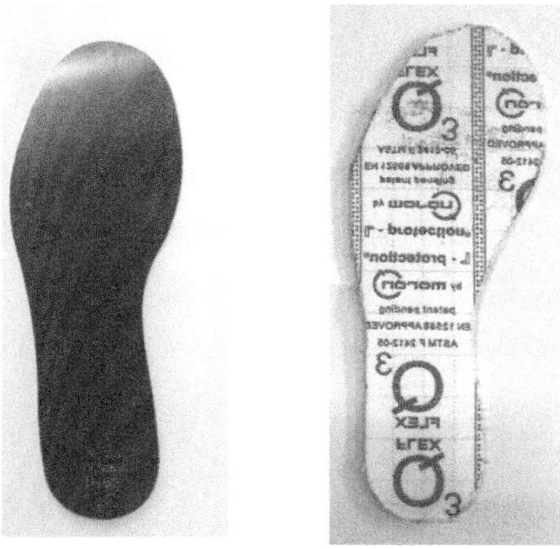

Figure 1. Left—metallic insert; right—Kevlar insert.

new innovative materials to be applied in this component. In addition to the tests mentioned, visits were made to different producers of safety shoes in order to acquire information and knowledge about the manufacturing and assembly process of this type of footwear.

In the design area, a new method will be used, which uses reverse engineering techniques for the development of new components from preexisting ones (Silva, et al., 2013). This methodology was tested in a brand new toe cap for safety shoes and also turn easier and quicker the development of a new product, boosting this improvement.

3 CONCLUSIONS

In the past years, the large number of accidents at work continued to result in serious injuries to a large percentage of workers. The European Union has created some strategies aiming to make workplaces safer.

Focused on the increase of comfort and safety of the existing safety footwear, an analysis of this type of footwear and the penetration resistant inserts was made in this research. This analysis allowed to conclude that there are some failures that could be solved if this component was more comfortable and lightweight, but at the same time more resistant.

For a correct optimization, several tests were performed on penetration resistant inserts samples. With an innovative methodology based in reverse engineering we aimed to develop a new insert that would increase the level of comfort, safety and reliability of safety shoes.

ACKNOWLEDGEMENTS

This work was financed l through founding from the Vale Inovação project n°2002/24148 (QREN).

REFERENCES

Aleven & W., A.A., 1994. Puncture Resistant Insole for Safety Footwear. Canada, US 5285583.

Battaglia, L., 2003. *Metalic Insert for Working and Safety Shoes*. Italia, EP 1 354526 A1.

Commission of the European Communities, 2007. *Communication from the commission to the European Parliament, the Council, the European Economic and Social Committee and the Committee of the Regions*. Brussels.

Communities, E., 2004. *Statistical analusis of soci-economic costs of accidents at work in the European Union*. Luxembourg.

Costa, S.L., Silva, J.V., Peixinho, N. & Mendonça, J.P., 2013. Innovative geometric redesign of safety footwear components using a reverse engineering approach. *ASME*.

EUROSTAT http://epp.eurostat.ec.europa.eu/(2013).

Goldcher, A. & b, D.A., 2005. Chaussures de sécurité, de protection et de travail. *EMC-Podologie*, pp. 12–23.

Harrison, F.A., 1999. *Puncture-Resistant and Impact Resistant Safety Shoe Insert*. Canada, US 5996257.

INE. *Instituto Nacional de Estatistica*. http://www.ine.pt/ (2013).

Okayasu, K., 1991. *Shoe Insole*. Estados Unidos, Patente US 5001848.

Sartor, L., Callegari, M. & Montemurro, A., 2011. *Insole Having Puncture-resistant Properties for Safety Footwear*. Estados Unidos, US 8082685 B2.

Silva, J.V. et al., 2013. Sustainable Reverse Engineering Methodology Assisting 3D Modeling of Footwear Safety Metalic Components. *ASME*.

Occupational Safety and Hygiene II – Arezes et al. (eds)
© 2014 Taylor & Francis Group, London, ISBN 978-1-138-00144-2

Crawler excavator track repair: Track fastening—noise characterization and personal protection

N.F. Couto, J. Castelo Branco & J. Santos Baptista
Research Laboratory on Prevention of Occupational and Environmental Risks (PROA/LABIOMEP), Faculty of Engineering, University of Porto, Porto, Portugal

ABSTRACT: The heavy machinery maintenance activity sector is characterized by the occurrence of intense noise during the working shift. This noise evaluation in the workplace was conducted to assess the noise emission during the task of removing tracks from an excavator, and to characterize in detail the nature of the risk in the sense of controlling the workers exposure to occupational noise. The measurements were made according to both the Portuguese and international framework, a narrow band analysis was made to the noise using the 1/1 octave band analysis method. In the measurements the presence of high intensity noise levels was showed with frequencies above 1 *kHz*. It was recognized the advantage of using a double protection system consisting in the simultaneous use of earplugs and earmuffs, granting a continuous protection implementing the obligation of earplugs use and making earmuffs use compulsory. According to these findings, this practice became mandatory.

1 INTRODUCTION

One of the most relevant problems of the workers' health in their occupational environment is high level noise exposure, particularly because the damage is not noticed in the short term. Quite the contrary this problem often occurs after a long-term (Miguel 2010). Noise-Induced Hearing Loss (*NIHL*) caused by occupational noise exposure is the second most frequent cause of hearing loss in the adult population being estimated that worldwide about 30 million workers are exposed to insecure noise levels in the workplace (Śliwinska-Kowalska, *et al.* 2006).

A study conducted in the American State of Tennessee demonstrated that most of the metallic construction company workers were exposed to dangerous levels of noise pressure (above 87 dB(A)) during their working activities (Mohammadi 2008).

This work aims at characterizing a heavy equipment maintenance workshop in the occupational hazard variable noise, particularly in the maintenance of tracks from crawler excavators during the fastening tasks using electric fastener tools.

2 MATERIALS AND METHODS

2.1 *Methodological approach*

A case-study of a crawler excavator maintenance workshop was conducted in order to understand in detail each part of the whole productive process.

A survey was conducted and delivered to the workers in order to identify the workstations with the greatest noise exposure, according to the workers point of view. That information was completed resorting to a noise pre-evaluation made in the workplace in a sense of determining the several noise pressure values (L_{Aeq}) for each task presented in the workplace.

Finally a task was chosen by the intensity of the noise emitted and a data collection was made with an 8:00 h day's work measurement, placing the sound-meter within the area of the noisiest task. A detailed analysis of this data was made, complemented with the determination of the 8:00 h working noise exposure, according to the NP EN ISO 9612:2011 standard.

The sound-meter was calibrated according to the standard and the correction value was 0,3 dB within the value of ±1,5 dB determined by the manufacturer.

Three 8:00 h measurements were conducted; the results are presented in Table 1.

The expanded uncertainty, associated to a probability of 95%, was determined using the C3 and C4 sections in the NP EN ISO 9612: 2011 standard. The value was 4,7 dB.

The uncertainty value for the noise levels was 2,57 dB within the limits determined by the standard (≤3,5 dB) Thus more measurements were not needed.

Table 1. Daily measurements (8:00 h) in the track repair section.

Measurement	Duration (h)	L_{Aeq} $dB(A)$	$LCpeak$ $dB(C)$
1	8	86,0	129,5
2	8	88,7	133,0
3	8	87,4	129,3

2.2 *Equipment*

The equipment used was a class 1 sound-meter *Solo*, model 01*dB*. The *fast response* mode was selected and the *A* filter was used to process the data in a way to characterize the noise level the workers were exposed to.

2.3 *The track repair section*

The chosen section to evaluate was the track repair section. In this section of the company the tasks that take place are as follows:

- Reception of the excavator tracks carried by a diesel forklift;
- Track placement over a roller table where the tracks are moved along the bolt fastening plant;
- Unscrewing of the worn-down tracks;
- Bolt removing from the track, and track link separation from the track shoes;
- Damaged track parts substitution;
- New tracks screwing;
- Full repaired track removing from the section by a diesel forklift.

3 NOISE MEASUREMENT RESULTS

Tasks—fastening: This task presents a noise that comprises a significant number of successive impacts (Figure 1).

The equipment used in the fastening operations is the same used to screw and unscrew the bolts from the tracks, however, in the unscrewing task, a greater number of impacts was verified when compared with the screwing task due to the driller bearing spring work, by means of successive impacts. The number of impacts needed to unscrew the tracks is greater because of the presence of debris in both bolts and nuts requiring more strength causing a greater *SPL*. The unscrewing task signal is presented in Figure 2.

Within the time domain, the noise of the fastening operations is intermittent ($T > 1$ s) with an impulsive character.

In terms of spectrum, the values are showed in Figure 3.

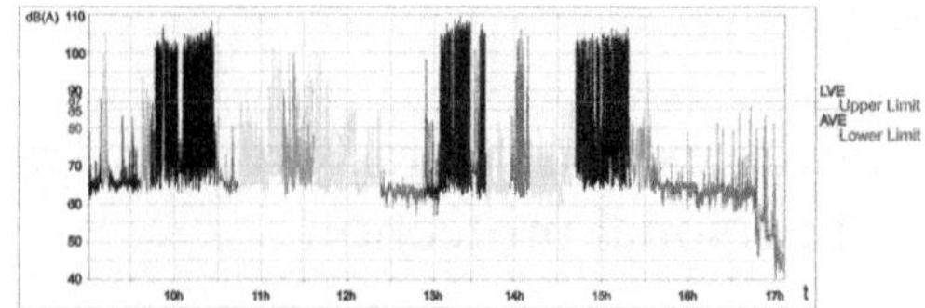

Figure 1. Sound wave of the 8 *h* noise produced in the track repair section.

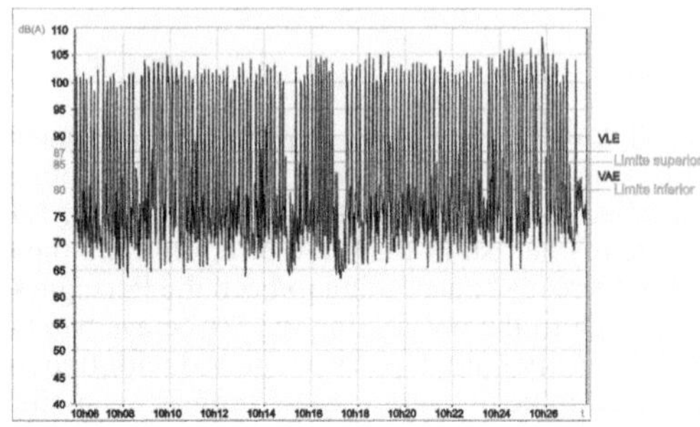

Figure 2. Unscrewing task: Characterization of the signal in terms of Sound Pressure Level (SPL)[1].

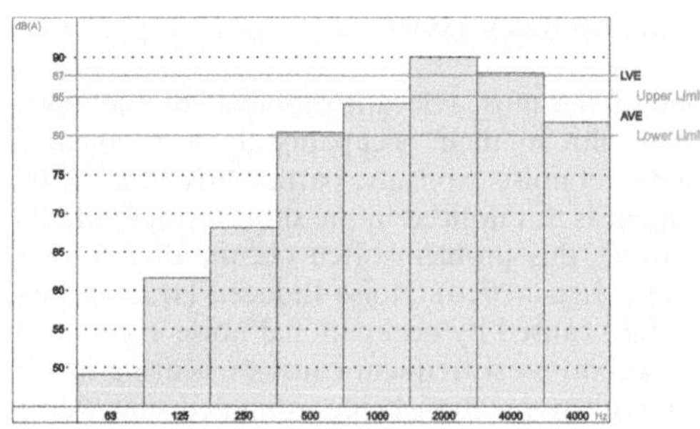

Figure 3. Bolt fastening—spectral characteristic by 1/8 band of frequency.

The noise incidence presented in the spectrum is between 0.5 and 8 kHz, being predominantly a high frequency noise.

3.1 *Noise spectral characteristic in 1/8 octave band analysis*

By frequency band, the noise sample collected in the 8:00 h presented the values listed in the Table 2.

A great influence of the bolt fastening task in the working shift noise is showed in the octave band

1. LVE: Limit Value of Exposure according to the Portuguese framework (*Decreto-lei* 182/2006).
AVE: Action Value of Exposure according to the Portuguese framework (*Decreto-lei* 182/2006).

Table 2. *SPL* values by octave band central frequency.

63 Hz	125 Hz	250 Hz	500 Hz	1000 Hz	2000 Hz	4000 Hz	8000 Hz
73,6 dB(A)	72,3 dB(A)	73,0 dB(A)	74,1 dB(A)	80,3 dB(A)	84,6 dB(A)	82,5 dB(A)	79,4 dB(A)

Table 3. *PPA* in use attenuation by frequency band.

		Attenuation (*M*)							S. deviation (*S*)						
PPA	Type	125 Hz	250 Hz	500 Hz	1000 Hz	2000 Hz	4000 Hz	8000 Hz	125 Hz	250 Hz	500 Hz	1000 Hz	2000 Hz	4000 Hz	8000 Hz
A	Earmuff	16,5 dB(A)	20,7 dB(A)	31,4 dB(A)	33,7 dB(A)	34,5 dB(A)	38,4 dB(A)	32,7 dB(A)	2,4 dB(A)	2,3 dB(A)	1,8 dB(A)	3,3 dB(A)	2,2 dB(A)	3,1 dB(A)	4,7 dB(A)
T	Earplug	31,4 dB(A)	28,8 dB(A)	32,5 dB(A)	33,8 dB(A)	35,6 dB(A)	39,3 dB(A)	41,9 dB(A)	7,3 dB(A)	8,9 dB(A)	8,1 dB(A)	7,3 dB(A)	4,3 dB(A)	6,0 dB(A)	5,0 dB(A)

frequency signal analysis. The median/average time spent in this operation during a shift is of about 3:00 *h*. When comparing the values presented in table 2 it is possible to see the prevalence of higher *SPL* values precisely in the same frequencies where the fastening task shows higher sound pressure levels (Figure 3).

4 PERSONAL PROTECTIVE EQUIPMENT

In order to guarantee in a continuous way the workers non exposure to noises that can induce danger to their health it's important to prescribe Personal Protective Equipment (*PPA*) that takes into account not only the *SPL* value during the 8:00 h day's work, but also the spectral characteristic of the noise signal. The *PPA* devices currently in use by the company are described in Table 3.

For determining the global attenuation provided by the use of the *PPA* present in Table 3, the 1/1 octave band analysis method was used as prescribed by the Portuguese framework, and the NP EN ISO 9612: 2011 with a confidence level of 98%.

In a way to assure a high level of protection to the workers, the measurement with the highest *SPL* value described in Table 1 was chosen.

The total value of attenuation that the *PPA* in use provides to the workers is described in the Table 4.

According to the NP EN 458:2006 standard, in order to achieve a *satisfactory* rate, the workers should be exposed to occupational *SPL* greater than 70 dB(A) and inferior or equal to 75 dB(A).

That classification is achieved with the earmuffs; however with the earplugs, the attenuation puts the workers exposed to *SPL* levels within the *acceptable* classification within the standard prescriptions (75 dB(A) $< L \leq$ 80 dB(A) and 65 dB(A) $< L \leq$ 70 dB(A)). The goal was to find a solution to not only grant a continuous level of protection to the workers but also to accomplish the ideal values of attenuation rated by the NP EN 458:2006 standard as *satisfactory* (70 dB(A) $< L \leq$ 75 dB(A)).

In a working context it was observed that the workers use earmuffs while fulfilling their tasks and by the time they finish a particular task, a

Table 4. Workers noise exposition values when using the PPA.

PPA	Type	$L_{Aeq.}$	*LAeqEf.*
A	Earmuff	88,7 dB(A)	74,0 dB(A)
T	Earplug	88,7 dB(A)	67,1 dB(A)

Table 5. Central frequency attenuation of the 5 different earplugs in study.

PPA	Type	Attenuation (M)							S. deviation (S)						
		125 Hz	250 Hz	500 Hz	1000 Hz	2000 Hz	4000 Hz	8000 Hz	125 Hz	250 Hz	500 Hz	1000 Hz	2000 Hz	4000 Hz	8000 Hz
T*	Earplug (silicone)	31,4 dB(A)	28,8 dB(A)	32,5 dB(A)	33,8 dB(A)	35,6 dB(A)	39,3 dB(A)	41,9 dB(A)	7,3 dB(A)	8,9 dB(A)	8,1 dB(A)	7,3 dB(A)	4,3 dB(A)	6,0 dB(A)	5,0 dB(A)
T1	Earplug (foam)	5,7 dB(A)	20,8 dB(A)	21,2 dB(A)	24,0 dB(A)	16,7 dB(A)	14,4 dB(A)	29,9 dB(A)	0,2 dB(A)	0,1 dB(A)	0,2 dB(A)	0,3 dB(A)	0,4 dB(A)	0,1 dB(A)	0,6 dB(A)
T2	Earplug (wool)	19,8 dB(A)	35,0 dB(A)	30,2 dB(A)	27,8 dB(A)	38,6 dB(A)	36,3 dB(A)	49,4 dB(A)	4,5 dB(A)	2,3 dB(A)	1,7 dB(A)	1,9 dB(A)	0,2 dB(A)	0,5 dB(A)	0,4 dB(A)
T3	Earplug (silicone)	19,4 dB(A)	34,7 dB(A)	40,8 dB(A)	41,3 dB(A)	40,5 dB(A)	45,6 dB(A)	47,3 dB(A)	2,6 dB(A)	3,2 dB(A)	6,6 dB(A)	4,5 dB(A)	6,2 dB(A)	5,4 dB(A)	6,0 dB(A)
T4	Earplug (foam)	19,6 dB(A)	33,4 dB(A)	27,4 dB(A)	35,4 dB(A)	43,8 dB(A)	40,0 dB(A)	42,9 dB(A)	4,7 dB(A)	1,9 dB(A)	1,4 dB(A)	1,0 dB(A)	0,1 dB(A)	0,1 dB(A)	2,6 dB(A)

* Currently in use by the company workers.

natural tendency of removing the PPA in use was showed, exposing the workers to the noise produced in the neighboring, putting themselves into a risky situation.

In that sense and bearing in mind the specific characteristics of the noise it should:

- Be made compulsory in the workplace a full-time utilization of earplugs due to the efficiency shown regarding impact noises. The plugs should be made in porous material (instead of silicone) to achieve a better efficiency preserving the communication speech intelligibility (M.R., et al. 2000; Kvaløy, *et al.* 2010; Abel, *et al.* 2011; e Norin, *et al.* 2011).
- Be previously made an inquiry to the workers regarding the PPA utilization comfort, in order to assure an 8:00 h permanent PPA use (Toivonen, et al. 2002; Gerges 2012).
- Due to the incidence of high frequency noise (>1.500 Hz) passive protective devices must be used (Buck., *et al.* 2004).
- In the tasks that present aggravated risk ($L_{Aeq.}$ > 85 dB(A)), the simultaneous use of earplugs and earmuffs should be compulsory in order to gain additional ear protection and remove the workers from the risk zone every time they finish a specific task (Gallagher, et al. 2010 e Behar, et al. 1999).

As previously mentioned, the permanent earplug use in the studied work sections is recommended once the noise produced in one work area exposes the workers in the contiguous working zone. In that sense, a group of combined ear protection devices previously studied by Behar et al. (1999) were evaluated in this particular work context, aiming at finding an alternative solution to the *PPA* presently in use by the company.

The first step was to determine which earplug was the best to be recommended to the workers regarding the noise in study. A set of calculations was made using the attenuation of 5 earplugs to determine the device with better performance according to NP EN 458:2006. The attenuation by central frequency is showed in Table 5 and the attenuation to the occupational noise is presented in Table 6.

Table 6. Earplug attenuation values regarding the noise in study.

PPA	$L_{Aeq.}$	*LAeqEf.*
T	88,7 dB(A)	67,1 dB(A)
T1	88,7 dB(A)	74,0 dB(A)
T2	88,7 dB(A)	73,8 dB(A)
T3	88,7 dB(A)	76,3 dB(A)
T4	88,7 dB(A)	73,9 dB(A)

Table 7. Double protection central frequency attenuation.

PPA	Type	Attenuation (*M*)							S. deviation (*S*)						
		125 Hz	250 Hz	500 Hz	1000 Hz	2000 Hz	4000 Hz	8000 Hz	125 Hz	250 Hz	500 Hz	1000 Hz	2000 Hz	4000 Hz	8000 Hz
T2 + A1	Earplug + earmuff	17,9 dB(A)	45,6 dB(A)	47,3 dB(A)	32,2 dB(A)	46,5 dB(A)	57,3 dB(A)	62,7 dB(A)	4,2 dB(A)	3,0 dB(A)	2,6 dB(A)	2,0 dB(A)	0,6 dB(A)	2,6 dB(A)	2,8 dB(A)
T2 + A2	Earplug + earmuff	16,1 dB(A)	44,7 dB(A)	46,3 dB(A)	31,6 dB(A)	47,0 dB(A)	56,6 dB(A)	61,5 dB(A)	3,7 dB(A)	3,6 dB(A)	2,8 dB(A)	2,1 dB(A)	0,5 dB(A)	3,0 dB(A)	1,0 dB(A)
T2 + A3	Earplug + earmuff	20,1 dB(A)	49,4 dB(A)	47,9 dB(A)	31,3 dB(A)	46,9 dB(A)	58,9 dB(A)	62,4 dB(A)	3,2 dB(A)	3,9 dB(A)	2,8 dB(A)	1,3 dB(A)	0,8 dB(A)	2,7 dB(A)	2,3 dB(A)
T2 + A4	Earplug + earmuff	19,5 dB(A)	47,9 dB(A)	49,5 dB(A)	37,4 dB(A)	48,5 dB(A)	64,1 dB(A)	66,9 dB(A)	3,6 dB(A)	3,5 dB(A)	2,7 dB(A)	1,7 dB(A)	0,7 dB(A)	0,7 dB(A)	2,2 dB(A)
T4 + A1	Earplug + earmuff	19,0 dB(A)	42,7 dB(A)	45,4 dB(A)	36,0 dB(A)	47,2 dB(A)	59,7 dB(A)	62,3 dB(A)	0,7 dB(A)	2,0 dB(A)	1,0 dB(A)	1,5 dB(A)	0,4 dB(A)	0,6 dB(A)	1,3 dB(A)
T4 + A2	Earplug + earmuff	19,1 dB(A)	42,4 dB(A)	45,5 dB(A)	35,8 dB(A)	47,2 dB(A)	59,5 dB(A)	63,6 dB(A)	0,9 dB(A)	2,3 dB(A)	0,8 dB(A)	1,6 dB(A)	0,6 dB(A)	1,7 dB(A)	0,9 dB(A)
T4 + A3	Earplug + earmuff	21,3 dB(A)	45,3 dB(A)	46,4 dB(A)	36,2 dB(A)	48,0 dB(A)	61,4 dB(A)	64,3 dB(A)	1,6 dB(A)	0,9 dB(A)	1,0 dB(A)	0,7 dB(A)	0,8 dB(A)	1,5 dB(A)	0,5 dB(A)
T4 + A4	Earplug + earmuff	22,1 dB(A)	43,7 dB(A)	46,1 dB(A)	35,7 dB(A)	48,1 dB(A)	60,2 dB(A)	64,3 dB(A)	0,9 dB(A)	4,4 dB(A)	1,5 dB(A)	1,2 dB(A)	0,4 dB(A)	2,9 dB(A)	0,7 dB(A)

Table 8. Double protection attenuation values regarding the noise in study.

PPA	$L_{Aeq.}$	*LAeqEf.*
T2 + A1	88,7 dB(A)	74,0 dB(A)
T2 + A2	88,7 dB(A)	74,0 dB(A)
T2 + A3	88,7 dB(A)	73,8 dB(A)
T2 + A4	88,7 dB(A)	73,8 dB(A)
T4 + A1	88,7 dB(A)	73,7 dB(A)
T4 + A2	88,7 dB(A)	73,7 dB(A)
T4 + A3	88,7 dB(A)	73,7 dB(A)
T4 + A4	88,7 dB(A)	73,6 dB(A)

The attenuation data show two important figures; the silicon earplugs have greater attenuation ratio but they also have grater standard deviation values; according to the NP EN 458:2006 standard, the T1, T2 and T4 earplugs have rated *satisfactory*.

Despite having registered the highest attenuation value, the silicone earplugs do not fulfill the standard requirements. The foam and wool earplugs (porous and sound-absorbing material) have showed to be more appropriate to the noise in study.

Following the double *PPA* protection logic, a series of calculations were made using this three devices associated to four different earmuffs. For the attenuation values of the combined devices the study conducted by Behar et al. (1999) was considered. The two best performing devices (T2 and T4) were considered. The attenuation by central frequency is showed in Table 7 and the attenuation to the occupational noise is presented in Table 8.

The set of protection devices that came closer to the average value between the optimal interval determined by the NP EN 458:2006 standard of 70 dB(A) $< L \leq$ 75 dB(A) was the one composed by T4 earplug plus the A4 earmuff, with a global attenuation value of 15.1 dB(A).

The T4 earplug, when used alone, has showed an attenuation value of 14.8 dB(A), therefore the increase in attenuation given by the simultaneous use of earmuffs and earplugs in this case study was of 0.3 dB(A).

5 CONCLUSIONS

The predominance of high frequency noise with a high *SPL* value characterizes the noise collected in this study.

Another issue worth emphasizing is the fact that the work produced during the accomplishment of one specific task exposes the workers of the neighboring tasks to the noise emitted, putting them in a potential risky situation.

It was demonstrated that by using the earplugs recommended in this study, the workers are protected from the noise and it was verified that using the earplugs together with the earmuffs the degree of attenuation increased 0,3 dB(A), achieving a classification of *Satisfactory* according to the NP EN 458:2006 Portuguese standard.

It was also verified that the earplugs made of porous material achieved more satisfactory attenuation regarding the noise in study, with a strong prevalence in high frequencies.

ACKNOWLEDGEMENTS

The authors are grateful to the Master of Science in Health & Safety (*MESHO*) of the Faculty of Engineering of the University of Porto (*FEUP*) for supporting in the development and international disclosure of this study.

BIBLIOGRAPHY

Abel, Sharon M; Nakashima, Ann; Saunders, Douglas 2011. *Speech understanding in noise with integrated in-ear and muff-style hearing protection systems*. Noise & Health, 13, 378–384.

Arezes, P.M. 2002. *Percepção do Risco de Exposição Ocupacional ao Ruído*. Braga: Universidade do Minho.

Behar, A., & Kunov, H. 1999. *Insertion loss from using double protection*. Applied Acoustics, 57, 375–385.

Gallagher, Hilary L.; Bjorn, Valerie S.; McKinley, Richard L. 2010. *Proceedings of Meetings on Acoustics*. Acoustical Society of America, 11, 1–13.

Gerges, S.N. 2012. *Earmuff comfort*. Applied Acoustics, 73(10), 1003–1012.

K., B., & V., Z.-J. 2004. *Active Hearing Protection Systems and Their Performance*. RTO-EN-HFM, 111, 3–22.

Kvaløy, Olav; Berg, Tone; Henriksen, Viggo 2010. *A Comparison Study of Foam versus Custom Silicone Earplugs Used as Part of an Intelligent Electronic Hearing Protector System*. International Journal of Acoustics and Vibration, 15, 151–155.

Mahendra, Prashanth K.V. & Venugopalachar, S. 2011. *The possible influence of noise frequency components on the health of exposed industrial workers - A review*. Noise & Health, 13,16–25.

Miguel, A.S. 2010. *Manual de Higiene e Segurança do Trabalho (11 ed.)*. Braga: Porto Editora.

Mohammadi, G. 2008. *Hearing conservation programs in selected metal fabrication industries*. Applied Acoustics, 69, 287–292.

Norin, Julie A.; Emanuel, Diana C.; Letowski, Tomasz R. 2011. *Speech Intelligibility and Passive, Level-Dependent Earplugs*. Ear & Hearing, 32, 642–649.

Śliwinska-Kowalska, M., Dudarewicz, A., Kotylo, P., Zamyslowska-szmytke, E., Pawlaczyk-Łuszczynska, M., & Gajda-Szadkowska, A. Individual Susceptibility to Noise-Induced Hearing Loss: Choosing An Optimal Method of Restrospective Classification of Workers Into Noise-Susceptible And Noise-Resistant Groups. International Journal of Occupational Medicine and Environmental Health. 26 de Setembro de 2006, Vol. 19, pp. 235–245.

Occupational Safety and Hygiene II – Arezes et al. (eds)
 ISBN 978-1-138-00144-2

Assessment of the performance of globe thermometers under different environmental conditions

D.A. Quintela, A.R. Gaspar & A.R. Raimundo
ADAI-LAETA, Department of Mechanical Engineering, University of Coimbra, Pólo II, Coimbra, Portugal

A.V.M. Oliveira
Department of Mechanical Engineering, Coimbra Institute of Engineering, Polytechnic Institute of Coimbra, Rua Pedro Nunes, Quinta da Nora, Coimbra, Portugal

D. Cardoso
Department of Mechanical Engineering, University of Coimbra, Pólo II, Coimbra, Portugal

ABSTRACT: This paper addresses the performance of globe thermometers under different environmental conditions. Two 150 mm standard globes from different constructors and one 50 mm globe are considered. The assessment of the response times, the evaluation of the steady-state temperature values obtained with each sensor and the effect of the air velocity on the globe temperature are the main goals. The results of the response times suggest that the values usually stated in the literature are not met and that longer periods must be considered. Significant differences obtained between the 150 and 50 mm sensors show the influence of the diameter of the globe. The effect of the air velocity on the globe temperature clearly shows its influence: in the actual environmental conditions, the globe temperature decreases with the increase of the air velocity.

1 INTRODUCTION

The assessment of thermal stress is widely believed to be a difficult, expensive and time-consuming task (Budd, 2001). Many reasons play a role in such scenario and since that assessment requires the measurement of physical parameters of the environment their correct evaluation is undoubtedly one of the main causes. One of the most important physical parameters of the environment is the globe temperature (t_g) and therefore its accurate evaluation must be carefully addressed. It is usually measured with a temperature sensor placed inside a globe, which can have different diameters. The globe temperature takes into account the heat transfer by radiation and convection between the globe and the environment, thus considering the mean radiant temperature (t_r), the air temperature (t_a) and the air velocity (v_a). This parameter has several applications both in calculation of heat stress indices and in the estimation of other physical variables. For the former, the WBGT index is one of the most used and well-known heat stress indices all over the world. However, its applicability across a wide range of potential scenarios and environments is limited because of the inconvenience of measuring t_g (Epstein and Moran, 2006). In many circumstances, measuring t_g is cumbersome and impractical (NIOSH, 1986; Moran and Pandolf, 1999). In the case of the estimation of other physical variables the mean radiant temperature is probably the most used and ISO 7726 (1998) states that the accuracy of such estimation depends on the accuracy of the measurement of the physical parameters involved, thus depends on t_g. Moreover, the assessment of the globe temperature is also an important issue in many activities, either indoors (professional kitchen; glass, ceramic and mining industries; firefighting …) or outdoors (military, road and building construction, agriculture, firefighting …). Hence, its correct measurement must be duly considered by occupational hygienists and health and safety professionals.

This paper focuses the performance of globe thermometers of different diameters (50 and 150 mm) under different environmental conditions. The assessment of the response times, the evaluation of the temperature values obtained with three different globes and the effect of the air velocity on the globe temperature are considered.

2 MATERIALS AND METHODS

The study of the performance of globe thermometers represents the main objective of the present work. Therefore, the definition of globe temperature and its measurement specifications are firstly addressed followed by a brief introduction of the WBGT index and by the definition of the mean radiant temperature. This section ends with the presentation of the experimental setup and measuring equipment.

2.1 *Globe temperature*

The globe temperature is the temperature indicated by a temperature sensor placed in the centre of a globe (ISO 7243, 1989). If we look to the measuring specifications proposed by different organizations small differences can be found. Juang and Lin (2007) address this issue, but since the present paper focuses the WBGT index adopted by ISO, the recommendations of this organization are listed in Table 1.

2.2 *WBGT index*

The WBGT index combines the measurement of the natural wet-bulb temperature (t_{nwb}) and the globe temperature (t_g) and in some situations the measurement of a basic parameter, the air temperature (t_a) (dry bulb temperature). Inside buildings and outside without solar load, the combination that better adapts to the human response is given by (Dukes-Dobos e Henschel, 1971, 1973):

$$WBGT = 0.7 \times t_{nwb} + 0.3 \times t_g \tag{1}$$

For outdoors with solar load the calculation of WBGT is:

$$WBGT = 0.7 \times t_{nwb} + 0.2 \times t_g + 0.1 \times t_a \tag{2}$$

The WBGT value is thus dependent on the value of t_g.

Table 1. Characteristics of the globe temperature sensor according to ISO 7243, 1989.

Diameter	150 mm
Mean emission coefficient	0.95 (matt black globe)
Thickness	As thin as possible
Measuring range	20 to 120°C
Accuracy of measurement:	
Range: 20 to 50°C	±0.5°C
Range: 50 to 120°C	±1.0°C

2.3 *Mean radiant temperature*

The Mean Radiant Temperature (t_r) is the uniform temperature of an imaginary enclosure in which radiant heat transfer from the human body is equal to the radiant heat transfer in the actual non-uniform enclosure (ISO 7726, 1998).

The t_r can be measured and/or estimated by different ways but for the present paper the estimation of t_r from t_g is of significance. ISO 7726 states that, in order to obtain an approximate value of the t_r, simultaneous measurements of globe and air temperatures and air velocity must be performed. The accuracy of such estimation depends on the type of environment being considered and on the accuracy of the measurement of the mentioned physical parameters. According to ISO 7726, equations 3 and 4 can be used to estimate t_r in natural and forced convection, respectively:

$$\bar{t}_r = \left[\left(t_g + 273\right)^4 + \frac{0{,}25 \times 10^8}{\varepsilon} \times \left(\frac{|t_g - t_a|}{D} \right)^{1/4} \times \left(t_g - t_a\right) \right] - 273 \tag{3}$$

$$\bar{t}_r = \left[\left(t_g + 273\right)^4 + \frac{1{,}1 \times 10^8 \times v_a^{0,6}}{\varepsilon \times D^{0,4}} \times \left(t_g - t_a\right) \right]^{1/4} - 273 \tag{4}$$

These equations put in evidence the role of the globe temperature in the estimation of t_r and the possibility to use different globe diameters. Therefore, the consequences of using globes smaller than the standard have to be carefully assessed.

2.4 *Experimental setup*

The tests were carried out from May 17th to June 17th in a wide laboratory of the University of Coimbra with 7 m height that is partially built underground where the environmental conditions do not vary significantly. These were continuously monitored with a temperature and humidity sensor from Testo (175-H2, ref. 0563 1758). During the 5 days necessary to perform the tests required for the present research the air temperature ranged between 20.5 and 22.8°C. The mean value of the relative humidity was 58%.

Three different globes were used. A 50 mm globe (G_50) made of a 0.3 mm cooper plate with an air temperature sensor from Testo (ref. 0613 1711) placed in the center and connected to the data-logger 175–T2 (ref. 0563 1755); a 150 mm globe (G_150BK) made of a 0.4 mm copper plate from Brüel & Kjær pertaining to the WBGT sensor (ref. MM0030) and; a globe of 150 mm (G_150T) from Testo (ref. 0554 0670) connected to the data-logger

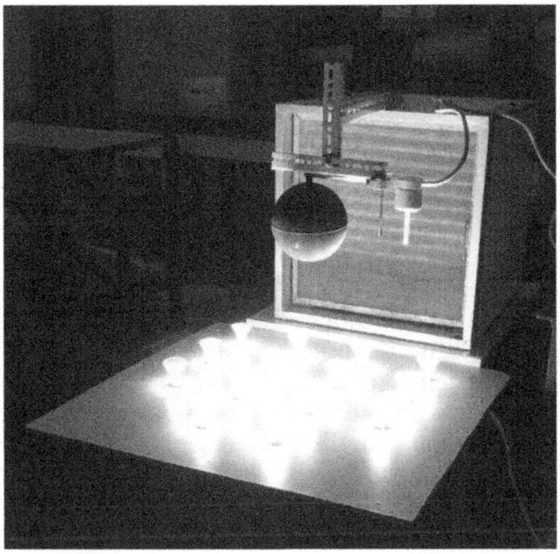

Figure 1. Experimental setup with the 14 lights on and with the WBGT sensor (ref. MM0030) from Brüel & Kjær.

Testo 400 (ref. 0563 4001). In this case the thickness of the copper plate is unknown.

To ensure the desired conditions in terms of air velocity, a helical fan from Sodeca (type 90/50) was used. Four mean air velocities were considered for the tests: approximately 0.35, 0.94, 1.68 and 2.43 m/s. This parameter was measured with an anemometer from Testo (ref. 0635 9340) connected to the data-logger Testo 445 (ref. 0560 4450). Additionally, to obtain an uniform flow in the neighborhood of the globes two metallic filters were placed in front of the fan. Both the helical fan and the filters were mounted in a wood box enclosure.

In order to simulate a significant radiative heat source 14 lamps from Philips, type 13117, each with a power of 150 W, were attached to a wood panel (0.68×0.75 m^2). This heat source was positioned at three different distances (L) from the center of the globes (25, 45 and 65 cm). A picture of the experimental setup is shown in Figure 1.

For the mentioned distances a total of 45 tests were performed considering natural and forced convection. Since ISO standard 7726 (1998) states that the stabilization period of a 150 mm globe is about 20 min to 30 min, according to the physical characteristics of the globes and the environmental conditions, a period of 60 min was selected for each trial to assure that the required stabilization period was always achieved. The acquisition rate was set to 1 minute in every sensor.

3 RESULTS AND DISCUSSION

In order to show a typical graphical representation of the results, Figure 2 illustrates the time

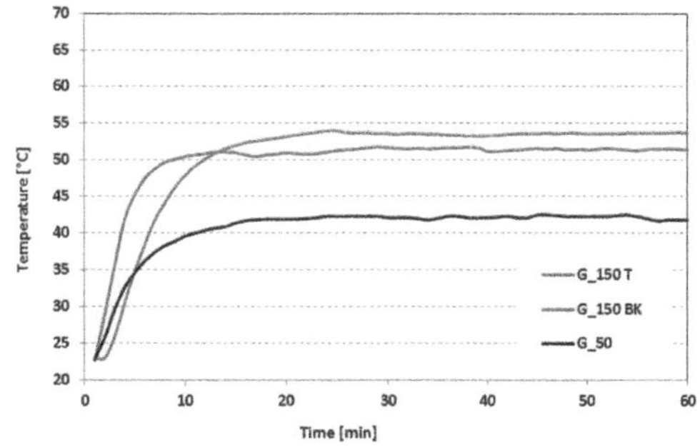

Figure 2. Example of the time evolution of the globe temperatures.

evolution of the globe temperatures in the test carried out in natural convection and for L = 45 cm. Figure 2 shows that the duration of the experiment was adequate, a scenario that is also true for all the other trials. In addition, the figure of each trial, combined with the Excel data sheet, enabled the definition of the final temperature (t_f), i.e., the equilibrium temperature of each sensor.

Table 2 characterizes the thermal conditions of the tests by presenting the value of t_g for each sensor.

As it might be expected, Table 2 shows that the highest values correspond to the lowest distance between the sensor and the heat source. For the case of the 50 mm globe temperature, the results show that the temperature varied from 27.1 to 53.5°C. The corresponding values for the 150 mm sensors

Table 2. Thermal conditions of the tests.

		Final temperature (t_f) [°C]		
Trial	Position	G_50	G_150BK	G_150T
Natural convection	25 cm	53.5	66.8	68.1
	45 cm	42.2	51.5	53.6
	65 cm	34.1	41.7	42.2
V_a = 0.35 m/s	25 cm	38.2	54.5	56.9
	45 cm	37.2	48.0	48.3
	65 cm	30.5	38.5	40.8
V_a = 0.94 m/s	25 cm	34.1	52.1	53.1
	45 cm	32.2	46.9	47.4
	65 cm	29.6	38.0	38.7
V_a = 1.68 m/s	25 cm	31.5	46.7	46.5
	45 cm	31.1	40.4	40.8
	65 cm	29.8	36.7	36.7
V_a = 2.43 m/s	25 cm	30.1	43.8	43.2
	45 cm	28.3	37.8	37.7
	65 cm	27.1	33.2	33.2
RANGE	25 cm	30.1–53.5	43.8–66.8	43.2–68.1
	45 cm	28.3–42.2	37.8–51.5	37.7–53.6
	65 cm	27.1–34.1	33.2–41.7	33.2–42.2

were 33.2 and 68.1°C. The 50 mm globe presents lower values than the 150 mm globes and among these the globe from Testo has higher values in the majority of the cases. Finally, for each sensor the temperature decreases with the air velocity.

3.1 *Response time*

The thermal inertia of a temperature sensor is an important matter whenever a temperature measurement is foreseen. The response time is thus a requirement that must be duly considered. ISO 7726 (1998) states that the response time is in practice the time after which the quantity being measured can be considered to be sufficiently close to the exact figure for the quantity to be measured. The response time does not depend solely on the intrinsic characteristics of the sensor (mass, surface area, presence of a protective shield) but also on the environmental conditions (air velocity, thermal radiation, ...). ISO 7726 (1998) recommends that the conditions under which the response times are obtained should be specified (*vd.* Table 2). The response times of the sensors defined as the period necessary to reach t_f are listed in Table 3 In addition, for the present purposes, the time necessary to reach the value that corresponds to 90% of t_f is also presented. Moreover, ISO 7726 (1998) states that measurements should not be made before a period equal to at least 1.5 times the response time (90%) of the probe has elapsed and that to perform a measurement it is necessary to wait, as a minimum, for the time equivalent to the response time.

Table 3. Response time of the sensors.

Trial	Position	G_50 t_f	G_50 90% × t_f	G_150BK t_f	G_150BK 90% × t_f	G_150T t_f	G_150T 90% × t_f
		Response time [min]					
Natural convection	25 cm	25	9	44	8	33	11
	45 cm	23	8	27	6	22	12
	65 cm	21	6	36	6	28	9
V_a = 0.35 m/s	25 cm	20	5	26	5	47	10
	45 cm	27	4	36	6	27	9
	65 cm	15	4	32	5	36	8
V_a = 0.94 m/s	25 cm	23	4	25	5	36	9
	45 cm	17	4	30	6	40	9
	65 cm	33	2	12	4	20	8
V_a = 1.68 m/s	25 cm	12	3	36	6	26	9
	45 cm	14	2	46	6	29	6
	65 cm	18	2	15	3	20	7
V_a = 2.43 m/s	25 cm	15	3	24	4	29	6
	45 cm	21	2	8	4	31	6
	65 cm	32	1	46	6	24	5

For the 50 mm globe, the response time varied between 12 and 33 minutes. The corresponding range for the 150 mm globes was 8 and 46 minutes and 20 and 47 minutes, respectively for the globes from Brüel & Kjær and Testo. The results put in evidence the faster response time of the 50 mm globe. Furthermore, the globe from BK, generally shows a faster response time particularly if the 90%. t_f values are considered.

The present results show that the response time of the 150 mm globes may be greater than the period of 20–30 min usually stated. In some cases the final temperature is seen to be reached after 40 minutes. In addition, it is important to take into account that the highest globe temperature in the present study was 68.1°C, thus even higher response times are to be expected in the case of environments with higher globe temperatures.

3.2 *Difference between the 150 and 50 mm globes*

Table 4 lists the temperature differences between the two 150 mm globes and the 50 mm globe and between the two globes of 150 mm.

The results show that the greater differences between the 150 and 50 mm globes correspond to the higher temperatures, which is to say to the lower distances to the heat source. In the case of the difference G_150T-G_50 and for the lower distance between the globes and the radiative source, the

Table 4. Temperature differences between the 50 and 150 mm globes.

Trial	Position	G_150T-G_50	G_150BK-G_50	G_150T-G_150BK
		Temperature differences (ΔT) [°C]		
Natural convection	25 cm	14.6	13.3	1.3
	45 cm	11.4	9.3	2.1
	65 cm	8.1	7.6	0.5
V_a = 0.35 m/s	25 cm	18.7	16.3	2.4
	45 cm	11.1	10.8	0.3
	65 cm	10.3	8.0	2.3
V_a = 0.94 m/s	25 cm	19.0	18.0	1.0
	45 cm	15.2	14.7	0.5
	65 cm	9.1	8.4	0.7
V_a = 1.68 m/s	25 cm	15.0	15.2	–0.2
	45 cm	9.7	9.3	0.4
	65 cm	6.9	6.9	0.0
V_a = 2.43 m/s	25 cm	13.1	13.7	–0.6
	45 cm	9.4	9.5	–0.1
	65 cm	6.1	6.1	0.0
RANGE	25 cm	13.1–19.0	13.3–18.0	2.4–(–0.6)
	45 cm	9.4–15.2	9.3–14.7	2.1–(–0.1)
	65 cm	6.1–10.3	6.1–8.4	2.3–0.0

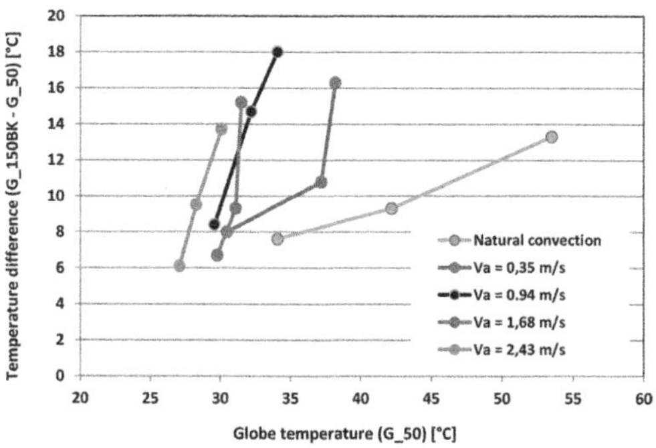

Figure 3. Temperature difference G_150BK-G_50 as a function of the temperature of the 50 mm globe.

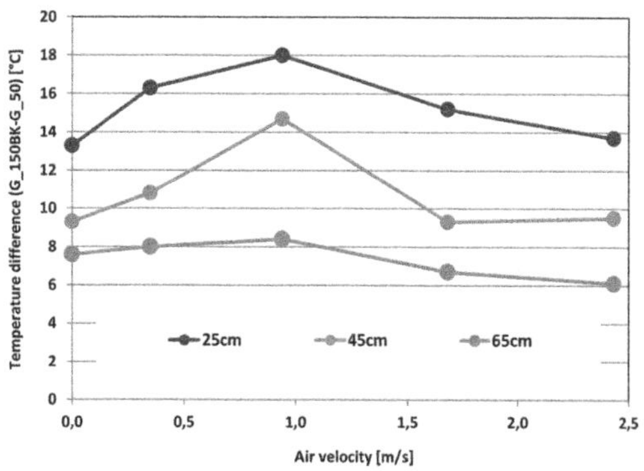

Figure 4. Influence of the air velocity on the temperature difference G_150BK—G_50.

temperature difference ranged between 13.1 and 19.0°C while in the case of the G_150BK-G_50 the corresponding variation was between 13.3 and 18.0°C. In either case the temperature difference is very significant. If the distance between the heat source and the globes is increased, i.e., the globe temperatures decrease, the corresponding extreme values decrease.

Figure 3 represents the temperature difference G_150BK-G_50 as a function of the temperature of the 50 mm globe and it shows that, for each velocity, the 50 mm globe temperature differences between the lowest and highest distances decrease significantly when the air velocity grows.

In natural convection this difference is 19.4°C while for the highest air velocity (2.43 m/s) the corresponding value is 3.0°C. An identical scenario is obtained for both 150 mm globes. Thus, whenever the temperature is changed, the air velocity increase tends to reduce the temperature difference.

3.3 *Effect of the air velocity*

Figure 4 puts in evidence the effect of the air velocity by showing the variation of the temperature difference (G_150BK-G_50) as a function of the air velocity. In fact this figure represents a different perspective of Figure 3 and shows that the difference G_150BK-G_50 obtained between L = 25 and 65 cm is 5.7°C in natural convection and 7.6°C for the higher air velocity (2.43 m/s). However, the highest difference (G_150BK-G_50 for the 25 and 65 cm distances) is 9.6°C and corresponds to an air velocity of approximately 1.0 m/s.

In case of the Testo globe the scenario is similar and the corresponding highest value is 9.9°C. Furthermore, it is worth mentioning that, for each distance between the heat source and the globes and as Figure 5 clearly puts in evidence, the highest difference always corresponds to the air velocity of approximately 1 m/s. For the case of the BK globe the highest differences (G_150BK-G_50) are 18.0, 14.7 and 8.4°C, respectively for the 25, 45 and 65 distances.

4 CONCLUSION

The present paper focuses the assessment of the globe temperature which accurate measurement depends on several features.

The response time results suggest that the response times usually stated in the literature are not met and that longer periods must be considered. Thus, the stabilization period of 20–30 minutes commonly adopted is not sufficient.

The influence of the diameter of the globe is addressed with a globe of 50 mm. The results show significant differences between this globe and the standard. It should be noted that the smaller the diameter of the globe, the greater the effect of the air temperature and air velocity, thus causing a reduction in the accuracy of the measurement of t_r (ISO 7726, 1998). In addition, the use of smaller globes leads to the measurement of lower globe temperatures.

The influence of the thickness of the copper plate of the globe is only briefly addressed. Despite not knowing the thickness of the copper plate of the globe from Testo, this sensor is heavier than the one from BK. This is probably the reason for the differences obtained in terms of temperature values and response times between the two globes.

The study of the effect of the air velocity on the globe temperature clearly shows a globe temperature decrease with the increase of the air velocity. In particular, the effect of the air velocity on the difference between the 150 and 50 mm globes and the "transition" that takes place around 1 m/s have to be further studied in order to fully understand such phenomena.

The main conclusion of the present paper is that the measurement of the globe temperature depends on many conditions. Therefore, its accurate measurement should be considered carefully. The present study represents a first approach to this matter that is actually being further addressed with complementary developments, namely by considering additional globe sensors with different thicknesses and different environmental conditions.

REFERENCES

Budd, G.M. 2001. Assessment of thermal stress—the essentials. *Journal of Thermal Biology* 26: 371–37.

Dukes-Dobos, F. & Henschel, A. 1971. The modification of the WBGT index for establishing permissible heat exposure limits in occupational work. HEW, USPHE, NIOSH, TR-69.

Dukes-Dobos, F. & Henschel, A. 1973. Development of Permissible Heat Exposure Limits for Occupational Work. ASHRAE Journal 9: 57.

Epstein, Y. & Moran, D. 2006. Thermal comfort and the heat stress indices. *Industrial Health* 44: 388–398.

ISO 7243, 1989. Hot Environments—Estimation of the Heat Stress on Working Man, Based on the WBGT—index (Wet bulb globe temperature). International Standard, 2nd Edition, *International Organization for Standardization* (ISO), Genève.

ISO 7726, 1998. Ergonomics of the thermal environment—Instruments for measuring physical quantities. International Standard, 2nd Edition. *International Organization for Standardization* (ISO), Genève.

Juang, Y. & Lin, Y. 2007. The effect of thermal factors on the measurement of wet bulb globe temperature. *Journal of Occupational Safety and Health* 15: 191–203.

Moran, D.S. & Pandolf, K.B. 1999. Wet bulb globe temperature (WBGT)—to what extent is GT essential. Aviat Space Environ Med 70: 480–484.

NIOSH, 1986. Criteria for a recommended standard: occupational exposure to hot environment. DHHS (NIOSH) Publication N° 86–113, 101–10, *National Institute for Occupational Safety and Health*, Washington DC.

Occupational Safety and Hygiene II – Arezes et al. (eds)
© 2014 Taylor & Francis Group, London, ISBN 978-1-138-00144-2

Different transposed legislation analysis in safety matter for construction sites: Three study cases

C. Reis
Universidade de Trás-os-Montes e Alto Douro, Vila Real, Portugal

C. Oliveira
Instituto Politécnico de Viana do Castelo, Escola Superior de Tecnologia e Gestão, Portugal

M.A. Araújo Mieiro & C. Santos
Universidade de Trás-os-Montes e Alto Douro, Vila Real, Portugal

ABSTRACT: The construction industry is characterized as a high-risk activity, being one of the biggest responsible for the deaths due to occupational accidents. Therefore, it's necessary to prevent and to adopt specific measures in order to improve the working conditions and reduce the accident rate. The knowledge of occupational risks, coupled with proper prevention, in which includes training and information, is the responsibility of all the interveners, whether they are technical, employers, employees, owners and public entities. The publication of the Council Directive 92/57/EEC on 24th June set the minimum Safety and Health requirements at temporary or mobile construction, as well as the figures of Safety Coordinator in Project and at Construction phase. Being a publication of the European Union, its members must to leap over for his internal law, adapting it to the reality of each Country. Its questioned how far the transpositions between three countries, are compatible within each other, having as base the original document? Considered as an object of study, the Council Directive 92/57/EEC, the Portuguese transposition, the Decree-law n.° 273/2003, 29th October, the Italian transposed the Decree-law n. 81, 9th April 2008 and the Real Decree 1627/1997, 24th October, from Spain, in order to compare the different transposed legislation between this three countries.

1 INTRODUCTION

1.1 *Directive 92/57/EEC*

The publication of the Council Directive 92/57/EEC of 24th June 1992, usually called "Construction site Directive", follows the article 16 from Directive 89/391/EEC of the Council of 12th June 1989 and, establishes the minimum requirements of the Health and Safety applicable to temporary or mobile construction. Nevertheless the applicability of the Directive, it's emphasized that it does not cover the activities of drilling and extraction in the extractive industries.

The current version (Directive 92/57/EEC) includes the changes conferred by the Directive 2007/30/EC of the European Parliament, the Council of 20th June 2007 and the Corrigendum, JO L 33 of 1993/02/09, Note that at the time of its initial publication, the State-Members were: Germany, Belgium, Denmark, Finland, France, Greece, Holland, England, Ireland, Italy, Luxembourg, Portugal and Sweden.

With its publication, for the first time, the Safety Coordinator regulates the safety and health matter during the project realization and the construction phases. Both have as their main functions, to ensure the adequate safety environment, whether at the level of technical definitions of the project, either at the level of the development of the work on the construction site.

1.2 *Decree-Law n. 273/2003 of 29th October*

On 29th October 2003, is published the Decree-Law n. 273/2003. This document revokes the Decree-Law n.155/1995 of 1st July and, continues to transpose into national law of Directive n. 92/57/EEC and its updates. This document is based on two fundamental objectives:

1. Take the philosophy consecrated of the General Principles of Prevention (GPP) to the building design act, in particular regarding architectural options and techniques to materialize, as well in the prevention of conception, allowing greater

efficiency in the configuration of the safety and health at work;
2. Strengthen the coordination between the different interveners, from the preparation of the project design, and also during the execution of the work, to boost the articulation and intervention sequences, contemplating the different requirement to planning the health and safety at work under a constructive undertaking.

Thus, the SCP (safety coordinator in project phase) is defined as a singular or collective person that performs, during the project elaboration, the coordination tasks in safety and health matter and, may also participate in the preparation of the negotiation process of the work and other preparatory acts execution of the work regarding the health and safety field. The SCP is nominated by the contractor whenever:

1. The design work is produced by more than one subject, since their architectural and technical options involves some complexity choices;
2. Intervening two or more companies at same time, including entity performer and subcontractors. Their specific responsibilities are to ensure that the author's project have always in mind, the PGP, collaborate with the contractor, always regarding safety and health matter, prepare the Health and Safety Plan (HSP) or, if prepared by another person designated by the contractor, proceed to the technical validation;
3. The works are subject to special risks, or have the need for prior notice opening construction site.

The SCC (safety coordinator in construction phase) is defined as a singular or collective person that executes, during the construction, the coordination regarding the safety and health matter. Their responsibilities are:

1. Give the proper support to the contractor in the elaboration and actualization of the prior notice;
2. Assessing the development and alterations of the HSP, and if necessary, propose to the performer entity, the appropriate changes in order to validate technically;
3. Analyze the suitability of the procedures safety sheets, and, if necessary, propose to the performer entity appropriate changes;
4. Verify the activities coordination of the companies and independent workers that are involved in the construction site, in order to prevent occupational risks;
5. Promote and ensure compliance with the HSP and other obligations of the performer entity subcontractors and the independent workers, in particular regarding the organization of the construction site, the emergency system, the constraints existing on the construction site and in the surrounding area, the work involving special risks, the special construction processes, the activities that may be incompatible in time or space, and the system of communication between the intervenient of the construction site;
6. Coordinate the control of the correct application of the work methods, in order to have influence on the safety and health field;
7. Promote mutual dissemination of information concerning occupational risks and its prevention among all those involved in the building site;
8. Register the coordination activities in terms of safety and health at work, in the work book under the legal regime. Failing that, it must be in accordance with a system of appropriate records, which should be established for the construction work;
9. Ensure that the entity performer take the necessary measures so that the access to the building site is restricted to authorized persons;
10. Regularly inform the contractor on the result of the evaluation of the existent health and safety on the building site;
11. Inform the contractor about the responsibilities of his role, under the present law;
12. Analyze the causes of the major accidents that may occur in the building site;
13. Integrate in the technical compilation, the arising elements from the execution of the works in their absence.

Whenever in the execution of the work, exists one of the following conditions: a total period of work exceeding 30 days and the simultaneous use of more than 20 employees; a execution period exceeding 500 days, corresponding to the sum of working days performed by each worker, the owner of the construction shall notify the competent authority in Portugal (ACT—Authority of Working Conditions). This communication and updates should be supported by the SCC and must identify all the characteristics of the contracts, the execution time, and the identification of all the interveners. The executing entity must place the prior communication and its updates, in a visible way on the building site.

1.3 *Real Decree 1627/1997 24th of October*

Subsequent to the publication of the Directive 92/57/EEC of 24th June, from the need for its transposition into the Spanish legislative and, the provisions of Article 6 from the Law 31/1995 of 8th November, 1995, is published the Real Decree 1627/1997 of 24th October, which defines the minimum requirements for safety and health applicable to the constructions, excluding extractive industries in open air, and/or subterranean and soundings.

Please, note that the Real Decree 1627/1997 was amended by the following publications: Real Decree 2177/2004 of 12th November, Real Decree 604/2006 of 19th May, Real Decree 1109/2007 of 24th August, and Real Decree 337/2010 of 19th March.

The Real Decree 1627/1997 declares the need for a safety coordinator in the project phase (as well as the application of the General Principles of Prevention), and at the construction phase, with the functions mentioned in the Decree-Law 273/2003 (Portugal). Although there is the similarity between these two documents, the document HSP have a complement called study of health and safety. This study shall always be executed whenever the execution of the contracted work, including already the project design, has a value equal or greater than €450,759.08, whenever there is a need for prior communication for the building site and, whenever the work includes construction of tunnels, galleries, underground works and dams. This study is done by the safety coordinator at the project stage. When the study of health and safety it's not required, it's necessary to prepare the base study of safety and health, in which it's prepared by a technical competent and subsequently validated by the safety coordinator in project. This study should include the following requirements:

1. Identification of occupational risks that can be avoided, indicating their technical solutions;
2. Occupational risks that can not be eliminated, specifying thus, preventive measures in order to minimize them;
3. Provide specific measures for the identified works that has special risks;
4. Complement with useful information to work day-to-day as well as for the subsequent work.

The prior communication for the building sites its responsibility of the executing entity, when it's foreseeable that the execution of the works involves one of the following situations:

1. Whenever it's expected that the period of use of employees exceeds 30 days, and that at some point, are more than 20 workers simultaneously;
2. When the volume of labour work estimated, i.e., the sum of days of work of all the workers exceeds 500 days.

The prior communication includes the study of health and safety, which will be permanently available to the ITSS and technical of the specialized organisms in safety and health issues on the competent Public Administrations.

1.4 *Decree-Law n. 81, 2008, Italy*

The General Principles of Prevention, published in Decree-Law n. 626 of 19th September 1994, particularly in the Article 3:

1. Assess the risks to health and safety;
2. Eliminate the risks in relation to the knowledge of technical progress and, when this is not possible, minimize them;
3. Reduce risks at the source;
4. Program aimed to prevent a complex that integrates consistently prevent the production company technical and organizational as well as the influence of the work environment;
5. Replace what is dangerous, for something that it's not, or less dangerous;
6. Comparison with the ergonomic principles in the conception of the workplaces, the choice of equipment and the definition of working methods and production, also to alleviating monotonous and repetitive work;
7. Priority for collective preventive measures, instead individual preventive measures;
8. Limit for a minimum workers who may or may be exposed to risks;
9. Limit the use of chemical, physical and biological agents during the work;
10. Monitor the workers health when exposed to special risks;
11. Avoid the workers exposure to the risk, for reasons inherent for each person;
12. Provide hygiene measures;
13. Provide collective and individual protection measures;
14. Provide emergency measures to be taken in case of first aid, fire fighting and evacuation of workers and serious and immediate danger;
15. The use of warning and safety signs;
16. Regular maintenance of facilities equipment, machinery, with special reference to safety devices, accordance with the manufacturers instructions;
17. Information, training, consultation and participation of workers and their representatives on issues relating to safety and health at work;
18. Provide appropriate instructions to the workers.

These basic principles already consider the rigor that Italy assigns to the themes of Health and Safety at Work.

On 14th August of 1996 was published the Decree-Law n. 494, which transposed into Italian law the Council Directive 92/57/EEC. After 12 years, was published the Decree-Law n. 81, 2008, revoking the Legislative Decree n. 494, and compiling a single document-directive for safety and health matter. This perspective is defined for temporary or mobile construction, whenever:

1. Simultaneous presence of several companies and the volume is less than 200 man per day;
2. Works that involves special risks; The SCP is responsible for preparing the HSP and Safety Coordination, which shall include risk assessment, procedures, preparations and necessary equipment for the fulfilment of good safety standards, as well as the costs and the preparation of a dossier containing useful information for the prevention and protection against risks exposed to workers;

   The SCC is particularly responsible for:

1. Checking if there are coordination and control;
2. The application of the HSP;
3. Adjust the appropriateness of the HSP document, organizing the participants of the undertaking;
4. Apply the disposition agreements between the social partners;
5. Notify the contractor or supervisor, using nonconformities;
6. Suspend the work if there are serious and imminent dangers.

Both coordinators, SCP and SCC, should meet the following requirements:

1. Be graduated in Engineering, Architecture, Geology, Agricultural Sciences and Forestry and having recognition by the employers or contractors of the experience of at least 1 year in the construction sector;
2. Be graduated in Engineering, Architecture and having recognition by the employers or contractors of the experience of at least 2 year in the construction sector;
3. Project designers, Industrial experts, Agricultural and having recognition by employers or contracting experience at least 3 years in the construction industry. Nevertheless, the appointment of these coordinators does not exempt the contractor or supervisor of their legal responsibilities.

Regarding the opening of the construction site, the Prior Notice shall be sent by the contractor or supervisor for the Local Health Unit and the Provincial Labour whenever:

1. There are more than one performer entity;
2. The only company has a working volume of not less than 200 men per days.

This document must be posted on the construction site and must be available to all the intervenient.

2 CONCLUSIONS

The European Directives publication has allowed the standardization of bases that must respect the discipline of Safety, Hygiene and Health at Work. This standardization, in the other hand, when transposed into national legislation for each country, is adapted for each of their realities.

In this work, it's possible to conclude that there are similarities in the laws of the three countries, namely the guarantee of safety conditions in the workplace, specifically in a building site, its responsibility of all involved and it's included in all consulted diplomas.

Regarding the health and safety plan (HSP), highlight to the Spanish legislation, for the differentiation of the type of Study of Safety and Health and Study Base of Safety and Health, for the overall value of the construction work, including the project design, and for the type of work, which turns out to require the technician to adjust the document to the type of project.

The General Principles of Prevention are common to all the analyzed diplomas and are undoubtedly a work base and an asset to the Health and Safety Occupational promotion across the Europe.

REFERENCES

Decreto-Lei n. 155/1995 de 1 de Julho. Diário da República n. 150 – I Série – A. Ministério do Emprego e da Segurança Social. Lisboa.

Decreto-Lei n. 273/2003 de 29 de Outubro. Diário da República n. 251/2003 – I Série – A. Ministério da Segurança Social e do Trabalho. Lisboa.

Decreto-Lei n. 81, 9 Abril 2008. Ministério delLavoro, della Salute e delle Politiche Sociali.

Directiva 92/57/CEE de 24 de Junho. JO L 245 de 26.8.1992, p. 6–22. Conselho das Comunidades Europeias. Luxemburgo.

Ley 31/1995, de 8 de Novembro. BOE n. 269 10-11-1995. Jefatura del Estado. Prevención de Riesgos Laborales. Madrid.

Real Decreto 1627/1997 de 24 de Outubro. BOE n. 256 25–10–1997. Ministério de Presidência. Madrid.

Real Decreto 337/2010 de 19 de Março. BOE n. 71 de 23/3/2010. Ministério de Trabajo e Inmigración. Madrid.

Pinto, A. (2005). Manual de Segurança, Construção, Conservação e Restauro de Edifícios. 1ª Edição – 2ª Reimpressão, Edições Sílabo, Lda. Lisboa.

Occupational Safety and Hygiene II – Arezes et al. (eds)
© 2014 Taylor & Francis Group, London, ISBN 978-1-138-00144-2

Preventive measures in the execution of metallic structures in Portugal

C. Reis
Universidade de Trás-os-Montes e Alto Douro, Vila Real, Portugal

C. Oliveira
Instituto Politécnico de Viana do Castelo, Escola Superior de Tecnologia e Gestão, Portugal

M.A. Araújo Mieiro
Universidade de Trás-os-Montes e Alto Douro, Vila Real, Portugal

T.M. Macedo Felgueiras
Vila Real, Portugal

ABSTRACT: The choice of this theme took into account the need to develop a Prevention Model appropriate to the execution of Steel Structures on building sites, in order to reduce the number of accidents existence in this type of work. The intention with the systematization of concepts was to detach some extremely important terms and definitions for the correct perception of the dissertation. It was made an exhaustive research on the existing literature about the subject under investigation, in order to enrich knowledge on this theme and understand how information is organized throughout the process of transformation of structures that are subject of study. It was on the other hand, developed a questionnaire to understand how steel construction companies manage their motivations and difficulties in implementing the concept of Safety at Work and act at that level. With such information were asked a few companies and the results were analyzed. The ultimate objective of this research work focused on obtaining a Prevention Model, this has been developed from tables of identification and risk prevention, adapted to the circumstances of each stage, in assembling on site of steel structures.

1 INTRODUCTION

1.1 *Building safety*

As is generally known, the sector of the construction is the one that presents the higher rates of labour accidents, mainly due to their specificities. The occupational safety, particularly in construction, has gradually evolved in a positive way over the years. Historically, safety as a synonym for the prevention of accidents has evolved in a crescent way, which consists in a growing number of factors and activities from the first actions for damages (injuries), to a broader concept which is sought to prevent all situations that cause unwanted effects to work. There's important authors in this field, such as John Gambatese, as well as López-Arquillos, who studied the risks in similar construction works. Both defend that the formwork activities are associated with a high frequency of accidents and injuries (Gambatese et al. 2013), thereby justifying this study research.

1.2 *Metallic structures*

It's relevant the growing demand for metallic structures that have been considered structural solutions in the construction sector in Portugal. As can be seen from Figure 1, it appears that since 2002 there has been a considerable increase in the production of metal construction work in Portugal. It can be seen a relative increase in the production of this type of construction, in the period of time ranging from the years 2002 to 2009, except for 2005, where there was a slight fall in the number of tons produced the material under investigation. To highlight the fact that the values of the years 2008 and 2009 have been obtained on an estimated basis, accordingly the data collected on Agência Europeia para a Segurança e Saúde no Trabalho.

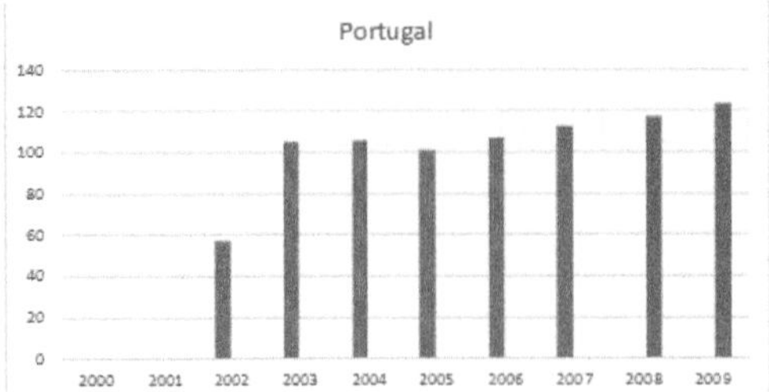

Figure 1. Production of metallic construction work in Portugal.

2 METHODOLOGY

2.1 *Study questionnaire*

This study research consists of a development of a questionnaire directed to some construction companies such as BySteel, SA, Frisomat Portugal, Martifer Metallic Constructions and Tegopi, SA. The objective of this research is to study, among other cause of accidents, risks, procedures and preventive measures, safeguarding the safety at work. I highlighted that, in 15 construction companies, only the four previously mentioned, were receptive to respond to this questionnaire. It is also worth to mention that these data were provided by those in charge in the safety matter of each company. This sample would be faithfully representative, if indeed encompassed a larger number of construction companies, not abstaining from the conclusions which are possible to take after this study.

The questionnaire has a set of questions that were asked to the companies responsible for the manufacture and assembly of metal structures in order to understand their concerns regarding health and safety matter in construction.

1. What are the most common risks when implementing these structures?
2. What types of welding are used for their manufacture?
3. What are the most common risks during transportation of the metallic structures?
4. What are the main causes of accidents during the transportation of structures to the construction site?
5. What types of safety procedures your company have during the transportation of the structures to the construction site?
6. What types of work platforms used for assembly?
7. What types of welding are used for assembly?
8. What are the most common risks during the assembly phase in the construction site?
9. What are the main causes of accident at work during the assembly of metal structures?
10. What types of safety procedures your company have for the assembly in this type of structures in the construction site?

3 CONCLUSIONS

It's possible to elaborate an analysis/interpretation of the results based on the proposed questions in the questionnaire, and for which can be drawn conclusions that are of particular interest to the topic of this work.

Questions related to the fabrication of metallic structures.

It was possible to determine by the companies that were under study that the risk involving falling material stand out from other risks mentioned by the companies, regarding to the risks most likely to occur in the process of manufacturing of this kind of structures.

Regarding types of welding used in the factories, it's clear that the semi-automatic welding (MIG7MAG) and the submerged arc are the most used as observable in Figure 3.

Questions related to the transportation of metallic structures.

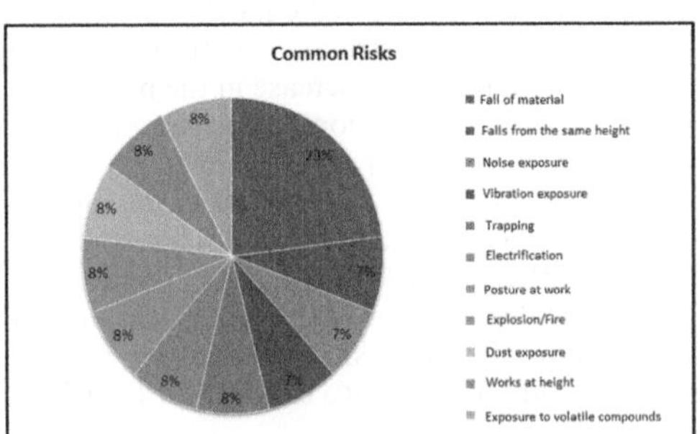

Figure 2. Common risks on the manufacture of metal structures.

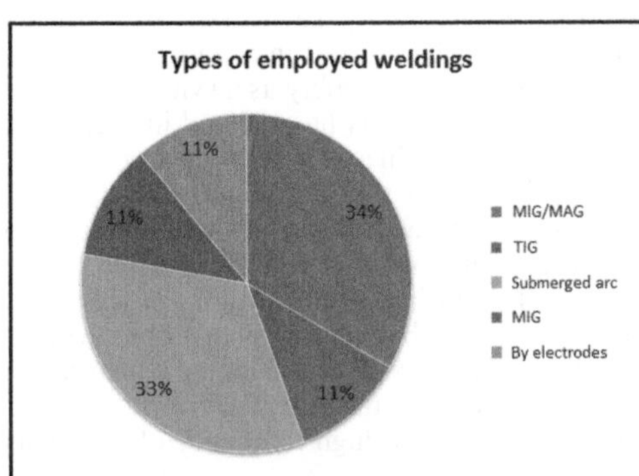

Figure 3. Types of welding works employed on the manufacture of metal structures.

Taking into account the graphical results, it appears that the falling material is presented as the most frequent risk, among several mentioned by the companies at the transportation stage to the construction site.

Regarding the main causes of accidents, the companies presented several causes as can be seen from Figure 5, although it should be noted the material poorly conditioned as the main cause.

As can be seen from Figure 6, there are several types of safety procedures taken by the companies, during the transportation of the metal structures to the building site. Still, it's important to refer that the proper packaging and the material signalled were the procedures most mentioned.

Questions related to the assembly stage in the construction site.

Regarding the type of platforms work used in the assembly of the metallic structures, the companies answered several types as shown in Figure 7, with a particular focus on movable platforms lift instead of the fixed types, as would be expected in this type of work of metallic construction.

About the types of welding types used in the assembly phase of metallic structures in the construction site, it was observed that the welding semi-automatic MIG/MAG assumes a greater preponderance in relation to the others.

In the Figure 9, it appears that the risks that involves a higher probability of occurrence at this

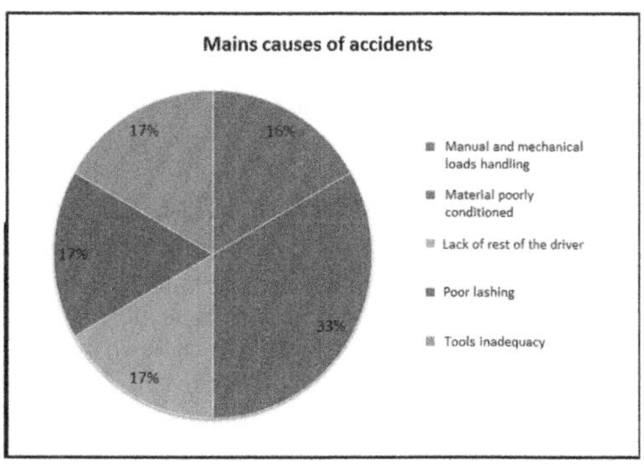

Figure 4. Common risks at the transportation phase.

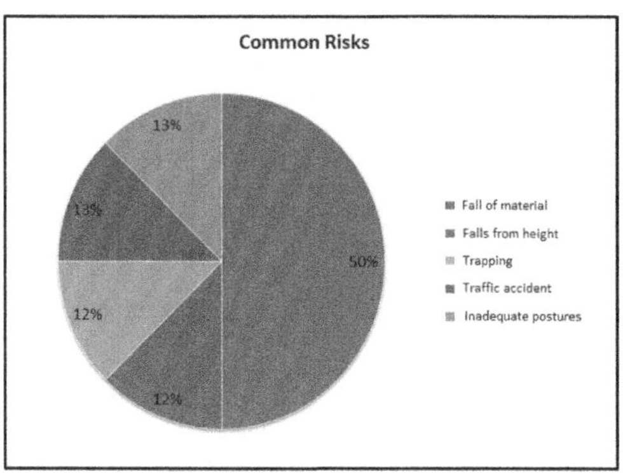

Figure 5. Main causes of accidents on the transportation phase.

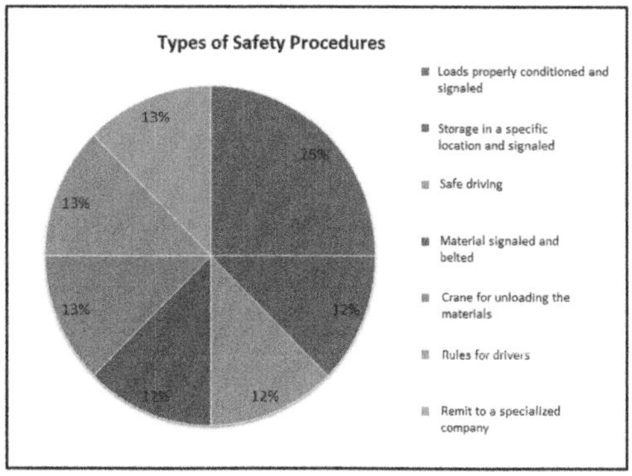

Figure 6. Types of safety procedures during the transportation phase.

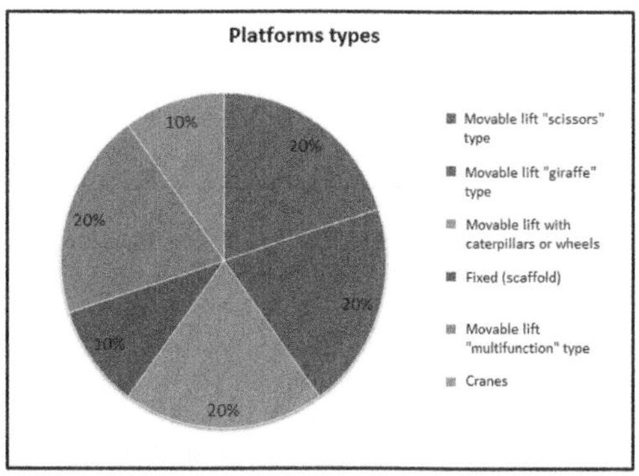

Figure 7. Types of working platforms used in assembly.

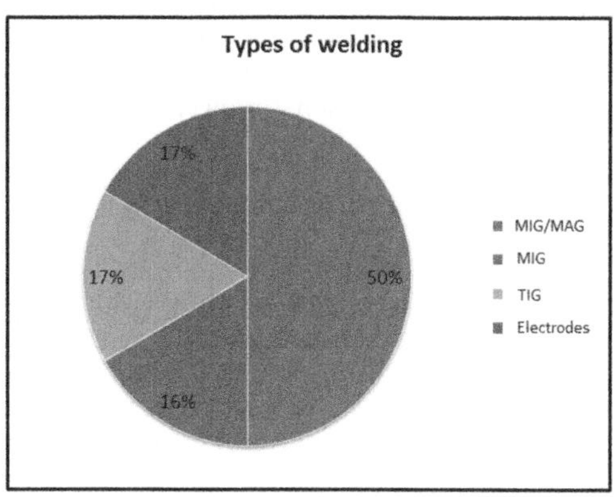

Figure 8. Types of welding used in the assembly phase of metallic structures, in the construction site.

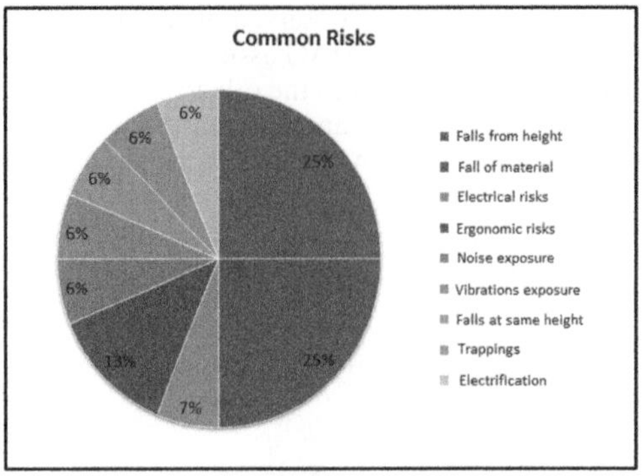

Figure 9. Most common risks during the assembly of the metallic structures, in the construction site.

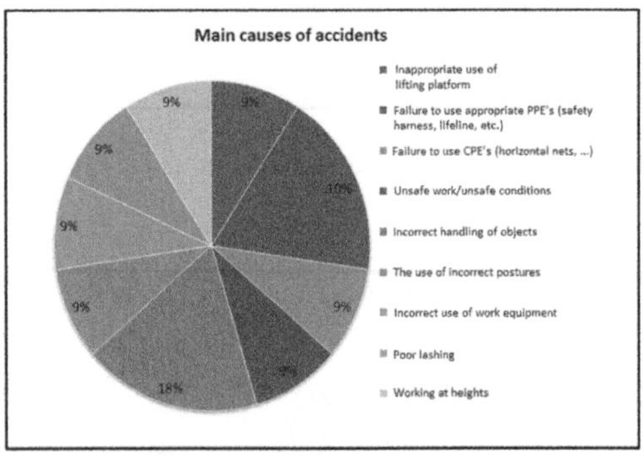

Figure 10. Main causes of accidents at work during the assembly phase of metallic structures.

stage of the assembly process in the construction site, are those that are associated with falls from height and falling material. With a lower incidence, but nevertheless with a greater importance of all others mentioned, may also be mentioned ergonomic risks.

In relation to the main incidence factor of accidents in the assembling of metal structures, several causes were pointed by the companies as shown in Figure 10. It's relevant to point the non proper application of the personal protective equipment (PPE's) and improper handling of objects, emphasizing that the number of cases of study do not permit to have a great accuracy what will eventually occur in reality.

According to Figure 11, it can be concluded that, the compliance with the Health and Safety Plan was the preventive instrument referred by the companies as the main procedure to take into account. Nevertheless, it can also be mentioned the provision of specific training, the use of protection devices, inspection/maintenance of machinery, the existence of means of fire fighting, the existence of safety animators and noise assessment.

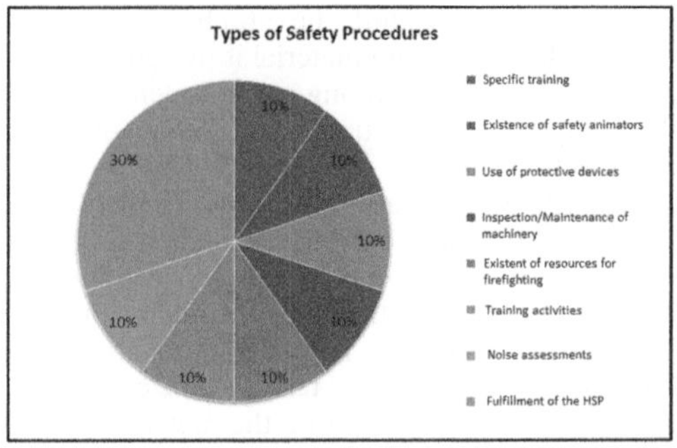

Figure 11. Types of safety procedures during the assembling phase.

Taking into account the obtained results, it is emphasized to the fact that only four inquiries were answered, which of course made unfeasible another rigor in the results interpretation.

However, it is noteworthy that the author Reis (2011), in a study about 709 accidents analysis in construction, verified that when it came to accidents caused by falls from height, many of them were at the time of steel structures installation and execution, and also for the non-use or incorrect use of the personal and collective protective equipment. This same author, also verified that the cause of construction accidents dominated the inadequate use of the work platforms. Also according to Reis (2011), the causes that contribute to accidents, like falling from height, are mainly the absence/non-use of protective equipment (collective and/or individual) and also, the poor training or lack of checking the condition of the material. That fact is also confirmed by the four companies interviewed, and possibly the answers would be similar if there were more companies answering to this questionnaire.

REFERENCES

Agência Europeia para a Segurança e Saúde no Trabalho. 2012. http://osha.europa.eu/pt/.

Felgueiras, T.M. 2011. Medidas preventivas na execução de estruturas metálicas. Vila Real.

López-Arquillos, A. Rubio-Romero, J.C. Gibb, A.G.F. Gambatese, J.A. 2013. Safety risk assessment for vertical concrete formwork activities in civil engineering construction.

Miguel, A. 2010. Manual de Higiene e Segurança do Trabalho. Porto Editora, 11ª Edição. Porto.

Reis Cristina (2011) - Análise dos acidentes na construção/ Analysis the construction accidents. International Symposium Occupational Safety and Hygiene - SHO 2011 Guimarães 10 e 11 de Fevereiro de 2011. ISBN: 978-972-99504-7-6, Book in 1 volume, 720 pages. Pages 368 a 372.

Occupational Safety and Hygiene II – Arezes et al. (eds)
© 2014 Taylor & Francis Group, London, ISBN 978-1-138-00144-2

Preventive measures in rehabilitation of buildings—facades shoring

C. Reis
Universidade de Trás-os-Montes e Alto Douro, Vila Real, Portugal

C. Oliveira
Instituto Politécnico de Viana do Castelo, Escola Superior de Tecnologia e Gestão, Portugal

M.A. Araújo Mieiro & V. Pereira
Universidade de Trás-os-Montes e Alto Douro, Vila Real, Portugal

ABSTRACT: The rehabilitation of building consists in restitute some capabilities to the building, that until then had lost. This procedure is extremely important, but also difficult and delicate since most of the times the degradation (of the building) is very pronounced. The present work focuses on the analysis of the shoring of the facades that were made in a building located in the council of Braga, during its rehabilitation work, and the preventive measures that are inherent in this type of intervention. The state of the walls and its structures and sometimes the ignorance of execution/procedure of the work could bring undesired risks in terms of occupational safety during the rehabilitation of buildings.

1 INTRODUCTION

1.1 *Rehabilitation in Portugal*

The housing need, which go against the degradation and abandonment of numerous fires entered in urban centers, make urban renewal/reconstruction an imperative issue in Portugal. However, much of this rehabilitation work has no project, is insufficiently prepared and involves construction companies with poor experience to perform such intervention.

The rehabilitation of buildings are mostly carried out by micro and small companies, that tends to give less importance to quality, environment and occupational safety and health. Both the owners of the work and the construction companies have serious difficulties in meet and comply all legal and regulatory frameworks that are currently in force, particularly in terms of quality of the materials and technologies to employ. It's important to refer that most of the times the records of the building are not trustworthy and/or credible, which compromise the rehabilitation and maintenance to run safely. The variety and unpredictability of this type of work requires construction companies with skills in diverse technologies, in order to obtain quality end products, made in safety and meeting the customer expectations.

1.2 *Framework*

In general, rehabilitation has the particularity of acting on an object built and with an aspect abandoned, but not necessarily in ruins. It's a process that restores the building, skills until then had lost. This is extremely important, but difficult to implement due to degradation of the building in most cases. The advances that take place continuously, whether in the field of materials, either in structures design, are thought primarily to the direct application of new concepts. However, the old buildings, plus the equity value they present, occupy a significant area in many historic city centres, often in poor condition, so it's urgent to proceed with its redevelopment with appropriate interventions.

2 METHODOLOGY

2.1 *Study case*

Work visits were made to the old CTT (mail post) building in Braga, Portugal. This work was being performed by the construction company PERI, in which allowed the assistance to the development of this study research. Note that this construction was already in an advanced state, so it was not possible to verify the shoring of the facades.

2.2 *Work Procedures with Special Risks (WPSR)*

This document is intended to establish the minimum conditions of safety to obey during the assembly of mettalic structures for the containment of facades. It's the obligation of the performer entity that these conditions are fulfilled. In case of non-compliance, it should be reported to the person responsible for the work by the safety coordination (as established in Decree-Law n. 273/2003 of 29th October).

The equipment used are varied, from the multi-loader, to angle grinders. For an effective safety during the construction process, a metal structure to contain the facades was mounted on the exterior of the building, with locks on doors and windows areas, so that the wall do not collapse into the interior of the building.

Being a rehabilitation work, and is in the historic center, one of the constraints would be the automobile circulation arising on the site, adding to the movement of machinery inherent of the work, increasing the risks for those who circulate and works on the site, including collision, fall height, falling materials and trampling. A detailed list of preventive measures was emitted in order to minimize or eliminate the risks mentioned above.

The assembling of the metallic structures by itself, is a complex and lenghty process and, involves serious risks for those who perform this tasks, such as falling from heights, trapping, cutting and crushing. For a greater individual and collective protection, it was necessary to to implement the PPE (Personal Protection Equipment) and CPE (Colective Protection Equipment). The implemented PPE is especially the protection helmet, steel toe boots, reflective vest, safety harness, mechanical protection gloves, googles and ear protection. As for the CPE, it was used the temporary signage, signage tape, lifeline, guardrails and physical barriers.

For each task inherent to the work, safety procedures sheets was implemented, in which describes the most frequent risks, the main causes of risk and the appropriate preventive measures.

Figure 1. Side elevation from Av. Liberdade with shoring.

2.3 *Preventive measures in the act of shoring up the facades*

The preservation of the facades is becoming into a more frequent way to meet the development needs, offering to the owner, an internal structure suitable to the needs of the users without creating a major impact on the urban environment in which it is inserted. These facades are in most cases resistant structural walls, and during the rehabilitation, there must be a great care about the current state of these walls and even conduct some repairs if needed, in order to avoid the collapse of the structure. A shoring work, carried out in a deficiently way, has the potential to allow the collapse of the building with disastrous consequences and probably, with deaths. It's important to highlight the crucial importance to follow the safety procedures, in order to have a continuous and, above of all, a safe work.

For each task/action to perform, it's necessary to elaborate the possible risks and their respective preventive measures to combat them. Below, is presented a list of risks and therefore, preventive measures for each task:

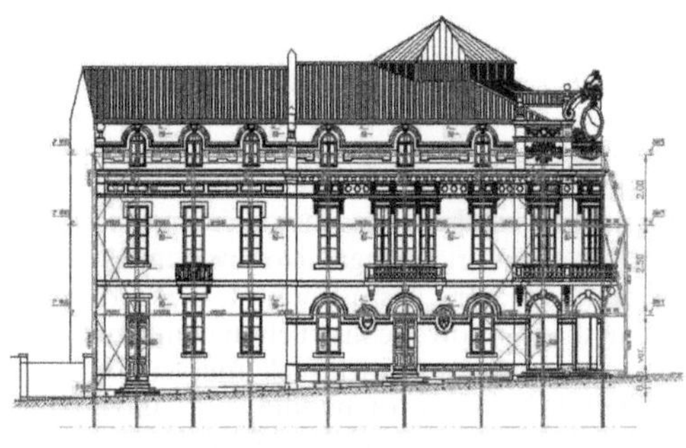

Figure 2. Side elevation design from Av. Liberdade with shoring.

Figure 3. Side elevation from street Dr. Gonçalo Sampaio with shoring.

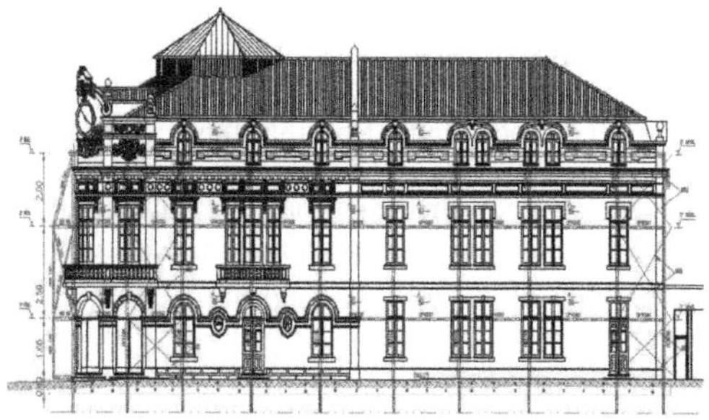

Figure 4. Side elevation design from street Dr. Gonçalo Sampaio with shoring.

Constraints risks:

1. Collision;
2. Falling from heights;
3. Falling materials;
4. Trampling.

Preventive measures:

1. The construction site will be prohibited through the placement of physical barriers and ribbon signaling with red and white stripes, to prevent the passage of vehicles and persons outside the service, as well to restrain the acess to the paths used for the movement of machinery.

Risks in the transport and packaging material:

1. Crushing;
2. Trampling;
3. Falling materials;
4. Cuttings.

Preventive measures:

1. The transportation should be done by the shortest possible route, in order to minimize excessive movement of the load;
2. The pieces should be well supported on beams and will be handled with the aid of straps or chains with safety latches on hooks;
3. The materials will be unloaded and placed in an area close to the assembly site, to avoid the probability of occurrence of accidents due to moving parts and machinery;
4. The elements stored in a vertical position must be leaned against a resistent element, under wooden crossties, with an angle that ensures stability and in order not to damage the engaging elements.

Risks in traffic areas and intervention:

1. Trampling;
2. Falling materials;
3. Cuttings.

Preventive measures:

1. The area where the work proceeds must be unimpeded, regularized and compacted, to allow the proper movement of machinery and workers;
2. The whole area around and under the assembly area will be prohibited and signaled voncerning the circulation and permanence of the workers;
3. Do not stay in range of the machines while in operation;
4. Proceed with the cut of all the existent infrastructure (water, electricity, gas, etc.).

Risks when handling pieces:

1. Falling from heights;
2. Trapping;
3. Falling materials;
4. Cuttings.

Preventive measures:

1. When moving pieces through the use of lifting equipment, it'll be necessary to use two support points in order to minimize possible oscillations and possible falling materials;
2. The pieces will be handled with the aid of polyester fiber straps;
3. In the pillars case, its positioning is accomplished using only one fixation point, and therefore should be an extra careful with the permanence of people nearby the pieces movement;
4. It's not permitted the permanence of unnecessary personnel in the proximity to the moving pieces;
5. Each element must be moved at least by two men. One is leading the element using a guide rope and the second worker is maneuvering from the lifting equipment;
6. When placed correctly in the site, it should proceed to the final assembly of the element, before disconnecting it from the point of suspension and without releasing the guide ropes;
7. Workers who have been assigned to the task of receiving pieces of metal in places with risk of falling from height, must wear safety harness, tying to structural fixed and safe elements;
8. Whenever it's not possible to avoid manual handling, the employer must take measures that attenuate the hardship of work and avoid the risks, including the taken routes with loads should be as shorten as possible;
9. Hand loads weighing more than 30 kg in occasional operations or 20 kg in frequent operations, difficult to handle, must be moved at least by two workers.

Risks in the assembly:

1. Falling from heights;
2. Trapping;

3. Falling materials;
4. Cuttings;
5. Trampling.

Preventive measures:

1. Access to workplaces only through the safe zones: stairs, walkways with guardrails and platforms;
2. If the work is performed in the absence of collective protection, workers should use safety harness secured to fixed resistant elements;
3. The placement of the pieces in the correct work place should be done using a multi-loader;
4. Operations must be performed synchronously. The heavy elements have a large inertia, so a slight oscillation is enough to knock down a man;
5. If some element starts to rotate about itself, it's necessary to control using the guide ropes. It should be strictly forbidden to directly use the body to its control;
6. The works should be suspended when there's wind exceeding 40 km/h, pluviosity or heavy fog;
7. In high places, without collective protection and where there are places where it can engage the safety harness, it must be installed safelines properly secured and resistant to fixed points;
8. The staff access to the structure for the realization of the screwing operation and other work related to the assembly, will be done using a platform lift with guardrails;
9. Workers who move themselfs in the lift platform, must be equipped with safety harness attached to a fixed point of the structure;
10. To assist in the positioning of assembled pieces at height, it shall be used ropes attached in the extremities to allow the workers to carry out the steering elements from the floor;
11. The components of the metallic structure can only be disconneceted from the lifts means, after being properly anchored, or permanently fixed.

2.4 *Constructive metallic phasing*

Below are presented the various stages of the construction process which involves assembling of the shoring system in welded metallic structures:

1. Analysis of the conditions of integrity of the facades and, installing and zeroing the targets to put on the facades as defined in the Plan of Instrumentation and Observation;
2. Confirmation of the geometry and elevation level of the facade foundations to preserve;
3. Implementation of micropiles foundation of the metallic structure to contain facades. The eventual drilling of the burried structural elements for the installation of micropiles, shall be performed by means of core drilling. This elements should have a sufficient sealing length, at the level of rocky massif, so that they can transmit to the ground by lateral attrition, the loads from the walls and the provisional metallic structure. To achieve this objective, it should be adopted cement grout with appropriate characteristics and, filling injection and sealing should be executed with dual shutter and non-return valves;
4. Execution of the repaving beams and the pier heads exterior of micropiles, foundation of the metallic structure to contain the façade that faces the street Gonçalo Sampaio. Assembly of the provisional metal structure;
5. Manual phased demolition of the inner structure (according to the project);
6. Realization, if necessary, injection of consolidation and reinforcement of the facade to preserve, in particularly in areas where will be executed the connections to the provisional metallic structure and, to the head pier of the micropiles foundation;
7. Work platform execution, in order to allow the circulation within the demolished zone, from the equipment for the execution of the repaving beam and the containment walls under the existing facade;
8. Execution of tie rods prestressed that will allow the sewing massifs, interior na exterior, to the facade wall to preserve. The sealing of the tie rods should be held with cement grout with appropriate characteristics;
9. Excavation and peripheral containment under the façade;
10. Implementation of the metallic structure containment of the wall façade to preserve, and the deactivation of prestressed tie rods;
11. Demolition of the reinforced concrete elements, whose geometric not compatible with the solutions provided in the projects of the remaining specialties, and to determine the occupation of the public road, in particular, the pier heads of the micropiles foundation of the metallic structure.

3 CONCLUSIONS

In Portugal, the conservation and rehabilitation of old buildings is becoming more relevant, which reflects in a growing area in the construction sector. It's important to ensure that the works that are associated with this intervention and, within these, the shoring facades, are carried out in order to ensure safety, whether from the point of view of the preservation of its architecture, but more importantly, ensure the safety of human life.

During the shoring of the facades, there must be a special attention to the current state of the walls and, if necessary, carry out some repairs to prevent the collapse of the structure, when it's proceeding the shoring of the facades, with disastrous consequences, may even be fatal. A shoring poorly built, has the potential to permit the collapse part of the structure, making it vital that this kind of intervention have to be designed by structural engineers or other competent professionals.

With this study, is expected to be able to transmit to all the interveners involved in the rehabilitation works, especially when these interventions requires shoring of facades, how important is the safety issue, and take special attention to the preventive measures that are inherent in this kind of work, in order to allow a safety construction, since Portugal is the EU countries with the highest rates of accidents at work in the construction sector.

REFERENCES

Appleton, J. 2003. Reabilitação de Edifícios Antigos – Patologias e tecnologias de intervenção, Edições Orion, Amadora.

Dos Reis, C.M. 2007. Melhoria da eficácia dos Planos de Segurança na redução dos acidentes na construção – Dissertação de Doutoramento em Engenharia Civil, Faculdade de Engenharia da Universidade do Porto, FEUP.

Gaspar, José de Freitas; Gerardo, Ana Margarida de Abreu. 2005. Informação Técnica – Segurança e Saúde no Trabalho: Construção: Manual de Prevenção; ISHST, Lisboa.

Gonçalves Pereira, V.H. 2010. Medidas preventivas na reabilitação de edifícios antigos – Escoramento de paredes. Vila Real.

Hume, Ian. 1997. Scaffolding and Temporary Works for Historic Building – The Building Conservation Directory.

Silva, V.C. 2007. Reabilitação Estrutural de edifícios Antigos – Técnicas Pouco Intrusivas, Geocorpa/Argumentum, Lisboa.

Occupational Safety and Hygiene II – Arezes et al. (eds)
© 2014 Taylor & Francis Group, London, ISBN 978-1-138-00144-2

Assessment of the health and safety plan on Vila Nova de Gaia county

C. Reis
University of Trás-os-Montes and Alto Douro, Vila Real, Portugal

C. Oliveira
Instituto Politécnico de Viana do Castelo, Escola Superior de Tecnologia e Gestão, Viana do Castelo, Portugal

M.A. Araújo Mieiro & P. Freitas
University of Trás-os-Montes and Alto Douro, Vila Real, Portugal

ABSTRACT: The developed study took in consideration the analysis of the quality of 36 safety and health plans in relation to municipal constructions taking place in Vila Nova de Gaia, in order to analyze the professional risk prevention instruments not only by observing what has been specified by the Portuguese legislation but also by what the best exercise of civil constructions advises. The applied methodology was based in the definition of a safety and health plan with a certain internal structure and then the sample plans were analyzed by comparing with the model plan. The results were frankly favourable and cheerful. Never the less it was registered an absence of important requirements defined in decree nr 273/2003, from the 29th October, especially in small dimension and less complex construction sites. As a consequence of the developed work stood out the need of a much larger awareness of the importance of safety and health plans, in respect to the technicians that create this professional risk prevention instruments, so this is what is proposed: not only a bigger exigency by the municipal entity that approves the enterprise licensing, on the observation of the legislation that rules the plan elaboration, but also, a control more active of the Work Condition Authority, in the applying area of the professional risk prevention measures.

1 INTRODUCTION

Historically, the work safety and health theme has been occupying a relevant role in governmental politics of industrialized countries, as a consequence of social movements and the evolution of philosophical and politic thought, accordingly Cabral, F. (2011). The national states and the international organizations have been adopting social guide-lines in order to normalize the work conditions. The research work presented here focused on the analysis of the safety and health plans quality, on the project stage, referring to civil construction work in Vila Nova de Gaia council, in Portugal, using the ruling legislation, specifically the Decree nr. 273/03, October 29th.

2 METHODOLOGY

The developed study was based on the references for the elaboration of a safety and health plan specified on L.M. Alves Dias and M. Santos Fonseca literary work: "Plano de Segurança e Saúde na Construção."

The collecting of data lasted nine months and 36 safety and health plans referring to the years from 2005 and 2010 were analysed. This documents were provided by the City Council of Vila Nova de Gaia.

For the gathering of data was elaborated a questionnaire subdivided on two chapters:

1. Type of organization of the safety and health:
   - Face sheet;
   - Preparation and presentation sheet;
   - Actualizations/corrections sheet;
   - Distribution sheet;
   - Signings sheet;
   - Global index;
2. Elements to incorporate on the safety and health program:
   - Descriptive note that includes: objective definition, previous communication, appropriate regulation, functional organogram, work schedule, work accident insurance and other,

timeline for the enterprise execution, methods and constructive processes.
- Characterization of the enterprise that includes: global characteristics, work qualities map, work program, workmanship chronogram, project of the building site, list of special work risks and list of dangerous materials.
- Actions for prevention of risks that comprehends: action plan on the conditioning existing in the local, signalization and circulation plan of the building site, collective protection plan, individual protection plan, utilization and control of the equipment of the building site, inspection and prevention plan, workers health plan, registration of accidents and disaster numbers plan, workers formation and information plan, visitors plan, emergency plan.

This way, was verified if the safety and health plan had or not the several items from each of the two chapters, having then delivered a classification at the end of each chapter, in function of the number of items included on each plan.

The classifications scale varied between "Very Weak" and "Very Good". Each obtained grade resulted of an arithmetic average, in which each chapter that included all items received a 100% grade, corresponding to "Very Good". In case of only half of the elements were present, the grade would be of 50%, corresponding to "Average". The valuation process was kept of in between grades, following the formula:

$$Grade = \frac{n}{nt} \times 100$$

where: n = quantity of verified items; and nt = quantity of items for chapter.

Following this valuation criteria, was considered the following scale of evaluation for the 2 chapters that should have a place on the safety and health plan:

"Very Weak" = 0%–25%
"Weak" = 26%–49%
"Average" = 50%–70%
"Good" = 71%–85%
"Very Good" = 86%–100%

Table 1. Evaluation considerations.

	Number of items	Deliberation (%)
Health plan organization	6	10
Descriptive document	8	28
Undertaking characterization	7	24
Risk prevention actions	11	38

As for the individual evaluation of each safety and health plan, has been standardized a specified weight for each one of the 2 chapters. It was considered that the chapter referent to the type of organization on the safety plan had a more formal point of view than substantial so was valuated with 10%. The second chapter, elements to include on the plan, was valuated with 90% proportionally divided in function of the number of items that made part of each one of the 3 elements.

This way, the next table stands out the given considerations:

This way the final evaluation of each safety and health plan was a result of the considered average obtained following this next formula:

$$x = OP \times 10\% + MD \times 28\% + CE \times 24\% + AP \times 38\%$$

$$Grade = \frac{n}{nt} \times 100$$

where: x = considered average; OP = safety plan organization; MD = Descriptive notes; CE = enterprise characterization; and AP = Risk prevention actions.

The average formula was created by Freitas, P.E. (2013), with the intention to have numerical final results regarding the HSP evaluation.

3 DATA ANALYSES

In relation to the Health and Safety Plan (HSP) organization chapter, was verified that 24 of the 36 analyzed plans, 77% of the sample, had a classifications varying between "Average" and "Good", as can be seen in Figure 1.

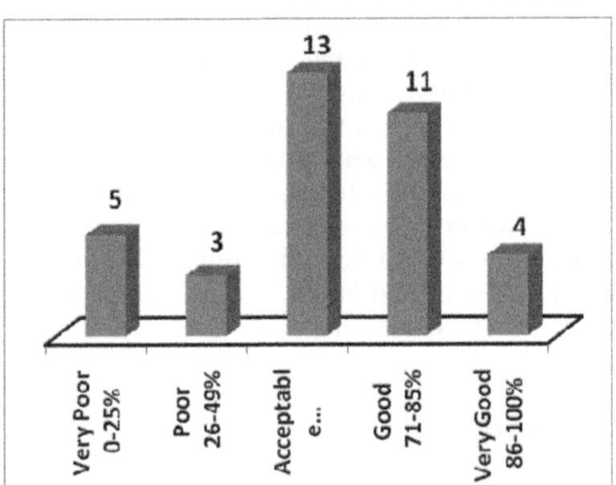

Figure 1. Evaluation of the "HSP organization".

The items with higher accomplishment percentage were the face sheet and global index. On the other hand, preparation and approval sheet registered a low presence on the elaboration of the analyzed plans. It also stands out that 4 plans fully followed the defined requirements for this chapter.

As for the descriptive notes, the global evaluation was satisfactory, as 26 plans followed most of the 8 requirements that shaped this safety and health plan element.

It is worth noticing that 4 plans presented all the items that were being analyzed, obtaining maximum score. The functional organogram and the stages requirements of the enterprise execution registered the lowest number of occurrences: 14 out of 36. The accident insurance item was observed in 28 plans, reflecting the commitment of the lead technicians as for the workers protection. As for the previous communication, 4 plans did not include it, as the ruling legislation does not demand the observance of this item, because of the fact that these were constructions with execution limits inferior to 30 days, and employed less than 20 workers.

With the inclusion of the enterprise characterization element on the elaboration of the safety and health plan, it is looked into describe with detail the building to construct, having in sight a rigorous perdition of the safety and health risks on the stage of project, and also on construction stage. Globally 64% of the sample satisfactory verified the seven requirements of this element. The chronogram of workmanship was absent in 19 plans. It stands out that the jobs with a special risk list had a frequency of 86%, being present in 31 plans as shown in Figure 2.

From the registered observations, was clear the technician worries for the following of the items who shape this element.

As for the risk prevention actions, it should be said that this will be as effective as better the enterprise characterization is. The number of actions to have in consideration in the risk prevention varies in direct relation to the dimension and complexity of each construction. As a matter of fact, it was verified that the obtained results in risk prevention actions are linked to the ones reached in the enterprise characterization. This way, 83% of the studied sample fulfilled the requirements of this element, present in a safety and health plan, as 17 plans had a "Good" or "Very Good" classification. Worth noticing that the highest frequency items were: emergency plan–92%; individual protection plan–89% and inspection and prevention plan–78%. Just 14 plans presented visitor plan. In the final and global analyses to the group of samples and studied characteristics for each safety and health plan, was concluded that 28 plans, corresponding to 78%, had a positive grade, 19 of them frankly favourable with evaluations of "Good" or "Very Good". The remaining 8 plans registered visible fails in their elaboration as seen in Figure 3.

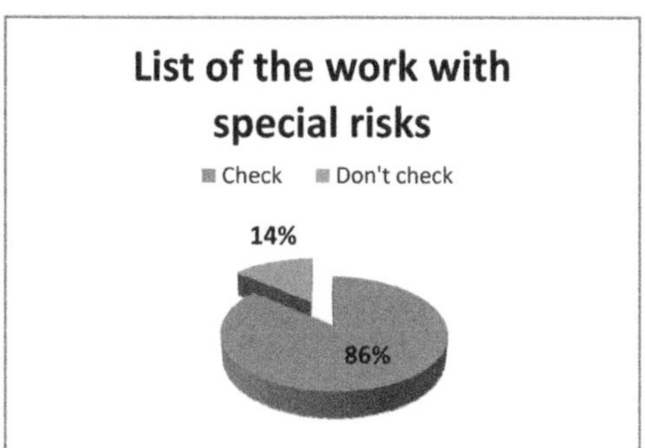

Figure 2. List of the work with special risks.

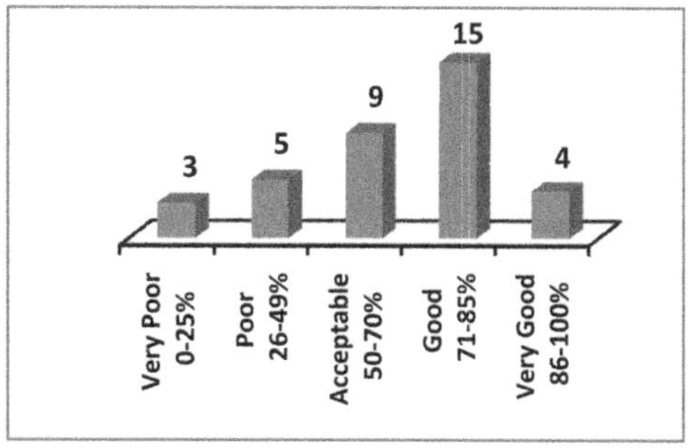

Figure 3. Final evaluation of the HSP.

4 CONCLUSIONS

The purpose of this research work was to make an analysis of the Health and Safety Plans quality, relating to the Municipality of Vila Nova de Gaia works, taking into account the methodology presented in this article. Although the evaluation of the Health and Safety Plans was satisfactory, it is possible to draw the following conclusions:

- Regarding the assessment of the safety and health plan organization, it appears that 24 of the 36 HSP (67%) had a score between the "Average" and "Good", 8 obtained unfavourable rating, and the remaining 4 had a "Very Good" score. It was observed that the preparation and approval sheet was verified only in 12 safety and health plans. In others research works on the same topic, but in other county, Alijó (Reis & Calisto, 2012) and Vila Real (Pinto & Reis,

2012), it was noted that this same item had less score, comparing with the other items;

- As for the evaluation of the safety and health plans descriptive document, it was verified that 72% of the required items had a satisfactory presentation, and the remaining 28% had lower ratings, being the functional organization chart and the stages of implementation item, that were absent in more than half. Comparing to the work of the authors Reis & Calisto and Pinto, it appears that the assessment is negative and the missing documents are the prior notice and the applicable regulation;
- As for the characterization of the undertaking, 64% safety and health plans presented the required items in a satisfactory way, and the remaining had a "Poor" rating. The manpower timeline was the element most missing, while the works list with the special risks was verified in 86% of cases. In the Reis 2012 and Pinto 2012 works, presented very poor ratings in this chapter, with minimal information regarding the general characteristics of the undertaking.
- Regarding the risk prevention actions, it was verified that 83% of the safety plans showed a positive evaluation and, only 17% of the plans had a negative evaluation. However, it should be highlighted that the highest scoring item in all referred studies, are the emergency plan and the individual protections plan.

Even thought that most of the safety and health plans referring to municipal constructions on the council of Vila Nova de Gaia obtained a satisfactory evaluation, it was noticed the absence of important requirements defined by Decree nr. 273/2003, October 29th that should be part of a safety plan. This fact was particularly evident in smaller dimension and complexity constructions. Although that the number of deadly work accidents in Portugal has been decreasing over the last decade, is clear that the construction enterprises and public constructions planning and execution should follow with more commitment and more rigorously, not only safety, health and hygiene referring legislation but also apply wisely opinions and ideas recommended in published studies by specialists in this area. The duty of knowledge and application of the ruling legislation to each economic situation is part National Right principle.

The conducted analysis, it was emphasized the need for a greater awareness of the technical who elaborate the safety and health plans, in order to avoid applying the same model to different works. Each project has its own characteristics, which distinguishes it from the others. It is noted that this was perceptible in other health and safety plans evaluated in Alijó and Vila Real counties, presenting this document only to comply with the legislation in force.

From the effectuated analyses, stood out the need of a bigger awareness form the technicians who elaborate the safety and health plans, in order to avoid applying the same plan model to different constructions, as each enterprise has own features that makes it different from the others. In these cases, unlike what happened in the municipality of Vila Nova de Gaia, the assessment of the health and safety plans was negative, perhaps because in this case it was evaluated the safety plans relating to applications for permits and not the city council works.

By one side, bigger exigency by the municipal entity that approves the licensing of an enterprise, as for the observance of the legislation the rules the elaboration of plans, and by another side, a more active supervision, the Work Conditions Authority, in the area of application of the professional risk prevention measures.

With a good planning and respect for the safety and health principles, the labour accidents are eliminated or reduced, the workers become motivated, productivity rises and health expenses decrease.

REFERENCES

Cabral, F. (2011). Segurança e Saúde do Trabalho—Manual de Prevenção de Riscos Profissionais. Lisboa. Verlag Dashofer.

Dias, L.M.A. Fonseca, M.S. (1996). Plano de Segurança e Saúde na Construção. Lisboa. Instituto Superior Técnico, Departamento de Engenharia Civil.

Dos Reis, C.M. (1998). Análise Económica da Segurança na Construção—Estudo de Alguns Casos. Porto. Faculdade de Engenharia do Porto. Dissertação para a obtenção de Grau de Mestre.

Freitas, P.E. (2013). Dissertação de Mestrado em Engenharia Civil—Análise da elaboração de planos de segurança e saúde no concelho de Vila Nova de Gaia. Vila Real. Universidade de Trás-os-Montes e Alto Douro.

Gonelha, L.M. Saldanha, R.A. (2006). Segurança, Higiene e Saúde no Trabalho em Estaleiros de Construção—Decreto-Lei nº 273/03, de 29 de Outubro, anotado e comentado (2ª ed.). Lisboa. Associação Forum Mercados Públicos.

Pinto D.; Reis C (2012). Análise de Planos de Segurança e Saúde/Analysis of Health and Safety plans. International Symposium Occupational Safety and Hygiene—SHO 2012 - Guimarães 9 e 10 de Fevereiro de 2012. ISBN 978-972-99504-8-3. Book in 1 volume, 506 pages. 348 a 350.

Reis C.; Calisto F. (2012). Análisis de la calidad de los planos de seguridad en Alijo—Portugal. XI Congresso Internacional de Prevención de Riesgos Laborales, ORP 2012, 23 a 25 de Maio de 2012, Bilbão, Espanha, Editado em CD ISBN: 978-84-615-7900-6.

Occupational Safety and Hygiene II – Arezes et al. (eds)
© 2014 Taylor & Francis Group, London, ISBN 978-1-138-00144-2

Stakeholder's perspectives on accident causes in rehabilitation and maintenance work in Portugal and Brazil

J.P. Couto
Department of Civil Engineering, University of Minho, Campus of Azurém, Guimarães, Portugal

E.R. Kohlman Rabbani
Civil Engineering Master's Program, University of Pernambuco, Recife, Pernambuco, Brazil

J. Duarte
Department of Civil Engineering, University of Minho, Campus of Azurém, Guimarães, Portugal

F.J.B. Santos
Civil Engineering Master's Program, University of Pernambuco, Recife, Pernambuco, Brazil

ABSTRACT: Construction accidents are one of the main indicators of the claims recorded at work all over the world. Identification of the causes and factors related to work accidents is essential for a proper analysis of risk and the consequent adoption of procedures and measures to achieve more effective prevention. Focus on rehabilitation and maintenance has increased worldwide and it is in this kind of construction that accidents are more frequent, principally motivated by a lack of planning and adequate projects that consider workers safety during the building life cycle. Studies have shown that the planning and project phases are essential in guaranteeing occupational safety during the maintenance and rehabilitation stages. Based on studies done in Portugal and Brazil during the past few years, this work puts forward the stakeholder's perspective on the main reasons for observed safety noncompliance and suggestions for procedures to improve these types of activities.

1 INTRODUCTION

The construction sector is considered by many to be one of the major pillars of a country's economy, but also one of the most dangerous industries (Enshassi, 1996). The policies adopted to increase productivity in the sector vary from country to country and depend on the strategy of each company. However, occupational safety and health have been widely recognized as two of the most influential aspects in companies' overall performance and have gradually come to no longer be seen as a luxury, but a necessity in order to avoid losses, injuries or even deaths (Abdul-Rashid et al., 2007).

Due to intense construction over the last decades, there has been a certain neglect of issues concerning construction quality (Araújo, 2009), which now has repercussions at the onset of the appearance of anomalies and premature building degradation. Thus, the need to invest more and more in repair and rehabilitation has grown, which, in the current economic environment, seems to be a viable alternative to new construction. In this type of work, the safety risks are still high, principally caused by poor Occupational Safety and Health (OSH) management skills related to lack of knowledge on how to intervene to prevent accidents, lack of prevention through design mechanisms/procedures, and poor manpower qualifications (Melo Filho et al., 2012).

In order to better understand the state of safety management for Rehabilitation and Maintenance Projects in Portugal and Brazil, this study tries to compare the results of two studies carried out independently in those countries and understand if main findings are corroborating the international literature. The fist one falls within the scope of a master thesis in sustainable construction and rehabilitation developed at the University of Minho, where a questionnaire was given to 57 actors, including owners, designers and contractors in the field of rehabilitation, with the objective of identifying their perspectives on the key factors that affect safety in rehabilitation and possible solutions for better safety management. It was concluded that the lack of technical expertise in organization and planning, lack of communication, and lack of skilled labor appear to be the main factors related to flaws in safety coordination and management of this type of work.

At the same time, research was carried out in Brazil by the University of Pernambuco in the development of a master thesis (Mélo Filho, 2009) in the area of occupational safety during maintenance work, with the objective of evaluating how Occupation Safety and Health noncompliance at eight building maintenance sites in Recife, Pernambuco, were related to project deficiencies and which improvements could be effectively introduced for safety management. The case study was conducted through on-site visits, informal interviews with the engineer responsible for the maintenance work, and analysis of the building projects and safety plans. It was found that the planning and design of a building are vital to ensuring worker safety during the maintenance phase. Deficiencies in the buildings' architectural design, irregularities in safety procedures, and a lack of OSH knowledge by managers responsible for contracting and monitoring maintenance companies were observed.

2. LITERATURE REVIEW

2.1 *Maintenance and rehabilitation*

According to Marcelli (2007), in civil engineering, maintenance is a set of necessary and indispensable measures for guaranteeing the proper functioning, conservation, and safety of a building's equipment and installations, of whatever type or size, resulting in a set of preventive and corrective actions whose purpose is to preserve the satisfactory fulfillment of the functions for which the building and its components were designed in order to ensure the desired service life of a building.

Post-work intervention activities in construction can be divided into maintenance activities and improvement activities, each having different goals. According to Xenos (1998), maintenance activities aim to maintain the original condition and operating performance by restoring any deterioration of these conditions, while improvement activities are those that aim to improve the original operating conditions, performance, and reliability through changes to the original design.

The term rehabilitation is used as a post-work intervention that aims to solve buildings pathologies or when there is a need to improve the functions of existing buildings. Rehabilitation offers important advantages as follows: it is ecologically smarter than demolition and new construction; helps maintain the streetscape and local architecture and it is often economically a more favorable option.

Both maintenance and rehabilitation works include similar risk situations, including work at heights, electrical hazards, ergonomic issues, and OSH management issues.

2.2 *Safety in international construction*

Internationally, there are a substantial number of studies aimed at clarifying the reasons related with safety performance. In Palestine, through an inquiry of 32 construction managers, Enshassi (1996) sought to assess the opinion of safety interveners, having concluded that the use of safety methods provides benefits, such as the reduction of accident costs, improvements in human relations, and consequent increase in productivity (Enshassi, 1996). For the construction managers, awareness of safety issues is directly related to a worker's age and experience, insofar as, through the results obtained, it may be concluded that the amount of accidents tends to decrease from the age of 30 onwards, demonstrating that more experience in performing a task lowers the risk of accidents.

Another study on the topic was carried out by Adbdul-Rashid et al. (2007), with the objective of identifying the main factors affecting safety at large projects in Egypt, using a questionnaire survey conducted on large construction contractors selected from the Egyptian Federation for Construction and Building Contractors (EFCBC). This study was based on a bibliographical research that made it possible to identify a set of 63 factors, grouped in 12 categories. In table 1, the

Table 1. Factors affecting safety performance in international literature (Abdul-Rashid et al., 2007).

Factors affecting safety performance
Implementation of safety management system in accordance with legislation;
Compliance with occupational safety and health legislation, codes and standards;
Definition of safety responsibility;
Development of safety policy;
Provision of safe working environment;
Development of emergency plans and procedures;
Development of safety committee;
Effective accident reporting;
High line management commitment;
Active supervisor's role;
Active personal role;
Understanding and implementation of safety management system;
Understanding and participation in occupational health and safety management system;
Quality of subcontractors;
Understanding and implementation of safety procedures;
Carrying out work in a safe and professional manner;
Type and method of construction;
Management's attitude towards safety;
Monetary incentives;
Disciplinary action.

main factors considered in the questionnaire are presented.

The authors concluded that the factors considered most relevant by companies were the need for awareness by the company's management and project managers regarding the implementation of a safety management system, and the need for frequent monitoring.

Fang et al. (2004) also carried out a study to analyze the factors of safety management in construction works in China. The main factors considered to be most influential in safety management were:

- Safety inspection;
- Safety meeting;
- Safety regulation enforcement;
- Safety education;
- Safety communication;
- Safety cooperation;
- Management-worker relationship;
- Safety resources.

In 2008, Aksorn & Hadikusumo carried out a study where they identified the success factors influencing safety in the performance of Thai construction projects. Through bibliographical research, they identified a set of 16 critical factors for success (CFS), and implemented a questionnaire to 80 actors involved in medium or large scale construction projects, with the purpose of determining which of the 16 critical factors of success were the most influential in safety management. The results obtained are presented in table 2, ordered from most to least influential.

On the basis of the results presented, it may be concluded that the most influential factor in work safety of Thai construction is the support given to work management. Moreover, Aksorn & Hadikusumo (2008) consider that these 16 critical success factors may be grouped into 4 sets, namely: worker involvement; safety prevention and control system; safety arrangement; and management commitment.

Table 2. Critical factors of success (Aksorn & Hadikusumo, 2008).

Critical Factors of Success (CFS)	
1	Management support;
2	Appropriate safety education and training;
3	Teamwork;
4	Clear and realistic goals;
5	Effective enforcement scheme;
6	Personal attitude;
7	Program evolution;
8	Personal motivation;
9	Delegation of authority and responsibility;
10	Appropriate supervision;
11	Safety equipment acquisition and maintenance;
12	Positive group norms;
13	Sufficient resource allocation;
14	Continuing participation of employees;
15	Good communication;
16	Personal competency;

2.3 *Prevention though Design—PtD*

Rwamamara et al. (2010), when interviewing Swedish designers, revealed the need to consider and develop tools for assessing work safety in the building project planning stage. However, designers have expressed that it would be uncommon to charge construction companies for this. The authors cite several studies that justify the investment in the planning stage, in view of the economic benefits related to increased productivity due to elimination of rework and reducing uncertainty caused by lack of proper communication among the parties involved.

According to Cooke et al. (2008), to improve safety outcomes in the construction industry, architects and design engineers should run a full risk assessment for each project prepared.

The PtD initiative is based on the premise of being an efficient way to prevent and control risks, injuries, illnesses, and fatalities at construction sites through actions taken while still in the conceptual and project stages (Gambatese et al., 2008). More specifically, according to Toole et al. (2011), PtD means among other concepts, to explicitly consider the safety of construction workers [including any post-work interventions] during the design of a project and during the constructability review process.

3 PERCEPTON OF THE CAUSES AFFECTING SAFETY PERFORMANCE IN REHABILITATION

As part of a study carried out at the University of Minho, entitled "Optimization of Management Rehabilitation Projects" (Araújo, 2009), a national survey was conducted, with the main objective of analyzing the main causes for failures in the management of rehabilitation work.

Based on the literature previously mentioned, a set of 15 causes deemed the best fit for this type of project were selected. A survey interview was conducted with 57 participants in this type of activity, including 10 owners, 20 designers, and 27 contractors, with management and leadership positions, in order to figure out the causes for most hazardous conditions, and failures, and poor safety performance.

Table 3 presents a summary matrix with the main causes (ranking) for each group of participants, which allows analyzing and comparing the views of different groups on the subject under study in Portugal (Araújo, 2009).

Through the analysis of table 3, it can be concluded that it is practically consensual that the difficulties associated with the use of labor-intensive, low-skilled workmanship are the main reason for the non-fulfillments in safety. Although with a lower degree of agreement between groups, the lack of a technical phase of work planning and organization also appears as one of the most prevalent reasons. There is equally the importance that contractors attach to the coordination and communication in the work, which certainly results in a more effective knowledge of its importance and consequences. The lack of research and observation of the target area by the contractor was also included among the main reasons, having this reason recorded a relative agreement among the various respondents. Beyond the analysis of the main causes of safety failures at rehabilitation worksites, a set of six measures to facilitate the achievement of safety objectives were presented to respondents. Respondents were asked to validate and rank the measures' importance in fulfilling safety conditions. The measures considered the most important were:

- Qualification of contractors, designers and technical staff in general;
- Diagnosis of the safety conditions of the building;
- Implementation of Health and Safety Plans (HSP) specific to each type of work;

Table 3. Ranking of the main causes for safety uncompliances by group of participants (Araújo, 2009).

	Ranking		
Causes affecting safety performance	Work owners	Designers	Contractors
Little qualified and specialized workmanship	1	2	1
Shortage of technicians in the phase of work planning and organization	2	6	4
Poor communication and coordination between the various actors in the work	–	4	2
Inadequate inspection of the workplace by the contractor	3	3	5

- Coordination of safety and efficiency of the planning and design phase.

4 SAFETY NON COMPLIANCE AT BUILDING MAINTENANCE WORKSITES

The research developed in Brazil (Melo Filho, 2009) presents the condition of specific building maintenance worksites in the city of Recife, with regard to their compliance with OSH regulations, and identifies key gaps in the buildings' design that can generate maintenance risks.

The study was limited to the civil engineering maintenance work, which included: low-voltage electrical maintenance; plumbing maintenance (cold water, cisterns, and sewer); façade maintenance; and reinforced concrete structure maintenance.

Eight building maintenance work sites were visited between October 2008 and February 2009, and analyzed using a check list and interview with the constructor manager. The building ages chosen for the study varied between 10 and 25 years, in order to maintain certain homogeneity across the sample, while at the same time allowing for differentiation between newer and older buildings.

The checklist sought to obtain:

- Information about the maintenance company;
- Characteristics of the worksite (location, building size and age, maintenance activity being performed);
- Documentation of OSH required by Brazilian law (safety programs/plans such as PCMAT and PCMSO according to Brasil (2013));
- Inspection of the physical facilities during maintenance work (seeking to identify design flaws in the installation and accessibility of its components.); and
- Compliance with NR18, Brazilian OSH-specific regulating standard for construction during the execution of maintenance activities.

This instrument was applied at worksites by observing the on-site work conditions; interviewing those directly responsible for the maintenance work (e.g. responsible engineer) who validated the information, sometimes in the presence of the building condominium managers; and analyzing the building design and safety plans and procedures used by the contractors and the ones required by the condominium managers.

Flaws were detected in eight of the architectural projects analyzed, which directly hindered maintenance operations. Some of the principal ones were: difficulty accessing the façade due to architectural details of the building; absence or insufficient quantity of anchoring accessories; electrical installations with difficult accessibility; and poor access to plumbing installations (Melo Filho et al., 2012).

These design flaws may indicate the absence or inefficiency of safety analysis during the project design phase of these buildings. Facilities with numerous hazards in the maintenance phase were detected, which could be avoided with the inclusion of specific safety devices in the construction phase of the work, if they had been planned for during the project phase. Improvement in the quality of projects was found, however, when comparing older buildings with newer ones (evident mainly in electrical installations).

With regard to the maintenance company safety procedures, a lack of qualified professionals to implement safety actions was found, with several breaches of OHS laws occurring, as well as ineffective application of protection measures. In the buildings visited, the following was found: absence of system for fastening and supporting suspended scaffolding structures, use of a single point of anchorage for attachment of equipment and cables for seatbelts, use of a counterweight system with granular solid anchorage; installation of ropes that rub against the parapet or constructive detail of the building without guarding the cable against wear; improper installation of scaffolding, with excessive deviation from the façade; and/or improvised equipment installation.

It was concluded that investment in the planning phase of maintenance work is an important measure of guaranteeing work quality, as well as of preventing of risks from occupational accidents.

5 CONCLUSIONS

Maintenance and Rehabilitation works are, naturally, an added advantage for the prevention and preserving of the country's heritage. However, it seems equally obvious that these should not be developed to the detriment of the safety conditions of those involved. The costs with accidents are considerably higher than the costs involved in measures to foster safety. However, even with these arguments, some construction, maintenance and rehabilitation companies are still not concerned about this problem or apply preventive measures, thus concurring to the fact that construction is one of the industries that contributes most to accidents in Portugal and Brazil.

The studies carried out independently in those countries presented similar safety problems (also confirmed by other international studies presented in section 2.2) such as: lack of qualification of contractors, designers and technical staff in general, poor communication and coordination among several stakeholders, shortage of technicians planning and organization phase, and design flaws that may indicate the absence or inefficiency of safety analysis during the project design phase or depreciation of safety issues on the part of the designers. The studies showed that the planning and project phases are essential in guaranteeing occupational safety during maintenance and rehabilitation interventions, corroborating, therefore, other international literature (see section 2.3). The most influential management procedures suggested by Fang (2004), which include safety inspection (monitoring), meetings, OSH regulation enforcement, education, communication, cooperation and resources are also essential to guarantee acceptable levels of safety during maintenance and rehabilitation works.

REFERENCES

Abdul-Rashid, I., Bassioni, H. & Bawazeer, F. 2007. Factors affecting safety performance in large construction contractors in Egypt. In: Boyed, D. (Ed), Procs *23rd Annual ARCOM Conference,* 3–5, September 2007, Belfast, UK, ARCOM—Association of Researches in Construction Management, 661–670.

ACT. 2012. Authority for the conditions of the work, Cabinet of strategy and planning. Accessed October 2012, available at: http://www.gep.msss.gov.pt/estatistica/acidentes/at2009sintese.pdf.

Aksorn, T. & Hadikusumu, B.H.W. 2008. Critical success factores influencing safety program performance in Thai construction projects. *Safety Science 46,* Asian Institute of Institute Technology, Pathumthani, Thailand, 713–715.

Araújo, J.D. 2009. Optimization of the management of Rehabilitation projects. *Dissertation of master's degree in Sustainable Construction (in Portuguese).* University of Minho, Guimarães.

Brazil. 2013. *Normas Regulamentadoras.* Accessed September 2013, available at: http://portal.mte.gov.br/legislacao/normasregulamentadoras-1.htm.

Cooke, T., Lingard, H., Blismas, N. & Stranieri, A. 2008. The development and evaluation of a decision support tool for health and safety in construction design. *Engineering, Construction and Architectural Management* 15 (4): 336–351.

Edwards, D.J., Holt, G.D., Harris, F.C. 1998. Predictive maintenance techniques and their relevance to construction plant. *Journal of Quality in Maintenance Engineering* 4 (1): 25–37.

Enshassi, A. 1996. Factors Affecting Safety on Construction Projects. *Department of Civil Engineering, IUG,* Gaza Strip, Palestine, 14–17.

Fang, D.P., Xie, F., Hung, X.Y. & Li, H. 2004. Factor analysis-based studies on construction workplace safety management in China. *Journal of Project Management,* Vol. 22. No 1, 43–49.

Gambatese, J.A., Behm, M. & Rajendran, S. 2008. Design's role in construction accident causality and prevention: Perspectives from an expert panel. *Safety Science*, 46(6): 675–691.

Gomide, T.L.F. 2006. Técnicas de inspeção e manutenção predial. São Paulo: Editora Pini.
Marcelli, M. 2007. *Sinistros na construção civil: causas e soluções*. São Paulo: Editora Pini.
Mélo Filho., E. 2009. Operation, use and maintance manuals adequacy to occupational safety and health regulations. *Master's Dissertation in Civil Construction (in Portuguese)*. Unviersity of Pernambuco. Recife: UPE, Brazil.
Mélo Filho, E., Kohlman Rabbani, E.R., Barkokébas Junior, B. 2012. Occupational safety evaluation at high-rise building maintenance worksites. *Produção* 22 (4): 817–830.
Rwamamara, R. et al. 2010. Using visualization technologies for design and planning of a healthy construction workplace. *Construction Innovation: Information, Process, Management*. 10 (3): 248–266.
Toole, T. et al. 2011. Prevention through Design: An important aspect of social sustainability. International Conference on Sustainable Design and Construction (ICSDC). *ASCE Proceedings*. Kansas City, Missouri: ASCE.
Xenos, H.G. 1998. *Gerenciando a manutenção produtiva*. Belo Horizonte: Indg Tecnologia e Serviços Ltda.

Occupational Safety and Hygiene II – Arezes et al. (eds)
© 2014 Taylor & Francis Group, London, ISBN 978-1-138-00144-2

Evaluation of indoor thermal environment of a manufacturing industry

L. Teixeira
DEGEI, IEETA, University of Aveiro, Aveiro, Portugal

M. Talaia
DFIS, CIDTFF, University of Aveiro, Aveiro, Portugal

M. Morgado
DEGEI University of Aveiro, Aveiro, Portugal

ABSTRACT: From the sustainability point of view, in an industrial context, it is desirable to achieve acceptable thermal sensation with the minimum energy. Companies should enhance ergonomic study of thermal comfort to improve its working environment. This paper aims to present a study to improve the thermal conditions in a manufacturing metalworking industry, following a methodology that identifies the workplaces thermally more vulnerable, and then evaluates the thermal comfort of these places and the thermal perception of its occupants. Data concerning the air temperature and relative humidity were collected using the measuring instrument '*Testo 435-4*', and were analyzed using *EsConTer* and *THI* indexes. The results allowed the identification of the most critical workstations in order to conduct a more focused study on these zones, helping the Department of Health and Safety at Work to define strategies to improve the working environment based on more accurate facts.

1 INTRODUCTION

"Work" is a concept that can be described as an activity created by humans that involves the production of a material, the execution of a task and the provision of a service, with the purpose of obtaining a particular economic result (profit) or social utility (Pinto, 2009). Ergonomics is the scientific area concerned with the understanding of the interactions among humans and other systems, in order to maximize human well-being and overall system performance. Thus, in the industrial context, ergonomics emerges as a science that aims to seek the adaptation of work to the worker in order to promote the stimulation and satisfaction of occupants.

Nowadays the playing field of ergonomics is extremely broad, being applied in different areas, from the physical stress and the mental processes to environmental factors that may affect the individual comfort and the public health (Rodrigues, 2007). Regarding the context of intervention, ergonomic measures can be taken in various contexts, in particular in industry, schools, hospitals, transportation, and construction, among others (Meles, 2012).

Actually ergonomics is an area of study that supports the business strategy of companies, keeping them competitive. However, due to the lack of knowledge, oftentimes the managers do not associate ergonomics to the efficiency of the organization nor believe that strategies based on concepts of ergonomics can contribute to increased productivity.

In industrial context, thermal environment and thermal comfort are important areas in the ergonomics field. Thermal comfort has been defined by Hensen (1991) as "a state in which there are no driving impulses to correct the environment by the behavior". American Society of Heating, Refrigerating and Air-Conditioning Engineers (ASHRAE) international standards (ASHRAE 55, 2004) and ISO standard 7730 (ISO 7730, 2005) define thermal comfort as: "the condition of mind which expresses satisfaction with the thermal environment". Djongyang *et al.* say that "comfort thermal occurs when body temperatures are held within narrow ranges, skin moisture is low, and the physiological effort of regulation is minimized". Nevertheless, this definition may be considered ambiguous and depends on several circumstances, such as the place where the human is, the reason why he is in that place, the seasons, among others.

Thermal comfort is a field studied by different professionals in order to understand the reactions of individuals and their adaptation to different thermal environments (Emmanuel, 2005). Particularly in the manufacturing industry, studies related to thermal comfort are very important, since it affects the working conditions, well-being, safety and health of the people who work in these environments. The literature reports several studies that have shown relations between the thermal comfort, thermal environment, public health and productivity in workplaces (Pinto, 2009; Félix *et al.*, 2010; Bluyssen *et al.*, 2011).

Besides the human component, other aspects related to energy consumption are also important in this area (Castilla *et al.*, 2011). Understand the behaviour and interaction of the occupants with the environment they are exposed is a crucial aspect to reducing energy consumption and emissions of carbon dioxide, and consequently to reach well-being and thermal comfort of the worker (Liu *et al.*, 2012; Daum *et al.*, 2011).

Rodrigues (2007) refers that an analysis of the thermal environment associated with the principles of industrial engineering is based on three fundamentals vectors: (i) one is related to public health, ensuring the thermal comfort of the occupants in their workspace; (ii) another is related to human performance which depends on physical robustness, i.e., an individual with thermal comfort has better performance in carrying their tasks, and (iii) the third vector concerns the energy conservation in the space.

In accordance with the main purpose of a manufacturing industry, studies of thermal comfort are fundamental to continuous improvement of organizations and consequently to increase the productivity of its members.

The thermal environment is an area of ergonomics that study a corrective and reactive intervention, namely the interaction between the workers and the thermal environment where they operate, in order to evaluate and improve comfort, safety and well-being, decreasing fatigue and increasing productivity. It was with the studies of Fanger (1972) that the study of the thermal environment has gained momentum and more followers in the research world.

According to Felix *et al.* (2010), air quality and thermal comfort in indoor environments can affect health and well-being of its occupants.

The literature shows that the analytical methods, most used to study the thermal environment, based on thermal indices with pre-defined limits, sometimes not take into account the feelings of its workers (Leal and Neves, 2013). Indeed, the evaluation of thermal comfort and the environment of a workplace must take into account two types of methods: (i) objective methods, consisting of analytical methods based on thermal indices, and (ii) subjective methods, which evaluates the thermal sensations of the individual in order to understand his reaction.

Starling *et al.* (2012) analyzed the environment and the level of thermal stress of a sector with five workstations in a company and concluded that the standards or normative parameters do not consider individual preferences and therefore, environments classified as "correct" do not necessarily guarantee the thermal comfort to its occupants.

Leite *et al.* (2013) showed that the workplace is defined as the set of surroundings that affects the administrative and operational activities and determine, in large part, the productivity coupled with the quality of the work produced. The relationship between the worker and the workplace should be understood as "win-to-win" relationship, to guarantee all the necessary conditions for the well-being of individuals, in order to ensure the effective performance of their tasks. DeRango (2003) stated that ergonomic interventions lead to decreased physical effort and increase productivity and Starling *et al.* (2013) consider that a comfortable thermal environment is essential to the well-being and efficiency of the performance of employees.

However, it should be stated that each individual may have a different response to the same thermal environment, due to their individual susceptibility and degree of acclimatization (Djongyang *et al.*, 2012). In these circumstances it is assumed that each individual might have different needs of thermal comfort.

The study by Daum *et al.* (2011) revealed that the individuals exposed to highly control indoor air conditioning systems, are twice more sensitive to temperature variations than individuals who are exposed to environments naturally ventilated by opening or closing windows and/or replacement of the type of clothing.

The authors of this work believe that the best results of workers may be associated with the better conditions of work, where the study of the thermal comfort is included. The identification of vulnerable areas within an industry is a key step to improve the thermal environment, making possible the implementation of strategies in line with existing conditions.

The study reported in this article aims to improve the thermal conditions in a manufacturing metalworking industry, following a methodology that identifies the workplaces thermally more vulnerable, and then evaluates the thermal comfort of these places and the thermal perception of its occupants, in order to ascertain the dynamics of the thermal environment.

2 MATERIALS AND METHODS

This study was carried out in a Portuguese manufacturing metalworking industry and aims to

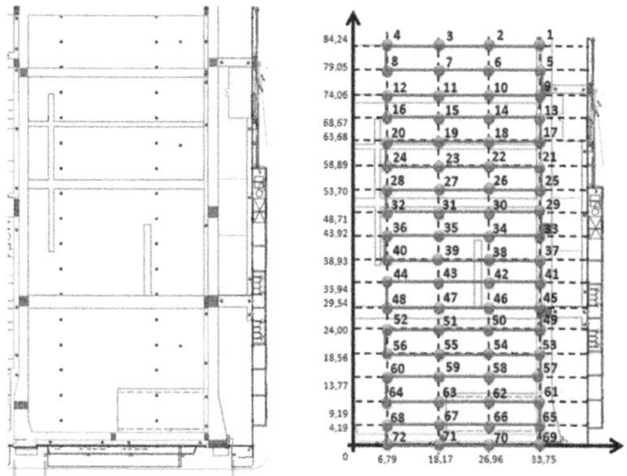

Figure 1. Layout and observation points in industrial sector studied.

improve the thermal conditions following a methodology that identify the workplaces more vulnerable, evaluates the thermal comfort of workstations and thermal perception of occupants.

To identify critical areas, the points of observation and data collection were marked and identified. As shown in Figure 1, 72 points of observation were identified. At each point, air temperature and relative humidity data were collected (using the measurement instrument '*Testo* 435-4') in two instants, one in the morning (10:30 am) and another in the afternoon (15:00).

The authors decided to collect only these two parameters since there are the most used to apply the chosen thermal indexes, used to analyze the result and identify the most vulnerable areas: (i) the T*emperature Humidity* Index (*THI*) (Nieuwolt, 1977) and (ii) *EsConTer* Index, a new index developed by Talaia and Simões (2009). This last was presented in several international conferences, and the index makes the thermal comfort one intuitive and easier interpretation tool of thermal sensation, based on a color scale, and in line with 7-point ASHRAE thermal sensation *scale* (ASHRAE 55, 2004).

Regarding the *THI* index, it was initially developed by Thom (1959) using the temperature of the wet bulb T_{wn} (°C) and the air temperature T (°C), and later, in order to facilitate the implementation and evaluation of the index, Nieuwolt (1977) modified it, replacing the temperature of the wet bulb by humidity relative humidity (%). The calculation formula for the *THI* is:

$$THI = 0{,}7T + T(RH/500) \quad (1)$$

where T = air temperature (°C); and RH = relative humidity (%).

The *THI* index was selected and used due to its simplicity of application.

Table 1. Thermal sensation based on *THI* [adapted by Talaia *et al.* (2013)].

THI	Thermal sensation
$THI < 8$	too cool
$8 \le THI < 21$	sun needed for comfort
$21 \le THI < 24$	COMFORTABLE
$24 \le THI < 26$	wind needed for comfort
$26 \le THI$	too hot

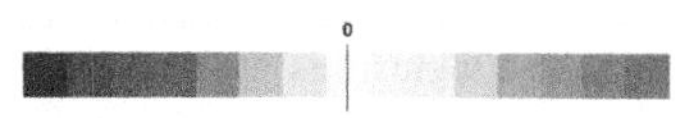

Figure 2. Color scale to thermal sensation.

The *THI* values for thermal sensation were adapted by Talaia *et al.* (2013) as shown in Table 1.

Regarding the *EsConTer* index (Talaia and Simões, 2009), its calculation formula is based on a color scale (Es), using data related to the thermal comfort sensation of occupants (Con) and it is classified as a thermal scale (Ter). The calculation formula is:

$$EsConTer = -3.75 + 0.103(T + T_w) \quad (2)$$

where T = air temperature (°C); and T_w = temperature of the wet bulb (°C).

The occupants answer the thermal sensation in a color scale from '−3' (dark blue color—thermal sensation: very cold) to '+3' (dark red color—thermal sensation: very hot), as shown in Figure 2.

Five workstations were identified as the most critical, and in these, the thermal sensations of its occupants were evaluated, using the *THI* and *EsConTer* indexes.

3 RESULTS AND DISCUSSION

Based on the '*termohigrometros*' data, registered during the morning and the afternoon, different graphs that show the dynamics of pattern temperature and relative humidity were generated, by using a *MatLab* program. Also, in the industrial sector studied, dynamic maps were generated, based on values for the *THI* and *EsConTer* indexes.

The values obtained with the application of *THI* and *EsConTer* indexes reveal a significant correlation coefficient (0.964), which raises the same considerations for both indices.

Optionally, we present results for one day (May 6, 2013).

Figure 3 and Figure 4 show the pattern obtained for the *EsConTer* and *THI* indices respectively, based on the same set of data.

Figure 3 shows the existence of zones vulnerable to thermal stress, corresponding to the highest values of the *EsConTer* index. Based on the highest values of the thermal index, it is possible to identify the existing critical workstations.

Figure 3 shows the existence of zones vulnerable to thermal stress, corresponding to the highest values of the *EsConTer* index, being possible to identify the existing critical workstations.

Figure 4 shows the pattern obtained when applying the *THI* index using the same data. The pattern identified is similar to that shown in Figure 3, indicating the same conclusions.

Figure 5 and Figure 6 present magnified images of Figure 3 and Figure 4 respectively, illustrating more vulnerable area, and their representations show an excellent agreement between them, when applied the *EsConTer* and *THI* indexes. The black circles in representations identify the critical workstations.

The critical workstations were studied using a colour scale of thermal sensation applied in two shifts. Values of air temperature T (°C) and the temperature of the wet bulb T_{wn} (°C) were registered. This set of values allowed calculating the *EsConTer* index for each workplace.

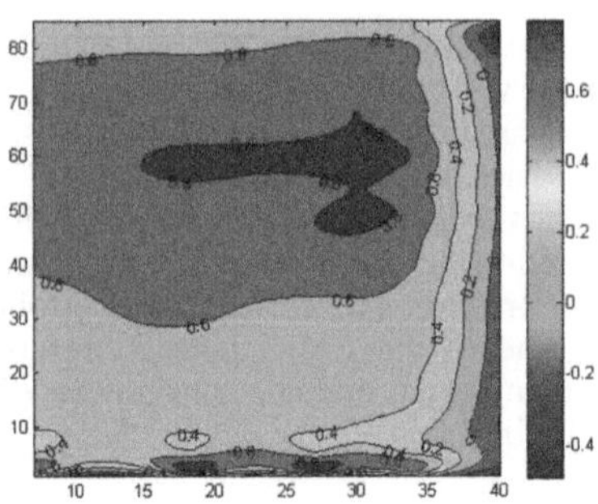

Figure 3. Pattern thermal sensation based on *EsConTer* (May 6, 2013).

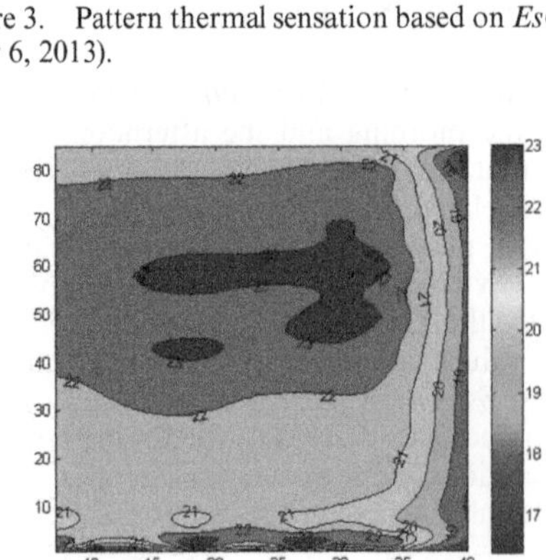

Figure 4. Pattern thermal sensation based on *THI* (May 6, 2013).

Figure 7 shows, for the morning work shift, the values of thermal perception of workers and the resulting values, after to apply the *EsConTer* index, indoor and outdoor the industrial sector.

The observation of Figure 7 shows the values of thermal sensation provided by *EsConTer*, the bold circles represent the values indoor of the industrial sector, while the gray circles represent the values outside. The findings shows that the thermal environment outside is slightly cold and the thermal environment indoor is slightly warm. This result is expected due to the presence of free energy sources.

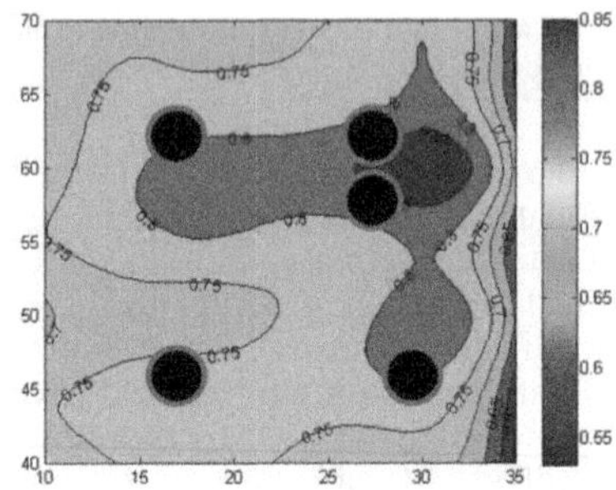

Figure 5. *EsConTer* and critical workstations.

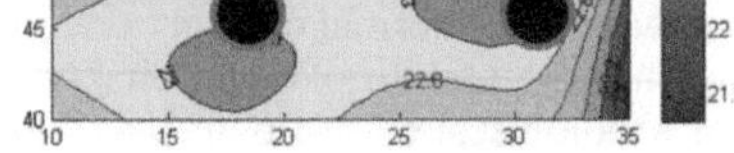

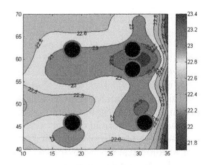

Figure 6. *THI* and critical workstations.

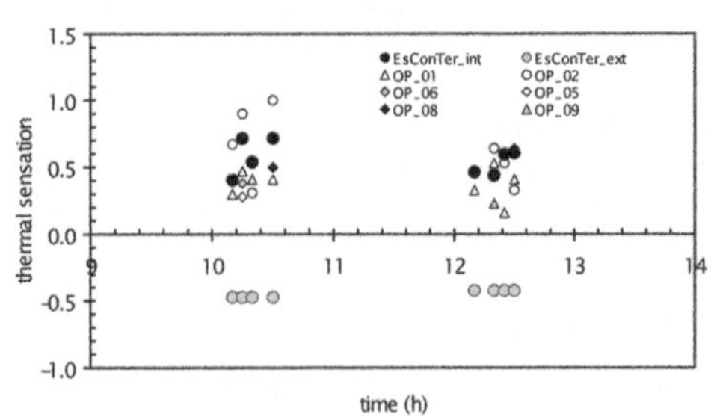

Figure 7. Values of the *EsConTer* and thermal sensation of workers—morning work shift.

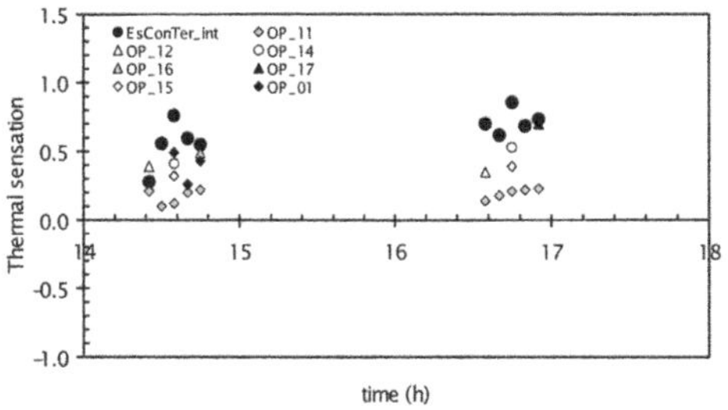

Figure 8. Values of the *EsConTer* and thermal sensation of workers—afternoon work shift.

In the same representation, each worker is encoded with the symbol OP_i, where *i* identify the worker. The representation of Figure 7 shows, unequivocally, that workers have a thermal sensation which is in agreement with the predicted values for the indoor of the industrial sector in a vulnerable zone.

Figure 8 shows, for the afternoon work shift, the values of thermal perception of workers and the resulting values, for the *EsConTer* index.

Similar to Figure 7, the observation of Figure 8 shows the values of thermal sensation provided by *EsConTer*, where the bold circles represents the indoor values of the industrial sector, and the values obtained with the response of the workers concerning their thermal sensations.

During the afternoon the thermal environment indoor remained slightly warm.

The worker coded with OP_{11} (female gender) has values of thermal sensation lower than the ones expected, in agreement with the findings obtained by Talaia and Alves (2011).

The representation of the Figure 8 shows that workers have a thermal sensation which is very close to the value indicated by the *EsConTer* index.

4 CONCLUSION

The results of this study show that it is possible to know an industrial sector with thermal discomfort, by means of usable tools and methodologies.

In the industrial sector studied, different dynamic maps with thermal standards were drawn in order to identify the most vulnerable areas. Additionally, the data about thermal sensations indicated by workers on a colour scale, were also obtained and analysed. The results allowed identifying the critical zone of the industrial company, in terms of thermal environment. Additionally, the results of this study proved to be very useful for the Department of Health and Safety at Work, as they were able to define strategies to improve the working environment based on accurate facts.

The usage of the *EsConTer* ant *THI* indexes showed the existence of an excellent correlation coefficient, either one can be used.

Also as a conclusion of this study, and based on the interpretation of the findings arising from the application of each index, which were concordant, it is possible to suggest for further studies the usage of the *EsConTer* index.

REFERENCES

Ashrae 55 (2004). *Thermal environmental conditions for human occupancy*. American Society of Heating, Refrigerating and Air-Conditioning Engineers - ISBN/ISSN: 1041–2336.

Bluyssen, P., Aries, M., Dommelen, P. (2011). Comfort of workers in office buildings: The Europian HOPE project. *Building an Environment*. 46, 280–288.

Castilla. M., Álvarez. J.D., Berenguel. M., Rodríguez. F., Guzmán. J.L., Pérez. M., (2011). A comparison of thermal comfort predictive control strategies, *Energy and Buildings*, 43(10), 2737–2746.

Daum, D., Haldi, F., Morel, N. (2011). A personalized measure of thermal comfort for building controls. *Building and Environment*. 46, 3–11.

DeRango, K. (2003). Office Workers' Productivity Enhanced by Ergonomics. *Employment Research Newsletter*. 10(3), 1-3.

Djongyang, N., Tchinda, R., Njomo, D. (2012). Thermal comfort: A review. *Renewable and Sustainable Energy Reviews*. 14, 2626–2640.

Emmanuel, R. (2005). Thermal comfort implications of urbanization in warm-humid city: the Colombo Metropolitan Region (CMR), Sri Lanka. *Building and Environment*. 40, 1591–1601.

Fanger, P. (1972). *Thermal Comfort*. McGraw-Hill (2nd ed): New-York.

Felix, V.; Moura, D.; Pereira, M, L.; Tribess, A. (2010) A. Evaluation of thermal comfort in surgical environments using Fanger's method and equivalent temperatures. *Ambiente Construído*, 10 (4), 69–78.

Hensen J. (1991). On the thermal interaction of building *structure and heating and ventilating system*, Technische Universiteit Eindhoven (PhD thesis).

ISO 7730 (2005). *Ergonomie des ambiances thermiques—Détermination analytique et interprétation du confort thermique par le calcul des indices PMV et PPD et par des critères de confort thermique local*. Switzerland, International Standardisation Organisation, Geneva, Suisse.

Leal, A., Neves, M., (2013). Study of Thermal Hot Environments: Contribution to a Technical Assessment. *Occupational Safety and Hygiene*. Editors Arezes et al. Taylor & Francis Group, London, 187–188.

Leite, F., Dias, E., Martins, L. (2013). Analysis of the ambience of the tasks of the cook and kitchen helper of a private hospital in the metropolitan area Recife. *Occupational Safety and Hygiene*. Editors Arezes et al. Taylor & Francis Group, London, 193–194.

Liu, J., Yao, R., Wang, J., Li, B. (2012). Occupants' behavioural adaptation in workplaces with non-central heating and cooling systems. *Applied Thermal Engineering*. 35, 40–54.

Meles, B. (2012). *Ergonomia Industrial e Conforto Térmico em postos de trabalho*. Universidade de Aveiro (Dissertation in Portuguese).
Nieuwolt, S. (1977). *Tropical climatology*. London: Wiley.
Pinto, A. (2009). *Análise ergonómica dos postos de trabalho com equipamentos dotados de visor em centros de saúde da administração regional de saúde do centro*. Universidade de Coimbra (Dissertation in Portuguese).
Rodrigues, F. (2007). *Conforto e Stress Térmico: uma Avaliação em Ambiente Laboral*, Universidade de Aveiro (Dissertation in Portuguese).
Starling, T., Mendonça, V., Alsina, O., Monteiro, L. (2013). Ergonomic Analysis at a Working Station Located in a Technology and Information Management Company from Sergipe, Brazil, Based on Temperature and Termic Stress Level. *Occupational Safety and Hygiene*. Editors Arezes et al. Taylor & Francis Group, London, 403–404.
Talaia, M., Meles, B. & Teixeira, L. (2013). Evaluation of the Thermal Comfort in Workplaces – a Study in the Metalworking Industry. *Occupational Safety and Hygiene*. Editors Arezes et al. Taylor & Francis Group, London, 473–477.
Talaia M. & Alves, J. (2011). A Condução e o Conforto Térmico na Segurança Rodoviária – Estudo de Percepção. *Proceedings 6° Cong° Luso – Moçamb° de Eng^a e 3° Cong° de Eng^a de Moçambique* (artigo CLME'2011_2806 A, 12 páginas). Maputo, 29 Agosto–2 Setembro.
Talaia, M. & Simões, H. (2009). *Índices PMV e PPD na Definição da "performance" de um Ambiente*. Livro de atas do V Encontro Nacional de Riscos e I Congresso Internacional de Riscos, Coimbra, de 29 a 31 de Maio.
Thom, E.C. (1959). The Discomfort Index, *Weatherwise*: 57–60.

Occupational Safety and Hygiene II – Arezes et al. (eds)
 ISBN 978-1-138-00144-2

Analysis of correlation between unsafe actions and unsafe conditions in the constitution of occupational accidents—case study

F.F. Sousa, L.B. Silva & E.L. Souza
Universidade Federal da Paraíba, Paraíba, Brazil

ABSTRACT: This paper aims to describe the relationship of occupational factors within the operational environment in the constitution of incidents and accidents in the production sector of a food factory. The field of research was chosen due to the fact that food industry lies in the second position of notifications and fines in public audit tied to accidents in Brazil in 2012. Therefore, the data were collected from January 2012 to August 2013. The obtaining of the results through regression modeling, $\alpha = 0.05$, demonstrated a relation among variables, showing the increase of accidents and incidents as well as the number of occurrences of unsafe actions of the employees, and the unsafe conditions intrinsic to the work environment. This representative relation reveals that safe and healthy environments contribute to the preventive behavior of the employees.

1 INTRODUCTION

It is very common in industrial activities, occupational accidents being composed by some non-conformities of the physical installation of equipments, or basically by unsafe actions adopted by some officials. However, as Rollenhagem et al (2010) mentions, it is still possible to find organizations that devote little attention to investigate these accidents.

According to Tavares and Abdolhamidzadeh et al (2010), the model of accident at work that defines quite well this situation is the "domino" effect, as the trajectory of an accident within an organization indicates that the risks pass through existing defenses due to unsafe actions committed by individuals and/or unsafe conditions intrinsic to the workplace, which sequentially causes accidents from minor to the most serious.

To Darbra, Palacios and Casal (2010), based on the "domino" theory, a relatively small accident can initiate a sequence of events that cause damage over a much larger area and can lead to much more serious consequences. Usually, this is called "domino" effect. This theory was developed in 1959 by Heinrich, showing five events that lead to the injury of the worker (IIDA, 2005): personality, human failures, causes of accidents (unsafe conditions and unsafe actions), accident, and injury.

According to Iida (2005), prevention based on this theory, should be done by removing the causes of accidents, to thereby, avoid the spread of the fall of the "dominoes". Though this phenomenon can occur in any industrial facility, congested plants such as offshore platforms or process plants where processing equipment and control systems are in close proximity, are especially prone to these types of primary event (DARBRA; PALACIOS; CASAL, 2010).

According to Reason (2008), the gaps found in the defenses against accidents at work occur for two reasons, described by the model of "Swiss Cheese" shown in Figure 1:

a) Active failures are the unsafe actions committed by people who are in straight contact with the system. They take a variety of forms: slips, lapses, mistakes and procedural flaws.
b) Latent conditions are the inevitable faults within the system. They arise from decisions made by the designers, constructors, procedure writers and top level management, which are characterized as unsafe conditions.

These decisions may be mistaken, but it does not have to be. All of these strategic decisions have potential for introduction of pathogenic agents in the system. Latent conditions have two types of adverse effects: can result in circumstances that can cause error in the workplace (for instance, time pressure, lack of staff, inadequate equipment, fatigue and inexperience) and can create long-lasting holes or weaknesses in the defenses (alarms unreliable, unworkable procedures, design and deficient constructions, etc.). Latent conditions, as

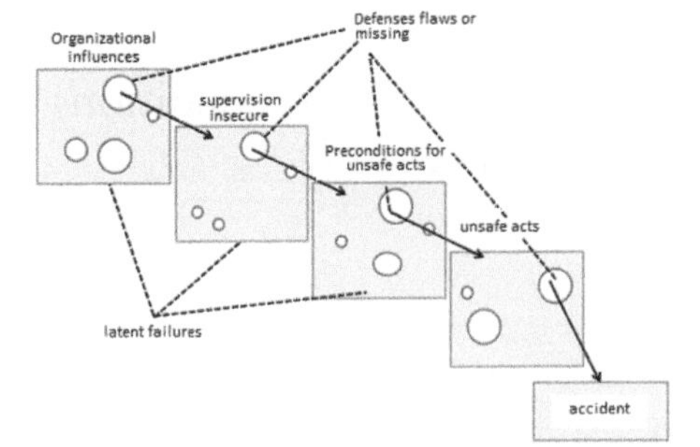

Figure 1. "Swiss Cheese" model, showing how the defenses and barriers can be penetrated by a hazard, causing damage.

the term suggests, can remain latent in the system for many years before they combine with active failures, creating an opportunity for of an accident (REASON, 2008).

To support the investigation of the causes of accidents in a work environment, some authors have already used statistical tools to assist in solving some problems. Martín *et al.* (2009) used Bayesian networks with the aim of analyzing the circumstances surrounding work tasks performed by auxiliary equipments (ladders, scaffolding, etc), which can result in falls; Malyshkina, Mannering, and Tarko (2009) applied processes of negative binomial Poisson in order to study the frequency of traffic accidents when driving cars; Elvik (2011) used operational criteria of causality with the multivariate statistical models developed to identify sources of systematic variation in counting accidents; Wang, Quddus and Ison (2011) applied a multivariate mixture of two stages to predict accident frequency at their severity levels; whereas Pines, Lemesch and Grafstein (1992) applied a linear regression model to highlight the trend of accidents in workplaces, in which states that the linear regression procedure is a useful tool for the evaluation and comparison of statistical trends of accidents during a relatively short period of time.

Considering the importance of the risks present in the workplace that impact decisively on the generation of accidents, this article aims to evaluate the relationship between unsafe actions and unsafe conditions in the constitution of labor incidents and accidents, covering the sector of production of a food factory.

2 METHODOLOGY

This work was conducted in a food factory located in the city of João Pessoa, Paraíba State, Brazil, specifically in the sector of production of 600 ml packages. The occurrences of accidents and incidents of work were collected during the period of January 2012 to August 2013. Each occurrence was specified by the team through direct investigation in the workplace. Every occurrence and its consequence were analyzed from the point of view of an unsafe condition or unsafe action. This information fueled the system of security management of the company, BOOK SSO (Occupational Health and Safety). These data were analyzed descriptively, and using mathematical modeling was studied the possible relations between the variables accidents and incidents, unsafe actions and unsafe conditions with the support of the software R.

3 RESULTS

The Table 1 displays the monthly occurrence of accidents and incidents reported and investigated during the period from January 2012 to August 2013, as well as the records of unsafe actions committed by employees of the production line, and unsafe conditions identified through the monitoring of security that is performed by the supervision staff.

The values of the descriptive measures are shown in Table 2. In this table is seen that the average number of occurrences of accidents is much

Table 1. Indicators of labor accidents and incidents.

Months (2012 a 2013)	Unsafe actions (N°)	Unsafe conditions (N°)	Accidents + incidents occurred (N°)
Jan/12	4	12	26
Feb/12	9	16	45
Mar/12	22	20	69
Apr/12	15	40	59
May/12	32	63	69
Jun/12	24	40	80
Jul/12	33	60	113
Aug/12	18	56	78
Sept/12	17	49	77
Oct/12	50	80	234
Nov/12	30	60	122
Dec/12	55	59	102
Jan/13	35	13	107
Feb/13	33	23	125
Mar/13	20	56	172
Apr/13	130	115	438
May/13	91	130	477
Jun/13	110	112	363
Jul/13	65	130	367
Aug/13	40	121	264

Table 2. Descriptive measures.

Descriptive measures	Unsafe actions	Unsafe conditions	Accidents and incidents
Minimum	4.00	12.00	26.00
First quartile	19.50	35.75	75.00
Third quartile	32.50	57.50	110.00
Median	41.65	62.75	169.40
Average	51.25	88.00	241.50
Maximum	130.00	130.00	477.00
Standard Deviation	33.76	39.65	138.70

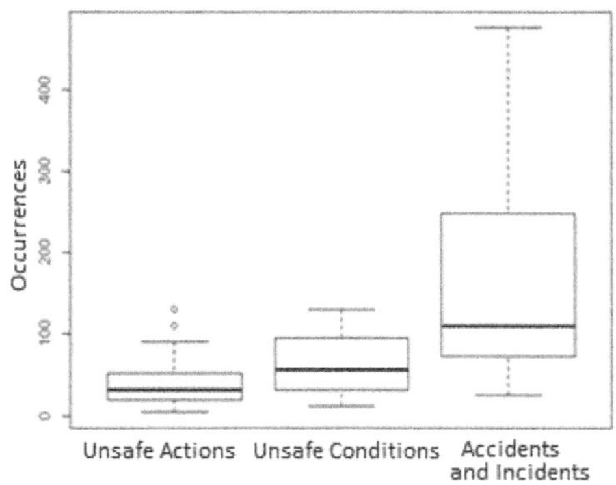

Figure 2. Variability in the number of unsafe actions, unsafe conditions and accidents/incidents.

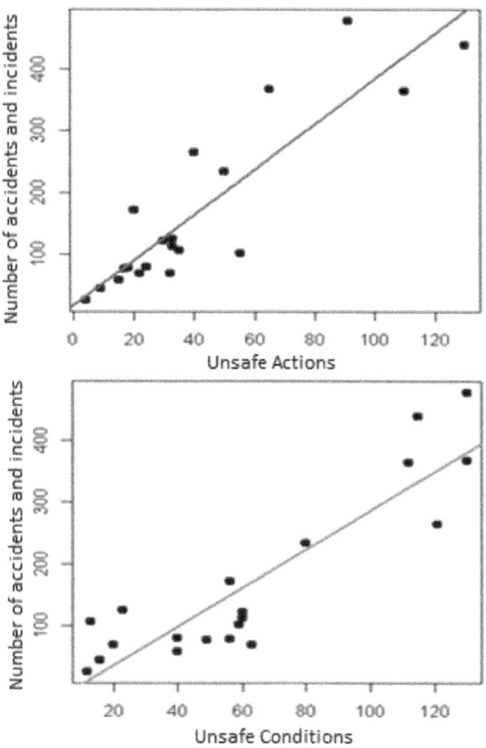

Figure 3. Upward trend in the number of accidents and incidents.

higher than the unsafe actions and unsafe conditions. This difference between the values of central tendency is also shown in the box plot (Figure 2).

The box plot provides a view of the variability in the number of unsafe actions, unsafe conditions and accidents/incidents. The box indicates the limits (first and third quartile) in which concentrate 50% of the records. The limits outside the box indicate the values in which are more than 98% of the observations recorded. Values out of these limits are considered atypical occurrences, being those periods where the number of occurrences is much higher than the others.

The graph of dispersion, depicted in Figure 3 indicates an upward trend in the number of accidents and incidents proportionally to the increase of unsafe actions and unsafe conditions.

Through the linear regression models (1) and (2) presented in Figure 3, it is possible to analyze the relationships among the variables: accidents and incidents; unsafe actions and unsafe conditions

$$AI = 16,6895 + 3,6653\ \text{AtI} \quad (1)$$

For every AtI there is an average increase of approximately 3.6 in AI. As $R^2 = 0,796$ (*p_value*, 0,05), then 79.60% of the variability of the average AI is due to AtI.

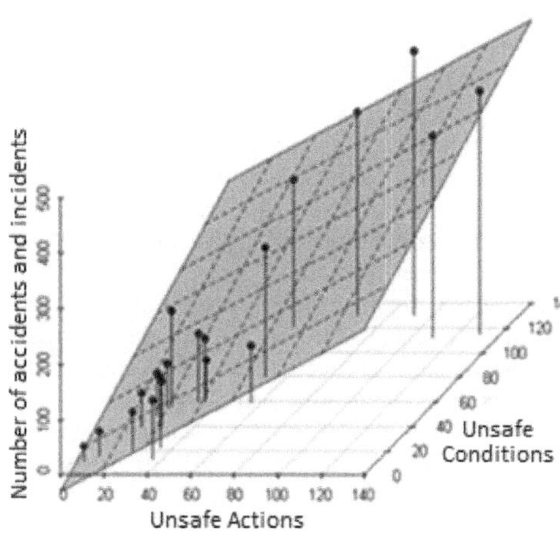

Figure 4. Upward trend in the number of accidents and incidents: Dispersion in three dimensions.

$$AI = -26,6966 + 3.242.CI \quad (2)$$

In each CI there is an average increase of approximately 3,2 in AI. As $R^2 = 0,7979$ (*p_value*, 0,05), then 79.79% of the average variability of AI is due to IC.

A dispersion graph in three dimensions, Figure 4, makes it clear the upward trend on the number of accidents with the increase of unsafe conditions along with unsafe actions. It is also observed a trend in the number of unsafe actions as unsafe conditions are more representative.

The model depicted in Figure 4 can be written as:

$$AI = -28,0209 + 1,7743.CI + 2,0656AtI$$

$$R^2 = 0,9018, \text{p-value} = 2.721e^{-9} < 0,05$$

4 CONCLUSION

Some food industries have on the center of their production, technological innovations which enable the production of products in a more feasible time. The investments in "*news ICT*" reduce the distance between the consumer and producer; facilitate the logistics system, as well as provide better optimization of the direct and indirect costs of the production processes. However, all of these innovations still do not make possible, reduce the risks that permeate the worker when performing their activities. This observation may be reflected in this article, as follows:

1. The average number of occurrences of accidents is much higher than the unsafe actions and unsafe conditions.
2. There is a trend of increase in the number of accidents and incidents to be proportional to the increase of unsafe actions and unsafe conditions—it is believed that such circumstances allow the appearance of several acts committed by the worker himself.
3. There is a positive and representative correlation between unsafe actions and unsafe conditions in case of occurrence of accidents and incidents in the production sector.
4. There is a positive and representative correlation between unsafe actions and the increase of occurrences of unsafe conditions. This relation confirms the importance of a good and salubrious work environment (appropriate equipment to the activities of the workers and their anthropometry; good building facilities, respecting the national and international regulatory standards).
5. 90% of accidents and incidents were due to the acts and unsafe conditions.

REFERENCES

ABDOLHAMIDZADEH, BAHMAN. ABBASI, TASNEEM. RASHTCHIAN, D. ABBASI, S.A. 2010. A new method for assessing domino effect in chemical process industry. *Journal of Hazardous Materials.* 182 (3): 416–426.

DARBRA, R.M. PALACIOS, ADRIANA. CASAL, JOAQUIM. 2010. Domino effect in chemical accidents: Main features and accident sequences. *Journal of Hazardous Materials.*183(3): 565–573.

ELVIK, RUNE. 2011. Assessing causality in multivariate accident models. *Accident Analysis & Prevention.* 43(1): 253–264.

IIDA, I. 2005. Ergonomia: projeto e produção. São Paulo: *Edgard Blucher*. 3: 225.

MARTÍN, J.E, RIVAS, T. MATÍAS, J.M. TABOADA, J. ARGÜELLES, A. 2009. A Bayesian network analysis of workplace accidents caused by falls from a height. *Safety Science.* 47(2): 206–214.

MALYSHKINA, NATALIYA V. MANNERING, FRED L. TARKO, ANDREW P. 2009. Markov switching negative binomial models: An application to vehicle accident frequencies. *Accident Analysis & Prevention.* 41(2): 217–226.

PINES, A. LEMESCH, C. GRAFSTEIN, O. 1992. Regression analysis of time trends in occupational accidents (Israel, 1970–1980). *Safety Science.* 15(2): 77–95.

REASON, JAMES. 2008. The human contribution: unsafe acts, accidents, and heroic recoveries. Great Britain. *MPG Books Ltd, Bodmin 10: 201.*

ROLLENHAGEN, CARL. WESTERLUND, JOAKIM. LUNDBERG, JONAS. HOLLNAGEL, ERIK. 2010. The context and habits of accident investigation practices: A study of 108 Swedish investigators. *Safety Science.* 48(7): 859–867.

TAVARES, J.C. 2010. Noções de prevenção e controle de perdas em segurança do trabalho. 8 Ed. São Paulo: *SENAC* 5: 231.

WANG, CHAO. QUDDUS, MOHAMMED A. ISON, STEPHEN G. 2011. Predicting accident frequency at their severity levels and its application in site ranking using a two-stage mixed multivariate model. *Accident Analysis & Prevention.* 43(6): 1979–1990.

Occupational Safety and Hygiene II – Arezes et al. (eds)
© 2014 Taylor & Francis Group, London, ISBN 978-1-138-00144-2

Proposal for evaluation of heat stress on the activity of vehicular patrolling by military police in Paraíba—Brazil

E.G.S. de Medeiros, F. de A. Gomes Filho & L.B. da Silva
Federal University of Paraíba, João Pessoa, Paraíba, Brazil

ABSTRACT: Considering the climate of the northeast region of Brazil and type of clothing used by police officers, it can be inferred that either the characteristics of the vehicle, as the use of uniforms should interfere with the thermal comfort of the agents. The methodology proposed in this study for assessing thermal environment in vehicles used to carry out the service vehicular patrolling by military police in Paraíba (Brazil) consists on the use of ISO 14505—"Ergonomics of the thermal environment—Evaluation of thermal environments in vehicles" which regards definitions of terms and procedures concerning the ergonomic evaluation of thermal environments in specific vehicles. Data processing will be done by means of descriptive analysis from measures of central tendency and the construction of mathematical models to evaluate the relationships between variables being studied based on regressive models.

1 INTRODUCTION

The constant interaction between man and his working environment can cause stress on individuals, be they physical or psychological. Moreover, it is known that favorable environments have positive impact on the workers. By staying long periods in inadequate environments, workers may experience symptoms such as discomfort, fatigue, pain, irritation, nervous disorders, stress and others (GUIMARÃES, 2004; MARCHI, 2007; PARSONS, 2000).

A working environment increasingly common in the XXI century is the automobile. This is used by drivers, taxi drivers, people who need to move continuously while performing their work activities and even the armed forces, as in the case of vehicular patrolling by the Military Police. Hense (2004) defines thermal comfort as "a state in which no impulses are directed to correct the sensation provided by the environment through the man's behavior." On the other hand, the American Society of Heating, Refrigerating and Air-Conditioning Engineers raise the delimitation of thermal comfort as "that condition of mind which is expressed satisfaction with the thermal environment".

In general, it can be said that the thermal comfort occurs when the body temperature does not vary abruptly, the rate of skin moisture is low and physical effort for temperature regulation is minimized. Moreover, the behavioral attitudes of individuals influence the comfort (DJONGYAN; TCHINDA; NKOMO, 2012).

The state of Paraiba is located in the Northeast region presenting a tropical and semi-arid climate with annual temperature average of 30°C (MARENGO, 2012). Iida (2005) mentions the fact that the climate, humidity and more incisively temperature influence the performance of workers, with documented evidence through studies in industry and laboratories. Such influences relate to productivity and increase the risks of accidents.

The attention regarding occupational health professionals working in public safety, such as military police, has been minority (FRAGA, 2005). For the author, few technical studies on the scope of these professionals have been developed regarding the management of Occupational Health and Safety. Vasconcelos (2007) states that the working conditions imposed on police officers may interfere with the quality of service, depending on the configuration of these conditions.

Thus, thermal comfort is considered one of the factors that may interfere with the working conditions of police officers who perform vehicular patrolling due to reasons mentioned above.

Considering the climate of the northeast region of Brazil and the type of clothing used by police officers when performing their service, it can be inferred that either the characteristics of the vehicle, as the use of uniforms, including bullet proof vest, should interfere with the thermal comfort of the agents.

2 POLICING AND THERMAL COMFORT

The situations which officers are submitted during the fulfillment of their workday require speed, perception, memory, prudence, observation, concentration, physical and psychological precision, besides exposing individuals to unfavorable situations (RODRIGUEZ-AÑEZ, 2003).

Moreover, usually the police are equipped with necessary instruments for the fulfillment of their workday (which may be 12 or 24 hours), as weapons, handcuffs, truncheon, bullet proof vest, among others, and only the vest weighs around 1.6 a 2.6 kg (SIMÕES, 2012). All these factors contribute negatively to physical, mental and thermal comfort of the agents of this type of work.

The Military Police of the State of Paraíba (PMPB) provides on its website the regulation related to the uniforms of the police. For the services of vehicular patrolling, focused activity of this study, the uniform required is the following (GOVERNO DA PARAÍBA, 2012):

- Black beret;
- Shoulder sleeves in black for officers and warrant officers with matching embroidered insignia;
- Embroidered insígnia with silver line on escutcheon fabric in black on both sleeves.
- Windbreakers gray in ripstop, long sleeve;
- Inner shirt, half sleeve in white mesh in transit units and training schools, and black for other units;
- Cover vest in black;
- Black nylon belt with black buckle;
- Black garrison belt. White in transit units;
- Gray pants or breeches in ripstop;
- Black socks;
- Black boots and shoelaces.

This entire garment directly interferes on the thermal comfort of individuals, as it increases the level of heat resistance, important variable related to the levels of thermal balance. Despite these factors, it was not found in recent literature research studies conducted on the thermal comfort considering the activities of the police officers and vehicular patrolling.

3 THERMAL COMFORT IN VEHICLES

Concomitantly to transformations who are suffering human society, environments where man is inserted also have been changing in order to make them more comfortable adapting them to the needs of its occupants and therefore it is becoming more common the development of efficient, sustainable and cost-effective systems to control such environments (MOURA, 2007; PEREIRA e ALCOBIA, 2006).

For these authors, with the increasing use of cars by people for various purposes, such as walking or even as a work tool, concerns about the questions of safety and comfort have grown increasingly matching to the performance and cost concerns carried by the buyers, showing as crucial items on the part of the one who is acquiring the vehicle.

Vehicular thermal comfort not only helps reduce the stress caused to its occupants, but also avoid other disturbances such as blurred vision (fogging effect), contributing to the safety of the driver and other occupants. Also, excessive exposure of individuals to heat causes the elevation of body temperature, sleepiness and may lead to physical collapse (ALAHMER *et al*, 2011; STEEN, 2012).

It can be said that the temperature inside a vehicle is an important factor in the occurrence of traffic accidents. A thermally comfortable environment results in an increase of the attention of the driver and improves the performance of the driver. Also ensures greater safety in different road conditions (FARZANEH e TOOTOONCHI, 2008).

ISO 14505—"Ergonomics of the thermal environment—Evaluation of thermal environments in vehicles" regards definitions of terms and procedures concerning the ergonomic evaluation of thermal environments in specific vehicles.

The standard is divided into three parts, where the first states the principles and methods for the determination of thermal stress, the second section presents how to determine the equivalent temperature and the third part regards about the evaluation of thermal comfort using the subjective response of individuals.

In particular, part 3 of ISO 14505, which is titled "Evaluation of thermal comfort using human subjects", provides the methods for the analysis of thermal comfort environments using subjects. It is said that these methods are quite appropriate because although subjective allow a direct assessment. Furthermore, it can be used together with the thermal indices valid for vehicles for the evaluation.

These methods quantify the responses of individuals through the development of questionnaires with subjective scales established by ISO 14505-3 (2007), which are based on psychological factors relevant to the thermal environment in which these individuals will be inserted.

These responses are given considering each part of the human body, so that you can get a preview of the body segments where individuals have greater discomfort.

ISO 14505-3 (2007) also suggests, as supplement to the subjective evaluation, the use of other methods: the objectives, those equipped with instruments to measure physical or mental conditions, and behavioral methods, consisting in the observation and interpretation of human behavioral aspects.

4 MATERIALS AND METHODS

The methodology proposed in this study for assessing thermal environment in vehicles used to carry out the service vehicular patrolling by military police in Paraiba State has two moments.

The first part will take place when the collection of anthropometric measurements, before the execution of the focus activity. This step aims at the acquisition of anthropometric data of individuals that may be related to later thermal sensations in their corporeal sessions collected through the questionnaire contained in ISO 14505-3. The anthropometric variables are divided into two parts. The first is with de individual standing:

- Stature;
- Shoulder width.

The second part considerate the individual seated:

- Head height from the seat;
- Shoulder height from seat;
- Popliteal height;
- Hip width, sitting.

Besides that, it will also be collected age, sex and weight from the individuals.

The second part will take place in three areas: gathering environmental variables; gathering personal variable and subjective assessment. Data collection of environmental variables such as air temperature, mean radiant temperature, air velocity and relative humidity of air will be through the use of two meters heat stress TGD-400 (Instrutherm) as shown in Figure 1.

One of the heat stress equipment shall be locate inside the vehicle and the other in an outdoor environment, so that it may be possible to capture the data of thermal variables at the same moment that performs the collection of subjective data through questionnaire ISO 14505-3 (2007).

The purpose of the achievement of these data is the possibility of constructing mathematical models so that obtain mathematical relationships between thermal variables indicated inside the vehicle as well as the external environment, along with the subjective responses individual about thermal comfort.

Personal variables were determined by the ISO/TS 14505-1 (2007) and ISO 9920 (2007). The metabolism was defined as 75 W/m^2 and the vestment as 1.6 clo.

Finally, for the collection of subjective data, shall be applied a questionnaire which will be collected data relating to ISO 14505-3, as explained before.

Data processing will be done by means of descriptive analysis from measures of central tendency, and the construction of mathematical and computational models to evaluate the relationships between variables being studied, based on regressive models.

Figure 1. Heat stress measurer.
Source: Criffer (2013).

5 FINAL CONSIDERATIONS

Some considerations on the measurement of human responses used to quantify thermal comfort should be recommended. It is important to emphasize that the questionnaires used must be standardized subjective scales established by ISO 14505-3 (2007) so that results can be compared with those of other studies. Scales are used to classify both the overall comfort and for specific areas of the body.

Furthermore the ratings given previously by others should not be presented to individuals in evaluation thereby avoiding distortions by those who will perform the test.

Yet, it is essential to carry out a pilot test and the training of staff who will collect the data.

Despite the simplicity of data collection the way the scales are presented can influence the results. Therefore, careful translation of scales and cultural aspects should be considered.

The conditions under which the test shall be carried out to will determine in-vehicle comfort evaluation. However it will hardly be possible to reproduce real conditions identically. Therefore, the application of laboratory testing allows greater control over the variables where it is possible to adjust them more easily in order to analyze the situation of interest.

On the other hand, in the field, it becomes a more difficult task, since the range of possible variables to be controlled decreases considerably. In this case ISO 14505-3 (2007) highlights the importance of researchers being properly attentive during the course of the test in order to identify the factors that influence the thermal comfort.

The major disadvantage of the study in the laboratory is the non-reproduction of unusual facts and/or unexpected, unlike what happens in the field.

Finally, the analysis of the results involves qualitative and quantitative aspects. It is necessary to process the data and the application of statistical tests to support, for example identifying the relationship of the variables and the relevance of the data collected.

Also must be interpreted the responses and behavior of people. The results generally express trends that should be discussed before you reach a more accurate conclusion.

REFERENCES

Alahmer A.; Omar, M.; Mayyas, A.R.; Qattawi, A. 2012. Analysis of vehicular cabins' thermal sensation and comfort state, under relative humidity and temperature control, using Berkeley and Fanger models. *Building and Environment* 48: 146–163.

Alahmer, A.; Mayyas, Ahmed; Mayyas, Abed A.; Omar, M.A.; Shan, D. 2011. Vehicular thermal comfort models; a comprehensive review. *Applied Thermal Engineering* 31 (6): 995–1002.

ANSI/ASHRAE. *Standard 55:* Thermal Environment Conditions for Human Occupancy, 2010.

Criffer. *TGD-300:* Medidor de stress térmico (IBUTG). Disponível em: < http://www.criffer.com.br/locacoes/locacao_termometro_globo.php>. Acesso em 4 Mai 2013.

Djongyang, N.; Tchinda, R.; Njomo, D. 2012. Estimation of some comfort parameters for sleeping environments in dry-tropical sub-Saharan Africa region, *Energy Conversion and Management*. 58: 110–119.

Farzaneh, Y.; Tootoonchi, A.A. 2008. Controlling automobile thermal comfort using optimized fuzzy controller. *Applied Thermal Engineering*, 28:1906–1917.

Fraga, C.K. *A Polícia Militar Ferida*: Da violência visível à invisibilidade da violência nos acidentes em serviço. Tese de Doutorado apresentada ao Programa de Pós-Graduação em Serviço Social da Pontifícia Universidade Católica do Rio Grande do Sul. Porto Alegre: PUCRS, 2005.

Governo da Paraíba. *Regulamento de uniformes da polícia militar da Paraíba.* Disponível em: <http://www.pm.pb.gov.br/pagina.php?conteudo = informacoes&s ub = inf_rupm> Acesso em 20 abr. 2012.

Guimarães, L.B.M. *Ergonomia de Processo*. Ed. FEENG, vol. 1, Porto Alegre, RS, 2004. 31 p.

Hense, J.L.M. *On the thermal interaction of building structure and heating and ventilating system.* PhD thesis, Technische Universiteit Eindhoven; 1991.

Iida, I. *Ergonomia*: projeto e produção. 3 ed., São Paulo, Editora Edgard Blücher Ltda, 2005. 450 p.

International Organization for Standardization. *ISO 14505-3*. Ergonomics of the thermal environment—Vehicles – Part 3: Evaluation of thermal comfort using human subjects. Geneve, ISO:2006.

Marchi, S.R. Análise da Influência da Cor no Potencial de Aproveitamento da Luz Natural no Ambiente Construído. 2007. Dissertação (Mestrado em Engenharia Mecânica) – Programa de pós-graduação em Engenharia Mecânica, Universidade do Paraná, Curitiba.

Marengo, J.A. Possíveis impactos da mudança de clima no Nordeste. *Revista Eletrônica de Jornalismo Científico.* http://www.comciencia.br/comciencia/?section = 8& edicao = 22&id = 248. Acessado em 30 abr 2012.

Moura, M.B.B. de. *Aprimoramentos em sistema de climatização veicular para melhoria de condições ambientais de cabine e redução no consumo de combustível.* Trabalho de Conclusão de Curso apresentado à Escola Politécnica da Universidade de São Paulo para obtenção do título de Mestre Profissional em Engenharia Automotiva. São Paulo, 2007.

Parsons, K.C. 2000. Environmental ergonomics: a review of principles, methods and models. *Applied Ergonomics*, 31:581–594.

Pereira, C.J.O.; Alcobia, J. *Ergonomia ambiental em veículos.* Tese de doutorado em Ciências de Engenharia Mecânica (Aerodinâmica), 2006.

Rodriguez-Añez, C.R. *Sistema de avaliação para a promoção e gestão do estilo de vida saudável e da aptidão física relacionada à saúde de policiais militares.* Tese apresentada ao Programa de Pós-Graduação em Engenharia de Produção da Universidade Federal de Santa Catarina. Florianópolis: 2003. 143 p. Disponível em: <http://teses.eps.ufsc.br/defesa/pdf/7030.pdf> Acesso em 20 abr. 2012.

Simões, M.C. *Formulação de um repositor hidroeletrolítico para o trabalho físico ostensivo de policiais militares, adaptado as variações climáticas de Florianópolis.* Tese apresentada ao Programa de Pós-Graduação em Engenharia de Produção da Universidade Federal de Santa Catarina. Florianópolis, 2003. 271p. Disponível em: <http://teses.eps.ufsc.br/defesa/pdf/3818.pdf> Acesso em 20 abr. 2012.

Steen, R. *Preparing for safe winter travel.* The American National Red Cross. Disponível em: <http://www.redcross.org> Acesso em 07 Mai 2012.

Vasconcelos, I.C. *Estudo ergonômico do colete à prova de balas utilizado na atividade policial.* Dissertação (Mestrado) – Universidade Estadual Paulista, Bauru, 2007.

Occupational Safety and Hygiene II – Arezes et al. (eds)
 ISBN 978-1-138-00144-2

Take the risk by others: Working conditions and health of workers of private security

P. Costa
Faculty of Engineering, University of Porto, Rua Dr. Roberto Frias, Porto, Portugal

L. Cunha
Portuguese Catholic University, Rua Diogo de Botelho, Porto, Portugal

ABSTRACT: The present study aims to analyze the working conditions and workers health effects in the Private Security. We chose a quantitative methodology by applying the Work and Health Survey (INSAT) as well as a qualitative approach by conducting semi-structured interviews. The results showed that the perception of employees towards risk pointed to the exposure to various types of risk inherent in the performance of their activities including physical risks, psychosocial, organizational and also relational constraints. Globally, they contributing to enhance and exacerbate the health problems of exposed workers and also suggest the need for intervention in the working conditions of these workers. In this sense, it is necessary to join efforts, so that the managers of companies and those responsible for monitoring the health of these workers develop more contextualized interventions in this activity.

1 INTRODUCTION

In recent years, the Private Security (PS) sector has gained visibility due to the current economic conjuncture, which potentiated the increase in criminality and, by consequence, in insecurity, due to other factors linked to the evolution of society (Poiares, 2008). Among these factors, may highlight the growth of urban areas and the emergence of large commercial surfaces. According to Rodrigues (2011, p.83), the main causes of PS' growth are "*changes in the use of urban space and the movement of people, the expertise of the crime and the emergence of new forms of crime and perceptions of violence and increasing sense of insecurity.*"

The growing role of the Private Security Activity (PSA) in most European countries stems from the paradigm shift in the field of public security. Thus, it has found a significant increase in the privatization of risk, as well as the transfer of functions, which up to now was exclusively performed by public entities. However this reality has been changing and now is being carried out by private PS companies (IGAI, 2010). In Portugal, the expansion of PSA, appears in consequence of the State's inability to satisfy the need to achieve a collective security (Rodrigues, 2011). Due to this weakness, companies and some private individuals search for alternative ways to public security, resorting to the services of PS' companies. Currently, "*individuals are called up to assume responsibilities to perform essential tasks of the State*" (Gonçalves, 2005, p.16). The difficulties which the Forces and Public Security Services are struggling lead to an increase of requests to PS companies that, year after year, see the growth of their areas of competence (Rodrigues, 2011).

The PSA, in a global dimension, is in constant evolution since the seventies. Furthermore, in the Portuguese case, is being assumed a relevant importance, not only by sustained growth, but also by the creation of jobs, as has been observed in the last years. According to the Relatório Anual da Segurança Privada (RASP), in 2011 were active (bound by contract of employment with a PS company) 40.287 vigilants and 112 licensed PS entities providing services (MAI, 2012).

The PS sector is characterized by subcontracting and there are considerable rotation rates. These factors put surveillance professionals in a very particular situation, because when the contract between the requesting entity and the PS company ceases, the vigilant is in a very weak and uncertain position about the employment status. When the contract is awarded to another PS company, the vigilant often does not know if will be relocated to the new company and in what conditions, or if they will remain with their current job. These tensions could become worse if the uncertainty remains for a long period of time (CoESS, 2004). When the changes in the job's world suggest an increased insecurity and precarious jobs (Benach *et al.*, 2007;

Ferrie *et al.*, 2008), it is essential to be aware of the risks that this situation also brings up.

PS workers also carry out their activity in different work contexts throughout their professional life and, therefore, they are exposed to the inherent risks in their activity or in the activity developed in the customer's premises (CoESS, 2004). Workers who perform their surveillance functions in places with a large flow of people, including train and subway stations, health centers, hospitals, etc., are more susceptible to be exposed to the risk of aggression (CoESS, 2004). Moreover, workers that perform inspection, control and authority activities, do the public outreach, do isolated work and do the handling and transportation of valuables (Freitas, 2011), are also more susceptible to physical and verbal abuses. According to the Relatório Anual de Segurança Interna (RASI), in Portugal, in 2011 there were 47 robberies of cars transporting money (GSG, 2011). However, there was a decrease in 2012 in comparison to 2011, recording only 26 robberies (GSG, 2012). Despite the decreasing thefts, it does not mean that the risk to the workers has declined once they have to manage the risk of attempted robbery. For this reason, our concern focuses mainly on the private venture that has no relation with statistical indicators, since the data shown are related to situations of accomplished acts and do not have consideration to the situations that were not accomplished but were the target of attempted robbery, where the risk had to be managed by the employee.

This study aims, so to contribute, to a better understanding of working conditions and the effects on the health of PS workers, taking into account their point of view of the protagonists of this work activity. Labor and health relations are complex and require careful and detailed analysis, so as comprehend them. According to Gollac *et al.*, (2000, p.23), "*labor and health relations are neither univocal nor instant.*" In this context it's necessary to understand and appreciate more closely the features of the working conditions that interfere in the various dimensions of health, but not just the work-related diseases and the ones officially recognized in the list of occupational diseases (Gollac *et al.*, 2000). However, it is evident that, even without a diagnosed pathology, some health problems may cause a decrease in the work productivity and profitability with important repercussions in economics (Alavinia *et al.*, 2009). The intervention in occupational health, calls for a more comprehensive than explanatory approach (Volkoff, 2002), which emphasizes aspects such as observing, understanding and understanding health (Honoré, 2002). Thus, it is import ant to allow a new perspective on the health at work, mainly more focused on the experiences and perceptions of the worker. The effects of work conditions about the health and well-being, in fact, seem so obvious, that transmit the idea that the majority workers have a satisfactory health even being exposed to various occupational risks. However, this is not the truth because their pathologies are not visible but they exist and could be serious.

Consequently, it seems that there is a great difficulty in relating health problems with job, and that exposure to occupational risks, follows a more health oriented line, diagnosing the disease, establishing a relationship with the exposure to these risks, resulting in a classical approach to the concept of health as absence of disease itself (Barros-Duarte *et al.*, 2007).

2 MATERIALS AND METHODS

It was used a quantitative approach through the implementation of the survey INSAT (Barros-Duarte *et al.*, 2010) and was used and a qualitative approach by conducting semi-structured interviews, which proved effective in gathering information specifically the perceptions of workers related to their work and health. The INSAT fits in epidemiological studies and has been built from the contribution of several European surveys, developed particularly in France and Portugal. The first version the INSAT appeared in 2007 (Barros-Duarte *et al.*, 2007) the version has been updated in 2010 according to the developed experience based on its application in different sectors of activity (Barros-Duarte et al., 2010). This instrument aims to evaluate the relationship between working conditions and health of workers. The INSAT is structured in a logical coherent, integrated and sequential way of work and its health effects. In this sense, the INSAT is divided into seven main groups: (I) The work; (II) Conditions and job characteristics; (III) Conditions of life out of work; (IV) Training and work; (V) Health and work; (VI) My health and my work; (VII) My health and my wellbeing. The application of INSAT was conducted person by person, individually and the survey lasted about 40 minutes. From the application of INSAT survey, it was noticed that there were issues that were not sufficiently developed. Therefore, there was the need to conduct interviews with the participants, in order to enrich the information gathered through this instrument. To do so, a semi structured interview guide was designed to allow the collection of information relevant to this study. The script of the interview consists of four main parts, which cover the following aspects: (I) Socio demographic and career; (II) Changes to the level of the sector of activity and work; (III) Impacts of perceived work activity; (IV) Proposals for improving transformation. The interviews were conducted after the implementation of the surveys to the

Table 1. Identification of participants in interviews.

Identification	Age	Gender
A.C.	51	M
A.T.	37	M
P.G.	42	M
M.V.	40	M
S.A.	34	M
C.C.	38	F

workers in an individual context, with an average duration of 45 minutes. Simultaneously, and during all the process of data collection, observations of the work situation were verified with the aim of gathering additional information that could complement the information provided by the workers. It should also be verified that both the application of surveys and the performance interviews were conducted in the schedule of work activity, taking into account the working hours of these workers and their availability. After collecting the data, the SPSS program (Statistical Package for the Social Sciences), version 21, proceeded with a descriptive statistical analysis and inductive issues of INSAT. The most important points of the interviews were analyzed in order to complement the information collected through the surveys.

The target population under study is composed by 766 workers of a company that provides from private security services located at Oporto's delegation city. The sample consists in 34 workers randomly selected (30 males and 4 females). Although, only 6 participants of these were selected to participate in the structured interviews (five males and one female), that aimed to explore some issues that have not been sufficiently explored in the INSAT questionnaire, as the Table 1 shows. For the interviews, were selected the participants that showed more knowledge of their activity, in order to provide a useful contribution to understand the data.

3 RESULTS AND DISCUSSIONS

The participants in this study were mostly male (88.2%) and only 11.8% were female. These workers carry out their activity in six specific work contexts, being the oil and rail transport sectors the most significant, representing 35.3% and 29.4% of the total sample, respectively. The results also showed that the majority are permanent workers (61.8%), despite this kind of work to bring a lot of job insecurity for these workers. This situation may be explained by the Collective Bargaining Agreement the professional surveillance, that does not provide the professional transition from current PS company to the new company (BTE, 2011).

Table 2. Physical and organizational constraints.

Risk factor	Participants exposed (%)	Degree of discomfort (%)
Exposed to the heat and/or the intense cold weather	58.8	50.0
Remain many hours standing up	55.9	47.0
Sleep at unusual hours	61.8	61.7
Lie in bed after midnight	73.5	55.9
Exceed the normal working hours	70.6	50.0
Frequently "jump" or have a short meal or skip making a break	67.7	52.9

Therefore, the fear and insecurity were reported by study participants. It was also found that the majority of employees working in a full time basis (94.1%) have rotating shifts (85.3%). These, as it is a rule, change their work schedule every week, on the basis of three shifts practiced by the company (morning shift, afternoon shift and night shift). The Table 2 shows that the participants of this study are subjected to various types of risks during the performance of their activities.

In the exercise of their activity, it is susceptible to occur variations of temperature, and exposure to adverse temperature conditions (58.8%), and it becomes painful for 50% of workers. The fact that they often perform the outdoor service, also contributes to increase thermal discomfort. According to the interviewed P.G. ["*Here we pass through tremendous thermal shock, it's horrible, it's horrible* ..."]. Workers indicate cold, heat and discomfort as factors that can negatively affect the performance of their duties. Still, according to the same participant, cold ["*we feel in the flesh, in the bones, in the body ... after a few years of work, we were worn out* ..."]. The participants consider that the uniforms are not suitable for thermal conditions of the premises where they perform their activity. As the participant A.T., refered ["*the uniforms are inappropriate to the service and the place, because here there are sudden temperature changes. Suddenly it's hot and suddenly it's cold*"]. The fact that the uniforms are not suitable for thermal conditions can cause problems for the health of exposed workers. The uniforms used are not only an imposition of the PS company, but also a legal imposition, since the law of PS defines the obligation of vigilant to be properly uniformed (approved uniforms).Workers spend many hours standing (55.9%) in the same position and/or in constant movement causing inconvenience and discomfort to 47% of the participants. This was most visible in refinery facilities and railway stations, because the demands of the activity by these customers contribute to the fact that the vigilants have to spend all the working hours standing up. Workers also refered that the shoes that are part of their uniforms are discomfortable. They are not suitable for walking or comfortable for long periods of time. The participant A.C., states that ["*we are forced to walk with shoes created for a reception, and are forced to walk with them eight hours in a station, to do kilometers, up and down the stairs ... the vigilant are uncomfor*table *with this situation*"]. And yet according with the participant P.G., the fact that spending long hours standing causes them discomfort and other health problems, including pain in the ["*ankles, legs, knees, thighs and hip at the end of the day, and all this added over the days and years ... we are exhausted ...*"]. It was further observed that the vigilants have to sleep at unusual hours (61.8%); lie in bed after midnight (73.5%) and frequently "jump" or have a short meal or skip making a break (67.7%), knowing that they work alone. The organizational constraints that cause the most discomfort in workers are the fact that they sleep at unusual hours (61.7%) and lie in bed after midnight (55.9%), as a result of different shifts that employees perform in their work activities. From all the participants, 70.6% reported that they extend the normal working hours, because the service must be permanently maintained by a vigilant, as it was provided in the contract between the PS company and the contractor. This constraint becomes painful for 50% of participants. Finally, workers complain in having a short or haven't a break (52.9%), because the According to the provisions of the CBA and the Labour Code, the work schedule is exercised continuously and doesn't predict breaks, even for meals. According to the interviewed P.G., the contract that the PS company celebrates with its customers does not include breaks. For this reason, during the eight hours of work in a day, the vigilant does not make one break at least for a small meal. The same participant stated that ["*customers, some of them more sensitive, will agree to our requests for at least a 15 minutes break during the eight hours service to eat a sandwich ...*"].

The participants have to confront themselves with situations of tension and conflict in relations with the public (85.3%), are therefore exposed to risk of verbal abuse (82.3); physical assault (76.5%) and risk of intimidation (76.5%). One of the aspects mentioned by workers as negative was the exposure to the abuse by the public and it was highlighted in interviews, ["*I've often been verbal assaulted, not physically, because I do not allow*", participant M.V.]. The responses to INSAT proves that statement and demonstrated that 64.7% of workers feel uncomfortable by being exposed to the risk of physical assault and 52.9% to the risk of verbal abuse, as table 3 shows.

The interviewed A.T., said that he was victim physical assault and even had to receive hospital treatment, ["*twice was stabbed with a sharp object ... that left marks*"]. The same participant reported that, ["*verbal abuse affected more than the physical aggression*"]. The respondent P.G., who has been the victim of verbal abuse several times says, ["*Of course this does not leave marks, but the fact is that when such a case occurs... I can go home and not sleep thinking about the situation ...*"]. The mark that this worker refers, are not physical but are mainly marks that affect psychologically. As reported by the interviewed A.C., ["*it is rare a day that a vigilant does not suffer attempts of aggression ... it can be both verbal and physical ...*"]. Workers may be affected by constraints of different nature, not only by the physical demands of the job, but also by the feeling of annoyance by being exposed to certain factors that may weaken them in various ways: feel exploited (29.4%), do not wish their children to perform this work (47%), not being able to perform this work at the age of 60 (29.4%) and feel insatisfied with their work (26.4%). These results require a better understanding of the reasons for insatisfaction, feeling that they are being exploited and the prospect of failure to pursue the current role. The feeling of insatisfaction and exploitation is due to the fact that these professionals do not have prospects for the development of their career, ["*a critical point are the professional categories ... is not motivating for those that are working in the industry for 20 years and suddenly arrives a new vigilant to which I have to pass all the information that I have acquired over these 20 years and reach the end of the month and I am paid equally ...*" interviewed M.V.]. The participant C.C. believes that ["*we should have more rights ... we do not have the best conditions ... I do not see any progress in this direction, and also do not see progress in terms of career...*"]. The constraints of the work, according

Table 3. Relational constraints.

Risk factor	Participants exposed (%)	Degree of discomfort (%)
Have to confront themselves with situations of tension and conflict in relations with the public	85.3	50.0
Exposed to a risk of verbal abuse	82.3	52.9
Exposed to a risk of physical assault	76.5	64.7
Exposed to a risk of intimidation	76.5	20.6

to perception by workers, contributed to enhance and aggravate some health problems, such as: headache (20.6%), back pain (23.5%) and sleep problems (20.6%). Other infra pathologic problems were reported by workers, in particular: perception that the days seem never end, so they are always tired (mentioned by 100% of participants affected); have to wake up early and have sleeplessness (77.8%); sleep poorly overnight (80%); take a long time to fall asleep (75%); spend most of their time awake and have pain while standing (75%); easily lose their heads (50%); have difficulty to stand up for a long period of time (50%); feel nervous and tense (30%).

The results led to a better understanding of the relationship between working conditions and health of workers and allowed to concluding beyond the logic of health as absence of disease, looking at a different perspective, taking into account both the less visible conditions of employment and employment problems, that workers face and are often causing suffering and discomfort (Barros-Duarte *et al.*, 2007). The authors reaffirm that the health problems infra pathological and its relation to the work should be, increasingly, a concern for the technical safety and health at work and for the employers.

4 CONCLUSIONS

PS workers are exposed to various types of risks inherent to the professional development of their activity. For this reason, it is necessary to consider the specific context in which they exercise their activity to understand better and act on the risks they are exposed. In this sense, it is expected that managers, business leaders and those responsibles for health surveillance of these workers can develop interventions anchored in the actual occupation, integrating the point of view of the protagonists of these work situations. Finally, it would be important to examine better the conditions of employment of professional surveillance, because this dimension also interferes directly with the health and well-being of workers.

The results are limited to a small number of employees. In this sense, in future studies would be an asset the integration of a larger number of participants in order to develop a study that portrays the reality of PSA at a sectoral level.

ACKNOWLEDGMENT

The authors would like to thank Master in Occupational Safety and Hygiene Engineering (MESHO) of the Faculty of Engineering of the University of Porto (FEUP), all the support in the development and international dissemination of this work.

REFERENCES

Alavinia, S.M., Molenaar, D. & Burdorf, A. (2009). *Productivity loss in the workforce: associations with health, work demands and individual characteristics.* American Journal of Industrial Medicine, 52, 49–56.

Barros-Duarte, C., Cunha, L. & Lacomblez, M. (2007). *Inquérito de Saúde e Trabalho: uma proposta metodológica para a análise dos efeitos das condições de trabalho sobre a saúde.* Laboreal, 3, (2) 54–62.

Barros-Duarte, C. & Cunha, L. (2010). *Inquérito de Saúde e Trabalho: outras questões, novas relações.* Laboreal, 6, (2) 19–26.

Benach, J. & Muntaner, C. (2007). *Precarious employment and health: developing a research agenda.* Journal of Epidemiology and Community Health, 61, 276–277.

Boletim do Trabalho e Emprego (BTE) (2011). *Contrato Coletivo de Trabalho entre a Associação das Empresas de Segurança*, (17) 1486–1507.

Confederation of European Security Services (CoESS) (2004). *Preventing occupational hazards in the private security sector.* Bruxelles.

Ferrie, J., Westerlund, H., Virtanen, M., Vahtera, J., Kivimäki, M. (2008). *Flexible labor markets and employee health. Scandinavian* Journal of Work, Environment and Health; 34(6), 98–110.

Freitas, L.C. (2011). Manual de Segurança e Saúde no Trabalho (2ª ed.). Lisboa, Edições: Sílabo.

Gabinete do Secretário Geral (GSG) (2011). *Relatório Anual de Segurança Interna.* Sistema de Segurança Interna. Lisboa.

Gabinete do Secretário Geral (GSG) (2012). *Relatório Anual de Segurança Interna.* Sistema de Segurança Interna. Lisboa.

Gollac, M. & Volkoff, S. (2000). Les conditions de travail. Paris. Editions La Découverte.

Gonçalves, P. (2005). *Entidades Privadas com Poderes Públicos—O exercício de poderes públicos de autoridade privadas com funções administrativas.* Coleções de Teses. Lisboa, Editora: Almedina.

Honoré, B. (2002). *A saúde em projeto.* Tradução—Ilda Cristina d`Espiney. Loures: Lusociência.

Inspeção Geral da Administração Interna (IGAI) (2010). *Guia de Fiscalização e de Investigação de Segurança Privada.* Lisboa.

Ministério da Administração Interna (MAI) (2012). *Relatório Anual de Segurança Privada.* Conselho de Segurança Privada. Lisboa.

Poiares, N. (2008). *Novos Horizontes para a Segurança Privada. Estudos em Homenagem ao Professor Doutor Artur Anselmo.* Coimbra, Editora: Almedina.

Rodrigues, N.P.G. (2011). *A segurança privada em Portugal. Sistemas e Tendências.* Coimbra, Editora: Almedina.

Volkoff, S. (2002). *Des comptes à rendre: usages des analyses quantitatives en santé au travail pour l'ergonomie.* Noisy-le-Grand: Centre d'Etudes de l'Emploi.

Occupational Safety and Hygiene II – Arezes et al. (eds)
© 2014 Taylor & Francis Group, London, ISBN 978-1-138-00144-2

Occupational noise exposure in gyms of county of Coimbra

S. Seco, J. Almeida, J. Figueiredo & A. Ferreira
Instituto Politécnico de Coimbra, ESTeSC, Coimbra Health School, Saúde Ambiental, Portugal

ABSTRACT: Occupational noise is one of the most important risk factors for workers and is also the most common physical agent found in the workplace. Thus, it's imperative to evaluate the noise exposure of professionals working in gyms and try to assess its impact on the health and well-being of workers. The test was composed of 4 gyms, 22 workers, with noise measurements made during all programs available. Questionnaires were applied to all professionals, about personal information, professional and health condition data. The professionals fulfilled a weekly schedule with all activities performed by them in the gyms under study and the respective exposure times. More than 50% of the individuals were exposed to noise levels which are harmful to their health, reaching in some cases up to 84,2 dB(A). According to professionals, none had hearing problems, however some symptoms of discomfort associated with noise exposure were identified: stress, irritability, and others.

1 INTRODUCTION

The population's health and well being are indispensable factors for the good social life. However, the technological development has caused very negative effects, namely the noise factor (Medeiros & Goldenberg, 1999; Arezes & Miguel, 2002).

According to the definition, noise is the "unwanted and uncomfortable" sound (Arezes & Miguel, 2002) and therefore, it can be unpleasant and/or intolerable for the human being (Pereira, 2009; APA, 2013). But these feelings are quite subjective, as they depend on several factors, such as the features of the noise itself; the time of exposure and the individual susceptibility—sex, age, living style, emotional mood and others (Harger & Barbosa-Branco, 2004; Costa, 2009).

The exposure to the noise in the working places is one of the most important risk factors for workers (Costa, 2009; Pereira, 2009; Mergulhão, 2009) and also the most common physical agent in the working area.

Nevertheless, the workers' routines may lead them to get used to this agent's presence and consequently, they stop noticing its existence in the working place (Pereira, 2009; Mergulhão, 2009). The exposure to intense noise levels conditions the aptitude to hear and communicate which may cause multiple damages in the auditory apparatus: acoustic trauma (Arezes, 2002; Arezes & Miguel, 2002; Pereira, 2009) and temporary and/or permanent hearing losses leading to the humming feeling (Arezes, 2002; Mergulhão, 2009; Mendes, 2011). That exposure also has consequences at the social, psychological and/or physiological extra auditory levels, associated to the stress and tiredness (Arezes, 2002; Costa, 2009; Pereira, 2009)—irritability, lack of motivation, bad mood, apathy, sleeping disturbs, among others which cause a general tiredness (Bernardi, et al, 2006; Pereira, 2009; Mergulhão, 2009). The blood system, as well as the vestibular, gastrointestinal, hormonal, respiratory and reproductive system may also suffer from some disturbances, even if the noise levels are quite low (Medeiros & Goldenberg, 1999; Costa, 2009; Pereira, 2009; Mergulhão, 2009).

The national legislation (Decree-Law nº 182/2006, 6th September, transposed by Directive nº 2003/10/CE, of the European Parliament and of the Council of 6th February 2003) establishes limit values of personal exposure and also establishes the superior action and inferior action values, from which it is absolutely necessary to take some measures, stressing also a number of measures to be taken every time those values are reached or exceeded. It is also needed in the range of this legal structure, a suitable control of the workers' health, namely through the audiometric exams.

The goal of this investigation is to evaluate the workers' exposure to the noise created in the gyms, as well as to understand its impact in their health and well being.

2 MATERIAL AND METHODS

This research was conducted in the gyms of the county of Coimbra and it belongs to the Level II, descriptive-correlative, observational and from the transversal cutting type. The studied sample was composed by the gyms belonging to the county of Coimbra and respective professionals. The sample was: 4 gyms and 22 workers.

The data collecting was done between November 2012 and January 2013 and it was divided into three stages. On the first stage it was collected data using a sound meter Brüel & Kjær, model 2260 and software BZ7210 and BZ7815 to evaluate the noise. Two measurements were made during each lesson, in each gym and also another measurement in the same spaces outside the lesson period. In the bodybuilding and cardiofitness rooms two measurements were also done, resulting in 131 measurements, during between nearly 5 to 10 minutes, always performing a sound meter calibration before each measurement series. The measurements were done having the reference of the methodology included in the Decree-Law nº 182/2006, from 6th September, transposed by Directive nº 2003/10/CE, of the European Parliament and of the Council of 6th February 2003. The equipment was placed as near the professionals as possible. The noise evaluation took place in a normal functioning period but it was not possible to do measurements in the empty bodybuilding and cardiofitness rooms. The measured analytical parameters were the equivalent continuous sound level ($L_{Aeq,T}$) and the peak sound pressure level ($L_{C,pico}$), to find out the daily personal exposure ($L_{EX,8h}$) to the noise and also the weekly personal exposure ($\bar{L}_{EX,8h}$). The second stage was the answering to a questionnaire to gather information related to the workers. The questionnaires were validated by preliminary application to a different population, and were subsequently applied to the target population. The questionnaires were applied after the institutions authorization, with guaranteed anonymity and confidentiality of the data collected for the study. This research is of purely academic interest, showing no economic or financial interests. It was not possible to do any preliminary audiogram during sampling. The third stage was the delivery of a timetable aimed at retrieving detailed information about every activity developed each day of the week, per worker, in the four gyms, as well as the exact duration of each activity, so it could be possible to find out the daily personal exposure to the noise.

The data analyzing was made using the software Noise Explorer Type 7815 and the statistic software IBM SPSS, version 21.0. To check the studied hypotheses, it was used the tests: t-Student for independent samples; t-Student for 1 sample; Analysing the variation to I factor for independent samples and tests of multiple comparison Games-Howell.

Understanding the statistic tests was possible using a significance level $\alpha = 0.05$ with a confidence interval of 95%. For a significant α ($\alpha \leq 0.05$), differences or associations between the groups were observed. However, for $\alpha > 0.05$, the observed associations or differences were not considered statistically significant.

3 RESULTS

The present study involved 22 professionals, from 4 gyms, 14 (63,4%) were males and 8 (36,4%) were females. For the calculus of the workers personal exposure to the noise, there was a sample of 20 people: 13 men and 7 women.

It was known that 8 (36,4%) of those individuals had professional activities outside these gyms and 7 of them are men. It was found out that the average age was 31 ± 7,21 years old, and both sexes presented very similar values. Men generally performed their job for a longer time than women, around 8 ± 8,11 years while the women presented the average number of 5 ± 2,30 years. It was also found out that these professionals worked around 7 ± 2,63 hours a day and the female worked around 8 ± 0,41 hours a day, compared to the male group (6 ± 2,89 hours a day). In the specific case of the sample for the measurement of the $L_{EX,8h}$, the medium age was still 31 years old. On average, the professionals worked 5.4 hours a day and the male group was the one which worked more (5,6 hours a day) compared to the 5,4 hours a day for the female group. As for the disturbances in health it was found out that the whole sample stressed they didn't have any auditory pathology. But the feeling of discomfort was reported by 13.6% of the individuals. It was found that there aren't average statistically significant differences about the number of years spent doing the job or about the age, related to the discomfort (p-value > 0,05). However, in average, the people who said they suffered from daily discomfort were those who had been working in this kind of activity for fewer time (5 ± 1,15 years) and also the youngest ones (29 ± 1). The practiced modalities were grouped according to their similarities (1, 2, 3, 4 and 5). In the group 1 there were those demanding more rhythm; the group 2 focused on body specific parts; the group 3 was related to the practice of flexibility, posture and relaxing; in the group 4 were those activities which traditionally have very intensive music types; and in group 5 there were the bodybuilding and cardiofitness which, according to the gyms, were practiced in the same space or in different spaces. It was observed the existence of statistically significant differences (p-value < 0.0001) between the average values of equivalent continuous sound level ($L_{Aeq,T}$) and the different groups of activities practiced by the professionals. The groups 1 and 4 revealed the highest average values of $L_{Aeq,T}$:

90.74 ± 3.64 dB(A) and 89.34 ± 2.57 dB(A) respectively. The activities where the average values of $L_{Aeq,T}$ were the lowest, belonged to the group 3 (66.10 ± 10.46 dB(A)). The average value of $L_{Aeq,T}$ of every group of activities was 81.36 ± 11 dB(A). It was also noticed the existence of differences between the average values of $L_{Aeq,T}$ found for each group of activities. There are differences between the groups 1 and 3, having that difference been significantly superior in group 1 (d = 24.64 dB(A)), in average. Comparing the groups 3 and 4, it was also observed the existence of statistically significant differences between the average values of $L_{Aeq,T}$ got for each of them (p-value ≤ 0.0001), having those differences been significantly inferior for the group 3 (d = −23.24 dB(A)). The Graphic 1 supports the previous analysis, showing the variation of $L_{Aeq,T}$ values, according to the group of developed activities and showing that the groups whose equivalent continuous sound levels got nearer were 1 and 4, as the group 3 is the one that stays further.

It was observed the existence of average differences between the values of $L_{C,pico}$ got in each of the practiced modalities and the inferior action value determined by the law (135 dB(C)). However, all those differences were significantly inferior to the value of the inferior action (p-value ≤ 0.0001), except in the case of the Located modality which really had an inferior average difference, but it was not significant (p-value > 0.05). It was also registered the existence of statistically significant differences (p-value < 0.0001) between average values of $L_{C,pico}$ of the different activity groups. The groups 1 and 4 were once again, those which revealed the highest average value: 105.06 ± 1.99 dB(C) and 103.83 ± 3.02 dB(C) respectively. The activities where the average values were the lowest belonged to the group 3: 77.49 ± 7.90 dB(C) and the average value of $L_{C,pico}$ of all the groups of activities was 94.53 ± 11.62 dB(C).

It was observed a behavior pattern similar to the one of the $L_{Aeq,T}$ values. Therefore, it was observed the existence of differences between the groups 1 and 3 (being that difference significantly higher in group 1 (d = 27.56 dB(C)) and between the groups 3 and 4, having those differences been significantly lower for group 3 (d = −26.34 dB(C). The Graphic 2 shows the $L_{C,pico}$ variation, according to the group of developed activities, and those which got nearer, as far as the peak sound pressure is concerned, kept being the groups 1 and 4. However, it was observed that even with a Confidence Interval (CI) of 95%, none of those values exceeded the 135 dB(C), having been found the lowest difference of 29.94 dB(C), related to that value.

Being 80 dB(A) the inferior action value determined for the levels of daily personal exposure ($L_{EX,8h}$) or weekly personal exposure ($L_{EX,8h}$ ≥ 80 dB(A)), it was observed that the weekly average of the daily personal exposure to the sound, in the studied sample, was 78.84 dB(A) (p-value > 0.05), not exceeding the value of the inferior action, which is established by law. Although there were differences between this value and the inferior action value (d = ±6.90dB(A)) the calculated average value for the workers was inferior 1.16 dB(A) related to the reference value.

It was observed that around 50% of the individuals were exposed to values under 80 dB(A), and in some cases their exposure slightly exceeded that value ($\overline{L}_{EX,8h}$ ≤ 80,11 dB(A)). The remaining 50% were exposed to noise values above the inferior action value—25% between 80.11 dB(A) and 83.28 dB(A) and the remaining 25% reached an exposure of 84.2 dB(A), having the superior action value never been reached (Graphic 3).

Concerning the number of weekly hours spent by the professionals practicing several modalities, it was observed that the inquired professionals who practiced for more time revealed a higher $\overline{L}_{EX,8h}$,

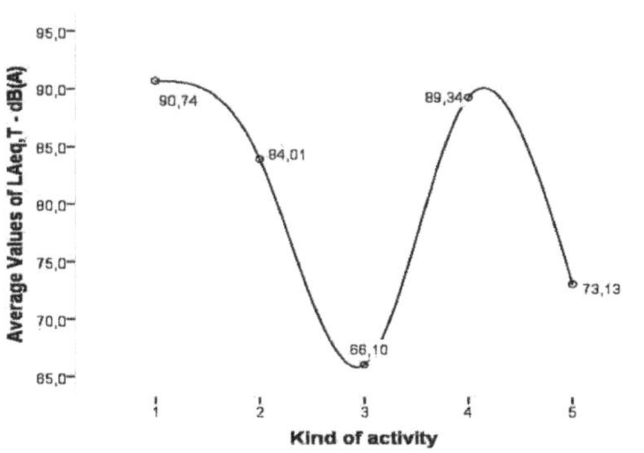

Graphic 1. Variation of $L_{Aeq,T}$ average values between the several groups of activities.

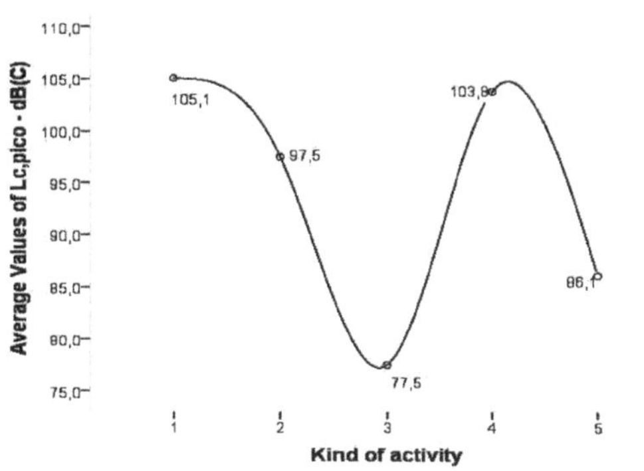

Graphic 2. Variation of the $L_{C,pico}$ average values between the activity groups.

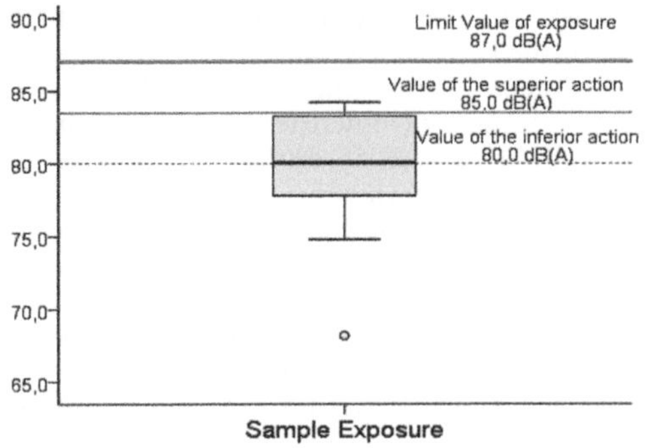

Graphic 3. Variation of the $\overline{L}_{EX,8h}$ values for the studied professionals, related to the legal values.

although it wasn't significant (p-value > 0,05). It was also observed the existence of difference between that exposure, as about the sex, the inferior action value, not only with men (d = −1,02dB(A)) but also with women (d = −1,41dB(A)). Although the differences weren't statistically significant (p-value > 0,05), it was revealed a higher exposure for men. It was observed that less than 50% of men had a $\overline{L}_{EX,8h}$ inferior to 80 dB(A)—less than 25% of them reached the 78,53 dB(A) and the remaining reached levels of 80,75 dB(A) and all the others were above this value. On women, it was observed that 50% didn't reach the 80 dB(A), although they have reached very near values ($\overline{L}_{EX,8h} \leq 79{,}83$ dB(A)) and on the remaining 50%, 25% reached 83,28 dB(A).

4 DISCUSSION

The exposure to sound in the gyms revealed itself quite high for many studied professionals ($L_{EX,8h} \geq$ 80 dB(A)) and the noise is the responsible agent for the degenerative hearing function in many individuals (Arezes & Miguel, 2002; Dias, 2006; CE, 2007). However, all the professionals informed they didn't have any hearing problems, even those who have been working there for a longer time, but in general, deafness is a disease which develops very slowly and it is difficult to be aware of (CE, 2007). On the opposite, it was identified, although in a small percentage of workers (13,6%) some symptoms of discomfort, which are frequently associated to the exposure to high levels of noise: stress, general tiredness, apathy, difficulties in focusing; irritability and sleeping disturbances (Harger & Barbosa-Branco, 2004; CE, 2007; Costa, 2009). The reason why the professionals who reported these symptoms were, on average, the youngest (in age and years of service) may be related to habituation to noise by older workers and/or workers who already work in this business for more years (CE, 2007; Mergulhão, 2009; Pereira, 2009).

The average $L_{Aeq,T}$ highest values were got by the modalities with a higher sound volume, associated with activities having more rhythm and demanding people's interaction, as well as much moving in the room. Some of those examples were STEP, Body Pump and Body Combat belonging to groups 1 and 4, respectively. In group 2, the activities had an intensive physical effort, but the activities were made without needing to move very much through the room, which may have contributed to the difference of the $L_{Aeq,T}$ average value, compared to the other groups. In the case of the activities of bodybuilding and cardiofitness (group 5), the values depended on the number of people in the room and the used equipment. The activities of flexibility, posture and relaxing, where a background music is used and which activities are made in a small space by each person, with soft movements, were those which revealed the lowest $L_{Aeq,T}$ values, being their practice (if compared to the other activity groups and having the same number of exposure hours) the less damaging for the professionals' health and that practice also got more monitors' adhesion.

Some of the noise measurements were conducted during the period of less affluence (holidays). Otherwise, it would be logical that the activity group 4, the one with more adhesion could be the noisiest, due to its conditions: very strong movements, music with a strong rhythm, much moving through all the room and a very strong interaction among people.

According to the Decree-Law nº 182/2006, from 6th September, the levels of peak sound pressure used as risk indicators to evaluate the effects of the exposure to brief and very intensive sounds (impulsive noise) (CE, 2007) must not exceed the 135 dB(C) and it was observed that the law was taken into consideration in every studied activity. The activities of STEP and FUNDANCE got the highest $L_{Aeq,T}$ average values, Body Pump and Body Combat belonging to the activity group 4 had a higher average $L_{C,pico}$ value, reaching close to the 135 dB(C). This can be explained by the fact that they are very intensive activities, with louder music and there was much interaction between the participants. As all the values were situated beyond the reference value ($L_{C,pico}$), the professionals' auditory function disturbance can be observed only at a long term (CE, 2007; Albizu, Lacerda, Gonçalves, Heupa, & Costa Junior, 2012).

The Portuguese legislation, refers two more physical parameters used as risk indicators: $L_{EX,8h}$ and $\overline{L}_{EX,8h}$, which enable the valuation of the long term exposure to the noise, as the daily exposure to noise was different along the week, the present research investigated the professionals' weekly personal exposure. More than 50% of the individuals were exposed to noise levels superior to 80 dB(A) and consequently it is recommended to eliminate or, if not possible, to reduce the exposure level to that agent. Knowing those figures, the effects on the auditory function are more serious (CE, 2007). The studied professionals' individual exposure was quite near the superior action value ($\bar{L}_{EX,8h}$ = 85 dB(A)), reaching the 84,2 dB(A). The employer might take some measures such as: rearranging the work timetables according to the type of the developed activities and their distribution along the week, installing sound limiting devices in the sound devices; changing the columns position; informing the professionals about the consequences of their exposure to noise and recommend to lower the sound volume; and correct maintenance of the used equipment. The action to eliminate the exposure to noise must primarily be focused on the source, promoting the collective protection (CE, 2007). Men were those who were mostly exposed to noise. This situation could be worse, as they pointed out to have other professional activities, thus, their personal exposure to occupational noise could have been even higher.

5 CONCLUSIONS

The workers' exposure to noise in the studied gyms exceeded in more than 50% of the cases, the inferior action value determined by the national law, and in some situations, the values were very near the superior action value, reaching the 84,2 dB(A). On the other hand, the level of the peak sound pressure was not exceeded in any case. Some workers referred to feel some discomfort symptoms, generally associated with the exposure to noise. This research strengthened the importance of assessing the noise in this activity, as it was possible to conclude that although practicing sport is a healthy activity, it can be damaging, when it is practiced in closed spaces.

In spite of the gathered conclusions, this inquiry also had some limitations: The measurements were partly made during holidays, with a smaller attendance, which probably influenced the measurements results; The measurements of noise, in each gym, were not made with the same professionals and therefore, the noise levels may differ according to the professionals.

The present data showed that the professionals' exposure to noise was mostly above the inferior action value. With such kind of study it would be possible to contribute to the improvement of these professionals' life quality, through the identification of the factors which determine an excessive exposure to noise. Suitable measures could also be considered, avoiding damages in the auditory function. Therefore, we propose a development of similar researches, reaching a higher sample of gyms and professionals.

REFERENCES

APA (2013). *Som, ruído e incomodidade.* Agência Portuguesa do Ambiente. Retrieved January 10, 2013, from http://www.apambiente.pt/index.php?ref=16&subref=86&sub2ref=529.

Albizu, E., Lacerda, A., Gonçalves, C., Heupa, A. & Costa Junior, H. (2012). *Audiological findings among workers from Brazilian food industry exposed to continuous and impulsive noise.* SHO 2012—Occupational Safety and Hygiene, 7–13.

Arezes, P. (2002). *Percepção do risco de exposição ocupacional ao ruído.* Escola de Engenharia da Universidade do Minho.

Arezes, P. & Miguel, A. (2002). *A exposição ocupacional ao ruído em Portugal.* Riscos Ocupacionais, 20(1), 61–69.

Bernardi, A., Fiorini, A., Costa, E., Ibanez, R. & Serra, T.R.R. (2006). *Saúde dos trabalhadores; 5 Protocolos de Complexidade Diferenciada—Perda auditiva induzida por ruído.* (M. da Saúde, Ed.) (1a ed., pp. 1–40). Brasília.

CE (2007). *Guia indicativo de boas práticas para a aplicação da Directiva 2003/10/CE "Ruído no trabalho".* Comissão Europeia (pp. 1–169). Luxemburgo.

Costa, H. (2009). *Exposição ao ruído ocupacional e sua repercussão na saúde dos trabalhadores da empresa CMP-Maceira.* Universidade de Coimbra.

Dias, A. (2006). *Associação entre perda auditiva induzida pelo ruído e zumbidos.* Cad. Saúde Pública, Rio de Janeiro, 22(1), 63–68.

Harger, M. & Barbosa-Branco, A. (2004). *Efeitos auditivos decorrentes da exposição ocupacional ao ruído em trabalhadores de marmorarias no distrito federal.* Rev Assoc Med Bras, 50(4), 396–399.

Medeiros, L. & Goldenberg, M. (1999). *Ruído: efeitos extra-auditivos no corpo humano.* Centro de Especialização em Fonoaudiologia clínica.

Mendes, A. (2011). *Ruído ocupacional em ambiente industrial.* Faculdade de Engenharia da Universidade do Porto.

Mergulhão, F. (2009). *Ruído ocupacional e sua perceção pelos alunos de medicina dentária.* Universidade Fernando Pessoa.

Pereira, A. (2009). *Avaliação da exposição dos trabalhadores ao ruído (análise de casos).* Universidade do Minho.

Occupational Safety and Hygiene II – Arezes et al. (eds)
 ISBN 978-1-138-00144-2

Ergonomic service quality of the elderly on the example of the financial market

M. Butlewski, A. Misztal & E. Tytyk
Poznan University of Technology, Chair of Ergonomics and Quality Management, Poznań, Poland

D. Walkowiak
Financial Sector, Poznań, Poland

ABSTRACT: The growing services market results in an increased interest in quality control methods in respect to intangible products of businesses such as services. Service quality requirements are even greater, because the demographic structure of Europe is changing due to the prolonging life expectancy and growing percentage of seniors in the population. Companies, such as those in the banking market, with a number of complex financial products, should increasingly respond to the needs of these elderly customers. In order to respond to this demand, on the basis of SERVQUAL methods and ergonomic guidelines, a new tool of assessing the ergonomic quality of financial services for the elderly has been created—ERGO-SERVQUAL. Surveyed customers assess their expectations prior to service and the experienced service quality and the degree of importance of specified factors. The research results determine the needs and expectations for the ergonomic quality of financial services.

1 INTRODUCTION

Currently, quality is a socio-economic problem. It applies to every country and every sphere of human life. Quality is both rational and irrational in nature. One can make a very parameterized description of it or leave a qualitative description (Mazur 2013). In many cases, one can speak not only of product quality but also of the quality of services, processes, or even quality of life or performed work (Misztal & Butlewski 2012). Quality management can be a path to achieving business excellence by setting goals and consistently striving to achieve them, but in the case of its abandonment it does not yield any added value (Jasiulewicz-Kaczmarek 2012).

With the right development of quality of the offered services, companies are able to obtain and maintain a competitive advantage in the market and meet the needs of a wide range of customers. Especially in the case of micro-businesses, which in Poland amount to 1.67 million, the quality of service is a feature that allows to survive on the market (Więcek-Janka 2008). Tools to manage service quality also identify critical characteristics and needs of users and consequently help to build lasting relationships with customers. In addition, it is worth noting that the best information technology, the most efficient prudential rules or the largest capital will mean little when the organization of the service provider will not be focused on the primacy of quality and will not demarcate a strategy to achieve a high quality of service (Opolski 2000).

Service quality focused on elderly people is due to the fact that they constitute an ever-increasing group. Because of the processes occurring during aging, elderly people have more specialized, but often hidden, needs. This is why the quality of services should be directed through ergonomic requirements which, because of their specificity, are ideal for shaping the environment to the needs of users.

2 BACKGROUND

2.1 *The need for a development of quality for the elderly*

Currently in Poland there are nearly 20% of people over 60 years of age. In the year 2035 the percentage of those over 65 years of age will reach 38%, and for all of Europe this factor, the so-called demographic load, will achieve a level of 43% (CSO 2009). The estimated life expectancy of people over 65 years (until recently the retirement age) is 17.3 years for men and 20.9 years for women (IIASA 2012).

On the other hand, it ought to be noted that aging is associated with changes such as (Duda 2013), decreasing adaptation to various physical, biological and psychosocial loads, decreasing organ reserves, increasing morbidity, and rising mortality. The changes of a physiological and psychological nature also cause personality changes, the course of which is largely defined by elusive factors such as a self-assessment of personal performance and satisfaction with life (Halicka 2004, Steuden 2011). These changes will have an impact on the demand for quality of products and services.

The declining abilities of some elderly people do not mean that they should be isolated and deprived of the benefit of various goods and services. On the contrary, research shows that mental workload and the need to solve problems have a beneficial effect on the welfare of elderly workers and their ability to work (Schooler et al. 1998). What is more, appropriate working conditions can significantly reduce the risk of memory problems and dementia, even after the completion of professional activity (Frisoni 1993). This means that more and more interest will have to be spent on adapting the environment to the needs of the people, including appropriate ergonomic development of service quality.

2.2 *From quality to service quality*

Quality is quite difficult to define because of its ambiguity. Particular definitions of it depend on the context in which it is used. According to Kolman (2009) quality is the degree to which a set of requirements is met and whose total satisfaction means to achieve a state of relative perfection; Hamrol & Mantura (1998) equate quality with an unlimited space of n-dimensional features, and as for the quality of the object, recognize a set of attributes belonging to it. The biggest influence on the shaping of modern achievements in knowledge in regard to quality management, in terms of theory and practice, is attributed to Deming, Juran, Crosby, and Ishikawa.

The beginnings of interest in service quality date back to the 1970s and 1980s. Very often the matter of service quality is addressed in publications relating to marketing, mainly because of the practical-economic result of the undertaken activities. On the basis of research, the following definition of service quality was developed: "Quality of service is a measure that indicates the extent to which the implementation of a specific provision meets the expectations of the customer. Providing high quality services means meeting the expectations of customers" (Parasuraman et al. 1985). Quality of service is defined as the ability to meet customer needs, which means the implementation of customer's expectations or even exceeding them (Rogoziński 2005). The resulting models of service quality imply its evaluation by a comparison of the expected and the experienced quality and often refer to its functional and technical aspects and also to elements associated with its image and marketing (Gummesson 1993).

The evaluation of service quality must be regarded in the context of earlier definitions associated with the identification of customer expectations and also needs to determine what criteria the customer takes into consideration. The biggest impact on the determination of these criteria had studies undertaken in 1985 by Berry, Parasuraman, and Zeithaml. As a result of the studies, ten basic criteria have been singled out for assessing the quality of services that can be applied across all sectors: (1) physical attributes; (2) reliability; (3) sensibility; (4) competence; (5) politeness; (6) credibility; (7) security; (8) availability; (9) communication skills, (10) understanding (Mazur 2002). A subsequent analysis of the above study showed partial overlap of the mentioned criteria. However, further work led to the construction of a tool by which service providers can anticipate customers' needs.

2.3 *What is ergonomic quality?*

Ergonomic quality is a category that describes the relation of a user, his/her specific and individualized conscious and unconscious needs, with the surrounding technical environment (Butlewski 2011). Ergonomic quality is subjective, but can be accurately defined within the range of its individual characteristics by means of specialized measurements (Branowski et al. 2013). This quality can be seen both from the point of view of a simple system of one human-technical object, as well as from complex technical-social systems, where man takes on several roles, and is both the creator and the recipient of the effect of his or her work (Butlewski & Tytyk 2012). In such a wider perspective, ergonomic quality is connected with the third generation ergonomics also known as macroergonomics (Pacholski 2000). It is also directly linked to quality management processes taking place in modern companies (Golas & Mazur 2008). Thus, the concepts of quality and ergonomics are interrelated and striving to achieve excellence in the company (the highest and unattainable level of quality) must also include ergonomic requirements (Jasiulewicz-Kaczmarek et al. 2013).

Ergonomic quality influences productivity (Butlewski & Tytyk 2008). Increased work efficiency through an improved ergonomic quality takes place at the same or lower human cost. One can also achieve the same profit without consequences (Górny 2012). Categories of ergonomic quality are particularly useful in the description of requirements for communities of people with special needs, for instance the elderly and people with disabilities, and this creates the opportunity to improve their quality of life in general (Misztal & Butlewski 2012). Therefore, it is advisable to transfer ergonomic criteria into the realm of methods of assessment of service quality to complete the methods with valid and perhaps not known criteria.

2.4 *Tools for assessing service quality*

Testing the quality of service is a more complex process than testing the quality of products. This is due to the fact that the characteristics of services significantly differ from the characteristics of products and the assessment of service quality is more subjective. Besides, it is often difficult to apply quantitative parameters during the measurement of service quality. An assessment of the level of service quality provided by different service providers depends on: (1) objective factors, (2) subjective factors, (3) reference point, (4) mood and (5) requirements on the part of the evaluator (Stoma 2012).

A review of methods assessing service quality points to their large variety (Ladhari 2008). However, we can distinguish three main groups: (1) multi-attributive, (2) event testing and (3) complaint assessment (Stoma 2012). The first group contains methods of measurement carried out from the perspective of the customer, and the overall assessment of service quality is the sum of individual assessments of specific quality attributes. Multi-attributive methods include, among others, SERVQUAL (Parasuraman et al. 1993), Penalty-Reward Factor and Vignette method. A large part of the methods is a modification of the existing ones, as a result, one can achieve methods for: evaluating service quality provided in historic houses—HISTOQUAL (Frochot & Hughes 2000), measuring service quality within education sector—HEdPERF (Higher Education PERFormance) (Abdullah 2005), measuring service quality delivered by online shopping websites—E-QUAL (Parasuraman et al. 2005), and the Key Quality Characteristics Assessment for Hospitals—KQCAH (Sower et al. 2001). On the basis of SERVQUAL, SERVPERF (Service Performance) was created (Cronin & Taylor 1994). It was a counterproposal that denied the relevance of measuring customer expectations in relation to service. The variety of methods indicates a large discrepancy in the approach to describe service quality.

3 RESEARCH METHODOLOGY

3.1 *Conception of a method to assess the ergonomic quality of financial services*

A review of models and methods for quality assessment resulted in the choice of the SERVQUAL method, despite some of its shortcomings and dysfunctions (Brown 1993), because this method was initially created to assess service quality in the banking market, and thus served as an appropriate foundation. Traditional SERVQUAL method defines service quality assessment as the difference between the expected and the experienced by the customer level of service quality. The total picture of the perceived service quality is thus the comparison of the perception of experienced services with prior expectations and an assessment of the importance of the given groups of criteria.

The method's 5 basic categories of ratings have been incorporated: (1) tangibles (e.g., physical appearance, proper conditions), (2) reliability (e.g. dependable and accurate service), (3) empathy (e.g. responding to individual needs of each customer), (4) assurance (e.g. qualification and knowledge of service provider) and (5) responsiveness (e.g. willingness to help, fast response). However, they have been supplemented with ergonomic requirements that followed mainly the following guidelines: (A) customers should fully understand the service provider; (B) documents must be legible and understandable, and as part of this: the use of phrases in documents ought to be understandable to customers, there should be an appropriate font size, adequate space for the signature, providing most relevant (key) information in the most visible way, the use of short paragraphs that make it easier to read and understand; (C) appropriate conditions during service, including an increased amount of time, conditions ensuring concentration, adequate lighting, no excessive noise, comfortable chairs, a comfortable position while completing documents, sufficient space at a desk, freedom of movement of arms and legs, etc.; (D) availability of facilities for people with reduced mobility, accessible parking, the ability to deliver services in the client's home; (E) adequate social facilities, access to toilets, drinks; (F) safety and confidentiality while filling out documents. The above mentioned ergonomic guidelines supplemented mainly the two basic SERVQUAL categories: reliability and responsiveness of service. The created questionnaire tool, included questions directed at each of the 5 categories, for which the respondent was asked to rate each category prior to receiving service (expectation) and directly after the service (experience). A scale of 0–5 was used, with 5 indicating the expected or experienced service quality to be ideal. The respondents were also to indicate the weight (importance) they assign to each category by distributing 100 points between the 5 categories.

The resulting method can be called ERGO-SERVQUAL to highlight its ergonomic component.

3.2 *Presentation of test procedures*

The first step of research was an analysis of literature in the field of assessing the quality of financial services and guidelines connected with design for elderly people. Then the SERVQUAL method was selected and was complemented with its ergonomic criteria. As a result of this, the first, draft version of ERGO-SERVQUAL was achieved, and it was tested on a preliminary group to determine the criteria that were the most relevant, important to service quality. The recipients were to assess the importance of specified service quality factors and their service quality expectations. On this basis, the final version of ERGO-SERVQUAL was created. After its improvement, it was used to study the ergonomic quality of service in a company providing short-term loans. In the making of the tool the following assumptions are adopted: (1) the developed sets of statements are based on criteria of quality, including ergonomic quality, and covering the specifics of financial services; (2) the number and type of dimensions were kept in the same state as the original SERVQUAL instrument, since there was no impediment to their application; (3) theses in the instrument were constructed in such a way as to permit an assessment of the ergonomic quality in financial services made by customers using this type of service. The study was conducted on a group of 50 clients, in the company offering loans from 117 to 707 euro for a period from 16 to 52 weeks (the maximum loan amount is close to the median gross wage in Poland, which is net of less than 500 euro) (CSO 2012). The research sample is characteristic of the offered financial services where the majority of respondents (68%) is more than 50 years old. No decision was made on the statistical analysis of the differences between the group of older and younger recipients, due to the small sample size of the latter, along with the difficulty in separating the effects of education and sex of respondents. Besides that, there was no significant difference between age groups.

3.3 *Results*

The results were determined from the subtraction of the expected value from the experienced value assigned to each category of service quality and the multiplication by the weight assigned to the category (e.g. [3–5] × 20) and averaged from all responses. In extreme cases, if one of the categories had received all of the points or weight or the difference between the experienced and the expected quality was also maximal, the implicit score would be 400.

The biggest lack of quality was noticed within the the following criteria: (1) adjusting services to client's needs (–26.91); (2) safety and confidentiality during service (–15.55); (3) social infrastructure (–11.70); (4) confidentiality in filling out documents (–10.61); (5) suitable working hours (–8.97); (6) short paragraphs in the contract (–8.88); (7) clarity of terms used in three sub-categories: checking understanding of the language used by employees (–7.99), using understandable terms (–6.22) and intelligibility of the staff's statements (–3.55).

The most positive quality was rated according to the following criteria: (1) convenient place of service realization (17.58); (2) aesthetic look of documents (11.99); (3) opportune time for the service (8.97); (4) speed and efficiency of service (8.37); (5) a sufficient amount of time (no pressure) (6.22).

4 DISCUSSION

In general, in 48% of criteria, the experienced value exceeded the expected value, in 45% it was lower than expected, and in 7% it was at the same level. Unfortunately, it wasn't possible to identify the reasons for the low feeling of quality in terms of the overall criterion-adjusting services to meet the needs of the client. They can be associated with specific features of the services provided where the consequences of obtaining a certain amount of money are distributed in time and those who take a loan are "forced" to accept the offered service due to their current life situation. An analysis of needs shows that the customers of these types of financial services, short-term loans, are mainly older adults, thus is it especially important to adapt the ergonomic service quality to their needs, because it reflects their psycho-physical needs.

5 CONCLUSION

The applied method used to evaluate the ergonomic quality of financial services, called ERGO-SERVQUAL, allowed to identify areas for improvement, also in terms of ergonomic criteria that have not been previously found in the original version. Customers have indicated the need for a greater sense of security and confidentiality, and also for a greater adjustment of intelligibility of the used phrases and concepts and an opportunity to use social facilities. Unfortunately, the quality of the research sample did not allow for examination of the relationship between needs and quality features differentiating clients such as level of education, age or gender. Such studies are planned for the future.

The changes introduced in the company after the presentation of research results made it possible to increase the experienced quality of services, as evidenced by a preliminary analysis of the quality currently experienced by the customers.

REFERENCES

Abdullah, F. (2005) HEdPERF versus SERVPERF: The quest for ideal measuring instrument of service quality in higher education sektor Quality Assurance in Education, 13 (4), pp. 305–328.

Branowski B., Pacholski L., Rychlik M., Zabłocki M., Pohl P., (2013) Studies on a New Concept of 3D Data Integration about Reaches and Forces of a Disabled Person on a Wheelchair (CAD Methods in Car and Market Ergonomics), Human Factors and Ergonomics in Manufacturing & Service Industries 23 (4) pp. 255–266 (2013).

Brown, T.J., Churchill Jr., G.A., Peter, J.P. (1993) Improving the measurement of service quality Journal of Retailing, 69 (1), pp. 127–139.

Butlewski M., Tytyk E., (2008) The method of matching ergonomic non-powered hand tools to maintenance tasks for the handicapped. 2nd International Conference on Applied Human Factors and Ergonomics, Las Vegas, Nevada, USA, Conference Proceedings, Edited by Waldemar Karwowski and Gavriel Salvendy.

Butlewski M., Tytyk E., (2012) The assessment criteria of the ergonomic quality of anthropotechnical mega-systems; pp. 298–306, [in]: Advances in Social and Organizational Factors, Edited by Peter Vink, CRC Press, Taylor and Francis Group, Boca Raton, London, New York, ISBN 978-1-4398-8.

Butlewski M. (2011), Ergonomic quality as a means of providing powered hand tools' safety, SAFETY OF THE SYSTEM: HUMAN—TECHNICAL OBJECT–ENVIRONMENT, Wyd. Politechnika Częstochowska.

Cronin J.J., Taylor S.A., (1994), SERVPERF versus SERVQUAL: Reconciling Performance-Based and Perceptions-Minus-Expectations Measurement of Service Quality, op.cit., pp. 125–131.

CSO (Central Statistical Office). 2009. Population projection for Poland 2008–2035 [Internet]. Warszawa (Poland): GUS; [cited 2013 July 30]. Available from: http://www.stat.gov.pl/cps/rde/xbcr/gus/PUBL_L_prognoza_ludnosci_na_lata2008_2035.pdf.

CSO (Central Statistical Office). 2012. Average monthly salary in the national economy in the years 1950–2012 [Internet]. Warszawa (Poland): CSO; [cited 2013 July 30]. Available from: http://www.stat.gov.pl/gus/5840_1630_PLK_HTML.htm.

Duda K., (2013), The aging process [in], Marchewka A., Dąbrowski Z., Żołądź J.A., (2013) Physiology of aging: Prevention and Rehabilitation/red. nauk. Publishing house PWN, Warszawa (in Polish).

Frisoni GB, Rozzini R, Bianchetti A, Trabucchi M., (1993) Principal lifetime occupation and MMSE score in elderly persons. J Gerontology Nov;48(6): 310–314.

Frochot, I., Hughes, H. (2000) HISTOQUAL: The development of a historic houses assessment scale Tourism Management, 21(2), pp. 157–167.

Gołaś H., Mazur A., (2008), Macroergonomic aspects of a quality management system, [in:] Macroergonomic paradigms of Management, Jasiak A. [red.] to honour 40 years scientific activity of Professor Leszek Pacholski, Poznan University of Technology Editorial Board, pp. 161–170.

Górny, A. (2012), Ergonomics in the formation of work condition quality. Work: A Journal of Prevention, Assessment and Rehabilitation, 41, pp. 1708–1711.

Gummesson E., (1993) Quality Management in Service Organizations, ISQA International Service Quality Association, New York, NY.

Halicka M. (2004), Life satisfaction of the elderly: a study of the theoretical and empirical, Białystok (in Polish).

Hamrol A., Mantura W., (1998) Quality management. Theory and practice, Publishing house: PWN, Warszawa—Poznań (in Polish).

IIASA (International Institute for Applied Systems Analysis). 2012. European demographic data sheet 2012 [Internet]. Laxenburg (Austria): IIASA; [cited 2013 July 30]. Available from: http://www.iiasa.ac.at.

Jasiulewicz-Kaczmarek M. Misztal A., Butlewski M., The holons model of quality improvement in SMEs, 2013 (in press).

Jasiulewicz-Kaczmarek M., (2012) Socio-technical integrity in maintenance activities, In: Vink P. (eds.): Advances in Social and Organizational Factors, CRC Press 2012, pp. 582–592.

Kolman R., (2009) Qualitology. Knowledge of the various areas of quality, Publishing house Placet, Warszawa (in Polish).

Ladhari, R. (2008) Alternative measures of service quality: A review; Managing Service Quality, 18 (1), pp. 65–86.

Mazur J., (2002) Marketing Management Services, Publishing house Diffin, Warszawa (in Polish).

Mazur, A., (2013) Application of Fuzzy Index to Qualitative and Quantitative Evaluation of the Quality Level of Working Conditions, Springer Berlin Heidelberg 978-3-642-39475-1.

Misztal A., Butlewski M., (2012) Life improvement at work, Wyd. PP, Poznań, ISBN: 978-83-7775-177-0.

Opolski K., (2000) Quality management in financial services, [in]: Managing service quality in banks and insurance companies, Ed. Garczarczyk J., AE w Poznaniu, Poznań (in Polish).

Pacholski, L., (2000) Macroergonomic paradox of entrepreneurship and economic renewal, in: Ergonomics for the new millennium, Human Factors and Ergonomics Society, pp. 185–188, Santa Monica—San Diego CA.

Parasuraman, A., Berry, L.L., Zeithaml, V.A., (1993) More on improving service quality measurement Journal of Retailing, 69 (1), pp. 140–147.

Parasuraman, A., Zeithaml, V.A., Malhotra, A., (2005) E-S-QUAL a multiple-item scale for assessing electronic service quality; Journal of Service Research, 7 (3), pp. 213–233.

Parasuraman, A. Zeithaml V.A., Berry L.L. (1985); A Conteptual Model of Service Quality and Its Implications for Future Research, Journal of Marketing, Vol. 49.

Rogoziński K., (2005) Quality of service in the horizon axiological, Quality Problems, no. 1, (in Polish).

Sower, V., Duffy, J., Kilbourne, W., Kohers, G., Jones, P. (2001) The dimensions of service quality for hospitals: Development and use of the KQCAH scale Health Care Management Review, 26 (2), pp. 47–59.

Steuden S. (2011) Psychology of aging and old age. PWN. Warszawa, (in Polish).

Stoma M., (2012) Models and methods for measuring the quality of services, Publishing house POLIHYMNIA, Lublin, (in Polish).

Więcek-Janka, E., Quality of service in the microenterprises, Economics and Organization of Enterprises (2008), pp. 90–94, (in Polish).

Occupational Safety and Hygiene II – Arezes et al. (eds)
© 2014 Taylor & Francis Group, London, ISBN 978-1-138-00144-2

Indoor Air Quality (IAQ) in the living rooms in homes for the elderly—case study in the county of Pombal

C. Carvalho, A. Ferreira & J.P. Figueiredo
College of Health Technology of Coimbra, Portugal

ABSTRACT: The concern with IAQ in buildings comes from the fact that nowadays the people spend more time inside buildings, and are exposed to the action of numerous pollutants. The study aims to evaluate the IAQ in Elderly Homes of the Council of Pombal, and the fulfillment of conditions imposed by decree-law, as well as the perception of the elderly compared to this same air quality. The maximum and average values of CO_2 recorded have exceeded the value established in the current legislation. The Hr, the Homes U and Z have revealed higher values than the established by legislation. Perceived symptoms were fatigue, dizziness and vomiting that demonstrated higher representation. Was the rural with the highest concentration of pollutants of CO_2, T and PM_{10}. The urban environment values of Hr and VOC's were higher.

1 INTRODUCTION

With the passage of time has been witnessing significant changes in urban and rural centers. These changes, resulting primarily from the increase in population concentration and overloads the infrastructure of these centers, without a commitment to deliver quality to urban inhabitants (Borrego, 2009; Costa, 2005). Nowadays, communities recognize that outdoor air pollution has serious health effects, since the air is increasingly contaminated (Ministry of Environmental). However, there is not the same awareness of the risks associated with poor air quality inside buildings, which turns out to be harmful, since it is contaminated also, directly and/or indirectly by air pollution. Due to this large imbalance has been a particular attention on the issue of quality of indoor air, because it is not only the existence of pollutants. But also the comfort level and the perception that each does the quality of the air we breathe (Bernardes, 2009; Verdelhos, 2011; Nabais, 2004). Sick Building Syndrome (SBS) is the building itself that instead of ensuring the welfare and preserve the health of its occupants, is the generator of different types of health problems (Carmo, 1999). The SBS is characterized by five types of symptoms: mucous membrane irritation of eyes, nose and throat; neuropsychiatric disorders (fatigue, headaches, confusion, dizziness); skin problems (itching, dry skin); asthma-like symptoms and, finally, unpleasant taste and odor (Verdelhos, 2011; Silva, C., 2005; Santos et al., 1992). In a general sense, the IAQ can be defined as one that does not affect the welfare of the occupants of the space in question (Bernardes, 2009; Sanguessuga, 2012). In a more technical definition, when the IAQ be achieved thereby meet the three basic requirements that govern human occupation of space: thermal environment acceptable, maintaining normal concentrations of expired gases CO_2; removal of pollutants or their dilution in order to maintain their levels within acceptable limits for health and odor control which may create nuisance (Bernardes, 2009). One of the most susceptible populations and with a weakened immune system are the elderly, and also with a higher prevalence of chronic diseases and respiratory problems. This age range is about 19–20 h/day indoors making it more susceptible to health complications associated with indoor air pollution (Project GERIA, 2012). It is then necessary to have an awareness regarding IAQ and, for the preventive and corrective measures that contribute to the proper management of the same buildings, because although air is a natural infinite, that is, with which there need be no concern in terms of quantity, needs to be preserved, since the maintenance of its quality becomes imperative (Sanguessuga, 2012; Santos, 2005). The study aims to evaluate the IAQ in Elderly Homes of the Council of Pombal, and the fulfillment of conditions imposed by decree-law, as well as the perception of the elderly compared to this same air quality because this age group is so vulnerable.

2 MATERIAL AND METHODS

The research study was developed in 2012/2013. The study was conducted in the municipality of Pombal. The Level II study, descriptive correlational, the cross sectional cohort. The type of sampling was not probabilistic, and the technique of convenience sampling. The sample represents all the homes of that municipality. The sample was composed of three nursing homes in the county of Pombal and by their seniors, of whom 33 responded to the questionnaires, being the total sample was 60. Data collection was conducted in two stages, the first visits to nursing homes with the aim of evaluating the physical space to be monitored. The questionnaires were administered by their family. The selection of the participants focused on those who were residents of the study areas (homes). The questionnaire consisted of two parts, the first relating to the personal information of the respondent and the second part allowed review, beyond the knowledge of the users about IAQ, the existence or absence of SBS. The second step was to evaluate the IAQ in the selected space, the living room of every nursing home, using the analytical evaluation of chemical compound as CO_2, CO, PM_{10}, and VOC's, and indicators of thermal comfort as, T and Hr, being the detection limit of the equipment VOC's: 0,00–2000/0–50000 mg/m^3; CO_2: 0–5000 ppm; PM: 0–200 mg/m^3; CO: 0–1000 ppm; T: 0,0–60,0 graus C and Hr: 0,0–100,0%.

The site selected was the living room of nursing homes, as this represents the place where the elderly stay longer. Measurements of chemical and physical parameters were carried out over a period of 5 minutes according to technical standarts (NT-SCE-02), and the result of measurement an average of three measurements made at the same site simultaneously. The measurements were carried out abroad preceded those performed inside spaces with the same sampling time according to the standard technique. IAQ measurements were taken at two different periods, before the elders occupy the lounges and during its occupation by placing the equipment in the most central position possible at each site, at approximately the height of the airways of users, the seated position. To carry out the measurements was taken as the reference NT-SCE-02. It was still necessary to resort to a portable device for the measurement of IAQ: brand and model Quest Technologies EVM-7. It is considered the Decree-Law n.° 79/2006, Decree-Law n.° 80/2006 and ISO 7730. Was used software SPSS version 21 for Windows. The interpretation of statistical tests were based on a significance level of p-value = 0.05 with Confidence Interval (CI) of 95%. This investigation has exclusively academic interest, subtracting any financial or economic interest.

3 RESULTS

The study sample was made up of the living rooms of nursing homes and the elderly present in them. It is important to mention that the 33 seniors who responded to the questionnaire 78.8% were female and mean age was 84 ± 7 years. Of the three homes studied, one was located in the urban area (home Z) and the remaining in the country (home X and Y). After application of tools for collecting data preset was made comparing the maximum and average analytical estimates of concentrations, CO_2, VOC's and PM_{10} in indoor air of several nursing homes with values established by decree-law, which is shown in the Board 1.

At the level of particulate matter 10 (PM_{10}) values estimated in different homes had levels below the limit set by decree-law (<0.15 mg/m^3). With respect to the pollutant CO_2 was found that homes X and Y revealed values (average and maximum) averaged over similar and in some cases highly significantly ($p < 0.05$) for the Home X. However, the Home Z expressed in average values (average and

Board 1. Concentration of polluants recorder in each nursing home, compared with the references values (with the occupation of the living room).

Home	N	$\bar{x} \pm s$	p	Reference value
Home X				
VMáx	3	1269,33 ± 41,07	0,007	984 ppm
VAverage	3	1208,33 ± 42,90	0,012	CO_2
Home Y				
VMáx	9	1726,78 ± 298,99	<0,0001	
VAverage	9	1649,22 ± 289,42	<0,0001	
Home Z				
VMáx	3	967,33 ± 125,739	0,840	
VAverage	3	901,67 ± 132,50	0,394	
Home X				
VMáx	3	2561,42 ± 517,88	0,022	600 ug/m^3
VAverage	3	2281,22 ± 150,19	0,003	COV's
Home Y				
VMáx	9	2393,20 ± 791,83	<0,0001	
VAverage	9	2190,88 ± 452,75	<0,0001	
Home Z				
VMáx	3	4747,98 ± 1187,22	0,026	
VAverage	3	3692,87 ± 815,77	0,022	
Home X				
VMáx	3	0,07 ± 0,008	0,003	0,15 mg/m^3
VAverage	3	0,06 ± 0,06	0,001	PM_{10}
Home Y				
VMáx	9	0,08 ± 0,04	<0,0001	
VAverage	9	0,08 ± 0,01	<0,0001	
Home Z				
VMáx	3	0,06 ± 0,01	0,001	
VAverage	3	0,05 ± 0,001	<0,0001	

Legend: N = Number of measurements, df = degree of freedom, x = mean, s = standard deviation; test: T-student for one sample.

maximum) lower, but not significantly ($p > 0.05$) below the legal limit (<984 ppm).

Regarding the pollutant VOC's, found that the three homes in the study revealed values (average and maximum) on average similar, but Home Z expressed in average values (maximum and average) strongly and significantly higher ($p < 0.05$). However, the three homes had reference values above the legal limit (600 ug/m^3). It is important to note that the pollutant CO, could not be assessed due to their results may be null. Of the three the sampled households, the Home X and Y had higher values than the reference values with respect to the maximum value of CO_2 and medium, although higher amounts may be due to the fact that Home Y have a large number of occupants. In the Z home, although not exceed the reference value, had a maximum value of CO_2 very close to the reference value.

There was also concern verify if there were significant differences in the amount of pollutants, according to the time points (before and after the room is occupied by the elderly) can be seen that there were statistically significant differences for the parameters CO_2, VOC's, PM_{10} and Hr.

Regarding the values of the parameters evaluated in the living room before and with the occupation of the elderly, it can be seen in Board 2, that the parameters CO_2, VOC's, T e Hr, the maximum, average and parameter PM_{10} values average during the occupation of the living room.

On average values (maximum and mean) of the pollutant CO_2 when the occupants, are not suitable, that is, they are above the reference value (984 ppm).

With respect to the average temperature was suitable when the living met with the occupants. For the values (maximum and average) of VOC's, it can be seen that on average they were not suitable.

Given that the Hr values stipulated in the legislation are 30% to 70% on average values of this, when the living room is occupied, they are not adequate.

Board 2. Comparison of the concentration of pollutants before the living room be occupied and the living room occupied by elderly.

Parameter	Height measurement	$\bar{x} \pm s$	p
CO (ppm)			
VMáx	Before busy	0,60 ± 0,83	0,294
	With occupants	0,33 ± 0,49	
CO_2 (ppm)			
VMáx	Before busy	522,1 ± 386,82	<0,001
	With occupants	1483,4 ± 398,19	
VAverage	Before busy	430,53 ± 286,54	<0,001
	With occupants	1411,5 ± 389,24	
COV's (ug/m^3)			
VMáx	Before busy	2018,6 ± 493,95	0,019
	With occupants	2897,8 ± 1232,7	
VAverage	Before busy	1836,74 ± 271,1	0,005
	With occupants	2509,35 ± 769,4	
PM_{10} (mg/m^3)			
VMáx	Before busy	0,075 ± 0,01	0,719
	With occupants	0,073 ± 0,03	
VAverage	Before busy	0,06 ± 0,01	0,017
	With occupants	0,07 ± 0,02	
Temp (°C)			
VMáx	Before busy	19,03 ± 2,24	0,062
	With occupants	20,53 ± 1,98	
VAverage	Before busy	18,83 ± 2,37	0,073
	With occupants	20,33 ± 2,05	
HR (%)			
VMáx	Before busy	70,75 ± 3,62	0,120
	With occupants	72,68 ± 2,93	
VAverage	Before busy	69,39 ± 3,22	0,032
	With occupants	71,87 ± 2,79	

Legend: df = degree of freedom, x = average, s = standard deviation; test: T-student for one sample.

Given that the three homes are located in different geographical areas (rural and urban), was of interest to observe in which the areas were recorded higher values of these parameters.

The existence of statistically significant differences between the geographical location of the Homes and the parameters CO_2, VOC's, PM_{10} and Hr ($p \leq 0.05$), is seen in Board 3.

The rural areas showed the highest values of CO_2, PM_{10} and T. The urban environment was significantly higher in VOC's and Hr.

Board 3. Concentration of these parameters in different homes depending on the geographic location (with the occupation of the living room).

Parameter	$\bar{x} \pm s$	p
CO_2 (ppm)		
VMáx		
Rural	1612,4 ± 328,83	<0,0001
Urban	967,3 ± 125,74	
VAverage		
Rural	1539,0 ± 317,83	0,001
Urban	901,7 ± 132,50	
COV's (ug/m³)		
VMáx		
Rural	2435,3 ± 714,53	0,001
Urban	4747,9 ± 1187,22	
VAverage		
Rural	2213,5 ± 393,51	<0,0001
Urban	3692,9 ± 815,77	
PM_{10} (mg/m³)		
VMáx		
Rural	0,08 ± 0,03	0,383
Urban	0,06 ± 0,01	
VAverage		
Rural	0,08 ± 0,02	0,011
Urban	0,05 ± 0,001	
Temp (°C)		
VMáx		
Rural	21,39 ± 0,90	<0,0001
Urban	17,07 ± 0,76	
VAverage		
Rural	21,21 ± 1,03	<0,0001
Urban	16,83 ± 0,75	
HR (%)		
VMáx		
Rural	70,04 ± 3,72	0,013
Urban	76,43 ± 0,25	
VAverage		
Rural	68,50 ± 2,97	0,001
Urban	75,87 ± 0,21	

Legend: N = number of measurements; df = degree of freedom, x = average, s = standard deviation; test: T-student for one sample.

Note that the values of CO does not appear in the table because it is not possible to analyze, since no values were obtained in the urban area.

Regarding the perception of the elderly face IAQ, after analysis of the questionnaires obtained the results presented below.

78.8% of the elderly found that air quality does not affect your health. With regard to thermal comfort, 30 of 33 elderly, this is, 90.9% indicated feeling good in the living room. With respect to the season, 69.2% indicated that the symptoms/diseases worsened depending on the seasons, this result was observed in Board 4. As to gender, both the majority of females (84.6%) and male (57.1%) indicated that air quality does not affect your health. In turn, the same applies to thermal comfort, female gender (92.3%) and male (85.7%) indicated that they felt comfortable. The elderly households indicated feeling certain symptoms. The symptom/disease with the highest expression were the headaches, dizziness and vomiting with a representative of 61.5% of participants, followed by breathing difficulties with an occurrence of 46.2%. In the event where he effected the relationship between the risk of the presence of symptoms and the occurrence of values of pollutants above the reference values, it was found that as the maximum values of the

Board 4. Frequency of comfort indices by gender.

	Gender		
	Female	Male	Total
	n (%)	n (%)	n (%)
Indoor air quality affects health			
No	22 (84,6)	4 (57,1)	26 (78,8)
Yes	4 (15,4)	3 (42,9)	7 (21,2)
Total	26 (100,0)	7 (100,0)	33 (100,0)
Thermal comfort			
No	2 (7,7)	1 (14,3)	3 (9,1)
Yes	24 (92,3)	6 (85,7)	30 (90,9)
Total	26 (100,0)	7 (100,0)	33 (100,0)
Go outside the building			
No	5 (71,4)	2 (33,3)	7 (53,8)
Yes	2 (28,6)	4 (66,7)	6 (46,2)
Total	7 (100,0)	6 (100,0)	13 (100,0)
Absence symptoms exterior building			
No	2 (100,0)	3 (100,0)	5 (100,0)
Yes	0 (0,0)	0 (0,0)	0 (0,0)
Total	2 (100,0)	3 (100,0)	5 (100,0)
Symptoms aggravate with the station of the year			
No	2 (28,6)	2 (33,3)	4 (30,8)
Yes	5 (71,4)	4 (66,7)	9 (69,2)
Total	7 (100,0)	6 (100,0)	13 (100,0)

Legend: n = number of elders.

Board 5. Symptoms/diseases perceived by elderly households.

	Symptoms		
	Absence of symptoms	Presence of symptoms	Total
Home X			
n	10	7	17
% Line	58,8%	41,2%	100,0%
% Total	30,3%	21,2%	51,5%
Home Y			
n	10	6	16
% Line	62,5%	37,5%	100,0%
% Total	30,3%	18,2%	48,5%
Total			
n	20	13	33
% Line	60,6%	39,4%	100,0%
% Total	60,6%	39,4%	100,0%

Legend: n = number of elders; $X^2 = 0,047$; degrees of freedom = 1; p-value = 0,829; test: X^2 of independence.

pollutants CO_2, and VOC's seniors in Homes X and Y expressed symptoms were on the presence of risk, since figures were above the value set in current decree-law.

The manifestation of symptoms and/or diseases, Home X, Y, was greater than 20%, as shown in the Board 5. Note that the Home Z is not represented in the previous Board, since the health problems of the elderly did not allow them to have a correct perception of reality.

4 DISCUSSION

The average CO_2 concentration was exceeded in the maximum and average at Home X and Home Y compared with the reference value. In the Z home, on average, the maximum amount of pollutant CO_2 was recorded 967,33 ppm, thus approaching the reference value. Also the Home Z, the average value was registered polluting CO_2 this result may be caused by reduced number of occupants. With respect to the amounts of VOC's, it can be seen that the concentration exceeds the reference value in all homes. With regard to values of Hr, them values recorded in the three homes exceeded the reference value considered good for human comfort. Analysis of the values of T, it was found that also the Home X and Z was the only ones who always expressed values below those stipulated in the current legislation. In view of these results it can be concluded that there is a risk for the health of the occupants from Home Z with respect to T and Hr felt. It was obtained higher values on average during the occupation by the elderly, for the parameters CO_2, VOC's, T, Hr and PM_{10}. Two homes belong to the rural areas, Homes X and Y and Home Z to the urban environment. We obtained higher values of CO_2, PM_{10} and T in rural areas and values of VOC's and Hr in urban areas. CO_2 is a colorless and odorless gas, its concentration in the air inside the space evaluation can, under certain reasons, it can give a good indication of the rate of ventilation, it is generated on the inside especially by human metabolism (APA, 2009). This idea is enhanced by the X and Y homes, as are the homes with more occupants. However, the locations of the measurements verified the existence of heat sources in operation, the presence of the kitchen next to the living room and the presence of visitors, are also important factors that can influence the levels of CO_2. The high concentration of VOCs in the air inside can result from many factors, namely the outside air, since the road transport is a light source such compounds (Quadros et al., 2010; Strausz, 2001), the introduction of new products and materials of construction, finishing, decoration, furniture and cleaning (Bernardes, 2009; Strausz, 2001). A major factor to refer was the moment that have developed measurements, these were held at Christmas, thus causing increased concentration of this pollutant and causing the air to become more contaminated, thus impairing the health of its occupants. The amounts of VOCs of the Home Z are higher, eventually to be also located in an urban area where car pollution is more intense compared to rural areas. For a building to be considered sick, it is necessary to ensure that at least 20% of its occupants manifest symptoms such as irritation of the mucous membranes, fatigue, headaches, confusion, dizziness, skin affections and sensitivity to odors and that they cease of the show when he moves to the outside of the building (Gioda et al., 2003; Gauer et al., 2003). Both at Home X, Y like home, reveal that over 20% of older perception symptoms inside the building, however perceived symptoms do not disappear when the seniors are away from it. But this condition is not observed in any of the homes, so it can be concluded that none of the homes has SBS. Perceived symptoms, the headaches, dizziness and vomiting were the symptoms with higher relief, ie, 61.5% reported these symptoms. Studies conducted indicate that inhalation of elevated CO_2 levels can cause headaches, nausea, dizziness, and cardiovascular effects and also can act as a simple asphyxiant and cause an irritant effect on airways (Santos et al., 1992; Ganer et al., 2003; Shirmer et al., 2003), also levels elevated VOC's can cause headaches. The elderly was asked whether the effects of air quality can affect your health. Generally seniors who reported having symptoms argue that IAQ may not affect health. You can also check that the elderly who perception headaches, dizziness and vomiting (61.5%) 62.5% stated that the quality of the air can't affect human health.

5 CONCLUSIONS

The people spend more time inside the buildings and to the growth of microorganisms, the use of cleaning products, the presence of polluting materials and equipment, the very human occupation and poor ventilation and air exchange are some of the contributors to the liberators of pollutants and harmful to human health (Borrego, 2009; Bernardes, 2009). The Homes X and Y, those who reported higher levels of pollutant CO_2, must make several changes to the organizational and structural. VOC's levels are above the reference value in the three homes, thus should be taken to reduce the levels of this pollutant. One limitation to the study was not to have controlled the type of ventilation in existing physical spaces evaluated and relate them with indicators of IAQ. One suggestion to help clean the air in the living room is the acquisition of plant speciesn (Strausz, 2001). Therefore, and due to the small number of studies, deepening the theme of IAQ becomes very important to maintain the link between environmental agents and the effects on human health remains a challenge for Environmental Health and the various professionals this area of intervention, which play a predominant role, which contributes to the improvement of IAQ problems in buildings, providing health and comfort to its occupants.

REFERENCES

APA, Agencia Portuguesa do Ambiente—Laboratório de Referência do Ambiente, 2009. *Qualidade do Ar em Espaços Interiores—Um guia técnico.*

Bernardes, A. *Análise Dos Métodos De Auditoria À Qualidade Do Ar Interior—Rsece.* Universidade de Aveiro, 2009.

Borrego, C., Costa, A., Valente, J., Lopes, M., & Miranda, A.I. (Fevereiro de 2009). *Qualidade do ar nas zonas urbanas.* Revista Indústria e Ambiente, pp. 16–19.

Carmo, A.T. (1999). *Qualidade do Ar Interno.* [Online] [Citação: 15 de Novembro de 2012]. http://www.labeee.ufsc.br/sites/default/files/disciplinas/Racine%20-%20IAQ.pdf.

Costa J. *Qualidade do Ar Interior e Conforto Térmico: Um estudo em espaços de estacionamento em Natal/RN com tipologias arquitectónicas diferenciadas*: Universidade Federal do Rio Grande do Norte; 2005.

Decreto-Lei n.º 79/2006 de 4 de Abril. Diário da Républica n.º 67/2006—I Série A, Lisboa: Ministério das Obras Públicas, Transportes e Comunicações. 2006.

Decreto-Lei n.º 80/2006 de 4 de Abril. Diário da Républica n.º 67/2006—I Série A, Lisboa: Ministério das Obras Públicas, Transportes e Comunicações. 2006.

Gauer, M.A, Szymanski, M.S.E., Pian, L.B., Schirmer, W.N (2008) *A poluição do ar em ambientes internos.*

Gioda, A., Aquino Neto, F.R. *Considerações sobre estudos de ambientes industriais e não-industriais no Brasil: uma abordagem comparativa.* Caderno da Saúde Pública, v. 19, n. 5, p. 1389–1387, 2003.

ISO 7730:2005—Ergonomics of the thermal environment—Analytical determination and interpretation of thermal comfort using calculation of the PMV and PPD indices and local thermal comfort criteria. Switzerland: s.n., 2005.

Ministério do Ambiente. Qualidade do Ar Interior. Agência Portuguesa do Ambiente [Online] [Citação: 23 de Outubro de 2012]. http://www.apambiente.pt/index.php?ref=16&subref=82&sub2ref=319&sub3ref=338.

Nabais R. *Odors Prevention in the Food Industry.* 2000:75–104. Natureza ANdCd. Poluição do Ar Interior Ameaça Saúde Pública. In: Quercus DNd, ed. Lisboa 2004.

Nota Técnica NT-SCE-02. Metodologias para auditorias periódicas da QAI em edifícios de serviços existentes no âmbito do RSECE. 2009.

Projecto GERIA—*Estudo Geriátrico dos Efeitos na Saúde da Qualidade do Ar Interior em Lares da 3ª Idade de Portugal*, Instituto Nacional de Saúde; 2012.

Quadros, M.E., Lisboa, H.M., (2010). *Controle da Poluição Atmosférica. Capítulo IX. Qualidade do Ar Interno.* [Online] [Citação: 3 de Junho de 2013]. http://www.lcqar.ufsc.br/adm/aula/Capitulo%209%20 Ar%20Interno.pdf.

Sanguessuga, M. *Síndroma dos Edifícios Doentes.* Escola Superior de Tecnologia da Saúde de Lisboa, 2012.

Santos, U., Rumel, D., Martarello, N., Ferreira, C., & Matos, M. (1992). *Síndrome dos Edifícios Doentes em Bancários.*

Santos, J.P.C.M. *Avaliação Experimental dos Níveis de Qualidade do Ar Interior em Quartos de Dormir* [dissertação]. Lisboa: Faculdade de Ciências e Tecnologias da Universidade Nova de Lisboa, 2008.

Schirmer, W.N., Pian, L.B., Szymanski, M.S.E., Gauer, M.A. (2008). *A poluição do ar em ambientes internos e a síndrome dos edifícios doentes.* Universidade Estadual do Centro-Oeste. [Online] [Citação: 23 de Maio de 2013]. http://www.scielo.br/pdf/csc/v16n8/a26v16n8.pdf.

Silva, Fernando Costa. Portal de Saúde Pública, (2005). [Online] [Citação: 5 de Novembro de 2012]. http://www.saudepublica.web.pt/05-promocaosaude/054-SOcupacional/SED.htm.

Sterling, T., Collett, C., & Rumel, D. (1991). *A epidemiologia dos "edifícios doentes".* Revista de Saúde Pública, pp. 56–63.

Strausz, M. *Análise de um acidente fúngico na Biblioteca Central de Manguinhos: um caso de Síndrome do Edifício Doente.* [Online] 2001. [Citação: 4 de Junho de 2013.] http://teses.icict.fiocruz.br/pdf/strauszmcm.pdf.

Verdelhos, V.M.M. (2011). *Caracterização da Qualidade do Ar Interior em Espaços Públicos com Permissão de Fumar.* [Online] [Citação: 20 de Outubro de 2012]. https://estudogeral.sib.uc.pt/handle/10316/19975.

Occupational Safety and Hygiene II – Arezes et al. (eds)
© 2014 Taylor & Francis Group, London, ISBN 978-1-138-00144-2

Occupational exposure noise—case study of a quarter of firefighters in center region

I. Gaminha, J. Almeida, H. Simões, J. Figueiredo & A. Ferreira
IPC, ESTeSC, Coimbra Health School, Coimbra, Portugal

ABSTRACT: Noise exposure is currently one of the occupational hazards with higher expression in the workplace. The noise is considered as the physical agent unwholesome more common in the workplace, being characterized as a factor of greater prevalence of the origins of occupational diseases. Professionals from a fire station are constantly exposed to risks and occupying a prominent place is noise exposure. We conducted a study level II descriptive—correlational, cross-cutting, comparing later with Decree-Law n° 182/2006, of 6 September. Measurements were performed at 11 locations in the building and 3 different vehicles. We analyzed records of occurrence, between the years 2009 and 2012, and calculated the Daily Personal Exposure. There were also calculated maximum exposure times in the car (needed to reach the Exposure Limit Value). Questionnaires were distributed in order to understand if the professionals considered their workplace noisy and cumbersome and detected some kind of symptoms. It was concluded that the noise that was felt in the administrative areas and in cars was nuisance and that some professionals had symptoms of habituation to noise. It was also concluded that the awakening by alarm action in dormitories, causes discomfort and disruption in the professionals.

1 INTRODUCTION

Noise is a factor of physical risk defined as any unwanted sound in the environment (Lee et al., 2005) with a growing trend with occupational agent, which will eventually be aggravated if measures are not taken and established structured programs for its reduction. In most cases, the sound level varies with time, it is necessary to clarify a relationship between the level and duration (Miguel, 2002).

One of the important effects of noise is on sleep disorders. This should be uninterrupted for physiological and mental functioning correctly (Berglund et al., 1999). Some of these professionals rest during the shift in facilities intended for that (dormitories). These are awaked with high and sudden sound intensities due to emergency alerts that happen. The effects of noise on duration and quality of sleep, under experimental conditions have been studied in detail with the aid of polysomnography (Antunes, 2009). The sense of hearing it's the most efficient form of waking up. When eyes are closed, optical stimulus are reduced, while the sense of hearing is only slightly misunderstand during sleep, keeping its primary function of alarm system. There is a significant increase from insomnia related to occupational noise due to intense noise intolerance, nervousness and irritation (Kroemer & Grandjean, 2005). Firefighters are exposed to many dangers (NIOSH, 2013). Apart from the noise from sirens and alarms, firefighters are also exposed to numerous chemicals and combustion products that may have ototoxic effects and aggravate a situation of hearing loss (Lees, 1995). It has recently been shown to trigger apoptosis and cell injury due to oxidation caused by the presence of excess free radicals formed by sound stimulation or by exposure to certain chemical agents. Studies of substances and conditions that protect cochlear hair cells against harmful noise and chemicals have been conducted (Oliveira, 2001). These professionals represent a population whose noise exposure data are difficult to obtain due to environmental factors, labor standards and the unpredictability of emergency responses (Kelly & Schwennker, 2013).

In addition to the fire department, the administrative sector is also affected by the noise from the emergency alarms. Jobs that require mental concentration or where language comprehension is important are occupations sensitive to noise, and even if the noise level is comparatively low can be disruptive (Huang, 2011).

The main objective of the study was to understand what sound levels practitioners of the barracks are exposed during their occupation. In addition to the daily personal exposure, we sought to verify that the noise was considered cumbersome and if workers detected some type of symptoms.

2 MATERIAL AND METHODS

This study was part of an investigation that aimed to quantify the noise exposure of workers in barracks firefighters. It was investigated the occupational noise exposure in these individuals, both administrative and firefighters.

The study was Level II descriptive, correlational, cross-cutting. The sample was defined as the type of non-probabilistic sampling and regarding the technique, accidental or convenience. The total study population was composed of 105 professionals from a fire station of the Central Region, and the sample was composed of 79 professionals. Data collection was conducted between February and April of 2013.

The institution's records collected from 2009 to 2012, inclusive, intended to understand and quantify the average number of exits and kilometers, made during the years referred to the trend and realize the same, in order to estimate a greater or lesser exposure, during the several months of the year.

Questionnaires were also applied in order to collect information related to noise exposure and health professionals, including: data social biographical, noise disturbance in performing tasks, annoyance by noise exposure in relation to used vehicles, the use of berths barracks, signs and symptoms detected over the state of health.

Regarding the measurement and interpretation of data, had to be based on the Decree-Law n° 182/2006 of 6 September, compared to the value of Lower Exposure Action Values (LEA), Upper Exposure Action Values (UEA) and Exposure Limit Values (ELV), values reported by the World Health Organization as the tolerable limit for the human ear (Berglund et al., 1999). The Brazilian Norm 10.152 establishes noise levels of acoustic comfort, setting the maximum decibel to be adopted in certain places, like administration rooms, and the values considered in the "Handbook of Ergonomics—Adapting the work to the man" for small offices and silent) (Lees, 1995). For firefighters calculated the minimum exposure time required to achieve the benchmarks in the legislation. Measurements were performed in 11 locations in the building (technical office, principal's office, secretaries, office of previous plans, the office of Head of Service, the office of command, logistics section and the operating room) and 3 different vehicles (INEM—Ambulance—Mercedes Benz 2001; VLCI—light vehicle firefighting—Land Rover 2010; VUCI—vehicle urban firefighting—Mercedes Benz 1998). All professionals classes are exposed to noise from emergency alarms triggered every occurrence requested. The alarms were categorized into three types: Alarm 1, designated by INEM—National Institute of Medical Emergency (emergencies within the emergency health and legal conflicts, poisoning, sudden illness, falls and trauma, burns, childbirth, drowning and evacuation of a doctor and assaults and rape, suicide), Alarm 2, called services (activities like cleaning of roads, water supply, door opening, closing, towing, tree fall, fallings, damages due to the fall of electric devices); and the Alarm 3, called General (fires, accidents, dangerous materials transport, among others). Measurements were performed on each type of situation (with the share of each type of alarm). The number of alerts triggered to advertise events is performed at least twice. Thus, for calculation purposes, the average number of alarms triggered corresponded to twice the average number of outputs.

After data collection, they were treated with the software Noise Explorer Type 7815 and the statistical software SPSS, version 20.0. We adopted a significance level of 5% (0.05) for the application of statistical tests.

3 RESULTS/DISCUSSION

We analyzed the monthly mean values obtained during the four years analyzed. In Figure 1 you can check the average number of outputs, per month and per type of alarm, and understand to what periods workers are more exposed to noise from alarms. It was found that for the Alarm 1 and Alarm 3 the variation of the average output over months was not significant (p-value > 0.05).

Analyzing the frequency of Alarm 1, it was verified that August was the month that got a lower average number of outlets (152 ± 25.6, making a daily average 5 outputs). However, medical emergencies in the months February, May and October tend to increase. October stood out that, on average, had 208 ± 36.3 sorties per month, giving a daily average of approximately 7 outputs. In the case of the events associated with the second alarm, the change in monthly average number of outputs was significant (p-value < 0.05). It was found that the average number of emergencies was equal to 255 ± 41.5 events per month for a total daily average of 8 outputs. Stood out in January that, on average, achieved an average total of 295 ± 85.3 outputs.

Emergencies associated with the Alarm 3 had an average number of occurrences equal to 113 ± 24.1 (daily average of 4 outputs). In such instances, stood out in July with a monthly average of 139 ± 31.7 outputs. Mentioned that there was still months with a monthly average of less than 91 ± 20.8 outputs. The values obtained per year for three types of alarm which may comprise up this variation. It was found that there was a significant change (p-value > 0.05) between the average annual level is output. It was found that, in general, the average number of occurrences, during analyzed years, declined. However, from 2011 to 2012, the emergencies associated with the Alarm 1 and, from 2009 to 2010, characterized by the emergency Alarm 2, increased slightly.

Regarding the number of kilometers traveled it was found that the monthly variation tendency wasn't significant (p-value > 0.05). Regarding the Alarm 1, it was found that February, May and October are the months in which more kilometers were effected, with an average of 3428 ± 559, 3367 ± 760 and 3306 ± 693, respectively. In total, the vehicles had an average of 3012 kilometers per month, equivalent to an average total of approximately 100 kilometers daily. Although the category with the highest number of occurrence, there was not the most kilometers logged. Still, November, January and February, walked up 3090 ± 525, 3347 ± 941 and 3416 ± 725 km, respectively. With regard to the second alarm, the total average per month was 3090 ± 681 kilometers (equivalent on average to 103 kilometers per day). Alarm 3 is the category that was traveled a greater number of kilometers. The average total number of kilometers made these emergencies was 3536 ± 1242 (a total of 118 km daily). Emphasize the months June and September that, on average, traveled more kilometers.

Concerning the results of the average number of kilometers driven each year, it was found that, for alarms 2 and 3, there was a significant variation between the average values obtained kilometers (p-value > 0.05). However, the highlight was the 3 to Alarm, the year 2010 to 2011, increased the number of kilometers made by approximately 8.8%.We point out that for this alarm in 2011 were covered, on average, 3745 ± 1283 km (equivalent to a daily average of 125 km).

Based on measurements of sound levels in the three types of vehicles, the values obtained of L_{Aeq} was 93.3 dB(A), 96.6 dB(A) and 103.2 dB(A) for vehicles VUCI, EMT and VLCI, respectively. On Table 2 are described the maximum intervals time permissible of exposure. Thus, we highlight the vehicle VLCI that with the siren, two minutes were sufficient for the worker to reached a $L_{EX,8h}$ of 80 dB(A). To achieve a $L_{EX,8h}$ of 85 dB(A) and 87dB(A) would suffice 7 and 11 minutes of exposure to noise, respectively. The car INEM obtained a value of L_{Aeq} of 96.6 dB(A), reaching a maximum allowable time of exposure to noise of 10, 30 and 50 minutes to reach values of LEA, UEA and ELV, respectively. As the vehicle VUCI the L_{Aeq} is significantly lower, however, the lower action value was achieved with 20 minutes of exposure to noise by the sound of vehicle (Table 1).

Table 1. Maximum permissible noise exposure in vehicles.

			Maximum allowable time (minutes)		
Vehicles	Year	LAeq dB(A)	LEA 80 dB(A)	UEA 85 dB(A)	ELV 87 dB(A)
VUCI	1998	93,3	20	70	110
INEM	2001	96,6	**10**	30	50
VLCI	2010	**103,2**	**2**	7	**11**

The exposure of these professionals to noise depends on the number of occurrences and, in part, on the number of kilometers performed on emergency vehicles.

The results of the measurements made in several places in the quarter are shown in Table 1. It was found that where it reached values higher noise levels were Registry (Level 1) and the cabinet previous plans (Level 0), a value of $L_{EX,8h}$ equal to 63.8 dB(A) and 64 94 dB(A), respectively. Importantly, there were no values below Exposure Personal Daily 53.12 dB(A) in places where the noise was quantified. This is explained by the structural conditions of the building and organization, because these two sites are closer to the speakers that emit the alert, present in the building.

When we quantified the noise, it was found that the values achieved for the Daily Personal Exposure evaluated in different locations in the building were below the reference values set out in legislation (LEA = 80 dB(A); UEA = 85 dB(A), ELV = 87 dB(A)). However, when compared with the reference values for the Acoustic Comfort (35 dB(A)–45 dB(A)), in rooms of Directors, present in the Brazilian Norm 10.152, the values of $L_{ex,8h}$ in local administration were analyzed always higher (ABNT, 1998). Compared with other references which considered that the usual noise and very small office systems must be between the 40 dB(A) and 45 dB(A), it was found that the values $L_{ex,8h}$ were also above recommended. Jobs that require mental concentration or where language comprehension is important occupations are "overly sensitive", and even though the level is comparatively low may be disturbing (Antunes, 2009). Reading problems, attention, problem solving and memory are the cognitive effects of greater importance associated with noise exposure (ABNT, 1998). However, as regards the tolerable limit of human hearing sound, the values obtained were below the limit value tolerable by the human ear (65 dB(A)) (Berglund et al., 1999).

It should be noted that the noise that is felt at the touch of the alert, although high, has a very short duration. Through the analysis of questionnaires, we found that most professionals Technical Assistants considered their workplace as being noisy and disruptive in performing tasks. It should be noted that any sound deemed undesirable by the occupier of the workplace can be considered as noise (ABNT, 1998). The level of speech lies between 55 and 65 dB(A). To have a good communication it is necessary that the noise level is 6 dB(A) below this range. Already above 60 dB(A) is necessary to raise your voice to enable communication (Kjellberg, 1994).

Class Professional Firefighters, could not create a standard relative to Personal Exhibition Night because the job is not fixed, and the time they remain in each space of the barracks is unknown and highly variable. In the analyzed vehicles, these professionals are exposed to high noise levels caused by siren. It was noticed that for VLCI car, which in turn is a very used car, seven minutes were enough noise exposure to achieve the value of Action Superior. However, the fireman to noise exposure is intermittent. This depends on the intensity and duration to which the worker is exposed (Shung & Chu, 2012). It is important that firefighters are sensitized to activate the sirens in the car when absolutely necessary, not producing or being exposed to noise unnecessarily. It was also concluded that the awakening by alarm action in dormitories, causes discomfort and disruption in the professionals. Regarding the questionnaires, there was a significant association between the perceptions of the workplace be noisy due to the Professional Class of workers in the barracks (p-value > 0.05). However, it should be noted that 60.8% of workers consider their workplace noisy.

When asked if the noise was disturbing the performance of work tasks 46.8% of professionals considered that the noise disturbs performing tasks.

As for the noise detected inside the vehicles, nearly 50% of subjects considered the noise emitted by the vehicle sirens annoyance. Of those who considered the noise nuisance in the execution of activities, most considered disturbing in terms of their cognitive abilities (concentration and reasoning), at physiological and psychological (stress, anxiety, heart rate) and the level of interpersonal communication. On the other hand, some professionals said that the noise was a risk inherent in their profession and they were used to.

Not all firefighters who considered their workplace noisy reported that this troubled them in performing tasks, some arguing that the noise was a risk inherent in their profession and are accustomed. It is not yet defined the extent to which people are accustomed to noise, always depends on external circumstances and other subjective factors internal, it is not possible to generalize (Kroemer & Grandjean, 2005). The change of cognitive and physiological and psychological workers are exposed to noise while the vehicles were also mentioned by firefighters, including driving, thinking and concentration and stress and anxiety that can lead to these professionals act in a thoughtless way. It was also shown the difficulty of transmitting and receiving the voice message, which reduces intelligibility and increases the fatigue state, and irritability.

With regard to the symptoms detected by the respondents, it was found that the "noise annoyance", "fatigue upon waking", "sleep disorders" and "Mood Disorders" are the most often mentioned by professionals 32.9%, 29.1%, 36.7% and 32.9% affirmative response, respectively. There was a considerable number of workers (n = 26) affected by this criterion of discomfort. Other symptoms like "Difficulties in hearing", "Lack of Concentration" and "irritation" were mentioned by 17.7%, 16.5% and 24.1% of respondents. "Headaches", "Ringing" (tinnitus) and "Tremors" were mentioned symptoms for about 20% of the professionals.

The relationship between the variables Symptoms and Time Job was not statistically significant (p-value > 0.05). However, it should be noted that the level of perception, with the exception of "irritation", "dizziness" and "lack of concentration", there was a slight decrease in the average career span of individuals who responded affirmatively to this condition. Of the 16.5% who said that the noise was distracting, the average time Job (in years) was 15.1 ± 6.4, with the remaining 63 Professional being 17.3 ± 6.9.

For variables symptoms and age, there was a statistically significant association in symptoms "dizziness" and "Gastrointestinal problems" (p-value < 0.05). It was found that workers that had these symptoms were significantly older than those that not suffered from such health problems. Age is a biological factor with a role in the effect of increasing deafness (Sliwinska-Kowalska & Dudarewicz, 2006).

Already on symptoms "Lack of Concentration" and "irritation", the average age tended to decrease slightly.

The younger and with less years in the profession, when exposed to noise, reflect easily lack of concentration and this may again translate than the other are more accustomed to perform work activities exposed to noise. Also refers to the 62% of individuals who considered distracting noise in the wards, 39.2% who said they had had some kind of symptom wake up this way.

In these situations the warning alarm is triggered in the same way and loudness. In the wards, it was found that the noise values of L_{Aeq} reached between 75 dB(A) and 85 dB(A). Noise can either awaken completely people or leave them in a state of semi-consciousness (Kroemer & Grandjean, 2005). Thus, it is arguable the psychological state of these professionals to wake up and leave for an emergency.

The present study was developed with many limitations, including the inability to create standards for the Exposition Personal Daily workers, since the number of occurrences and the number of kilometers made, the number of night occurrences (where you can not trigger the siren vehicles), and the time that firefighters spend at each location is always uncertain.

4 CONCLUSION

It was concluded that exposure to noise from professionals inside the barracks, triggered by alarms, during the period analyzed, did not reach noise levels sufficient to cause health effects on workers. However, through questionnaires and compared with some references, it could be concluded that the noise that is felt inside the barracks, when triggered alarms is upsets and discomfort in performing tasks. It was also concluded that some professionals have a clinical state of habituation to the noise to which they are exposed.

Exposure to noise in the vehicles also vary with the type of driver. It is important that firefighters are sensitized to activate the sirens in the car when absolutely necessary, not producing or being exposed to noise unnecessarily. It was also concluded that the awakening by alarm action in dormitories, causes discomfort and disruption in the professionals.

The present study was developed with many limitations, including the inability to create standards for the Daily Noise Exposure Level, since the number of occurrences and the number of kilometers made, the number of night occurrences (where you can not trigger the siren vehicles), and the time that firefighters spend at each location is always uncertain.

In a perspective of future studies, it is important to deepen the exposure of firefighters to noise in the vehicle, as it is the most worrying situation, the use of these professionals dosimeters (as it does not have a position or fixed duration and thus it would be easier study their noise exposure) and audiometric examinations in the universe studied in order to understand if there are already in fact professionals with pathologies at the hearing.

REFERENCES

ABNT—Associação Brasileira de Normas Técnicas. Nível de ruído para conforto acústico. NB—101152, 1998.

Antunes, M. *Exposição ao ruído de baixas frequências em meio ocupacional vs Repercussões na Qualidade do Sono*. Mestrado em Ciências do Sono. Faculdade de Medicina de Lisboa; 2009.

Berglund, B., Lindvall, T., Schwela, D. *Guidelines for Community Noise*. World Health Organization, Geneva, 1999.

Huang K., 2011, *A study on effects of thermals, luminous, and acoustic environments on indoor environmental comfort in offices*, Building and Environment 49, 304–309, 2011.

Kelly S., Schwennker C., *Firefighter Noise Exposure During Training Activities and General Equipment Use*; Journal of Occupational and Environmental Hygiene, 10: 116–121; 2013.

Kjellberg A. *Noise in the Office: Part II—The scientific basis for the guide*. International Journal of Industrial Ergonomics, 1994.

Kroemer, K., Grandjean, E. *Manual de Ergonomia—Adaptando o Trabalho ao Homem*. 5ª edição—Edições Sílabas; 2005.

Lee L., Kim B., Kim J.-H., Kim J., *Analysis of occupational noise for the healthy life according to the job characteristics*; Health Vol. 4, No. 10, 897–903, 2012.

Lees, P. *Combustion products and other firefighter exposures*. Occupational Med 10(4):691–707, 1995.

Miguel, A. *Manual de Higiene e Segurança do Trabalho*. 12ª edição; 2012. Porto Editora.

NIOSH. *Promoting Hearing Health among Fire Fighters*; DHHS (National Institute for Occupational Safety and Health) Publication No. 2013–142; 2013.

Oliveira, J. *Prevenção e proteção contra perda auditiva induzida pelo ruído*. In: Nudelmann, A. Pair—Perda Auditiva Induzida pelo Ruído: volume II. Rio de Janeiro: Revinter, 2001.

Shung I., Chu I. *Hearing effects from intermittent and continuous noise exposure in a study of Korean factory workers and firefighters*. BMC Public Health, 2012.

Sliwinska-Kowalska M, Dudarewicz A. *Individual susceptibility to noise-induced hearing loss: choosing an optimal method of retrospective classification of workers into noise-susceptible and noise-resistant groups*. International Journal of Occupational and Environmental Health; 19(4): 235–245, 2006.

Occupational Safety and Hygiene II – Arezes et al. (eds)
 ISBN 978-1-138-00144-2

Risk factors associated with musculoskeletal symptoms in footwear sewing workers

L. Afonso & M.E. Pinho
Faculty of Engineering, University of Porto, Porto, Portugal

P.M. Arezes
University of Minho, Minho, Portugal

ABSTRACT: Beyond the lack of studies on this issue, particularly in Portugal, the relationship between the prevalence of musculoskeletal symptoms and many work-related risk factors is still very unclear. Therefore, in order to contribute to a better understanding of WMSD risk factors, this study aims to analyze the association between the prevalence of musculoskeletal symptoms in sewing workers of the footwear industry and some of the work-related risk factors for the development of WMSD. Both direct observation and a questionnaire survey were performed, and SPSS was used in statistical data analysis. Some of the individual, occupational, environmental and organizational/psychosocial risk factors were found to be associated with the reported symptoms, but those were distinct in the two companies under study. The results emphasize the multifactorial nature of WMSD and the need for the implementation of new and more effective ergonomic prevention programs, more centered on the identified risk factors.

1 INTRODUCTION

Due to the present socio-economic context and globalization, the increasing workload and quality demands are threatening workers' health and leading to high productivity losses and tremendous financial costs for both individuals and society. Particularly the impact on workers' musculoskeletal health seems to be even more concerning.

The growing importance of this issue has led institutions and organizations such as European Commission (EC, 2010) to consistently identify Musculoskeletal Disorders (MSD) as a priority in the prevention of safety and health at the companies level.

In fact, according to the statistical data from the Bureau of Labour Statistics (2012), in 2011, MSD were responsible for 33% of all accidents and illnesses related to labour absenteeism in the United States of America (USA). Meanwhile, in Europe, according to Health and Safety Executive (2012), countries like Austria, Germany and France show a high number of lost working days due to Work-related Musculoskeletal Disorders (WMSD). The same source refers that WMSD are one of the most common causes of work absenteeism due to illness in the United Kingdom while in France, according to the European Agency for Safety and Health at Work (EU-OSHA, 2010) in the year 2006, WMSD led to 7 millions of lost working days, which represents a cost of 710 million Euro for the companies. Likewise, Woolf & Pfleger (2003) referred that the expenses with the productivity and wages losses due to WMSD correspond to 2.4% and 1.3% of the gross domestic product of Canada and USA, respectively. The socio-economic impact also seems clear in the manufacturing industry (EC, 2010). In Portugal, the footwear industry represents an important part of the manufacturing industry and is responsible for a growing amount of the national exports. However, the traditional taylorist organization system, which is characterized by risk factors recognized in several studies as predominant for the onset of WMSD, is still adopted by most companies (Todd *et al.*, 2008). The sewing sector is referred to as one of the sectors where the workplaces are at a higher risk (Aghili *et al.*, 2012; Roquelaure *et al.*, 2004).

In this context, this study aims to check if there are statistically significant associations between the prevalence of Musculoskeletal Symptoms (MSS) reported by the workers of the footwear industry sewing sector and the studied risk factors. The study of the MSS arises as a predictor of subsequent MSD in the studied populations, therefore constituting an important step towards the WMSD prevention (Smith *et al.*, 2009).

In the last decades, physical factors for WMSD have been widely studied in the literature. On the contrary, psychosocial factors have been neglected and there is still a lot to explore. For that reason, they were privileged in this study while, although have been registered, some of the physical factors were not subjected to a specific analysis in this investigation.

2 MATERIALS AND METHODS

2.1 *Subjects*

This study was conducted in the sewing sector of two companies (A and B) of the footwear industry, placed in the municipality of Felgueiras. From a global population of 130 female workers, samples (34 in company A and 32 in company B) were selected through pre-established inclusion and exclusion criteria. The inclusion criterion was: individuals who had been performing duties in the company for at least one year, who were present on the day of the companies' visit and who agreed to participate in the study. All the identified individuals who did not fulfil the inclusion criterion as well as those presenting musculoskeletal problems, which were not related to the job, were excluded. All the participants were informed about the objectives of the study and that all information collected would be treated as strictly confidential and anonymous.

2.2 *Methods*

For the characterization of the activity carried out by the workers of the sewing sector, the guide for the ergonomic analysis of workstations by the Finnish Institute of Occupational Health (FIOH), translated and adapted into Portuguese by University of Minho (Gomes da Costa, 2004), was used. For data collection on the musculoskeletal symptoms reported by the workers as well as on the potential risk factors, a questionnaire was built up based on 3 validated questionnaires: Nordic Musculoskeletal Questionnaire in its translated and validated version for the Portuguese population (Mesquita *et al.*, 2010) to evaluate health aspects; Dutch Musculoskeletal Questionnaire—extended version (Hildebrandt *et al.*, 2001) to evaluate socio-demographic and work aspects and Copenhagen Psychosocial Questionnaire—medium size version (Kristensen *et al.*, 2005) to evaluate psychosocial aspects.

Work-related MSS were evaluated in terms of pain perception or discomfort of the respondents, whenever they were present in the last 12 months for at least a week (Sluiter *et al.*, 2001), in the following body regions: neck, thoracic region, lumbar region, hips/thighs, knees, ankles/feet, shoulders, elbows and wrists/hands.

SPSS® v.21.0 software was used in data statistical analysis. The Chi-Square independence test was performed to look for statistically significant associations between the prevalence of the reported MSS and the observed risk factors. For those associations, the Relative Risk (RR) was performed whenever possible, while Odds Ratio (OR) was used in the remaining cases. The corresponding 95% confidence intervals (95% CI) were also estimated.

2.3 *Procedures*

In December 2012 the activity of sewing workers was characterized through direct observation and the support of the FIOH ergonomic analysis guide.

Subsequently, a pre-test of the constructed questionnaire was conducted in 10 of the study participants. Then, during January 2013, the questionnaire was applied to all the selected workers by means of a structured interview. The duration of each interview was approximately 8 minutes.

3 RESULTS AND DISCUSSION

Herein, are presented and discussed the statistically significant results found in the study of the association between the prevalence of MSS (presented in the paper entitled "Prevalence of WMSD in the sewing sector of two companies of the footwear industry") and the studied work-related socio-demographic (section 3.1), occupational (section 3.2), environmental (section 3.3) and organizational/psychosocial (section 3.4) risk factors.

3.1 *Socio-demographic factors*

The final sample consisted of 34 workers from company A and 32 from company B, which mean response rates of 80% and 38%, respectively.

The workers are aged between 22 and 55 years old and, in average, they are 38 ± 8 years old (39 ± 8 for company A and 38 ± 8 for company B).

Table 1 shows the statistically significant associations between the prevalence of MSS and the socio-demographic factors.

In company A, the number of workers under 40 years old reporting shoulders symptoms is significantly greater than expected (OR = 5.000; CI 95%: 1.030–24.279). This result seems to contradict the results of other studies which found an association between the prevalence of MSS and the increase in age of the sewing workers (Aghili *et al.*, 2012; Wang *et al.*, 2009). On the other hand, others have found a higher prevalence of

Table 1. Association between the prevalence of MSS and socio-demographic factors.

Risk factor	Region of the body	p
Company A		
Age	Shoulders	0.038*
Marital status	MSS 12 months	0.048**
	Lower back	0.046**
Company B		
Education level	Ankles/feet	0.024**

*Chi-square independence test; **Fisher's exact test.

MSS on shoulders and neck in younger workers (Roquelaure *et al.*, 2012). This situation may be explained by the workers' inexperience (Roquelaure *et al.*, 2012), though it is worth noting that the percentage of workers with less than 20 years of experience is the same in both companies and that, in company B, the prevalence of MSS was not associated with age. Another likely reason is the so called "healthy worker effect", that is, the possible bias on results that may be due to the fact that the healthier workers remain employed longer (Benavides *et al.*, 2006). Although the literature refers a cause and effect relationship between workers' age and MSS, this relationship is not linear, which suggests that there may be other factors involved (Guo *et al.*, 2004).

Despite the fact that the majority of the participants are married, it can be observed that in company A, the risk of the single or widow workers (p = 0.048; RR = 4.348; CI 95%: 1.366–13.889) reporting symptoms in the last 12 months is more than four times that found for those who are married or who live together. Similar conclusion was made by Kaergaard & Andersen (2000), who found a higher risk of neck and shoulders symptoms in single seamstresses. On the contrary, in company B, the risk of workers with an education level up to the 4th grade (p = 0.048; RR = 8.800; CI 95%: 1.122–69.036) reporting ankles/feet symptoms is almost nine times the risk of those with a higher education level. These data seem to confirm the findings of Ozturk & Esin (2011), who reported a higher prevalence of MSS in workers with low education levels. This is likely due either to the increased difficulty of workers with low education levels to understand working instructions (Costa *et al.*, 2009) or to the unclear transmission of information to the workers. However, the type of pedal used in company B may also have influenced the development of the symptoms in those regions, since the pedal of the machine is actuated with only one foot, while in company A, the pedal is actuated with both feet simultaneously, thus eliminating the static load on one foot.

3.2 *Occupation related factors*

Statistically significant associations between the perceived MSS and the occupational risk factors were only found in company A (Table 2), where the risk of reporting elbows symptoms is higher in workers with experience up to 20 years (p = 0.017; RR = 1.800; CI 95%: 1.045–3.101) than in more experienced workers. Also Roquelaure *et al.* (2012) found that longer job experience was associated with a lower MSS prevalence rate. On the contrary, Zhang *et al.* (2011) referred that the higher the number of years of experience, the higher the probability of developing MSS.

The results of these studies seem to indicate that workers with less years of work in the same job may have more difficulties in performing their work, which may lead to inadequate postures while carrying out their activities. However, a work experience up to 20 years may not necessarily imply little experience. This situation in company A may have been worsened by the lack of a elbows support, which, unlike company B, was not provided to the workers.

In what concerns the employment contract type, data indicate that the risk of workers with a fixed-term contract reporting MSS in the last 12 months is 5.6 times (p = 0.021; RR = 5.618; CI 95%: 1.923–16.393) that found for permanent workers. The results of the current study seem to support the findings of other authors (Benavides *et al.*, 2006; Roquelaure *et al.*, 2012), which reported a higher prevalence of MSS in the wrists/hands region of temporary workers when compared to the permanent workers. This fact is possibly due to the fear that the operators with a fixed-term contract have of becoming unemployed (Benavides *et al.*, 2006). Besides that, workers with less professional experience are more frequently exposed to time constraints (Roquelaure *et al.*, 2012). In company A, the prevalence of neck MSS is also statistically associated with the duration of the working rest breaks (p = 0.020), and the percentage of workers reporting neck MSS was higher in those enjoying rest breaks lower than 10 minutes than in those having longer rest breaks. In fact,

Table 2. Association between the prevalence of MSS and occupation related factors.

Risk factor	Region of the body	p**
Company A		
Work experience	Elbows	0.017
Contract type	MSS 12 months	0.021
Rest breaks	Neck	0.020

**Fisher's exact test.

the insufficient rest breaks during work are being pointed out as a risk factor for the development of MSS in sewing workers, mainly in the neck and shoulders (Wang *et al.*, 2007; Zhang *et al.*, 2011). This situation may indicate that, in company A, the duration of rest breaks may not be enough for workers to recover from the effort of the used muscles.

3.3 *Environmental factors*

In company A, the risk of workers reporting MSS in the lumbar region (p = 0.052; RR = 2.160; CI 95%: 1.152–4.051), in neck (p = 0.006; RR = 2.778; CI 95%: 1.500–5.145), and in shoulders (p = 0.033; RR = 3.333; CI 95%: 1.342–8.281) is higher for those who perceive to be exposed to extreme temperatures. In fact, during the observation of the working environment in both companies, the room temperature seemed to be relatively higher in company A than in company B. In this context, it is likely that the symptoms reported by the workers in those body regions may be explained by a significant increase of temperature while carrying out their tasks. Similar results were found by Gold *et al.* (2009), which referred that the exposure to extreme temperatures was associated with the prevalence of MSS. However, furthermore, the seat used in company A is not adjustable, forcing workers to perform the activity in an uncomfortable position which may explain a higher prevalence of symptoms in those regions (Rempel *et al.*, 2007; Wang *et al.*, 2008). On the other hand, the workers of company B who consider to be exposed to poor lighting conditions show a risk of reporting MSS in the neck region of 6.5 times (p = 0.034; RR = 6.500; CI 95%: 1.375–30.731), and in the lumbar region of 13 times (p = 0.015; RR = 13.000; CI 95%: 1.621–104.247) the risk of those who do not consider to be exposed to that risk factor. In fact, Parimalam *et al.* (2007) referred that the prevalence of back pain in sewing workers might be associated with inadequate positions due to poor lighting. It is important to highlight that the sewing activity is a precision work; therefore, when the lighting is inadequate, the workers' visual fatigue may predispose to inadequate positions, such as the flexion of the neck and back (Institut National de Recherche et de Sécurité, 2011). In this context, it was observed that, although there was a greater amount of natural light in company B, the artificial lighting levels could be lower, especially due to the fact that the luminaire support was too high, forcing workers to lean forward to improve their vision of the work plane and, thus, contributing to the development of MSS in the referred regions.

3.4 *Organizational and psychosocial factors*

Table 3 shows the statistically significant associations between the prevalence of MSS and the psychosocial factors.

In company A, the workers who perceive a low leadership recognition and support show a risk of reporting symptoms in the last 12 months 2.4 times the risk of those who do not perceive that risk factor (p = 0.001; RR = 2.396; CI 95%: 1.116–5.142). Similar results were found by Kaergaard & Andersen (2000) for the regions of the neck and shoulders of sewing workers. In addition, Canjuga *et al.* (2010) referred that the high social support by the managers was also associated with less absences from work in the short run. Besides that, the workers who perceive not being able to easily express their opinions and feelings at work show a risk of reporting symptoms in the last 12 months (p = 0.028; RR = 1.579; CI 95%: 1.031–2.419) that is 1.6 times the risk of those who perceive to be able to do it. In fact, Wang *et al.* (2005) observed a statistically significant association between the low work control and the prevalence of MSS in the sewing sector. By their side, in a study conducted in office workers, Van den Heuvel *et al.* (2005) found an association between the development of the carpal tunnel syndrome and low work control.

In company B, the prevalence of MSS in the last 12 months is associated with the workers' perception of working very fast (p = 0.023). Moreover, the workers who consider themselves exposed to a very fast work pace have a chance of reporting wrists/ hands symptoms (p = 0.041; OR = 9.800; CI 95%: 1.036–92.696) that is 9.8 times that of those who do not consider to be exposed to that risk factor. These results corroborate what was found by other authors in studies conducted in sewing operators, which found that the prevalence of MSS in the wrists is associated to the fast work

Table 3. Association between the prevalence of MSS and organizational/psychosocial factors.

Risk factor	Region of the body	p
Company A		
Low leadership recognition	MSS 12 months	0.001**
Difficulty of expression at work	MSS 12 months	0.028*
Company B		
Very fast work pace	MSS 12 months	0.023**
	Wrists/hands	0.041**

*Chi-square independence test; **Fisher's exact test.

pace (Ozturk & Esin, 2011; Wang *et al.*, 2009). This situation may also have been aggravated by the fact that the sewing workers make many repetitive movements with wrists/hands (Sarder *et al.*, 2006; Sealetsa & Thatcher, 2011).

4 CONCLUSIONS

The risk factors associated with the reported MSS were distinct in the two companies, which seems to indicate the multifactorial nature of WMSD. This situation demonstrates that, in general, the prevalence of MSS in both companies is due to the adoption of the taylorist production system, which continues to prevail in the Portuguese footwear industry, as well as other important individual and psychosocial work aspects. However, despite the interesting conclusions made, the results must be carefully interpreted because of the reduced sample size.

This exploratory study seems to point out to the need for the implementation of a prevention ergonomic program focused on the identified risk factors, as well as for more research about the influence of each risk factor individually on the prevalence of MSS. In this context, it would be interesting to carry out an extensive study, including a large sample of companies and including also workers from the remaining sectors of this particular economic activity in order to be able to extrapolate the results with greater reliability to the studied populations.

ACKNOWLEDGMENTS

The authors would like to thank Master in Occupational Safety and Hygiene Engineering (MESHO), of the Faculty of Engineering of the University of Porto (FEUP), all the support in the development and international dissemination of this work.

REFERENCES

Aghili, M., *et al.* (2012). Evaluation of musculoskeletal disorders in sewing machine operators of a shoe manufacturing factory in iran. *International Journal of Industrial Ergonomics, 62*(3).

Benavides, F.G., *et al.* (2006). Associations between temporary employment and occupational injury: What are the mechanisms? *Occupational and Environmental Medicine, 63*(6), 416–421. doi: 10.1136/oem.2005.022301.

BLS. (2012). Nonfatal occupational injuries and illnesses requiring days away from work, 2011 (pp. 32). Washington D.C: Bureau of Labor Statistics.

Canjuga, M., *et al.* (2010). Correlates of short-and long-term absence due to musculoskeletal disorders. *Occupational Medicine-Oxford, 60*(5), 358–361.

Costa, L.d.C.M., *et al.* (2009). Prognosis for patients with chronic low back pain: Inception cohort study. *BMJ: British Medical Journal, 339*.

EC. (2010). *Health and safety at work in Europe (1999–2007)*. Belgium: Publications Office of the European Union.

EU-OSHA. (2010). OSH in figures: Work-related musculoskeletal disorders in the EU—Facts and figures (pp. 179). Luxembourg: European Agency for Safety and Health at Work.

Gold, J., *et al.* (2009). Skin temperature in the dorsal hand of office workers and severity of upper extremity musculoskeletal disorders. *International Archives of Occupational and Environmental Health, 82*(10), 1281–1292. doi: 10.1007/s00420-009-0450-5.

Gomes da Costa, L. (2004). *Guia para a análise ergonómica de postos de trabalho*. Universidade do Minho.

Guo, H.-R., *et al.* (2004). Prevalence of musculoskeletal disorder among workers in Taiwan: A nationwide study. *Journal of Occupational Health, 46*(1), 26–36.

Hildebrandt, V., *et al.* (2001). Dutch Musculoskeletal Questionnaire: Description and basic qualities. *Ergonomics, 44*(12), 1038–1055.

HSE. (2012). Annual Statistics Report (pp. 23). London: Health and Safety Executive.

INRS. (2011). Les Troubles musculosquelettiques du membre supérieur: Gide pour les préventeurs. Paris: Institut National de Recherche et de Sécurité pour la Prévention des Accidents du Travail et de Meladies Professionnelles.

Kaergaard, A. & Andersen, J.H. (2000). Musculoskeletal disorders of the neck and shoulders in female sewing machine operators: Prevalence, incidence, and prognosis. *Occupational and Environmental Medicine, 57*(8), 528–534. doi: 10.1136/oem.57.8.528.

Kristensen, T.S., *et al.* (2005). The Copenhagen Psychosocial Questionnaire-a tool for the assessment and improvement of the psychosocial work environment. *Scandinavian Journal of Work, Environment & Health*, 438–449.

Mesquita, C., *et al.* (2010). Portuguese version of the standardized Nordic musculoskeletal questionnaire: Cross cultural and reliability. *Journal of Public Health, 18*(5), 461–466. doi: 10.1007/s10389-010-0331-0.

Ozturk, N. & Esin, M.N. (2011). Investigation of musculoskeletal symptoms and ergonomic risk factors among female sewing machine operators in Turkey. *International Journal of Industrial Ergonomics, 41*(6), 585–591. doi: 10.1016/j.ergon.2011.07.001.

Parimalam, P., *et al.* (2007). Knowledge, attitude and practices related to occupational health problems among garment workers in Tamil Nadu, India. *Journal of Occupational Health, 49*(6), 528–534.

Rempel, D.M., *et al.* (2007). A randomized controlled trial evaluating the effects of new task chairs on shoulder and neck pain among sewing machine operators—The Los Angeles garment study. *Spine, 32*(9), 931–938.

Roquelaure, Y., *et al.* (2004). Surveillance program of neck and upper limb musculoskeletal disorders: Assessment over a 4 year period in a large company. *Annals of Occupational Hygiene, 48*(7), 635–642. doi: 10.1093/annhyg/meh054.

Roquelaure, Y., *et al.* (2012). Working in temporary employment and exposure to musculoskeletal constraints. *Occupational Medicine-Oxford, 62*(7), 514–518.

Sarder, M., *et al.* (2006). Ergonomic workplace evaluation of an Asian garment-factory. *Journal of Human Ergology, 35*(1–2), 45.

Sealetsa, O.J. & Thatcher, A. (2011). Ergonomics issues among sewing machine operators in the textile manufacturing industry in Botswana. *Work-a Journal of Prevention Assessment & Rehabilitation, 38*(3), 279–289.

Sluiter, J.K., *et al.* (2001). Criteria document for evaluating the work-relatedness of upper-extremity musculoskeletal disorders. *Scandinavian Journal of Work, Environment & Health*, 1–102.

Smith, C.K., *et al.* (2009). Psychosocial factors and shoulder symptom development among workers. *American Journal of Industrial Medicine, 52*(1), 57–68.

Todd, L., *et al.* (2008). Health survey of workers exposed to mixed solvent and ergonomic hazards in footwear and equipment factory workers in Thailand. *Annals of Occupational Hygiene, 52*(3), 195–205.

Van den Heuvel, S.G., *et al.* (2005). Psychosocial work characteristics in relation to neck and upper limb symptoms. *Pain, 114*(1), 47–53.

Wang, P.C., *et al.* (2005). Work organization and work-related musculoskeletal disorders for sewing machine operators in garment industry. *Annals of epidemiology, 15*(8), 655.

Wang, P.C., *et al.* (2007). Work-organisational and personal factors associated with upper body musculoskeletal disorders among sewing machine operators. *Occupational and Environmental Medicine, 64*(12), 806–813.

Wang, P.C., *et al.* (2008). A Randomized controlled trial of chair interventions on back and hip pain among sewing machine operators: The Los Angeles garment study. *Journal of Occupational and Environmental Medicine, 50*(3), 255–262. doi: 10.1097/JOM.0b013e318163869a.

Wang, P.C., *et al.* (2009). Self-reported pain and physical signs for musculoskeletal disorders in the upper body region among Los Angeles garment workers. *Work-a Journal of Prevention Assessment & Rehabilitation, 34*(1), 79–87.

Woolf, A.D. & Pfleger, B. (2003). Burden of major musculoskeletal conditions. *Bulletin of the World Health Organization, 81*(9), 646–656.

Zhang, F., *et al.* (2011). Quantify work load and muscle functional activation patterns in neck-shoulder muscles of female sewing machine operators using surface electromyogram. *Chinese Medical Journal, 124*(22), 3731.

Occupational Safety and Hygiene II – Arezes et al. (eds)
© 2014 Taylor & Francis Group, London, ISBN 978-1-138-00144-2

Environmental risks: A systematic review

B. Barkokébas Jr., E.M.G. Lago & F.M. Cruz
University of Pernambuco, Recife, Brazil

H.C. Albuquerque Neto
Doctoral Program in Occupational Safety and Healthy, Porto, Portugal

ABSTRACT: The diffusion of technology arising of the industrial revolution spurred industrial processes as well as showed a number of risks present in the workplace. Thus, the study has aimed to conduct a systematic review of the environmental risks, leaning on the risks of accident, chemical, biological, physical and ergonomic. The research was conducted in 26 portals scientific journals, following the assumptions of the PRISMA statement. In total 1169 publications were found, however, at most three publications met the inclusion criteria, and of these only two pored over the case study. It was concluded that six risk agents are common between the two publications being distributed in four different risks. Moreover, the risk of accidents is not addressed in detail making it a vacancy in the study. Finally, this article is relevant due to the small number of publications available, showing a large field for future studies.

1 INTRODUCTION

One of the challenges for job security in the new millennium are the issues of occupational health and safety (Jayaprakash, 2003) which have demanded the synergistic efforts of several countries (Aneziris et al., 2010), among which environmental risks emerge as critical issues for most rulers (Youssef et al., 2011). Currently, numerous human activities, which are linked by different levels of complexity, cause most of the risks (Bohm and Pfister, 2005) which have deleterious effects on the physical and mental health of human beings (Chen et al., 2013).

In this context, the experts in health and safety at work search for a place of work which is safe and healthy and is free from risks, among which we can mention: accidents, physical, chemical, biological and exposure to ergonomic problems (Froneberg, 2006). Thus, the work undertaken in contemporary literature which outlines these risks deserves attention since it is very relevant to the diagnosis of different risks in a single working environment. Therefore, the present study aims to conduct a systematic review of environmental risks, focusing on the risks of accidents, physical, chemical, biological and ergonomic, highlighting the characteristics of the causative agents and the prevailing trends in this field.

2 METHODOLOGICAL PROCEDURES

In order to achieve the proposed goal, the literature was reviewed in a systematic way, covering current topics which are relevant, and checking the data and information that arises from an analysis of environmental risks. The study was based on online surveys of different search portals which are integrated in the MetaLib tool from Ex Libris®. According to Zhonghua and Ling (2013), the MetaLib® streamlines the discovery process, presenting users with content from multiple providers of information, using a clear and familiar interface.

From the above, scientific journals that had undergone a systematic peer review process were selected. This process has been used for decades in order to ensure the quality, readability and relevance of manuscripts, whilst filtering poorer studies, as well as to ensure a fair and impartial assessment of a manuscript (Bunner and Larson, 2012). Moreover, another quality factor was considered when it was decided to select only papers, and exclude editorials, letters to the editor, conferences, special issues for conferences, opinion pieces for articles and magazines from the study. As such, all the 26 portals studied are shown in Table 1.

In order to investigate the scientific articles dealing with environmental risks, four combinations of keywords were used, as these would be located in the publication's title. Given this, the search for the articles was divided into four stages as outlined in Table 2.

Looking at Table 2, it can be seen that the keywords alternate between the terms risk(s)/hazard(s) and environmental/occupational due to conceptual variations in the literature. These concepts are very similar, however, and depending on the perspective

Table 1. Portals where the searches of scientific publications were made.

Types	Name of the portals surveyed
Scientific magazines	ACM Digital Library, ACS Journals, Annual Reviews, ASME Digital Library, BioMed Central Journals, Cambridge Journals Online, ASCE, Directory of Open Access Journals (DOAJ), Emerald, Geological Society of America (GSA), Highwire Press, IEEE Xplore, Taylor and Francis, IOPscience Journals, MetaPress, nature.com, Oxford Journals, Royal Society of Chemistry, SAGE, SciELO, Science Magazine, ScienceDirect, SIAM, SpringerLink, The Chronicle of Higher Education e Wiley Online Library

Table 2. Procedures for the undertaking of the research.

Stages	Keyword	Time period of publication	Period of the research
1	"Environmental risk(s)"	2000 to 2013	03/07/3013 to 23/07/2013
2	"Environmental hazard(s)"		
3	"Occupational risk(s)"		
4	"Occupational hazard(s)"		

of the study, they can be misinterpreted, resulting in a survey of all the terms. According to (Ulbig et al., 2010) the scientific community interprets "risk" and "hazards" as completely different concepts, but empirical studies by the same authors showed that lay people do not know how to differentiate the concepts, which means that the conceptual difference is not always explicit. Given this, studies by Raybould and Cooper (2005) state that the hazard of a substance or action is its ability to cause harm and risk and is estimated by combining the hazard and exposure measurements. The authors also clarify that exposure is the likelihood of being exposed to a hazard.

A similar phenomenon occurs with "environmental" and "occupational", and many risks in the literature include both terms. According to Semple (2005), occupational exposure is usually measured using simple and well validated techniques, whilst environmental exposures require much more sensitive instruments, making them more difficult to evaluate.

In defining the keywords, it was decided that the publications should have been written in the last

Table 3. Procedures for the exclusion of articles.

Stages of screening	Criteria for exclusion
1	Identify which articles are repeated
2	Identify which articles are not accessible to the reader
3 (optional stage)	Identify which articles are written in a language in which the authors are not proficient

Table 4. Procedures for the eligibility of the articles.

Stages of screening	Criteria for eligibility
1	Identify which articles do not address chemical, physical, biological, ergonomic and accidents in their scopes

eleven years (except the year 2013, as this is ongoing) and it is these which form the core of this current research. Subsequently, two exclusion quantitative criteria were adopted for the screening of articles, whose purpose was to select only those publications that are fit for a systematic reading. The exclusion criteria are briefly presented in Table 3.

It should be noted that the articles written in languages that were not English, Portuguese, Spanish, German and French, would be removed because the authors are fluent in only those languages just cited. There is, however, a further exclusion criterion which is the criterion for the eligibility of the articles, which is a criterion which ensures minimal homogenity in the description of methods and results. Thus, unlike the exclusion criteria, the eligibility criterion requires a reading of the publications. The procedures used are shown in Table 4.

It needs to be clarified that the decision to be made is the same for the exclusion criteria, given that if an article does not have the basic elements of eligibility, it must be discarded from further analysis. In performing all the steps of screening, we sought a method/procedure that would guarantee greater integrity in the design process of the systematic review. Therefore, we adopted the PRISMA statement (Preferred Reporting Items for Systematic reviews and Meta-Analyses) which, according to Urrútia and Bonfill (2010), is a tool to help improve the clarity and transparency of publications for systematic reviews.

3 RESULTS

After the completion of the search for scientific articles, 1169 in a universe of publications were found from combinations of the keywords identified. This detailed discrimination process is shown in Table 5.

Later, the process of deleting items was initiated, as can be seen in Table 6, adopting the exclusion criteria displayed in Table 3.

From Table 6 it appears that few articles are repeated, which can be explained by the terms in the keywords used for searching. As explained in the methodological procedures, there is disagreement over the meaning of the terms, which means the key words are distributed amongst different studies. Some publications were unavailable for download, making it impossible to read them, or confirm their findings. Furthermore, all the publications were found within the language proficiencies of the authors, which meant the additional step for screening the items was not required. In short, it appears that most publications were able to meet the eligibility criterion, which is outlined in Table 7.

Table 7 shows that only three publications address the five risks that from the focus of this article. Furthermore, in order to succinctly show the processes of the systematic review methodology, the flow of information through different the different stages is outlined, as proposed by Moher et al. (2010) and shown in Figure 1.

This number (3) is much less than the number of works which were screened (1054), which attests to the fact that most current publications focus on

Table 5. Number of publications found.

Stages	Keyword	Number of publication
1	"Environmental risk(s)"	784
2	"Environmental hazard(s)"	107
3	"Occupational risk(s)"	194
4	"Occupational hazard(s)"	84
Total of publications		1169

Table 6. Procedures for the exclusion of articles and the publications excluded.

Stages of screening	Criteria for exclusion	Number of publication
1	Identify which articles are repeated	48
2	Identify which articles are not accessible to the reader	67
Total of works excluded		115
Total number of works meeting the eligibility criteria		1054

Table 7. Procedures for the eligibility of the articles and the number of publications excluded.

Stage of screening	Criteria for eligibility	Number of publication
1	Identify which articles do not address chemical, physical, biological, ergonomic and accidents in their scopes	1051
Total number of works which meet this criteria		1054
Total number of works which meet the objectives of the article		3

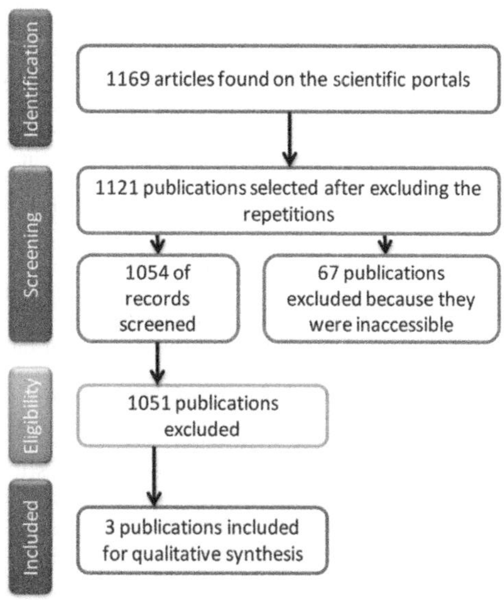

Figure 1. Flow of information through the different phases of a systematic review.

studies dedicated to a particular or pre-defined group of risks. However, a consolidated assessment of the risks which occur in the work environment is essential in the search for improvements in occupational safety for the worker. Thus, the works which were found should be discussed systematically, whilst investigating the variables that compose them. As such, Table 8 presents the eligible articles and highlights their general characteristics.

All three works relate to the field of health and were designed in Brazil. One of the possible causes for this is that Brazil has an Environmental Risk Prevention (PPRA) Program regulated by Norma Regulamentadora 9 (BRASIL, 1994) of the Ministry of Labour and Employment (MTE). This program aims to preserve the health and integrity of the employees, and works for the minimization

Table 8. Information relating to the eligible publications.

Author(s) and year	Criteria for eligibility	Type of article	Country in which the study was undertaken
Alencar (2005)	Pharmaceutical industry	Case study	Brazil
Silva and Zeitoune (2009)	Hemodialysis unit	Case study	Brazil
Volquind et al. (2013)	Anesthesiology practice	Literature review	Brazil

Table 9. Risk agents recorded the two selected works.

Risks	Similar approaches to Alencar (2005) and Silva & Zeitoune (2009)
Physical	Exposure to noise and variations in temperature
Chemical	Contact with and handling of chemical products
Biological	Exposure to microorganisms
Ergonomic	Repetitive movements, difficultly with mobility, inadequate lighting
Accidents	There is no thorough detailing of the risk of accidents

of the five risks that are the objects of this study. Another factor to be considered is that the work of Volquind et al. (2013) is a review of the literature and does not address the risks from an empirical aspect. Given this, we excluded this work on the systematization of risk agents, given that these would be described in the literature. Therefore, Table 9 presents the main risk agents that are presented in the other two works.

As can be observed in Table 9, ergonomic risks are those that have greater relevance in these studies and there are no detailed exemplifications of the risks of accidents in either work. This highlights the need for a thorough study of this area, given that the risk of accidents can be a serious threat to the lives of workers. Moreover, the articles demonstrate the productive process in question, which is using questionnaires to ask employees about the risks which prevail in their sectors. This will assist the work of safety expert at the time of diagnosing such risks.

4 FINAL CONSIDERATIONS

4.1 *Conclusions*

The ability of mankind to assess risks intuitively and manage them has been critical to the survival of humanity and its evolution (Jardine et al., 2003) (Moffett, 2012), but there are many risks that are difficult to perceive and easily understand (Yamashita, 2009). Therefore, holistic research on the various risks of a work activity is a major contribution to the full physical health of the worker.

It can promote measures for improving the minimization of possible risks. From a global perspective, it can be shown that the most frequent work related risks are chemical, ergonomic, biological and physical accidents. In this sense, this paper aims to conduct a systematic review of recent publications that address these risks, which can then be applied to the occupational or environmental fields. In order to do this this, we used the PRISMA Statement as a methodological tool to guide the research, and scientific journal portals available in MetaLib® for research publications 1169 studies were initially obtained from the survey of search portals, but only 3 were eligible in that they contained the five environmental risks that are the focus of this article.

Moreover, of these three articles, we identified one which is a review of the literature, which therefore makes it impossible to characterize the risks in the empirical study. This means that in only two of the studies was a characterization of the risks made. In view of the small number of eligible publications identified amongst the large number found, one can discern a trend in the literature which focuses on individual risks and their consequences. On the other hand, there is a gap in the literature of studies which address the risks as a whole, such as multidisciplinary studies. Another important factor in the two works found is the lack of exemplification of potential causative agents of accidents, which raises doubts about what risks exist within the work activity. Finally, the articles are more concerned with detailing the process of productive labor and quantifying the number of workers at risk.

4.2 *Limitations of the study and opportunities for future work*

The basis of the study is the search for articles in scientific journals through checking the keywords listed in the title of the publication. However, this method proved to be a limiting factor, and should be addressed in future studies. For instance, new searches could be made through academic databases and search engines, as well as screening the abstracts as well as the publication titles. Based on these results, the fact that the eligible works were all undertaken in Brazil, does not preclude the existence of other work relevant to the topic, which were not selected. Therefore, it is advisable not to use only Brazilian search portals for a new survey of the publications.

REFERENCES

Alencar, J.R.B. 2005. Riscos ocupacionais na fabricação de medicamentos: análise de uma indústria localizada no Nordeste brasileiro. *Revista Brasileira de Saúde Ocupacional,* 32, 49–67.

Aneziris, O.N., Papazoglou, I.A. & Doudakmani, O. 2010. Assessment of occupational risks in an aluminium processing industry. *International Journal of Industrial Ergonomics,* 40, 321–329.

Bohm, G. & Pfister, H. 2005. Consequences, morality, and time in environmental risk evaluation. *Journal of Risk Research,* 8, 461–479.

Brasil 1994. Norma Regulamentadora 9. Brasil: Ministério do Trabalho e Emprego.

Bunner, C. & Larson, E. 2012. Assessing the quality of the peer review process: Author and editorial board member perspectives. *American Journal of Infection Control,* 40, 701–704.

Chen, J., Chen, S. & Landry, P.F. 2013. Migration, environmental hazards, and health outcomes in China. *Social Science & Medicine,* 80, 85–95.

Froneberg, B. 2006. National and International Response to Occupational Hazards in the Healthcare Sector. *Annals of the New York Academy of Sciences,* 1076, 607–614.

Jardine, C., Hrudey, S., Shortreed, J., Craig, L., Krewski, D., Furgal, C. & Mccoll, S. 2003. Risk Management Frameworks for Human Health and Environmental Risks. *Journal of Toxicology and Environmental Health, Part B,* 6, 569–718.

Jayaprakash, K. 2003. Acquired Methaemoglobinemia (Met Hb) in Goldsmiths—A Hitherto Unobserved Occupational Hazard. *Indian Association of Occupational Health,* 7.

Moffett, D.B. 2012. Environmental Hazards and Risk Assessment in Primary Care. *American Journal of Lifestyle Medicine,* 6, 342–346.

Moher, D., Liberati, A., Tetzlaff, J., Altmane, D.G. & Group, T.P. 2010. Preferred reporting items for systematic reviews and meta-analyses: The PRISMA statement. *International Journal of Surgery,* 8.

Raybould, A. & Cooper, I. 2005. Tiered tests to assess the environmental risk of fitness changes in hybrids between transgenic crops and wild relatives: the example of virus resistant Brassica napus. *Environmental Biosafety Research,* 4, 127–140.

Semple, S. 2005. Occupational Medicine. *Assessing Occupational and Environmental Exposure,* 55, 419–424.

Silva, M.K.D. & Zeitoune, R.C.G. 2009. Riscos ocupacionais em um setor de hemodiálise na perspectiva dos trabalhadores da equipe de enfermagem. *Escola Anna Nery,* 13, 279–286.

Ulbig, E., Hertel, R.F. & Böl, G.F. 2010. Evaluation of Communication on the Differences between "Risk" and "Hazard". *In:* OFFICE, F.I.F.R.A.P. (ed.). Berlin.

Urrútia, G. & Bonfill, X. 2010. Declaración PRISMA: una propuesta para mejorar la publicación de revisiones sistemáticas y metaanálisis. *Medicina clínica,* 135, 507–511.

Volquind, D., Bagatini, A., Monteiro, G.M.C., Londero, J.R. & Benvenutti, G.D. 2013. Occupational Hazards and Diseases Related to the Practice of Anesthesiology. *Revista Brasileira de Anestesiologia,* 63, 227–232.

Yamashita, H. 2009. Making invisible risks visible: Education, environmental risk information and coastal development. *Ocean & Coastal Management,* 52, 327–335.

Youssef, M.A., Omer, A.A., Ibrahim, M.S., Ali, M.H. & Cawlfield, J.D. 2011. Geotechnical investigation of sewage wastewater disposal sites and use of GIS land use maps to assess environmental hazards: Sohag, upper Egypt. *Arabian Journal of Geosciences* 4, 719–733.

Zhonghua, Y. & Ling, H. 2013. An investigate MetaLib usage by Chinese students in Loughborough University. *Journal of Theoretical and Applied Information Technology,* 48, 1403–1409.

Occupational Safety and Hygiene II – Arezes et al. (eds)
ISBN 978-1-138-00144-2

Prevalence of WMSD in the sewing sector of two companies of the footwear industry

L. Afonso & M.E. Pinho
Faculty of Engineering, University of Porto, Porto, Portugal

P.M. Arezes
University of Minho, Minho, Portugal

ABSTRACT: The few available studies indicate a high prevalence of WMSD in footwear industry, particularly in the sewing sector. Since there is a lack of studies on this issue, especially in Portugal, this paper aims to investigate the prevalence of WMSD among workers of the sewing sector in two companies of the footwear industry. Observations in the workplace were carried out and, further, a questionnaire survey was performed. The results show that wrists/hands (42%), neck (32%), lower-back (30%) and shoulders (23%) were the most affected by musculoskeletal symptoms, although neck and back have been the only to show statistically significant differences between both companies. The study has contributed to the awareness of the prevalence of WMSD in the sewing sector of the footwear industry and point out to the need of more studies focusing either on the prevalence of symptoms or on risk factors for WMSD in this sector.

1 INTRODUCTION

In recent decades, there have been profound changes in the working conditions, which are associated with the economic globalization and the technological innovation (Uva *et al.*, 2008). These changes show visible consequences on the workers' health, as well as in their productivity (Meerding *et al.*, 2005). Among those consequences, Work-Related Musculoskeletal Disorders (WMSD) have become the most common form of occupational diseases worldwide, either in developed countries or in developing countries (Mody & Brooks, 2012).

According to the European Survey on New and Emerging Risks, carried out in 2009 by the European Agency for Safety and Health at Work (2009), WMSD were identified as one of the main preoccupations in what concerns Occupational Health and Safety in the European companies. Besides the associated suffering, they are responsible for high socio-economic costs, both for the governments of each country and for the society in general (Brooks, 2006). In the European Union member states, the percentage of lost days at work seems to be high in several fields of the economic activity. According to Health and Safety Executive (2012), around 61% of people, whose main health problem related to work is a musculoskeletal one, were on sick leave in 2007. In the United States, the annual costs related to the compensation for WMSD were estimated at around 45 to 50 billion dollars (Denis *et al.*, 2008). In Portugal, although the occupational disease is mentioned in the occupational diseases list and despite their notification is mandatory, the data about the prevalence of WMSD is scarce, especially in some specific economic activities, such as the footwear industry. Partly, this may be due to the difficult to establish a clear and precise diagnosis of the pathology (Cunha-Miranda *et al.*, 2010), which very likely may lead to a underestimation of its prevalence in the working population.

Due to the scarce information about the WMSD in the footwear industry, this paper aims (1) to assess and compare the prevalence of the Musculoskeletal Symptoms (MSS) in the workers of the sewing sector of two Portuguese companies of the footwear industry, (2) to identify some work related risk factors and (3) to highlight the activity performed in both companies. In this context, the study of the MSS arises as a predictor of subsequent musculoskeletal disorders in the studied population, therefore constituting an important step towards the WMSD prevention (Smith *et al.*, 2009).

2 MATERIALS AND METHODS

2.1 *Subjects*

The studied population consists of 130 sewing sector workers of the two footwear industry companies (A and B) from Felgueiras municipality. The participants' selection was based on the following inclusion criterion: individuals who had been performing duties in the company for at least one year, who were in the company on the day in which the survey was carried out and who agreed to participate in it. All the identified individuals who did not fulfil the inclusion criterion, as well as those presenting MSS that were not related to the job, were excluded. The final sample consisted of 66 females (34 from company A and 32 from company B). All the participants were informed about the objectives of the study and that all the information collected would be treated as strictly confidential and anonymous.

2.2 *Methods*

The instruments used for the data collection were a guide for the ergonomic analysis of workstations, published by the Finnish Institute of Occupational Health (FIOH), translated and adapted into Portuguese by the University of Minho (Gomes da Costa, 2004) and a questionnaire to gather information on the MSS reported by the workers, which was built, partly based on three validated questionnaires: Nordic Musculoskeletal Questionnaire in its translated and validated version for the Portuguese population (Mesquita *et al.*, 2010) to evaluate health aspects; Dutch Musculoskeletal Questionnaire—extended version (Hildebrandt *et al.*, 2001), to evaluate socio-demographic and work aspects and Copenhagen Psychosocial Questionnaire—medium size version (Kristensen *et al.*, 2005), to evaluate psychosocial aspects.

Work-related MSS were evaluated in terms of pain perception or discomfort of the respondents, whenever they were present in the last 12 months for at least a week (Sluiter *et al.*, 2001), in the following body regions: neck, thoracic region, lumbar region, hips/thighs, knees, ankles/feet, shoulders, elbows and wrists/hands. All the participants with symptoms in a particular region answered the question about the limitations in the performance of normal activities during the last 12 months due to the reported symptoms. An ordinal scale (Mesquita *et al.*, 2010) for the classification of the pain intensity in the different body regions (scale from 0 to 10, in which 0 represents the absence of pain and 10 represents the maximum pain) was included. The participants have also answered a question about the absence from work due to the reported symptoms and when they were worsened.

SPSS® v.21.0 software was used in data statistical analysis. The Chi-Square independence test, the t-Student and Mann-Whitney test were performed to look for statistically significant differences between the two companies regarding to risk factors of the activity under study. The Relative Risk (RR) was performed whenever possible to do so for all the statistically significant relationships. A significance level of 5% was used.

2.3 *Procedures*

In December 2012 the activity of sewing workers was characterized through direct observation and the support of the FIOH ergonomic analysis guide.

Subsequently, in order to make sure that the questions translated from English into Portuguese were easy to understand, a pre-test of the constructed questionnaire was conducted in 10 of the study participants. Then, during January 2013, the questionnaire was applied to all the selected workers by means of a structured interview. The duration of each interview was approximately 8 minutes.

3 RESULTS AND DISCUSSION

3.1 *Characterisation of the sample*

An overall response rate of 52.3% (80% in company A and 38% in company B) was reached.

The fact that the population under study was exclusively composed by women may have had significantly exacerbated the prevalence of the reported MSS (Aghili *et al.*, 2012). On the contrary, Nag *et al.*, (2010) did not find any significant differences in the prevalence of the MSS between genders.

The ages of the participants are between 22 and 55 years old and the workers' age average is 39 ± 8 years old in company A and 38 ± 8 years old in company B. Analysing the ages of the participants, it was observed that in both companies the average age was similar, which suggests a population of almost middle-aged individuals. It should be noted that the prevalence of MSS tends to increase with age in the sewing sector (Aghili *et al.*, 2012; Wang *et al.*, 2007).

The participants show an average body mass index slightly higher in company A (26 ± 4 kg/m^2) than in company B (25 ± 4 kg/m^2). However, this difference is not statistically significant. Some authors point out that the individuals, particularly female, with anthropometric characteristics that are far from the average population, tend to have greater probability of developing WMSD (Uva *et al.*, 2008).

In what concerns education, in both companies the workers did not study beyond the 9th grade. According to Ozturk & Esin (2011), the prevalence of symptoms was more common in low-educated workers. Besides, no statistically significant differences were found between companies

A and B for any of the studied socio-demographic characteristics.

In what concerns labour activity, two variables distinguish, in a statistically significant manner in the two companies: time in the company in years ($p = 0.028$) and break time in min/day ($p < 0.001$). The probability of the workers having less than 5 years seniority in the company B (RR = 1.754; IC 95%: 1.050–2.929) is almost twice that of the workers of company A. This result suggests that company B may be more recent than company A.

On the other hand, the probability of the workers of company A taking breaks shorter than 10 min (RR = 3.889; IC 95%: 1.862–8.124) is almost four times that of the workers of company B, where a longer break can be observed (>10 min). According to Uva *et al.*, (2008), regular rest periods ease the work overload and allow the muscles to recover from the effort associated to an intense work pace. In this context, insufficient breaks during work in company A may have contributed to the development of the reported MSS.

3.2 *Prevalence of musculoskeletal symptoms*

Table 1 shows that 76% of the workers reported symptoms (79% in company A and 72% in company B). These results are similar to those obtained by other authors such as Nag *et al.*, (2010), which reported a prevalence of 79% for the weavers, while an higher prevalence value (83%) was found for office workers by Parimalam *et al.*, (2007). However, in the sewing sector, Wang *et al.*, (2007) and Roquelaure *et al.*, (2004) registered relatively lower prevalences (58% and 39%, respectively).

It should be noted that the influence probability of the confounding variables (physical exercise, heavy household chores, other working and extra-working activities), which could interfere in the prevalence of the reported symptoms were taken into account. The short time spent in physical activities by workers of both companies may be one of the risk factors for MSS, though Brooks (2006) considers that it is just a prevention factor and not a WMSD risk factor. Besides, the workers with MSS not related with the job were excluded of the study.

Table 1. Characterization of musculoskeletal reported symptoms by body region.

MSS n (%)	Overall 66 (100)	Company A 34 (51)	Company B 32 (49)
MSS 12 months	50 (76)	27 (79)	23 (72)
Neck	21 (32)	16 (47)	5 (16)
Shoulders	15 (23)	11 (32)	4 (13)
Elbows	14 (21)	9 (26)	5 (16)
Wrists/hands	18 (42)	13 (38)	15 (47)
Thoracic region	4 (6)	4 (12)	0 (0)
Lumbar region	20 (30)	16 (47)	4 (13)
Hips/thighs	6 (9)	2 (6)	4 (13)
Knees	14 (21)	10 (29)	4 (13)
Ankles/feet	10 (15)	5 (15)	5 (16)

Among the workers who reported symptoms, the most affected body regions are wrists/hands (42%), neck (32%), lumbar region (30%) and shoulders (23%). In company A, the most prevalent regions are neck and the lumbar region (47%), wrists/hands (38%) and shoulders (32%). On the other hand, in company B only the wrists/hands have a high prevalence (47%). Similarly to the current study, Kalınkara *et al.*, (2012) referred the back (30.5%), neck (27.0%) and shoulders (15.6%) as the most prevalent regions. In turn, Ozturk & Esin (2011) registered relatively higher prevalence rates, and the most affected regions were torso (62.5%), neck (50.5%), shoulders (50.2%) and the lumbar region (23.9%) of sewing workers. Still, some studies reported significantly lower prevalence rates. For instance, Warnakulasuriya *et al.*, (2012) registered lower prevalence rates for the lumbar region (19%), wrists/hand (16.8%), elbows (15.7%), shoulders (15%), knees (15%) and neck (14%).

The comparison between the results of the two companies shows that the prevalence of the MSS is higher for the upper limbs, and for company A in most of the regions of the body, except for the wrists/hands, hips/thighs and ankles/feet in which the prevalence was higher for company B. Nonetheless, statistically significant differences between the two companies were only found for neck ($p = 0.006$) and the lumbar region ($p = 0.002$). Thus, the risk of the workers of company A reporting neck symptoms is almost twice (RR = 1.905; IC 95%: 1.239–2.929) that of the workers of company B. Similar conclusion was made in what concerns the risk of symptoms in the lumbar region (RR = 2.044; IC 95%: 1.341–3.117).

Taking into account the results of the current study, the high prevalence of MSS in some of the regions of the body may be explained by the way the workers of the two companies do their jobs. Direct observation of the work performance showed that workers spend their entire journey in a sitting position, and adopt extreme positions of the back, neck and shoulders for long periods of time, because the work is demanding and requires a lot of concentration. These positions are referred by several studies as responsible for the high prevalence of MSS in those regions (Zhang *et al.*, 2011). Moreover, workers were sitting in seats without backs or wheels, which might worsen their adopted posture, causing them to lean forward in a more pronounced way, aggravating the symptoms (Todd *et al.*, 2008). This situation is reinforced by the perception of 37% of the workers, who consider that the reported symptoms are due to the fact that they work for long periods of time in a sitting position. The probability of the workers attributing their perceived pain or discomfort to the long working periods in a sitting position is higher in company A ($p = 0.029$; RR = 1.737; IC 95%: 1.075–2.806) than in company B. On the other hand, in both companies, the symptoms in wrists/hands may be connected with having to perform highly repetitive movements (Wang *et al.*, 2007). This may be supported by the fact that more than half of the workers of both companies reported repetitive movements as the most frequent cause for their symptoms. In fact, it was observed that the work cycles are short (0.5–5 min) and that the work is performed with hands placed at a high level (above elbow height). Conversely, insufficient breaks in both companies (Wang *et al.*, 2007) as well as the use of non-ergonomic scissors to cut off rows surplus, may also have contributed to the high prevalence of symptoms (Ozturk & Esin, 2011; Todd *et al.*, 2008).

WMSD are often associated to high levels of in-capacity due to the pain. In the current study, the maximum intensity (10) of pain was perceived for most of the regions in which the prevalence is high, namely the lumbar region for company A and the wrists/hands for company B. However, no significant differences were found in the reported pain intensity between both companies. Lower levels of pain intensity were found by other authors (Ozturk & Esin, 2011; Wang *et al.*, 2010), although caution is required, since the method of registration of the pain intensity used in those studies was different which may compromise the comparison of results. However, Wang *et al.*, (2007) referred that the prevalence of pain, moderate to severe, for the shoulders and neck was of 24% while for the distal end it was of 16%. These results can be explained by the insufficient breaks, as previously mentioned. The high values found for the pain intensity may also, in a certain way, be explained by the low education level of the workers of both companies, since Dorner *et al.*, (2011) mentions a relationship between the education level and intensity of perceived pain.

The regions whose symptoms interfere most in the activities carried out by the workers are shoulders (88%), elbows (83%) and wrists/hands (74%), although no statistically significant differences have been found between both companies for any of the body regions. Conflicting data were found by Côté *et al.*, (2008) in a working population in which a greater interference was found for the neck.

Only 9% of the workers were absent from work due to their MSS. Among the workers of company A who were absent from work (12%), half referred the lumbar region symptoms and 2 workers of company B (6%) stated that it was due to the symptoms reported in the elbows and wrists/hands regions. However, no statistically significant differences were found between both companies. The absenteeism rate found seems to be in accordance with the results found by Saidu *et al.*, (2011), in which 7.4% of the participants were absent from work due to the reported symptoms. The low absenteeism rates may, eventually, be due to the difficulty that many health professionals have in the diagnosis of musculoskeletal disorders (Cunha-Miranda *et al.*, 2010).

3.3 *Environmental conditions of the workplace*

The workplace environment condition with more complaints is the existence of extreme temperatures (38%). A statistically significant difference between both companies ($p = 0.049$) was found and the probability of workers reporting the exposure to extreme temperatures is higher in company B (RR = 1.640; IC 95%: 1.012–2.657) than in company A. Extreme temperatures may be related to the development of WMSD (Gold *et al.*, 2009), however, extreme temperatures were not observed in both companies.

Natural light was predominant in company B when compared to company A. This situation may explain the higher trunk flexion of company A workers to improve their vision, which can be the cause of the reported neck and lumbar regions MSS (Parimalam *et al.*, 2007).

3.4 *Organisational and psychosocial conditions of the workplace*

The most reported risk factors are the low recognition by managers (64%), the difficulty to express feelings and work opinions (59%) and the very fast work pace (56%). Comparing both companies, the same situation was observed although only for the very fast work pace a statistically significant difference ($p = 0.003$) has been found. The chance of company B participants reporting the perception of working very fast (RR = 2.351; IC 95%: 1.245–4.440) is more than twice the registered in company A. Also Wang *et al.*, (2010) found an association between the prevalence of MSS in neck and shoulders regions and the intense work pace in sewing workers.

The difficulty to express work opinions had also been referred as a risk factor for the development of MSS. In fact, Wang *et al.*, (2005) observed a statistically significant connection between the low control over work and the prevalence of MSS in the neck and shoulders of the sewing workers.

In what concerns the low recognition by the managers, Van den Heuvel *et al.*, (2005) reported a connection between the carpal tunnel syndrome and the low social support by the supervisors, in office workers. On the other hand, Bongers *et al.*, (2002) state that a high social support from colleagues and managers play an important role in the injury recovery.

4 CONCLUSIONS

This study demonstrated a high prevalence of MSS among the sewing sector workers, mainly in the upper part of the body. The workers from company A were the most affected by the symptoms. The pain intensity was also very high, but only 9% of the workers were absent from work due to the MSS.

More studies focusing on the specific working conditions of the footwear industry and the potential risk factors for the development of WMSD are needed.

ACKNOWLEDGMENTS

The authors would like to thank Master in Occupational Safety and Hygiene Engineering (MESHO), of the Faculty of Engineering of the University of Porto (FEUP), all the support in the development and international dissemination of this work.

REFERENCES

Aghili, M., *et al.* (2012). Evaluation of musculoskeletal disorders in sewing machine operators of a shoe manufacturing factory in iran. *International Journal of Industrial Ergonomics, 62*(3).

Bongers, P.M., *et al.* (2002). Are psychosocial factors, risk factors for symptoms and signs of the shoulder, elbow, or hand/wrist?: A review of the epidemiological literature. *American Journal of Industrial Medicine, 41*(5), 315–342. doi: 10.1002/ajim.10050.

Brooks, P.M. (2006). The burden of musculoskeletal disease: A global perspective. *Clinical Rheumatology, 25*(6), 778–781. doi: 10.1007/s10067-006-0240-3.

Côté, P., *et al.* (2008). The burden and determinants of neck pain in workers. *European Spine Journal, 17*, 60–74.

Cunha-Miranda, L., *et al.* (2010). Prevalence of rheumatic occupational diseases: Proud study. *Ata Reumatológica Portuguesa, 35*(2), 215–226.

Denis, D., *et al.* (2008). Intervention practices in musculoskeletal disorder prevention: A critical literature review. *Applied Ergonomics, 39*(1), 1–14.

Dorner, T.E., *et al.* (2011). The impact of socio-economic status on pain and the perception of disability due to pain. *European Journal of Pain, 15*(1), 103–109.

EU-OSHA. (2009). Novos riscos emergentes para segurança e saúde no trabalho (pp. 24). Luxemburgo: Agência Europeia para a Segurança e Saúde no Trabalho.

Gold, J., *et al.* (2009). Skin temperature in the dorsal hand of office workers and severity of upper extremity musculoskeletal disorders. *International Archives of Occupational and Environmental Health, 82*(10), 1281–1292. doi: 10.1007/s00420-009-0450-5.

Gomes da Costa, L. (2004). *Guia para a análise ergonómica de postos de trabalho*. Universidade do Minho.

Hildebrandt, V., *et al.* (2001). Dutch Musculoskeletal Questionnaire: Description and basic qualities. *Ergonomics, 44*(12), 1038–1055.

HSE. (2012). Annual Statistics Report (pp. 23). London: Health and Safety Executive.

Kalınkara, V., *et al.* (2012). Anthropometric measurements related to the workplace design for female workers employed in the textiles sector in Denizli, Turkey. *Eurasian journal of anthropology, 2*(2), 102–111.

Kristensen, T.S., *et al.* (2005). The Copenhagen Psychosocial Questionnaire-a tool for the assessment and improvement of the psychosocial work environment. *Scandinavian journal of work, environment & health*, 438–449.

Meerding, W.J., *et al.* (2005). Health problems lead to considerable productivity loss at work among workers with high physical load jobs. *Journal of Clinical Epidemiology, 58*(5), 517–523. doi: 10.1016/j.jclinepi.2004.06.016.

Mesquita, C., *et al.* (2010). Portuguese version of the standardized Nordic musculoskeletal questionnaire: Cross cultural and reliability. *Journal of Public Health, 18*(5), 461–466. doi: 10.1007/s10389-010-0331-0.

Mody, G.M. & Brooks, P.M. (2012). Improving musculoskeletal health: Global issues. *Best Practice & Research in Clinical Rheumatology, 26*(2), 237–249. doi: 10.1016/j.berh.2012.03.002.

Nag, A., *et al.* (2010). Gender differences, work Stressors and musculoskeletal disorders in weaving industries. *Industrial Health, 48*(3), 339–348.

Ozturk, N. & Esin, M.N. (2011). Investigation of musculoskeletal symptoms and ergonomic risk factors among female sewing machine operators in Turkey. *International Journal of Industrial Ergonomics, 41*(6), 585–591. doi: 10.1016/j.ergon.2011.07.001.

Parimalam, P., *et al.* (2007). Knowledge, attitude and practices related to occupational health problems among garment workers in Tamil Nadu, India. *Journal of Occupational Health, 49*(6), 528–534.

Roquelaure, Y., *et al.* (2004). Surveillance program of neck and upper limb musculoskeletal disorders: Assessment over a 4 year period in a large company. *Annals of Occupational Hygiene, 48*(7), 635–642. doi: 10.1093/annhyg/meh054.

Saidu, I.A., *et al.* (2011). Prevalence of musculoskeletal injuries among factory workers in Kano Metropolis, Nigeria. *International Journal of Occupational Safety and Ergonomics, 17*(1), 99–102.

Sluiter, J.K., *et al.* (2001). Criteria document for evaluating the work-relatedness of upper-extremity musculoskeletal disorders. *Scandinavian journal of work, environment & health*, 1–102.

Smith, C.K., *et al.* (2009). Psychosocial factors and shoulder symptom development among workers. *American Journal of Industrial Medicine, 52*(1), 57–68. doi: 10.1002/ajim.20644.
Todd, L., *et al.* (2008). Health survey of workers exposed to mixed solvent and ergonomic hazards in footwear and equipment factory workers in Thailand. *Annals of Occupational Hygiene, 52*(3), 195–205.
Uva, A., *et al.* (2008). Lesões Músculoesqueléticas Relacionadas com o Trabalho—Guia de orientação para a prevenção. *Direcção Geral da Saúde.*
Van den Heuvel, S.G., *et al.* (2005). Psychosocial work characteristics in relation to neck and upper limb symptoms. *Pain, 114*(1), 47–53.
Wang, P.C., *et al.* (2010). Follow-up of neck and shoulder pain among sewing machine operators: The Los Angeles garment study. *American Journal of Industrial Medicine, 53*(4), 352–360. doi: 10.1002/ajim.20790.
Wang, P.C., *et al.* (2007). Work-organisational and personal factors associated with upper body musculoskeletal disorders among sewing machine operators. *Occupational and Environmental Medicine, 64*(12), 806–813. doi: 10.1136/oem.2006.029140.
Wang, P.C., *et al.* (2005). Work organization and work-related musculoskeletal disorders for sewing machine operators in garment industry. *Annals of epidemiology, 15*(8), 655.
Warnakulasuriya, S.S.P., *et al.* (2012). Musculoskeletal pain in four occupational populations in Sri Lanka. *Occupational Medicine-Oxford, 62*(4), 269–272. doi: 10.1093/occmed/kqs057.
Zhang, F., *et al.* (2011). Quantify work load and muscle functional activation patterns in neck-shoulder muscles of female sewing machine operators using surface electromyogram. *Chinese Medical Journal-Beijing, 124*(22), 3731.

Occupational Safety and Hygiene II – Arezes et al. (eds)
© 2014 Taylor & Francis Group, London, ISBN 978-1-138-00144-2

Industrial hygiene and safety of work in a manufacturing industry profiles aluminum

D.F. Santos, T.S. Januario, M.B.G. Santos, S.S. Justino & A.S.R. Mayer
Federal University of Campina Grande, Campina Grande, Paraíba, Brazil

ABSTRACT: Environmental conditions are essential for the effective implementation of labor activities, accordingly, assess the associated agents and scenarios likely to risks and accidents is a major factor in the current conduct of management and production processes. The study aims to determine a diagnosis of simplified environmental work conditions associated with a metal industry specializing in manufacturing aluminum profiles. Therefore, we carried out measurements of the main environmental aspects, namely: noise, temperature and lighting in two different sectors of the enterprise (administrative and anodizing), verifying the quality of the work environment and the existence of risk more evident for workers. Was applied cheklist regarding security conditions in the scope of work, as well as questionnaires about the usability of personal protective equipment. Upon review, it was identified that the sector anodizing has a greater propensity to agents and risks environmental and infers greater susceptibility to accidents, unsafe and unhealthy conditions.

1 INTRODUCTION

Industrial production increasingly requires efficiency and commitment from employees, who in some cases are subjected to intense work routines and comprehensive driving technological processes involving machinery without proper training, inadequate working conditions for the activity the level of working overtime too high. Through this perspective, the worker is subject to develop serious problems in the short and long term, in terms of occupational health and accidents in the work.

Work safety refers to science that operates in the prevention of work accidents related risk factors operating (SALIBA, 2010). The fundamentals of workplace safety often referred to situations where unsafe acts or not aggregated to unsafe conditions originate from accidents that can affect the condition of the labor employed on a temporary or permanent basis. The occupational hygiene aims not only to identify, assess and control risks later on labor activities, but provide alternatives to examine the scope of these risks not only from the aspect of occupational diseases, as well as through the issues inherent to the comfort and well-being in this context.

According to the Brazilian Aluminum Association (Abal), the segment of the metallurgical industry admits relevance in economic and industrial scope, since it has high rates of growth in productivity and market expansion. However, this performance is a significant parallel to unsafe working conditions, often inhumane, subjecting workers to hazards sectoral effective. In addition, there is a lack of knowledge about the actual risks, the lack of effective training, the non-recognition of the need for proper use of Personal Protective Equipment (PPE), the lack of an infrastructure work that enables a routine work more befitting with the activities of each employee.

From the assimilation of the need for a more comprehensive assessment of this segment, this study aimed to conduct a simplified diagnosis of occupational safety and health on the working conditions of a metallurgical industry specializing in manufacturing aluminum profiles, in the city of Campina Grande—PB—Brazil. It emphasized identification of environmental agents inherent in two specific sectors of business, measurement and comparison of these parameters, the provision of checklists on assessed risks and the level of usability of PPE suitable for the job in question, as well as suggestions for combat the discontinuities with Brazilian law.

2 BIBLIOGRAPHIC REVISION

Every work activity infers environmental risks, especially those inherent in the industry. A legal instrument that allows for the promotion of workplace healthier and safer consists Program Risk and Accident Prevention, the PPRA, which is predicted by Norm No. 9 (1978). Establishes the preservation of health and physical integrity of workers, through the anticipation, recognition, evaluation and control of the occurrence of the risks existing or that may be inherent in the work environment (SZABÓ JR, 2012).

Two other regulatory standards are essential for designing security diagnostics and occupational hygiene: NR 15 on Unhealthy Activities and Operations, which defines the activities subject to unsanitary conditions, the limits of tolerance and the evaluation criteria (SZABÓ JR, 2012), and NR 17 which deals with ergonomics, determining the basic fundamentals that can make routines work more comfortable and safe in order to avoid the incidence of occupational diseases.

The aluminum industry, according to Abal (2011) holds a 0.7% of GDP and 3.2% of industrial GDP, also admitted a production of aluminum plates of 507 700 tonnes in year 2011. Thus, the sector is important for the growth of the domestic industry. Allied to this scenario are the working conditions susceptible to environmental risks evident as inadequate acoustic conditions, high temperatures, insufficient lighting and exposure to chemicals.

The acoustic aspects is reflected throughout the body and not only your hearing, loud noise environments with permanent and generate a range of negative effects, such as altered mood, ability to concentrate, inferences on metabolism and risk of cardiovascular problems, and hearing loss often irreversible. Excessive heat is introduced by the activities and the equipment used in the processes, as well as the characteristics of the local environment. Luminous conditions need to consider the environment, dimensions, features, colors and predominant age workers, and can effectively undermine the health of workers, and affect your productivity. According to Silva et al. (2013), a good lighting design promotes wellness for employees and enables an execution of the work more enjoyable.

From this, the study aims to examine the working conditions relating to professionals in the metallurgical industry for the manufacture of aluminum profiles, assessing the impact of the acoustic aspects, thermal and luminance on the work environment, and consequently on the health and integrity worker.

3 METHODOLOGICAL ASPECTS

3.1 *Characterization of the enterprise*

The work was performed in the metallurgical industry working in the production of aluminum and aluminum alloys in primary form and the transformation of aluminum plates for aluminum profiles. Located in Campina Grande—PB—Brazil admits two branches and has 96 employees, also has the construction industry as the main market area. Its production process includes the following steps: receiving and storage of raw materials, transportation of raw materials, mat cutting, extrusion, stretching, oven aging, anodizing, drying, and packaging, storing and shipping.

The research was focused on two sectors of the organization: the administrative sector and the sector of anodizing. The administrative sector works with the coordination of production services and other services related to the administration. It consists of five workers: one administrative assistant, two in charge of production and technical safety. The same is arranged in an area of 19.26 m^2 (4, 63 m × 4, 16 m), has a ceiling height of 3 meters, tile floors, plaster ceiling and lighting is basically artificial.

The sector comprises anodizing the servicesdanodizing of aluminum profiles as a way to ensure maximizing the resistivity of the aluminum-nio as well as providing color to the profiles (matte black, natural or white). This sector has fifteen employees in effective local implementation of the activities, the production hall is built in masonry, with a ceiling height of 8 meters, cement floor, metal structure covered with translucent resin tiles, natural and artificial lighting and ventilation natural.

3.2 *Characterization of research*

Aimed at meeting the research objectives were addressed four main methodological aspects: literature, quantification of agents acoustic, thermal and luminance sectors evaluated, qualitative identification of environmental risks through the application of check sheets safety inspection (cheklist) and perceived level of usability of PPE by workers' parties of both sectors.

Regarding the literature was used as a query parameter to primary and secondary sources from books recognized the area, articles and scientific and technical journals, as well as sites related theme. In addition, the literature search was combined with field research, in order to provide data and results of more concise.

Regarding the quantitative approach, we used experimental procedures through the use of the device thermo—hydro—decibel—lux meter, referring to Instrutherm brand, model THDL-400. This instrument consists of a device that implements planned measurements aspects inherent noise (decibel) levels of illuminance (lux) and temperature (°C). The same was properly calibrated. Measurements relating to noise levels and temperature were carried out in the cycle of 8 hours of work runs, corresponding to a shift of production, in order that the operation of such a process requires specification. The intervals between the measured noise and temperature both sectors accounted for 30 minutes between each measurement. Regarding the examination of aspects luminance, measurements took place in the morning and afternoon and 1 hour intervals were between medictions.

The measurement procedure was basically similar for both sectors. The measurement was carried out involving noise as follows: there is a unit near the ear of the worker checked the result, it was noted the measured and evaluated in accordance with Annex 1 of the NR 15. The temperature was measured at the time of employment of employees within the cycles previously specified, and for reasons of operational limitations was not feasible to measure the index of thermal sensation. Therefore, we used only the ambient temperature as a parameter evaluating the results by NR 17. Compared to lighting, set up four points effective measurement (p1, p2, p3 and p4), where the measurements obtained were analyzed according to NBR 5382 (1985) under the premise of regular field work with luminaries centralized, effective the average luminance of the sectors in focus. To quantify the ideal luminance were used three factors: age, speed and precision and reflectance of the background. Established the sum of the scores found, considering the signal based on NBR 5413 (1992). After the measurements was structured a brief analysis across sectors in order to identify the highest incidence of the acoustic aspects, thermal and luminance in each workplace.

Settled yet check sheets safety inspection (checklists) as a way to evaluate the work environment in the aspect of unsafe acts and conditions, based on the specifications contained in regulatory standards No. 9 (Programme for the Prevention of Accidents and Hazards—PPRA), No. 15 (Unhealthy Activities and Operations) and No. 17 (Ergonomics). The aspects evaluated in check sheets were PPE, buildings, facilities and services in electricity, transport and storage of materials, machinery and equipment, fire protection and safety signs.

With regard to the indices usability of Personal Protective Equipment (PPE), applied a questionnaire divided into three main blocks of questions, the first correspondent identification of accidents in routine labor of the worker, the awareness of the use and the importance of applying them, as well as the verification of the availability of such PPE for the organization in question. The second phase of evaluation includes verification of workers who use these devices effectively, the existence of training to optimize the use of these identification and discomfort regarding use. The last segment evaluation covers the frequency of use by workers evaluated. We applied a total of 20 questionnaires between administrative sectors and anodizing.

4 ANALYSES OF RESULTS

4.1 *Quantification and analysis of risk agents*

The quantitative approach allowed the identification and measurement of noise levels, thermal and luminance company evaluated. Tables 1 and 2 show the results for the measurements corresponding to each sector evaluated.

Regarding the measurement of noise levels, held readings in one shift production, and using as parameter the Annex 1 of the NR 15, which sets the maximum permissible tolerance limit for a working day of 8 hours the noise level of 85 dB, it is inferred that the noise level in both sectors is evaluated according coma s standard specifications.

Table 1. Quantification of risk agents—administrative sector.

Risk agent	Maximum value	Minimum value	Average value	Standard	Standardized
Noise	65,6 dB	39,7 dB	56,1 dB	85 dB	NR 15
Temperature	24,9 °C	21,7 °C	23,1 °C	20 °C e 23 °C	NR 17
Lighting	275 lux	215 lux	247,75 lux	500 lux	NBR 5413

Table 2. Quantification of risk agents—anodizing industry.

Risk agent	Maximum value	Minimum value	Average value	Standard	Standardized
Noise	83,8 dB	60,1 dB	73,3 dB	85 dB	NR 15
Temperature	28,6 °C	22,1 °C	25,2 °C	20 °C e 23 °C	NR 17
Lighting	287 lux	185 lux	238,1 lux	500 lux	NBR 5413

Thus, it appears that the noise pattern identified in routine both sectors provides favorable conditions for work and admits acoustic comfort, not focusing unsanitary. Note that the sector holds adonizaçao noise levels very close to the limit of tolerance provided in Rule, a fact that shows the highest incidence of acoustic aspects in the work environment. The exposure time of workers to approximate indexes can also, over time, contribute to possible occupational illnesses, injuries mainly in the hearing aid.

In appearance temperature, the administrative sector keeps track of such a variable through the use of an air-conditioning enabling environment holds the ideal temperature for thermal comfort. Identified an average temperature of 23.1 °C. According to NR-17, specifically subsection 17.5.2., the index temperature must be between 20 ° C and 23 °C. Thus, it can be inferred that the administrative environment admits thermal conditions consistent with the activities, allows certain level of comfort to the workers and not offering unhealthy conditions. Regarding the anodizing industry average temperature was 25.2 °C, above the level specified in the standard, due to the characteristics of the work environment and the type of activity performed. Thus, it appears that the industry anodizing submit their workers to unfavorable thermal conditions, which cause discomfort and discomfort which can cause possible scenarios unsanitary work and occupational diseases.

Regarding the lighting aspect of the management sector, the artificial luminance of the same is made by means of a recessed luminaries type trough with two fluorescent lamps. Despite the existence of a window, the scene lacks the natural lighting, as it relates to another area of the company. The fact that the colors of the walls and roof are clear raises the level of reflectance environment. The anodizing industry has mainly natural and artificial lighting due to wide doors of the shed; moreover, still holds light walls. Both sectors have inadequate levels of luminance for the activities performed, according to NBR 5413. Such evidence notes the effectiveness of an unhealthy work environment that can cause illnesses over the years, depending on the time of exposure to this inadequacy, and requires greater effort and attention of workers in the conduct of its activities.

4.2 *Analysis of check sheets safety inspection (checklist) on environmental conditions*

Regarding the analysis of the checklist applied it was found that there is a requirement for use of PPE in most sectors of the enterprise, except for the administrative sector. The company offers the same as provided in NR 6, but does not evaluate the adequacy of the routine work of each sector. The maintenance and cleaning of PPE are characterized as regular, but not effectively in this type of practice. The company's buildings are in good condition, as well as stairs and ramps. There is no evidence of cracks or leaks and drains accessible or broken glass panes.

In covering electrical installations both the company and the machinery and equipment are adequate to NR 10. All electrical design is grounded and isolated so that there is no accident involving liquid. The sockets are protected, all wiring is built and the tools used for this purpose services are electrically isolated. As for the transport and storage of cargoes, materials are stocked respecting the distance of 0.50 m (standard) and forklifts meet the specifications as smooth running, maintenance and use, according to NR 11.

The machinery meets the minimum distance spacing of 0.60 m to 0.80 m and there are specific markings for hallways and storage areas, as NR 12. The drive devices can be connected and disconnected by the sector worker and others in an emergency. In relation to flammable fuels and both are properly allocated by the specifications of NR 20, moreover, there is the danger identifying both the containers as the storage location.

On the issue of fire safety, it was found that runners and outputs meet maximum width of 1.20 m and which is the direction of the external ports. The firefighting equipment are accessible to workers, and fire extinguishers are within 1.60 m above floor level, according to NR 23. The signs are well visualized, following the NR 26, in size and suitable locations.

4.3 *Usability of personal protective equipment*

It was found that 20% of the total estimated indicated the occurrence of any event or accident inherent in the employment context, which are inherent to the sector anodizing. In its entirety, the workers of both sectors said they consider important to use and agree to apply them in the performance of work procedures. Such a scenario infers a relative change in the outlook of such workers on the effective contribution that PPE bring work routines, so it appears that there is a process of awareness of employees regarding the usability and efficiency of PPE.

Another aspect evaluated was the realization of the provision of personal protective equipment appropriate for each function. It was identified that the main PPE used in the workplace are: air-purifying respirators, masks, safety helmets, gloves, boots and earmuffs clamshell or insertion. Only the sector anodizing necessarily requires the

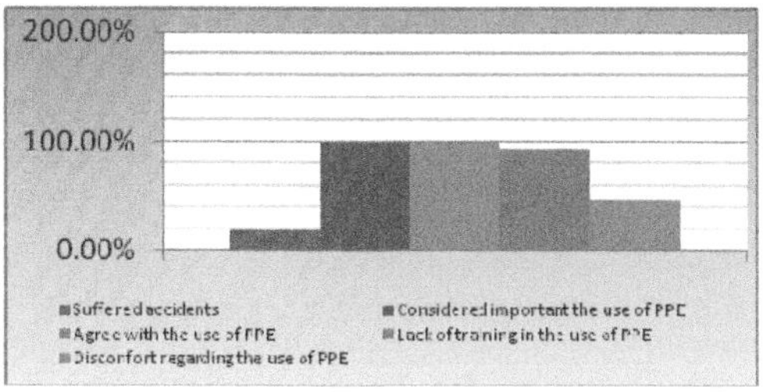

Figure 1. Identification of the usability of PPE.

use of PPE, but you cannot exempt the administrative sector of occurrences arising from the lack of this type of equipment.

As the anodization process emits gases and works directly with chemicals such tools are essential to worker safety. There is effectively the distribution of PPE among employees, but there is no study on the suitability of these instruments the activities, so a total of 86.67% found the PPE employees as appropriate to its operations.

It was found that 86.67% of workers admit analyzed the use of PPE, a percentage considerably, given the rates of risks relating to the sector. Although most workers admit that uses the IPE, 93.33% of them said they had not received any specific training on the use of such equipment, which represents a significant percentage, since improper use can cause damage effectively dangerous health and integrity of the worker, as well as enhance the action of environmental agents inherent to the sector.

Another aspect considered was the identification of discomfort in regard to the use of PPE. The results were very relevant, where almost half of the workers in the industry admits that the use of such equipment is a nuisance in the execution of their work activity. The discomfort indicated refers primarily to the heat generated by the use of respirators, masks, gloves and helmets, the tightness that some PPE offer, both in relation to the head as the ear of the worker. This finding of discomfort influences the conduct of work and employee performance. A possibility of easing this discomfort labor could be the addition of proper training in the use of PPE and an educational policy of the company that aims to educate the essentiality of the use of such equipment. Although all workers affirm the importance and willingness to use the PPE, we need to encourage, teach, ratify these instruments exist for the work routine more enjoyable and safe, and may prevent substantial damage to workers' health.

The last segment of evaluation regarding the use of PPE covers the frequency of use by workers evaluated. Given the results, it was found that the frequency of use is acceptable, about 60% of employees say always use the PPE, but by the environmental risks inherent in the industry anodizing, it was expected that the frequency of use was more effective. Approximately 26.67% admit to using PPE's sometimes during routine work, justified response, according to the workers, by the rush to operationalize the production process or by forgetting them. Only 13.33% have no frequency of use, the main reason of this fact is related to the discomfort associated with the use of PPE, where even some of them aware of the risks to which they are subjected, not choose to continue using the equipment that can prevent accidents or more serious illnesses future.

5 CONCLUSION

From the analysis of environmental risks, it can be concluded that for both sectors of the company, relating to noise ratios are appropriate to the work routine and the norm, it is only necessary to show the industry that infers anodizing levels close to the limit comfort, being more effective control measures. Regarding the temperature, the rate of thermal discomfort admitted sectors, highlighting the need for more accurate assessments. Luminous conditions are below standard specified in the standard, for the two sectors evaluated, that is dangerous especially regarding the handling of chemicals by workers in the sector anodizing. It would be essential to a new lighting plan based on the specifications of the standard.

The conditions for the check sheet identified a company concerned with the working conditions offered to its employees by providing the minimum requirements in the standard. Regarding the usability of PPE showed the importance of the awareness of the use, but still lacks the training to promote the proper use of them.

Thus, workers in the anodizing are more prone to hazards and environmental agents, especially as regards to chemicals, excessive heat routine and inadequate lighting for driving activity. As the values of the agents evaluated in this sector are at higher levels or even above the limits of the standard, leading to unsanitary. The administrative sector is more effective when it comes to issues of health and safety, and submits their workers to less risk and less unhealthy conditions.

REFERENCES

Brazilian Association of aluminum. Abal, 2011.

Brazilian association of technical standards. Nbr 5413—interior luminance. Rio de Janeiro: Abnt, 1992.

Brazilian association of technical standards. Nbr 5382—Verification of interior luminance. Rio de Janeiro: Abnt, 1985.

Ministry of Labour, and Employment of Brazil. List of national regulatory standards (6, 9, 10, 11, 12, 15, 17, 20 and 23). Official Gazette of Brazil, Brasilia.

Saliba, T.M. Basic course in safety and occupational hygiene. 3rd Ed São Paulo: LTr, 2010.

Silva, J.M.N., et al. Proposal for lighting design of a metallurgical based on the determination of the level of average illuminance. In: Occupational Safety and Hygiene, 2013. p. 369–370.

Szabó A.M Jr,. Manual of Hygiene and Occupational Medicine/Adalberto Júnior Mohai Szabó. 4. ed. current. —Sao Paulo: Rideel, 2012.

Occupational Safety and Hygiene II – Arezes et al. (eds)
© 2014 Taylor & Francis Group, London, ISBN 978-1-138-00144-2

Ergonomic method for the implementation of occupational safety systems

M. Butlewski & M. Sławińska
Poznan University of Technology, Chair of Ergonomics and Quality Management, Poznań, Poland

ABSTRACT: The implementation of solutions that improve occupational safety is an extremely important task, not only because of the social aspect of occupational safety but also because of the direct and indirect costs carried by the lack of safety. Therefore, it is important to ensure that the efficiency of safety systems is as large as possible, and this is directly related to the so-called human factor. The following article has characterized factors responsible for the efficiency of safety systems from the ergonomics standpoint and on the basis of an analysis of literature regarding the topic has proposed a method for evaluating the effectiveness of safety systems.

1 INTRODUCTION

Occupational safety systems play a key role in shaping the policies of each company. The safety of employees and property is a good, which when lacking can result in significant direct costs and immense indirect costs. Hence, it is important that the implemented systems and safety devices perform their role and achieve high efficiency, which largely depends on the human factor. Therefore, ergonomic design and system diagnostics can be very effective tools in this respect where the point of reference is the human factor. Man is considered to be the weakest part of the system, but one has to remember that it is in him/her that most of the responsibility lies.

An ergonomic approach creates opportunities for analysis and design of any situation in which humans act, as its primary function is to strive for maximum efficiency of man with minimal human cost (Górny 2012, Butlewski & Tytyk 2008). This is because the purpose of the ergonomic design of the human-technical object system is to ensure that a worker has optimal operating conditions in which he can make full use of his skills and professional qualifications. The ergonomic design of work systems puts the emphasis on the organization of elements and relationships in such a way that the planned method of work, such as at the work position of industrial processes diagnosis, is the result of a thorough knowledge of human relations with the environment and with the technical knowledge of the possible states of the system, for which an employee's tasks are planned. The choice of the operator's action mode should also take into account a range of ergonomic factors including features that determine the efficiency of a particular service provider, such as age and health.

Safety does not only depend on methods of work of a relatively simple ergonomic design (Tytyk 2001), but also on production procedures aimed at increased resistance to interference (Mrugalska & Kawecka-Endler 2012). Studies also show that the third generation ergonomics called macroergonomics (Pacholski 2000, Jasiak & Misztal 2004), which is associated with quality management processes taking place in modern companies (Gołaś & Mazur 2008), may allow for the achievement of significant improvements in safety by improving performance and obtaining a "safety culture" (Kleiner 1999). The need for a methodical-ergonomic approach results from the fact that by modifying the workplace, in order to improve the safety conditions, some tools are used, which at the same time change the current functioning conditions and along with the positive effect of increased safety within a specified dimension may inadvertently have a negative impact on work safety (usually in another area or dimension). It is therefore necessary to create appropriate strategies to use technical objects, in which the assessment of the final effects will refer to the human factor, and not stop just at the efficiency of the device. This strategy, however, must be consistent with other dimensions of business excellence such as maintenance, quality, environment, and logistics (Jasiulewicz-Kaczmarek 2012) and also have financial justification, because an investment in safety also has to pay off (López-Alonso 2013). It is therefore necessary to apply a methodological approach to the problem of safety that can be effectively based on ergonomic principles and can achieve maximum effects.

2 BACKGROUND

2.1 *The need for safety systems and devices*

The growing competition and safety requirements lead to the elimination of any inconsistency, and any errors. This requirement applies to both the characteristics of the products quality as well as the safety of workers. Depending on the type of industry and approved level of considerations, the analyzed factors may be: rolling direction, normalized bending radius and punch to die clearance, in case of low alloy steel by typical manufacturing processes (Vatter 2013), as well as behavior and attitudes of employees so as to influence individuals and groups toward safety (Bowen 2013).

Another factor that heightens the need for complex safety systems is increasingly complex technical objects, such as hydroelectric or nuclear power plants, whose range of destruction justifies deep safety analysis while taking into account all risk factors (Ren 2011). Safety systems in the case of large industrial facilities have been successfully functioning for years. Operators are provided with continuous information about critical states of the observed factors on the basis of mathematical algorithms. (Critical Function Monitoring System) (Harmon 1983). Systems of this type are characteristic of strategic objects, but it seems reasonable that some of them can also be used for less spectacular projects, because their effects are certainly tangible, as evidenced by the annual number of accidents at work in Poland—94 000, of which more than 400 result in death within six months after the accident (Partnership for prevention 2012). Even more dramatic evidence of this need is the number of fatal road accidents, which makes it necessary to systematically analyze the circumstances in which we make mistakes that are risking lives (Watson 2003).

The need for implementation of safety systems is reflected by studies which have shown that complex modern production systems, which include all aspects of functioning of an enterprise, often do not fulfill the most common requirements of the safety standards (Mattila et al. 1996). The increase in complexity of system design also results in the need to focus on safety issues and to regard them in a systematic way (Butlewski 2012, Sławińska 2008).

2.2 *Characteristics of safety systems and devices*

Safety systems are significant because of their potential, which is represented by different types of resources, capacities and capabilities (Lozownicka-Stupnicka 2000): (1) potential for structure integration VS (t), dependent on the quality and intensity of the relation between elements of the system (due to the purposes of its operation)—potential for system creation, (2) resources of components of VEL(t) system (3) human potential VL (t), (4) biotic potential of ecosystems VB (t), (5) informative potential VJ (t), (6) technical potential VTE (t), (7) energy potential VEN (t), (8) substantial potential VM (t), (9) regulatory, control potential VR (t), (10) economic potential VE (t), (11) time resources VT (t), and others. The overall capacity of the system or object is a function of all of them.

The introduction into a system of a new element during the operation of the technical object is associated with a risk-prone situation of safety failure. This is due to the fact that in the work system, the level of safety depends on its relative dynamics, which consist of the characteristics of a temporary flow of risks and a corresponding defensive action flow of the system (referred to as adaptive capacity). That is why there is a need to monitor the state of the system in terms of the abovementioned resources which by changing their values may change the dynamic flow of the system thereby reducing safety.

2.3 *The causes of failure and the human factor in safety systems*

Safety systems have a tendency to fail, which becomes the subject of a detailed analysis and examination procedure to eliminate such danger in the future. In most situations the direct cause of failure may be damage to the system, its exposure to unexpected load, or fault of the employee.

Despite the great variety of technical events that cause failures, it should be noted that it is the human who is the weakest element of the technical—social system and he is the most unreliable element. Depending on the type of industry, the human factor is recognized as the cause of up to 80% of failures. For the aerospace industry, the human factor is at fault in over 75% of the accidents out of which 70% (overall 52.5%) are operators' mistakes—of the cabin crew (Wiegmann & Shappell 1999). However, according to an American study human error in aviation is between 21% and 67%, where the lower limit was determined for a group of minor accidents (Rash et al. 2006). Diversity of the results also arises from the fact that the concept of human error is very wide and can include both obvious situations like exceeded safety limit, which is estimated to 43.5% of all safety system failures (Saurin et al. 2005), as well as human errors in the design of systems, which is a category that adds additional percentages against the human factor.

The proportion of the human causes in system failures is dependent on many factors that one cannot control, but much of it can be identified and reduced. Modeling of safety led to the identification of four major sources of determining the reduction in the number of accidents (Guastello 1989): (1) ambient danger and hazard levels, (2) variables that impact upon human performance capability, (3) initial accident rates, and (4) the mathematical function that interrelates those variables. Among them—2, is an ergonomic super-category covering the general determinants of human efficiency. A more specific categorization of human causes of accidents can be the categorization performed during an analysis of plane crashes, where the most common was (Xiaoli 2004): air traffic control supervising and control, unsafe crew behavior and its preconditions, confusing air traffic control command—the primary cause of crew decision-making error, mishear and forgetting or misremembering—the dominant form of crew perception error, lack of crew resource management skills, especially, the lack of crew cooperation and effective crosscheck. These categories, with the exception of names, are universal for all jobs where the safety of the whole system is dependent on the operator's actions.

Despite such a significant unreliability it is the human—operator that is treated as a superordinate constituent, having a supervisory function, which should be the last to function effectively. It is therefore apparent that the most unreliable part of the safety system is, in the case of operator systems, the final chance to ensure safety. It also shows the importance of our topic, because even small progress in improving the function of the operator influences the safety of the entire system.

3 ERGONOMIC FEATURES OF SAFETY SYSTEMS

The standard ergonomic features of safety systems are the operator's environment as well as the characteristics of control and signaling elements. Early on it was noted that in the case of strategically important safety systems it is essential to ensure that the console environment is designed to minimize fatigue, eyestrain, and discomfort by optimizing light fixtures, as well as minimizing noises made by fans or footsteps and harmonizing colors and brightness throughout the control room (Takeda et al. 1985). However, the work environment itself is not enough, so procedures were created that allowed to make decisions in high stress situations in an ergonomic way. An example of such an approach are 'soft' procedures that are used in the nuclear industry (Reynes & Beltranda 1990).

The system of operating technical objects, which includes the prevention of states of safety threats, takes up ergonomic activities in the following aspects: (1) analysis and documentation of operational (or analogical) technical system; (2) identification of important factors relating to mental overload in the method of work; (3) modeling of exposure time to ergonomic factors of automated work processes; (4) simulation of the situational context in use of technical facilities. In addition, apart from the above mentioned steps in safety management of a work system, it is necessary to obtain empirical data about situational conditions of task overload, and thus determine when and why the job requirements exceeded or approached the limit of human potential. This need arises from the fact that regardless of the kinds of elements with which a man creates relationships, in ergonomic design there is one and the same problem—the need to assess the level of environmental—technical adaptation to psychophysical capabilities of man. While searching for ergonomic solutions to ensure reduction in the level of human failure, one should have the data that will allow to undertake design actions.

4 METHOD OF ERGONOMIC ANALYSIS OF SAFETY SYSTEMS

The method of ergonomic analysis of work done by an operator in the safety system is based on the method of reconfiguration of the chain of events (Sławińska 2011), where ergonomic knowledge resources are obtained, and they support ergonomic design of processes using technical objects. This results in a pragmatic model of the use of technical means, that is presented in the language of the decision-maker of the given fragment of the modeled reality (Legutko 1999). The proposed method consists of 10 stages, preceded by a definition of the goals of the proceedings. The particular stages have been characterized below:

Stage 1. Recognition of the nature of cooperation of the employee with the technical environment, and as a result producing data on the technological and organizational factors of the work system. At this stage, an analysis of photographs from a working day is to be conducted. The analysis ought to be performed in terms of relationships in the chain of events and the determining possible deficits in system resources.

Stage 2. Identification of potentially dangerous events, including an estimate of their effects (e.g., analysis of the signals unnoticed when operator is busy, etc.). This analysis may be conducted on the basis of rules in an occupational risk assessment, where factors such as the risk scale and the probability of a given situation arising will need to be taken into consideration.

Stage 3. Identification of ergonomic factors, which may affect the efficiency of the operator. At this stage it is essential to consider factors that depend on the operator's specific features (age, sex, character) and environment (specific states of environmental elements).

Stage 4. Search for possible solutions related to the operator—technical object system. Depending on the estimated level of load—an analysis aimed at finding technical elements that reduce the requirements set by the system on the operator or at a creative search for methods of support of the limited psychophysical capacity of a human. In search of these, some heuristic techniques can be used (Butlewski 2013), since in theory, this research area is unlimited. It can refer to any work system environment.

Stage 5. Defining expectations of the planned method of work. The fifth stage may be joined with the fourth step, when the assumptions will not correspond to feasible solutions.

Stage 6. Determination of requirements of the planned information system, for example, a specific course of action during interaction with the information system under extreme conditions. At this stage, one must make the assessment of the maximum load and determine the maximum border potential (VS_{max}), as well as a minimal border potential of the system (VS_{min}). At the same time, appropriate indicators and measures of the level of load are applied.

Stage 7. Simulation of the working conditions of the system and testing of the synergistic effect—in various courses of tasks done by the employee in extreme, difficult, and typical situations in order to determine the operation of the system in the context of a difficult situation. Particularly important for the evaluation of the ergonomics of an interaction process is an experiment involving transitions between states of the system (e.g., from normal to emergency).

Stage 8. Determination of the ergonomically optimal system situation for the initially adopted specific design criteria.

Stage 9. Estimating the cost of adverse effects of human behavior and aiming to design conditions conducive to adequate behavior. The purpose of this step is to ensure a low level of stress, with the coinciding large responsibility for the work process.

Stage 10. Recognition of the efficiency of ergonomic reengineering, thus the determination whether the carried out procedure showed adequate effects in relation to the costs.

The proposed procedure should be repeated with every change in the safety system because any change can cause unpredictable consequences. For an example of the proposed method see Table 1.

Table 1. Example of method implementation.

Stage No.	Method implementation effects and examples of safety system development
1	Relationship of operator with the technical environment—e.g., determination of when the operator's behavior affects the security of the system
2	Risk assessment of the system—e.g., determination of consequences of a certain reaction by an anonymous system operator
3	Ergonomic hazards identification—determination of factors reducing the efficiency of the safety system due to the physical and psychological well-being of a particular operator—e.g., hidden symptoms that prevent a quick response
4	Determination of technical ways to limit the unreliability of the human factor—e.g., use of automatic recognition of critical situations
5	Determination of ways to limit the unreliability of the human factor by methods of work—e.g., use of additional breaks
6	List of streams of information along with an indication of their maximum intensity—e.g., number of information provided to the operator to which he can react
7	Event simulation scenarios and determination of the minimum potential at which the safety system maintains efficiency—e.g., number of operators required
8	Defined situation of ergonomically optimal load on the operator—e.g., list of exclusions—lack of quick application startup/restart
9	Cost analysis of human error in the safety system—e.g., lack of response to an emergency signal
10	Comparison of the costs incurred due to changes with the potential costs set out in stage 9

5 CONCLUSION

Ergonomic actions may be the way to obtain greater safety, especially in the case of operator systems. However, each technical tool that ensures safety must be supported by organizational factors, to achieve a sufficiently high level of safety. Elements which are very important in the workplace constitute error management climate and safety communication, which significantly improve safety at work (Casey & Krauss 2013).

Despite the differences in percentage, studies shows a significant, several-tens percent share of the human factor in the number of caused accidents and failures, which means that it is highly justified to perform actions in order to reduce this level.

Because safety is attributable to time and space, in which the system itself exists, ergonomic redesigns of operational processes of technical objects is a way to address specific system features. The proposed framework of ergonomic application can provide the appropriate level of safety in an operational system involving the optimal investment in exploitation. In this structure ergonomic requirements play an important role in determining the choice of specified solutions of safety devices. A lack of compliance with ergonomic operating criteria or a lack of the introduction of ergonomic modifications decides on the lack of adequate use of the system or its withdrawal from service.

REFERENCES

Bowen, M., The problem with people: The complex nature of human behavior, (2013) Society of Petroleum Engineers—SPE Americas E and P Health, Safety, Safety, and Environmental Conference 2013, pp. 546–550.

Butlewski, M., The issue of product safety in contemporary design. in: Safety of the system, Technical, organizational and human work safety determinants. Red. Szymon Salamon. Wyd. PCzęst. Częstochowa 2012. ISBN 978-83-63500-13-9, ISSN 1428-1600.

Butlewski, M., Tytyk, E., The method of matching ergonomic non-powered hand tools to maintenance tasks for the handicapped. 2nd International Conference on Applied Human Factors and Ergonomics, Las Vegas, Nevada, USA, 2008; Conference Proceedings, Edited by Waldemar Karwowski and Gavriel Salvendy (CD ROM).

Butlewski, M., Heuristic methods aiding ergonomic design, (2013) Lecture Notes in Computer Science (including subseries Lecture Notes in Artificial Intelligence and Lecture Notes in Bioinformatics), 8009 LNCS (PART 1), pp. 13–20.

Casey, T.W., Krauss, A.D., The role of effective error management practices in increasing miners' safety performance, (2013) Safety Science, 60, pp. 131–141.

Gołaś, H., Mazur, A., Macroergonomic aspects of a quality management system, [in:] Macroergonomic paradigms of Management, Jasiak A. [red.] to honour 40 years scientific activity of Professor Leszek Pacholski, Poznan University of Technology Editorial Board, 2008, pp. 161–170.

Górny, A. (2012), Ergonomics in the formation of work condition quality. Work: A Journal of Prevention, Assessment and Rehabilitation, 41, pp. 1708–1711.

Guastello, S.J., Catastrophe modeling of the accident process: Evaluation of an accident reduction program using the Occupational Hazards Survey, (1989) Accident Analysis and Prevention, 21 (1), pp. 61–77.

Harmon, D.L., Critical function monitoring system algorithm development., (1983) IEEE Transactions on Nuclear Science, NS-31 (1), pp. 862–867.

Jasiak, A., Misztal, A., Macroergonomics and maroergonomic design, Publishing house Politechniki Poznańskiej, Poznań 2004, ISBN 83-7143-471-5. (in polish).

Jasiulewicz-Kaczmarek, M., (2012), Socio-technical integrity in maintenance activities, In: Vink P. (eds.): Advances in Social and Organizational Factors, CRC Press 2012, pp. 582–592.

Kleiner, B.M., Macroergonomic analysis and design for improved safety and quality performance., (1999) International journal of occupational safety and ergonomics: JOSE, 5 (2), pp. 217–245.

Legutko, S., Fundamentals of machine operation, Publishing house Politechnika Poznańska, Poznan (1999). (in polish).

López-Alonso, M., Ibarrondo-Dávila, M.P., Rubio-Gámez, M.C., Munoz, T.G. The impact of health and safety investment on construction company costs, (2013) Safety Sciencc, 60, pp. 151–159.

Łozownicka-Stupnicka, T., Risk assessment and risk in complex systems man—Technical object—Environment, Series: "Sanitary and Water Engineering", Politechnika Krakowska, Kraków 2000.

Mattila, M., Perälä, M., Vannas, V., Flexible manufacturing systems, compliance to the safety standards, (1996) International Journal of Advanced Manufacturing Technology, 12 (1), pp. 60–65.

Mrugalska, B., Kawecka-Endler, A. Practical application of product design method robust to disturbances, Human Factors and Ergonomics in Manufacturing and Service Industries.—2012, Vol. 22, Iss. 2, pp. 121–129.

Pacholski, L., Macroergonomic paradox of entrepreneurship and economic renewal, in: Ergonomics for the new millennium, Human Factors and Ergonomics Society, pp. 185–188, Santa Monica—San Diego CA, 2000.

Partnership for prevention [Internet.] 2012. Warszawa: CIOP; [cited 2013 July 24]. http://www.ciop.pl/28802.html.

Rash, C.E., LeDuc, P.A., Manning, S.D., Human Factors in U.S. Military Unmanned Aerial Vehicle Accidents, Advances in Human Performance and Cognitive Engineering Research, Volume 7, 2006, pp. 117–131.

Ren, Q.-W., Theory and methods of high arch dam's entire failure under disaster conditions, (2011) Gongcheng Lixue/Engineering Mechanics, 28 (SUPPL. 2), pp. 85–96.

Reynes, L., Beltranda, G., A computerized control room to improve nuclear power plant operation and safety, (1990) Nuclear Safety, 31 (4), pp. 504–513.

Saurin, T.A., Formoso, C.T., Cambraia, F.B., Analysis of a safety planning and control model from the human error perspective, (2005) Engineering, Construction and Architectural Management, 12 (3), pp. 283–298.

Sławińska, M., Ergonomics of automated systems, Publishing house Politechniki Poznańskiej, (2008), (in polish).

Sławińska, M., Ergonomic reengineering of automated exploitation processes of technological equipment (ZUT), Dissertations No. 462, Publishing house Politechniki Poznańskiej, (2011), (in polish).

Takeda, S., Akiyama, A., Katoh, T., Kudo, K., Man-machine interface of TRISTAN., (1985) IEEE Transactions on Nuclear Science, NS-32 (5), 1985 p.

Thornton, C.H., Failure statistics categorized by cause and generic class., (1985) pp. 14–23.

Tytyk, E., Projektowanie ergonomiczne. PWN, Warszawa-Poznań, 2001.

Vatter, P.H., Hildering, S., Tsoupis, I., Merklein, M., Development of a damage prediction system for bending and cutting of high strength steels, (2013) Key Engineering Materials, 554–557, pp. 2479–2486.

Watson, G.S., Papelis, Y.E., Chen, L.D., Transportation safety research applications utilizing high-fidelity driving simulation, (2003) Advances in Transport, 14, pp. 193–202.

Wiegmann, D.A., Shappell, S.A., Human error and crew resource management failures in naval aviation mishaps: A review of U.S. Naval safety center data, 1990–96, (1999) Aviation Space and Environmental Medicine, 70 (12), pp. 1147–1151.

Xiaoli, L., Classified statistical report on 152 flight incidents of less than separation standard occurred in china civil aviation during 1990–2003, (2004) Progress in Safety Science and Technology V. 4: Proceedings of the 2004 International Symposium on Safety Science and Technology, (A), pp. 166–171.

Occupational Safety and Hygiene II – Arezes et al. (eds)
 ISBN 978-1-138-00144-2

A review of effects from exposure to static magnetic fields in magnetic resonance imaging

V.M. Silva
Programa Doutoral em Segurança e Saúde Ocupacionais, Faculdade de Engenharia da Universidade do Porto (FEUP), Porto, Portugal
Unidade de Ressonância Magnética, Serviço de Radiologia—Centro Hospitalar São João, EPE (CHSJ), Portugal
Área Técnico-Científica de Ciências Morfológicas, Escola Superior de Tecnologia da Saúde do Porto (ESTSP), Instituto Politécnico do Porto, Porto, Portugal

I.M. Ramos
Faculdade de Medicina da Universidade do Porto, Porto, Portugal

ABSTRACT: Nowadays, magnetic resonance imaging is widely accepted, worldwide. Every day, all human beings are exposed to earth magnetic field, but this field is not perceptible. With the introduction of magnetic resonance in Medicine, in 1980s, a new trend of exposure to magnetic fields was introduced. Static magnetic field used in magnetic resonance imaging is very strong and bring hazards and effects to workers who operate with them and to patients who do magnetic resonance exam. There are mechanical and biological risks and effects due to static magnetic field in magnetic resonance. Mechanical risks are those who are best known, such as attraction, deflection and torque. Biological effects include nausea, dizziness, vertigo, headaches, head ringing, magnetophosphenes, metallic taste, visual disturbances and cognitive and motor effects (like short-time memory, concentration problems, working memory and attention cognitive domains). It is of vital importance a well knowledge of static magnetic field effects and risks for a better health and safety for both workers and patients.

1 INTRODUCTION

All human beings are constantly exposed to the magnetic field created by earth, which is not usually perceptible by people because it is weak and does not interfere with their lives (Feychting, 2005). Earth magnetic field, measured in Gauss (G) is approximately 0,3 G (or $3*10^{-5}$ Tesla—T) in equator and 0,7 G (or $7*10^{-5}$ T) at the poles. Nowadays, most magnets used in clinical practice have Static Magnetic Field (SMF) strengths above 0,5 T (Crook, 2009).

On 19th century, the introduction of strong electromagnets on certain industries has occurred. On 20th century, ElectroMagnetic Fields (EMF) were introduced in Medicine (1980s) with Magnetic Resonance Imaging (MRI) equipments for better diagnosis and care (Coskun, 2010; Schenck, 2005). A new degree of human exposure to EMF was introduced, emerging new challenges, paradigms, hazards and effects.

In the last decade, many advances on MRI brought new and more sophisticated equipments, allowing better accuracy on diagnosis. With these improvements, SMF strengths become higher, increasing effects associated. In clinical use, 3 T equipments are used, but on investigation field equipments with 7 T SMF strength are used (Cavon, 2007; Crook, 2009; Glover, 2007).

For its operation, MRI requires (Coskun, 2010; Crook, 2009; De Wilde, 2007; Hartwig, 2009; Marshall, 2007; McRobbie, 2012):

1. A SMF with a given field strength (from 0,2 T to 3 T in clinical use);
2. Time-Varying magnetic Field Gradients (TVGF), with a variable intensity and slew rate;
3. RF fields, which works as a transmitter and receiver system.

Each of these fields has associated risks and effects. In this paper, only effects related to SMF will be reported. It is important to distinguish risk from effect. A risk is the possibility, high or low, that somebody may be harmed by the hazard brought, in this case, by MRI equipment. Effects are health alterations from the exposure to a type of source.

Some of health effects are well known and described on literature, but others are less common and not well explored and studied. No technology is absolutely safe, and workers who operate MRI equipments must be aware of that fact.

2 AIM

This paper has as major objective to make a survey of effects and risks associated to SMF used in MRI. SMF international guidelines addressed to MRI will be shown, as well as some ways to prevent some effects.

3 METHODOLOGY

Some search databases were used to detect important and relevant studies, articles and reviews published after 2005: MetaLib of Exlibris, Pubmed/Medline, SpringerLink. A wide range of combined keywords were used, such as: MRI, EMF, occupational exposure, SMF, health effects, risks and safety. Only published studies with relevant conclusions and a good background description were eligible. Brief communications and not published data were excluded. In addition international reports, guidelines and directives from recognized entities, such as: Institute of Electrical and Electronics Engineers (IEEE), European Commission (EC) and International Commision on Non-Ionizing Radiation Protection (ICNIRP) on EMFs and MRI exposures were considered.

4 EFFECTS FROM EXPOSURE TO SMF IN MRI

According to Crook et al., several investigations were developed on how SMFs interact with human tissues. The research has been carried out for over a century (Crook, 2009).

SMF strength used in clinical and research MRI is growing due to better image quality for better accuracy on diagnosis. So, risks and effects associated to SMF due to this fact will increase. SMF in MRI is always switched on and workers must be aware of that.

Risks and of MRI SMF can be classified as: biological and mechanical.

Some of biological risks are already known and included: nausea, dizziness, vertigo, headaches, head ringing, magnetophosphenes, and metallic taste (Chakeres, 2005; Coskun, 2010; Crook, 2009; McRobbie, 2012; Möller, 2008; Schenck, 2005; Wang, 2008). Visual disturbances and cognitive and motor effects (e.g. short-time memory, concentration problems, working memory and attention cognitive domains) are not well studied and need more investigation. For example, some authors consider to exist some cognitive effects related to exposure to SMF (de Vocht, 2006), but other authors consider them absent (Atkinson, 2007; Lepsien, 2012; Schlamann, 2010).

All symptoms are reversible and transitory, not long-term and cease after few minutes after exposure tm SMF or when the subject walks away from near the magnet (principal source of SMF) (Feychting, 2005; McRobbie, 2012; Yamaguchi-Sekino, 2011). These sensory effects are thought to arise as consequence of movement near the SMF, namely by head motion and other rapid movements.

When a SMF is applied to biological tissues, ionic currents are present and a net force is applied to the moving ions. These forces, called magneto hydrodynamic, will interact with liquids existing in human body, such as (Chakeres, 2005; Schenck, 2005):

- Blood vessels: blood is considered as a conductive medium, causing a magneto hydrodynamic effect which produces a voltage across the vessel. This biological effect is nonpathologic and include an elevation of T-wave in ElectroCardioGram (ECG), when occurs the maximal flow rate of blood (ventricular contraction);
- Endolymphatic tissues in inner ear: it is thought to be the source for sensations of vertigo and nausea upon exposure to SMFs.

Several studies were performed to perceive some effects related to exposure to SMF in MRI. In 2012, van Nierop et al. performed a double-blind randomized crossover study on 31 healthy volunteers where they were asked to make shake head movements before every neurocognitive tasks and neurocognitive functioning on attention, concentration and visuospatial orientation were affected (van Nierop, 2012).

In 2007, Glover et al. made an experimental study where volunteers experienced vertigo-like sensations when moving from a SMF to another (Glover, 2007).

Metallic taste was another symptom studied, in 2007, by Cavon et al., where half of volunteers had experienced metallic taste after making movements near a SMF of 7 T (Cavon, 2007).

Workers are exposed to SMFs and they can perceive some of the health effects stated above and others, such as: vertigo, nausea, dizziness, and difficulties on concentration, headaches, metallic taste and sleep disorders. As stated above, all symptoms are transitory and reversible.

In workers, adverse effects due to SMFs are caused by significant electrical currents induced in the head when they move rapidly in a powerful SMF near the entrance of the MRI equipment room or near MRI magnet.

Mechanical risks are well known and, sometimes, are the only effects and risks that workers and patients are aware of. Attraction of ferromagnetic materials towards the magnet is one hazard and can be named by projectile or missile effect and anything or anyone between the path of the object and the magnet can be struck (Fig. 1). There are some documented fatalities, according to Crook et al. Any ferromagnetic object placed near the magnet can experience this force and acceleration. It is proportional to the product of the SMF strength with the TVGF (Crook, 2009; Gowland, 2005; Hartwig, 2009; Marshall, 2007).

The effect of SMF on medical metallic implants and devices can cause deflection or malfunction of the implanted device. This can result on serious damages on patients, causing severe injuries or in fatalities (Crook, 2009; Marshall, 2007; Schenck, 2005). Any ferromagnetic device positioned in a SMF experiences forces which have a tendency to move it in space and torques.

As stated above, implant medical devices can suffer torsion and deflection by influence of a high SMF and the induced RF energy can generate a located energy deposition near the device producing a local temperature elevation (Bassen, 2006; Hartwig, 2009; Marshall, 2007). It is very important to know the nature of the medical device and its components and if those are compatible with a MRI equipments. Standardized tests for testing implantable devices compatibility increase safety of patients and support both the MRI users and devices manufacturers (Schaefers, 2006). With the growing of non-communicable diseases and injuries (Habib, 2010), the number of MRI scans of patients with medical implants will increase (Schaefers, 2006).

SMF is not just confined to the scanning room, where the patient is located. Magnetic field goes beyond operating room, through SMF lines. One aspect to take into account is that fact that the higher SMF strength, the greater will be the scope

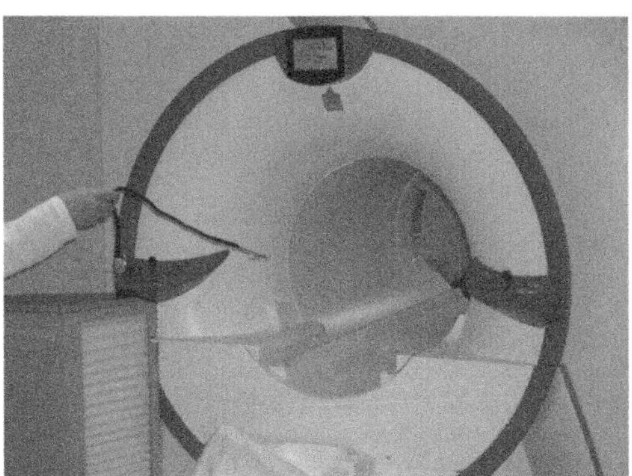

Figure 1. Attraction force caused by a high static magnetic field. A ferromagnetic material (a stethoscope) is attracted toward the magnet.

of SMF lines. An important line to be taken into account is the line for persons with cardiac pacemaker (magnetic flux density (B) > 0,5 mT), up to the distance no longer than about 5 meters from magnet (Karpowicz, 2006). It is of vital importance to label this area to avoid incidents with patients with pacemaker.

Information panels must be located where there SMF may interfere with ferromagnetic devices, indicating the field strength and what kind of objects and medical devices are in risk of incompatibility with MRI equipment. SMF is never off, so it is very important the knowledge of SMF potential effects and hazards.

Workers operating MRI scanners are the main occupational group highest exposed to EMFs in Medicine. There are occupational limits of exposure for the three main sources: i) SMF; ii) TVGF; ii) RF fields. Three international commissions and institutes guidelines and standards were taken into account: IEEE, ICNIRP and International Electrotechnical Commission (IEC)—IEC/EN 60601-2-33 safety standard for MR system.

In 2004, European Commission (EC) established a directive—2004/40/EC—on the minimum health and safety requirements regarding the exposure of workers to the risks arising from physical agents, such as EMFs. This directive brought new paradigms on occupational exposure from MRI environments (European Commission, 2004). Employers must assess occupational EMFs exposure levels and if there are areas where limits may exceed, they have to be well indicated and working procedures may be altered to restricted access, receiving all necessary information about the potential risks and hazards.

ICNIRP recommends that occupational exposure of the head and trunk should not exceed a spatial peak magnetic flux density of 2 T, except in certain applications. When the exposure is restricted to limbs, exposures till 8 T are acceptable (International Commision on Non-Ionizing Radiation Protection, 2009). IEC operates a three-tier system of limits (International Electrotechnical Commission, 2010; McRobbie, 2012). It recommends an exposure limit of 4 T both to trunk and head and limbs. IEEE has lower spatial peak magnetic flux density limits: 0.5 T (The Institute of Electric and Electronics Engineers, 2005) (Table 1).

There are also permissible occupational exposure values to SMF of a whole working day (Table 2).

Some studies were developed to assess exposure of MRI staff to SMF. The majority of those studies were performed after EC launched the 2004/40 Directive.

On those studies, SMF exposure did not exceed legal action values (Bradley, 2007; Fuentes, 2008; Groebner, 2011; Kännälä, 2009). But, certain

Table 1. SMF limits for occupational exposure (all values peak).

International Commission/ Institute	Trunk and head (T): ceiling value	Limbs (T): ceiling value
ICNIRP	2 (for specific work applications exposure up to 8 T can be justified)	8
IEC	4	4
IEEE	0.5 (500 mT)	0.5 (500 mT)

Table 2. Permissible occupational exposure to SMF of a whole working day.

International Commission/Institute	Whole body (action value)	Limbs (action value)
ICNIRP	200 mT	Not fixed
IEC	200 mT	Nor fixed
IEEE	500 mT	500 mT

measures may be taken to reduce to a minimum SMF exposure, trying to avoid the appearance of symptoms associated to SMF.

Another situation that workers must be aware of is pregnancy. Many studies were undertaken to show if pregnancy is not a limitation for MRI. The majority of laboratory studies and investigations showed no evidence of injury during pregnancy and no substantial increase on negative outcomes from different sources of magnetic fields used in MRI, namely SMF (DeWilde, 2005).

Pregnant MRI workers are authorized to work around MRI environments throughout all stages of pregnancy, but only in acceptable and limited actions and zones. In the majority of imaging centers they are not permitted to work around ultra-high SMFs (e.g. 3 T) and to enter in the equipment room, even in lower SMF, except in emergency situations. Given options to pregnant MRI staff should be given when working in these environments (Crook, 2009).

5 CONCLUSIONS

MRI is an imaging technique widely accepted by medical community and it is spreading worldwide, increasing the number of MRI exams performed.

EMFs exposures in MRI environments have potential effects and hazards to patients and workers. These risks and effects are originated from three sources: a) SMF; b) TVGF and c) RF fields. All of these fields have different effects that MRI staff and others professionals must be aware of.

SMFs used in MRI are increasing its magnitude and the frequency of effects associated may increase, such as nausea, vertigo, dizziness, headache, magnetophosphenes, visual disturbances, cognitive functioning, magneto-hydrodynamic forces and others. Hazards, such as induced forces (torque, deflection and projectile effect) have to be taken in care because of MRI interaction with medical devices and implants and metallic objects.

There are international guidelines and standards that regulate the limit and action values of the different sources of SMF that take in care both workers and patients. Standardized safety procedures and a better knowledge on potential risks of MRI increase the safety of patients and workers and should support MRI workers on their clinical practice.

Knowing potential risks and effects of SMF exposure in MRI, workers can have different behaviors:

- Avoid rapid movements of head and body near MRI equipment, which is the main source of SMF;
- Reduce to a minimum the entries in operating room, where MRI equipment is installed;
- Explain to patients and other workers from others departments some of the mechanical risks associated to SMF;
- Knowing the potential movements that induce some symptoms, try to avoid them, adopting other postures and movements;
- Inform patients to walk and lift slowly to avoid symptoms.

So, it is of vital importance a well knowledge of static magnetic field effects and risks for a better health and safety for both workers and patients.

REFERENCES

Atkinson, I.C., Renteria, L., Burd, H., Pliskin, N.H., Thulborn, K.R., (2007). Safety on human MRI at static fields above the FDA 8 T guideline: sodium imaging at 9.4 T does not affect vital signs or cognitive ability. *Journal of Magnetic Resonance Imaging, 26*(1222–1227).

Bassen, H., Kainz, W., Mendoza, G., Kellom, T., (2006). MRI-induced heating of selected thin wire metallic implants—laboratory and computational studies—findings and new questions raised. *Minimally Invasive Therapy, 15*(2), 76–84.

Bradley, J.K., Nyekiova, M., Price, D.P., Lopez, L.D., Crawley, T. (2007). Occupational Exposure to Static and Time-varying Gradient Magnetic Fields in MR Units. *Journal of Magnetic Resonance Imaging, 26*, 1204–1209.

Cavon, I.D., Glover, P.M., Bowtell, R.W., Gowland, P.A., (2007). Tresholds for Perceiving Metallic Taste at High Magnetic Field. *Journal of Magnetic Resonance Imaging, 26*, 1357–1361.

Chakeres, D.W., de Vocht, F., (2005). Static magnetic field effects on human subjects related to magnetic resonance imaging systems. *Progress in Biophysics and Molecular Biology, 87*, 255–265.

Coskun, O. (2010). Magnetic resonance imaging and safety aspects. *Toxicology and Industrial Health, 27*(4), 307–313.

Crook, N., Robinson, L., (2009). A review of the safety implications of magnetic resonance imaging at field strengths of 3 Tesla and above. *Radiography, 15*, 351–356.

de Vocht, F., Stevens, T., van Wendel-de-Joode, B., Engels, H., Kromhout, H., (2006). Acute Neurobehavioral Effects of Exposure to Static Magnetic Fields: Analysis of Exposure-Response Relations. *Journal of Magnetic Resonance Imaging, 23*, 291–297.

De Wilde, J.P., Grainger, D., Price, D.L., Renaud, C., (2007). Magnetic resonance imaging safety issues including an analysis of recorded incidents within the UK. *Progress in Nuclear Magnetic Resonance Spectroscopy, 51*, 37–48.

DeWilde, J.P., Rivers, A.W., Price, D.L., (2005). A review of the current use of magnetic resonance imaging and safety implications for the fetus. *J Progresse Biophys Mol Biol, 87*, 335–353.

European Commission. (2004). Directive 2004/40/EC of the European Parliament and of the Council of 29 April 2004 on the minimum health and safety requirements regarding the exposure of workers to the risks arising from physical agents (electromagnetic fields). *Official Journal of the European Union L159*(1–26).

Feychting, M. (2005). Health effects of static magnetic fields—a review of the epidemiological evidence. *Progress in Biophysics and Molecular Biology, 87*, 241–246.

Fuentes, M.A., Trakic, A., Wilson, S.J., Crozier, S., (2008). Analysis and Measurements of Magnetic Field Exposures for Healthcare Workers in Selected MR Environments. *IEEE Transactions on Biomedical Engineering, 55*(4), 1355–1364.

Glover, P.M., Cavin, I., Qian, W., Bowtell, R., Gowland, P.A., (2007). Magnetic-field-induced vertigo: a theoretical and experimental investigation. *Bioelectromagnetics, 28*(5), 349–361.

Gowland, P.A. (2005). Present and future magnetic resonance sources of exposure to static fields. *Progress in Biophysics and Molecular Biology, 87*, 175–183.

Groebner, J., Umathum, R., Bock, M., Krafft, A., Semmler, W., Rauschenberg, J., (2011). MR safety: simultaneous B0, dθ/dt, and dB/dt measurements on MR-workers up to 7 T. *Magn Reson Mater Phy, 24*, 315–322.

Habib, S.H., Saha, S., (2010). Burden of non-communicable disease: Global overview. *Diabetes & Metabolic Syndrome: Clinical research & Reviews, 4*, 41–47.

Hartwig, V., Giovannetti, G., Vanello, N., Lombardi, M., Landini, L., Simi, S., (2009). Biological Effects and Safety in Magnetic Resonance Imaging: A Review. *International Journal of Environmental Research and Public Health, 6*, 1778–1798.

International Commision on Non-Ionizing Radiation Protection. (2009). ICNIRP Guidelines on Limits of Exposure to Static Magnetic Fields. *Health Physics, 96*(4), 504–514.

International Electrotechnical Commission. (2010). Medical electrical equipment—Part 2-33: Particular requirements for the basic safety and essential performance of magnetic resonance equipment for medical diagnosis (IEC 60601-2-33). Geneva.

Kännälä, S., Toivo, T., Alanko, T., Jokela, K., (2009). Occupational exposure measurements of static and pulsed gradient fields in vicinity of MRI scanners. *Physics in Medicine and Biology, 54*, 2243–2257.

Karpowicz, J. (2006). Health Risk Assessment of Occupational Exposure to a Magnetic Field From Magnetic Resonance Imaging Devices. *International Journal of Occupational Safety and Ergonomics (JOSE), 12*(2), 155–167.

Lepsien, J., Müller, K., von Cramon, D.Y., Möller, H.E., (2012). Investigation of Higher-Order Conginitive Functions During Exposure to a High Static Magnetic Field. *Journal of Magnetic Resonance Imaging, 36*, 835–840.

Marshall, J., Martin, T., Downie, J., Malisza, K., (2007). A Comprehensive Analysis of MRI Research Risks: In Support of Full Disclosure. *Can. J.. Neurol. Sci., 34*, 11–17.

McRobbie, D.W. (2012). Occupational Exposure in MRI. *The British Journal of Radiology, 85*, 293–312.

Möller, H.E., von Cramon, D.Y., (2008). Survey of risks related to static magnetic fields in ultra high MRI. *Rofo, 180*(4), 293–301.

Schaefers, G., Melzer, A., (2006). Testing methods for MR safety and compatibility of medical devices. *Minimally Invasive Therapy, 15*(2), 71–75.

Schenck, J.F. (2005). Physical interactions of static magnetic fields with living tissues. *Prog Biophys Mol Biol., 87*(2–3), 185–204.

Schlamann, M., Voigt, M.A., Maderwald, S., Bitz, A.K., Kraff, O., Ladd, S.C., Ladd, M.E., Forsting, M., Wilhelm, H., (2010). Exposure to High-Field MRI Does Not Affect Conginite Function. *Journal of Magnetic Resonance Imaging, 31*, 1061–1066.

The Institute of Electric and Electronics Engineers. (2005). IEEE standard for safety levels with respect to human exposure to radio frequency electromagnetic fields, 3 kHz to 300 GHz—description (pp. C95.91-199). New York.

van Nierop, L. (2012). Effects of magnetic stray fields from a 7 Tesla MRI scanner on neurocognition: a double-blind randomised crossover study. *Occup Environ Med, 69*(10), 759–766.

Wang, H., Trakic, A., Liu, F., Crozier, S., (2008). Numerical Field Evaluation of Healthcare Workers When Bending Towards High-Field MRI Magnets. *Magnetic Resonance in Medicine, 59*, 410–422.

Yamaguchi-Sekino, S., Sekino, M., Ueno, S., (2011). Biological Effects of Electromagnetic Fields and Recently Updated Safety Guidelines for Strong Magnetic Fields. *Magn Reson Med Sci, 10*(1), 1–10.

Occupational Safety and Hygiene II – Arezes et al. (eds)
© 2014 Taylor & Francis Group, London, ISBN 978-1-138-00144-2

Ergonomic model of hotel service quality for the elderly and people with disabilities

M. Butlewski
Poznan University of Technology, Chair of Ergonomics and Quality Management, Poznan, Poland

J. Jabłońska
Wroclaw University of Technology, The Faculty of Architecture, Wroclaw, Poland

ABSTRACT: The needs of the elderly and people with disabilities (E&D) are slowly starting to be noticed in the tourism sector, probably because of the purchasing power potential of these two increasingly large groups of customers. Despite this improvement, people with disabilities continue to face a number of barriers that prevent them from being satisfied with the use of hotel services, even if these are somewhat tailored for them, because they are more complex (Shaw & Coles 2004) and require a number of subjective factors to be taken into account, which are partly unconscious and hence not articulated. Therefore, the identified criteria have been characterized in the ergonomic quality model of hotel services that takes into account dimensions of different needs (accessibility, informativity, service) along with the formal requirements of ergonomic design.

1 INTRODUCTION

Designing for groups with special needs, such as the elderly and people with disabilities is a postulate tackled for many years, also in scientific literature. Within the framework of this theme, a number of approaches and methodologies have been developed, including universal design, inclusive design, design-for-all, barrier-free design, accessible design (Goldsmith 2000, Meyer et al. 1998, Branowski & Zabłocki 2006). Yet, widely used solutions still only to some extent address the needs of a wide range of users, by limiting to the needs of wheelchair users, not including the less visible but not less important needs (Eichhorn et al. 2007). This trend is particularly evident in the case of the tourism and hotel industry.

The full and conscious adjustment of hotels is important not only from a moral point of view, which assumes equality of all social groups in regard to access to public facilities, but also it is economically justified. Elderly people are becoming a more and more powerful group of consumers who want to enjoy a range of goods which are available only to people without disabilities. In a similar, and sometimes even worse, situation are people with disabilities. They currently constitute the largest group of consumers, which is neglected by the market of tourism services (Burnett 2001). There are 27 million people with disabilities in the European Union, which is a large but untapped potential on the tourist services market (Bizjak et al. 2011). A change of the fixed habits of the elderly, even in the case of such ailments as chronic pain, can have a positive impact on their well-being and health. Studies show that people exposed to such an indisposition, after an increase in tourist activity, changed the perception of pain sensation and to a lesser extent considered it as an aggravating factor (Alizadeh-Fard & Sarpoolaky 2012). Unfortunately, the level of activity, especially of people with disabilities, in Poland is at a very low level—it is estimated that only 1% of them is systematically involved in or does any form of sport or recreation (Bednarczuk 2008), and the reasons for this occurrence must be found in the absence of a suitably shaped environment. Hence, there is a need to develop a set of requirements, which should be satisfied by construction works to meet the needs of the less efficient part of society.

2 BACKGROUND

2.1 *Needs of the elderly and people with disabilities in the field of hotel services*

Aging is associated with changes such as (Duda 2013): decreased ability to adapt to a variety of physical, biological and psychosocial loads, reduced reserves of the organs, organ dysfunction which distort functioning, for example night rest, physiological needs and daily rhythm (Marchewka et al. 2013). Features of personality also change

(Halicka 2004, Steuden 2011). With age the ability to memorize decreases and spatial orientation worsens (Lewandowski 2013). These changes are partly consistent with the restrictions, which are typical for people with disabilities that are best characterized by the ICF disability classification. Within its framework the most important limitations relating to areas have been identified (ICF 2001): (1) communicating—receiving messages (d310–d329) and creating messages (d330–349), (2) moving and manipulating objects (d430–d449), (3) walking and other means of mobility (d450–d469), (4) taking care of oneself (d510–d599). An analysis of literature sources allowed to collect the needs of E&D in the context of hotel services in five categories, which have been referred in Table 1.

People with disabilities and the elderly not only expect a hotel room to be adapted to their needs, but also all other services available to clients such as spa, gym, pool (Navarro & Andreu 2013). It is therefore important that the assumed model of ergonomic quality of hotel services describes all the possible range of services, not just the minimum level that allows only to passively spend the night.

2.2 *Legal requirements for public utilities*

When considering hotels, it should be pointed out that these facilities, which offer among others accommodation services, and commercial and tourist functions (e.g. SPA and Wellness), do not always,

Table 1. Characteristics of the E&D needs in the context of hotel services.

T.	Characteristics of the need	Authors
General	Satisfaction with overall tourist service	Kozak 2001
	Fulfillment of needs	Tian-Cole and Crompton 2003
	Safety—in the case of a fire or other emergency which requires evacuation do not leave people with disabilities as last	Navarro & Andreu 2013
	The possibility to choose a room as other customers, rather than the assignment of several adapted ones	Navarro & Andreu 2013
Access	Physical access	Darcy 1998
	Sensory access	Darcy 1998
	Communication access	Darcy 1998
Informative	Information about the availability of objects at the destination	McKercher et al. 2003; Shaw et al. 2005; Stumbo and Pegg 2005; Yau et al. 2004
	Credibility of rich information	Cavinato and Cuckovich 2002
	Data features: richness and reliability of information, appropriate travel information sources, communication and customer oriented services	Yau et al. 2004
	The use of information regarding the availability of the main news channels (no stigma)	Cheng 2002; Burnett and Bender 2001; Darcy 1998
	Accessible information provision (availability of information for people with different ranges and levels of disability)—in particular for the elderly in terms of the websites	Pühretmair 2007
Service	Experience in serving people with disabilities	Navarro & Andreu 2013
	Empathy	Misztal & Butlewski 2012
	Taking seriously observations and needs	Navarro & Andreu 2013
	Self-service-independence by freeing oneself from the role of an "object" of care and overcoming doubts as well as building one's own confidence	Blichfeldt & Nicolaisen 2011
Other	The need to assist children with disabilities when spending holiday	Kim & Lehto 2013
	The need for cooperation in the decisions concerning the interior and decorations for people who stay for a longer period of time in hotels for seniors	Knight et al. 2010
	The need to limit the amount of available food due to the excessive consumption of the elderly	Reis & Vanhoni 2009
	Elderly people have a problem with urinary incontinence and quite reluctantly ask for any help in this regard	Mitteness 1995

Table 2. Model of ergonomic qualities.

Characteristic of a need	Architectural parameter	Variable parameter
Satisfaction	Safety, hygiene and comfort of a facility	General quality of an architectural project Project quality of object adjustment to different needs
	Standard of service	Thematic training
Fulfillment of needs	Implementation of all the primary and secondary needs at the hotel by every guest*	Appropriate scope and type of the rooms in accordance with their function
	Services	Tailoring an offer to the possibility of service implementation
Safety	Adjustment of general communication routes for the evacuation of all visitors	Properly developed essential elements to ensure safety**
	All senses warning systems	Systems reliability
	Development of procedures and instruction manuals in case of danger of fire	Congruence of instruction to reality Completeness of instructions
	Guides i.c. of danger	Thematic training
Wide selection of services	Adapted rooms in each standard offered by a given hotel or all rooms adapted	Quality of the project to adapt the rooms and bathrooms to different needs**
Physical access	Form and shape of the architectural solutions	Quality of the project to adapt the form and shape of architectural solutions to different needs**
Sensory access	Material and technical solutions	Quality of the project to adapt material and technical solutions to different needs**
Communication access	All senses communication broadcast messages system	Reliability of the systems
Information about the availability of facilities in the destination	Marketing	Completeness of information regarding the adapted object, reported to the public Information access
Accessible & credibility of rich information	Method of presenting an offer	The quality of the presented information regarding the adapted object reported to the public
Information characteristics richness and reliability	Visual identification system that is accessible and comprehensible to all the guests	Quality of the project of visual identification system
	A concierge service	Thematic training
Information on the main information channels	Development of tourism information system	Reliability of the systems
Experience in serving people with disabilities	Training for staff	Thematic training
Empathy & serious consideration	Proper selection of employees by staff	Thematic training
Self-service-independence	Appliances and equipment by guests with smaller disability	Project quality of adjusting equipment, appliances for different needs**
	Workshops for people with disabilities	Thematic training
Support children with disabilities while spending holidays	Rooms, environment and equipment tailored to the needs of children with disabilities	Quality of the project adapted to the rooms, environment and equipment to meet the needs of children with disabilities**
	Qualified staff to work with children with disabilities	Congruence of eligibility to requirements Quality of qualifications
Co-participation in design for people staying longer in hotels	Provide visitors with "corner" of creative work	Project quality adjusted to the "corner" of creative work**
	Collaboration with architects, interior designers, decorators	Willingness to work Effectiveness of cooperation
Nutrition	Adjusting the food court to special cooking	Quality of the project**
	Cooperation of guests and catering staff	Willingness to work Effectiveness of cooperation
Special problems e.g. Urinary incontinence	Training for staff	Flexibility in answer the needs
	Training for staff	Thematic training

*What stated the offer presented to the customer.
**The variable parameter that will be greatly expanded in the following part of the research.

in the light of Polish law, fulfill regulation requirements for public facilities. While often the regulations of hotels and public facilities overlap, due to the regulatory duality inconsistencies are created, leading to situations in which hotels are not properly adjusted to meet the needs of the E&D.

Architectural requirements for public facilities refer mainly to "technical conditions" (OJ No. 75, item. 690) such as horizontal or vertical mobility. They dictate: the use of passenger lifts from ground floor, or for low buildings, at least one entrance, which enable a person with disabilities to have access to all public spaces (e.g. meeting rooms, dining rooms), and each floor, should have a room that meets all sanitary standards. Facilities are also required to provide: customized parking spaces, proper placement of the front door, ramp with landings, elevators, well-engineered stairs and railings.

Requirements regarding hotels (OJ No. 22, item 169) include, a number of adapted rooms, requirements for devices located in public areas (buttons, switches and communicators), properly adapted furniture, and possibilities to operate electronic devices from bed (e.g. the nurse call installation), as well as the use of handles in the bathroom.

In view of the above, it is concluded that the existing administrative regulations in Poland take into account people with disabilities and comprehensively specify information related to public areas. However, it seems that they are fairly general and do not raise many architectural issues that users with limited psychophysical possibilities encounter during the use of hotel service.

3 RESEARCH METHODOLOGY

The investigation aims to develop a model of ergonomic quality with regard to hotel service quality for users with special needs. In order to achieve the aim it was necessary to do a broad research of literature sources concerning the studied group with ICF International Classification of Functionality and attractions available to the clients of hotels. Finally, by analyzing the collected information it was possible to build a model of ergonomic quality regarding hotel services, shown in Table 2. For the purposes of analysis the category of ergonomic quality was used, because it describes the relation of the user, his specific and individual needs, both conscious and unconscious with the surrounding technical environment (Butlewski 2011). Though it is subjective, it may be subject to thorough determination with respect to several of its characteristics through specialized measurements (Branowski et al. 2013).

For the purposes of this study and in order to develop a model, the following concepts can be defined: architectural parameter, variable (architectural) parameter, and the scope of the variable (architectural) parameter. By architectural parameter can be understood a part of a building or its fragment which constitutes a separated entity that contains its own characteristics, regardless of its dimensions, for example handrail, countertop, floor, upholstery. The variable parameter is assumed to be individual for a given item and objectively measurable or describable feature of exterior or interior architecture, interior design, furniture and appliances, such as length or shape of a handrail, the height placement of the upper countertop surface, floor texture, color, upholstery, etc. The scope of the variable is assumed to be an interval in which the parameter can be modified, e.g. height of a step or the presence or absence of a variable, such as nurse call signal: present/not present.

4 ERGONOMIC QUALITY MODEL OF HOTEL SERVICES FOR E&D

In view of the mentioned dimensions, and characteristics of the needs and the rules relating to Polish construction law with the relevant implementing regulations, it is concluded that it is not possible to precisely describe every solution for possible disabilities and limitations. An unlimited number of combinations, situations, and the continuous development of the disease registry, as well as the artistic-engineering character of an architectural entity as a work protected by copyright, is conducive to determine a model of ergonomic quality based on variables defined in ranges. As diagnostic categories accepted are the abovementioned in Table 1 characteristics of needs. The physical strain is assumed to be the skeletal muscular effort: in overcoming architectural barriers, handling and use of equipment and facilities, efficient movement around the site; and for psychological: mental effort that is put into the execution of the above mentioned activities, and any disruptions of this process caused by external factors (Tytyk 2001). One should also note the presence of long or shortwave stress associated with the abovementioned activities and a stay outside the usual place of residence. The model describes ergonomic qualities relating to architectural parameters categorized on the basis of the reported needs.

5 DISCUSSION

The ergonomic quality parameters derived as a result of the conducted analysis, will form the foundation of future research, a survey of the elderly and people with disabilities in regard to their hotel stay expectations. The future assessment may be carried out within the scope of the answers: YES, NO, point or percentage, depending on the type of architectural item being evaluated. The development of an evaluation system and carrying out the tests will be conducted within the framework of the next part of the study. The degree of importance the found ergonomic quality parameters actually have regarding spending a night at a hotel will be able to be determined. The results of the analysis will give hotels knowledge of the actual needs and expectations of the elderly and people with disabilities and what areas they need to improve on.

6 CONCLUSION

Supporting the activity of the elderly and people with disabilities is an action which brings not only positive effects to them, but also to their closer and further environment. The road to this support must be the creation of appropriate conditions that can be obtained through the use of models of ergonomic quality. Understanding the criteria of ergonomic quality of hotel services by policy makers may allow for breaking down both physical an mental barriers on the side of service providers (ignorance, lack of empathy), and also on the side of the recipients, and, therefore, people with disabilities and the elderly whose excessive passivity exposes them to a chronic syndrome, a sense of powerlessness—Seligman's helplessness syndrome (Lee 2012).

REFERENCES

Alizadeh-Fard, S., Sarpoolaky, M.K., The effect of tourism experience on elderly with chronic pain, (2012) Current Issues in Hospitality and Tourism Research and Innovations—Proceedings of the International Hospitality and Tourism Conference, IHTC 2012, pp. 491–495.

Bednarczuk, G., Directions of development in sports organizations for the disabled in Poland, (2008) Postepy Rehabilitacji, 22 (3), pp. 35–42. (in polish).

Bizjak, B., Knežević, M., Cvetrežnik, S., Attitude change towards guests with disabilities. Reflections From Tourism Students, (2011) Annals of Tourism Research, 38 (3), pp. 842–857.

Blichfeldt, B.S., Nicolaisen, J., Disabled travel: Not easy, but doable, (2011) Current Issues in Tourism, 14 (1), pp. 79–102.

Branowski B., Pacholski L., Rychlik M., Zabłocki M., Pohl P., (2013) Studies on a New Concept of 3D Data Integration about Reaches and Forces of a Disabled Person on a Wheelchair (CAD Methods in Car and Market Ergonomics), Human Factors and Ergonomics in Manufacturing & Service Industries, 23 (4), pp. 255–266 (2013).

Branowski B., Zabłocki M., (Creation and contamination of design principles and construction principles in the design for people with disabilities, (in): Ergonomics of the product. Ergonomic principles of products design), 2006, ISBN: 83-7143-238-0 (in polish).

Burnett, J., and H. Bender Baker, 2001 Assessing the Travel-Related Behaviors of the Mobility-Disabled Consumer. Journal of Travel Research 40(1):4–11.

Burnett, J.J., Baker, H.B., Assessing the travel-related behaviors of the mobility-disabled consumer, (2001) Journal of Travel Research, 40 (1), pp. 4–11.

Butlewski M., Ergonomic quality as a means of providing powered hand tools' safety, safety of the system: human—technical object–environment, 2011, Wyd. Politechnika Częstochowska.

Cavinato, J., and M. Cuckovich 1992 Transportation and Tourism for the Disabled: An Assessment. Transportation Journal 31(3):46–53. Cheng 2002.

Darcy S., 1998 Anxiety to Access: Tourism Patterns and Experiences of New South Wales People With a Physical Disability. Sydney: Tourism New South Wales.

Darcy, S., and P. Daruwalla, 1999 The Trouble with Travel: People with Disabilities and Travel. Social Alternatives, 18(1):41–46. Kozak 2001.

Duda K., Aging process (in), Marchewka A., Dąbrowski Z., Żołądź J.A., (2013) Physiology of aging: prevention and rehabilitation, PWN, Warszawa 2013. (in polish).

Eichhorn, V, Miller, G., Michopoulou, E., Buhalis D. (2007) Enabling Disabled Tourists? Accessibility Tourism Information Schemes—Annals of Tourism Research—http://epubs.surrey.ac.uk/1090/.

Goldsmith S. 2000, Universal design, Architectural Press, Oxford-Auckland-Boston.

Halicka M., Life satisfaction of elderly people: a theoretical-empirical study Białystok 2004, (in polish).

International classification of functioning, disability and health (ICF) [Internet]. 2013. WHO; [cited 2013 July 22]. Available from: http://www.who.int/classifications/icf/en.

Internet-based System of legal acts. Ordinance of the Minister of Infrastructure on the technical specifications, which should correspond to the buildings and their position of 12 April 2002. 2002 O.J. (L 75) 690. [cited 2013 Sep 9]. Available from: http://isap.sejm.gov.pl/ (in polish).

Internet-based System of legal acts. Appendix 1 to regulation of the Minister of economy and labour dated 19 August 2004, "requirements for equipment, qualifications of the staff and and the scope of the provided services, including food service for hotels' legal status. 2006 O.J. (L 22) 169. (consolidated version) [cited 2013 Sep 12]. Available from: http://isap.sejm.gov.pl/ (in polish).

Kim, S., Lehto, X.Y, Leisure travel of families of children with disabilities: Motivation and activities, (2013) Tourism Management, 37, pp. 13–24.

Knight, C., Haslam, S.A., Haslam, C., In home or at home? How collective decision making in a new care facility enhances social interaction and wellbeing amongst older adults, (2010) Ageing and Society, 30 (8), pp. 1393–1418.

Lee, B.K., Agarwal, S., Kim, H.J., Influences of travel constraints on the people with disabilities' intention to travel: An application of Seligman's helplessness theory, (2012) Tourism Management, 33 (3), pp. 569–579.

Lewandowski M.H., Central and peripheral nervous system-physiological aging and prevention (in), Marchewka A., Dąbrowski Z., Żołądź J.A., (2013) Physiology of aging: prevention and rehabilitation PWN, Warszawa 2013. (in polish).

Marchewka A., Dąbrowski Z., Żołądź J.A., (2013) Physiology of aging: prevention and rehabilitation, PWN, Warszawa 2013 (in polish).

McKercher, B., Packer T., Yau M., Lam P., (2003) Travel Agents as Facilitators or Inhibitors of Travel: Perceptions of People with Disabilities. Tourism Management 24(4):465–474.

Meyer, Beth, Rogers, Wendy A., Mead, Sherry E., Guidelines for age-inclusive design, (1998) Proceedings of the Human Factors and Ergonomics Society.

Misztal A., Butlewski M., Life improvement at work, Wyd. PP, Poznań 2012, ISBN: 978-83-7775-177-0.

Mitteness, L.S., Barker, J.C., Finlayson, E., Residential managers' experience with urinary incontinence in elderly tenants, (1995) Journal of Applied Gerontology, 14 (4), pp. 408–425.

Navarro, S., Andreu, L., (2013) Value co-creation among hotels and disable customers: a qualitative study of the participants′ view, GIKA Conference 2013.

Pühretmair, F., It's Time to Support Accessible Tourism, ENTER 2007, Research Track CD of the Proceedings, Springer Verlag, Ljubljana, Slovenia (2007).

Reis F., M., Vanhoni R., P. Nutritional quality of the dinner consumed by elderly people staying in a hotel in Balneário Camboriú, Brazil (2009) Revista Chilena de Nutricion, 36 (2), pp. 120–127.

Shaw, G., C. Veitch, Coles T., (2005) Access, Disability, and Tourism: Changing Responses in the United Kingdom. Tourism, Review International 8(3):167–176.

Shaw, G., Coles, T., Disability, holiday making and the tourism industry in the UK: A preliminary survey (2004) Tourism Management, 25 (3), pp. 397–403.

Steuden S. Psychology of aging. PWN. Warszawa 2011 (in polish).

Stumbo, N., Pegg S., (2005) Travellers and Tourists with Disabilities: A Matter of Priorities and Loyalties. Tourism Review International 8(3):195–209. Stumbo and Pegg 2005.

Tian-Cole, S., Crompton J., (2003) A Conceptualization of the Relationship between Service Quality and Visitor Satisfaction, and their Links to Destination Selection. Leisure Studies 22(1):65–80. Yau et al 2004.

Tytyk E., Ergonomic design PWN, Warszawa—Poznań, 2001 (in polish).

Yau, M., McKercher B., Packer T., (2004) Travelling with a Disability—More than an Access Issue. Annals of Tourism Research 31(4):946–960.

Occupational Safety and Hygiene II – Arezes et al. (eds)
© 2014 Taylor & Francis Group, London, ISBN 978-1-138-00144-2

Safety in the work environment of a company in the textile in Campina Grande-PB—Brazil

K.R. Araujo, V.K.O. Almeida, M.B.G. Santos & R.I. Gomes
Universidade Federal de Campina Grande, PB, Brazil

ABSTRACT: Several factors affect the performance and health of workers in the textile sector, highlighting the risks of physical agents, chemical, ergonomic and accidents. This study aims to identify and propose solutions to the risk situations in a clothing industry, emphasizing the luminal, acoustic and thermal insulation. Were conducted noise measurements, illuminance, temperature and relative humidity in two sectors, as well as checklists were applied in the workplace. Through the measurements, were identified nonconformities regarding to the safety signs and the importance and usability or Personal Protective Equipment (PPE). It was concluded that the adequacy aspects related to environmental conditions, through obedience regulatory standards, can contribute to the improvement of unsafe conditions and worker wellbeing, provided they are taken into account the suggestions made. As a result of this deployment, we could contribute in reducing work accidents and occupational diseases.

1 INTRODUCTION

Work safety is a major issue for companies, given that the practice of training and constant monitoring of the regulations governing the conditions of a more secure environment, improves performance and increases worker productivity and product quality.

Investing in job security is not common in companies. Usually companies are reactive and act only when something unpleasant occurs, such as when some employee falls ill due to their exposure to risks in the workplace, therefore, the practice of continuous improvement of safety through prevention is essential for the company.

The main objective of preparing this evaluation project is to analyze if they are nonconforming with legislation luminal aspects, noise and heat of a clothing company grants the city of Campina Grande, Paraiba, Brazil. Whereas, the specific objectives are: to compare the levels of noise, illumination, humidity and temperature, sectors and administrative mounting bags, are in agreement with what is considered the level of comfort, identify whether the proper personal protective equipment are being used by workers, as well as identify the risks, at the work environment, by the use of a checklist.

2 TEXTILE SECTOR

The textile industry is very broad and consists of various stages of production that are interrelated. According to the National Confederation of Industry (CNI), the Brazilian textile sector is divided into 4 groups of processes or production stages, they are: spinning, weaving, processing and garment. The supply chain in this segment can be further divided into: segment producer and supplier of fiber, the manufacturing segment of yarn, woven and knitted fabrics and finished goods segment that encompasses the making.

Taking into account the supply chain activities performed in the segment producer and supplier are strongly supported by automation of the production process, while the segment of finished goods is a large presence of micro and small businesses that are distributed throughout the country, making it so what else employs labor within this sector.

Thus, it is unlikely that during the cooking process can not occur any work accident, or even that it caused some occupational disease of workers in this sector.

2.1 *Agentes of risks*

Regarding the risks present in this sector Santo et al. (2009) points out that among the numerous risks which are subject workers of a textile mention: physical hazards (noise, vibration, illuminance and heat), chemical hazards (dust, hazardous substances and dyes); mechanical risk (cutting equipment accidents and falls); ergonomic risk (poor posture, repetitive movements and physical effort), among others that cause harm to health.

For greater prevention and greater caution in this regard is important to have a better interaction between the company and the worker focusing on knowledge of the aspects that cause occupational diseases, in order to remove them. The worker needs to know to identify, control and eliminate such risks and participate in actions that seek better health at work, they need to both training and full support of the employer.

Environmental agents are elements or substances in the environment that surrounds the human being, that when agents found above the tolerance limits cause damage to people's health. The productive sector is more influenced by physical and chemical agents, while the administrative sectors and human resources, and the influence it can have these same agents can be derived from the productive sector, depending on the location of these two sectors, still has a strong influence of ergonomic risk. For this purpose, in the textile industry has physical risk for the presence of a bad light, noise and heat. Besides, there is chemical risk through the glue residue and its strong aroma.

With regard to lighting, natural light would be the ideal for any environment, but due to the variation of the position of the sun relative to Earth the incidence of light rays also vary. Thus artificial lights are used, which if they generate a poor lighting can cause eye injuries, eyestrain, risk of accidents, strobe effect and consequently the drop in productivity.

As for the noise, is one of the most common physical agents in environments, facilities and labor activities. Generally, noise is commonly defined as an unwanted sound. The noise requires special attention due to its impact on the human body. According to Saliba (2010, p.191), "the noise contributes to gastrointestinal disorders and disorders related to the nervous system," as acceleration pulse and elevated blood pressure, as well as unwanted effects to the hearing aid. Examples of these effects undesirable: rupture of the eardrum, from exposure to noise from explosions or violent impact, and deafness resulting from exposure to high levels of noise.

The heat is present in whatever working environment, it is derived from solar radiation or machines overheat. According to Souza (2003) apud Spillere & Furtado (2007, p.12) in hot times of the year, in certain geographic areas, the actual solar heat can worsen the situation of the industrial worker. The kind of activity and the clothes used by the workers also have an important influence to the heat comfort, as heavier the activity the greater the amount of heat produced by the body and aid in heat exchange required between the body and the environment. Among the consequences of elevated body temperature are heat exhaustion, dehydration, heat cramps, heat shock and heat stroke. As with all risk environments control measures are carried out primarily at source, making use of air-conditioned and exhaust system, and finally at the worker, making use the adequate PPE and limiting his exposure time.

In NR-15 are treated the factors that influence the heat, and among them presents up moisture from the air. The very low humidity bothers the worker and hampers their perspiration, on the other hand, high humidity also creates an environment conducive to the growth of fungi, bacteria and viruses that might contaminate the workers.

3 METHODOLOGY

For the realization of this work was done a literature search on the environmental aspects which expose workers to risks and are causative agents of occupational diseases. Instruments were applied as a checklist of safety inspection and a questionnaire covering the usability and importance of personal protective equipment, and finally, we performed a quantitative assessment of risk agents, noise, illumination and heat. From this, and according to the norms, 15:17 regulating the conditions under which the work environment must be considered to be a comfortable and risk-free, were analyzed and compared two sectors of the company, the administrative sector and industry assembling the bags.

The check list for safety inspection was applied to the company, on September 2, 2013. The checklist applied aimed to identify the current state of the company on all matters: personal protective equipment—PPE; buildings, facilities and services in electricity, machinery and equipment, fire protection, safety signs, because the agents were considered relevant to this study, whose discontinuities can lead to trigger hazards or risks for employees. Still, from the data obtained by applying the checklist was prepared a risk map for the company.

A questionnaire containing closed questions was applied through informal interview with two officials of the assembly sector and 3 responsible for the administrative sector of the company, was registered on September 2, 2013 and addresses the usability of PPE, according to NR 6, taking into consideration the comfort and frequency of use.

The quantitative evaluation was performed from template "*in loco*" granted by the business owner. During the visit held on August 30, 2013 from 7:00 to 17:00, we collected measurements of risk agents: noise, illumination, temperature and humidity. Measurements were performed with the apparatus, thermo—hydro—decibel—Luximeter, the manufacturer and model Instrutherm THDL-400.

In this unit it is possible to perform measurements about those risks already mentioned, these measurements were performed in the morning and afternoon.

For purposes of analysis of noise in the environment, measurements were made near the ear of the worker in both sectors, following cycles of 1 hour and 15 minutes between each starting from 7 o'clock in the morning and finish at 17 pm. Been averaged the results in order to compare them with the values established by NR 15 establishing the maximum exposure time for a given noise level.

The illuminance measurements were analyzed in four (4) points p1, p2, p3, p4—in every job that were studied, which were distributed as proposed by NBR 5382 (1985) in the model 4.2 for field work regularly with central light fixture. The measurements also followed cycles with intervals of 1 hour and 15 minutes between each measurement. After the measurements were made the arithmetic mean for the parameter P collected for each sector, which were the values of the average illuminance of the sectors.

Temperature measurements were made in environmental jobs, initially with 15-minute intervals, with the first measurement to seven (7) hours, however, it was noticed that the temperature had little variation in very short time intervals, and thus adopted, temperature measurements every 1 hour and 15 minutes, using the same standard measurement of other environmental agents.

The ambient humidity was measured by following cycles with intervals of 1 hour and 15 minutes between each measurement and were conducted in the two target sectors of the study. For comparison purposes specified in the law, was used NR-17, which establishes the minimum humidity that must be present in the environment.

4 RESULTS AND DISCUSSION

4.1 *Description of company*

The company is located in the neighborhood of the Upper White, Campina Grande—PB. Has 18 employees, 10 men and 8 women, aged 20 to 55 years, with no failures, distributed in gas cutting, montage, collage, sewing and administration. The working hours in the company is from 7:00 to 17:00. The factory floor has an area of about 141 (one hundred forty-one) square meters of covered area, right foot height of four (4) feet in masonry construction common, with walls and ceiling painted white floor in gray, the granilite.

This topic will present and discuss the results obtained from what was outlined in the methodology described in the previous item.

4.2 *Occupational and health and safety at work*

Was identified in the company that the PPE required for worker protection are earplugs, masks and gloves. The manager of the company provides these PPE as set out in item 6.3 of NR-6, but not all employees use them. As for the buildings, it was observed that the floor of the workplace presents no discontinuities that hinder the movement of people, the stairs at the site is in perfect condition to be used, the floors above ground have fall protection, no cracks, open drains or broken glass panes.

The facilities and services in electricity are prevented from shocks, all the manufactures installation is grounded and suspended without chance of a wire contact with water, there are some accessories that increase the number of outputs of the sockets, but all are protected. The machinery and equipment meet the required spacing of 0.60 m to 0.80 m, and drive devices have stop located in areas that are not hazardous in the machine and can be switched on and off by the operator and can be switched off in case of emergency by any other person other than the operator.

As for fire protection was identified that the outlet openings do not yet have the required width so that employees can get out in case of fire, they do not open towards the exterior of the site and which are not clearly marked by signs or illuminated signs indicating the direction of the exit. The portable fire-fighting equipment, fire extinguishers, are readily accessible, easy viewing and in places where there is less likelihood of fire block access, there are no signs indicating the firefighting equipment. A fire extinguisher is water, type A, to combat protections for wood and paper, and the other is powder, chemical type C, ideal for fighting fires in electrical equipment, fire extinguishers are within the expiration date. As for safety signs, yet there is no type of signaling.

4.3 *Importance and usability of PPE*

The areas evaluated showed the need for employees to use PPE's, which according to these NR6 should be offered by the employer with a certificate of approval—CA—and ensured the quality for use. The questionnaire included the following results for the usability of PPE sectors.

For the initial assessment was identified that the PPE is made available by the employer, but is not offered training for any industry because it identified that only one person from each industry uses their facilities, and only the assembly sector use PPE frequently and uses all. The other people do not use protective equipment, because they feel uncomfortable, and also, misunderstand the risks they are exposed to, when not using them. They feel pain or discomfort during or after labor activity and some believe that these symptoms are from lack of use of PPE.

Since it is possible to conclude that there is availability of PPE in the company, but what they lack is knowledge on the part of workers and delivery of proper training for the use becomes constant.

4.4 *Agents hazardous environments*

4.4.1 *Lighting*

From the illuminance values obtained in the two sectors of the company were taken the maximum and minimum values in order to compare them with the values normalized such information is in Table 1.

The data collected were the result of conditions found in the company as to the measurements there has been no change in the environment.

For comparison with NBR 5413, which imposes the minimum illuminance, average or maximum setting for each sector in accordance with aspects of workers' age, the importance of speed and accuracy of the activity that is performed and the reflectance of the background task. To do so we have to mount and administrative sectors, the value of the sum of that to the administrative sector and for the assembly sector is characterized as average illuminance. Thus, the value of illuminance required should be 1000 Lux and 500 Lux for the administrative sector and assembly respectively.

Therefore, comparing the values obtained in measurements and values required by the standard, it is observed that the values present in the environment are far below what it should be, being insufficient. Should this value to both the type of lamps used and gutters, as presented to weather conditions on the day measurements were taken. The deficiency of illumination at the site causes the ciliary muscle to work harder to capture as much light, thereby causing eye fatigue in workers.

4.4.2 *Noise*

The noise present in the environment consistent with the classification of intermittent noise. The results obtained for the two sectors, administrative and assembly, from the measurements are presented in Table 2.

Given the values found in noise measurements, we evaluate that in none of the times at which the measurements were made, was not exceeded the noise limit stipulated for daily exposure of 8 hours (85 db). It was noticed that in the administrative sector the noise level was lower when compared with the industry's mounting bags due to its location as the assembly sector of the bags is nearest the compressor, causing major source of noise.

Although the findings are within the tolerance limits for exposure of 8 hours per day according to the NR-15, they are beyond the value set by the NR -17 is 65 dB as the level of acoustic comfort. These values might generate difficulty concentrating, and hearing loss lifelong labor of the worker. For the administrative sector, a way to reduce the intensity of noise, would put an acoustic barrier (wall), isolating the sector, as in the assembly sector, the simplest solution would be through the use of PPE (hearing protectors).

4.4.3 *Temperature*

Through Table 3, the information can be seen in relation to temperature variation for each studied sector.

Comparing the temperature variation between the sectors, taking into account the same time of measurement can be observed that the values of the assembly sector, generally are somewhat higher than in the administrative. This is due to the existence of a window in the administrative sector and to be higher than the assembly sector.

In relation to the tolerance limits for the temperature, we can use as a reference in 5.2 NR17, where it recommends for comfort conditions, the temperature is between 20 °C (twenty) and 23 °C (twenty three degrees centigrade) Thus it is clear that at all times measured, temperatures exceeded the limit established by the NR 17. However, it was not possible to analyze the issues of unhealthiness through the WBGT index (wet bulb globe temperature and natural), due to lack of equipment, which would still need to measure the heat radiation from the globe thermometer.

Table 1. Illuminance levels.

Sectors	Maximum value	Minimum value	Average	Standardized	Norm
Administrative	123	81	106	1000	NBR 5413
Mounting	93	72	78	500	

Source: Data obtained from the survey (2013).

Table 2. Values of noise levels.

Sectors	Maximum value	Minimum value	Average	Standardized	Norm
Administrative	73	41	60	85	NR 15
Mounting	82	42	71		

Source: Data obtained from the survey (2013).

Table 3. Values of temperature.

Sectors	Maximum	Minimum	Average	Standardized	Norm
Administrative	27,6	25,1	26,5	20°C a 23°C	NR17
Mounting	27,9	25,8	26,8		

Source: Data obtained from the survey (2013).

4.4.4 *Moisture*

The standard that deals with the moisture present in workplaces is NR 17, where on your point 17.5.2 is the relative humidity should not be below 40% (forty percent). Given that, relative humidity activities considered unhealthy are those that are developed in areas flooded or waterlogged, and which have excessive moisture.

In the company, the humidity measurements in the two sectors studied showed 82.3% of relative humidity. This measure is due to the climatic conditions of the day in which the measurements were made, the day found himself humid with rain showers in your course. Thus accordance with the measure is what is suggested by NR 17. In return, proves ideal for acting the proliferation of fungi and bacteria thus increasing the chances of occurrence of colds and flu for example.

5 CONCLUSION

During this work, it became clear that the company employees understand the risks they are facing, but in an informal and superficial way, not knowing the real consequences of an unsafe environment. Workers are more likely to neglect situations involving their lives and that of their colleagues.

The clarity of the existing risks in the work environment, as well as the importance of using PPE and consequences of illumination aspects, sound and heat when in violation of the rules that regulate, allows the company and its employees to take preventive measures to avoid accidents and occupational diseases.

Thus, we suggest the implementation of knowledge strategies, such as lectures and training in order to clarify the risk map of the company and benefits to health and physical integrity of the worker to use his protective gear. It is vital to provide workers plus protective equipment, provide constant information so that they can really become aware of their quality of life at work, despite the risks in this environment.

REFERENCES

Brazilian Association of Technical Standards Nbr (5382): Iluminance Interior. Rio de Janeiro. 1991.

Brazilian Association of Technical Standards Nbr (5413): Verificatiom of Iliminance. Rio de Janeiro. 1985.

Brazil. Ministry of Labour. Ordinance No. 3214 of June 8, 1978—NR 6, NR 15 and NR 17. Official Gazette of the Federative Republic of Brazil, Brasilia.

Saliba, Tuffi Messiah. Basic Course in Occupational Health and Safety. 3. Ed. Sao Paulo: LTR, 2010.

Service Industry. National Department. Technical Assessment of Environmental Agents: Manual Sesi Brasilia: Sesi 2007.

Occupational Safety and Hygiene II – Arezes et al. (eds)
© 2014 Taylor & Francis Group, London, ISBN 978-1-138-00144-2

Exposure assessment to ionizing radiation in the manufacture of phosphate fertilizer

J.G. Estevez
Instituto Piaget, Portugal

F.O. Nunes
Instituto Superior de Engenharia de Lisboa, Lisboa, Portugal

ABSTRACT: Present study develops and implements a specific methodology for the assessment of health risks derived from occupational exposure of workers to ionizing radiation in the fertilizer manufacturing industry. Negative effects on the health of exposed workers are identified, according to the types and levels of exposure to which they are subject, namely an increase of the risk of cancer even with long term exposure to low level radiation. Ionizing radiation types, methods and measuring equipment are characterized. The methodology developed in a case study of a phosphate fertilizer industry is applied, assessing occupational exposure to ionizing radiation caused by external radiation and the inhalation of radioactive gases and dust.

1 INTRODUCTION

Natural sources of radiation are present in several workplaces such as spas, caves, mines or other environmental conditions which work together towards the existence of significant doses of radiation.

Various industrial processes and activities where natural radioactive isotopes are handled and the presence of radon can also be identified.

Council Directive 96/29/EURATOM, of 13 May, provides basic safety standards for the protection of the health of workers and the general public against the dangers arising from ionizing radiation.

1.1 *Effects on health*

The health effects resulting from exposure to low doses of ionizing radiation are often observable only several years after exposure, making it difficult to establish a clear cause-effect and hence determine exposure limits that are safe (Snashall & Patel, 2003).

According the toxicological profiles for radium and radon established by the Agency for Toxic Substances and Disease Registry, the formation of cataracts is induced when a radiation dose is applied to the lens of the eye. Cataracts caused by radiation can take months to years to appear.

Exposure to low level radiation increases the risk of cancer, depending on several factors, namely, the amount and type of radiation, the individual doses received, the age and sex of the person, their individual susceptibility, if exposure occurred during a short or a long period of time and the presence of other substances which can enhance the effect inductor (ICRP, 1990).

Such evidence may be important due to the possible carcinogenic effects caused by very low doses of ionizing radiation, such as the emission of alpha particles by radon and its radioactive daughters (Little, 2000).

The problem of radon is referred to in the Commission Recommendation 90/143/EURATOM, of 21 February, regarding the protection of populations against their indoor exposure, as follows: "Exposure to radon is not a new phenomenon and epidemiological studies of various groups of minors exposed to elevated concentrations at work have revealed an excess of lung cancer deaths."

This phenomenon may be exacerbated if the worker is also exposed to other contaminants, yielding synergistic effects resulting from exposure to two or more contaminants. The synergistic effect between exposure to ionizing radiation and exposure to toxic chemicals such as, tobacco smoke, asbestos, fumes, vapors and leaches is often referred to in scientific literature (IARC, 1993).

1.2 *Types of exposure*

Gamma radiation may enter the human body only through the skin, but exposition to alpha radiation occurs by inhalation of radioactive dusts (including radon) and through wounds.

Alpha radiation penetration through the airways can be very significant in dusty environments with radioactive particles, with the resulting risk depending greatly on the size and quantity of particles in the air.

A high concentration of radon in the air is significantly hazardous to human health. The descendants of the nuclear decay of radon (polonium, bismuth and lead) are associated with dust particles entering the respiratory system, which emit high-energy alpha particles. Through the exposure to radionuclides attached to dust in the air, mainly by breathing such dust, depending on the size of the dust, the radionuclides may be deposited in the worker's lungs and implement a high-energy alpha radiation dose to that lung tissue.

The phosphates used in Portugal, usually from Tunisia, Syria, Morocco and Senegal, can contain significant concentrations of uranium which rely heavily on specific locations where they are extracted and are not systematically controlled from the radiological point of view on their arrival. These phosphates are handled, processed, stored and still generate huge amounts of industrial wastes (phosphogypsum) containing high concentrations of radioactive isotopes. Worker exposure occurs in areas of mining, transportation, loading and unloading of the raw material, processing industries and waste produced and are accumulated close to the industrial units.

2 MATERIALS AND METHODS

To estimate the total radiation dose to which workers may be exposed additive effects of external gamma radiation dose, inhalation and ingestion of radioactive dust and radon and its radioactive progeny should be considered.

The gamma spectrometry have been performed in the laboratory of ITN (Instituto Tecnológico e Nuclear). The activity of radionuclides present measured in Bq/kg.

2.1 *Laboratory measurement of radiological activity of raw materials*

With gamma spectrometry present radioactive nuclides and their activities are identified. The steps are as follows:

1. Collect samples of phosphate rock in different factory sites which were kept in plastic bags.
2. Subsequently, already in laboratory, passage of samples into sealed glass vials.
3. 30-day waiting until the sample obtains balance between the radiological ^{226}Ra and their descendants.

The presence of a significant radiological activity in the phosphate rock may suggest levels of radioactive contamination all along the manufacturing process, particularly in receiving tank, filtration equipment, pipes and conveyors.

2.2 *Measurement of gamma radiation field*

Measuring the radiation field in terms of gamma radiation using a radiation meter is important since it is a less visible danger as the effects of ionizing radiation at low doses are usually observable only several years after exposure, therefore it is often difficult to establish a clear cause-effect relationship.

The dose rate of external gamma radiation in mSv/h, should be measured in various workplaces using a meter ionizing radiation. Several measurements on different days should be performed in every workplace. The results obtained should be processed in terms of mean values and standard deviations shown in plant.

Alternatively, the radiation field measuring of the external radiation dose range may also be characterized, using a digital dosimeter sensitive to gamma radiation. This dosimetry should be performed over a significant period of time (hours or days).

A total of 96 measurements of external gamma radiation was conducted. The equipments used were a meter ionizing radiation ALNOR, model RDS 120 and an individual dosimeter gamma Automess, model ADOS.

2.3 *Measuring the concentration of dust in the air*

The method used to determine the concentration of dust in the air is based on gravimetric analysis (determination of mass by weighing) of the dust filters before and after collection. The weight difference of the filter before and after the collection of dust gives us the weight of dust trapped in the filter and indirectly a measure of the concentration of dust in the air expressed as mg/m^3.

To assess occupational exposure to total particulate pumps, air intake must be calibrated to 1 L/min with 2 hours of measurement. For aerosol particles (separated by an aluminum cyclone) vacuum pumps must be calibrated to 2.5 L/min at 2.5 hour measurements.

To determine the concentration of dust in the air, total dust and respirable dust in work areas were measured through the use of 4 SIDEKICK air suction pumps.

The level of dust concentration in the air provides a rough indication of the amount of radionuclides derived from phosphate rock in the air. The dust in the air may constitute a danger to the health of workers due to the possibility of inhalation, ingestion and skin absorption of radioactive elements present in these dusts. A significant danger to the health of exposed workers results from alpha radiation emission by these radioactive dusts, highly energetic and able to destroy the internal tissues not protected by human skin.

Furthermore, if the workplace has a high concentration of dust, radon present in the air will generate other radioactive elements in their natural process of radioactive decay, namely polonium, bismuth and lead. These radioactive elements are associated with respirable dust in the air and are inhaled and can be deposited throughout the respiratory system. Alpha radiation associated with these radioactive elements can cause irreversible damage to the tissues of the lungs, including the development of neoplasms.

2.4 *Measuring the concentration of radon*

Radon is a radioactive inert gas produced by the natural decay of uranium that is present in phosphate rock, which accounts for a significant dose of radiation to which workers may be exposed. Generally, radon escapes into the open air and is quickly diluted in the atmosphere, remaining only at low concentrations, and therefore not of great concern. However, in enclosed or poorly ventilated spaces, as existing in the industry studied, radon can accumulate reaching high concentrations that may be harmful to the health of workers.

Initially, fast measurements of radon concentration in the atmosphere were made in areas where higher concentrations of the radioactive gas were predicted and simultaneously had a significant occupancy of workers. These measurements were performed using a radon detection equipment, DURRIDGE, model RAD7, direct reading and equipped with air drying filter with sampling periods of 5 minutes.

In a second phase, measurements were taken using more stringent nuclear track detectors. The six detectors used were installed and collected after 29 days of exposure in the selected workplace and subsequently sent to the ITN to determine the concentration of radon by counting the number of micro-holes produced in the film detector (LA 115 Kodak).

2.5 *Conversion of radon concentration for a dose of radon*

To account for the effects of the sum of external radiation and gamma radiation from radon, it becomes necessary to convert radon concentration (Bq/m^3) into radon effective dose (Sv), ie, equivalent dose of radiation that has the same destructive effect on biological tissues as an equal dose of gamma radiation. Regarding the conversion factor to be used, there are still some uncertainties, with conversion data values significantly different for epidemiological analysis and for dosimetric analysis. The value specified in the legislation is 1.4 Sv per $J \cdot h \cdot m^3$ which corresponds to about 6.7 nSv per $Bq \cdot h \cdot m^{-3}$ while the conversion factor recommended (UNSCEAR, 2000) is 9 nSv per $Bq \cdot h \cdot m^{-3}$ and takes into account the quality factors.

2.6 *Application of numerical simulation models in the design phase*

The radiological protection against external radiation may be achieved through technical measures: reducing the amount of radioactive material present, placing shields between the radiation source and the workers exposed as much as possible or removing the radioactive sources. It can also be reduced through organizational measures, reducing the exposure time of workers to radioactive sources in order to limit the total dose received.

These variables can be simulated on the computer via a calculation tool that incorporates an appropriate model (World Information Service on Energy, 2013) for the propagation of external gamma radiation, and considers the distances between the radioactive sources and workers exposed shields, and the estimated time of worker exposure to these radiations. One can thus predict exposure levels to which workers may be exposed by simulating various scenarios. For example, in the design phase or remodeling of a plant, exposure levels can be minimized, simulating different layouts and types of protections and thus choosing the best options to ensure maximum protection of workers against ionizing radiation.

As an example, considering the geometry shown in Figure 1, it should be noted that the dose rate values obtained by computer simulation

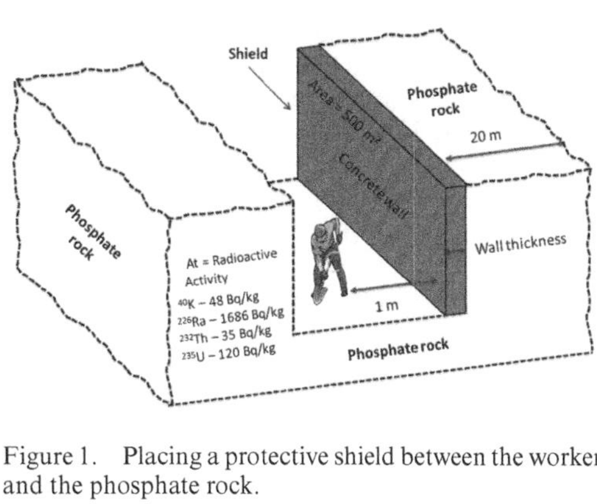

Figure 1. Placing a protective shield between the worker and the phosphate rock.

of the model were very similar to values obtained in field measurements performed. Thus, it can be concluded that the model used to predict the rate of external radiation dose range can provide suitable results, enabling its use as a tool for prediction in the phase design of new installations where the risk of radioactive contamination is present.

Still by way of example, to implement the use of a shield (Fig. 1), the construction of a concrete wall is considered, a structure to be very suitable from the standpoint of radiation protection, ease of construction and associated costs. Next, a series of computer simulations have been performed to compute the effectiveness of this shield as a function of wall thickness for the case of Senegal phosphate.

For a display area of 500 m^2, 1 m distance between the radioactive source and receiver, with a thick layer of 20 m phosphate rock, variation of the thickness of the concrete wall is simulated, verifying that a protection of 90% is achieved at a thickness of concrete wall of about 13 cm and a protection of 95% at a wall thickness of about 19 cm. The results thus show the use of this type of screen to be feasible from both a technical and an economical point of view.

3 RESULTS AND DISCUSSION

3.1 *Radiological activity of phosphate rock*

By gamma spectrometry performed in the ITN laboratory, the amount of radioactive nuclides present in the various samples used as raw material in the manufacture of fertilizers were determined.

Thus, radiological activity was measured with samples of phosphate rock derived from various sources, namely, Tunisia, Syria, Morocco and Senegal in order to identify their radiological differences.

The measured values showed that the samples of phosphate rock have a significant radiological activity, varying significantly with its provenience. Hence, for example, phosphate rock from Senegal has a radioactivity of the radionuclide ^{226}Ra of approximately 1700 Bq/kg (over four times that of Tunisian phosphate rock with 400 Bq/kg, relationship that also holds approximately in the case of the radionuclide ^{235}U).

3.2 *Dose rate of external gamma radiation*

Several measurements of external gamma radiation in various workplaces were conducted using a ionizing radiation meter.

The measured values indicate that in shipping and handling areas of dry phosphates, annual dose of radiation exposure of some workers can reach 2.97 mSv/year.

3.3 *Dose of radiation from radon*

In the case of the phosphate rock analyzed in laboratory a strong presence of ^{226}Rd, radon parent, was found. A high concentration of radon in enclosed spaces where there was a significant storage of phosphate rock or their products derivatives is, then, expected to occur.

The values found in the tunnel of the conveyor inlet phosphate rock (472 Bq/m^3) and in the control room of the industrial process (461 Bq/m^3) actually exceed the reference value legally established for existing buildings (400 Bq/m^3), getting the value found in the elevator of phosphate rock in floor one (291 Bq/m^3) above the reference value for new buildings (200 Bq/m^3) and the values of the other sites below this value.

3.4 *Concentration of dust in the workplace*

The measured values only exceed the limits established for inert dust bins in the area of raw materials (28.3 mg/m^3 for total dust). However, in the case of dust with a significant content of radioactive isotopes, significant risk associated with inhalation, ingestion and absorption of this dust can be identified.

3.5 *Estimate of total dose*

To estimate the total radiation dose to which workers may be exposed the following factors should be considered:

1. Activity of radionuclides present in the dust in the air (measured values).
2. Dust concentration in the air in workplaces (measured values).
3. Average amount of air inhaled by workers and the appropriate conversion factor (from the literature).
4. External dose of gamma radiation present in the workplace (measured values).
5. Concentration of radon in working areas (measured values).
6. Number of hours of exposure to various types of contamination (obtained by interviews and observation).
7. Type of respiratory protection used (breathing mask type, filter type, efficiency, operating conditions and proper placement, duration of use).

It is also important to consider the concentration of radon in confined spaces such as tunnels, tanks, warehouses, machinery and other poorly ventilated places which normally have high concentrations. These sites are rarely occupied by workers in the normal operation but can be filled by workers in maintenance operations, including cleaning and repairing equipment.

The total dose of radiation to which the worker is exposed can be calculated from the sum of the different doses: inhalation, ingestion and absorption through wounds of dust with radionuclides, inhalation of radon and its short half-life descendants and external gamma radiation.

According to (Birky et al., 1998), to estimate the radiation dose by inhalation of radioactive dusts to which workers may be exposed, it can be assumed that the average value of air ventilated by workers is 35 L/min with a maximum value of 85 L/min for hard work and 13 L/min for light work. It would also be necessary to take into account the protection factor introduced by the use of masks against dust and mists covering your nose and mouth with a maximum protection factor of ten.

In this study, only the contributions of radiation doses received from external gamma radiation and inhalation of radon were considered, since the radiation doses by inhalation and absorption through the skin and wounds are difficult to determine accurately and it is assumed that their contribution to the total dose is less significant.

The calculations relative to the estimated total dose received by workers (excluding the contribution to the radiation dose received by inhalation of radioactive dusts and absorption through skin and wounds) show that this radiation level can be significantly dangerous to the health of workers, reaching a maximum value of 7.7 mSv/year (Fig. 2).

3.6 *Calculation of the increased risk of cancer*

To estimate the risk of developing cancer due to exposure to low doses of ionizing radiation (stochastic effects) it is considered that there is an

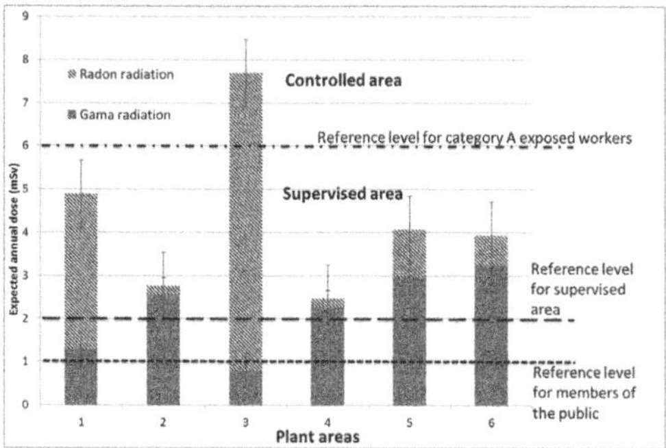

Figure 2. Expected annual dose by classified areas.

increased risk of fatal cancer of 0.04 per Sievert per year for exposed workers (ICRP, 1990).

Performing the calculation for dose values estimated for the various areas of the facility, one may obtain the respective percentage of increased cancer risk throughout life. The values obtained confirm that workplaces most at risk are those areas of the control room, next to carriers of phosphate rock powder and the grinding zone and sieve, which coincide with areas of high dustiness and poor ventilation (Estevez, 2012).

4 CONCLUSIONS

High levels of radioactivity caused by radioactive radionuclides ^{226}Ra and ^{235}U were found in the phosphate rock. The one from Senegal presented more than four times the phosphate rock from Tunisia.

There were very significant levels of dust concentration resulting from the phosphate rock, which cannot and must not be considered simply as inert dusts. In this particular case of the phosphate rock dust with a significant content of radioactive isotopes that emit ionizing radiation, significant risks to the health of workers associated with inhalation, ingestion, and absorption through the skin and unprotected wounds are identified.

Some workers may receive an annual dose exceeding 6 mSv/year by the cumulative effect of exposure to external gamma radiation, alpha radiation from radon and its progeny.

In workplaces with poor ventilation and high dustiness, the most significant component of radiation dose observed corresponds to radon and its radioactive progeny. A significantly increased risk was found in maintenance operations involving exposure to fine dust-cleaning equipment and bag filters of vacuum systems.

Thus, based on the identification and assessment of risks in each particular installation, the control measures of exposure to develop (technical, organizational and individual protection) should be directed towards the maintenance of the levels of exposure according to the ALARA principle (As Low As Reasonably Achievable).

Special attention must be granted to the use of adequate personal protective equipment to avoid exposures associated with inhalation, ingestion and absorption through skin in workplaces with significant levels of radioactive dust concentration, such as control room, areas next to carriers of phosphate rock powder and the grinding zone and sieve.

REFERENCES

Birky, B.K., Tolaymat, T. and Warren, B.C. 1998. Evaluation of Exposure to Technologically Enhanced Naturally Occurring Radioactive Materials (TENORM) in the Phosphate Industry. *Florida Institute of Phosphate Research*, USA.

Estevez, J. Gil. 2012. Protección de los trabajadores frente al riesgo de exposición a radiaciones ionizantes en la industria de fosfatos en Portugal. *Tesis doctoral. Departamento de Ciencias Biomédicas de la Universidad de Léon.*

IARC—International Agency for Research on Cancer 1993. Monographs on the evaluation of the carcinogenic risk of chemicals to humans. Some naturally occurring substances. *Food items and constituents, heterocyclic aromatic amines and mycotoxins.* Lyon.

ICRP—International Commission on Radiological Protection 1990. Recommendations of the International Commission on Radiological Protection. *Publication 60.*

Little, John B. 2000. Radiation carcinogenesis. *Carcinogenesis: Integrative Cancer Research.* Vol. 21 (3), pp. 397–404.

Snashall, D. and Patel, D. 2003. ABC of Occupational and Environmental Medicine. 2nd Edition. *BMJ Publishing Group*, London.

UNSCEAR—United Nations Scientific Committee on the Effects of Atomic Radiation 2000. *Sources and Effects of Ionizing Radiation.* New York.

World Information Service on Energy. (2013, June 14). WISE Uranium Project. Retrieved from http://www.wise-uranium.org:http://www.wise-uranium.org/rdcx.html.

Occupational Safety and Hygiene II – Arezes et al. (eds)
 ISBN 978-1-138-00144-2

Shiftwork experience: Worker's vision of its impacts

I.S. Silva, J. Prata, A.I. Ferreira & A. Veloso
School of Psychology, University of Minho, Braga, Portugal

ABSTRACT: Shiftwork, especially the one performed during the night, has been associated with health problems (e.g. sleep and digestive problems, tiredness). Besides effects on health, this work schedule also has been associated with additional troubles in the family and social life. Although the negative impact is the more emphasized perspective in the studies, this work schedule may, however, under some personal circumstances, represent advantages compared to the standard work schedule. In this study, we are shown the main impacts (positive and negative) put into perspective by 184 workers in relation to four work schedules (morning, afternoon, night, and rotating system) based on their own work experience. The data, of a qualitative nature, was achieved when we asked approximately 1,400 shift workers in the industrial sector in six companies from the north of Portugal. Specifically, the data is related to the comments that the workers spontaneously gave about their own experience. Generally speaking, the results indicate that each work schedule is put into perspective as having negative and positive impacts, even if in an asymmetric mode (e.g., at the level of health, the shifts involving night work are evaluated mainly negatively while the afternoon shift is viewed as advantageous). The results also suggest that the variables influencing the experience of shift work are many and diverse.

1 INTRODUCTION

The organisation of working hours constitutes one of the key elements of working conditions and employment relationship (Parent-Thirion et al., 2007), having a strong impact beyond the working context.

Shiftwork is one possibility, amongst others, of the organisation of the working time. Typically, this modality refers to the sequence of working teams (i.e. shifts) (Thierry & Jansen, 1998) heading to an extension of the working period, including its continuation up to 24 hours per day, 7 days a week. According to the Fifth European Working Conditions Survey—2010, 17% of the active population of the European Union worked on shiftwork (Eurofound, 2012).

Shiftwork, especially if it includes the night shift, has been associated with several negative effects on the employees' health. Its impact may be felt on 1) the quality and quantity of sleeping (e.g. Akerstedt, 2003; Costa, 1996); 2) the digestive system (e.g. Costa, 1996, 1997; Knutsson, 2003); 3) the cardiovascular system (e.g. Boggild & Knutsson, 1999); 4) the psychological well-being (e.g. Cole et al., 1990; Wedderburn, 2000); 5) women's reproductive health (e.g. Nurminen, 1998; Mozurkewich et al., 2000) and, more recently, there is a certain evidence that it may constitute a risk factor for developing oncologic disease, namely breast cancer (Hansen, 2001; Megdal et al., 2005). Even though the association between shiftwork and health effects is complex and multifaceted (e.g. Smith et al., 2003; Totterdell, 2005), the disruption of circadian rhythms is commonly referred to as a potential major factor explaining this negative impact.

Besides health, the difficulties experienced by shift workers can also be felt on the familial and conjugal realm (e.g., organisation of domestic life, marital and parental relationships) as well as on their social lives (e.g., difficulty with engaging in social activities) (e.g., Gadbois, 2004; Wedderburn, 2000). In this case, difficulties occur as a result of the existing conflict between working hours and the temporal organisation of society regarding the privileged periods dedicated to family and social life, which tend to happen at the end of the day or over the weekend (Baker et al., 2003).

Even though the dominant perspective tends to consider the negative impact of shiftwork in terms of health, family and social life, this way of organizing the working hours may bring advantages to the employees when compared to the standard schedule (09:00 am–05:00 pm): more free time, easiness of reconciliation with family life and economic rise (Silva, 2008; West et al., 2012).

2 STUDY'S AIMS

The current study helps, to illustrate the complexity of shiftwork, trying to understand the diverse aspects associated with it in the perspectives of the employees themselves and taking into account the type of shift system. Particularly, the study features the analysis of the workers' comments on their experiences as shift workers. These comments were made as the employees were undertaking a survey on the effects of shift work regarding health, family and social life. At the end of the survey, there was a blank space in which the employees could freely make any "comments or suggestions related to the experience of being a shift-working employee". The free nature of the comments not only provides a strong contribution to the understanding of shift work effects, but also emphasizes those effects and their relation with personal circumstances.

The data used for this study, derived only from the analysis of the open question.

3 METHODOLOGY

Of the 1,386 workers that participated in the undertaking of the survey, 236 spontaneously made "comments or suggestions", for a 17% participation rate. These workers may have been working on a fixed shift (morning, afternoon or night, Monday through Friday) or on a continuous rotating system. Typically, the start and end times of fixed shifts were as follows: morning (06:00 am–02:00 pm), afternoon (02:00 pm–10:00 pm) and night (10:00 pm–06:00 am). In the rotating system, the direction of the rotation was clockwise and at the rate of seven workdays, with two or four days off. The six companies that participated in data collection are industrial (mainly from the textile industry) located in the North of Portugal.

In each of the six companies the questionnaire was applied after the consent of the respective Directors or the Human Resources Department, and the procedures for data collection were negotiated on a case by case basis between each company and the research team. In five of the six companies, the questionnaires were given to the Human Resources Department, who then gave them to employees. Employees would return them in sealed envelopes addressed to the research team. Each questionnaire was accompanied with an introduction letter which explained the aims of the study, the voluntary nature of the participation, the procedures for returning the questionnaires, as well as the contact information for the research team. In the sixth case, the data was collected through personal visits from the researchers. Subsequently, the workers inserted the questionnaires in a closed box. In collection procedures, anonymity and confidentiality of the data was assured.

The questionnaires contained questions of different natures, namely: socio-demographic data, regarding the family situation; issues related to physical and psychological health; satisfaction with working schedules; and the impact of working hours on family and social life. At the end of the questionnaire there was an open question, which consisted of space where the workers could make any comments and/or suggestions related to their experience as shift workers, if they wished. As mentioned, the present study is based on the analysis of the data collected in this open question through content analysis.

All answers were subject to content analysis (Bardin, 2009). First, the answers were analyzed by three authors according to a previously-defined categorization system, which included five general categories: "positive aspects" of shiftwork, "negative aspects" of shiftwork, "positive and negative aspects" of shift work, "general remarks" (which refers to suggestions/recommendations about the implementation of shiftwork within the organisations), and "irrelevant" (when the comment did not fit the scope of the study). The three independent categorisations were compared and one could observe the absence of agreement (i.e. at least one of the evaluations was different than the others) in 7.6% of the cases. These cases were discussed jointly until a clarification was reached and a respective agreement was found. The discussion of the agreements and disagreements and further analysis of the content resulted the refinement of the initial categorisation system. In total, there were defined 4 categories: "positive aspects" of shiftwork; "negative aspects" of shiftwork; "irrelevant", and "general comments".

As it can be seen on Table 3, the categories "positive aspects" and "negative aspects" were subdivided into 7 subcategories (work/life interface; family life; economic aspects; general; adaptation; organisational aspects; and health). At this stage, the comments regarding the category "irrelevant" (n = 52) were excluded.

Thus, the 236 participations resulted in 364 comments as one answer could refer to appreciations regarding more than one shift and/or considering distinct dimensions (e.g. health, family) of a certain shift. After excluding the comments in the category "irrelevant," the authors had a total of 312 comments, involving 184 workers for effective analysis.

Table 1 presents a brief characterisation of these workers, given their work schedule at the time the data was collected. Most workers were men (60.3%), average age 37.3 years (SD = 10.7), are married (71.7%) and about half of the sample (52.7%) have children aged up to 18 years.

Table 1. Characterisation of shift workers who made comments in relation to their work schedule.

Shift	Morning (n=46)		Afternoon (n=57)		Night (n=61)		Rotating (n=20)		Total (N=184)*	
Variable	n	%	n	%	n	%	n	%	n	%
Gender										
Man	17	9.2	32	17.4	46	25	16	8.7	111	60.3
Woman	29	15.8	25	13.6	15	8.2	4	2.2	73	39.7
Age										
M**	38.7		35.9		36.7		38.5		37.4	
(SD)***	(11)		(10.6)		(13)		(8.1)		(10.7)	
Status										
Single	10	5.4	15	8.2	14	7.6	–	–	39	21.2
Married	35	19.	40	30.3	43	23.7	14	7.6	132	71.7
Widow	–	–	1	0.54	–	–	–	–	1	0.5
Separated	1	0.5	1	0.5	4	2.2	–	–	6	2.3
Children										
Yes****	27	14.7	29	15.7	31	16.8	10	5.4	97	52.6

*Given the presence of missing values on the "Status" variable, N obtained in this case is less than the sample size (N = 184); **M (Mean); ***SD (Standard Deviation); ****Children aged up to 18 years.

Table 2. Frequency of positive and negative aspects depending on shifts.

Shift	Morning		Afternoon		Night		Rotating	
Variable	n	%	n	%	n	%	n	%
Positive aspects	35	60.3	50	54.9	41	41.8	9	24.3
Negative aspects	23	39.7	41	45.1	57	58.2	28	75.7

4 PRESENTATION AND DISCUSSION OF THE RESULTS

Table 2 provides the response frequencies achieved in the "positive aspects" and "negative aspects" categories according to the four types of work schedules: morning shift, afternoon, night, and rotating system.

The proportion of positive and negative aspects reported in each shift is different, whether it includes night work or not. In shifts without night hours (i.e., the morning and afternoon shift), the number of positive aspects reported is higher than the negative. In shifts where the work schedule requires night work (i.e., night and rotating shift), the pattern is reversed. This is particularly highlighted in the rotating shift.

After this general analysis, Table 3 presents a more detailed analysis, including the frequency of subcategories listed in each shift (i.e., the interface work-life, health, family life, etc.). Finally, for each case, we present an excerpt of typical responses in the ruling subcategory, namely, in the case of the morning shift, the answer presented in the "positive aspects" aims to exemplify the subcategory "Interface work/life" and so on—(*"In fact we get up very early but we have the whole afternoon free for any other tasks"*). Beyond the categories related to "positive aspects" and "negative aspects," it was also considered the category "general remark" (with 31 answers), which globally referred to suggestions/recommendations in the implementation by the shift work organizations—(*"All workers should work in the shifts of their preference"* or *"Workers who work in continuous shifts should not work more than 10 years in a particular shift or should change when asked"*) are examples of responses that were included there.

Globally, the results are in agreement with the literature. The shifts that require night working (in a fixed or alternate timetable) are viewed by shift workers as having more negative than positive aspects. Another aspect that emerges from the results presented in Table 2 and is less common in the literature is the references to positive aspects, yet asymmetrically between shifts. Finally, the obtained results also suggest a significant diversity

Table 3. Frequency of responses by categories and subcategories "positive aspects" and "negative aspects" according to the shifts.

Morning shift (06:00 am–02:00 pm)	
Category	
Positive aspects ($n = 35$) (e.g.: *"It's true that we get up very early, but we have the whole afternoon free for other tasks."*) (e.g.: *"... working in shifts allows us to organise life at home, and do other things like go to the bank, to the doctor, without having to miss work ..."*)	Negative aspects ($n = 23$) (e.g.: *"In my opinion, working the morning shift is very tiring. Still, it is one of the few shifts that allows us to have some time for the family, even with some sacrifice."*) (e.g.: *"The morning shift—6:00 am to 2:00 pm—is a bit uncomfortable because we have to get up too early ..."*)
Subcategory	Subcategory
– Work/life interface ($n = 12$) – Family life ($n = 9$) – Economic aspects ($n = 4$) – General ($n = 4$) – Adaptation ($n = 3$) – Organisational aspects ($n = 2$) – Health ($n = 1$)	– Health ($n = 15$) – Family life ($n = 4$) – Economic aspects ($n = 3$) – Organisational aspects ($n = 1$)
Afternoon shift (02:00 pm–10:00 pm)	
Category	
Positive aspects ($n = 50$) (e.g.: *"It is a shift that allows well-being, health and enough free time to do a bit of everything."*) (e.g.: *"My shift gives me more free time to do the housework. Besides, I think it's better for health."*)	Negative aspects ($n = 41$) (e.g.: *"My shift leaves me little time for the family."*) (e.g.: *"Working in shifts deprives us of family time ..."*)
Subcategory	Subcategory
– Work/life interface ($n = 21$) – Health ($n = 12$) – Family life ($n = 7$) – Adaptation ($n = 3$) – Organisational aspects ($n = 3$) – General ($n = 3$) – Economic aspects ($n = 1$)	– Family life ($n = 16$) – Organisational aspects ($n = 8$) – Health ($n = 5$) – Work/life interface ($n = 5$) – Economic aspects ($n = 4$) – General ($n = 3$)
Night (10:00 pm–06:00 am)	
Category	
Positive aspects ($n = 41$) (e.g.: *"The night shift is my favorite only because it is well-paid."*) (e.g.: *"I like working this shift because it gives me plenty of free time to do other things I like."*)	Negative aspects ($n = 57$) (e.g.: *"Work at night is very tiring and harmful to health ..."*) (e.g.: *"Considering the effort of working at night, we are underpaid."*)
Subcategory	Subcategory
– Economic aspects ($n = 11$) – Work/life interface ($n = 10$) – Adaptation ($n = 6$) – Health ($n = 5$) – Organisational aspects ($n = 3$) – Family life ($n = 3$) – General ($n = 3$)	– Health ($n = 20$) – Economic aspects ($n = 15$) – Family life ($n = 10$) – Organisational aspects ($n = 7$) – Adaptation ($n = 2$) – General ($n = 2$) – Work/life interface ($n = 1$)

(*Continued*)

Table 3. (*Continued*).

Rotating shifts	
Category	
Positive aspects ($n = 9$) (e.g.: *"The only good thing about working in shifts is that you don't need to miss work to solve personal problems ...."*) (e.g.: *"I would like to work on a fixed shift without loss of pay"*)	Negative aspects ($n = 28$) (e.g.: *"... the rotating is the worst in terms of meals and rest..."*) (e.g.: *"These work schedules deprive us of social lives and affects our mood in a bad way."*)
Subcategory	Subcategory
– Work/non work interface ($n = 4$) – Economic aspects ($n = 3$) – Health ($n = 1$) – Organisational aspects ($n = 1$)	– Health ($n = 12$) – Family life ($n = 8$) – Economic aspects ($n = 4$) – Organisational aspects ($n = 2$) – Work/life interface ($n = 2$)

of aspects associated with each shift (health, family, social, economic or organisational).

Considering the specific results of each shift, it turns out that the positive aspects associated with the morning shift are located mainly at the interface between work and private lives, including household responsibilities. Contributing to this appraisal, the amount of free time this schedule allows will surely stand, as indicated by the research done by Silva (2008). Nevertheless, the free time linked to this schedule is achieved at the expense of a work schedule starting early, which may have a negative impact on health, especially in terms of sleep (reduction of sleep and troubles in waking up). Such troubles on health (sleep) linked to the morning shift are particularly felt when the shift starts very early (e.g., Kecklund et al., 1997 Åkerstedt, 2003).

Regarding the afternoon shift, it is observed that the shift workers indicate dimensions of social/family nature both in positive and negative terms. Such dichotomy suggests that the way in which the shift is evaluated is dependent on the personal circumstances of each worker (e.g., the spouse's work schedule), even though the research indicates that the periods related to the late afternoon and the weekends are most appreciated from the family and social viewpoint (Baker et al., 2003; Gadbois, 2004). We must also stress that this is the only shift where the subcategory "health" gathers more positive than negative references comparatively. This positive assessment is mainly due to the non-interference of this work schedule with the repose.

Contrasting with the assessment of the afternoon shift, the shifts that include nights (in a fixed or switched way) are evaluated negatively on health, especially in terms of sleep and well-being. This assessment is consistent with the empirical evidence in this field; such troubles are mainly related to the need of reversing the normal cycle of activity-repose (Smith et al., 2003; Silva, 2012). In fact, the studies with shift workers that involve rotating shifts (e.g., Parkes, 2002) and night work (e.g., Silva, 2008) indicate that the quality of the daytime sleep is not comparable to the night time sleep; this tends to be shorter and of lower quality. From the understood advantages of addressing these schedules, especially the night shift, we emphasize the work-life interface with more free time and the economic benefits associated with the schedule; these benefits aim to recognize, on behalf of the legislator, that such schedules have increased demands compared to the standard ones.

5 FINAL CONSIDERATIONS

In general, the presented study suggests that the variables susceptible of influencing the experience of working in shifts are many and varied. In fact, in addition to variables of a more individual nature (where the impact on health stands out), the reports of the workers also emphasize variables of social/organisational and contextual nature (where the interface work/life and family life stands out) tend to be less considered in the study of shiftwork.

REFERENCES

Åkerstedt, T. (2003). Shift work and disturbed sleep/wakefulness. *Occupational Medicine*, 53, 89–94.

Baker, A., Ferguson, S., & Dawson, D. (2003). The perceived value of time: Controls versus shiftworkers. *Time & Society*, 12(1), 27–39.

Bardin, L. (2009). *Análise de conteúdo (L'Analyse de Contenu)* (4ª ed.). Lisboa: Edições 70, LDA.

Boggild, H., & Knutsson, A. (1999). Shift work, risk factors and cardiovascular disease. *Scandinavian Journal of Work and Environmental Health*, 25 (2), 85–99.

Cole, R.J., Loving, R.T., & Kripke, D.F. (1990). Psychiatric aspects of shiftwork. *Occupational Medicine: State of the Art Reviews*, 5(2), 301–314.

Costa, G. (1996). The impact of shift and night work on health. *Applied Ergonomics*, 27(1), 9–16.

Costa, G. (1997). The problem: Shiftwork. *Chronobiology International*, 14(2), 89–98.

Eurofound (2012), *Fifth European Working Conditions Survey*, Publications Office of the European Union, Luxembourg.

Gadbois, C. (2004). Les discordances psychosociales des horaires postés: Questions en suspens. *Travail Humain*, 67(1), 63–85.

Handy, J. (2010). Maintaining family life under shiftwork schedules: A case study of a New Zealand petrochemical plant. *New Zealand Journal of Psychology*, 39(1), 29–37.

Hansen, J. (2001). Light at night, shiftwork, breast cancer risk. *Journal of the National Cancer Institute*, 93(20), 1513–1515.

Kecklund, G., Åkerstedt, T., & Lowden, A. (1997). Morning work: Effects of early rising on sleep and alertness. *Sleep*, 20(3), 215–223.

Knutsson, A. (2003). Health disorders of shift workers. *Occupational Medicine*, 53, 103–108.

Mozurkewich, E.L., Luke, B., Avni, M., & Wolf, F.M. (2000). Working conditions and adverse pregnancy outcome: A meta-analysis. *Obstetrics and Gynecology*, 95, 623–635.

Nachreiner, F. (1998). Individual and situational determinants of shiftwork tolerance. *Scandinavian Journal of Work and Environmental Health*, 24(Suppl. 3), 35–42.

Nurminen, T. (1998). Shift work and reproductive health. *Scandinavian Journal of Work and Environmental Health*, 24(Suppl. 3), 28–34.

Parent-Thirion, A., Macías, F., Hurley, J., & Vermeylen, G. (2007). *Fourth european working conditions survey*. Luxembourg: Office for Official Publications of the European Communities.

Parkes, K.R. (2002). Age, smoking, and negative affectivity as predictors of sleep patterns among shiftworkers in two environments. *Journal of Occupational Health Psychology*, 7(2), 156–173.

Silva, I.S. (2008). *Adaptação ao trabalho por turnos*. Dissertação de Doutoramento em Psicologia do Trabalho e das Organizações. Braga, Universidade do Minho.

Silva, I.S. (2012). *As condições de trabalho no trabalho por turnos. Conceitos, efeitos e intervenções*. Lisboa: Climepsi Editores.

Smith, C.S., Folkard, S., & Fuller, J.A. (2003). Shiftwork and working hours. In J.C. Quick & L.E. Tetrick (Eds.), *Handbook of occupational health psychology* (pp.163–183) (2nd ed.).Washington, DC: American Psychological Association.

Thierry, H., & Jansen, B. (1998). Work time and behaviour at work. In P.J.D. Drenth, H. Thierry & C.J. de Wolff (Eds.), *Handbook of work and organizational psychology* (Vol. 2: Work Psychology) (2nd ed., pp. 89–119). East Sussex: Psychology Press.

Totterdell, P. (2005). Work schedules. In J. Barling, E.K. Kelloway & M.R. Frone (Eds.), *Handbook of work stress* (pp. 35–62). Thousand Oaks: Sage Publications.

Wedderburn, A. (2000) (Ed.). *Shiftwork and health*. European Studies on Time. Dublin: European Foundation for the Improvement of Living and Working Conditions.

West, S., Mapedzahama, V., Ahern, M., & Rudge, T. (2012). Rethinking shiftwork: Mid-life nurses making it work. *Nursing Inquiry*, 19(2), 177–187, doi: 10.1111/j.1440-1800.2011.00552.x.

Occupational Safety and Hygiene II – Arezes et al. (eds)
 ISBN 978-1-138-00144-2

Ergonomic conditions of work bus urban driver and bus collector

B. Barkokébas Jr.
Civil Engineering Postgraduate Program, University of Pernambuco, Brazil

A.B.B.H. Pinto & L.B. Martins
Department of Design, Federal University of Pernambuco, Recife, PE, Brazil

E.M.G. Lago & F.M. da Cruz
Civil Engineering Postgraduate Program, University of Pernambuco, Brazil

B.M. Guimarães
Department of Design, Federal University of Pernambuco, Recife, PE, Brazil

ABSTRACT: In this article we sought to address the issue public transport system, focusing on the element buses, more precisely in the professionals who use it: driver and collector. With a sample of 185 professionals and using the methodology of intervention proposed by Moraes and Mont'Alvão (2003) allowed the identification of working conditions in the sector of Urban Passenger Transport in the Metropolitan Region of Recife.

1 INTRODUCTION

Public transport has a great importance in day-to-day lives, whose main representative is the city bus. Drivers and collectors are agents that, a brief analysis, make the interface between the organization of road public transport urban passenger with society, interfering with the sense of security and well being of the community hall.

The behavior of these operators is of fundamental importance to the development of this activity, since failures may result in job losses endangering both society and the workers themselves.

Errors in the work of the collector can cause conflicts with passengers and financial losses, while errors in the work of the driver may cause accidents and endanger people's lives (Prange, 2010).

According Zanelato and Oliveira (2013), the bus driver is the professional driving vehicles of public companies, private and interstate, which triggers commands such as gait and direction, steering the vehicle by a predetermined route, in accordance with the laws of transit, in order to carry passengers to their destinations.

Costa (2006) reports that the jobs of drivers and collectors on trams began, and over the years, public companies were being privatized and they went to work for private companies. Thus, the author criticizes stating that despite the labor achievements, these professionals are still subjected to a sacrificial work, with the presence of noise above the limit, overload the musculoskeletal system due to repetitive movements and maintaining sitting posture for long period, high temperature, heavy traffic and vehicle vibration which can lead to dysfunction of any type of organism.

Lima et al (2003) corroborates this line of thought, saying that, like any worker, the driver and collector are exposed to a number of adverse conditions that could endanger health: cardiovascular, musculoskeletal problems, hemorrhoids, gastrointestinal problems, respiratory problems and stress.

According to ABHO (Brazilian Association of Occupational Hygienists) can classify risks working as environmental hazards (physical, chemical and biological) because they propagate the environment and safety hazards (mechanical, ergonomic) because they are static or due to unsuitability of the environment to man.

We define environmental risk under paragraph 9.1.5 of the NR-9 as: "... the physical, chemical and biological environments in existing work, due to its nature, concentration or intensity and exposure time are capable of causing damage to workers' health."

Thus, the working conditions have major influence on workers' earnings, which confirms the need for a study on the working conditions that aims at identifying solutions and tools that contribute to these professionals perform their tasks with the lowest risk of accidents and damage to health, with more comfort and well-being, thus increasing work efficiency.

This research proposes to survey and analyze the working conditions of these professionals so that they can intervene in safety and quality of life using the methodology proposed by Mont'Alvão & Moraes (2003).

2 METHODOLOGY

2.1 *Intervention methodology proposed by Moraes and Mont'Alvão*

To search for ergonomic analysis of a job we adopted the methodology of Intervention proposed by Moraes e Mont'Alvão (2003).

According to the authors the method × task × System Human Machine consists of five phases: Phase Assessment, Diagnosis Phase, the design stage; Phase Ergonomic Validation and Testing, and Phase Detailing Ergonomic and Optimization. However, for this research to be conducted in the metropolitan area of Recife, the first two major phases are considered sufficient to achieve goal of knowing what the real working conditions of these professionals, therefore, will be explained below in advance of what makes these two phases.

2.1.1 *Phase 1—assessment ergonomic*

This initial scanning is considered where problems are mapped in a broader sense. The assessment phase is divided into two phases: the systematization and questioning. In this large phase techniques are used as unsystematic observations in the workplace, with photographic and video records and interviews with users.

The systematization according Mont'Alvão and Moraes (2003), corresponds to the presentation of system models that characterize the serial position within the macro system, its hierarchical ordering, which corresponds to systems that are above and below the target system; expansion identifying who are the parallel systems, modeling and communication, in other words, as the system communicates with man.

The questioning is divided into three phases, namely: recognition of facts, development of problem and formulation of the problem. Recognition of the problem is the initial phase of questioning, where problems are identified that are more apparent in the first contact being parsed. The development of the problem is the selection and classification in detail the different aspects found from an observation further.

This stage ends with ergonomic advice are suggestions for improvements that are considered the first research hypothesis, there are ergonomic recommendations, since they can not be defined solutions to initiate projects.

2.1.2 *Phase 2—diagnose ergonomic*

According Mont'Alvão and Moraes (2003), is the Diagnose phase deepening problems found, sorted and prioritized in the previous phase. This phase is primarily based on Task Analysis which corresponds to the analysis, synthesis, interpretation, evaluation and processing requirements of the task in the light of knowledge and theories on human characteristics. Diagnosis is made up of four subphases, they are, macroergonômica analysis, behavioral analysis task; profile and voice operators and task analysis.

Macroergonômica analysis is the analysis of hierarchical communications company, workers and work organization.

Behavioral analysis of the task is the detailed study of the task, and it inserts: the activities, decision making, drives, oral and gestural movement, materials movement, postures and positions of the human body.

The profile and voice of the operators is the stage where collecting the opinions, complaints, reviews and suggestions from employees.

The analysis of the physical environment of the task is data collection physical environment, or illumination, noise, vibration, and temperature. Ends up the second major step, with ergonomic recommendations in relation to the environment, arrangement of jobs, task schedule.

Ergonomics has, in its scope, a variety of methodologies that address the job. Among them, the proposed intervention Ergonomizadora Mont'Alvão and Moraes (2003), was the basis for the choice of methods and techniques used in research into the working conditions of bus drivers and collectors of the Metropolitan Region of Recife.

2.2 *Sample size*

The target population for this research is the drivers and collectors of urban transport in the metropolitan area of Recife, a total of 19 urban public transport companies are used as a sample. It was observed that of the nineteenth, one uses the microbus, it made the driver also could play the role of the collector. The total sample is 185 professionals representing 5 collectors and 5 drivers of each company, noting that one of them has only driver.

2.3 *Methods and techniques for developing the chosen methodology*

The methods and techniques used in this study to collect data on the behavior of the driver and collector of urban transportation in the Metropolitan Region of Recife were:

First **observations** unsystematically not participating in the main bus terminal and the selected rows. In the second stage, diagnosis ergonomic, it is more objective observations, still unsystematic and not participating, but with some planning, using photographic records and filming.

Became non-structured focused **interviews** with coordinators/tax for each line studied to draw a profile of drivers and collectors.

Evaluated the job of the driver and collector through **cheklist**, based on checklists that help you quickly identify problems that recur in jobs.

We used the **questionnaire** for this study was divided into four sections: personal data; desktop: bus terminal and bus; conditions of physical and cognitive, and work organization. It was concluded by the **diagram Corlett**, with the object of mapping the presence of pain/discomfort perceived by the respondent.

3 RESULTS

3.1 *Results found in the assessment*

3.1.1 *Workplace and activity bus driver*

The job of a bus driver is composed of the urban public chair that can provide height adjustment, front distance with respect to proximity to the wheel and lateral distance. The steering wheel, gear rates which can be manual or automatic and pedals are also elements that compose this post. The control panel provides this professional information such as speed, fuel quantity and features buttons that light triggered the lights inside the vehicle. Like the dashboard, the door opening lever are also part of the driver's use of tools. And last but not least, are the internal mirrors located on the front and rear of the bus's interior and exterior mirrors located on both sides of the bus.

The activity bus driver consists of urban collective actions carried out simultaneously, repeatedly and over a long period of time and without large numbers of breaks. These are the actions to press the clutch, drive and change gears, accelerate and brake when necessary, turn the steering wheel positioning the vehicle in the desired direction; look mirrors to observe passengers and other vehicles; trigger buttons is to open doors or turn on lights, watch the control panel to learn the technical conditions of the vehicle, observe signs and traffic signs; safely transport passengers and efficient and carries all the value charged during labor to where it accounted.

3.1.2 *Workplace and activity bus collector*

The job of a bus collector urban collective comprises the chair that can provide height adjustment of the seat and backrest of the back, the foot support when there is, the cash drawer, where it is stored all the money raised during the work, the ratchet and ratchet release button, means of organizing and counting the number of passengers were transported during that turn, and is formed by the validator, which according to the Consortium Greater Recife, is a device that reads and transmits the card data VEM (Electronic Valley Metropolitan) to the central through wireless technology.

The activity of urban public bus collectors also constituted by actions carried out simultaneously, repeatedly and over a long period of time and without large numbers of breaks. These are the actions collect and receive money crossings; spend the change when necessary; observe whether the passenger is using the card correctly COMES; trigger the release button ratchet passage of each passenger; give information when they are asked, the passenger is inside/outside the bus or the driver; assist the bus driver; preventing or relieving action of walking inside the vehicle; responsible for maintaining order on the bus when necessary; responsible for informing the end of the workday the exact number registered at the turnstile, so as to present the company collects the amount, if it fails to amount, it is the responsibility of the collector to make up the difference.

3.2 *Some problems founded*

3.2.1 *The driver*

The highest prevalence of pain in the **right shoulder** to the left compared to bus drivers may be related to the constant need to shift. Furthermore, it is observed postures when changing gear, such as flexion and abduction of the right shoulder. Over time, the adoption of these movements can generate biomechanical overload the musculoskeletal structures may lead to myofascial pain syndromes, inflammation in tendons and bursae causing acute pain in the efforts of the upper limb. The initial framework can evolve to chronicity resulting in constant pain and loss of strength in the upper limb (Kapandji, 1990).

Meanwhile, the prevalence of pain in the **lumbar spine** may be related to vibration, with the extended sojourn in sitting posture and postures. Over time, the sustaining such a position can lead to complaints of back pain, early degeneration of the intervertebral discs and the emergence of hernias disk in lumbar region. These postures are harmful to employees by the fact much of the tension produced by them are transmitted to the spine (Magee, 2005). Also, overloading can generate tension in lumbar region and the intervertebral discs that are thin and somewhat resistant to pressure, favoring the occurrence of rupture disc (Kapandji, 1990).

Furthermore, the prevalences of **lower limb** pain may be related to the permanency of the seated posture throughout the working day and the flexion—extension of the lower limbs when using pedals bus. This situation can cause compression in the region of the thigh, and consequently, the blood vessels in the lower limbs (Mairiaux, 1992), altering blood flow in the region and facilitating the onset of muscle fatigue and pain.

The prevalence of pain in the **neck** in the drivers may be related to vibration and awkward postures encountered while performing the task. The need for attention and tension during the execution of work activities makes drivers remain in forward head, which can cause increased level of static tension of suboccipital muscles along with all the extensor muscles of the cervical spine to support the head (Viel, Esnault, 2000), which over time can cause neck pain, degeneration of intervertebral discs and disc hernias in the cervical region.

In this context, the prevalence of pain in the driver's **right arm** may be related to the large amount of flexion-extension movements of the elbow during the act of changing the march of the bus. Thus, over time and the long working hours can increase the biomechanical overload in the region, causing muscular fatigue, complaints of pain around the right arm and inflammation of the tendons that arm.

The bus driver on the left—Improper posture of head, spine and trunk musculature. Tilt anterior spinal flexion exacerbated associated with a depression of the neck and shoulder. Tendency to a forward head, with tension of the back muscles superiorly and a kyphosis of the thoracic spine.

The bus driver on the right—driver with trunk rotation, left arm with shoulder abduction, forearm supination and palmar handgrip. Right upper limb adduction of the shoulder, forearm pronated and wrist in radial deviation.

3.2.2 *The collector*

With respect to collectors, the prevalence of **lower limb** pain may be related to permanency in the sitting posture, absence of support for the feet, or not use the support, and inadequate postures of flexion of knees. This situation can cause excessive compression on the posterior region of the knee causing difficulty in venous return and compression of the sciatic nerve. This led to leg swelling and nerve compression symptoms like numbness or tingling in the legs (Chaffin, 2001).

The prevalence of pain in the **lumbar spine** may be related to vibration, prolonged permanence in sitting posture and postures. Over time, the sustaining such a position can lead to complaints of back pain, early degeneration of the intervertebral discs and the emergence of hernias disk in **lumbar** region. This situation is detrimental to the worker, since overcharging may generate tension in the lumbar region and the intervertebral discs that are thin and somewhat resistant to pressure, favoring the occurrence of rupture disc (Kapandji, 1990).

The prevalence of pain in the **neck** in the collectors may be related to vibration and awkward postures found during work. The association of these risk factors for long periods of time can cause biomechanical overloads in the region and lead to complaints of pain and tissue changes important, for example, early degeneration of the intervertebral discs and the emergence of cervical disc hernias.

The collector on the left—The maintenance of knee flexion will lead to increased internal pressure on this joint and the contact between the bony parts, generating biomechanical overload in this region.

The collector on the right—combinations of arm movements: right shoulder adduction is semiflexion with elbow (this time presents a compensatory movement of the scapula and a slight rotation of the shoulder girdle to the opposite side). While the contralateral limb delivers payback to the passenger with a radial deviation of the wrist.

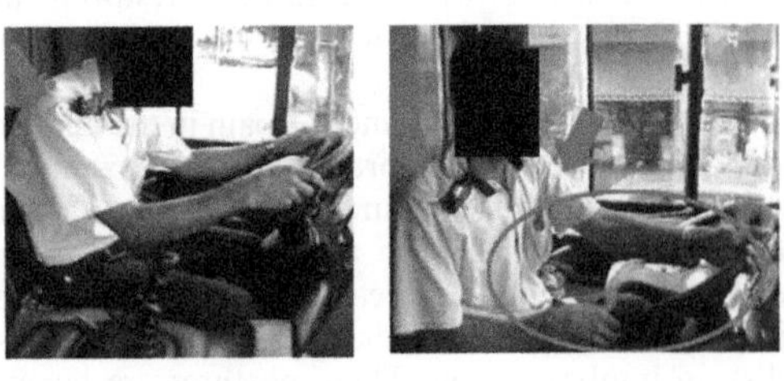

Figure 1. The bus driver.

Figure 2. The bus collector.

4 CONSIDERATIONS

With respect to risk assessment of ergonomic workstations driver and collector, can be identified:

The **driver** during the execution of their activities, spends a lot of time sitting with little time for breaks can cause circulatory impairment. The act of driving requires many twists trunk (side slopes associated with rotations of the trunk), with repetitive movements of the upper limb, which may cause postural problems. The engine next to the job hinders the passage causing the driver needs to elevate his legs to get as much out of his job, showing a dimensional problem of the job. Another issue to be studied is the effect of the sun, or during rain, often need to adopt awkward postures or compensatory measures to enable them to perform the task. Thus, the activity of the driver needs to maintain maximum attention, because any carelessness can cause an accident, besides being bombarded with various information such as traffic lights, the cars around you, pedestrian walkway, drive the bus, request entry/descent passengers, in other words, lots of information that must be processed instantly and coded so that it can respond quickly and safely.

The collectors adopts postures at rest or with anterior tilt arm raised above the shoulder to sustain and maintain balance. Often adopts a relaxed position with the support over the sacral region which can lead to back pain and overload the musculoskeletal system and may cause postural problems. During the execution of work activities, uses very repetitive motion clamp and handle with wrist movements, and movements associated trunk which can lead to the appearance of nervous disorders such as the carpal tunnel syndrome. How to spend much time with the legs dangling, if you have not or do not use the footrest, there may also be circulatory disorders. His job requires allow easy access due to the position of the arms of the chair, on some buses analyzed, there are retractable, and the position of the ratchet and cassette; since some collectors shorter stature need not dwell on the drawer so he can record the journey in electronic turnstile, showing dimensional problems. Thus, it is an activity that has great mental burden since it deals with money, because the collector is constantly required to receive and spend the change correctly, and often have to give information to users. It is also a job that requires the worker remains in a constant state of alert due to probable burglary/theft.

Organizational problems, both for the driver and for the collector, take into account breaks between trips, work rhythms, monotony, time to meet the route, which often comes all intertwined, because at the moment there is a traffic jam travel delays hence the stress increases, the breaks tend to decrease the pace of work becomes more pronounced and recovered by the controlling terminal becomes larger.

REFERENCES

Associação Brasileira de Normas Técnicas—ABNT. Normas Brasileiras NBR 15570—Transporte—Especificações técnicas para fabricação de veículos de características urbanas para transporte coletivo de passageiros. ABNT, 2009.

Chaffin, D.B., Anderson, G.B.J. e Martin, B.J. Biomecânica ocupacional. Belo Horizonte: Ergo, 2001.

Contandriopoulos, Andre-Pierre; Champagne, François, Potvin, Louise, et. al; Souza, Silvia Ribeiro de, Trad. Saber preparar uma pesquisa. 3 ed. São Paulo: Hucitec; Rio de Janeiro: Abrasco, 1999.

Costa, E.A.V.G. da. Estudo dos Constrangimentos Físicos e Mentais Sofridos pelos motoristas de ônibus urbano na cidade do Rio de Janeiro. Dissertação. Universidade Pontifícia Católica—PUC—RIO, 2006.

Kapandji, I.A. Fisiologia articular. vol 3: tronco e coluna vertebral. 5a ed. São Paulo: Manole. 1990.

Lakatos, E.M.. Fundamentos de metodologia científica/Marina de Andrade Marconi, Eva Maria Lakatos.—7. ed.—São Paulo: Atlas 2009.

Lima, A.A. et al. Estudo ergonômico do posto de trabalho do motorista de ônibus circular da linha Grajaú-Centro. Monografia de curso de Especialização. Pontifícia Universidade Católica do Rio de Janeiro, PUC/JR, 2010.

Mairiaux, Ph. Polígrafo curso Ergonomia, UCL—Universidade Católica de Louvain: Cap. V: A postura de trabalho, Cap. VI: Concepção do posto de trabalho, Bélgica, 1992.

Ministério do Trabalho e Emprego. Norma Regulamentadora do Ministério do Trabalho—NR 17—Ergonomia. Disponível em: http://portal.mte.gov.br/legislacao/normas-regulamentadoras-1.htm.

Moraes, Anamaria de; Mont'alvão, Cláudia. Ergonomia: Conceitos e Aplicações. 2 ed., Rio de Janeiro: 2AB editora, 2000.

Pádua, Elisabeth M.M. de. Metodologia da Pesquisa: Abordagem Teórico-prática/Elisabete Matallo Marchesini de Pádua.—10 ed. rev. e atual.—Campinas, SP: Papirus, 2004.

Prange, A.P.L. Quem dá mais, cobra mais! uma análise das normas antecedentes do ofício de motorista de ônibus em um contexto específico. Revista Estudos Pesquisas em Psicologia. Rio de Janeiro, v. 11, n. 2, 2010.

Viel, E.; Esnault, M. Lombalgias e Cervicalgias da Posição Sentada: conselhos e exercícios. São Paulo: Manole, 2000.

Zanelato, L.C.; Oliveira, L.C. Fatores Estressantes Presentes no Cotidiano dos Motoristas de Ônibus Urbano. 2003 Retirado do site: http://www.sepq.org.br/sitesipeq/pdf/poster1/08.pdf., em jun. 2013.

Occupational Safety and Hygiene II – Arezes et al. (eds)

A study of associations of occupational accidents to number of employees, and to hours worked

P.H. Marques
ISLA—Instituto Superior de Gestão e Administração, Santarém, Portugal
UNIDEMI—R&D Unit, Faculdade de Ciências e Tecnologia, Universidade Nova de Lisboa, Portugal

J. Atouguia, F.H. Marques, C. Palhais, A.R. Pinto, L.A. Silva, M.M. Silva, M.G. Andrade & P. Torres
ISLA—Instituto Superior de Gestão e Administração, Santarém, Portugal

V. Jesus
ISEGI—Instituto Superior de Estatística e Gestão de Informação, Univ. Nova de Lisboa, Lisboa, Portugal

ABSTRACT: This article is the result of a 4 hours class, in a MSc course, aiming to the group production of a paper. The object of study was the variation of occupational accidents with number of employees, hours worked and average duration of workday. Scientific goal was searching correlations between variables. Learning objective was demonstrating steps of scientific method, and the educational objective was breaking down prejudices about the research process. Starting from a brief literature review, research hypotheses were formulated. Subsequently, statistical methods were applied to a data set of occupational accidents with sick leave, average number of employees, hours worked per year, and average duration of the workday, of a large company, during two decades. Results showed statistically significant and strong direct associations of accidents to hours worked, as well as to number of employees. As to the association of accidents to average duration of workday, it appeared to be statistically significant, but of weak intensity. This essay also enabled to interpolate and extrapolate, with proper caution, accidents expected in a given workload. The skills developed become useful to research, as well as to manage labour risk prevention.

1 INTRODUCTION

1.1 *The formative context*

This article was based on a statistical laboratory class of 4 hours, of MSc in Management of Occupational Risks Prevention, for subsequent group writing of a scientific paper. The articulation of tasks assigned to 9 coauthors allowed the article to be completed in the correspondent 36 hours of work.

The scientific objective was to assess correlations between variables. The learning objective was to demonstrate the steps of the scientific method, and the educational objective was to break down prejudices about the difficulty of scientific research.

It started with a brief literature review about the object of study—the variation of occupational accidents with number of employees and with hours worked, as well as with the average duration of the workday.

From what has been learned from this (just enough to support the study) review, research hypotheses were formulated.

After that, descriptive and confirmatory statistics were performed with the available data—the occupational accidents with sick leave, average number of employees, hours worked per year, and average duration of the workday, of a company, during 20 years in a row.

Findings were discussed and (with proper caution) interpolated and extrapolated—estimating accidents expected in a given workload.

Finally, participants were asked to identify the skills developed that are useful to investigate, as well as to manage labour risk prevention.

1.2 *Background*

In this study, an accident at work with sick leave, is one resulting in incapacity for work that lasts longer than 3 consecutive days. These accidents may result in death or injury, sometimes permanently disabling, not only for work but also for a good quality of life.

The European Union statistics highlight the human cost and social consequences of labour accidents. Information on accidents at work are of great relevance for countries and organizations to better understand the importance of occupational health and safety. Thus, statistical data are essential for the prevention of accidents and a starting point to work safely (Hämäläinen, Saarela & Takala, 2009). According to *Gabinete de Estratégia e Planeamento* (GEP, 2008) (Office of Strategy and Planning of the Portuguese Government), statistical information on accidents at work aims to further and more rigorous analysis of accidents and should lead to more effective prevention measures. *«Statistics are the most common method of risk analysis, allowing the safety specialist effective knowledge of labour accidents and subsequent control prioritization of various risks »* (Miguel, 2000). Thus, literature recognizes that statistical analysis of occupational accidents and their results disclosure can contribute to its reduction.

Management of occupational risk prevention involves technical skills of risks evaluation and control, but also requires management skills, such as statistical analysis of accidents phenomena.

Statistical analyzes with different quality, allow different learning potential, as exemplified by four studies performed during this century, that are summarized below.

In a MSc research (Panzer, 2004) held in a metallurgical company in the Brazilian city of Porto Alegre, using only descriptive statistics, hours of work and accidents were recorded over two years. There were monthly fluctuations both in accidents and hours of work (with greater or lesser extension of overtime daily work), and these fluctuations were out of phase with each other. Investigator interpreted that work accidents were not related with overtime work because the rise and fall of both of them did not occur in the same month. However, looking at study data one can see, in several cases, that a month with an increase in hours of overtime was followed by growth of occupational injuries in the following month. By itself, this observation raises doubt—if this research had tested with inferential statistics the association between the mentioned variables, could have it reached a different conclusion?

In a review by (Salminen, 2010) concerning 8 studies from various countries presenting accident rates for 8 hours daily work compared with rates for 10, 12 and more than 12 hours, it was concluded that occurrence of injuries from accidents at work was:

- 41% higher for the journey of 10 hours per day, compared to 8 daily hours;
- 14% higher for 12 hours of daily work, compared to 8 daily hours;
- 98% higher when working over 12 hours a day, compared to 8 daily hours (in 3 studies).

On the other hand, in a publication of *Direction de l 'Animation de la Recherche, des Études et des Statistiques* (DARES), a survey of medical monitoring of exposure to occupational hazards, between 1994 and 2010, in France, revealed there was lower frequency of accidents during night work, but with a higher severity rate. Further concluded that increasing weekly working had a direct effect on the accident rate. On average, the incidents were 2.4 times more frequent when the weekly work was over 39 hours. In addition, there was a 40% increase in the risk of accidents, when working without interruption, due to the absence of the 48-hour weekly rest (DARES, 2012).

The latter study was done exclusively with descriptive statistics, but managed to evidence these results, due to the tendency of a large amount of data collected for 17 years. Similarly to this approach, during the 1st edition of the MSc in Management of Occupational Risks Prevention, at *ISLA Santarém*, a training of theoretical research addressed the relationships of occupational accidents and the average number of employees and of occupational accidents with total of hours worked per year, applying exclusively descriptive statistics over a large volume of data collected for 19 years. It has been concluded that, under those study conditions, there appeared to exist direct associations of the number of work accidents with the average number of employees and hours worked per year, but also that this conclusion would require validation through confirmatory statistics (Baptista *et al.*, 2012).

2 RESEARCH QUESTIONS

This work intends to respond to research challenges left by the previous study (Baptista *et al.*, 2012), which had examined almost all the same data—with just a year of records less than those currently available—now checking the conjectures made by those authors, about the possible relationships between variables.

This study uses confirmatory statistics to highlight relationships between accidents and other variables, which may or may not be intuitive. This intended to investigate correlations between the number of accidents, number of employees, hours worked per year and the average duration of working day. For this porpuse, the following hypotheses were formulated:

- $H1_{null}$: Work accidents do not vary with average number of employees;
- $H1_{research}$: Work accidents vary with average number of employees;
- $H2_{null}$: Work accidents do not vary with hours worked per year;

- H2 $_{research}$: Work accidents vary with hours worked per year;
- H3 $_{null}$: Work accidents do not vary with the average workday duration;
- H3 $_{research}$: Work accidents vary with the average workday duration.

3 METHODOLOGY

Methodologically, the research featured as descriptive and confirmatory statistics, focusing on quantitative data collection and analysis using the statistical tool SPSS Statistics Version 20 to test the previously formulated hypotheses.

The information processed consisted of publicly available data at official reports (former *Balanço Social* and current *Relatório Único*) of a large Portuguese company not interest in being identified. For sufficiently representative sampling of accidents at work, all considered data complied with that concept of work accident resulting in incapacity for work longer than 3 consecutive days. Data was collected during 20 consecutive years (from 1993 to 2012). Purposely, accidents with no sick leave were not considered, due to lack of guarantees of being so consistently reported over the years as accidents resulting in more than 3 days sick leave.

The studied variables were:

- "accidents with sick leave"—defined by the total work accidents that resulted in incapacity for work for a period exceeding 3 consecutive days, in each of studied years;
- "average number of employees"—defined by the weighted average number of workers, in each of studied years;
- "hours worked per year"—defined by the total hours worked by all the workers, in each of studied years;
- "average workday duration"—defined by the average daily hours worked per individual, in each of studied years.

After descriptive statistic analysis of variables behavior, we performed confirmatory statistic analysis, to test the association between "accidents with sick leave" and each of the other variables, adopting a significance level of 5% (confidence interval of 95%).

4 RESULTS

Figure 1 shows the behavior of each variable over time, converted into the same order of magnitude.

Despite annual fluctuations, one can see that the exponential trends of "accidents with sick leave", "average number of employees" and "hours worked per year", seem to coincide in decreasing. The smaller fluctuations of "average workday duration" are hardly noticeable on the scale represented, and do not appear to have a comparable behavior to accidents.

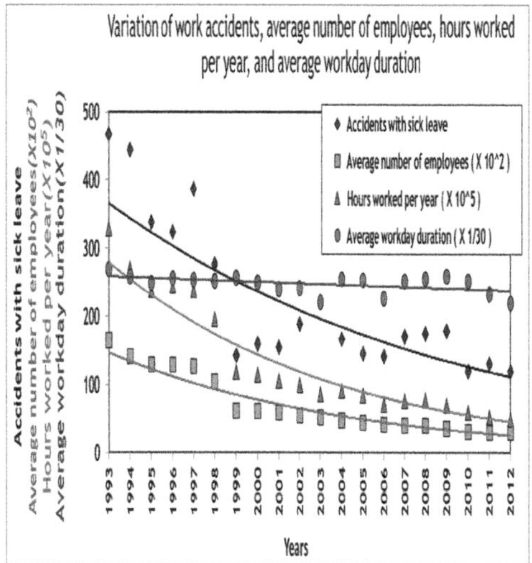

Figure 1. Variation of work accidents, average number of employees, hours worked per year, and average workday duration, over 20 years.

Table 1. SPSS output of Pearson test.

	Hours worked per year	Average number of employees	Average workday duration
Accidents with sick leave			
Pearson correlation	0.959	0.959	0.491
Sigma (2-tailed)	0.000	0.000	0.028

Table 1 shows results of Pearson test for the linear correlation of variable "accidents with sick leave" with each of the other variables. In all cases, the Sigma 2-tailed is less than the 5% significance adopted. This means that the null hypothesis (independence of variables) can be rejected, so one can accept all the research hypotheses H1, H2 and H3—that is, there are statistically significant associations between the dependent variable and each independent variable.

For a confidence interval of 99%, test values reveal very strong correlations between the dependent variable "accidents with sick leave" and each of independent variables "average number of employees" and "hours worked per year".

As for the correlation between the dependent variable "accidents with sick leave" and independent "average workday duration", it can be seen that correlation is weaker.

The behavior of the dependent variable "accidents with sick leave" with variation of each of independent variables is represented in Figures 2, 3 and 4.

5 DISCUSSION

The coincidence of decreasing trends, visible in the descriptive statistics, allow only to suspect of direct associations of yearly accidents to employees and to hours worked.

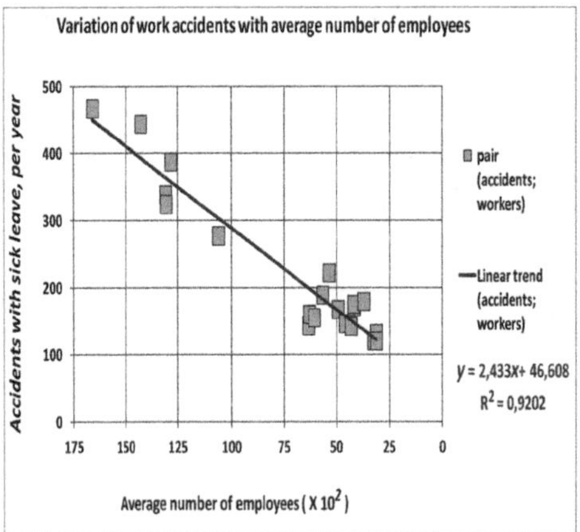

Figure 2. Variation of work accidents with average number of employees.

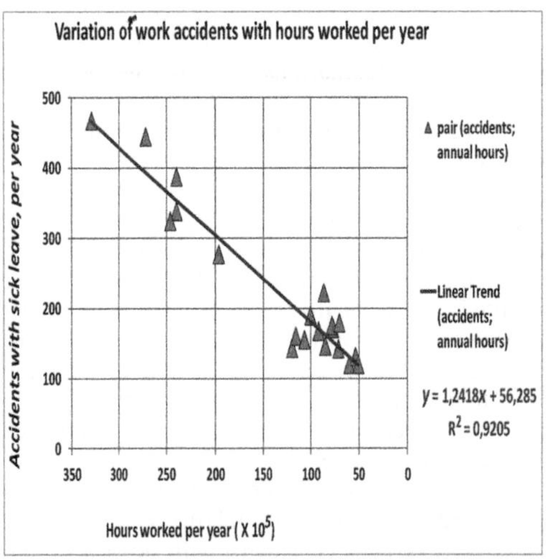

Figure 3. Variation of work accidents with hours worked per year.

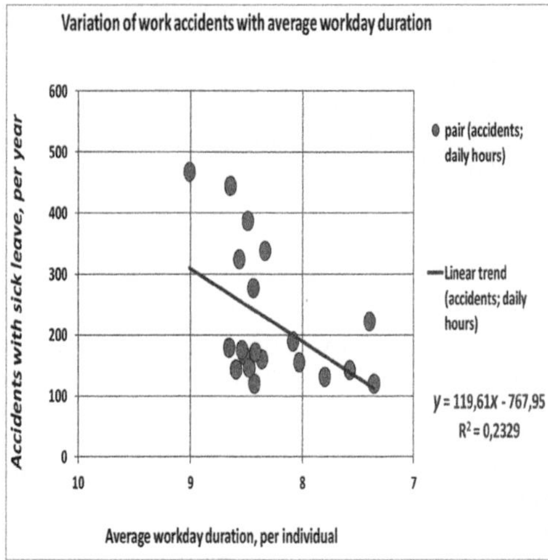

Figure 4. Variation of work accidents with average workday duration.

On the other hand, through the association tests, it can be stated that accidents tend to vary directly with the number of workers and hours worked per year. Linear trends of direct correlations observed between the average number of employees and accidents, as well as between hours worked annually and accidents, are very strong (Pearson correlation = 0.959 and a goodness of fit of $R^2 = 0.92$). Such trends allow to interpolate and extrapolate the expected number of accidents due to a given workload—with appropriate reservations, considering the levels of analysis that were studied (the level of an organization), of the study period (two decades) and if all other risk factors keep constant.

In the case of association test of accidents to the average duration of the working day, this suggests a direct variation of accidents with the duration of the working day. Covariance of these variables can also be represented by linear trend line, but with a weak correlation (Pearson correlation = 0.491 and a goodness of fit of $R^2 = 0.23$). This result converges with the ones founded in literature (Salminen, 2010) (DARES, 2012), but is not as evident as in those studies. Although present results are consistent with an increase in propensity of workers to suffer accidents, as the duration of working day is prolonged and fatigue is accumulated, this effect is not that much shown in the present study, maybe due to a smaller range of the average daily working hours—in this case, ranging from just more than 7 to less than 9.5 hour working day.

Taking into account the population size and the period studied, it is legitimate to extrapolate these results to comparable conditions. It follows that, when a greater workload is expected, then, to prevent accidents increase, risk control measures must be strengthened. However, these results only allow speculation to other levels of analysis (individual, region, country or world economy), or for other periods (daily, weekly, monthly, annual, secular or millennial)—because, in significantly different conditions, the relationships may differ.

In future studies, it would be interesting to compare in detail, accidents occurred during periods of overtime and those affecting workers in normal work journey, all other things being equal.

6 CONCLUSION AND FINAL COMMENTS

Study conditions confirm that accidents with sick leave vary directly with the independent variable "average number of employees" and also with "hours worked per year"—these last variables having strong predictive value over work accidents. The later variable also might directly depend on "average workday duration" but, in this case, the covariance follows a trend of weak correlation—meaning that the length of the working day has no sufficient predictive value of work accidents. Hence the need for studies of accidents with larger amplitudes of workday hours.

If it is proven that the overtime work causes more accidents, what should be done to prevent it? An obvious answer would be to have the sufficient number of workers for more hours of work, without being necessary individual extra time of work. This implies a resilient response to increased workload. It might be achieved by a pool of workers able to perform the functions occasionally, that could be recruited among workers who have left the organization or who have retired.

By continuing a previous study about almost the same data, this paper innovated by confirmatory statistical analysis and the study of another association.

Pedagogical and didactic aspects remained valued, because this accessible exercise, done in a MSc class, served simultaneously to:

- break down prejudices about the difficulty of producing scientific work, by showing that it is not an impossible task, you do not need endless time, that does not necessarily imply trial or only isolated individual work;
- demonstrate how to meet fundamental requirements of scientific work, such as the representativeness of data (using reliable data in sufficient quantity), the explicit objectives (with clear and well-formulated hypotheses), the methodology robustness (applying well-established and appropriate to data statistical analysis), the conclusions being limited to evidence found, and the reproducibility of the study.

Thus, the skills developed in this study proved to be useful for the production of scientific knowledge, in general, and to management of occupational risks prevention, in particular. This approach exemplifies how professionals who manage risks at work can contribute to decision-making based on evidence resulting from organizational learning that is possible by analyzing the data of their own organizations.

REFERENCES

Baptista, F.D., Marques, F.H., Nunes, L.M., David, M.S., Silva, S.M., Marques, P.H., 2012. Relationships between occupational accidents and the number of employees, and between occupational accidents and total of hours worked, In: Neves *et al.* (Eds), *Vertentes e Desafios da Segurança 2012*, Leiria, 25–26 oct 2012: 30–37.

DARES, 2010. L'Évolution des Risques Professionnels dans le Secteur Privé entre 1994 et 2010: premiers résultats de l'enquête SUMER. *Dares Analyses*, 23: 1–10.

Gabinete de Estratégia e Planeamento, 2008. *Acidentes de Trabalho—2006*. Lisboa: Coleção Estatística. Retrieved from GEP: http://www.gep.msss.gov.pt/.

Hämäläinen, P., Saarela, L., & Takala, J., 2009. Global trend according to estimated number of occupational accidents and fatal work-related diseases at region and country level. *Journal of Safety Research*, 40: 125–239.

Miguel, A.S., 2000. *Manual de Higiene e Segurança do Trabalho*. Porto: Porto Editora—5th edition.

Panzer, R.A., 2004. Correlação existente entre horas extras e acidentes de trabalho, queixas de dores osteomusculares, absenteísmo e produtividade em uma empresa metalúrgica de Porto Alegre, nos anos 2002 e 2003. *Trabalho de Mestrado*. Porto Alegre, Brasil: Universidade Federal do Rio Grande do Sul: 112.

Salminen, S., 2010. Shift Work and Extended Working Hours as Risk Factors for Occupational Injury. *The Ergonomics Open Journal*, 3: 14–18.

Occupational Safety and Hygiene II – Arezes et al. (eds)
© 2014 Taylor & Francis Group, London, ISBN 978-1-138-00144-2

3D human thermoregulation model: Bioheat transfer in tissues and small vessels

J.C. Guedes, J. Santos Baptista & J.M. Soeiro de Carvalho
Research Laboratory on Prevention of Occupational and Environmental Risks (PROA/LABIOMEP), Faculty of Engineering, University of Porto, Portugal

ABSTRACT: There are several models that can predict human physiological response to heat and cold environment. However there are only a few models that simulate and predict the temperature distribution all over the body, considering the individual differences and transient conditions. The main goal of this study is to present part of a whole body, three-dimensional thermoregulation model that simulates biological heat transfer in tissues and small vessels. The development of the 3D model begins by solving the bioheat equation in one dimension and then adapting it to a more complex problem. The final 3D model test shows that the model behaves as expected.

1 INTRODUCTION

1.1 *Context*

In occupational health and safety it is important to predict the risk of exposure to environmental hazards, especially when the exposure meets human tolerance limits.

The ergonomics of thermal environment studies the effect of a combination of environmental physical variables on human thermal response. In spite of the several studies and patterns developed to predict thermal comfort or discomfort (Moran, Shitzer, & Pandolf, 1998; Fiala, Lomas, & Stohrer, 1999; Malchaire, 2006), only a few combine the prediction of thermo-regulation of the whole body model with expousure to extreme hot conditions, concerning the individual differences (Havenith, 2001), real-time results (Yokota et al., 2008), transient environments and three-dimensional heat diffusion (Gonzalez, 2004).

1.2 *Model presentation*

This article presents a small part of a whole body thermo-physiological model that intends to predict human physiological response to extreme heat conditions. The model can be divided into two main modules that treat, respectively, the passive and the active system of the human thermoregulation.

The passive system includes the modeling of the thermo-physical properties of the human body as blood, vessels, tissues, body size and geometry; and considers the important role of circulatory system in heat dissipation.

The active system includes the mechanisms that allow the body to adapt by itself to stressful hot environments (such as sweating and vasodilation).

The module now presented is the basis of the passive system. It allows predicting heat dissipation throughout any part of the body considering the geometry of the segment and its composition (i.e. takes into account the size and shape of the body element and its internal constitution as well as the distribution of the various kinds of tissue). It also allows calculating the spatial over time distribution of the temperature in different body parts, considering environmental exposure and body condition at exposure time.

2 MATHEMATICAL EXPOSITION OF THE MODEL

2.1 *Differential equation for modeling tissue and small vessels*

Heat transfer through tissue and small vessels is modeled by classical heat diffusion equations and considers the properties and geometry of tissue of different body parts.

In the inner layers the heat conduction through the tissues follows the heat conservation laws, being the bioheat equation, presented by Pennes (1948), an evolution of the general equations that describe heat conduction and heat diffusion.

So the Equation 1 (bioheat equation), as used in this work, is the main module of the heat transfer between the inner and outer and between the

Table 1. Summary of the classification of physical behavior of the equations of heat conduction problems [adapted from: Versteeg & Malalasekera (2007)].

Problem type	Equation type	Prototype	Conditions	Solution
Equilibrium problem	Elliptic	$D\nabla^2\phi + S = 0$	Initials	Closed domain and stable
Dynamic problem with heat dissipation	Parabolic	$\partial\varphi/\partial t = D\nabla^2\varphi + S_\varphi$	Boundary and initials	Open domain and stable

tissues that form the human body (bone, muscle, fatness and skin) (Albuquerque Neto, 2010):

$$\rho_t c_t \frac{\partial T_t}{\partial t} = k_t \nabla^2 T_t + \hat{\dot{V}}_{sv}\rho_{bl}c_{bl}(T_{ar} - T_t) + \hat{q}_t \tag{1}$$

where ρ_t is the specific weight of the tissue in $Kg \cdot m^{-3}$; c_t is the specific heat of the tissue in $J \cdot kg^{-1}.°C$; T_t is tissue temperature in $°C$; k_t is the thermal conductivity of the tissue in $W \cdot m^{-1} \cdot °C$; $\nabla^2 T_t$ is temperature gradient over the tissue in $°C \cdot m^{-1}$; $\hat{V}_{sv}$ is the arterial blood flux that flows throughout each body part in; ρ_{bl} is $m^3 \cdot m^{-3} \cdot s$ the specific weight of the blood in $Kg \cdot m^{-3}$; c_{bl} is the specific heat of the blood in $J \cdot kg^{-1}.°C$; T_{ar} is arterial blood temperature in $°C$; $\hat{q}_t$ is the tissues internal heat production in $W \cdot m^{-3}$.

The temperature gradient form depends on the geometry of the body element and the number of dimensions considered. The gradient in three-dimensional heat diffusion using orthogonal coordinates, assumes the standard form:

$$\frac{\partial^2 T_t}{\partial x^2} + \frac{\partial^2 T_t}{\partial y^2} + \frac{\partial^2 T_t}{\partial z^2} \tag{2}$$

where x,y,z are the orthogonal coordinates in m.

From Equation 2 we can get temperature gradient for one or two-dimensional diffusion, by eliminating the element corresponding to the coordinates to which heat conduction is null.

2.2 *General energy conservation equation*

Physical, chemical and biological systems and processes that change both in space and time are modeled by PDEs that remarkably assume a similar form (eq. 3) Guyer et al. (2009):

$$\frac{\partial \phi}{\partial t} = H(\phi, \mu_i) \tag{3}$$

where ϕ represents the variable of interest that evolves over time and considering other influencing state variables μ_i. The general energy conservation equation can be defined by each combination of the transient, convection, and diffusion or source terms.

$$\left[\frac{\partial \rho\phi}{\partial t}\right]_{transient} + \left[\nabla(\vec{u})\phi\right]_{convection} = \left[\nabla(D\nabla\phi)^n\right]_{diffusion} + \left[S_\phi\right]_{source} \tag{4}$$

The bioheat equation comes from the energy conservation equation (eq. 4). It represents a particular case where the convection term is null. The terms of Equation 1 can be classified as follows:

- $\rho_t c_t\ (\partial T_t/\partial t)$—represents the transient term of the equation, this is, the time evolving temperature;
- $k_t \nabla^2 T_t$—is the spatial diffusion term, that represents the temperature variation over the geometries;
- $\hat{\dot{V}}_{sv}\rho_{bl}c_{bl}(T_{ar} - T_t) + \hat{q}_t$—is the source term, responsible for the endogenous heat production and heat provided to the element through blood flow.

The steady-state form of the equation can be set by equaling the transient term to zero. So the solution of the bioheat equation can assume an elliptical form, if the solution does not depend on time, otherwise it takes the form of a parabolic equation. Table 1 resumes the practical aspects to take into account when solving each of the two types of problem.

3 MODEL SOLUTION

3.1 *Program implementation*

The model solution was implemented using a high level computer language (python) and open n source software to allow a sustainable development of the model and studying more complex mechanisms of thermal diffusion.

3D meshes were created using Gmsh assisted by a finite volume method solver using python—FiPy. Results are presented by plotting 4D graphs over time using a 3D Scientific Data Visualization and Plotting software, MayaVi.

3.2 *Description of the simulated element*

The body element simulated is considered a simple geometrical element divided into 4 layers, from inside out respectively, bone tissue, muscle, fat and skin. This division is considered, by some authors, to be detailed enough to produce good results and simple enough to implement (Ferreira & Yanagihara, 2009).

The values used to build a body element structure (the thickness of the different layers and the size of the body element structure) come from Werner & Buse (1988). Thermo-physical properties, specific weight of the tissues, specific heat and thermal conductivity were atributed to each layer according to Werner & Buse (1988), Ferreira & Yanagihara (2009) and Albuquerque Neto (2010). The internal basal heat generation came also from Werner & Buse (1988) with the exception of bone tissue that was corrected in accordance with Albuquerque Neto (2010). Tissue blood flow at rest for each layer came from Werner & Buse (1988) too. All the mentioned values are listed on Table 2.

Other assumptions took into account standard values:

- Arterial temperature was set at 37°C;
- Initial temperatures of the tissues were: 36.5°C for bone; 36.0°C for muscle, 34.0°C for fat and 33.0°C for skin.

The model also keeps common boundary conditions that intend to simulate the body part exposed to a specific uniform environment. As in real exposure the outer skin is in contact with hot air, in the model boundary condition the exterior faces of the mesh were set to 35°C in 1D algorithm and 40°C in the 3D algorithm.

3.3 *Development and testing methodology*

The implementation of the model started by modeling the simplest case (heat diffusion through a cylinder with four layers), solving the heat equation in one dimension. Next step was to include the effect of the different thermo-physical properties

Table 2. Summary of tissues' thermal properties and radius of body part [adapted from Ferreira & Yanagihara (2009) and Werner & Buse (1988)].

Tissue	Radius	ρ_t	c_t	k_t	$\hat{V}_{SV}$	$\hat{q}_t$
Bone	0.0268	1357	1700	0.75	0	0
Muscle	0.0540	1085	3800	0.51	483	501
Fat	0.0571	920	2300	0.21	398.7	4
Skin	0.0582	1085	3680	0.47	362	368
Blood	–	1059	3850	–	–	–

*All the presented values are in SI units.

of the four layers. After it, program behavior was tested and calculations were compared with the expected results.

Once the 1D model completed and tested, we began to develop the 3D model, based on it. The entire program was developed concerning a modular architecture. This kind of architecture will make easier future upgrades to the model.

Several conditions were tested to see if the boundary and initial conditions generate the expected diffusion results (longitudinal diffusion and radial diffusion in two directions) according to thermodynamics laws.

Finally the 3D model behavior was validated according to literature by testing the influence of:

- internal heat production;
- arterial temperature;
- tissue vascularization;
- outer temperature of contact;
- initial tissue temperatures.

4 RESULTS

4.1 *1D heat diffusion program*

All the tests were performed for 3600 seconds with a time step of 20 seconds. To compare with infinite exposure the Steady-State (SS) result was also calculated.

The 1D program uses the values presented in 3.2 paragraph. However, to test the behavior of the model, we changed one variable at a time.

Figure 1 shows the model results in extreme conditions, when the arterial blood temperature was set to 0,0°C. From the inner center of the considered

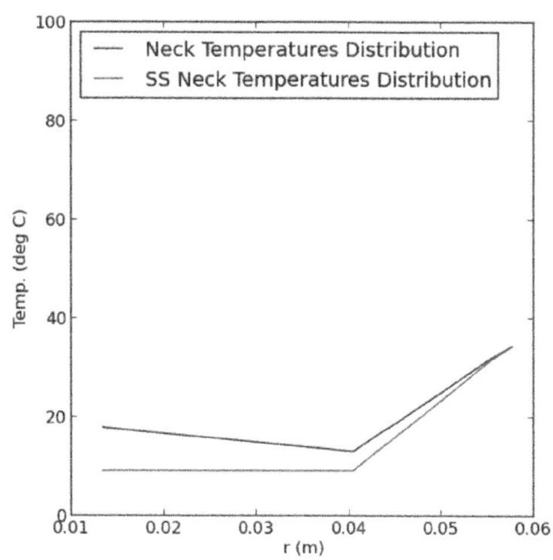

Figure 1. One dimension heat diffusion considering a 0,0°C arterial blood temperature. (SS means steady-state).

body element (r = 0 m) to the outside (r = 0.0582 m), the temperature values decrease over time with the exception of the outer face which remains at 35°C (the considered air temperature that keeps the exterior faces at constant temperature). The black curve shows the temperatures in the body component after one hour. After this time lapse, temperatures in the bone and muscle were below 20°C and could go to under 10°C (Table 3).

When the arterial blood temperature is changed to 100°C, the model responds as expected. The inner temperature of the tissues increases, but the outer temperature remains similar to the imposed boundary condition (Table 4).

By increasing internal heat production ($\hat{q}_t$) 100 fold, internal temperature values increase as well, according to the expected results, however not as much as by changing arterial blood temperature (T_{ar}). This shows that the model is much more sensitive to changes in arterial blood temperature as mentioned by Pennes (1948) and Albuquerque Neto (2009) (Table 5).

Table 3. Temperature distribution over the body element with 0,0°C arterial blood temperature (T_{ar}).

Tissue	Radius	t = 3600s	Steady-state
Bone	0.01340	17.9790	9.2050
Muscle	0.04040	13.0900	9.2050
Fat	0.05555	32.1170	31.7810
Skin	0.05765	34.4280	34.3670

*Radius is presented in meters and temperature in °C.

Table 4. Temperature distribution over the body element with 100,0°C of arterial blood temperature (T_{ar}).

Tissue	Radius	t = 3600s	Steady-state
Bone	0.01340	68.0584	83.43913
Muscle	0.04040	76.6108	83.43913
Fat	0.05555	40.43432	41.02473
Skin	0.05765	36.07776	36.18416

*Radius is presented in meters and temperature in °C.

Table 5. Temperature distribution over the body element with internal heat production ($\hat{q}_t$) increased 100 times.

Tissue	Radius	t = 3600s	Steady-state
Bone	0.0134	48,98686197	55,16724198
Muscle	0.0404	52,41970119	55,16724198
Fat	0.05555	36,58258168	36,82014851
Skin	0.05765	35,32605937	35,36887498

*Radius is presented in meters and temp. in °C.

The model showed sensitivity to the increase and decrease of blood flow.

If the arterial blood temperature is high and blood flow is also high, the temperature distribution in tissues tends to increase. If the arterial blood temperature is low and blood flow is also low, the temperature distribution in tissues tends to decrease.

The direction of the heat flux also changes with the change of the boundary condition. If the outer faces are kept at 35°C the heat flows from the right to the left (as seen above), otherwise if the boundary condition is changed to a value much below the core temperature the result is the opposite.

4.2 *3D heat diffusion program*

The 3D model was tested at one dimensional diffusion in z, x and y coordinates.

Figure 2 shows the z coordinate heat diffusion by applying a constant hot temperature at the top face of the cylindrical mesh. As expected that produces a unidirectional heat flow, seeing fastest changes in more conductive tissues (the inner bone and muscle physical volumes) that easily reach higher temperatures.

The heat flux also changes direction when boundary condition change from the top faces to the bottom faces of the mesh.

The distribution variation of the temperature absolute values has a similar behavior to the 1D model. The temperature reaches higher values increasing the arterial temperature than with a higher internal heat production. Also vascularization leads to higher or lower temperatures depending if the temperature of the arterial blood is higher or lower than the initial values.

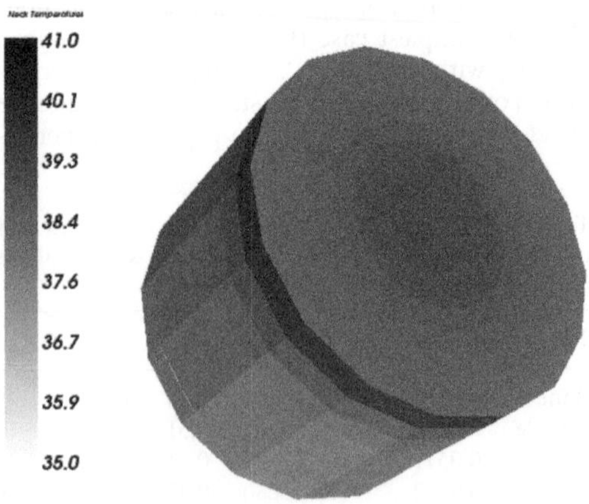

Figure 2. Three dimensional heat diffusion along z coordinate.

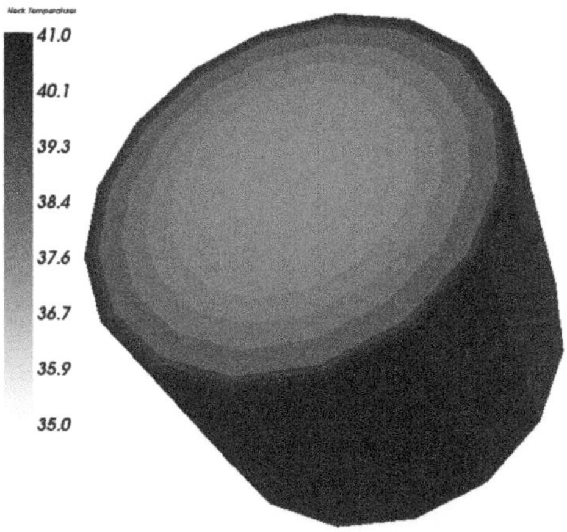

Figure 3. Heat flux from outside-in over the radial direction.

In Figure 3 the boundary condition kept the exterior faces of the body element that contact with the air at a constant 40°C temperature. The result is a unidirectional heat flow from outer cells to inner cells in radial direction. Remember that initial temperatures of the tissues are equal in all the tissue volume layers. So the result are circular isometric temperature curves.

5 CONCLUSIONS

The 3D model behaves as expected considering the thermo dynamical rules, is sensible enough to variable changes and respects literature conditions.

Although the tests only consider unidirectional heat diffusion, the model is prepared to be fully tested considering:

- heat diffusion in 3 directions;
- individual characteristics of human body;
- more complex boundary conditions;
- more complex initial conditions.

The program scheme follows a modular structure, easily adapted to new conditions. The use of high level programming language and open source software will allow free further improvement of the model and use in particular cases.

After the whole body system is completed and validated, the program will be upgraded to consider the active system (thermoregulation mechanisms) and a whole body energy expenditure model. Several clothing models can also be easily adapted to the program.

This model is detailed enough to be used to predict local discomfort issues and its effects on the whole body. Issues with outdoor or indoor radiant heat can also be included. A further possibility to integrate this model with common CFD software is being studied.

With a complete 3D model of the human body that predicts physiological response to environment and physical exertion we will be able to test human physical performance, suitability of environment to a specific task, time exposure limits to extreme temperatures, harder and safer training programs to the adaption to extreme environmental conditions.

REFERENCES

Albuquerque Neto, C. (2010). Modelo integrado dos sistemas térmico e respiratório do corpo humano. Tese de Doutorado, Escola Politécnica, Universidade de São Paulo, São Paulo. Recuperado em 2013-09-20, de http://www.teses.usp.br/teses/disponiveis/3/3150/tde-28022011-124824/.

Ferreira, M.S., & Yanagihara, J.I. (2009). A transient three-dimensional heat transfer model of the human body. *International Communications in Heat and Mass Transfer, 36*(7), 718–724. doi: http://dx.doi.org/10.1016/j.icheatmasstransfer.2009.03.010.

Fiala, D., Lomas, K.J., & Stohrer, M. (1999). A computer model of human thermoregulation for a wide range of environmental conditions: the passive system. [Research Support, Non-U.S. Gov't]. *J Appl Physiol, 87*(5), 1957–1972.

Gonzalez, R.R. (2004). SCENARIO revisited: comparisons of operational and rational models in predicting human responses to the environment. *Journal of Thermal Biology*, 29(7–8), 515–527. doi: http://dx.doi.org/10.1016/j.jtherbio.2004.08.021.

Guyer, J.E., Wheeler, D., & Warren, J.A. (2009). "FiPy: Partial Differential Equations with Python," *Computing in Science & Engineering* 11(3) pp. 6–15, doi:10.1109/MCSE.2009.52, http://www.ctcms.nist.gov/fipy.

Havenith, G. (2001). Individualized model of human thermoregulation for the simulation of heat stress response. [Research Support, Non-U.S. Gov't]. *J Appl Physiol*, 90(5), 1943–1954.

Malchaire, J.B. (2006). Occupational heat stress assessment by the Predicted Heat Strain model. [Research Support, Non-U.S. Gov't Review] *Ind. Health*, 44(3), 380–387.

Moran, D.S., Shitzer, A., & Pandolf, K.B. (1998). A physiological strain index to evaluate heat stress. *Am J Physiol*, 275(1 Pt 2), R129–134.

Pennes, H.H. (1948). Analysis of tissue and arterial blood temperatures in the resting human forearm. *Journal of Applied Physiology*, 1(2), 93–122.

Werner, J., & Buse, M. (1988). Temperature profiles with respect to inhomogeneity and geometry of the human body. *Journal of Applied Physiology*, 65(3), 1110–1118.

Yokota, M., Berglund, L., Cheuvront, S., Santee, W., Latzka, W., Montain, S., Moran, D. (2008). Thermoregulatory model to predict physiological status from ambient environment and heart rate. *Comput Biol Med*, 38(11–12), 1187–1193. doi: http://dx.doi.org/10.1016/j.compbiomed.2008.09.003.

Occupational Safety and Hygiene II – Arezes et al. (eds)
© 2014 Taylor & Francis Group, London, ISBN 978-1-138-00144-2

Work organization and the interfaces with occupational stress

J.C.S. Pessoa
Universitary Center of João Pessoa, João Pessoa, Paraiba, Brazil

L.S. Nascimento Júnior
Federal University of Rio Grande do Norte, Health Science College, Santa Cruz, Rio Grande do Norte, Brazil

M.L.C. Santos
Federal Institute of Paraíba, João Pessoa, Paraiba, Brazil

ABSTRACT: This study aimed to analyze the relationship between the work organization in a call center and occupational stress. Therefore, we seek to identify the main factors which cause occupational stress, investigating its signs and symptoms, relate it to the physical and organizational situations presents in the work environment of the subjects. Individual interviews were conducted, with an application of a form planned in advance, and the technique of direct observation. With these results, it was established that there are ergonomic and organizational deficiencies, and the presence of signs and symptoms of the condition under study. Soon, it was realized that the participants are exposed to certain occupational stress triggers, whether due to direct and/or indirect physical load, highlighting the need for ergonomic suitability for the job activity, as is located in a reserved environment and that there is organizational strategies that enhance human capital.

1 INTRODUCTION

Nowadays, the stress, which is a process of human adaptation to the demands of the environment, has shown a negative connotation—distress—in different work environments, making the individual susceptible to errors and accidents, hindering their productivity, ability to work and quality of life. It can be considered that stress is a summation of the responses of the organism to cope with stressors.

The results of these reactions are related to the way the individual react depending on your mental, physical and emotional balance (Fiori, 1997). The tendency to investigate and correlate stress with the organizational aspects of the work is due to the fact of human spend most of their time in this environment, and the factors that compose it, have a great influence on their satisfaction and quality of life.

In this context, this paper seeks to clarify how the working conditions of the operators in a call center can be considered triggering sources of occupational stress, trying to relate the organization of work, characterized by promoting monotony, repetitiveness and fatigue with the triggering of the distress in this professional who works on risky situations for your voice, your body and your mind (Carrer et al., 2012; Constancio et al., 2012).

The performance of any work activity involves understanding the interaction and interdependence of at least three aspects: physical, cognitive and emotional, that through the analysis of neurotic syndrome of telephone operators, studies by Le Guillant in 1956, it was possible find that work demands may develop headaches, tinnitus, hissing and obsessive thoughts about work, as the repetition of stereotyped fragments of speeches, and changes in sleep and mood (Añez et al., 2005).

With this, it is required the applicability of the knowledge of ergonomics in order to identify alternatives that reduce useless sacrifices imposed on workers, and create work environments that are more cooperative and motivators. Therefore, this study has significant importance since it provides an opportunity to raise awareness as to the necessary actions for its prevention and treatment in the academic and scientific areas of the various of human knowledge.

2 METHODOLOGY

The research was comprised of officials from central call office of the Federal Center of Technological Education of Paraíba—CEFET/PB—Brazil, with a total of three operators. This universe was defined by the criteria of workers acceptance and was constituted as follows: indication of a focus group that was related to the research topic, and he was exposed to possible triggers of occupational stress.

For data collection, we have used the form, interviews and direct observation. The form was administrated to each operator in the actual work environment through the interview technique. The form was characterized by objective and subjective questions, divided into five parts: (a) occupation and work organization, (b) assessment of environmental conditions, (c) postures during work, (d) general health and (e) stress symptoms. The development of this instrument was based from information provided by the authors: Oliveira (1991), Goldberg (1996), Deliberato (2002) and Lipp (2005).

During the form application, was also allowed direct observation of the completion of the surveyed work activities.

The collected data were treated qualitatively and quantitatively, but is important to say that only the item relating to general health made exclusive use of the qualitative approach.

3 RESULTS AND DISCUSSION

3.1 *Characterization of the subject*

The research was formed of three employees of the central office of the Federal Institute of Paraiba (IFPB), all of them are female, aged bounded to 33.3% between the ages of 20–29 years, and 66.7% between the ages of 30–39 years, with a minimum age of 24 years and maximum age of 38 years.

It was observed that two of the employees are in this function work time ranging from 1–10 years, while one works 11–20 years. Denoting the long period that operators are exposed to stress factors relating to both the desktop and the activity itself, making all physical and chemical changes, used by the body as an adaptation strategy against the stressor, are inadequately maintained for a long period of time, which may lead to chronic stress, thus making the organism more susceptible to infections and physical and/or mental diseases (Rossi et al., 2010).

3.2 *Evaluation of occupation and work organization*

According to the observation made, the basic activity developed by the operators of this call center is to make and answer external and internal calls through one telecommunication equipment that is placed at a table with rectangular and whose commands are operated by audible or visual signals. To carry out the activity, operators, use headphone, constantly, perform movements of extension and flexion of the fingers on the commands.

These actions were performed in three shifts, as follows: the morning shift responsibility of an operator. For the afternoon, an employee responds to the needs of work and the night shift is fulfilled by another employee, making a workload of five hours a day, five days a week by telephone, with no practice of extra hour.

It was noted that there are no breaks at work, however they have the possibility to leave the job when affected by physiological needs.

Regarding the autonomy in the execution of actions, two workers reported not own it. According to Rosh (1996 apud Montoro, 1998), this activity is considered as one of the functions subjected to stress, since it has many responsibilities, but few chances in the processes of decision making and control.

Of the total, two considered their activity as monotonous. The monotony, according to Iida (2005), is considered a stressor, due to job dissatisfaction, especially in organizations where the work pace is steady.

The fact is that the activity of operator calls is considered how a standardized work, overly controlled and repeatable, because according Wells (2000), this term is used to describe a quick manual act with little rest between movements. The repeatability of the shares has its effects accented when operators are impossible to express spontaneous behaviors. As noted by Dejours (1992, p.101) *"it is mandatory, somehow, that the attendant repress its intentions, its inactivity, its language, in other words, its personality"*.

The searches also stated that they perform, in addition to telephone functions, the receptionist function, and this overwhelms the quality performance of the function. As France and Roberts (1999), this situation can be considered as conflicting roles, since the professional has to perform more than one role in a given situation.

3.3 *Evaluation of environmental conditions*

In a general way, the operators said that there is noise in the workplace, interfering directly on the workload of these, occurring either by lack of acoustic treatment in the workplace, as the large concentration of people in the industry alongside, passing to be a disturbing element in the execution of the activity. About it, Grandjean (2005) states that intellectual works that require mental concentration or activity in which the understanding of the conversation is important, belonging to "professions sensitive to noise." Therefore, the presence of noise in the workplace can cause impaired concentration, feelings of discomfort and stress, since it has physical and/or psychological effects.

Regarding factor lighting, all telephone operators stated that the lighting is comfortable for the execution of the activity. The lighting of the workplace comes from two sources of artificial light and this situation is favored by the fact that the work environment is defined not by walls, but by windows that let in some ways, the presence of natural light during the day.

The subjects considered the temperature appropriated by the presence of adjustable cooling system environment. The temperature of the job must be related to the activity performed, since excessive heat has an effect on employee productivity, causing tiredness, drowsiness, reduced attention and increasing the chances of failure (Grandjean, 2005).

3.4 *Evaluation of working postures adopted in performing the activity*

About body postures, four items were analyzed. Regarding the fixed posture, two of respondents said they keep them. To speak about the practice of movements of torso twists, two employees do it. The third item was rate trunk flexion in sitting posture, two operators work sometimes as an option to perform alternating posture, and just one reported that actually performs this kind of posture to perform the activity. Finally, with respect to cervical flexion, it was found that one employee perform, along with that one, for each, do not achieve sometimes only held during the task.

Wells (2000) and Nascimento & Moraes (2000) argue that the sitting posture associated with the work computer, generally require a longer stay in the body static posture, repetitive movements, and for longer period of time carry out and can result in fatigue, pain/muscle disorders, postural changes and/or circulatory. Still they can adopt a poor posture due to the existing facilities or the lack of postural awareness, being common to see people sitting incorrectly.

3.5 *Assessment of overall health*

According to the data collected through the survey form, it was found that the telephonists from the call center at IFPB have impairment of their physical and mental health, according to Table 1, showing that occupational stress affects the human being in all aspects, interfering in their relationship to work, their family relationships and their performance in other activities. Añez et al. (2005) argue that, to carry out any work activity, there must be an interaction and interdependence of aspects: physical, cognitive and emotional.

3.6 *Evaluation of signs and symptoms of stress*

According to data collected by the form, referring to the symptoms of stress, it was found that all telephone operators have signs and symptoms of stress. In relation to the physical symptoms, it was found that two of the interviewed had dry mouth, fatigue, dizziness and insomnia, 100% had muscle tension and one of them had wheezing. Santos (2004) says that sleep disorders appear in workers as a result of concerns about the performance of some task, pressure head or even the constant image of the company, even outside.

With regard to cognitive symptoms, it was found that one of the total interviewed presents memory loss, mental fatigue and a greater likelihood for errors and accidents, and two had loss of humor and about the emotional symptoms studied showed that all of them have anxiety and two employees present irritability. Finally, with regard to behavioral symptoms, it was found that two of them had changes in appetite. But it was also possible to notice symptoms relating to the relationship, as mistrust, corresponding to one of the three.

According to Santos (2004), irritability at work corresponds to a form of psychic wear, due to fatigue in the workplace, which usually affects the

Table 1. Impairments at physical and/or mental health in telephone operators.

Impairment at physical/mental health	Subject A	Subject B	Subject C
Ability to concentrate	Yes	Yes	Yes
Insomnia caused by worries	No	Sometimes	Yes
More time to execute the tasks	Yes	Yes	Yes
Loss of interest in daily activities	Yes	Sometimes	Yes
Facility to get along with others	Yes	Yes	Yes
Ability to make decisions	Sometimes	Yes	Yes
Constantly tired	No	Yes	Yes
Inability to overcome difficulties	No	No	No
Excessive activities	No	Yes	No
Nervous and always tense	No	Yes	Yes

home, making hard the family relationship. In their studies, Santos & Vianna (2011) found higher percentages of 60% by relating work activities with the presence of signs and symptoms of stress in telemarketers.

Realize that these professionals are exposed to stressors that encompass the physical, psychological and cognitive factors, that because that any work activity has at least three aspects: physical, cognitive and emotional, which are closely linked and that the overhead of either an overload promotes the other two. However, it is important to emphasize personal vulnerability factor that must be considered, because people react differently when exposed to the same stressors.

4 CONCLUSION

Occupational stress corresponds to a failure of the process of human adaptation, developed due to changes in organizational factors, ergonomic and biological, corporately, resulting in signs and symptoms. Accordingly, due to the complexity of this process, it is difficult to diagnose without a full assessment of the workplace and occupational action performed. In addition, it is also important to ascertain the non-occupational activities aiming to detect the contributions in the development of this pathology.

As the objective of this research was to investigate the working conditions that can trigger occupational stress in the call center operators at IFPB, using the form developed by the researchers from the literature. It was established that the development of pathology in the study was related to occupational factors as the repetitiveness of the task performed, dissatisfaction with their homework, no breaks at work, and was still registered the presence of internal and external noise, and ergonomic factor as inappropriate. As for the bio-psychosocial factors, susceptibility was observed, inherent in human beings, for the development of fatigue, monotony and lack of autonomy.

All these factors were featured by literature on this subject, and those can interfere with health, safety and worker comfort.

It was concluded that the employees are exposed to certain triggers of occupational stress, since the activity performed is regarded as an employment situation of risk, either by direct physical load (maintaining the same posture for long periods etc.), or by loading physical indirect, resulting from somatization of anxiety and everyday stress. It is important to note that these factors trigger the physical stress and/or mental impairment that interfere directly on the productivity and quality of the task performed, and the quality of life of employees.

The analyzed data in the study allowed reaffirm the importance for a deeper understanding of the occupational stress pathology, which, directly or indirectly, has been part of everyday life for many workers. This assertion is based on the finding that the human being is in the process of constant adaptation to new technologies, thus seeking to upgrade themselves to a market, increasingly restrictive and competitive in order to ensure their survival.

Thus, is important to denote that it would be more appropriate for the job of these operators were located in a more reserved environment, allowing a better employee performance, without the physical and/or mental overload. In addition, organizational strategies can be used in order to promote job enrichment, correction factors in the physical environment and to implement effective management of interpersonal factors.

In conclusion, is suggested that future studies aimed at monitoring the changes proposed by this study as also the studies of this disease in relation to other jobs.

REFERENCES

Añez, C.R.R.; David. E.; Lobo, M. 2005. *Ergonomia, estresse e trabalho*. Available in http: //www.saudeetrabalho.com.br/download_2/estresse-ciro-anez.doc. Acess in 11.12.2012.

Carrer, P. et al. 2012. Risk Factors and health status among operators at a large call center in Italy. Proceedings of *30th International Congress on Occupational Health*—Icoh.

Constancio, S. et al. 2012. Body aches in call center operators and the relationship with voice use during work activities. *Rev Soc Bras Fonoaudiol*. 17(4):377–384.

Dejours, C. 1992. *A Loucura do Trabalho: estudo da Psicopatologia do Trabalho*. 5. ed. São Paulo: Cortez—Oboré.

Deliberato, P.C.P. 2002. *Fisioterapia Preventiva: fundamentos e aplicações*. São Paulo. Editora Manole.

Fiori, A.M. 1997. Stress Ocupacional. *Revista CIPA*. São Paulo, p. 40–49.

França, A.C.L. & Rodrigues, A.L. 1999. *Stress e trabalho: guia básico com abordagem psicossomática*. São Paulo: Atlas.

Goldberg, D.P. 1996. *Questionário de saúde geral de Goldberg: manual técnico QSG—adaptação brasileira*. São Paulo: Casa do Psicólogo.

Grandjean, E. 2005. *Manual de ergonomia: adaptando o trabalho ao homem*. 4. ed. Porto Alegre: Artes Médicas.

Iida, I. 2005. *Ergonomia projeto e produção*. 2. ed. São Paulo: Edgar Blucher.

Lipp, M. *Inventário de sintomas e efeitos do estresse*. Available in: www.estresse.com.br. Access in: 12.01.2012.

Montoro, O.C.P. 1998. *O estresse na comunicação do telemarketing.* Monografia de conclusão do Curso de Voz. CEFAC—Centro de Especialização em Fonoaudiologia Clínica. São Paulo-SP.

Nascimento, N.M. & Moraes, R.A. 2000. *Fisioterapia nas empresas: saúde x trabalho.* Rio de Janeiro: Taba Cultural.

Oliveira, C.R. 1991. Lesão por Esforços Repetitivos (LER). *Revista brasileira de saúde ocupacional.* n. 73. v. 19, p. 59–85.

Rossi, A.M.; Quick, J.C.; Perrewé, P.L. 2009. Stress e qualidade de vida no trabalho: o positivo e o negativo. São Paulo: Atlas.

Santos, R.V. 2004. Psicopatologia do trabalho. *Revista CIPA*, São Pulo, p. 24–32.

Santos, A.C. & Vianna, M.I. 2011. Prevalence of stress reaction among telemarketers and psychosocial aspects related to occupation. *J epidemiol community health* vol. 65.

Wells, R. Relação dos distúrbios osteomusculares com o trabalho. In: Ranney, D. 2000. *Distúrbios Osteomusculares Crônicos relacionados ao Trabalho.* São Paulo: Rocca, p. 68–85.

Occupational Safety and Hygiene II – Arezes et al. (eds)
© 2014 Taylor & Francis Group, London, ISBN 978-1-138-00144-2

Key elements to effective emergency management in a school complex

F.H. Marques
ISLA—Instituto Superior de Gestão e Administração, Santarém, Portugal

M.C. Neves
ISLA—Instituto Superior de Línguas e Administração, Leiria, Portugal

P.H. Marques
Universidade Europeia, Laureate International Universities, Lisboa, Portugal
UNIDEMI—R&D Unit, Faculdade de Ciências e Tecnologia, Universidade Nova de Lisboa, Portugal

ABSTRACT: This descriptive study aims to analyze the implementation of Fire Self-Protection Measures (which become mandatory, in Portugal, since 2009) during a five-year period in a large school complex occupied by 3,500 persons. The methodology is based on qualitative research, namely non-participant observation, check-lists, questionnaires, photographs and videos. Since 2009 until 2013, it was provided general training for students and school staff, as well as specific training for safety staff. Additionally, fire drills were conducted annually in order to test occupants regarding fire fighting, first aid and evacuation. The lessons learned from each drill allowed improvements in subsequent emergency response. The results of this study show that occupants' emergency performance has increased due to an effective emergency management. The school emergency management has proven to go beyond legal requirements and, by setting a good example both in school and community, it enhances the safety of students and their families.

1 INTRODUCTION

On the 1th January 2009, it became mandatory, in Portugal, new legislation about fire safety in buildings (Decree-Law 220/2008 and Building Fire Safety Technical Regulation 1532/2008). These new regulations established for the first time Fire Self-Protection Measures (FSPM), which applies not only to new but also existing buildings, that are evaluated according to fire risk level and, since then, are classified in twelve use types according to their utilization/function: residential, parking, administrative, educational, hospitals/nursing homes, shows/public meetings, hotels/restaurants, commercial/transport, sports/leisure centres, museums/galleries, libraries/archives and factories/warehouses.

There are four fire risk levels: first level means low risk; second means medium risk; third means high risk; and fourth means very high risk. The criteria for classifying the fire risk level include, among others, the number of total occupants, number of occupants with disabilities and height of the building.

The current building fire safety regulation, besides closing the gaps of the previous dispersed legislation, has also introduced new concepts and significant changes, by setting out the duties and responsibilities regarding building construction (Castro & Abrantes, 2009). In accordance with the new safety procedures, fire drills must be done on a periodic basis (except for buildings classified in low fire risk level) and companies/organizations must provide specific fire safety training for the safety staff, as well as general fire safety training to prepare all building occupants for an emergency. Another novelty is that schools must also carry out evacuation training program for students.

As matter of fact, an emergency plan must be a "living document" (Linaza, 2006) which demands constant updating and it should be functional, realistic, practical, operational, user-friendly and understandable. Moreover, fire drills must be conducted in order to test the effectiveness of the emergency plan, as well as the organization's ability and readiness to handle a fire emergency (Ramos, 2011).

1.1 *Aim of the study*

This paper aims to summarize a descriptive study about the implementation of FSPM, during a six-year period from 2008 to 2013, in a large preschool and school complex in the North of Portugal.

The objectives of this descriptive study are:

- To analyze the implementation of FSPM, during a six-year period from 2008 to 2013;
- To evaluate effectiveness of the fire training and response readiness of safety staff;
- To evaluate effectiveness of the fire drills and behaviors of all school occupants.

1.2 *Object of the study*

The buildings construction (that took place in the middle of the twentieth century) happened before fire safety regulation and didn't consider some aspects like means of escape, fire-resisting construction materials and access routes for the Fire Department.

The school complex consists of 10 buildings, most of them of medium-height, from nursery school (children aged from 2 years) to high school (students aged up to 18 years) and is catalogued as use type IV—educational facilities (in accordance with Portuguese Decree-Law 220/2008).

Nowadays it's occupied by 500 employees and 3,000 children/students and due to two risk factors, the number of occupants (more than 1,500) and building height (more than 9 meters), it's classified in a very high risk category (in accordance with Portuguese Decree-Law 220/2008).

People with disabilities and other special needs like children aged between 2 and 5 were also included in the emergency-planning process. In fact, the emergency planner should take into account not only the building fire protection and fire alarm system, but also the physical and psychological characteristics of the occupants (Lopes, 2008).

Considering the complexity of ancient buildings (some of them with long walking distance and inadequate number of escape routes and exits), as well as persons with mobility impairment and the great numbers of people that can be inside buildings, it is rather difficult to organize a quick and safe evacuation, unless the school board is engaged in planning and practicing for emergency on a regular basis.

The implementation of FSPM in this school compound followed several stages, the first being a technical fire safety analysis, a fire safety record system and the elaboration of an Emergency Plan (all done in 2008). We were able to know thoroughly the facilities and make a flexible Emergency Plan based on the outcome of fire risk assessment and consisting of procedures adapted to local conditions.

In the second stage, from 2009 until 2013, we carried out on a regular basis evacuation and fire safety training, as well as properly conducted fire drills.

Each year, since 2009, all students and school staff had general fire safety training that included didactic movies and leaflets about emergency evacuation (see Photo 1).

It is interesting to point out that young people may be the best means of introducing to their families and local community notions of safety and prevention. As a matter of fact, in Portugal there are still indicators of "*low safety culture*" due to insufficient safety training and so priority must be given to develop education programs in this area, in particular for the younger (Castro & Abrantes, 2009). It is essential that students have a basic understanding of fire safety and prevention best practices in order to bring change to society.

During these last 5 years, it was also conducted practical training on an annual basis for safety staff about emergency management, as well as firefighting with real flame and fire extinguishers (see Photos 2, 3 & 4). The training also consisted of evacuation drills which enabled the key staff to familiarize with procedures for emergency preparedness and response.

It is important to emphasize that school board provided refresher training program once a year, which exceeds the requirements of building fire safety regulation, because it does not establish

Photo 1. Awareness session for students.

Photo 2. Staff training on evacuation.

Photo 3. Training on emergency management.

Photo 4. Training fire fighting with real flame.

periodicity of training, only setting that training must be conducted at the beginning of the school year.

Between 2009 and 2013 it was possible to carry out 5 fire drills on the assumption that a real emergency existed. Fire drills were conducted annually (accordingly to the building fire safety regulation) and all school occupants (about 3,500) had to evacuate.

2 METHODOLOGY

This descriptive study is being developed in a MSc in Management of Occupational Risk Prevention and the methodology is based on qualitative research, namely non-participant observation.

In order to evaluate the effectiveness of emergency planning we collected data during fire drills by gathering check-lists (filled in by observers and evaluators), post-fire drill questionnaires (filled in by safety staff), photographs and videos (made by the students of Audio-Visual courses). All fire drills were recorded and evaluated.

3 RESULTS AND DISCUSSION

The 5 fire drills that were carried out between 2009 and 2013 were adequately planned and evaluated, in order to test emergency procedures, identify vulnerabilities and find out what should be improved for a better response to a fire emergency.

In the first fire drill, that took place in 2009, there was no intervention of external emergency services and school building's safety staff had to conduct an emergency evacuation in order to make a self-assessment of its emergency response capacity. The main goals were: to ensure that all exit routes and stairways were not obstructed and could be used during evacuation; to test the working conditions and effectiveness of all fire and emergency equipments; and to test if safety staff was able to conduct their duties successfully.

From the second drill onwards, emergency services like Fire Department, Police Department, Civil Protection and Medical Emergency Services also participated, and so they were able to address the difficulties they would have to overcome if a real hazard would happen in a large school complex such as this—involving the evacuation of thousands of occupants, some of them with special needs (see Photos 5 & 6).

The fire drills planning included meetings to discuss the script with safety staff and emergency services, in order to ensure that evacuation exercises were done properly and taken seriously by all.

During fire drills, observers and evaluators (most of them external) were placed at strategic points—inside and outside the buildings—to look for good and bad practices, and they were told to not intervene directly during evacuation exercise. They were also previously asked to identify opportunities for improvement, by filling in a check-list and then mentioned them at the debriefing meeting after the drill.

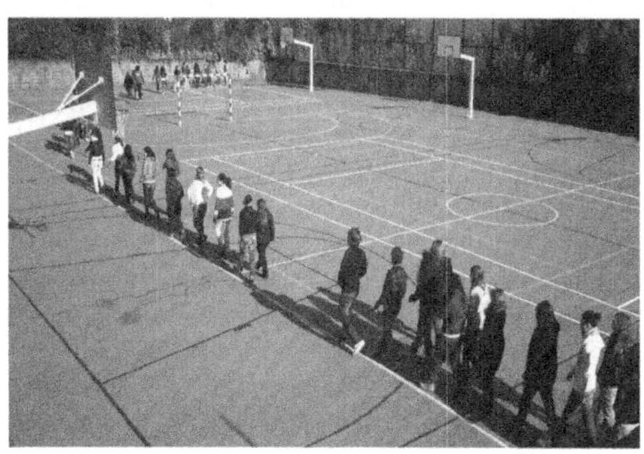

Photo 5. Evacuation of some occupants to one of the meeting points.

Photo 6. Occupants assembled in another meeting point and confirmation of attendance, to convey information to the Fire Department.

Photo 7. Students pretending to be injured.

Post-fire drill debriefing meetings were also held to gather information from participants, observers and evaluators, encouraging them to give feedback, in order to evaluate the effectiveness of the evacuation procedures and to find out what improvements could be done. All activities conducted during fire drill and emergency evacuations were recorded and documented and subsequently, evaluation reports were done for each fire drill, based on the information, photos and videos collected, which became an essential tool for monitoring the emergency management.

During fire drills there was simulation of smoke and fire-fighting with extinguishers. From the second drill onwards there was also participants playing the role of injured people in need of help to be evacuated from the buildings (see Photo 7).

The presence of smoke made the emergency scenario more believable (see Photo 8) and enabled occupants to follow safety procedures that they learned during the fire and evacuation training. It is noteworthy that smoke can reduce visibility to

Photo 8. Real smoke created on purpose to simulate a fire outbreak.

near zero and may cause several difficulties like the possibility of people becoming trapped or slowed down in a fire and can even cause panic (Coelho, 2010).

Each year, the degree of complexity of emergency scenarios has been increased, by placing new challenges to the school complex and external emergency services, with hypothetical disasters and multiple eventualities that have a probability of happening.

Usually, emergency planning takes into consideration less catastrophic scenarios, but one should also consider disasters of great gravity, in which the hazard impact would be so devastating that it would become infeasible the return to premises, after evacuation, and it would compel all the occupants of the school perimeter. Undoubtedly it is also important to work up scenarios for huge hazard impact and it may be necessary to extract all occupants on foot or by bus to a refuge outside the school perimeter (Alexander, 2002). Therefore the custody and welfare of children and students would then be a difficult issue to handle by the school and would need to be prepared in advance, because children and most students could not be permitted to travel on their own to home.

Taking into consideration this new challenge, the fifth fire drill (carried out at the beginning of the school year 2013/2014) integrated a huge hazard impact—the fire near a tank of liquefied petroleum gas, with risk of Boiling Liquid Expanding Vapor Explosion (BLEVE)—thus, becoming the first exercise imposing the need for extraction of occupants. This huge high risk of explosion (BLEVE) has forced the extraction of about 700 out of the 3,500 evacuated occupants, walking by foot and re-assembling in a safe place far from the school complex.

This time, extraction was partial and done by foot, but in future even more complex scenario, requiring complete extraction of occupants to refuges outside the school complex, then it must be done by bus or other external means of transportation. The biggest challenge would probably be when children and students are released to parents or adults with authorization to collect them, like other family members, babysitters and housekeepers. In a real life situation, parents and other adults in charge of children/students and even local community cooperation, will be crucial.

4 CONCLUDING REMARKS AND PROSPECTS FOR PROGRESS

Certain organizations, due to their size, complexity and characteristics of occupants, require specific solutions to overcome the difficulties of an emergency planning (Alexander, 2002).

In Portugal, it is seems that certain schools are better prepared for emergency than buildings of another use types, maybe because some companies/organizations are not aware that they are required to implement FSPM or regard it without the desirable commitment, as they do not have the responsibility to protect children and teenagers.

There are several reasons why fire drills are important: firstly, they are an opportunity to practice evacuation procedures and to make sure that all staff is familiar with them; secondly, they are also useful for testing escape routes, exits and fire alarm system.

During these five fire drills that were held in the school complex, we were able to verify that almost all occupants, including students, followed evacuation procedures and there was no real injuries resulting from the exercises. Moreover, safety staff has shown increased concern in implementing the recommendations for improvements that were highlighted at the debriefings meetings after each drill.

As a matter of fact, it was possible to learn from fire drills' results, the following lessons: telephone and cell phone communications among safety staff must be improved; all escape routes and exits must be unlocked and unobstructed; evacuees should use the correct evacuation route to leave the floor/building; all occupants must be evacuated and should not carry cumbersome or heavy objects; the next goal for each building occupants should be to reduce evacuation time; the five meetings point pre-defined in the planning process were not enough (this was realized during the first fire drill and, since then, it was established the existence of six meeting points).

The implementation of the various stages of emergency management in this school, that were beyond legal requirements, including regular training and fire drills being conducted based on a script and debriefings, were the key elements to an effective emergency response and to achieve more complex emergency planning in the long term.

In all fire drills, results have been accomplished, taking into consideration the growing exposure of occupants to more complex emergency scenarios, like, for instance, their extraction after evacuation.

Despite the size, the complexity and the fact that school compound was built before the new legislation, preliminary findings shows a successful emergency management.

This school complex, by setting a good example, can thus help families and local community become more accustomed to the notion that emergency planning is essential and that everyone has a roll to play in disaster prevention and mitigation of consequences.

ACKNOWLEDGEMENT

Recognition is due to the school complex *Colégio das Caldinhas* for rendering this paper possible, by providing photos, videos all the resources necessary to project implementation.

REFERENCES

Alexander, D.E, 2002. *Principles of Emergency Planning and Management*. England: Terra Publishing.

Castro, C.F, & Abrantes, J.B, 2009. *Manual de Segurança contra Incêndios em Edifícios*. Sintra: Escola Nacional de Bombeiros.

Coelho, A.L, 2010. *Incêndio em Edifícios*. Amadora: Edições Orion.

Linaza, L.M, 2006. *Elaboración de un Plan de Emergencia en la Empresa*. Madrid: FC Editorial.

Lopes, N.C, 2008. *Gestão de Emergência: Processos de Evacuação*. Lisboa: Verlag Dashofer.

Ministério da Administração Interna, 2008. Decreto-Lei n.° 220/2008. *Diário da República*, 1.ª série, n.° 220, de 12 de Novembro de 2008.

Ministério da Administração Interna, 2008. Portaria n.° 1532/2008. *Diário da República*, 1.ª série, n.° 250, de 29 de Dezembro de 2008.

Ramos, P, 2011. Organização de Simulacros—pode ser a fingir, mas é para levar a sério. *Proteger*, 13: 39–43.

Occupational Safety and Hygiene II – Arezes et al. (eds)
© 2014 Taylor & Francis Group, London, ISBN 978-1-138-00144-2

A study on the reality of Portuguese companies about work health and safety

A. Bastos
CICCOPN—Vocational Training Centre for the Building and Public Works Industry of the North, Maia, Portugal

J. Sá
Higher School of Business Sciences, Polytechnic Institute of Viana do Castelo, Valença, Portugal

O. Silva
Lusófona University of Porto, Portugal

M.C. Fernandes
CEPESE—Centro de Estudos da População, Economia e Sociedade, Porto, Portugal

ABSTRACT: This article aims to study the behaviour of Portuguese companies in the field of Occupational Health and Safety at Work, over the past three years. This study was based on a questionnaire made available to a set of enterprises geographically distributed across the country and with various sectors of activity. This study revealed that a large proportion of companies have organized systems of OHS but also that the number of hours of training given to workers who suffered injuries in the last three years was clearly low. This research also allowed us to analyze the relationship between accidents and the different sectors of activity, as well as the costs of these accidents, showing that the best option is still prevention.

1 INTRODUCTION

The daily presence, in the media, of news referring to work-related accidents has been making our society more and more conscious of this unfortunate reality. Agents with responsibility in the area of Occupational Health and Safety (OHS) have been developing multiple campaigns to raise awareness to the several types of work-related accidents and occupational diseases which occur in many sectors of activity. The legal obligation that companies organize their OHS services constituted a major step towards improving working conditions.

Companies' certification according to Standard OSHA 18001 is also of the utmost importance for improving working conditions, since it is a way for companies to organize their Health and Safety systems. There are many advantages to this implementation, but Fernández-Muñiz et al. (2012) emphasize its added value for enhancing the lines of communication between members of the organization for the transmission of information to workers in matters relating to OHS.

The business world includes many different areas of activity, and this sometimes makes it harder to organize efficient OHS services, meaning such services that have no accidents or occupational diseases. This fact has been stimulating Health and Safety technicians to adopt a more proactive attitude, constantly looking for adequate solutions to problems they are faced with and implementing prevention in companies. The multiple publications dealing with OHS, published either by Autoridade para as Condições de Trabalho (ACT) or by scholars or researchers in the field, have been a major contribution as source of information, both for companies and for workers. The experience gained by the OHS technicians throughout the years is a central factor not to be discarded; on the contrary, these professionals have contributed heavily to the adoption of OHS good practices in the companies they work for. According to Nold and Bochmann (2010), only by combining the knowledge acquired through technical publications with the years-long experience will OHS technicians be able to organize OHS services capable of preventing work-related accidents by improving working conditions, thus allowing companies to increase their productivity.

The implementation of organized OHS services in companies sometimes faces problems relating to a resistance to change from the management, due to the fact that no gain is achieved in the very short-term. Rundmo and Hale (2003) refer that there are managers who consider the involvement of people in prevention takes too long to bring about results. Among those managers, some go even further and consider that if their workers respect the safety rules the production goals become harder to achieve.

This situation undoubtedly hampers the OHS technicians' actions. Bastos et al. (2013), in their article, prove otherwise, showing that the lack of investment in OHS training generates losses higher than the investment needed for that training. According to Fernández-Muñiz et al. (2012), investing in the improvement of working conditions has a positive impact on customers' satisfaction, since it enhances the image they have of the company, thus increasing its competitiveness. The occurrence of work-related accidents usually entails interruptions in production, which implies delays in the customers' delivery deadlines. Lai et al. (2011) state that the OHS management plays a very important role in keeping deadlines with customers, and stress that by doing this the company improves its customers' satisfaction. In that sense, organizing OHS services may be a big help in achieving that goal. Cui et al. (2013) consider it possible to improve OHS in organizations through the adoption of safe behaviours. This necessary change in behaviour may be achieved by developing the workers' cognitive process when it comes to work-related health and safety issues.

The secret to success is, without a question, getting the workers involved in the activities performed by the organized OHS services, since one of the main goals these services aim to achieve is a change in the workers' attitude towards adopting safe behaviours. O'Toole (2002) mentions that in those companies where the management is committed to OHS, this situation has a very positive influence in the workers. It is thus crucial that the management takes their responsibility and assures the necessary conditions for the promotion of OHS in the organization. According to Haines et al. (2001), OHS programmes adopted are more likely to succeed when the implementation counts on the workers' commitment.

Companies face numerous challenges when it comes to safety in the workplace, and it is up to the Safety Technicians to identify hazardous situations and to organize such activities that respond to them in the best way possible. As an example, Kinnersley and Roelen (2007) call attention to the fact that often the design of machines, equipment, worksites and other elements can be held accountable for a considerable number of work-related accidents and incidents in companies. From a different perspective, Villanueva and Garcia (2011) stress the more advanced age, shift work and the temporary workforce as factors that increase the risk workers are exposed to. For these situations, the involvement of OHS services is extremely important, namely in terms of training and informing the workers. Specifically in terms of training, Luria and Yagil (2010) recommend that the one aimed at temporary workers be different from the one directed at effective workers, and they defend that training for the temporary workforce should be based on actions of short duration in more specific areas.

These are some of the issues organizations have to deal with on a daily basis and which, taking into consideration all specificities, must be taken into account when implementing organized Occupational Health and Safety services.

2 STUDY PRESENTATION

The study we present here was based on an online survey sent to companies from several sectors of activity through the social networks.

We obtained responses from 900 companies in the sectors of Industry, Construction, Services, Agriculture and Solidarity, from which we gathered 613 valid answers.

This research aimed at finding out whether companies were well organized in terms of Occupational Health and Safety, in order to face the work-related accidents which happened in the last 3 years, as well as a way of prevention.

These answers showed a clear predominance of the Construction and Services sectors of activity, with 47% and 39.6% of the accidents, respectively, as can be seen from Figure 1.

As for regions, most of the companies (68.7%) are localized in the North region of the country, as Figure 2 shows. This result is consistent with the INE data that point to a supremacy of companies in the northern region.

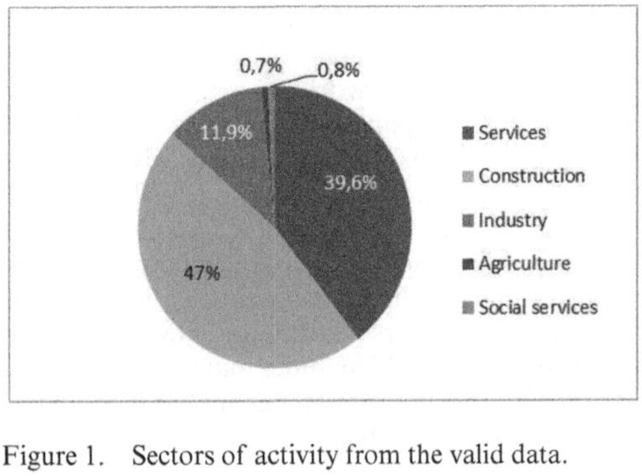

Figure 1. Sectors of activity from the valid data.

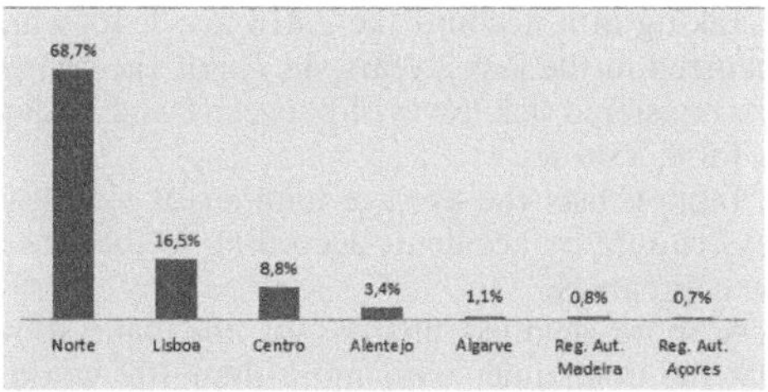

Figure 2. Regions—NUTS II.

When considering the size of the companies taking part in this study, there was a balance in the sample, and companies with a workforce of between 51 and 100 were the least represented, as can be seen in Table 1.

In terms of the type of Occupational Health and Safety services used, most companies (51.2%) use their Own Services and only a very small number use Shared Services (6.2%). It is noteworthy that 8.8% of the companies have no organized OHS service.

Data analysis from this set of companies showed that in the last 3 years 305 companies registered no work-related accident, whereas 308 suffered accidents in their work activities.

Of the companies without accidents, 47.9% belong to the Services sector and 48.5% have between 1 and 10 workers. As to the existence of OHS services, only 14.1% of the companies claim not to have such a service, as Table 3 shows.

Out of the 308 case studies of companies where work-related accidents occurred, 49% belong to the Construction sector, and a larger percentage of the accidents happened in companies with over 100 workers (45.8%) and in companies with 11 to 50 workers (26.3%). However, a detailed analysis of the following table reveals a percentage of 62.4% of accidents happening in larger companies, that is, with over 50 workers.

As for the OHS Services, only 3.6% of the companies where accidents occurred didn't have one, whereas most of the companies (61.7%) use their own services.

Portuguese legislation, specifically Decree-law no. 102/2009, of 10th September, states that "*The employer shall provide safety and health at work ...* and may adopt the modalities of internal service, joint service or external service".

The internal service is part of the structure of the company and is dependent on the employer. It is mandatory for establishments that have at least 400 employees or establishments that expose a minimum of 30 workers to high-risk activities.

The shared service is a common agreement among various companies and establishments owned by companies that are not in the same group or are required to establish internal service.

On the other hand, the external service is characterized as one which is developed by another entity which, through a contract with the employer, carries out health and safety at work.

The study sample is characterized by a set of companies of different size, as shown in Table 4, especially for those companies with more than

Table 1. Size of companies: number of workers.

No. of workers	Frequency	%
1 to 10	183	29,9
11 to 50	183	29,9
51 to 100	67	10,9
Over 100	180	29,4
Total	613	100,0

Table 2. Companies' OHS services.

	Frequency	%
OHS services		
None	54	8.8
Own	314	51.2
Outsourced	207	33.8
Shared	38	6.2
Total	613	100.0

Table 3. OHS services in companies.

	Companies with no accidents		Companies with accidents	
	No.	%	No.	%
OHS services				
None	43	14.1	11	3,6
Own	124	40.7	190	61,7
Outsourced	123	40.3	84	27,3
Shared	15	4.9	23	7,5
Total	305	100.0	308	100

Table 4. Company size: number of workers.

No. of workers	Frequency	%
1 to 10	35	11.4
11 to 50	81	26.3
51 to 100	51	16.6
Over 100	141	45.8
Total	308	100.0

100 workers (45.8%) and with a number of employees between 11 and 50 (26.3%).

Of the 35 firms that employ between 1 and 10 people, 4 have no health and safety service organized and have a frequency of 65 accidents without deadly consequences. Of these companies, 74.3% belong to the Construction sector.

With a number of workers between 11 and 50, we registered 81 companies, 70.4% of which belong to the Construction sector. Of these, 3 did not have any type of work health and safety service, 39 have internal services, 33 external services and 6 shared services. There were 296 accidents, one of which resulted in death.

Regarding the number of companies whose size is between 51 and 100 employees, 51 responded, 2 of which did not have any type of health and safety system organized, whereas 34 use internal services, 14 external services and only 1 has a shared service. The Construction sector is the most widely represented (58.8% of the total) and it was found that in the three years under review the 505 accidents resulted in 3 deaths.

As for companies with more than 100 workers, we recorded 1952 accidents, of which 9 resulted in death. Of these companies, 2 have no health and safety service, 113 have internal services, 15 external services and 11 use shared services. Of these companies, 50.4% belong to the services sector.

Table 5. Average number of days of sick leave of the victims.

	Frequency	%
Days of sick leave		
Up to 7	57	18.5
8 to 30	150	48.7
31 to 60	64	20.8
61 to 90	12	3.9
More than 90	23	7.5
Didn't answer	2	0.6
Total	308	100.0

Taking into account the 2,818 accidents which occurred in the last 3 years, 48.7% of the companies registered sick leaves of between 8 and 30 days, as Table 5 shows.

Table 6 lists the average number of sick leave days caused by accidents according to the size of the companies.

A more detailed analysis of the data shows that the companies with more than 100 workers are those with the highest number of accidents (45.4%). Analyzing the severity of accidents at work (number of days of absence), it appears that companies with more than 100 workers show a tendency for less severe accidents. As we can see, 43.9% of the accidents with a maximum of 7 days of leave refer to companies with more than 100 workers and the same happens with accidents causing between 8 and 60 days of leave. However, when observing workplace accidents in companies with more than 100 workers and leaves between 61 and 90 days, it appears that the percentage is reduced to 33.3%. As for work-related accidents causing over 90 days of leave, the percentage decreases further to 13%.

Of the total number of accidents, 22 were unfortunately fatal, 19 of which in the Construction sector, with the other 3 occurring in the Services sector.

Besides the victims' personal damage, companies report a set of very significant negative consequences directly relating to work-related accidents, namely:

- Need to use additional hours of work to comply with the clients' requests (38.3%);
- Delays in the delivery of the clients' requests (38%);
- Admission of new workers to replace the ones who suffered accidents (23.1%);
- Compensation to the victims (18.5%);
- Workers' permanent loss of capacities, resulting from previous work-related accidents (16.9%).

Still other consequences, such as "Fines" (2.3%), "Cancellation of orders by the clients" (1.3%) and

Table 6. Average number of days of sick leave according to companies' size.

No. of workers	1 to 10	11 to 50	51 to 100	>100	Total	**% companies with accidents > 100**
Up to 7 days	6	17	9	25	57	**43.9%**
8 to 30 days	18	30	27	75	150	**50.0%**
31 to 60 days	5	20	7	32	64	**50.0%**
61 to 90 days	1	5	2	4	12	**33.3%**
over 90 days	5	9	6	3	23	**13.0%**
Total	35	81	51	139	306	
%	11.4%	26.5%	16.7%	45.4%	100%	

"Compensations to clients due to delays and/or cancellation of requests" (1.3%), although less mentioned, should not be overlooked.

The training of employees is a subject of great importance in the prevention of occupational accidents and diseases, so an important part of this study aimed at knowing the average number of hours of training in OHS attended by the employees in the last three years. The results collected indicate that 89% of injured workers attended less than 25 hours of training and, of these, 47.1% had no training provided by the company. From this we can infer that the few hours of training (25 in three years) and their absence must surely have contributed to the accidents that occurred in the companies checked. The percentage of injured workers who attended between 25 and 50 hours of OHS training was 10.4%, and those who attended more than 50 hours correspond to 0.6%, which leads us to conclude that workers which attend more hours of training are less prone to accidents.

The more educated and informed workers are about the risks and the preventive measures associated with the tasks they perform in their day-to-day, the more they tend to adopt attitudes that may contribute to the improvement of existing conditions and reduce the number of accidents.

3 CONCLUSIONS

Workers have the right to perform their work in conditions that safeguard their health and their safety. One of the ways companies can ensure working conditions and safety is by using OHS systems according to the size and activities that they perform. We can conclude from this study that approximately 91.2% of the companies comply with this legal requirement. It is important that the inspection body (ACT) acts accordingly towards companies that stubbornly resist to organizing OHS services.

Whenever an employee is to perform a task, he/she must demonstrate OHS competence, so his/her training must not be neglected, taking into account the risks associated with their work, thus ensuring their physical integrity. The lack of training demonstrated by injured workers, the target of this study, quite clearly presents a high percentage (89%) of workers with little or no training. Unfortunately the consequence for 22 workers was death.

It is therefore imperative to invest more in occupational health and safety training!

REFERENCES

Bastos, A., Sá, J., Silva, O., (2013). The Importance of Training for Preventing Occupational Risks. *CICOT 2013—2nd Working Conditions International Congress, (ISBN: 978-989-97762-6-5),* Civeri Publishing, pp. 308–312.

Cui, L., Fan, D., Fu, G., Zhu, C.J. (2013). An integrative model of organizational safety behaviour. *Journal of Safety Research*, 45, 37–46.

Fernández-Muñiz, B., Montes-Peón, J.M., Vázquez-Ordás, C.J. (2012). Safety climate in OHSAS 18001-certified organisations: Antecedents and consequences of safety behaviour. *Accident Analysis and Prevention*, 45, 745–758.

Haines III, V.Y., Merrheim, G., Roy, M. (2001). Understanding Reactions to Safety Incentives. *Journal of Safety Research*, 32, 17–30.

Kinnersley, S., Roelen, A. (2007). The contribution of design to accidents. *Safety Science*, 45, 31–60.

Lai, D.N.C., Liu, M., Ling, F.Y.Y. (2011). A comparative study on adopting human resource practices for safety management on construction projects in the United States and Singapore. *International Journal of Project Management*, 29, 1018–1032.

Luria, G., Yagil, D. (2010). Safety perception referents of permanent and temporary employees: Safety climate boundaries in the industrial workplace. *Accident Analysis and Prevention*, 42, 1423–1430.

Nold, A., Bochmann, F. (2010). Examples of evidence-based approaches in accident prevention. *Safety Science*, 48, 1044–1049.

O'Toole, M. (2002). The relationship between employees' perceptions of safety and organizational culture. *Journal of Safety Research*, 33, 231–243.

Rundmo, T., Hale, A.R. (2003). Managers' attitudes towards safety and accident prevention. *Safety Science,* 41, 557–574.

Villanueva, V., Garcia, A.M. (2011). Individual and occupational factors related to fatal occupational injuries: A case-control study. *Accident Analysis and Prevention*, 43, 123–127.

Occupational Safety and Hygiene II – Arezes et al. (eds)
© 2014 Taylor & Francis Group, London, ISBN 978-1-138-00144-2

Influence of rotating shiftwork in health and performance of air traffic controller: A review

J.C.F. Siqueira, L.B. da Silva, M.S. Negreiros & E.L. de Souza
Department of Manufacturing Engineering, Federal University of Paraiba, Paraiba, Brazil

M.G. Andrade Filho
ICMC, Sao Paulo University, Sao Paulo, Brazil

ABSTRACT: Air Traffic Control is considered to be one of the most stressful occupations due to short time response demands and the great responsibility involved in the activity. For these reasons, many studies have been focused in improving the air traffic controller performance in order to help the prevention of aviation incidents and accidents. The objective of this article was centered in analyzing the results of rotating shifts over the health and performance of the air traffic controller by means of a literature review. Data from CAPES/MEC-Brazil, SciELO and Science Direct from the last 30 years were used through the key-words "air traffic control" and "shiftwork" or "shift work". 18 out of 89 articles that address the effects of rotating shifts over the health and performance of the air traffic controller were chosen. The analysis detected a great concern with the psychological and physiological factors of the ATC and its association with the workload, even when different methodologies were adopted by the authors. Results show that rotating shifts affect negatively the laborer. Counterclockwise shifts seem to cause a larger negative impact, and night shifts, especially with light workloads, directly affect the attention level and time of reaction, resulting in a greater possibility of incidents and accidents. Further ergonomic studies should be realized to determine the influence of environmental factors, (such as light and sound) associated to organizational factors (such as work schedule) over the health and performance of different sector controllers (ACC, APP and TWR). It is also important to evaluate physiological parameters during the shift rotations in which the construction of a mathematical model may help the analysis. These combined measures should be useful to result in a better decision making process over the work schedule organization of the controllers, especially if considered the intensification of domestic and international flights.

1 INTRODUCTION

The human being, as a diurnal species, has its active rhythm in daytime; which is the optimal work period, and its normal rest period (sleep) during the night. Abrupt deviations from this normal rhythm of work and rest can lead to problems, such as accidents, incidents or health problems, acute or chronic, because the biological clock adapts slowly and is different for each individual (Arendt 2010).

Due to modern society's growing demand for 24 hours continuous services, night time jobs and alternating shifts have become a reality for many workers such as the ones involved in the security, transport and health industries (Natale et al. 2003).

In a report about the priorities for occupational safety and health research in Europe 2013–2020, published by European Agency for Safety and Health at Work, the concern with globalization and the changing world of work, included well-being at work, work-related stress, chronic diseases and health conditions (European Agency for Safety and Health at Work 2013).

Linked to the transport industry and included in this type of need are the Air Traffic Controllers (ATC), whose profession emerged after World War II, in response to a significant increase in aviation safety demands in order to help pilots with the navigation, rational use of the controlled airspace and ultimately to prevent incidents and accidents with aircrafts. His main function is to keep a safe, regular and efficient air traffic flow, through information processing (Ministério da Defesa, Brasil, 2009). In reason of the life dealing responsibilities and short time decision making demands, the air traffic controller has been classified as one of the most stressful professions. Besides that, the work schedule has been acknowledged as a contributing factor to acute and chronic health problems, and also to stress and its associated errors (Edwards et al. 2012).

Air traffic control organization is divided in three different departments, which work in different environments and have distinct geographic and jurisdictional functions. (Ministério da Defesa Brasil, 2009): 1) Area Control Center (ACC): Provides service to the controlled flights above 14000 feet in the areas under its jurisdiction. Uses for this purpose sophisticated radars and VHF radios; 2) Approach Control (APP): Provides service to the controlled flights arriving or departing from one or more airports and to flights realized between 2000 and 14000 feet in the areas under its jurisdiction. Its controllers are responsible for radar vigilance, level change of aircraft with different performances, communications with pilots via radio. They use computers and radios to monitor the flight progress; 3) Control Tower (TWR): Provides service to the aerodrome area and flights below 2000 feet, uses specific frequency, limited to communication between tower and landed aircrafts or authorized vehicles in the maneuver area. The control is visual with the use of radios to further the instructions, but computers are also used to monitor the flights (Brink & Nel 2009; Vargas & Guimarães 2007; Ribas et al. 2011; Carvalho & Ferreira 2012).

Staring from the perception of ATC importance to the air transport industry safety and the recognition of the need of studies about the changing of the types of work organization, emerged the idea to research the influence of the shiftwork in the health and performance of these professionals. Within this objective this article aims, by means of a literature review, to reflect about the implications of rotating shiftwork in the physical and mental performance and health of air traffic controllers.

2 METHODS

A bibliographic research of articles published and present in the database of the Coordenação de Aperfeiçoamento de Pessoal de Nível Superior/Ministério da Educação—Brazil (CAPES/MEC), Scientific Electronic Library Online (SciELO) and Science Direct, from the last 30 years were used through the key-words "air traffic control" and "shiftwork" or "shift work". After the initial research, the criteria selection for the articles was: 1) written in English, Portuguese or Spanish; 2) focus on rotating shift work in air traffic control; 3) fully available. Then, the selected articles were chronologically cataloged, considering country of origin and number of air traffic controllers that participated in the original researches.

3 RESULTS

After the initial research, only 18 from 89 articles were selected. The others were excluded for not meeting the criteria of focus on rotating shift work in air traffic control. The selected items that treat the effects of rotating shifts over the health and performance of the air traffic controller were chosen and cataloged in Table 1, presenting author(s), year of publication, country, and sample size of ATC (and categories).

4 DISCUSSION

In every research analyzed there was a clear concern with the physiological and psychological factors and their relation to the workload even when different methodologies were applied. Variations at all work shifts' parameters were identified making it harder to establish which work organization type would be more suitable for the ATCs to perform their activities.

Studies concerned on work effects are very variable even when related to a single profession, the

Table 1. Selected articles characteristics.

Author	Year	Country	ATC
Folkard & Condon	1987	17 countries	422 (ACC, APP, TWR)
Costa	1993	Italy	20
McAdaragh	1995	USA	997
Schroeder et al.	1998	USA	52
Sega et al.	1998	Italy	80 ATC e 240 CG
Stoynev & Minkova	1998	Bulgaria	60
Weikert & Johansson	1999	Sweden	36 incidents
Rocco et al.	2000	USA	6,753
Corradini & Cacciari	2002	Italy	18 (7 TWR, 11 APP)
Cruz & Boquet	2003	USA	28
Natale et al.	2003	Italy	18
Mélan et al.	2007	France	15
Santhi et al.	2007	USA	18
Signal & Gander	2007	New Zealand	28 (ACC)
Signal et al.	2009	New Zealand	28
Orasanu et al.	2011	USA	179 (95 ACC, 41 APP, 43 TWR)
Pruchnicki et al.	2011	USA	1 accident
Fricke-Ernst	2011	Netherlands	189

ATC: Air traffic Controller; ACC: Area Control Center; APP: Approach Control; TWR: Control Tower; CG: control group.

work shifts are very diversified since the shift work is acknowledge as being like any other "out of normal hours job" that includes: fixed night shifts, overtime hours and rotating shifts. Some work schedule factors, such as shift rotating direction, changing schedule speed between different shifts, duration of the shift and the full cycle of shifts as well as day off distribution may influence fatigue, performance, safety and well being of the worker, especially if not considering the circadian cycle of the ATC worker (Sallinen & Kecklund 2010; Karlson et al. 2009).

Corradini & Cacciari (2002); Cruz & Boquet (2003); Schroeder et al. (1998); Signal & Gander (2007); Stoynev & Minkova (1998); Costa (1993) analyzed the ATCs' physical and/or mental performance. In their research they used instruments such as the: Multiple Task Performance Battery (MTPB), French paper-and-pencil version of Thayer's activation–deactivation adjective checklist, and Psychomotor Vigilance Task (PVT), in order to register accidents/incidents, communication efficiency with the pilots and psychological testing. Furthermore they verified the alert levels, selective awareness, mental performance, chronobiological pattern, work perception and stress.

According to Costa (1993), the alterations found in 6 day cycles of counterclockwise rotation, regarding the chronobiological pattern of each individual, were compatible with those of other workers. On the other hand, Schroeder et al. (1998) research pointed towards a worsening in fatigue and alert levels with rotating shifts, without alterations however, in 8 to 10 hours shifts. In a comparative study between clockwise and counterclockwise rotation shifts, no different effects over ATCs complex tasks performance or surveillance were detected. (Cruz & Boquet 2003). However, Signal & Gander (2007) observed that the sleepiness reduces when counterclockwise rotation is used and that the best performance on the night shift is directly related to a larger period of sleep on the previous night. Natale et al. (2003) also point to the fact that working in counterclockwise shifts result in a constant reduction of sleepiness on the following morning shift, resulting on the next night shift to be initiated with a sleep deficit. Stoynev & Minkova (1998) pointed that fast time rotation generated less physical, psychological and social problems than the other schedules. Orasanu et al. (2011) indicated that the brain's sleep and circadian systems interact in a non linear manner e are constantly defied by long days, irregular and unpredictable shifts, high workload and other environmental factors.

According Fricke-Ernst (2011), the level of attention was significantly reduced higher the sleepiness perceived by the ATC, though the results of applied attention test were higher than of others occupational groups.

By request of the North American Congress, Della Rocco et al. (2000) realized a survey through the internet about the fatigue factor over the ATCs and its relation with the shift patterns used and the rotating practices. Related to the fatigue and psychological health, it was observed that rotating shift schedules are worst than straight shifts. Counterclockwise rapidly rotating schedules with midnights, presented itself as the most fatigue inducing ones, with worsening of the sleep quality, more gastrointestinal symptoms and workers with higher body mass index in comparison with Counterclockwise rapidly rotating schedules without midnights or Straight 5 s (4–5 days on the same shift before change). No difference in Cardiovascular Scale from the Physical Health Questionnaire of the ATCs was detected on the different schedules.

It is important to observe that the ATCs' performance was worst on the night shift. Santhi et al. (2007) pointed out loss of awareness on the night shift, along with a decline in the accuracy and slowing of the time response due to the loss of sensibility and increase of errors, although the low workload also contributes to lowering of awareness level. Mélan et al. (2007) research shows that worst scores were obtained during the night shift and the beginning of the morning. This performance reduction during the night shifts has also as a contributing factor the low workload as demonstrated by the communication efficiency reduction between ATCs and pilots (Corradini & Cacciari 2002), and also by the larger number of accidents due to work routine failure on these periods (Weikert & Johansson 1999).

The workload of the ATCs can be measured in a variety of ways. The most used was the number of aircrafts being controlled at a given period (Corradini & Cacciari 2002; Costa 1993), which proved itself a reliable index for this purpose (Collet et al. 2009). Another method used was NASA-Task Load Index (Schroeder et al. 1998), which relates physical, mental and time demands, performance, effort (both physical and mental), and frustration level.

Besides these instruments, during the Flight 5191 crash analysis, Pruchnicki et al. (2011) suggested that the use of mathematical modeling could be useful in retrospective accident analysis (if historical data is available) and also prospectively by identifying fatigue risk factor associated to the workload in different shifts. These authors realized a retrospective flight accident analysis combining: 1) sleep and circadian cycle reconstruction; 2) mathematical modeling, predicting historical performance of sleep and circadian cycle; and 3) behavioral evidence of fatigue at the 30 minutes period prior to the accident, in accordance to the data collected by the agencies responsible for flight accident investigations. Even though it was possible to identify the presence of fatigue, it was not possible to detect a direct relation between fatigue and the accident.

To verify acute and chronic effects in health, physiologic data such as temperature, blood pressure, heart rate and urine samples were collected in addition to the detection of specific problems, such as coronary diseases and sleep disturbances.

Folkard & Condon (1987) detected that the greater the number of successive night shifts at the schedule, the greater the possibility of night shift paralysis.

Regarding the systemic blood pressure, there were no differences during the work shifts with rotating hours compared to a control group with men of the general population (Sega et al. 1998). Stoynev & Minkova (1998) results also showed small influence of the rapid clockwise rotation over the heart rate, blood pressure and temperature. Though, McAdaragh (1995) found that almost half of all controllers working at full-time are afflicted with circadian disruption at least 50% of the time.

A series of studies performed with other types of professionals have associated a number of pathologies to the work shifts, particularly cardiovascular system dysfunctions. Many of the studies that compare the effects of shift work are realized with health and emergency professionals. The differences encountered by the researchers between the ATCs and other professionals may be related to the task performed, working environment or to the countermeasures adopted to compensate the night shifts. Signal et al. (2009) in a research about the napping effects during the night shifts detected that, despite the poor quality and the short time, the nap periods increased the awareness and performance of the ATCs.

In a literature review about the effects of work shifts on workers in general, Smith & Eastman (2012) analized the mechanisms that affect the workers in this type of work organization. They may be useful in a comparative evaluation between ATCs and other professionals to identify possible similarities and differences. These mechanisms were divided by the authors in three different moments:

1. Misalignment between the internal circadian clock and the tasks performed (work, sleep and feeding). The biological rhythm's objective is to line up the physiological functions of the species with the environment and it influences hormones, behaviors, cognitive functions, metabolism, cell reproduction, apoptosis and stress responses (Arendt, 2010). Acute and chronic health problems increase at the same measure that the individual works desynchronized with his biological clock. (Arendt 2010; Karlson et al. 2009).
2. Sleep Deprivation (chronic or partial): the circadian rhythm phase changes occur due to *zeitgebers*, being light the main one. The time of sleep, however, also has an influence in non photic *zeitgebers* like exercise, social activities and food ingestion. Daytime sleeping is shorter and of poor quality compared with nighttime sleep (Arendt 2010). Sleep privation allied with circadian desynchronism has been known as a factor for obesity, metabolic syndrome, diabetes and glycolic intolerance, along with an increase in error and accidents.
3. Melatonin suppression by light exposure during nighttime: The melatonin hormone or nocturnal hormone is the main factor to promote humoral break for the seasonal organization of the day and the circadian rhythm. Its secretion can be suppressed by illumination, with suppression possible to be detected at a 30–50 lux rate and the maximum suppression at 1000–2000 lux. The melatonin suppression is associated to hazardous effects on health and a fast rise in awareness and temperature, although this mechanism is not totally clarified. (Arendt 2010).

These above referred mechanisms may affect each different sector controller in distinct ways.

The studies listed in Table 1 were realized with ATCs in specific groups or in all of them. However comparisons between the different categories were not made. Due to the different environmental conditions and instruments applied to each control sector the results obtained may not be true for all the ATCs.

Vargas & Guimarães (2007), note that the Approach Control is considered the most critical from the cognitive point of view because of the diversity of tasks. Their conclusion is important to avoid the risk of generalization with the use of the obtained data to all types of controllers and also to avoid the creation of mathematic models for risk factor analysis on the performance and health of the ATC that do not consider the different environment, tasks and workload of the control sectors.

Although the researched articles presented diverse methodologies, it was possible to identify the influence of the night shift on the tasks performed by the ATCs, with impacts on awareness reduction and fatigue elevation. These factors can lead to an increase in aviation incidents and accidents with reflections on the ATCs' health.

5 CONCLUSION

The researchers emphasize in their articles that the shift work brings a series of side effects for the

ATC, such as: lowering in the level of awareness, slower time response, precision decline, problems with pilots communication, privation of sleep and fatigue. However in the articles researched, the relation between shift rotation and the development of diseases was not explored due to the different methodologies applied and the non utilization of a specific category of controller as subject.

Further ergonomic studies should be realized to determine the influence of environmental factors, (such as light and sound) associated to organizational factors (such as work schedule) over the health and performance of different sector controllers (ACC, APP and TWR). It is also important to evaluate physiological parameters during the shift rotations in which the construction of a mathematical model may help the analysis. These combined measures should be useful to result in a better decision making process over the work schedule organization of the controllers, especially if considered the intensification of domestic and international flights.

REFERENCES

Arendt, J., 2010. Shift work: coping with the biological clock. *Occupational medicine (Oxford, England)*, 60(1), pp.10–20.

Brink, E. & Nel, P., 2009. *The relationship between occupational stress, emotional intelligence and coping strategies in air traffic controllers.* Stellenbosch University.

Carvalho, P.V.R. de & Ferreira, B., 2012. Modeling activities in air traffic control systems: antecedents and consequences of a mid-air collision. *Work*, 41, pp.232–239.

Collet, C., Averty, P. & Dittmar, A., 2009. Autonomic nervous system and subjective ratings of strain in air-traffic control. *Applied ergonomics*, 40(1), pp.23–32.

Corradini, P. & Cacciari, C., 2002. The Effect of Workload and Workshift on Air Traffic Control: A Taxonomy of Communicative Problems. *Cognition, Technology & Work*, 4(4), pp.229–239.

Costa, G., 1993. Evaluation of workload in air traffic controllers. *Ergonomics*, 36(9), pp.1111–1120.

Cruz, C. & Boquet, A., 2003. Clockwise and counter-clockwise rotating shifts: effects on vigilance and performance. *Aviation, space, and environmental medicine*, 74(6), pp.606–614.

Della Rocco, P. et al., 2000. *Shiftwork and Fatigue Factors in Air Traffic Control Work: Results of a Survey*, Washington, DC, United States of America.

Edwards, T. et al., 2012. Factor interaction influences on human performance in air traffic control: The need for a multifactorial model. *Work*, 41, pp.159–166.

European Agency for Safety and Health at Work, 2013. *Priorities for occupational safety and health research in Europe: 2013–2020* K. Sas & A. Suarez, eds., European Agency for Safety and Health at Work.

Folkard, S. & Condon, R., 1987. Night shift paralysis in air traffic control officers. *Ergonomics*, 30(9), pp.1353–1363.

Fricke-ernst, C., Kluge, A. & Kötteritzsch, A., 2011. Comparison of controller attention decrease during different break patterns in night shifts. *Proceedings of the Human Factors and Ergonomics Society Annual Meeting*, 55, pp.1195–1199.

Karlson, B. et al., 2009. Effects on sleep-related problems and self-reported health after a change of shift schedule. *Journal of occupational health psychology*, 14(2), pp.97–109.

Mcadaragh, R.M., 1995. Humand circadian rhythms and the shift work practices of air traffic controllers. *The Journal of Aviation/Aerospace Education & Research*, 5(3), pp.7–15.

Mélan, C., Galy, E. & Cariou, M., 2007. Mnemonic Processing in Air Traffic Controllers: Effects of Task Parameters and Work Organization. *The International Journal of Aviation Psychology*, 17(4), pp.391–409.

Ministério da defesa, 2009. *Regras do ar e serviços de tráfego aéreo*, Brasília/DF.

Natale, V., Martoni, M. & Cicogna, P., 2003. Effects of circadian typology on sleep-wake behavior of air traffic controllers. *Psychiatry and clinical neurosciences*, 57(5), pp.539–541.

Orasanu, J. et al., 2011. Work Schedules and Fatigue Management Strategies in Air Traffic Control (ATC). *Proceedings of the Human Factors and Ergonomics Society Annual Meeting*, 55(1), pp.1–5.

Pacetti, G.A. et al., 2011. Análise do controle autonômico de controladores de tráfego aéreo por meio da variabilidade cardíaca. *ConScientiae Saúde*, 10(4), pp.689–695.

Pruchnicki, S. a, Wu, L.J. & Belenky, G., 2011. An exploration of the utility of mathematical modeling predicting fatigue from sleep/wake history and circadian phase applied in accident analysis and prevention: the crash of Comair Flight 5191. *Accident analysis and prevention*, 43(3), pp.1056–1061.

Ribas, V.R. et al., 2011. Hematological and immunological effects of stress of air traffic controllers in northeastern Brazil. *Revista brasileira de hematologia e hemoterapia*, 33(3), pp.195–201.

Sallinen, M. & Kecklund, G., 2010. Shift work, sleep, and sleepiness—differences between shift schedules and systems. *Scandinavian Journal of Work, Environment & Health*, 36(2), pp.121–133.

Santhi, N. et al., 2007. Acute Sleep Deprivation and Circadian Misalignment Associated with Transition onto the First Night of Work Impairs Visual Selective Attention. *PloS one*, 2(11), p.e1233 (1–9).

Schroeder, D.J., Rosa, R.R. & Witt, L.A., 1998. Some effects of 8- vs. 10-hour work schedules on the test performance/alertness of air traffic control specialists. *International Journal of Industrial Ergonomics*, 21, pp.307–321.

Sega, R. et al., 1998. Ambulatory blood pressure in air traffic controllers. *American journal of hypertension*, 11(2), pp.208–212.

Signal, T.L. et al., 2009. Scheduled napping as a countermeasure to sleepiness in air traffic controllers. *Journal of sleep research*, 18(1), pp.11–19.

Signal, T.L. & Gander, P.H., 2007. Rapid counterclockwise shift rotation in air traffic control: effects on sleep and night work. *Aviation, space, and environmental medicine*, 78(9), pp.878–885.

Smith, M.R. & Eastman, C.I., 2012. Shift work: health, performance and safety problems, traditional countermeasures, and innovative management strategies to reduce circadian misalignment. *Nature and science of sleep*, 4, pp.111–132.

Stoynev, A.G. & Minkova, N.K., 1998. Effect of forward rapidly rotating shift work on circadian rhythms of arterial pressure, heart rate and oral temperature in air traffic controllers. *Occupational medicine*, 48(2), pp.75–79.

Vargas, C.V. De & Guimarães, L.B. de M., 2007. *Análise ergonômica do trabalho do controle de aproximação aérea (APP) em situação real*. Universidade Federal do Rio Grande do Sul.

Weikert, C. & Johansson, C.R., 1999. Analysing Incident Reports for Factors Contributing to Air Traffic Control Related Incidents. *Proceedings of the Human Factors and Ergonomics Society Annual Meeting*, 43(20), pp.1075–1079.

Occupational Safety and Hygiene II – Arezes et al. (eds)

 ISBN 978-1-138-00144-2

Perception of risk inherent in the working conditions of women agricultural workers in northeastern Brazil

F.D. Sá, A.G. Magalhães & L.S. Nascimento Júnior
Federal University of Rio Grande do Norte, Health Science College of Trairi, Santa Cruz, Rio Grande do Norte, Brazil

D.S. Dantas
Federal University of Rio Grande do Norte, Department of Public Health, Natal, Rio Grande do Norte, Brazil

ABSTRACT: Agriculture represents an important economic activity in Brazil, especially in the Northeast region. In your job, women that execute agricultural activities are exposure to many risks that could affect direct your health condition. The perception of risk by the worker is a good way to combat and eliminate them. In this sense, this study seeks to identify the perception of women that works in agricultural activity about the risk inherent in their working conditions.

1 INTRODUCTION

Agricultural associations in Brazil work as a cooperative joint ownership and in their majority have management democratically controlled. According to Food and Agriculture Organization of the United Nations (FAO, 2013) being the agriculture the main source of income and employment in rural areas, where most of the population is poor and lives in a situation of food insecurity, these cooperatives become a vital role in supporting small farmers—men and women—and marginalized groups, provide market opportunities for small farmers, training in natural resource management, access to information, technology, innovation and extension services.

This study had how objects an association of women agricultural workers in northeastern Brazil, where climatic conditions are extremely unfavorable, with long periods of drought and has a population with low levels of Human Development (HDI).

In addition to the vulnerability related to climatic factors, these women work in the riverfront and are subject to various risks inherent in the scenario where they perform their occupational activities. They are exposed to mechanical aggressors, because use various tools and handling machines, saws, scythes, machetes, to other agents of a physical nature such as solar radiation, lightning, extreme temperatures, heat and noise, chemicals agents and fertilizer for soil correction, pesticides, medicines for veterinary and biological, as bitten by venomous animals, bacteria and viruses in the care of the earth and factors related to the organization of work how long hours of work, intensive work cycles, related to the different stages of production, relations subaltern perpetuated since the days of slavery, among others (Dias & Mendes, 1999).

Faced with the diversity of work situations and processes in the rural sector and the lack of accurate information about conditions of work and health, becomes difficult to draw the profile of risk exposure and occupational health or presence of illness on these workers. In this context, the evaluation of the risk perception becomes a relevant tool.

According to Wiedemann (1993), the perception of risk is defined as the "ability to interpret a situation of potential damage to the health or life of the person or others, based on previous experiences and their extrapolation to a future time, varying since a vague opinion to a firm conviction". For the author, the perception of risk is based primarily on images and beliefs and has roots, in a lesser extent, in some previous experience, per example, accidents that a happened before, the knowledge of past disasters and relation with information about the probability of an accident occurring.

Studies about perception of risks arise from the decade of 70/80 as an important counterpoint to the utilitarian perspective of techniques risk analysis, based on knowledge of engineering, toxicology, economics and agricultural sciences and that didn't include the beliefs, fears and restlessness of the communities involved (Gomes & Freitas, 1997).

In this sense, this study seeks to identify the perception of these workers about the risk inherent in their working conditions, as well as the measures taken to minimize or eliminate such risks.

2 METHODOLOGY

During the field research, initially was conducted a diagnosis of the local situation, where were observed the environmental characteristics, the people involved in labor and yours social relations, as well as the organization of tasks, among other aspects.

After this observation, was possible to structure an interview form, becoming the instrument of data collection. The questionnaire consisted in four blocks: (1) identification of socio-demographic data, (2) identification of occupational hazards, such as: environmental hazards, ergonomic, physical, chemicals, biological and accidents involving the environment and the nature of tasks performed, (3) questions about preventive actions towards risks or accidents; and (4) the occurrence of work accidents in recent months.

To construct the sample of the study were used the answers of 19 women that works in an agricultural association located in a municipality in Rio Grande do Norte, located inside of Northeast—Brazil.

When the registration and collection of data was finished, all the information obtained in the interviews was transferred to spreadsheets in MS Excel to conduct simple descriptive statistics.

3 RESULTS E DISCUSSION

The sample was formed by women, with age between 19–55 years, with average of 36,5 years, that execute activities of planting, care, watering and harvest of agricultural products. They made these tasks, on average, for 9 years and earning less or equal to 255,00 BRL. This profile isn't different of others women that works with agricultural, because according to data from the National Household Sample Survey (IBGE, 2009), women residing in rural areas in Brazil begin to work very early, 51, 8% enter in active life between 10 and 14 years of age and are extremely underpaid and undervalued in the Northeast. These data indicate a reality of life and work with strong potential to produce deleterious effects on general health and occupational health at these women.

When they are questioned about the presence of situations that provide some kind of risk in their work, the samples was unanimous in asserting that question. They are distributed in Table 1.

Table 1. Distribution of frequency for the presence of occupational risks.

	Frequency		Percentage (%)	
Occupational risk	Yes	No	Yes	No
Venomous animals	13	6	68,5	31,5
Sun rays	16	3	84	16
Handling of cutting tools	7	12	36	64
Contact with chemicals products	9	10	47	53
Handling of inadequate tools	5	14	26	74
Falls	8	11	42	58
Handling of heavy tools	8	11	42	58
Posturas inadequadas	17	2	40,6	59,4
Others	4	15	20,8	79,2

The majority of them mentioned the adoption of wrong or uncomfortable postures during the action of her tasks how occupational risks, making about 90% of the total sample. About this, Leite et al. (2007) reports the presence of bad positions of the body or the exercise of the labor activities with inadequate corporal arrangements like source of prejudice to the agricultural worker, because can provide the development of chronicle or acute pathologies with occupational nexus. This same author appoints the application of methods that search determinate bad postures during the work, to improve your work conditions, yet these professionals present variable musculoskeletal problems due to inappropriate positioning.

This result was approximated with those who correlated the sun rays as a possible risk for burns and skin diseases (84%). Was large too the number of workers who cited the possibility of attack by venomous animals as a risk present in your work activity and such situation was presented by 68.5% of the total sample.

These data call attention because Iida (2009) describes the exposure to sunlight for long periods, without breaks and the adoption of appropriate fluid and caloric supply, as a trigger for health problems such as muscle cramps, syncopes, heat exhaustion, premature aging and skin cancer. In addition, Silva (2005) says that is very common in Brazilian agricultural work the presence of biological hazards, including accidents with venomous animals that have no direct relationship with the work, but as the way how is executed, being common snakebite, arachnidism and scorpion sting and accidents with animals with the urticant hair in contact with skin.

Next, was cited as occupational risk the contact with chemicals (47%), which Dias (2006) classifies them as chemical hazards, covering agents to fix or fertilize the soil and pesticides. Also appeared the possibility of falls or handling of heavy equipment (with 42%, each one) and the manipulation of cutting tools (36%). Fehlberg et al. (2001) in their study found a prevalence of approximately 30% of the accidents to be caused by handling of equipment and tools among the agricultural population.

In a smaller proportion, with 26% of the sample, was mentioned by farmers the manipulation of tools inadequate or poorly maintained. Still, few farmers cited risks as the presence of spines in some plants, handling electrical materials, the possibility of contracting skin diseases and aggressions by co-workers, giving a percentage of 5.2% each.

Curiously, the use of pesticides, occupational risk more prevalent in agriculture according to the World Health Organization and more discussed in the literature both nationally and internationally, had his report less frequent among this sample of farmers (WHO, 2000; Rainbard & O'Neil, 1995). In this case it should be noted that the starting point of any study of risk perception is how the interpretation of a lay person—understood here as one who didn't acquire specific knowledge in your life about the object in question life—to a particular hazard, differs of the interpretation by an "expert" (Wiedemann, 1993). The risk perception of the population is generally quite different from that of the experts, especially scientists. Their interpretations are based much more on their own beliefs and convictions than on facts and empirical data elements that form the basis for the construction of risk perception of technicians and scientists. Therefore, the low reference this risk may be due to the sparing use of pesticides, as well as the negligence of the risk inherent in these products.

When asked about performing actions in their work aimed at reducing or eliminating accidents, 58% said they do some kind of intervention in routine work to make it safer.

First was mentioned the use of Personal Protective Equipment (PPE) for 63% of them. Among the PPE's appeared as frequently used by these farmers, the use of long clothes (66.6%), hat (50%), sunblock (41.7%), boots (33.3%), gloves (16, 7%) and apron (8.4%).

Recena & Caldas (2008) found different data, because among the farmers who participated in their research, many said they didn't make use of PPE, claiming no use for personal reasons and not for the lack of knowledge about their effectiveness. However, the same authors showed the same PPE as most cited by those who said to use them (hat, long clothes, boots and gloves). On the use of sunscreen, Penha et al. (2007) cites this measure, as well as adoption and hydration breaks as a preventive resource for the emergence of problems resulting from exposure of the skin to sunlight.

Another measure mentioned by the sample was the participation in meetings/activities that promote the improvement of health and work conditions, this amount corresponded to 79% of the total. The intervention guided by popular education for health is presented by Peres (1999) as an effective way to that the labor practices carried out by farmers to become healthy and safe as they make them knowledgeable of the risks and hazards to health and safety, making them accountable for their actions to avoid such situations.

Penha et al. (2008) adds that training and programs for improvement of agricultural jobs should be designed and implemented in an interdisciplinary approach between professionals from agricultural, health and ergonomics. In our research, the women mentioned that they were submitted to the educational interventions conducted by institutions like the Municipal Department of Agriculture, Universities and EMATER (Enterprise Technical Assistance and Rural Extension), which is the government body that seeks to promote and enable the rural labor, especially the small landholdings, agricultural cooperatives and family subsistence farming.

Finally, were mentioned the practice of medical consultations periodically as preventive measures. This total was 43%. These consultations reported refer to the constant access to units of primary health care in Brazil represented by the Family Health Strategy and denote self-health care on an overall vision, not necessarily tied to periodic assessment of occupational health.

When they are asked about having suffered some sort of accident or near accident on his work in recent months, the sample was distributed unanimously to say no.

These figures with low frequency of accidents are presents in similar studies, Fehlberg et al. (2001) found only 11% of casualties among their respondents and Faria (1997) recorded 10% of accidents involving farmers.

4 CONCLUSION

We can conclude that the women that execute agricultural work in this cooperative present a clear perception of the risks to which they are subjected by their working condition. The risks most prevalent were for accidents and ergonomic, being this last most frequently reported. We emphasize the reporting of low frequency risk for use of pesticides, which goes against the literature.

The workers were also worried about mitigate this exposure with simple measures, which have access and are concerned to participate in educational activities for improvement of working conditions and occupational thus perform follow doctor.

Though in the presence of scarcity of resources and frequent risks is notable the efforts of governmental and university to provide training and minimum fostering these workers.

Although the sample was limited and in a singular location of region and studies relevant to that category are still scarce, the results can't be generalized to all the community farm. Thus, it is suggested new studies are developed for more consistent outcomes.

REFERENCES

Dias, E.C. 2006. Condições de vida, trabalho, saúde e doença de trabalhadores rurais no Brasil. In: Pinheiro, T.M.M. *Saúde do Trabalhador Rural.* RENAST: Brasília.

FAO. 2013. World Food and Agriculture statistical yearbook. Rome.

Felhberg, M.F. et al. 2001. Prevalence and associated factors to rural occupational accidents. *Rev Saúde Pública.* 35(3): 269–275.

Gomez C.M.; Freitas, C.M. 1997. Análise de riscos tecnológicos na perspectiva das ciências sociais. *Hist Ciênc Saúde Manguinhos* 3: 485–504.

Iida, I. 2005. *Ergonomia—Projeto e produção.* 2ª ed. São Paulo: Edgard Blücher.

Leite, B.R.B. et al. 2007. Importância da ergonomia e segurança do trabalho na melhoria das condições de trabalho do trabalhador canavieiro. *Proceedings of XVII ENEGEP—Encontro Nacional de Engenharia de Produção.* Foz do Iguaçu—PR—Brazil.

Mendes, R.; Dias, E.C. 1999. Saúde dos trabalhadores. In: Rouquayrol, M.Z.; Almeida Filho, N. *Epidemiologia e saúde.* Rio de Janeiro: MEDSI, p. 431–458.

Rainbard, G.; O'Neil, D. 1995. Occupational disorders affecting agricultural workers in tropical developing countries: results of a literature review. *Applied Ergonomics* v.26, p.187–193.

Recena, M.C.P. & Caldas, E.D. 2008. Risk perception, attitudes and practices on pesticide use among farmers of a city in Midwestern. *Rev Saúde Pública.* 42(2): 294–301.

Silva, J.M. et al. 2005. Pesticides and work: a dangerous combination for the Brazilian agricultural workers health. *Ciênc. saúde coletiva.* v.10 n.4 Rio de Janeiro.

WHO—World Health Organization. 1990. Public health impact of pesticides used in agriculture. Geneva: UNEP.

Wiedemann, P.M. 1993. *Introduction risk perception and risk communication.* Jülich: Programme Group Humans; Environment, Technology (MUT): Research Centre Jülich.

Occupational Safety and Hygiene II – Arezes et al. (eds)
© 2014 Taylor & Francis Group, London, ISBN 978-1-138-00144-2

Self-rated health, social factors and life style among farmers in Brazil

A.G. Magalhães, L.S. Nascimento Júnior & F.D. Sá
Federal University of Rio Grande do Norte, Health Science College of Trairi, Santa Cruz, Rio Grande do Norte, Brazil

D.E. Souza
Brazilian Institute of Geography and Statistics, Natal, Rio Grande do Norte, Brazil

E.S.R. Viana
Department of Physiotherapy, Federal University of Rio Grande do Norte, Natal, Rio Grande do Norte, Brazil

D.S. Dantas
Department of Public Health, Federal University of Rio Grande do Norte, Natal, Rio Grande do Norte, Brazil

ABSTRACT: The purpose of this study was to determine the Self-Related Health (SRH) and associated factors among women that work with agriculture in Rio Grande do Norte State, in northeast Brazil. An investigation was conducted from secondary data about 18 women, where the SRH was compared with the socio-economics conditions showed by the sample. In the results, it's possible to see that women without access to leisure activities, with alcoholism habits had a higher proportion of self-reported poor health.

1 INTRODUCTION

Agriculture is a human activity, which includes a number of different tasks and occupies a huge number of people worldwide. Estimates indicate that the northeastern in Brazil is the region that concentrates the highest number of people living in the countryside, which corresponds to 26.49% of the population that lives there, according to data from the Brazilian Institute of Geography and Statistics (IBGE, 2010). Studies reveal that the agricultural sector offers high risk of producing health problems related to occupational exposures. However, in Brazil, face to the lack of official records and the shortage of population studies on the topic, the health problems among rural workers are not well established (Faria, 2005).

The interaction of different situations during the work provides conditions more or less favorable to health problems. The evaluation of these aspects and their influence on health workers has advanced considerably, but its measurement still appears as a challenge for occupational epidemiology. The current reality is demanding of the researchers involved with the issue of occupational health and safety a differentiated approach to the hegemonic view, which conceives the appears of disease after biological manifestations of individual, linked to causal nexus linked to conditions of occupational risk.

Dejours (1986), disseminates the concept of health related to psychosomatic associating physical illness whereby is happening on the mental level. Although relevant and frequently mentioned, the relationship between perceived health and the relationship to the work situation has been insufficiently researched and applied (Martinez, 2002). The systematic results about the relationship between perceived health and work be used as subsidies in the design, implementation and evaluation of preventive and corrective measures on the situation of life and work for the promotion and protection of workers' health.

The Self-Rated Health (SRH) is a subjective measure that combines components physical, emotional and satisfaction with life, being considered a good predictor of health status, because is associated with progressive worsening of health, with a strong association with mortality, with other morbidity and socio-demographics' factors (Fonseca et al., 2008, Bluestein & Rutledge, 2006; Goldberg et al., 2001; Dowd & Does, 2007; Peres, 2010). Comprehended as a marker of inequalities between subgroups of populations, simple to use and used in investigations of population for high reliability and validity (Peres, 2010). Health conditions are influenced by the conditions of life, so precarious social conditions cause lower life expectancy and higher levels of illnesses. The older women, the poorest and least educated considered their health status as fair or poor. The greater the number of self-reported morbidities, the greater the proportion of individuals with negative self-rated health, and the effect of comorbidities was higher among females (Fonseca, 2008; Peres, 2010).

In this way, the simple fact of being a woman in almost all societies, even in the most developed, is related to heavier load of work, less pay and consequently greater poverty. They are also aimed at less qualified jobs and are not considered or they aren't paid by your housework and the care of the children and the home (Brasil, 2010).

In respect to rural women workers, public policies are currently implemented by the Brazilian government, seeking intervene in this reality through actions that can generate economic autonomy, individualization of rights, stimulating a sovereign insertion of women in the family, the economy and citizen participation. (Di Sabbato et al., 2009).

Nonetheless, few studies have been conducted on the health of rural populations, so this study sought to examine the self-reported health of women that works in agriculture of a northeastern Brazilian city and social variables associated in a group of women living in Santa Cruz—RN, inside the Brazilian northeast.

2 METHODS

The present study was developed from secondary data from the Project PET—Health and Vigilance, conducted in Santa Cruz—RN—Brazil. Initially, a cluster sampling was performed, using as reference families registered in family health program of the city. From this strategy 637 women aged 10–49 years were interviewed. This population included women aged 18 years and were rural workers, performing tasks in agriculture, bringing the number of eighteen women in the sample of this study.

The questionnaire, that was applied, included questions about socio-demographic and economic factors and SRH. The socio-demographic and economic factors included were age, educational level (years of study), race (white, brown and black), marital status (single, married/cohabiting, divorced, widowed), amount of children, smoking, alcohol consumption, average monthly income, amount of goods (TV, radio, refrigerator, mobile phone, computer, microwave, motorcycle, dvd player and car) and hours of work per day.

Neighborhood characteristics (perception of safety in the neighborhood, illumination, community mobilization, quality of health services, leisure and quality of air) evaluated on a five-point scale, ranging from excellent to poor (excellent, good, fair, poor and very poor).

Self-Rated Health (SRH) was assessed by a five-point scale, ranging from excellent to poor, through the three questions:

1. How do you evaluate your health? (SHR current)
2. How do you evaluate your health compared to other people your age? (SHR compared to people of the same age)
3. How do you evaluate your health compared to a year ago? (SHR compared to a year ago)

For statistical analysis, the sociodemographic and economic factors were dichotomized. Continuous variables (age, educational level, amount of children, average monthly income, amount of goods and hours of work per day) were classified according to the median. Race was dichotomized into white e nonwhite (brown and black) and marital status into single and married. SHR current, SHR compared to people of the same age and SHR compared to a year ago were dichotomized into good (Excellent, good and fair) and precarious (poor and very poor).

All variables were normally distributed, through the Shapiro-Wilk test. To analyze the sociodemographic and economic factors and three category of SRH, descriptive statistical was used. To check the associations between SRH e other variables the prevalence rate was calculated and prevalence proportions were compared through the chi-square test. For cases below five, however, Fisher's test was considered. Statistical analysis was conducted in SPSS 20.0. A p-value < 0.05 indicates a statistically significant result. The research protocol was approved by the Ethics Committee of the Federal University of Rio Grande do Norte.

3 RESULTS

In Table 1, socio-demographic and economic factors data are presented.

The average age of workers women was 33 years, most had five or fewer years of formal education (61.1%) of non-white race (72.2%), married (83.3%), with monthly earnings equal to or less than 255,00 BRL, with three or fewer children (66.7%), worked an average of five hours or less per day (88.9%).

About this profile, a study conducted in the same Brazilian state was found, by Rebouças and Lima (2013), which found the same average age

Table 1. Socio-demographics and economics variables from the sample.

Variable	N	Fr (%)
Age		
≤33 years	9	50%
>33 years	9	50%
Years of study		
≤5 years	11	61.1%
>5 years	7	38.9%
Race		
White	5	27.8%
Non-white (brown and black)	13	72.2%
Marital status		
Single	3	16.7%
Married	15	83.3%
Average monthly income		
≤255,00 BRL	12	66.7%
>255,00 BRL	6	33.3%
Number of children		
≤3 children	12	66.7%
>3 children	6	33.3%
Smoker		
Yes	3	16.7%
No	15	83.3%
Alcoholism		
Yes	3	16.7%
No	15	83.3%
Number of properties		
≤5	12	66.7%
>5	6	33.3%
Average work time daily		
≤5 hours	16	88.9%
>5 hours	2	11.1%

for farmers. Still, demographics data pointed to the low education of these professionals, with 39% who never went to school and 43% who attended a few years, and most (58%) of these are concentrated in the northeast region (IBGE, 2009).

Stropasolas (2004) to discuss the values of women working in agriculture says that, following a common pattern for different activities, women who study more tend to marry less or later and have fewer children when compared to those do not study or attending less school.

When was analyzed the perception of self-reported current health 66.7% evaluated it as good and 33.3% as poor, 83.3% evaluated their health compared to people of the same age as good and 66.7% rated it as good the own health compared for a year, as shown in Table 2.

As presented in Table 3, among the socio-demographics and economics variables, as well as the neighborhood characteristics, showed increased risk for a poor perception of health, the fact of to be single (PR = 1.67, CI = 1.10; 2:52); alcoholism (RP = 1.67, CI = 1.10; 2:52) and self-assessment of leisure as poor (PR = 15:00, CI = 1.22, 185.20). In contrast possess few assets represented decreased risk for poor health (PR = 00:10, CI = 0:10; 0.97.) The other variables were not statistically associated with the outcome of SRH.

Table 2. Self-rated among women who works in agriculture.

Variable	Poor	Good
Self-Reported Health (SRH)	33,3%	66,7%
SRH compared to people in the same age	16,7%	83,3%
SAR compared with the own health for a year	33,3%	66,7%

4 DISCUSSION

The vulnerable situation of women against some diseases and causes of death has more association with his condition of discrimination in society than to biological factors, in addition, women live longer than men, however, get sick more often (Brasil, 2010). The results presented here are according to these proposals and also with the general situation of the rural female population in Brazil, indicated by population studies, as these data point to the strong presence of women in agricultural work, making almost half of the workforce in this economic segment (Butto & Dantas, 2011).

According to data from the National Household Sample Survey (IBGE, 2009), among women residing in rural areas in Brazil 53.3% don't complete the primary school and among those who declare themselves how non-white the situation is even more precarious because 10.4% are illiterate between 30 and 39 years. These women begin to work very early, 51, 8% enter in active life between 10 and 14 years of age and are extremely underpaid in the Northeast, where they earn on average 205,00 BRL, with 91.8% earning up to one wage (678,00 BRL).

In addition, a significant percentage of the women residing in rural Brazil, 30.7% are not remunerated or only produce for self-consumption (46.7%). Still, the data point to the low education of these professionals, with 39% who didn't go to school and 43% attended few years, and the majority (58%) of these are concentrated in the northeast (DIEESE, 2011). These data are consistent with the profile drawn in the present study, in which the time of the study sample was 0–11 years of study, with an average 4.7 years and with an average monthly income of less than half an one wage, demonstrating a reality of life and work with

Table 3. Prevalence and variables associated with self-rated health in women who work in agriculture in a northeastern Brazilian city.

Variable	SRH precarious (%)	SRH good (%)	RP (IC95%)
Years of study			
≤5 years	66.7	66.7	1.00
>5 years	33.3	33.3	(0.12–7.99)
Race			
White	33.3	25.0	1.50
Non-white	66.7	75.0	(0.18–12.77)
Marital status			
Single	16.7	25.0	1.67
Married	83.3	75.0	(1.10–2.52)
Average monthly income			
≤255,00 BRL	50.0	75.0	0.33
>255,00 BRL	50.0	25.0	(0.04–2.63)
Number of children			
≤3 children	66.7	66.7	1.00
>3 children	33.3	33.3	(0.12–7.99)
Smoker			
Yes	33.3	8.3	5.50
No	66.7	91.7	(0.38–78.57)
Alcoholism			
Yes	16.7	25.0	1.67
No	83.3	75.0	(1.10–2.52)
Average work time daily			
≤5 hours	83.3	33.3	10.00
>5 hours	16.7	66.7	(0.85–117.0)
N° of properties			
≤5	33.3	83.3	0.10
>5	66.7	16.7	(0.10–0.97)
Quality of the health assessment			
Poor	33.3	33,3	1.00
Good	66.7	66,7	(0.12–7.99)
Security			
Poor	50.0	41,7	1.40
Good	50.0	58,3	(0.19–10.03)
Iluminatiom			
Poor	33.3	8,3	5.50
Good	66.7	91,7	(0.38–78.57)
Community mobilization			
Poor	83.3	50	5.00
Good	16.7	50	(0.44–56.62)
Leasure			
Poor	83.3	25	15.00
Good	16.7	75	(1.22–185.20)
Air quality			
Poor	33.3	41,7	0.70
Good	66.7	58,3	(0.09–5.43)

strong potential to produce deleterious effects in the state of general and occupational health in this population.

In the present study, there was a high prevalence of poor SRH among women who composed the sample (33.3%), these were mostly single and alcoholics and was associated with having fewer recreational opportunities. These data are higher than found in the study about the profile lifestyle and leisure habits of industrial workers, which showed that the highest proportion of workers who reported a negative perception of their own health were women (18.4%) and married (17.3%) (Nahas, 2009). Asfar et al. (2007) confirms the above and added that women get sick more than men, and that this disparity occurs in most developing countries.

Stropasolas (2004) to discuss the values of women working in agriculture says that these workers marry early and have children very early, because they see the formation of a new family following the patterns displayed by their predecessor's family. If this standard does not happen, the woman sees herself in a situation of conflict and anxiety to marry, favoring psychological distress and depression.

Regarding the leisure activity, Bicalho et al., (2010) in his study on the prevalence of individuals active in the field of leisure showed that women were three times less active. Corroborating the results presented by Nahas (2009) observed thatwill that the proportion of workers classified as inactive during leisure time were mostly women (60.6%), this relationship can be evidenced with the foregoing in this research, where the sample study, women who reported their leisure with poor had fifteen times more chances of poor SRH. About it, Stropasolas (2011) speaking about farmers saw that recreational activities are male privilege, and as is knew that lack of leisure activities favor the appearance of illness and decreased quality of life, the privation of which results in negative effects on perception of health among women farmers.

Still, on the use of alcohol and their relationship to the health of farm workers, Farias et al. (2000) found that the few numbers of drinkers could justify the few hospitalizations between the farmer populations studied by him.

In contrast, few possess represented decreased risk for poor health (PR = 00:10, 95% CI = 0:10; 0.97), corroborating the idea presented by Buss & Pellegrini (2007) where health and quality of life is not dependent on amount of access to material things, but the lack of opportunity in the pursuit of healthy practices or improve the feeling of well-being in any environment in which we are embedded.

5 CONCLUSION

The study contributed to the discussion of gender and work in relation to perceived health in an important part of Brazilian women, rural workers. Although the intense process of urbanization that occurred in Brazil in the last century, still persists in some regions significant portion of the population living and working in rural areas, with the largest contingent in the Northeast. In this group, the results showed that women without access to leisure activities, who drank alcohol regularly and married, had a higher proportion of self-reported poor health. However, evidence to indicate, similar to studies with urban women, that can determine marital status married worse perceived health, possibly due to the multiple journey, besides the role of primary provider of family care, especially in rural so that new perspectives need to be directed to the study population, either by health professionals or public policy makers.

REFERENCES

Asfar, T. et al. 2007. Self-rated health and its determinants among adults in Syria: a model from the Middle East. *BMC Public Health* 7: 177.

Bicalho, P.G. et al. 2010. Adult physical activity levels and associated factors in rural communities of Minas Gerais State, Brazil. *Revista de Saúde Pública* 44.5: 884–893.

Bluestein, D. & Rutledge C.M. 2006. Perceived health and geriatric risk stratification. *Canadian Family Physician*, vol. 52.

Brasil. Ministério da Saúde. Secretaria de Gestão Estratégica e Participativa. Departamento de Apoio à Gestão Participativa e ao Controle Social. 2010. *Saúde da mulher: um diálogo aberto e participativo/* Ministério da Saúde, Secretaria de Gestão Estratégica e Participativa, Departamento de Apoio à Gestão Participativa e ao Controle Social.—Brasília: Editora do Ministério da Saúde.

Buss, P.M. & Pellegrini, A. 2007. A saúde e seus determinantes sociais. *PHYSIS: Rev. Saúde Coletiva* 17(1): 77–93.

Butto, A.; Dantas, I (org.). 2011. Autonomia e cidadania: políticas de organização produtiva para as mulheres no meio rural/Brasília: Ministério do Desenvolvimento Agrário.

Calnan M., Wadsworth E., May M., Smith A., Wainwright D. 2004. Job strain, effort—reward imbalance, and stress at work: competing or complementary models? *Scand J Publ Health* 32(2): 84–93.

Dejours C. 1986. Por um novo conceito de saúde. *Revista Brasileira de Saúde Ocupacional,* São Paulo 14(54): 7–11.

DIEESE—Departamento Intersindical de Estatística e Estudos Socioeconômicos. 2011. Anuário das mulheres brasileiras. DIEESE: São Paulo.

Di Sabato, A. & Melo, H.P. 2009. Gênero e Trabalho Rural. *Estatísticas Rurais e Economia Feminista.* MDA: Brasília.

Dowd J.B. & Does A.Z. 2007. The predictive powers of self-rated health for subsequent mortality risk vary by socioeconomic status in the US? *International Journal of Epidemiology* 36: 1214–1221.

Faria, N.M.X. et al. 2000. Processo de produção rural e saúde na serra gaúcha: um estudo descritivo. *Cad. Saúde Pública* 16(1): 115–128.

Faria, N.M.X. 2005. A saúde do trabalhador rural. Tese de Doutorado. Programa de Pós- Graduação em Epidemiologia. Universidade Federal de Pelotas. Pelotas.

Fonseca, S.A. et al. 2008. Percepção de saúde e fatores associados em industriários de Santa Catarina, Brasil. *Cad. Saúde Pública, Rio de Janeiro* 24(3): 567–576.

Goldberg, P. et al. 2001. Longitudinal study of associations between perceived health status and self-reported diseases in the French Gazel cohort, *J Epidemiol Community Health* 55: 233–238.

IBGE—Instituto Brasileiro de Geografia e Estatística. 2009. *Pesquisa Nacional por Amostra de Domicílios—PNAD*.

Lundberg, O. & Manderbacka, K. 1996. Assessing reliability of a measure of self-rated health. *Scand J Soc Med* 24(3): 218–224.

Martinez, M.C. 2002. *As relações entre a satisfação com aspectos psicossociais no trabalho e a saúde do trabalhador.* Dissertação de Mestrado, Faculdade de Saúde Pública, Universidade de São Paulo, São Paulo.

Nahas, M.V. 2009. *Estilo de vida e hábitos de lazer dos trabalhadores das indústrias brasileiras: relatório geral.* Brasília: SESI.

Rebouças, M.A. & Lima, V.L.A. 2013. Caracterização socioeconômica dos agricultores familiares produtores e não produtores de mamão irrigado na agrovila canudos, ceará mirim (RN). *Holos* 79–95.

Siegrist, J. et al. 2004. The measurement of effort-reward imbalance at work: European comparisons. *Soc Sci Med* 58(8): 1483–499.

Stropasolas, W.L. 2004. O valor do casamento na agricultura familiar. *Estudos Feministas* 12(1).

Peres, M.A. et al. 2010. Auto-avaliação da saúde em adultos no Sul do Brasil. *Revista Saúde Pública* 44(5): 901–911.

Occupational Safety and Hygiene II – Arezes et al. (eds)
© 2014 Taylor & Francis Group, London, ISBN 978-1-138-00144-2

Prevalence of chronic pain and associated factors in administrative assistants of a Brazilian university

D.S. Dantas
Department of Public Health, Federal University of Rio Grande do Norte, Natal, Rio Grande do Norte, Brazil

A.G. Magalhães, L.S. Nascimento Júnior & F.D. Sá
Federal University of Rio Grande do Norte, Health Science College of Trairi, Santa Cruz, Rio Grande do Norte, Brazil

G.B. Nascimento & C.H. Moreira
Department of Psysiotherapy, State Universidade of Paraiba, Campina Grande, Paraíba, Brazil

ABSTRACT: In the literature, there are many definitions and classifications assigned for pain, but the consensus is that this is a subjective sensation, which highlights the need for that a multidimensional evaluation would be made, not just focused on intensity. Pain, be acute or chronic, will bring several changes of sleep, appetite and cravings in the individual's life and can, in severe cases, incapacitating the individual to work and active social life. Psychological or biological aspects, and more specifically the exercise of the work, are considered as sources of pain. As regards the labor activity, factors such as demands, deadlines, repetitive postures and inadequate workstation, which require postural compensations of employees, contributing for the appearance of pain. The present study aimed to evaluate the prevalence of chronic pain in administrative assistants and associated factors.

1 INTRODUCTION

Pain, in respect to discomfort, is dependent of the intensity, but can cause some or a lot of suffering, impairing their usual activities, such as work, sleep, leisure, sexuality and social life, promoting a decrease in quality of life.

According to the International Association for the Study of Pain (IASP), pain is defined "as a subjective experience unpleasant sensory and emotional associated with actual or potential injury of the tissue, when described in terms of such damage, which was experienced by almost all people" (Carvalho, 1999).

Kreling (2006) says that, acute or chronic, the pain, generally, leads the individual to manifest symptoms such as changes in sleep patterns, appetite and libido, manifestations of irritability, changes in energy, impaired concentration and restrictions on the ability for family, professional and social activities. Consequently, the individual who has chronic pain, the persistence of the pain, extends and strengthens these symptoms.

To Iida (2005) in addition to the requirements of the task, often the worker takes inadequate postures due to poor design of machinery, equipment and workplace, which if kept for a long time, can cause bodily pain, fatigue, occupational diseases and absenteeism. Pain syndromes resulting from work, to Oliveira (2000), include: headache, back pain, and Work-Related Musculoskeletal Disorders (WRMD's).

Frutuoso (2006) adds that a negative view of work by the worker may potentiate the disease. Dejours (1994) states that long hours working or work on rotating shifts cause irritability and despondency, triggering relationship problems from family life, contributing to the onset of pain syndromes.

Although pain is highly subjective, understanding its mechanisms and/or processes, and the development of tools to assess and measure reliably the perception of pain, including its multidimensionality, are essential to better control and manage it.

In this sense this study aimed to determine the prevalence of chronic pain and associated variables among office workers of a Brazilian higher educational institution, as well as evaluates the pain in a multidimensional context.

2 METHODOLOGY

Was conducted a census study, together with the 96 workers who perform the function of administrative assistant in the State University of Paraiba—Campus I, located in Campina Grande—Paraiba—Brazil. The inclusion criterion was to be in full exercise of the work activities and consent to participate.

The data collection was performed by trained interviewers and consisted of the application of a form of interview. The protocol was composed by two parts: the first part was developed exclusively for this research in order to characterize the sample and collect relevant socio-demographics data, handedness and occupational data. At the end, asked the participant about the presence of any pain that persists for more than six months, characterizing the pain like chronicle (Scopel et al., 2007; IASP, 2008). The second part of the protocol, which is composed by a body diagram, a numeric categorical scale and McGill Pain Questionnaire (MPQ), aimed to evaluate the characteristics of pain and was only answered by the workers who gave that question an affirmative answer.

The body diagram is an instrument in which the worker must indicate with symbols specific according to the intensity, the region which has pain with time equal or bigger than six months. The Numerical Categorical Scale (NCE) was used together with the body diagram and quantifies the intensity of pain using numbers. We used a 11-point scale, from 0 to 10, where zero is no pain and 10 the worst possible pain. The values were grouped in low pain (1 to 3), moderate pain (4 to 6) and severe pain (7 to 10) (Andrade et al., 2006).

We used a version translated by Pimenta & Teixeira (1996), of the McGill Pain Questionnaire (MPQ). The MPQ is a multidimensional scale used around the world, "that provides quantitative measures of pain, making possible statistical treatment", that is composed of 78 descriptors or words on pain divided into twenty subgroups and four major groups: sensory-discriminative, affective-emotional, cognitive-evaluative and miscellaneous.

According to Silva and Ribeiro Filho (2006), the pain score was proposed by Kremer et al. in 1982 and is obtained by the summation of the intensity values of the descriptors for each dimension and then sharing that value added the total score in that dimension, which will generate values ranging from zero, indicating that the patient did not select any descriptor of a given dimension to describe the pain, and one, indicating that the patient selected all descriptors ordinations higher a given dimension to describe their pain. These scores were obtained in the total and for each one of the four areas.

For statistical analysis we used the Kolmogorov-Smirnov test, face to the need of to prove the normality of the data. Then, they were analyzed by this descriptive statistics and inferential. For employment of association tests, the variables were dichotomized. In the case of numerical variables were dichotomized from these average values. To verify the association of socioeconomics and occupational factors with the occurrence of pain was calculated the prevalence ratio. And finally, to verify differences in the scores of different areas of McGill according to sex, we used the t-test.

Statistical analysis was conducted in SPSS 20.0. A p-value < 0.05 indicates a statistically significant result. The research protocol was approved by the Ethics Committee of the State University of Paraíba.

3 RESULTS

The study sample was composed mainly of women (51.0%), with a predominance of 20–29 years (32.3%), right-handed (91.7%), married/consensual marriage (49.0%), with teaching complete medium (44.8%). Regarding the occupational characteristics, we found that workers spend most time sitting (64.6%), with weekly workload of 40 hours of work (97.9%) and time in the profession for less than five years (37.5%).

Among the workers studied, the prevalence of chronic pain was 57.3%. Comparatively analyzing workers with and without chronic pain can we perceive differences in socio-demographic characteristics of those (Table 1).

Workers with chronic pain were mostly female (65.5%), aged 40–59 years (65.4%), right-handed (92.7%), married/consensual marriage (54.5%), with higher education (47.3%), remained seated most of the time (67.3%), with weekly workload of 40 hours (96.2%) and working in the profession are less than five years (29.2%).

Of the variables analyzed in a comparative way, there was a statistically significant difference between workers with and without chronic pain, in relation to gender ($p = 0.001$), age ($p = 0.048$) and time occupying the profession ($p = 0.012$). Therefore, chronic pain was mostly reported by women aged over 40 years and length of service exceeding 20 years.

When analyzing the prevalence ratio for the occurrence of chronic pain, it was observed that protective factors were: being male (PR = 12:24, CI: 0:10; 0:58), have aged less than 40 years old (RP = 0:45, CI 0:20, 1:04) and working in the profession for less than 16 years (PR = 00:36, CI 00:15, 0.82).

The average of places with pain was 1.68 ± 0.91, and as showed at Table 2, the most appointed region was the thoracic column, followed by back, sacral and coccygeal and the majority classified the intensity how moderate.

Table 1. Characteristics of the population according to the presence of chronic pain.

Variables	With chronic pain n = 55	Without chronic pain n = 41	p*
Sex			
Male	19 (34.5)	28 (68.3)	0.001
Female	36 (65.5)	13 (31.7)	
Age			
20–29 years	12 (21.8)	18 (43.9)	0.048
30–39 years	6 (10.9)	4 (9.7)	
40–49 years	18 (32.7)	10 (24.4)	
50–59 years	18 (32.7)	7 (17.1)	
60 years or more	1 (1.9)	2 (4.9)	
Handedness			
Right-handed	51 (92.7)	37 (90.2)	0.663
Left-handed	4 (7.3)	4 (9.8)	
Civil state			
Single	19 (34.7)	17 (41.5)	0.205
Married/consensual marriage	30 (54.5)	17 (41.5)	
Divorced	3 (5.4)	3 (7.2)	
Widower	3 (5.4)	4 (9.8)	
Level of education			
High school	25 (45.4)	18 (43.9)	0.880
Graduate	26 (47.3)	16 (39.0)	
Post-graduate	4 (7.3)	7 (17.1)	
Predominant position			
Sitting most of the time	37 (67.3)	25 (61.0)	0.833
Sitting and standing in equal periods	1 (1.8)	1 (2.4)	
Standing most of the time	17 (30.9)	15 (36.6)	
Working hour per week			
≤40 hours	53 (96.2)	41 (100)	–
>40 hours	2 (3.8)	0	
Time occupying the profession			
Until 5 years	16 (29.2)	20 (48.8)	0.012
06–10 years	3 (5.4)	1 (3.4)	
11–20 years	3 (5.4)	7 (17.1)	
21–30 years	22 (40.0)	8 (19.5)	
Over 30 years	11 (20.0)	5 (12.2)	

* Calculated by X^2.

The McGill pain questionnaire permits realize a score from zero to one for each domain. Using this analysis, we calculated the average score and workers with pain were grouped according to gender. As shown in Table 3, there are significant differences between the domains sensory-discriminative, affective-emotional, evaluative and total score between the sexes. Furthermore, we can learn that for both sexes, the highest scores were reached in the areas of evaluation, followed by sensory-discriminative domain.

Table 2. Localization and intensity of chronic pain at participating workers.

Variables	Percentage (%)
Numbers of locals with pain	
1	52.7
2	31.0
3	12.7
4	1.8
5	1.8
Locals	
Cervical region	18
Thoracic region	25
Lumbar, sacral e coccyx region	21
Shoulder and upper limb	16
Lower limb	14
Head, face and mouth	6
Intensity by the NCE	
Low pain (1 a 3)	11.1
Moderate pain (4 a 6)	53.7
Severe pain (7 a 10)	35.2

Table 3. Standardized scores for the different dimensions in the McGill Questionnaire.

Dimensions in the McGill Questionnaire	Sex		
	Male (n = 19)	Female (n = 35)	p* value
Sensory-discriminative	0.31	0.59	0.001
Affective-emotional	0.22	0.46	0.002
Evaluativ	0.44	0.75	0.004
Miscellany	0.29	0.43	0.102
Total	0.30	0.54	0.001

* Calculated by T-student test.

4 DISCUSSION

In this sample the high incidence of chronic pain seen in workers corroborates studies Kreling (2006), which found a prevalence of chronic pain in 61.4% of workers, and by Panazzolo et al. (2007).

When comparing the demographic profile of workers with chronic pain with the total sample, we realize that there are no big differences. Just a fact that should be highlighted is the largest report of chronic pain in women, given that resembles that found by Verhaak et al. (1998), that after review of nine studies on the prevalence of chronic pain was observed in two of these the frequency of occurrence of chronic pain was the same for men and women and in seven, namely the majority, the pain was more frequent in women in any of studies reported experiencing greater in men.

Kreling (2006), in his study notes some factors that could explain the higher incidence of pain in women among them we can mention the hormonal changes intrinsic to the female menstrual cycle, which would make the woman more sensitive, consequently increase their perception of pain, and cultural character that allows a woman to express her pain, but this reproach expression in man, since as parameter insensitivity associate virility individuals aged below 40 years had protection for the development of pain, showing that events chronobiological are related to that experience.

In addition, it was perceived that individuals with a shorter time in the profession had lower pain reports than those with longer experience, which may be explained by the cumulative exposure to ergonomic work-related risk factors, and establishes the causal link with the occupation.

The percentage of reference of only one location of pain is similar to Kreling (2006) studies in which the 505 workers studied, 66% reported pain happens more than six months only at one place and the average number of places was 1.32. Still on the same study, the reference sites of pain are different, because in this study, the most quoted site was head, face and mouth (26.7%), followed by the lumbar spine (19.4%).

The site thirst checked in our sample corroborates the findings of Dellaroza, Pepper and Matsuo (2007) that in a study with 451 elderly servers in Londrina—Paraná—Brazil, observed the dorsal region (21.7%) as a local thirst most cited for chronic pain.

Correlating activity time with the function performed by workers and the work environment observed during the visits, is conjectured that the higher reported pain in the thoracic, lumbosacral and coccygeal region is derived from the inadequacy in the ergonomic design of furniture which are not allow a backboard suitable for the column, neither the maintenance of the worker in appropriate posture, requiring the worker to remain for a long period in trunk flexion and forward head for better viewing and handling, whether the monitor, either objects on the desktop.

Authors such as Marras (2001) and Schneider et al. (2005) summarize the problem of back pain is related to static or repetitive work, tasks that require frequent bending and rotation of the spine and the number of times that the same approach is adopted in a short time, in this case as inappropriate furniture imposes the adoption of these postures erroneous, possibly it is related to the factors causing the pain.

In a more psychological reading, Marras (2001) and Schneider et al. (2005) believe that the frames of chronic back pain problems such termination in career, breaking of family relationships, personal isolation and depression, among others.

Considering that a good part of workers spend a good time on the computer doing repetitive exercises typing, we could expect a higher incidence of pain in the wrist and upper limbs, consistent with carpal tunnel syndrome, but it is believed due to the wide dissemination of the syndrome and its prevention in the media and considering the high level of education of the sample, the workers have some care to adopt preventive measures which might decrease the occurrence of pain in these locations.

The moderate intensity, more obtained in this study, contraries the studies of Dellaroza et al. (2007), which reported with higher intensity (50%) the pain, which corresponds to the range of one to three in the NCE.

According to Silva and Ribeiro Filho (2006), MPQ assesses pain in three dimensions: sensory, affective and evaluative. Following this line, in most part of the workers, the sensory dimension showed the highest scores, noting that the mechanical, thermal and spatial properties of the pain are more present in the painful experience of the tension, fear, dread, fears and neurovegetative responses, fact does not exclude the affective properties in painful experience that is without a doubt a subjective experience.

Analyzing the scores obtained in the areas sensory-discriminative and affective-emotional, it is clear that the participants of both sexes achieved higher scores in the first. Therefore, for the sample studied, pain was related with questions neurovegetative more than psychological.

Finally, a comparative analysis between the genders reinforces the difference in pain expression between men and women, once they have obtained higher scores in all areas analyzed.

5 CONCLUSION

It is concluded that chronic pain showed high prevalence among the workers studied, and mainly affects women. Beyond gender, age and working time in the profession were factors that contributed to the onset of chronic pain.

Another important factor was greater contribution the sensory-discriminative component to the painful experience.

REFERENCES

Andrade, F.A. et al. 2006. Mensuração da dor no idoso: uma revisão. *Rev Latino-am Enfermagem*, v. 14 n. 2, março-abril, 2006.

Carvalho, M.M.M.J. 1999. Prefácio. In: ________(Org.) *Dor:* Um estudo multidisciplinar. São Paulo: Summus. p. 7–8.
Dejours, C. et al. 1994. *Psicodinâmica do Trabalho*. São Paulo: Atlas, 1994.
Dellaroza, M.S.G. et al. 2007. Prevalence and characterization of chronic pain among the elderly living in the community. *Cad. Saúde Pública*, Rio de Janeiro, v. 23 n. 5, p. 1151–1160.
Frutuoso, J.T. 2006. *Mensuração de aspectos psicológicos presentes em portadores de dor crônica relacionada ao trabalho*, 2006. Tese (Doutorado) Programa de Pós-Graduação em Engenharia de Produção, Universidade Federal de Santa Catarina, Florianópolis.
IASP—International Association for Study of Pain. *PAIN*. Disponíble at: <www.iasp-pain.org> Access in 15/05/2012.
Iida, I. 2005. *Ergonomia Projeto e produção*. 2ª Ed. São Paulo: Edgar Blucher.
Kreling, M.C.G.D. et al. 2006. Prevalence of chronic pain in adult workers. *Rev Bras Enferm* v. 59 n.4, p. 509–513.
Marras, W.S. 2001. Spine biomechanics, government regulation, and prevention of occupational low back pain. *The Spine Journal* s.1, p. 163–165.
Mccaffery, M. & Ferrel, B. 1997. Nurses' knowledge of pain assessment and management: how much progress have we made? *J Pain Symptom Manage*. v. 14 n. 3.
Oliveira, J.T. 2000. Aspectos comportamentais da síndrome de dor crônica. *Arquivos de Neuro-psquiatria*, v.58 n. 2-A, p. 360–65.
Panazzolo, D. et al. 2007. Dor crônica em idosos moradores do conjunto cabo frio, cidade de londrina/pr . *Rev. Dor* v. 8 n. 3.
Pimenta, C.I. 1999. Fundamentos teóricos da dor e de sua avaliação. In: Carvalho, M.M.J. (Org). *Dor:* um estudo multidisciplinar. São Paulo: Summus, p. 31–46.
Pimenta, C.A.M. & Teixeira, M.J. 1996. Questionário de dor McGILL: proposta de adaptação para a língua portuguesa. *Rev.Esc.Enf. USP* v. 30 n. 3 p. 473–483.
Schneider, S. et al. 2005. Workplace stress,lifestyle and social factors as correlates of back pain: a representative study of the German working population. *Int Arch Occup Environ Health* s. 78, p. 253–269.
Scopel, E. et al. 2007. Medidas de Avaliação da dor. *Revista Digital—Buenos Aires* v. 11 n. 105.
Silva, J.A. & Ribeiro Filho, N.P. 2006. *Avaliação e Mensuração de dor:* Pesquisa, teoria e prática. Ribeirão Preto: Funpec-editora.
Verhaak, P.F.M. et al. 1998. Prevalence of chronic benign pain disorder among adults: a review of the literature. *Pain* s. 77, p. 231–239.

Occupational Safety and Hygiene II – Arezes et al. (eds)
 ISBN 978-1-138-00144-2

Occupational risks in a workplace of a metallurgical in Paraíba—Brazil

J.G. Silva, R.M.A. Lima, M.B.G. Santos, I.F. Araújo & E.S.P. Sales
Federal University of Campina Grande, Campina Grande, Paraíba, Brazil

ABSTRACT: Environmental conditions at work can significantly impair sensory and psychological aspects of the employees, affecting the well-being and quality of services. This paper aims to collect data to verify aspects of thermal comfort, acoustic and ergonomic in a job Oxipira machine that is automated cutting and drilling sheet metal. We conducted a literature research about comfort, highlighting key aspects of national legislation. Quantitative data were collected using specific equipment. Qualitative assessments were obtained from the direct observation of the production process. The activity is designed to sitting position, but the operators alternate standing posture. It was observed above the noise level acceptable for the purpose of comfort, as the effective temperature and the relative humidity did not reveal inadequacies. The air velocity was below the recommended for air renewal. The activity object of the study revealed some inadequacies and so were some recommendations.

1 INTRODUCTION

In any production process, concern for comfort is also essential because the environmental conditions of the working environment can significantly affect the sensory and psychological aspects of who uses it, decreasing the yield capacity, thus justifying the current preoccupation with association between work environment and comfort conditions present in it.

Iida (2005) emphasizes that the adverse environmental conditions such as extreme temperatures, noise pollution and vibrations cause discomfort, increasing the risk of accidents, which may cause considerable damage to health.

There are many variables that may be associated with environmental comfort in a work environment, e.g. noise, lighting, temperature, humidity, purity and air velocity, radiation, physical effort, type of clothing, among others. Each one represents an important part in the welfare of workers and the quality of services. Health problems often are related to one or more of these variables of comfort, as can also be linked to changes in the individual order, social and technical (Silva et al., 2002).

This work aims to verify aspects of thermal, acoustic and ergonomic comfort in a specific job, where the machine Oxipira is installed, which is automated cutting and drilling of sheet metal in a metallurgical company. To do so, the agents relevant to the physical activity developed based on the criteria defined in the Regulatory Standards NR-15 and NR-17-01 in NHO and NBR10152.

2 CHARACTERIZATION OF THE COMPANY

Nowadays many modern buildings are steel structured, this initial positioning enabled the studied company to a pioneering and exponential growth, resulting in great partnership at the beginning of the year 2011, with a group of investors.

Located in the industrial district of Queimadas—PB, it has an area of 85.000 m^2, with 14,000 m^2 of built area subdivided into manufacturing, painting and offices and 20,000 m^2 for the stock and shipping.

It is a company specialized in manufacturing and assembly works in the sectors of oil and gas, power plants, buildings with multiple floors (commercial or residential), sheds (industrial or commercial), shopping centers, distribution centers and logistics, supermarkets, airports, terminals roads, railways, bridges, flyovers and footbridges. Having production capacity of around one thousand (1,000) tons a month.

2.1 *The equipment Oxipira*

The Oxipira is a machine used for cutting and drilling of metal plates with different thicknesses. Presents a rugged and durable enabling its use in the industrial environment with maximum working cycle. It has low operating cost and maintenance, which contributes to the quality of the parts produced, resulting in cost-effective once the final cost of a part is low and quality is high. The Oxipira is capable of performing detailed cuts, allowing for greater exploitation of raw material which impacts on cost reduction. The handling is carried out by means of a computerized interface, without the need for manual adjustments to the thickness, shape and size of the cut piece.

2.2 *Ergonomic features*

This analysis refers to a diagnosis ergonomic, based on Regulatory Standard NR-17 (2007), making necessary a study of the workplace in order to detect occupational risk factors capable of providing subsidies for the ergonomic solutions company, adapting the legislation.

Grandjean (1998) states that the practical goal of ergonomics is the adaptation of the workplace, instruments, machines, timetables, environmental requirements of the human being. The achievement of such goals, the industrial level, provides improvement in labor income and income from human effort.

Therefore, the ergonomic analysis studies work circumstances seeking to adapt it to the man from the analysis of environmental and organizational conditions, among them, there are the thermal and acoustic effects, seeking to reveal improved practice tasks with comfort, health, safety and effectiveness.

2.3 *Aspects thermal*

Thermal comfort is inserted into the environmental comfort, where also part visual comfort acoustic comfort and air quality. Since this is a set of variables that directly interfere in human performance.

Lamberts & Xavier (2002) defines thermal comfort as a state of mind that expresses satisfaction with the environment. The body conditions also affect this feeling; this feeling is so subjective, as it differs from person to person and varies according to weight, metabolism, amount of body fat, type of activity performed, among others.

According to Fanger (1970), as the thermal comfort involves environmental variables and also subjective or personal variables, it is not possible that a group of people subject to the same environment at the same time, all of it is satisfied with the same thermal conditions, due to individual characteristics of people.

A comfortable environment is one that triggers metabolic heat production, which remains stable heat exchange with the environment. The environment not being in a situation of stability, there may be occasions where the harsh heat exchange establishes a risk to the welfare of the individual, because even the human body possessing mechanisms of thermoregulation, cannot maintain the internal temperature constant and consistent.

2.4 *Acoustic aspects*

The acoustics aims to examine the elements of sound and its relationship to the human senses, in order to make minimal unfavorable conditions, such as noise. Having the goal of maximally reduce the noise causing hearing damage and control the interference to enable a good understanding between the person that plays the sound and the one who will listen to it.

Noise-induced hearing loss continues to be one of the most prevalent occupational conditions and occurs across a wide spectrum of industries. Occupational hearing loss is preventable through a hierarchy of controls, which prioritize the use of engineer controls over administrative controls and personal protective equipment (Kirchner, 2012). When significant noises can interfere with complex tasks and simple as they may confound the worker and mitigate the effects of monotony and repetitiveness enabling the rise of occupational diseases and accidents at work.

The parameters used to assess the level of noise are present in the Norm 15—Unhealthy Activities and Operations (1978) which determines the exposure limits as a function of working time.

2.5 *Regulatory standards*

2.5.1 *NR -17*

This Norm aims to establish parameters for adapting working conditions to the psychophysiological characteristics of workers, in order to provide maximum comfort, safety and efficient performance. Include aspects related to lifting, transporting and unloading of materials, to furniture, equipment and environmental conditions of the job and the organization of work itself.

2.5.2 *NR 15*

This Norm comments on the activities, operations and unhealthy implications. For evaluation purposes will only be used to Table 1 of Annex 01 of NR, which shows the maximum allowable limits during the workday.

2.5.3 *NBR 10152*

This Standard sets noise limits compatible with the acoustic comfort in different environments.

2.5.4 *NHO 01*

This Technical Standard is to establish criteria and procedures for the assessment of occupational exposure to noise, which implies potential risk of occupational deafness. It was took into account only the topic related to evaluation.

3 MATERIALS AND METHODS

We performed a literature search of comfort and its thermal, acoustic and ergonomic characteristics, taking into account the main aspects involved in the national legislation. Taking as reference the query structuring NR17 (2007) establishing parameters that allow the adaptation of working conditions to the psychophysiological characteristics of workers, in order to provide maximum comfort, safety and efficient performance, it was also considered aspects related to NR 15 (2011) that discusses the unhealthy activities and operations and their implications, and NBR 10152 determining noise limits compatible with the acoustic comfort. To determine the assessment procedures, NHO 01 was consulted.

Quantitative data were collected through direct research through observation and the use of specific equipment for each measurement was obtained the necessary analyzes and conclusions.

Noise levels were measured with the aid of digital decibel meter multifunctional, being positioned near the auditory zone of the worker. The effective temperature and air velocity and relative humidity were measured with a portable digital anemometer. Measurements were taken with the instrument located near the worker so that there barriers that could jeopardize the legitimacy of the data obtained. The measurements were performed during three consecutive days of April during the evening hours, which are carried out at intervals of 5 minutes when done using the decibel meter and 60 minutes when done using the anemometer. Qualitative assessments were made based on the observation of the activity in the workplace.

3.1 *Materials used*

The variables that make up the thermal environment, air temperature, Tar (°C), relative humidity, RH (%), were collected simultaneously by the instrument Psychrometer—AN—4870 portable anemometer manufacturer ICEL, as well as the noise level (dB) was collected by use of a Thermo-Hygro—decibel meter, light meter portable model THDL—400 manufacturer Instrutherm.

4 ANALYSIS OF RESULTS

4.1 *Ergonomic features*

The activity performed at the workstation machine Oxipira includes discontinuously manual transport of loads, alternating between machine operation, inspection and cargo transportation to final disposal and storage. This transport is made by employees from the work stations.

The transport of metal plates to the machine Oxipira is via muncks and physical effort performed by the worker is compatible with his strength and ability to not compromise his health or safety, but he does not receive satisfactory training or instructions regarding methods of working that should be used, in order to safeguard their health and prevent accidents.

The work station offers a work area easy to reach and a good display by the worker. It is planned to sitting position, but the operators alternate this posture with the standing position, it becomes necessary for the execution machine adjustment, inspection and transportation. The height and distance of the panel provide operators the required distance from the eyes to the field work. The chair and panels provide workers able to maintain proper posture, facilitating the visualization and execution of activities, thus meeting the criteria established ergonomic comfort by NR17.

The seat used by machine operators has dimensional features that allow proper positioning and movement of the body segments. It has adjustable height to the height of the operators, so the seat can meet different anthropometric measures, also allowing good accommodation for the workers who will use this seat in another turn.

The chair conformation has not used in the base which allows the worker's posture changes, providing relief from the tension on the muscles supporting. By owning rounded front edge and back with slightly adapted form to the body, allow less compression on the back of the thighs and lower back protection. From the aspects observed, one can see that the seat meets the basic requirements of comfort, offering employees good working conditions.

The machine Oxipira has footrest at the proper height for operators, enabling switching position of the feet, making the implementation of activities less stressful and more comfortable. But there are no supports in place suitable for reading documents, being performed on the control panel or even just using the hand as support, causing operators to drive frequent neck and eyestrain.

The machine used in the workplace study offers no mobility conditions sufficient to allow adjustment of the screen's ambient lighting or adjusting the keyboard according to the task to be performed. The distances between eye-screen and eye-keyboard are practically the same, since they are attached on the same basis. The distance eye-document varies with the position of the hands of the workers with the document or the place where they will allocate this role, since there is not a fixed base for supporting the same. Hindering activities that involve reading documents for typing.

The designs used to base the cut are difficult readability and understanding, as well as having small characters, for the interpretation of the data requires a level of advanced technical knowledge.

The performance evaluation for purposes of compensation does not take into consideration only the productivity, ensuring that employees do not need to commit excesses effort to enable damage to their health. There is no rest break predefined, but the workers have the freedom to leave of absence and go drink water and/bathroom. And at times there are stops activity when there is waiting to reach a new part in the process or when it is not necessary inspection and the machine is in operation. Not overloading operators.

4.2 *Acoustic aspects*

The activity studied shows no correlation or equivalence with those listed in NBR 10152, the sound level was varied between 60.1 dB and 103.5 dB depending on the thickness of the plate to be cut (As shown in Fig. 1) and it is unacceptable for purposes of comfort, since it is considered comfort zone when the environment has a level of down to 65 dB (A).

4.3 *Thermal aspects*

The effective temperature in the studied area can be considered ideal in order to not exceed the maximum thermal comfort stipulated by NR17, between 20°C and 23°C, and decreased by 0.5°C less than the minimum.

In Figure 2 we can observe the values of thermal sensation effective temperature during the days of measurement.

The air velocity did not exceed 0.75 m/s, maximum value estimated by NR17. The value measured during the whole process was 0.0 m/s thus revealing the inadequacy regarding the renewal of the

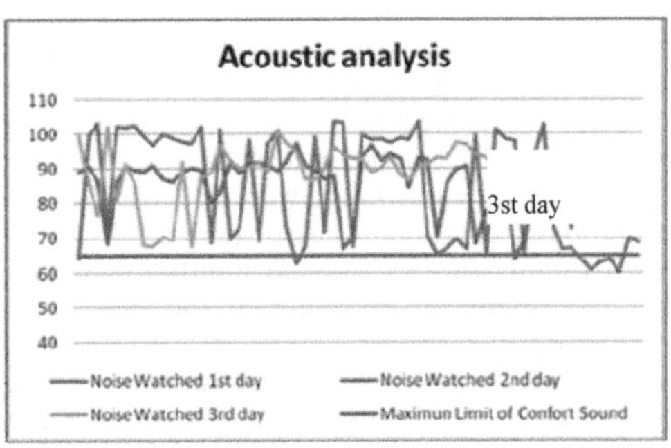

Figure 1. Acoustic analysis.
Source: Authors own (2012).

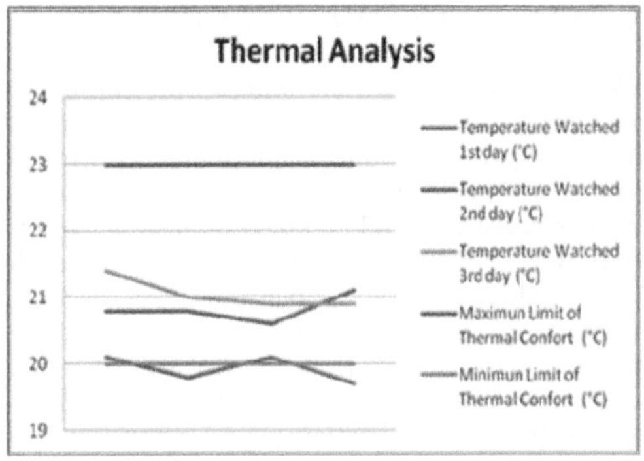

Figure 2. Thermal aspects.
Source: Authors own (2012).

air in which the operators are submitted, they are exposed to an environment with no ventilation either natural or artificial.

The average relative humidity found in the environment varied according to the days of observation, but most were recorded lower rates or very close to 50%, considered ideal for a work environment that offers comfort to the worker.

4.4 *Luminance aspects*

The lighting factory was only analyzed using criteria as perceived by researchers. Therefore, no measurements were made.

From means of compliance was noted that the environment where workers perform their activities has no good natural lighting which is unstable, since it changes according to the climate and weather and time, even with artificial perception researchers is insufficient.

Points of artificial light that there are very distant from each other, having a small capacity of luminance, since there are few points of light right foot and the work environment is high. Therefore, the lighting is not evenly distributed causing shadows and glare which also result from variation in natural lighting, which may adversely affect the performance of the inspection and safety of workers.

As no measurements were made with technical equipment, we cannot say whether the job meets the minimum levels of luminance established in NBR 5413.

5 SUGGESTIONS FOR IMPROVEMENT AND CONCLUSIONS

As conclusions of the analysis, job analysis could identify some shortcomings related to ergonomics, unsatisfactory conditions of acoustic comfort.

As for the comfort level on the temperature of the gas machine Oxipira, was considered sufficient, though, still can make improvements in this regard.

Some recommendations are suggested, based on the established diagnosis system dysfunctions man task, which aims to improve the working conditions of the gas studied and, consequently, improve product quality and increase productivity.

As for the thermal aspects recommends the inclusion of means of natural ventilation and artificial, enabling the cooling of the environment favoring the heat exchange medium with the worker. To this end it is suggested that implements ventilators and windows to facilitate the entry of air.

With regard to luminance aspects using artificial lighting uniformly distributed which is indicated to decrease the glare, and shadows on the environment, avoiding possible accidents caused by poor lighting.

The noise level observed in the survey, although high is not responsible for damages to workers because they use protective equipment ear—shell. Thus making the noise level near desirable for the acoustic comfort of the individual. It is not possible to enclose their sources in the machine, considering that there is no possibility to make changes to the same physical structure and the materials that constitute it. The machine is already away from other work stations, avoiding the spread in other ways than that in which it is embedded. It is possible to implement other automations to the process, thus reducing the time of exposure of the worker to the machine and muscle overload worker.

Regarding safety, it is proposed to adapt the equipment used, such as the implementation of a system of pulleys to suspend the material, thus preventing accidents caused by falling equipment and personnel, besides the use of benches for support papers/documents and placement of support brackets for the arms of workers as a way to reduce the problems related to muscle overload responsible for the occurrence of MSDs and fatigue.

It is suggested for the job the adoption of rotation operators, as well as physical activities stretching, to reduce the effects of fatigue of daily activities.

It is important not to cease the company's interest in the welfare and health of its employees, as observed through security policies and organization.

Where it is present routines of periodic examinations. It is advised that, in addition, for the pursuit of employee awareness through lectures on themes related to the environmental hazards they are exposed to and train them in the use and maintenance of personal protective equipment.

REFERENCES

Brazilian Association of Technical Standards. NBR 10152: Noise level for acoustic comfort. Rio de Janeiro, 1987.

Brazilian Association of Technical Standards. Standard Occupational Hygiene 01: Assessment of occupational noise exposure. 2001.

Fanger, P.O. Thermal Comfort. New York: McGraw- Hill Book Company, 1970.

Grandjean, E. Ergonomics Handbook. 4th ed. Publisher Bookman. Porto Alegre, 1998.

IIDA, I. Ergonomics, design and production. Sao Paulo: Edgard Blücher, 2005.

Kirchner, D. Bruce Kirchner, et al. Occupational Noise-Induced Hearing Los. ACOEM Task Force on Occupational Hearing Loss. American College of Occupational and Environmental Medicine, 2012.

Lamberts, R. Xavier, A.A.P. Thermal Comfort and Heat Stress. Laboratory for Energy Efficiency in buildcations. Florianopólis: UFSC, 2002.

MTE, Ministry of Labour and Employment of Brazil. List of regulatory Standards (15 and 17). Available at: <http://portal.mte.gov.br/legislacao/normas-regulamentadoras-1.htm>. Accessed August 10, 2013.

Silva, L.B. et al. Comparative analysis between a theoretical model and windchill declared by workers in environments with VDT. National Congress of Engineering Mechanics, 2002 João Pessoa: ABCM, 2002.

Occupational Safety and Hygiene II – Arezes et al. (eds)
© 2014 Taylor & Francis Group, London, ISBN 978-1-138-00144-2

One-way coupling between a CFD model and a transient thermal model of the human body

S. Teixeira, R.F. Oliveira, N. Rodrigues, A.S. Miguel & J.C. Teixeira
School of Engineering, University of Minho, Guimarães, Portugal

J. Santos Baptista
Research Laboratory on Prevention of Occupational and Environmental Risks (LABIOMEP/CIGAR), Faculty of Engineering, University of Porto, Porto, Portugal

ABSTRACT: For the same environment, thermal perception varies with people and satisfaction with the occupational working place has a great influence on work performance and productivity. There are several methods for solving problems of thermal comfort, including computer simulation of the thermal system comprising the Human Body—Clothing—Environment and also CFD (Computational Fluid Dynamics) techniques used to analyze the HVAC (Heating, Ventilating, and Air Conditioning) systems, particularly in the vicinity of the human body. In the current study, a CFD model was coupled with a thermoregulatory model of the human body to describe the fluid flow, heat and mass transfer between the ventilation air and a human manikin inside a room. The computational model solves the heat, mass and momentum conservation equations in the computation domain using a finite volume discretization method in the ANSYS © environment. The results can later be used to calculate the PMV-PPD index.

1 INTRODUCTION

In industrialized countries, people spend most of their time inside buildings where a wide range of environmental parameters are set within strict limits. The correct calculation and optimization of these parameters are very important because they will dictate the local comfort sensation which, in turn, affects the productivity and working performance of each person in the same environment (Djongyang et al. 2010).

The sensation of thermal comfort is dependent on the heat balance between human body and its surrounding environment and this balance can be achieved acting in the common four parameters usually related to indoor climatic conditions (air temperature, air velocity and relative humidity and, mean radiant temperature) and the two parameters related to the human body (its metabolic heat production and clothing insulation). Therefore, the building of a comfortable environment depends on a proper design of the equipment and installations, a suitable acclimatization and the correct clothing selection (Parsons 2003).

HVAC systems are often used to obtain the required interior air quality whether to guarantee adequate hygienic levels or to provide thermal comfort. These systems generate the necessary rates of air renewal of the building, its pressurization, and the control of its temperature and humidity. Its correct simulation will help building designers to achieve a comfortable working place.

There are several ways of handling thermal comfort issues. One of the most promising is based upon the use of computational simulation of the human body thermal system—clothing—environment. This method coupled with the continuous development of Computational Fluid Dynamics techniques (CFD) makes the modeling of HVAC systems for velocity field and temperature distribution particularly at the vicinity of the human body possible.

This system presents several difficulties due to the complex geometry of the human body and its thermo-physiologic properties. Nonetheless, it is possible to obtain data with an acceptable level of accuracy that can be compared with real information (Balocco & Lio 2011, Ho et al. 2009, Kilic & Sevilgen 2008).

This study investigates the influence of the environment in the human body and in a one-way coupling the influence of this heat source in the environment. For this purpose, a CFD model is developed using ANSYS® and integrated with a thermoregulatory model of the human body to simulate the environmental conditions at a working place (Teixeira et al. 2010).

2 MATHEMATICAL MODELS

There are practical, fast and cost effective ways to predict fluid behavior in complex situations, including CFD models (Versteeg & Malalasekera 1995). Such models include a set of equations that govern the conservation of mass, momentum and energy transfer to compute fluid flow distribution and energy transport in the domain.

The CFD simulation process was carried out using the ANSYS™ Workbench platform to develop the mesh, run the CFD solver and the post-processing of the results. The configuration of the Fluent™ included the energy model for the thermal exchanges between all boundaries and the species model for a simplified air mixture (H_2O, N_2 and O_2).

For the turbulence modeling, the standard k-ε with wall functions was selected due to its stability and precision of numerical results. To model radiation heat transfer, the Surface-to-Surface radiation (S2S) type model was implemented, modeled by a discrete beam approach. Solutions are obtained iteratively and the convergence is accepted when the residuals for mass, velocity, pressure and temperature are below 1E-05.

The solution algorithm scheme was the SIMPLE for the pressure-velocity coupling. The Standard method was used for the spatial discretization of the pressure while the Second Order Upwind was used for the momentum, k, ε, H_2O, O_2 and energy equations. The simulation was carried out in steady state, assuming constant the input data and boundary conditions.

The implementation of the CFD model requires the proper characterization of the human body. For this, the interaction between the body and the environment (fluid inside the room and walls temperature) is determined by a thermoregulatory model. In this, the body is divided into 16 distinct parts. Part 1, which represents the lungs-heart where metabolic heat is generated, is considered as a unique system while all the other parts divided into three layers: core, shell and skin.

Each layer has a uniform temperature and within each part exchanges are between the core and the shell and between this and the skin. Each one of the 15 parts (2–16) of human body can be covered with up to two layers of clothing. When the body part is naked, heat exchanges to the surrounding air by convection, radiation and evaporation. When the body part is covered, i.e, when the body is clothed, skin exchanges heat with the first layer of clothes and the last layer exchanges by convection and radiation with the air. The model also includes the moisture diffusion through the cloth fabrics.

A main program solves all the balance equations numerically, using the Runge-Kutta-Merson method. This method was chosen by its simplicity and robustness in solving this kind of differential equations. The initial temperature values in all body parts and layers (core, shell, skin and cloth) and the water vapor concentration at the cloth layers in all covered body parts are given and can be introduced by the user. A Fortran computer code was written and details of the model can be found elsewhere (Teixeira et al. 2010).

3 CFD MODEL SET UP

The implementation of the models is initiated by the detailed design of the room, including the presence of the human model, using the SolidWorks™ software. The room has the overall maximum dimensions of 7.3 m × 3.5 m × 3.4 m.

A representation of the surface boundaries is shown in Figure 1, where it is possible to identify the inlet in the ceiling and the outlet in the lower part of the door located at the corner of the room. The wall opposite to the air outlet is a continuous boundary that represents a glass window that is fully closed (right hand side of Fig. 1).

Some assumptions were implemented in the original geometry: the shape of the air inlet section shape was assumed to be rectangular although the actual air entrance in the room had a grille. The same simplification also was introduced in the outflow area in the lower part of the door. In both, the velocity vectors were taken at an angle to compensate for the leaned grille.

In addition, there were three lamps and two electrical heaters in the room. Two of the lamps were positioned longitudinally 0.5 m from the ceiling and a third was located perpendicular to those in a small recess into the ceiling. One heater is fixed in the wall next to the window, while the other (a portable unit) is close to the door. Both are on the same side wall.

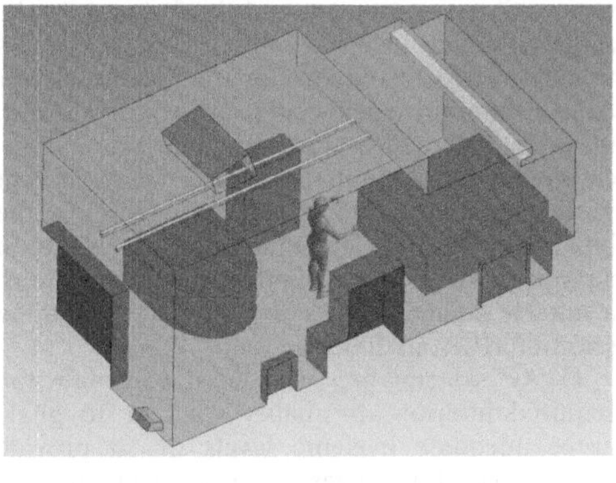

Figure 1. Representation of the room and its main boundaries (colored).

The boundaries of various furniture elements, grouped by the material type, were also included in the physical model. The manikin model was positioned in the center of the room as shown in Figure 1. Because the human model was divided into nine boundaries (head, torso, belly, arms, forearms, hips, legs, feet and hands), it was possible to assign a different material and temperature to each one.

Using the ANSYS™ software for meshing, which is integrated in the Workbench platform, the 3D tetrahedral non structured grid was obtained. Using the refinement tools for the boundary walls and advanced size function for the proximity and curvature, resulted in a set of elements ranging in size from 4.5 to 400 mm. A suitable grid (grid independent) containing 947,521 nodes and 5,214,299 elements was obtained (Fig. 2).

The inlet boundary values were measured using a Bruel & Kjaer weather station. This is a device with various sensors that allows for the measurement of surface and air temperature, air velocity and relative humidity in a confined environment. The inlet airflow was defined as a velocity inlet condition with a turbulence intensity that is a function of the Reynolds number.

In order to define the boundary conditions for the human manikin, a preliminary solution for the velocity field was obtained assuming the dummy as a simple adiabatic wall. All the other boundary conditions (walls, heat sources, inlet velocity and outflows) were already defined.

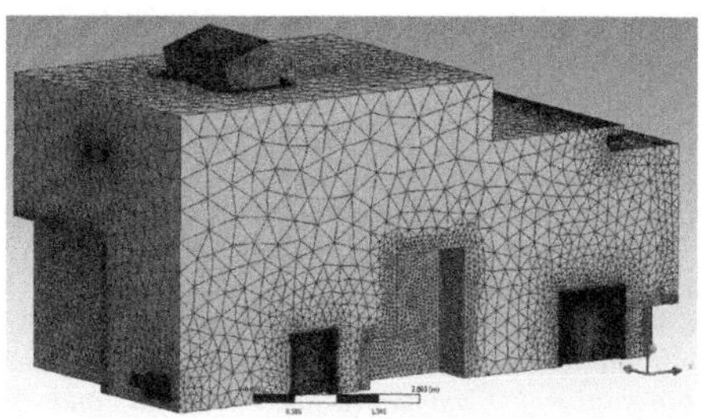

Figure 2. Representation of the computational mesh used in the simulation.

Table 1. Boundary conditions for the human dummy.

Location	Boundary type	Material	Temperature
Torso and belly	Wall	Cloth	30.74 °C
Arms and forearms	Wall	Cloth	28.83 °C
Hips	Wall	Cloth	27.77 °C
Legs	Wall	Cloth	29.74 °C
Feet	Wall	Cloth	28.55 °C
Head	Wall	Skin	34.28 °C
Hands	Wall	Skin	35.22 °C

The post-processing of the CFD results simulation provided the temperature field surrounding the human dummy. It was calculated by averaging the data over a volume centered in the dummy. For this purpose, a 1 m radius sphere was assumed. The average temperature surrounding the dummy was then used as an input to the thermoregulatory model of the human body. In the study, only the hands and head of the dummy were not clothed and the other parts were modeled with a normal clothing for office work. The metabolic rate was taken as 152 W.

The results for the different parts of the human body were calculated after a stabilization period (1 hour), as listed in Table 1. These data were used as the temperature boundary condition at the human dummy for Fluent™ final simulation. Subsequently, and considering a 65% relative humidity for the human skin, the H_2O mass fraction was calculated in the air at the vicinity of the human dummy. The O_2 concentration (22.82%) and the humidity (65%) were assumed constant and all the other configuration parameters are kept as described previously.

4 RESULTS AND DISCUSSION

Figures 3 and 4 show the contours for the absolute velocity field in the two perpendicular planes bisecting the dummy.

The results show the strong influence of the human body as a thermal source. It is clearly observed a strong upwards draft (thermal plume) that overrides the mechanical ventilation flow into the room. As a consequence of this new pattern, the furniture on the right side of Figure 3 induces a recirculation zone.

The same patterns can be observed in Figure 4. The upwards flow along the body yields flow separation near the dummy's right shoulder. This non symmetry pattern is also due to the location of the heating elements on that side of the manikin.

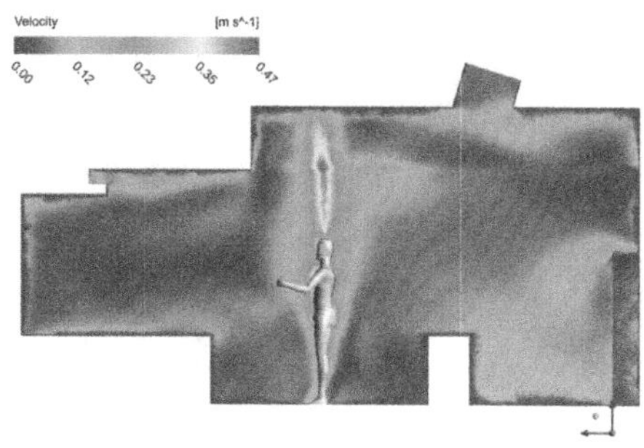

Figure 3. Velocity distribution at a xz plane.

The lighting sources in Figure 4 act as bluff bodies inducing flow separation on the upwards draft.

The temperature field shows a pattern similar to that of the velocity magnitude. The thermal plume around the human body is clearly identified in Figures 5 and 6, which results in a thermal stratification near the top of the room.

The interaction between this warm flux and the inlet flow of fresh air, results in a reorientation of this flux to right of side of the room and the presence of a layer of cool fluid near the ceiling surface. Overall temperature gradients of approximately 15 C are observed inside the room. The cooler layer in the vicinity of the ceiling flows downwards along the side walls (Fig. 6) in counter current to the upwards flow induced by the heat sources (heating elements and human body).

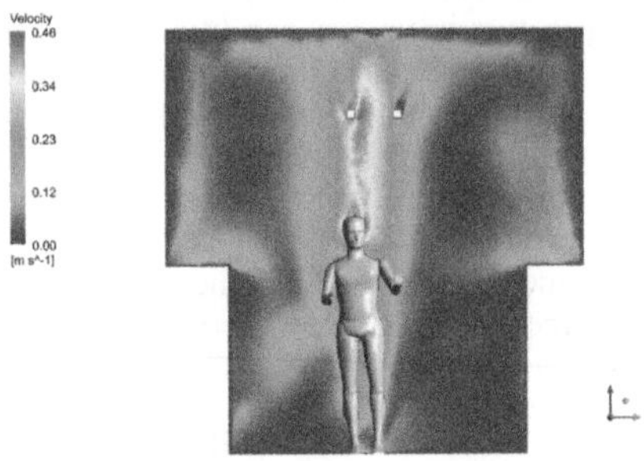

Figure 4. Velocity distribution at a yz plane.

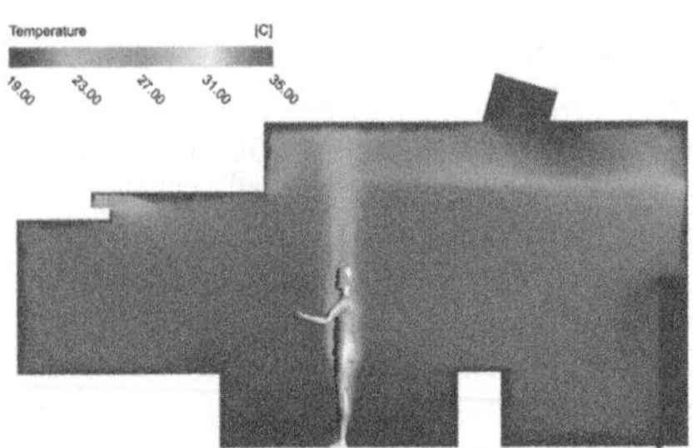

Figure 5. Temperature distribution at a xz plane.

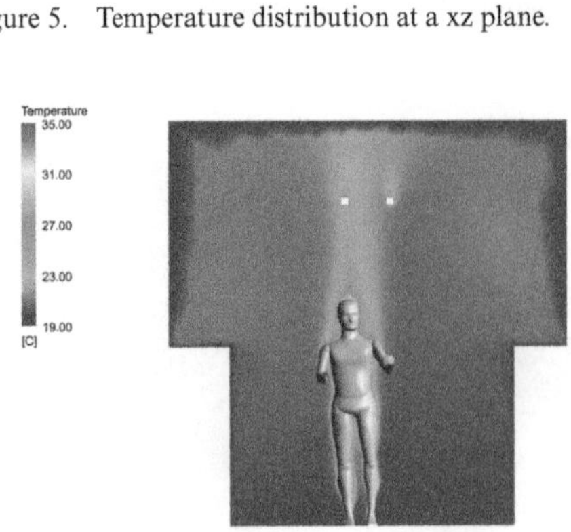

Figure 6. Temperature distribution at a yz plane.

In brief, it can be stated that the air surrounding the human dummy is heated and drives an upwards hot air plume from the human body.

5 CONCLUSIONS

The present work reports the use of a CFD model coupled with a thermoregulation model to compute the thermal evaluation of a working place. The boundary conditions were based on experimental measurements of temperature, air velocity and humidity at the inlet air ventilation supply.

The results seem very satisfactory and it can be concluded that the use of computational fluid dynamics techniques is a very suitable tool for the evaluation of thermal comfort (Rodrigues et al. 2013). The inclusion of detailed data regarding the human surface (temperature and humidity) as a function of the metabolic rate and clothing is paramount to the accuracy of the calculations.

REFERENCES

Balocco, C. & Lio, P. 2011. Assessing ventilation system performance in isolation rooms. *Energy and Buildings* 43(1): 246–252.

Djongyang, N. Tchinda, R. & Njomo, D. 2010. Thermal comfort: A review paper. *Renewable and Sustainable Energy Reviews* 14(9): 2626–2640.

Ho, S.H. Rosario, L. & Rahman, M.M. 2009. Three-dimensional analysis for hospital operating room thermal comfort and contaminant removal. *Applied Thermal Engineering* 29(10): 2080–2092.

Kilic, M. & Sevilgen, G. 2008. Modelling airflow, heat transfer and moisture transport around a standing human body by computational fluid dynamics. *International Communications in Heat and Mass Transfer* 35(9): 1159–1164.

Parsons, K.C. 2003. *Human Thermal Environments*. 2nd ed. Taylor & Francis.

Rodrigues, N. Oliveira, R.F. Teixeira, S.F.C.F. Miguel, A.S. Teixeira, J.C. & Baptista, J.S. 2013. Thermal comfort evaluation of an operating room through CFD methodology. In P. Arezes, J.S. Baptista, M.P. Barroso, P. Carneiro, P. Cordeiro, N. Costa, R.B. Melo, A.S. Miguel, & G. Perestrelo (eds.), *Occupational Safety and Hygiene*: 411–416. London, UK: CRC Press—Taylor & Francis.

Teixeira, S. Leão, C.P. Neves, M. Arezes, P. Cunha, A. & Teixeira, J.C. 2010. Thermal comfort evaluation using a CFD study and a transient thermal model of the human body. In J.C.F. Pereira & A. Sequeira (eds.), *V European Conference on Computational Fluid Dynamics* Lisbon: ECCOMAS.

Versteeg, H.K. & Malalasekera, W. 1995. *An introduction to computational fluid dynamics: the finite volume method.* Harlow, England: Longman.

Occupational Safety and Hygiene II – Arezes et al. (eds)
© 2014 Taylor & Francis Group, London, ISBN 978-1-138-00144-2

The importance of ergonomics analysis in prevention of MSDs: Exploratory study in Swedwood-Portugal

C. Coelho
ULP—Universidade Lusófona do Porto, Porto, Portugal

P. Oliveira
ULP—Universidade Lusófona do Porto, Porto, Portugal
ISLA, Instituto Superior de Línguas e Administração de Leiria, Leiria, Portugal

E. Maia
ISLA, Instituto Superior de Línguas e Administração de Leiria, Leiria, Portugal

R. Rangel
Serviço de Química e Toxicologia Forenses do Instituto Nacional de Medicina Legal e Ciências Forenses, I.P., Portugal
CENCIFOR—Centro de Ciências Forenses, Portugal

M. Dias-Teixeira
REQUIMTE, Instituto Superior de Engenharia do Politécnico do Porto, Porto, Portugal
CIEG—Centro de Investigação em Economia e Gestão, Universidade Lusófona de Humanidades e Tecnologias, Lisboa, Portugal
ISLA, Instituto Superior de Gestão e Administração, Santarém, Portugal

ABSTRACT: Studying the relationship between work and health involves a scrupulous identification of occupational risk factors, as well as the positive or negative impact on employees. This study was developed in-Swedwood Portugal. Company has its own production lines in jobs that require intervention in order to reduce the risk of developing MSDs. To quantify the level of risk and ergonomic elaboration of an action plan of preventive and corrective measures to be implemented in the company proceeded to the application of ergonomic evaluation methodology and analysis of anthropometric data. The research was conducted in three phases: (1) application of methodology Checklist OSHA, allowing you to prioritize more detailed evaluations and adapt the methodology of level 2, (2) workload assessment with the methodology RULA (classification integrated risk of MSDs) and Equation NIOSH (calculation of the recommended weight limit). Results suggest that it is imperative to ascertain and make immediate modifications in 65% of jobs. Is need to implement preventive and corrective measures, using tools of ergonomics and engineering, such as interventions based on anthropometry, organization of working time, changes to working methods, reorganization of production layout and equipment introduction aid the implementation of tasks.

1 INTRODUCTION

Ergonomics is the scientific study of the work-place conditions and job demands regarding its adaptation the capabilities of the working population. The goal of ergonomics is to develop, through the operation of several scientific subjects, a better adaptation of man to technology and work environments. It provides the overall analysis of work situations, promoting their improvement and thereby guarantying the best working conditions for employees, taking also into account the efficiency of the system (Montmollin M. and Darses F., 2011). Comprises a set of actions, including the study and transformation work conditions, helping to optimize the relationship between the physical environment, work processes and the effort expended by man. All actions are carried out in accordance with the characteristics of the individual and the type of relationship that is established regarding the interface individual vs. machine (Bento A., 2007), work organization and improvement of working conditions (Iida I. 2000). As through detailed analysis of all tasks can reflect on whether human activity (Montmollin M. and Darses F., 2011), the modification is recommended to improve the health status of workers. In the industrial sector, contributes to improve the efficiency, reliability and quality of operations (Iida I., 2000). In an initial stage of an ergonomic evaluation of an industrial unit must be distinguished between prescribed and the real work. The prescribed work comprises the goals set by the company to carry out a particular activity (e.g. number of parts to be produced; defects to be avoided) and the actual work is carried out concretely in the production unit independently of the performances held (Rabardel P. et al., 1998) (Montmollin M. and Darses F., 2011). The non-fulfillment of the goals set by management may lead to human errors. This assessment provides information regarding the etiological mismatch between the prescribed work and real work, such as the human idiosyncrasies, work organization, the binomial man-machine and the context of the preparation of the activities. Observation is the preferred method by Ergonomist. All the movements, gestures and glances are observed; the use of technical devices and the interactions between people are also analyzed. Significant information is collected such as the production methods, errors and defects aiming the knowledge of the strategies developed by the workers. Additionally, physical or chemical parameters are recorded and, sometimes very faithfully, demonstrate the worker's intervention.

1.1 *Work related musculoskeletal injuries*

In the industrial setting, working conditions rarely meet the needs of the workers; therefore it urges to explore the interactions between man and work, intervening to prevent potential negative impacts, such as the appearance of musculoskeletal injuries related to work (MSDs). This type of injuries are a frequent consequence of occupational diseases and have been increasing with the implementation of new methods and models of work organization consisting, in nowadays, in an important issue in Health and Safety at Work. Being defined as a set of inflammatory and degenerative dysfunctions of the muscoskeletal system, it has its origin on occupational risk factors, such as repeatability, overload, posture maintained rhythms of intense work, exposure to vibration, manual handling, static or inadequate postures, insufficient breaks, low temperatures and organizational factors, high demands and low job control (Uva A.S. et al., 2008) (Bernard B.P., 1997, Forde M.S. et al., 2002, Malchaire J.B. et al., 2001). Individual factors such as age, sex, muscle strength, anthropometric characteristics also contribute to the development of risk factors (Armstrong T.J. et al., 1993, Punnett L. and Wegman D.H., 2004). Affecting one or more regions of the body, with the neck and upper limbs being the most affected areas, occurs in the course of professional activity with repetitive movements, postures maintained and manual handling of load (Baxter P. et al., 2010).

1.2 *Method of risk analysis MSDs-Rapid Upper Limb Assessment (RULA)*

Regarding the prevention of MSDs, the Portuguese legislation in force does not specify any method for risk assessment. The RULA method, seeks to assess the risk of musculoskeletal disorders of the upper limbs (McAtamneyand L., Corlett N., 1993). This method is used to assess posture, strength and movements associated with sedentary tasks, such as the use of computers, manufacturing or other where the worker is sitting or standing without walking. It must be used as a first analysis to assess the level of exposure of the upper limbs to risk factors such as posture, muscle contraction, static repetition and strength and to determine the factors that contribute most to the risk associated with the task. Establish, for each zone, ranges of posture and describes a score in accordance with the overload level. Similarly, evaluates whether the job is static (postures maintained for more than a minute) or repeated (frequency of movement of the segments ≥ 4 per minute) and the force or load requirements. Its application consists in recording different working postures, classified by as coring system, using diagrams of body posture and tables that assess the risk of exposure to factors of external load.

2 METHODS

2.1 *Sample characterization*

This work was performed at Swedwood Portugal, registered with the CAE activity 31091—Manufacture of Wooden Furniture for other purposes. Its facilities occupy a total area of roughly 370,000 m^2, of which about 175 000 m^2 correspond to a covered area and the remaining as an open area, divided into 4 blocks. Daily, the factory manufactures more than 30,000 pieces of furniture for the IKEA Group. The plant operates 5 days/week, 24 hours/day, in three shifts (8 hours each). It has about 1300 employees (47% women and 53% men), 50% of which come from a radius of 10 Kmand 80% live within a radius of 20 km. There are different production technologies in this complex business. This study focuses on the analysis of the production process of the Board on frame. This sector is divided into 5 areas (Cutting, Frames and Coldpress; Edgebanding & Drill; lacquering and Packing), but in this study will only cover two areas (Cutting and Frames). Each area has different tasks of the overall production process.

In the Cutting area, the melamine and HDF board is cut according to the specifications of the final product. This process is realized through an integrated automatic machines (machines cutting discs, rollers mats, etc.). In the Frames section, more elaborate cuts are made to the pieces from the Cutting, supplements are mounted (Frame) via a hot glue system. HDF boards are then mounted to the produced Frame.

2.2 *Ergonomic assessment tools*

This research used the following two approaches to assess the MSD risks at the Cutting and Frames area.

1. Baseline risk identification of ergonomic factors (OSHA Checklist)
   This checklist includes questions on working postures of the back and neck, arms and hands, legs during sitting and standing tasks. It also gives examples of the type of action at a technical, organizational and individual level that can be put in place to prevent or reduce the risks caused by awkward postures.
   The main risk factors evaluated by the OSHA Checklist tool are force, posture, repetitiveness, and work duration. Tasks associated with other risk factors, such as exposure to vibration, mechanical pressure, and low temperature, should be referred to professional personnel for further analyses.
   The results show that the OSHA Checklist is an effective and rapid screening instrument to monitor the potential ergonomic risk for upper arm and the results from this checklist in examining the neck, shoulder and back symptoms were identical with that of the questionnaire survey.
2. Rapid Upper Limb Assessment (RULA)
   This method was applied to identify postural stress of upper limbs. The risk is calculated into scores and classified into four action levels. A RULA sheet consists of body posture diagrams and scoring tables. Based on the RULA method, the human body is divided into two parts, which are part A for Arm and Wrist analysis while part B for Neck, Trunk and Leg Analysis.

Table 1. RULA: Score and indication.

Score	Indication
1 and 2	Acceptable posture
3 and 4	Changes are recommended
5 and 6	Changes are soon required
7	Changes are immediately required

   A scoring system is used to assign scores at every step, depending on the body position, with the higher scores for more awkward postures. RULA method is widely used in ergonomic field.

The ergonomic risk evaluation process included a formal meeting and a tour of the shop floor (onsite inspection). In the meeting, the chief inspector first introduced the ergonomic risk evaluation process and then dialogued with the employers and major crewmembers to gather an overview of the work and the key problems experienced by the studied company. During the tour, researchers collected field data of specific tasks by talking with individual operators, taking photographic and video evidence of manual materials handling tasks, and measuring task demands and physical dimensions of inspected workstations. OSHA Checklist and RULA tools that enabled the quantification of ergonomic risk were then used for explaining potential causes of MSDs. Some weeks later, the researchers drafted and sent the company an inspection report with improvement suggestions, and a follow up meeting was arranged afterward to discuss the efficacy of the ergonomic interventions.

2.3 *Ergonomic practices: operation statement*

In cutting area, during the task 1 (Fig. 1) the worker's receipt of material from the cutting and forwarding the same to Foil, using a manual trolley. Pieces are forwarded to the trolley through automatic rolls and then the employee takes the trolley up to the destination rolls. The cargo withdrawal from the trolley is performed manually resorting to the operator's labor. In the task 2 (Fig. 2) (set of two tasks

Figure 1. Transport of the in parts manual car for the foil.

Figure 2. Stations receiving the slats PAUL.

performed by two employees), the operator receives the pieces from the PAUL rollers. The task consists in the reception and orientation of parts on the *baseboard*. The *baseboard* height is adjustable with a sensor that adjusts the height depending on the height of the load. In task 3 (Fig. 2) (set of two tasks performed by two employees), the operator receives and supports the pieces orientation on the *baseboard*. The height of the *baseboard* is adjustable via a sensor, which adjusts the height depending on the height of the load. The operator receives the slats above the PAUL's output elevator.

In Frames area, the task 1 (Fig. 3) is performed by an operator who is collecting the car carrying the slats, positioned next to the operators that perform the frames assembly, the operator takes the car to the supply area, where the slats needed for mounting the frame that is being produced are unloaded, in the trolley and then back, forwarding the car to the mounting area of the frames. The task 2 (Fig. 3) is performed by mounting the frames with hot glue. The mounting of this piece is performed by an operator who has at his side (right or left depending on the job) the slats, which are placed at its disposal by another operator in a trolley. Also, underneath the assembly panel are containers from where pieces for the assembly are taken. There is also a stand where to rest the hot glue gun in order to facilitate access to it. The assembly is held in pairs, i.e. two pieces at a time. The assembly of this

Figure 3. Frames supply to assembly stations and frames assembly.

pieces performed by a collaborator throughout the entire work shift with an average production of 89 pieces/hour worked per collaborator.

3 RESULTS AND DISCUSSION

This study conducted an ergonomic evaluation to reduce the identified risk factors associated with the developing of MSDs in a Cutting and Frames Area. Based on the results of our ergonomic assessment, the worktable was redesigned to improve work postures for the tasks in this areas.

In Table 2 is a summary of the assessment made in the area of cutting and suggesting: in task 1, that the trolley replacement for mechanical cart maneuvered by the worker; in task 2 (Fig. 2), a plan of rotation of workers (decreasing exposure time); gymnastics; changes to the machine-the position in which the employee receives the slats requires an impractical approach, in this sense is proposed a change to the machine in order to improve working posture; in task 3, a Plan of rotation of the workers (decreasing exposure time); gymnastics, training and awareness; changes to the machine-the position in which the employee receives the slats requires an impractical approach, regarding this situation is proposed that the machine is modified in order to improve working posture.

Table 3 is a summary of the assessment made in the Frames area and suggests: in task 1, gymnastics, training and awareness; limiting the number of pieces by handle; changes to the trolley-adaptation to anthropometric measures of the Portuguese population; in task 2 suggested: labor gymnastics, training and awareness; rotating workers Plan (decreased exposure time); adaptation of the piece to the worker it's proposed to adopt measures in the sense that the distribution of frames to assemble remade according to the height of employees, i.e., for the assembly of smaller frames, this should be done by workers smaller stature. For

Table 2. Summary of evaluation of tasks performed in cutting area.

Task	Anatomic region	Posture	Muscles	Strength	Result	Final result	Observations
1*	Arm/forearm/pulse/pulse rotation	5	0	3	8	7	Level of action 4*
	Neck/torso/legs	5	0	3	8		
2*	Arm/forearm/pulse/pulse rotation	4	1	0	5	7	
	Neck /torso/ legs	5	1	0	6		
3*	Arm/forearm/pulse/pulse rotation	5	1	0	6	7	
	Neck /torso/ legs	5	1	0	6		

*Task 1—Transport with car parts manual for the foil.
Task 2—First place of receipt to the slats PAUL.
Task 3—Second post of receipt to the slats PAUL.
Action level 4—It is imperative to conduct investigations and immediate modifications.

Table 3. Summary of evaluation of tasks performed in area frames.

Task	Anatomic region	Posture	Muscles	Strength	Result	Final result	Observations
1*	Arm/forearm/pulse/pulse rotation	5	1	1	7	7	Level of action 4*
	Neck/torso/legs	5	1	1	7		
2*	Arm/forearm/pulse/pulse rotation	7	1	0	8	7	
	Neck/torso/legs	4	1	0	5		

*Task 1—Frames supply to assembly stations.
Task 2—Frames assembly.
Action level 4—It is imperative to conduct investigations and immediate modifications.

larger frames, the assembly should be carried out by employees of higher stature; Work platforms placement (elevation of the worker).

4 CONCLUSION

With this study was possible to intervene in reducing the risk for MSDs in two ways. On one hand, intervening in the workplace through the layout reorganization, adapting the machinery, the equipment and tools used, among others. But when these changes were not enough, it was possible to intervene at the level of employee turnover and implementation of programs of gymnastics.

Although the proposed measures do not stand training as a key factor, this is a tool of great importance as regards the promotion of safety in the workplace. There are various movements made by workers who may be regarded as unnecessary, eliminating these movements through training and awareness result in a significant gain reduction MSDs development. The training, however, is not the solution to all problems, especially if one has in mind specific aspects such as workers' age, their seniority in the company, the type of functions or even their perception of risk. The differences found in various jobs, pointed out that the training of workers for the perception of risk need to be designed for each situation and not to be carried out a generic training would suit all situations.

A properly designed job, takes advantage of human capabilities, as it considers the limitations and power efficiency of the system, if this is not achieved, the system performance is limited, and it can pose a risk to the worker.

This consideration becomes relevant due to the increasing complexity of systems that often leads man near their limits. The need of the perfect knowledge of the physical characteristics of workers becomes evident, therefore, to consider the machinery and equipment as an extension of the worker himself so that it operates with maximum efficiency and comfort, this is only possible if the design the future user is analyzed and considered.

REFERENCES

Armstrong T.J., Buckle P., Fine L.J., Hagberg M., Jonsson B., Kilbom A., Kuorinka I.A., Silverstein B.A., Sjogaard G. & Viikari-juntura E.R. 1993. A conceptual model for work-related neck and upper-limb musculoskeletal disorders. *Scandinavian Journal of Work, Health & Environment,* 19: 73–84.

Baxter P., Aw T.C., Cockcroft A., Durrington P. & Harrington J.M. 2010. *Hunter's diseases of occupations,* London, CRC Press.

Bento A. 2007. Ergonomia: Contributos para a Gestão de Recursos Humanos. IN Caetano A. & Vala J. (Eds.) *Gestão de Recursos Humanos Contextos, Progressos e Técnicas.* Lisboa, Editora RH.

Bernard B.P. 1997. *Musculoskeletal disorders and workplace factors: a critical review of epidemiologic evidence for work-related musculoskeletal disorders of the neck, upper extremity, and low back,* Cincinnati, OH, US Department of Health and Human Services.

Forde M.S., Punnett L. & D.H., W. 2002. Pathomechanisms of work-related musculoskeletal disorders: conceptual issues. *Ergonomics,* 45: 619–630.

Iida I. 2000. *Ergonomia—Projeto e produção,* São Paulo, Editora Edgard Blucher LTDA.

Malchaire J.B., Cock N. & S., V. 2001. Review of the factors associated with musculoskeletal problems in epidemiological studies. *International archives of occupational and environmental health,* 74: 79–90.

Mcatamney L. & Corlett N. 1993. RULA: a survey method for the investigation of work-related upper limb disorders. *Applied Ergonomics,* 24: 91–99.

Montmollin M. & Darses F. 2011. *A Ergonomia,* Lisboa, Instituto Piaget.

Punnett L. & Wegman D.H. 2004. Work-related musculoskeletal disorders: the epidemiologic evidence and the debate. *Journal of Electromyography and Kinesiology,* 14: 13–23.

Rabardel P., Carlin N., Chesnais M., Lang N. & Pascal M. 1998. *Ergonomie: concepts et methods,* Toulouse, Octares Editions.

Uva A.S., Carnide F., Serranheira F., Miranda L.C. & M.F., L. 2008. *Lesões Musculoesqueléticas Relacionadas com o Trabalho—Guia de Orientação para a Prevenção,* Lisboa, Direcção-Geral de Saúde—Programa Nacional Contra as Doenças Reumáticas.

Occupational Safety and Hygiene II – Arezes et al. (eds)
© 2014 Taylor & Francis Group, London, ISBN 978-1-138-00144-2

Environmental risks in a company in the food sector in Campina Grande, PB

L.M.S. Pereira, P.F. Veríssimo, M.B.G. Santos & R.J. Silva
Federal University of Campina Grande, Campina Grande, Paraíba, Brazil

ABSTRACT: This work aims to identify and propose measures to control the risk of environmental agents and the production process in a food company in the small town of Campina Grande—PB, Brazil. The work environment was characterized by means of observations of the production process and measurements of environmental variables related to noise, lighting, temperature and humidity. The major problems encountered were related to noise and illuminance.

1 INTRODUCTION

In a world where, every day, are increasing discoveries and technological innovations, the dissemination of information on the prevention of accidents and occupational diseases becomes decisive for the quality of life in the workplace and respect the integrity of the worker is valued. Temperature, humidity, lighting, noise are factors that influence and affect worker performance and can have negative repercussions for your health. And the consequences caused by poor working conditions is not restricted to the affected worker and can interfere with your performance in your mood, in their relationships with colleagues, in their family life, creating burden for the other workers, who have also are pressed and inserted in the same production process with characteristics (time, quality, need, etc.) quite peculiar (Alevato, 2009).

Thus, the main objective of this work is to identify the aspects of acoustic, thermal and luminance of a food company in the city of Campina Grande—PB, Brazil and propose solutions to the discontinuities in accordance with current legislation.

2 THEORETICAL

The increase in the number of accidents and fatalities resulting from work are poor and insecure working environments coupled with lack of oversight by the competent bodies. Compliance with the rules of safety and health at work—the Regulatory Standards of the Ministry of Labor and Employment (NRs)—is a challenge for most companies that generally prioritize the improvement of product quality at the expense of better environmental conditions work (Lacerda et al., 2005).

In general, workers are more prone to injuries and illnesses caused by psychological pressure that is submitted. This is the profile of workers in food industries, whose activities require care accented on the control of products, which can cause them to wear emotional, physical and psychological. Such factors may directly influence the pace of production, is the incidence of occupational diseases or by accidents.

In relation to this, we started to see the conditions of comfort working environments of industries producing food as factors that may contribute to the reduction of physical and emotional exhaustion of workers and consequently reduce rates of accidents in this sector which currently has the highest rates recorded, with 8.91% of cases (BPA, 2007).

There are many variables that may be present in the work environment. As environmental comfort, for example, are associated with the following: noise, lighting, temperature, physical exertion, among others. Each represents an important part in the welfare of workers and the quality of services. Health problems often are related to one or more of these variables of comfort, as can also be linked to changes in the individual order, social and technical (Silva et al., 2002).

In the food industry, a number of professional activities puts labor environments that present thermal conditions than those to which the human body is normally accustomed to. These professionals are exposed to intense heat or cold, which can seriously compromise your health. The man who works in high temperature environments suffer from fatigue, your income decreases, errors occur in perception and reasoning and appear serious psychological disorders that can lead to exhaustion and prostration (Vieira, 1997).

The low temperatures, in turn, have an influence on motor skills. Hands when exposed to cold, have impairments of touch and movement of the joints, making the work slower and may increase errors and accidents (Vieira, 1997).

The noise level is another very important parameter to be considered in food industries, given that they are present causing different sources such disorder. The presence of noise in the workplace can cause damage to hearing and even deafness. (Tomaz et al., 2000).

Inadequate lighting can cause eye strain, stress and even depression, the temperature may influence mood, well-being and performance of tasks; air quality affects the well-being and health, noise can cause irritability and lack concentration, and the inadequacy of jobs ergonomics results in a decrease in performance (Bogo, 2009).

The physical and environmental conditions in a given workspace, resulting in one way or another, impacts on health and well-being of worker. According to Iida (2005), environments with excessive heat, noise and vibration constitute sources of stress at work, because they cause discomfort, increase the risk of accidents and can cause significant damage to health.

Therefore, the work environment where the HVAC systems, lighting and sound, if controlled, can contribute to the efficiency and effectiveness of the tasks performed in the workplace, as well as promoting greater comfort for workers (Silva, 2001).

3 DESCRIPTION OF THE COMPANY AND PRODUCTION PROCESS

The study is now in the food industry, producing fruit pulp, founded in 1992, located in the city of Campina Grande, Brazil. The company currently has seven employees, five men and two women, with shift working Monday to Friday from 7:30 to 17:30 with meal break from 11:30 to 13:30, the distributed entres sectors of administration, reception, cleaning, sorting, weighing, pulping, bottling and storage, as described below:

- Administration: department responsible for managing the company.
- Reception: industry responsible for the control of customer orders, as well as the request for suppliers of fruit.
- Weighing of fruits: The weighing must be done in two stages, one when it receives the fruits of the suppliers to have a value of the quantity of fruits and other after selection, to have a real value of the amount of fruit being processed.
- Selection: In the selection, the staff observe the fruit characteristics, such as, dents, and degree of maturation, then go to the selected fruit cleaning. In the production process of pulp fruit selection is an essential step, because any and all fruit nonstandard—rotten, bruised or green—can interfere with the final flavor of the pulp.
- 1st Wash: After selection, the fruits are taken to a tank where it is appropriate washing and disinfection—with 220 ppm chlorine—at this stage the fruits are immersed in chlorinated water tank, where they wait 3 min. so that impurities such as sand fruit is retained. The washing process or wet cleaning is to be done with bacteriologically pure water and treated.
- 2nd Wash: After the 1st wash fruits are forwarded to another tank of dechlorinated water, where they remain for another 3 min.
- Pulping and Refinement: In this step, preparing the pulp of fibrous material, in addition to achieve reduction of particle size, then the refiner is coupled to depulper which have the role of leach pulp processed. The total time of the fruit in the process of pulping is 1 hour, taking into account the capacity of the machine is 2500 kg per hour.
- Potting: Subsequently, the process proceeds to the pulp where they will be filling properly packaged and chilled in cold designed for this purpose.
- Storage and Freezing: The pulps are stored in casks and placed in inventory (cold room).

4 METHODOLOGY

To achieve the objective of this research, we used four methodological aspects:

- Literature Review through consultations in primary and secondary sources, ie books, journal articles, technical and scientific, as well as consultation to sites related to the topic discussed.
- Identification Qualitative Risk by applying check sheets (checklist) in two sectors of the company (reception and pulping), it dealt with the topics: buildings, provision of machinery and equipment, fire protection, safety signs.
- Identification profile design as the usability and importance of PPE. Informal interviews were conducted with six employees and face where the questions were asked and recorded by a member of the group.
- In the quantitative evaluation, the variables that make up the thermal environment, temperature, Ta (°C), relative humidity, RH (%), were collected simultaneously by the instrument Psychrometer Anemometer Portable AN-4870 Manufacturer ICEL, to your specifications technical shown in Table 1, as well as the noise level (dB) was collected by use of a thermo-Hygro-decibel

Table 1. Technical specifications of portable anemometer psychrometer-NA-4870 manufacturer ICEL.

Escala	Valores	Resolução	Exatidão
Temperature (°C)	0 à 70	0,1	±0,6
Relative humidity (%)	0 à 100	0,1	±3,0
Air velocity (m/s)	0 à 35	0,1	±3,0

Table 2. Technical specifications of Thermo-Hygro-decibel meter-luximeter THDL-400 manufacturer Instrutherm.

Escala	Valores	Resolução	Exatidão
Sound pressure dB (A)	35–130	0,1	±3,5
Brightness (Luz)	0–20000	0,1	±5,0

meter-luximeter portable THDL-400 model Instrutherm manufacturer, whose technical characteristics are shown in Table 2.

Data were collected every 15 minutes for five days in two sectors: in the reception and the pulping industry, thus making a comparison between them.

The noise measurements were performed with the sensor positioned as close as possible to the auditory zone according to Norm 15 (NR 15), which comes from unhealthy activities and operations.

For lighting, set up four points effective measurement (p1, p2, p3, p4), where the measurements found were analyzed according to NBR 5382 (1985) for field work regularly with central light fixture. Then, averages were made of the values found from the degree of luminance which concerned workers. Measurements of illuminance levels in the fields of work has been carried out by positioning the base of a horizontal photocell within certain industries, obtaining the illuminance levels in lux. We established three factors for quantification ideal age, speed, accuracy and background reflectance, according to NBR 5413/92, in which the values are set to medium and minimum illuminance into service for artificial lighting indoors, where they performed activities industry.

Verification of the temperature was carried out according NR 17, which deals with the setting parameters that allow adjustment of the working conditions of workers psychophysiological characteristics. Remained in the place where the worker, the equipment was positioned close to even with waiting 15 minutes to stabilize the measurement.

Already measurements of ambient humidity were conducted in both sectors with 15 minute break between each.

5 RESULTS AND DISCUSSIONS

5.1 *Qualitative assessment*

We observed that the building floor, stairs, ramps, corridors and traffic areas, there the risk of slipping and feature disagreement with NR 8 establishing minimum technical requirements that are observed in the buildings, to ensure comfort and safety for who work there. Well with no material or structure is used for fall protection, such as guardrails or support to go up or down stairs. However, there are no cracks in walls, open drains or broken glass panes.

Regarding the provisions of the machines (despolpadeiras machine, filling and funnel) and equipment, realize that they are in compliance with the required spacing of 0.60 m to 0.80 m, have drive and stopping devices located in areas that are not dangerous in machine and can be turned on and off by the operator can be disabled in an emergency by any other person other than the operator, fully compliant with the NR 12, which deals with the protection and measures to ensure the health and integrity of the worker in machinery.

As for fire protection was identified that the outlet openings have the required width (1.20 m) so that employees can exit in case of fire according to NR 23, however, openings, exits and passageways are not clearly marked by signs or signal lights indicating the direction of the output being true to the NR 26, which deals with safety signs.

The portable fire-fighting equipment are located within easy access, easy viewing and in places where there is less likelihood of fire block access, there are no signs indicating the fire fighting equipment. As for safety signs, yet there is no type of signaling.

5.2 *Identification profile design as the usability and importance of PPE*

Regarding Personal Protective Equipment (PPE), we found that approximately 20% of employees did not use them, because they feel uncomfortable and irrelevant. The PPE is provided free by the manager of the company, as required by NR 6, which deals with the individual devices used by the employee, for the protection of risks likely to threaten the safety and health at work. However, they do not receive specific training regarding the use of this equipment, since improper use can cause damage effectively dangerous to health and integrity of the worker.

It was possible to identify the pulping industry, it is necessary to use earplugs, taking into account that the site has a high noise level, but 20% of employees do not use, there is also the need to use boots, with a view that the floor is slippery and can cause falls. It is also essential to use gloves, masks and headdresses because it is a food company. However, gloves are not used by employees.

In the sector of receipt no need to use PPE, since the industry is not near noise zone.

5.3 *Quantitative analysis of environmental risks*

5.3.1 *Noise*

According to NR 15, the maximum allowable daily exposure to one shift of 8 hours of work is 85 dB. In the pulping industry average noise level was 98 dB exceeding this maximum level, is shown in Table 3.

Therefore, employees are exposed to high noise levels, thus are subject to occupational diseases. The high amount of noise hinders communication, concentration, attention, redimento at work may contribute to accident occurrences. We can control the noise level using three distinct ways: control at source, and control the trajectory, when the two previous controls do not show appreciable reductions, control the receiver (person).

The control at the source can be done through new projects, maintenance, insulation structure, reducing the surface area of vibrant settings. Control the trajectory is done by enclosing total, partial, acoustical barriers. The control in the receiver is done with the use of headphones, which can be a model shell, or insert. Other means of control in the receiver are: training, education (awareness) and reduced exposure time.

In the sector of the reception, it was identified that the noise is below the tolerance limits set by NR 15, is shown in Table 4.

5.3.2 *Lighting*

Tables 5 and 6 present data relating to lighting industry pulping and reception.

Table 3. Noise level in the pulping industry.

Agents risk	Minimum	Médium	Maximum	Standardized
Noise (dB)	95	98	101	85

Source: Data obtained from the survey (2013).

Table 4. Noise level in the sector reception.

Agents Risk	Minimum	Médium	Maximum	Standardized
Noise (dB)	52	61	64	85

Source: Data obtained from the survey (2013).

Table 5. Level lighting industry pulping.

Agents risk	Minimum	Médium	Maximum	Standardized
Lighting (Lux)	90,3	98,3	105,7	750

Source: Data obtained from the survey (2013).

Table 6. Level lighting industry reception.

Agents Risk	Minimum	Médium	Maximum	Standardized
Lighting (Lux)	160	165	168	750

Source: Data obtained from the survey (2013).

The lighting of the site is predominantly artificially (CFLs), with little contribution of daylighting in some sectors. In general lighting company provides disability sectors, since, in addition to the inadequate amount of LUX, according Abreu, Spinelli, Zanardi (2009) and Silva Junior (2005), the bulbs have a protection that greatly decreases the intensity light, especially in areas involved with food.

According to the measurements, the pulping sector showed the lowest level of illuminance to average 98 LUX, since the sector was receiving an average of 165 LUX, when the recommended would be around 750 LUX.

Therefore, one can conclude that the conditions of illuminance are insufficient to guarantee comfort and safety while performing activity. Ie, the type of lighting used in the workplace must be related to the type of activity that is performed. Besides being distributed and diffuse uniformly, the lighting must be designed and installed to avoid glare, annoying reflections, shadows and contrasts excessive.

5.3.3 *Ambient temperature*

In Tables 7 and 8 ambient temperatures can be seen in the sectors of pulping and reception.

With regard to thermal conditions may be observed that in the area of the pulping temperature conforms to NR 17, whereas the effective temperature is between 20°C and 23°C. In the sector of the reception there is disagreement with the standard, because the average was around 24°C.

5.3.4 *Humidity*

The norm that regulates the moisture present in the workplace is NR 17, is that the relative humidity should not be below 40% (forty percent).

Table 7. Ambient temperature sector pulping.

Agents risk	Minimum	Médium	Maximum	Standardized
Anbient temperature (°C)	22	23	23	20 a 23

Source: Data obtained from the survey (2013).

Table 8. Temperatura ambiente no setor de recepção.

Agents risk	Minimum	Médium	Maximum	Standardized
Anbient temperature (°C)	24	25	25	20 a 23

Source: Data obtained from the survey (2013).

Given that, relative humidity activities considered unhealthy are those that are developed in areas flooded or waterlogged, and which have excessive moisture.

In the company, the humidity measurements in the two sectors studied showed 82.3% relative humidity of the air, thus presenting accordance with what is suggested by the NR 17.

6 CONCLUSION

After evaluation at the company, it can be concluded comfort conditions of the work environment in two sectors, pulping and reception.

- As for the building, it was noticeable that there is a disagreement with the NR 8, which establishes minimum technical requirements that must be observed in the buildings, to ensure comfort and security for those who work in it.
- Machines (despolpadeiras, potting machine, hopper) are in accordance with the NR 12, which deals with the protection and measures to ensure the health and integrity of the worker operating machinery.
- As for fire protection was identified that the outlet openings have the required width (1.20 m) as the NR 23, however, openings, exits and passageways are not clearly marked by signs or traffic lights indicating the direction of the output being true to the NR 26, which deals with safety signs.
- As for safety signs, yet there is no type of signaling.
- Regarding Personal Protective Equipment (PPE), we found that approximately 20% of employees did not use them, because they feel uncomfortable and irrelevant.
- The sector with the highest risk was the pulping, since environmental variables (noise, temperature and lighting) were all above the regulated limit.
- In the area of reception the only disagreement was found with respect to the low level of illuminance, in disagreement with NBR5413.

REFERENCES

Abreu E.S.; Spinelli, M.G.N.; Pinto, A.M.S. 2009. *Gestão de Unidades de Alimentação e Nutrição: um modo de fazer.* São Paulo.

Alevato, H.; Araújo, M.E.G. 2009. *Gestão, Organização e Condições de trabalho.* V Congresso Nacional de Excelência em Gestão—Anais. Niterói.

Anuário Brasileiro de Proteção, 2004. Novo Hamburgo: MPF Publicações.

Bogo, A.J.; Souza, C.R. 2009. *Avaliação pós-ocupação em edifícios de escritório objetivando aferição do nível de satisfação dos usuário.* VI Encontro Nacional de Conforto no Ambiente Construído, São Pedro SP.

Iida, I. *Ergonomia: Projeto e Produção.* 1990. São Paulo: Edgar Blücher.

Lacerda, L.D.; Santelli, R.E.; Campos, R.C. 2005. *Metais em sedimentos.* In: II Workshop do Projeto de Caracterização e monitoramento Ambiental da Bacia Potiguar. Natal, 15 p.

Ministério do Trabalho e Emprego. 2004. Disponível em: http://www.portal.mte.gov.br/legislacao/. Acesso em: 10 set 2013.

Silva, L.B.; Coutinho, A.S.; Másculo, F.S.; Xavier, A.A.P.; Fialho, F.A.P. 2002. *Análise comparativa entre um modelo teórico e a sensação térmica declarada por trabalhadores em ambientes com VDT.* João Pessoa: ABCM.

Tomaz, A.F.; Rodrigues, C.L.P.; Másculo, F.S. 2000. *Avaliação das condições ambientais e organização do trabalho do sub-setor de lanternagem de uma empresa de transporte urbano na cidade de João Pessoa.* Revista Principia, N° 8, Ano 4.

Vieira, S.D.G. 1997. *Estudo de caso: análise ergonômica do trabalho em uma empresa de fabricação de móveis tubulares.* Dissertação (Mestrado em Enge. de Produção)—Universidade Federal de Santa Catarina, Florianópolis,

Occupational Safety and Hygiene II – Arezes et al. (eds)
© 2014 Taylor & Francis Group, London, ISBN 978-1-138-00144-2

Recent application of infrared thermography in work-related musculoskeletal disorders

A. Seixas
Universidade Fernando Pessoa, Porto, Portugal
LABIOMEP, CIAFEL, Faculty of Sport, University of Porto, Portugal

R. Vardasca & J. Gabriel
IDMEC-FEUP, LABIOMEP, Faculty of Engineering, University of Porto, Portugal

J.P. Vilas-Boas
LABIOMEP, CIFI 2D, Faculty of Sport, University of Porto, Portugal

ABSTRACT: The burden of work-related musculoskeletal disorders is an important problem in the occupational setting. New techniques to quantify the exposure to risk factors are important to understand possible pathophysiological mechanisms that may lead to the development of these disorders. Existing methods have known limitations and new approaches are required. Work-related musculoskeletal disorders have been associated with blood flow impairments. Skin temperature is influenced by blood flow and might be an interesting parameter in their management. Thermography is a simple way to record skin temperature and produces a thermal map of the evaluated body part. A literature review was conducted in order to analyze the recent applications of thermal imaging in the management of work-related musculoskeletal disorders.

1 INTRODUCTION

Injuries affecting a structure of the musculoskeletal system are often called "Musculoskeletal Disorders". According to several authors these disorders may develop in a sudden or insidious onset and can be responsible for functional impairment and symptom provocation for short periods or the whole lifetime (Bernard, 1997, Sanders & Dillon, 2006, Sanders & Stricoff, 2006, Woolf & Pfleger, 2003, Mody & Brooks, 2012). If the musculoskeletal disorder has been induced or aggravated by work or its circumstances the term Work-Related Musculoskeletal Disorder (WRMD) is adequate (Schneider et al., 2010, Luttmann et al., 2003).

The functional capacity of the musculoskeletal system can be assessed through numerous methods in order to identify the risk leading to musculoskeletal disorders, to monitor the effects of occupational interventions and research. The most common used instruments are questionnaires, functional tests and observational methods. While using questionnaires data collection costs are low, several parameters of exposure can be determined simultaneously and all risk factors are assessed with a similar approach but these instruments require better validity testing research (Barrero et al., 2009). Wind et al. (2005) in a systematic review identified 13 questionnaires and 14 functional tests and concluded that only three questionnaires showed high levels of validity and reliability and that none of the functional tests had high level of validity and reliability. Observational methods may be useful but differences were found while using different observational methods on the same target (Takala et al., 2010).

WRMD have a high impact in either individuals and society, the burden associated has been extensively discussed in the literature. The need for research in new technologies that allow a better understanding of these disorders and objective outcomes is crucial to reduce the impact of WRMD.

2 INFRARED THERMAL IMAGING

Temperature is a physiological health status indicator as the presence of pathology may affect local thermal balance, increasing or decreasing skin temperature through blood flow regulation. This principle supports the use of thermography, a measurement tool applied to images of temperature distribution of the skin that is simple, non-invasive, non-ionizing and objective measurement. Several studies relate the appearance of musculoskeletal disorders with disturbances in blood flow, especially in repetitive strain injuries (Pritchard et al., 1999, Gold et al., 2010, Brunnekreef et al., 2006).

Thermal imaging has been used for research, to study diseases where skin temperature is an indicator of the underlying tissues status or where blood flow is increased or decreased and can be applied as a diagnostic procedure or as an outcome measure (Ring & Ammer, 2012).

2.1 *Equipment*

Several imaging modalities work within the electromagnetic spectrum but mainly allow access to anatomical information.

Thermal cameras are used in thermography to capture and monitor the amount of heat dissipated by infrared radiation and to produce infrared thermograms, which are images of temperature distribution of the target.

It was in the 1960s that the impact of temperature distribution measurements from thermograms began (Ring, 1995) however equipment and examination protocols have evolved significantly in the last decade (Ring & Ammer, 2012). The latest generation of high thermal and spatial resolution cameras improved the potential of thermography and small, but meaningful, variations in thermal patterns can be identified and assessed.

2.2 *Standardization requirements*

As a method to measure skin temperature, thermography must meet criteria of measurements like validity, reliability, sensitivity and responsiveness. In order to meet these criteria the examiner must be aware of the sources of variability of measurements performed. The object or subject being studied, the imaging system, the position of the subject or object during image capture and the environment conditions (temperature, humidity and air flow) are common sources of errors.

Plassmann et al. (2006) proposed a series of simple tests for quality assurance in thermal imaging. The use of external sources of reference temperatures is important for systematic calibration checks.

Thermal patterns are often represented in false colour scales where different colours represent different temperature values. In medicine, the use of a rainbow false colour scale has been recommended, since at human eye has better discrimination. The temperature colour scale used at acquisition should be displayed in the thermogram since its absence makes the image poorly defined as the range of temperatures is essential to correctly understand the image and to allow future comparisons (Ring & Ammer, 2012).

The concern with standardization procedures has been increasing in the literature and guidelines have emerged regarding equipment preparation, subject preparation, body positioning, examination environment conditions, image recording, region of interest definition and evaluation of thermograms (Ammer & Ring, 2008, Ammer, 2008, Schwartz, 2006).

3 RECENT APPLICATIONS OF THERMAL IMAGING IN THE MANAGEMENT OF WORK-RELATED MUSCULOSKELETAL DISORDERS

Thermal imaging has not been extensively used in occupational medicine in the management of WRMD but some studies have attempted to use the technology as diagnostic tool and as a means to study risk factor exposure in order to better understand the pathophysiological mechanisms leading to injury development.

3.1 *Thermal imaging as a diagnostic tool in the occupational setting*

Three studies were conducted to study the potential use of thermography to identify groups of patients with WRMD.

Gold et al. (2004) studied office workers with WRMD and healthy controls simulating a keyboard typing activity. The subjects with WRMD were distributed in two groups according to the presence of cold hands induced by keyboard use and thermograms of the dorsal hand skin temperature were obtained at baseline, 0–2 minutes, 3–5 minutes and 8–10 minutes after typing. The authors found three distinct temperature patterns during the 9 minutes typing activity followed by an observation period of 10 minutes and concluded that infrared thermography was able to discriminate between the three groups of subjects. The same group (Gold et al., 2009), in another study simulating a similar task aimed to establish the suitability of using dorsal hand skin temperature as an indicator of WRMSD in office workers and addressed the reproducibility of thermal measurements. Symptomatic and asymptomatic office workers and controls were evaluated the conclusions pointed skin temperature of the dorsal hands measured by infrared thermography as a reliable measurement to determine the severity of WRMD in office workers.

Mohamed et al. (2011) studied two groups of pianists, one with pain related to piano playing and one without associated pain, focusing in hand, forearm and arm skin temperatures. Thermograms were captured at baseline, immediately after each of the three piano-playing activities with increasing difficulties, 15 minutes after and 30 minutes after the last piano exercise. The authors found that the experimental design conduced to meaningful results and that the temperature of the hands, but not the temperature of the forearm and arm, was significantly higher in the group with pain related to piano playing.

More studies are needed in order to establish thermal imaging as a useful diagnostic tool but the literature that has been published on the subject is promising.

3.2 *Thermal imaging as a sensitive technique to assess risk factors and task demands on the musculoskeletal system*

Several studies have been recently published contributing to understand the pathophysiological mechanisms of injury development.

Gold et al. (2010) aimed to establish the correlation between the skin temperature of the dorsal hand of office works and relative blood flow measured by near infrared spectroscopy. Thermographic evaluation followed the methodology previously published by the research group (Gold et al., 2004, Gold et al., 2009). The authors found a moderate correlation between relative blood flow and the temperature of the hand during the 10 minutes following a typing task. Skin temperature and relative blood flow were influenced by the typing speed and despite reasonably correlated both parameters were found to be highly variable between subjects. Individuals typing more than 50 words per minute evidenced faster overall post-typing temperature decrease towards the baseline values.

Govindu and Babski-Reeves (2012) analyzed skin temperature over the muscles in the thenar eminence in 12 participants simulating a pipetting task. The effects of pipette volumes, solvent viscosity and gender on thermal parameters and subjective rating of discomfort of the thumb thenar muscles were analyzed. Thermal parameters were not sensitive to solvent viscosity and pipetting volume but were correlated with subjective ratings of discomfort. An increase in thermal parameters and discomfort ratings was observed. The rate of temperature change may be more adequate to describe the impact of task demands on muscles over time and may predict the amount and change in discomfort experienced during the task.

Camargo et al. (2012) simulated a textile industry activity with repetitive movements of the wrist. The activity was emulated for 3 hours and 30 minutes and involved movements like reaching, taking, dropping and others. Only two subjects were studied, and the behavior of wrist temperatures in both cases was similar, increasing during the first hour and a half of work and declining over the next 2 hours.

Barker et al. (2006) and Bertmaring et al. (2008) applied thermal imaging to assess shoulder overhead activities. Barker et al. (2006) evaluated the effects of task parameters on middle deltoid and trapezius. Thermography was a sensitive method to detect changes in task parameters, working at 33% duty cycle and lower work height resulted in higher temperatures and increased rates of temperature change. At the 50% work cycle the temperatures were lower. The temperatures at 67% work cycle were expected to be even lower but the opposite was true. The authors advanced the accumulation of waste products in the muscle and lower levels of blood flow due to exertion as possible explanations.

Another study (Bertmaring et al., 2008) proposed to quantify surface temperature changes in the anterior deltoid and evaluate the efficacy of thermography as an assessment tool. Two work loads and two shoulder angles were evaluated and surface temperature, discomfort ratings and endurance time were assessed during overhead static exertions until exhaustion. The work loads were 15% and 30% maximum voluntary contraction since blood flow has been cited to begin at 20% of maximum voluntary contraction and 90° and 115° shoulder angles were chosen as they represented overhead postures previously published. In this study, shoulder angle affected temperature rates of change. According to the authors, lower temperature slopes may present increased risk of injury due to lower blood flow levels. Working at a lower shoulder angle allows more blood to be distributed to working muscles resulting in faster rates of change. The results related to the exertion levels were unexpected since the deltoid thermal readings were not influenced by them, possibly because the exertion levels spanned the 20% cited in the literature as cut point to reduced blood flow.

Using infrared thermography as a diagnostic and/or assessment complementary tool has several advantages. Being a non-contact, non-invasive and fast method allows real time dynamic temperature monitoring without interfering with the subject being monitored, even when monitoring large areas. In opposition to other imaging modalities whose focus is anatomical information, the focus of infrared thermography is the physiology of the human being, associated with the microcirculation and autonomous nervous system. The use of false colour coded thermograms allows faster and easier subjective analysis. The fact that it is a non-ionizing method, recording the natural radiation emitted from the human body surface, makes it suitable for repeated and prolonged measurements. When assessing the microvascular system, the great advantage of thermography, compared with laser doppler, is that it is a faster method, able to measure a larger area, although more research is needed for reaching the same sensitivity.

Caution, however, is needed when using this imaging modality. Well controlled environment and evaluation protocols are needed since several sources of variability of thermal images may be present, arising from the subject being evaluated, the imaging system, the image capture protocol and/or the environmental conditions such as temperature, humidity and air flow.

4 CONCLUSION

Despite several limitations in the previously mentioned studies, attention should be placed in thermal imaging as an assessment tool in occupational setting.

Epidemiological studies are needed in large cohorts of workers to establish objective criteria of early development of WRMD, reducing the associated burden.

Longitudinal studies should be conducted in order to test the suitability of this physiological parameter in the workplace to evaluate the severity of WRMD. Workers at risk to develop musculoskeletal disorders, working with different exertion levels should be monitored regularly in order to understand the possible relationship between the rate of temperature changes and the onset and severity of WRMD. The correlation with subjective parameters like perceived pain, perceived effort and discomfort ratings should also be studied.

Some research has been done to understand the relation between thermal readings and other physiological parameters, like blood flow, but more studies are needed in order to fully understand this relation. Studies focusing on the relation between thermal readings and muscle activity, measured by surface electromyography, are highly demanded. Using the objective criteria of surface electromyography to establish the state of muscle fatigue, the thermal characterization of the process of muscle fatigue could be known and more information would be added about the effect of muscle activity on the rate of temperature change.

Research is still required in order to structure specific evaluation protocols and guidelines, allowing thermography to become an even more reliable diagnostic tool but, more importantly, a more reliable assessment tool to be used in the field.

Further research is needed in order to clarify the possible relationship between muscle activity, thermal readings and task demands.

The relationship between working experience and temperature is also to be fully understood. Different groups of workers with varying work experiences should be compared.

More studies focusing on the reliability of thermal imaging are highly required.

REFERENCES

Ammer, K. (2008) The Glamorgan Protocol for recording and evaluation of thermal images of the human body. Thermology international, 18, 125–144.

Ammer, K. & Ring, E. (2008) Standard procedures for infrared imaging in medicine. In Diakides, N. & Bronzino, J. (Eds.) Medical Infrared Imaging. Boca Raton: CRC Press.

Barker, L.M., Hughes, L.E. & Babski-Reeves, K.L. (2006) Efficacy of using thermography to assess shoulder loads during overhead intermittent work. Proceedings of the Human Factors and Ergonomics Society Annual Meeting. SAGE Publications.

Barrero, L.H., Katz, J.N. & Dennerlein, J.T. (2009) Validity of self-reported mechanical demands for occupational epidemiologic research of musculoskeletal disorders. Scandinavian journal of work, environment & health, 35, 245.

Bernard, B.P. (1997) Musculoskeletal Disorders and Workplace Factors—A Critical Review of Epidemiologic Evidence for Work-Related Musculoskeletal Disorders of the Neck, Upper Extremity, and Low Back. Cincinnati: U.S.: National Institute for Occupational Safety and Health; Center for Disease Control and Prevention.

Bertmaring, I., Babski-Reeves, K. & Nussbaum, M.A. (2008) Infrared imaging of the anterior deltoid during overhead static exertions. Ergonomics, 51, 1606–1619.

Brunnekreef, J.J., Oosterhof, J., Thijssen, D.H., Colier, W.N. & Van Uden, C.J. (2006) Forearm blood flow and oxygen consumption in patients with bilateral repetitive strain injury measured by near-infrared spectroscopy. Clinical physiology and functional imaging, 26, 178–184.

Camargo, C., Ordorica, J., De La Vega, E., Olguín, J., López, O. & López, J. (2012) Analysis of temperature on the surface of the wrist due to repetitive movements using sensory thermography. Work: A Journal of Prevention, Assessment and Rehabilitation, 41, 2569–2575.

Gold, J.E., Cherniack, M. & Buchholz, B. (2004) Infrared thermography for examination of skin temperature in the dorsal hand of office workers. European journal of applied physiology, 93, 245–251.

Gold, J.E., Cherniack, M., Hanlon, A., Dennerlein, J.T. & Dropkin, J. (2009) Skin temperature in the dorsal hand of office workers and severity of upper extremity musculoskeletal disorders. International archives of occupational and environmental health, 82, 1281–1292.

Gold, J.E., Cherniack, M., Hanlon, A. & Soller, B. (2010) Skin temperature and muscle blood volume changes in the hand after typing. International Journal of Industrial Ergonomics, 40, 161–164.

Govindu, N.K. & Babski-Reeves, K.L. (2012) Thermographic assessment of the thenar thumb muscles during pipetting. International Journal of Human Factors and Ergonomics, 1, 268–281.

Luttmann, A., Jäger, M., Griefahn, B. & Caffier, G. (2003) Preventing Musculoskeletal Disorders in the Workplace, India: World Health Organization.

Mody, G.M. & Brooks, P.M. (2012) Improving musculoskeletal health: Global issues. Best Practice & Research Clinical Rheumatology, 26, 237–249.

Mohamed, S., Frize, M. & Comeau, G. (2011) Assessment of piano-related injuries using infrared imaging. Engineering in Medicine and Biology Society, EMBC, 2011 Annual International Conference of the IEEE. IEEE.

Plassmann, P., Ring, E. & Jones, C. (2006) Quality assurance of thermal imaging systems in medicine. Thermology international, 16, 10–15.

Pritchard, M., Pugh, N., Wright, I. & Brownlee, M. (1999) A vascular basis for repetitive strain injury. Rheumatology, 38, 636–639.

Ring, E. (1995) The history of thermal imaging. In Ammer, K. & Ring, E. (Eds.) The Thermal Image in Medicine and Biology. Wien: Uhlen-Verlag.

Ring, E. & Ammer, K. (2012) Infrared thermal imaging in medicine. Physiological measurement, 33, R33.

Sanders, M. & Dillon, C. (2006) Diagnosis of Work-Related Musculoskeletal Disorders. In Karwowski, W. (Ed.) International Encyclopedia of Ergonomics and Human Factors, Second Edition—3 Volume Set. Kentucky: CRC Press.

Sanders, M. & Stricoff, R. (2006) Rehabilitation of Musculoskeletal Disorders. In Karwowski, W. (Ed.) International Encyclopedia of Ergonomics and Human Factors, Second Edition—3 Volume Set. Kentucky: CRC Press.

Schneider, E., Irastorza, X.B. & Copsey, S. (2010) OSH in Figures: Work-related Musculoskeletal Disorders in the EU-Facts and Figures. OSH in figures. Luxemburg: Office for Official Publications of the European Communities.

Schwartz, R. (2006) Guidelines for neuromusculoskeletal thermography. Thermol Int, 16, 5–9.

Takala, E.-P., Pehkonen, I., Forsman, M., Hansson, G.-Å., Mathiassen, S.E., Neumann, W.P., Sjøgaard, G., Veiersted, K.B., Westgaard, R.H. & Winkel, J. (2010) Systematic evaluation of observational methods assessing biomechanical exposures at work. Scandinavian journal of work, environment & health, 3–24.

Wind, H., Gouttebarge, V., Kuijer, P.P.F. & Frings-Dresen, M.H. (2005) Assessment of functional capacity of the musculoskeletal system in the context of work, daily living, and sport: a systematic review. Journal of Occupational Rehabilitation, 15, 253–272.

Woolf, A.D. & Pfleger, B. (2003) Burden of major musculoskeletal conditions. Bull World Health Organ, 81, 646–56.

Occupational Safety and Hygiene II – Arezes et al. (eds)
© 2014 Taylor & Francis Group, London, ISBN 978-1-138-00144-2

Ergonomic work analysis of swing arm cutting machine by RULA method

A.C. Mendes, J. Moreira, R. Maia & M.A. Gonçalves
ESEIG, Vila do Conde, Portugal

ABSTRACT: This work consists in evaluating ergonomic workstation—Swing Arm Cutting Machine (SACM) from a footwear industry. The evaluation method adopted was the methodology of RULA (Rapid Upper Limb Assessment) using the Nordic Musculoskeletal Questionnaire (NMQ). There were identified critical factors related to posture of the operator arising from the SACM design and job requirements. From the results of the ergonomic evaluation were defined corrective and preventive measures to minimize the risks of exposure to musculoskeletal injuries.

1 INTRODUCTION

As a result from the high demands of today's market, companies must worry increasingly about the health of its employees, so that they produce more and better. In this context, the ergonomics are framed as a participant in the process, by adapting the work to the human being through various methods, such as Ergonomic Work Analysis to a job.

According to Motta (2009) by offering better working conditions, ergonomics reduces fatigue and "stress" and consequently the increase in well-being and productivity of employees.

In the footwear industry, the SACM workstation, implies that the worker is standing all day and makes efforts, causing significant pain on neck, shoulders and dorsal region.

Based in this fact, this paper aims to evaluate the ergonomic risk of this workstation, applying the RULA (Rapid Upper Limb Assessment) method, identify and compare the possible symptoms associated with movements performed using the Nordic Musculoskeletal Questionnaire (NMQ) and define corrective and preventive measures to minimize the risks of exposure to musculoskeletal injuries.

2 LITERATURE REVIEW

Developed by McAtamney and Corlett and published in 1993 issue of the journal Applied Ergonomics, the RULA is used in ergonomics investigations of workplaces where postures, forces and muscle activities have been shown to contribute do work-related upper limb disorders (Lueder, 1996).

This method evaluates posture, strength and upper limb movements, such as the neck, back and arms, forearms and wrists associated with sedentary tasks which could contribute to the onset of musculoskeletal injuries (David, 2005; Espinoza, 2011).

The four main applications of RULA are:

- Risk assessment musculoskeletal, usually as part of a comprehensive ergonomic research;
- Comparison of effort between musculoskeletal design of current workstation and modified;
- Evaluate outcomes such as productivity or equipment compatibility;
- Advise employees about musculoskeletal risks created by different working postures.

Basically, this method comprises three steps (McAtamney & Corlett, 1993):

- Selection of posture or postures for evaluation;
- Evaluation of working postures by using a grid of points, diagrams of body parts and tables;
- Conversion of these scores on four measures proposed.

This technique addresses ergonomic risk results from a score of 1 to 7 where higher scores mean higher levels of perceived risk, as can be seen in Table 1.

A low score on the RULA method does not guarantee, however, that the workplace is free of ergonomic risks, as well as a high score does not guarantee that a severe problem exists. The method was developed to detect postures work or risk factors that deserve more attention (Lueder, 1996).

One of the advantages of this method is that it is not necessary to use specialized equipment and its application does not interfere in the work situation (Marran & Karwowski, 2006).

Table 1. Level of Intervention for the results of the RULA method (Adapt from McAtamney & Corlett, 1993).

Levels	Score	Results
Level 1	1 or 2 points	Posture acceptable, if not maintained or repeated for extended periods of time.
Level 2	3 or 4 points	Posture that needs investigation and changes may be necessary.
Level 3	5 or 6 points	Stance to investigate and change quickly.
Level 4	7 points or more	Posture investigating and change urgently.

The NMQ was published by Kuorinka in 1987, with the purpose of standardizing the measurement of reported musculoskeletal symptoms and thus facilitates comparison of results across studies, allowing comparison of low back, neck, shoulder and general complaints for use in epidemiological studies. The tool was not developed for clinical diagnosis (Crawford, 2007).

The authors do not indicate this questionnaire as a basis for clinical diagnosis, but for the identification of musculoskeletal disorders and as such may be an important diagnostic tool environment or job.

The instrument consists of multiple-choice or binary for the occurrence of symptoms in different anatomical regions in which they are most common. The respondent shall report the occurrence of symptoms 12 months and considering the 7 days preceding the interview, as well as to report the occurrence of withdrawal from daily activities in the last year (Pinheiro, 2009).

The NMQ can be used as a self-administered questionnaire or as an interview. However, significantly higher frequencies of musculoskeletal problems were reported when the questionnaire was administered as part of a focused study on musculoskeletal issues and work factors than when administered as part of a periodic general health examination (Andersson, 1987).

3 METHODOLOGY

On this method there were considered five steps:

a. Analysis of the Workplace—Collection of data on the functional structure of the job of SACM, operator target analysis, equipment and characterization of the workplace, assessment of thermal comfort, brightness and noise.
b. Questionnaire on Musculoskeletal symptoms—Distribution of the NMQ to determine the symptoms, completed by four employees working in SACM, three men and a woman aged 18 to 45 year.
c. Application Method RULA—With the aid of the videos were observed residence times of different combinations of positions of the trunk, arms, legs, wrists and neck and applied the RULA method, using a spreadsheet;
d. Definition of correction and prevention measures, by analyzing the results obtained by the method of RULA and the NMQ.

4 EVALUATION OF ERGONOMIC WORKSTATION: SACM

The chosen workplace was SACM, from Cutting section. In this section various pieces of leather are cut or other materials that will be transformed and joined together in later sections.

In the case of leather, the various parts to be cutted are affected by, some characteristics as strain, rigidity, and porosity. Thus, for example, should be cut in the upper area of leather with the best features, which better presentation and confers resistance to most exposed area of the shoe. In this activity, the worker spends most of his time standing, operating with SACM.

The described task is to operate the SACM, which is put on the leather and cutting tools so that when pressing these split into several skin flaps with the exact format that is intended. This activity requires constant efforts of the upper limbs, which justifies the use of the method of postural analysis RULA.

4.1 *Working environment*

Physical space: The worker has a surface and cubing enough according to art.° 8 da Portaria 53/71, 3th February, amends by Portaria 702/80, de 22nd September—General regulations and safety occupational hygiene in industrial establishments.

Level Lighting: Natural lighting comes from window openings at the top of the pavilion, while the artificial lighting is generally from 8 type fluorescent luminaires, 40 watts each, existing a fluorescent lamp of 60 W, above the machine. In the luminance evaluation of the light meter equipped with a LP model luminance sensor, obtaining an average value of luminance of 857 lux, being within the range recommended by the ISO 8995:2002—Lighting of indoor work places.

Thermal environment: The temperature of the work environment was pleasant in the day that were done filming and analysis of the work place, however the worker complained that the winter period, with the gate open there are air currents and thermal discomfort. A measurement carried out, in loco showed a value of 23° C and a relative humidity of 50%.

Noise: The assessment of noise exposure was based on Decreto-Lei 182/2006, of 6th September in which they are defined: $L_{EX,8h}$—daily level for personal exposure of workers to noise and L_{CPICO}—maximum peak sound pressure level. It was used an Integrating sound level meter, accuracy class 2, equipped with microphone model. The values average were $L_{EX,8h} = 75.2$ dB(A) and $L_{CPICO} = 109.4$ dB(C), both below the lower action levels, so there is no risk of noise exposure to SACM workers.

4.2 *Division activities in various postures*

4.2.1 *Activity 1*

This activity consists of the placement of the leather on the workbench and in their preparation for the cut, as is visible in Figure 1. It was divide into the following postures:

- Posture 1: put the leather on the workbench;
- Posture 2: smoothing leather as much as possible;
- Posture 3 and 4: wrap leather material in the form of roll;
- Posture 5: place the leather in the cutting position.

4.2.2 *Activity 2*

In this activity, cutting tools are placed over the leather on the workbench, identified as posture 6 and is represented in Figure 2.

4.2.3 *Activity 3*

The third activity consists on cutting the leather, identified in Figure 3. This activity is divided into four postures:

- Postures 7 and 8: pull the cutting machine;
- Posture 9: put the machine in the desired position, trying to center the machine with mold;
- Posture 10: pressing machine which works as a kind of press, making the cut.

Figure 1. Place the leather on the workbench and prepair for the cut (activity 1).

Figure 2. Cutting tools is placed over the leather (activity 2).

Figure 3. Cutting the leather (activity 3).

Figure 4. Remove the leather from the cutting tool (activity 4).

4.2.4 *Activity 4*

In this activity the worker release the machine that will return to its initial position and remove the leather from the cutting tool while idealizes where will put the same or another cutting tool next, to maximize the utilization of leather. This activity is observable in Figure 4, which identifies the posture 11.

4.2.5 *Activity 5*

Activity 5 represents the blending and storage of cut pieces, as shown in Figure 5. This activity was divided in two postures.

Figure 5. Blending and storage of cut pieces (activity 5).

- Posture 12: add a certain amount of cut pieces and count the pieces of the lot;
- Posture 13: organized on a shelf beside her.

On average, activity 1 is repeated every 10 minutes during 20 seconds. Activities 2, 3 and 4, during these 10 minutes, successively repeated in 10 seconds cycles. Finally, the activity 5, like the first, is repeated 10 in 10 minutes taking about 15 seconds.

4.3 *RULA application*

These photos provided the analysis of the angles between the segments of the body and the characteristics of each posture by the RULA method.

The results are shown into two groups (Table 2):

- Group A—Group A—Evaluation of the Neck, Trunk and Feet;
- Group B—Assessment of Arms, Forearms and Fists.

4.4 *Analysis of results*

Based on the value related to the overall score can be promoted by evaluating corrective actions on the findings in accordance with the Table 2.

The results of method RULA show that any of the postures obtained five activity score 1 or 2, so, there was no posture which is fully acceptable if it was maintained for long periods. All of the positions reported results that deserve anthropometric investigation.

During activities 1, 2, 4 and 5 only one posture showed different final result of 3 or 4 points, which means that these activities must be investigated to detected possible changes. In these 4 activities only posture 2 of Activity 1, where the employee tries to straighten up the leather in order to prepare it for cutting, earned a score of 5 points, which reflects the need to be investigated and a possible change soon. This difference in the remaining point postures due to a bent posture laterally of the trunk and neck.

Table 2. Results of RULA method application for the post of work: SACM.

Activity	Posture	Group A	Group B	Final result
Activity 1	Posture 1	5	3	4 points
	Posture 2	4	5	5 points
	Posture 3	3	3	3 points
	Posture 4	4	3	3 points
	Posture 5	3	2	3 points
Activity 2	Posture 6	2	4	4 points
Activity 3	Posture 7	4	5	5 points
	Posture 8	6	6	7 points
	Posture 9	3	5	4 points
	Posture 10	6	7	7 points
Activity 4	Posture 11	2	4	4 points
Activity 5	Posture 12	3	3	3 points
	Posture 13	4	1	3 points

The third activity, which is cutting the leather, was the activity that demonstrated the need for research and immediate change. The activity was divided into four postures and presented a posture with 4 points, other with 5 and other two with 7, which represents well this need. The two highest scores occurred when the worker pull the machine to the cutting posture and in the phase of the pressure to make the cut (postures 8 and 10), which is justified because it is a repetitive action and a load from 2 to 10 kilograms. Also target of an investigation and possible change soon is the posture of the worker to put yourself in posture to pull the machine (posture 7). A score of 5 points obtained in this task, it must be because it is a repetitive action and the score of the arms, neck and trunk.

So we can conclude that in cutting activity, posture where the employee focuses the machine into position (posture 9) was the one that obtained the lowest score, with 4 values, which means however that is also target of an investigation to detect possible changes.

The results obtained in the Nordic questionnaire indicate consequences associated with this working place are shown in Table 3.

The dorsal region is the most affected, there 75% of employees indicate that they felt pain at this place in the last 12 months and 50% of workers in the last 7 days.

Also with a high incidence, are pains in the shoulders and neck, with 50% of workers with symptoms in the last 12 months, and even originated preventing the completion of other tasks.

In the remaining areas of the upper body, at least one of the surveyed workers had pain in the last 12 months.

Table 3. Results of the NMQ.

Zone body	Symptoms in the last 12 months (%)	Impediment to perform normal activities because of this problem in the last 12 months (%)	Symptoms in the last 7 days (%)
Neck	50	25	0
Shoulders	50	25	50
Elbows	25	0	25
Forearm	25	0	0
Wrists/hands/fingers	25	25	25
Dorsal	75	0	50
Lower back	25	0	0
Hips and/or thighs	0	0	0
Knees	0	0	0
Ankles	25	0	25

At the bottom of the human body are not present complaints, with the exception of the female sex worker noted that having foot pain occasionally. This may be related to the fact that these workers have shorter stature than other workers and physical effort resulting higher being.

5 PREVENTIVE AND CORRECTIVE MEASURES

The ergonomic analysis of the workplace concluded that this job causes MSDs, focusing on the upper body, especially the lower back, neck and shoulders, related mostly to activity 3, which involves cutting leather.

Consequently it is pertinent activated a program of prevention of MSDs, acting at the level of knowledge among workers in the correct use of the SACM, avoiding excessive displacements of the swing arm involving rotation and inclination of the trunk (posture 8) as well as the inclination of the trunk and neck when lowering the swing arm (posture 10).

It also appears that this job entails more significant risks to the feminine gender, as the SACM is dimensioned for males, with a work plan that seems to favor individuals with heights exceeding 170 cm, and should be performed an anthropometric study of it and realized all the changes needed to adapt the workplace to each worker. For example, can and should be provided a platform for the people of smaller stature.

Besides the intervention at the level of the SACM, it is extremely important to carry out training and awareness of employees to the correct postures.

Regarding the organization of work, and once it comes to repetitive work, we propose to conduct studies on the introduction of frequent breaks of shorter duration, which will allow greater recovery of tissues and anatomical structures involved, and the rotation of posts job.

In order to minimize episodes of pain and absenteeism, we also propose the realization of labor gymnastics.

6 CONCLUSION

Evaluating ergonomic workstation using RULA method it was possible identify the more critical positions in SACM of the footwear industry, with are mostly associated with the third activity, cutting the leather.

The assessment and analysis of the results obtained in the NMQ showed that none of the activities identified in the studied approach is acceptable, and may be the cause of the symptoms associated with MSDs. Further investigation extended over time, using more companies and considering the factor age and gender should be taken into account in a future study.

As a conclusion of this study, we proposed corrective measures body and suitability of the job task in order to minimize the risk of exposure of workers to MSDs and adjust the positions of the worker on the this workplace.

With this study, we can conclude that it is extremely important to use methods of postural analysis in various human activities, to act and prevent work-related upper limb disorders.

REFERENCES

Andersson et al. 1987. The importance of variations in questionnaire administration. *Appl Ergon* 18: 229–232.

Crawford, J.O. 2007. The Nordic Musculoskeletal Questionnaire. *Occupacional Medicine* 57: 300–301.

David, G.C. 2005. Ergonomic methods for assessing exposure to risk factors for work-related musculoskeletal disorders. *Occupational Medicine*. 55: 190–199.

Espinoza, L.A. & Nájera, J.M. 2011. Postural load on the personnel of a Costal Rican university and impliocations for occupational health. *Cuadernos de Investigación UNED*, Vol. 3(1): 59–62.

Kuorinka, I. et al. 1987. Standardized Nordic questionnaires for the analysis of musculoskeletal symptoms. *Appl Ergon* 18: 233–237.

Lueder, R. 1996. A Proposed RULA for Computer Users. *Proceeding of the Ergonomics Summer Workshop*, UC Berkeley Center for Occupational & Envirомental Health Continuing Education Program, San Francisco.

Marras, W.S & Karwowski, W. 2006. *Fundamentals and Assessment Tools for Occupational Ergonomics*. 2. ed. CRC Press.

McAtamney, L. & Corlett, E.N. 1993. A survey method for the investigation of work-related upper limb disorders. *Applied Ergonomics* 24(2): 91–99.

Motta, F.V. 2009. Avaliação ergonômica de postos de trabalho no setor de pré-impressão de uma indústria gráfica. Monografia submetida à coordenação de curso de engenharia de produção da Universidade Federal de Juiz de Fora.

Pinheiro, F.A. et al. 2002. Validação do Questionário Nórdico de Sintomas Osteomusculares como medida de morbidade. *Revista de Saúde Pública*. Vol. 36(3): 307–312.

Stanton, N. et al. 2005. *Handbook of Human Factors and Ergonomics Methods*. CRC Press.

Occupational Safety and Hygiene II – Arezes et al. (eds)
 ISBN 978-1-138-00144-2

Evaluation of sidewalks on a busy avenue in a Brazilian metropolis

A.N.J. Mandel Lins
Occupational Safety Engineering Graduate Program, University of Pernambuco, Recife, Brazil

A. Mandel Lins
Undergraduate and Graduate Program in Law and Public Administration, Grupo Ser Educacional, Recife, Brazil

E.R. Kohlman Rabbani
Civil Engineering Graduate Program (PEC), University of Pernambuco, Recife, Brazil

ABSTRACT: This paper investigates the compliance of sidewalks to Technical Standard NBR 9050 on an avenue in Recife, a large Brazilian city, where property owners are responsible for sidewalk construction and maintenance (Law 16292 and Decree 20604). Diverse factors may influence public discourse, so in addition to measuring walkability, this study aims to identify solutions for sidewalk compliance according to the actual perception of interviewees. For this, free observation, individual semi-structured interviews, questionnaires, and focus groups were conducted, utilizing microanalysis, enunciation analysis, and categorization. The respondents gave a grade of 3.51 (on a scale of 0 to 10), but they did not feel obligated to build or maintain their own sidewalks. When building them, they do not contract qualified professionals and fail to comply with legal requirements. In the short and medium term, walkability will not improve if the government does not take responsibility for building and maintaining sidewalks.

Keywords: sidewalks, walkability, quality

1 INTRODUCTION

According to the definition adopted by the Brazilian Traffic Code—CTB (Brazil, 1997), the sidewalk is "part of the street, usually separated at a different level, not intended for vehicle circulation, reserved for pedestrian traffic and, where possible, the placement of street furniture, signs, vegetation and other items."

As can be seen, the definition adopted by the CTB pavement includes some of the elements necessary for the development of a safe and comfortable walk in a space not destined for motor vehicles. However, the pedestrian does not walk only on sidewalks. For this reason, studies whose scope covers other factors, such as the infrastructure of plazas and parks, intersection ramps, leveling, and signage at vehicle crossings, and others. All this and more is covered by the concept of "walkability," whose understanding is crucial to the context of this article.

The concept of walkability covers the concept of the sidewalk. It is old, and is related to the quality of life of the pedestrian (Bradshaw, 1993). For Santos & Cruz (2008), this concept can be described by the quality of spaces designed for walking that provide pedestrians, with or without mobility difficulties, good conditions of access to the various parts of the city.

The first known published work on the measurement of walkability was Bradshaw (1993). This author created 10 categories of *a priori* values to measure walkability in the streets of the neighborhood where he lived in Ottawa, Canada. His article explained little about the technical infrastructure of the place, referring primarily to the application of a questionnaire survey to determine what users thought about the issue. However, the idea of creating an official index of walkability was brilliant because it encouraged debate, revealing problems of public administration on this issue to society in general.

Based on this initiative, other researchers began to measure walkability, performing technical surveys of sidewalks and other elements related to infrastructure, as well as applying opinion surveys (e.g., ADB, 2011; Kubat, 2007; Rutz, Merino, Prado, 2007; Santos & Cruz, 2008).

Works such as Bradshaw (1993), researched the views of pedestrians, but did not perform an in-depth technical survey of infrastructure conditions. On the other hand, there are studies such as Santos & Cruz (2008), which provide important technical surveys of the condition of sidewalks and other infrastructure-related elements, but do not study what the pedestrian thinks about it.

Evaluating walkability is always important, even if this is done on an avenue whose sidewalks are visibly in poor condition. The research, however, is not limited to this. Its principal objective is to identify, based on the actual perception of respondents, solutions for effective compliance of sidewalks with technical standard NBR 9050, taking into consideration that the responsibility for construction and maintenance of sidewalks lies with property owners (Law 16292 and Decree 20604 of Recife).

2 METHODOLOGY

A "case study"[1] was chosen to be applied at three locations along Avenida Cons. Aguiar in the city of Recife/Pernambuco/Brazil. The area was selected because it has a high income level and therefore a large amount of taxes are collected (based on oral information).[2] A form was elaborated based on the category codes proposed by Bradshaw (1993) to measure walkability. These codes or meanings were identified through micro-analysis (Strauss, Corbin, 2008). A semi-structured interview was processed using analysis of enunciation (Bardin, 2011).

Ten items were chosen and assigned weights for calculating a final average grade. These items were selected from Technical Standard NBR 9050 (ABNT, 2004), the Recife Municipal Law No. 16,292 of 29 January 1997, and the Recife Municipal Decree No. 20,604 of 20 August 2004 (see Table 1).

The weight of each item was established according to the opinion of the research subjects, who had the choice of weights ranging from 1 to 3.

Table 1. Choice of assessment items and their weights.

Evaluated items	Weight
Pavement condition	3
Access	2
Sidewalk rest areas	2
Ramps	2
Widths (throughway and service zones)	1
Obstacles (throughway zone)	3
Maintenance (work areas)	2
Crosswalk	2
Pedestrian flow	3
Signage	1

The research subjects were: (a) a former president of the Urbanization Company of Recife[3] (URB), regarded as a technical expert in the area, (b) 30 pedestrians, 15 men and 15 women, of whom 1 woman and 2 men had difficulties walking, but did not require the use of equipment (wheelchair, crutches, etc.), (c) 4 participants in a focus group detailed in the next section.

In order to achieve richer and more reliable data, a combination of four techniques of data collection was used.[4] This procedure is known as "multifaceted observation" because it combines more than three sampling techniques on the same object.

Before initiating data collection in the case study environment, a test was used to choose the most appropriate equipment for field collection (e.g., type of camera and tape measure). The test also provided input to carry out some adjustments in the research form, and the preparation of the focus group and semi-structured interview.

The test was held on 29 August 2013 on Herculano Bandeira Avenue, which is located about three kilometers away from the section of road selected for the case study. This test area was chosen because of its many similarities with the area of analysis.

Details on the collection and processing of data are discussed in the next sections.

3 DATA COLLECTION AND PROCESSING

3.1 *Technical questionnaire*

Sidewalk technical data were gathered from observations made during the data collection, based on technical and legal requirements for sidewalks. These are presented in Table 2. Photographic records were made to illustrate situations encountered at the research sites. Some of these records are shown in Photographs 1–4 presented in this article.

The technical data are shown in Table 2.

The weighted average obtained from the technical questionnaire was 2.67 on a scale of 0 to 10 and is detailed in Figure 1.

3.2 *Free observation*

The "free observation" was performed over a two-day period: 30–31 August 2013. Difficulties walking

[1]A Case Study is not a method in itself, but a specific approach strategy for collecting, analyzing, and interpreting data (Yin, 2001).

[2]Declarations of the Fiscal Auditor of the Treasury of Pernambuco, Mr. Ricardo Ralinoe, and the Fiscal Auditor of the City of Recife, Mr. Soares de Carvalho Junior Elísio on 10 September 2013.

[3]The URB is a Public Firm of Recife, responsible for building construction, urban development, and public services (Recife, 2005).

[4]According to Denzin (1989), the combination of techniques enhances the research and allows for the study of the same object from different angles.

Table 2. Technical data.

Qualitative evaluation	Weight	Grade
Pavement condition: almost the entire sidewalk pavement has broken surfaces, holes, gaps and obstacles. There is no use of tactile paving and none meet the criteria established by law.	3	1
Access: almost all access to buildings and parking areas fails to meet standards.	2	1
Sidewalk rest areas: there were no rest areas found on the sidewalks in the researched areas.	2	0
Ramps: almost all ramps intended for vehicular access had slope and width irregularities.	2	1
Widths (throughway and service zones): there are no sections with less than 1.5 m width; however, there is no separation between service and throughway zones.	1	8
Obstacles (throughway zone): as there is no separation between lanes, obstacles of all kinds are found blocking the open lane.	3	1
Maintenance (work areas): there is no regular maintenance and where work areas exist, they do not prioritize pedestrians.	2	1
Crosswalk: almost all crosswalks do not allow adequate sight distance neither for pedestrians nor drivers.	2	3
Pedestrian flow: insignificant to affect the travel speed of passersby in the sections studied.	3	10
Signage: signs found do not obey thc rules and in many cases are non-existent.	1	1

Photograph 1. Obstacles present on sidewalk.

Photograph 2. Faded crosswalks.

were observed in almost all sections analyzed with the presence of obstacles (photograph 1), faded crosswalks (photograph 2), damaged wheelchair access ramps (photograph 3) and damaged pavement (photograph 4). The free observation served to make comparisons between the results determined by other methods of data collection and analysis.

Photograph 3. Damaged wheelchair ramps.

Photograph 4. Damaged pavement.

x 3 x 2 x 2 x 2 x 1 x 3 x 2 x 2 x 3 x 1

$$\frac{1+1+0+1+8+1+1+3+10+1}{3+2+2+2+1+3+2+2+3+1}=\frac{56}{21}=2{,}67$$

Figure 1. Weighted average from technical questionnaire.

3.3 *Semi-structured interview*

The interview was conducted on 2 September 2013 with a former President of the Urbanization Company of Recife. The technique known as "Analysis of enunciation" (Bardin, 2011)[5] was used. The interview helped identify solutions for effective compliance to sidewalk standards. The central category of this categorization was "legitimacy," describing as solutions: "construction and maintenance of all sidewalks under the responsibility of the government." On 6 September 2013 an email containing questions about the same items technical (subsection 3.1) was sent. The e-mail was answered on 11 September 2013, resulting in a weighted average of 3.86 (See Fig. 2).

3.4 *Questionnaires*

The questionnaires contained ten open questions and were conducted by two researchers with 30 respondents on two different occasions:

Tuesday, 3 September 2013, in front of the Avenida Cons. Aguiar, no. 2775; Thursday, 5 September 2013, between the corner of Avenida Cons. Aguiar and Rua Artur Muniz, and Rua Vicência; and Saturday, Brazilian Independence Day, 7 September 2013, at the corner of Rua Barão de Souza Leão and Avenida Cons. Aguiar, where it intersects with Rua Padre Bernardino Pessoa.

The technique of "focused interview" was used when applying the questionnaires forms (Merton, 1987). The forms took, on average, 15 minutes to complete. Open-ended questions were asked, followed by more specific questions. At the end, grades were requested for the technical questionnaires items (subsection 3.1) and the value 1–3 for the degree of importance each item, with one = somewhat important, two = important, and three = very important. For the analysis and interpretation, "categorization (open, axial, selective, and central)" was used with tools such as "flip flop" and "mini-frameworks" (Strauss & Corbin, 2008). The calculation of the weighted average of the questionnaires forms returned a value of 4.05, which is shown in Figure 3.

The central category from questionnaires forms was the "legitimacy," describing as solutions:

$$\frac{\overset{\times 3}{2} + \overset{\times 2}{1} + \overset{\times 2}{1} + \overset{\times 2}{1} + \overset{\times 1}{9} + \overset{\times 3}{5} + \overset{\times 2}{1} + \overset{\times 2}{5} + \overset{\times 3}{10} + \overset{\times 1}{3}}{3+2+2+2+1+3+2+2+3+1} = \frac{81}{21} = 3{,}86$$

Figure 2. Weighted average of semi-structured interview.

$$\frac{\overset{\times 3}{4} + \overset{\times 2}{3} + \overset{\times 2}{2} + \overset{\times 2}{3} + \overset{\times 1}{10} + \overset{\times 3}{5} + \overset{\times 2}{0} + \overset{\times 2}{0} + \overset{\times 3}{10} + \overset{\times 1}{2}}{3+2+2+2+1+3+2+2+3+1} = \frac{85}{21} = 4{,}05$$

Figure 3. Weighted average from questionnaires forms.

$$\frac{\overset{\times 3}{2} + \overset{\times 2}{2} + \overset{\times 2}{1} + \overset{\times 2}{1} + \overset{\times 1}{8} + \overset{\times 3}{5} + \overset{\times 2}{1} + \overset{\times 2}{3} + \overset{\times 3}{9} + \overset{\times 1}{1}}{3+2+2+2+1+3+2+2+3+1} = \frac{73}{21} = 3{,}48$$

Figure 4. Weighted average from focus group.

"construction and maintenance of all sidewalks under the responsibility of the government, and to recover the costs from the owners who are able to pay."

3.5 *Focus group*

According to Riccio & Ruediger (2006, p. 151), focus groups are characterized by the "possibility of real-time intervention in the course of analyzing and comparing the perceptions of the participants, their similarities, and contradictions regarding a topic or group of topics related to the subject of the research."

The focus group occurred during the holiday of 7 September 2013 at 8 am with four employees of a cafeteria located at Rua Carlos Pereira Falcão, on the corner with Avenida Cons. Aguiar. The four participants all lived in neighborhoods further away and walked, on average, 400 meters down the sidewalks of Avenida Cons. Aguiar until reaching their place of work. The activity lasted one hour and 10 minutes. While one researcher directed the focus group, the other took notes of the discussions. The weighted average calculated for the focus group was 3.48 points (see Fig. 4).

The central category from the focus group was "legitimacy," describing the following solutions: "strict supervision, construction, and maintenance of all public sidewalks with recovery of costs from owners who can afford it."

4 FINAL CONSIDERATIONS

The literature on scientific methodology was reviewed. With this, it was possible to employ different techniques that returned somewhat different assessments in the "grade" category. However, all techniques pointed to the same central category (Strauss & Corbin, 2008, p. 146) of "legitimacy." According to this core category, the regulation that requires homeowners to build their own sidewalks is not legitimate, or is not accepted by society. Several value categories point to this same conclusion. The main ones are: (a) the respondents pay taxes, but

[5]Bardin (2011, p. 218) teaches that communication has two languages: the hidden and what society wants to hear.

feel that the cost-benefit relationship is disadvantageous (Sayeg, 2003), (b) the respondents do not feel obliged to build their driveways because they have to pay for them, but cannot decide on the placement of bus stops, trees, payphones, mailboxes, posts and other furniture on their sidewalks (Barros, 2013), and (c), the obligation to pay for the sidewalk came to the property owner in a "top down" manner, as described by Tenório (2004). That, according to Saravia (2006), creates legitimacy and effectiveness issues for public policies.

The results point to a difficulty in contextualizing the benefits of walkability for those who pay taxes and build their driveways. The solution to reach a consensus is through debate (Fleury, 2006).

The weighted average obtained from the technical questionnaire was the lowest (2.67). This is due to three factors: (a) exemption; (b) questionnaire of the situation refined *in loco*; (c) the results of the technical questionnaire were obtained from different areas.

The weighted average closest to the technical questionnaires average was that from the focus group (3.48), revealing that debate frees people, causing them to grow politically (Aron, 1976). The overall average from all types of evaluations was 3.51.

Based on this study, three possible alternatives are proposed: (a) The government should build sidewalks and make them in compliance with standards. Maintenance would also be the responsibility of the Government, except in cases of damage caused by third parties. (b) The government, in cases where the property owner has failed to build an adequate sidewalk, must build it and cover the costs through specific taxes. (c) The government should promote periodic inspections and grant deadlines for compliance. In cases where this is not done, the government must build the sidewalks and cover the cost administratively and/or judicially.

The users' perception is that the compliance of sidewalks with legal standards will only be effective upon the initiative of the government. This indicates that the issue revolves around the "decision making" process.

REFERENCES

ABNT. 2005. NBR 9050. Acessibilidade à edificações, mobiliário, espaços e equipamentos urbanos.

ADB. 2011. *Walkability and Pedestrian Facilities.* Accessed on 12 september 2013, available at Asian Develop. Bank: http://www.adb.org/publications/walkability-and-pedestrian-facilities-asian-cities-state-and-issues.

Aron, R. 1976. *Essai sur les libertés.* Paris: Calmann-Lévy.

Bardin, L. 2011. *Análise de conteúdo.* São Paulo: Edições 70.

Barros, C., Monteiro, P., Mafra, H., Lima, E., Moss, M., Ferreira, C., & Araripe, V. 2013. Recife Enxerido. Recife: C.B.

Bradshaw, C. 1993. Creating and using a rating system for neighborhood walkability. Colorado.

Brazil. 1962. *Lei Federal n.4.150.* Accessd on 27 August 2013, available at Câmara dos Deputados: http://www2.camara.leg.br/legin/fed/lei/1960–1969/lei-4150–21-novembro-1962-353924-publicacaooriginal-1-pl.html.

Brazil. 1997. *Lei Federal n. 9.503.* Accessed on 28 August 2013, available at Pres. República: http://www.planalto.gov.br/ccivil_03/leis/l9503.htm.

Capra, F. 2005. *As conexões ocultas* (4 ed.). (E.P.-C. Ltda, Ed., & M.B. Cipolla, Trad.) São Paulo: Cultrix.

Denzin, N.K. 1989. *The research act: a theoretical introduction to sociological methods* (3 ed.). Prentice Hall.

Fleury, S. 2006. *Democracia, descentralização e desenvolvimento: Brasil & Espanha.* (F.G. Vargas, Ed.) Rio de Janeiro: FGV.

Flick, U. 2009. *Introdução à pesquisa qualitativa* (3 ed.). (J.E. Costa, Trad.) Porto Alegre: Artmed.

Kubat, A.S. 2007. *6th International Space Syntax Symposium ISTANBUL 2007.* Accessed on 12 September 2013, available at Istanbul Technical University: http://www.cevse.itu.edu.tr/english/index.html.

Merton, R.K. 1987. Focussed Interview and Focus Groups. *Public Opinion Quarterly, 51*, 550–566.

Recife. 2011. *Prefeitura do Recife.* Accessed on 28 August 2013, available at Prefeito sanciona lei: http://www.recife.pe.gov.br/noticias/imprimir.php?codigo=176875.

Riccio, V., & Ruediger, M.A. Grupo focal: método e análise simbólica da organização da sociedade. In: Vieira, M.M. & Zouain, D.M. Pesquisa qualitativa em administração. 2 ed. Rio de Janeiro: FGV, 2006, p. 151–172.

Rutz, N., Merino, E., & Prado, F.H. 2007. *16 Congresso Brasileiro de Transporte e Trânsito.* Accessed on 13 September 2013, available at CBTU: http://www.cbtu.gov.br/estudos/pesquisa/antp_16congr/resumos/arquivos/antp2007_206.pdf.

Santos, E.C., & Cruz, I.A. 2008. Recuperação da área central com base no aumento do índice de caminhabilidade. *da Vinci / Universidade Positivo. Núcleo de ciências exatas e tecnológicas., 5*(1), pp. 21–49.

Saravia, E., & Ferrarezy, E. 2006. *Políticas Públicas (org).* Brasília: ENAP.

Sayeg, R.N. 2003. *Sonegação tributária e complexidade.* (R. electron, Ed.) Accessed on 19 July 2012, available at RAE eletrônica: http://dx.doi.org/10.1590/S1676-56482003000100010.

Strauss, A., & Corbin, J. 2008. *Pesquisa qualitativa: técnicas e procedimentos para o desenvolvimento de teoria fundamentada* (2 ed.). Porto Alegre: Artmed.

Tenório, F. 2004. *Tem razão a administração? Ensaios de teoria organizacional* (2 ed.). Ijuí: Unijuí.

Thiry-Cherques, H. 2009. Saturation in qualitative research: empirical sizing estimation. *PMKT*(3), 20–27. Accessed on 13 June 2012, available at PMKT: http://www.revistapmkt.com.br/Portals/9/Edicoes/Revista_PMKT_003_02.pdf.

Vergara, S. C. 2005. *Métodos de pesquisa em administração.* São Paulo: Atlas.

Yin, R.K. 2001. *Estudo de caso: planejamento e métodos* (2 ed.). Porto Alegre: Bookman.

Occupational Safety and Hygiene II – Arezes et al. (eds)
© 2014 Taylor & Francis Group, London, ISBN 978-1-138-00144-2

Occurrence of pain in metals industry workers and its relation to work

F. Torres, F.S. Menezes & A.A.P. Xavier
UTFPR—Federal University of Technology of Paraná, Ponta Grossa, Brazil

ABSTRACT: The aim of this study was to check the occurrence of pain among workers of a metals industry in Brazil and its relation to the work performed. The data collection was carried out through the application of three instruments: a questionnaire with individual questions and Nordic Musculoskeletal Questionnaire. The sample was composed by 90 employees. According to the results, 63% of the workers frequently or always feel some kind of pain, 32% say they rarely feel any pain, and only 5% do not mention any pain. The body segment with the highest incidence of pain was the lumbar region. The results also show that the greater the length of performance in the same function, the higher the occurrence of pain. So the pain-related musculoskeletal symptoms are present in the every day's lives of the individuals surveyed, and may even evolve over the years to RSI/WRMSD.

1 INTRODUCTION

The Repetitive Strain Injury (RSI) or Work-Related Musculoskeletal Disorders (WRMSD) currently reach different categories of workers (Brasil, 2012; Mendes, 2004), and can be defined as a group of diseases that affect muscles, tendons, nerves and vessels of the upper and lower limbs and that is directly related to the requirements of the tasks, physical environments and with the work organization (Chiavegato Filho; Pereira JR, 2004).

Several factors shape RSI/WRMSDs as a major socioeconomic and public health problem nowadays in Brazil: the magnitude of the prevalence of the affected population, the wide range of impacted economic sectors, the clinical complexity with the high cost of therapies, and absenteeism from work for temporary or permanent disability (Neves; Nunes, 2009).

Pain is often reported in the literature as one of the major causes of disability, and is considered a health problem not only for the physical aspect but also for the great socioeconomic impact and consequent impairment quality of life, as well as for lost workdays and high medical costs (Chalot, 2006; Marras, 2000; Iguti, Hoehne, 2006).

The group of metalworkers fits perfectly as an example of what the division of labor has produced because they perform their activities in sessions, doing exactly the same thing every day, in the same way, with the same repetitive efforts, some for even long years (Lima, 1997).

Thus, it becomes important to know the profile of the metalworkers affected by WMSDs with the objective of previewing the onset of these disorders, identifying their triggers, and also establishing the need for prevention and health recovery, minimizing the occurrence of new cases, and therefore providing effective and quick return of workers to their works and activities of everyday life.

Therefore, this study aimed at identifying and characterizing the symptoms related to musculoskeletal pain/discomfort presented by the workers of a metals industry in the city of Ponta Grossa, Brazil, and associating these symptoms with occupational variables.

2 METHODS

This study was characterized by a quantitative descriptive cross-sectional research, and was approved in the Ethics Committee of Federal University of Technology—Paraná by protocol number: 16029613.9.0000.5547.

The site of this research was an Industry of the metal-mechanic branch, in the city of Ponta Grossa, Brazil, chosen for its accessibility and interest in the research development. Comprising approximately 1076 employees distributed in 29 sectors, among them the painting sector, the object of this study due to an internal demand presented by the company.

To collect data, scheduled visits according to the availability of the company were done. For better accuracy of data regarding information registered from each questionnaire, it was opted for the method of structured interviews with up to 8 workers in the interview room at a time Ninety employees were interviewed, who firstly answered a questionnaire about individual data, consisting of objective closed questions, and then the Nordic Musculoskeletal Questionnaire (QNSO), adapted, which provided basic data for the research.

Data were processed and analyzed using descriptive statistics, which aims to sort, summarize and describe the collected data (Piccoli, 2006). The results were presented in form of graphs, represented numerically in percentage.

3 RESULTS

The average age was 31 years old, with a minimum age of 19 years and a maximum of 59 years. The age group that prevailed in the study was between 19 and 25 years, corresponding to 38% of the participants.

According to the function performed in the company, most workers are in the Production Assistant function, with 42% of workers, followed by the Painter function, with 33% of workers, being the only female participating in the study in the function of Technician in Chemistry.

Concerning working time, the majority of the employees work for more than a year in the company, accounting for 63% of all the sector workers, with an average of 40 months working in the function, being the minimum three months and the maximum 384 months.

Regarding the occurrence of musculoskeletal symptoms in the studied workers, it was identified through the QNSO that 63% of the workers often or always feel some kind of pain/discomfort, 32% claim they rarely feel any pain/discomfort, and only 5% did not relate any kind of pain/discomfort (Fig. 1).

By associating the working time with the occurrence of pain, it was found that 40% of the workers with less than 6 months of work reported feeling frequently or always some sort of pain/discomfort in some region of the body, among those between 6 months and one year of work, 61% reported feeling often or always some sort of pain/discomfort, and 68% of those working for more than a year in the same function reported feeling often or always any pain/discomfort.

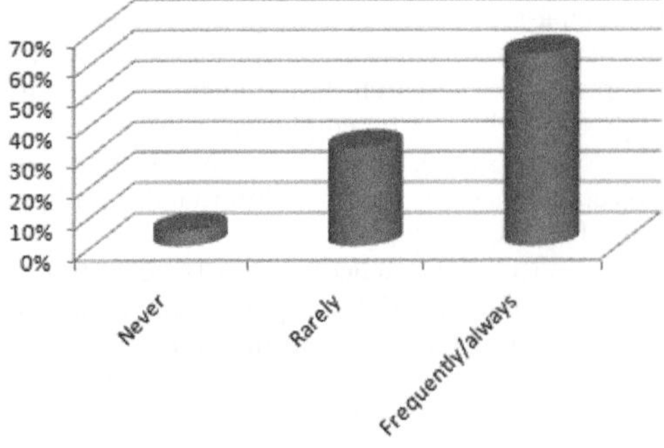

Figure 1. Occurrence of musculoskeletal symptom according to its frequency.

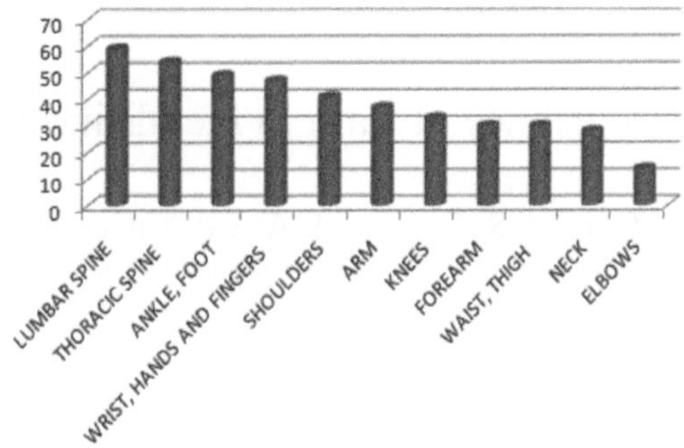

Figure 2. Occurrence of pain according to the body region.

It is observed in Figure 2 that the highest incidence of pain was in the lumbar spine region with 59 subjects, followed by the thoracic spine region with 54 cases, ankle and foot with 49, wrist, hands and fingers with 47, shoulders with 41, arm with 37, knees with 33, forearm and Waist with 30, neck with 28, and elbows with 14, respectively. Considering these numbers, it was observed that the most part of workers indicated presence of pain in different body segments.

About the relation of the musculoskeletal pain/discomfort symptoms to the function performed in the company, it was found that some of the workers reported that frequently or always felt pain/discomfort in some part of the body as follows: in the Painter function, 77% and Production Assistant, 63% of them, being the most affected area the lumbar spine.

4 DISCUSSION

The results showed that the injured workers were relatively young. Reis et al. (2000) had studied 565 injured workers in Brazil and there was a predominance of patients younger than 40 years. The data from this study also concur with other research that found the distribution of people with RSI/WMSD met by Salim (2003) with the predominance of cases in the age group 30–39 years.

Industrial employees tend to suffer more from episodes harmful. Santana et al. (2012) assessed the causes of injuries at work in 1,283,442 young subjects who were temporarily away from work in 2006 in Brazil. Of these, over 40% were employees of industries. According to Rhee et al. (2013), industrial workers have more occupational injuries to any worker in other sectors.

With regard to the occurrence of musculoskeletal symptoms in the study, it was found that over 90% of respondents in this study reported feeling some kind of musculoskeletal symptoms. Of these, mostly were indicated presence of pain in different body segments.

Workers in the metallurgical industry in Brazil, found that 75.2% of the workers studied reported some type of musculoskeletal symptoms in the last twelve months (Picoloto; Silveira, 2008). Another study examined 461 workers in an industry of trucks in England, checking through QNSO that 79% of workers had some kind of musculoskeletal problem in the last year, agreeing with the present study (Hussain, 2004). Worst results to those were found in 5654 workers in the production line of eight aluminum industries in Norway. The QNSO was used and it was found that 94% of workers (ranging from 87% to 95%) had pain in one or more parts of body (Morken, et al., 2002).

By associating the working time with the occurrence of pain in this study, it was found that as longer the work as greater the occurrence of pain/discomfort felt 'often' or 'always' (score of QNSO) by workers. A similar study that analyzed 256 workstations in an industry of cars manufactured by a global company, to find that older workers still complain of symptoms of lumbar spine, despite low demand imposed by their current jobs, indicating cumulative strain (Landau, et al., 2008).

Analyzing the anatomical location, the lumbar region was the most prevalent in the present study. Similar results were found in a study of 155 workers belonging to four factories in various sectors of production (food, metallurgical, pharmaceutical and automobile assembly) where it was found that the overall prevalence of pain was in lumbar region (Schierhout; Myers; Bridger, 1993).

A cross-sectional study with 3,479 front-line workers from 60 factories in China, which also was used the QNSO. The body regions most affected were the lower back, neck, shoulders and upper back. The results were similar to that found in the present study (Yu, et al., 2012).

As regards the role played by the worker, the painter was the function most affected, where 77% of these workers reported some type of pain. Of these, 80% of them declared to be the lumbar spine pain the higher symptom. In a study of 28 painters of 20 timber industries in Sweden, it was found a prevalence of 46% for the lower back, and this region was the highest percentage among workers, agreeing with the present study. These authors also did a comparison of this prevalence with a previous study, conducted in 278 industrial workers in wood and metal workers and engineers in 2884, that found similar findings (Bjoring; Hagg, 2000). Therefore, the physical demands associated with the painting work, like squatting and the static postures sustained by a long time, appears to be a causal factor of pain in industrial labor.

5 CONCLUSIONS

The activities in industrial environments are generally characterized by the presence of biomechanical factors as handling and transportation of cargo, and the use of force, causing physical effort and usually high repetitiveness.

In the studies presented here, there is a high incidence of musculoskeletal symptoms in the industry workers, including those in the metals industry. The region of largest occurrence of musculoskeletal symptoms, in most cases, was the spine, particularly the lumbar region. These results emphasize the importance of the lumbar region to work, as a region of support of the body, often ends up being used improperly as a result of poor posture and cargo transportation.

It was found that the musculoskeletal symptoms related to pain in the analyzed functions are present in the daily lives of the surveyed individuals causing pain and discomfort and may even, over the years, evolve to RSI/WRMDs.

Considering the investigated scenario, it was identified the need to promote orientations, ergonomic analysis and preventive actions in order to promote a better way for the worker to develop his/her activities with the objective of eliminating or minimizing the incidence of musculoskeletal disorders.

REFERENCES

Bjoring, G. & Hagg, G.M. 2000. Musculoskeletal exposure of manual spray painting in the woodworking industry—an ergonomic study on painters. *International Journal of Industrial Ergonomics* 26:603–614.

Brasil. Ministério da Saúde. 2012. *Dor relacionada ao Trabalho.* 2012. *Protocolos de Complexidade diferenciada.* Brasília, DF. Disponível em: http://bvsms.saude.gov.br/bvs/publicacoes/dor_relacionada_trabalho_ler_dort.pdf. Acesso em: 15 dezembro.

Chalot, S.D. et al. 2006. Prevalência de dor e qualidade de vida na população das unidades básicas de saúde do município de Embu-SP. *Anais da 58ª Reunião Anual da SBPC*—Florianópolis, SC—Julho/.

Chiavegato Filho, & Pereira J.R. 2004. LER/DORT: multifatorialidade etiológica e modelos explicativos. *Interface—Comunic., Saúde, Educ.* 8(14): 149–162, set.2003-fev.

Hussain, T. 2004. Musculoskeletal symptoms among truck assembly workers. *Occup Med* 54:506–512.

Iguti, A.M. & Hoehne, E.L. 2003. Lombalgias e trabalho. *Rev Bras Saúde Ocup* 28:78–87.
Landau, K. et al. 2008. Musculoskeletal disorders in assembly jobs in the automotive industry with special reference to age management aspects. *International Journal of Industrial Ergonomics* 38:561–576.
Lima, M.E.A. et al. 1997. LER/DORT–Lesões por Esforços Repetitivos, Dimensões Ergonômicas e Psicossociais. Belo Horizonte: Ed. Health.
Marras, W.S. 2000. Occupacional low back disorder causation and control. *Ergonomics* 43:880–902.
Mendes, R. 2004. *Patologia do Trabalho.* 2a ed. Rio de Janeiro: Atheneu.
Morken, T. et al. 2002. Effects of a training program to improve musculoskeletal health among industrial workers—effects of supervisors role in the intervention. *International Journal of Industrial Ergonomics* 29:95–99.
Neves, R.F. & Nunes, M.O. 2009. Incapacidade, cotidiano e subjetividade: a narrativa de trabalhadores com LER/ DORT. Interface. *Comunicação saúde educação* 13(30): 55–66, jul./set.
Piccoli, J.C.J. 2006. *Normatização para trabalhos de conclusão em educação física*. Canoas: Ulbra.
Picoloto, D. & Silveira, E. 2008. Prevalência de sintomas osteomusculares e fatores associados em trabalhadores de uma indústria metalúrgica de Canoas—RS. *Ciência & Saúde Coletiva* 13(2):507–516.
Reis, R.J., Pinheiro, T.M.M., Navarro, A., Martin M.M. 2000. Perfil da demanda atendida em ambulatório de doenças profissionais e a presença de lesões por esforços repetitivos. *Rev Saúde Pública* 34(3):292–298.
Rhee, K.Y., Choe, S.W., Kim, Y.S. et al. 2013. The Trend of Occupational Injuries in Korea from 2001 to 2010. *Safety and Health at Work* 4(1):63–70.
Salim, C.A. 2003. Doenças do Trabalho. Exclusão, segregação e relações de gênero. *São Paulo em Perspectiva* 17(1):11–24.
Santana, V., Villaveces, A., Bangdivala, S.I. et al. 2012. Workdays Lost Due to Occupational Injuries Among Young Workers in Brazil. *American Journal of Industrial Medicine* 55(10):917–925.
Schierhout, G.H., Myers, J.E., Bridger, R.S. 1993. Musculoskeletal pain and workplace ergonomic stressors in manufacturing industry in South Africa. *International Journal of Industrial Ergonomics* 12: 3–11.
Yu, W. et al. 2012. Work-related injuries and musculoskeletal disorders among factory workers in a major city of China. *Accident Analysis and Prevention* 48:457–463.

Occupational Safety and Hygiene II – Arezes et al. (eds)
© 2014 Taylor & Francis Group, London, ISBN 978-1-138-00144-2

A proposal for the use of serious games in occupational safety

J.R. Jordan & G. Letti
QualiAll, Universidade do Planalto Catarinense, Lages, SC, Brazil

T.L. Pinto
TDA Sistemas, Florianópolis, SC, Brazil

ABSTRACT: A expressive amount of accidents occur daily in businesses worldwide. It is estimated that 4% of all global gross domestic product is intended for issues of removals, production interruptions, medical expenses and workers' compensation. However, training in work safety currently used by companies worldwide, apparently have not been able to significantly reduce the rate of accidents, or increase the level of learning and assimilation of content related to the safety at work of their workers. The objective of this work is to present a proposal to develop an educational environment using serious games as a training tool in industrial environments and services with electricity, thereby increasing the assimilation of content and consequently reduce accidents in the workplace.

1 INTRODUCTION

According to the ILO (2010), in the world around 6300 people die each day as a result of accidents or occupational diseases. Annually, more than 2.3 million deaths occur as result of 317 million occupational accidents. This means that every second, a worker suffers an occupational accident. In addition to the irreparable human losses, there is a serious economic consequence: 4% of all global gross domestic product is spent on issues of removals, production interruptions, medical expenses and workers' compensation. The reasons to explain the high number of occurrences of accidents are the most diverse, involving failures take on projects of work systems, equipment, tools, deficiency in maintenance processes of the various component parts of the work. Human factors stand out as the major cause of labor-related accidents, including psychosocial characteristics of the worker, negative attitudes towards the activities preventers, aspects of personality, lack of attention, among others. One can add the deficiency in the educational programs of qualification in occupational safety applied in industries. Thus, workers often face occupational safety training as something boring or unnecessary. The alteration of this situation, with the imperious and consequent reduction of accidents in the workplace, necessarily involves training and qualification of employees. Studies by several researchers (Abdelhamid & Everett, 2000; Tam & Fung, 2011; Toole, 2002; Tse, 2005), suggest that the lack or inadequate training in safety is an important factor in the high rate of accidents in the industry. Thus, the adoption of a program of efficient safety training that can improve the performance of safety by preventing accidents, is regarded as an important and effective method for improving the occupational health and safety (Abdelhamid & Everett, 2000; Dong et al. 2009; Halperin & McCann, 2004; Lee & Halpin, 2003; Wallen & Mulloy, 2006). Other researchers (Hoffmann and Stetzer, 1998) and (Topf, 2000) suggest that the appropriation of the cognitive process to respect for the job safety, coupled with a process of efficient training and communication, offer an excellent opportunity for employees to assimilate the necessary knowledge, effectively increasing the performance in the processes of occupational safety requirements.

2 LEARNING SYSTEMS

2.1 *Fundamentals*

According to Trajano et al. (2008), learning systems are used to facilitate the transfer of knowledge to an individual or group of individuals. Developed software are made to integrate themselves knowledge that will be acquired by those who use them. Thus, when the interaction occurs, the user can assimilate the knowledge passed through the system and then apply it. The (Fig. 1), extracted from Trajano et al. (2008) shows schematically this interaction.

2.2 *Conceptualizing the serious games*

Building on the great success that video games have obtained, researchers soon considered taking advantage of this popularity, the creation of tools

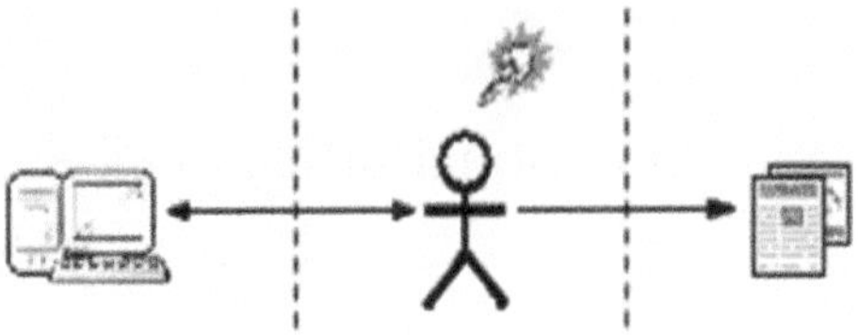

Figure 1. Learning demonstration.

to support learning. These are known as Serious Games, where the game world is merged with the educational content required entry. According to Alvarez (2007) apud Mouaheb et al. (2012), the game is a serious, "computer application, Which AIMS to combine aspects of both the serious, but not limited to, teaching, learning, communication, or Further information with entertainment from the spring game. Such an association has intended to depart from mere entertainment." To Michael & Chen (2006), the simplest definition for serious games is "games that do not have entertainment or fun as primary purpose." This definition also includes educational games and simulators, games for disclosure of commercial products (*advergames*), military applications games and even for games for health management training in large corporations. Thus, taking advantage of the inherent characteristics of this type of application, such as interactivity, the user can absorb knowledge so almost unconscious. During the course of interaction the player ends up developing the skills necessary to achieve the goal set by the game's storyline. And in addition, also learn exercises and content that you want to spend. Due to the active participation of the player helps to bring you closer to the subject to be "studied", which also contributes greatly facilitate learning. When the user gets closer to the problem he can understand it better, and with it, solve it in a better way (Trajano et al. 2008).

According to Prensky (2011), there are at least a dozen reasons for the use of serious games in training:

- They are a form of entertainment;
- Have rules that allow structure;
- Have goals that facilitate motivation;
- Are interactive that require them to work;
- Are you provide adaptive evolution;
- Have results and feedback that enable learning;
- Have several levels of win that provide satisfaction and self-confidence;
- Have problem solving that enhances our creativity;
- Have it simulates interaction and social groups;
- Games are form of play. They enable intense and passionate participation;

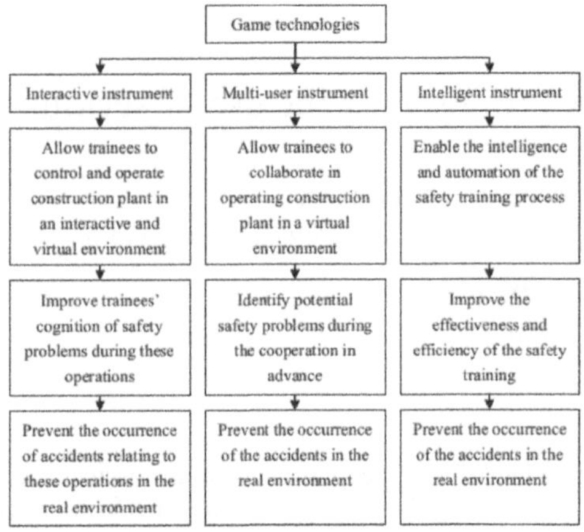

Figure 2. Framework of applying game technology.

- Have Representation and History and it generates excitement;
- Have Conflict, Competition, Challenge, Opposition. And that creates adrenaline.

2.3 *Serious game and training*

Serious games are part of a broad class of applications training, from the medical to the field of psychology. The technology is based on the content and implemented through a variety of channels such as visual technology, digital technology, simulation technology and operational technology multi-user (Guo et al. 2012). These authors propose a framework for the use of that called Game Technology (Fig. 2).

2.4 *Serious game: Main features*

According to Mouaheb et al. (2012), the main features of the serious game is:

- An object teaching priorities: the serious game is a learning process;
- A means of entertainment in parallel;
- A technology of information and communication: serious games are an application of video game technologies;
- It targets multiple learning objectives: to teach, train, educate, heal;
- It applies in almost every field: education, vocational training, health, defense, politics, advertising.

3 PROPOSAL OF A SERIOUS GAME FOR WORKING WITH ELECTRICITY

The proposed game called Game NR 10, aims to support the teaching-learning relationship in work safety in the electricity area. More specifically, this serious game implements a script that describes the steps to be performed while doing electrical maintenance/operational procedures, and as part of the test defines when these steps are planned and then executed, and how much work, time and resources will be needed. Therefore, any testing strategy must incorporate test planning, test case design, test execution, data collection and evaluation of the resulting data (Pressman, 2011).

The environment is composed of internal and external scenarios, where the employee must perform routine tasks, combining operational procedures with safety procedures, so that assimilate these procedures fast, simple and durable. The game begins with the arrival of the worker at the factory and the first scenario is the locker room, where you should select properly protective equipment (stage dress-up) (Fig. 3).

After selecting properly their personal safety equipment and wear them, the employee will be able to move to the next level of the game, where you must find and solve a given problem in the environment (which simulates a deposit (Fig. 4) A company). In this part of the game, the worker must relate cause and effect, and the solution was allowed to take on several attempts. The number of

Figure 3. Dress up room.

Figure 4. Storehouse.

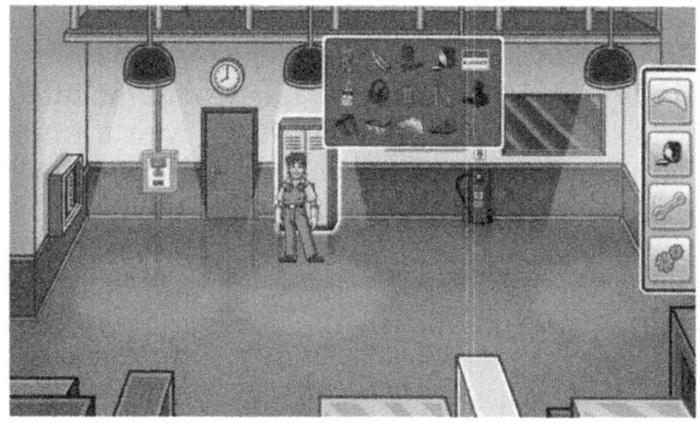

Figure 5. Signs and lock-out.

attempts to solve the problem will be filed and will appear in the accompanying report. To complete this step, it requires developer knowledge in electricity, occupational health and safety, as well as notions of organization and cleanliness.

By owning complex operations, life-threatening and tangible and intangible assets, companies need highly qualified and with full knowledge of the actions to be taken to control and system operation and safety. Not admitting operational errors or rash action, it may cause loss of life, and improper shutdowns affecting thousands of people and causing damage to the company.

Thus, the game presents situations where the employee should identify dangerous situations, isolate the workplace through signs and lock out equipment and mainly learn to correctly select the equipment to be used in each service type (Fig. 5).

4 EVALUATION OF LEARNING

The evaluation of learning is an important element in the game NR 10 that was considered, as a way to check the level of learning of the worker. The taxonomy of the assessment can be divided into three major groups: assessment, written, oral and practical assessment. To Michael and Chen (2006), there are important features in games that are responsible for evaluating the learning player (employee) with respect to the game. One is known as a tutorial, and consists of a learning system that aims to teach the player to play the game. Tutorials consist of a series of instructions that must be followed step by step instructions for new, more complex can be passed. Also there is the scoring system. The points system basically makes it clear to the player what tasks are important in the game. The score of the player's performance can be viewed similarly to score on a written test, in the sense that it must determine what are the key points that most influence the learning content safety.

5 DISCUSSION

Thirty workers were selected and were randomly divided into two groups of fifteen, with a workload of 40 hours. The first group was trained through the traditional system: classroom and an instructor. The second group was trained using the game NR 10. After training, were evaluated according to the metrics of speed, concentration and number of errors. At the end of the course, we applied a test containing 100 questions relating to training in electrical safety. The group that was trained with the serious game NR 10 had an increase of correct answers 40% greater than the first group, with a response rate 55% better.

6 CONCLUSIONS

The serious game NR 10 was used as a tool in the training of safety operations with electricity. The characteristics of the site and the situation were analyzed and the risks inherent in the study were identified. The game offers the employee a virtual environment that allows him to study different situations within the industry. At the same time, provides the worker with sufficient knowledge to help them identify potential safety issues. The main advantage of the game NR 10 is that it allows workers to study and practice the operating procedures and safety requirements of the plant in a virtual environment that resembles the real environment with the same design characteristics. The workers that were trained with the serious game NR 10 had an increase of correct answers 40% greater than the first group, with a response rate 55% better.

REFERENCES

Abdelhamid, T., Everett, J., 2000. *Identifying root causes of construction accidents*. Academy of Management Journal 41, 644–657.

Guo, H. Li H., Chana, G., Skitmorec, M. 2012. *Using game technologies to improve the safety of construction plant operations.* Accident Analysis and Prevention 48, 204–213.

Halperin, K.M., McCann, M., 2004. *An evaluation of scaffold safety at construction*. Health and Safety, Pittsburgh, PA.

Hoffmann, D.A., Stetzer, A., 1996. *A cross-level investigation of factors.* Journal of Risk.

ILO. *Emerging risks and new patterns of prevention in a changing world of work*. Genebra, 2010.

Lee, S., Halpin, D.W., 2003. *Predictive tool for estimating accident risk*. Journal of Risk.

Michael, D., Chen, S. *Serious Games: Games That Educate, Train, and Inform*. Course Technology PTR, 2006.

Mouaheb, H., Fahli, A., Moussetad, M., Eljamalic, S. 2012. *The serious game: what educational benefits?* Procedia—Social and Behavioral Sciences 46, 5502–5508.

Prensky, M., 2011. *Digital game based learning*. New York: McGraw-Hill.

Pressman, R. *Engenharia de software: uma abordagem profissional,* AMGH, (2011). 7ed.

Tam, V.W.Y., Fung, I.W.H., 2011. *Tower crane safety in the construction industry: A Hong Kong study*. Safety Science 49 (2), 208–215.

Toole, M., 2002. *Construction site safety roles.* Journal of Construction Engineering and Management 128 (3), 203–210.

Topf, M.D., 2000. *General next?* Occupational Hazards 62, 49–50.

Trajano, B.M., Souza, A.L., Santos, K.F. *Criação de Sistemas de Aprendizagem Educacional Utilizando Jogos Sérios na Unidade do CEFET de Recife.* III Congresso de Pesquisa e Inovação da Rede Norte Nordeste de Educação TecnológicaFortaleza—CE—2008.

Tse, S., 2005. *Study of the Impact of Site Safety Cycle on Safety Performance of Contractors in Hong Kong*, The University of Hong Kong, Hong Kong Special Administrative Region.

Wallen, E.S., Mulloy, K.B., 2006. *Computer-based training for safety: comparing methods with older and younger workers.* Journal of Safety Research 37 (5), 461–467.

Occupational Safety and Hygiene II – Arezes et al. (eds)
 ISBN 978-1-138-00144-2

Summer parties in Algarve—DJs and audience noise exposure

V. Rosão
SCHIU, Vibration and Noise Engineering, Faro, Portugal

R. Constantino
SWORK, Health and Safety at Work, Faro, Portugal

ABSTRACT: The legislative requirement to limit noise exposure of workers is one important achievements of the humanity progress, reflected by European Directives. It is widely accepted that the most important acoustic parameter of the noise exposure is *Daily Noise Exposure Level* ($L_{EX,8h}$, dB(A)). The initial concern of noise exposure was extended, over time, to less obvious work places, such as Music. Recognizing the potential occupational hazards in this sector, Directive 2003/10/EC establishes the obligation to develop a *Code of Conduct*. The present study was undertaken within Algarve, a very important tourism region, where many summer DJs parties are conducted, aimed to access the noise exposure in both DJs and Audience areas. Based on the results of the present study, we can observe that DJs and Audience are often exposed to extremely high noise level and according to this fact several recommendations are suggested to minimize their consequences.

1 INTRODUCTION

The legislative requirement to limit noise exposure of workers is one of the important achievements of the progress of humanity and has a history of implementation, in Europe, of about 30 years, since the publication of the European Directives 80/1107/EEC (OJEC, 1980) and 86/188/EEC (OJEC, 1986). This legislation was revised by Directive 2003/10/EC (OJEU, 2003a).

Since the beginning the essential acoustic parameter for the quantification of noise exposure is called *Daily Noise Exposure Level* ($L_{EX,8h}$) (dB(A): reference sound pressure of 20 µPa), which corresponds to an energetic average of noise levels over time, penalized (higher value) when the daily exposure of the worker is more than 8 hours and benefited (lower value) when the worker's daily exposure is less than 8 hours. The energetic average of noise levels is usually called *A-weighted Equivalent Continuous Sound Pressure Level* and represented by $L_{Aeq,T}$, so we can write (OJEU, 2003a):

$$L_{EX,8h} = L_{Aeq,T} + 10\log\left(\frac{T}{8h}\right) \quad (1)$$

where T is the time, in hours, of the daily noise exposure of the worker.

In 1986 legislation (OJEC, 1986) the limit of the noise exposure of workers was $L_{EX,8h}$ = 90 dB(A) and in 2003 legislation (OJEU, 2003a), currently in force, the limit is $L_{EX,8h}$ = 87 dB(A).

Over time there has been, naturally, an evolution of the quantitative limits referred, as well as an evolution of sensitivity of the population, in general, and of the specialist in this matter in particular. This development meant that the initial concerns centered on noise exposure of workers in heavy industry it was extending, over time, to another type of activities where common sense did not recognize, usually, the existence of risks associated with noise exposure, e.g., in Schools (EASH, 2013) and Music (Behar, 2006).

One of the major problems concerning music activities is that many musicians and entertainment workers, such as DJs, regard themselves as self-employed people or freelancers, which are not covered by general Health and Safety at Work Directive (OJEU, 1989) as well as noise exposure legal obligations. Recognizing the professional noise exposure of these workers and also the potential occupational hazards, Directive 2003/10/EC (OJEU, 2003a) was established. According to this directive and for the case of the music and entertainment sectors, should be established as an obligation, the development, until 2008, of a *Code of Conduct* for these sectors.

The European Communities developed the *Code of Conduct* referred in the form of a chapter (Chapter 8) in the reference (EC, 2008a), which has a Portuguese version (EC, 2008b). The UK has developed the *Code of Conduct* in the form of a single document more extensive (HSE, 2008).

Other aspect of the development which has take place in this matter relates to the emergence of rules limiting not only the noise exposure of workers but also the noise exposure of the audience of musical performances. For example, the German Standard DIN 15905-5 (GS, 2007) states that at any point accessible to the audience the *A-weighted Equivalent Continuous Sound Pressure Level*, every half hour ($L_{Aeq,30\,min}$), cannot exceed 99 dB(A). This limitation is more permissive, and possibly better adjusted, than the limit of a maximum value, with the time weighting *Fast* ($L_{FAeq,Max}$) of 100 dB(A), established in the *Guidelines for Community Noise* (WHO, 1999). Note that, although this kind of limitation may have problems associated with the individual liberties of the people, there are studies (PINCHE, 2005) that show its necessity, because they show that a good portion of adolescents visitors of discotheques consider that music is usually too loud and uncomfortable and prefer to listen the music not so loud.

2 CHARACTERIZATION CARRIED

2.1 *Summer DJs parties in algarve*

The Algarve (south of Portugal) is a region very popular, especially in summer, for national and international tourists, largely due to having a great number of high quality beaches along its 150 km of coastline. Fun offers, in this region, are many and varied, and lately (some years ago) begun to be fashionable the Summer DJs Parties, where the noise levels are usually high.

As explained in the INTRODUCTION, the music and entertainment sectors have a special consideration of the legislation currently in force in Europe (OJEU, 2003a). For this reason and because there are no known studies about noise exposure in Summer DJs Parties, in Algarve, it was considered appropriate to access the typical noise exposure associated. Have been considered two areas of characterization: one area next to DJs and other area in the Audience (in the middle of the dance floor).

Once the DJs are usually Freelancers, the legal limits (OJEU, 2003a) are not directly applicable, as they applies to Workers, and the definition of Worker in accordance with the Framework Directive 89/391/EEC (OJEC, 1989) is: *any person employed by an employer, including trainees and apprentices but excluding domestic servants.*

Thus, in the absence of legal enforcement for DJs, should be their own the first interested in knowing the noise exposure that they are usually subject and find the best ways to minimize it, since any affectation of they hearing mean, sure, an affectation of they career. Be noted, in this regard, what is established in Council Recommendation 2003/134/EC (OJEU, 2003b).

2.2 *Objective*

The aim of this work is to obtain quantitative data about noise exposures, in DJs area and in Audience, in typical Summer DJs Parties, in Algarve, and analyze and discuss the results and the main problems associated.

2.3 *Equipment and measurements*

Measurements were made near the DJs in 16 Parties (P1-P16) and in the Audience (in the middle of the dance floor) in 9 Parties (P1-P9). The equipment used was a Class 1 Integrator Sound Level Meter. In each Party, measurements were made every half hour, between 00:00 and 06:00 am. Measurement dates were as follows (month-day), all in 2013:

- DJs: P1: 07-28; P2: 07-31; P3: 08-01; P4: 08-02; P5: 08-03; P6: 08-04; P7: 08-06; P8: 08-07; P9: 08-08; P10: 08-09; P11: 08-10; P12: 08-11; P13: 08-12; P14: 08-15; P15: 08-16; P16: 08-17.
- Audience: P1: 08-15; P2: 08-16; P3: 08-18; P4: 08-19; P5: 08-21; P6: 08-22; P7: 08-23; P8: 08-24.

All parties were featured outdoors, so the values obtained are not representatives of Parties in closed spaces. Equipment has been fixed, near the DJ, to the support structure of the mixer (1 meter high and 1 meter away from the DJ's ear), and in the middle of the dance floor fixed to structure (3 meters high).

2.4 *Results*

In Tables 1 and 2 are shown the results obtained, for *A-weighted Equivalent Continuous Sound Pressure Level*, every half hour ($L_{Aeq,30\,min}$), and total (6h) energetic average, in the area of DJs. In Table 3 are shown the results obtained in the Audience (in the middle of the dance floor).

The results higher than 99 dB(A), that exceed what is established in German Standard DIN 15905-5 (GS, 2007), are identified in bold.

2.5 *Daily noise exposure for DJs*

To determine the *Daily Noise Exposure Level* ($L_{EX,8h}$), with special interest for workers (DJs included, despite the particularities associated with freelancers), given the legal limits applicable, it is also necessary to know the typical daily exposure time.

Table 1. A-weighted continuous equivalent level near DJ (Part 1).

	$L_{Aeq,30\,min}$ [dB(A)]							
	Parties							
Half-hours	P1	P2	P3	P4	P5	P6	P7	P8
00:00	75	88	81	75	73	84	**100**	92
00:30	86	91	84	76	84	85	**100**	94
01:00	91	95	90	81	88	87	**101**	97
01:30	94	96	92	85	92	93	**106**	**107**
02:00	95	**101**	93	89	93	96	**104**	**108**
02:30	96	99	93	92	96	97	**105**	**107**
03:00	**101**	**101**	**101**	97	99	97	**103**	**103**
03:30	**101**	99	**104**	**100**	**106**	97	**105**	**106**
04:00	**102**	99	**103**	**100**	**107**	97	**105**	**103**
04:30	**103**	99	**104**	**102**	**106**	98	**104**	**104**
05:00	**101**	97	**103**	**102**	**104**	97	**105**	**103**
05:30	**102**	97	**105**	99	**102**	**100**	**100**	**100**
6h average	99	98	**101**	98	**102**	96	**104**	**104**

Table 2. A-weighted continuous equivalent level near DJ (Part 2).

	$L_{Aeq,30\,min}$ [dB(A)]							
	Parties							
Half-hours	P9	P10	P11	P12	P13	P14	P15	P16
00:00	84	85	85	87	86	95	91	77
00:30	95	89	89	93	91	**100**	94	88
01:00	**102**	89	96	94	91	**100**	99	95
01:30	**106**	90	97	95	93	**103**	**101**	**101**
02:00	**101**	90	99	98	94	**104**	**100**	**104**
02:30	**103**	90	97	97	97	**108**	**111**	**101**
03:00	**102**	92	98	98	97	**111**	**115**	**108**
03:30	**108**	93	98	**100**	94	**113**	**116**	**109**
04:00	**105**	95	97	99	97	**110**	**116**	**110**
04:30	**104**	93	96	98	93	**109**	**116**	**111**
05:00	**104**	94	97	99	57	**110**	**116**	**112**
05:30	**103**	92	96	100	44	**112**	**110**	**112**
6h average	**104**	92	97	98	94	**109**	**113**	**108**

Table 3. A-weighted continuous equivalent level in the middle of the dance floor.

	$L_{Aeq,30\,min}$ [dB(A)]								
	Parties								
Half-hours	P1	P2	P3	P4	P5	P6	P7	P8	P9
00:00	71	84	80	86	96	78	85	81	91
00:30	83	90	89	93	96	83	87	88	94
01:00	88	90	93	93	97	87	90	90	99
01:30	89	90	95	94	96	89	92	91	**101**
02:00	92	89	95	96	98	92	92	92	**100**
02:30	89	88	96	95	97	92	91	93	**111**
03:00	91	91	96	95	**105**	96	95	99	**115**
03:30	92	92	96	94	**103**	92	94	98	**116**
04:00	92	90	97	92	**103**	92	91	98	**116**
04:30	90	90	96	94	**100**	93	92	98	**116**
05:00	89	89	95	93	95	91	92	98	**116**
05:30	93	89	97	93	93	92	91	99	**110**

Once the legislation is historically based on the application to industrial activities, where the noise is typically stable and regular, every working day, it makes more sense to talk about a single value of daily noise exposure. In the case of activities where noise can change from day to day and from season to season (summer, winter), such as DJs, it makes less sense to talk about a single value of daily noise exposure.

Since, even in industries, there may be activities with noise change from day to day, the law (OJEU, 2003a) allows, in these cases, the determination of a weekly average value ($L_{EX,w}$), using the following expression (IOS, 1990):

$$L_{EX,w} = 10\log\left[\frac{1}{5}\sum_{i=1}^{m} 10^{\frac{L_{EX,8h,i}}{10}}\right] \quad (2)$$

where m is the number of working days in the week and $L_{EX,8h,i}$ is de *Daily Noise Exposure Level* in each of these days. This expression is equivalent to:

$$L_{EX,w} = L_{Aeq,T} + 10\log\left(\frac{T}{8\text{h}\times 5}\right) \quad (3)$$

where T is the total of working hours during the week.

Equation (1) and equation (3) assume a logarithmic (base 10) variation of noise exposure as a function of the ratio (T/T_0) between the effective duration of exposure (T, in hours) and the reference duration of exposure (T_0, in hours; for daily average T_0 = 8h and for weekly average $T_0 = 8 \times 5 = 40$h).

The possibility of using an annual average is not provided in the legislation (OJEU, 2003a), although we believe to be appropriate to write the following expression for the calculation of the annual average ($L_{EX,a}$), with greater potential of applicability to cases of greater annual variability of noise exposure, as is the case for DJs:

$$L_{EX,a} = L_{Aeq,T} + 10\log\left(\frac{T}{8\text{h}\times 22\times 11}\right) \quad (4)$$

Table 4. Daily Noise Exposure Level calculated for DJs.

		$L_{EX,8h}$			$L_{EX,a}$ [dB(A)]						
		ap			ap						
Parties	$L_{Aeq,6h}$	100	90	80	70	60	50	40	30	20	10
P10	**92**	**91**	**89**	87	**84**	**81**	77	72	66	58	47
P13	**94**	**93**	**91**	**89**	86	**83**	79	74	68	60	49
P6	**96**	**95**	**93**	**91**	**88**	**85**	**81**	76	70	62	51
P11	**97**	**96**	**94**	**92**	**89**	86	**82**	77	71	63	52
P2,4,12	**98**	**97**	**95**	**93**	**90**	87	83	78	72	64	53
P1	**99**	**98**	**96**	**94**	**91**	**88**	**84**	79	73	65	54
P3	**101**	**100**	98	96	93	90	86	**81**	75	67	56
P5	**102**	**101**	**99**	**97**	**94**	**91**	87	**82**	76	68	57
P7,8,9	**104**	**103**	**101**	**99**	**96**	**93**	**89**	**84**	78	70	59
Av.	**105**	**104**	**102**	**100**	**97**	**94**	**90**	**85**	79	71	60
P16	**108**	**107**	**105**	**103**	**100**	**97**	**93**	**88**	**82**	74	63
P14	**109**	**108**	**106**	**104**	**101**	**98**	**94**	**89**	**83**	75	64
P15	**113**	**112**	**110**	**108**	**105**	**102**	**98**	**93**	87	79	68

ap: Annual percentages of noise exposure. Av.: Average. P2,4,12: P2, P4 and P12; P7,8,9: P7, P8 and P9.

In the above equation are assumed as reference 22 days of work per month and 11 months of work per year, which means T_0 = 1936h.

Table 4 shows the values of $L_{EX,a}$ calculated for the different Parties characterized (see Tables 1 and 2), and the overall energetic average of all Parties, order by increasing L_{Aeq}, for different annual percentages (T/T_0) of noise exposure (it was assumed no exposure to noise in time supplementary to the percentage indicated and a common working week with 5 days):

- 100%: about 48 weeks per year of exposure.
- 90%: about 43 weeks per year of exposure.
- 80%: about 39 weeks per year of exposure.
- 70%: about 34 weeks per year of exposure.
- 60%: about 29 weeks per year of exposure.
- 50%: about 24 weeks per year of exposure.
- 40%: about 19 weeks per year of exposure.
- 30%: about 14 weeks per year of exposure.
- 20%: about 9 weeks per year of exposure.
- 10%: about 5 weeks per year of exposure.

In bold and on dark gray background are values higher than 99 dB(A) (values extremely high), in bold and on light gray background are values greater than 87 dB(A) (*Exposure Limit Value* of legislation (OJEU, 2003a)), in no bold and on dark gray background are values greater than 85 dB(A) (*Upper Exposure Action Value* of legislation (OJEU, 2003a)) and in bold and no background are values greater than 80 dB(A) (*Lower Exposure Action Value* of legislation (OJEU, 2003a)).

Note that the analysis of the values of Table 4 should be done with some care, since it is using an annual average not provided in legislation. For example, the International Standard ISO 9612 (IOS, 2009) indicates that the calculations should be made using the so-called *Nominal Day*, or *Nominal Week* (there are no references to a *Nominal Year*), of noise exposure, and this day/week should correspond to a typical day/week (which in some way, in terms of annual average, may be deducted from the values of Table 4) or for security, the highest day/week of noise exposure. The values in Table 4 must be understood as being indicative (and somehow lower bounds), for cases of non-occurrence of a total noise exposure in a year (1936 hours of noise exposure per year). For security, should be used the values associated with the day/week of greater noise exposure, regardless of higher or lesser percentage of annual exposure. Very probably the most suitable exposure value to consider, in each case (for each DJ), should be between the values of Table 4 and the values of the day/week with maximum exposure, but only audiometric tests, prolonged and directed to DJs, can provide evidence about the most appropriate exchange rate to consider for DJs (see, e.g, the chapter 3.3 of reference (USDHHS, 1998) for a more detailed discussion about the best exchange rate). Note also that in some cases and in some instances the DJs use headphones with very high sound intensity, which can significantly increase the noise exposure.

3 CONCLUSIONS AND RECOMMENDATIONS

The noise exposure is one of the most important risks related to the music and entertainment activities, affecting not only professionals but also the audience itself. In an occupational perspective hazards should not be ignored because they can inflect a variety of health and safety risks to workers, such as, hearing loss and other physiological disorders (increasing blood pressure) (HSE, 2008). It is crucial to develop studies quantifying noise exposure of their activities, providing important information for regulations purposes and reviews.

The results in Table 3 show that, in the Audience, only in 2 Parties (P5, P9) was exceeded the limit value of $L_{Aeq,30\,min} = 99$ dB(A) (GS, 2007). Note that, for the Audience, the measurements were made in the middle of the dance floor. For Audience zones closer to the DJs, or closer to main loudspeakers, the values should tend to the values presented in Tables 1 and 2.

The results in Tables 1 and 2 show that, in the Audience closer to DJ, only in 3 Parties (P10, P11, P13) values were not higher than 99 dB(A).

The results in Table 4 show that, in the DJs area, on all Parties characterized was exceeded the *Exposure Limit Value* ($L_{EX,8h} = 87$ dB(A)) applicable to workers and established by legislation (OJEU, 2003a), and in 8 Parties (P3, P5, P7, P8, P9, P16, P14 and P15) was exceeded the value of 99 dB (A), which is considered very high. Table 4 shows too that, for the average value of all Parties, is required annual percentage of noise exposure less than or equal to 40% (less than or equal to about 19 weeks per year) for $L_{EX,a}$ to be less than 87 dB(A) (*Exposure Limit Value* of legislation).

Taking into account the values obtained and the explained above, it is confirmed that the noise levels occurring in Summer DJs Parties, in Algarve, are indeed very high. This underscores the need to do something, for the protection of users and for the protection of workers, in particular DJs.

We hope that, having regard to the quantification performed by this work, the people involved become more awake to the effective need of some type of intervention, because the values obtained show a high probability of hearing loss for workers exposed. Note that, in conversations with some people involved in this kind of Parties, is a common observation that, most DJs, have already significant hearing loss.

Note also that one of the measures recommended in the *Code of Conduct* ((EC, 2008a & EC, 2008b) is the existence, in performances with amplified music, of a greater number of loudspeakers, distributed around the enclosure, each of them emitting lower noise levels, instead of few loudspeakers, located near the stage, emitting higher noise levels to cover the whole area of the enclosure. In all Parties characterized in this work, the loudspeakers stand on the stage, emitting from there to the whole area. Another common type of measure for these cases, is the possibility, for workers and users, to visualize in real-time and on a screen, or similar, the values of noise levels during the Parties, which also was not observed in any of the Parties characterized. Only was observed characterization of noise levels in order to check compliance with the legal noise limits (environmental law) in the Houses in the neighborhood of the Parties.

In order to promote more safety and healthy workplaces, a number of recommendations should be taken in to account, considering not only the freelancer/self-employed workers, but also the employer or contractor, acoustic and health and safety at work professionals.

The major recommendation relies on the fact that contractor and freelancers must work together to achieve the reduction of noise risks. The contractor has the obligation to promote a safety environment for their workers, through risk assessment and implementation of collective or individual measures. When the risks are high, the contractor should manage a prioritized action plan.

Whenever it is possible, the collective measures, as noise control, equipment specification and position should be the first to implement since they protect a high number of people (workers and audience). On the other way freelancers should be aware of the regulations of level noise exposure and also their health consequences. The hearing protection should be implemented when collective measures are not sufficient to reduce or eliminate the noise risk. Point out that some of the DJs observed used hearing protectors, typically with a noise attenuation of approximately 6 dB, which, according to the values obtained is clearly insufficient, if not implemented further action.

It is strongly recommended that freelancer DJs required Health Surveillance for themselves and regularly hearing checks, that prevents hearing damages identifies early signs of hearing loss. Additionally, noise level exposure assessments are also strongly recommended to these professionals, providing the key information to implement preventing measures.

In all of these matters the acoustic and health and safety professionals should play one of the most important role, making not only the required measuring studies, but also advising about the technical problems.

In addition some information sessions must conducted, for both contractors and workers, explaining the main issues concerning this subject.

ACKNOWLEDGMENTS

The authors thank all the people who contributed to the realization of the measurements, especially to Nuno Lourenço, Rui Lopes and Zélia Ramos.

REFERENCES

Behar, A. & Wong, W, & Kunov, H. 2006. Risk of Heraing Loss in Orchestra Musicians: Review of Literature. Medical Problems of Pergorming Atists, 21: 164–168.

Brockt, G. 2010. Control of noise exposure for employees in the music sector. Proceedings of 20th International Congress on Acoustics (ICA 2010). Sydney, Australia.

European Agency for Safety and Health (EASH) 2013. Occupational safety and health and education: a whole-school approach.

European Communities (EC), 2008a. Non-binding guide to good practice for the application of Directive 2003/10/EC "Noise at work".

European Communities (EC), 2008b. Guia indicativo de boas práticas para a aplicação da directiva 2003/10/CE "ruído no trabalho".

German Standard (GS) 2007. DIN 15905-5. Event-Technology—Sound Engineering—Part 5: Measures To Prevent the Risk of Hearing Loss of the Audience by High Sound Exposure of Electroacoustic Sound Systems.

Healthy and Safety Executive (HSE), 2008. Sound Advice: Control of noise at work in music and entertainment.

International Organization for Standardization (IOS) 1990. ISO 1999. Acoustics: Determination of occupational noise exposure and estimation of noise-induced hearing impairment.

International Organization for Standardization (IOS) 2009. ISO 9612. Acoustics: Determination of occupational noise exposure: Engineering method.

Official Journal of the European Communities (OJEC). 1980. L327. Council Directive 80/1107/EEC, of 27 November, on the protection of workers from the risk related to exposure to chemical, physical and biological agents at work.

Official Journal of the European Communities (OJEC). 1986. L137. Council Directive 86/188/EEC, of 12 May, on the protection of workers from the risk related to exposure to noise at work.

Official Journal of the European Communities (OJEC). 1989. L391. Council Directive 89/391/EEC, of 12 June, on the introduction of measures to encourage improvements in the safety and health of workers at work.

Official Journal of the European Union (OJEU), 2003a. L42. Directive 2003/10/EC, of the European Parliament and of the Council, of 6 February, on the minimum health and safety requirements regarding the exposure of workers to the risks arising from physical agents (noise).

Official Journal of the European Union (OJEU), 2003b. L53. Council recommendation 2003/134/EC, of 18 February, concerning the improvement of the protection of the health and safety at work of self-employed workers.

Policy Interpretation Network on Children's Health and Environment (PINCHE), 2005. Final Report Exposure Workpackage 1. European Work Group.

United States Department of Health and Human Services (USDHHS) 1998. Criteria for a recommended standard: occupational noise exposure: revised criteria 1998.

World Health Organization (WHO) 1999. Guidelines for Community Noise.

Occupational Safety and Hygiene II – Arezes et al. (eds)
© 2014 Taylor & Francis Group, London, ISBN 978-1-138-00144-2

Occupational risks of accident at a work: A case study in footwear industry in Paraíba—Brazil

M.É. Laurindo Sousa, M. Lira, B.J. Farias, S. Brito Lira & M.B. Gama Santos
Federal University of Campina Grande, PB, Brazil

ABSTRACT: This work will focus on the analysis of two sectors of the production department of a company footwear from the city of Campina Grande, Brazil-Paraíba the chafradeira the sewing industry and cutter insoles in cutting industry. The research was made possible through the measurement of the noise, heat and illuminance of work environment soon after it was made comparisons between the sectors and its possible risks. From addition was analyzed the usability of Individual Protection Equipment (IPE) and aspect of comfort, thus ending with the check list and analysis of results.

1 INTRODUCTION

Aspects that are fleeing from human comfort there are in great majority of factories and industries today, these factors, in addition to causing stress, may cause a risk of occupational accidents in more various employees of any existing industry in the company. As an example of these aspects we can cite the noise factor, thermal and lighting.

The objective of this study lies in the identification of risk agents promoters of work accidents and occupational diseases and in the proposition of improvements aimed at control of these risks. Have knowledge about the risks typical of work accidents that have occurred in a footwear company, related to thermal environments, bright and acoustic together with a check-list related to security and usability of EPI.

2 CONCEPTS OF WORK ACCIDENTS

The concept of an accident at work, emphasizing the thermal environments, bright and acoustic, even being widely commented, is different. For the NB18 (Brazilian Standard of Register of Accidents), the accident at work is "an unforeseen occurrence and undesirable, instant or not, related to the exercise of labor, which resulted in, or could result personal injury." (ABNT, 2000).

For Costella, 1999, these accidents can be caused by many reasons, among them: characteristics of the workers themselves (carelessness, inattention, unprepared and inability), hostile environment and dangerous (older machines and dangerous, lack of maintenance and heavy work and unhealthy) and the lack of a prevention program (unconcern with the security on the part of companies and workers).

3 FOOTWEAR INDUSTRY

Over the past four decades, Brazil has played an important role in the history of shoes. The largest country in Latin America is one of the principal who has the manufacturing of leather, which is the third position in the ranking of the largest producers, they still have great importance in the participation of women's shoes that combines quality at competitive prices.

The Paraiba registered one of the largest growths that occurs between January and July 2010, about 50 million pairs of shoes. And that the total value of exports was 15 million pairs, an increase of 10% compared to the same period of 2009. Today, the main destinations are the latin American countries, especially Argentina and Panamá.

Approximately 400 producers of footwear, formal and informal, are installed in Paraiba, 50% in the city of Campina Grande, one of the largest footwear from the northeast. But it is in the metropolitan area of the capital, João Pessoa, that many companies were created for the account of a series of benefits, such as reduction of taxes.

Retailers, in turn, have a greater opportunity for growth, not only by the increase of sales premises, employment and credit for the families, but also by reducing the rush caused by the elevation of the consumption of cars, electronics and home appliances, driven by the reduction of IPI.

4 METHODOLOGY

For the fulfilment of the objective of this research, were used four methodological aspects:

- Bibliographic Review: through consultations in primary and secondary sources, i.e. in books, articles, as well as, consulting the various sites related to item addressed.
- Qualitative Identification of environmental risks: by applying sheets of verification (check list) of safety inspection, in two sectors of the company (cutting and sewing).
- Profile Identification of design regarding the usability of IPE: the instrument of data collection for this methodological aspect was the questionnaire applied during informal interviews.
- Qualitative Aspects: diagnosis acoustic, thermal and luminous of sectors. The data collection was carried out during the days 22, 23 and august 26, 2013, to measure the sound pressure (dB), environment (Â°C) and iluminancia (lux). It was used the instrument of type End-higro-decibel-luxmeter, portable model THDL-1300, having as a manufacturer to Instrutherm Dec-405 sound level meter, to measure the temperature was used the instrument digital anemometer model AN 4870 of ICEL.

5 RESULTS AND DISCUSSIONS

5.1 *Characterization of the company*

The company was created in 1992, and produces handbags, footwear and leather belts. The production is divided in two locations in which produce handbags and shoes.

Account with the collaboration of 52 employees, distributed among the sectors administrative, financial, design and production. And has an average daily production of about 200 shoes and 12 scholarships, but the maximum production revolves around 300 shoes and 15 scholarships.

The work shifts are divided into: morning (Monday to Thursday and Saturday from 7:00 am to 11:00 am) and afternoon (Monday to Thursday from 1:00 pm to 6:00 pm, and on Saturdays from 1:00 pm to 3:00 pm).

In a general way, the sector of production in this company account with 5 sectors: storehouse, cut (where it locates the nibbler sock), sewing (is the chanfradeira), assembly and shipping. Cut (where cut and mold parts and products where staff saw) seam (where the sewing machine and chamfering ir) Assembly (responsible for the assembly and moutin of footwear, as well as the finish or the final product, such as milling) and transport (which is responsible for delivery to the final consumer through retail).

5.2 *Analysis of results*

As For the qualitative identification of environmental risks we obtained the following results:

- Buildings: Through the interviews and direct observation, it was possible to identify unsafe conditions in manufacturing environment, such as the existence of boxes with products that are currently being processed in the corridors of movement, which hinders the passage of workers.
- Facilities and Services in Electricity: The electrical networks are already installed above the jobs and next the bulbs, thus avoiding direct contact of the wires with the worker, as the NR 10—Safety in electrical installations and services.
- Transport, Handling, Storage and Handling of Materials: Because this is a manual process and small material, the production is not fast. In This way there is the creation of inventories in these sectors, not being required to use trans-paleteiras, forklifts or some type of vehicle load or transport, as the NR 17—Ergonomics.
- Machines and Equipment: It was observed that there are few machines (only 2) in sectors here clarified, and that they comply with yes the minimum spacing of 0.60 m to 0.80 m of distance between them, as the NR 12—Safety at work on machines and equipment.

  Fire Protection: in spite of their own departments, there is the presence of the extinguisher of any kind, they are located close to the site and in easy viewing, according to NR 23—Protection against fires.
- Safety Signs: In departments observed there are markings of passage, this being a way of signalling. There are also signs of warning and/or guidance regarding the use of PPE's, as called for in the NR—5 Internal Commission for the Prevention of Accidents.

5.3 *EPI's*

As For the aspects of identification of the design and usability of IPE, analyzing the perception of the respondents regarding the concepts relating to safety at work, it was observed that, at the operational level, it was difficult conceptualizing security. However, the same were related to daily life, situations of insecurity.

The management offers the IPE and puts warnings near the machines alerting the workers about the use of the same, but these do not usually follow the tips. Working without protection and without adequate clothing, being more exposed to accidents at work. This is also why there is not part of the supervisors adequate supervision and rigid on the correct use of PPEs and on the prevention of such accidents from the use of them.

For this reason, it is clear the role of basic education on the use of protective equipment for the prevention of accidents, which may occur through the training and skills listed, not only in the admission of this company. For the aspects of project identification regarding the usability of the IPE, analyzing the perception of respondents about the concepts relating to safety at work, it was observed that, at the operational level, it was difficult to conceptualize security. However, the same related to the everyday situations of insecurity.

The administration provides the warnings IPE and places near workers notice machines on its use, but these do not usually follow the tips. Work without protection and without appropriate clothing, being more exposed to accidents. According to a survey of the regional organization and Employment of Paraiba (SRTE-PB), between 2008 and 2009, the area more dangerous to work the footwear industry. With 25,000 jobs created, the sector was responsible for 26.6% of all accidents in the state. In both years, the PB-SRTE recorded 2,680 accidents, with 714 left in the footwear industry. The boilers are usually cause burns, cuts and hands even crushing, which in some cases, leads to the death or disability of the employee.

During the implementation of the operating cycle of the chamfer the noise level maximum was of 83.2 dB, the value below allowed by Brazilian legislation 2836 by NR 15 Annex 1, which is 85 dBA in a journey of seven hours of daily. The noise level equivalent was influenced by rotation. Thus, he offers short-term risk for the employee from this station. However, it should be noted the necessity of the use of protective equipment, and, thus, prevent future damage of the auditory system of the employee, with the realization of constant monitoring.

In the field of inner soles maximum level was 96.8 dB, exceeding the maximum allowed.

This value observed, it is considered critical, demanding an immediate action, either by isolating the source or with the use of proper IPE. Thus, the use of this type of equipment does not cause more serious problems of the auditory system of the worker, which may lead to loss of hearing temporary or permanent. Thus, there is also a need for constant monitoring.

The thermal conditions, we Obtained.

It can be observed that the maximum working temperature was 34°C and the minimum of 29.3°C, with an average of 26.8°C in sector of chanfraria. In the station of cut in strips insoles, the maximum temperature was 33.2°C and 28.2°C more low, thus obtaining an average of 30.8 degrees C. In both sectors, it was found that they are above the default authorized, found in NR 17, which says that it should be between 20 and 23°C.

Thus, it is necessary to consider the implementation of a suitable ventilation system to the location, by forming the analysis of constant temperature, so that you can control it. And, if possible, perform the insulation from the heat source.

About lighting have.

As regards the aspects of illuminance in the field of work, it was observed that the climatic values recommended for the tour operator, when the work is carried out by the hands and with the operator seated, allow the variation in air temperature between 19 to 22°C, with relative humidity between 40 to 70% and air velocity of 0.10 m s^{-1}. When the operation requires that the operator work in, the temperature must be from 2 to 4°C lower and the air velocity above 0.10 m s^{-1}, in relation to the completion of the operation sitting (Delgado, 1991).

As can be seen from the data of Table 3, it has to draw up a new lighting system in such jobs, considering the factors to have an adequate lighting (of the type and amount of light/lamp, distribution, maintenance and optimal color for each sector). Since both are activities that stretch the vision and require attention, there is a need for a better lighting, avoiding problems of vision and some accidents, such as cuts and amputation of fingers/hands.

The noise level on the first day of measurements in chanfradeira obtained an average of 78.74, on the basis of day average was 77.6 and in the last day of medition was of 79.94. The topper sock had rates greater than noise, given that on the first day the mean was 90.72, already on the second day was 91.52 and the third day was an average of 92.94.

In relation to the thermal conditions it was observed that the sector of the chamfer showed an

Table 1. Data on noise.

Setors	Maximum value	Minumum value	Average	Standardized	Norm
Chafering	83,2 dB	71,2 dB	71,9 dB	85 dB	NR 15
Cutter Insole	96,8 dB	82 dB	88,1 dB		

Table 2. Data on temperature.

Sectors	Maximum value	Minumum value	Average	Standardized	Norm
Chamfering	34,0°C	29,3°C	26,8°C	20°C a 23°C	NR 17
Cutter insole	33,2°C	28,2°C	30,8°C		

Table 3. Data on lighting.

Sectors	Maximum value	Minumum value	Average	Standardized	Norm
Chamfering	132 lux	97 lux	104 lux	1000 lux	NBR 5413
Cutter insole	98 lux	84 lux	79 lux	500 lux	

average of 31.4°C in first, 32.44°C on the second day and 32.3°C in third-ro. Already on the mower to sock the first day was an average of 30.9°C, the second day 31,04°C and the third 31.1°C. It can be observed that the sector of chamfer showed a higher temperature than the industry of cutting insole.

6 CONCLUSION

From these analyzes it can be said that in the noise factor job cutting of sock got worse number, already in relation to the temperature, it was seen that the two jobs are analyzed above allowed and recommended, and in relation to the lighting, the two sectors are outside the recommended.

There are still aspects related to the security of the environment that need to be improved, and in a general way it may be said that the workers of the two jobs are harmed, because didn't work sitting, the other has difficulties in getting by there are boxes on the floor, and the two work with noise levels, temperatures and seasonal illuminations like planets. When you deal with the usability of the IPE, it was seen that no uses all the recommended equipment.

REFERENCES

Costella, M.F. Analysis of Occupational Accidents and Diseases Occurring in Construction Activity in Rio Grande do Sul in 1996 and 1997. Porto Alegre - UFRGS. Master's Dissertation—Graduate Course in Civil Engineering, Federal University of Rio Grande do Sul 149p, 1999.

Delgado, L.M. El agricultural tractor features utilización y. Madrid: Laboreo Solotractor, 1991. 235p.

Goncalves, Edwar Abreu. Safety and health at work in 1200 (one thousand two hundred) questions and answers. Ltr Sao Paulo, 1996.

Iida, I. Ergonomics: Design and production. Sao Paulo: Edgard Blucher, 1990. 465p.

Iida, I.; Wierzzbicki, H.A.J. Ergonomics. Lecture notes. Sao Paulo: EPUSP, 1978. 282p.

MICHEL, Osvaldo. Accidents at work and occupational diseases. Sao Paulo: LTR publisher, 2000. Available in: Accessed: 09/09/2013 9:00 ace; Available at: http://portal.mte.gov.br/data/files/8A7C816A2E7311D1012FE5B50DCD522C/nr_08_atualizada_2011.pdf. Accessed: 09/05/2013 16:00 ace; Available at: http://portal.mte.gov.br/data/files/8A7C812D311909DC0131678641482340/nr_05.pdf; Accessed: 09/09/2013 to 14:00 pm.

Occupational Safety and Hygiene II – Arezes et al. (eds)
 ISBN 978-1-138-00144-2

Identification and proposition of preventive measures for environmental risks on a t-shirts factory in Campina Grande, Brazil

T.F. Chaves, M.B.G. Santos & L.F. Monteiro
Federal University of Campina Grande, Campina Grande, Paraíba, Brazil

ABSTRACT: This article intended to identify risks relating to thermal environments, lights and acoustic present in two sectors of a t-shirts factory in the city of Campina Grande, proposing suggestions for control. The methodology used was developed from direct observation and questionnaires and checklists applied with officials from the company's manufacturing sector. The data collection aimed to verify the working conditions in relation to physical condition and occupational safety, by legislation that is regulated through Regulatory Standards (RS/1978). From the results, it was found that some problems regarding the failure to specific legislation, such as the failure to use Personal Protective Equipment (PPE), Collective Protection Equipment (CPE) and lack of safety signs.

1 INTRODUCTION

Facing a currently a worrying scenario for statistics of occupational accidents, stating that in Brazil in the year 2012 were recorded 711.164 accidents at work, data that puts Brazil in fourth place in the ranking of the International Labour Organisation which lists the places with higher cases of occupational accidents. The number of workers who die on the job is also very high in Brazil, data from the Federal Government of Brazil show that in 2011, 2.884 deaths were recorded.

In such a context, there are several studies that address occupational health and safety of work, wherein by methods appropriate preventive, studying the causes of accidents, allowing the adoption of technical measures that aim to control or reduce the occurrence of workplace accidents.

In this sense, it can be affirmed that the safety of the work is a science of fundamental importance in the management of a company or factory, having the safety of the work the best way to preserve the physical and mental integrity of the worker. Once, it aims to identify the causative agents of these problems, quantify its extent and to take measures for its reduction or extinction.

Thus were created in Brazil by the Ministry of Labor and Employment (MLE) in 1988, the Regulatory Standards (RS), which aims to guide and determine how the shares should be made to prevent environmental risks physical. Among these standards, we can highlight the Regulatory Standards (RS)-9, which addresses the Program Environmental Risk Prevention—PRSP, aiming to preserve the health and physical integrity of workers, through the anticipation, recognition, evaluation and control of the occurrence environmental risks exist or those that may exist in the workplace.

Another standard widely used in industry is the Regulatory Standards (RS)-17/1978 that emphasizes ergonomics aspects in the workplace, analyzing the adaptation of the workplace to man, can improve various activities on behalf of the worker. Also noteworthy as other allied standards peculiar to the protection of the worker and the Regulatory Standards (RS)-6/1978 RS-15/1978. The Regulatory Standards (RS)-6/1978 defines protection and responsibilities of the use of PPE. Thereby, the use of these, stands out for its efficiency and effectiveness. Already Regulatory Standards (RS)-15, defines the unhealthy activities, the tolerance limits and the criteria for their evaluation.

With this in mind, this research aims to identify the risks relating to thermal environments, lights and noise present in two sectors of a t-shirt factory in the Brazil, proposing suggestions for control.

2 ENVIRONMENTAL RISKS

Environmental risks are considered the agents physical, chemical, biological, ergonomic and mechanical accidents that may bring or cause harm to workers' health in the workplace, due to its nature, concentration, intensity and time of exposure.

In this context, it is known that every worker is exposed to risks in your workplace. Thus, it is necessary the study, identification and prevention of these risks, seeking soften or eliminate accidents and occupational diseases of production systems. Since it, fatalities as the removal and loss of workers directly reflect productivity and profitability of the company.

2.1 *Physical hazards*

The physical agents of hazards can be defined as the various types of energy to which the worker is exposed while performing their activities. Among them, we highlight mainly the noise, extreme temperatures and inadequate lighting.

2.2 *Noise*

Noise is one of the main physical agents present in the workplace, in various types of installations or professional activities. For their enormous occurrence and since the health effects of exposed individuals are considerable, is a major focus of attention of hygienists and professionals facing the health and safety of workers, according to the Manual of Assessment Techniques SESI Environmental Agents.

It is known that noise can be classified into three types: continuous, intermittent and impact. The first mentioned, which is one level of 3 dB sound pressure varies during a long period (over 15 minutes) of observation. Already intermittent noise stands out for being one whose sound pressure level varies up to 3 dB for short periods (less than 15 minutes and 0,2 seconds). Regarding the noise impact, it has high intensity and short duration, lasting less than 1 second.

It is also important to point out that the excess exposure to a particular type of noise can lead the employee to submit nervous fatigue, high blood pressure, changes in heart rate and breathing, and irritability.

2.3 *Heat*

The heat can be defined as the energy in transit across the boundary of a system provided there is a temperature difference between the system and environment. Since the mains mechanisms of heat transfer are: conduction, convection and irradiation.

A workplace that presents high temperatures can cause many problems to the developer, among them are: sunlight, heat prostration, heat cramps, cataracts, dehydration and skin rashes and stress. The Regulatory Standards (RS)-17 indicates that the range of comfort in a workplace that is expected in the range of 20 ° C and 23 ° C.

2.4 *Lighting*

Assessments lighting aims to quantify the luminance in workplaces, seeking its subsequent comparison with the minimum values established by Brazilian legislation, as well as provide general recommendations to achieve the adequacy of lighting conditions the activities developed there.

In this sense, the evaluation of the lighting in the workplace is based on measurement of luminance level, which according to NBR 5413 is defined as the limit of the ratio of the luminous flux received by the surface around a point considered to the surface area when this goes to zero.

In measuring the luminance should make use of the procedures established by the NBR 5382, because it presents the way in which they must scan the internal illumination light that is originated from the general environment. To establish minimum luminance suited to environments we use the NBR 5413, which will have varying values taking into account the age of the worker, speed and precision and reflectance of the background task.

Therefore, it is understood as an enlightenment "aggressive agent" of vision trigger certain illnesses. But still, when it is inadequate, and, in most cases, the inadequacy refers to the deficiency of enlightenment, we can notice some consequences, such as: greater eyestrain and generally higher risk of accident, lower productivity/quality, environment psychologically negative.

3 METHODOLOGY

3.1 *Description of company*

The company under study is located in Campina Grande—Brazil and has operating in the market for women's clothing, its production is geared to the demands of the national market. The company is part of the textile sector and its cool form is a limited partnership formed by two partners.

The businesses has a production volume with 400 t-shirts for monthly and feature a staff of 10 employees, divided between the sectors of design, pattern maker, sewing and screen printing. Where, they all work in the routine Commonly 8 hours per day divided into two shifts (morning and afternoon) and including rest breaks.

3.2 *Methodological procedures*

For the development of this paper was were performed: literature review, through aspects of the legislation, books and article for the theoretical foundation of the same; case studies in a company in the field of women's clothing in Campina Grande—Brazil, in which stood the risks present in two sectors (sewing and management office); application of check-list and questionnaires to identify usability of PPE by employees of the sewing industry and the entrepreneurs and the measurements of noise levels, luminance and temperature to verify compliance.

3.3 *Instrumentation*

The data collection was carried out during the days 29th June 2013 to 05th July 2013, with 30-minute intervals throughout the work shift, for the variables noise, temperature and luminance.

For measurement of noise, in dB, and the luminance in lux, the instrument used was the type Thermo-hygro-decibel-portable light meter, model THDL—400, the producer Instrutherm. For the measurement of the temperature in °C, the instrument was used digital anemometer model AN 4870 Manufacturer ICEL.

Through the machine settings, it is known that the accuracy for noise is ± 3.5 dB, to the temperature variable is ± 0.6 ° C and the luminance accuracy is ± 5.0 lux.

3.4 *Description of work environment*

The research was conducted in two areas of the company. The first sector is responsible for sewing and the second sector by the management of the organization.

The first to be featured is the sewing sector, which is located in a covered area of 12 m^2, where only two people working. The sewing sector it is a masonry construction in good condition and physical structure and is composed of dark walls (illumination with few entries), and it is considered a hot workplace by workers and has a dark floor (darkening the workplace, further hindering performance activities).

The workplace does not have windows in their physical arrangement, having as light input just a few door and glass tiles and there is presence of artificial lighting (lamps, luminaries, and flashlight). These observations reflect in a bad light and ventilation.

The management sector, is an area of 9 m^2 covered area, where only two employees working. This has a physical structure in good condition, walls and floors clear. Have inputs of light and air and the presence of artificial lighting, as lamps and fixtures. So basically we can say that its structure provides a good worker comfort.

4 RESULTS AND DISCUSSION

To assess levels maximum, average and minimum of noise, temperature and luminance in the two sectors of the company under study, we used data from measurements that led to the values shown in Tables 1 and 2.

4.1 *Noise*

As for the noise, it was observed that the maximum value found in the sewing industry was 78 dB and the management sector was 66 dB. Regarding the minimum noise found in the sewing sector we find the value of 70 dB and the management sector was 53 dB. Given these values, the amplitude of the sectors are respectively 8 dB at the sewing sector and 13 dB in the management sector. According to these data, it is noticed that the noise levels of the two sectors are within the tolerance limit for an 8 hour per day of continuous work, according to RS-15, since, the level of noise with the maximum allowable load of 8 hours daily is 85 dB at a frequency of continuous exposure.

4.2 *Heat*

With respect to temperature, it was observed that the higher values were found in the sewing sector, which was 24,5°C. In the management sector the highest value observed was 21,3°C. Already the lowest values of heat found in sewing sector was

Table 1. Quantification of risk agents—sewing sector.

Risk agent	Maximum value	Minimum value	Average value	Standard
Noise	78 dB	70 dB	74 dB	85 dB
Temperature	24,5°C	22,8°C	23,7°C	20°C–23°C
Lighting	80,6 lux	247,3 lux	167,8 lux	500 lux

Table 2. Quantification of risk agents—management sector.

Risk agent	Maximum value	Minimum value	Average value	Standard
Noise	66 dB	53 dB	60 dB	85 dB
Temperature	21,3°C	19,8°C	20,3°C	20°C–23°C
Lighting	72,4 lux	165,7 lux	128,5 lux	500 lux

22,8°C and the management sector was 19,8°C. Given these values, it can be seen that the amplitude values are considerably lower, with 1,7°C in the sewing sector and 1,5°C in the management sector. The requirements of Regulatory Standards (RS)-17, which regulates the range of thermal comfort in a workplace as being between 20 and 23°C, it appears that the management sector have temperatures in compliance. However, the sewing sector is out of compliance of thermal comfort, since it presents temperature above 23°C.

4.3 *Lighting*

As for lighting, the equipment used for measuring this variable was the light meter. And NBR 5382 was used to define the average luminance. With the number of rows of luminaires with M = 2 and the number of luminaires per row and N = 3. For the sewing sector, the calculated average luminance was 137,80 lux for measurements made at 08:00 am.

However, as there is a strong influence of sunlight on the desktop, the measurements were repeated at 11:00, with average luminance of 162,75 lux. For the management sector, with M = 2 and N = 6, the average level of luminance for 08:00 am was 90,67 lux and at 11:00 am was 98,15 lux.

In this sense, it is known that according to Regulatory Standards (RS)-15, all workplaces must have adequate lighting, whether artificial or natural. According to the Brazilian Association of Technical Standards (ABNT) NBR 5413 assumes the values of minimum luminance in workplaces. Realizes the need to increase the level of luminance, in view of the possible consequences that workers may have, such as eye diseases and visual fatigue.

4.4 *Profile design workers when the use of PPE*

In both sectors studied in company, 100% of employees say the company never provided the PPE and they also had no knowledge or awareness of the importance of use protective equipment, and further confirmed and emphasized that it was never used in any protection tool in the production process of the company. So before this undesirable outcome, it was not possible to make a correct analysis of the usability and comfort of PPE in the company. However, once informed about what it was and the importance of use of PPE, 80% of employees have recognized the importance of using them and said they believe in the efficiency of PPE. And in this scenario, the entrepreneurs have pledged to provide employees free PPE appropriate to the risk, in perfect condition and working order. In addition to requiring the use thereafter, guide and train workers on the proper use and perform periodic maintenance and cleaning thereof.

5 CONCLUSION

The only variable where both sectors were disabled were the luminance levels, however, the sewing sector still showed lower levels. Therefore, it is suggested to be supplied the deficiency in the lighting industry continues to maintain the lamps and the addition of a greater number of fixtures in the sector so that it can achieve the proper levels for good lighting in the type of environment studied.

Therefore, some control measures can be adopted to prevent these problems, such as the reduction of the work activity, blowing fresh air or exhaust gases hot, reflective barriers and reduction of moisture; regarding staff, acclimatization, periodic medical examinations, and provision of water and salt and caster staff. Furthermore, it also indicates the use of PPE in the production system of the company. Aiming to minimize or eliminate the risk of accidents at work present in sectors studied.

In all, the objective was achieved, as the results were satisfactory and one can see the importance and necessity of study of such variables. And also the aim is for the company to deploy the measures suggested to be improved working conditions of both sectors studied, thus increasing performance and decreasing the production occupational problems.

Finally, it is important to clarify that the purpose of this undeniable work concerns the prevention of occupational health. The important thing is to use our simple knowledge to propose some kind of improvement to the worker. Without forgetting that the proposed improvements will benefit not only the employee but also the employer and the entire organization.

REFERENCES

Almeida, L.A; Santos, M.B.G; Montei-Ro, L.F. 5RD SEPRONe—2010. Diagnosis of Hy-giene, Health and Safety in a Microenterprise Sector Footwear. Maceió, AL, Brasil.

Brazilian association of technical standards. Nbr 5382-Verification of interior luminance. Rio de Ja-neiro: Abnt, 1978.

Coelho, F. SESI/DN, 2007. Technical asses-sment of environmental agents: Manual SESI. Brasí-lia, DF. Brazil.

Goncalves, E.dwar A. breu. 2nd ed. LTr, 1995. General Introductory. Notes technical-legal safety and occupational medicine. 2nd ed. Sao Paulo, SP, Brazil. LTr, 1995.

Ministry of labour and employment. Regulatory Standards.

Saliba, T.M. 3rd ed. LTr 2010. Basic course in safety and occupational hygiene. São Paulo, SP, Brazil.

Szabó Júnior, A.M. 4th. ed. current. Manual of Hygiene and Occupational Medicine.

Occupational Safety and Hygiene II – Arezes et al. (eds)
© 2014 Taylor & Francis Group, London, ISBN 978-1-138-00144-2

Noise pollution in a shopping center of Konya (Turkey)

A. Alaş
Department of Biology, A.K. Education Faculty, Necmettin Erbakan University, Meram, Konya, Turkey

Ç. Çiftçi Ulusoy
Department of City and Regional Planning, Necmettin Erbakan University, Meram, Konya, Turkey

M. Shahriari
Department of Industrial Engineering, Engineering-Architecture Faculty, Necmettin Erbakan University, Meram, Konya, Turkey

Ş.N. Alaş
Department of Statistics, Sciences Faculty, Selçuk University, Selçuklu, Konya, Turkey

ABSTRACT: Konya is one of the 16 metropolis of Turkey. Population of Konya is more than one millions. Survey area is one of the big shopping centers in the city. Elevated shopping center noise levels potentially affect communication, stress levels, aggressive behavior in customers and working people during the day. The purpose of this research study was to characterize the environmental noise inside a shopping center in Konya (Turkey). Data was collected in decibels (dB) at six locations in the shopping center over 3 different time periods on Sunday and Wednesday as crowded and normal work days respectively. Recorded data was analyzed and compared across locations and over time, in the light of World Health Organization (WHO) and Turkish noise standards. The 16.0 version of SPSS software were used for analyses of factorial variance. The results showed that the average sound levels varied between 62.25 (±3.56) and 74.36 (±3.53) dB on Wednesday and Sunday respectively. The highest rate of noise concerns with the last hour of the work day.

1 INTRODUCTION

Noise is generally defined as unwanted sound. In daily usage, noise refers to an unwanted annoying auditory input. Noise pollution today is one of the main forms of urban environmental pollution and is responsible for negative impacts that are harmful to the environment and the quality of life of the population (WHO, 2003).

High sound levels not only affect the verbal quality of communication but also contribute to serious problems in the intellectual development of students, such as impaired learning, writing and speaking difficulties, limitations in reading comprehension and development of vocabulary (Berglund, Lindvall & Schwela,1990).

Noise in its various forms has become part of the modern life. A large number of citizens in modern environments are daily exposed to background noise. Noise is not only subjectively annoying but it can also have an adverse effect on hearing and health in general. However, in the light of recent evidence, even such levels of noise that do not cause damage to the peripheral hearing can have a persistent effect on speech brain function, perception, and attention control, affecting the individual's behavior (Kujala & Brattico 2009).

According to researchers, the disturbing presence of noise in modern society is a well-known form of environmental pollution.

Elevated shopping center noise levels potentially affect communication, stress levels, aggressive behavior in customers and working people during the day. The purpose of this research study was to characterize the environmental noise inside a shopping center in Konya (Turkey).

1.1 *Study area*

Old and modern city Konya, with a population of about 1.200 million, is the center of Turkey. It is also one of the 16 metropolis of Turkey.

Study area is one of the big shopping centers in the city. Inside of this center, there are amusement center, cinemas, fast-food restaurants, various shops, hypermarket, electronic markets etc.

A lot of visitors visit this shopping center for shopping and to have a great time different times of a week.

1.2 *Collecting data and analyses method of data*

Data was collected in decibels (dB) by using sound level meter at six locations in the shopping center over 3 different time periods on Sunday and Wednesday as crowded and normal work days respectively. Recorded data was analyzed and compared across locations and over time, in the light of World Health Organization (WHO) and Turkish noise standards. The 16.0 version of SPSS software were used for analyses of factorial variance.

According to the standards given by World Health Organization the noise level should not be more than 70 dB in shopping centers and the similar places (WHO, 1995).

2 RESULTS AND DISCUSSION

Figure 1 shows the sources and effects of noises in the shopping center.

Sound was measured in the area of the study two days a week; one ordinary day (Wednesday) and one day during the weekend (Sunday) which is more crowded.

Sound level in different stations and different hours in the study area were showed in Figure 2.

The results showed that the average sound levels varied between 62.25 (±3.56) and 74.36 (±3.53) dB on Wednesday and Sunday respectively. There were significant differences between stations and hours (<0.01).

As can be seen in Figure 2, the noise level is different with regards to the stations and the hours as well.

According to the results, sound level is the highest level at 19.00 p.m. This is after working time.

Obviously, excessive noise can affect people health in different ways. In case the exposure to

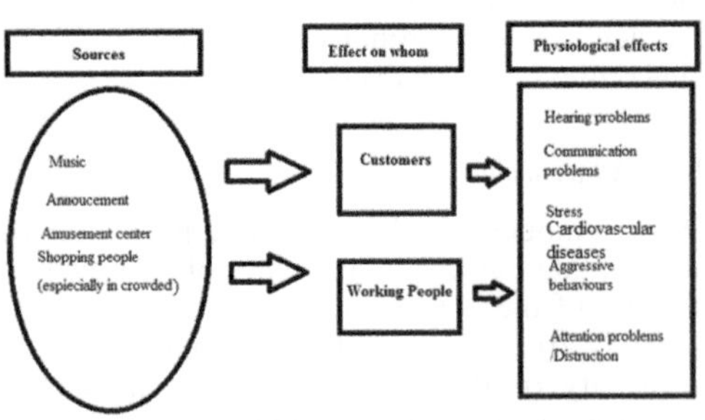

Figure 1. Sources and effects of the noises in the shopping center, Konya-Turkey.

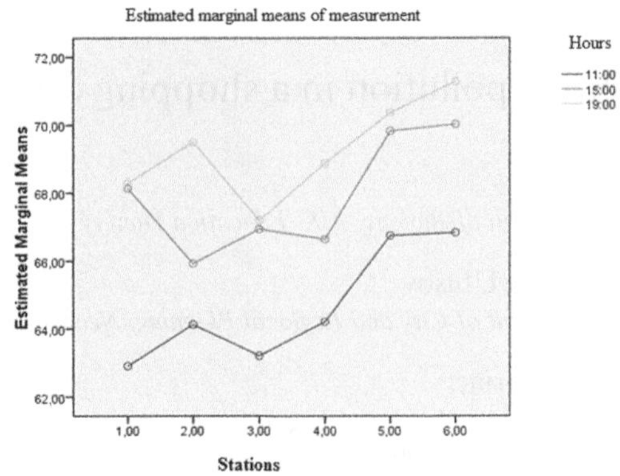

Figure 2. Sound level in different stations and different hours.

that noise is frequent enough and if the noise levels are high enough, it could reduce hearing ability, could have cardiovascular effects, could cause stress (depending on quality of the sound). Noise also interferes with communication on the people who are exposed to noise in their workplace and has effect on their performance and their productivity. Loud noise can prevent people both from communicating with one another. Not only is the noise level important, equally so is the duration of time that the people are exposed to the noise. Generally, continued exposure to noise above 80 dB(A) will result in some hearing loss over the time (Edworthy, 1997).

Babisch, (2002), has proposed a noise effects reaction scheme as shown in Figure 3 (WHO, 2009).

According to epidemiological studies, sound/noise is a psychosocial stressor that activates the sympathetic and endocrine system. Acute noise effects do not only occur at high sound levels in occupational settings, but also at relatively low environmental sound levels when, more importantly, intended activities such as concentration, relaxation or sleep are disturbed (WHO, 2009).

Results of this study show that in the light of literature knowledge, health of working people under high noise condition in this shopping center have been faced at risks.

In contrast to many other environmental problems, noise pollution continues to grow and it is accompanied by an increasing number of complaints from people exposed to the noise. The growth in noise pollution is unsustainable because it involves direct, as well as cumulative, adverse health effects (WHO, 1999).

It also adversely affects future generations, and has socio-cultural, esthetic and economic effects.

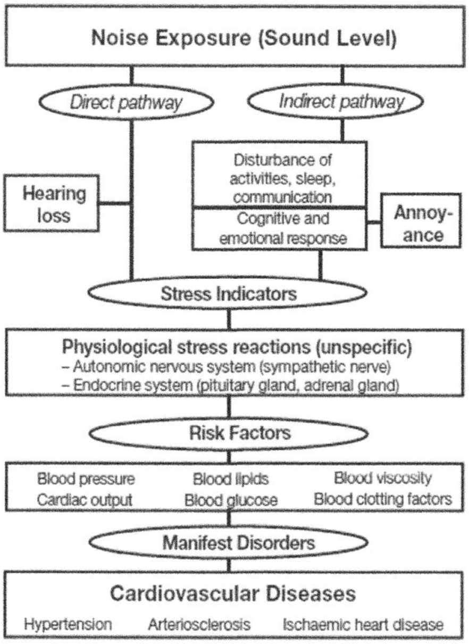

Figure 3. Noise effects reaction scheme (WHO, 2009).

Noise pollution affects both health and behavior. Unwanted sound (noise) can damage psychological health. Noise pollution can cause annoyance and aggression, hypertension, high stress levels, hearing loss, sleep disturbances, and other harmful effects (Rosen & Olin, 1965; Field, 1993).

Noise pollution also is a cause of annoyance. On the other hands, there is fairly consistent evidence that loud noise also increases aggressive behavior in individuals predisposed to aggressiveness (WHO, 1995).

According to our results in this research, we can say that people may be negatively affected from high level sounds during visiting and shopping. This is supported by Rosen & Olin, (1965); Field, (1993); Edworthy, (1997); Kujala & Brattico, (2009) and WHO, (2009).

Continual noise from various sources may not necessarily directly affect our hearing or our physical health, but it may well affect the quality of life and our general well-being. For this reason, some noise preventive measures can be applied in this shopping center.

1. Music sounds must be cut down inside of this center
2. More acoustic materials can be used inside of this shopping center
3. Cautionary signals must be used for people at the high level sound places
4. People must prefer normal work days than crowded days for visiting and shopping
5. Working people informed and educated about hazardous of high level sounds
6. Working interval must be rearranged for workers to protect their health
7. Workers must be controlled from medical examination time to time
8. Periodically, sound level measurements in shopping centers can be done by Ministry of Environment and City Planning.

3 CONCLUSIONS

A study carried out in Konya, Turkey to investigate the noise level in a shopping center. The results showed that the noise level in the shopping center is higher than expected and the standards. This could create a serious health problem specially on the people who work there and also those who spend their time in the shopping center. Different ways could be proposed to minimize the noise level in the area under the study informing people about the problem, improving the structure of shopping center by using acoustic materials, regular control of the sound and the health of the people who work in this shopping center. The study should be extended to other shopping centers and the places where are crowded.

REFERENCES

Berglund, B., Lindvall, T., & Schwela, D.H. (Eds.). 1990. *Guidelines of community noise*. Stokholm University, Karolinska Institute, Stockholm.

Edworthy, J. 1997. Noise and its effects on people: An overview. *Intern. J. Environmental Studies,* (51) 335–344.

Field, J.M. 1993. Effect of personal and situational variables upon noise annoyance in residential areas, *Journal of the Acoustical Society of America*, 93: 2753–2763.

Kujala, T. & Brattico, E. 2009. Detrimental noise effects on brain's speech functions. *Biological Psychology*, (81) 135–143.

Rosen, S. & P. Olin, P. 1965. Hearing Loss and Coronary Heart Disease, *Archives of Otolaryngology*, 82:236.

WHO. 1995. *Guidelines for community noise*. Edited by Birgitta Berglund Thomas Lindvall Dietrich H Schwela. Stockholm University and Karolinska Institute. 21pp.

WHO. 2003. World Health Organization. *Résumé D'orientation Des Directives De I'oms Relatives Au Bruit Dans I'environmental.*

WHO. 2009. *Night noise guidelines for europe*. WHO Regional Office for Europe Scherfigsvej 8. DK-2100 Copenhagen Ø, Denmark. 184pp.

Occupational Safety and Hygiene II – Arezes et al. (eds)

Contributions to the design of Emergency Management Intelligent Systems

M.J. Simões-Marques
CINAV—Portuguese Navy, Lisbon, Portugal

I.L. Nunes
Faculdade de Ciências e Tecnologia, Campus de Caparica, Portugal
UNIDEMI—Unidade de Investigação e Desenvolvimento em Engenharia Mecânica e Industrial, Portugal

ABSTRACT: Emergency Management in the context of Disaster Relief operations places problems that are at the cross roads of many knowledge domains and involve many actors. Emergency Management is a complex process which calls for decision support tools. This paper discusses the characteristics Emergency Management Intelligent Systems should offer, and presents a fuzzy multicriteria methodology which can be used in the development of such systems.

1 INTRODUCTION

Emergency Management (EM) goal is to reduce vulnerability to hazards and respond to disasters. EM is a complex process that is at the cross roads of many knowledge domains and demands the coordination of a large number of actors with different cultures, aims and views of the world that are asked to combine efforts, and who are faced with challenges of multidisciplinary nature. Naturally the development of decision support systems to assist emergency managers in this context is also a big challenge. EM is a core element of Disaster Relief operations, which aims to provide efficient and effective response to multiple and often conflicting needs in situations of scarce resources, considering several complementary functional elements, such as Health, Supply, Transportation or Construction. In all these elements decision-making relates to the basic questions What, Where, When, Who, Why, How, How Much. These questions become particularly difficult to answer in critical situations, where the timeliness and complexity of the decision-making process is especially sensitive, and resources are usually very limited (Simões-Marques, 2005).

Briefly, the main goals of an Emergency Management Intelligent System (EMIS) are: providing EM decision support; ensuring a reliable and flexible network, including mobile components, to support information sharing; serving the needs of the various actors of the decision process, including compilation of information to produce a standardized, integrated and consolidated picture on the status of incidents and on the usage of resources for a given area of interest; and providing dynamic advice to decision-makers regarding courses of action. Standardizing, integrating and consolidating information requires eliciting knowledge in problem domain (e.g., ontologies to characterize the types of incidents and the resources useful to the potential areas of emergency response (Galton and Worboys, 2011)). These activities are usually referred as Knowledge Engineering (Turban et al., 2010). An effective EM requires the adoption of robust criteria for evaluating the degree of incident response urgency and the adequacy of allocating a given resource to that incident. Examples of factors to consider in assessing the suitability of the resources to assign to a specific incident are skills/capabilities, vicinity, and availability. An EMIS should also provide advice on courses of actions, which must adapt dynamically to context evolution. Obviously the adoption of the recommended actions is not mandatory, and the emergency manager has to validate system's recommendations. Nevertheless, the use of common robust advice tools in a distributed/collaborative environment ensures predictability and coherence of the parallel and concurrent decision-making processes, which facilitates responders' unity of effort.

Since an EMIS must be scalable and flexible, the area covered by the system must be adapted to the needs of decision-makers. In an organization based on a hierarchy of emergency operation centers some workstations must be dedicated to strategic level decision-making (i.e., broad scope and large data granularity), and others to tactical level decision-making (focused on restricted areas of intervention). Naturally, the larger the area covered the greater the complexity of the integrated information and lesser the degree of detail that is possible to apprehend. Thus, a general coordination center is concerned with the overall picture, analyzing where the "hot spots" are and making a macro management of resources (e.g., moving available means to places where they are scarce). On other hand, a local decision-maker is concerned with individual incidents, in real time, and making discrete allocations of the resources. Therefore, an EMIS should support transparently different levels of decision-making; however, the amount and the way data is presented to each user profile should be adjusted to the specific needs. To collect, integrate and share EM information it is necessary a network and terminals with features suited to the roles of managers and responders. The main coordination centers may meet their functional requirements in fixed installations; however, local command posts and responders on the ground need to have portable terminals which ensure greater mobility (e.g., portable computers, tablets or smartphones).

The next sections introduce the main features of the SINGRAR model (Simões-Marques, 1999) which was used in the implementation of a distributed EMIS expert system installed on board of ships and that, in the last decade, received continuous improvement. It proceeds discussing the application of this model to the context of Disaster Relief operations, and illustrates the use of the underlying methodology in an EM scenario.

2 SINGRAR MODEL

2.1 *Overview*

Despite experienced professionals can discard computer assisted decision support for many daily tasks, these systems are a major asset and provide competitive advantage for training and simulation, and specially for real life EM on complex, stressful and lasting crisis situations, where humans tend to fail their judgement (a issue commonly addressed as bounded rationality (Simon, 1955)). Complex Disaster Relief scenarios like man-made disasters, natural disasters or humanitarian assistance operations may be given as examples where the need for support becomes evident.

In fact, SINGRAR is a methodology developed for implementing distributed EM expert systems. SINGRAR is the Portuguese acronym for Integrated System for Priority Management and Resource Assignment. The model, developed by the first author, was initially implemented in an EMIS for the Portuguese Navy to assist EM onboard warships.

The SINGRAR model was designed to be flexible and scalable, adapting to the characteristics of the Universe of Discourse and the scope of the EM problem at hand (e.g., naval and industrial applications; local and regional scope).

A system based on this model (a) provides a common platform for the compilation of incident status, (b) supports the assessment of priorities of alternative lines of action (e.g., incident response, infrastructure repair) based on context (e.g., incident types), (c) supports the resource assignment process, and (d) provides advice regarding, for instance, responders engagement to specific incidents.

Due both to the complexity of the evaluation and advice problems handled by the SINGRAR model and to the vagueness of most of the data under consideration, a fuzzy logics approach was selected.

2.2 *Fuzzy multiple attribute decision making*

There is a lack of EM intelligent decision support tools. There are strong reasons for this. First, these decision-making processes are extremely complex due to the high number of parameters under consideration. Second, it is not easy to code and compute common vague concepts used by humans, such as "*many casualties*", "*large incident*", "*fast response*" or "*important asset*". Even if the terms of human language are constrained by some formalism, it is hard to handle statements such as "*asset A, which is very important to respond to threat X, is degraded*" or "*asset B, which is fundamental to respond to Y-type incidents, is committed to a Z-type incident*". These problems are largely increased in multi-domain scenarios (e.g., security, rescue, damage control, supply chain, reconstruction, health) like the ones that characterize Disaster Relief operations. Classical set theory and Boolean logics present serious limitations to manipulate data that presents such ill-defined outlines.

The complexity of the evaluation problem handled and the vagueness of the data processed by the SINGRAR model, led to the selection of an approximate reasoning approach based on a Fuzzy Multiple Attribute Decision Making (FMADM) methodology ((Chen and Hwang, 1992, Zimmermann, 2001). The Fuzzy Set Theory (FST) (Zadeh, 1965) is a generalization of the classical set theory that provides a way to incorporate the vagueness inherent to phenomena whose information is qualitative and provides a strict mathematical framework that allows its study with some precision and accuracy. To better understand the rationale behind the model SINGRAR can be compared to a rule based priority management inference process. In the incident response assessment context the basic rule is:

IF *impact₁ is high* **AND** *severity₁ is high* **OR**
... OR
impact$_n$ is high **AND** *severity$_n$ is high*
THEN *response priority is high*

The different factors considered can include, for instance, human health, infrastructures or environment. The individual impact and severity factors considered in this rule may have to be assessed based on finer attributes, which reason about information with smaller granularity. This decomposition of inference process can be extended to as many levels as required to characterize the situation at hand with the desired degree of detail.

Obviously, due to the high number of foreseeable incident types and to the high complexity of interdependencies, defining crisp/Boolean rules for each relevant combination (defined in the impacts × severity domain) would be almost impossible. The approach used in the SINGRAR model is based on a fuzzy quantification of the degree of truth of each statement in the condition side of the rule, followed by its recursive aggregation by means of fuzzy operators. In fact, the judgment conveyed by the previous rule is numerically computed using the following formulation, whose result expresses the degree of truth of "*response priority is high*":

$$\mu_P = \bigcup_{i=1}^{n}(\mu_{I_i} \otimes \mu_{S_i}) \qquad (1)$$

where μ_P = truth degree of the conclusion "*response priority is high*"; μ_{Ii} = truth degree of the ith condition "*impact is high*"; μ_S = truth degree of the ith condition "*severity is high*"; $\cup$ = fuzzy union operator (*t-conorm*); and $\otimes$ = fuzzy intersection operator (*t-norm*).

Both conditions and conclusion are quantified in the interval [0, 1], where 0 means no priority, impact or severity and 1 means the highest priority, impact, or severity. Intermediate values represent different degrees of priority, impact, or severity.

The model can be further improved by assigning weights to the different decision factors, incorporating their preference or importance in the decision. Thus, the previous formulation can evolve to:

$$\mu_P = \bigcup_{i=1}^{n} w_i(\mu_{I_i} \otimes \mu_{S_i}) \qquad (2)$$

where w_i = weight assigned to the ith impact factor.

When several incidents converge in the same situation the global priority assessment may require the aggregation of the several partial priorities. This can result in following formulation:

$$\mu_P = \bigcup_{j=1}^{m} \mu_{P_j} \qquad (3)$$

where μ_P = global assessment of the conclusion "*response priority is high*"; μ_{P_j} = partial assessment of the conclusion "*response priority is high*" considering the jth micro-incident, computed using (2); $\cup$ = fuzzy union operator (*t-conorm*).

The sophistication of the model results also of a careful selection of fuzzy aggregation operators that model desired effects, such as synergy of factors.

In the model the membership (or truth) degrees can be defined either by means of linguistic variables or continuous membership functions. The relation between antecedents and consequents of the inference process (i.e., between the conditions (IF side) and the conclusion (THEN side) of the rules) is defined by means of fuzzy relations.

Detailed descriptions of the SINGRAR and on the basics of FST can be found in (Simões-Marques, 1999, Simões-Marques and Pires, 2003) applied to naval context, and in (Simões-Marques and Nunes, 2013) to EM.

Considering that the inference chain can have several levels the evaluation of the criteria can be a rather complex task. In fact the SINGRAR model considers several operational and technical factors related, for instance, with selection and prioritisation of a set of tasks the emergency response system is able to perform, taking into consideration different potential incident scenarios and the characteristics of the Universe with which the EM response system is interacting. Incident response priority and resource assignment assessment processes both follow a similar approach where a ranking process sorts priority levels evaluated by a rating process. The scheme of the SINGRAR FMADM model combining incident response prioritization and resource assignment processes is depicted in Figure 1, where the rating and ranking processes are shown only for the priority evaluation phase.

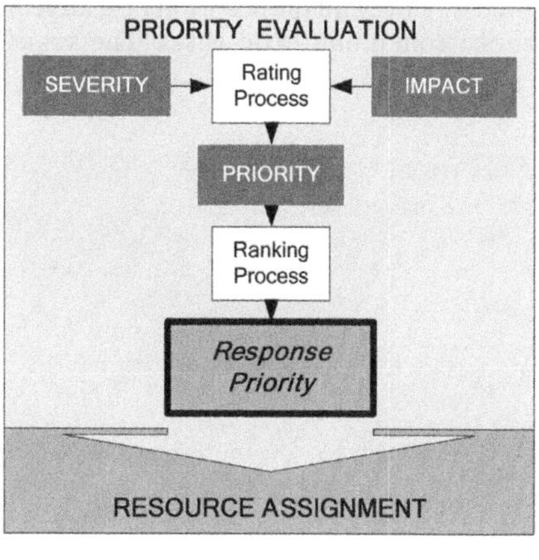

Figure 1. Scheme of the SINGRAR FMADM model.

Whenever possible the application of the FMADM model to the decision-support in a specific context requires the previous parameterization of the Knowledge Base to the geographical and organizational context and of the Inference Engine to the business rules applicable in such context. This improves the accuracy of the decision-support provided. Nevertheless the basic framework to adopt in emergency management can be quite generic. For those interested (Simões-Marques and Nunes, 2013) offers further perspective about fuzzy relations that can be used to assess the utility of the assets to assign in an incident considering, for instance, the relevant relationships between responders, incident types, incident impact types, geographical context or coordination responsibility. SINGRAR model was meant for implementation in an expert system shell which allows the parameterisation of the Knowledge Base and the User Interface for different problem contexts. Further to computing response priority and resource assignment an expert system also offers recommendations and explanations.

Considering the Human Factor, Usability is a very important issue that affects the effectiveness and success of systems, which is particularly critical in complex systems, and when accuracy and timeliness is decisive. Therefore, Usability of EMIS is of utmost relevance. (Nunes and Simões-Marques, 2013) offers a perspective on how such systems can contribute to effectively support decision-makers.

3 EMERGENCY MANAGEMENT SCENARIO

This section illustrates the use of the SINGRAR model in a fictitious EM scenario (Fig. 2).

A terrorist attack to a merchant vessel loaded with hazardous materials resulted in an explosion followed by a fire and the emission of a large plume of smoke containing toxic gases. The vessel was moored in a harbour, close to large urban areas. A strong response force was required because some main infrastructures and services are being directly affected by the heavy smoke plume and the population is also at risk. Thus, local EM Coordination Centre took control of the situation. In complement of the local fire department, the harbour master was engaged, and maritime fire fighting capabilities were mobilized, consisting of a couple of tugs equipped with high capacity water cannons, and other means to control a potential spill of toxic materials. The harbour master also interdicted the navigation of commercial traffic in the harbour and mobilized the local shuttle ferries to support the evacuation of populations in the affect area. The incident occurred near a bridge which is a main road and railway infrastructure. The heavy smoke caused road accidents, forcing traffic authorities to close the highway. Train circulation was also stopped.

The dense plume of smoke spread and the population in the area had to be evacuated, particularly an existing hospital. This required the activation of civil protection lodging plans and the call of Armed Forces CBRN expert teams to provide transportation and other logistic support.

This scenario is used to illustrate the application of the model in rating and ranking the response to this incident compared to other concurrent incidents. For maintaining the example simple, only a minor portion of the inference chain will be considered, and the evaluation of the incident will be based on linguistic variables (*severity* and *impact*) defined as discrete fuzzy sets:

$$severity = 1/extreme + 0.75/high + 0.5/medium + 0.25/low + 0/null$$

$$impact = 1/extreme + 0.75/high + 0.5/medium + 0.25/low + 0/null$$

As mentioned, the model can weigh the importance of the impact factors. In this example the weights are also defined using a linguistic variable:

$$importance = 1/high + 0.9/medium + 0.8/low$$

Let's assume the impact factors considered for assessing the "terrorist attack" incident are the ones presented in Table 1. The table shows the linguistic classification assigned to each factor, as well as the corresponding linguistic weight classification.

To compute the incident response priority (using formula (2)) is required the selection of a fuzzy intersection operator to use in this aggregation. Considering its characteristics, the *product* function presents an adequate behavior. This aggregation requires also the selection of a fuzzy union operator. The most basic fuzzy union operator is the *max* function, which will be used in this

Figure 2. Scenario of a complex EM situation.

Table 1. Classification of the severity factors and weights for "terrorist attack" incident.

	Factor		
	Human health	Infrastructure	Environment
Impact	*Extreme*	*High*	*High*
Severity	*Extreme*	*Low*	*High*
Importance	*High*	*Medium*	*High*

Table 2. Example of a list of incident priorities.

Incident	Rating	Ranking
Terrorist attack	1	1st
Forest fire	0.75	2nd
Car chain accident	0.68	3rd

example. Considering the membership degrees for each term of the linguistic variables results:

$$\bigcup_{i=1}^{n} w_i(\mu_{I_i} \otimes \mu_{S_i})$$
$$= \max(1 \times 1 \times 1, 0.9 \times 0.25 \times 0.75, 1 \times 0.75 \times 0.75) = 1$$

Recalling that the conclusion is quantified in the interval [0, 1], where 0 means no priority, and 1 means the highest priority, the "terrorist attack" incident is considered as having the highest priority.

Whenever there are several incidents to respond to, individual incident priority assessments can be compared in order to rank the assignment of resources. Table 2 reflects a scenario where the "terrorist attack" incident priority is rank ordered against two other incidents. As one can observe this incident is the first in terms of response priority, considering that the "forest fire" and "car chain accident" incidents have lower priority ratings.

Ranking response priorities is critical to the second stage of the EM process, which is assigning resources. Unfortunately there is no room in this paper for addressing this stage thoroughly. Nevertheless, one can note that the assessment of resources needed has to consider, among other factor the type of response activities required, the geographical context, and the skills of the different potential available responders. It becomes self-evident the need for decision-support in complex Disaster Relief operations (like the one presented in the scenario above) where the response coordination involves several entities that should cooperate to improve the effectiveness and efficiency of their collective effort. This goal can be achieved if all entities share a common operational picture that allows gaining situational awareness, and use common tools that help anticipating the courses of action of other stakeholders involved in the process. The use of common tools also helps breaking cultural and procedural barriers, contributing to strengthen the maturity of the interaction among groups of responders and to elevate the level of global preparedness to face critical situations. The use of common EM Intelligent Systems is undoubtedly a step in that direction.

4 CONCLUSIONS

Once some level of understanding about concurrent emergency incidents is attained, decisions have to be made regarding the more effective distribution of the response according to the identified needs. One goal of the EM decision-making process is setting intervention and support priorities. A second goal is levelling the resources, avoiding both excesses and deficits. Multi-Attribute Decision Making methodologies help dealing with this type of problems. However most of the information available in critical situations is vague or imprecise. Fuzzy Logics provides a coherent mathematical framework to deal with uncertainty and imprecision. Thus, the use of FMADM methodologies is an adequate approach to help managing priorities and allocating finite amounts of resources. The paper offers a brief description SINGRAR model considering its application to support EM. The model was developed to dynamically manage priorities based on situational parameters and select the most adequate resources to assign considering the specifics of the crisis context. An EMIS based on this model is able to ensure shared situational awareness through a common operational picture, and uniform and coherent recommendations regarding lines of action and resource assignment, contributing to the desired unity of effort. An EMIS developed following a user-centred design can effectively support the complexity of the decision-making process. The EM scenario illustrated a small scale yet complex incident where an EMIS is useful and described some relationships that can be considered to support the decision process, contributing to a coordinated, effective and efficient emergency response.

ACKNOWLEDGEMENTS

This work was funded by the Portuguese Navy.

REFERENCES

Chen, S.-J. & Hwang, C.-L. 1992. *Fuzzy Multiple Attribute Decision Making. Methods and Applications*, Springer Verlag.

Galton, A. & Worboys, M. 2011. An Ontology of Information for Emergency Manag. 8th Int ISCRAM Conf. Lisbon.

Nunes, I.L. & Simões-Marques, M. 2013. SINGRAR Usability Study. In Marcus, A. (Ed.) Design, User Experience, and Usability. Design Philosophy, Methods, and Tools. LNCS 8012, Springer Berlin Heidelberg.

Simões-Marques, M. & Nunes, I.L. 2013. A Fuzzy Multicriteria Methodology to Manage Priorities and Resource Assignment in Critical Situations. In Zeimpekis, V., Ichoua, S. & Minis, I. (eds.) Humanitarian and Relief Logistics. Springer New York.

Simões-Marques, M.J. 1999. Sistema de Apoio à Decisão para a Gestão de Prioridades de Reparação de Equip.s e Afectação de Recursos, em Navios, em Situação de Combate. MSc Thesis, FCT—UNL.

Simões-Marques, M.J. 2005. Contributos para a aplicação das TI na gestão de emergência. As missões navais de apoio a populações vítimas de catástrofe como Case Study Análise e Gestão de Riscos, Segurança e Fiabilidade (Vol. 1).

Simões-Marques, M.J. & Pires, F.J. 2003. SINGRAR—A fuzzy distributed expert system to support Command and Control activities in Naval environment. European J. of Operations Research, 145, 343–362.

Simon, H.A. 1955. A behavorial model of rational choice. Quarterly Journal of Economics, 69: 99–118.

Turban, E., Sharda, R. & Delen, D. 2010. Decision Support and Business Intelligence Systems, Prentice Hall.

Zadeh, L.A. 1965. Fuzzy sets. Inf. and Control, 8: 338–353.

Zimmermann, H.-J. 2001. Fuzzy Set Theory and Its Applications, Kluwer Academic Publishers.

Occupational Safety and Hygiene II – Arezes et al. (eds)
 ISBN 978-1-138-00144-2

User-centered design of the interface prototype of a BI intelligence mobile application

Gonçalo Ferreira
Faculdade de Ciências e Tecnologia, Campus de Caparica, Portugal

Bruno Mourão
Sonae, Portugal

Isabel L. Nunes
Faculdade de Ciências e Tecnologia, Campus de Caparica, Portugal
UNIDEMI, Unidade de Investigação e Desenvolvimento em Engenharia Mecânica e Industrial, Portugal

ABSTRACT: User-Centered Design (UCD) is an Ergonomic approach for the development of human computer interfaces. The recent explosion of mobile touchscreen equipments, such as smartphones and tablets, brings very interesting opportunities but also presents some challenges in the domain of UCD. This paper discusses the different phases (users' needs identification, definition of functional requirements, prototype design and testing) of the use of a UCD approach in the development of a Business Intelligence mobile application. The prototype assessment was done based on the usability attributes (efficiency, effectiveness and satisfaction).

1 INTRODUCTION

Nowadays smartphones and tablets are playing a key role in industries, e.g., in the retail industry, they are being used in activities such as barcode changing, order processing and stock location (Roche, 2012). However, turning processes and activities simpler and easier doesn't happen merely by introducing this type of devices. Instead, is due mainly to the usability of the applications, designed for improving users' experience and making their tasks easier. Having usability in consideration in every stage of product development is essential to obtain an easy to use product, capable of improving the user's productivity. However, when it comes to multitouch interfaces, there isn't much body of knowledge on usability, perhaps due to the young nature of the devices. Despite this, there is a vast literature regarding guidelines and good practices, which, coupled with the guidelines provided by the creators of smartphones and tablets operative systems can somewhat ensure usability for touch/multitouch systems (Simões-Marques & Nunes, 2012). It's in this context that the study presented in this article emerged. A retail company decided to provide their store managers with a solution that would grant them mobility. This job requires people to be in constant motion, moving between the storefront, the warehouse and the office. Traditionally, they were forced to go to a specific place, such as their office, in order to access business information (through a wide array of applications) necessary for them to make decisions, thus the importance of developing a mobile Business Intelligence (BI) application. The main goal of this study was the development, evaluation and testing of 2 interface prototypes, one for smartphone and the other for tablet, of a mobile touchscreen BI application for store managers at a retail company, which grants them mobility, while ensuring access to management information supporting decision-making and also means to detect problems and inefficiencies. The complete study can be found in (Ferreira, 2013).

2 INTERFACE PROTOTYPES DEVELOPMENT

The process of developing any product capable of satisfying its target user needs requires steps that should be taken in a structured way. Generally, a product is only positively accepted by the target user if it has usability. Ideally, usability engineering should accompany the entire lifecycle of the product, with important activities happening even before designing the actual user interface (Nielsen, 1993).

2.1 *Methodology*

The methodology to develop the BI application followed a User-Centered Design (UCD) approach and consisted of 4 phases: user needs assessment, system functional requirements definition, interface prototypes development and usability testing and evaluation. Each phase encompassed various activities. Their length ranged from only a few hours to more than a month.

2.2 *User needs assessment*

In order to collect the user needs, a sample of 7 store managers was selected based on their availability and interest in this particular subject. All the store managers were male with an average age of 43 years (34–52 years) and the working experience varied from 6 to 20 years, averaging at 13 years. After having the sample defined, a set of individual interviews took place. These interviews lasted for about 1 hour each and occurred in a casual way, within the users' working hours and in their working place. The method used for these interviews varied based on the product in development, but the goal was always to find existing or latent needs (Kraft, 2012). Another purpose of the interviews was to present the project goal. The users were asked: (1) what Key Performance Indicators (KPI) they classified as important to access while they're away from office, (2) which level of importance they assign to each of those KPI, (3) what applications they currently access to retrieve the said KPI and (4) which tasks they need to complete on a daily basis. After the interviews, the users gave a guided tour of their workplace (namely the storefront) while, at the same time, describing some tasks that usually arise on a daily basis. These tours were very important to gain a better understanding of the users' working environment, how the application would be used and of the company's processes. These insights played an important part at this point. By observing routines and procedures it was possible to confront users with eventual needs that they didn't raise in the first place. Tours' information was added to the information collected from the interviews. A total of 29 needs were collected. Then they were listed, in a structured way (e.g. "access the workplace's video feed" or "analyze KPI 1") and coded with a number.

2.3 *System functional requirements definition*

After collecting, listing and organizing user needs, they were ranked by priority order. During this process some user needs were discarded and the list was shortened from 29 to 20. An analysis which compared the constraints of not having the needs satisfied by a mobile BI application versus the benefits of having the needs satisfied by the application was performed. This analysis was very important because it allowed identifying current weaknesses and opportunities for future improvements (Nielsen, 1993).

The next step involved translating user needs (i.e., information content) into detailed system functional requirements, addressing how to provide that information (what users' were going to see, which buttons they were going to be able to press, etc.), with the aim of helping the future interface design. A meeting with all users, validated the entire prioritization process along with each of the 20 user needs. By promoting a group discussion where users could express their opinions while hearing others' opinions, it was almost guaranteed that every single decision regarding the content/functionalities of the BI application would satisfy the majority. The array of changes proposed by users and approved by everyone, allowed shortening the list from 20 to 6 needs. This change may seem too drastic, however, only 5 needs were completely dropped while the others were just reclassified and would still be part of the mobile BI application.

2.4 *Preparations for the interface design*

Before starting to design the 2 prototypes, research was performed looking for similar BI applications, suitable prototyping software, defining the prototyping approach and acknowledging the differences between iOS and Android user experience.

2.5 *Interface design*

There were 2 criteria behind the choice of which screens were going to be designed. First of all, there was the need to showcase all the different functionalities and sections of the application. In addition, by following a scenario approach, several screens were designed to later provide testing subjects with logical paths to follow in the Cognitive Walkthrough (CW). A total of 199 screens were designed (122 for smartphone and 77 for tablet). Naturally, the tablet interface prototype took considerably less screens because more information could be presented in a single screen.

3 USABILITY EVALUATION AND TESTING

3.1 *Heuristic evaluation*

Heuristic evaluation can be performed on interface prototypes, which makes it a suitable tool for finding usability problems in the early stages of product development. Ideally, heuristic evaluation should be performed by several evaluators in order to find different usability problems (Nielsen, 1993). However, in this study, only one person subjected both interface prototypes to a heuristic evaluation. A few studies were done, with the purpose of developing a list of usability heuristics to have in mind while designing the mobile applications. 16 heuristics were collected, with the aim of covering most aspects of the interfaces prototype (10 heuristic from Nielsen, 1995; 4 heuristic from Gong & Tarasewich, 2004; 1 heuristic from Sjöberg, 2005 and from Haywood & Reynolds, 2008). The heuristic evaluation revealed that the interfaces were in compliance with 14 of the 16 heuristics, which resulted in 3 changes: creation of one additional screen for the interfaces with the aim of alerting users when they enter an invalid product code; a complete overhaul of the tab bar icons; and a slight modification in the magnifying glass icon.

3.2 *Cognitive walkthrough*

A set of 6 representative tasks were defined with the purpose of truthfully recreate the usage of the BI application on users' daily routines. The tasks were defined focusing on procedures that could be realistically recreated such as analyzing certain values, charts and notifications. Each task had a similar difficulty level (all of the tasks took approximately the same number of steps and amount of time), and since the CW was going to be performed on interface prototypes (participants weren't going to interact with any real data or access any real functionalities), in practice, the tasks consisted in users finding a particular screen. Also, participants usually learn the product as their experience with it grows. So in order to compensate for their experience with the interfaces, the order of tasks was randomized. In other words, each participant was going to perform the tasks in a different order. This procedure is called "Counterbalancing" (Tullis & Albert, 2008).

Since there were 2 interface prototypes, participants were going to be subjected to 2 CWs (one for each prototype) with a 2 to 3 weeks interval between them, in order to avoid interference on the results of the second CW from their previous experience with the first CW. Therefore, 8 participants were selected all male with an average age of 42 years (34–52 years) and their working experience varies from 6 to 20 years, averaging 13 years. It's a sample larger than the "magic number 5" which guarantees a higher probability of finding usability problems while hoping to reduce the dispersion of the results. The output of the CW study comes in the form of a set of performance as well as self-reported usability metrics along with participants' comments. The metrics collected were the following: Task success; Task time; Number of errors; Efficiency; Self-Reported (After-Scenario Questionnaire and System Usability Scale). The participants were monitored while performing the tasks but their interaction with the monitor was kept to a minimum, only giving a few minor hints whenever the participants seemed lost, misread the task or forgot about its purpose. After 2 or 3 weeks participants that finished the 1st CW were contacted again for the 2nd testing session.

3.3 *Analysis and discussion of results*

3.3.1 *Task success*

All the tasks performed on both prototypes were completed successfully by all the participants. This indicates that both interface prototypes were designed in a very intuitive way.

3.3.2 *Task time*

This metric can be displayed in various ways; perhaps the most common way is to present the average amount of time participants spent on each task. However, since time data is typically skewed, it is more appropriate to present the geometric mean. Also, to account for the potential variability across participants' times (some took longer than normal while others were very fast) it is important to display confidence intervals to show the variability in the data (Tullis & Albert, 2008).

Analyzing the results from both prototypes' testing, it can be concluded that participants took approximately the same amount of time for successfully completing tasks. Exception was task 3, which took considerably more time completing in the smartphone prototype. Task 3 asks participants to analyze a certain chart. While in the tablet interface the charts are always displayed side-by-side with the tables and thus, always visible, on the smartphone interface participants had to access the charts by pressing the "Chart" button. The existence of one more button is enough to increase participants' cognitive effort, forcing them to think where they should press to access the chart.

3.3.3 *Errors*

When analyzing errors, it is important to know how many errors participants made while trying to complete each task. In this case, 2 types of errors were defined: wrong choice in a menu or list (Type A) and taking a wrong set of actions (Type B). It's important to establish this difference because some errors (Type A) may be more severe in terms of usability than others (Type B).

Additionally to knowing how many errors were made on each task, it was also important to know the percentage of participants that committed those errors because there may have been participants who did more than one error while others did. These results, stands out for the high number of errors happening on task 6. Specifically, 7 Type B errors in the tablet prototype and 5 errors (2 Type A and 3 Type B) in the smartphone prototype. This could indicate a severe problem in terms of usability in the "Notifications" section but it did not. First of all, all the Type B errors for this task happened because participants didn't click on the appropriate element within the correct section. In order to open any list of notifications within the "Notifications" section, participants have to press the blue circular button on the right of the respective row. However, most participants assumed they could access the list by simply clicking either on the title or the red indicator. After realizing that that action didn't trigger anything, they immediately clicked on the correct button. It's safe to assume that users will quickly understand how to open the notifications after failing their first attempt.

3.3.4 *Efficiency*

In order to analyze efficiency, the number of actions (mouse clicks) taken by the participants on each task was counted. Since all tasks were defined to take approximately the same amount of effort, by presenting the average number of actions a participant took to complete a task it is possible to find which tasks require the most amount of effort (Tullis & Albert, 2008). Also, it is important to establish comparisons between the ideal number of actions to complete each task (previously defined) and the actual participants' performance. It is noticeable that the average number of actions only exceeds the ideal number of clicks by approximately 1 click for all the tasks. This can be explained by the small number of mistakes done by participants. The most frequent type of errors was Type B which usually meant the addition of one more click to the ideal number of clicks. Whenever participants would choose a wrong section (Type A error) they would quickly notice it and return to the right path. Regarding the dispersion of the results, task 3 of the tablet prototype has the most dispersion (confidence interval of 2.1 clicks). This is due to some participants taking an exceedingly high number of actions (7 clicks) when compared to the majority of the remaining participants (4 clicks). Task 3 could be completed in 2 ways, through KPI 2a section (6 clicks) or through KPI 2b section (4 clicks). The participants who followed the KPI 2a section path took more actions and thus contributed to increasing the dispersion of the results. In sum, having a 100% successful task rate in all the tasks for both prototypes was a good presage for the following results. Both the task time and efficiency metrics revealed that participants took reasonable time and number of actions to complete the tasks and the errors metric enabled to find some minor usability problems. Performance-wise it can be concluded that both interface prototypes had good usability.

3.3.5 *Self-reported metrics*

The type of self-reported data collected after each task was the "After-Scenario Questionnaire" (ASQ) which was developed to be used immediately after a scenario completion, where a scenario is a collection of related tasks (Lewis, 1991). The ASQ consists of a set of 3 statements accompanied by a 7-point rating scale of "strongly disagree" to "strongly agree" and they focus on fundamental areas of usability (Tullis & Albert, 2008). Being statements designed to be presented after an entire scenario, each refers to the tasks as a whole. Therefore, a little adaptation had to be done in order for the statements to be presented to the participants after each task. The final statements, presented were the following: a) I am satisfied with the ease of completing this task; b) I am satisfied with the amount of time it took to complete this task. Each statement was accompanied by a 5-point rating scale of "strongly disagree" to "strongly agree". This is a modification concerning the original ASQ, but a study, assessing the impact of different Likert scales on participants' answers, concluded that both scales are comparable (Dawes, 2008). Therefore, having a 5-point or 7-point rating scale doesn't have a significant impact on participants' answers.

Concerning self-reported metrics, there was one more questionnaire delivered to participants after completing the entire set of tasks. The aim of this questionnaire was to assess participants' overall perception of usability regarding the interface prototypes. The said questionnaire is called the System Usability Scale (SUS), a reliable and low-cost usability scale that consists of 10 statements (Brooke, 1996), with the odd-numbered statements worded positively and the even-numbered statements worded negatively (Lewis & Sauro, 2009). Each statement is accompanied by a 5-point rating scale of "strongly disagree" to "strongly agree". The statements presented to participants were the following: 1) I think that I would like to use this application frequently. 2) I found the application unnecessarily complex. 3) I thought the application was easy to use. 4) I think that I would need the support of a technical person to be able to use this application. 5) I found the various functions in this application were well integrated. 6) I thought there was too much inconsistency in this application. 7) I would imagine that most people would learn to use this application very quickly. 8) I found the application very cumbersome to use. 9) I felt very confident using the application. 10) I needed to learn a lot of things before I could get going with this application.

The choice of SUS was also due to its simplicity and reduced number of statements. There are other questionnaires like QUIS (Chin, Diehl & Norman, 1988) or SUMI (Kirakowski, 1994) which are more thorough but the higher number of statements may cause some discomfort on participants who may not have the available time to answer so many questions. Besides, the SUS was developed with the intention of being a tool that could quickly and easily collect users subjective rating of a product's usability (Bangor, Kortum & Miller, 2008), which was exactly what was needed for this study. A study, that analyzed various SUS scores, indicates that the SUS is a highly robust and versatile tool for usability professionals (Bangor, Kortum & Miller, 2008).

3.3.6 *After-scenario questionnaire*

After collecting the participants' answers for the 2 statements presented to them after each task, the results were converted to a numeric scale (from 1 to 5). Averaging the results of both statements gives an overall view of the participants' satisfaction concerning their experience with the prototypes. From these results it can be concluded that the participants were extremely satisfied with the usability of both prototypes. The lowest satisfaction rating was 4.63 for the tablet prototype and the highest rating (5.0) was given to several tasks for both prototypes.

3.3.7 *Single Usability Metric (SUM)*

The SUM intends to summarize 4 variables (satisfaction, task time, task success and errors) into one single usability score for each task (Sauro & Kindlund, 2005). The data entered per participant and task was the following: Participant satisfaction (the average of the 2 ASQ answers); Task completion; N. of errors recorded on task; and Task time. After entering data, it's required to specify the confidence level and the number of error opportunities for each task. The confidence level chosen was 95% as it was the level used for all calculations. Regarding the number of error opportunities, it wasn't easy to assign a truthful value because it's somewhat subjective and hard to define what constitutes an error opportunity. It was assumed that 5 error opportunities for each task would be a reasonable value. After entering all the data, the SUM score was calculated. Each of the 6 tasks had a very good SUM score for both prototypes. The lowest overall score was task 6 for the tablet which scored 91.8%.

3.3.8 *System usability scale*

The application of the SUS questionnaire allow to assess participants' perceived usability of their entire experience with both prototypes. When analyzing the results, it is important to recall that its purpose is to provide a single combined rating. According to a study where 50 studies' average SUS scores were compiled, a SUS score higher than 80% can be considered pretty good (Tullis & Albert, 2008). Both prototypes obtained very good SUS scores; the tablet prototype scored 97.81% while the smartphone prototype scored 99.06%. Therefore there is no doubt they received very positive scores.

4 CONCLUSIONS

It is clear that having applied the UCD philosophy proved very beneficial for the development of 2 interface prototypes with high usability that satisfy user needs. Centering all the product development activities on users improves the chance of satisfying their needs. In this case, the initial meetings allowed understanding user needs as well as their working environment which translated into an accurate user needs definition. Also, being familiarized with the applications users recurred to was a very important step to accurately define the functional requirements and design the interfaces in a familiar way to users. The contribution of users in the interface design phase also helped in designing better interfaces. All of these factors contributed to such good results from the metrics collected as well as the ASQ, SUS and SUM scores. Resorting to usability evaluation and testing activities allowed a more detailed inspection of the interfaces, discovering more usability problems. In sum, the work described in this study can be considered successful. All the initial proposed objectives were successfully accomplished in the expected time and all the parts involved (the author, the company and the users) were satisfied with the results. Hopefully, the prototypes developed as a consequence of this work will inspire the developers of the BI application and the design of the real interfaces will closely resemble the one proposed here.

ACKNOWLEDGEMENTS

This work was funded by *Fundação para a Ciência e a Tecnologia* under the scope of project *PEst-OE/EME/UI0667/2011*.

REFERENCES

Bangor, A., Kortum, P.T. & Miller, J.T. 2008. An Empirical Evaluation of the System Usability Scale. Int J of Human–Computer Interaction, 24: 574–594.

Brooke, J. 1996. SUS—A quick and dirty usability scale. P.W. Jordan, B. Thomas, I.L. McClelland & B. Weerdmeester, Usability Evaluation in Industry. Taylor & Francis.

Chin, J.P., Diehl, V.A. & Norman, K.L. 1988. Development of an Instrument Measuring User Satisfaction of the Human-Computer Interface. Proc. SIGCHI Conf. on HF in Computing Systems: 213–218. Washington, D.C.: ACM Press/Addison-Wesley Publishing Co.

Cornett, C. 2012. Similarities and differences in iOS and Android UX design. In http://www.archer-group.com/2012/.

Dawes, J. 2008. Do data characteristics change according to the number of scale points used? An experiment using 5-point, 7-point & 10-point scales. Int J Mark Res, 50: 61–77.

Ferreira, G.L. 2013. User-centered design of the interface prototype of a business intelligence mobile application. MSc thesis, in http://hdl.handle.net/10362/10034.

Gong, J. & Tarasewich, P. 2004. Guidelines for handheld mobile device interface design. Proc 2004 DSI Annual Meeting: 3751–3756. Boston, Massachusetts.

Haywood, A. & Reynolds, R. 2008. Usability guidelines | Touchscreens, in http://experiencelab.typepad.com/files/design-guidelines-touchscreens-1.pdf.

Kirakowski, J. 1994. The Use of Questionnaire Methods for Usability Assessment, in http://sumi.ucc.ie/sumipapp.html.

Kraft, C. 2012. User Experience Innovation. Apress.

Lewis, J.R. 1991. Psychometric evaluation of an after-scenario questionnaire for computer usability studies: The ASQ. SIGCHI Bulletin, 23: 78–81.

Lewis, J.R. & Sauro, J. 2009. The Factor Structure of the System Usability Scale. Proc. 1st Int Conf on Human Centered Design: 94–103. Heidelberg, Germany: Springer-Verlag.

Nielsen, J. 1993. Usability Engineering. Academic Press, Inc.

Nielsen, J. 1995. 10 Usability Heuristics, in http://www.nngroup.com/articles/ten-usability-heuristics/.

Roche, M. 2012. How Smartphones are now being used in major industries, in http://www.dokisoft.com/how-smartphones-are-now-being-used-in-major-industries/.

Sauro, J. & Kindlund, E. 2005. A Method to Standardize Usability Metrics. Proc. SIGCHI Conf on Human Factors in Comput. Syst.: 401–409. Portland, Oregon: ACM Press.

Simões-Marques, M. & Nunes, I.L. 2012. Usability of Interfaces. In I.L. Nunes (ed), Ergonomics—A Systems Approach. InTech.

Sjöberg, S. 2005. A Touch Screen Interface for Point-Of-Sale Applications in Retail Stores. Umeå, Sweden: Umeå Univ.

Tullis, T. & Albert, B. 2008. Measuring the User Experience: Col., Anal., & Pres Usability Metrics. Morgan Kaufmann.

Occupational Safety and Hygiene II – Arezes et al. (eds)
© 2014 Taylor & Francis Group, London, ISBN 978-1-138-00144-2

Occupational noise exposure during pregnancy: A systematic review

M.A. Bastos, S.C.A. Duarte & R.M.S. Araújo
Doctor Ship Program of Engineering School in Porto University, Porto City, Portugal

C. Ramalho
Doctor Specialized in Obstetrics and Gynecology and Professor of Medical School in Porto University, Porto City, Portugal

J. Torres da Costa
Professor of Medical School in Porto University Porto City, Portugal

ABSTRACT: The concern about women professional/environmental place with noise exposure and the human reproduction. This article shows a systematic review to identify studies that presents the theme of noise exposure in professional/environmental place during pregnancy and reproductive health. The systematic review searched among published articles in1990 until 2013. The screening was made with methodological procedures from PRISMA. It resulted in 35 (thirty-five) articles. Among those 9 (nine) will be included in the final part the analysis of review. Some studies present divergencies in the association to exposure noise in a professional/environmental place. Others associate the chronic noise exposure to the congenital abnormalities in fetus: Pregnancy hypertension. The selected articles used information from patients chart, interviews and questionary. The theme revealed a diversity in studies, although they needed to concern bringing a contribution and clarify the information given. The methodological choice has a great importance for the new studies to aggregate relevant data to the area.

1 INTRODUCTION

The emerging markets situation reveal a reaction of developing and growth, with an increase in short and long range on the sector of production and services. Consequently the labor Market presents an increasing demand for skilled workers. Although, some productive sectors are affected by the lack of those professionals.

Allied to this fact, statistic data shows that the livelihood in the family has changed and women are increasing in that position on their families.

The reflections of world financial crisis arrive to families homes and impose to them the search of financial support. Those information were given by IBGE (Brazilian institute of geography and statistics), that reveals 37,4% of Brazilian families has the woman as the livelihood (IBGE 2012). On the other hand the importance of women on labor Market shows their economic power in other countries. Statics from 2012 presents that Brazil has 43% of the woman in the labor market, in Canada 46%, Australia 45,5%, China 44,5%, Europe 45%, EUA 43%, Japan 42% and Argentina 40% (Statistics from 2012).

The demand for skilled workers and the concern for more careful procedures, requires attention and speial cares, provided women new job opportunities.

The new familiar structure has changed. The classic role of men and women has changed among the "traditional Family". Nonetheless, these facts don't hazard families, but the functional changes between the livelihoods in families are growing (Simões and Hashimoto 2012).

The professional daily actions require attention and it needs a safe environment. It encourages de raising attention for the methodological procedures that allows an analysis of safety and health of the workers during pregnancy. At all costs, the studies have shown a few or no harm to the fetus or pregnant women on what they regular do as an activity. All the harms are related to standing up, bending over, rasing and long work jorney, nonetheless those results are not concluded for bodily risks at work specially for pregnant women. (Salihu, Myers, and August 2012).

The European agency for health and safety at work, says that the pregnant worker who was exposed to loud noises in her work, may put her fetus in risk. (EU-OSHO 2012).

To Magann et al. (2005), the previews studies has failed because of the differences in the analyzed population, of its bounds and variables that could get in the way of the goals from the analysis.

The excessive exposition to the noise has been related to deaf, hypertension, cardiac ischemic disease, opposite performance at work, increasing the aggressive behavior and risk of premature birth (Magann et al. 2005).

This systematic review has a goal of pointing out the studies that reveal the consequences of noise exposure professional/environmental during pregnancy. It was applied a methodology that allowed to organize the results and correlate it to the hearing ability of a children or fetus during pregnancy. Also to analyze the quality of procedures applied in each experience told in the selected articles.

2 METHODOLOGY

The strategy to the analysis was made for a research on an electronic data base:

ACM Digital Library, ACS Journals, Acad. S. Complete, AGRICOLA Articles, Annual Reviews, BioMed Central Journals, Biomedical, Business S. Complete, Energy C. (DOE), High wire Press, IEE Xplore, Inf. (T. and Francis), Medline, MetaPress, Nature.com, Oxford Journals, PsycArticles, Pubmed, ScienceDirect, SCOPUS, Web of Science e Zentralblatt MATH. The words that were used on the research were: "noise", "pregnancy", "gestational malformations", "miscarriage", "gestational risk", "preterm birth", "gestational noise level", "delivery complications", "newborn behavior".

To point out the articles that were published, was used some of the key words, through the Metalib.

For the research among studies about noise exposure during pregnancy in professional/environmental place, the words pointed out made some matches listed on the following Chart 1.

Those matches resulted in seven key word matches, in which were identified one by one of the published texts, based on criteria of inclusion and exclusion.

The researches were made through the integrated advanced form in scientific magazines and the data base from FEUP library. For the inclusion some criteria were applied.

Chart 1. Key word match.

Match	Key word
1	"Noise and pregnancy"
2	"Noise and gestational malformations"
3	"Noise and miscarriage"
4	"Noise and gestational risk"
5	"Noise and preterm birth"
6	"Gestational noise level and delivery complications"
7	"Gestational noise level and newborn behavior"

2.1 *Criteria applied for exclusion*

The selected texts on the base were listed in an electronic worksheet. For that was used Microsoft Excel.

Was not made any restriction as far as the population studied. On the first step were excluded the articles under the criteria: repeated ones; unidentified author; unidentified year; not published in English and before 1990. Afterwards the ones that didn't have an abstract were excluded.

2.2 *Criteria applied for inclusion*

These articles became acceptable due to its abstract reading, therefor to identify the goals with clear and acceptable for inclusion, even the methodology for the consequences of the exposure to noise during pregnancy.

Thereafter, to define the quality of the article was made criteria for analysis of the whole text, to be acceptable in the group of selected articles (Urrutia and Bonfill 2010). Those were the selected ones, they included the theme and identified the evidences, or even, the consequences of noise exposure in pregnant women.

After the acceptance step. Important information was researched on those articles towards the theme, such as:

"country; activity; sample; goals; methodology; period of research; volume of the noise in which the pregnant women are exposed to (dB); hours of exposure; attached factors as incidence; type of statistics analysis".

By the end of it the selected articles for review were analyzed by specific criteria and systematic. Those methodological steps were executed according the procedures and criteria specified by the PRISMA methodology (Preferred Reporting Items for Systematic Reviews and Meta-Analyses) (Urrutia and Bonfill 2010).

3 RESULTS

The selected articles are presented in the Chart 2, and was pointed out by the key word match on Chart 1.

From 3.126 articles included, 6 passed through analysis and are quantified on Chart 3 by its source. During the selection of articles that contain the reviewed theme, was perceptible the reason in which the articles were excluded, because of not being on the theme or not having the defined period of time, or even for having an unknown author or unknown year, in which were excluded 3.120 articles. The range of their ages was between 20 and 40 years old. The volume of the noise was between 80 and 90 dB.

Chart 2. Selected articles by word matches—identifying step.

	Amount of publications Conjuntos de palavras chaves						
Source	1	2	3	4	5	6	7
ACM Digital Library	–	–	–	4	–	–	–
ACS Journals	–	–	–	20	–	6	–
Acad. S. Complete	10	–	1	–	2	1	–
AGRICOLA Articles	1	–	–	–	–	–	–
Annual Reviews	–	–	–	7	–	3	–
BioMed Central Journals	–	13	1	–	62	22	–
Biomedical	–	1	–	8	–	–	–
Business S. Complete	1	–	–	–	–	–	–
Energy C. (DOE)	500	–	–	–	–	–	–
High wire Press	–	5	2	–	25	8	–
IEE Xplore	–	17	4	–	2	–	–
Inf. (T. and Francis)	–	117	185	133	126	57	71
Medline	–	4	–	–	–	–	–
MetaPress	–	–	–	1	5	–	–
Nature.com	–	39	34	173	182	37	–
Oxford Journals	–	–	–	–	–	4	–
PsycArticles	222	4	12	23	63	8	–
Pubmed	239	–	14	–	3	–	–
Science Direct		–	–	–	1	–	–
SCOPUS	296	2	–	10	32	–	–
Sem identificação	–	–	–	60	–	–	–
Web of Science.	214	2	7	18	2	–	–
Zentralblatt MATH	–	–	–	–	1	–	–
Amount of matches	1.483	204	260	457	505	146	71
Σ Pubications from identifying step						3.126	

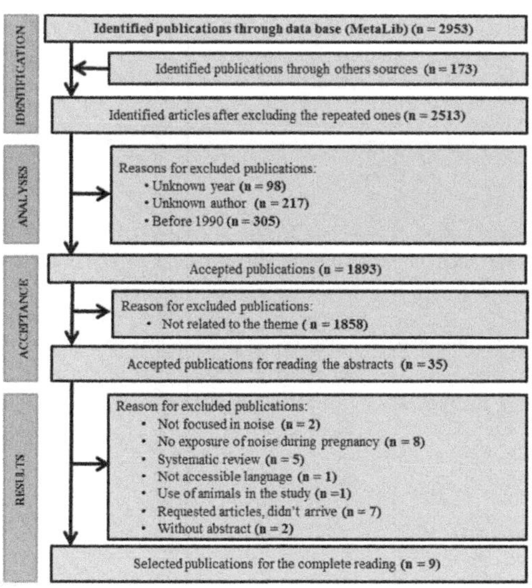

Picture 1. Flowchart of systematic review from publications.

Chart 3. Amount of selected publications.

Source	Amount
High wire Press	1
PubMed	3
SCOPUS	4
Web of Science	1
Total	9

The nine fully analyzed articles are presented summed up on the Charts 4 and 5. In general all of them present the consequences during pregnancy; however the opinions are not the same.

4 DISCUSSION

The articles that were accepted present a significant variable methodology. Those information were taken from Charts 4 and 5 which reveal the range of works and variety of data. The data are

Chart 4. Information from the selected publications on systematic review (Parte 1).

Author/year	Country/ period	Sample	Socio-economics	Education	Age range	Childbirth	Marital status	Smokers	Use of alcohol	Activities
Hartikainen, A.L. et al., 1994	Filand/ 1983–1987	111 grávidas exposta e 181 não expostas	33% low	ND	<20 a 35 years	1 a 3	73 a 86%	31 a 35%	23 a 26%	>% services, industry.
Bendokiene, I.; Grazuleviciene, R., & Dedele, A., 2011	Kaunas, Lituânia 2007–2009	3,121 pregnant women (Voluntary)	Low: 26 a 31% Medium: 51 a 57% High: 15 a 17%	Primary: 2 a 8% Secondary: 37 a 44% University: 47 a 59%	20– 45 years old	>1: 71 a 78%	Married: 78 a 86%	2 a 5%	49 a 51% ≥ 2 week: 1 a 4%	varied
Rocha, E.B.; Azevedo, M.F. & Ximenes Filho, J.C., 2005	Fortaleza, Brazil August 2002 to june 2003	80 children 45 controle 35 G. study	Low: 1–2 payment	ND	Children 0 a 6 months	ND	ND	ND	ND	ND
Magann, E.F. et al., 2005	California. EUA	814	ND	ND	ND	Range: 1 a 4	ND	ND	ND	ND
Croteau, A.; Marcoux, S. & Brisson, C.; 2008	Canadá, jan 1997 to 7 de Mar 1999	Cases (n = 1.242) and control (n = 4.513)	ND	Study: 42% > 14 years Control: 36% > 14 years	20–40 years	Range: 1 a 3	Married: 96%	NA	NA	NA
Niemtzow, R.C., 1993	Review		ND	ND	ND	ND	ND			Industrial
Wu, T. et al., 1996	Taiwan/1991	200	High and middle	School years	17–40	ND	ND	3%	5,5%	ND
Zhang et al., 1992	Changhai	1875			20–35	Varies	ND	ND	ND	ND
Hrubá, D., 1999	Six count. European 1990–1992	3897	ND	ND	ND	ND	ND	Evaluation	ND	Industrial

Obs: ND: Not Defined; NA: Not Associated; A: Associated.

Chart 4. Information from the selected publications on systematic review (Parte 2).

Author/year	Evaluation of the exposure	Analyzed parameters	Including criteria	Type of noise/range	The use of ear protector	Understand the exposure	Noise exposure
Hartikainen, A.L. et al., 1994	Interview/ questionaries'	Heavy work Psycological burden Standing up shifts Noise louder Vibration Temperature	Mother age, parity work condition	48% das > 89 dB L_{Aeq} (8 h) 60% exposed to vibrations	>80% of your work time e 39% of them more then 95%	90% don't consider their job involves noise exposure	Occupational
Bendokiene, I.; Grazuleviciene, R. & Dedele. A., 2011	Interview/ questionaries'	Hypertension	Mother age, work condition	Low, ≤50 dB(A); moderate, 51–60 dB(A); and high, ≥61 dB(A).	ND	ND	Occupational and environmental
Rocha, E.B.; Azevedo, M.F. & Ximenes Filho, J.C., 2005	Interview/ questionaries'	ND	ND	Mother: 80 a 90 dB	ND	ND	Occupational
Magann, E.F. et al., 2005	Interview/ questionaries'	Standing up	Active, no medical concerns	85 dB	ND	ND	Occupational
Croteau, A.; Marcoux, S. & Brisson, C., 2008	Interview/ by phone	Chronic noise congenital anomaly		ND	ND	ND	Occupational
Niemtzow, R.C., 1993	Revision	Noise exposure	Ensemble	Ensemble	ND	ND	General
Wu, T. et al., 1996	Densímetro + questionaries'	ND	Mother weight, the sex, pregnancy period	ND	ND	ND	Amplified music 12%
Zhang, et al., 1992	Questionaries'	Baby weight Childbirth	Mother weight, the sex, pregnancy period	ND	ND	ND	General
Hrubá, D., 1999	Questionaries'	Noise and others		ND	ND	ND	Ocupacional

Obs: ND: Not Defined; NA: Not Associated; A: Associated.

Chart 5. Information from the selected publications on systematic review.

Author/year	Premature childbirth	Congenital abnormalities	Hypertension	Under weight	Fetus deaf	Fetus: intellectual development	Exposure to noise
Hartikainen, A.L. et al., 1994	A	Abnormalities 3% (considered low) NA	NA	A	ND	ND	ND
Bendokiene, I.; Grazuleviciene, R. & Dedele. A., 2011	ND	ND	82 a 87%	ND	ND	ND	Subject of the study
Rocha, E.B.; Azevedo, M.F. & Ximenes Filho, J.C., 2005	ND	ND	ND	ND	NA	ND	ND
Magann, E.F. et al., 2005	A	ND	ND	ND	ND	ND	ND
Croteau, A.; Marcoux, S. & Brisson, C., 2008	NA	ND	NA	NA	ND	ND	ND
Niemtzow, R.C., 1993	ND	ND	ND	ND	ND	ND	85 a 95 dB
Wu, T. et al., 1996	ND	ND	ND	NA Countryside area	ND	ND	Traffic noise around home 22%
Zhang, et al., 1992	ND	ND	ND	A	ND	A	ND
Hrubá, D., 1999	ND	ND	ND	ND	ND	ND	ND

Obs: ND: Not Defined; NA: Not Associated; A: Associated.

not heterogeneous. They point to data fragility, because they present evidences and not any conclusive result. It highlights the necessity in future studies to be done, so it would include a bigger portion of the population and for a much longer period of investigation. Hohmann et al. (2013) It suggests applying objective measurements, the propagation of a noise map and its measure tool for noise volume, that is during the following 12 years. For more conclusive results.

Para Hartikainen et al. (1994), the loss of the hearing is undeniable and for life a life time when exposed to a loud noise. This exposure takes to stress in human beings, in which is clear disorders in communication, work performance and sleep.

Besides inducing peripheral vasoconstriction and physiological reactions in a short period of time, on the systems: neurologic, respiratory and vegetative, endocrinologic.

The articles, on its majority were applied by questionary on pregnant women and a controlled group. Under this context, the articles show the following evidences, as congenital abnormalities in fetus on a low percentage (Hartikainen et al. 1994), tendency to premature delivery (Magann et al. 2005); do not associate this with pregnancy hypertension (Haelterman et al. 2007) and (Croteau, Marcoux, and Brisson 2007). Never the less, this review showed that hypertension presented its effects on 82 pregnant women enterviewed that were exposed to the noise (Bendokiene, Grazuleviciene, and Dedele 2011).

The variance of results falls on the range that envolves the studies and it can influence on its causes presented as effect on pregnant women exposed to noise. To Niemtzow (1993), says that between 85 to 95 dB can be tolerated by a fetus in the uterus. Zhang. Cal & Lee (1992), in their article shows that the exposure to the noise in a work envirомnt during pregnancy can cause fetus death.

On studies, Hrubá, Kubla. & Tyrlik (1999) gives the results, about noise, they've had the influence on much parameters to analyse, such as exposure to chemical products (solvents and pesticides). Was not possible to avaluate clearly the risks of noise exposure in a work enviromnt during pregnancy. Tha datas were presented on the Charts 4 and 5.

5 CONCLUSION

The systematic review presented what was an object of study until today, even as achieved results and its limits presented by authors. Allowed to put on a map the results in a way of contributing to the development and definition of its methodology to be applied for the research that I will prepare within this theme.

REFERENCES

Bendokiene, I., R. Grazuleviciene, and A. Dedele. 2011. "Risk of hypertension related to road traffic noise among reproductive-age women." *Noise Health* no. 13 (55):371–7. http://www.ncbi.nlm.nih.gov/pubmed/22122952. doi: 10.4103/1463-1741.90288.

Croteau, A., Marcoux, S., & Brisson, C. 2007. Work activity in pregnancy, preventive measures, and the risk of preterm delivery. *Am J Epidemiol* no. 166 (8):951–65. http://www.ncbi.nlm.nih.gov/pubmed/17652310. doi: 10.1093/aje/kwm171.

EU-OSHO, Agência Europeia para a Segurança e Saúde no Trabalho. 2012. *Facts 57: The impact of noise at work*. https://osha.europa.eu/en/publications/factsheets/57.

Haelterman, E., Marcoux, S., Croteau, A. & Dramaix M. 2007. Population-based study on occupational risk factors for preeclampsia and gestational hypertension. *Scandinavian Journal of Work, Environment & Health* no. 33 (4):304–317. doi: 10.5271/sjweh.1147.

Hartikainen, A.L., Sorri, M., H. Anttonen, R. Tuimala & Laara, E. 1994. "Effect of occupational noise on the course and outcome of pregnancy." *Scandinavian Journal of Work, Environment & Health* no. 20 (6):444–450. doi: 10.5271/sjweh.1376.

Hohmann, C., L. Grabenhenrich, Y. de Kluizenaar, C. Tischer, J. Heinrich, C.M. Chen, C. Thijs, M. Nieuwenhuijsen, and T. Keil. 2013. "Health effects of chronic noise exposure in pregnancy and childhood: a systematic review initiated by ENRIECO." *Int J Hyg Environ Health* no. 216 (3):217–29. http://www.ncbi.nlm.nih.gov/pubmed/22854276. doi: 10.1016/j.ijheh.2012.06.001.

Hrubá. D., Kubla L. & Tyrlik, M. 1999. Occupational risk for human re´roproduction: Elaspac study no. 4: 210–215. Central European Journal of Public Helth.

IBGE, Instituto Brasileiro de Geografia e Estatística. 2012. Pesquisa Nacional por Amostra de Domicílios (PNAD). Relatório 2011. tp://ftp.ibge.gov.br/Trabalho_e_Rendimento/Pesquisa_Nacional_por_Amostra_de_Domicilios_anual/2011/Sintese_Indicadores/sintese_pnad2011.pdf.

Lercher, P., G.W. Evans, M. Meis, and W.W. Kofler. 2002. Ambient neighbourhood noise and children's mental health.pdf.

Magann, E.F., S.F. Evans, S.P. Chauhan, T.E. Nolan, J. Henderson, J.H. Klausen, J.P. Newnham, and J.C. Morrison. 2005. The effects of standing, lifting and noise exposure on preterm birth, growth restriction, and perinatal death in healthy low-risk working military women. *J Matern Fetal Neonatal Med* no. 18 (3):155–62. http://www.ncbi.nlm.nih.gov/pubmed/16272037. doi: 10.1080/14767050500224810.

Niemtzow, R.C. 1993. Loud noise and pregnancy. 1993. no. 158:10–12. Melitary Medicine.

Salihu, H.M., J. Myers, and E.M. August. 2012. Pregnancy in the workplace. *Occup Med (Lond)* no. 62 (2):88–97. http://www.ncbi.nlm.nih.gov/pubmed/22355087. doi: 10.1093/occmed/kqr198.

Simões, F.I.W., and F. Hashimoto. 2012. Mulher-mercado de trabalho e as configurações familiares do século XX. *Revista Vozes dos Vales da UFVJM: Publicações Acadêmicas, no 02, ano I. Belo Horizonte/MG—Brasil. Out 2012*. Accessed 9 de abril 2013. http://www.ufvjm.edu.br/site/revistamultidisciplinar/files/2011/09/Mulher-mercado-de-trabalho-e-as-configura%C3%A7%C3%B5es-familiares-do-s%C3%A9culo-XX_fatima.pdf.

Statistics, U.S. Bureau of Labor. 2012. *Labor Market*. http://www.bls.gov/fls/chartbook/2012/section2.pdf.

Urrutia, G., and X. Bonfill. 2010. PRISMA declaration: a proposal to improve the publication of systematic reviews and meta-analyses. *Med Clin (Barc)* no. 135 (11):507–11. http://www.ncbi.nlm.nih.gov/pubmed/20206945.

Zimmerman, E., and A. Lahav. 2013. Ototoxicity in preterm infants: effects of genetics, aminoglycosides, and loud environmental noise. *J Perinatol* no. 33 (1):3–8. http://www.ncbi.nlm.nih.gov/pubmed/22878560. doi: 10.1038/jp.2012.105.

Zhang. M., Cal, W. & Lee, D.J. 1992. Occupational hazards and pregnancy outcomes no. 2:397–408. American Journal of Industrial Medicine.

Occupational Safety and Hygiene II – Arezes et al. (eds)
© 2014 Taylor & Francis Group, London, ISBN 978-1-138-00144-2

Author index

Printed and bound by CPI Group (UK) Ltd, Croydon, CR0 4YY
18/10/2024
01776219-0016